1984
ANNUAL BOOK OF ASTM STANDARDS

Iron and Steel Products

VOLUME
01.03 Steel — Plate, Sheet, Strip, Wire

Includes standards of the following committee:

A-1 on Steel, Stainless Steel, and Related Alloys

Publication Code Number (PCN): 01-010384-02

 1916 Race Street/Philadelphia, PA 19103

Editorial Staff

Managing Standards Editor:
  Roberta A. Priemon
Associate Standards Editors:
  Joan L. Cornillot
  Dorothy F. Savini
Senior Assistant Editors:
  Sharon R. Powell
  Sharon L. Kauffman
  Margaret C. Sherlock
Assistant Editors:
  Sarajane H. Hart
  Joyce A. Barton
  Kenneth W. O'Brien
  Deborah Paz
  Andrea L. Urian
Editorial Assistants:
  Eileen D. Cannon
  Leigh Anne Loux
  Robin E. Craig
Senior Indexer:
  Harris J. Shupak
Secretary:
  Faith C. McKinney

Vice-President, Publications and Marketing:
  Robert L. Meltzer

Copyright © by AMERICAN SOCIETY FOR TESTING AND MATERIALS 1984
Library of Congress Catalog Card Number: 40-10712

Printed in Easton, MD, U.S.A.

Foreword

ASTM

ASTM, founded in 1898, is a scientific and technical organization formed for "the development of standards on characteristics and performance of materials, products, systems, and services; and the promotion of related knowledge." It is the world's largest source of voluntary consensus standards.

The Society operates through more than 140 main technical committees with 1944 subcommittees. These committees function in prescribed fields under regulations that ensure balanced representation among producers, users, and general interest participants.

The Society currently has 28,800 active members, of whom approximately 17,400 serve as technical experts on committees, representing 82,000 units of participation.

Membership in the Society is open to all concerned with the fields in which ASTM is active. A membership application with detailed information may be found at the back of this volume. Additional information may be obtained from Member, Committee, and Sales Services, ASTM, 1916 Race St., Philadelphia, Pa. 19103.

1984 Annual Book of ASTM Standards

The 1984 *Annual Book of ASTM Standards* consists of 66 volumes, divided among 16 sections, of which this volume is one. It contains formally approved ASTM standard specifications, test methods, classifications, definitions, and practices, and related material such as proposals. These terms are defined as follows in the Regulations Governing ASTM Technical Committees:

Categories:

standard—as used in ASTM, a document that has been developed and established within the consensus principles of the Society and that meets the approval requirements of ASTM procedures and regulations.

Discussion—The term "standard" serves in ASTM as an adjective in the title of documents, such as test methods or specifications, to connote specified consensus and approval. The various types of standard documents are based on the needs and usages as prescribed by the technical committees of the Society.

proposal—a document that has been approved by the sponsoring committee for publication for information and comment prior to its consideration for adoption as a standard.

Discussion—Complete balloting procedures are not required for proposals and no designation is assigned to them.

emergency standard—a document published by the Society to meet a demand for more rapid issuance of a specific standard document.

Discussion—The Executive Subcommittee of the sponsoring committee must recommend the publishing of an emergency standard and the Committee on Standards must concur in the recommendation. Emergency standards are not full consensus documents because they are not submitted to Society ballot.

Types:

The various types of ASTM documents are to provide a flexibility of form, communication, and

FOREWORD

usage for both the technical committees and the myriad users of ASTM documents. The type of ASTM document that is developed and titled is based on the technical content and intended use, not on the degree of consensus achieved. The three categories of ASTM documents (standard, emergency standard, and proposal) can be of the following forms and types:

classification—a systematic arrangement or division of materials, products, systems, or services into groups based on similar characteristics such as origin, composition, properties, or use.

guide—a series of options or instructions that do not recommend a specific course of action.

Discussion—Whereas a practice prescribes a general usage principle, a guide only suggests an approach. The purpose of a guide is to offer guidance, based on a consensus of viewpoints, but not to establish a fixed procedure. A guide is intended to increase the awareness of the user to available techniques in a given subject area and to provide information from which subsequent evaluation and standardization can be derived.

practice—a definitive procedure for performing one or more specific operations or functions that does not produce a test result. (Compare *test method*.)

Discussion—A practice is not a downgraded test method. Examples of practices include procedures for conducting interlaboratory testing programs or other statistical procedures; for writing statements on sampling or precision and accuracy; and for selection, preparation, application, inspection, necessary precautions for use or disposal, installation, maintenance, and operation of testing equipment.

specification—a precise statement of a set of requirements to be satisfied by a material, product, system, or service that indicates the procedures for determining whether each of the requirements is satisfied.

Discussion—It is desirable to express the requirements numerically in terms of appropriate units together with their limits.

terminology—a document comprising definitions of terms; descriptions of terms; explanations of symbols, abbreviations, or acronyms.

test method—a definitive procedure for the identification, measurement, and evaluation of one or more qualities, characteristics, or properties of a material, product, system, or service that produces a test result. (Compare *practice*.)

A new edition of the Book of Standards is issued annually. Each volume contains all actions approved by the Society at least six months before the issue date. Later actions are included wherever this is possible without delaying the appearance of the volume. In addition, a list of new and revised standards received too late for processing is appended to the end of each table of contents just before going to press. New and revised standards approved by the Society between the annual appearances of any given volume are made available as separate copies. The 1984 edition of the Book of Standards comprises approximately 54,500 pages and includes over 7100 ASTM standards.

Purpose and Use of ASTM Standards

An ASTM standard represents a common viewpoint of those parties concerned with its provisions, namely, producers, users, consumers, and general interest groups. It is intended to aid industry, government agencies, and the general public. The use of an ASTM standard is purely voluntary. It is recognized that, for certain work or in certain regions, ASTM specifications may be either more or less restrictive than needed. The existence of an ASTM standard does not preclude anyone from manufacturing, marketing, or purchasing products, or using products, processes, or procedures not conforming to the standard. Because ASTM standards are subject to periodic review and revision, those who use them are cautioned to obtain the latest revision.

Consideration of Comments on ASTM Standards

An ASTM standard is subject to revision at any time by the responsible technical committee and must be reviewed every five years and if not revised, either reapproved or withdrawn. Your comments

FOREWORD

are invited either for revision of any standard or for additional standards and should be addressed to ASTM Headquarters. Your comments will receive careful consideration at a meeting of the responsible technical committee, which you may attend. If you feel that your comments have not received a fair hearing you should make your views known to the ASTM Committee on Standards, 1916 Race St., Philadelphia, Pa. 19103.

Using the Annual Book of ASTM Standards

The standards are assembled in each volume in alphanumeric sequence of their ASTM designation numbers except for Volumes 11.01, 11.02, and 05.04, which are assembled by subject matter. Volume 06.03 is assembled first by committee, then in alphanumeric sequence. Each volume has two tables of contents—one, a list of the standards in alphanumeric sequence of their ASTM designations; the other, a list of the standards classified according to subject. A subject index of the standards in each volume appears at the back of each volume.

Availability of Individual Standards

Each ASTM standard is available as a separate copy from ASTM. Special quantity prices and discounts for members can be obtained from Sales Service. When ordering, give ASTM standard designation and year of issue, title, quantity desired, and shipping instructions.

Obsolete Editions

This new edition of the *Annual Book of ASTM Standards* makes last year's edition obsolete. Each volume of the *Annual Book of ASTM Standards* is replaced annually because of additions of new standards and significant revisions in existing standards. On the average, about 30 % of each volume is new or revised. For practical purposes, therefore, it is not wise to use obsolete volumes. However, for teaching purposes, these outdated volumes might be useful.

Precautionary Caveat

In January 1983, the Board of Directors approved the inclusion of the following precautionary caveat:

This standard may involve hazardous materials, operations, and equipment. This standard does not purport to address all of the safety problems associated with its use. It is the responsibility of whoever uses this standard to consult and establish appropriate safety and health practices and determine the applicability of regulatory limitations prior to use.

Inclusion of the caveat is required in test methods, specifications (where test methods are detailed other than by reference), practices, and guides. Implementation of the caveat will be phased in as new, revised, and reapproved standards are approved by the Society.

Disclaimer

The American Society for Testing and Materials takes no position respecting the validity of any patent rights asserted in connection with any item mentioned in these standards. Users of these standards are expressly advised that determination of the validity of any such patent rights, and the risk of infringement of such rights, are entirely their own responsibility.

ANNUAL BOOK OF ASTM STANDARDS
Listed by Section and Volume

Section 1—Iron and Steel Products
Volume 01.01 Steel Piping, Tubing, and Fittings
  01.02 Ferrous Castings; Ferroalloys; Shipbuilding
  01.03 Steel Plate, Sheet, Strip, and Wire
  01.04 Structural Steel; Concrete Reinforcing Steel; Pressure Vessel Plate and Forgings; Steel Rails, Wheels, and Tires
  01.05 Steel Bars, Chain, and Springs; Bearing Steel; Steel Forgings
  01.06 Coated Steel Products

Section 2—Nonferrous Metal Products
Volume 02.01 Copper and Copper Alloys
  02.02 Die-Cast Metals; Light Metals and Alloys
  02.03 Electrical Conductors
  02.04 Nonferrous Metals—Nickel, Lead, and Tin Alloys, Precious Metals, Primary Metals; Reactive Metals
  02.05 Metallic and Inorganic Coatings; Metal Powders, Sintered P/M Structural Parts

Section 3—Metals Test Methods and Analytical Procedures
Volume 03.01 Metals—Mechanical Testing; Elevated and Low-Temperature Tests
  03.02 Metal Corrosion, Erosion, and Wear
  03.03 Metallography; Nondestructive Tests
  03.04 Magnetic Properties and Magnetic Materials; Metallic Materials for Thermostats and for Electrical Resistance, Heating, and Contacts
  03.05 Chemical Analysis of Metals; Sampling and Analysis of Metal Bearing Ores
  03.06 Emission Spectroscopy; Surface Analysis

Section 4—Construction
Volume 04.01 Cement; Lime; Gypsum
  04.02 Concrete and Mineral Aggregates
  04.03 Road and Paving Materials; Traveled Surface Characteristics
  04.04 Roofing, Waterproofing, and Bituminous Materials
  04.05 Chemical-Resistant Nonmetallic Materials; Vitrified Clay and Concrete Pipe and Tile; Masonry Mortars and Units; Fiber-Cement Products; Precast Concrete Products
  04.06 Thermal Insulation; Environmental Acoustics
  04.07 Building Seals and Sealants; Fire Standards; Building Constructions
  04.08 Natural Building Stones; Soil and Rock
  04.09 Wood

Section 5—Petroleum Products, Lubricants, and Fossil Fuels
Volume 05.01 Petroleum Products and Lubricants (I) D 56–D 1660
  05.02 Petroleum Products and Lubricants (II) D 1661–D 2896
  05.03 Petroleum Products and Lubricants (III) D 2891–latest; Catalysts
  05.04 Test Methods for Rating Motor, Diesel, and Aviation Fuels
  05.05 Gaseous Fuels; Coal and Coke

Section 6—Paints, Related Coatings, and Aromatics
Volume 06.01 Paint—Tests for Formulated Products and Applied Coatings;
  06.02 Paint—Pigments, Resins and Polymers
  06.03 Paint—Fatty Oils and Acids, Solvents, Miscellaneous; Aromatic Hydrocarbons; Naval Stores

Section 7—Textiles
Volume 07.01 Textiles—Yarns, Fabrics, and General Test Methods
  07.02 Textiles—Fibers, Zippers

LISTED BY SECTION AND VOLUME

Section 8—Plastics
Volume 08.01   Plastics (I) C 177–D 1600
        08.02   Plastics (II) D 1601–D 3099
        08.03   Plastics (III) D 3100–latest
        08.04   Plastic Pipe and Building Products

Section 9—Rubber
Volume 09.01   Rubber, Natural and Synthetic—General Test Methods; Carbon Black
        09.02   Rubber Products, Industrial—Specifications and Related Test Methods; Gaskets; Tires

Section 10—Electrical Insulation and Electronics
Volume 10.01   Electrical Insulation (I)
        10.02   Electrical Insulation (II)
        10.03   Electrical Insulating Liquids and Gases; Electrical Protective Equipment
        10.04   Electronics (I)
        10.05   Electronics (II)

Section 11—Water and Environmental Technology
Volume 11.01   Water (I)
        11.02   Water (II)
        11.03   Atmospheric Analysis; Occupational Health and Safety
        11.04   Pesticides; Resource Recovery; Hazardous Substances and Oil Spill Response; Waste Disposal; Biological Effects and Environmental Fate

Section 12—Nuclear, Solar, and Geothermal Energy
Volume 12.01   Nuclear Energy (I)
        12.02   Nuclear (II), Solar, and Geothermal Energy

Section 13—Medical Devices
Volume 13.01   Medical Devices

Section 14—General Methods and Instrumentation
Volume 14.01   Molecular and Mass Spectroscopy; Chromatography; Resinography; Temperature Measurement; Microscopy; Computerized Systems
        14.02   General Test Methods, Nonmetal; Statistical Methods; Appearance of Materials; Hazard Potential of Chemicals; Particle Size Measurement; Thermal Measurements; Laboratory Apparatus; Metric Practice; Durability of Nonmetallic Materials

Section 15—General Products, Chemical Specialities, and End Use Products
Volume 15.01   Refractories; Manufactured Carbon and Graphite Products; Activated Carbon
        15.02   Glass; Ceramic Whitewares; Porcelain Enamels
        15.03   Space Simulation; Aerospace Materials; High Modulus Fibers and Their Composites
        15.04   Soap; Polishes; Cellulose; Leather; Resilient Floor Coverings
        15.05   Engine Coolants; Halogenated Organic Solvents; Industrial Chemicals
        15.06   Adhesives
        15.07   End Use Products
        15.08   Fasteners
        15.09   Paper; Packaging; Flexible Barrier Materials; Business Copy Products

Section 00—Index
Volume 00.01   Subject and Alphanumeric Index

ANNUAL BOOK OF ASTM STANDARDS
Listed by Subjects

SUBJECT	VOLUME
Acoustics, Environmental	04.06
Activated Carbon	15.01
Adhesives	15.06
Aerosols	15.09
Aerospace Industry Methods	15.03
Aggregates	04.02
Aluminum and Aluminum Alloys	02.02
Amusement Rides and Devices	15.07
Appearance of Materials	14.02
Aromatic Hydrocarbons and Related Chemicals	06.03
Atmospheric Analysis	11.03
Biological Effects and Environmental Fate	11.04
Building Constructions	04.07
Building Seals and Sealants	04.07
Building Stones	04.08
Business Copy Products	15.09
Carbon Black	09.01
Carbon Products, Manufactured	15.01
Cast Iron	01.02
Catalysts	05.03
Cellulose	15.04
Cement	14.01
Hydraulic	14.01
Rubber	09.01
Ceramic Materials	15.02
Ceramic Whitewares	15.02
Ceramics for Electronics	10.04
Porcelain Enamel	15.02
Chemical Analysis of Metals	03.05
Chemical-Resistant Nonmetallic Materials	04.05
Chemicals, Industrial	15.05
Chromatography	14.01
Clay and Concrete Pipe and Tile	04.05
Closures	15.09
Coal and Coke	05.05
Compatibility and Sensitivity of Materials in Oxygen-Enriched Atmospheres	14.02
Computerized Systems	14.01
Concrete and Mineral Aggregates	04.02
Concrete Products, Precast	04.05
Concrete Reinforcing Steel	01.04
Consumer Products	15.07
Coolants, Engine	15.05
Copper and Copper Alloys	02.01
Corrosion, Metal	03.02
Criteria for the Evaluation of Testing and Inspection Agencies	14.02
Die-Cast Metals	02.02
Ductile Iron	01.02

SUBJECT	VOLUME
Durability of Nonmetallic Materials	14.02
Electrical Conductors	02.03
Electrical Contacts	03.04
Electrical Insulating Materials	10.01, 10.02, 10.03
Electrical Protective Equipment for Workers	10.03
Electronics	10.04, 10.05
Erosion and Wear	03.02
Evaluating Testing and Inspection Agencies	14.02
Fasteners	15.08
Fatigue	03.01
Fences	01.06
Ferroalloys	01.02
Ferrous Castings	01.02
Fiber-Cement Products	04.05
Filtration	14.02
Fire Standards	04.07
Flexible Barrier Materials	15.09
Food Service Equipment	15.07
Footwear, Safety and Traction for	15.07
Forensic Sciences	13.01
Fracture Testing	03.01
Gaseous Fuels	05.05
Gaskets	09.02
Glass	15.02
Graphite Products, Manufactured	15.01
Gypsum	04.01
Halogenated Organic Solvents	15.05
Hazardous Substances and Oil Spill Response	11.04
Hazard Potential of Chemicals	14.02
High Modulus Fibers and Composites	15.03
Index (for all parts)	00.01
Industrial Chemicals	15.05
Iron Castings	01.02
Knock Test Manual	05.04
Laboratory Apparatus	14.02
Leather	15.04
Lime	04.01
Magnesium and Magnesium Alloys	02.02
Magnetic Properties	03.04
Malleable Iron	01.02
Masonry Units	04.05
Meat and Poultry	15.07
Medical and Surgical Materials and Devices	13.01
Metallic and Inorganic Coatings	02.05
Metallography	03.03
Metal Powders	02.05
Metals, Chemical Analysis	03.05

LISTED BY SUBJECTS

SUBJECT	VOLUME
Metals, Effect of Temperature on Properties	03.01
Metals, Physical and Mechanical Testing	03.01
Metric Practice	14.02
Microscopy	14.01
Mortars	04.05
Naval Stores	06.03
Nickel and Nickel Alloys	02.04
Nondestructive Testing	03.03
Nonferrous Metals, General	02.04
Nonmetals, General Test Methods	14.02
Nuclear Materials	12.01, 12.02
Occupational Health and Safety	11.03
Oil Spill Response, Hazardous Substances	11.04
Ores, Metal Bearing, Sampling and Analysis	03.05
Packaging	15.09
Paint and Related Coatings and Materials:	
Fatty Oils and Acids, Solvents, Miscellaneous	06.03
Pigments, Resins and Polymers	06.02
Tests for Formulated Products and Applied Coatings	06.01
Paper	15.09
Particle Size Measurement	14.02
Pesticides	11.04
Petroleum Products and Lubricants	05.01, 05.02, 05.03, 05.04
Plastics	08.01, 08.02, 08.03
Plastic Pipe and Building Products	08.04
Polishes	15.04
Porcelain Enamel	15.02
Pressure Vessel Plate and Forgings	01.04
Products Liability Litigation, Technical Aspects of	14.02
Protective Clothing	15.07
Protective Coating and Lining Work for Power Generation Facilities	06.01
Protective Equipment, Electrical, for Workers	10.03
Radioisotopes and Radiation Effects	12.02
Reactive and Refractory Metals	02.04
Refractories	15.01
Resilient Floor Coverings	15.04
Resinography	14.01
Resource Recovery	11.04
Road and Paving Materials	04.03
Roofing, Waterproofing, and Bituminous Materials	04.04
Rubber	09.01, 09.02
Security Systems and Equipment	15.07
Sensory Evaluation of Materials and Products	15.07
Shipbuilding	01.02
Sintered P/M Structural Parts	02.05
Soap	15.04
Soil and Rock	04.08
Solar Energy Conversion	12.02
Space Simulation	15.03
Spectroscopy	03.06, 14.01
Sports Equipment and Facilities	15.07
Statistical Methods	14.02
Steel:	
Bars	01.05
Bearing Steel	01.05
Bolting	01.01, 15.08
Castings	01.02
Chain	01.05
Concrete Reinforcing	01.04
Fasteners	15.08
Forgings	01.04, 01.05
Galvanized	01.06
Piping, Tubing, and Fittings	01.01
Plate, Sheet, and Strip	01.03
Pressure Vessel Plate and Forgings	01.04
Rails, Wheels, and Tires	01.04
Springs	01.05
Stainless Steel	01.01, 01.02, 01.03, 01.04, 01.05
Structural Steel	01.04
Wire	01.03
Surface Analysis	03.06
Surgical Materials and Devices	13.01
Temperature Measurement	14.01
Textiles:	
Fabrics	07.01
Felts	07.01
Fibers	07.02
Floor Coverings	07.01
General Test Methods	07.01
Tire Cords	07.01
Yarns	07.01
Zippers	07.02
Thermal Measurements	14.02
Thermal Insulation	04.06
Thermocouples	14.01
Thermostats for Electrical Resistance, Heating, and Contacts	03.04
Tires	09.02
Traveled Surface Characteristics	04.03
Vacuum Cleaners	15.07
Vitrified Clay Pipe	04.05
Waste Disposal	11.04
Water	11.01, 11.02
Wood	04.09

Contents

1984 ANNUAL BOOK OF ASTM STANDARDS, VOLUME 01.03

Standards Relating to Steel Plate, Sheet, Strip, and Wire

A complete Subject Index begins on p. 737

Listed below are those standards included in this book and those standards that appeared previously that have been discontinued within the past five years. Since the standards in this book are arranged in alphanumeric sequence, no page numbers are given in this contents.

In the serial designations prefixed to the following titles, the number following the dash indicates the year of adoption as standard or, in the case of revision, the year of last revision. Thus, standards adopted or revised during the year 1984 have as their final number, 84. A letter following this number indicates more than one revision during that year, that is 84a indicates the second revision in 1984, 84b, the third revision, etc. Standards that have been reapproved without change are indicated by the year of last reapproval in parentheses as part of the designation number, for example, (1984). A superscript epsilon indicates an editorial change since the last revision or reapproval—$\epsilon1$ for the first change, $\epsilon2$ for the second change, etc.

§A 109 – 83	Specification for Steel, Carbon, Cold-Rolled Strip
§A 109M – 77	Specification for Steel, Carbon, Cold-Rolled Strip [Metric]
A 164 – 71	Specification for Electrodeposited Coatings of Zinc on Steel (Discontinued 1980‡—Replaced by B 633)
§A 167 – 82	Specification for Stainless and Heat-Resisting Chromium-Nickel Steel Plate, Sheet, and Strip
§A 176 – 83	Specification for Stainless and Heat-Resisting Chromium Steel Plate, Sheet, and Strip
§A 177 – 82	Specification for High-Strength Stainless and Heat-Resisting Chromium-Nickel Steel Sheet and Strip
§A 227/A 227M – 83	Specification for Steel Wire, Cold-Drawn for Mechanical Springs
§A 228/A 228M – 83	Specification for Steel Wire, Music Spring Quality
§A 229/A 229M – 83	Specification for Steel Wire, Oil-Tempered for Mechanical Springs
A 230/A 230M – 83	Specification for Steel Wire, Oil-Tempered Carbon Valve Spring Quality
A 231/A 231M – 83	Specification for Chromium-Vanadium Alloy Steel Spring Wire
A 232 – 77	Specification for Chromium-Vanadium Alloy Steel Valve Spring Quality Wire
§†A 240 – 83	Specification for Heat-Resisting Chromium and Chromium-Nickel Stainless Steel Plate, Sheet, and Strip for Pressure Vessels
§A 262 – 81	Recommended Practices for Detecting Susceptibility to Intergranular Attack in Stainless Steels
†A 263 – 82	Specification for Corrosion-Resisting Chromium Steel Clad Plate, Sheet, and Strip
†A 264 – 82	Specification for Stainless Chromium-Nickel Steel Clad Plate, Sheet, and Strip
†A 265 – 81a	Specification for Nickel and Nickel-Base Alloy Clad Steel Plate
§A 313 – 81	Specification for Chromium-Nickel Stainless and Heat-Resisting Steel Spring Wire
A 345 – 75 (1979)	Specification for Flat-Rolled Electrical Steels for Magnetic Applications
§A 366 – 72 (1979)	Specification for Steel, Carbon, Cold-Rolled Sheet, Commercial Quality
A 368 – 82	Specification for Stainless and Heat-Resisting Steel Wire Strand

§ Approved for use by agencies of the Department of Defense and, if indicated on the standard, replaces corresponding Federal or Military document.

† Adopted by or under consideration for adoption by the Boiler and Pressure Vessel Committee of the American Society of Mechanical Engineers. The ASME Boiler and Pressure Vessel Code Specifications are identical with or based upon these ASTM Specifications.

‡ Although this standard has been officially withdrawn from Society approval, a brief description is included for information only.

CONTENTS

§†A 370 – 77ϵ2	Methods and Definitions for Mechanical Testing of Steel Products
A 380 – 78	Recommended Practice for Cleaning and Descaling Stainless Steel Parts, Equipment, and Systems
A 401 – 77 (1983)	Specification for Chromium-Silicon Alloy Steel Spring Wire
A 407 – 77 (1983)	Specification for Steel Wire, Cold-Drawn, for Coiled-Type Springs
A 407M – 80	Specification for Steel Wire, Cold-Drawn, for Coiled-Type Springs [Metric]
§†A 412 – 82	Specification for Stainless and Heat-Resisting Chromium-Nickel-Manganese Steel Plate, Sheet, and Strip
†A 414/A 414M – 83	Specification for Carbon Steel Sheet for Pressure Vessels
A 417 – 74 (1980)ϵ1	Specification for Steel Wire, Cold-Drawn, for Zig-Zag, Square-Formed, and Sinuous-Type Upholstery Spring Units
A 417M – 80	Specification for Steel Wire, Cold-Drawn, for Zig-Zag Square-Formed and Sinuous-Type Upholstery Spring Units [Metric]
§A 424 – 80	Specification for Steel Sheet for Porcelain Enameling
A 457 – 82	Specification for Hot-Worked, Hot-Cold-Worked, and Cold-Worked Alloy Steel Plate, Sheet, and Strip for High Strength at Elevated Temperatures
A 478 – 82	Specification for Chromium-Nickel Stainless and Heat-Resisting Steel Weaving Wire
§†A 480/A 480M – 83a	Specification for General Requirements for Flat-Rolled Stainless and Heat-Resisting Steel Plate, Sheet, and Strip
§A 492 – 82	Specification for Stainless and Heat-Resisting Steel Rope Wire
§A 493 – 82a	Specification for Stainless and Heat-Resisting Steel for Cold Heading and Cold Forging—Bar and Wire
§A 505 – 78	Specification for General Requirements for Steel Sheet and Strip, Alloy, Hot-Rolled and Cold-Rolled
§A 506 – 73 (1980)	Specification for Steel Sheet and Strip, Alloy, Hot-Rolled and Cold-Rolled, Regular Quality
§A 507 – 73 (1980)	Specification for Steel Sheet and Strip, Alloy, Hot-Rolled and Cold-Rolled, Drawing Quality
§A 510 – 82	Specification for General Requirements for Wire Rods and Coarse Round Wire, Carbon Steel
§A 510M – 82	Specification for General Requirements for Wire Rods and Coarse Round Wire, Carbon Steel [Metric]
A 544 – 82	Specification for Steel Wire, Carbon, Scrapless Nut Quality
§A 545 – 82	Specification for Steel Wire, Carbon, Cold-Heading Quality, for Machine Screws
§A 546 – 82	Specification for Steel Wire, Medium-High-Carbon, Cold-Heading Quality, for Hexagon-Head Bolts
§A 547 – 82	Specification for Steel Wire, Alloy, Cold-Heading Quality, for Hexagon-Head Bolts
§A 548 – 82	Specification for Steel Wire, Carbon, Cold-Heading Quality, for Tapping or Sheet Metal Screws
§A 549 – 82	Specification for Steel Wire, Carbon, Cold-Heading Quality, for Wood Screws
§A 555 – 80	Specification for General Requirements for Stainless and Heat-Resisting Steel Wire
§A 568 – 83	Specification for General Requirements for Steel, Carbon and High-Strength Low-Alloy Hot-Rolled Sheet and Cold-Rolled Sheet
§A 568M – 77	Specification for General Requirements for Steel, Carbon and High-Strength Low-Alloy Hot-Rolled Sheet and Cold-Rolled Sheet [Metric]
§A 569 – 72 (1979)	Specification for Steel, Carbon (0.15 Maximum, Percent), Hot-Rolled Sheet and Strip, Commercial Quality
§A 570 – 79	Specification for Hot-Rolled Carbon Steel Sheet and Strip, Structural Quality
§A 580 – 83	Specification for Stainless and Heat-Resisting Steel Wire
§A 581 – 80	Specification for Free-Machining Stainless and Heat-Resisting Steel Wire
A 604 – 77 (1982)	Method for Macroetch Testing of Consumable Electrode Remelted Steel Bars and Billets
§A 606 – 75 (1981)	Specification for Steel Sheet and Strip, Hot-Rolled and Cold-Rolled, High-Strength, Low-Alloy, with Improved Atmospheric Corrosion Resistance
A 607 – 83	Specification for Steel Sheet and Strip, Hot-Rolled and Cold-Rolled, High-Strength, Low-Alloy Columbium and/or Vanadium
§A 611 – 82	Specification for Steel, Cold-Rolled Sheet, Carbon, Structural
§A 619/A 619M – 82	Specification for Steel Sheet, Carbon, Cold-Rolled, Drawing Quality
§A 620/A 620M – 82	Specification for Steel Sheet, Carbon, Cold-Rolled, Drawing Quality, Special Killed
§A 621/A 621M – 82	Specification for Steel Sheet and Strip, Carbon, Hot-Rolled, Drawing Quality
§A 622/A 622M – 82	Specification for Steel Sheet and Strip, Carbon, Hot-Rolled, Drawing Quality, Special Killed
A 628 – 74	Specification for Tool-Resisting Composite Steel Plates for Security Applications (Discontinued 1983‡)
§A 635 – 81	Specification for Hot-Rolled Carbon Steel Sheet and Strip, Commercial Quality, Heavy-Thickness Coils (Formerly Plate)

CONTENTS

A 635M – 81	Specification for Hot-Rolled Carbon Steel Sheet and Strip, Commercial Quality, Heavy-Thickness Coils (Formerly Plate) [Metric]
§A 659 – 72 (1979)	Specification for Steel, Carbon (0.16 Maximum to 0.25 Maximum, Percent), Hot-Rolled Sheet and Strip, Commercial Quality
§A 666 – 82	Specification for Austenitic Stainless Steel, Sheet, Strip, Plate, and Flat Bar for Structural Applications
A 679 – 77 (1983)	Specification for Steel Wire, High Tensile Strength, Hard-Drawn, for Mechanical Springs
§A 680/A 680M – 81	Specification for Steel, High-Carbon, Strip, Cold-Rolled Hard, Untempered Quality
§A 682 – 77 (1983)	Specification for General Requirements for Steel, High-Carbon, Strip, Cold-Rolled, Spring Quality
§A 682M – 77 (1983)	Specification for General Requirements for Steel, High-Carbon, Strip, Cold-Rolled, Spring Quality, [Metric]
§A 684/A 684M – 81	Specification for Steel, High-Carbon, Strip, Cold-Rolled Soft, Untempered Quality
§A 693 – 82	Specification for Precipitation-Hardening Stainless and Heat-Resisting Steel Plate, Sheet, and Strip
§A 700 – 81	Recommended Practices for Packaging, Marking, and Loading Methods for Steel Products for Domestic Shipment
A 702 – 81	Specification for Steel Fence Posts and Assemblies, Hot-Wrought
A 708 – 79	Recommended Practice for Detection of Susceptibility to Intergranular Corrosion in Severely Sensitized Austenitic Stainless Steel
A 713 – 77 (1983)	Specification for Steel Wire, High-Carbon Spring, for Heat-Treated Components
A 715 – 81[ϵ1]	Specification for Steel Sheet and Strip, Hot-Rolled, High-Strength, Low-Alloy, with Improved Formability
A 749 – 83	Specification for General Requirements for Steel, Carbon and High-Strength, Low-Alloy Hot-Rolled Strip
§A 749M – 83	Specification for General Requirements for Steel, Carbon and High-Strength, Low-Alloy Hot-Rolled Strip [Metric]
A 751 – 82	Methods, Practices, and Definitions for Chemical Analysis of Steel Products
A 752 – 83	Specification for General Requirements for Wire Rods and Coarse Round Wire, Alloy Steel
A 752M – 83	Specification for General Requirements for Wire Rods and Coarse Round Wire, Alloy Steel [Metric]
A 763 – 83	Practices for Detecting Susceptibility to Intergranular Attack in Ferritic Stainless Steels
A 764 – 79	Specification for Steel Wire, Carbon, Drawn, Galvanized and Galvanized at Size for Mechanical Springs
A 793 – 81	Specification for Rolled Floor Plate, Stainless Steel
A 794 – 82	Specification for Steel, Carbon (0.16 Maximum to 0.25 Maximum %), Cold-Rolled Sheet, Commercial Quality)
A 805 – 82	Specification for Cold-Rolled Carbon Steel Flat Wire
A 812/A 812M – 83	Specification for Steel Sheet, Hot-Rolled, High-Strength, Low-Alloy for Welded Layered Pressure Vessels
B 670 – 80	Specification for Precipitation-Hardening Nickel Alloy (UNS N07718) Plate, Sheet, and Strip for High-Temperature Service
§E 437 – 80	Specification for Industrial Wire Cloth and Screens (Square Opening Series)
§E 454 – 80	Specification for Industrial Perforated Plate and Screens (Square Opening Series)
E 527 – 83	Practice for Numbering Metals and Alloys (UNS)
§E 674 – 80	Specification for Industrial Perforated Plate and Screens (Round Opening Series)

RELATED MATERIAL

	PAGE
ASTM Metric Practice (Excerpts) (E 380)	719
Index	737
ASTM Membership Application Blank	

List by Subjects

1984 ANNUAL BOOK OF ASTM STANDARDS, VOLUME 01.03

STEEL PLATE, SHEET, STRIP, AND WIRE

A complete Subject Index begins on p. 737

Since the standards in this book are arranged in alphanumeric sequence, no page numbers are given in this list by subjects.
The standards listed in italics are related documents included for information only and do not appear in this volume.

STEEL PLATE, SHEET, STRIP

Specifications for:

§A 666 – 82	Austenitic Stainless Steel, Sheet, Strip, Plate, and Flat Bar for Structural Applications
†A 414/A 414M – 83	Carbon Steel Sheet for Pressure Vessels
†A 263 – 82	Corrosion-Resisting Chromium Steel Clad Plate, Sheet, and Strip
A 345 – 75 (1979)	Flat-Rolled Electrical Steels for Magnetic Applications
§†A 480/A 480M – 83	Flat-Rolled Stainless and Heat-Resisting Steel Plate, Sheet, and Strip, General Requirements for
§†A 240 – 83	Heat-Resisting Chromium and Chromium-Nickel Stainless Steel Plate, Sheet, and Strip for Pressure Vessels
§A 177 – 82	High-Strength Stainless and Heat-Resisting Chromium-Nickel Steel Sheet and Strip
§A 635 – 81	Hot-Rolled Carbon Steel Sheet and Strip, Commercial Quality, Heavy-Thickness Coils (Formerly Plate)
A 635M – 81	Hot-Rolled Carbon Steel Sheet and Strip, Commercial Quality, Heavy-Thickness Coils (Formerly Plate) [Metric]
§A 570 – 79	Hot-Rolled Carbon Steel Sheet and Strip, Structural Quality
A 457 – 82	Hot-Worked, Hot-Cold-Worked, and Cold-Worked Alloy Steel Plate, Sheet, and Strip for High Strength at Elevated Temperatures
†A 265 – 81a	Nickel and Nickel-Base Alloy Clad Steel Plate
B 670 – 80	Precipitation-Hardening Nickel Alloy (UNS N07718) Plate, Sheet, and Strip for High-Temperature Service
§A 693 – 82	Precipitation-Hardening Stainless and Heat-Resisting Steel Plate, Sheet, and Strip
§†A 412 – 82	Stainless and Heat-Resisting Chromium-Nickel-Manganese Steel Plate, Sheet, and Strip
§A 167 – 82	Stainless and Heat-Resisting Chromium-Nickel Steel Plate, Sheet, and Strip
§A 176 – 83	Stainless and Heat-Resisting Chromium Steel Plate, Sheet, and Strip
†A 264 – 82	Stainless Chromium-Nickel Steel Clad Plate, Sheet, and Strip
§A 568 – 83	Steel, Carbon and High-Strength Low-Alloy Hot-Rolled Sheet and Cold-Rolled Sheet, General Requirements
§A 568M – 77	Steel, Carbon and High-Strength Low-Alloy Hot-Rolled Sheet and Cold-Rolled Sheet, General Requirements [Metric]
A 749 – 83	Steel, Carbon and High-Strength, Low-Alloy Hot-Rolled Strip, General Requirements for
§A 749M – 83	Steel, Carbon and High-Strength, Low-Alloy Hot-Rolled Strip, General Requirements for [Metric]
§A 366 – 72 (1979)	Steel, Carbon, Cold-Rolled Sheet, Commercial Quality

§ Approved for use by agencies of the Department of Defense and, if indicated on the standard, replaces corresponding Federal or Military documents.

† Adopted by or under consideration for adoption by the Boiler and Pressure Vessel Committee of the American Society of Mechanical Engineers. The ASME Boiler and Pressure Vessel Code Specifications are identical with or based upon these ASTM Specifications.

LIST BY SUBJECTS

Specifications for:

§A 109 – 83	Steel, Carbon, Cold-Rolled Strip
§A 109M – 77	Steel, Carbon, Cold-Rolled Strip [Metric]
§A 569 – 72 (1979)	Steel, Carbon (0.15 Maximum, Percent), Hot-Rolled Sheet and Strip, Commercial Quality
A 794 – 82	Steel, Carbon (0.16 Maximum to 0.25 Maximum %), Cold-Rolled Sheet, Commercial Quality
§A 659 – 72 (1979)	Steel, Carbon (0.16 Maximum to 0.25 Maximum, Percent), Hot-Rolled Sheet and Strip, Commercial Quality
§A 680/A 680M – 81	Steel, High-Carbon, Strip, Cold-Rolled Hard, Untempered Quality
§A 684/A 684M – 81	Steel, High-Carbon, Strip, Cold-Rolled Soft, Untempered Quality
§A 682 – 77 (1983)	Steel, High-Carbon, Strip, Cold-Rolled, Spring Quality, General Requirements for
A 682M – 77 (1983)	Steel, High-Carbon, Strip, Cold-Rolled, Spring Quality, General Requirements for [Metric]
§A 611 – 82	Steel, Cold-Rolled Sheet, Carbon, Structural
A 702 – 81	Steel Fence Posts and Assemblies, Hot-Wrought
§A 507 – 73 (1980)	Steel Sheet and Strip, Alloy, Hot-Rolled and Cold-Rolled, Drawing Quality
§A 505 – 78	Steel Sheet and Strip, Alloy, Hot-Rolled and Cold-Rolled, General Requirements for
§A 506 – 73 (1980)	Steel Sheet and Strip, Alloy, Hot-Rolled and Cold-Rolled, Regular Quality
§A 621/A 621M – 82	Steel Sheet and Strip, Carbon, Hot-Rolled, Drawing Quality
§A 622/A 622M – 82	Steel Sheet and Strip, Carbon, Hot-Rolled, Drawing Quality, Special Killed
A 607 – 83	Steel Sheet and Strip, Hot-Rolled and Cold-Rolled, High-Strength, Low-Alloy Columbium and/or Vanadium
§A 606 – 75 (1981)	Steel Sheet and Strip, Hot-Rolled and Cold-Rolled, High-Strength, Low-Alloy, with Improved Atmospheric Corrosion Resistance
A 715 – 81[c1]	Steel Sheet and Strip, Hot-Rolled, High-Strength, Low-Alloy, with Improved Formability
§A 619/A 619M – 82	Steel Sheet, Carbon, Cold-Rolled, Drawing Quality
§A 620/A 620M – 82	Steel Sheet, Carbon, Cold-Rolled, Drawing Quality, Special Killed
§A 424 – 80	Steel Sheet for Porcelain Enameling
A 812/A 812M – 83	Steel Sheet, Hot-Rolled, High-Strength, Low-Alloy for Welded Layered Pressure Vessels
A 793 – 81	Steel, Stainless, for Rolled Floor Plate
A 284 – 81	Low and Intermediate Tensile Strength Carbon-Silicon Steel Plates for Machine Parts and General Construction (see Vol 01.04)
A 283 – 81	Low and Intermediate Tensile Strength Carbon Steel Plates, Shapes, and Bars (see Vol 01.04)
F 56 – 82	Stainless Steel Sheet and Strip for Surgical Implants (see Vol 13.01)
F 139 – 82	Stainless Steel Sheet and Strip for Surgical Implants (Special Quality) (see Vol 13.01)
A 328 – 81	Steel Sheet Piling (see Vol. 01.04)

STEEL BOLTING MATERIALS

The standards covering steel bolting materials are published in Volumes 01.01 and 15.08 of the Annual Book of ASTM Standards

STEEL FOR CONCRETE REINFORCEMENT AND PRESTRESSED CONCRETE

The standards covering steel bar and wire used for concrete reinforcement and prestressed concrete are published in Volume 01.04 of the Annual Book of ASTM Standards

STEEL WIRE

Specifications for:

§A 313 – 81	Chromium-Nickel Stainless and Heat-Resisting Steel Spring Wire
A 478 – 82	Chromium-Nickel Stainless and Heat-Resisting Steel Weaving Wire
A 401 – 77 (1983)	Chromium-Silicon Alloy Steel Spring Wire
A 231/A 231M – 83	Chromium-Vanadium Alloy Steel Spring Wire
A 232 – 77	Chromium-Vanadium Alloy Steel Valve Spring Quality Wire
§A 581 – 80	Free-Machining Stainless and Heat-Resisting Steel Wire
§A 493 – 82a	Stainless and Heat-Resisting Steel for Cold Heading and Cold Forging—Bar and Wire
§A 492 – 82	Stainless and Heat-Resisting Steel Rope Wire
§A 580 – 83	Stainless and Heat-Resisting Steel Wire
§A 555 – 80	Stainless and Heat-Resisting Steel Wire, General Requirements for
A 368 – 82	Stainless and Heat-Resisting Steel Wire Strand
A 805 – 82	Steel Flat Wire, Cold-Rolled Carbon
§A 547 – 82	Steel Wire, Alloy, Cold-Heading Quality, for Hexagon-Head Bolts

LIST BY SUBJECTS

Specifications for:

§A 545 – 82	Steel Wire, Carbon, Cold-Heading Quality, for Machine Screws
A 548 – 82	Steel Wire, Carbon, Cold-Heading Quality, for Tapping or Sheet Metal Screws
§A 549 – 82	Steel Wire, Carbon, Cold-Heading Quality, for Wood Screws
A 764 – 79	Steel Wire, Carbon, Drawn Galvanized at Size for Mechanical Springs
A 544 – 82	Steel Wire, Carbon, Scrapless Nut Quality
A 407 – 77 (1983)	Steel Wire, Cold-Drawn, for Coiled-Type Springs
A 407M – 80	Steel Wire, Cold-Drawn, for Coiled-Type Springs [Metric]
§A 227/A 227M – 83	Steel Wire, Cold-Drawn for Mechanical Springs
A 417 – 74 (1980)[e1]	Steel Wire, Cold-Drawn, for Zig-Zag, Square-Formed, and Sinuous-Type Upholstery Spring Units
A 417M – 80	Steel Wire, Cold-Drawn, for Zig-Zag Square-Formed and Sinuous-Type Upholstery Spring Units [Metric]
A 713 – 77 (1983)	Steel Wire, High-Carbon Spring, for Heat-Treated Components
A 679 – 77 (1983)	Steel Wire, High Tensile Strength, Hard-Drawn, for Mechanical Springs
§A 546 – 82	Steel Wire, Medium-High-Carbon, Cold Heading Quality, for Hexagon-Head Bolts
§A 228/A 228M – 83	Steel Wire, Music Spring Quality
A 230/A 230M – 83	Steel Wire, Oil-Tempered Carbon Valve Spring Quality
§A 229/A 229M – 83	Steel Wire, Oil-Tempered for Mechanical Springs
A 752 – 83	Wire Rods and Coarse Round Wire, Alloy Steel, General Requirements for
A 752M – 83	Wire Rods and Course Round Wire, Alloy Steel, General Requirements for [Metric]
§A 510 – 82	Wire Rods and Coarse Round Wire, Carbon Steel, General Requirements for
§A 510M – 82	Wire Rods and Coarse Round Wire, Carbon Steel, General Requirements for [Metric]
B 502 – 70 (1980)	*Aluminum-Clad Steel Core Wire for Aluminum Conductors, Aluminum-Clad Steel Reinforced (see Vol 02.03)*
B 416 – 81	*Concentric-Lay-Stranded Aluminum-Clad Steel Conductors (see Vol 02.03)*
B 415 – 81	*Hard-Drawn Aluminum-Clad Steel Wire (see Vol 02.03)*
B 501 – 69 (1979)	*Silver-Coated Copper-Clad Steel Wire for Electronic Application (see Vol 02.03)*
F 55 – 82	*Stainless Steel Bar and Wire for Surgical Implants (see Vol 13.01)*
F 138 – 82	*Stainless Steel Bars and Wire for Surgical Implants (Special Quality) (see Vol 13.01)*
A 421 – 80	*Uncoated Stress-Relieved Steel Wire for Prestressed Concrete (see Vol 01.04)*
A 416 – 80	*Uncoated Seven-Wire Stress-Relieved Steel Strand for Prestressed Concrete (see Vol 01.04)*
A 185 – 79	*Welded Steel Wire Fabric for Concrete Reinforcement (see Vol 01.04)*
B 500 – 72 (1982)	*Zinc-Coated (Galvanized) and Aluminum-Coated (Aluminized) Stranded Steel Core for Aluminum Conductors, Steel Reinforced (ACSR) (see Vol 02.03)*

Metallic Coated Steel Products

The standards covering metallic coated steel products (including galvanized and aluminized steel, tin mill products, and other coated steel products) are published in Volume 01.06 of the Annual Book of ASTM Standards.

INDUSTRIAL SIZING SCREENS

Specifications for:

§E 674 – 80	Industrial Perforated Plate and Screens (Round Opening Series)
§E 454 – 80	Industrial Perforated Plate and Screens (Square Opening Series)
§E 437 – 80	Industrial Wire Cloth and Screens (Square Opening Series)

SURGICAL IMPLANTS

Specifications for:

F 55 – 82	*Stainless Steel Bar and Wire for Surgical Implants (see Vol 13.01)*
F 138 – 82	*Stainless Steel Bars and Wire for Surgical Implants (Special Quality) (see Vol 13.01)*
F 56 – 82	*Stainless Steel Sheet and Strip for Surgical Implants (see Vol 13.01)*
F 139 – 82	*Stainless Steel Sheet and Strip for Surgical Implants (Special Quality) (see Vol 13.01)*

GENERAL TEST METHODS

Test Methods for:

A 751 – 82	Chemical Analysis of Steel Products, Methods, Practices, and Definitions for
A 604 – 77 (1982)	Macroetch Testing of Consumable Electrode Remelted Steel Bars and Billets
§†A 370 – 77[e2]	Mechanical Testing of Steel Products, Methods and Definitions for

LIST BY SUBJECTS

Test Methods for:

E 112 – 82	Average Grain Size of Metals (see Vol 03.03)
E 10 – 78	Brinell Hardness of Metallic Materials (see Vol 03.01)
A 255 – 67 (1979)	End-Quench Test for Hardenability of Steel (see Vol 01.05)
E 140 – 83	Hardness Conversion Tables for Metals (Relationship Between Brinell Hardness, Vickers Hardness, Rockwell Hardness, Rockwell Superficial Hardness, and Knoop Hardness) (see Vol 03.01)
E 23 – 82	Notched Bar Impact Testing of Metallic Materials (see Vol 03.01)
E 3 – 80	Preparation of Metallographic Specimens (see Vol 03.03)
E 59 – 78 (1982)	Sampling Steel and Iron for Determination of Chemical Composition (see Vol 03.05)
E 8 – 83	Tension Testing of Metallic Materials (see Vol 03.01)

Practices for:

A 380 – 78	Cleaning and Descaling Stainless Steel Parts, Equipment, and Systems
A 763 – 83	Detecting Susceptibility to Intergranular Attack in Ferritic Stainless Steels
§A 262 – 81	Detecting Susceptibility to Intergranular Attack in Stainless Steels
A 708 – 79	Detection of Susceptibility to Intergranular Corrosion in Severely Sensitized Austenitic Stainless Steel
E 527 – 83	Numbering Metals and Alloys (UNS)
§A 700 – 81	Packaging, Marking, and Loading Methods for Steel Products for Domestic Shipment
E 178 – 80	Dealing with Outlying Observations (see Vol 14.02)
E 709 – 80	Magnetic Particle Examination (see Vol 03.03)
A 400 – 69 (1982)	Selection of Steel Bar Compositions According to Section (see Vol 01.05)
E 177 – 71 (1980)	Use of the Terms Precision and Accuracy as Applied to Measurement of a Property of a Material (see Vol 14.02)

Definitions of Terms Relating to:

E 44 – 83	Heat Treatment of Metals (see Vol 03.03)
E 6 – 83	Methods of Mechanical Testing (see Vol 03.01)

METRIC PRACTICE

Standard for:

§E 380 – 82	Metric Practice (Excerpts) (see Related Material section)

Designation: A 109 – 83

Standard Specification for
STEEL, CARBON, COLD-ROLLED STRIP[1]

This standard is issued under the fixed designation A 109; the number immediately following the designation indicates the year of original adoption or, in the case of revision, the year of last revision. A number in parentheses indicates the year of last reapproval. A superscript epsilon (ϵ) indicates an editorial change since the last revision or reapproval.

This specification has been approved for use by agencies of the Departments of Defense and for listing in the DoD Index of Specifications and Standards.

NOTE—Table 10 was corrected editorially and the designation changed on Aug. 30, 1983.

1. Scope

1.1 This specification covers cold-rolled carbon steel strip in cut lengths or coils, furnished to closer tolerances than cold-rolled carbon steel sheet, with specific temper, with specific edge or specific finish, and in sizes as follows:

Width, in.	Thickness, in.
Over ½ to 23¹⁵⁄₁₆	0.2499 and under

1.2 Cold-rolled strip is produced with a maximum of the specified carbon not exceeding 0.25 percent.

1.3 This specification does not include the product in narrow widths known as cold-rolled sheet, slit from wider widths, Specification A 568 for General Requirements for Steel, Carbon and High-Strength Low-Alloy Hot-Rolled Sheet and Cold-Rolled Sheet,[2] nor does it include cold-rolled carbon spring steel.

NOTE 1—A complete metric companion to Specification A 109 has been developed—A 109M; therefore, no metric equivalents are presented in this specification.

2. Applicable Documents

2.1 *ASTM Standards:*
A 370 Methods and Definitions for Mechanical Testing of Steel Products[2]
A 700 Practices for Packaging, Marking, and Loading Methods for Steel Products for Domestic Shipment[2]
E 8 Methods for Tension Testing of Metallic Materials[3]
E 30 Methods for Chemical Analysis of Steel, Cast Iron, Open-Hearth Iron and Wrought Iron[4]

3. Descriptions of Terms

3.1 *Carbon Steel* is the designation for steel when no minimum content is specified or required for aluminum, chromium, cobalt, columbium, molybdenum, nickel, titanium, tungsten, vanadium, zirconium or any other element added to obtain a desired alloying effect; when the specified minimum for copper does not exceed 0.40 percent or when the maximum content specified for any of the following elements does not exceed the percentage noted: managanese 1.65, silicon 0.60, or copper 0.60.

3.1.1 In all carbon steels small quantities of certain residual elements unavoidably retained from raw materials are sometimes found which are not specified or required, such as copper, nickel, molybdenum, chromium, etc. These elements are considered as incidental and are not normally determined or reported.

3.2 *Cold Reduction* is the process of reducing the thickness of the strip at room temperature. The amount of reduction is greater than that used in skin-rolling (see 3.7).

3.3 *Annealing* is the process of heating to and holding at a suitable temperature and then cooling at a suitable rate, for such purposes as reducing hardness, facilitating cold working, producing a desired microstructure, or obtain-

---

[1] This specification is under the jurisdiction of ASTM Committee A-1 on Steel, Stainless Steel and Related Alloys, and is the direct responsibility of Subcommittee A01.19 on Sheet Steel and Strip.
Current edition approved Aug. 30, 1983. Published November 1983. Originally published as A 109 – 26 T. Last previous edition A 109 – 81.
[2] *Annual Book of ASTM Standards*, Vol 01.03.
[3] *Annual Book of ASTM Standards*, Vol 03.01.
[4] *Annual Book of ASTM Standards*, Vol 03.05.

A 109

ing desired mechanical, physical, or other properties.

3.3.1 *Box Annealing* involves annealing in a sealed container under conditions that minimize oxidation. The strip is usually heated slowly to a temperature below the transformation range, but sometimes above or within it, and is then cooled slowly.

3.3.2 *Continuous Annealing* involves heating the strip in continuous strands through a furnace having a controlled atmosphere followed by a controlled cooling.

3.4 *Normalizing* is heating to a suitable temperature above the transformation range and then cooling in air to a temperature substantially below the transformation range. In bright normalizing the furnace atmosphere is controlled to prevent oxidizing of the strip surface.

3.5 *Temper* is a designation by number to indicate the hardness as a minimum, as a maximum, or as a range. The tempers are obtained by the selection and control of chemical composition, by amounts of cold reduction, by thermal treatment, and by skin-rolling.

3.6 *Dead Soft* refers to the temper of strip produced without definite control of stretcher straining or fluting. It is intended for deep drawing applications where such surface disturbances are not objectionable.

3.7 *Skin-Rolled* is a term denoting a relatively light cold rolling operation following annealing. It serves to reduce the tendency of the steel to flute or stretcher strain during fabrication. It is also used to impart surface finish, or affect hardness or other mechanical properties, or to improve flatness (shape).

3.8 *Finish* refers to the degree of smoothness or luster of the strip. The production of specific finishes requires special preparation and control of the roll surfaces employed.

4. Ordering Information

4.1 Orders for material to this specification shall include the following information, as necessary, to describe adequately the desired product:

4.1.1 Quantity,
4.1.2 Name of material (cold-rolled carbon steel strip),
4.1.3 Condition (oiled or not oiled),
4.1.4 Temper (Section 7),
4.1.5 Edge (Section 8),
4.1.6 Finish (Section 9),
4.1.7 Dimensions,
4.1.8 Coil size requirements (Section 12.2),
4.1.9 ASTM designation and date of issue,
4.1.10 Copper-bearing steel, if required,
4.1.11 Application (part identification or description),
4.1.12 Cast or heat analysis (request, if required), and
4.1.13 Special requirements, if required.

NOTE 2—A typical ordering description is as follows: 20,000 lb Cold-Rolled Strip, Oiled, Temper 4, Edge 3, Finish 3, 0.035 in. by 9 in. by coil, 5000 lb max, coil weight, 16-in. ID ASTM A 109 dated ———, for Toaster Shells.

5. Manufacture

5.1 The steel shall be made by the open-hearth, basic-oxygen, or electric-furnace process.

5.2 Cold-rolled carbon steel strip is normally produced from rimmed, capped, or semi-killed steel. When required, special killed steel may be specified, and aluminum is normally used as the deoxidizer.

5.3 Cold-rolled carbon strip is manufactured from hot-rolled descaled coils by cold reducing to the desired thickness on a single stand mill or on a tandem mill consisting of several single stands in series. Sometimes an anneal is used at some intermediate thickness to facilitate further cold reduction or to obtain desired temper and mechanical properties in the finished strip. An anneal is used at final thickness for the production of Temper 4 or Temper 5.

6. Chemical Requirements

6.1 *Cast or Heat Analysis (Formerly Ladle Analysis)*—Each cast of heat of steel shall be analyzed by the manufacturer to determine the percentage of carbon, manganese, phosphorus, sulfur, and copper (when specified). This analysis shall conform to the requirements shown in Table 1. When requested, cast analysis shall be reported to purchaser or his representative.

6.2 *Product, Check, or Verification Analysis* may be made by the purchaser on the finished material.

6.2.1 Capped or rimmed steels are not technologically suited to product analysis due to the nonuniform character of their chemical composition and therefore, the tolerances in Table 2 do not apply. Product analysis is appropriate on these types of steel only when misapplication is apparent or for copper when copper steel is specified.

6.2.2 For steels other than rimmed or capped, when product analysis is made by the purchaser, the chemical analysis shall not vary from the limits specified by more than the amounts in Table 2. The several determinations of any element shall not vary both above and below the specified range.

6.3 For referee purposes, if required, Methods E 30 shall be used.

6.4 For applications where cold-rolled strip is to be welded, care must be exercised in selection of chemical composition, as well as mechanical properties, for compatibility with the welding process and its effect on altering the properties.

7. Temper and Bend Test Requirement

7.1 Cold-rolled carbon strip specified to temper numbers shall conform to the Rockwell hardness requirements shown in Table 3.

7.2 Bend-test specimens shall stand being bent at room temperatures as required in Table 4.

7.3 All mechanical tests are to be conducted in accordance with Methods and Definitions A 370.

8. Edge

8.1 The desired edge number shall be specified as follows:

8.1.1 *Number 1 Edge* is a prepared edge of a specified contour (round or square), which is produced when a very accurate width is required or when an edge condition suitable for electroplating is required, or both.

8.1.2 *Number 2 Edge* is a natural mill edge carried through the cold rolling from the hot-rolled strip without additional processing of the edge.

8.1.3 *Number 3 Edge* is an approximately square edge, produced by slitting, on which the burr is not eliminated. Normal coiling or piling does not necessarily provide a definite positioning of the slitting burr.

8.1.4 *Number 4 Edge* is a rounded edge produced by edge rolling either the natural edge of hot-rolled strip or slit-edge strip. This edge is produced when the width tolerance and edge condition are not as exacting as for No. 1 edge.

8.1.5 *Number 5 Edge* is an approximately square edge produced from slit-edge material on which the burr is eliminated usually by rolling or filing.

8.1.6 *Number 6 Edge* is a square edge produced by edge rolling the natural edge of hot-rolled strip or slit-edge strip. This edge is produced when the width tolerance and edge condition are not as exacting as for No. 1 edge.

9. Finish

9.1 The finish is specified normally as one of the following:

9.1.1 *Number 1 or Matte (Dull) Finish* is a finish without luster, produced by rolling on rolls roughened by mechanical or chemical means. This finish is especially suitable for lacquer or paint adhesion, and is beneficial in aiding drawing operations by reducing the contact friction between the die and the strip.

9.1.2 *Number 2 or Regular Bright Finish* is a finish produced by rolling on rolls having a moderately smooth finish. It is suitable for many requirements, but not generally applicable to bright plating.

9.1.3 *Number 3 or Best Bright Finish* is generally of high luster produced by selective rolling practices, including the use of specially prepared rolls. Number 3 finish is the highest quality finish commonly produced and is particularly suited for bright plating. The production of this finish requires extreme care in processing and extensive inspection.

10. Dimensional Tolerances

10.1 The dimensional tolerances shall be in accordance with the following:

Tolerances for	Table Number
Thickness, in.	5
Width, in.	6, 7, 8
Length, in.	9
Camber, in.	10

11. Workmanship

11.1 Cut lengths shall have a workmanlike appearance and shall not have imperfections of a nature or degree for the product, the grade, and the description ordered that will be detrimental to the fabrication of the finished part.

11.2 Coils may contain some abnormal imperfections which render a portion of the coil unusable since the inspection of coils does not afford opportunity to remove portions containing imperfections as in the case with cut lengths.

12. Packaging

12.1 Unless otherwise specified, the strip shall be packed and loaded in accordance with

A 109

Practices A 700.

12.2 When coils are ordered it is customary to specify a minimum or range of inside diameter, maximum outside diameter, and a maximum coil weight, if required. The ability of manufacturers to meet the maximum coil weights depends upon individual mill equipment. When required, minimum coil weights are subject to negotiation.

13. Marking

13.1 As a minimum requirement, the material shall be identified by having the manufacturer's name, ASTM designation, weight, purchaser's order number, and material identification legibly stenciled on the top of each lift or shown on a tag attached to each coil or shipping unit.

14. Inspection

14.1 When purchaser's order stipulates that inspection and tests (except product analysis) for acceptance on the steel be made prior to shipment from the mill, the producer shall afford the purchaser's inspector all reasonable facilities to satisfy him that the steel is being produced and furnished in accordance with the specification. Mill inspection by the purchaser shall not interfere unnecessarily with the manufacturer's operation. All tests and inspection (except product analysis) shall be made at the place of manufacture unless otherwise agreed.

15. Rejection and Rehearing

15.1 Unless otherwise specified, any rejection shall be reported to the producer within a reasonable time after receipt of material by the purchaser.

15.2 Material that is reported to be defective subsequent to the acceptance at the purchaser's works shall be set aside, adequately protected, and correctly identified. The producer shall be notified as soon as possible so that an investigation may be initiated.

15.3 Samples that are representative of the rejected material shall be made available to the producer. In the event that the producer is dissatisfied with the rejection, he may request a rehearing.

TABLE 1 Cast or Heat Analysis

Element	Composition, percent	
	Temper No. 1, 2, 3	Temper No. 4, 5
Carbon, max	0.25	0.15
Manganese, max	0.60	0.60
Phosphorus, max	0.035	0.035
Sulfur, max	0.04	0.04
Copper, when copper steel is specified, min.	0.20	0.20

TABLE 2 Tolerances for Product Analysis

Element	Limit or Maximum of Specified Element Percent	Tolerance	
		Under Minimum Limit	Over Maximum Limit
Carbon	to 0.15, incl	0.02	0.03
	over 0.15 to 0.25, incl	0.03	0.04
Manganese	to 0.60, incl	0.03	0.03
Phosphorus		...	0.01
Sulfur		...	0.01
Copper		0.02	...

A 109

TABLE 3 Temper and Hardness Requirement

Temper	Thickness, in.	Rockwell Hardness Minimum	Rockwell Hardness Maximum (approx.)
No. 1 (hard)	0.070 and Over	B84	...
	Under 0.070 to 0.040, incl	B90	...
	Under 0.040 to 0.025, incl	30T76	...
	Under 0.025	15T90	...
Softer Tempers[A]			
No. 2 (half-hard)	0.040 and Over	B70	B85
	Under 0.040 to 0.025, incl	30T63.5	30T73.5
	Under 0.025	15T83.5	15T88.5
No. 3 (quarter-hard)	0.040 and Over	B60	B75
	Under 0.040 to 0.025, incl	30T56.5	30T67
	Under 0.025	15T80	15T85
No. 4[B] (skin-rolled)	0.040 and Over	...	B65
	Under 0.040 to 0.025, incl	...	30T60
	Under 0.025	...	15T82
No. 5[B] (dead-soft)	0.040 and Over	...	B55
	Under 0.040 to 0.025, incl	...	30T53
	Under 0.025	...	15T78.5

[A] Rockwell hardness values apply at time of shipment. Aging may cause slightly higher values when tested at a later date.

[B] Number 4 and 5 temper may sometimes be ordered with a carbon range of 0.15–0.25 %. In each instance the maximum hardness requirement is established by agreement.

TABLE 4 Temper and Bend Test Requirement

Temper	Bend Test Requirement
No. 1 (hard)	Not required to make bends in any direction.
No. 2 (half-hard)	Bend 90 deg across[a] the direction of rolling around a radius equal to that of the thickness.
No. 3 (quarter-hard)	Bend 180 deg across[a] the direction of rolling over one thickness of the strip and 90 deg in[a] the direction of rolling around a radius equal to the thickness.
No. 4 (skin-rolled)	Bend flat upon itself in any direction.
No. 5 (dead-soft)	Bend flat upon itself in any direction.

[a] To bend "across the direction of rolling" means that the bend axis (crease of the bend) shall be at right angle to the length of the strip. To bend "in the direction of rolling" means that the bend axis (crease of the bend) shall be parallel with the length of the strip.

TABLE 5 Thickness Tolerances[a] Cold-Rolled Carbon Steel Strip
(Coils and Cut Lengths)

Specified Thickness, in.		Thickness Tolerances, Over and Under, in.							
		Specified Width, in.							
Under	To and incl	Under 1 to ½, excl	Under 3 to 1, incl.	3 to 6, incl	Over 6 to 9, incl	Over 9 to 12, incl	Over 12 to 16, incl	Over 16 to 20, incl	Over 20 to 23 15/16, incl
0.250	0.200	0.003	0.004	0.0045	0.0045	0.005	0.0055	0.0055	0.0055
0.200	0.161	0.002	0.0035	0.004	0.004	0.0045	0.0045	0.005	0.005
0.161	0.100	0.002	0.002	0.003	0.003	0.003	0.0035	0.0045	0.005
0.100	0.069	0.002	0.002	0.0025	0.003	0.003	0.0035	0.0035	0.0035
0.069	0.050	0.002	0.002	0.0025	0.0025	0.0025	0.003	0.003	0.003
0.050	0.040	0.002	0.002	0.0025	0.0025	0.0025	0.0025	0.0025	0.0025
0.040	0.035	0.002	0.002	0.002	0.002	0.002	0.002	0.002	0.002
0.035	0.032	0.0015	0.0015	0.002	0.002	0.002	0.002	0.002	0.002
0.032	0.029	0.0015	0.0015	0.0015	0.002	0.002	0.002	0.002	0.002
0.029	0.026	0.001	0.0015	0.0015	0.002	0.002	0.002	0.002	0.002
0.026	0.023	0.001	0.001	0.001	0.0015	0.0015	0.002	0.002	0.002
0.023	0.020	0.001	0.001	0.001	0.0015	0.0015	0.0015	0.0015	0.0015
0.020	0.013	0.00075	0.00075	0.00075	0.001	0.001	0.0015	0.0015	0.0015
0.013	0.009	0.00075	0.00075	0.00075	0.001	0.001	0.001	0.001	0.001
0.009	0.007	0.00075	0.00075	0.00075	...	...	...	...	...
0.007	...	0.0005	0.0005	0.0005	...	...	...	...	...

[a] Measured ⅜ in. or more in from the edge on 1 in. or wider; and on narrower than 1 in., at any place between the edges.

TABLE 6 Width Tolerances of Edge Numbers 1, 4, 5, and 6

Edge Number	Specified Width, in.	Specified Thickness, in.	Width Tolerance, Over and Under, in.
1	Over ½ to ¾, incl	0.0938 and Under	0.005
1	Over ¾ to 5, incl	0.125 and Under	0.005
4	Over ½ to 1, incl	0.1875 to 0.025, incl	0.015
4	Over 1 to 2, incl	0.2499 to 0.025, incl	0.025
4	Over 2 to 4, incl	0.2499 to 0.035, incl	0.047
4	Over 4 to 6, incl	0.2499 to 0.047, incl	0.047
5	Over ½ to ¾, incl	0.0938 and Under	0.005
5	Over ¾ to 5, incl	0.125 and Under	0.005
5	Over 5 to 9, incl	0.125 to 0.008, incl	0.010
5	Over 9 to 20, incl	0.105 to 0.015, incl	0.010
5	Over 20 to 23 15/16, incl	0.080 to 0.023, incl	0.015
6	Over ½ to 1, incl	0.1875 to 0.025, incl	0.015
6	Over 1 to 2, incl	0.2499 to 0.025, incl	0.025
6	Over 2 to 4, incl	0.2499 to 0.035, incl	0.047
6	Over 4 to 6, incl	0.2499 to 0.047, incl	0.047

TABLE 7 Width Tolerances for Number 2 (Mill) Edge

Specified Width, in.	Width Tolerance, Over and Under, in.
Over ½ to 2, incl	1/32
Over 2 to 5, incl	3/64
Over 5 to 10, incl	5/64
Over 10 to 15, incl	3/32
Over 15 to 20, incl	⅛
Over 20 to 23 15/16, incl	5/32

TABLE 8 Width Tolerances for Number 3 (Slit) Edge

Specified Thickness, in.	Width Tolerances, Over and Under, in.				
	Specified Width, in.				
	Over ½ to 6, incl	Over 6 to 9, incl	Over 9 to 12, incl	Over 12 to 20, incl	Over 20 to 23¹⁵⁄₁₆, incl
Over 0.160 to 0.2499, incl	0.016	0.020	0.020	0.031	0.031
Over 0.099 to 0.160, incl	0.010	0.016	0.016	0.020	0.020
Over 0.068 to 0.099, incl	0.008	0.010	0.010	0.016	0.020
Over 0.016 to 0.068, incl	0.005	0.005	0.010	0.016	0.020
Up to 0.016, incl	0.005	0.005	0.010	0.016	0.020

TABLE 9 Length Tolerances

Specified Width, in.	Length Tolerances Over, No Tolerance Under, in.		
	Specified Length, in.		
	24 to 60, incl	Over 60 to 120, incl	Over 120 to 240, incl
Over ½ to 12, incl	¼	½	¾
Over 12 to 23¹⁵⁄₁₆, incl	½	¾	1

TABLE 10 Camber Tolerances, Cold-Rolled Carbon Steel Strip

(Coils and Cut Lengths Applicable to all Types of Edges)

NOTE 1—Camber is the deviation of a side edge from a straight line. The standard for measuring this deviation is based on any 8-ft length. It is obtained by placing an 8-ft straightedge on the concave side and measuring the maximum distance between the strip edge and the straightedge.

NOTE 2—For strip less than 8 ft, tolerances are to be established in each instance.

NOTE 3—When the camber tolerances shown in Table 10 are not suitable for a particular purpose, cold-rolled strip is sometimes machine straightened.

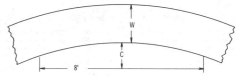

W = width of strip, in.
C = camber, in.

Width, in.	Standard Camber Tolerance, in.
Over ½ to 1½, incl	½
Over 1½ to 23¹⁵⁄₁₆	¼

Flatness Tolerances

It has not been practicable to formulate flatness tolerances for cold-rolled carbon steel strip to represent the wide range of widths and thicknesses and variety of tempers produced in coils and cut lengths.

APPENDIX

X1. GENERAL INFORMATION AND METALLURGICAL ASPECTS

X1.1 Mechanical Properties

X1.1.1 Table X1.1 shows the approximate mechanical properties corresponding to the five commercial tempers of cold-rolled carbon steel strip. This table is presented as a matter of general information. The limits of tensile strength, etc., are not intended as criteria for acceptance or rejection unless specifically agreed to by the manufacturer when accepting the order. The exact processing by different manufacturers will naturally vary slightly, so that absolute identity cannot be expected in their commercial tempers of cold-rolled strip.

X1.2 Identified Part

X1.2.1 Cold-rolled carbon steel strip can be furnished in the various tempers to make an identified part provided the fabrication of the part is compatible with the grade and temper of the steel specified. Proper identification of parts may include visual examination, prints or descriptions, or a combination

A 109

of these. It is the general experience that most identified parts can be satisfactorily produced from one of the tempers. There are applications or requirements that necessitate additional controls or limit the choice of processing methods. For most end part application only one kind of mechanical test requirement is normally employed. This test requirement is generally the Rockwell hardness test.

X1.3 Rockwell Scales and Loads

X1.3.1 Various scales and loads are employed in Rockwell testing, depending on the hardness and thickness of the strip to be tested. It is common practice to make the Rockwell hardness test at a point midway between the side edges on a single thickness only. There is some overlapping among the different scales, but the best scale to use in any given case is the one which will give the maximum penetration, without showing undue evidence of impression on the undersurface and without exceeding B100 or its equivalent on the dial. The use of a lighter load results in a loss of sensitivity, while a heavier load leads to a loss in accuracy. If the Rockwell ball is flattened by using it on a hard sample, it should be replaced, otherwise the subsequent readings will be affected. A tolerance for check testing, of two Rockwell points on the B scale below the minimum and above the maximum of the range specified, is commonly allowed to compensate for normal differences in equipment. It is recommended that hardness numbers be specified to the same scale as that to be used during testing.

X1.4 Aging Phenomenon

X1.4.1 Although the maximum ductility is obtained in steel strip in its dead soft (annealed last) condition, such strip is unsuited for many forming operations due to its tendency to stretcher strain or flute. A small amount of cold rolling (skin-rolling) will prevent this tendency, but the effect is only temporary due to a phenomenon calling aging. The phenomenon of aging is accompanied by a loss of ductility with an increase in hardness, yield point, and tensile strength. For those uses in which stretcher straining, fluting, or breakage due to aging of the steel is likely to occur, the steel should be fabricated as promptly as possible after skin-rolling. When the above aging characteristics are undesirable, special killed (generally aluminum killed) steel is used.

TABLE X1.1 Approximate Mechanical Properties for Various Tempers of Cold-Rolled Carbon Strip

NOTE—These values are given as information only and are not intended as criteria for acceptance or rejection.

Temper	Tensile Strength,[a] psi (MPa)	Elongation in 2 in. for 0.050 in. Thickness of Strip,[b] percent	Remarks
No. 1 (hard)	90 000 ± 10 000 (620) ± (70)	...	A very stiff, cold-rolled strip intended for flat blanking only, and not requiring ability to withstand cold forming.
No. 2 (half-hard)	65 000 ± 10 000 (450) ± (70)	10 ± 6	A moderately stiff cold-rolled strip intended for limited bending.
No. 3 (quarter-hard)	55 000 ± 10 000 (380) ± (70)	20 ± 7	A medium soft cold-rolled strip intended for limited bending, shallow drawing and stamping.
No. 4 (skin-rolled)	48 000 ± 6 000 (330) ± (40)	32 ± 8	A soft ductile cold-rolled strip intended for deep drawing where no surface strain or fluting is permissible.[c]
No. 5 (dead-soft)	44 000 ± 6 000 (300) ± (40)	39 ± 6	A soft ductile cold-rolled strip intended for deep drawing where stretcher strains or fluting are permissible.[c] Also for extrusions.

[a] Tensile properties are based on the standard tension-test specimen for sheet metals, see Fig. 7 of Methods E 8.
[b] Elongation in 2 in. varies with thickness of strip. For Temper No. 5, dead-soft temper, the percentage of elongation = 41 + 10 log "t" (t = thickness, in.). Other tempers vary in a similar way.
[c] See X1.4 for *Aging Phenomenon*.

The American Society for Testing and Materials takes no position respecting the validity of any patent rights asserted in connection with any item mentioned in this standard. Users of this standard are expressly advised that determination of the validity of any such patent rights, and the risk of infringement of such rights, are entirely their own responsibility.

This standard is subject to revision at any time by the responsible technical committee and must be reviewed every five years and if not revised, either reapproved or withdrawn. Your comments are invited either for revision of this standard or for additional standards and should be addressed to ASTM Headquarters. Your comments will receive careful consideration at a meeting of the responsible technical committee, which you may attend. If you feel that your comments have not received a fair hearing you should make your views known to the ASTM Committee on Standards, 1916 Race St., Philadelphia, Pa. 19103.

Metric

Designation: A 109M – 77

An American National Standard

Standard Specification for
STEEL, CARBON, COLD-ROLLED STRIP [METRIC][1]

This standard is issued under the fixed designation A 109M; the number immediately following the designation indicates the year of original adoption or, in the case of revision, the year of last revision. A number in parentheses indicates the year of last reapproval. A superscript epsilon (ϵ) indicates an editorial change since the last revision or reapproval.

This specification has been approved for use by agencies of the Department of Defense and for listing in the DoD Index of Specifications and Standards.

1. Scope

1.1 This specification covers cold-rolled carbon steel strip in cut lengths or coils, furnished to closer tolerances than cold-rolled carbon steel sheet, with specific temper, with specific edge or specific finish, commonly available by size as follows:

Width, mm	Thickness, mm
Through 600	Through 6

1.2 This specification does not include the product in narrow widths known as cold-rolled sheet, slit from wider widths, Specification A 568M, for General Requirement for Steel, Carbon and High-Strength Low-Alloy Hot-Rolled Sheet and Cold-Rolled Sheet [Metric],[2] nor does it include cold-rolled carbon spring steel, Specification A 682M, for General Requirements for Steel, High-Carbon, Strip, Cold-Rolled, Spring Quality [Metric].[2]

1.3 This specification covers only metric (SI) units and is not be used or confused with U.S. customary units.

NOTE—This metric specification is the equivalent of Specification A 109, and is compatible in technical content.

2. Applicable Documents

2.1 *ASTM Standards:*
A 370 Methods and Definitions for Mechanical Testing of Steel Products[2]
A 700 Practices for Packaging, Marking, and Loading Methods for Steel Products for Domestic Shipment[2]
E 8 Methods for Tension Testing of Metallic Materials[3]
E 350 Methods for Chemical Analysis of Carbon Steel, Low-Alloy Steel, Silicon Electrical Steel, Ingot Iron, and Wrought Iron[4]

2.2 *Military Standards:*
MIL-STD-129 Marking for Shipment and Storage[5]
MIL-STD-163 Steel Mill Products, Preparation for Shipment and Storage[5]

2.3 *Federal Standards:*
Fed. Std. No. 123 Marking for Shipments (Civil Agencies)[5]
Fed. Std. No. 183 Continuous Identification Marking of Iron and Steel Products[5]

3. Descriptions of Terms

3.1 *Carbon Steel* is the designation for steel when no minimum content is specified or required for aluminum, chromium, cobalt, columbium, molybdenum, nickel, titanium, tungsten, vanadium, zirconium, or any other element added to obtain a desired alloying effect; when the specified minimum for copper does not exceed 0.40 % or when the maximum content specified for any of the following elements does not exceed the percentage noted: manganese 1.65, silicon 0.60, or copper 0.60.

3.1.1 In all carbon steels small quantities of certain residual elements unavoidably retained from raw materials are sometimes found which are not specified or required, such as copper, nickel, molybdenum, chromium, etc. These elements are considered as

---

[1] This specification is under the jurisdiction of ASTM Committee A-1 on Steel, Stainless Steel and Related Alloys, and is the direct responsibility of Subcommittee A01.19 on Steel Sheet and Strip.
Current edition approved Nov. 3, 1977. Published January 1978.
[2] *Annual Book of ASTM Standards*, Vol 01.03.
[3] *Annual Book of ASTM Standards*, Vol 03.01.
[4] *Annual Book of ASTM Standards*, Vol 03.05.
[5] Available from Naval Publications and Forms Center, 5801 Tabor Ave., Philadelphia, Pa. 19120.

A 109M

incidental and are not normally determined or reported.

3.2 *Cold Reduction* is the process of reducing the thickness of the strip at room temperature. The amount of reduction is greater than that used in skin-rolling (see 3.7).

3.3 *Annealing* is the process of heating to and holding at a suitable temperature and then cooling at a suitable rate, for such purposes as reducing hardness, facilitating cold working, producing a desired microstructure, or obtaining desired mechanical, physical, or other properties.

3.3.1 *Box Annealing* involves annealing in a sealed container under conditions that minimize oxidation. The strip is usually heated slowly to a temperature below the transformation range, but sometimes above or within it, and is then cooled slowly.

3.3.2 *Continuous Annealing* involves heating the strip in continuous strands through a furnace having a controlled atmosphere followed by a controlled cooling.

3.4 *Normalizing* is heating to a suitable temperature above the transformation range and then cooling in air to a temperature substantially below the transformation range. In bright normalizing the furnace atmosphere is controlled to prevent oxidizing of the strip surface.

3.5 *Temper* is a designation by number to indicate the hardness as a minimum, as a maximum, or as a range. The tempers are obtained by the selection and control of chemical composition, by amounts of cold reduction, by thermal treatment, and by skin-rolling.

3.6 *Dead Soft* refers to the temper of strip produced without definite control of stretcher straining or fluting. It is intended for deep drawing applications where such surface disturbances are not objectionable.

3.7 *Skin-Rolled* is a term denoting a relatively light cold rolling operation following annealing. It serves to reduce the tendency of the steel to flute or stretcher strain during fabrication. It is also used to impart surface finish, or affect hardness or other mechanical properties or to improve flatness (shape).

3.8 *Finish* refers to the degree of smoothness or luster of the strip. The production of specific finishes requires special preparation and control of the roll surfaces employed.

4. Ordering Information

4.1 Orders for material to this specification shall include the following information, as necessary, to describe adequately the desired product:

4.1.1 Quantity,

4.1.2 Name of material (cold-rolled carbon steel strip),

4.1.3 Condition (oiled or not oiled),

4.1.4 Temper (Section 7),

4.1.5 Edge (Section 8),

4.1.6 Finish (Section 9),

4.1.7 Dimensions,

4.1.8 Coil size requirements (12.3),

4.1.9 ASTM designation and date of issue,

4.1.10 Copper-bearing steel, if required,

4.1.11 Application (part identification or description),

4.1.12 Cast or heat analysis (request, if required), and

4.1.13 Special requirements if required.

NOTE 1 — A typical ordering description is as follows: 10 000 kg Cold-Rolled Strip, Oiled, Temper 4, Edge 3, Finish 3, 0.80 mm by 300 by coil. 2500 kg max, coil weight, 400 mm ID ASTM A 109M dated_____, for Toaster Shells.

5. Manufacture

5.1 Cold-rolled carbon steel strip is normally produced from rimmed, capped, or semi-killed steel. When required, special killed steel may be specified, and aluminum is normally used as the deoxidizer.

5.2 Cold-rolled carbon strip is manufactured from hot-rolled descaled coils by cold reducing to the desired thickness on a single stand mill or on a tandem mill consisting of several single stands in series. Sometimes an anneal is used at some intermediate thickness to facilitate further cold reduction or to obtain desired temper and mechanical properties in the finished strip. An anneal is used at final thickness for the production of Temper 4 or Temper 5.

6. Chemical Requirements

6.1 *Cast or Heat Analysis (Formerly Ladle Analysis)* — Each cast of heat of steel shall be analyzed by the manufacturer to determine the percentage of carbon, manganese, phosphorus, sulfur, and copper (when specified). This analysis shall conform to the requirements shown in Table 1. When requested, cast analysis shall be reported to the purchaser or his representative.

6.2 *Product, Check, or Verification Analysis (Formerly Check Analysis)* may be made by the purchaser on the finished material.

6.2.1 Non-killed steels (such as capped or rimmed) are not technologically suited to product analysis due to the nonuniform character of their chemical composition, and therefore the tolerances in Table 2 do not apply. Product analysis is appropriate on these types of steel only when misapplication is apparent or for copper when copper steel is specified.

6.2.2 For steels other than non-killed (rimmed or capped), when product analysis is made by the purchaser, the chemical analysis shall not vary from the limits specified by more than the amounts in Table 2. The several determinations of any element shall not vary both above and below the specified range.

6.3 For referee purposes, if required, Methods E 350 shall be used.

7. Temper and Bend Test Requirements

7.1 Cold-rolled carbon strip specified to temper numbers shall conform to the Rockwell hardness requirements shown in Table 3.

7.2 Bend tests shall be conducted at room temperature and bent to the requirements shown in Table 4.

7.3 All mechanical tests are to be conducted in accordance with Methods and Definitions A 370.

7.4 It is recommended that hardness numbers be specified in the same scale as that which will be used in testing the strip.

8. Edge

8.1 The desired edge number shall be specified as follows:

8.1.1 *Number 1 Edge* is a prepared edge of a specified contour (round or square), which is produced when a very accurate width is required or when an edge condition suitable for electroplating is required, or both.

8.1.2 *Number 2 Edge* is a natural mill edge carried through the cold rolling from the hot-rolled strip without additional processing of the edge.

8.1.3 *Number 3 Edge* is an approximately square edge, produced by slitting, on which the burr is not eliminated. Normal coiling or piling does not necessarily provide a definite positioning of the slitting burr.

8.1.4 *Number 4 Edge* is a rounded edge produced by edge rolling either the natural edge of hot-rolled strip or slit-edge strip. This edge is produced when the width tolerances and edge condition are not as exacting as for No. 1 edge.

8.1.5 *Number 5 Edge* is an approximately square edge produced from slit-edge material on which the burr is eliminated usually by rolling or filing.

8.1.6 *Number 6 Edge* is a square edge produced by edge rolling the natural edge of hot-rolled strip or slit-edge strip. This edge is produced when the width tolerance and edge condition are not as exacting as for No. 1 edge.

9. Finish

9.1 The finish is specified normally as one of the following:

9.1.1 *Number 1 or Matte (Dull) Finish* is a finish without luster, produced by rolling on rolls roughened by mechanical or chemical means. This finish is especially suitable for lacquer or paint adhesion, and is beneficial in aiding drawing operations by reducing the contact friction between the die and the strip.

9.1.2 *Number 2 or Regular Bright Finish* is a finish produced by rolling with rolls having a moderately smooth finish. It is suitable for many requirements, but not generally applicable to bright plating.

9.1.3 *Number 3 or Best Bright Finish* is generally of high luster produced by selective rolling practices, including the use of specially prepared rolls. Number 3 finish is the highest quality finish commonly produced and is particularly suited for bright plating. The production of this finish requires extreme care in processing and extensive inspection.

10. Dimensional Tolerances

10.1 The dimensional tolerances shall be in accordance with the following:

Tolerances, mm	Table Number
Thickness	5
Width	6, 7, 8
Length	9
Camber	10

11. Workmanship

11.1 Cut lengths shall have a workmanlike appearance and shall not have imperfections of a nature or degree for the product, the grade, and the description ordered that will be detrimental to the fabrication of the finished part.

11.2 Coils may contain some abnormal imperfections that render a portion of the coil unusable since the inspection of coils does not afford an opportunity to remove portions containing imperfections as in the case with cut lengths.

12. Packaging

12.1 Unless otherwise specified, the sheet and strip shall be packaged and loaded in accordance with Practices A 700.

12.2 When specified in the contract or order, and for direct procurement by or direct shipment to the government, when Level A is specified, preservation, packaging, and packing shall be in accordance with the Level A requirements of MIL-STD-163.

12.3 When coils are ordered it is customary to specify a minimum or range of inside diameter, maximum outside diameter, and a maximum coil weight, if required. The ability of manufacturers to meet the maximum coil weights depends upon individual mill equipment. When required, minimum coil weights are subject to negotiation.

13. Marking

13.1 As a minimum requirement, the material shall be identified by having the manufacturer's name, ASTM designation, weight, purchaser's order number, and material identification legibly stenciled on top of each lift or shown on a tag attached to each coil or shipping unit.

13.2 When specified in the contract or order, and for direct procurement by or direct shipment to the government, marking for shipment, in addition to requirements specified in the contract or order, shall be in accordance with MIL-STD-129 for military agencies and in accordance with Fed. Std. No. 123 for civil agencies.

13.3 For Government procurement by the Defense Supply Agency, strip material shall be continuously marked for identification in accordance with Fed. Std. No. 183.

14. Inspection

14.1 When the purchaser's order stipulates that inspection and test (except product analysis) for acceptance on the steel be made prior to shipment from the mill, the manufacturer shall afford the purchaser's inspector all reasonable facilities to satisfy him that the steel is being produced and furnished in accordance with the specification. Mill inspection by the purchaser shall not interfere unnecessarily with the manufacturer's operation. All tests and inspection (except product analysis) shall be made at the place of manufacture unless otherwise agreed.

15. Rejection and Rehearing

15.1 Unless otherwise specified, any rejection shall be reported to the manufacturer within a reasonable time after receipt of material by the purchaser.

15.2 Material that is reported to be defective subsequent to the acceptance at the purchaser's works shall be set aside, adequately protected, and correctly identified. The manufacturer shall be notified as soon as possible so that an investigation may be initiated.

15.3 Samples that are representative of the rejected material shall be made available to the producer. In the event that the manufacturer is dissatisfied with the rejection, he may request a rehearing.

A 109M

TABLE 1 Cast or Heat Analysis

Element	Composition, %	
	Temper No. 1, 2, 3	Temper No. 4, 5
Carbon, max	0.25	0.15
Manganese, max	0.60	0.60
Phosphorus, max	0.035	0.035
Sulfur, max	0.04	0.04
Copper, when copper steel is specified, min	0.20	0.20

TABLE 2 Tolerances for Product Analysis

Element	Limit or Maximum of Specified Element Percent	Tolerance	
		Under Minimum Limit	Over Maximum Limit
Carbon	to 0.15, incl	0.02	0.03
	over 0.15 to 0.25, incl	0.03	0.04
Manganese	to 0.60, incl	0.03	0.03
Phosphorus		...	0.01
Sulfur		...	0.01
Copper		0.02	...

TABLE 3 Temper and Hardness Requirement

Temper	Thickness, mm		Rockwell Hardness	
	Over	Through	Minimum	Maximum (approx.)
No. 1	1.8	...	B84.0	...
(hard)	1.0	1.8	B90.0	...
	0.6	1.0	30T76.0	...
	...	0.6	15T90.0	...
Softer Tempers[A]				
No. 2	1.0	...	B70.0	B85.0
(half-hard)	0.6	1.0	30T63.5	30T73.5
	...	0.6	15T83.5	15T88.5
No. 3	1.0	...	B60	B75
(quarter-hard)	0.6	1.0	30T56.5	30T67.0
	...	0.6	15T80.0	15T85.0
No. 4[B]	1.0	...	...	B65
(skin-rolled)	0.6	1.0	...	30T60.0
	...	0.6	...	15T82.0
No. 5[B]	1.0	...	...	B55
(dead-soft)	0.6	1.0	...	30T53.0
	...	0.6	...	15T78.5

[A] Rockwell hardness values apply to special killed steels and also rimmed or semi-killed steels at time of shipment only. Aging of these latter steels may cause slightly higher values when tested at a later date.

[B] Number 4 and 5 temper may sometimes be ordered with a carbon range of 0.15 to 0.25 %. In each instance the maximum hardness requirement is established by agreement.

TABLE 4 Temper and Bend Test Requirements

Temper	Bend Test Requirement
No. 1 (hard)	Not required to make bends in any direction.
No. 2 (half-hard)	Bend 90 deg across[A] the direction of rolling around a radius equal to that of the thickness.
No. 3 (quarter-hard)	Bend 180 deg across[A] the direction of rolling over one thickness of the strip and 90 deg in[A] the direction of rolling around a radius equal to the thickness.
No. 4 (skin-rolled)	Bend flat upon itself in any direction.
No. 5 (dead-soft)	Bend flat upon itself in any direction.

[A] To bend "across the direction of rolling" means that the bend axis (crease of the bend) shall be at right angle to the length of the strip. To bend "in the direction of rolling" means that the bend axis (crease of the bend) shall be parallel with the length of the strip.

A 109M

TABLE 5 Thickness Tolerances[A] Cold-Rolled Carbon Steel Strip Ordered to Nominal Thickness
(Coils and Cut Lengths)

Specified Nominal Thickness, mm		Thickness Tolerance, Over and Under, for Specified Width, mm				
Over	Through	To 75, incl	Over 75 to 150, incl	Over 150 to 300, incl	Over 300 to 450, incl	Over 450 to 600, incl
...	0.2	0.015	0.015	...	...	...
0.2	0.3	0.02	0.02	...	...	...
0.3	0.4	0.02	0.02	0.03	0.04	0.04
0.4	0.5	0.02	0.02	0.03	0.04	0.04
0.5	0.6	0.03	0.03	0.04	0.04	0.04
0.6	0.7	0.04	0.04	0.05	0.05	0.05
0.7	0.8	0.04	0.04	0.05	0.05	0.05
0.8	0.9	0.04	0.05	0.05	0.05	0.05
0.9	1.0	0.05	0.05	0.05	0.05	0.05
1.0	1.5	0.05	0.07	0.07	0.08	0.08
1.5	2.0	0.05	0.07	0.07	0.09	0.09
2.0	3.0	0.05	0.08	0.08	0.12	0.13
3.0	4.0	0.05	0.08	0.08	0.12	0.13
4.0	5.0	0.09	0.10	0.10	0.13	0.13
5.0	6.0	0.10	0.12	0.12	0.14	0.14

[A] Measured 10.0 mm or more in from the edge on 25.0 mm or wider; and at any place between the edges on narrower than 25.0 mm.

TABLE 6 Width Tolerances of Edge Numbers 1, 4, 5, and 6, Cold-Rolled Carbon Steel Strip

Edge No.	Specified Width, mm		Specific Thickness, mm		Width Tolerance, mm
	Over	Through	min	max	Over and Under
1	...	200	...	3.0	0.13
4	...	25	0.6	5.0	0.38
4	25	50	0.6	6.0	0.65
4	50	150	1.0	6.0	1.20
5	...	100	...	3.0	0.13
5	100	500	0.4	3.0	0.25
5	500	600	0.6	2.0	0.38
6	...	25	0.6	5.0	0.38
6	25	50	0.6	6.0	0.65
6	50	150	1.0	6.0	1.20

TABLE 7 Width Tolerance for Edge Number 2 (Mill), Cold-Rolled Carbon Steel Strip

Specified Width, mm		Width Tolerance, mm
Over	Through	Over and Under
...	50	0.8
50	100	1.2
100	200	1.6
200	400	2.5
400	500	3.0
500	600	4.0

TABLE 8 Width Tolerances for Edge Number 3 (Slit), Cold-Rolled Carbon Steel Strip

Specified Width, mm		Width Tolerance, Over and Under, mm, for Specified Thickness, mm			
Over	Through	To 1.5, incl	Over 1.5 to 2.5, incl	Over 2.5 to 4.5, incl	Over 4.5 to 6.0, incl
...	100	0.13	0.20	0.25	0.40
100	200	0.13	0.25	0.40	0.50
200	300	0.25	0.25	0.40	0.50
300	450	0.40	0.40	0.50	0.80
450	600	0.50	0.50	0.50	0.80

A 109M

TABLE 9 Length Tolerances, Cold-Rolled Carbon Steel Strip

Specified Width, mm		Tolerance Over Specified Length (No Tolerance Under), mm		
		Specified Length, mm		
Over	Through	600 to 1500, incl	Over 1500 to 3000, incl	Over 3000
...	300	10	15	20
300	600	15	20	25

TABLE 10 Camber Tolerances, Cold-Rolled Carbon Steel Strip

(Coils and Cut Lengths Applicable to all Types of Edges)

NOTE 1—Camber is the deviation of a side edge from a straight line. The standard for measuring this deviation is based on any 2000-mm length. It is obtained by placing an 2000-mm straightedge on the concave side and measuring the maximum distance between the strip edge and the straightedge.

NOTE 2—For strip less than 2000 mm, tolerances are to be established in each instance.

NOTE 3—When the camber tolerances shown in Table 9 are not suitable for a particular purpose, cold-rolled strip is sometimes machine straightened.

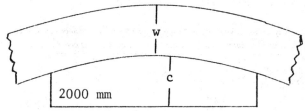

W = Width of strip, mm
C = Camber, mm

Over	Width, mm	Standard Camber Tolerance, mm
	Through	
...	50	10
50	600	5

Flatness Tolerances
It has not been practicable to formulate flatness tolerances for cold-rolled carbon steel strip to represent the wide range of widths and thicknesses and variety of tempers produced in coils and cut lengths.

APPENDIX

X1. GENERAL INFORMATION AND METALLURGICAL ASPECTS

X1.1 Mechanical Properties

X1.1.1 Table X1.1 shows the approximate mechanical properties corresponding to the five commercial tempers of cold-rolled carbon steel strip. This table is presented as a matter of general information. The limits of tensile strength, etc., are not intended as criteria for acceptance or rejection unless specifically agreed to by the manufacturer when accepting the order. The exact processing by different manufacturers will naturally vary slightly, so that absolute identity cannot be expected in their commercial tempers of cold-rolled strip.

A 109M

X1.2 Identified Part

X1.2.1 Cold-rolled carbon steel strip can be furnished in the various tempers to make an identified part provided the fabrication of the part is compatible with the grade and temper of the steel specified. Proper identification of parts may include visual examination, prints or descriptions, or a combination of these. It is the general experience that most identified parts can be satifactorily produced from one of the tempers. There are applications or requirements that necessitate additional controls or limit the choice of processing methods. For most end part application only one kind of mechanical test requirement is normally employed. This test requirement is generally the Rockwell hardness test.

X1.3 Rockwell Scales and Loads

X1.3.1 Various scales and loads are employed in Rockwell testing, depending on the hardness and thickness of the strip to be tested. It is common practice to make the Rockwell hardness test at a point midway between the side edges on a single thickness only. There is some overlapping among the different scales, but the best scale to use in any given case is the one that will give the maximum penetration, without showing undue evidence of impression on the undersurface and without exceeding B100 or its equivalent on the dial. The use of a lighter load results in a loss of sensitivity, while a heavier load leads to a loss in accuracy. If the Rockwell ball is flattened by using it on a hard sample, it should be replaced, otherwise the subsequent readings will be affected. A tolerance for check testing of two Rockwell points on the B scale below the minimum and above the maximum of the range specified is commonly allowed to compensate for normal differences in equipment. It is recommended that hardness numbers be specified to the same scale as that to be used during testing.

X1.4 Aging Phenomenon

X1.4.1 Although the maximum ductility is obtained in steel strip in its dead soft (annealed last) condition, such strip is unsuited for many forming operations due to its tendency to stretcher strain or flute. A small amount of cold rolling (skin-rolling) will prevent this tendency, but the effect is only temporary due to a phenomenon called aging. The phenomenon of aging is accompanied by a loss of ductility with an increase in hardness, yield point, and tensile strength. For those uses in which stretcher straining, fluting, or breakage due to aging of the steel is likely to occur, the steel should be fabricated as promptly as possible after skin-rolling. When the above aging characteristics are undesirable, special killed (generally aluminum killed) steel is used.

TABLE X1.1 Approximate Mechanical Properties for Various Tempers of Cold-Rolled Carbon Strip

NOTE—These values are given as information only and are not intended as criteria for acceptance or rejection.

Temper	Tensile Strength[A], MPa	Elongation in 50 mm for 1.2-mm Thickness of Strip,[B] %	Remarks
No. 1 (hard)	620 ± 70	...	A very stiff, cold-rolled strip intended for flat blanking only, and not requiring ability to withstand cold forming.
No. 2	450 ± 70	10 ± 6	A moderately stiff cold-rolled strip intended for limited bending.
No. 3 (quarter-hard)	380 ± 70	20 ± 7	A medium soft cold-rolled strip intended for limited bending, shallow drawing and stamping.
No. 4 (skin-rolled)	330 ± 40	32 ± 8	A soft ductile cold-rolled strip intended for deep drawing where no surface stain or fluting is permissible.[C]
No. 5 (dead-soft)	300 ± 40	39 ± 6	A soft ductile cold-rolled strip intended for deep drawing where stretcher stains or fluting are permissible.[C] Also for extrusions.

[A] Tensile properties are based on the standard tension-test specimen for sheet metals. See Fig. 7 of Methods E 8.
[B] Elongation in 50 mm varies with thickness of strip. For Temper No. 5, dead-soft temper, the percentage of elongation = 41 + 10 log "t" (t = thickness, mm). Other tempers vary in a similar way.
[C] See X1.4 for *Aging Phenomenon*.

The American Society for Testing and Materials takes no position respecting the validity of any patent rights asserted in connection with any item mentioned in this standard. Users of this standard are expressly advised that determination of the validity of any such patent rights, and the risk of infringement of such rights, are entirely their own responsibility.

This standard is subject to revision at any time by the responsible technical committee and must be reviewed every five years and if not revised, either reapproved or withdrawn. Your comments are invited either for revision of this standard or for additional standards and should be addressed to ASTM Headquarters. Your comments will receive careful consideration at a meeting of the responsible technical committee, which you may attend. If you feel that your comments have not received a fair hearing you should make your views known to the ASTM Committee on Standards, 1916 Race St., Philadelphia, Pa. 19103.

Last ASTM Designation: A 164 – 71

Standard Specification for
ELECTRODEPOSITED COATINGS OF ZINC ON STEEL

This specification covers requirements for electroplated zinc coatings on steel articles that are required to withstand corrosion. Three types of coatings are covered: namely, Type GS, Type LS, and Type RS.

Formerly under the jurisdiction of Committee B-8 on Metallic and Inorganic Coatings, this specification was discontinued in 1980 and replaced by Specification B 633, for Electrodeposited Coatings of Zinc on Iron and Steel.[1]

---

[1] *Annual Book of ASTM Standards*, Vol 02.05.

Designation: A 167 – 82

Standard Specification for
STAINLESS AND HEAT-RESISTING CHROMIUM-NICKEL STEEL PLATE, SHEET, AND STRIP[1]

This standard is issued under the fixed designation A 167; the number immediately following the designation indicates the year of original adoption or, in the case of revision, the year of last revision. A number in parentheses indicates the year of last reapproval. A superscript epsilon (ϵ) indicates an editorial change since the last revision or reapproval.

This specification has been approved for use by agencies of the Department of Defense and for listing in the DoD Index of Specifications and Standards.

1. Scope

1.1 This specification covers stainless and heat-resisting chromium-nickel steel plate, sheet, and strip.

1.2 The values stated in inch-pound units are to be regarded as the standard.

2. Applicable Documents

2.1 *ASTM Standards:*
A 370 Methods and Definitions for Mechanical Testing of Steel Products[2]
A 480/A 480M Specification for General Requirements for Flat-Rolled Stainless and Heat-Resisting Steel Plate, Sheet, and Strip[2]
A 751 Methods, Practices, and Definitions for Chemical Analysis of Steel Products[2]
E 527 Practice for Numbering Metals and Alloys (UNS)[2]

3. Description of Terms

3.1 Plate, sheet, and strip as used in this specification are described as follows:

3.1.1 *Plate*—Material ³⁄₁₆ in. (4.76 mm) and over in thickness and over 10 in. (254 mm) in width.

3.1.2 *Sheet*—Material under ³⁄₁₆ in. (4.76 mm) in thickness and 24 in. (609.6 mm) and over in width.

3.1.3 *Strip*—Material under ³⁄₁₆ in. (4.76 mm) in thickness and under 24 in. (609.6 mm) in width.

4. General Requirements for Delivery

4.1 Material furnished under this specification shall conform to applicable requirements of the current edition of ASTM Specification A 480/A 480M.

5. Ordering Information

5.1 Orders for material under this specification shall include the following information:

5.1.1 Quantity (weight or number of pieces),
5.1.2 Name of material (stainless steel),
5.1.3 Form (plate, sheet, or strip),
5.1.4 Dimensions,
5.1.5 Type (Table 1),
5.1.6 Edge, strip only (see Specification A 480),
5.1.7 Mechanical properties (Section 8),
5.1.8 Finish (see Specification A 480/ A 480M); specify whether one or both sides are to be polished.
5.1.9 ASTM designation and date of issue, and
5.1.10 Additions to the specification or special requirements, and
5.1.11 Restrictions (if desired) on methods for determining yield strength (see footnote C of Table 2).

---

[1] This specification is under the jurisdiction of ASTM Committee A-1 on Steel, Stainless Steel and Related Alloys and is the direct responsibility of Subcommittee A01.17 on Flat Stainless Steel Products.
Current edition approved Oct. 29, 1982. Published December 1982. Originally published as A 167 – 35 T. Last previous edition A 167 – 81a.
[2] *Annual Book of ASTM Standards*, Vol 01.03.

A 167

NOTE 1—A typical ordering description is as follows: 200 pieces, stainless steel sheets, 0.060 in. (1.52 mm) by 48 in. (1.219 m) by 120 in. (3.048 m), Type 304, No. 2D Finish. ASTM A 167 dated ____.

6. Process

6.1 The steel shall be made by one or more of the following processes: electric-arc, electric-induction, or other suitable processes.

7. Chemical Requirements

7.1 The steel shall conform to the requirements as to chemical composition specified in Table 1.

7.2 Methods and practices relating to chemical analysis required by this specification shall be in accordance with Methods, Practices, and Definitions A 751.

8. Mechanical Requirements

8.1 The material shall conform to the mechanical property requirements specified in Table 2.

9. Special Tests

9.1 If any special tests are required that are pertinent to the intended application of the material ordered, they shall be as agreed upon between the seller and the purchaser.

10. Test Specimens

10.1 Tension test specimens shall be taken from the finished material and shall be selected either in the longitudinal or transverse direction. The tension test specimen shall conform to the appropriate Sections 7, 8, or 9 of Methods and Definitions A 370.

NOTE 2—For plate material up to and including ¾ in. (19.05 mm) thick, the sheet type tension test specimen described in Section 8 (Fig. 4) of Methods and Definitions A 370 is also permitted.

10.2 Hardness tests may be made on the grip ends of the tension test specimen before they are subjected to the tension test.

11. Number of Tests

11.1 In the case of sheet, strip, or plate produced in coil form, two or more hardness tests (one from each end of the coil); one bend test, when required; and one or more tension tests shall be made on specimens taken from each coil. If the hardness difference between the two ends of the coil exceeds 5HRB, or equivalent, tensile properties must be determined on both ends. In the case of sheet, strip, or plate produced in cut lengths, one tension and one or more hardness tests shall be made on each 100 or less sheets, strips, or plates of the same gage and heat produced under the same processing conditions.

12. Test Methods

12.1 *Yield Strength*:

12.1.1 The yield strength shall be determined by the offset method as described in Methods and Definitions A 370. The offset shall be 0.2 % of the gage length of the specimen.

12.1.2 Between the yield strength and fracture of the specimen, the cross head speed of the testing machine shall be between ⅛ in. (3.18 mm) and ½ in. (12.70 mm)/in. (25.4 mm) of gage length/min.

13. Permissible Variations in Dimensions

13.1 Unless otherwise specified in the purchase order, material shall conform to the permissible tolerances shown in Specification A 480.

13.2 *Sheet, Strip, and Plate*—Material with No. 1 finish may be ground to remove surface defects, provided such grinding does not reduce the thickness or width at any point beyond the permissible variations in dimensions.

14. Workmanship

14.1 The material shall be of uniform quality consistent with good manufacturing and inspection practices. The steel shall have no defects of a nature or degree, for the grade and quality ordered, that will be detrimental to the stamping, forming, machining, or fabrication of finished parts.

15. Inspection

15.1 Inspection of the material by the purchaser's representative at the producing plant shall be made as agreed upon by the purchaser and the seller as part of the purchase order.

16. Certification

16.1 Upon request of the purchaser in the contract or order, the producer's certification that the material was manufactured and tested in accordance with this specification together with a report of the test results shall be furnished at the time of shipment.

17. Rejection

17.1 Unless otherwise specified, any rejection based on tests made in accordance with

A 167

this specification shall be reported to the seller within 60 working days from receipt of the material by the purchaser.

18. Rehearing

18.1 Samples tested by the purchaser in accordance with this specification that represents rejected material shall be retained for three weeks from the date of the notification to the seller of the rejection. In case of dissatisfaction with the results of the tests, the seller may make claim for a rehearing within that time.

TABLE 1 Chemical Requirements[A]

UNS Designation[B]	Type	Carbon	Manganese	Phosphorus	Sulfur	Silicon	Chromium	Nickel	Molybdenum	Other Elements
S30100	301	0.15	2.00	0.045	0.030	1.00	16.00–18.00	6.00–8.00	...	...
S30200	302	0.15	2.00	0.045	0.030	1.00	17.00–19.00	8.00–10.00	...	...
S30215	302B	0.15	2.00	0.045	0.030	2.00–3.00	17.00–19.00	8.00–10.00	...	...
S30400	304	0.08	2.00	0.045	0.030	1.00	18.00–20.00	8.00–10.50	...	...
S30403	304L	0.030	2.00	0.045	0.030	1.00	18.00–20.00	8.00–12.00	...	...
S30453	304LN	0.030	2.00	0.045	0.030	1.00	18.00–20.00	8.00–12.00	...	N 0.10–0.16
S30500	305	0.12	2.00	0.045	0.030	1.00	17.00–19.00	10.50–13.00	...	...
S30800	308	0.08	2.00	0.045	0.030	1.00	19.00–21.00	10.00–12.00	...	...
S30900	309	0.20	2.00	0.045	0.030	1.00	22.00–24.00	12.00–15.00	...	...
S30908	309S	0.08	2.00	0.045	0.030	1.00	22.00–24.00	12.00–15.00	...	...
S30940	309Cb	0.08	2.00	0.045	0.030	1.00	22.00–24.00	12.00–16.00	...	Cb + Ta 10 × C min; 1.10 max
S31000	310	0.25	2.00	0.045	0.030	1.50	24.00–26.00	19.00–22.00	...	...
S31040	310Cb	0.08	2.00	0.045	0.030	1.50	24.00–26.00	19.00–22.00	...	Cb + Ta 10 × C min; 1.10 max
S31008	310S	0.08	2.00	0.045	0.030	1.50	24.00–26.00	19.00–22.00	...	Cu 0.50–1.00; N 0.180–0.220
S31254	...	0.020	1.00	0.030	0.010	0.80	19.50–20.50	17.50–18.50	6.00–6.50	...
S31600	316	0.08	2.00	0.045	0.030	1.00	16.00–18.00	10.00–14.00	2.00–3.00	...
S31640	316Cb	0.08	2.00	0.045	0.030	1.00	16.00–18.00	10.00–14.00	...	Mo 2.0–3.0; N 0.10 max; Cb + Ta 10 × C min; 1.10 max
S31603	316L	0.030	2.00	0.045	0.030	1.00	16.00–18.00	10.00–14.00	2.00–3.00	...
S31653	316LN	0.030	2.00	0.045	0.030	1.50	16.00–18.00	10.00–14.00	2.00–3.00	N 0.10–0.16
S31635	316Ti	0.08	2.00	0.045	0.030	1.00	16.00–18.00	10.00–14.00	...	Mo 2.0–3.0; N 0.10 max; Ti 5 × (C + N) min; 0.70 max
S31700	317	0.08	2.00	0.045	0.030	1.00	18.00–20.00	11.00–15.00	3.00–4.00	...
S31703	317L	0.030	2.00	0.045	0.030	1.00	18.00–20.00	11.00–15.00	3.00–4.00	...
S32100	321	0.08	2.00	0.045	0.030	1.00	17.00–19.00	9.00–12.00	...	N 0.10 max; Ti 5 × (C + N) min; 0.70 max
S32109	321H	0.04–0.10	2.00	0.045	0.030	1.00	17.00–19.00	9.00–12.00	...	N 0.10 max; Ti 4 × (C + N) min; 0.70 max
S34700	347	0.08	2.00	0.045	0.030	1.00	17.00–19.00	9.00–13.00	...	Cb + Ta 10 × C min; 1.10 max
S34800	348	0.08	2.00	0.045	0.030	1.00	17.00–19.00	9.00–13.00	...	Cb + Ta 10 × C min; 1.10 max; Ta 0.10 max; Co 0.20 max
S38100	XM-15	0.08	2.00	0.030	0.030	1.50–2.50	17.00–19.00	17.50–18.50	...	

[A] Maximum unless range or minimum is indicated.
[B] New designation established in accordance with ASTM E 527 and SAE J 1086, Recommended Practice for Numbering Metals and Alloys (UNS).

TABLE 2 Mechanical Property Requirements

UNS Designation[A]	Type	Tensile Strength, min		Yield Strength, min[B,C]		Elongation in 2 in. or 50 mm, min, %	Hardness, max[D]	
		ksi	MPa	ksi	MPa		Brinell	Rockwell B
S30100	301	75	515	30	205	40.0	202	92
S30200	302	75	515	30	205	40.0	202	92
S30215	302B	75	515	30	205	40.0	217	95
S30400	304	75	515	30	205	40.0	202	92
S30403	304L	70	485	25	170	40.0	183	88
S30453	304LN	75	515	30	205	40.0	202	92
S30500	305	70	485	25	170	40.0	183	88
S30800	308	75	515	30	205	40.0	183	88
S30900	309	75	515	30	205	40.0	217	95
S30940	309Cb	75	515	30	205	40.0	217	95
S30908	309S	75	515	30	205	40.0	217	95
S31000	310	75	515	30	205	40.0	217	95
S31040	310Cb	75	515	30	205	40.0	217	95
S31008	310S	75	515	30	205	40.0	217	95
S31600	316	75	515	30	205	40.0	217	95
S31640	316Cb	75	515	30	205	40.0	217	95
S31603	316L	70	485	25	170	40.0	217	95
S31653	316LN	75	515	30	205	40.0	217	95
S31635	316Ti	75	515	30	205	40.0	217	95
S31700	317	75	515	30	205	35.0	217	95
S31703	317L	75	515	30	205	35.0	217	95
S32100	321	75	515	30	205	40.0	217	95
S32109	321H	75	515	30	205	40.0	217	95
S34700	347	75	515	30	205	40.0	183	88
S34800	348	75	515	30	205	40.0	183	88
S38100	XM-15	75	515	30	205	40.0	217	96
S31254	...	94	650	44	300	35.0	220	96

[A] New designation established in accordance with ASTM E 527 and SAE J 1086, Recommended Practice for Numbering Metals and Alloys (UNS).
[B] See 12.1.
[C] Yield strength shall be determined by the offset method at 0.2 % limiting permanent set in accordance with Methods A 370. Unless otherwise specified (see 5.1.11), an alternative method of determining yield strength may be based on a total extension under load of 0.5 %.
[D] Either Brinell or Rockwell B hardness is permissible.

The American Society for Testing and Materials takes no position respecting the validity of any patent rights asserted in connection with any item mentioned in this standard. Users of this standard are expressly advised that determination of the validity of any such patent rights, and the risk of infringement of such rights, are entirely their own responsibility.

This standard is subject to revision at any time by the responsible technical committee and must be reviewed every five years and if not revised, either reapproved or withdrawn. Your comments are invited either for revision of this standard or for additional standards and should be addressed to ASTM Headquarters. Your comments will receive careful consideration at a meeting of the responsible technical committee, which you may attend. If you feel that your comments have not received a fair hearing you should make your views known to the ASTM Committee on Standards, 1916 Race St., Philadelphia, Pa. 19103.

Designation: A 176 – 83

Standard Specification for
STAINLESS AND HEAT-RESISTING CHROMIUM STEEL PLATE, SHEET, AND STRIP[1]

This standard is issued under the fixed designation A 176; the number immediately following the designation indicates the year of original adoption or, in the case of revision, the year of last revision. A number in parentheses indicates the year of last reapproval. A superscript epsilon (ϵ) indicates an editorial change since the last revision or reapproval.

This specification has been approved for use by agencies of the Department of Defense and for listing in the DoD Index of Specifications and Standards.

1. Scope

1.1 This specification covers stainless and heat-resisting chromium steel plate, sheet, and strip available in a wide variety of surface finishes.

1.2 The values stated in inch-pound units are to be regarded as the standard.

2. Applicable Documents

2.1 *ASTM Standards:*
A 370 Methods and Definitions for Mechanical Testing of Steel Products[2]
A 480/A 480M Specification for General Requirements for Flat-Rolled Stainless and Heat-Resisting Steel Plate, Sheet, and Strip[2]
A 751 Methods, Practices, and Definitions for Chemical Analysis of Steel Products[2]
E 527 Practice for Numbering Metals and Alloys (UNS)[2]

3. Descriptions of Terms

3.1 Plate, sheet, and strip as used in this specification are described as follows:

3.1.1 *Plate*—Material 3/16 in. (4.76 mm) and over in thickness and over 10 in. (254 mm) in width.

3.1.2 *Sheet*—Material under 3/16 in. (4.76 mm) in thickness and 24 in. (609.6 mm) and over in width.

3.1.3 *Strip*—Material under 3/16 in. (4.76 mm) in thickness and under 24 in. (609.6 mm) in width.

4. General Requirements for Delivery

4.1 Material furnished under this specification shall conform to the applicable requirements of the current edition of Specification A 480/A 480M.

5. Ordering Information

5.1 Orders for material under this specification shall include the following information:

5.1.1 Quantity (weight or number of pieces),
5.1.2 Name of material (stainless steel),
5.1.3 Form (plate, sheet, or strip),
5.1.4 Dimensions,
5.1.5 Type (Table 1),
5.1.6 Edge, strip only (see Specification A 480/A 480M),
5.1.7 Mechanical properties (Section 9),
5.1.8 Finish (see Note 3 and Specification A 480/A 480M), specify whether one or both sides are to be polished.
5.1.9 ASTM designation and date of issue,
5.1.10 Restrictions (if desired) for determining yield strength (see footnote B of Table 2), and
5.1.11 Additions to the specification or special requirements.

NOTE 1—A typical ordering description is as follows: 200 pieces, stainless steel sheets, 0.060 (1.52 mm) by 48 in. (1219 mm) by 120 in. (3048 mm), Type 410 No. 2B Finish, ASTM A 176 dated _____.

---

[1] This specification is under the jurisdiction of ASTM Committee A-1 on Steel, Stainless Steel and Related Alloys and is the direct responsibility of Subcommittee A01.17 on Flat Stainless Steel Products.
Current edition approved July 29, 1983. Published November 1983. Originally published as A 176 – 35 T. Last previous edition A 176 – 82a.
[2] *Annual Book of ASTM Standards*, Vol 01.03.

NOTE 2—The 409 grade is a titanium-stabilized type normally produced for applications such as heat and oxidation resistance, where smoothness and uniformity of finish are not of particular importance. It is generally produced with a hot- or cold-rolled, annealed and pickled finish. Because its surface appearance differs somewhat from a No. 1 or 2D finish obtained on the other steels of Table 1, it is not customary nor necessary to use finish numbers when ordering this steel.

6. Process

6.1 The steel shall be made by one or more of the following processes: electric-arc, electric-induction, or other suitable commercial processes.

7. Chemical Requirements

7.1 The steel shall conform to the requirements as to chemical composition specified in Table 1.

7.2 Methods and practices relating to chemical analysis required by this specification shall be in accordance with Methods, Practices, and Definitions A 751.

8. Mechanical Requirements

8.1 The material shall conform to the mechanical property requirements specified in Table 2.

9. Bending Requirements

9.1 The bend test specimen shall withstand cold bending through the angle specified in Table 2 without cracking on the outside of the bent portion.

10. Special Tests

10.1 If any special tests are required that are pertinent to the intended application of the material ordered, they shall be as agreed upon between the seller and the purchaser.

11. Test Specimens

11.1 Tension test specimens shall be taken from finished material and shall be selected either in the longitudinal or transverse direction. The tension test specimens shall conform to the appropriate Sections 7, 8, or 9 of Methods and Definitions A 370.

NOTE 3—For plate material up to and including ¾ in. (19.05 mm) thick, the sheet-type tension test specimen described in Section 8 (Fig. 4) of Methods and Definitions A 370 is also permitted.

11.2 Bend test specimens shall conform to Section 14 of Methods and Definitions A 370.

11.3 Hardness tests may be made on the grip ends of the tension test specimen before they are subjected to the tension test.

12. Number of Tests

12.1 In the case of sheet, strip, or plate produced in coil form, two or more hardness tests (one from each end of the coil); one bend test, when required; and one or more tension tests shall be made on specimens taken from each coil. If the hardness difference between the two ends of the coil exceeds 5HRB, or equivalent, tensile properties must be determined on both ends. When material is produced in sheet, strip, or plate form, one tension, one bend and one or more hardness tests shall be made on each 100 or less sheets, strips, or plates of the same gage and heat produced under the same processing conditions.

13. Test Methods

13.1 *Yield Strength*:

13.1.1 The yield strength shall be determined by the offset method as described in Methods and Definitions A 370. The offset shall be 0.2 % of the gage length of the specimen.

13.1.2 Between the yield strength and fracture of the specimen, the cross-head speed of the testing machine shall be between ⅛ in. (3.18 mm) and ½ in. (12.70 mm)/in. (25.4 mm) of gage length/min.

13.2 *Bend Test*:

13.2.1 Material up to and including ⅜ in. (9.52 mm) in thickness shall be bent over a single piece of flat stock of the same thickness as the material tested, allowing the test material to form its natural curvature.

NOTE 4—The bend may be made over a diameter equal to the thickness of the test material.

13.2.2 Material over ⅜ in. (9.52 mm) in thickness shall be bent over two pieces of flat stock, each being of the same thickness of the material tested, allowing the test material to form its natural curvature.

NOTE 5—The bend may be made over a diameter equal to two times the thickness of the test material, or over a single piece of flat stock twice the thickness of the test material.

13.2.3 The axis of the bend shall be parallel to the direction of rolling.

14. Permissible Variations in Dimensions

14.1 Unless otherwise specified in the purchase order, material shall conform to the permissible tolerances shown in Specification A 480/A 480M.

14.2 *Sheet, Strip, and Plate*—Material with No. 1 finish may be ground to remove surface defects, provided such grinding does not reduce the thickness or width at any point beyond the permissible variations in dimensions.

15. Workmanship

15.1 The material shall be of uniform quality consistent with good manufacturing and inspection practices. The steel shall have no defects of a nature or degree, for the grade and quality ordered, that will be detrimental to the samping, forming, machining, or fabrication of finished parts.

16. Inspection

16.1 Inspection of the material by the purchaser's representative at the producing plant shall be made as agreed upon between the purchaser and the seller as part of the purchase order.

17. Certification

17.1 Upon request of the purchaser in the contract or order, the producer's certification that the material was manufactured and tested in accordance with this specification together with a report of the test results shall be furnished at the time of shipment.

18. Rejection

18.1 Unless otherwise specified, any rejection based on tests made in accordance with this specification shall be reported to the seller within 60 working days from receipt of the material by the purchaser.

19. Rehearing

19.1 Samples tested by the purchaser in accordance with this specification that represent rejected material shall be retained for three weeks from the date of the notification to the seller of the rejection. In case of dissatisfaction with the results of the tests, the seller may make claim for a rehearing within that time.

A 176

TABLE 1 Chemical Requirements[A]

UNS Designation[B]	Type	Composition, %								
		Carbon	Manganese	Phosphorus	Sulfur	Silicon	Chromium	Nickel	Nitrogen	Other Elements
S40300	403	0.15	1.00	0.040	0.030	0.50	11.50–13.00	0.60	...	...
S40500	405	0.08	1.00	0.040	0.030	1.00	11.50–14.50	0.60	...	Al 0.10–0.30
S40900	409	0.08	1.00	0.045	0.045	1.00	10.50–11.75	0.50	...	Ti 6 × C min, 0.75 max
S41000	410	0.15	1.00	0.040	0.030	1.00	11.50–13.50	0.75	...	...
S41008	410S	0.08	1.00	0.040	0.030	1.00	11.50–13.50	0.60	...	...
S41050	...	0.040	1.00	0.045	0.030	1.00	10.50–12.50	0.60–1.10	0.10	...
S42900	429	0.12	1.00	0.040	0.030	1.00	14.00–16.00	0.75	...	...
S43000	430	0.12	1.00	0.040	0.030	1.00	16.00–18.00	0.75	...	...
S44200	442	0.20	1.00	0.040	0.040	1.00	18.00–23.00	0.60	...	...
S44400	...	0.025	1.00	0.040	0.030	1.00	17.50–19.50	1.00	0.035	Mo 1.75–2.50 Ti + Cb 0.20 + 4 (C + N) min, 0.80 max
S44600	446	0.20	1.50	0.040	0.030	1.00	23.00–27.00	0.75	0.25	...
S44626	XM-33	0.06	0.75	0.040	0.020	0.75	25.0–27.0	0.50	0.04	Mo 0.75–1.50 Ti 0.20–1.00 and 7 (C + N) min Cu 0.20 max
S44627	XM-27	0.010[C]	0.40	0.020	0.020	0.40	25.00–27.50	0.50	0.015[C]	Mo 0.75–1.50 Cu 0.20 max Ni + Cu 0.50 max Cb 0.05–0.20
S44635	...	0.025	1.00	0.040	0.030	0.75	24.5–26.0	3.5–4.5	0.035	Mo 3.5–4.5 Ti + Cb 0.20 + 4 (C + N) min, 0.80 max
S44660	...	0.030	1.00	0.040	0.030	1.00	25.0–28.0	1.0–3.50	0.040	Mo 3.0–3.50 Ti + Cb = 0.20 – 1.00 and 6 (C + N) min
S44700	...	0.010	0.30	0.025	0.020	0.20	28.0–30.0	0.15	0.020[D]	Mo 3.5–4.2 Cu 0.15 max
S44800	...	0.010	0.30	0.025	0.020	0.20	28.0–30.0	2.0–2.5	0.020[D]	Mo 3.5–4.2 Cu 0.15 max
S44735	...	0.030	1.00	0.040	0.030	1.00	28.00–30.00	1.00	0.045	Mo 3.60–4.20 Ti + Cb 0.20–1.00 Ti + Cb 6 (C+N) min

[A] Maximum unless range or minimum is indicated.
[B] New designation established in accordance with ASTM E 527 and SAE J1086, Recommended Practice for Numbering Metals and Alloys (UNS).
[C] Produce analysis tolerance over the maximum limit for carbon and nitrogen shall be 0.002 %.
[D] Carbon + nitrogen = 0.025 max.

TABLE 2 Mechanical Test Requirements

UNS Designation	Type	Tensile Strength, min		Yield Strength,[A,B]		Elongation in 2 in. or 50 mm, min, %	Hardness, Max[C]		Cold Bend, Deg[D]
		ksi	MPa	ksi	MPa		Brinell	Rockwell B	
S40300	403	70	485	30	205	25.0[E]	183	88	180
S40500	405	60	415	25	170	20.0	183	88	180
S40900	409	55	380	30	205	22.0[E]	...	80	180
S41000	410	65	450	30	205	22.0[E]	217	95	180
S41008	410S	60	415	30	205	22.0[E]	183	88	180
S41050	...	60	415	30	205	22.0	183	88	180
S42900	429	65	450	30	205	22.0[E]	183	88	180
S43000	430	65	450	30	205	22.0[E]	183	88	180
S44200	442	75	515	40	275	20.0	217	95	180
S44400	...	60	415	40	275	20.0	217	95	180
S44600	446	75	515	40	275	20.0	217	95	135
S44626	XM-33	68	470	45	310	20.0	217	95	180
S44627	XM-27	65	450	40	275	22.0[E]	190	90	180
S44635	...	90	620	75	515	20.0	270	27[F]	180
S44660	...	80	550	55	380	18.0	241	100	180
S44700	...	80	550	60	415	20.0	...	20[F]	not reqd
S44800	...	80	550	60	415	20.0	...	20[F]	not reqd
S44735	...	80	550	60	415	18.0	...	25[E]	180

[A] See 13.1.1.
[B] Yield strength shall be determined by the offset method at 0.2 % limiting permanent set in accordance with Methods and Definitions A 370. Unless otherwise specified (see 5.1.10), an alternative method of determining yield strength may be based on a total extension under load of 0.5 %.
[C] Either Brinell or Rockwell B hardness is permissible.
[D] See 13.2.
[E] Material 0.050 in. (1.27 mm) and under in thickness shall have a minimum elongation of 20.0 %.
[F] Rockwell C scale.

The American Society for Testing and Materials takes no position respecting the validity of any patent rights asserted in connection with any item mentioned in this standard. Users of this standard are expressly advised that determination of the validity of any such patent rights, and the risk of infringement of such rights, are entirely their own responsibility.

This standard is subject to revision at any time by the responsible technical committee and must be reviewed every five years and if not revised, either reapproved or withdrawn. Your comments are invited either for revision of this standard or for additional standards and should be addressed to ASTM Headquarters. Your comments will receive careful consideration at a meeting of the responsible technical committee, which you may attend. If you feel that your comments have not received a fair hearing you should make your views known to the ASTM Committee on Standards, 1916 Race St., Philadelphia, Pa. 19103.

Designation: A 177 – 82

Standard Specification for
HIGH-STRENGTH STAINLESS AND HEAT-RESISTING CHROMIUM-NICKEL STEEL SHEET AND STRIP[1]

This standard is issued under the fixed designation A 177; the number immediately following the designation indicates the year of original adoption or, in the case of revision, the year of last revision. A number in parentheses indicates the year of last reapproval. A superscript epsilon (ϵ) indicates an editorial change since the last revision or reapproval.

This specification has been approved for use by agencies of the Department of Defense for listing in the DoD Index of Specifications and Standards.

1. Scope

1.1 This specification covers high-strength stainless and heat resisting chromium-nickel steel sheet and strip.

1.2 The values stated in inch-pound units are to be regarded as the standard.

2. Applicable Documents

2.1 *ASTM Standards:*
A 370 Methods and Definitions for Mechanical Testing of Steel Products[2]
A 480/A 480M Specification for General Requirements for Flat-Rolled Stainless and Heat-Resisting Steel Plate, Sheet, and Strip[2]
A 751 Methods, Practices, and Definitions for Chemical Analysis of Steel Products[2]

3. Descriptions of Terms

3.1 Sheet and strip as used in this specification apply to the following:

3.1.1 *Sheet*—Material under ³⁄₁₆ in. (4.76 mm) in thickness and 24 in. (609.6 mm) and over in width.

3.1.2 *Strip*—Cold-rolled material ³⁄₁₆ in. (4.76 mm) and under in thickness and under 24 in. (610 mm) in width.

4. General Requirements for Delivery

4.1 Material furnished under this specification shall conform to the applicable requirements of the current edition of Specification A 480/A 480M.

5. Ordering Information

5.1 Orders for material under this specification shall include the following information:

5.1.1 Quantity (weight and number of pieces),
5.1.2 Name of material (stainless steel),
5.1.3 Condition (cold-rolled),
5.1.4 Form (sheet or strip),
5.1.5 Dimensions,
5.1.6 Edge (strip only) (see Specification A 480/A 480M),
5.1.7 Type,
5.1.8 ASTM designation, and
5.1.9 Additions to specification or special requirements.

NOTE 1—A typical ordering description is as follows: 200 pieces, high-strength stainless steel sheets, 0.060 by 48 by 120 in., Type 301, ½-hard temper, tensile strength 150 000 psi min, ASTM A 177.

6. Process

6.1 The steel shall be made by one or more of the following processes: electric-arc, electric-induction, or other suitable commercial processes.

7. Chemical Requirements

7.1 The steel shall conform to the following chemical composition:

---

[1] This specification is under the jurisdiction of ASTM Committee A-1 on Steel, Stainless Steel and Related Alloys and is the direct responsibility of Subcommittee A01.17 on Flat Stainless Steel Products.
Current edition approved Oct. 29, 1982. Published December 1982. Originally published as A 177 – 35 T. Last previous edition A 177 – 80.
[2] *1983 Annual Book of ASTM Standards*, Vol 01.03.

A 177

Element	Composition, %
Carbon, max	0.15
Manganese, max	2.00
Phosphorus, max	0.045
Sulfur, max	0.030
Silicon, max	1.00
Chromium	16.00–18.00
Nickel	6.00–8.00

7.2 Methods and practices relating to chemical analysis required by this specification shall be in accordance with Methods, Practices, and Definitions A 751.

8. Finish

8.1 The finish of sheet and strip in each of the tempers specified in Table 1 is that obtained by the cold rolling necessary to develop the required mechanical properties.

9. Mechanical Requirements

9.1 *Tensile Properties*—The material shall conform to the mechanical properties specified in Table 1.

9.2 *Bending Properties*—The bend test specimens shall withstand cold bending through the angles specified in Table 2 and Table 3, without cracking on the outside of the bent portion.

10. Test Specimens

10.1 Tension and bend test specimens shall be taken from finished material and shall be selected in the transverse direction, except in the case of strip under 9 in. (228.6 mm) in width, in which case tension test specimens shall be selected in the longitudinal direction.

10.1.1 The axis of the bend shall be parallel to the direction of rolling.

11. Number of Tests

11.1 In the case of sheet and strip produced in coil form, two or more hardness tests (one from each end of the coil); one bend test, when required; and one or more tension tests shall be made on specimens taken from each coil. If the hardness difference between the two ends of the coil exceeds 5HRB, or equivalent, tensile properties must be determined on both ends. When material is produced in sheet or strip form, at least one tension test and one bend test shall be made on each 100 or less sheets or strips of the same gage and heat produced under the same processing conditions.

12. Test Methods

12.1 *Yield Strength*:

12.1.1 The yield strength shall be determined by the offset method as described in Methods and Definitions A 370. The limiting permanent offset shall be 0.2 % of the gage length of the specimen. An alternative method of determining yield strength may be used, based on the following total extensions under load:

Yield Strength min, psi	Yield Strength, min, MPa	Total Extension Under Load in 2-in. or 50-mm Gage Length, in. (mm)
75 000	515	0.0098 (0.25)
110 000	530	0.0125 (0.32)
135 000	655	0.0144 (0.37)
140 000	675	0.0148 (0.38)

12.1.2 The requirements of this specification for yield strength will be considered as having been fulfilled if the extension under load for the specified yield strength does not exceed the specified values.

NOTE 2—The value obtained in this manner should not, however, be taken as the actual yield strength for 0.2 % permanent set.

12.2 *Tension Test*—Between yield and fracture of the test specimen, the tests shall be conducted at a constant strain rate between ⅛ in./in (0.125 mm/mm)/min and ½ in./in. (0.500 mm/mm)/min, inclusive, or at a crosshead speed that will give a strain rate within this range. For the purposes of this specification, the rate of strain may be determined by a strain rate pacer, indicator, or controller, or by dividing the unit elongation by the elapsed time from yield strength to fracture.

12.3 *Bend Test*—Specimens shall be bent around a diameter equal to the bend factor, shown in Table 2 and Table 3, times the nominal thickness of the test specimen.

13. Workmanship

13.1 The material shall be of uniform quality consistent with good manufacturing and inspection practices. The steel quality shall be satisfactory for the production or fabrication of finished parts.

14. Inspection

14.1 Inspection of the material by the purchaser's representative at the producing plant shall be made as agreed upon by the purchaser and the seller as part of the purchase order.

A 177

15. Certification

15.1 Upon request of the purchaser in the contract or order, the producer's certification that the material was manufactured and tested in accordance with this specification together with a report of the test results shall be furnished at the time of shipment.

16. Rehearing

16.1 Samples tested by the purchaser in accordance with this specification that represent rejected material shall be retained for 3 weeks from the date of the notification to the seller of the rejection. In case of dissatisfaction with the results of the tests, the seller may make claim for a rehearing within that time.

TABLE 1 Tensile Strength Requirements

Temper	Tensile Strength, min		Yield Strength,[A] min		Elongation in 2 in. or 50 mm, min, %	
	ksi	MPa	ksi	MPa	Up to 0.015 in. (0.38 mm) in thickness	Over 0.015 in. (0.38 mm) in thickness
¼ hard	125	860	75	515	25	25
½ hard	150	1035	110	760	15	18
¾ hard	175	1205	135	930	10	12
Full hard	185	1275	140	965	8	9

[A] See 12.1.

TABLE 2 Bend Test Requirements, Free Bend

Temper	Thickness 0.050 in. (1.27 mm) and Under		Thickness Over 0.050 in. (1.27 mm) Through 0.187 in. (4.75 mm)	
	Bend Angle, deg	Bend Factor[A]	Bend Angle, deg	Bend Factor[A]
¼ hard	180	1	90	2
½ hard	180	2	90	2
¾ hard	180	3	...	...
Full hard	180	4	...	...

[A] See 12.3.

TABLE 3 Bend Test Requirements, Controlled bend (V Block)

Temper	Thickness 0.050 in. (1.27 mm) and Under		Thickness Over 0.050 in. (1.27 mm) Through 0.187 in. (4.75 mm)	
	Bend Angle, deg	Bend Factor[A]	Bend Angle, deg	Bend Factor[A]
¼ hard	135	2	135	3
½ hard	135	4	135	4
¾ hard	135	6	...	...
Full hard	135	6	...	...

[A] See 12.3.

The American Society for Testing and Materials takes no position respecting the validity of any patent rights asserted in connection with any item mentioned in this standard. Users of this standard are expressly advised that determination of the validity of any such patent rights, and the risk of infringement of such rights, are entirely their own responsibility.

This standard is subject to revision at any time by the responsible technical committee and must be reviewed every five years and if not revised, either reapproved or withdrawn. Your comments are invited either for revision of this standard or for additional standards and should be addressed to ASTM Headquarters. Your comments will receive careful consideration at a meeting of the responsible technical committee, which you may attend. If you feel that your comments have not received a fair hearing you should make your views known to the ASTM Committee on Standards, 1916 Race St., Philadelphia, Pa. 19103.

Designation: A 227/A 227M – 83

Standard Specification for
STEEL WIRE, COLD-DRAWN FOR MECHANICAL SPRINGS[1]

This standard is issued under the fixed designation A 227/A 227M; the number immediately following the designation indicates the year of original adoption or, in the case of revision, the year of last revision. A number in parentheses indicates the year of last reapproval. A superscript epsilon (ϵ) indicates an editorial change since the last revision or reapproval.

1. Scope

1.1 This specification covers two classes of round cold-drawn steel spring wire having properties and quality for the manufacture of mechanical springs that are not subject to high stress or requiring high fatigue properties and wire forms.

1.2 The values stated in either SI (metric) units or inch-pound units are to be regarded separately as standard. The values stated in each system are not exact equivalents; therefore, each system must be used independent of the other.

2. Applicable Documents

2.1 *ASTM Standards:*
A 370 Methods and Definitions for Mechanical Testing of Steel Products[2]
A 510 Specification for General Requirements for Wire Rods and Coarse Round Wire, Carbon Steel[3]
A 510M Specification for General Requirements for Wire Rods and Coarse Round Wire, Carbon Steel [Metric][3]
A 700 Practices for Packaging, Marking, and Loading Methods for Steel Products for Domestic Shipment[2]
A 751 Methods, Practices, and Definitions for Chemical Analysis of Steel Products[2]
E 29 Recommended Practice for Indicating Which Places of Figures Are to Be Considered Significant in Specified Limiting Values[4]
E 380 Metric Practice[4]

2.2 *American National Standard:*
B32.4M Preferred Metric Sizes for Round, Square, Rectangle, and Hexogon Metal Products[5]

2.3 *Military Standard:*
MIL-STD-163 Steel Mill Products, Preparation for Shipment and Storage[6]

2.4 *Federal Standard:*
Fed. Std. No. 123 Marking for Shipment (Civil Agencies)[6]

3. Ordering Information

3.1 Orders for material under this specification shall include the following of each item:
3.1.1 Quantity (mass),
3.1.2 Name of material (cold-drawn steel mechanical spring wire) and class (Table 2),
3.1.3 Wire diameter (Table 2),
3.1.4 Packaging (Section 13),
3.1.5 Cast or heat analysis report, if requested (Section 5),
3.1.6 Certification or test report, or both, if specified (Section 12), and
3.1.7 ASTM designation and date of issue.

NOTE 1—A typical ordering description is as follows: 15 000 kg Cold-Drawn Mechanical Spring Wire, Class I, Size 5.00 mm in 700-kg coils to ASTM A 227M dated ——, or for non-SI units, 30 000 lb Cold-Drawn Mechanical Spring Wire, Class I, Size 0.207 in. diameter in 500-lb coils to ASTM A 227 dated ——.

---

[1] This specification is under the jurisdiction of ASTM committee A-1 on Steel, Stainless Steel, and Related Alloys and is the direct responsibility of Subcommittee A01.03 on Steel Rod and Wire.
Current edition approved July 29, 1983. Published November 1983. Originally published as A 227 – 39 T and A 227M – 80. Last previous edition A 227 – 77 and A 227M – 80.
[2] *Annual Book of ASTM Standards*, Vol 01.04.
[3] *Annual Book of ASTM Standards*, Vol 01.03.
[4] *Annual Book of ASTM Standards*, Vol 14.02.
[5] Available from American National Standards Institute, 1430 Broadway, New York, N. Y. 10018.
[6] Available from Naval Publications and Forms Center, 5801 Tabor Ave., Philadelphia, Pa. 19120.

A 227/A 227M

4. Manufacture

4.1 The steel may be made by any commercially accepted steel-making process. The steel may be either ingot cast or strand cast.

4.2 The finished wire shall be free of detrimental pipe and undue segregation.

4.3 The wire shall be cold drawn to produce the desired mechanical properties.

5. Chemical Composition

5.1 The steel shall conform to the requirements for chemical composition prescribed in Table 1.

5.2 *Cast or Heat Analysis*—Each cast or heat of steel shall be analyzed by the manufacturer to determine the percentage of elements prescribed in Table 1. This analysis shall be made from a test specimen preferably taken during the pouring of the cast or heat. When requested, this shall be reported to the purchaser and shall conform to the requirements of Table 1.

5.3 *Product Analysis*—An analysis may be made by the purchaser from finished wire representing each cast or heat of steel. The chemical composition thus determined, as to elements required or restricted, shall conform to the product analysis requirements specified in Table 10 of Specification A 510 or A 510M.

5.4 For referee purposes, Methods, Practices, and Definitions A 751 shall be used.

6. Mechanical Properties

6.1 *Tension Test:*

6.1.1 *Requirements*—The material as represented by tension test specimens shall conform to the requirements prescribed in Table 2A or Table 2B.

6.1.2 *Number of Tests*—One test specimen shall be taken for each ten coils or fraction thereof, in a lot. Each cast or heat in a given lot shall be tested.

6.1.3 *Location of Tests*—Test specimens shall be taken from either end of the coil.

6.1.4 *Test Method*—The tension test shall be made in accordance with Methods and Definitions A 370.

6.2 *Wrap Test:*

6.2.1 *Requirements*—The material as represented by the wrap test specimens shall conform to the requirements specified in Table 3A or Table 3B. Wrap test on wires over 8.5 mm or 0.312 in. in diameter is not applicable. Since the conventional methods will not accommodate over 8.5 mm or 0.312 in., an alternative test procedure may be agreed upon between purchaser and producer.

6.2.2 *Number of Tests*—One test specimen shall be taken for each ten coils, or fraction thereof, in a lot. Each cast or heat in a given lot shall be tested.

6.2.3 *Location of Test*—Test specimens shall be taken from either end of the coil.

6.2.4 *Test Method*—The wrap test shall be made in accordance with Methods and Definitions A 370, Supplement IV.

7. Dimensions and Permissible Variations

7.1 The permissible variations in the diameter of the wire shall be as specified in Table 4A or Table 4B.

8. Workmanship and Appearance

8.1 *Workmanship*—The wire shall not be kinked or improperly cast. To test for cast, a few convolutions of wire shall be cut from the coil and placed on a flat surface. The wire shall lie flat on itself and not spring up excessively nor show a wavy condition.

8.1.1 Each coil shall be one continuous length of wire, properly coiled and firmly tied. Welds made prior to cold drawing are permitted.

8.2 *Appearance*—The surface shall be smooth and free of defects such as seams, pits, die marks, and other defects tending to impair the use of the wire for springs. Any additional surface requirements must be negotiated at the time of entry of the order.

9. Retests

9.1 If any test specimen exhibits obvious defects or shows the presence of a weld, it may be discarded and another specimen substituted.

10. Inspection

10.1 Unless otherwise specified in the contract or purchase order, the manufacturer is responsible for the performance of all inspection and test requirements specified in this specification. Except as otherwise specified in the contract or purchase order, the manufacturer may use his own or any other suitable facilities for the performance of the inspection and test requirements unless disapproved by the purchaser at the time the order is placed. The purchaser shall have the right to perform any of

A 227/A 227M

the inspections and tests set forth in this specification when such inspections and tests are deemed necessary to assure that the material conforms to prescribed requirements.

11. Rejection and Rehearing

11.1 Unless otherwise specified, any rejection based on tests made in accordance with this specification shall be reported to the manufacturer as soon as possible so that an investigation may be initiated.

11.2 The material must be adequately protected and correctly identified in order that the manufacturer may make a proper investigation.

12. Certification

12.1 When specified in the purchase order or contract, a manufacturer's or supplier's certification shall be furnished to the purchaser that the material was manufactured, sampled, tested, and inspected in accordance with this specification and has been found to meet the requirements. When specified in the purchase order or contract, a report of the test results shall be furnished.

13. Packaging, Marking, and Loading for Shipment

13.1 The coil mass, dimensions, and the method of packaging shall be agreed upon between the manufacturer and purchaser.

13.2 The size of the wire, purchaser's order number, ASTM specification number, heat number, and name or mark of the manufacturer shall be marked on a tag securely attached to each coil of wire.

13.3 Unless otherwise specified in the purchaser's order, packaging, marking, and loading for shipments shall be in accordance with those procedures recommended by Practices A 700.

13.4 *For Government Procurement*—Packaging, packing, and marking of material for military procurement shall be in accordance with the requirements of MIL-STD-163, Level A, Level C, or commercial as specified in the contract or purchase order. Marking for shipment of material for civil agencies shall be in accordance with Fed. Std. No. 123.

TABLE 1 Chemical Requirements

Element	Composition, %
Carbon	0.45–0.85[A]
Manganese	0.30–1.30[B]
Phosphorus, max	0.040
Sulfur, max	0.050
Silicon	0.15–0.35

[A] Carbon in any one lot may not vary more than 0.13 %.
[B] Manganese in any one lot may not vary more than 0.30 %.

A 227/A 227M

TABLE 2A Tensile Requirements, SI Units[A]

Diameter,[B] mm	Class I Tensile Strength, MPa min	Class I Tensile Strength, MPa max	Class II Tensile Strength, MPa min	Class II Tensile Strength, MPa max
0.50	1960	2240	2240	2520
0.55	1940	2220	2220	2500
0.60	1920	2200	2200	2480
0.65	1900	2180	2180	2460
0.70	1870	2140	2140	2410
0.80	1830	2100	2100	2370
0.90	1800	2070	2070	2340
1.00	1770	2040	2040	2310
1.10	1740	2000	2000	2260
1.20	1720	1980	1980	2240
1.40	1670	1930	1930	2180
1.60	1640	1880	1880	2120
1.80	1600	1840	1840	2080
2.00	1580	1810	1810	2040
2.20	1550	1780	1780	2010
2.50	1510	1730	1730	1960
2.80	1480	1700	1700	1920
3.00	1460	1680	1680	1900
3.50	1420	1630	1630	1840
4.00	1380	1590	1600	1700
4.50	1350	1550	1550	1750
5.00	1320	1510	1510	1700
5.50	1300	1490	1490	1670
6.00	1280	1470	1470	1650
6.50	1250	1440	1440	1630
7.00	1220	1410	1410	1600
7.50	1200	1390	1390	1580
8.00	1190	1370	1370	1550
9.00	1160	1340	...	...
10.00	1130	1310	...	...
11.00	1110	1280	...	...
12.00	1090	1260	...	...
14.00	1050	1210	...	...
16.00	1010	1170	...	...

[A] Tensile strength values for intermediate diameters may be interpolated.
[B] Preferred sizes. For a complete list, refer to ANSI B32.4M, Preferred Metric Sizes for Round, Square, Rectangle, and Hexagon Metal Products.

TABLE 2B Tensile Requirements, Inch-Pound Units[A]

Diameter, in.	Class I Tensile Strength, ksi min	Class I Tensile Strength, ksi max	Class II Tensile Strength, ksi min	Class II Tensile Strength, ksi max
0.020	283	323	324	364
0.023	279	319	320	360
0.026	275	315	316	356
0.029	271	311	312	352
0.032	266	306	307	347
0.035	261	301	302	342
0.041	255	293	294	332
0.048	248	286	287	325
0.054	243	279	280	316
0.062	237	272	273	308
0.072	232	266	267	301
0.080	227	261	262	296
0.092	220	253	254	287
0.106	216	248	249	281
0.120	210	241	242	273
0.135	206	237	238	269
0.148	203	234	235	266
0.162	200	230	231	261
0.177	195	225	226	256
0.192	192	221	222	251
0.207	190	218	219	247
0.225	186	214	215	243
0.250	182	210	211	239
0.312	174	200	201	227
0.375	167	193	194	220
0.438	165	190	191	216
0.500	156	180	181	205
0.562	152	176	177	201
0.625	147	170	171	294

[A] Tensile strength values for intermediate diameters may be interpolated.

TABLE 3A Wrap Test Requirements, SI Units

Diameter, mm	Mandrel Size Class I	Mandrel Size Class II
0.50 to 4.0, incl	1X[A]	2X
Over 4.0 to 8.0, incl	2X	4X

[A] For 1X mandrel, wire may be wrapped on itself.

TABLE 3B Wrap Test Requirements, Inch-Pound Units

Diameter, in.	Mandrel Size Class I	Mandrel Size Class II
0.020 to 0.162, incl	1X[A]	2X
Over 0.162 to 0.312, incl	2X	4X

[A] For 1X mandrel, wire may be wound on itself.

A 227/A 227M

TABLE 4A Permissible Variations in Wire Diameter, SI Units[A]

Diameter, mm	Permissible Variations, plus and minus, mm	Permissible Out-of Round, mm
To 0.70, incl	0.02	0.02
Over 0.70 to 2.00, incl	0.03	0.03
Over 2.00 to 9.00, incl	0.05	0.05
Over 9.00	0.08	0.08

[A] For purposes of determining conformance with this specification, all specified limits are absolute as defined in Recommended Practice E 29.

TABLE 4B Permissible Variations in Wire Diameter, Inch-Pound Units[A]

Diameter, in.	Permissible Variations, plus and minus, in.	Permissible Out-of-Round, in.
0.020 to 0.028, incl	0.0008	0.0008
Over 0.028 to 0.075, incl	0.001	0.001
Over 0.075 to 0.375, incl	0.002	0.002
Over 0.375 to 0.625, incl	0.003	0.003

[A] For purposes of determining conformance with this specification, all specified limits are absolute as defined in Recommended Practice E 29.

The American Society for Testing and Materials takes no position respecting the validity of any patent rights asserted in connection with any item mentioned in this standard. Users of this standard are expressly advised that determination of the validity of any such patent rights, and the risk of infringement of such rights, are entirely their own responsibility.

This standard is subject to revision at any time by the responsible technical committee and must be reviewed every five years and if not revised, either reapproved or withdrawn. Your comments are invited either for revision of this standard or for additional standards and should be addressed to ASTM Headquarters. Your comments will receive careful consideration at a meeting of the responsible technical committee, which you may attend. If you feel that your comments have not received a fair hearing you should make your views known to the ASTM Committee on Standards, 1916 Race St., Philadelphia, Pa. 19103.

Designation: A 228/A 228M – 83

Standard Specification for
STEEL WIRE, MUSIC SPRING QUALITY[1]

This standard is issued under the fixed designation A 228/A 228M; the number immediately following the designation indicates the year of original adoption or, in the case of revision, the year of last revision. A number in parentheses indicates the year of last reapproval. A superscript epsilon (ε) indicates an editorial change since the last revision or reapproval.

This specification has been prepared for use by agencies of the Department of Defense and for listing in the DoD Index of Specifications and Standards.

1. Scope

1.1 This specification covers a high quality, round, cold-drawn steel music spring quality wire, uniform in mechanical properties, intended specially for the manufacture of springs subject to high stresses or requiring good fatigue properties.

1.2 The values stated in either SI (metric) units or inch-pound units are to be regarded separately as standard. The values stated in each system are not exact equivalents; therefore, each system must be used independently of the other.

2. Applicable Documents

2.1 *ASTM Standards:*
A 370 Methods and Definitions for Mechanical Testing of Steel Products[2]
A 510 Specification for General Requirements for Wire Rods and Coarse Round Wire, Carbon Steel[3]
A 510M Specification for General Requirements for Wire Rods and Coarse Round Wire, Carbon Steel [Metric][3]
A 700 Practices for Packaging, Marking, and Loading Methods for Steel Products for Domestic Shipment[2]
A 751 Methods, Practices, and Definitions for Chemical Analysis of Steel Products[2]
E 29 Recommended Practice for Indicating which Places of Figures are to be Considered Significant in Specified Limiting Values[4]
E 380 Metric Practice[4]

2.2 *Military Standard:*
MIL-STD-163 Steel Mill Products, Preparation for Shipment and Storage[5]

2.3 *Federal Standard:*
Fed. Std. No. 123, Marking for Shipment (Civil Agencies)[5]

3. Ordering Information

3.1 Material furnished under this specification shall conform to the applicable requirements of the current edition of either Specifications A 510M or A 510.

3.2 Orders for material under this specification shall include the following information for each ordered item:

3.2.1 Quantity (mass),

3.2.2 Name of material (music steel spring wire),

3.2.3 Dimensions (Table 1 and Section 7),

3.2.4 Finish (see 8.2),

3.2.5 Packaging (Section 13),

3.2.6 Heat analysis report, if requested (see 5.2),

3.2.7 Certification or test report, or both, if specified (Section 12), and

3.2.8 ASTM designation and year of issue.

NOTE—A typical metric ordering description is as follows: 2500 kg Music Spring Wire, 1.40 mm diameter, phosphate coated in 25 kg coils to ASTM A228 dated ___, or for inch-pound units, 5000 lb Music Spring Wire, 0.055 in. diameter, phosphate coated in 50 lb coils to ASTM A228 dated ___.

---

[1] This specification is under the jurisdiction of ASTM Committee A-1 on Steel, Stainless Steel, and Related Alloys and is the direct responsibility of Subcommittee. A01.03 on Steel Rod and Wire.
Current edition approved July 29, 1983. Published November 1983. Originally published as A 228 – 39 T. Last previous edition A 228 – 77.
[2] *Annual Book of ASTM Standards*, Vol. 01.04.
[3] *Annual Book of ASTM Standards*, Vol. 01.03.
[4] *Annual Book of ASTM Standards*, Vol. 14.02.
[5] Available from Naval Publications and Forms Center, 5801 Tabor Ave., Philadelphia, PA 19120.

4. Manufacture

4.1 The steel may be made by any commercially accepted steel-making process. The steel may be either ingot cast or strand cast.

4.2 The finished wire shall be free from detrimental pipe and undue segregation.

4.3 The wire shall be cold drawn to produce the desired mechanical properties.

5. Chemical Requirements

5.1 The steel shall conform to the requirements for chemical composition prescribed in Table 2.

5.2 *Heat Analysis*—Each heat of steel shall be analyzed by the manufacturer to determine the percentage of elements prescribed in Table 2. This analysis shall be made from a test specimen preferably taken during the pouring of the heat. When requested, this shall be reported to the purchaser and shall conform to the requirements of Table 2.

5.3 *Product Analysis*—An analysis may be made by the purchaser from finished wire representing each heat of steel. The chemical composition thus determined, as to elements required or restricted, shall conform to the product analysis requirements specified in Table 10 of Specification A 510M or A 510.

5.4 For referee purposes, Methods A 751 shall be used.

6. Mechanical Requirements

6.1 *Tension Test:*

6.1.1 *Requirements*—The material as represented by tension test specimens shall conform to the requirements prescribed in Table 1.

6.1.2 *Number of Tests*—One test specimen shall be taken from each end of every coil or from the top or outside end of each reel or spool.

6.1.3 *Test Method*—The tension test shall be made in accordance with Methods A 370.

6.2 *Coiling Test:*

6.2.1 *Requirements*—The coiling test shall be applied only to sizes 2.6 mm or 0.105 in. and smaller in diameter.

6.2.2 *Number of Tests*—Specimens taken from each end of every coil or from the top or outside end of each reel or spool shall be tested for conformance.

6.2.3 *Test Method*—A length of wire as shown in Table 3 shall be closed wound on an arbor three to three and one-half times the diameter of the wire. The closed coil shall be stretched so that it sets to approximately three times its original length. The wire so tested shall show a uniform pitch with no splits or fractures.

7. Dimensions and Permissible Variations

7.1 The permissible variations in the diameter of the wire shall be as specified in Table 4.

8. Workmanship, Finish, and Appearance

8.1 *Workmanship*—The wire shall not be kinked or improperly cast. To test for cast, one convolution of wire shall be cut from the coil, reel, or spool and placed on a flat surface. The wire shall not spring up nor show a wavy condition. Wire below 4.0 mm or 0.156 in. diameter shall lie flat while wire diameters larger than the above shall lie substantially flat.

8.1.1 Each coil, reel, or spool shall be one continuous length of wire, properly coiled and firmly tied.

8.1.2 Welds made prior to cold drawing are permitted. If unmarked welds are unacceptable to the purchaser, special arrangements should be made with the manufacturer at the time of purchase.

8.2 *Finish*—Music wire is supplied with many different types of finish such as bright, phosphate, tin, and others. Finish desired should be specified on purchase orders.

8.3 *Appearance*—The surface shall be smooth and free from defects such as seams, pits, die marks and other defects tending to impair the use of the wire for springs. Any additional surface requirements must be negotiated at the time of entry of the order.

9. Retests

9.1 If any test specimen exhibits obvious defects or shows the presence of a weld, it may be discarded and another specimen substituted.

10. Inspection

10.1 Unless otherwise specified in the contract or purchase order, the manufacturer is responsible for the performance of all inspection and test requirements specified in this specification. Except as otherwise specified in the contract or purchase order, the manufacturer may use his own or any other suitable facilities for the performance of the inspection and test requirements unless disapproved by the purchaser at the time the order is placed. The purchaser shall have the right to perform any of the inspections and tests set forth in this specification when such inspections and tests are deemed necessary to assure that the material conforms to prescribed requirements.

11. Rejection and Rehearing

11.1 Unless otherwise specified, any rejection based on tests made in accordance with these specifications shall be reported to the manufacturer as soon as possible so that an investigation may be initiated.

11.2 The material must be adequately protected and correctly identified in order that the manufacturer may make a proper investigation.

12. Certification

12.1 When specified in the purchase order or contract, a producer's or supplier's certification shall be furnished to the purchaser that the material was manufactured, sampled, tested, and inspected in accordance with this specification and has been found to meet the requirements. When specified in the purchase order or contract, a report of the test results shall be furnished.

13. Packaging, Marking, and Loading for Shipment

13.1 The coil, reel or spool mass, dimensions, and the method of packaging shall be agreed upon between the manufacturer and purchaser.

13.2 The size of the wire, purchaser's order number, ASTM specification number, heat number, and name or mark of the manufacturer shall be marked on a tag securely attached to each coil, reel or spool of wire.

13.3 Unless otherwise specified in the purchaser's order, packaging, marking, and loading for shipments shall be in accordance with those procedures recommended by Practices A 700.

13.4 *For Government Procurement:*

13.4.1 Packaging, packing, and marking of material for military procurement shall be in accordance with the requirements of MIL-STD-163, Level A, Level C, or commercial as specified in the contract or purchase order. Marking for shipment of material for civil agencies shall be in accordance with Fed. Std No. 123.

A 228/A 228M

TABLE 1 Tensile Requirements

SI Units

Diameter, mm[A,B]	Tensile Strength, MPa		Diameter, mm[A,B]	Tensile Strength, MPa	
	min	max		min	max
0.10	3000	3300	0.90	2200	2450
0.11	2950	3250	1.00	2150	2400
0.12	2900	3200	1.1	2120	2380
0.14	2850	3150	1.2	2100	2350
0.16	2800	3100	1.4	2050	2300
0.18	2750	3050	1.6	2000	2250
0.20	2700	3000	1.8	1980	2220
0.22	2680	2980	2.0	1950	2200
0.25	2650	2950	2.2	1900	2150
0.28	2620	2920	2.5	1850	2100
0.30	2600	2900	2.8	1820	2050
0.35	2550	2820	3.0	1800	2000
0.40	2500	2750	3.2	1780	1980
0.45	2450	2700	3.5	1750	1950
0.50	2400	2650	3.8	1720	1920
0.55	2380	2620	4.0	1700	1900
0.60	2350	2600	4.5	1680	1880
0.65	2320	2580	5.0	1650	1850
0.70	2300	2550	5.5	1620	1820
0.80	2250	2500	6.0	1600	1800

Inch-Pound Units

Diameter, in.[A]	Tensile Strength, ksi		Diameter, in.[A]	Tensile Strength, ksi	
	min	max		min	max
0.004	439	485	0.055	300	331
0.005	426	471	0.059	296	327
0.006	415	459	0.063	293	324
0.007	407	449	0.067	290	321
0.008	399	441	0.072	287	317
0.009	393	434	0.076	284	314
0.010	387	428	0.080	282	312
0.011	382	422	0.085	279	308
0.012	377	417	0.090	276	305
0.013	373	412	0.095	274	303
0.014	369	408	0.100	271	300
0.015	365	404	0.102	270	299
0.016	362	400	0.107	268	296
0.018	356	393	0.110	267	295
0.020	350	387	0.112	266	294
0.022	345	382	0.121	263	290
0.024	341	377	0.125	261	288
0.026	337	373	0.130	259	286
0.028	333	368	0.135	258	285
0.030	330	365	0.140	256	283
0.032	327	361	0.145	254	281
0.034	324	358	0.150	253	279
0.036	321	355	0.156	251	277
0.038	318	352	0.162	249	275
0.040	315	349	0.177	245	270
0.042	313	346	0.192	241	267
0.045	309	342	0.207	238	264
0.048	306	339	0.225	235	260
0.051	303	335	0.250	230	255

[A] Tensile strength values for intermediate diameters may be interpolated.
[B] Preferred sizes. For a complete list, refer to ANSI B32.4, Preferred Metric Sizes for Round, Square, Rectangle and Hexagon Metal Products.

TABLE 2 Chemical Requirements

Element	Composition, %
Carbon	0.70–1.00
Manganese	0.20–0.60
Phosphorus, max	0.025
Sulfur, max	0.030
Silicon	0.10–0.30

TABLE 3 Closed Coil Lengths vs Wire Diameter

SI Units

Closed Coil, Length, mm, min	Wire Diameter, mm
50	to 0.75, incl
75	Over 0.75 to 1.00, incl
100	Over 1.00 to 1.25, incl
125	Over 1.25

Inch-Pound Units

Closed Coil Length, in., min	Wire Diameter, in.
2	to 0.030, incl
3	Over 0.030 to 0.040, incl
4	Over 0.040 to 0.050, incl
5	Over 0.050

TABLE 4 Permissible Variations in Wire Diameter[A]

SI Units

Diameter, mm	Permissible Variations, plus and minus, mm	Permissible Out-of-Round, mm
to 0.25, incl	0.005	0.005
Over 0.25 to 0.70, incl	0.008	0.008
Over 0.70 to 1.50, incl	0.010	0.010
Over 1.50 to 2.00, incl	0.013	0.013
Over 2.00	0.03	0.03

Inch-Pound Units

Diameter, in.	Permissible Variations, plus and minus, in.	Permissible Out-of-Round, in.
0.004 to 0.010, incl	0.0002	0.0002
Over 0.010 to 0.028, incl	0.0003	0.0003
Over 0.028 to 0.063, incl	0.0004	0.0004
Over 0.063 to 0.080, incl	0.0005	0.0005
Over 0.080 to 0.250, incl	0.001	0.001

[A] For purposes of determining conformance with this specification, all specified limits are absolute as defined in Recommended Practice E 29.

The American Society for Testing and Materials takes no position respecting the validity of any patent rights asserted in connection with any item mentioned in this standard. Users of this standard are expressly advised that determination of the validity of any such patent rights, and the risk of infringement of such rights, are entirely their own responsibility.

This standard is subject to revision at any time by the responsible technical committee and must be reviewed every five years and if not revised, either reapproved or withdrawn. Your comments are invited either for revision of this standard or for additional standards and should be addressed to ASTM Headquarters. Your comments will receive careful consideration at a meeting of the responsible technical committee, which you may attend. If you feel that your comments have not received a fair hearing you should make your views known to the ASTM Committee on Standards, 1916 Race St., Philadelphia, Pa. 19103.

Designation: A 229/A 229M – 83

Standard Specification for
STEEL WIRE, OIL-TEMPERED FOR MECHANICAL SPRINGS[1]

This standard is issued under the fixed designation A 229/A 229M; the number immediately following the designation indicates the year of original adoption or, in the case of revision, the year of last revision. A number in parentheses indicates the year of last reapproval. A superscript epsilon (ϵ) indicates an editorial change since the last revision or reapproval.

1. Scope

1.1 This specification covers two classes of oil-tempered steel spring wire intended especially for the manufacture of mechanical springs and wire forms.

1.2 The values stated in either SI (metric) units or inch-pound units are to be regarded separately as standard. The values stated in each system are not exact equivalents; therefore, each system must be used independent of the other.

2. Applicable Documents

2.1 *ASTM Standards:*
A 370 Methods and Definitions for Mechanical Testing of Steel Products[2]
A 510 Specification for General Requirements for Wire Rods and Coarse Round Wire, Carbon Steel[3]
A 510M Specification for General Requirements for Wire Rods and Coarse Round Wire, Carbon Steel [Metric][3]
A 700 Practices for Packaging, Marking, and Loading Methods for Steel Products for Domestic Shipment[2]
A 751 Methods, Practices, and Definitions for Chemical Analysis of Steel Products[2]
E 29 Recommended Practice for Indicating Which Places of Figures Are to Be Considered Significant in Specified Limiting Values[4]
E 380 Metric Practice[4]

2.2 *American National Standard:*
B32.4M Preferred Metric Sizes for Round, Square, Rectangle, and Hexagon Metal Products[5]

2.3 *Military Standard:*
MIL-STD-163 Steel Mill Products, Preparation for Shipment and Storage[6]

2.4 *Federal Standard:*
Fed. Std. No. 123 Marking for Shipment (Civil Agencies)[6]

3. Ordering Information

3.1 Material furnished under this specification shall conform to the applicable requirements of the current edition of either specification A 510 or A 510M.

3.2 Orders for material under this specification shall include the following information for each ordered item:

3.1.1 Quantity (mass or weight),
3.1.2 Name of material (oil-tempered steel mechanical spring wire) and class (Table 2),
3.1.3 Dimensions (Section 7),
3.1.4 Chemical composition (Table 1), if required,
3.1.5 Packaging (Section 12),
3.1.6 Cast or heat analysis report, if desired (see 5.2),
3.1.7 Certification or test report, or both, if specified (Section 12), and
3.1.8 ASTM designation and date of issue.

---

[1] This specification is under the jurisdiction of ASTM Committee A-1 on Steel, Stainless Steel, and Related Alloys, and is the direct responsibility of Subcommittee A01.03 on Steel Rod and Wire.
Current edition approved July 29, 1983. Published November 1983. Originally published as A 229 – 39 T. Last previous edition A 229 – 77.
[2] *Annual Book of ASTM Standards*, Vol 01.04.
[3] *Annual Book of ASTM Standards*, Vol 01.03.
[4] *Annual Book of ASTM Standards*, Vol 14.02.
[5] Available from American National Standards Institute, 1430 Broadway, New York, N. Y. 10018.
[6] Available from Naval Publications and Forms Center, 5801 Tabor Ave., Philadelphia, Pa. 19120.

A 229/A 229M

NOTE 1—A typical metric ordering description is as follows: 10 000 kg oil-tempered steel mechanical spring wire, Class I, 8.00 mm diameter, in 250-kg coils to ASTM A 229M dated ——, or for inch-pound units, 20 000 lb oil-tempered steel mechanical Spring Wire, Class I 0.315 in. diameter, in 500-lb coils to ASTM A 229 dated ——.

4. Manufacture

4.1 The steel may be made by any commercially accepted steel-making process. The steel may be either ingot cast or strand cast.

4.2 The finished wire shall be free of detrimental pipe and undue segregation.

4.3 The wire shall be hardened and tempered to produce the desired mechanical properties.

5. Chemical Composition

5.1 The steel shall conform to the requirements for chemical composition prescribed in Table 1.

5.2 *Cast or Heat Analysis*—Each cast or heat of steel shall be analyzed by the manufacturer to determine the percentage of elements prescribed in Table 1. This analysis shall be made from a test specimen preferably taken during the pouring of the cast or heat. When requested, this shall be reported to the purchaser and shall conform to the requirements of Table 1.

5.3 *Product Analysis*—An analysis may be made by the purchaser from finished wire representing each cast or heat of steel. The chemical composition thus determined, as to elements required or restricted, shall conform to the product analysis requirements specified in Table 10 of Specification A 510 or A 510M.

5.4 For referee purposes, Methods, Practices, and Definitions A 751 shall be used.

6. Mechanical Properties

6.1 *Tension Test:*

6.1.1 *Requirements*—The material as represented by tension test specimens shall conform to the requirements prescribed in Table 2A or Table 2B.

6.1.2 *Number of Tests*—One test specimen shall be taken for each ten coils or fraction thereof, in a lot. Each cast or heat in a given lot shall be tested.

6.1.3 *Location of Tests*—Test specimens shall be taken from either end of the coil.

6.1.4 *Test Method*—The tension test shall be made in accordance with Methods and Definitions A 370.

6.2 *Wrap Test:*

6.2.1 *Requirements*—Wire 4.00 mm or 0.162 in. and smaller in diameter shall wind on itself as an arbor without breakage. Larger diameter wire, up to and including 8.00 mm or 0.312 in. shall wind, without breakage, on a mandrel twice the wire diameter. Wrap test on wires over 8.00 mm or 0.312 in. diameter is not applicable.

6.2.2 *Number of Tests*—One test specimen shall be taken for each ten coils, or fraction thereof, in a lot. Each cast or heat in a given lot shall be tested.

6.2.3 *Location of Test*—Test specimens shall be taken from either end of the coil.

6.2.4 *Test Method*—The wrap test shall be made in accordance with Methods and Definitions A 370, Supplement IV.

7. Dimensions and Permissible Variations

7.1 The permissible variations in the diameter of the wire shall be as specified in Table 3A or Table 3B.

8. Workmanship and Appearance

8.1 The surface of the wire as received shall be free from rust and excessive scale. No serious die marks, scratches, or seams may be present.

8.2 The wire shall be uniform in quality and temper and shall not be wavy or crooked.

8.3 Each coil shall be one continuous length of wire, properly coiled and firmly tied. Welds made prior to cold drawing are permitted.

9. Retests

9.1 If any test specimen exhibits obvious defects or shows the presence of a weld, it may be discarded and another specimen substituted.

10. Inspection

10.1 Unless otherwise specified in the contract or purchase order, the manufacturer is responsible for the performance of all inspection and test requirements specified in this specification. Except as otherwise specified in the contract or purchase order, the manufacturer may use his own or any other suitable facilities for the performance of the inspection and test requirements unless disapproved by the purchaser at the time the order is placed.

A 229/A 229M

The purchaser shall have the right to perform any of the inspections and tests set forth in this specification when such inspections and tests are deemed necessary to assure that the material conforms to prescribed requirements.

11. Rejection and Rehearing

11.1 Unless otherwise specified, any rejection based on tests made in accordance with this specification shall be reported to the manufacturer as soon as possible so that an investigation may be initiated.

11.2 The material must be adequately protected and correctly identified in order that the manufacturer may make a proper investigation.

12. Certification

12.1 When specified in the purchase order or contract, a manufacturer's or supplier's certification shall be furnished to the purchaser that the material was manufactured, sampled, tested, and inspected in accordance with this specification and has been found to meet the requirements. When specified in the purchase order or contract, a report of the test results shall be furnished.

13. Packaging, Marking, and Loading for Shipment

13.1 The coil mass, dimensions, and the method of packaging shall be agreed upon between the manufacturer and purchaser.

13.2 The size of the wire, purchaser's order number, ASTM specification number, heat number, and name or mark of the manufacturer shall be marked on a tag securely attached to each coil of wire.

13.3 Unless otherwise specified in the purchaser's order, packaging, marking, and loading for shipments shall be in accordance with those procedures recommended by Practices A 700.

13.4 *For Government Procurement*—Packaging, packing, and marking of material for military procurement shall be in accordance with the requirements of MIL-STD-163, Level A, Level C, or commercial as specified in the contract or purchase order. Marking for shipment of material for civil agencies shall be in accordance with Fed. Std. No. 123.

TABLE 1 Chemical Requirements

Element	Composition, %
Carbon	0.55–0.85
Manganese	0.30–1.20[A]
Phosphorus, max	0.040
Sulfur, max	0.050
Silicon	0.15–0.35

[A] Generally 0.80/1.20 % manganese for diameter 5.00 mm or 0.192 in. and larger; 0.30/0.90 % for diameters less than 5.00 mm or 0.192 in. The choice of composition shall be optional with the manufacturer unless the purchaser definitely specifies otherwise.

A 229/A 229M

TABLE 2A Tensile Requirements, SI Units[A]

Diameter,[B] mm	Class I Tensile Strength, MPa		Class II Tensile Strength, MPa	
	min	max	min	max
0.50	2050	2250	2230	2450
0.55	2020	2220	2220	2440
0.60	2000	2200	2210	2430
0.65	1950	2150	2190	2410
0.70	1950	2150	2170	2390
0.80	1900	2100	2140	2360
0.90	1850	2050	2100	2320
1.00	1800	2000	2060	2280
1.10	1780	1980	2030	2240
1.20	1750	1950	2000	2210
1.40	1700	1900	1950	2150
1.60	1650	1850	1900	2100
1.80	1620	1820	1860	2060
2.00	1600	1800	1820	2020
2.20	1580	1780	1790	1990
2.50	1550	1750	1750	1950
2.80	1520	1720	1710	1900
3.00	1500	1700	1690	1880
3.50	1450	1620	1640	1830
4.00	1400	1580	1600	1780
4.50	1380	1550	1560	1740
5.00	1350	1520	1520	1700
5.50	1320	1500	1500	1680
6.00	1300	1480	1480	1660
7.00	1280	1450	1450	1630
8.00	1250	1430	1430	1610
9.00	1220	1400	1410	1590
10.00	1200	1380	1400	1580
11.00	1180	1350	1380	1560
12.00	1150	1320	1360	1540
14.00	1120	1300	1340	1520
16.00	1120	1300	1320	1500

[A] Tensile strength values for intermediate diameters may be interpolated.
[B] Preferred sizes. For a complete list, refer to ANSI B32.4M, Preferred Metric Sizes for Round, Square, Rectangle, and Hexagon Metal Products.

TABLE 2B Tensile Requirements, Inch-Pound Units

Diameter,[A] in.	Class I Tensile Strength, MPa		Class II Tensile Strength, MPa	
	min	max	min	max
0.020	293	323	324	354
0.023	289	319	320	350
0.026	286	316	317	347
0.029	283	313	314	344
0.032	280	310	311	341
0.035	274	304	305	335
0.041	266	296	297	327
0.048	259	289	290	320
0.054	253	283	284	314
0.062	247	277	278	308
0.072	241	271	272	302
0.080	235	265	266	296
0.092	230	260	261	291
0.106	225	255	256	286
0.120	220	250	251	281
0.135	215	240	241	266
0.148	210	235	236	261
0.162	205	230	231	256
0.177	200	225	226	251
0.192	195	220	221	246
0.207	190	215	216	241
0.225	188	213	214	239
0.244	187	212	213	238
0.250	185	210	211	236
0.312	183	208	209	234
0.375	180	205	206	231
0.438	175	200	201	226
0.500	170	195	196	221
0.562	165	190	191	216
0.625	165	190	191	216

[A] Tensile strength values for intermediate diameters may be interpolated.

TABLE 3A Permissible Variations in Wire Diameter, SI Units[A]

Diameter, mm	Permissible Variations, plus and minus, mm	Permissible Out-of Round, mm
To 0.70, incl	0.02	0.02
Over 0.70 to 2.00, incl	0.03	0.03
2.00 to 9.00, incl	0.05	0.05
Over 9.00	0.08	0.08

[A] For purposes of determining conformance with this specification, all specified limits are absolute as defined in Recommended Practice E 29.

TABLE 3B Permissible Variations in Wire Diameter, Inch-Pound Units[A]

Diameter, in.	Permissible Variations, plus and minus, in.	Permissible Out-of-Round, in.
0.020 to 0.028, incl	0.0008	0.0008
Over 0.028 to 0.075, incl	0.001	0.001
Over 0.075 to 0.375, incl	0.002	0.002
Over 0.375	0.003	0.003

[A] For purposes of determining conformance with this specification, all specified limits are absolute as defined in Recommended Practice E 29.

The American Society for Testing and Materials takes no position respecting the validity of any patent rights asserted in connection with any item mentioned in this standard. Users of this standard are expressly advised that determination of the validity of any such patent rights, and the risk of infringement of such rights, are entirely their own responsibility.

This standard is subject to revision at any time by the responsible technical committee and must be reviewed every five years and if not revised, either reapproved or withdrawn. Your comments are invited either for revision of this standard or for additional standards and should be addressed to ASTM Headquarters. Your comments will receive careful consideration at a meeting of the responsible technical committee, which you may attend. If you feel that your comments have not received a fair hearing you should make your views known to the ASTM Committee on Standards, 1916 Race St., Philadelphia, Pa. 19103.

Designation: A 230/A 230M – 83

Standard Specification for
STEEL WIRE, OIL-TEMPERED CARBON VALVE SPRING QUALITY[1]

This standard is issued under the fixed designation A 230/A 230M; the number immediately following the designation indicates the year of original adoption or, in the case of revision, the year of last revision. A number in parentheses indicates the year of last reapproval. A superscript epsilon (ϵ) indicates an editorial change since the last revision or reapproval.

1. Scope

1.1 This specification covers the highest quality of round carbon steel spring wire, uniform in quality and temper, intended especially for the manufacture of valve springs and other springs requiring high-fatigue properties.

1.2 The values stated in either SI (metric) units or inch-pound units are to be regarded separately as standard. The values stated in each system are not exact equivalents; therefore, each system must be used independent of the other.

2. Applicable Documents

2.1 *ASTM Standards:*
A 370 Methods and Definitions for Mechanical Testing of Steel Products[2]
A 510 Specification for General Requirements for Wire Rods and Coarse Round Wire, Carbon Steel[3]
A 510M Specification for General Requirements for Wire Rods and Coarse Round Wire, Carbon Steel [Metric][3]
A 700 Practices for Packaging, Marking, and Loading Methods for Steel Products for Domestic Shipment[2]
A 751 Methods, Practices, and Definitions for Chemical Analysis of Steel Products[2]
E 29 Recommended Practice for Indicating Which Places of Figures Are to Be Considered Significant in Specified Limiting Values[4]
E 380 Metric Practice[4]

2.2 *American National Standard:*
B32.4M Preferred Metric Sizes for Round, Square, Rectangle, and Hexagon Metal Products[5]

2.3 *Military Standard:*
MIL-STD-163 Steel Mill Products, Preparation for Shipment and Storage[6]

2.4 *Federal Standard:*
Fed. Std. No. 123 Marking for Shipment (Civil Agencies)[6]

3. Ordering Information

3.1 Material furnished under this specification shall conform to the applicable requirements of the current edition of either specification A 510 or A 510M.

3.2 Orders for material under this specification shall include the following information for each ordered item:

3.1.1 Quantity (mass or weight),
3.1.2 Name of material (oil-tempered carbon steel valve spring quality wire),
3.1.3 Dimensions (Section 8),
3.1.4 Chemical composition (Table 1), if required,
3.1.5 Packaging (Section 14),
3.1.6 Cast or heat analysis report, if desired (see 5.2),
3.1.7 Certification or test report, or both, if specified (Section 13), and
3.1.8 ASTM designation and date of issue.

---

[1] This specification is under the jurisdiction of ASTM Committee A-1 on Steel, Stainless Steel, and Related Alloys, and is the direct responsibility of Subcommittee A01.03 on Steel Rod and Wire.
Current edition approved July 29, 1983. Published November 1983. Originally published as A 230 – 39 T. Last previous edition A 230 – 77.
[2] *Annual Book of ASTM Standards*, Vol 01.04.
[3] *Annual Book of ASTM Standards*, Vol 01.03.
[4] *Annual Book of ASTM Standards*, Vol 14.02.
[5] Available from American National Standards Institute, 1430 Broadway, New York, N. Y. 10018.
[6] Available from Naval Publications and Forms Center, 5801 Tabor Ave., Philadelphia, Pa. 19120.

NOTE 1—A typical metric ordering description is as follows: 20 000 kg oil-tempered carbon steel valve spring quality wire, 6.00 mm diameter in 125-kg coils to ASTM A 230M dated ——, or for inch-pound units, 40 000 lb oil-tempered carbon steel valve spring quality wire, 0.250 in. diameter in 350-lb coils to ASTM A 230 dated ——.

4. Manufacture

4.1 The steel may be made by any commercially accepted steel-making process. The steel may be either ingot cast or strand cast.

4.2 The finished wire shall be free of detrimental pipe and undue segregation.

4.3 The wire shall be hardened and tempered to produce the desired mechanical properties.

5. Chemical Composition

5.1 The steel shall conform to the requirements for chemical composition prescribed in Table 1.

5.2 *Cast or Heat Analysis*—Each cast or heat of steel shall be analyzed by the manufacturer to determine the percentage of elements prescribed in Table 1. This analysis shall be made from a test specimen preferably taken during the pouring of the cast or heat. When requested, this shall be reported to the purchaser and shall conform to the requirements of Table 1.

5.3 *Product Analysis*—An analysis may be made by the purchaser from finished wire representing each cast or heat of steel. The chemical composition thus determined, as to elements required or restricted, shall conform to the product analysis requirements specified in Table 10 of Specification A 510 or A 510M.

5.4 For referee purposes, Methods, Practices, and Definitions A 751 shall be used.

6. Mechanical Properties

6.1 *Tension Test:*

6.1.1. *Requirements*—The material as represented by tension test specimens shall conform to the requirements prescribed in Table 2A or Table 2B.

6.1.2 *Number of Tests*—One test specimen shall be taken for each five coils, or fraction thereof, in a lot. Each cast or heat in a given lot shall be tested.

6.1.3 *Location of Tests*—Test specimens shall be taken from either end of the coil.

6.1.4 *Test Method*—The tension test shall be made in accordance with Methods and Definitions A 370.

6.2 *Wrap Test:*

6.2.1 *Requirements*—Wire 4.00 mm or 0.162 in. and smaller in diameter shall wind on itself as an arbor without breakage. Larger diameter wire, up to and including 6.50 mm or 0.250 in. shall wind, without breakage, on a mandrel twice the diameter of the wire.

6.2.2 *Number of Tests*—One test specimen shall be taken for each five coils, or fraction thereof, in a lot. Each cast or heat in a given lot shall be tested.

6.2.3 *Location of Test*—Test specimens shall be taken from either end of the coil.

6.2.4 *Test Method*—The wrap test shall be made in accordance with Methods and Definitions A 370, Supplement IV.

7. Metallurgical Properties

7.1 *Decarburization:*

7.1.1 *Requirements*—Transverse sections of the wire properly mounted, polished, and etched shall show no completely decarburized (carbon-free) areas when examined at a magnification of 100 diameters. Partial decarburization shall not exceed a depth of 0.040 mm or 0.0015 in.

7.1.2 *Number of Tests*—One test specimen shall be taken for each five coils, or fraction thereof, in a lot.

7.1.3 *Location of Tests*—Test specimens may be taken from either end of the coil.

7.2 *Surface Condition:*

7.2.1 The surface shall be smooth and free of defects such as seams, pits, die marks, and other defects tending to impair the fatigue value of the springs. Any additional surface requirements must be negotiated at the time of entry of the order.

7.2.2 *Number of Tests*—One test specimen from each end of every coil shall be tested for conformance to the provisions of 7.2.1.

7.2.3 *Test Method*—The surface shall be examined after etching in a solution of equal parts of hydrochloric acid and water that has been heated to approximately 80°C for a sufficient length of time to remove up to approximately 1 % of the diameter of the wire. Examination shall be made using 10× magnification.

8. Dimensions and Permissible Variations

8.1 The permissible variations in the diameter of the wire shall be as specified in Table 3A or Table 3B.

8.2 *Number of Tests*—One test specimen from each end of the coil shall be tested for conformance (see 8.1).

9. Workmanship and Appearance

9.1 The wire shall be uniform in quality and temper and shall not be wavy or crooked.

9.2 Each coil shall be one continuous length of wire, properly coiled, and firmly tied.

9.3 No welds are permitted in the finished product and any welds made during processing must be removed

10. Retests

10.1 If any test specimen exhibits obvious defects or shows the presence of a weld, it may be discarded and another specimen substituted.

11. Inspection

11.1 Unless otherwise specified in the contract or purchase order, the manufacturer is responsible for the performance of all inspection and test requirements specified in this specification. Except as otherwise specified in the contract or purchase order, the manufacturer may use his own or any other suitable facilities for the performance of the inspection and test requirements unless disapproved by the purchaser at the time the order is placed. The purchaser shall have the right to perform any of the inspections and tests set forth in this specification when such inspections and tests are deemed necessary to assure that the material conforms to prescribed requirements.

12. Rejection and Rehearing

12.1 Unless otherwise specified, any rejection based on tests made in accordance with this specification shall be reported to the manufacturer as soon as possible so that an investigation may be initiated.

12.2 The material must be adequately protected and correctly identified in order that the manufacturer may make a proper investigation.

13. Certification

13.1 When specified in the purchase order or contract, a manufacturer's or supplier's certification shall be furnished to the purchaser that the material was manufactured, sampled, tested, and inspected in accordance with this specification and has been found to meet the requirements. When specified in the purchase order or contract, a report of the test results shall be furnished.

14. Packaging, Marking, and Loading for Shipment

14.1 The coil mass, dimensions, and the method of packaging shall be as agreed upon between the manufacturer and purchaser.

14.2 The size of the wire, purchaser's order number, ASTM specification number, heat number, and name or mark of the manufacturer shall be marked on a tag securely attached to each coil of wire.

14.3 Unless otherwise specified in the purchaser's order, packaging, marking, and loading for shipments shall be in accordance with those procedures recommended by Practices A 700.

14.4 *For Government Procurement*—Packaging, packing, and marking of material for military procurement shall be in accordance with the requirements of MIL-STD-163, Level A, Level C, or commercial as specified in the contract or purchase order. Marking for shipment of material for civil agencies shall be in accordance with Fed. Std. No. 123.

TABLE 1 Chemical Requirements

Element	Composition, %
Carbon	0.60–0.75[A]
Manganese	0.60–0.90[A]
Phosphorus, max	0.025
Sulfur, max	0.030
Silicon	0.15–0.35

[A] Carbon and manganese may be varied by the manufacturer at his discretion, provided the mechanical properties specified are maintained and provided the purchaser does not specifically stipulate otherwise.

A 230/A 230M

TABLE 2A Tensile Requirements, SI Units

Diameter, mm	Tensile Strength, MPa		Reduction of Areas, min, %
	min	max	
Less than 1.50	1700	1850	...
1.50 to 2.50, incl	1650	1800	...
Over 2.50 to 3.50, incl	1600	1750	40
Over 3.50 to 4.00, incl	1580	1720	40
Over 4.00 to 4.80, incl	1550	1700	40
Over 4.80 to 5.50, incl	1520	1680	40
Over 5.50 to 6.50, incl	1500	1650	40
Over 6.50	1450	1600	40

TABLE 2B Tensile Requirements, Inch-Pound Units

Diameter, in.	Tensile Strength, ksi		Reduction of Area, min, %
	min	max	
Less than 0.062	245	265	...
0.062 to 0.092, incl	240	260	...
Over 0.092 to 0.128, incl	235	255	40
Over 0.128 to 0.162, incl	230	250	40
Over 0.162 to 0.192, incl	225	245	40
Over 0.192 to 0.225, incl	220	240	40
Over 0.225 to 0.250, incl	215	235	40
Over 0.250	210	230	40

TABLE 3A Permissible Variations in Wire Diameter, SI Units[A]

Diameter, mm	Permissible Variations, plus and minus, mm	Permissible Out-of Round, mm
Less than 1.50	0.01	0.01
1.50 to 2.50, incl	0.02	0.02
Over 2.50 to 3.50, incl	0.03	0.03
3.50 to 4.50, incl	0.04	0.04
Over 4.50	0.05	0.05

[A] For purposes of determining conformance with this specification, all specified limits are absolute as defined in Recommended Practice E 29.

TABLE 3B Permissible Variations in Wire Diameter, Inch-Pound Units[A]

Diameter, in.	Permissible Variations, plus and minus, in.	Permissible Out-of-Round, in.
Less than 0.062	0.0005	0.0005
0.062 to 0.092, incl	0.0008	0.0008
Over 0.092 to 0.148, incl	0.001	0.001
Over 0.148 to 0.177, incl	0.0015	0.0015
Over 0.177	0.002	0.002

[A] For purposes of determining conformance with this specification, all specified limits are absolute as defined in Recommended Practice E 29.

TABLE 4 Preferred Diameters for Steel Wire[A], mm

1.6	2.8	4.5
1.8	3.0	4.8
2.1	3.2	5.0
2.2	3.5	5.5
2.4	3.8	6.0
2.5	4.0	6.5
2.6	4.2	

[A] For a complete list of preferred sizes, refer to ANSI B32.4M, Preferred Metric Sizes for Round, Square, Rectangle, and Hexagon Metal Products.

The American Society for Testing and Materials takes no position respecting the validity of any patent rights asserted in connection with any item mentioned in this standard. Users of this standard are expressly advised that determination of the validity of any such patent rights, and the risk of infringement of such rights, are entirely their own responsibility.

This standard is subject to revision at any time by the responsible technical committee and must be reviewed every five years and if not revised, either reapproved or withdrawn. Your comments are invited either for revision of this standard or for additional standards and should be addressed to ASTM Headquarters. Your comments will receive careful consideration at a meeting of the responsible technical committee, which you may attend. If you feel that your comments have not received a fair hearing you should make your views known to the ASTM Committee on Standards, 1916 Race St., Philadelphia, Pa. 19103.

Designation: A 231/A 231M – 83

Standard Specification for
CHROMIUM-VANADIUM ALLOY STEEL SPRING WIRE[1]

This standard is issued under the fixed designation A 231/A 231M; the number immediately following the designation indicates the year of original adoption or, in the case of revision, the year of last revision. A number in parentheses indicates the year of last reapproval. A superscript epsilon (ϵ) indicates an editorial change since the last revision or reapproval.

1. Scope

1.1 This specification covers round chromium-vanadium alloy steel spring wire having properties and quality intended for the manufacture of springs used at moderately elevated temperatures. This wire shall be either in the annealed and cold-drawn or oil-tempered condition as specified by purchaser.

1.2 The values stated in either SI (metric) units or inch-pound units are to be regarded separately as standard. The values stated in each system are not exact equivalents; therefore, each system must be used independent of the other.

2. Applicable Documents

2.1 *ASTM Standards:*
A 370 Methods and Definitions for Mechanical Testing of Steel Products[2]
A 700 Practices for Packaging, Marking, and Loading Methods for Steel Products for Domestic Shipment[2]
A 751 Methods, Practices, and Definitions for Chemical Analysis of Steel Products[2]
A 752 Specification for General Requirements for Wire Rods and Coarse Round Wire, Alloy Steel[3]
E 29 Recommended Practice for Indicating Which Places of Figures Are to Be Considered Significant in Specified Limiting Values[4]
E 380 Metric Practice[4]

2.2 *American National Standard:*
B32.4M Preferred Metric Sizes for Round, Square, Rectangle, and Hexagon Metal Products[5]

2.3 *Military Standard:*
MIL-STD-163 Steel Mill Products, Preparation for Shipment and Storage[6]

2.4 *Federal Standard:*
Fed. Std. No. 123 Marking for Shipment (Civil Agencies)[6]

3. Ordering Information

3.1 Orders for material under this specification shall include the following information for each ordered item:
3.1.1 Quantity (mass),
3.1.2 Name of material (chromium-vanadium alloy steel wire),
3.1.3 Wire diameter (Table 2),
3.1.4 Packaging (Section 14),
3.1.5 Cast or heat analysis report (if requested) (5.2),
3.1.6 Certification or test report, or both, if specified (Section 13), and
3.1.7 ASTM designation and date of issue.

NOTE 1—A typical ordering description is as follows: 20 000 kg oil-tempered chromium-vanadium alloy steel wire, size 6.00 mm in 150-kg coils to ASTM A 231M dated ——, or for inch-pound units, 40 000 lb oil-tempered chromium-vanadium alloy steel spring wire, size 0.250 in. in 350-lb coils to ASTM A 231 dated ——.

4. Manufacture

4.1 The steel may be made by any commercially accepted steel–making process. The steel may be either ingot cast or strand cast.

---

[1] This specification is under the jurisdiction of ASTM Committee A-1 on Steel, Stainless Steel, and Related Alloys, and is the direct responsibility of Subcommittee A01.03 on Steel Rod and Wire.
Current edition approved July 29, 1983. Published November 1983. Originally published as A 231 – 39 T. Last previous edition A 231 – 77.
[2] *Annual Book of ASTM Standards*, Vol 01.04.
[3] *Annual Book of ASTM Standards*, Vol 01.03.
[4] *Annual Book of ASTM Standards*, Vol 14.02.
[5] Available from American National Standards Institute, 1430 Broadway, New York, N. Y. 10018.
[6] Available from Naval Publications and Forms Center, 5801 Tabor Ave., Philadelphia, Pa. 19120.

A 231/A 231M

4.2 The finished wire shall be free from detrimental pipe and undue segregation.

5. Chemical Composition

5.1 The steel shall conform to the requirements as to chemical composition specified in Table 1.

5.2 *Cast or Heat Analysis*—Each cast or heat of steel shall be analyzed by the manufacturer to determine the percentage of elements prescribed in Table 1. This analysis shall be made from a test specimen preferably taken during the pouring of the cast or heat. When requested, this shall be reported to the purchaser and shall conform to the requirements of Table 1.

5.3 *Product Analysis (formerly Check Analysis)*—An analysis may be made by the purchaser from finished wire representing each cast or heat of steel. The chemical composition thus determined, as to elements required or restricted, shall conform to the product (check) analysis requirements specified in Table 5 in Specification A 752.

5.4 For referee purposes, Methods, Practices, and Definitions A 751 shall be used.

6. Mechanical Properties

6.1 *Annealed and Cold Drawn*—When purchased in the annealed and cold-drawn condition, the wire shall have been given a sufficient amount of cold working to meet the purchaser's coiling requirements and shall be in a suitable condition to respond properly to heat treatment. In special cases the hardness, if desired, shall be stated in the purchase order.

6.2 *Oil Tempered*—When purchased in the oil-tempered condition, the tensile strength and minimum percent reduction of area, sizes 2.50 mm or 0.105 in. and coarser, of the wire shall conform to the requirements as shown in Table 2A or Table 2B.

6.2.1 *Number of Tests*—One test specimen shall be taken for each ten coils, or fraction thereof, in a lot. Each cast or heat in a given lot shall be tested.

6.2.2 *Location of Tests*—Test specimens shall be taken from either end of the coil.

6.2.3 *Test Method*—The tension test shall be made in accordance with Methods and Definitions A 370.

6.3 *Wrap Test*:

6.3.1 Oil tempered or cold drawn wire 4.00 mm or 0.162 in. and smaller in diameter shall wind on itself as an arbor without breakage.

Larger diameter wire up to and including 8.00 mm or 0.312 in. in diameter shall wrap without breakage on a mandrel twice the wire diameter. The wrap test is not applicable to wire over 8.00 mm or 0.312 in. in diameter.

6.3.2 *Number of Tests*—One test specimen shall be taken for each ten coils or fraction thereof, in a lot. Each cast or heat in a given lot shall be tested.

6.3.3 *Location of Test*—Test specimens shall be taken from either end of the coil.

6.3.4 *Test Method*—The wrap test shall be made in accordance with Supplement IV of Methods and Definitions A 370.

7. Metallurgical Properties

7.1 *Surface Condition:*

7.1.1 The surface of the wire as received shall be free of rust and excessive scale. No serious die marks, scratches, or seams may be present. Based upon examination of etched end specimen, seams shall not exceed 3.5 % of the wire diameter, or 0.25 mm or 0.010 in., whichever is the smaller as measured on a transverse section.

7.1.2 *Number of Tests*—One test specimen shall be taken for each ten coils or fraction thereof, in a lot. Each cast or heat in a given lot shall be tested.

7.1.3 *Location of Test*—Test specimens shall be taken from either or both ends of the coil.

7.1.4 *Test Method*—The surface shall be examined after etching in a solution of equal parts of hydrochloric acid and water that has been heated to approximately 80°C. Test ends shall be examined using 10× magnification. Any specimen which shows questionable seams of borderline depth shall have a transverse section taken from the unetched area, properly mounted and polished and examined to measure the depth of the seam.

8. Dimensions and Permissible Variations

8.1 The permissible variations in the diameter of the wire shall be as specified in Table 3A or Table 3B.

9. Workmanship and Appearance

9.1 *Annealed and Cold Drawn*—The wire shall not be kinked or improperly cast. To test for cast, a few convolutions of wire shall be cut loose from the coil and placed on a flat surface. The wire shall lie flat on itself and not spring

49

up nor show a wavy condition.

9.2 *Oil Tempered*—The wire shall be uniform in quality and temper and shall not be wavy or crooked.

9.3 Each coil shall be one continuous length of wire properly coiled. Welds made prior to cold drawing are permitted. If unmarked welds are unacceptable to the purchaser, special arrangements should be made with the manufacturer at the time of the purchase.

9.4 *Appearance*—The surface shall be smooth and free of defects such as seams, pits, die marks, and other defects tending to impair the use of the wire for springs. Any additional surface requirements must be negotiated at the time of entry of the order.

10. Retests

10.1 If any test specimen exhibits obvious defects or shows the presence of a weld, it may be discarded and another specimen substituted.

11. Inspection

11.1 Unless otherwise specified in the contract or purchase order, the manufacturer is responsible for the performance of all inspection and test requirements specified in this specification. Except as otherwise specified in the contract or purchase order, the manufacturer may use his own or any other suitable facilities for the performance of the inspection and test requirements unless disapproved by the purchaser at the time the order is placed. The purchaser shall have the right to perform any of the inspections and tests set forth in this specification when such inspections and tests are deemed necessary to assure that the material conforms to prescribed requirements

12. Rejection and Rehearing

12.1 Unless otherwise specified, any rejection based on tests made in accordance with this specification shall be reported to the manufacturer as soon as possible so that an investigation may be initiated.

12.2 The material must be adequately protected and correctly identified in order that the manufacturer may make a proper investigation.

13. Certification

13.1 When specified in the purchase order or contract, a manufacturer's or supplier's certification shall be furnished to the purchaser that the material was manufactured, sampled, tested, and inspected in accordance with this specification and has been found to meet the requirements. When specified in the purchase order or contract, a report of the test results shall be furnished.

14. Packaging, Marking, and Loading for Shipment

14.1 The coil mass, dimensions, and the method of packaging shall be as agreed upon between the manufacturer and purchaser.

14.2 The size of the wire, purchaser's order number, ASTM Specification number, heat number, and name or mark of the manufacturer shall be marked on a tag securely attached to each coil of wire.

14.3 Unless otherwise specified in the purchaser's order, packing, marking, and loading for shipments shall be in accordance with those procedures recommended by Practices A 700.

14.4 *For Government Procurement:*

14.4.1 Packaging, packing, and marking of material for military procurement shall be in accordance with the requirements of MIL-STD-163, Level A, Level C, or commercial as specified in the contract or purchase order. Marking for shipment of material for civil agencies shall be in accordance with Fed. Std. No. 123.

A 231/A 231M

TABLE 1 Chemical Requirements

Element	Analysis, %
Carbon	0.48–0.53
Manganese	0.70–0.90
Phosphorus	0.040 max
Sulfur	0.040 max
Silicon	0.15–0.35
Chromium	0.80–1.10
Vanadium	0.15 min

TABLE 2A Tensile Requirements, SI Units[A]

Diameter,[B] mm	Tensile Strength, MPa min	Tensile Strength, MPa max	Reduction of Area, min, %
0.50	2060	2260	c
0.55	2050	2240	c
0.60	2030	2220	c
0.65	2010	2200	c
0.70	2000	2160	c
0.80	1980	2140	c
0.90	1960	2120	c
1.00	1940	2100	c
1.10	1920	2080	c
1.20	1900	2060	c
1.40	1860	2020	c
1.60	1820	1980	c
1.80	1800	1960	c
2.00	1780	1930	c
2.20	1750	1900	c
2.50	1720	1860	45
2.80	1680	1830	45
3.00	1660	1800	45
3.50	1620	1760	45
4.00	1580	1720	40
4.50	1560	1680	40
5.00	1520	1640	40
5.50	1480	1620	40
6.00	1460	1600	40
6.50	1440	1580	40
7.00	1420	1560	40
8.00	1400	1540	40
9.00	1380	1520	40
10.00	1360	1500	40
11.00	1340	1480	40
12.00	1320	1460	40

[A] Tensile strength values for intermediate diameters may be interpolated.
[B] Preferred sizes. For a complete list, refer to ANSI B32.4M, Preferred Metric Sizes for Round, Square, Rectangle, and Hexagon Metal Products.
[C] The reduction of area test is not applicable to wire under 2.50 mm or 0.105 in. in diameter.

TABLE 2B Tensile Requirements, Inch-Pound Units[A]

Diameter, in.	Tensile Strength, ksi min	Tensile Strength, ksi max	Reduction of Area, min, %
0.020	300	325	c
0.032	290	315	c
0.041	280	305	c
0.054	270	295	c
0.062	265	290	c
0.080	255	275	c
0.105	245	265	45
0.135	235	255	45
0.162	225	245	40
0.192	220	240	40
0.244	210	230	40
0.283	205	225	40
0.312	203	223	40
0.375	200	220	40
0.438	195	215	40
0.500	190	210	40

[A] Tensile strength values for intermediate diameters may be interpolated.
[B] Preferred sizes. For a complete list, refer to ANSI B32.4M, Preferred Metric Sizes for Round, Square, Rectangle, and Hexagon Metal Products.
[C] The reduction of area test is not applicable to wire under 2.50 mm or 0.105 in. in diameter.

TABLE 3A Permissible Variations in Wire Diameter, SI Units[A]

Diameter, mm	Permissible Variations, plus and minus, mm	Permissible Out-of-Round, mm
To 0.70, incl	0.02	0.02
Over 0.70 to 2.00, incl	0.03	0.03
Over 2.00 to 9.00, incl	0.05	0.05
Over 9.00	0.08	0.08

[A] For purposes of determining conformance with this specification, all specified limits are absolute as defined in Recommended Practice E 29.

TABLE 3B Permissible Variations in Wire Diameter, Inch-Pound Units[A]

Diameter, in.	Permissible Variations, plus and minus, in.	Permissible Out-of-Round, in.
0.020 to 0.028, incl	0.0008	0.0008
Over 0.028 to 0.075, incl	0.001	0.001
Over 0.075 to 0.375, incl	0.002	0.002
Over 0.375 to 0.500, incl	0.003	0.003

[A] For purposes of determining conformance with this specification, all specified limits are absolute as defined in Recommended Practice E 29.

The American Society for Testing and Materials takes no position respecting the validity of any patent rights asserted in connection with any item mentioned in this standard. Users of this standard are expressly advised that determination of the validity of any such patent rights, and the risk of infringement of such rights, are entirely their own responsibility.

This standard is subject to revision at any time by the responsible technical committee and must be reviewed every five years and if not revised, either reapproved or withdrawn. Your comments are invited either for revision of this standard or for additional standards and should be addressed to ASTM Headquarters. Your comments will receive careful consideration at a meeting of the responsible technical committee, which you may attend. If you feel that your comments have not received a fair hearing you should make your views known to the ASTM Committee on Standards, 1916 Race St., Philadelphia, Pa. 19103.

ASTM Designation: A 232 – 77

Standard Specification for
CHROMIUM-VANADIUM ALLOY STEEL VALVE SPRING QUALITY WIRE[1]

This standard is issued under the fixed designation A 232; the number immediately following the designation indicates the year of original adoption or, in the case of revision, the year of last revision. A number in parentheses indicates the year of last reapproval. A superscript epsilon (ϵ) indicates an editorial change since the last revision or reapproval.

1. Scope

1.1 This specification covers the highest quality of round chromium-vanadium alloy steel valve spring wire, uniform in quality and temper, intended for the manufacture of valve springs and other springs requiring high-fatigue properties, when used at moderately elevated temperatures. The wire shall be either in the annealed and cold-drawn or oil-tempered condition as specified by the purchaser.

1.2 The values stated in inch-pound units are to be regarded as the standard.

2. Applicable Documents

2.1 *ASTM Standards:*
A 370 Methods and Definitions for Mechanical Testing of Steel Products[2]
A 700 Recommended Practices for Packaging, Marking, and Loading Methods for Steel Products for Domestic Shipment[3]
E 29 Recommended Practice for Indicating Which Places of Figures Are to Be Considered Significant in Specified Limiting Values[4]
E 30 Methods for Chemical Analysis of Steel, Cast Iron, Open-Hearth Iron, and Wrought Iron[5]
E 380 Standard for Metric Practice

3. Ordering Information

3.1 Orders for material under this specification shall include the following information for each ordered item:
3.1.1 Quantity (weight),
3.1.2 Name of material (chromium-vanadium alloy steel valve spring quality wire),
3.1.3 Dimensions (Section 8),
3.1.4 Packaging (Section 11),
3.1.5 Cast or heat analysis report, if desired,
3.1.6 Condition (Sections 1 and 6), and
3.1.7 ASTM designation and date of issue.

NOTE 1—A typical description is as follows: 40,000 lb valve oil tempered chromium-vanadium alloy steel valve spring quality wire, size 0.250 in. in 350-lb coils to ASTM A 232 dated ——.

4. Materials and Manufacture

4.1 The steel shall be made by the open hearth, basic-oxygen, or electric-furnace process.

4.2 A sufficient discard shall be made to ensure freedom from injurious piping and undue segregation.

5. Chemical Requirements

5.1 The steel shall conform to the requirements as to chemical composition specified in Table 1.

5.2 An analysis may be made by the purchaser from finished wire representing each heat of steel. The average of all the separate determinations made shall be within the limits specified in the Analysis column of Table 1. Individual determinations may vary to the extent shown in the

---

[1] This specification is under the jurisdiction of ASTM Committee A-1 on Steel, Stainless Steel and Related Alloys and is the direct responsibility of Subcommittee A01.03 on Steel Rod and Wire.
Current edition approved June 24, 1977. Published August 1977. Originally published as A 232 – 39 T. Last previous edition A 232 – 74.
[2] *Annual Book of ASTM Standards,* Vol 01.02.
[3] *Annual Book of ASTM Standards,* Vols 01.01, 01.03, 01.04, and 01.05.
[4] *Annual Book of ASTM Standards,* Vol 14.02.
[5] *Annual Book of ASTM Standards,* Vol 03.05.

A 232

Product Analysis Tolerance column, except that the several determinations of a single element in any one heat shall not vary both above and below the specified range. For referee purposes, Method E 30 shall be used.

5.3 The analysis of each heat of steel shall be furnished by the manufacturer, when requested, showing the percentages of the elements specified in Table 1.

6. Mechanical Requirements

6.1 *Temper Requirements:*

6.1.1 *Annealed and Cold Drawn*—When purchased in the annealed and cold-drawn condition, the wire shall have been given a sufficient amount of cold working to meet the purchaser's coiling requirements and shall be in a suitable condition to respond properly to heat treatment. In special cases the hardness, if desired, shall be stated in the purchase order.

6.1.2 *Oil Tempered*—When purchased in the oil-tempered condition, the tensile strength and minimum percent reduction of area (sizes 0.105 in. (2.67 mm) and coarser) of the wire shall conform to the requirements prescribed in Table 2.

6.1.3 *Number of Tests*—One test specimen shall be taken for each five coils, or fraction thereof, in a lot. Each cast or heat in a given lot shall be tested.

6.1.4 *Location of Tests*—Test specimens shall be taken from either end of the coil.

6.1.5 *Test Method*—The tension test shall be made in accordance with Methods A 370.

6.2 *Wrap Test:*

6.2.1 Wire, oil tempered or cold drawn, 0.162 in. (4.11 mm) and smaller in diameter shall wind on itself as an arbor without breakage. Larger diameter wire up to and including 0.312 in. (7.92 mm) in diameter shall wrap without breakage on a mandrell twice the wire diameter. The wrap test is not applicable to wire over 0.312 in. in diameter.

6.2.2 *Number of Tests*—One test specimen shall be taken for each five coils or fraction thereof, in a lot. Each cast or heat in a given lot shall be tested.

6.2.3 *Location of Test*—Test specimens shall be taken from either end of the coil.

6.2.4 *Test Method*—The wrap test shall be made in accordance with Supplement IV of Methods and Definitions A 370.

7. Metallurgical Requirements

7.1 *Decarburization:*

7.1.1 Transverse sections of the wire properly mounted, polished, and etched shall show no completely decarburized (carbon-free) areas when examined at a magnification of 100 diameters. Partial decarburization shall not exceed a depth of 0.001 in. (0.025 mm) on wire 0.192 in. (4.88 mm) and smaller or 0.0015 in. (0.038 mm) on larger than 0.192 in.

7.1.2 To reveal the decarburization more accurately in the untempered wire, the specimen shall be hardened and tempered before microscopical examination. Prior to hardening, the specimen shall be filed flat on one side enough to reduce the diameter at least 20 %. The subsequent mounted specimen shall show the flattened section, as well as the original wire edge. Any decarburization on this flattened section shall necessitate a new specimen for examination.

7.1.3 *Number of Tests*—One test specimen shall be taken for each five coils, or fraction thereof, in a lot.

7.1.4 *Location of Tests*—Test specimens may be taken from either end of the coil.

8. Permissible Variations in Dimensions

8.1 The permissible variations in the diameter of the wire shall be as specified in Table 3.

8.2 *Number of Tests*—One test specimen from each end of the coil shall be tested.

9. Surface Condition

9.1 The surface of the wire must be free of imperfections such as seams, pits, die marks, scratches, and other defects tending to impair the fatigue value of the springs.

9.2 *Number of Tests*—One test specimen from each end of every coil shall be tested for conformance to the provisions of 9.1.

9.3 *Test Method*—The surface shall be examined after etching in a solution of equal parts of hydrochloric acid and water that has been heated to approximately 175°F (80°C) for a sufficient length of time to remove up to approximately 1 % of the diameter of wire. Examination shall be made using 10× magnification.

10. Workmanship

10.1 *Annealed and Cold Drawn*—The wire shall not be kinked or improperly cast. To test

A 232

for cast, a few convolutions of wire shall be cut loose from the coil and placed on a flat surface. The wire shall lie flat on itself and not spring up nor show a wavy condition.

10.2 *Oil Tempered*—The wire shall be uniform in quality and temper and shall not be wavy or crooked.

10.3 Each bundle shall be one continuous length of wire, properly coiled and firmly tied.

10.4 No welds are permitted in the finished product and any welds made during processing must be removed.

11. Packaging, Marking, and Loading

11.1 The coil weight, coil dimensions, and method of packaging shall be agreed upon between the manufacturer and the purchaser.

11.2 The size of the wire, purchaser's order number, the ASTM specification number and condition (Section 1 and 6.1), heat number, and name or mark of the manufacturer shall be marked on a tag securely attached to each coil of wire.

11.3 When specified in the purchaser's order, packaging, marking, and loading for shipments shall be in accordance with those procedures recommended by Recommended Practices A 700.

12. Inspection

12.1 The manufacturer shall afford the inspector representing the purchaser all reasonable facilities to satisfy him that the material being furnished is in accordance with this specification. All tests (except product analysis) and inspections may be made at the place of manufacture prior to shipment, and shall be so conducted as not to interfere unnecessarily with the operation of the works.

13. Rejection and Rehearing

13.1 Unless otherwise specified, any rejection based on tests made in accordance with this specification shall be reported to the manufacturer within a reasonable length of time.

13.2 The material must be adequately protected and correctly identified in order that the producer may make a proper investigation.

TABLE 1 Chemical Requirements

	Analysis, %	Product Analysis Tolerance, %
Carbon	0.48–0.53	±0.02
Manganese	0.70–0.90	±0.03
Phosphorus	0.020 max	+0.005
Sulfur	0.035 max	+0.005
Silicon	0.20–0.35	±0.02
Chromium	0.80–1.10	±0.05
Vanadium	0.15 min	−0.01

ASTM A 232

TABLE 2 Tensile Requirements

Diameter, in. (mm)[A]	Tensile Strength, ksi (MPa)		Reduction of Area, min, %
	min	max	
0.020 (0.51)	300 (2070)	325 (2240)	[B]
0.032 (0.81)	290 (2000)	315 (2170)	[B]
0.041 (1.04)	280 (1930)	305 (2100)	[B]
0.054 (1.37)	270 (1860)	295 (2030)	[B]
0.062 (1.57)	265 (1830)	290 (2000)	[B]
0.080 (2.03)	255 (1760)	275 (1900)	[B]
0.106 (2.69)	245 (1690)	265 (1830)	45
0.135 (3.43)	235 (1620)	255 (1760)	45
0.162 (4.11)	225 (1550)	245 (1690)	40
0.192 (4.88)	220 (1520)	240 (1650)	40
0.244 (6.20)	210 (1450)	230 (1590)	40
0.283 (7.19)	205 (1410)	225 (1550)	40
0.312 (7.92)	203 (1400)	223 (1540)	40
0.375 (9.52)	200 (1380)	220 (1520)	40
0.438 (11.12)	195 (1340)	215 (1480)	40
0.500 (12.70)	190 (1310)	210 (1450)	40

[A] Tensile strength values for intermediate sizes may be interpolated.
[B] The reduction of area test is not applicable to wire under 0.105 in. (2.67 mm) in diameter.

TABLE 3 Permissible Variations in Wire Diameter

NOTE—For the purpose of determining conformance with this specification, all specified limits are absolute as defined in Recommended Practice E 29.

Diameter, in. (mm)	Permissible Variations, plus and minus, in. (mm)	Permissible Out-of-Round, in. (mm)
0.020 to 0.075 (0.51 to 1.90), incl	0.0008 (0.02)	0.0008 (0.02)
Over 0.075 to 0.148 (1.90 to 3.76), incl	0.001 (0.03)	0.001 (0.03)
Over 0.148 to 0.375 (3.76 to 9.52), incl	0.0015 (0.04)	0.0015 (0.04)
Over 0.375 to 0.500 (9.52 to 12.70), incl	0.002 (0.05)	0.002 (0.05)

The American Society for Testing and Materials takes no position respecting the validity of any patent rights asserted in connection with any item mentioned in this standard. Users of this standard are expressly advised that determination of the validity of any such patent rights, and the risk of infringement of such rights, are entirely their own responsibility.

This standard is subject to revision at any time by the responsible technical committee and must be reviewed every five years and if not revised, either reapproved or withdrawn. Your comments are invited either for revision of this standard or for additional standards and should be addressed to ASTM Headquarters. Your comments will receive careful consideration at a meeting of the responsible technical committee, which you may attend. If you feel that your comments have not received a fair hearing you should make your views known to the ASTM Committee on Standards, 1916 Race St., Philadelphia, Pa. 19103.

Designation: A 240 – 83

Standard Specification for
HEAT-RESISTING CHROMIUM AND CHROMIUM-NICKEL STAINLESS STEEL PLATE, SHEET, AND STRIP FOR PRESSURE VESSELS[1]

This standard is issued under the fixed designation A 240; the number immediately following the designation indicates the year of original adoption or, in the case of revision, the year of last revision. A number in parentheses indicates the year of last reapproval. A superscript epsilon (ϵ) indicates an editorial change since the last revision or reapproval.

This specification has been approved for use by agencies of the Department of Defense and for listing in the DoD Index of Specifications and Standards.

1. Scope

1.1 This specification[2] covers chromium, chromium-nickel, and chromium-manganese-nickel stainless and heat-resisting steel plate, sheet, and strip for pressure vessels.

1.2 Some steels covered by this specification, especially the chromium steels, because of their particular alloy content and specialized properties, may require special care in their fabrication and welding. Specific procedures are of fundamental importance, and it is presupposed that all parameters will be in accordance with approved methods capable of producing the desired properties in the finished fabrication. Satisfactory welded pressure vessels have been produced in some thicknesses of all these steels.

1.3 The values stated in inch-pound units are to be regarded as the standard.

2. Applicable Documents

2.1 *ASTM Standards*:
A 262 Practices for Detecting Susceptibility to Intergranular Attack in Stainless Steels[3]
A 370 Methods and Definitions for Mechanical Testing of Steel Products[4]
A 480 Specification for General Requirements for Flat-Rolled Stainless and Heat-Resisting Steel Plate, Sheet, and Strip[5]
A 751 Methods, Practices, and Definitions for Chemical Analysis of Steel Products[4]

3. Descriptions of Terms

3.1 Plate, sheet, and strip as used in this specification are:

3.1.1 *Plate*—Material $\frac{3}{16}$ in. (4.8 mm) and over in thickness and over 10 in. (254 mm) in width.

3.1.2 *Sheet*—Material under $\frac{3}{16}$ in. (4.8 mm) in thickness and 24 in. (609.6 mm) and over in width.

3.1.3 *Strip*—Material under $\frac{3}{16}$ in. (4.8 mm) in thickness and under 24 in. (609.6 mm) in width.

4. General Requirements for Delivery

4.1 Material furnished under this specification shall conform to the applicable requirements of the current edition of Specification A 480. In case of conflict, the requirements of this specification shall prevail.

5. Ordering Information

5.1 Orders for material under this specification shall include the information specified in Section 4 of Specification A 480.

6. Process

6.1 If a specific type of melting is required

---

[1] This specification is under the jurisdiction of ASTM Committee A-1 on Steel, Stainless Steel and Related Alloys and is the direct responsibility of Subcommittee A01.17 on Flat Stainless Steel Products.
Current edition approved July 29, 1983. Published November 1983. Originally published as A 240 – 40 T. Last previous edition A 240 – 82c.
[2] For ASME Boiler and Pressure Vessel Code applications see related Specification SA-240 in Section II of that Code.
[3] *Annual Book of ASTM Standards*, Vol 01.05.
[4] *Annual Book of ASTM Standards*, Vol 01.04.
[5] *Annual Book of ASTM Standards*, Vol 01.03.

by the purchaser, it shall be stated on the purchase order.

6.2 When specified on the purchase order, or when a specific type of melting has been specified on the purchase order, the material manufacturer shall indicate on the test report the type of melting used to produce the material.

7. Heat Treatment

7.1 *Austenitic Types:*

7.1.1 Except for the H types and S31254, the austenitic chromium-nickel steels (series 300, XM-15, and XM-21) shall be solution-annealed to meet the mechanical property requirements of this specification and, when specified on the purchase order, shall be capable of meeting the test for resistance to intergranular corrosion specified in 10.2. Solution-annealing shall consist of heating the material to a temperature of 1900°F (1040°C) minimum for an appropriate time followed by water quenching or rapidly cooling by other means.

7.1.1.1 For the H types, where cold working is involved in processing prior to annealing, the minimum solution-annealing temperature for Types 321H, 347H, and 348H shall be 2000°F (1095°C) and for Types 304H and 316H, 1900°F (1040°C). If the H types are hot-finished, the minimum solution-annealing temperatures for Types 321H, 347H, and 348H shall be 1925°F (1050°C) and for Types 304H and 316H, 1900°F (1040°C).

7.1.1.2 For S31254, solution-annealing shall consist of heating the material to a temperature of 1750 ± 25°F (955 ± 14°C) for an appropriate time followed by water quenching or rapidly cooling by other means.

7.1.1.3 The chromium-manganese-nickel types (201, 202, XM-17, XM-18, XM-19, XM-29, and XM-31) shall be solution-annealed to meet the mechanical property requirements of this specification and, when specified on the purchase order, shall be capable of meeting the test for resistance to intergranular corrosion specified in 10.2.

7.2 *Duplex Types:*

7.2.1 For T329 solution-annealing shall consist of heating the material to a temperature of 1750 ± 25°F (954 ± 15°C) for an appropriate time followed by water quenching or rapid cooling by other means.

7.2.2 S32550, S31200, and S31803 shall be solution annealed to meet the mechanical property requirements of this specification. Solution annealing shall consist of heating to a minimum temperature of 1900°F (1038°C) for an appropriate time followed by water quenching or rapidly cooling by other means.

7.3 *Martensitic and Ferritic Types:*

7.3.1 The chromium steels (series 400, XM-8, XM-27 and XM-33) shall be heat treated in such a manner as to satisfy all the requirements for mechanical and bending properties specified in Section 9.

8. Chemical Requirements

8.1 An analysis of each melt shall be made by the manufacturer to determine the percentages of the elements specified in Table 1. This analysis shall be made from a test specimen preferably taken during the pouring of the heat. The chemical composition thus determined shall be reported to the purchaser, or his representative, and shall conform to the requirements specified in Table 1.

8.2 When a product (check or verification) analysis is performed, the chemical composition thus determined may vary from the specified limits by the amounts shown in Table 1, Product Analysis Tolerances, of Specification A 480.

8.3 Methods and practices relating to chemical analysis required by this specification shall be in accordance with Methods, Practices, and Definitions A 751.

9. Mechanical Requirements

9.1 *Tensile Properties*—The material shall conform to the tensile property requirements specified in Table 2.

9.2 *Bending Properties*—The bend-test specimen from chromium steels 1 in. (25.4 mm) and less in thickness, shall withstand cold bending through the angle specified in Table 2 without cracking on the outside of the bent portion. Bend test on chromium steel over 1 in. (25.4 mm) in thickness shall be done when specifically required by the purchase order.

9.3 *Hardness*—The material shall conform to the hardness requirements as specified in Table 2.

10. Special Tests

10.1 If any special tests are required which are pertinent to the intended application of the material ordered, they shall be agreed upon between the seller and the purchaser.

10.2 The intergranular corrosion test, Practice E of Practices A 262, is not required unless it is specified on the purchase order. If so specified, it shall be performed on sensitized specimens of Types 304L, 304LN, 309Cb, 310Cb, 316Cb, 316L, 316LN, 316Ti, 317L, 321, 347, and 348, and in the case of other chromium-nickel steels (series 300) on specimens representative of the as-shipped material. When specified, all flat-rolled products of the chromium-nickel series (series 300) in thickness up to and including 2 in. (50 mm) nominal size shall be capable of passing the intergranular corrosion test in the as-shipped condition. In the case of heavier plates of types other than 304L, 304LN, 309Cb, 310Cb, 316Cb, 316L, 316LN, 316Ti, 317L, 321, 347, and 348 the applicability of this test shall be a matter for negotiation between the seller and the purchaser.

10.2.1 The H types are not normally subject to intergranular corrosion tests.

10.2.2 When intergranular corrosion tests are required on the chromium-manganese-nickel types (Types 201, 202, XM-17, XM-18, XM-19, XM-29, and XM-31), they shall be as agreed upon between the seller and purchaser.

10.2.3 When intergranular corrosion testing is specified for Types XM-27, XM-33, S44400 and S44635, the test shall be Practice E of Practices A 262, using samples prepared as agreed upon between the seller and the purchaser.

11. Test Specimens

11.1 Tension-test specimens shall be taken from finished material and shall be selected either in the longitudinal or transverse direction. The tension-test specimen shall conform to the appropriate Sections 7, 8, or 9 of Methods A 370.

NOTE 2—For plate material up to and including ¾ in. (19.05 mm) thick, the sheet-type tension specimen described in Section 8 (Fig. 4) of Methods A 370 is also permitted.

11.2 Transverse-bend test specimens from sheet and strip shall be the full thickness of the material and approximately 1 in. (25.4 mm) in width. The edges of the test specimen may be rounded to a radius equal to one half the thickness.

11.2.1 The width of the strip for which bend tests can be made is subject to practical limitations on the length of the bend-test specimen. For narrow strip the following widths can be tested:

Strip Thickness, in. (mm)	Minimum Strip Width and Minimum Specimen Length for Transverse Bend Test, in. (mm)
0.100 (2.54) and under	½ (12.70)
0.101 (2.57) to 0.140 (3.56), excl	1 (25.40)
0.140 (3.56) and over	1½ (38.10)

11.2.2 Bend-test specimens for sheet and strip may be of any suitable length over the minimum length.

11.3 Bend-test specimens taken from plates up to and including ½ in. (12.7 mm) in thickness shall be the full thickness of the material, of a suitable length, and between 1 in. (25.4 mm) and 2 in. (50.8 mm) in width. The corners of the cross section of the specimen may be broken with a smooth file, but no appreciable rounding of the corners may be permitted.

11.3.1 In the case of plates over ½ in. (12.7 mm) in thickness, bend test specimens, machined to 1 in. (25.4 mm) nominal width by ½ in. (12.7 mm) nominal thickness and at least 6 in. (152.4 mm) in length, may be used. One surface, to be the outside surface in bending, shall be the original surface of the plate; however surface preparation by light grinding is permitted. The edges may be rounded to a ¹⁄₁₆-in. (1.6-mm) radius. When permitted by agreement between the seller and the purchaser, the cross section may be modified to ½ in. (12.7 mm) nominal square.

11.4 Hardness tests may be made on the grip ends of the tension test specimens before they are subjected to the tension test.

12. Number of Tests

12.1 In the case of sheet, strip, or plate produced in coil form, two or more hardness tests (one from each end of the coil); one bend test, when required; and one or more tension tests shall be made on specimens taken from each coil. If the hardness difference between the two ends of the coil exceeds 5 HRB, or equivalent, tensile properties must be determined on both ends.

12.2 In the case of sheet, strip, or plate produced in cut lengths, one tension test; one bend test, when required; and one or more hardness tests shall be made on each 100 or less pieces of the same heat and nominal thickness rolled separately or continuously and heat treated within the same operating period, either as a lot or continuously.

NOTE 3—The term *continuously*, as applied to heat treatment, is meant to describe a heat-treating operation in which one cut length follows another through the furnace. Interspersement of different melts is permissible if they are of approximately the same nominal thickness and are heat treated in the same operating period and under the same conditions (time and temperature).

12.3 One corrosion test, when required, as described in Section 10 shall be selected from each heat and thickness subjected to the same heat treatment practice. Such a specimen may be obtained from specimens selected for mechanical testing.

13. Test Methods

13.1 The properties enumerated in this specification shall be determined in accordance with methods as specified in Specification A 480.

13.2 *Bend Tests*:

13.2.1 Material up to and including ⅜ in. (9.5 mm) in thickness shall be bent over a single piece of flat stock of the same thickness as the material tested, allowing the test material to form its natural curvature.

NOTE 4—The bend may be made over a diameter equal to the thickness of the test material.

13.2.2 Material over ⅜ in. (9.5 mm) in thickness shall be bent over two pieces of flat stock, of the same thickness as the material tested, allowing the test material to form its natural curvature.

NOTE 5—The bend angle may be made over a diameter equal to two times the thickness of the test material, or over a single piece of flat stock twice the thickness of the test material.

13.2.3 The axis of the bend shall be parallel to the direction of rolling.

13.3 Retests may be made in accordance with the provisions of Section 3 of Methods A 370.

14. Retreatment

14.1 If any specimens selected to represent any heat fail to meet any of the test requirements, the material represented by such specimens may be reheat-treated and resubmitted for test.

15. Permissible Variations in Dimensions

15.1 Unless otherwise specified in the purchase order, material shall conform to the permissible tolerances shown in Specification A 480.

15.2 *Sheet, Strip, and Plate*—Material with No. 1 finish may be ground to remove surface imperfections, provided such grinding does not reduce the thickness or width at any point beyond the permissible variations in dimensions.

16. Workmanship

16.1 The material shall be of uniform quality consistent with good manufacturing and inspection practices. The steel shall have no imperfections of a nature or degree, for the type and quality ordered, that will adversely affect the stamping, forming, machining or fabrication of finished parts.

17. Marking

17.1 Each sheet, strip, or plate shall be marked on one face, in the locations indicated below with the specification number and its suffix, type, test identification number, and the name or mark of the manufacturer. The characters shall be of such size as to be clearly legible. The marking shall be sufficiently stable to withstand normal handling. Unless otherwise specified by the purchaser, the marking, at the producer's option, may be done with: (*1*) marking fluid (if a specific maximum impurity limit of designated elements in the marking fluid is required by the purchaser, it shall be so stated on the purchase order), (*2*) low-stress blunt-nosed-continuous or low-stress blunt-nosed-interrupted-dot die stamp, (*3*) a vibratory tool with a minimum tip radius of 0.005 in. (0.1 mm), or (*4*) electrochemical etching.

17.1.1 Flat sheet, strip in cut lengths, and plate shall be marked in two places near the ends.

17.1.2 Sheet and strip in coils shall be marked near the outside end of the coil. The inside of the coil shall also be marked or shall have a tag or label attached and marked with the information of 17.1.

A 240

17.1.3 Material less than ¼ in. (6.4 mm) in thickness shall not be marked with die stamps.

17.2 The manufacturer's test identification number shall be legibly stamped on each test specimen, if to be shipped to the customer.

17.3 Marking for identification for U.S. Government procurement shall be in accordance with Specification A 480.

18. Inspection

18.1 Inspection of the material by the purchaser's representative at the producing plant shall be made as agreed upon by the purchaser and the seller as part of the purchase order.

19. Material Test Report

19.1 A report of the results of all tests required by this specification and the type of melting used when required by the purchase order shall be supplied to the purchaser.

20. Certification

20.1 Upon request of the purchaser in the contract or order, the producer's certification that the material was manufactured and tested in accordance with this specification together with the report of the test results shall be furnished at the time of shipment.

21. Rejection

21.1 Unless otherwise specified, any rejection based on tests made in accordance with this specification shall be reported to the seller within 60 working days from receipt of the material by the purchaser.

22. Rehearing

22.1 Samples tested by the purchaser in accordance with this specification that represent rejected material shall be retained for 3 weeks from the date of the notification to the seller of the rejection. In case of dissatisfaction with the results of the tests, the seller may make claim for a rehearing within that time.

ASTM A 240

TABLE 1 Chemical Requirements Composition, %[A]

Austenitic (Chromium-Nickel) (Chromium-Manganese-Nickel)

UNS Designation[B]	Type	Carbon[C]	Manganese	Phosphorus	Sulfur	Silicon	Chromium	Nickel	Other Elements
S20100	201	0.15	5.50–7.50	0.060	0.030	1.00	16.00–18.00	3.50–5.50	N 0.25 max
S20200	202	0.15	7.50–10.0	0.060	0.030	1.00	17.00–19.00	4.00–6.00	N 0.25 max
S30200	302	0.15	2.00	0.045	0.030	1.00	17.00–19.00	8.00–10.00	N 0.10 max
S30400	304	0.08	2.00	0.045	0.030	1.00	18.00–20.00	8.00–10.50	N 0.10 max
S30403	304L	0.030	2.00	0.045	0.030	1.00	18.00–20.00	8.00–12.00	N 0.10 max
S30453	304LN	0.030	2.00	0.045	0.030	1.00	18.00–20.00	8.00–12.00	N 0.10–0.16
S30409	304H	0.04–0.10	2.00	0.045	0.030	1.00	18.00–20.00	8.00–10.50	...
S30451	304N	0.08	2.00	0.045	0.030	1.00	18.00–20.00	8.00–10.50	N 0.10–0.16
S30500	305	0.12	2.00	0.045	0.030	1.00	17.00–19.00	10.50–13.00	...
S30908	309S	0.08	2.00	0.045	0.030	1.00	22.00–24.00	12.00–15.00	...
S30940	309Cb	0.08	2.00	0.045	0.030	1.00	22.00–24.00	12.00–16.00	Cb + Ta 10 × C min, 1.10 max
S31008	310S	0.08	2.00	0.045	0.030	1.50	24.00–26.00	19.00–22.00	...
S31040	310Cb	0.08	2.00	0.045	0.030	1.50	24.00–26.00	19.00–22.00	Cb + Ta 10 × C min, 1.10 max
S31254	...	0.020	1.00	0.030	0.010	0.80	19.50–20.50	17.50–18.50	Mo 6.00–6.50; Cu 0.50–1.00; N 0.180–0.220
S31600	316	0.08	2.00	0.045	0.030	1.00	16.00–18.00	10.00–14.00	Mo 2.00–3.00; N 0.10 max
S31603	316L	0.030	2.00	0.045	0.030	1.00	16.00–18.00	10.00–14.00	Mo 2.00–3.00; N 0.10 max
S31653	316LN	0.030	2.00	0.045	0.030	1.00	16.00–18.00	10.00–14.00	N 0.10–0.16; Mo 2.00–3.00
S31609	316H	0.04–0.10	2.00	0.045	0.030	1.00	16.00–18.00	10.00–14.00	Mo 2.00–3.00
S31635	316Ti	0.08	2.00	0.045	0.030	1.00	16.00–18.00	10.00–14.00	Ti 5 × (C + N) min, 0.70 max N 0.10 max Mo 2.0–3.0
S31640	316Cb	0.08	2.00	0.045	0.030	1.00	16.00–18.00	10.00–14.00	Cb + Ta 10 × C min, 1.10 max N 0.10 max Mo 2.0–3.0
S31651	316N	0.08	2.00	0.045	0.030	1.00	16.00–18.00	10.00–14.00	Mo 2.00–3.00; N 0.10–0.16
S31700	317	0.08	2.00	0.045	0.030	1.00	18.00–20.00	11.00–15.00	Mo 3.00–4.00; N 0.10 max
S31703	317L	0.030	2.00	0.045	0.030	1.00	18.00–20.00	11.00–15.00	Mo 3.00–4.00; N 0.10 max
S32100	321	0.08	2.00	0.045	0.030	1.00	17.00–19.00	9.00–12.00	Ti 5 × (C + N) min, 0.70 max N 0.10 max
S32109	321H	0.04–0.10	2.00	0.045	0.030	1.00	17.00–19.00	9.00–12.00	Ti 4 × (C + N) min, 0.70 max
S34700	347	0.08	2.00	0.045	0.030	1.00	17.00–19.00	9.00–13.00	Cb + Ta 10 × C min; 1.10 max
S34709	347H	0.04–0.10	2.00	0.045	0.030	1.00	17.00–19.00	9.00–13.00	Cb + Ta 8 × C min; 1.00 max
S34800	348	0.08	2.00	0.045	0.030	1.00	17.00–19.00	9.00–13.00	Cb + Ta 10 × C min; 1.10 max Ta 0.10 max Co 0.20 max
S34809	348H	0.04–0.10	2.00	0.045	0.030	1.00	17.00–19.00	9.00–13.00	Cb + Ta 8 × C min; 1.00 max Ta 0.10 max Co 0.20 max

TABLE 1 *Continued*

UNS Designation[B]	Type	Carbon[C]	Manganese	Phosphorus	Sulfur	Silicon	Chromium	Nickel	Other Elements
Austenitic (Chromium-Nickel) (Chromium-Manganese-Nickel)									
S38100	XM-15	0.08	2.00	0.030	0.030	1.50–2.50	17.00–19.00	17.50–18.50	...
S30452	XM-21	0.08	2.00	0.045	0.030	1.00	18.00–20.00	8.00–10.50	N 0.16–0.30
S30815	...	0.10	0.80	0.040	0.030	1.40–2.00	20.00–22.00	10.00–12.00	N 0.14–0.20 Ce 0.03–0.08
S21600	XM-17	0.08	7.50–9.00	0.045	0.030	1.00	17.50–22.00	5.00–7.00	Mo 2.00–3.00 N 0.25–0.50
S21603	XM-18	0.03	7.50–9.00	0.045	0.030	1.00	17.50–22.00	5.00–7.00	Mo 2.00–3.00 N 0.25–0.50
S20910	XM-19	0.06	4.00–6.00	0.040	0.030	1.00	20.50–23.50	11.50–13.50	N 0.20–0.40 Cb 0.10–0.30 V 0.10–0.30 Mo 1.50–3.00
S24000	XM-29	0.08	11.50–14.50	0.060	0.030	1.00	17.00–19.00	2.25–3.75	N 0.20–0.40
S21400	XM-31	0.12	14.00–16.00	0.045	0.030	0.30–1.00	17.00–18.50	1.00	N 0.35 min
Duplex (Austenitic-Ferritic)									
S32550	...	0.04	1.5	0.040	0.030	1.0	24.0–27.0	4.5–6.5	Cu 1.5–2.5 Mo 2.0–4.0 N 0.10–0.25
S31200	...	0.030	2.00	0.045	0.030	1.00	24.0–26.0	5.5–6.5	Mo 1.2–2.0 N 0.14–0.20
S31803	...	0.030	2.00	0.030	0.20	1.00	21.0–23.0	4.50–6.50	Mo 2.50–3.50 N 0.08–0.20
S32900	329	0.08	1.00	0.040	0.030	0.75	23.00–28.00	2.50–5.00	Mo 1.0–2.0
Ferritic or Martensitic (Chromium)									
S40500	405	0.08	1.00	0.040	0.030	1.00	11.50–14.50	0.60	Al 0.10–0.30
S40900	409	0.08	1.00	0.045	0.045	1.00	10.50–11.75	0.50	Ti 6 × C min; 0.75 max
S41000	410	0.15	1.00	0.040	0.030	1.00	11.50–13.50	0.75	...
S41008	410S	0.08	1.00	0.040	0.030	1.00	11.50–13.50	0.60	...
S41050	...	0.040	1.00	0.045	0.030	1.00	10.50–12.50	0.60–1.10	N 0.10 max
S42900	429	0.12	1.00	0.040	0.030	1.00	14.00–16.00	0.75	...
S43000	430	0.12	1.00	0.040	0.030	1.00	16.00–18.00	0.75	...
S43035	XM-8	0.07	1.00	0.040	0.030	1.00	17.00–19.00	0.50	Ti 0.20 + 4 (C + N) min; 1.10 max Al 0.15 max N 0.04 max
S44400	...	0.025	1.00	0.040	0.030	1.00	17.5–19.5	1.00	Ti + Cb 0.20 + 4 (C + N) min; 0.80 max Mo 1.75–2.50 N 0.035 max
S44627	XM-27	0.010[D]	0.40	0.020	0.020	0.40	25.00–27.50	0.50	Mo 0.75–1.50 Cb 0.05–0.20 Cu 0.20 max N 0.015 max[D] Ni + Cu 0.50 max
S44626	XM-33	0.06	0.75	0.040	0.020	0.75	25.00–27.00	0.50	Mo 0.75–1.50 Ti 0.20–1.00 7 × (C + N) min N 0.04 max Cu 0.20 max

A 240

TABLE 1 *Continued*

UNS Designation[B]	Type	Carbon[C]	Manganese	Phosphorus	Sulfur	Silicon	Chromium	Nickel	Other Elements
Ferritic or Martensitic (Chromium)									
S44635	...	0.025	1.00	0.040	0.030	0.75	24.5–26.0	3.5–4.5	Mo 3.5–4.5 Ti + Cb 0.20 + 4 (C + N) min; 0.80 max N 0.035 max
S44660	...	0.030	1.00	0.040	0.030	1.00	25.0–28.0	1.0–3.50	Mo 3.0–3.50 Ti + Cb = 0.20 – 1.00 and 6 (C + N) min N 0.040 max
S44700	...	0.010	0.30	0.025	0.020	0.20	28.0–30.0	0.15	Mo 3.5–4.2 (C + N) 0.025 max Cu 0.15 max N 0.020 max
S44800	...	0.10	0.30	0.025	0.020	0.20	28.0–30.0	2.0–2.5	Mo 3.5–4.2 (C + N) 0.025 max Cu 0.15 max N 0.020 max
S44735	...	0.030	1.00	0.040	0.030	1.00	28.00–30.00	1.00	Mo 3.60–4.20 N 0.045 max Ti + Cb 0.20–1.00 Ti + Cb 6 (C + N) min

[A] Maximum, unless range or minimum is indicated.

[B] New designation established in accordance with ASTM E 527 and SAE J 1086, Recommended Practice for Numbering Metals and Alloy (UNS).

[C] Carbon analysis shall be reported to nearest 0.01 % except for the low-carbon types, which shall be reported to nearest 0.001 %.

[D] Product (check or verification) analysis tolerance over the maximum limit for C and N in XM-27 shall be 0.002 %.

ASTM A 240

TABLE 2 Mechanical Test Requirements

UNS Designation	Type	Tensile Strength, min		Yield Strength[A] min		Elongation in 2 in. or 50 mm, min, %	Hardness, max[B]		Cold Bend[C]
		ksi	MPa	ksi	MPa		Brinell	Rockwell B	
Austenitic (Chromium-Nickel) (Chromium-Manganese-Nickel)									
S20100	201	95	655	38	260	40.0	...	...	...
S20200	202	90	620	38	260	40.0	...	...	...
S30200	302	75	515	30	205	40.0	202	92	not required
S30400	304	75	515	30	205	40.0	202	92	not required
S30403	304L	70	485	25	170	40.0	183	88	not required
S30453	304LN	75	515	30	205	40.0	202	92	not required
S30409	304H	75	515	30	205	40.0	183	92	not required
S30451	304N	80	550	35	240	30.0	183	92	not required
S30500	305	75	515	30	205	40.0	183	88	not required
S30908	309S	75	515	30	205	40.0	217	95	not required
S30940	309Cb	75	515	30	205	40.0	217	95	not required
S31008	310S	75	515	30	205	40.0	217	95	not required
S31040	310Cb	75	515	30	205	40.0	217	95	not required
S31254	...	94	650	44	300	35.0	220	96	not required
S31600	316	75	515	30	205	40.0	217	95	not required
S31603	316L	70	485	25	170	40.0	217	95	not required
S31653	316LN	75	515	30	205	40.0	217	95	not required
S31609	316H	75	515	30	205	40.0	217	95	not required
S31635	316Ti	75	515	30	205	40.0	217	95	not required
S31640	316Cb	75	515	30	205	30.0	217	95	not required
S31651	316N	80	550	35	240	35.0	217	95	not required
S31700	317	75	515	30	205	35.0	217	95	not required
S31703	317L	75	515	30	205	40.0	217	95	not required
S32100	321	75	515	30	205	40.0	217	95	not required
S32109	321H	75	515	30	205	40.0	217	95	not required
S34700	347	75	515	30	205	40.0	183	92	not required
S34709	347H	75	515	30	205	40.0	183	92	not required
S34800	348	75	515	30	205	40.0	183	92	not required
S34809	348H	75	515	30	205	40.0	183	92	not required
S38100	XM-15	75	515	30	205	40.0	217	95	not required
S30452	XM-21								
Sheet and Strip		90	620	50	345	30.0	241	100	not required
Plate		85	585	40	275	30.0	241	100	not required
S30815	...	87	600	45	310	40.0	217	95	not required
S21600	XM-17								
Sheet and Strip		100	690	60	414	40.0	...	100	not required
Plate		90	620	50	345	40.0	241	100	not required
S21603	XM-18								
Sheet and Strip		100	690	60	415	40.0	...	100	not required
Plate		90	620	50	345	40.0	241	100	not required
S20910	XM-19								
Sheet and Strip		120	825	75	515	30.0	241	100	not required
Plate		100	690	55	380	35.0	241	100	not required
S24000	XM-29								
Sheet and Strip		100	690	60	415	40.0	...	100	not required
Plate		100	690	55	380	40.0	...	100	not required
S21400	XM-31								
Sheet		125	860	70	485	40.0	...	...	not required
Strip		105	725	55	380	40.0	...	...	not required

TABLE 2 Continued

UNS Designation	Type	Tensile Strength, min ksi	Tensile Strength, min MPa	Yield Strength[A] min ksi	Yield Strength[A] min MPa	Elongation in 2 in. or 50 mm, min, %	Hardness, max[B] Brinell	Hardness, max[B] Rockwell B	Cold Bend °[C]
Duplex (Austenitic-Ferritic)									
S32550	...	110	760	80	550	15.0	297	32[E]	not required
S31200	...	100	690	65	450	25.0	220	...	not required
S31803	...	90	620	65	450	25.0	290	32[E]	not required
S32900	329	90	620	70	485	15.0	271	28[E]	not required
Ferritic or Martensitic (Chromium)									
S40500	405	60	415	25	170	20.0	183	88	180
S40900	409	55	380	30	205	20.0	...	80	180
S41000	410	65	450	30	205	20.0	217	95	180
S41008	410S	60	415	30	205	22.0[D]	183	88	180
S41050	...	60	415	30	205	22.0	183	88	180
S42900	429	65	450	30	205	22.0[D]	183	88	180
S43000	430	65	450	30	205	22.0[D]	183	88	180
S43035	XM-8	65	450	30	205	22.0	183	88	180
S44400	...	60	415	40	275	20.0	217	95	180
S44627	XM-27	65	450	40	275	22.0	190	90	180
S44626	XM-33	68	470	45	310	20.0	217	95	180
S44635	...	90	620	75	517	20.0	270	27[E]	180
S44660	...	80	550	55	380	18.0	241	100	180
S44700	...	80	550	60	415	20.0	...	20[E]	180
S44800	...	80	550	60	415	20.0	...	20[E]	180
S44735	...	80	550	60	415	18.0	...	25[E]	180

[A] Yield strength shall be determined by the offset method at 0.2 % limiting permanent set in accordance with Methods A 370. Unless otherwise specified (see 5.1.11), an alternative method of determining yield strength may be based on total extension under load of 0.5 %.
[B] Either Brinell or Rockwell B Hardness is permissible.
[C] See 9.2. Bend test not required for steels of the chromium-nickel series or for chromium steels thicker than 1 in. (2.54 mm).
[D] Material 0.050 in (1.27 mm) and under in thickness shall have a minimum elongation of 20.0%.
[E] Rockwell C scale.

The American Society for Testing and Materials takes no position respecting the validity of any patent rights asserted in connection with any item mentioned in this standard. Users of this standard are expressly advised that determination of the validity of any such patent rights, and the risk of infringement of such rights, are entirely their own responsibility.

This standard is subject to revision at any time by the responsible technical committee and must be reviewed every five years and if not revised, either reapproved or withdrawn. Your comments are invited either for revision of this standard or for additional standards and should be addressed to ASTM Headquarters. Your comments will receive careful consideration at a meeting of the responsible technical committee, which you may attend. If you feel that your comments have not received a fair hearing you should make your views known to the ASTM Committee on Standards, 1916 Race St., Philadelphia, Pa. 19103.

Designation: A 262 – 81

Standard Practices for
DETECTING SUSCEPTIBILITY TO INTERGRANULAR ATTACK IN AUSTENITIC STAINLESS STEELS[1]

This standard is issued under the fixed designation A 262; the number immediately following the designation indicates the year of original adoption or, in the case of revision, the year of last revision. A number in parentheses indicates the year of last reapproval. A superscript epsilon (ϵ) indicates an editorial change since the last revision or reapproval.

These practices have been approved for use by agencies of the Department of Defense and for listing in the DoD Index of Specifications and Standards.

1. Scope

1.1 These practices cover the following five tests:

1.1.1 *Practice A*—Oxalic Acid Etch Test for Classification of Etch Structures of Austenitic Stainless Steels (Sections 2 to 6, incl),

1.1.2 *Practice B*—Ferric Sulfate-Sulfuric Acid Test for Detecting Susceptibility to Intergranular Attack in Austenitic Stainless Steels (see Sections 7 to 13, incl),

1.1.3 *Practice C*—Nitric Acid Test for Detecting Susceptibility to Intergranular Attack in Austenitic Stainless Steels (see Sections 14 to 20, incl),

1.1.4 *Practice D*—Nitric-Hydrofluoric Acid Test for Detecting Susceptibility to Intergranular Attack in Molybdenum-Bearing Austenitic Stainless Steels (see Sections 21 to 27, incl), and

1.1.5 *Practice E*—Copper-Copper Sulfate-Sulfuric Acid Test for Detecting Susceptibility to Intergranular Attack in Austenitic Stainless Steels (see Section 28).

1.2 The following factors govern the application of these practices:

1.2.1 Susceptibility to intergranular attack associated with the precipitation of chromium carbides is readily detected in all five tests.

1.2.2 Sigma phase in wrought chromium-nickel-molybdenum steels, which may or may not be visible in the microstructure, can result in high corrosion rates only in nitric acid.

1.2.3 Sigma phase in titanium or columbium stabilized alloys, which may or may not be visible in the microstructure, can result in high corrosion rates in both the nitric acid and ferric sulfate-sulfuric acid solutions.

1.3 The oxalic acid etch test is a rapid method of identifying, by simple etching, those specimens of certain stainless steel grades which are essentially free of susceptibility to intergranular attack associated with chromium carbide precipitates. These specimens will have low corrosion rates in certain corrosion tests and therefore can be eliminated (screened) from testing as "acceptable."

1.4 The ferric sulfate-sulfuric acid test, the nitric acid test, and the nitric-hydrofluoric acid test are based on weight loss determinations and, thus, provide a quantitative measure of the relative performance of specimens evaluated. In contrast, the copper-copper sulfate-sulfuric acid test is based on visual examination of bend specimens and, therefore, classifies the specimens only as acceptable or non-acceptable.

1.5 In most cases either the 24-h copper-copper sulfate-sulfuric acid test or the 120-h ferric sulfate-sulfuric acid test, combined with the oxalic acid etch test, will provide the required information in the shortest time. All stainless grades listed in the accompanying table may be evaluated in these combinations of screening and corrosion tests, except those specimens of molybdenum-bearing grades (for example 316, 316L, 317, and 317L), which rep-

---

[1] These practices are under the jurisdiction of ASTM Committee A-1 on Steel, Stainless Steel and Related Alloys, and are the direct responsibility of Subcommittee A01.14 on Methods of Corrosion Testing.
Current edition approved July 31, 1981. Published November 1981. Originally published as A 262 – 43 T. Last previous edition A 262 – 79.

resent steel intended for use in nitric acid environments.

1.6 For AISI Grades 316, 316L, 317, and 317L only, the nitric-hydrofluoric acid test may be used to provide test results in 4 h.

1.7 The 240-h nitric acid test must be applied to stabilized and molybdenum-bearing grades intended for service in nitric acid and to all stainless steel grades which might be subject to end grain corrosion in nitric acid service.

1.8 Only those stainless steel grades are listed in Table 1 for which data on the application of the oxalic acid etch test and on their performance in various quantitative evaluation tests are available.

1.9 Extensive test results on various types of stainless steels evaluated by these practices have been published in Ref **(10)**.[2]

PRACTICE A—OXALIC ACID ETCH TEST FOR CLASSIFICATION OF ETCH STRUCTURES OF AUSTENITIC STAINLESS STEELS (1)

2. Scope

2.1 The oxalic acid etch test is used for acceptance of material but not for rejection of material. This may be used in connection with other evaluation tests to provide a rapid method for identifying those specimens which are certain to be free of susceptibility to rapid intergranular attack in these other tests. Such specimens have low corrosion rates in the various hot acid tests, requiring from 4 to 240 h of exposure. These specimens are identified by means of their etch structures which are classified according to the following criteria:

2.2 The oxalic acid etch test may be used to screen specimens intended for testing in Practice B—Ferric Sulfate-Sulfuric Acid Test, Practice C—Nitric Acid Test, Practice D—Nitric-Hydrofluoric Acid Test, and Practice E—Copper-Copper Sulfate-Sulfuric Acid Test.

2.2.1 Each practice contains a table showing which classifications of etch structures on a given stainless steel grade are equivalent to acceptable, or possibly nonacceptable performance in that particular test. Specimens having acceptable etch structures need not be subjected to the hot acid test. Specimens having nonacceptable etch structures must be tested in the specified hot acid solution.

2.3 The grades of stainless steels and the hot acid tests for which the oxalic acid etch test is applicable are listed in Table 2.

2.4 Extra low carbon grades, such as 304L, 316L, and 317L, are tested after sensitizing heat treatments at 1200 to 1250°F (650 to 675°C), which is the range of maximum carbide precipitation. These sensitizing treatments must be applied before the specimens are submitted to the oxalic acid etch test. The most commonly used sensitizing treatment is 1 h at 1250°F.

3. Apparatus

3.1 *Source of Direct Current*—Battery, generator, or rectifier capable of supplying about 15 V and 20 A.

3.2 *Ammeter*—Range 0 to 30 A (Note 1).

3.3 *Variable Resistance* (Note 1).

3.4 *Cathode*—A cylindrical piece of stainless steel or, preferably, a 1-qt (0.946-L) stainless steel beaker.

3.5 *Large Electric Clamp*—To hold specimen to be etched.

3.6 *Metallurgical Microscope*—For examination of etched microstructures at 250 to 500 diameters.

3.7 *Electrodes of the Etching Cell*—The specimen to be etched is made the anode, and a stainless steel beaker or a piece of stainless steel as large as the specimen to be etched is made the cathode.

3.8 *Electrolyte*—Oxalic acid, ($H_2C_2O_4 \cdot 2H_2O$), reagent grade, 10 weight % solution.

NOTE 1—The variable resistance and the ammeter are placed in the circuit to measure and control the current on the specimen to be etched.

4. Preparation of Test Specimens

4.1 *Cutting*—Sawing is preferred to shearing, especially on the extra-low carbon grades. Shearing cold works adjacent metal and affects the response to subsequent sensitization. Microscopical examination of an etch made on a specimen containing sheared edges, should be made on metal unaffected by shearing. A convenient specimen size is 1 by 1 in. (25 by 25 mm).

4.2 The intent is to test a specimen representing as nearly as possible the surface of the material as it will be used in service. Therefore the preferred sample is a cross section including the surface to be exposed in service. Only such

---

[2] The boldface numbers in parentheses refer to list of references found at the end of these practices.

surface finishing should be performed as is required to remove foreign material and obtain a standard, uniform finish as described in 4.3. For very heavy sections, specimens should be machined to represent the appropriate surface while maintaining reasonable specimen size for convenient testing. Ordinarily, removal of more material than necessary will have little influence on the test results. However, in the special case of surface carburization (sometimes encountered, for instance, in tubing or castings when lubricants or binders containing carbonaceous materials are employed) it may be possible by heavy grinding or machining to completely remove the carburized surface. Such treatment of test specimens is not permissible, except in tests undertaken to demonstrate such effects.

4.3 *Polishing*—On all types of materials, cross sectional surfaces should be polished for etching and microscopical examination. Specimens containing welds should include base plate, weld heat-affected zone, and weld metal. Scale should be removed from the area to be etched, by grinding to an 80 or 120-grit finish on a grinding belt or wheel without excessive heating, and then polishing on successively finer emery papers, No. 1, ½, ⅙, ⅔, and ⅜, or finer. This polishing operation can be carried out in a relatively short time since all large scratches need not be removed. Whenever practical, a polished area of 1 cm^2 or more is desirable. If any cross sectional dimension is less than 1 cm, a minimum length of 1 cm should be polished. When the available length is less than 1 cm, a full cross section should be used.

4.4 *Etching Solution*—The solution used for etching is prepared by adding 100 g of reagent grade oxalic acid crystals ($H_2C_2O_4 \cdot 2H_2O$) to 900 mL of distilled water and stirring until all crystals are dissolved.

4.5 *Etching Conditions*—The polished specimen should be etched at 1 A/cm^2 for 1.5 min. To obtain the correct current density:

4.5.1 The total immersed area of the specimen to be etched should be measured in square centimetres, and

4.5.2 The variable resistance should be adjusted until the ammeter reading in amperes is equal to the total immersed area of the specimen in square centimetres.

4.6 *Etching Precautions:*

4.6.1 Etching should be carried out under a ventilated hood. Gas, which is rapidly evolved at the electrodes with some entrainment of oxalic acid, is poisonous and irritating to mucous membranes.

4.6.2 A yellow-green film is gradually formed on the cathode. This increases the resistance of the etching cell. When this occurs, the film should be removed by rinsing the inside of the stainless steel beaker (or the steel used as the cathode) with an acid such as 30 % HNO_3.

4.6.3 The temperature of the etching solution gradually increases during etching. The temperature should be kept below 50°C by alternating two beakers. One may be cooled in tap water while the other is used for etching. The rate of heating depends on the total current (ammeter reading) passing through the cell. Therefore, the area etched should be kept as small as possible while at the same time meeting the requirements of desirable minimum area to be etched.

4.6.4 Immersion of the clamp holding the specimen in the etching solution should be avoided.

4.7 *Rinsing*—Following etching, the specimen should be thoroughly rinsed in hot water and in acetone or alcohol to avoid crystallization of oxalic acid on the etched surface during drying.

4.8 On some specimens containing molybdenum (AISI 316, 316L, 317, 317L) which are free of chromium carbide sensitization, it may be difficult to reveal the presence of step structures by electrolytic etching with oxalic acid. In such cases, an electrolyte of a 10 % solution of ammonium persulfate, $(NH_4)_2S_2O_8$, may be used in place of oxalic acid. An etch of 5 or 10 min at 1 A/cm^2 in a solution at room temperature readily develops step structures on such specimens.

5. Classification of Etch Structures

5.1 The etched surface is examined on a metallurgical microscope at 250× to 500× for wrought steels and at about 250× for cast steels.

5.2 The etched cross-sectional areas should be thoroughly examined by complete traverse from inside to outside diameters of rods and tubes, from face to face on plates, and across all zones such as weld metal, weld-affected zones, and base plates on specimens containing welds.

5.3 The etch structures are classified into the following types (Note 2):

5.3.1 *Step Structure (Fig. 1)*—Steps only between grains, no ditches at grain boundaries.

5.3.2 *Dual Structure (Fig. 2)*—Some ditches at grain boundaries in addition to steps, but no single grain completely surrounded by ditches.

5.3.3 *Ditch Structure (Fig. 3)*—One or more grains completely surrounded by ditches.

5.3.4 *Isolated Ferrite (Fig. 4)*—Observed in castings and welds. Steps between austenite matrix and ferrite pools.

5.3.5 *Interdendritic Ditches (Fig. 5)*—Observed in castings and welds. Deep interconnected ditches.

5.3.6 *End-Grain Pitting I (Fig. 6)*—Structure contains a few deep end-grain pits along with some shallow etch pits at 500×. (Of importance only when nitric acid test is used.)

5.3.7 *End-Grain Pitting II (Fig. 7)*—Structure contains numerous, deep end-grain pits at 500×. (Of importance only when nitric acid test is used.)

NOTE 2—All photomicrographs were made with specimens that were etched under standard conditions: 10 % oxalic acid, room temperature, 1.5 min at 1 A/cm^2.

5.4 The evaluation of etch structures containing steps only and of those showing grains completely surrounded by ditches in every field can be carried out relatively rapidly. In cases which appear to be dual structures, more extensive examination is required to determine if there are any grains completely encircled. If an encircled grain is found, the steel should be evaluated as a ditch structure. Areas near surfaces should be examined for evidence of surface carburization.

5.4.1 On stainless steel castings (also on weld metal) the steps between grains formed by electrolytic oxalic acid etching tend to be less prominent than those on wrought materials, or are entirely absent. However, any susceptibility to intergranular attack is readily detected by pronounced ditches.

5.5 Some wrought specimens, especially from bar stock, may contain a random pattern of pits. If these pits are sharp and so deep that they appear black (Fig. 7) it is possible that the specimen may be susceptible to end grain attack in nitric acid only. Therefore, even though the grain boundaries all have step structures, specimens having as much or more end grain pitting than that shown in Fig. 7 cannot be safely assumed to have low nitric acid rates and should be subjected to the nitric acid test whenever it is specified. Such sharp, deep pits should not be confused with the shallow pits shown in Figs. 1 and 6.

6. Use of Etch Structure Classifications

6.1 The use of these classifications depends on the hot acid corrosion test for which stainless steel specimens are being screened by etching in oxalic acid and is described in each of the practices. Important characteristics of each of these tests are described below.

6.2 *Practice B—Ferric Sulfate-Sulfuric Acid Test* is a 120-h test in boiling 50 % solution which detects susceptibility to intergranular attack associated primarily with chromium carbide precipitate. It does not detect susceptibility associated with sigma phase in chromium-nickel-molybdenum stainless steels (316, 316L, 317, 317L) which is known to lead to rapid intergranular attack only in certain nitric acid environments. It does not detect susceptibility to end grain attack which is also found only in certain nitric acid environments. The ferric sulfate-sulfuric acid test does reveal susceptibility associated with a sigma-like phase constituent in stabilized stainless steels, AISI 321 and 347.

6.3 *Practice C—Nitric Acid Test* is a 240-h test in boiling, 65 % nitric acid which detects susceptibility to rapid intergranular attack associated with chromium carbide precipitate and with sigma-like phase precipitate. The latter may be formed in molybdenum-bearing and in stabilized grades of austenitic stainless steels and may or may not be visible in the microstructure. This test also reveals susceptibility to end grain attack in all grades of stainless steels.

6.4 *Practice D—Nitric-Hydrofluoric Acid Test* is a 4-h test in a solution of 10 % nitric acid and 3 % hydrofluoric acid at 70°C. It is applicable only to molybdenum-bearing grades of austenitic stainless steels (AISI 316, 316L, 317, 317L) and detects only susceptibility to intergranular attack associated with chromium carbide precipitates. It does not detect susceptibility to intergranular attack associated with sigma phase or end grain corrosion, which, so far, are known to lead to rapid intergranular attack only in certain nitric acid environments.

6.5 *Practice E—Copper-Copper Sulfate-Sulfuric Acid Test* is a 24-h test in a boiling solution containing 16 % sulfuric acid and 6 % copper sulfate with the test specimen embedded in metallic copper shot or grindings, which detects susceptibility to intergranular attack associated with the precipitation of chromium-rich carbides. It does not detect susceptibility to intergranular attack associated with sigma phase, or end-grain corrosion, both of which have been observed to date only in certain nitric acid environments.

PRACTICE B—FERRIC SULFATE-SULFURIC ACID TEST FOR DETECTING SUSCEPTIBILITY TO INTERGRANULAR ATTACK IN AUSTENITIC STAINLESS STEELS (2)

7. Scope

7.1 This practice describes the procedure for conducting the boiling, 120-h ferric sulfate-50 % sulfuric acid test (Note 3) which measures the susceptibility of stainless steels to intergranular attack. The presence or absence of intergranular attack in this test is not necessarily a measure of the performance of the material in other corrosive environments. The test does not provide a basis for predicting resistance to forms of corrosion other than intergranular, such as general corrosion, pitting, or stress-corrosion cracking.

NOTE 3—See Practice A for information on the most appropriate of the several test methods available for the evaluation of specific grades of stainless steel.

7.1.1 The ferric sulfate-sulfuric acid test detects susceptibility to intergranular attack associated with the precipitation of chromium carbides in unstabilized austenitic stainless steels. It does not detect susceptibility to intergranular attack associated with sigma phase in austenitic stainless steels containing molybdenum, such as Types 316, 316L, 317, and 317L.

NOTE 4—To detect susceptibility to intergranular attack associated with sigma phase in austenitic stainless steels containing molybdenum, the nitric acid test, Practice C, should be used.

7.2 In stabilized stainless steel, Type 321 (and perhaps 347) the ferric sulfate-sulfuric acid test detects susceptibility associated with precipitated chromium carbides and with a sigma phase which may be invisible in the microstructure.

7.3 The ferric sulfate-sulfuric acid test may be used to evaluate the heat treatment accorded as-received material. It may also be used to check the effectiveness of stabilizing columbium or titanium additions and of reductions in carbon content in preventing susceptibility to rapid intergranular attack. It may be applied to wrought products (including tubes), castings, and weld metal.

7.4 Specimens of extra low carbon and stabilized grades are tested after sensitizing heat treatments at 1200 to 1250°F (650 to 675°C), which is the range of maximum carbide precipitation. The length of time of heating used for this sensitizing treatment determines the maximum permissible corrosion rate for such grades in the ferric sulfate-sulfuric acid test. The most commonly used sensitizing treatment is 1 h at 1250°F.

8. Rapid Screening Test

8.1 Before testing in the ferric sulfate sulfuric acid test, specimens of certain grades of stainless steels (see Table 3) may be given a rapid screening test in accordance with procedures given in Practice A, Oxalic Acid Etch Test for Classification of Etch Structures of Austenitic Stainless Steels. Preparation, etching, and the classification of etch structures are described therein. The use of etch structure evaluations in connection with the ferric sulfate-sulfuric acid test is specified in Table 3.

8.1.1 Corrosion test specimens having acceptable etch structures in the oxalic acid etch test will be essentially free of intergranular attack in the ferric sulfate-sulfuric acid test. Such specimens are acceptable without testing in the ferric sulfate-sulfuric acid test. All specimens having nonacceptable etch structures must be tested in the ferric sulfate-sulfuric acid test.

9. Apparatus

9.1 The apparatus (Note 6) is illustrated in Fig. 8.

9.1.1 A four-bulb Allihn or Soxhlet condenser with a 45/50 ground glass joint. Overall length: about 13 in. (330 mm), condensing section, 9½ in. (241 mm).

9.1.2 A 1-L Erlenmeyer flask with a 45/50

ground glass joint. The ground glass opening is somewhat over 1½ in. (38 mm) wide.

9.1.3 The glass cradle (Note 5) can be supplied by a glass-blowing shop. To pass through the ground glass joint on the Erlenmeyer flask, the width of the cradle should not exceed 1½ in., and the front-to-back distance must be such that the cradle will fit the 1⅓-in. (34-mm) diameter opening. It should have three or four holes to increase circulation of the testing solution around the specimen.

NOTE 5—Other equivalent means of specimen support, such as glass hooks or stirrups, may also be used.

9.1.4 Boiling chips must be used to prevent bumping.

9.1.5 A silicone grease[3] is recommended for the ground glass joint.

9.1.6 During testing, there is some deposition of iron oxides on the upper part of the Erlenmeyer flask. This can be readily removed, after test completion, by boiling a solution of 10 % hydrochloric acid in the flask.

9.1.7 A device such as an electrically heated hot plate which provides heat for continuous boiling of the solution.

9.1.8 An analytical balance capable of weighing to the nearest 0.001 g.

NOTE 6—No substitutions for this equipment may be used. The cold-finger type of condenser with standard Erlenmeyer flasks may not be used.

10. Ferric Sulfate-Sulfuric Acid Test Solution

10.1 Prepare 600 mL of 50 % (49.4 to 50.9 %) solution as follows:

10.1.1 **Caution**—Protect the eyes and use rubber gloves for handling acid. Place the test flask under a hood.

10.1.2 First, measure 400.0 mL of distilled water in a 500-mL graduate and pour into the Erlenmeyer flask.

10.1.3 Then measure 236.0 mL of reagent grade sulfuric acid of a concentration which must be in the range of 95.0 to 98.0 % by weight in a 250-mL graduate. Add the acid slowly to the water in the Erlenmeyer flask to avoid boiling by the heat evolved.

NOTE 7—Loss of vapor results in concentration of the acid.

10.1.4 Weigh 25 g of reagent grade ferric sulfate (contains about 75 % $Fe_2(SO_4)_3$) and add to the sulfuric acid solution. A trip balance may be used.

10.1.5 Drop boiling chips into the flask.

10.1.6 Lubricate ground glass joint with silicone grease.

10.1.7 Cover flask with condenser and circulate cooling water.

10.1.8 Boil solution until all ferric sulfate is dissolved (see Note 7).

11. Preparation of Test Specimens

11.1 A specimen having a total surface area of 5 to 20 cm^2 is recommended. Specimens containing welds should be cut so that no more than ½-in. (13-mm) width of base metal is included on either side of the weld.

11.2 The intent is to test a specimen representing as nearly as possible the surface of the material as used in service. Only such surface finishing should be performed as is required to remove foreign material and obtain a standard, uniform finish as specified. For very heavy sections, specimens should be machined to represent the appropriate surface while maintaining reasonable specimen size for convenience in testing. Ordinarily, removal of more material than necessary will have little influence on the test results. However, in the special case of surface carburization (sometimes encountered, for instance, in tubing or castings when lubricants or binders containing carbonaceous materials are employed) it may be possible by heavy grinding or machining to remove the carburized layer completely. Such treatment of test specimens is not permissible, except in tests undertaken to demonstrate such surface effects.

11.3 When specimens are cut by shearing, the sheared edges should be refinished by machining or grinding prior to testing.

11.4 All surfaces of the specimen, including edges, should be finished using No. 80 or 120 grit abrasive paper. If dry abrasive paper is used, polish slowly to avoid overheating. Sand blasting should not be used.

11.5 All traces of oxide scale formed during heat treatments must be thoroughly removed. Any scale which cannot be removed by grinding, for example, in stamped numbers, must be removed by immersing the specimen in concentrated nitric acid at about 200°F (93°C). (Residual oxide scale causes galvanic action and consequent activation in the test solution.)

---

[3] Dow Corning Stopcock Grease has been found satisfactory for this purpose.

11.6 The specimen should be measured including the inner surfaces of any holes and the total exposed area calculated.

11.7 The specimen should then be degreased and dried using suitable nonchlorinated agents, such as soap and acetone, and then weighed to the nearest 0.001 g.

12. Procedure

12.1 Place specimen in glass cradle and immerse in boiling solution.

12.2 Mark liquid level on flask with wax crayon to provide a check on vapor loss which would result in concentration of the acid. If there is an appreciable change in the level, the test must be repeated with fresh solution and a reground specimen.

12.3 Continue immersion of the specimen for a total of 120 h, then remove specimen, rinse in water and acetone, and dry.

12.4 Weigh specimen and subtract weight from original weight.

12.5 No intermediate weighings are usually necessary. The tests can be run without interruption for 120 h. However, if preliminary results are desired, the specimen can be removed at any time for weighing.

12.6 No changes in solution are necessary during the 120-h test periods.

12.7 Additional ferric sulfate inhibitor may have to be added during the test if the corrosion rate is extraordinarily high as evidenced by a change in the color of the solution. More ferric sulfate must be added if the total weight loss of all specimens exceeds 2 g. (During the test, ferric sulfate is consumed at a rate of 10 g for each 1 g of dissolved stainless steel.)

12.8 Several specimens may be tested simultaneously. The number (3 or 4) is limited only by the number of glass cradles that can be fitted into the flask.

13. Calculation and Report

13.1 The effect of the acid solution on the material shall be measured by determining the loss of weight of the specimen. The corrosion rates should be reported as inches of penetration per month (Note 8), calculated as follows:

Inches per month = $(287 \times W)/(A \times t \times d)$

where:
t = time of exposure, h,
A = area, cm^2,
W = weight loss, g, and
d = density, g/cm^3
for chromium-nickel steels, $d = 7.9$ g/cm^3
for chromium-nickel-molybdenum steels, $d = 8.00$ g/cm^3

NOTE 8—Conversion factors to other commonly used units for corrosion rates are as follows:
inches per month × 12 = inches per year
inches per month × 1,000 = mils per month
inches per month × 12,000 = mils per year
inches per month × 8350 × density = milligrams per square decimetre per day
inches per month × 34.8 × density = grams per square metre per hour
1.00 in.^{2} = 6.45 cm^2

PRACTICE C—NITRIC ACID TEST FOR DETECTING SUSCEPTIBILITY TO INTERGRANULAR ATTACK IN AUSTENITIC STAINLESS STEELS

14. Scope

14.1 This practice describes the procedure for conducting the boiling nitric acid test (**3**) as employed to measure the relative susceptibility of austenitic stainless steels to intergranular attack. The presence or absence of intergranular attack in this test is not necessarily a measure of the performance of the material in other corrosive environments; in particular, it does not provide a basis for predicting resistance to forms of corrosion other than intergranular, such as general corrosion, pitting, or stress-corrosion cracking.

14.2 The boiling nitric acid test may be used to evaluate the heat treatment accorded "as-received" material. It is also sometimes used to check the effectiveness of stabilizing elements and of reductions in carbon content in preventing susceptibility to rapid intergranular attack.

NOTE 9—Intergranular attack in nitric acid is associated with one or more of the following: (*1*) intergranular precipitation of chromium carbides, (*2*) sigma or transition phases in molybdenum-bearing grades, and (*3*) sigma phase constituents in stabilized grades. The boiling nitric acid test should not be used for extra low carbon molybdenum-bearing grades unless the material tested is to be used in nitric acid service. See Practice A, Oxalic Acid Etching Test, for information on the most appropriate of the several test methods available for the evaluation of specific grades of stainless steel.

14.3 Specimens of extra low carbon and stabilized grades are tested after sensitizing heat treatments at 1200 to 1250°F (650 to 675°C), which is the range of maximum carbide precipitation. The length of time used for this sensitizing treatment determines the maximum permissible corrosion rate in the nitric acid test. The most commonly used sensitizing treatment is 1 h at 1250°F.

14.4 This practice may be applied to wrought products (including tubes), castings, and weld metal of the various grades of stainless steel (Note 9).

15. Rapid Screening Test

15.1 Before testing in the nitric acid test, specimens of certain grades of stainless steel as given in Table 1 may be given a rapid screening test in accordance with procedures given in Practice A, Oxalic Acid Etch Test for Classification of Etch Structures of Austenitic Stainless Steels. The use of the etch structure evaluations in connection with the nitric acid test is specified in Table 4.

15.1.1 Corrosion test specimens having acceptable etch structures in the oxalic acid etch test will be essentially free of intergranular attack in the nitric acid test; such specimens are acceptable without testing in the nitric acid test. All specimens having nonacceptable etch structures must be tested in the nitric acid test.

16. Apparatus

16.1 *Container*—A 1-L Erlenmeyer flask equipped with a cold finger-type condenser, as illustrated in Fig. 9, is recommended.

NOTE 10—Two other types of containers have been employed in the past and may be used if agreed upon between the supplier and purchaser. One of these consists of a one-litre Erlenmeyer flask with a ground glass joint and equipped with a 30-in. (762-mm) reflux condenser; it has been shown that results obtained with a reflux condenser tend to be somewhat higher than with the cold finger-type condenser due to greater vapor loss. The second type of container is the so called multi-sample testing apparatus (4) which was designed to permit the testing of a large number of specimens simultaneously by providing for replacement of the acid in contact with the specimens several times per hour with redistilled acid. Because of the lesser accumulation of corrosion products in the testing solution, the rates obtained with the multi-sample tester are consistently lower than those obtained with the conventional apparatus; the differences are small on properly annealed or stabilized material which will show low rates in both types of test but can be very large for sensitized specimens. For research purposes or where results are to be compared directly, it is essential that the same type of apparatus be used for all tests.

16.2 *Specimen Supports*—Glass hooks, stirrups, or cradles for supporting the specimens in the flask fully immersed at all times during the test and so designed that specimens tested in the same container do not come in contact with each other.

16.3 *Heater*—A means for heating the test solutions and of keeping them boiling throughout the test period. An electrically heated hot plate is satisfactory for this purpose.

16.4 *Balance*—An analytical balance capable of weighing to at least the nearest 0.001 g.

17. Nitric Acid Test Solution

17.1 The test solution shall be 65 ± 0.2 weight % as nitric acid determined by analysis. This solution may be prepared by adding distilled water to concentrated nitric acid (reagent grade HNO_3, sp gr 1.42) (Note 11) at the rate of 108 mL of distilled water per litre of concentrated nitric acid.

NOTE 11—The nitric acid used should conform to the recommended specifications for analytical reagent chemicals of the American Chemical Society (5) as follows:

Nonvolatile matter, max, %	0.0005
Sulfate (SO_4), max, %	0.0002
Arsenic, max, %	0.000003
Chlorine, max, % about	0.00007
Heavy metals, max, % about lead and iron	0.0005 0.0001

In addition, the fluorine content shall not exceed 0.0001 % and phosphate (PO_4) shall not exceed 0.00002 %.

18. Preparation of Test Specimens

18.1 The size and shape of the specimen must be considered with respect to available facilities for accurate weighing and the volume of test solution to be used. Normally, the maximum convenient weight of specimen is about 100 g. Specimens containing welds should be cut so that no more than ½ in. (13 mm) width of base metal is included on either side of the weld. Furthermore, in the case of bar, wire, and tubular products, the proportion of the total area represented by the exposed cross section may influence the results. Cross-sectional areas in these products may be subject to end grain attack in nitric acid. The proportion of end grain in the specimen should therefore be kept low unless such surface is actually to be exposed in service involving nitric acid. When specimens of such products are being tested in research investigations, the ratio of the cross-sectional area exposed to the total area should be kept constant from test to test. For inspection tests, specimens cut from bars, wires, or tubes should be proportioned so that the areas of the exposed cross sections shall not exceed half the total exposed area of the specimen.

18.2 Special heat treatment of specimens prior to testing or the use of specimens which contain a weld may be specified.

18.3 When specimens are cut by shearing, the sheared edges should be refinished by machining or grinding prior to testing.

18.4 All surfaces of the specimen, including edges, should be finished using No. 80 or 120 grit abrasive paper. If dry abrasive paper is used, polish slowly to avoid overheating. Sandblasting should not be used.

18.5 The intent is to test a specimen representing as nearly as possible the surface of the material as it will be used in service. Only such surface machining should be performed as is required to remove foreign material and obtain a standard uniform finish as specified in 18.4. For very heavy sections, specimens should be machined to represent the appropriate surface while maintaining reasonable specimen size for convenience in testing. Ordinarily, removal of more material than necessary will have little influence on the test results. However, in the special case of surface carburization (sometimes encountered, for instance, in tubing or castings when lubricants or binders containing carbonaceous materials are employed), it may be possible by heavy grinding or machining to remove the carburized surface completely. Such treatment of test specimens is not permissible except in tests undertaken to demonstrate such surface effects.

18.6 The specimen should be measured including the inner surfaces of any holes and the total exposed area calculated.

18.7 The specimen should then be degreased and dried using suitable nonchlorinated agents, such as soap and acetone (Note 12), and then weighed to the nearest 0.001 g (see 16.4).

NOTE 12—The cleaning treatment described may be supplemented by immersing the specimen in nitric acid (for example, 20 weight % at 120 to 140°F (49 to 60°C) for 20 min), followed by rinsing, drying, and weighing. In the case of small-diameter tubular specimens which cannot be conveniently resurfaced on the inside, it is desirable to include in the preparation an immersion in boiling nitric acid (65 %) for 2 to 4 h using the same apparatus as for the actual test. The purpose of these treatments is to remove any surface contamination that may not be accomplished by the regular cleaning method and which may increase the apparent weight loss of the specimen during the early part of the test.

18.8 It is common practice to test only one specimen of each material or lot of material, as defined by those using the test for specification purposes. However, the use of at least two specimens for check purposes is recommended.

19. Procedure

19.1 Use a sufficient quantity of the nitric acid test solution to cover the specimens and to provide a volume of at least 125 mL/in.^{2} (20 mL/cm^2) of specimen surface. Normally, a volume of about 600 cm^3 is used.

19.2 The best practice is to use a separate container for each test specimen.

NOTE 13—For routine evaluations, it is acceptable to test as many as three specimens in the same container provided that they all are of the same grade and all show satisfactory resistance to corrosion. If more than one of the specimens tested in the same container fail to pass the test, it is necessary to retest all specimens in separate containers, since excessive corrosion of one specimen may result in accelerated corrosion of the other specimens tested with it. Excessive corrosion may often be detected by changes in the color of the test solution, and it may be appropriate to provide separate containers for such specimens without waiting until the end of the test period. A record should be made showing which specimens were tested together.

NOTE 14—If the multi-sample testing apparatus (see Note 10) is employed, a large number of specimens may be tested in the large container provided.

19.3 After the specimens have been placed in the acid in the container, pass cooling water through the condenser and bring the acid to a boil on the hot plate and then keep boiling throughout the test period (Note 15). After each test period, rinse the specimens with water and treat by scrubbing with rubber or a nylon brush under running water to remove any adhering corrosion products, after which they should be dried and weighed. Drying may be facilitated, if desired, by dipping the specimens in acetone after they are scrubbed.

NOTE 15—Care should be taken to prevent contamination of the testing solution, especially by fluorides, either before or during the test. Experience has shown that the presence of even small amounts of hydrofluoric acid will increase the corrosion rate in the nitric acid test. It is not permissible, for example, to conduct nitric-hydrofluoric acid tests in the same hood with nitric acid tests.

19.4 For most consistent results, the test should consist of five boiling periods of 48 h each (Note 16) with a fresh test solution being used in each period.

NOTE 16—For specification purposes, those experienced in the use of the test may, by mutual agreement, shorten the standard test to three 48-h boiling periods. However, if with this shorter test procedure the rate of attack in the third period should exceed that in either the first or second periods to some previously agreed-upon extent, then the test should be continued for a total of five periods. As an alternative, when the test is being used for inspection prior to approval of steel for shipment, a procedure may be agreed upon by the purchaser and the manufacturer whereby the material will be released for shipment following satisfactory performance in three 48-h boiling periods with final acceptance being dependent upon satisfactory performance in the longer test of five 48-h boiling periods. Also, by mutual agreement, a combination of one 48-h period and two 96-h periods (not necessarily in that order) instead of five 48-h test periods may be acceptable for routine evaluations.

20. Calculation and Report

20.1 *Calculation*—The effect of the acid on the material shall be measured by determining the loss of weight of the specimen after each test period and for the total of the test periods. Such weight-loss determinations should be made with the accuracy prescribed in 16.1.4. The corrosion rates are usually reported as inches per month (Note 17), calculated in the following rate of corrosion equation:

$$\text{Inches per month} = (287 \times W)/(A \times d \times t)$$

where:
t = time of exposure, h,
A = total surface area, cm^2,
W = weight loss, g, and
d = density of the sample, g/cm^3.

NOTE 17—Conversion factors to other commonly used units for corrosion rates are as follows:

inches per month × 12 = inches per year
inches per month × 1000 = mils per month
inches per month × 12 000 = mils per year
inches per month × 8350 × density = milligrams per square decimetre per day
inches per month × 34.8 × density = grams per square metre per hour
1.00 $in.^2$ = 6.45 cm^2

20.2 *Report*—Results should be reported for the individual periods, as well as the average for the three or five test periods.

PRACTICE D—NITRIC-HYDROFLUORIC ACID TEST FOR DETECTING SUSCEPTIBILITY TO INTERGRANULAR ATTACK IN MOLYBDENUM-BEARING AUSTENITIC STAINLESS STEELS (6)

21. Scope

NOTE 18—See Practice A for information on the most appropriate of the several test methods available for the evaluation of specific grades of stainless steel.

21.1 This practice describes the procedure for conducting the 70°C 4-h, 10 % nitric-3 % hydrofluoric acid test as employed to measure the susceptibility of molybdenum-bearing stainless steels to intergranular attack. The presence or absence of intergranular attack in this test is not necessarily a measure of the performance of the material in other corrosive environments. The test does not provide a basis for predicting resistance to forms of corrosion other than intergranular, such as general corrosion, pitting, or stress-corrosion cracking.

21.1.1 The 10 % nitric-3 % hydrofluoric acid test detects susceptibility to intergranular attack associated with the precipitation of chromium carbides in molybdenum-bearing stainless steels (Types 316, 317, 316L, and 317L). It does not detect susceptibility to intergranular attack associated with sigma phase in these same types of stainless steel.

21.2 The 10 % nitric-3 % hydrofluoric acid test may be used to evaluate the heat treatment accorded as-received material (Types 316 and 317 stainless steel). It may also be used to check the effectiveness of reduction in carbon content in preventing susceptibility to rapid intergranular attack (Types 316L and 317L). It may be applied to wrought products (including tubes), castings, and weld metal.

21.3 Specimens of extra low carbon grades (Types 316L and 317L) are tested after sensitizing heat treatments at 1200 to 1250°F (650 to 675°C), which is the range of maximum carbide precipitation. The length of time of heating used for this sensitizing treatment will generally be one or two hours. The most commonly used sensitizing treatment is 1 h at 1250°F.

22. Summary of Practice

22.1 *Types 316 and 317 Stainless Steel*—The material submitted for evaluation is tested in each of two conditions of heat treatment: (*1*) as-received (commercially annealed) and (2) laboratory-annealed for 1 h at 1900 to 2000°F (1040 to 1095°C), and water quenched.

22.2 *Types 316L and 317L Stainless Steel*—The material submitted for evaluation test is tested in each of two conditions of heat treatment: (1) a sensitized specimen, and (2) a baseline specimen, which usually is the as-received (commercially annealed) specimen. If the as-received specimen does not show a step structure in the oxalic acid etch test, a portion of it must be laboratory annealed to give a step structure and that sample used as the baseline specimen (see footnote c of Table 5).

23. Rapid Screening Test

23.1 Before testing in the 10 % nitric-3 % hydrofluoric acid test, specimens of the stainless steels as given in Table 5 may be given a rapid screening test in accordance with procedures given in Practice A, Oxalic Acid Etch Test for Classification of Etch Structures of Austenitic Stainless Steels. Preparation, etching, and the classification of etch structures are described therein. The use of etch-structure evaluations in connection with the 10 % nitric-3% hydrofluoric acid test is specified in Table 5.

23.1.1 Corrosion test specimens having acceptable etch structures in the Oxalic Acid Etch Test will be largely free of intergranular attack in the 10 % nitric-3% hydrofluoric acid test. Such specimens are acceptable without testing in the 10 % nitric-3 % hydrofluoric acid test. All specimens having nonacceptable etch structures must be tested in the 10 % nitric-3 % hydrofluoric acid test.

24. Apparatus

24.1 The apparatus is illustrated in Fig. 9.

24.1.1 *Test Cylinders* (Note 19)—The tests are conducted in cylinders of poly(vinyl chloride) (PVC) as shown in Fig. 10. The test cylinders can be made from 12-in. (305-mm) lengths of 1¼-in. PVC pipe (³⁄₁₆-in. (4.76 mm) wall) by either of the following two techniques: (1) one end of each pipe length is plugged with a disk of ³⁄₁₆-in. PVC sheet and the disk is then heat-welded in place with PVC filler rod, or (2) one end of each pipe length is plugged by solvent-welding a 1¼-in. NPS (Schedule 80) socket-type PVC cap onto the end of the pipe.

24.1.2 *Specimen Holders* (Note 19)—The specimens may be suspended in the test solution by means of either a specimen holder (as described in this paragraph) or a TFE-fluorocarbon string through an appropriate hole drilled in one end of the specimen. The specimen holders may be made in either of two ways: (1) a 1-in. (25-mm) length of TFE-fluorocarbon tubing (1-in. inside diameter) is drilled at one end to accommodate a ³⁄₁₆-in. (4.76-mm) TFE-fluorocarbon rod. The holders are flattened into an elliptical shape which is maintained by inserting the rod through the two holes and upsetting the ends of the rod with a hammer, or (2) a ½-in. NPS (Schedule 80) socket-type PVC cap is machined to reduce the outer diameter of the cap to 1¹⁄₁₆ in. (27.0 mm). Holes ¼ in. (6.35 mm) in diameter are then drilled in the bottom and sides of the machined PVC caps to allow free circulation of the test solution. Smaller holes are drilled at the top of the PVC cap to attach the TFE-fluorocarbon string. A loop of the same string is attached to each specimen holder and used to suspend it in the test cylinder.

NOTE 19—All poly(vinyl chloride) materials should be specified on the order as Schedule 80, rigid unplasticized normal impact PVC.

24.1.3 *Constant-Temperature Bath*—The desired solution test temperature of 70°C is obtained by placing the PVC cylinders within a rack in a constant-temperature water bath. The temperature of the bath is maintained at 72 to 73°C to offset the low thermal conductivity of the poly(vinyl chloride).

24.1.4 *10 % Nitric-3 % Hydrofluoric Acid Test Solution:*

24.1.4.1 **Caution**—The 10 % nitric-3 % hydrofluoric acid solution will cause severe burns if it comes into contact with the skin. Therefore, extreme care should be exercised in handling this solution. Rubber gloves should be worn. Spilled acid should immediately be washed from the skin with an excess of water and emergency first-aid treatment obtained.

24.1.4.2 A 10 % nitric-3 % hydrofluoric acid solution (by weight) is prepared by mixing 111 mL of 65 % nitric acid (sp gr 1.39), 54 mL of 48 % hydrofluoric acid (sp gr 1.16), and 784 mL of distilled water in a polyethylene carboy. Fresh test solution should be made up daily to avoid changes in concentration due to evaporation.

25. Preparation of Test Specimens

25.1 A specimen having a total surface area of 5 to 20 cm² is recommended. Specimens containing welds should be cut so that no more than a ½-in. (13-mm) width of base metal is included on either side of the weld.

25.2 The intent is to test a specimen representing as nearly as possible the surface of the material as it will be used in service. Only such finishing should be performed as is required to remove foreign material and obtain a standard, uniform finish as specified. For very heavy sections, specimens should be machined to represent the appropriate surface while maintaining reasonable specimen size for convenience in testing. Ordinarily, removal of more material than necessary will have little influence on the test results. However, in the special case of surface carburization (sometimes encountered, for instance, in tubing or castings when lubricants or binders containing carbonaceous materials are employed), it may be possible by heavy grinding or machining to remove the carburized layer completely. Such treatment of test specimens is not permissible.

25.3 When specimens are cut by shearing, the sheared edges should be refinished by machining or grinding prior to testing.

25.4 All surfaces of the specimen, including edges, should be finished using No. 80 or 120 grit abrasive paper. If dry abrasive paper is used, polish slowly to avoid overheating. Sand blasting should not be used.

25.5 The specimen should be measured and the total exposed area, including the inner surfaces of any holes, calculated in square centimetres.

25.6 The specimen should then be degreased and dried using suitable nonchlorinated agents, such as soap and acetone, and then weighed to the nearest 0.001 g.

26. Procedure

26.1 *Types of Test Specimens:*

26.1.1 *Types 316 and 317*—Test two specimens, one representing the as-received condition and one representing the laboratory-annealed condition. (The laboratory-annealed specimen must show a step structure in the oxalic acid etch test.)

26.1.2 *Types 316L and 317L*—Test two specimens, one representing the sensitized condition and the other a base-line specimen, which is usually the as-received (commercially annealed) specimen. If the as-received specimen does not show a step structure in the oxalic acid etch test, laboratory anneal a portion of it to give a step structure and use that sample as the base-line specimen (see footnote *c* of Table 5).

26.2 Fill the PVC test cylinders with 200 mL of 10 % nitric-3 % hydrofluoric acid solution and then heat in the constant temperature water bath until the solution temperature, as measured with a thermometer, is 70 ± 0.5°C.

NOTE 20—It is important that the test solution be maintained at 70 ± 0.5°C because small changes in solution temperature produce large changes in the corrosion rate.

26.3 When the solution temperature is 70 ± 0.5°C, lower the specimens into the solution by means of the specimen holders or string. After a 2-h exposure, remove the specimens from the test, wash in distilled water and acetone, dry, and weigh. Then expose the two specimens for an additional 2-h test period in fresh test solutions at 70 ± 0.5°C.

26.4 Test only one specimen in each cylinder. In addition, it is preferable to test simultaneously the two specimens representing each of the two conditions of heat treatment for each material evaluated.

27. Calculation and Report

27.1 *Calculation of Individual Corrosion Rates*—The effect of the acid solution on each of the two specimens of each material shall be measured by determining the loss of weight of the specimen. The average corrosion rate in inches per month (Note 21), based on the 4-h test exposure, is calculated for each specimen as follows:

$$\text{Inches per month} = (287 \times W)/(A \times t \times d)$$

where:
- t = time of exposure, h,
- A = area, cm²,
- W = weight loss, g, and
- d = density, g/cm³, for chromium-nickel-molybdenum steels, $d = 8.0$ g/cm³

27.2 *Calculations of Corrosion-Rate Ratios:*

27.2.1 *Types 316 and 317 Stainless Steel*—The ratio of the corrosion rate for the as-received specimen to the corrosion rate for the laboratory-annealed specimen is determined.

27.2.2 *Types 316L and 317L Stainless*

Steel—The ratio of the corrosion rate for the sensitized specimen to the corrosion rate for the as-received specimen is determined.

27.3 *Significance of Corrosion Rate Ratios*—A value of 1.5 or less for the above ratios indicates that the degree of intergranular attack in the 10 % nitric-3 % hydrofluoric acid test was not significant. A ratio of greater than 1.5 indicates that significant intergranular corrosion has occurred in the 10 % nitric-3 % hydrofluoric acid test.

NOTE 21—Conversion factors to other commonly used units for corrosion rates are as follows:

inches per month × 12 = inches per year
inches per month × 1000 = mils per month
inches per month × 12 000 = mils per year
inches per month × 8350 × density = milligrams per square decimetre per day
inches per month × 34.8 × density = grams per square metre per hour
1.00 in.^{2} = 6.45 cm^2

PRACTICE E—COPPER-COPPER SULFATE-SULFURIC ACID TEST FOR DETECTING SUSCEPTIBILITY TO INTERGRANULAR ATTACK IN AUSTENITIC STAINLESS STEELS (7) (8)

28. Scope

28.1 This practice describes the procedure by which the copper-copper sulfate-sulfuric acid test is conducted to determine the susceptibility of austenitic stainless steels to intergranular attack. The presence or absence of intergranular corrosion in this test is not necessarily a measure of the performance of the material in other corrosive media. The test does not provide a basis for predicting resistance to other forms of corrosion, such as general corrosion, pitting, or stress-corrosion cracking.

28.2 The copper-copper sulfate-sulfuric acid test indicates susceptibility to intergranular attack associated with the precipitation of chromium-rich carbides. It does not detect susceptibility associated with sigma phase. This test may be used to evaluate the heat treatment accorded as-received material. It may also be used to evaluate the effectiveness of stabilizing element additions (Cb, Ti, etc.) and reductions in carbon content to aid in resisting intergranular attack.

28.3 All wrought products and weld material of austenitic stainless steels can be evaluated by this test.

29. Rapid Screening Test

29.1 Before testing in the copper-copper sulfate-sulfuric acid test, specimens of certain grades of stainless steel (see Table 6) may be given a rapid screening test in accordance with the procedures given in Practice A (Sections 2 through 6). Preparation, etching, and the classification of etch structures are described therein. The use of etch-structure evaluations in connection with the copper-copper sulfate-sulfuric acid test is specified in Table 6.

29.1.1 Corrosion test specimens having acceptable etch structures in the oxalic acid etch test will be essentially free of intergranular attack in the copper-copper sulfate-sulfuric acid test. Such specimens are acceptable without testing in the copper-copper sulfate-sulfuric acid test. All specimens having nonacceptable etch structures must be tested in the copper-copper sulfate-sulfuric acid test.

30. Summary of Practice

30.1 A suitable sample of an austenitic stainless steel, embedded in copper shot or grindings, is exposed to boiling acidified copper sulfate solution for 24 h. After exposure in the boiling solution, the specimen is bent. Intergranular cracking or crazing is evidence of susceptibility.

31. Apparatus

31.1 A 1-L glass Erlenmeyer flask with a ground 45/50 glass joint and four-bulb Allihn condenser with 45/50 ground glass joint (as in 9.1.1 and 9.1.2 and Fig. 8) are required. A silicone grease is recommended for the ground glass joint.

31.2 *Specimen Supports*—An open glass cradle capable of supporting the specimens and copper shot or grindings in the flask is recommended.

NOTE 22—It may be necessary to embed large specimens, such as from heavy bar stock, in copper shot on the bottom of the test flask. A copper cradle may also be used.

31.3 *Heat Source*—Any gas or electrically heated hot plate may be utilized for heating the test solution and keeping it boiling throughout the test period.

32. Acidified Copper Sulfate Test Solution

32.1 Dissolve 100 g of copper sulfate

($CuSO_4 \cdot 5H_2O$) in 700 mL of distilled water, add 100 mL of sulfuric acid (H_2SO_4, cp, sp gr 1.84), and dilute to 1000 mL with distilled water.

NOTE 23—The solution will contain approximately 6 weight % of anhydrous $CuSO_4$ and 16 weight % of H_2SO_4.

33. Copper Addition

33.1 Electrolytic grade copper shot or grindings may be used. Shot is preferred for its ease of handling before and after the test.

33.2 A sufficient quantity of copper shot or grindings is to be used to cover all surfaces of the specimen whether it is in a vented glass cradle or embedded in a layer of copper shot on the bottom of the test flask.

33.3 The amount of copper used, assuming an excess of metallic copper is present, is not critical. The effective galvanic coupling between copper and the test specimen may have importance **(9)**.

33.4 The copper shot or grindings may be reused if they are cleaned in warm tap water after each test.

34. Specimen Preparation

34.1 The size of the sample submitted for test and the area from which it is to be taken (end or middle of coil, midway surface and center, etc.) is generally specified in the agreement between the purchaser and the seller. The testing apparatus dictates the final size and shape of the test specimen. The specimen configuration should permit easy entrance and removal through the neck of the test container.

34.1.1 Table 7 may be used as a guide to determine acceptable specimen sizes. There may be restrictions placed on specimen size by the testing apparatus.

34.1.2 Specimens obtained by shearing should have the sheared edges machined or ground off prior to testing. Care should be taken when grinding to avoid overheating or "burning." A "squared" edge is desirable.

34.2 Any scale on the specimens should be removed mechanically unless a particular surface finish is to be evaluated. Chemical removal of scale is permissible when this is the case. Mechanical removal of scale should be accomplished with 120-grit iron-free aluminum oxide abrasive.

34.2.1 Each specimen should be degreased using a cleaning solvent such as acetone, alcohol, ether, or a vapor degreaser prior to being tested.

34.3 All austenitic material in the "as-received" (mill-annealed) condition should be capable of meeting this test.

34.3.1 Specimens of extra low carbon and stabilized grades are tested after sensitizing heat treatments at 1200 to 1250°F (650 to 675°C), which is the range of maximum carbide precipitation. The most commonly used sensitizing treatment is 1 h at 1250°F. Care should be taken to avoid carburizing or nitriding the specimens. The heat treating is best carried out in air or neutral salt.

NOTE 24—The sensitizing treatment (1250°F) is performed to check the effectiveness of stabilized and 0.03 % maximum carbon materials in resisting carbide precipitation, hence, intergranular attack.

35. Test Conditions

35.1 The volume of acidified copper sulfate test solution used should be sufficient to completely immerse the specimens and provide a minimum of 50 mL/in.^{2} (8 mL/cm^2) of specimen surface area.

35.1.1 As many as three specimens can be tested in the same container. It is ideal to have all the specimens in one flask to be of the same grade, but it is not absolutely necessary. The solution volume-to-sample area ratio is to be maintained.

35.1.2 The test specimen(s) should be immersed in ambient test solution which is then brought to a boil and maintained boiling throughout the test period. Begin timing the test period when the solution reaches the boiling point.

NOTE 25—Measures should be taken to minimize bumping of the solution when glass cradles are used to support specimens. A small amount of copper shot (8 to 10 pieces) on the bottom of the flask will conveniently serve this purpose.

35.1.3 The time of the test shall be a minimum of 24 h unless a longer time is agreed upon between the purchaser and the producer. If not 24 h, the test time shall be specified on the test report. Fresh test solution would not be needed if the test were to run 48 or even 72 h. (If any adherent copper remains on the specimen, it may be removed by a brief immersion in concentrated nitric acid at room temperature.)

36. Bend Test

36.1 The test specimen shall be bent through 180° and over a diameter equal to the thickness of the specimen being bent (see Fig. 11). In no case shall the specimen be bent over a smaller radius or through a greater angle than that specified in the product specification. In cases of material having low ductility, such as severely cold worked material, a 180° bend may prove impractical. Determine the maximum angle of bend without causing cracks in such material by bending an untested specimen of the same configuration as the specimen to be tested.

36.1.1 Duplicate specimens shall be obtained from sheet material so that both sides of the rolled samples may be bent through a 180° bend. This will assure detection of intergranular attack resulting from carburization of one surface of sheet material during the final stages of rolling.

NOTE 26—Identify the duplicate specimen in such a manner as to ensure both surfaces of the sheet material being tested are subjected to the tension side of the 180° bends.

36.1.2 Samples machined from round sections or cast material shall have the curved or original surface on the outside of the bend.

36.1.3 The specimens are generally bent by holding in a vise and starting the bend with a hammer. It is generally completed by bringing the two ends together in the vise. Heavy specimens may require bending in a fixture of suitable design. An air or hydraulic press may also be used for bending the specimens.

36.1.4 Tubular products should be flattened in accordance with the flattening test, prescribed in ASTM Methods and Definitions A 370, for Mechanical Testing of Steel Products.[4]

36.1.5 When agreed upon between the purchaser and the producer, the following shall apply to austenitic stainless steel plates 0.1875 in. (4.76 mm) and thicker:

36.1.5.1 Samples shall be prepared according to Table 7.

36.1.5.2 The radius of bend shall be two times the sample thickness, and the bend axis shall be perpendicular to the direction of rolling.

36.1.5.3 Welds on material 0.1875 in. and thicker shall have the above bend radius, and the weld-base metal interface shall be located approximately in the centerline of the bend.

36.1.5.4 Face, root, or side bend tests may be performed, and the type of bend test shall be agreed upon between the purchaser and the producer. The bend radius shall not be less than that required for mechanical testing in the appropriate material specification (for base metal) or in ASME Code Section IX (for welds).

37. Evaluation

37.1 The bent specimen shall be examined under low (5 to 20×) magnification (see Fig. 12). The appearance of fissures or cracks indicates the presence of intergranular attack (see Fig. 13).

37.1.1 When an evaluation is questionable (see Fig. 14), presence or absence of intergranular attack shall be determined by metallographic examination of a longitudinal section of the specimen at a magnification of 100 to 250×.

NOTE 27—Cracking that originates at the edge of the specimen should be disregarded. The appearance of deformation lines, wrinkles, or "orange peel" on the surface, without accompanying cracks or fissures, should be disregarded also.

NOTE 28—Cracks suspected as arising through poor ductility may be investigated by bending a similar specimen which was not exposed to the boiling test solution. A visual comparison between these specimens should assist in interpretation.

[4] *Annual Book of ASTM Standards,* Vols 01.01, 01.02, 01.03, and 01.05.

REFERENCES

(1) For original descriptions of the use of etch structure classifications, see Streicher, M. A., "Screening Stainless Steels from the 240-h Nitric Acid Test by Electrolytic Etching in Oxalic Acid," *ASTM Bulletin,* No. 188, February 1953, p. 35; also "Results of Cooperative Testing Program for the Evaluation of the Oxalic Acid Etch Test," *ASTM Bulletin,* No. 195, January 1954, p. 63.

(2) For original description of ferric sulfate-sulfuric acid test, see Streicher, M. A., "Intergranular Corrosion Resistance of Austenitic Stainless Steels: A Ferric Sulfate-Sulfuric Acid Test," *ASTM Bulletin,* No. 229, April 1958, (*STP 95*), pp. 77–86.

(3) For original descriptions of the boiling nitric acid test, see Huey, W. R., "Corrosion Test for Research and Inspection of Alloys," *Transac-*

tions, *Am. Soc. Steel Treating,* Vol 18, 1930, p. 1126; also, "Report of Subcommittee IV on Methods of Corrosion Testing," *Proceedings,* Am. Soc. Testing Mats., Vol 33, Part I, 1933, p. 187.

(4) For details, see DeLong, W. B., "Testing Multiple Specimens of Stainless Steels in a Modified Boiling Nitric Acid Test Apparatus," *Symposium on Evaluation Tests for Stainless Steels, ASTM STP 93,* Am. Soc. Testing Mats., 1950, p. 211.

(5) See *Industrial and Engineering Chemistry,* Vol 17, 1925, p. 756; also, "A.C.S. Analytical Reagents. Specifications Recommended by Committee on Analytical Reagents," Am. Chem. Soc., March 1941.

(6) For original description of 10 % nitric-3 % hydrofluoric acid test, see Warren, D., "Nitric-Hydrofluoric Acid Evaluation Test for Type 316L Stainless Steel," *ASTM Bulletin,* No. 230, May 1958. pp. 45–56.

(7) The use of copper to accelerate the intergranular corrosion of sensitized austenitic stainless steels in copper sulfate-sulfuric acid was first described by H. J. Rocha in the discussion of a paper by Brauns, E., and Pier, G., *Stahl und Eisen,* Vol 75, 1955, p. 579.

(8) For original evaluation of the copper-copper sulfate-sulfuric acid test, see Scharfstein, L. R., and Eisenbrown, C. M., "An Evaluation of Accelerated Strauss Testing," *ASTM STP* 369, 1963, pp. 235–239.

(9) Subtle effects due to variations in copper surface areas, galvanic contact, condenser design, etc., are described by Herbsleb, G., and Schwenk, W., "Untersuchungen zur Einstellung des Redoxpotentials der Strausschen Lösung mit Zusatz von Mettalischem Kupfer," *Corrosion Science,* Vol 7, 1967, pp. 501–511.

(10) Brown, M. H., "Behavior of Austenitic Stainless Steels in Evaluation Tests for the Detection of Susceptibility to Intergranular Corrosion," *Corrosion,* Vol 30, January 1974, pp. 1–12.

TABLE 1 Application of Evaluation Tests for Detecting Susceptibility to Intergranular Attack in Austenitic Stainless Steels

NOTE 1—For each corrosion test, the types of susceptibility to intergranular attack detected are given along with the grades of stainless steels in which they may be found. These lists may contain grades of steels in addition to those given in the rectangles. In such cases, the acid corrosion test is applicable, but not the oxalic acid etch test.

NOTE 2—The oxalic acid etch test may be applied to the grades of stainless steels listed in the rectangles when used in connection with the test indicated by the arrow.

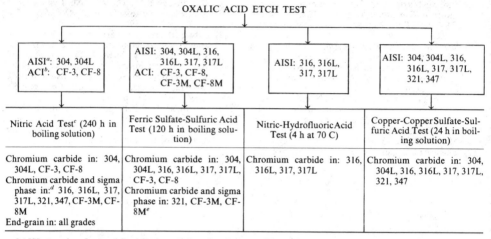

OXALIC ACID ETCH TEST

AISI[a]: 304, 304L ACI[b]: CF-3, CF-8	AISI: 304, 304L, 316, 316L, 317, 317L ACI: CF-3, CF-8, CF-3M, CF-8M	AISI: 316, 316L, 317, 317L	AISI: 304, 304L, 316, 316L, 317, 317L, 321, 347
Nitric Acid Test[c] (240 h in boiling solution)	Ferric Sulfate-Sulfuric Acid Test (120 h in boiling solution)	Nitric-Hydrofluoric Acid Test (4 h at 70 C)	Copper-Copper Sulfate-Sulfuric Acid Test (24 h in boiling solution)
Chromium carbide in: 304, 304L, CF-3, CF-8 Chromium carbide and sigma phase in:[d] 316, 316L, 317, 317L, 321, 347, CF-3M, CF-8M End-grain in: all grades	Chromium carbide in: 304, 304L, 316, 316L, 317, 317L, CF-3, CF-8 Chromium carbide and sigma phase in: 321, CF-3M, CF-8M[e]	Chromium carbide in: 316, 316L, 317, 317L	Chromium carbide in: 304, 304L, 316, 316L, 317, 317L, 321, 347

[a] AISI: American Iron and Steel Institute designations for austenitic stainless steels.
[b] ACI: Alloy Casting Institute designations.
[c] The nitric acid test may be also applied to AISI 309, 310, 348, and AISI 410, 430, 446, and ACI CN-7M.
[d] Must be tested in nitric acid test when destined for service in nitric acid.
[e] To date, no data have been published on the effect of sigma phase on corrosion of AISI 347 in this test.

TABLE 2 Applicability of Etch Test

	AISI Grade No.	ACI Grade No.
Practice B—Ferric Sulfate-Sulfuric Acid Test	304, 304L, 316, 316L, 317, 317L	CF-3, CF-8, CF-3M, CF-8M
Practice C—Nitric Acid Test	304, 304L	CF-8, CF-3
Practice D—Nitric Hydrofluoric Acid Test	316, 316L, 317, 317L	...
Practice E—Copper-Copper Sulfate-Sulfuric Acid Test	304, 304L, 316, 316L, 317, 317L, 321, 347	...

TABLE 3 Use of Etch Structure Classifications from the Oxalic Acid Etch Test with Ferric Sulfate-Sulfuric Acid Test

NOTE—Grades AISI 321 and 347 cannot be screened because these grades may contain a type of sigma phase which is not visible in the etch structure but which may cause rapid corrosion in the ferric sulfate-sulfuric acid test.

Grade	Acceptable Etch Structures	Nonacceptable Etch Structures[a]
AISI 304	Step, Dual, End Grain, I & II	Ditch
AISI 304L	Step, Dual, End Grain, I & II	Ditch
AISI 316	Step, Dual, End Grain, I & II	Ditch
AISI 316L	Step, Dual, End Grain, I & II	Ditch
AISI 317	Step, Dual, End Grain, I & II	Ditch
AISI 317L	Step, Dual, End Grain, I & II	Ditch
AISI 321	None	...
ACI CF-3	Step, Dual, Isolated Ferrite Pools	Ditch, Interdendritic Ditches
ACI CF-8	Step, Dual, Isolated Ferrite Pools	Ditch, Interdendritic Ditches
ACI CF-3M	Step, Dual, Isolated Ferrite Pools	Ditch, Interdendritic Ditches
ACI CF-8M	Step, Dual, Isolated Ferrite Pools	Ditch, Interdendritic Ditches

[a] Specimens having these structures must be tested in the ferric sulfate-sulfuric acid test.

A 262

TABLE 4 Use of Etch Structure Classification from Oxalic Acid Etch Test with Nitric Acid Test

NOTE—AISI 316, 316L, 317, 317L, 347, and 321 cannot be screened because these steels may contain sigma phase not visible in the etch structure. This may cause rapid intergranular attack in the nitric acid test.

Grade	Acceptable Etch Structures	Nonacceptable Etch Structures[a]
AISI 304	Step, Dual, End Grain I	Ditch, End Grain II
AISI 304L	Step, Dual, End Grain I	Ditch, End Grain II
ACI CF-8	Step, Dual, Isolated Ferrite Pools	Ditch, Interdendritic Ditches
ACI CF-3	Step, Dual, Isolated Ferrite Pools	Ditch, Interdendritic Ditches

[a] Specimens having these structures must be tested in the nitric acid test.

TABLE 5 Use of Etch Structure Classifications from the Oxalic Acid Etch Test with 10 % Nitric-3 % Hydrofluoric Acid Test

Grade	Condition of Heat Treatment	Acceptable Etch Structures	Nonacceptable Etch Structures[a]
AISI 316	As-received[b]	Step, Dual, End Grain I and II	Ditch
AISI 317	As-received[b]	Step, Dual, End Grain I and II	Ditch
AISI 316L	As-received[c]	Step, End Grain I and II	Dual, Ditch
	Sensitized	Step, Dual, End Grain I and II	Ditch
AISI 317L	As-received[c]	Step, End Grain I and II	Dual, Ditch
	Sensitized	Step, Dual End Grain I and II	Ditch

[a] Specimens having these structures must be tested in 10 % nitric-3 % hydrofluoric acid test.

[b] If the as-received specimen shows a ditch structure and the HNO_3-HF test is to be applied, then a duplicate specimen must be laboratory-annealed in order to produce a base-line specimen having a step structure. The laboratory-annealed specimen (step structure) along with the original as-received specimen (ditch structure) is subjected to the HNO_3-HF test. The final criterion is then the ratio of the corrosion rate for the as-received specimen to the corrosion rate for the laboratory-annealed specimen.

[c] When the oxalic acid etch test is used with the HNO_3-HF test for Types 316L and 317L, the as-received specimens should show a step structure (free from precipitated carbides). If it does not, a portion of the as-received specimen must be laboratory annealed to produce a base-line specimen having a step structure and that sample used for the HNO_3-HF test. This requirement is necessary because the base-line specimen must be completely free of intergranular attack in the HNO_3-HF test in order to obtain an indicative ratio of the corrosion rate of the sensitized specimen to the corrosion rate of the base-line specimen.

TABLE 6 Use of Etch Structure Classifications from the Oxalic Acid Etch Test with the Copper-Copper Sulfate-Sulfuric Acid Test

Grade	Acceptable Etch Structures	Nonacceptable Etch Structures[a]
AISI 304	Step, Dual, End Grain I and II	Ditch
AISI 304L	Step, Dual, End Grain I and II	Ditch
AISI 316	Step, Dual, End Grain I and II	Ditch
AISI 316L	Step, Dual, End Grain I and II	Ditch
AISI 317	Step, Dual, End Grain I and II	Ditch
AISI 317L	Step, Dual, End Grain I and II	Ditch
AISI 321	Step, Dual, End Grain I and II	Ditch
AISI 347	Step, Dual, End Grain I and II	Ditch

[a] Specimens having these structures must be tested in the copper-copper sulfate-sulfuric acid test.

TABLE 7 Sizes of Test Specimens

Type of Material	Size of Test Specimen
Wrought wire or rod:	
Up to ¼ in. in diameter, incl	Full diameter by 3 in. (min) long
Over ¼ in. in diameter	Cylindrical segment ¼ in. thick by 1 in. (max) wide by 3 to 5 in. long[a]
Wrought sheet, strip, plates, or flat rolled products:	
Up to 3/16 in. thick, incl	Full thickness by ⅜ to 1 in. wide by 3 in. (min) long
Over 3/16 in. thick	3/16 to ½ in. thick by ⅜ to 1 in. wide by 3 in. (min) long[b]
Tubing:	
Up to 1½ in. in diameter, incl	Full ring, 1 in. wide[c]
Over 1½ in. in diameter	A circumferential segment 3 in. (min) long cut from a 1-in. wide ring[d]

[a] When bending such specimens, the curved surface shall be on the outside of the bend.

[b] One surface shall be an original surface of the material under test and it shall be on the outside of the bend. Cold-rolled strip or sheets may be tested in the thickness supplied.

[c] Ring sections are not flattened or subjected to any mechanical work before they are subjected to the test solution.

[d] Specimens from welded tubes over 1½ in. in diameter shall be taken with the weld on the axis of the bend.

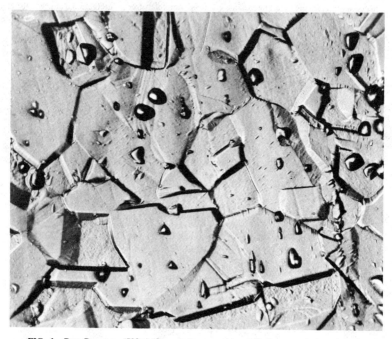

FIG. 1 Step Structure (500×) (Steps between grains, no ditches at grain boundaries)

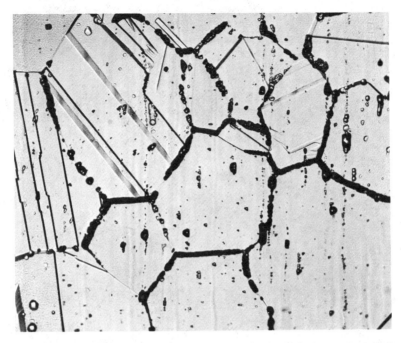

FIG. 2 Dual Structure (250×) (Some ditches at grain boundaries in addition to steps, but no one grain completely surrounded)

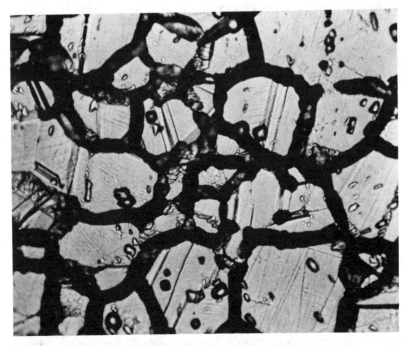

FIG. 3 Ditch Structure (500×) (One or more grains completely surrounded by ditches)

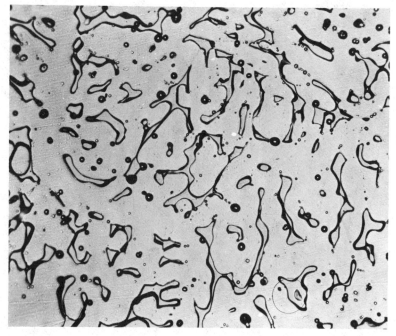

FIG. 4 Isolated Ferrite Pools (250×) (Observed in castings and welds. Steps between austenite matrix and ferrite pools)

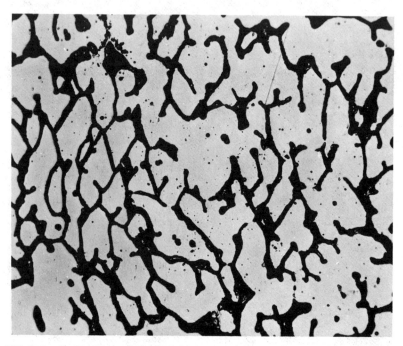

FIG. 5 Interdendritic Ditches (250×) (Observed in castings and welds. Deep interconnected ditches)

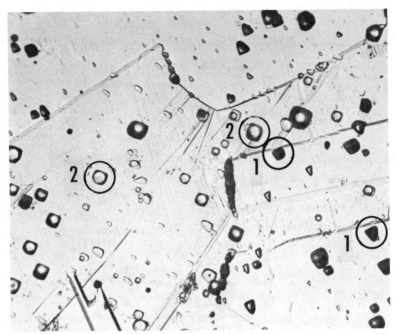

To differentiate between the types of pits, use a magnification of 500× and focus in the plane of etched surface. The pits which now appear completely black are end grain pits.

FIG. 6 End Grain Pitting I (500×) (A few deep end grain pits (see 1 in figure) and shallow etch pits (2))

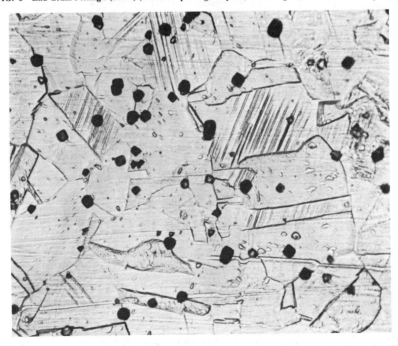

This or a greater concentration of end grain pits at 500× (using standard etching conditions) indicates that the specimen must be tested when screening is for nitric acid test.

FIG. 7 End Grain Pitting II (500×)

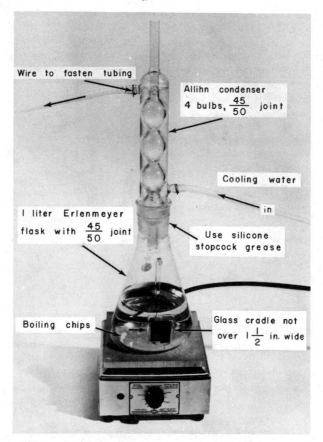

FIG. 8 Apparatus for Ferric Sulfate-Sulfuric Acid Test

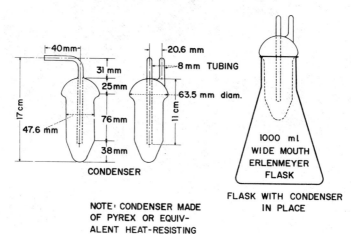

FIG. 9 Flask and Condenser for Nitric Acid Test

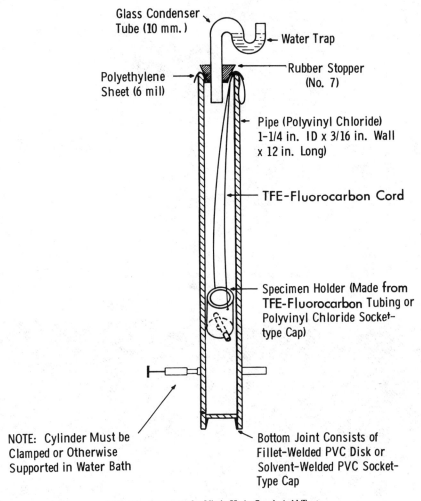

FIG. 10 Apparatus for Nitric-Hydrofluoric Acid Test

FIG. 11 A Bent Copper-Copper Sulfate-Sulfuric Acid Test Specimen

FIG. 12 Passing Test Specimen—View of the Bent Area. (20× magnification before reproduction)

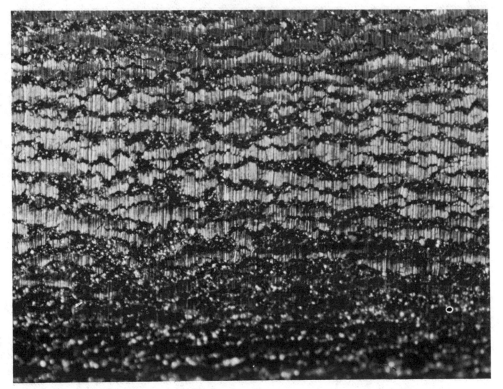

FIG. 13 Failing Test Specimen. (Note the many intergranular fissures. Bent area at 20× magnification before reproduction.)

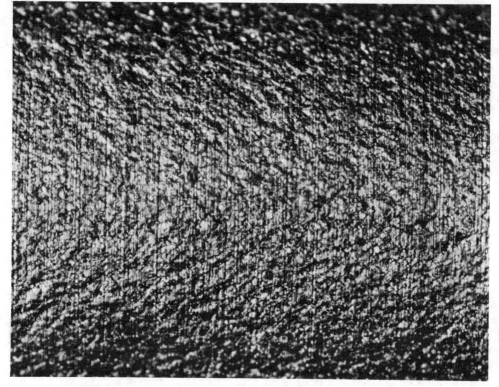

FIG. 14 Doubtful Test Result. (Note the traces of intergranular fissures and "orange-peel" surface. Bent area at 20× magnification before reproduction.)

The American Society for Testing and Materials takes no position respecting the validity of any patent rights asserted in connection with any item mentioned in this standard. Users of this standard are expressly advised that determination of the validity of any such patent rights, and the risk of infringement of such rights, are entirely their own responsibility.

This standard is subject to revision at any time by the responsible technical committee and must be reviewed every five years and if not revised, either reapproved or withdrawn. Your comments are invited either for revision of this standard or for additional standards and should be addressed to ASTM Headquarters. Your comments will receive careful consideration at a meeting of the responsible technical committee, which you may attend. If you feel that your comments have not received a fair hearing you should make your views known to the ASTM Committee on Standards, 1916 Race St., Philadelphia, Pa. 19103.

Designation: A 263 – 82

Standard Specification for
CORROSION-RESISTING CHROMIUM STEEL-CLAD PLATE, SHEET, AND STRIP[1]

This standard is issued under the fixed designation A 263; the number immediately following the designation indicates the year of original adoption or, in the case of revision, the year of last revision. A number in parentheses indicates the year of last reapproval. A superscript epsilon (ϵ) indicates an editorial change since the last revision or reapproval.

1. Scope

1.1 This specification[2] covers plate, sheet, and strip of a carbon steel or low-alloy steel base to which is integrally and continuously bonded on one or both sides a layer of corrosion-resisting chromium steel. The material is generally intended for pressure vessel use.

1.2 The values stated in inch-pound units are to be regarded as the standard.

2. Applicable Documents

2.1 *ASTM Standards*:
A 20/A 20M Specification for General Requirements for Steel Plates for Pressure Vessels[3]
A 240 Specification for Heat-Resisting Chromium and Chromium-Nickel Stainless Steel Plate, Sheet, and Strip for Pressure Vessels[3,4]

3. Description of Terms

3.1 This material is considered as single-clad or double-clad corrosion-resisting chromium-steel plate, sheet, or strip depending on whether one or both sides are covered.

3.2 The terms plate, sheet, and strip as used in this specification apply to the following:

3.2.1 *Plate*—Material ³⁄₁₆ in. (2.73 mm) and over in thickness and over 10 in. (254 mm) in width.

3.2.2 *Sheet*—Material under ³⁄₁₆ in. in thickness and 24 in. (609.6 mm) and over in width, material under ³⁄₁₆ in. in thickness and all widths and finishes of Nos. 3 to 8 inclusive, and

3.2.3 *Strip*—Cold-rolled material under 24 in. in width and ³⁄₁₆ in. and under in thickness.

4. General Requirements for Delivery

4.1 Material furnished under this specification shall conform to the applicable requirements of the current edition of Specification A 20.

5. Materials and Manufacture

5.1 *Process:*

5.1.1 The steel shall be made by the open-hearth, electric-furnace (with separate degassing and refining optional), or basic-oxygen processes, or by secondary processes whereby steel made from these primary processes is remelted using electroslag remelting or vacuum arc remelting processes.

5.1.2 The alloy cladding metal may be metallurgically bonded to the base metal by any method that will produce a clad steel that will conform to the requirements of this specification.

5.2 *Heat Treatment*—Unless otherwise specified or agreed upon between the purchaser and the manufacturer, all plates shall be furnished in the normalized or tempered condition, or both.

6. Thickness of Cladding Metal

6.1 The minimum thickness and tolerances on the thickness of the alloy cladding metal shall be agreed upon between the purchaser and the manufacturer.

---

[1] This specification is under the jurisdiction of ASTM Committee A-1 on Steel, Stainless Steel and Related Alloys, and is the direct responsibility of Subcommittee AO1.17 for Flat Stainless Steel Products.
Current edition approved July 30, 1982. Published September 1982. Originally published as A 263 – 43 T. Last previous edition A 263 – 79.
[2] For ASME Boiler and Pressure Vessel Code applications see related Specification SA-263 in Section II of that Code.
[3] *Annual Book of ASTM Standards*, Vol 01.04.
[4] *Annual Book of ASTM Standards*, Vols 01.03 and 01.04.

7. Chemical Requirements

7.1 The composite plate, sheet, or strip shall conform to any desired combination of alloy cladding metal and base metal as described in 7.2 and 7.3 and as agreed upon between the purchaser and the manufacturer.

7.2 *Alloy Cladding Metal*—The alloy cladding metal specified shall conform to the requirements as to chemical composition prescribed for the respective chromium steel in Specification A 240.

7.3 *Base Metal*—The base metal may be carbon steel or low-alloy steel conforming to the ASTM specifications for steels for pressure vessels. The base metal shall conform to the chemical requirements of the specification to which it is ordered.

8. Product Analysis

8.1 Product analyses may be required on the finished product only when the composite plate thickness is sufficient to permit obtaining drillings or millings without danger of contamination from the adjacent layer.

8.2 If product analysis is specified by the purchaser, it shall be made from drillings or millings taken from a broken test specimen. The chemical analysis thus determined shall conform to the requirements prescribed in Section 7.

9. Mechanical Requirements

9.1 *Plate:*

9.1.1 *Tensile Properties:*

9.1.1.1 For clad plates up to 1½ in. (38.1 mm), inclusive, maximum gage with base steel of 70 000 psi (485 MPa) tensile strength or less, the tensile properties as determined by a tension test on the composite plate shall be equal to the minimum and not more than 5000 psi (35 MPa) over the maximum prescribed in the specification for the base steel used.

9.1.1.2 For clad plates over 1½ in. gage or for clad plate with base steel of more than 70 000 psi minimum tensile strength or 40 000 psi (275 MPa) minimum yield strength, the tensile properties shall be determined by pulling a test of the base steel only, in which case the properties shall meet the requirements of the specifications for the base steel.

9.1.2 *Ductility*—Two bend tests of the composite plate shall be made, one with the alloy cladding in tension and the other with the alloy cladding in compression, to determine the ductility of the materials. On double-clad plates, the bend tests shall be made so that one specimen represents the alloy cladding in tension on one side while the other specimen represents the alloy cladding in tension on the opposite side. The bend test having the alloy cladding in tension shall be made according to and shall conform to the requirements prescribed in the specifications for the cladding metal. The bend test having the base metal in tension shall be made according to and shall conform to the requirements prescribed in the specifications for the base metal.

9.1.3 *Shear Strength*—When required by the purchaser, the minimum shear strength of the alloy cladding and base metals shall be 20 000 psi (140 MPa). The shear test, when specified, shall be made in the manner indicated in Fig. 1.

9.1.4 *Bond Strength*—As an alternative to the shear strength test provided in 9.1.2 and when required by the purchaser, three bend tests shall be made with the alloy cladding in compression to determine the quality of the bond. These bend tests shall be made in accordance with the specifications for the cladding metal. At least two of the three tests shall show not more than 50% separation on both edges of the bent portion. Greater separation shall be cause for rejection.

9.2 *Sheet and Strip:*

9.2.2 The bend test specimens of sheet and strip shall stand being bent cold, without cracking on the outside of the bent portion, through an angle of 180°.

9.2.3 The bend test specimens shall be bent over a single piece of flat stock of the same thickness as the material tested, allowing the test material to form its natural curvature. The axis of the bend shall be transverse to the direction of rolling.

NOTE 1—The bend may be made over a diameter equal to the thickness of the test material.

9.2.4 The bond between the alloy cladding and the base metal shall be ascertained by observation of the behavior of the composite sheet or strip when sheared with the alloy side down.

10. Test Specimens

10.1 *Plate:*

10.1.1 The tension test specimens from plate

shall conform to the requirements prescribed in the specifications for the base metal.

10.1.2 Bend test specimens shall be taken from the middle of the top of the plate as rolled, at right angles to its longitudinal axis.

10.1.3 When required by the purchaser, the shear test specimen shall be taken from a top or bottom corner of the plate as rolled, parallel to its longitudinal axis.

10.1.4 All tests shall be made on specimens in the same condition of heat treatment to which the composite plate is furnished.

10.1.5 For plates 1½ in. (38.1 mm) and under in thickness tension test specimens shall be full thickness of the material except as specified in 9.1.1.1 and 9.1.1.2; the bend test specimen shall be the full thickness of the material and shall be machined to the form and dimensions shown in Fig. 2, or may be machined with both edges parallel.

10.1.6 For plates over 1½ in. in thickness, tension tests shall be of the form shown in Fig. 3 and shall be of all base steel conforming to the requirements of the specification for the base steel.

10.1.7 For plates over 1½ in. in thickness the bend tests specimens need not be greater in thickness than 1½ in. but shall not be less than 1½ in. Specimens shall be of the form and dimensions shown in Fig. 2. In reducing the thickness of the specimen, both the alloy cladding and the base steel shall be machined so as to maintain the same ratio of clad metal to base steel as is maintained in the plate, except that the thickness of the clad metal need not be reduced below ⅛ in. (3.17 mm).

10.1.8 The sides of the bend test specimen may have the corners rounded to a radius not over ¹⁄₁₆ in. (1.58 mm) for plates, 2 in. (50.8 mm) and under in thickness, and not over ⅛ in. for plates over 2 in. in thickness.

10.2 *Sheet and Strip:*

10.2.1 Bend test specimens from sheet and strip shall be of the full thickness of the material and shall have a width equal to six times the thickness but not less than 1 in. (25.4 mm). The length shall be equal to the width of the sheet or strip but in no case longer than 12 in. (304.8 mm). The edges of the test specimen may be rounded to a radius equal to one half the thickness.

10.2.2 Any test specimen used for the determination of the minimum thickness of alloy cladding metal shall have distorted metal, due to burning or shearing, completely removed.

11. Number of Tests and Retests

11.1 *Plate:*

11.1.1 One or more tension tests, as required by the specifications for the base metal, one face bend test (alloy cladding in tension), one reverse bend test (alloy cladding in compression), and, when specified, one shear test or three bond bend tests shall be made representing each plate as rolled. Each specimen shall be in the final condition of heat treatment required for the plate.

11.1.2 If any test specimen shows defective machining or develops flaws, it may be discarded and another specimen substituted.

11.1.3 If the percentage of elongation of any tension test specimen is less than that specified in 9.1.1 and any part of the fracture is more than ¾ in. (19.1 mm) from the center of the gage length of a 2-in. (50.8-mm) specimen or is outside the middle third of the gage length of an 8-in. (203.2-mm) specimen, as indicated by scribe scratches marked on the specimen before testing, a retest shall be allowed.

11.2 *Sheet and Strip:*

11.2.1 On single-clad materials four bend tests shall be made of each lot of material. A lot shall consist of no more than the number of sheets or strips required to make 1000 lb (454 kg), but in no case more than 100 sheets or strips from the same melt of steel and specified thickness. On double-clad materials two bend tests shall be made on each lot of material.

11.2.2 On single-clad materials two of the bend tests shall be made with the alloy cladding metal in tension and two with the alloy cladding metal in compression.

11.2.3 On double-clad materials, the bend tests shall be made so that one specimen will represent the alloy cladding metal in tension on one side while the other specimen will represent the alloy cladding metal in tension on the opposite side.

12. Permissible Variations in Thickness and Mass

12.1 Composite plates, sheets, and strips shall conform to the dimensional and mass requirements prescribed in Tables 1, 2, 3, 4, and 5.

13. Workmanship and Finish

13.1 The material shall be free of injurious defects, shall have a workmanlike appearance, and shall conform to the designated finish.

13.2 Plate alloy surfaces shall be sandblasted, pickled, or, blast-cleaned and pickled.

13.3 The finish for the alloy surfaces of sheets and strips shall be as specified in the applicable sections of Specification A 240.

14. Repair of Cladding by Welding

14.1 Unless otherwise specified, the material manufacturer may repair defects in cladding by welding provided the following requirements are met:

14.1.1 Prior approval is obtained from the purchaser if the repaired area exceeds 3 % of the cladding surface.

14.1.2 The welding procedure and the welders or welding operators are qualified in accordance with Section IX of the ASME Code.

14.1.3 The defective area is removed and the area prepared for repair is examined by a magnetic particle method or a liquid penetrant method to ensure all defective area has been removed. Method of test and acceptance standard shall be as agreed upon between the purchaser and the manufacturer.

14.1.4 The weld will be deposited in a suitable manner so as to leave its surface condition equivalent in corrosion resistance to the alloy cladding.

14.1.5 The repaired area is examined by a liquid penetrant method in accordance with 14.1.3.

14.1.6 The location and extent of the weld repairs together with the repair procedure and examination results are transmitted as a part of the certification.

14.2 All repairs in Alloy Type 410 or repairs penetrating into the base steel shall be stress relieved to eliminate residual stresses.

15. Inspection

15.1 The manufacturer shall afford the inspector all reasonable facilities, without charge, to satisfy him that the material is being furnished in accordance with this specification. All tests (except product analysis) and inspection shall be made at the place of manufacture prior to shipment, unless otherwise specified, and shall be so conducted as not to interfere unnecessarily with the operation of the works.

16. Rejection and Rehearing

16.1 Unless otherwise specified, any rejection based on tests made in accordance with Section 8 shall be reported to the manufacturer within 5 working days from the receipt of samples by the purchaser.

16.2 Material which shows injurious defects subsequent to its acceptance at the manufacturer's work will be rejected, and the manufacturer shall be notified.

16.3 Samples tested in accordance with Section 8 that represent rejected material shall be preserved for three weeks from the date of the test report. In case of dissatisfaction with the results of the tests, the manufacturer may make claim for a rehearing within that time.

17. Report

17.1 The chemical analysis of the base metal shall be certified to the purchaser by the manufacturer.

17.2 The results of the tests in Section 9 shall be reported to the purchaser or his representative.

18. Marking

18.1 Except as specified in 18.2, the name or brand of the manufacturer, the manufacturer's test identification number, the class of the base steel, the type number of the alloy cladding metal, and the specified minimum tensile strength shall be legibly stamped on each finished single-clad plate in two places on the base steel side not less than 12 in. (304.8 mm) from the edges. The manufacturer's identification number shall be legibly stamped on each test specimen.

18.2 For double-clad material or for material under ¼ in. (6.35 mm) in thickness, the marking specified in 18.1 shall be legibly stenciled instead of stamped.

TABLE 1 Permissible Variations in Dimensions and Weight of Standard Sheet (No. 1 and No. 2 Finishes)

Specified Thickness, in. (mm)	Permissible Variations in Thickness, plus or minus, in. (mm)	
0.005 (0.127)	0.001 (0.025)	
0.006 to 0.007 (0.152 to 0.178)	0.0015 (0.038)	
0.008 to 0.016 (0.203 to 0.406)	0.002 (0.051)	
0.0161 to 0.026 (0.409 to 0.660)	0.003 (0.076)	
0.027 to 0.040 (0.686 to 1.016)	0.004 (0.102)	
0.041 to 0.058 (1.041 to 1.473)	0.005 (0.127)	
0.059 to 0.072 (1.499 to 1.829)	0.006 (0.152)	
0.073 to 0.083 (1.854 to 2.108)	0.007 (0.178)	
0.084 to 0.098 (2.133 to 2.489)	0.008 (0.203)	
0.099 to 0.114 (2.515 to 2.896)	0.009 (0.229)	
0.115 to 0.130 (2.921 to 3.302)	0.010 (0.254)	
0.131 to 0.145 (3.327 to 3.683)	0.012 (0.305)	
0.146 to 0.176 (3.708 to 4.470)	0.014 (0.356)	
Specified Width, in. (mm)[A]	Permissible Variations in Width, in. (mm)	
	Plus	Minus
Up to 42 (1066.8)	1/16 (1.59)	0
42 and Over (1066.8 and Over)	1/8 (3.18)	0
Specified Length, in. (mm)[A]	Permissible Variations in Length, in. (mm)	
	Plus	Minus
Up to 120 (3048)	1/16 (1.59)	0
120 and Over (3048 and Over)	1/8 (3.18)	0
Weight Permissible variations in weight apply only to polished finishes. The actual weight of any one item of an ordered thickness and size in any finish is limited in overweight by the following tolerances:	Estimated Permissible Variations, Actual Weight Over the Estimated Weight, %[B]	
Any item of 5 sheets or less, or any item estimated to weigh 200 lb (90.72 kg) or less	10	
Any item of more than 5 sheets, and estimated to weigh more than 200 lb (90.72 kg)	7.5	

[A] Sheet 0.131 in. (3.33 mm) and over in thickness, regardless of size, may have permissible variations of ± 1/4 in. (6.35 mm) in width and in length, respectively.

[B] There is no under variation in weight for No. 1 and No. 2 finishes, these finishes being limited in under variations only by the permissible variations in thickness. Polished sheets may actually weigh as much as 5 % less than the estimated weight. Estimated weight of the composite plates may be calculated using the following weights of the component materials:

	Weight per Square Foot for Material 1 in. in Thickness, lb
Steel	40.8
Chromium steel cladding	41.2

ASTM A 263

TABLE 2 Permissible Variations in Thickness of Cold-Rolled Strip

NOTE—Permissible variations in thickness are based on measurements taken ⅜ in. (9.53 mm) in from the edge on cold-rolled strip 1 in. (25.4 mm) or over in width and at any place on the strip on material less than 1 in. (25.4 mm) in width.

Specified Thickness, in. (mm)	Permissible Variations in Thickness, plus or minus, for Widths Given, in. (mm)									
	3⁄16 to ½ (4.76 to 12.7 mm)	½ to 1 (12.7 to 25.4 mm)	1 to 1½ (25.4 to 38.1 mm)	1½ to 3 (38.1 to 76.2 mm)	3 to 6 (76.2 to 152.4 mm)	Over 6 to 9 (152.4 to 228.6 mm)	Over 9 to 12 (228.6 to 304.8 mm)	Over 12 to 16 (304.8 to 406.4 mm)	Over 16 to 20 (406.4 to 508.0 mm)	Over 20 to 23 9⁄16 (508.0 to 608.1 mm)
0.249 to 0.161 (6.632 to 4.09), incl	0.002 (0.0508)	0.002 (0.0508)	0.003 (0.0762)	0.003 (0.0762)	0.004 (0.1016)	0.004 (0.1016)	0.004 (0.1016)	0.005 (0.1270)	0.006 (0.1524)	0.006 (0.1524)
0.160 to 0.100 (4.08 to 2.54), incl	0.002 (0.0508)	0.002 (0.0508)	0.002 (0.0508)	0.002 (0.0508)	0.003 (0.0762)	0.004 (0.1016)	0.004 (0.1016)	0.004 (0.1016)	0.005 (0.1270)	0.005 (0.1270)
0.099 to 0.069 (2.53 to 1.75), incl	0.002 (0.0508)	0.002 (0.0508)	0.002 (0.0508)	0.002 (0.0508)	0.003 (0.0762)	0.003 (0.0762)	0.003 (0.0762)	0.004 (0.1016)	0.004 (0.1016)	0.004 (0.1016)
0.068 to 0.050 (1.74 to 1.27), incl	0.002 (0.0508)	0.002 (0.0508)	0.002 (0.0508)	0.002 (0.0508)	0.003 (0.0762)	0.003 (0.0762)	0.003 (0.0762)	0.003 (0.0762)	0.004 (0.1016)	0.004 (0.1016)
0.048 to 0.040 (1.26 to 1.02), incl	0.002 (0.0508)	0.002 (0.0508)	0.002 (0.0508)	0.002 (0.0508)	0.003 (0.0762)	0.003 (0.0762)	0.003 (0.0762)	0.003 (0.0762)	0.004 (0.1016)	0.004 (0.1016)
0.039 to 0.035 (1.01 to 0.88), incl	0.002 (0.0508)	0.002 (0.0508)	0.002 (0.0508)	0.002 (0.0508)	0.0025 (0.0635)	0.003 (0.0762)	0.003 (0.0762)	0.003 (0.0762)	0.003 (0.0762)	0.003 (0.0762)
0.034 to 0.032 (0.87 to 0.81), incl	0.0015 (0.038)	0.0015 (0.038)	0.0015 (0.038)	0.0015 (0.038)	0.0025 (0.0635)	0.0025 (0.064)	0.0025 (0.064)	0.0025 (0.064)	0.003 (0.076)	0.003 (0.076)
0.031 to 0.029 (0.80 to 0.74), incl	0.0015 (0.038)	0.0015 (0.038)	0.0015 (0.038)	0.0015 (0.038)	0.002 (0.051)	0.0025 (0.064)	0.0025 (0.064)	0.0025 (0.064)	0.003 (0.076)	0.003 (0.076)
0.028 to 0.026 (0.73 to 0.66), incl	0.001 (0.025)	0.001 (0.025)	0.0015 (0.038)	0.0015 (0.038)	0.0015 (0.038)	0.002 (0.051)	0.002 (0.051)	0.002 (0.051)	0.0025 (0.064)	0.003 (0.076)
0.025 to 0.023 (0.65 to 0.58), incl	0.001 (0.025)	0.001 (0.025)	0.001 (0.025)	0.001 (0.025)	0.0015 (0.038)	0.002 (0.051)	0.002 (0.051)	0.002 (0.051)	0.0025 (0.064)	0.0025 (0.064)
0.022 to 0.020 (0.57 to 0.51), incl	0.001 (0.025)	0.001 (0.025)	0.001 (0.025)	0.001 (0.025)	0.0015 (0.038)	0.002 (0.051)	0.002 (0.051)	0.002 (0.051)	0.0025 (0.064)	0.0025 (0.064)

TABLE 2 Continued

NOTE—Permissible variations in thickness are based on measurements taken ⅜ in. (9.53 mm) in from the edge on cold-rolled strip 1 in. (25.4 mm) or over in width, and at any place on the strip on material less than 1 in. (25.4 mm) in width.

Specified Thickness, in. (mm)	Permissible Variations in Thickness, plus or minus, for Widths Given, in. (mm)									
	³⁄₁₆ to ½ (4.76 to 12.7 mm)	½ to 1 (12.7 to 25.4 mm)	1 to 1½ (25.4 to 38.1 mm)	1½ to 3 (38.1 to 76.2 mm)	3 to 6 (76.2 to 152.4 mm)	Over 6 to 9 (152.4 to 228.6 mm)	Over 9 to 12 (228.6 to 304.8 mm)	Over 12 to 16 (304.8 to 406.4 mm)	Over 16 to 20 (406.4 to 508.0 mm)	Over 20 to 23¹⁵⁄₁₆ (508.0 to 608.1 mm)
0.019 to 0.017 (0.50 to 0.43), incl	0.001 (0.025)	0.001 (0.025)	0.001 (0.025)	0.001 (0.025)	0.001 (0.025)	0.0015 (0.038)	0.0015 (0.038)	0.002 (0.051)	0.002 (0.051)	0.002 (0.051)
0.016 to 0.015 (0.42 to 0.38), incl	0.001 (0.025)	0.001 (0.025)	0.001 (0.025)	0.001 (0.025)	0.001 (0.025)	0.0015 (0.038)	0.0015 (0.038)	0.0015 (0.038)	0.002 (0.051)	0.002 (0.051)
0.014 to 0.013 (0.37 to 0.33), incl	0.001 (0.025)	0.001 (0.025)	0.001 (0.025)	0.001 (0.025)	0.001 (0.025)	0.0015 (0.038)	0.0015 (0.038)	0.0015 (0.038)	0.0015 (0.038)	0.0015 (0.038)
0.012 (0.30)	0.001 (0.025)	0.001 (0.025)	0.001 (0.025)	0.001 (0.025)	0.001 (0.025)	0.001 (0.025)	0.001 (0.025)	0.0015 (0.038)	0.0015 (0.038)	0.0015 (0.038)
0.011 (0.28)	0.001 (0.025)	0.001 (0.025)	0.001 (0.025)	0.001 (0.025)	0.001 (0.025)	0.001 (0.025)	0.001 (0.025)	0.0015 (0.038)	0.0015 (0.038)	0.0015 (0.038)
0.010 (0.25)	0.001 (0.025)	0.001 (0.025)	0.001 (0.025)	0.001 (0.025)	0.001 (0.025)	0.001 (0.025)	0.001 (0.025)	0.001 (0.025)	...	...
0.009 to 0.006 (0.23 to 0.15), incl	0.00075 (0.019)	0.00075 (0.019)	0.00075 (0.019)	0.00075 (0.019)	0.00075 (0.019)	...	...	...	...	...
Under 0.006 (Under 0.15)	0.0005 (0.013)	0.0005 (0.013)	0.0005 (0.013)	0.0005 (0.013)	0.0005 (0.013)	...	...	...	...	...

TABLE 3 Permissible Variations in Dimensions of Cold-Rolled Strip

Edge	Description	Specified Width, in. (mm)	Specified Thickness, in. (mm)	Permissible Variations in Width, plus or minus, in. (mm)[A]
No. 1	Round or Square Edge, Rolled	9/32 (7.14) or under	1/16 (1.59) or under	0.005 (0.125)
		Over 9/32 to 3/4 (7.14 to 19.05), incl.	3/32 (2.39) or under	0.005 (0.125)
		Over 3/4 to 5 (19.05 to 127.0), incl.	1/8 (3.18) or under	0.005 (0.125)
No. 5	Square rolled or filed, after slitting	9/32 (7.14) or under	1/16 (1.59) or under	0.005 (0.125)
		Over 9/32 to 3/4 (7.14 to 19.05), incl.	3/32 (2.39) or under	0.005 (0.125)
		Over 3/4 to 5 (19.05 to 127.0), incl.	1/8 (3.18) or under	0.005 (0.125)
		Over 5 to 9 (127.0 to 228.6), incl.	1/8 to 0.008 (3.18 to 0.203), incl.	0.010 (0.254)
		Over 9 to 20 (228.6 to 508.0), incl.	0.105 to 0.015 (2.667 to 0.381)	0.010 (0.254)
		Over 20 to 23 15/16 (508.0 to 608.0), incl.	0.080 to 0.023 (2.03 to 0.584)	0.015 (0.381)
No. 6	Square edge, rolled	1/2 (12.7) or under	3/16 (4.76) or under	0.010 (0.254)
		Over 1/2 to 15/16 (12.7 to 23.81), incl.	3/16 to 0.025 (4.76 to 0.635), incl.	1/64 (0.396)
		Over 15/16 to 2 (23.81 to 50.8), incl.	1/4 to 0.025 (6.35 to 0.635), incl.	1/32 (0.794)
		Over 2 to 6 (50.8 to 152.4), incl.	1/4 to 0.225 (6.35 to 5.72), incl.	3/64 (1.648)

Specified Thickness, in. (mm)	Permissible Variations in Width of No. 3 Edge Strip, in. (mm)			
	Up to 2 in. (50.8)	Over 2 to 6 in. (50.8 to 152.4)	Over 6 to 12 in. (152.4 to 304.8)	Over 12 to 23 15/16 incl. (304.8 to 608.0)
0.050 (1.27) or under	0.007 (0.178)	0.010 (0.254)	0.015 (0.381)	0.015 (0.381)
0.051 to 0.083 (1.28 to 2.11)	0.010 (0.254)	0.012 (0.305)	0.015 (0.381)	0.020 (0.508)
0.084 (2.12) or Over	0.012 (0.305)	0.015 (0.381)	0.018 (0.457)	0.20 (0.508)

Specified Length, ft (cm)	Permissible Variation in Length, plus, in. (mm)
Up to 5, incl	3/8 (9.53)
Over 5 (152.40) to 10 (304.8), incl	1/2 (12.7)
Over 10 (304.8) to 20 (609.6), incl	5/8 (15.88)
Over 20 (609.6) to 30 (914.4), incl	3/4 (19.05)
Over 30 (914.4) to 40 (1219.2), incl	1 (25.4)
Over 40 (1219.2) to 60 (1828.8), incl	1 1/2 (38.1)
Over 60 (1828.8) to 90 (2743.2), incl	2 (50.8)
Over 90 (2743.2) to 200 (6096.0), incl	2 1/2 (63.5)

Specified Width, in. (mm)	Permissible Camber[B]
1 1/2 and under (38.1 and under)	1/2 in. in any 8 ft length (12.7 mm in any 243.84 cm)
Over 1 1/2 to 23 15/16 (Over 38.1 to 608.0), incl	1/4 in. in any 8 ft length (6.35 mm in any 243.84 cm)

[A] These permissible variations are plus or minus. If variation is desired all one way, double the figure indicated.
[B] Camber shall be determined by placing an 8 ft (243.84 cm) straightedge against the concave side of the strip.

A 263

TABLE 4 Permissible Variations in Dimensions of Hot-Rolled Strip

Specified Thickness, Bwg or in. (mm)	Permissible Variations in Thickness, plus or minus, in. (mm)[A]			
	5 in. (127 mm) and Under in Width	Over 5 to 10, incl. (127 to 254 mm) in Width	Over 10 to 15, incl. (254 to 381 mm) in Width	Over 15 to 23¹⁵⁄₁₆, incl. (381 to 608 mm) in Width
No. 23 to No. 18, incl	0.003 (0.076)	0.004 (0.102)	. . .	. . .
No. 1 to No. 11, incl	0.004 (0.102)	0.005 (0.127)	0.006 (0.152)	0.007 (0.178)
No. 10 to 0.187 (4.75 mm), incl	0.005 (0.127)	0.006 (0.152)	0.007 (0.178)	0.008 (0.203)
0.188 to 0.249 (4.76 to 6.32 mm), incl	0.006 (0.152)	0.007 (0.178)	0.008 (0.203)	0.009 (0.229)

Specified Width, in. (mm)	Permissible Variation in Width, in. (mm)	
	Plus	Minus
Up to 1 (25.4) by No. 16 Bwg and heavier	¹⁄₃₂ (0.794)	¹⁄₃₂ (0.794)
Up to 1 (25.4) by No. 17 Bwg and lighter	¹⁄₃₂ (0.794)	¹⁄₃₂ (0.794)
Over 1 to 2 (25.4 to 50.8), incl	¹⁄₃₂ (0.794)	¹⁄₃₂ (0.794)
Over 2 to 5 (50.8 to 127), incl	³⁄₆₄ (1.911)	³⁄₆₄ (1.911)
Over 5 to 10 (127 to 254), incl	¹⁄₁₆ (1.588)	¹⁄₁₆ (1.588)
Over 10 to 15 (254 to 381), incl	³⁄₃₂ (2.381)	³⁄₃₂ (2.381)
Over 15 to 20 (381 to 508), incl	⅛ (3.175)	⅛ (3.175)
Over 20 to 23¹⁵⁄₁₆ (508 to 608), incl	⁵⁄₃₂ (3.969)	⁵⁄₃₂ (3.963)

Specified Length, ft. (cm)	Permissible Variations in Length, in. (mm)					
	Up to 3 in. (76.2 mm) in Width		Over 3 to 6 in. (76.2 to 152.4 mm) in Width		Over 6 in. (152.4 mm) in Width	
	Plus	Minus	Plus	Minus	Plus	Minus
Up to 5 (152.4)	¼ (6.35)	0	⅜ (9.53)	0	½ (12.7)	0
Over 5 to 10 (152.4 to 304.8), incl	⅜ (9.53)	0	½ (12.7)	0	¾ (19.1)	0
Over 10 to 20 (304.8 to 609.6), incl	½ (12.7)	0	⅝ (15.88)	0	1 (25.4)	0
Over 20 to 30 (609.6 to 914.4), incl	¾ (19.1)	0	¾ (19.1)	0	1¼ (31.8)	0
Over 30 to 40 (914.4 to 1219.2), incl	1 (25.4)	0	1 (25.4)	0	1½ (38.1)	0
Over 40 (1219.2)	1½ (38.1)	0	1½ (38.1)	0	1¾ (44.5)	0

Permissible Camber

Strip shall be as straight as practicable and not more than ¼ in. (6.35 mm) out of line in any 8 ft (243.8 cm) of length, with camber measurement taken on the concave edge.

[A] Thickness measurement shall be taken ⅜ in. (9.53 mm) from edge of strip. In the case of center measurements for thickness the permissible variations shall be that for the edge measurement plus the following:

  0.001 in. (0.025 mm) for width ½ to 1½ in. (12.77 to 38.1 mm)
  0.002 in. (0.051 mm) for width 1½ to 2½ in. (38.1 to 63.5 mm)
  0.003 in. (0.076 mm) for width 2½ to 5 in. (63.5 to 127.0 mm)
  0.004 in. (0.102 mm) for width 5 to 10 in. (127.0 to 254.0 mm)
  0.005 in. (0.127 mm) for width 10 to 15 in. (254.0 to 381.0 mm)
  0.006 in. (0.152 mm) for width 15 to 23¹⁵⁄₁₆ in. (381.0 to 608.1 mm)

A 263

TABLE 5 Permissible Variations in Thickness and Overweights of Plates

Thickness[A]

All plates shall be ordered to thickness and not to weight per square foot. No plates shall vary more than 0.01 in. (0.254 mm) under the thickness ordered, and the overweight of each lot[B] in each shipment shall not exceed the amount given in the table below.

Weight[C]

Specified Thickness, in. (mm)	Permissible Excess in Average Weight per Square Foot Plates for Widths Given in Inches (mm). Expressed in Percentage of Nominal Weight							
	Under 48 (1219.2)	48 (1219.2) to 60 (1524.0) excl	60 (1524.0) to 72 (1828.8) excl	72 (1828.8) to 84 (2133.6) excl	84 (2133.6) to 96 (2438.4) excl	96 (2438.4) to 108 (2743.2) excl	108 (2743.2) to 120 (3048.0) excl	120 (3048.0) to 132 (3352.8) excl	132 (3352.8) to 144 (3657.6) excl
3/16 (4.76) to 1/4 (6.35), excl	10.5	12.0	13.5	15.0	18.0	...	...	...	...
1/4 (6.35) to 5/16 (7.94), excl	9.0	10.5	12.0	13.5	15.0	18.0	21.0	24.0	28.5
5/16 (7.94) to 3/8 (9.52) excl	7.5	9.0	10.5	12.0	13.5	15.0	18.0	21.0	25.5
3/8 (9.52) to 7/16 (11.11), excl	7.0	7.5	9.0	10.5	12.0	13.5	15.0	18.0	22.5
7/16 (11.11) to 1/2 (12.7), excl	6.0	7.0	7.5	9.0	10.5	12.0	13.5	15.0	19.5
1/2 (12.7) to 5/8 (15.88), excl	5.5	6.0	7.0	7.5	9.0	10.5	12.0	13.5	16.5
5/8 (15.88) to 3/4 (19.05), excl	4.5	5.5	6.0	7.0	7.5	9.0	10.5	12.0	13.5
3/4 (19.05) to 1 (25.4), excl	4.0	4.5	5.5	6.0	7.0	7.5	9.0	10.5	12.0
1 (25.4) or Over	4.0	4.0	4.5	5.5	6.0	7.0	7.5	9.0	10.5

[A] Spot grinding is permitted to remove surface imperfections not to exceed 0.01 in. (0.254 mm) under the specified thickness.
[B] The term "lot" means all of the plates of each group width and each group thickness.
[C] The weight of individual plates shall not exceed the nominal weight by more than one and one-third times the amount prescribed in the above table.

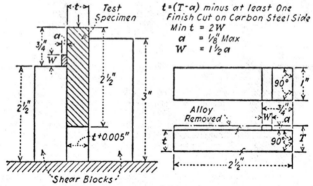

Metric Equivalents

in.	mm	in.	mm
0.005	0.127	1	25.4
1/8	3.17	2 1/2	64.5
3/4	19.1	3	76.2

FIG. 1 Test Specimen and Method of Making Shear Test of Clad Plate

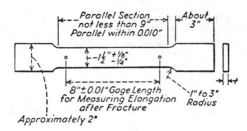

Metric Equivalents

in.	mm	in.	mm
0.01	0.254	1½	38.1
⅛	3.17	2	50.8
¼	6.35	3	76.2
1	25.4	8	203.2

NOTE 1—When necessary, it is permissible to use a narrower specimen, but in such a case the reduced portion shall be not less than 1 in. in width.

NOTE 2—Punch marks for measuring elongation after fracture shall be made on the flat or on the edge of the specimen and within the parallel section; either a set of nine punch marks 1 in. apart, or one or more sets of 8-in. punch marks may be used.

NOTE 3—The dimension t is the thickness of the test specimen as provided for in the applicable material specifications.

FIG. 2 Standard Rectangular Tension Test Specimen with 8-in. Gage Length

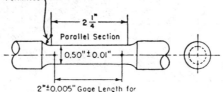

Metric Equivalents

in.	mm	in.	mm
0.003	0.076	⅜	9.53
0.005	0.127	0.50	12.7
0.01	0.254	2	50.8
⅛	3.18	2¼	64.5

NOTE—The gage length and fillets shall be as shown, but the ends may be of any shape to fit the holders of the testing machine in such a way that the load shall be axial. The reduced section shall have a gradual taper from the ends toward the center, with the ends 0.003 to 0.005 in. larger in diameter than the center.

FIG. 3 Standard Round Tension Test Specimen with 2-in. Gage Length

The American Society for Testing and Materials takes no position respecting the validity of any patent rights asserted in connection with any item mentioned in this standard. Users of this standard are expressly advised that determination of the validity of any such patent rights, and the risk of infringement of such rights, are entirely their own responsibility.

This standard is subject to revision at any time by the responsible technical committee and must be reviewed every five years and if not revised, either reapproved or withdrawn. Your comments are invited either for revision of this standard or for additional standards and should be addressed to ASTM Headquarters. Your comments will receive careful consideration at a meeting of the responsible technical committee, which you may attend. If you feel that your comments have not received a fair hearing you should make your views known to the ASTM Committee on Standards, 1916 Race St., Philadelphia, Pa. 19103.

Designation: A 264 - 82

Standard Specification for
STAINLESS CHROMIUM-NICKEL STEEL CLAD PLATE, SHEET, AND STRIP[1]

This standard is issued under the fixed designation A 264; the number immediately following the designation indicates the year of original adoption or, in the case of revision, the year of last revision. A number in parentheses indicates the year of last reapproval. A superscript epsilon (ϵ) indicates an editorial change since the last revision or reapproval.

1. Scope

1.1 This specification[2] covers plate, sheet, and strip of carbon steel or low-alloy steel base to which is integrally and continuously bonded on one or both sides a layer of stainless chromium-nickel steel. The material is generally intended for pressure vessel use.

1.2 The values state in inch-pound units are to be regarded as the standard.

2. Applicable Documents

2.1 *ASTM Standards*:
A 20/A 20M Specification for General Requirements for Steel Plates for Pressure Vessels[3]
A 240 Specification for Heat-Resisting Chromium and Chromium-Nickel Stainless Steel Plate, Sheet, and Strip for Pressure Vessels[3,4]
A 480 Specification for General Requirements for Flat-Rolled Stainless and Heat-Resisting Steel Plate, Sheet, and Strip[4]

3. Description of Terms

3.1 This material is considered as single-clad or double-clad corrosion-resisting chromium-nickel steel plate, sheet, or strip depending on whether one or both sides are covered.

3.2 The terms plate, sheet, and strip as used in this specification apply to the following:

3.2.1 *Plate*—Material ³⁄₁₆ in. (2.73 mm) and over in thickness and over 10 in. (254 mm) in width.

3.2.2 *Sheet*—Material under ³⁄₁₆ in. in thickness and 24 in. (609.6 mm) and over in width, material under ³⁄₁₆ in. in thickness and all widths and finishes of Nos. 3 to 8 inclusive, and

3.2.3 *Strip*—Cold-rolled material under 24 in. in width and ³⁄₁₆ in. and under in thickness.

4. General Requirements for Delivery

4.1 Material furnished under this specification shall conform to the applicable requirements of the current edition of Specification A 20.

5. Materials and Manufacture

5.1 *Process:*

5.1.1 The steel shall be made by the open-hearth, electric-furnace (with separate degassing and refining optional), or basic-oxygen processes, or by secondary processes whereby steel made from these primary processes is remelted using electroslag remelting or vacuum arc remelting processes.

5.1.2 The alloy cladding metal may be metallurgically bonded to the base metal by any method that will produce a clad steel which will conform to the requirements of this specification.

5.2 *Heat Treatment:*

5.2.1 Unless otherwise specified or agreed upon between the purchaser and the manufacturer, all austenitic stainless steel clad plates shall be given a heat treatment consisting of heating to the proper temperature for the solution of the chromium carbides in the cladding followed by individual air cooling. For base

---

[1] This specification is under the jurisdiction of ASTM Committee A-1 on Steel, Stainless Steel and Related Alloys, and is the direct responsibility of Subcommittee A01.17 on Flat Stainless Steel Products.
Current edition approved July 30, 1982. Published September 1982. Originally published as A 264 - 43 T. Last previous edition A 264 - 79.
[2] For ASME Boiler and Pressure Vessel Code applications see related Specification SA-264 in Section II of that Code.
[3] *Annual Book of ASTM Standards*, Vol 01.04.
[4] *Annual Book of ASTM Standards*, Vol 01.03.

metals of air-hardening low-alloy steels the above heat treatment shall be followed by a tempering treatment.

5.2.2 When plates over 1 in. (25.4 mm) in thickness are to be cold formed, the purchaser may specify that such plates be heat treated for grain refinement of the base metal.

6. Thickness of Cladding Metal

6.1 The minimum thickness and tolerances on the thickness of the alloy cladding metal shall be agreed upon between the purchaser and the manufacturer.

7. Chemical Requirements

7.1 The composite plate, sheet, or strip may conform to any desired combination of alloy cladding metal and base metal as described in 7.2 and 7.3 and as agreed upon between the purchaser and the manufacturer.

7.2 *Alloy Cladding Metal*—The alloy cladding metal specified shall conform to the requirements as to chemical composition prescribed for the respective chromium-nickel steel in Specification A 240.

7.3 *Base Metal*—The base metal may be carbon steel or low-alloy steel conforming to the ASTM specifications for steel for pressure vessels. The base metal shall conform to the chemical requirements of the specification to which it is ordered.

8. Product Analysis

8.1 Product analyses may be required on the finished product only when the composite plate thickness is sufficient to permit obtaining drillings or millings without danger of contamination from the adjacent layer.

8.2 If product analysis is specified by the purchaser, it shall be made from drillings or millings taken from the final piece or a broken test specimen. In order to avoid contamination by the base plate metal, cladding samples shall be taken from the test coupon by removal and discard of all the base metal plate material, plus 40% of the cladding thickness from the bonded side, not to exceed $\frac{1}{16}$ in. (1.588 mm). The material shall be chemically cleaned and millings shall be taken to represent the full cross-section of the remainder.

8.3 The results of the product analysis shall conform to the requirements of Section 7 with variations in limits permitted by Specification A 480.

9. Mechanical Requirements

9.1 *Plate:*

9.1.1 *Tensile Properties:*

9.1.1.1 For clad plates up to 1½ in. (38.1 mm), inclusive, maximum gage with base steel of 70 000 psi (485 MPa) tensile strength or less, the tensile properties as determined by a tension test on the composite plate shall be equal to the minimum and not more than 5000 psi (35 MPa) over the maximum prescribed in the specification for the base steel used.

9.1.1.2 For clad plates over 1½ in. gage, or for clad plate with base steel of more than 70 000 psi minimum tensile strength or 40 000 psi (275 MPa) minimum yield strength, the tensile properties shall be determined by pulling a test of the base steel only, in which case the properties shall meet the requirements of the specifications for the base steel.

9.1.2 *Ductility*—Two bend tests of the composite plate shall be made, one with the alloy cladding in tension and the other with the alloy cladding in compression to determine the ductility of the materials. On double-clad plates, the bend tests shall be made so that one specimen represents the alloy cladding in tension on one side while the other specimen represents the alloy cladding in tension on the opposite side. Bend tests shall be made according to and shall conform to the requirements prescribed in the specification for the base metal.

9.1.3 *Shear Strength*—When required by the purchaser, the minimum shear strength of the alloy cladding and base metals shall be 20 000 psi (140 MPa). The shear test, when specified, shall be made in the manner indicated in Fig. 1.

9.1.4 *Bond Strength*—As an alternative to the shear strength test provided in 9.1.3 and when required by the purchaser, three bend tests shall be made with the alloy cladding in compression to determine the quality of the bond. These bend tests shall be made in accordance with the specifications for the cladding metal. At least two of the three tests shall show not more than 50% separation on both edges of the bent portion. Greater separation shall be cause for rejection.

9.2 *Sheet and Strip:*

9.2.1 The bend test specimens of sheet and strip shall stand being bent cold through an angle of 180° without cracking on the outside of the bent portion. The bend test specimens shall be bent around a pin the diameter of which is equal to the thickness of the material. The axis of the bend shall be transverse to the direction of rolling.

9.2.2 The bond between the alloy cladding and the base metal shall be ascertained by observation of the behavior of the composite sheet or strip when sheared with the alloy side down.

10. Test Specimens

10.1 *Plate:*

10.1.1 The tension test specimens from plate shall conform to the requirements prescribed in the specifications for the base metal.

10.1.2 Bend test specimens, shall be taken from the middle of the top of the plate as rolled, at right angles to its longitudinal axis.

10.1.3 When required by the purchaser, the shear test specimen shall be taken from a top or bottom corner of the plate as rolled, parallel to its longitudinal axis.

10.1.4 All tests shall be made on specimens in the same condition of heat treatment to which the composite plate is furnished.

10.1.5 For plates 1½ in. (38.1 mm) and under in thickness tension test specimens shall be the full thickness of the material, except as specified in 9.1.1.1 and 9.1.1.2, and the bend test specimen shall be full thickness of the material and shall be machined to the form and dimensions shown in Fig. 2, or may be machined with both edges parallel.

10.1.6 For plates over 1½ in. in thickness, tension tests shall be of the form shown in Fig. 3 and shall be of all base steel conforming to the requirements of the specification for the base steel.

10.1.7 For plates over 1½ in. in thickness the bend test specimens need not be greater in thickness than 1½ in. but shall not be less than 1½ in. Specimens shall be of the form and dimensions shown in Fig. 2. In reducing the thickness of the specimen, both the alloy cladding and the base steel shall be machined so as to maintain the same ratio of clad metal to base steel as is maintained in the plate, except that the thickness of the clad metal need not be reduced below ⅛ in. (3.17 mm).

10.1.8 The side of the bend test specimen may have the corners rounded to a radius not over 1/16 in. (1.58 mm) for plates, 2 in. (50.8 mm) and under in thickness, and not over ⅛ in. for plates over 2 in. in thickness.

10.2 *Sheet and Strip:*

10.2.1 Bend test specimens from sheet and strip shall be of the full thickness of the material and shall have a width equal to six times the thickness but not less than 1 in. (25.4 mm). The length shall be equal to the width of the sheet or strip but in no case longer than 12 in. (304.8 mm). The edges of the test specimen may be rounded to a radius equal to one half the thickness.

10.3 Any test specimen used for the determination of the minimum thickness of alloy cladding metal shall have distorted metal, due to burning or shearing, completely removed.

11. Number of Tests and Retests

11.1 *Plate:*

11.1.1 One or more tension tests, as required by the specifications for the base metal, one face bend test (alloy cladding in tension), one reverse bend test (alloy cladding in compression), and, when specified, one shear test or three bond bend tests shall be made representing each plate as rolled. Each specimen shall be in the final condition of heat treatment required for the plate.

11.1.2 If any test specimen shows defective machining or develops flaws, it may be discarded and another specimen substituted.

11.1.3 If the percentage of elongation of any tension test specimen is less than that specified in 9.1.1 and any part of the fracture is more than ¾ in. (19.1 mm) from the center of the gage length of a 2-in. (50.8-mm) specimen or is outside the middle third of the gage length of an 8-in. (203.2-mm) specimen, as indicated by scribe scratches marked on the specimen before testing, a retest shall be allowed.

11.2 *Sheet and Strip:*

11.2.1 On single-clad materials four bend tests shall be made of each lot of material. A lot shall consist of no more than the number of sheets or strips required to make 1000 lb (454 kg), but in no case more than 100 sheets or strips from the same melt of steel and specified thickness. On double-clad materials two bend tests shall be made on each lot of material.

11.2.2 On single-clad materials two of the bend tests shall be made with the alloy cladding metal in tension and two with the alloy cladding metal in compression.

11.2.3 On double-clad materials, the bend tests shall be made so that one specimen will represent the alloy cladding metal in tension on one side while the other specimen will represent the alloy cladding metal in tension on the opposite side.

12. Permissible Variations in Thickness and Weight

12.1 Composite plates, sheets, and strips shall conform to the dimensional and weight requirements prescribed in Tables 1, 2, 3, 4, and 5.

13. Workmanship and Finish

13.1 The material shall be free of injurious defects, shall have a workmanlike appearance, and shall conform to the designated finish.

13.2 Plate alloy surfaces shall be sandblasted, pickled, or, blas-cleaned and pickled.

13.3 The finish for the alloy surfaces of sheets and strips shall be as specified in Specification A 240.

14. Repair of Cladding by Welding

14.1 Unless otherwise specified, the material manufacturer may repair defects in cladding by welding provided the following requirements are met:

14.1.1 Prior approval is obtained from the purchaser if the repaired area exceeds 3% of the cladding surface.

14.1.2 The welding procedure and the welders or welding operators are qualified in accordance with Section IX of the ASME Code.

14.1.3 The defective area is removed and the area prepared for repair is examined by a magnetic particle method or a liquid penetrant method to ensure all defective area has been removed. Method of test and acceptance standard shall be as agreed upon between the purchaser and the manufacturer.

14.1.4 The weld will be deposited in a suitable manner so as to leave its surface condition equivalent in corrosion resistance to the alloy cladding.

14.1.5 The repaired area is examined by a liquid penetrant method (see 14.1.3).

14.1.6 The location and extent of the weld repairs together with the repair procedure and examination results are transmitted as a part of the certification.

14.2 At the request of the purchaser or his inspector, plates shall be reheat treated following repair by welding.

15. Inspection

15.1 The manufacturer shall afford the inspector all reasonable facilities, without charge, to satisfy him that the material is being furnished in accordance with this specification. All tests (except product analysis) and inspection shall be made at the place of manufacture prior to shipment, unless otherwise specified, and shall be so conducted as not to interfere unnecessarily with the operation of the works.

16. Rejection and Rehearing

16.1 Unless otherwise specified, any rejection based on tests made in accordance with Section 8 shall be reported to the manufacturer within five working days from the receipt of samples by the purchaser.

16.2 Material that shows injurious defects subsequent to its acceptance at the manufacturer's work will be rejected, and the manufacturer shall be notified.

16.3 Samples tested in accordance with Section 8 that represent rejected material shall be preserved for 3 weeks from the date of the test report. In case of dissatisfaction with the results of the tests, the manufacturer may make claim for a rehearing within that time.

17. Report

17.1 The chemical analysis of the base metal shall be certified to the purchaser by the manufacturer.

17.2 The results of the tests in Section 9 shall be reported to the purchaser or his representative.

18. Marking

18.1 Except as specified in 18.2, the name or brand of the manufacturer, the manufacturer's test identification number, the class of the base steel, the type number of the alloy cladding metal and the specified minimum tensile strength shall be legibly stamped on each finished single-clad plate in two places on the base steel side not less than 12 in. (304.8 mm) from the edges. The manufacturer's identification number shall be legibly stamped on each test specimen.

18.2 For double-clad material or for material under ¼ in. (6.35 mm) in thickness, the marking specified in 18.1 shall be legibly stenciled instead of stamped.

ASTM A 264

TABLE 1 Permissible Variations in Thickness of Cold-Rolled Strip

NOTE—Permissible variations in thickness are based on measurements taken ⅜ in. (9.53 mm) in from the edge on cold-rolled strip 1 in. (25.4 mm) or over in width, and at any place on the strip on material less than 1 in. (25.4 mm) in width.

Specified Thickness, in. (mm)	Permissible Variations in Thickness, plus or minus, for Widths Given, in. (mm)									
	3/16 to 1/2 (4.76 to 12.7 mm)	1/2 to 1 (12.7 to 25.4 mm)	1 to 1½ (25.4 to 38.1 mm)	1½ to 3 (38.1 to 76.2 mm)	3 to 6 (76.2 to 152.4 mm)	Over 6 to 9 (152.4 to 228.6 mm)	Over 9 to 12 (228.6 to 304.8 mm)	Over 12 to 16 (304.8 to 406.4 mm)	Over 16 to 20 (406.4 to 508.0 mm)	Over 20 to 23 15/16 (508.0 to 608.1 mm)
0.249 to 0.161 (6.632 to 4.09), incl	0.002 (0.0508)	0.002 (0.0508)	0.003 (0.0762)	0.003 (0.0762)	0.004 (0.1016)	0.004 (0.1016)	0.004 (0.1016)	0.005 (0.1270)	0.006 (0.1524)	0.006 (0.1524)
0.160 to 0.100 (4.08 to 2.54), incl	0.002 (0.0508)	0.002 (0.0508)	0.002 (0.0508)	0.002 (0.0508)	0.003 (0.0762)	0.004 (0.1016)	0.004 (0.1016)	0.004 (0.1016)	0.005 (0.1270)	0.005 (0.1270)
0.099 to 0.069 (2.53 to 1.75), incl	0.002 (0.0508)	0.002 (0.0508)	0.002 (0.0508)	0.002 (0.0508)	0.003 (0.0762)	0.003 (0.0762)	0.003 (0.0762)	0.004 (0.1016)	0.004 (0.1016)	0.004 (0.1016)
0.068 to 0.050 (1.74 to 1.27), incl	0.002 (0.0508)	0.002 (0.0508)	0.002 (0.0508)	0.002 (0.0508)	0.003 (0.0762)	0.003 (0.0762)	0.003 (0.0762)	0.003 (0.0762)	0.004 (0.1016)	0.004 (0.1016)
0.048 to 0.040 (1.26 to 1.02), incl	0.002 (0.0508)	0.002 (0.0508)	0.002 (0.0508)	0.002 (0.0508)	0.003 (0.0762)	0.003 (0.0762)	0.003 (0.0762)	0.003 (0.0762)	0.004 (0.1016)	0.004 (0.1016)
0.039 to 0.035 (1.01 to 0.88), incl	0.002 (0.0508)	0.002 (0.0508)	0.002 (0.0508)	0.002 (0.0508)	0.0025 (0.0635)	0.003 (0.0762)	0.003 (0.0762)	0.003 (0.0762)	0.003 (0.0762)	0.003 (0.0762)
0.034 to 0.032 (0.87 to 0.81), incl	0.0015 (0.038)	0.0015 (0.038)	0.0015 (0.038)	0.0015 (0.038)	0.0025 (0.0635)	0.0025 (0.064)	0.0025 (0.064)	0.0025 (0.064)	0.003 (0.076)	0.003 (0.076)
0.031 to 0.029 (0.80 to 0.74), incl	0.0015 (0.038)	0.0015 (0.038)	0.0015 (0.038)	0.0015 (0.038)	0.002 (0.051)	0.0025 (0.064)	0.0025 (0.064)	0.0025 (0.064)	0.003 (0.076)	0.003 (0.076)
0.028 to 0.026 (0.73 to 0.66), incl	0.001 (0.025)	0.001 (0.025)	0.0015 (0.038)	0.0015 (0.038)	0.002 (0.051)	0.002 (0.051)	0.002 (0.051)	0.002 (0.051)	0.0025 (0.064)	0.003 (0.076)
0.025 to 0.023 (0.65 to 0.58), incl	0.001 (0.025)	0.001 (0.025)	0.001 (0.025)	0.001 (0.025)	0.0015 (0.038)	0.002 (0.051)	0.002 (0.051)	0.002 (0.051)	0.0025 (0.064)	0.0025 (0.064)
0.022 to 0.020 (0.57 to 0.51), incl	0.001 (0.025)	0.001 (0.025)	0.001 (0.025)	0.001 (0.025)	0.0015 (0.038)	0.002 (0.051)	0.002 (0.051)	0.002 (0.051)	0.0025 (0.064)	0.0025 (0.064)

TABLE 1 *Continued*

Permissible Variations in Thickness, plus or minus, for Widths Given, in. (mm)

Specified Thickness, in. (mm)	3/16 to 1/2 (4.76 to 12.7 mm)	1/2 to 1 (12.7 to 25.4 mm)	1 to 1 1/2 (25.4 to 38.1 mm)	1 1/2 to 3 (38.1 to 76.2 mm)	3 to 6 (76.2 to 152.4 mm)	Over 6 to 9 (152.4 to 228.6 mm)	Over 9 to 12 (228.6 to 304.8 mm)	Over 12 to 16 (304.8 to 406.4 mm)	Over 16 to 20 (406.4 to 508.0 mm)	Over 20 to 23 19/16 (508.0 to 608.1 mm)
0.019 to 0.017 (0.50 to 0.43), incl	0.001 (0.025)	0.001 (0.025)	0.001 (0.025)	0.001 (0.025)	0.001 (0.025)	0.0015 (0.038)	0.0015 (0.038)	0.002 (0.051)	0.002 (0.051)	0.002 (0.051)
0.016 to 0.015 (0.42 to 0.38), incl	0.001 (0.025)	0.001 (0.025)	0.001 (0.025)	0.001 (0.025)	0.001 (0.025)	0.0015 (0.038)	0.0015 (0.038)	0.0015 (0.038)	0.002 (0.051)	0.002 (0.051)
0.014 to 0.013 (0.37 to 0.33), incl	0.001 (0.025)	0.001 (0.025)	0.001 (0.025)	0.001 (0.025)	0.001 (0.025)	0.0015 (0.038)	0.0015 (0.038)	0.0015 (0.038)	0.002 (0.051)	0.002 (0.051)
0.012 (0.30)	0.001 (0.025)	0.001 (0.025)	0.001 (0.025)	0.001 (0.025)	0.001 (0.025)	0.001 (0.025)	0.0015 (0.038)	0.0015 (0.038)	0.0015 (0.038)	0.0015 (0.038)
0.011 (0.28)	0.001 (0.025)	0.001 (0.025)	0.001 (0.025)	0.001 (0.025)	0.001 (0.025)	0.001 (0.025)	0.001 (0.025)	0.0015 (0.038)	0.0015 (0.038)	0.0015 (0.038)
0.010 (0.25)	0.001 (0.025)	0.001 (0.025)	0.001 (0.025)	0.001 (0.025)	0.001 (0.025)	0.001 (0.025)	0.001 (0.025)	0.001 (0.025)	0.0015 (0.038)	0.0015 (0.038)
0.009 to 0.006 (0.23 to 0.15), incl	0.00075 (0.019)	0.00075 (0.019)	0.00075 (0.019)	0.00075 (0.019)	0.00075 (0.019)	...	...	...	...	...
Under 0.006 (Under 0.15)	0.0005 (0.013)	0.0005 (0.013)	0.0005 (0.013)	0.0005 (0.013)	0.0005 (0.013)	...	...	...	...	...

A 264

TABLE 2 Permissible Variations in Dimensions and Weight of Standard Sheet (No. 1 and No. 2 Finishes)

Specified Thickness, in. (mm)	Permissible Variations in Thickness, plus or minus, in. (mm)
0.005 (0.127)	0.001 (0.025)
0.006 to 0.007 (0.152 to 0.178)	0.0015 (0.038)
0.008 to 0.016 (0.203 to 0.406)	0.002 (0.051)
0.0161 to 0.026 (0.409 to 0.660)	0.003 (0.076)
0.027 to 0.040 (0.686 to 1.016)	0.004 (0.102)
0.041 to 0.058 (1.041 to 1.473)	0.005 (0.127)
0.059 to 0.072 (1.499 to 1.829)	0.006 (0.152)
0.073 to 0.083 (1.854 to 2.108)	0.007 (0.178)
0.084 to 0.098 (2.133 to 2.489)	0.008 (0.203)
0.099 to 0.114 (2.515 to 2.896)	0.009 (0.229)
0.115 to 0.130 (2.921 to 3.302)	0.010 (0.254)
0.131 to 0.145 (3.327 to 3.683)	0.012 (0.305)
0.146 to 0.176 (3.708 to 4.470)	0.014 (0.356)

Specified Width, in. (mm)[A]	Permissible Variations in Width, in. (mm)	
	Plus	Minus
Up to 42 (1066.8)	1/16 (1.59)	0
42 and Over (1066.8 and Over)	1/8 (3.18)	0

Specified Length, in. (mm)[A]	Permissible Variations in Length, in. (mm)	
	Plus	Minus
Up to 120 (3048)	1/16 (1.59)	0
120 and Over (3048 and Over)	1/8 (3.18)	0

Weight Permissible variations in weight apply only to polished finishes. The actual weight of any one item of an ordered thickness and size in any finish is limited in overweight by the following tolerances:	Estimated Permissible Variations, Actual Weight Over the Estimated Weight, %[B]
Any item of 5 sheets or less, or any item estimated to weigh 200 lb (90.72 kg) or less	10
Any item of more than 5 sheets, and estimated to weigh more than 200 lb (90.72 kg)	7.5

[A] Sheet 0.131 in. (3.33 mm) and over in thickness, regardless of size, may have permissible variations of ±1/4 in. (6.35 mm) in width and in length, respectively.

[B] There is no under variation in weight for No. 1 and No. 2 finishes, these finishes being limited in under variations only by the permissible variations in thickness. Polished sheets may actually weigh as much as 5 % less than the estimated weight. Estimated weight of the composite plates may be calculated using the following weights of the component materials:

	Weight per Square Foot for Material 1 in. in Thickness, lb
Steel	40.8
Chromium steel cladding	41.2

TABLE 3 Permissible Variations in Dimensions of Cold-Rolled Strip

Edge	Description	Specified Width, in. (mm)	Specified Thickness, in. (mm)	Permissible Variations in Width, plus or minus, in. (mm)
No. 1	Round or square edge rolled	9/32 (7.14) or under	1/16 (1.59) or under	0.005 (0.125)
		Over 9/32 to 3/4 (7.14 to 19.05), incl	3/32 (2.39) or under	0.005 (0.125)
		Over 3/4 to 5 (19.05 to 127.0), incl	1/8 (3.18) or under	0.005 (0.125)
No. 5	Square rolled or filed, after slitting	9/32 (7.14) or under	1/16 (1.59) or under	0.005 (0.125)
		Over 9/32 to 3/4 (7.14 to 19.05), incl	3/32 (2.39) or under	0.005 (0.125)
		Over 3/4 to 5 (19.05 to 127.0), incl	1/8 (3.18) or under	0.005 (0.125)
		Over 5 to 9 (127.0 to 228.6), incl	1/8 to 0.008 (3.18 to 0.203), incl	0.010 (0.254)
		Over 9 to 20 (228.6 to 508.0), incl	0.105 to 0.015 (2.667 to 0.381)	0.010 (0.254)
		Over 20 to 23 15/16 (508.0 to 608.0), incl	0.080 to 0.023 (2.03 to 0.584)	0.015 (0.381)
No. 6	Square edge, rolled	1/2 (12.7) or under	3/16 (4.76) or under	0.010 (0.254)
		Over 1/2 to 15/16 (12.7 to 23.81), incl	3/16 to 0.025 (4.76 to 0.635), incl	1/64 (0.396)
		Over 15/16 to 2 (23.81 to 50.8), incl	1/4 to 0.025 (6.35 to 0.635), incl	1/32 (0.794)
		Over 2 to 6 (50.8 to 152.4), incl	1/4 to 0.225 (6.35 to 5.72), incl	3/64 (1.648)

Permissible Variations in Width of No. 3 Edge Strip, in. (mm)[A]

Specified Thickness, in. (mm)	Up to 2 in. (50.8)	Over 2 to 6 in. (50.8 to 152.4)	Over 6 to 12 in. (152.4 to 304.8)	Over 12 to 23 15/16, incl (304.8 to 608.0)
0.050 (1.27) or under	0.007 (0.178)	0.010 (0.254)	0.015 (0.381)	0.015 (0.381)
0.051 to 0.083 (1.28 to 2.11)	0.010 (0.254)	0.012 (0.305)	0.015 (0.381)	0.020 (0.508)
0.084 (2.12) or Over	0.012 (0.305)	0.015 (0.381)	0.018 (0.457)	0.020 (0.508)

Specified Length, ft (cm)	Permissible Variations in Length, plus, in. (mm)
Up to 5, incl	3/8 (9.53)
Over 5 (152.40) to 10 (304.8), incl	1/2 (12.7)
Over 10 (304.8) to 20 (609.6), incl	5/8 (15.88)
Over 20 (609.6) to 30 (914.4), incl	3/4 (19.05)
Over 30 (914.4) to 40 (1219.2), incl	1 (25.4)
Over 40 (1219.2) to 60 (1828.8), incl	1 1/2 (38.1)
Over 60 (1828.8) to 90 (2743.2), incl	2 (50.8)
Over 90 (2743.2) to 200 (6096.0), incl	2 1/2 (63.5)

Specified Width, in. (mm)	Permissible Camber[B]
1 1/2 and under (38.1 and under)	1/2 in. in any 8 ft length (12.7 mm in any 243.84 cm)
Over 1 1/2 to 23 15/16 (over 38.1 to 608.0), incl	1/4 in. in any 8 ft length (6.35 mm in any 243.84 cm)

[A] These permissible variations are plus or minus. If variation is desired all one way, double the figure indicated.
[B] Camber shall be determined by placing an 8-ft (2.44-m) straightedge against the concave side of the strip.

TABLE 4 Permissible Variations in Dimensions of Hot-Rolled Strip

Specified Thickness, Bwg or in. (mm)	Permissible Variations in Thickness, plus or minus, in. (mm)[A]			
	5 in. (127 mm) and Under in Width	Over 5 to 10, incl (127 to 254 mm) in Width	Over 10 to 15, incl (254 to 381 mm) in Width	Over 15 to 23¹⁵⁄₁₆, incl (381 to 608 mm) in Width
No. 23 to No. 18, incl	0.003 (0.076)	0.004 (0.102)	...	...
No. 17 to No. 11, incl	0.004 (0.102)	0.005 (0.127)	0.006 (0.152)	0.007 (0.178)
No. 10 to 0.187 (4.75 mm), incl	0.005 (0.127)	0.006 (0.152)	0.007 (0.178)	0.008 (0.203)
0.188 to 0.249 (4.76 to 6.32 mm), incl	0.006 (0.152)	0.007 (0.178)	0.008 (0.203)	0.009 (0.229)

Specified Width, in. (mm)	Permissible Variation in Width, in. (mm)	
	Plus	Minus
Up to 2 (25.4 to 50.8), incl	¹⁄₃₂ (0.794)	¹⁄₃₂ (0.794)
Over 2 to 5 (50.8 to 127), incl	³⁄₆₄ (1.911)	³⁄₆₄ (1.911)
Over 5 to 10 (127 to 254), incl	¹⁄₁₆ (1.588)	¹⁄₁₆ (1.588)
Over 10 to 15 (254 to 381), incl	³⁄₃₂ (2.381)	³⁄₃₂ (2.381)
Over 15 to 20 (381 to 508), incl	⅛ (3.175)	⅛ (3.175)
Over 20 to 23¹⁵⁄₁₆ (508 to 608), incl	⁵⁄₃₂ (3.969)	⁵⁄₃₂ (3.969)

Specified Length, ft (cm)	Permissible Variations in Length, in. (mm)					
	Up to 3 in. (76.2 mm) in Width		Over 3 to 6 in. (76.2 to 152.4 mm) in Width		Over 6 in. (152.4 mm) in Width	
	Plus	Minus	Plus	Minus	Plus	Minus
Up to 5 (152.4)	¼ (6.35)	0	⅜ (9.53)	0	½ (12.7)	0
Over 5 to 10 (152.4 to 304.8), incl	⅜ (9.53)	0	½ (12.7)	0	¾ (19.1)	0
Over 10 to 20 (304.8 to 609.6), incl	½ (12.7)	0	⅝ (15.88)	0	1 (25.4)	0
Over 20 to 30 (609.6 to 914.4), incl	¾ (19.1)	0	¾ (19.1)	0	1¼ (31.8)	0
Over 30 to 40 (914.4 to 1219.2), incl	1 (25.4)	0	1 (25.4)	0	1½ (38.1)	0
Over 40 (1219.2)	1½ (38.1)	0	1½ (38.1)	0	1¾ (44.5)	0

Permissible Camber

Strip shall be as straight as practicable and not more than ¼ in. (6.35 mm) out of line in any 8 ft (2.44 m) of length, with camber measurement taken on the concave edge.

[A] Thickness measurement shall be taken ⅜ in. (9.53 mm) from edge of strip. In the case of center measurements for thickness the permissible variations shall be that for the edge measurement plus the following:
    0.001 in. (0.025 mm) for width ½ to 1½ in. (12.77 to 38.1 mm)
    0.002 in. (0.051 mm) for width 1½ to 2½ in. (38.1 to 63.5 mm)
    0.003 in. (0.076 mm) for width 2½ to 5 in. (63.5 to 127.0 mm)
    0.004 in. (0.102 mm) for width 5 to 10 in. (127.0 to 254.0 mm)
    0.005 in. (0.127 mm) for width 10 to 15 in. (254.0 to 381.0 mm)
    0.006 in. (0.152 mm) for width 15 to 23¹⁵⁄₁₆ in. (381.0 to 608.1 mm)

A 264

TABLE 5 Permissible Variations in Thickness and Overweights of Plates

Thickness[A]

All plates shall be ordered to thickness and not to weight per square foot. No plates shall vary more than 0.01 in. (0.254 mm) under the thickness ordered, and the overweight of each lot[B] in each shipment shall not exceed the amount given in the table below.

Weight[C]

Permissible Excess in Average Weight per Square Foot Plates for Widths Given in Inches (mm). Expressed in Percentage of Nominal Weight

Specified Thickness, in. (mm)	Under 48 (1219.2)	48 (1219.2) to 60 (1524.0) excl	60 (1524.0) to 72 (1828.8) excl	72 (1828.8) to 84 (2133.6) excl	84 (2133.6) to 96 (2438.4) excl	96 (2438.4) to 108 (2743.2) excl	108 (2743.2) to 120 (3048.0) excl	120 (3048.0) to 132 (3352.8) excl	132 (3352.8) to 144 (3657.6) excl
3/16 (4.76) to 1/4 (6.35), excl	10.5	12.0	13.5	15.0	18.0	...	...	...	...
1/4 (6.35) to 5/16 (7.94), excl	9.0	10.5	12.0	13.5	15.0	18.0	21.0	24.0	28.5
5/16 (7.94) to 3/8 (9.52), excl	7.5	9.0	10.5	12.0	13.5	15.0	18.0	21.0	25.5
3/8 (9.52) to 7/16 (11.11), excl	7.0	7.5	9.0	10.5	12.0	13.5	15.0	18.0	22.5
7/16 (11.11) to 1/2 (12.7), excl	6.0	7.0	7.5	9.0	10.5	12.0	13.5	15.0	19.5
1/2 (12.7) to 5/8 (15.88), excl	5.5	6.0	7.0	7.5	9.0	10.5	12.0	13.5	16.5
5/8 (15.88) to 3/4 (19.05), excl	4.5	5.5	6.0	7.0	7.5	9.0	10.5	12.0	13.5
3/4 (19.05) to 1 (25.4), excl	4.0	4.5	5.5	6.0	7.0	7.5	9.0	10.5	12.0
1 (25.4) or Over	4.0	4.0	4.5	5.5	6.0	7.0	7.5	9.0	10.5

[A] Spot grinding is permitted to remove surface imperfections not to exceed 0.01 in. (0.254 mm) under the specified thickness.
[B] The term "lot" means all of the plates of each group width and each group thickness.
[C] The weight of individual plates shall not exceed the nominal weight by more than one and one-third times the amount prescribed in the above table.

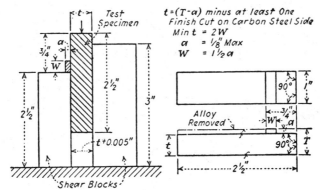

FIG. 1 Test Specimen and Method of Making Shear Test of Clad Plate

Metric Equivalents

in.	mm	in.	mm
0.005	0.127	1	25.4
1/8	3.17	2½	64.5
3/4	19.1	3	76.2

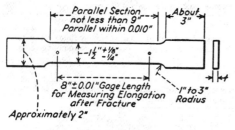
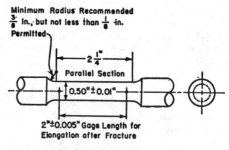

Metric Equivalents

in.	mm	in.	mm
0.01	0.254	1½	38.1
⅛	3.17	2	50.8
¼	6.35	3	76.2
1	25.4	8	203.2

NOTE 1–When necessary, it is permissible to use a narrower specimen, but in such a case the reduced portion shall be not less than 1 in. in width.

NOTE 2–Punch marks for measuring elongation after fracture shall be made on the flat or on the edge of the specimen and within the parallel section; either a set of nine punch marks 1 in. apart, or one or more sets of 8-in. punch marks may be used.

NOTE 3–The dimension *t* is the thickness of the test specimen as provided for in the applicable material specifications.

FIG. 2 Standard Rectangular Tension Test Specimen with 8-in. Gage Length

Metric Equivalents

in.	mm	in.	mm
0.003	0.076	⅜	9.53
0.005	0.127	0.50	12.7
0.01	0.254	2	50.8
⅛	3.18	2¼	64.5

NOTE—The gage length and fillets shall be as shown, but the ends may be of any shape to fit the holders of the testing machine in such a way that the load shall be axial. The reduced section shall have a gradual taper from the ends toward the center, with the ends 0.003 to 0.005 in. larger in diameter than the center.

FIG. 3 Standard Round Tension Test Specimen with 2-in. Gage Length

---

The American Society for Testing and Materials takes no position respecting the validity of any patent rights asserted in connection with any item mentioned in this standard. Users of this standard are expressly advised that determination of the validity of any such patent rights, and the risk of infringement of such rights, are entirely their own responsibility.

This standard is subject to revision at any time by the responsible technical committee and must be reviewed every five years and if not revised, either reapproved or withdrawn. Your comments are invited either for revision of this standard or for additional standards and should be addressed to ASTM Headquarters. Your comments will receive careful consideration at a meeting of the responsible technical committee, which you may attend. If you feel that your comments have not received a fair hearing you should make your views known to the ASTM Committee on Standards, 1916 Race St., Philadelphia, Pa. 19103.

Designation: A 265 – 81a

Used in USNRC-RDT standards

Standard Specification for
NICKEL AND NICKEL-BASE ALLOY CLAD STEEL PLATE[1]

This standard is issued under the fixed designation A 265; the number immediately following the designation indicates the year of original adoption or, in the case of revision, the year of last revision. A number in parentheses indicates the year of last reapproval. A superscript epsilon (ϵ) indicates an editorial change since the last revision or reapproval.

1. Scope

1.1 This specification[2] covers plate of a carbon steel or low-alloy steel base to which is integrally and continuously bonded on one or both sides a layer of nickel or nickel-base alloy. The material is generally intended for pressure vessel use.

1.2 The values stated in inch-pound units are to be regarded as the standard.

2. Description of Terms

2.1 This material is considered as single-clad or double-clad nickel or nickel-base alloy clad steel plate, depending on whether one or both sides are covered.

2.2 The term plate as used in this specification applies to material 3/16 in. (2.73 mm) and over in thickness, and over 10 in. (254 mm) in width.

3. Materials and Manufacture

3.1 *Process:*

3.1.1 The steel shall be made by the open-hearth, electric-furnace (with separate degassing and refining optional), or basic-oxygen processes, or by secondary processes whereby steel made from these primary processes is remelted using electroslag remelting or vacuum arc remelting processes.

3.1.2 The cladding metal may be metallurgically bonded to the base metal by any method that will produce a clad steel which will conform to the requirements of this specification.

3.2 *Heat Treatment*—Unless heat treatment is agreed upon by the purchaser and the manufacturer, composite plates will be furnished in the as-rolled condition.

4. Thickness of Cladding Metal

4.1 The minimum thickness and tolerances on the thickness of the cladding metal shall be agreed upon by the purchaser and the manufacturer.

5. Chemical Requirements

5.1 The composite plate may conform to any desired combination cladding metal and base metal as described in 5.2 and 5.3, and as agreed upon by the purchaser and the manufacturer.

5.2 *Cladding Metal*—The nickel or nickel-base alloy cladding metal specified shall conform to the requirements as to chemical composition prescribed for the respective metal in the following ASTM specifications:[3]

B 127, Nickel-Copper Alloy (UNS N04400) Plate, Sheet, and Strip,
B 162, Nickel Plate, Sheet, and Strip,
B 168, Nickel-Chromium-Iron Alloy (UNS N06600) Plate, Sheet, and Strip,
B 333, Nickel-Molybdenum Alloy (UNS N10001 and N10665),
B 409, Nickel-Iron-Chromium Alloy Plate, Sheet, and Strip,
B 424, Nickel-Iron-Chromium-Molybdenum-Copper Alloy (UNS N08825) Plate, Sheet, and Strip,
B 443, Nickel-Chromium-Molybdenum-Columbium Alloy (UNS N06625) Plate, Sheet, and Strip,
B 463, Chromium-Nickel-Iron-Molybdenum-Copper-Columbium Stabilized Alloy (UNS N08020) Plate, Sheet, and Strip and,
B 575, Low Carbon Nickel-Molybdenum-Chromium Alloy (UNS N10276 and N06455), and
B 582, Nickel-Chromium-Iron-Molybdenum-Copper Alloy (UNS-N06007).

5.3 *Base Metal*—The base metal may be carbon steel or low-alloy steel conforming to

---

[1] This specification is under the jurisdiction of ASTM Committee A-1 on Steel, Stainless Steel and Related Alloys, and is the direct responsibility of Subcommittee A01.17 on Flat Stainless Steel Products.
Current edition approved July 31 and Oct. 30, 1981. Published December 1981. Originally published as A 265 – 43 T. Last previous edition A 265 – 79.
[2] For ASME Boiler and Pressure Vessel Code applications see related Specification SA-265 in Section II of that Code.
[3] *Annual Book of ASTM Standards*, Vol 02.04.

the ASTM specifications for steels for pressure vessels. The base metal shall conform to the chemical requirements of the specification to which it is ordered.

6. Product Analysis

6.1 Product analyses may be required on the finished product only when the composite plate thickness is sufficient to permit obtaining drillings or millings without danger of contamination from the adjacent layer.

6.2 If product analysis is specified by the purchaser, it shall be made from drillings or millings taken from the final piece or a broken test specimen. In order to avoid contamination by the base plate metal, cladding samples shall be taken from the test coupon by removal and discard of all the base metal plate material, plus 40% of the cladding thickness from the bonded side, not to exceed 1/16 in. (1.588 mm). The material shall be chemically cleaned and millings shall be taken to represent the full cross section of the remainder.

7. Mechanical Requirements

7.1 *Tensile Properties*—The tensile properties as determined by a tension test on the composite plate shall be equal to or greater than the minimum requirements prescribed in the specifications for the base steel plate used.

7.1.1 For clad plates up to 1½ in. (38.1 mm), inclusive, maximum gage with base steel of 70 000 psi (485 MPa) tensile strength or less, the tensile properties as determined by a tension test on the composite plate shall be equal to the minimum and not more than 5000 psi (35 MPa) over the maximum prescribed in the specification for the base steel used.

7.1.2 For clad plates over 1½ in. gage, or for clad plate with base steel of more than 70 000 psi minimum tensile strength or 40 000 psi (275 MPa) minimum yield strength, the tensile properties shall be determined by pulling a test of the base steel only, in which case the properties shall meet the requirements of the specifications for the base steel.

7.2 *Ductility*—Two bend tests of the composite plate shall be made, one with the cladding metal in tension and the other with the cladding metal in compression, to determine the ductility of the materials. On double-clad plates, the bend tests shall be made so that one specimen represents the cladding metal in tension on one side while the other specimen represents the cladding metal in tension on the opposite side. Bend tests shall be made according to and shall conform to the requirements prescribed in the specifications for the base metal.

7.3 *Bond Strength*—When required by the purchaser, three bend tests shall be made with the cladding metal in compression to determine the quality of the bond. Bends shall be made in accordance with the specifications for the base metal. At least two out of the three tests shall show not more than 50% separation on both edges of the bent portion. Greater separation shall be cause for rejection.

7.4 *Shear Strength*—As an alternative to the bend test provided in 7.3 and when required by the purchaser, the minimum shear strength of the cladding and base metals shall be 20 000 psi (140 MPa). The shear test when specified, shall be made in the manner indicated in Fig. 1.

8. Test Specimens

8.1 The tension test specimens from plate shall conform to the requirements prescribed in the specifications for the base metal.

8.2 Bend test specimens shall be taken from the middle of the top of the plate as rolled, at right angles to its longitudinal axis.

8.3 When required by the purchaser, the shear test specimen shall be taken from a top or bottom corner of the plate as rolled, parallel to its longitudinal axis.

8.4 All tests shall be made on specimens in the same condition of heat treatment to which the composite plate is furnished.

8.5 For plates 1½ in. (38.1 mm) and under in thickness tension test specimens shall be the full thickness of the material except as specified in 7.1.1 and 7.1.2, and the bend test specimen shall be full thickness of the material and shall be machined to the form and dimensions shown in Fig. 2, or may be machined with both edges parallel.

8.6 For plates over 1½ in. in thickness tension tests shall be of the form shown in Fig. 3 and shall be of all base steel conforming to the requirements of the specification for the base steel.

8.7 For plates over 1½ in. in thickness the bend test specimens need not be greater in thickness than 1½ in. but shall not be less than

A 265

1½ in. Specimens shall be of the form and dimensions shown in Fig. 2. In reducing the thickness of the specimen, both the alloy cladding and the base steel shall be machined so as to maintain the same ratio of clad metal to base steel as is maintained in the plate, except that the thickness of the clad metal need not be reduced below ⅛ in. (3.17 mm).

8.8 The sides of the bend test specimen may have the corners rounded to a radius not over ¹⁄₁₆ in. (1.58 mm) for plates 1½ in. and under in thickness, and not over ⅛ in. for plates over 1½ in. in thickness.

8.9 Any test specimen used for the determination of the minimum thickness of cladding metal shall have distorted metal, due to burning or shearing, completely removed.

9. Number of Tests and Retests

9.1 One or more tension tests, as required by the specifications for the base metal, one face bend test (cladding metal in tension), one reverse bend test (cladding metal in compression), and, when specified, one shear test or three bond bend tests shall be made representing each plate as rolled. Each specimen shall be in the final condition of heat treatment required for the plate.

9.2 If any test specimen shows defective machining or develops flaws, it may be discarded and another specimen substituted.

9.3 If any part of the fracture takes place outside of the middle half of the gage length or in a punched or scribed mark within the reduced section, the elongation value obtained may not be representative of the material. If the elongation so measured meets the minimum requirements specified, no further testing is indicated, but if the elongation is less than the minimum requirements the test shall be discarded and a retest made.

10. Permissible Variations in Thickness and Weight

10.1 Composite plates shall conform to the thickness and weight requirements prescribed in Table 1.

11. Workmanship and Finish

11.1 The material shall be free of injurious defects, and shall have a workmanlike appearance.

11.2 Clad plates shall be furnished in the as-rolled condition, unless otherwise specified.

11.3 The clad surface may be supplied as-rolled, or as-rolled and sand blasted as specified.

12. Repair of Cladding by Welding

12.1 Unless otherwise specified, the material manufacturer may repair defects in cladding by welding provided the following requirements are met:

12.1.1 Prior approval is obtained from the purchaser if the repaired area exceeds 3% of the cladding surface.

12.1.2 The welding procedure and the welders or welding operators are qualified in accordance with Section IX of the ASME Code.

12.1.3 The defective area is removed and the area prepared for repair is examined by a magnetic particle method or a liquid penetrant method to ensure all defective area has been removed. Method of test and acceptance standard shall be as agreed upon by the purchaser and the manufacturer.

12.1.4 The weld will be deposited in a suitable manner so as to leave its surface conditon equivalent in corrosion resistance to the alloy cladding.

12.1.5 The repaired area is examined by a liquid penetrant method (see 12.1.3).

12.1.6 The location and extent of the weld repairs together with the repair procedure and examination results are transmitted as a part of the certification.

13. Inspection

13.1 The manufacturer shall afford the inspector, without charge, all reasonable facilities to satisfy him that the material is being furnished in accordance with this specification. All tests (except product analysis) and inspection shall be made at the place of manufacture prior to shipment, unless otherwise specified, and shall be so conducted as not to interfere unnecessarily with the operation of the works.

14. Rejection and Rehearing

14.1 Unless otherwise specified, any rejection based on tests made in accordance with Section 6, shall be reported to the manufacturer within 5 working days from the receipt of samples by the purchaser.

14.2 Material which shows injurious defects subsequent to its acceptance at the manufacturer's work will be rejected, and the manufacturer shall be notified.

14.3 Samples tested in accordance with Section 6, that represent rejected material shall be preserved for 3 weeks from the date of the test report. In case of dissatisfaction with the results of the tests, the manufacturer may make claim for a rehearing within that time.

15. Report

15.1 The chemical analysis of the base metal shall be certified to the purchaser by the manufacturer.

15.2 The results of the tests in Section 7, shall be reported to the purchaser or his representative.

16. Marking

16.1 Except as specified in 16.2, the name or brand of the manufacturer, the manufacturer's test identification number, the class of the base steel, the designation of the cladding metal, and the specified minimum tensile strength shall be legibly stamped on each finished single-clad plate in two places on the base steel side not less than 12 in. (304.8 mm) from the edges. The manufacturer's test identification number shall be legibly stamped on each test specimen.

16.2 For double-clad material or for material under ¼ in. (6.35 mm) in thickness, the marking specified in 16.1 shall be legibly stenciled instead of stamped.

TABLE 1 Permissible Variations in Thickness and Overweights of Plates Thickness[a]

NOTE—All plates shall be ordered to thickness and not to weight per square foot. No plates shall vary more than 0.01 in. (0.254 mm) under the thickness ordered, and the overweight of each lot[b] shall not exceed the amount given in the table below.[c]

Specified Thickness, in. (mm)	Permissible Excess in Average Weight per Square Foot of Plates for Widths Given in Inches (mm), Expressed in Percentage of Nominal Weight[d]								
	Under 48 (1219.2)	48 (1219.2) to 60 (1524.0) excl	60 (1524.0) to 72 (1828.8) excl	72 (1828.8) to 84 (2133.6) excl	84 (2133.6) to 96 (2438.4) excl	96 (2438.4) to 108 (2743.2) excl	108 (2743.2) to 120 (3048.0) excl	120 (3048.0) to 132 (3352.8) excl	132 (3352.8) to 144 (3657.6) excl	144 (3657.6) to 152 (3860.8) incl
³⁄₁₆ to ¼ (4.76 to 6.35), excl	5.25	6.00	6.75	7.50	9.00	11.50	...	...	...	...
¼ to ⁵⁄₁₆ (6.35 to 7.94), excl	4.50	5.25	6.00	6.75	7.50	10.50	12.00	13.00	14.25	...
⁵⁄₁₆ to ⅜ (7.94 to 9.52), excl	3.75	4.50	5.25	6.00	6.75	7.50	9.00	10.00	12.75	16.00
⅜ to ⁷⁄₁₆ (9.52 to 11.11), excl	3.50	3.75	4.50	5.25	6.00	6.75	7.50	8.50	11.25	15.00
⁷⁄₁₆ to ½ (11.11 to 12.7), excl	3.00	3.50	3.75	4.50	5.25	6.00	6.75	7.50	9.75	12.75
½ to ⅝ (12.7 to 15.88), excl	2.75	3.00	3.50	3.75	4.50	5.25	6.00	6.75	8.25	10.50
⅝ to ¾ (15.88 to 19.05), excl	2.25	2.75	3.00	3.50	3.75	4.50	5.00	5.50	6.25	6.75
¾ to 1 (19.05 to 25.4), excl	2.00	2.25	2.75	3.00	3.50	3.75	4.25	4.75	5.25	6.00
1 or Over (25.4 or Over)	2.00	2.00	2.25	2.75	3.50	3.50	3.75	4.25	4.75	5.25

[a] Spot grinding is permitted to remove surface imperfections not to exceed 0.01 in. (0.254 mm) under the specified thickness.

[b] The term "lot" means all of the plates of each group width and each group thickness.

[c] The weight of individual plates shall not exceed the nominal weight by more than one and one-third times the amount prescribed in the above table.

[d] Nominal weights for the composite plates shall be calculated using the following weights for the component materials:

	Density		Weight per Square Foot for Material 1 in. (25.4 mm) in Thickness, lb
	lb/in.³	g/cm³	
Steel	0.283	7.83	40.80
Nickel	0.321	8.88	46.22
Nickel-copper alloy	0.319	8.83	45.94
Nickel-chromium-iron alloy	0.307	8.49	44.21

This standard is subject to revision at any time by the responsible technical committee and must be reviewed every five years and if not revised, either reapproved or withdrawn. Your comments are invited either for revision of this standard or for additional standards and should be addressed to ASTM Headquarters. Your comments will receive careful consideration at a meeting of the responsible technical committee, which you may attend. If you feel that your comments have not received a fair hearing you should make your views known to the ASTM Committee on Standards, 1916 Race St., Philadelphia, Pa. 19103.

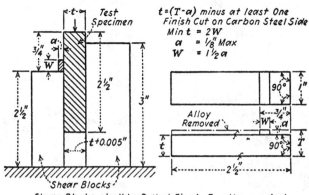

Shear Blocks shall be Bolted Firmly Together against Filler Piece which Provide Space 0.005" Wider than t of Specimen

Metric Equivalents

in.	mm	in	mm
0.005	0.127	1	25.4
1/8	3.17	2½	64.5
¾	19.1	3	76.2

FIG. 1 Test Specimen and Method of Making Shear Test of Clad Plate

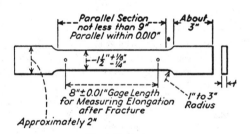

Metric Equivalents

in.	mm	in.	mm
0.01	0.254	1½	38.1
1/8	3.17	2	50.8
¼	6.35	3	76.2
1	25.4	8	203.2

NOTE 1—When necessary, it is permissible to use a narrower specimen, but in such a case the reduced portion shall be not less than 1 in. in width.

NOTE 2—Punch marks for measuring elongation after fracture shall be made on the flat or on the edge of the specimen and within the parallel section; either a set of nine punch marks 1 in. apart, or one or more sets of 8-in. punch marks may be used.

NOTE 3—The dimension t is the thickness of the test specimen as provided for in the applicable material specifications.

FIG. 2 Standard Rectangular Test Specimens with 8-in. Gage Length

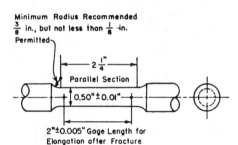

Metric Equivalents

in.	mm	in.	mm
0.003	0.076	3/8	9.53
0.005	0.127	0.50	12.7
0.01	0.254	2	50.8
1/8	3.18	2¼	64.5

NOTE—The gage length and fillets shall be as shown, but the ends may be of any shape to fit the holders of the testing machine in such a way that the load shall be axial. The reduced section shall have a gradual taper from the ends toward the center, with the ends 0.003 to 0.005 in. larger in diameter than the center.

FIG. 3 Standard Round Tension Tests Specimen with 2-in. Gage Length

The American Society for Testing and Materials takes no position respecting the validity of any patent rights asserted in connection with any item mentioned in this standard. Users of this standard are expressly advised that determination of the validity of any such patent rights, and the risk of infringement of such rights, are entirely their own responsibility.

Designation: A 313 – 81

Standard Specification for
CHROMIUM-NICKEL STAINLESS AND HEAT-RESISTING STEEL SPRING WIRE[1]

This standard is issued under the fixed designation A 313; the number immediately following the designation indicates the year of original adoption or, in the case of revision, the year of last revision. A number in parentheses indicates the year of last reapproval. A superscript epsilon (ϵ) indicates an editorial change since the last revision or reapproval.

This specification has been approved for use by agencies of the Department of Defense and for listing in the DoD Index of Specifications and Standards.

1. Scope

1.1 This specification covers stainless and heat-resisting steel round spring wire intended especially for the manufacture of springs.

1.2 The values stated in inch-pound units are to be regarded as the standard.

2. Applicable Documents

2.1 *ASTM Standards*:

A 555 Specification for General Requirements for Stainless and Heat-Resisting Steel Wire[2]

A 751 Methods, Practices, and Definitions for Chemical Analysis of Steel Products[3]

3. General Requirements for Delivery

3.1 In addition to the requirements of this specification, all requirements of the current edition of Specification A 555 shall apply.

4. Ordering Information

4.1 Orders for material under this specification shall include the following information:

4.1.1 Quantity (weight).

4.1.2 Name of material (stainless steel spring wire),

4.1.3 Finish (see 8.1),

4.1.4 Dimension (diameter),

4.1.5 Type designation (Table 1),

4.1.6 ASTM designation and date of issue, and

4.1.7 Special requirements.

NOTE—A typical ordering description is as follows:

2000 lb Stainless Steel Spring Wire, hard-drawn, bright finish, 0.032-in. diameter, in 100-lb 16-in. coils, Type 302 to ASTM A 313, dated _____.

5. Chemical Requirements

5.1 The steel shall conform to the requirements as to chemical composition prescribed in Table 1.

5.2 Methods and practices relating to chemical analysis required by this specification shall be in accordance with Methods A 751.

6. Condition

6.1 Types 302 Class 1, 304, 305, 316, 321, 347, and Grade XM-28 shall be cold drawn to produce the required mechanical properties.

6.2 Type 631, Type 302 Class 2, and Grade XM-16 shall be furnished in the cold-drawn condition ready for fabrication. Following fabrication Type 631 and Grade XM-16 shall be aged or precipitation hardened to produce their maximum strength properties. The tensile strengths to be obtained following the prescribed heat treatment are shown in Tables 4 and 5 for hardened wire. Type 302 Class 2 shall be stress relieved following fabrication and meet the requirements shown in Table 7. The nominal as-drawn tensile strengths are provided as a guide for this spring manufacturer.

7. Mechanical Requirements

7.1 *Tensile Properties*:

---

[1] This specification is under the jurisdiction of ASTM Committee A-1 on Steel, Stainless Steel and Related Alloys, and is the direct responsibility of Subcommittee A01.08 on Wrought Stainless Steel Products.

Current edition approved July 31, 1981. Published November 1981. Originally published as A 313 – 47 T. Last previous edition A 313 – 80a.

[2] *Annual Book of ASTM Standards*, Vol 01.03.

[3] *Annual Book of ASTM Standards*, Vol 01.01, 01.02, 01.03, 01.04, and 01.05.

7.1.1 Types 302 Class 1 and 304 shall conform to the requirements shown in Table 2.

7.1.2 Types 305, 316, 321, and 347 shall conform to the requirements shown in Table 3.

7.1.3 Type 631 shall conform to the requirements shown in Table 4 when heat treated 900°F (482°C) for 1 h and air cooled.

7.1.4 Grade XM-16 shall conform to the requirements shown in Table 5 when heat treated 850°F (454°C) for ½ h and air cooled.

7.1.5 Grade XM-28 shall conform to the requirements shown in Table 6.

7.1.6 Type 302 Class 2 shall conform to the requirements shown in Table 7.

7.2 *Coiling Tests*:

7.2.1 Wire 0.162 in. (4.11 mm) and smaller in diameter shall wind on itself as an arbor without breaking.

7.2.2 Wire larger than 0.162 in. (4.11 mm) in diameter shall wind without breaking on a mandrel having a diameter twice the diameter of the wire.

7.3 *Uniformity*:

7.3.1 In the as-cold drawn condition, a specimen coil shall be wound on an arbor of the size specified in Table 7 to form a tightly wound coil.

7.3.2 After winding, the specimen coil shall be stretched to a permanent set four times its as-wound length. After this treatment, the specimen coil shall show uniform pitch with no splits or fractures.

7.4 *Cast*—A loop or ring shall be cut from the bundle and allowed to fall on the floor. The wire shall lie flat and not spring up or show a wavy condition.

7.5 *Bend Test*—When specified in the purchase order, Types 302, 304, 305, 316, 321, and 347 shall be tested by the bend test. A piece not more than 10 in. (254 mm) long shall be selected from each test sample. These specimens shall be tested in a bending machine conforming substantially to Fig. 1. Bends shall be made at as nearly a uniform rate as possible, not exceeding 50 bends per minute, and in no case shall the speed be so great as to cause undue heating of the wire. The test specimen shall be bent back and forth through a total angle of 180 deg until failure occurs. Each 90-deg movement in either direction shall be counted as one bend. The wire shall withstand the minimum number of bends specified in Tables 2 and 3.

8. Finish

8.1 Stainless steel spring wire is supplied with different types of finish such as bright, copper, lead, oxide, and other.

SUPPLEMENTARY REQUIREMENTS

Unless otherwise specified in the purchase order, the following supplementary requirements shall apply when this specification is used in government procurement of Type 631 spring wire up to and including 0.162 in. (4.11 mm) in diameter.

S1. Wrapping Test

S.1.1 A wire specimen shall be wrapped five complete turns around a mandrel equal to the diameter of the wire without any surface breaks or cracks occurring in the wire. One specimen shall be taken from every ten coils in the lot.

S2. Surface Examination

S2.1 A wire specimen shall be etched electrolytically in a 75 % phosphoric acid solution with a current density of 1 A/in.^{2} for a sufficient time to remove up to 1 % of the diameter. After etching the surface of the wire specimen it shall be examined under a 10 power microscope for splits, seams, pits, die marks, scratches, or other imperfections tending to impair the fatigue resistance of springs. Appropriate higher magnification should be used for sizes below 0.125 in. (3.17 mm). Lubricating coatings, which are insoluble in acid etch solution, shall be removed before etching.

A 313

TABLE 1 Chemical Requirements

UNS Designation[B]	Type	Composition,[A] %							
		Carbon	Manganese	Phosphorus	Sulfur	Silicon	Chromium	Nickel	Other Elements
S 30200	302	0.15	2.00	0.045	0.030	1.00	17.00–19.00	8.00–10.00	N 0.10
S 30400	304	0.08	2.00	0.045	0.030	1.00	18.00–20.00	8.00–10.50	N 0.10
S 30500	305	0.12	2.00	0.045	0.030	1.00	17.00–19.00	10.50–13.00	
S 31600	316	0.08	2.00	0.045	0.030	1.00	16.00–18.00	10.00–14.00	Mo 2.00–3.00; N 0.10
S 32100	321	0.08	2.00	0.045	0.030	1.00	17.00–19.00	9.00–12.00	Ti 5 × C min
S 34700	347	0.08	2.00	0.045	0.030	1.00	17.00–19.00	9.00–13.00	Cb + Ta 10 × C min
S 17700	631	0.09	1.00	0.040	0.030	1.00	16.00–18.00	6.50– 7.75	Al 0.75–1.50
S 45500	XM-16	0.05	0.50	0.040	0.030	0.50	11.00–12.50	7.50– 9.50	Mo 0.50 max; Ti 0.80–1.40; Cu 1.50–2.50; Cb + Ta 0.10–0.50
S 24100	XM-28	0.15	11.00–14.00	0.060	0.030	1.00	16.50–19.00	0.50–2.50	N 0.20–0.45

[A] Maximum unless range is shown.
[B] New designations established in accordance with ASTM E 527 and SAE J1086, Recommended Practice for Numbering Metals and Alloys (UNS).

ASTM A 313

TABLE 2 Tensile Strength Requirements for Types 302 Class 1 and 304[A]

Diameter, in. (mm)	Bend Test Minimum Number of Bends	psi(MPa) min	psi(MPa) max
Up to 0.009 (0.23), incl		325 000 (2240)	355 000 (2450)
Over 0.009 (0.23) to 0.010 (0.25), incl		320 000 (2205)	350 000 (2415)
Over 0.010 (0.25) to 0.011 (0.28), incl		318 000 (2190)	348 000 (2400)
Over 0.011 (0.28) to 0.012 (0.30), incl		316 000 (2180)	346 000 (2385)
Over 0.012 (0.30) to 0.013 (0.33), incl		314 000 (2165)	344 000 (2370)
Over 0.013 (0.33) to 0.014 (0.36), incl		312 000 (2150)	342 000 (2360)
Over 0.014 (0.36) to 0.015 (0.38), incl		310 000 (2135)	340 000 (2345)
Over 0.015 (0.38) to 0.016 (0.41), incl		308 000 (2125)	338 000 (2330)
Over 0.016 (0.41) to 0.017 (0.43), incl		306 000 (2110)	336 000 (2315)
Over 0.017 (0.43) to 0.018 (0.46), incl		304 000 (2095)	334 000 (2300)
Over 0.018 (0.46) to 0.020 (0.51), incl		300 000 (2070)	330 000 (2275)
Over 0.020 (0.51) to 0.022 (0.56), incl		296 000 (2040)	326 000 (2250)
Over 0.022 (0.56) to 0.024 (0.61), incl		292 000 (2015)	322 000 (2220)
Over 0.024 (0.61) to 0.026 (0.66), incl	8	291 000 (2005)	320 000 (2205)
Over 0.026 (0.66) to 0.028 (0.71), incl	8	289 000 (1995)	318 000 (2190)
Over 0.028 (0.71) to 0.031 (0.79), incl	8	285 000 (1965)	315 000 (2170)
Over 0.031 (0.79) to 0.034 (0.86), incl	8	282 000 (1945)	310 000 (2135)
Over 0.034 (0.86) to 0.037 (0.94), incl	8	280 000 (1930)	308 000 (2125)
Over 0.037 (0.94) to 0.041 (1.04), incl	8	275 000 (1895)	304 000 (2095)
Over 0.041 (1.04) to 0.045 (1.14), incl	8	272 000 (1875)	300 000 (2070)
Over 0.045 (1.14) to 0.050 (1.27), incl	8	267 000 (1840)	295 000 (2035)
Over 0.050 (1.27) to 0.054 (1.37), incl	8	265 000 (1825)	293 000 (2020)
Over 0.054 (1.37) to 0.058 (1.47), incl	7	261 000 (1800)	289 000 (1990)
Over 0.058 (1.47) to 0.063 (1.60), incl	7	258 000 (1780)	285 000 (1965)
Over 0.063 (1.60) to 0.070 (1.78), incl	7	252 000 (1735)	281 000 (1935)
Over 0.070 (1.78) to 0.075 (1.90), incl	7	250 000 (1725)	278 000 (1915)
Over 0.075 (1.90) to 0.080 (2.03), incl	7	246 000 (1695)	275 000 (1895)
Over 0.080 (2.03) to 0.087 (2.21), incl	7	242 000 (1670)	271 000 (1870)
Over 0.087 (2.21) to 0.095 (2.41), incl	7	238 000 (1640)	268 000 (1850)
Over 0.095 (2.41) to 0.105 (2.67), incl	5	232 000 (1600)	262 000 (1805)
Over 0.105 (2.67) to 0.115 (2.92), incl	5	227 000 (1565)	257 000 (1770)
Over 0.115 (2.92) to 0.125 (3.17), incl	5	222 000 (1530)	253 000 (1745)
Over 0.125 (3.17) to 0.135 (3.43), incl	3	217 000 (1495)	248 000 (1710)
Over 0.135 (3.43) to 0.148 (3.76), incl	3	210 000 (1450)	241 000 (1660)
Over 0.148 (3.76) to 0.162 (4.11), incl	3	205 000 (1415)	235 000 (1620)
Over 0.162 (4.11) to 0.177 (4.50), incl	3	198 000 (1365)	228 000 (1570)
Over 0.177 (4.50) to 0.192 (4.88), incl	1	194 000 (1335)	225 000 (1550)
Over 0.192 (4.88) to 0.207 (5.26), incl	1	188 000 (1295)	220 000 (1515)
Over 0.207 (5.26) to 0.225 (5.72), incl	1	182 000 (1255)	214 000 (1475)
Over 0.225 (5.72) to 0.250 (6.35), incl	1	175 000 (1205)	205 000 (1415)
Over 0.250 (6.35) to 0.278 (7.06), incl	1	168 000 (1160)	198 000 (1365)
Over 0.278 (7.06) to 0.306 (7.77), incl	1	161 000 (1110)	192 000 (1325)
Over 0.306 (7.77) to 0.331 (8.41), incl	1	155 000 (1070)	186 000 (1280)
Over 0.331 (8.41) to 0.362 (9.19), incl	1	150 000 (1035)	180 000 (1240)
Over 0.362 (9.19) to 0.394 (10.00), incl	1	145 000 (1000)	175 000 (1205)
Over 0.394 (10.00) to 0.438 (11.12), incl	1	140 000 (965)	170 000 (1170)
Over 0.436 (11.12) to 0.500 (12.70), incl	1	135 000 (930)	165 000 (1150)
Over 0.500 (12.70)	1	130 000 (895)	160 000 (1105)

[A] When wire is specified in straightened and cut lengths, the minimum tensile strength shall be 90 % of the values listed in the table.

A 313

TABLE 3   Tensile Strength Requirements for Types 305, 316, 321, and 347[A]

Diameter, in. (mm)	Bend Test Minimum Number of Bends	psi(MPa) min	psi(MPa) max
Up to 0.009 (0.23)		245 000 (1690)	275 000 (1895)
Over 0.009 (0.23) to 0.010 (0.25), incl		245 000 (1690)	275 000 (1895)
Over 0.010 (0.25) to 0.011 (0.28), incl		240 000 (1655)	270 000 (1860)
Over 0.011 (0.28) to 0.012 (0.30), incl		240 000 (1655)	270 000 (1860)
Over 0.012 (0.30) to 0.013 (0.33), incl		240 000 (1655)	270 000 (1860)
Over 0.013 (0.33) to 0.014 (0.36), incl		240 000 (1655)	270 000 (1860)
Over 0.014 (0.36) to 0.015 (0.38), incl		240 000 (1655)	270 000 (1860)
Over 0.015 (0.38) to 0.016 (0.41), incl		235 000 (1620)	265 000 (1825)
Over 0.016 (0.41) to 0.017 (0.43), incl		235 000 (1620)	265 000 (1825)
Over 0.017 (0.43) to 0.018 (0.46), incl		235 000 (1620)	265 000 (1825)
Over 0.018 (0.46) to 0.019 (0.48), incl		235 000 (1620)	265 000 (1825)
Over 0.019 (0.48) to 0.020 (0.51), incl		235 000 (1620)	265 000 (1825)
Over 0.020 (0.51) to 0.022 (0.56), incl		235 000 (1620)	265 000 (1825)
Over 0.022 (0.56) to 0.024 (0.61), incl		235 000 (1620)	265 000 (1825)
Over 0.024 (0.61) to 0.026 (0.66), incl	8	235 000 (1620)	265 000 (1825)
Over 0.026 (0.66) to 0.028 (0.71), incl	8	235 000 (1620)	265 000 (1825)
Over 0.028 (0.71) to 0.032 (0.81), incl	8	235 000 (1620)	265 000 (1825)
Over 0.032 (0.81) to 0.036 (0.91), incl	8	235 000 (1620)	265 000 (1825)
Over 0.036 (0.91) to 0.041 (1.04), incl	8	235 000 (1620)	265 000 (1825)
Over 0.041 (1.04) to 0.047 (1.19), incl	8	230 000 (1585)	260 000 (1790)
Over 0.047 (1.19) to 0.054 (1.37), incl	8	225 000 (1550)	255 000 (1760)
Over 0.054 (1.37) to 0.062 (1.57), incl	7	220 000 (1515)	250 000 (1725)
Over 0.062 (1.57) to 0.072 (1.83), incl	7	215 000 (1480)	245 000 (1690)
Over 0.072 (1.82) to 0.080 (2.03), incl	7	210 000 (1450)	240 000 (1655)
Over 0.080 (2.03) to 0.092 (2.34), incl	7	205 000 (1415)	235 000 (1620)
Over 0.092 (2.34) to 0.105 (2.67), incl	5	200 000 (1380)	230 000 (1585)
Over 0.105 (2.67) to 0.120 (3.05), incl	5	195 000 (1345)	225 000 (1550)
Over 0.120 (3.05) to 0.148 (3.76), incl	3	185 000 (1275)	215 000 (1480)
Over 0.148 (3.76) to 0.166 (4.22), incl	3	180 000 (1240)	210 000 (1450)
Over 0.166 (4.22) to 0.177 (4.50), incl	3	170 000 (1170)	200 000 (1380)
Over 0.177 (4.50) to 0.207 (5.26), incl	1	160 000 (1105)	190 000 (1310)
Over 0.207 (5.26) to 0.225 (5.72), incl	1	155 000 (1070)	185 000 (1275)
Over 0.225 (5.72) to 0.250 (6.35), incl	1	150 000 (1035)	180 000 (1240)
Over 0.250 (6.35) to 0.312 (7.92), incl	1	140 000 (965)	170 000 (1170)
Over 0.312 (7.92) to 0.375 (9.53), incl	1	135 000 (930)	165 000 (1140)
Over 0.375 (9.53) to 0.500 (12.70), incl		130 000 (895)	160 000 (1105)
Over 0.500 (12.70)		125 000 (850)	155 000 (1970)

[A] When wire is specified in straightened and cut lengths, the minimum tensile strength shall be 90 % of the values listed in the table.

A 313

TABLE 4 Tensile Strength Requirements for Type 631[A]

Diameter, in. (mm)	Cold Drawn Condition C, psi(MPa) Nominal	Condition CH-900[B], psi (MPa)	
		min	max
0.010 (0.25) to 0.015 (0.38), incl	295 000 (2035)	335 000 (2310)	365 000 (2515)
Over 0.015 (0.38) to 0.020 (0.51), incl	290 000 (2000)	330 000 (2275)	360 000 (2480)
Over 0.020 (0.51) to 0.029 (0.74), incl	285 000 (1965)	325 000 (2240)	355 000 (2450)
Over 0.029 (0.74) to 0.041 (1.04), incl	275 000 (1895)	320 000 (2205)	350 000 (2415)
Over 0.041 (1.04) to 0.051 (1.30), incl	270 000 (1860)	310 000 (2135)	340 000 (2345)
Over 0.051 (1.30) to 0.061 (1.55), incl	265 000 (1825)	305 000 (2100)	335 000 (2310)
Over 0.061 (1.55) to 0.071 (1.80), incl	257 000 (1770)	297 000 (2050)	327 000 (2255)
Over 0.071 (1.80) to 0.086 (2.15), incl	255 000 (1760)	292 000 (2015)	322 000 (2220)
Over 0.086 (2.15) to 0.090 (2.18), incl	245 000 (1690)	282 000 (1945)	312 000 (2150)
Over 0.090 (2.18) to 0.100 (2.54), incl	242 000 (1670)	279 000 (1925)	309 000 (2130)
Over 0.100 (2.54) to 0.106 (2.69), incl	238 000 (1640)	274 000 (1890)	304 000 (2095)
Over 0.106 (2.69) to 0.130 (3.30), incl	236 000 (1625)	272 000 (1875)	302 000 (2080)
Over 0.130 (3.30) to 0.138 (3.50), incl	230 000 (1585)	260 000 (1795)	290 000 (2000)
Over 0.138 (3.50) to 0.146 (3.71), incl	228 000 (1570)	258 000 (1780)	288 000 (1985)
Over 0.146 (3.71) to 0.162 (4.11), incl	226 000 (1560)	256 000 (1765)	286 000 (1970)
Over 0.162 (4.11) to 0.180 (4.57), incl	224 000 (1545)	254 000 (1750)	284 000 (1960)
Over 0.180 (4.57) to 0.207 (5.26), incl	222 000 (1530)	252 000 (1740)	282 000 (1945)
Over 0.207 (5.26) to 0.225 (5.72), incl	218 000 (1505)	248 000 (1710)	278 000 (1915)
Over 0.225 (5.72) to 0.306 (7.77), incl	213 000 (1470)	242 000 (1670)	272 000 (1875)
Over 0.306 (7.77) to 0.440 (11.2), incl	207 000 (1425)	235 000 (1620)	265 000 (1825)
Over 0.440 (11.2) to 0.625 (15.88), incl	203 000 (1400)	230 000 (1585)	260 000 (1795)

[A] When wire is specified in straightened and cut lengths, the minimum tensile strength shall be 90 % of the values listed in the table.
[B] Aged at 900°F (482°C) for 1 h and air cooled.

TABLE 5 Tensile Strength Requirements for Grade XM-16[A]

Diameter, in. (mm)	Cold Drawn, psi(MPa) Nominal	Age Hardened[B], psi(MPa)	
		min	max
0.010 (0.25 to 0.040 (1.02), incl	245 000 (1690)	320 000 (2205)	350 000 (2415)
Over 0.040 (1.02) to 0.050 (1.27), incl	235 000 (1620)	310 000 (2135)	340 000 (2345)
Over 0.050 (1.27) to 0.060 (1.52), incl	225 000 (1550)	305 000 (2100)	335 000 (2310)
Over 0.060 (1.52) to 0.075 (1.90), incl	220 000 (1515)	295 000 (2035)	325 000 (2240)
Over 0.075 (1.90) to 0.085 (2.16), incl	215 000 (1480)	290 000 (2000)	320 000 (2205)
Over 0.085 (2.16) to 0.095 (2.41), incl	210 000 (1450)	285 000 (1965)	315 000 (2170)
Over 0.095 (2.41) to 0.110 (2.79), incl	200 000 (1380)	278 000 (1915)	308 000 (2125)
Over 0.110 (2.79) to 0.125 (3.17), incl	195 000 (1345)	272 000 (1875)	302 000 (2080)
Over 0.125 (3.17) to 0.150 (3.81), incl	190 000 (1310)	265 000 (1825)	295 000 (2035)
Over 0.150 (3.81) to 0.500 (12.7), incl	180 000 (1240)	260 000 (1795)	290 000 (2000)

[A] When wire is straightened and cut lengths, the minimum tensile strength shall be 90 % of the values listed in the table.
[B] Aged at 850°F (454°C) for ½ h and air cooled.

A 313

TABLE 6 Tensile Strength Requirements for Grade XM-28[A]

Diameter, in. (mm)	psi (MPa) min	psi (MPa) max
Up to 0.009 (0.23), incl	325 000 (2240)	355 000 (2450)
Over 0.009 (0.23) to 0.010 (0.25), incl	320 000 (2205)	350 000 (2415)
Over 0.010 (0.25) to 0.011 (0.28), incl	318 000 (2195)	348 000 (2400)
Over 0.011 (0.28) to 0.012 (0.30), incl	316 000 (2180)	346 000 (2385)
Over 0.012 (0.30) to 0.013 (0.33), incl	314 000 (2165)	344 000 (2370)
Over 0.013 (0.33) to 0.014 (0.36), incl	312 000 (2150)	342 000 (2360)
Over 0.014 (0.36) to 0.015 (0.38), incl	310 000 (2135)	340 000 (2345)
Over 0.015 (0.38) to 0.016 (0.41), incl	308 000 (2125)	338 000 (2330)
Over 0.016 (0.41) to 0.017 (0.43), incl	306 000 (2110)	336 000 (2315)
Over 0.017 (0.43) to 0.018 (0.46), incl	304 000 (2095)	334 000 (2305)
Over 0.018 (0.46) to 0.020 (0.51), incl	300 000 (2070)	330 000 (2275)
Over 0.020 (0.51) to 0.022 (0.56), incl	296 000 (2040)	326 000 (2250)
Over 0.022 (0.56) to 0.024 (0.61), incl	292 000 (2015)	322 000 (2220)
Over 0.024 (0.61) to 0.026 (0.66), incl	289 000 (1995)	319 000 (2200)
Over 0.026 (0.66) to 0.028 (0.71), incl	286 000 (1970)	316 000 (2180)
Over 0.028 (0.71) to 0.032 (0.81), incl	282 000 (1945)	312 000 (2150)
Over 0.032 (0.81) to 0.037 (0.94), incl	277 000 (1910)	307 000 (2120)
Over 0.037 (0.94) to 0.041 (1.04), incl	273 000 (1880)	303 000 (2090)
Over 0.041 (1.04) to 0.047 (1.19), incl	270 000 (1860)	300 000 (2070)
Over 0.047 (1.19) to 0.054 (1.37), incl	265 000 (1825)	295 000 (2035)
Over 0.054 (1.37) to 0.087 (2.21), incl	260 000 (1795)	290 000 (2000)
Over 0.087 (2.21) to 0.120 (3.05), incl	255 000 (1760)	285 000 (1965)
Over 0.120 (3.05) to 0.166 (4.22), incl	250 000 (1725)	280 000 (1930)
Over 0.166 (4.22) to 0.192 (4.88), incl	240 000 (1655)	270 000 (1860)
Over 0.192 (4.88) to 0.225 (5.72), incl	230 000 (1585)	260 000 (1795)
Over 0.225 (5.72) to 0.278 (7.06), incl	215 000 (1480)	245 000 (1690)
Over 0.278 (7.06) to 0.331 (8.41), incl	200 000 (1380)	230 000 (1585)
Over 0.331 (8.41) to 0.394 (10.00), incl	185 000 (1275)	215 000 (1480)
Over 0.394 (10.00) to 0.500 (12.70), incl	160 000 (1105)	190 000 (1310)

[A] When wire is specified in straightened and cut lengths, the minimum tensile strength shall be 85 % of the values listed in the table.

TABLE 7 Tensile Strength Requirement for Type 302 Class 2

Diameter, in. (mm)	psi (MPa) Cold Drawn Nominal	psi (MPa) Stress Relieved[A] min	psi (MPa) Stress Relieved[A] max
0.050 (1.27) to 0.160 (4.06), incl	290 000 (1998)	290 000 (2000)	340 000 (2343)

[A] Stress relieved at 800 to 850°F (427 to 454°C) for ½ h and air cooled.

TABLE 8 Arbor Diameter Size for Uniformity Test

Wire Diameter, in. (mm)	Arbor Diameter, in. (mm)
0.034 (0.86) and under	0.102 (2.59)
Over 0.034 (0.86) to 0.045 (1.14), incl	0.145 (3.68)
Over 0.045 (1.14) to 0.055 (1.40), incl	0.212 (5.38)
Over 0.055 (1.40) to 0.125 (3.17), incl	0.250 (6.35)
Over 0.125 (3.17) to 0.180 (4.57), incl	0.350 (8.88)

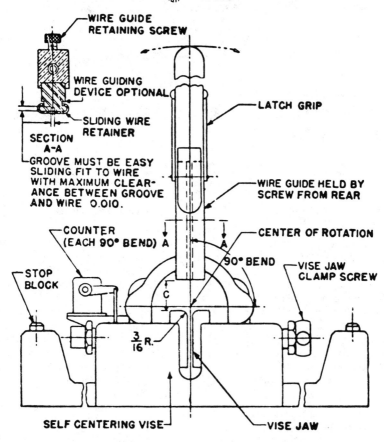

Diameter of Wire, in. (mm)	Clearance C ± 0.005, in. (mm)
Over 0.026 to 0.105 (0.66 to 2.67), incl	0.688 (17.48)
Over 0.105 to 0.162 (2.67 to 4.11), incl	0.813 (20.65)
Over 0.162 to 0.180 (4.11 to 4.57), incl	0.938 (23.83)

FIG. 1 Schematic Arrangement of Bending Machine

The American Society for Testing and Materials takes no position respecting the validity of any patent rights asserted in connection with any item mentioned in this standard. Users of this standard are expressly advised that determination of the validity of any such patent rights, and the risk of infringement of such rights, are entirely their own responsibility.

This standard is subject to revision at any time by the responsible technical committee and must be reviewed every five years and if not revised, either reapproved or withdrawn. Your comments are invited either for revision of this standard or for additional standards and should be addressed to ASTM Headquarters. Your comments will receive careful consideration at a meeting of the responsible technical committee, which you may attend. If you feel that your comments have not received a fair hearing you should make your views known to the ASTM Committee on Standards, 1916 Race St., Philadelphia, Pa. 19103.

Designation: A 345 – 75 (Reapproved 1979)

An American National Standard

Standard Specification for
FLAT-ROLLED ELECTRICAL STEELS FOR MAGNETIC APPLICATIONS[1]

This standard is issued under the fixed designation A 345; the number immediately following the designation indicates the year of original adoption or, in the case of revision, the year of last revision. A number in parentheses indicates the year of last reapproval. A superscript epsilon (ε) indicates an editorial change since the last revision or reapproval.

1. Scope

1.1 This specification covers general procedures for specifying requirements in the procurement and delivery of flat-rolled electrical steels for magnetic applications. When an applicable individual specification does not exist, this specification enables the user to order a suitable material to be supplied under controlled conditions with respect to magnetic quality, sampling, testing, packaging, etc., by specifying certain requirements on the purchase order and citing this specification.

1.2 Individual specifications that are in conformity with this specification are:

A 665 Specification for Flat-Rolled, Grain Oriented, Silicon-Iron, Electrical Steel, Fully Processed Types 27G053, 30G058, and 35G066[2]

A 677 Specification for Flat-Rolled, Nonoriented Electrical Steel, Fully Processed Types[2]

A 683 Specification for Flat-Rolled, Nonoriented Electrical Steel, Semiprocessed Grades[2]

A 725 Specification for Flat-Rolled, Grain-Oriented, Silicon Iron Electrical Steel, Fully Processed Types 27H076, 30H083, 35H094[2]

A 726 Specification for Cold-Rolled Carbon Steel Sheet, Magnetic Lamination Quality Types 1, 2, and 2S[2]

1.3 When an individual specification of 1.2 covers the specific magnetic characteristics required, it should be cited on the purchase order instead of this specification.

1.4 The values stated in inch-pound units are to be regarded as standard.

2. Applicable Documents

2.1 *ASTM Standards*:

A 34 Methods of Testing Magnetic Materials[2]

A 340 Definitions of Terms, Symbols, and Conversion Factors Relating to Magnetic Testing[2]

A 664 Recommended Practice for the Identification of Standard Electrical Steel Grades in ASTM Specifications[2]

A 700 Practices for Packaging, Marking, and Loading Methods for Steel Products for Domestic Shipment[3]

2.2 *Other Publications*:

American Iron and Steel Institute. *Steel Products Manual, Flat-Rolled Electrical Steel*[4]

3. Definitions of Terms and Symbols

3.1 The terms and symbols used in this specification are defined in Definitions A 340.

4. Ordering Information

4.1 Orders for material under this specification shall include as much of the following information as necessary to describe the desired material:

---

[1] This specification is under the jurisdiction of ASTM Committee A-6 on Magnetic Properties and the direct responsibility of Subcommittee A06.02 on Materials Specifications.
Current edition approved July 25, 1975. Published September 1975. Originally published as A 345 – 49 to replace A 310. Last previous edition A 345 – 70.
[2] *Annual Book of ASTM Standards*, Vol 03.04.
[3] *Annual Book of ASTM Standards*, Vol 01.03.
[4] Available from the American Iron and Steel Institute, 150 East 42nd St., New York, N. Y. 10017.

4.1.1 Specification A 345 or the individual specification number for the specification that shall govern.

4.1.2 Class of electrical steel, whether non-oriented or grain oriented, fully processed or semiprocessed, etc.

4.1.3 Core loss type number or standard grade designation (only if an individual specification is cited). The limiting value of the core loss or other magnetic property that shall control and all applicable test conditions and test methods shall be stated on the order if an individual specification is not cited.

4.1.4 Surface coating type.

4.1.5 Thickness, width and length, if in cut lengths instead of coils, for the ordered item.

4.1.6 Total weight of ordered item.

4.1.7 Limitations on coil size or lift weight.

4.1.8 End use. Whenever possible, state a single end use for the desired material. For instance, specify whether it is for flat punched or stamped laminations, sheared laminations, wound cores, formed cores, welded lamination cores, adhesive-bonded cores, etc. This will help the supplier to provide material with the most desirable physial characteristics for the user's fabricating practices.

4.1.9 Exceptions to the cited specification or a statement of special requirements.

5. Materials and Manufacture

5.1 Normally, these electrical steels are composed principally of iron with relatively small amounts of alloying elements such as silicon and aluminum. Other chemical elements are either in residual amounts or added in small amounts to improve fabrication. The manufacturer shall provide on request a statement of nominal chemistry being supplied.

5.2 The chemical composition and the method of manufacture shall not be unduly prescribed. Any restriction on the conditions of manufacture shall be negotiated between the manufacturer and the purchaser.

5.3 When changes in the manufacture of successive shipments of material due to changing technology are believed to increase the likelihood of adverse effects upon magnetic or fabrication performance in the specified end use, the manufacturer shall notify the purchaser before shipment is made so that he can be afforded an opportunity to evaluate the effects.

6. Magnetic Requirements

6.1 Electrical steels are normally purchased to specified maximum core loss requirements. The purchaser shall make clear to the supplier the limiting values of core loss required for the ordered material. The grain direction of the test specimen, whether as sheared or given a specific anneal, the test induction and frequency, the test method, and other information pertinent to the proper qualification of the material shall be specified.

6.2 When the desired end use imposes definite limits on other magnetic properties such as exciting current, permeability, coercive force, etc., the purchaser is responsible for so specifying on the order. The purchaser shall also state whether specific tests are required for these other magnetic properties or whether the specified characteristics are for informational purposes only.

7. Electrical Requirements

7.1 Electrical steels are normally provided with an electrical resistivity appropriate to the core-loss limit and the specified end use. If the electrical resistivity must be restricted, the limiting value shall be negotiated with the supplier.

7.2 The surface insulation resistance inherent in the processing of electrical steels for magnetic applications may differ widely with the class of electrical steel and the intended end use. Several types of applied coatings are available to attain different levels of insulation resistance as needed for critical requirements. These available core plate insulations are listed in Annex A1. If the inherent mill-processed surface lacks sufficient resistance for the user's purposes, the purchaser shall specify which of the available coatings shall be applied and any limiting value of the insulation quality.

8. Mechanical Requirements

8.1 Where requirements exist for ductility, lamination factor, tensile or yield strength, etc., that differ from those inherent in the usual product meeting the magnetic requirements, they should be specified along with any limiting values and the test methods and test conditions that apply.

9. Dimensions and Permissible Variations

9.1 *Thickness*—Electrical steels are normally

supplied in certain standardized decimal thicknesses for the various classes of electrical steel covered by the individual specifications listed in 1.2. The specified thickness should be one of the standardized decimal thicknesses whenever possible. Where the requirements of the end use indicate thicknesses that are lighter or heavier than those commonly offered, the manufacturer should be consulted by the purchaser and a thickness agreeable to both (and agreement on the corresponding effects on the magnetic requirements) should be negotiated.

9.2 *Thickness Variations*—The thickness supplied shall be as close as possible to the ordered decimal thickness. The variations with respect to the ordered thickness may differ appreciably with the class of electrical steel to be provided and the mill equipment normally used for its manufacture. The specified thickness tolerances should not be more stringent than required for satisfactory fabrication of the desired end product. The tolerances given in the AISI Steel Products Manual, *Flat-Rolled Electrical Steel*, represent normal commercial tolerances. For further details or requirements more stringent than the normal ones, the manufacturer should be consulted.

9.3 *Width Limitations*—Maximum widths that are available are limited by the width capability of the rolling and other steel-processing equipment used for the processing of the various classes of electrical steels. Narrower than economic rolled widths are usually provided as sub-widths slit from an economic width. Tolerances on the ordered width are dependent on the limitations imposed by the equipment required for the ordered width. The AISI Steel Products Manual or the supplier should be consulted for the normal tolerances than can be provided.

9.4 *Cut Lengths*—Material to be supplied as sheets or blanks is generally obtained by shearing from coils. The tolerances normally available may be determined by consulting the AISI Steel Products Manual, *Flat-Rolled Electrical Steel*, or the manufacturer of the desired material.

9.5 *Camber*—In cut lengths or coiled strip, the deviation of a side edge from a true straight line touching both ends of the side over a span of 96 in. (2438 mm) of length is normally limited to ¼ in. (6.35 mm) or a fraction thereof.

10. Workmanship and Finish

10.1 *Flatness*:

10.1.1 Adequately defining a limiting degree of flatness of electrical steel for commercial purposes is extremely difficult; therefore, no specific limits or methods of qualification for flatness evaluation are generally accepted. It is recognized that sharp waves and buckles are objectionable and that they should be avoided in the delivered material to an extent that will assure that it is suitable for fabrication of the intended end product.

10.1.2 The purchaser shall inform the supplier of any requirements for a degree of flatness more critical than that provided by the usual commercial manufacturing practices. Procedures for judging or evaluating the required degree of flatness shall be negotiated between the purchaser and the supplier.

10.2 *Surface Defects*—The surface shall be reasonably free of loose dust and essentially free of manufacturing defects such as holes, blisters, slivers, indentations, etc., which would interfere with its effective use in the intended application. Surface oxide and core plate coatings should be thin and tightly adherent.

11. Sampling

11.1 The manufacturer shall assign a number to each test lot for identification. The test lot shall conform to the requirements of Methods A 34 unless otherwise agreed between the purchaser and the supplier.

11.2 Samples shall be obtained after the final mill heat treatment or the final operation that may have a significant influence on the magnetic properties of the electrical steel to be supplied. One or both ends of the full width coil identified as the test lot shall be sampled in accordance with Methods A 34 or as designated in the individual specification or the purchase order.

12. Specimen Preparation

12.1 The required samples shall be made into Epstein test specimens for magnetic testing, and into suitable specimens for electrical or mechanical tests as required by Methods A 34 or by the test method cited in the individual specification or the purchase order. Care should be practiced to eliminate any bent, twisted, dented, highly burred, or improperly prepared pieces from the test specimen.

13. Test Methods

13.1 The required tests for core loss to determine the core-loss grade, and other tests when required, shall be in accordance with the desired test method of Methods A 34 or as designated in the individual specification or the purchase order.

13.2 The density of the material for testing purposes will vary according to the chemical composition of the material to be supplied. The proper test density shall be determined and used in the testing by the supplier in compliance with the requirements of Methods A 34.

14. Certification

14.1 The producer or supplier shall submit to the purchaser, as promptly as possible at the time of shipment, a certified report of the measured core loss values or other required test values to show that the material conforms to the individual specification or the purchase order. The test methods and applicable test conditions, including the test density, shall be clearly stated. The test report shall also carry the lot identification, purchase order number, and other information that is deemed necessary to identify the test results with the proper shipment and item.

15. Marking

15.1 Each package of coils or lift of cut lengths shall have firmly attached to it, outside its wrappings, a tag showing the purchaser's order number, specification number, grade designation, coating or surface type designation, thickness, width (and length if in sheet form), weight and test lot number. In addition, each wide coil shall have the specification number, grade designation, coating or surface type designation, thickness, width, weight, and test lot number marked on the outer surface of the coil itself. In a lift of narrow coils, each narrow coil in the package shall be tagged with the specification number, grade designation, coating or surface type designation, thickness, width, and test lot number.

16. Packaging

16.1 Methods of packaging, loading, and shipping, unless otherwise specified, shall correspond to Practices A 700.

ANNEX

A1. DESCRIPTION[5] OF CORE PLATE COATINGS

Core Plate Designation	Description
C-1	An organic enamel or varnish coating sometimes used for cores not immersed in oil. It enhances punchability and is resistant to ordinary operating temperatures. It will not withstand stress-relief annealing.
C-2	An inorganic insulation consisting of a glass-like film formed during the high-temperature annealing of electrical steel, particularly grain-oriented electrical steel. This insulation is intended for air-cooled or oil-immersed cores. It will withstand stress-relief annealing and has sufficient interlamination resistance for wound cores of narrow-width strip such as in distribution transformers. It is not intended for stamped laminations because it is abrasive to dies.
C-3	An enamel or varnish coating intended for air-cooled or oil-immersed cores. C-3 enhances punchability and is resistant to normal operating temperatures. It will not withstand stress-relief annealing.
C-4	Consists of a chemically treated or phosphated surface useful for air-cooled or oil-immersed cores. It will withstand stress-relief annealing in relatively neutral atmospheres.
C-5	An inorganic insulation similar to C-4 but with ceramic fillers added to increase the electrical insulation properties. C-5 can be used in air-cooled or oil-immersed cores and will endure stress-relief annealing.

NOTE—For applications where factors other than insulative are of primary importance, a thin coating of Core Plate C-1 or C-3 may be desirable. In such cases, the core plate designation should be suffixed by the letter A. (Example: Core Plate C-3A)

[5] Abstracted from the American Iron and Steel Institute, *Steel Products Manual*, SPMAA, *Flat-Rolled Electrical Steel*, July 1968.

The American Society for Testing and Materials takes no position respecting the validity of any patent rights asserted in connection with any item mentioned in this standard. Users of this standard are expressly advised that determination of the validity of any such patent rights, and the risk of infringement of such rights, are entirely their own responsibility.

This standard is subject to revision at any time by the responsible technical committee and must be reviewed every five years and if not revised, either reapproved or withdrawn. Your comments are invited either for revision of this standard or for additional standards and should be addressed to ASTM Headquarters. Your comments will receive careful consideration at a meeting of the responsible technical committee, which you may attend. If you feel that your comments have not received a fair hearing you should make your views known to the ASTM Committee on Standards, 1916 Race St., Philadelphia, Pa. 19103.

Designation: A 366 – 72 (Reapproved 1979)

An American National Standard

Standard Specification for
STEEL, CARBON, COLD-ROLLED SHEET, COMMERCIAL QUALITY[1]

This standard is issued under the fixed designation A 366; the number immediately following the designation indicates the year of original adoption or, in the case of revision, the year of last revision. A number in parentheses indicates the year of last reapproval. A superscript epsilon (ϵ) indicates an editorial change since the last revision or reapproval.

This specification has been approved for use by agencies of the Department of Defense and for listing in the DoD Index of Specifications and Standards.

1. Scope

1.1 This specification covers cold-rolled carbon steel sheet of commercial quality, in coils or cut lengths. This material is intended for exposed or unexposed parts where bending, moderate drawing, forming, and welding may be involved.

1.2 This specification is not applicable to Specification A 109, for Steel, Carbon, Cold-Rolled Strip.[2] Narrow widths cut from wide sheet are not strip, unless they qualify as strip because of thickness, special finish, special edge, or special temper.

1.3 The values stated in inch-pound units are to be regarded as the standard.

2. Applicable Documents

2.1 *ASTM Standards*:
A 370 Methods and Definitions for Mechanical Testing of Steel Products[2]
A 568 Specification for General Requirements for Steel, Carbon and High-Strength Low-Alloy Hot-Rolled Sheet and Cold-Rolled Sheet[2]

3. General Requirements for Delivery

3.1 Material furnished under this specification shall conform to the applicable requirements of the current addition of Specification A 568, unless otherwise provided herein.

4. Ordering Information

4.1 Orders for material under this specification shall include the following information, as required, to adequately describe the required material:

4.1.1 ASTM specification number and date of issue,

4.1.2 Name of material (cold-rolled sheet, commercial quality),

4.1.3 Copper-bearing steel (if required),

4.1.4 Finish (indicate unexposed (Class 2), or exposed (Class 1), matte (dull) finish, commercial bright, or luster as required),

4.1.5 Specify oiled or not oiled, as required,

4.1.6 Dimensions (thickness, width, and whether cut lengths or coils),

4.1.7 Coil size (must include inside diameter, outside diameter, and maximum mass),

4.1.8 Application (part identification and description),

4.1.9 Special requirements (if required),

4.1.10 Cast or heat analysis report (request, if required)

NOTE—A typical ordering description is as follows: ASTM A 366 – 72, Cold-Rolled Sheet, Commercial Quality, Class 1, Matte Finish, Oiled, 0.035 by 30 by 96 in. for Part No. 4560, Door Panel.

5. Chemical Requirements

5.1 The cast or heat analysis of the steel shall conform to the chemical composition shown in Table 1.

6. Mechanical Requirements

6.1 *Bend Test*—The material shall be capable of being bent, at room temperature, in

---

[1] This specification is under the jurisdiction of ASTM Committee A-1 on Steel, Stainless Steel and Related Alloys, and is the direct responsibility of Subcommittee A01.19 on Steel Sheet and Strip.
Current edition approved June 29, 1972. Published August 1972. Originally published as A 366 – 53 T. Last previous edition A 366 – 68.
[2] *Annual Book of ASTM Standards*, Vol. 01.03.

A 366

any direction through 180 deg flat on itself without cracking on the outside of the bent portion (see Section 14 of Methods and Definitions A 370).

6.2 *Hardness*—If no special flattening is required, sheet of this quality is not expected to exceed a hardness equivalent of Rockwell B60 at time of shipment.

6.3 Moderate deformations on identified parts are assessed by the use of the scribed square test as described by Specification A 568. Experience has shown that if the percent increase in area of any drawn portion of a satisfactory untrimmed part is 25 percent or less, commercial quality should give satisfactory performance. If it is more than 25 percent, drawing quality or drawing quality special killed should be specified.

6.4 If freedom from stretcher strains or fluting during fabrication is required, Class 1 finish should be specified and material should be effectively roller leveled immediately before use. Material so specified is subject to aging with elapsed time.

7. Certification and Reports

7.1 When requested, the producer shall furnish copies of a report showing test results of the cast or heat analysis. The report shall include the purchase order number, ASTM designation number, and the cast or heat number representing the material.

TABLE 1 Chemical Requirements

Element	Composition, percent
Carbon, max	0.15
Manganese, max	0.60
Phosphorus, max	0.035
Sulfur, max	0.040
Copper, when copper steel is specified, min	0.20

The American Society for Testing and Materials takes no position respecting the validity of any patent rights asserted in connection with any item mentioned in this standard. Users of this standard are expressly advised that determination of the validity of any such patent rights, and the risk of infringement of such rights, are entirely their own responsibility.

This standard is subject to revision at any time by the responsible technical committee and must be reviewed every five years and if not revised, either reapproved or withdrawn. Your comments are invited either for revision of this standard or for additional standards and should be addressed to ASTM Headquarters. Your comments will receive careful consideration at a meeting of the responsible technical committee, which you may attend. If you feel that your comments have not received a fair hearing you should make your views known to the ASTM Committee on Standards, 1916 Race St., Philadelphia, Pa. 19103.

Designation: A 368 – 82

Standard Specification for
STAINLESS AND HEAT-RESISTING STEEL WIRE STRAND[1]

This standard is issued under the fixed designation A 368; the number immediately following the designation indicates the year of original adoption or, in the case of revision, the year of last revision. A number in parentheses indicates the year of last reapproval. A superscript epsilon (ϵ) indicates an editorial change since the last revision or reapproval.

1. Scope

1.1 This specification covers stainless and heat-resisting steel wire strand composed of a multiplicity of round wires and suitable for use as guy wires, overhead ground wires, and similar purposes.

1.2 The values stated in inch-pound units are to be regarded as the standard.

2. Applicable Documents

2.1 *ASTM Standards:*
A 555 Specification for General Requirements for Stainless and Heat-Resisting Steel Wire[2]
A 751 Methods, Practices, and Definitions for Chemical Analysis of Steel Products[2]

3. General Requirements for Delivery

3.1 In addition to the requirements of this specification, all requirements of the current edition of Specification A 555 shall apply.

4. Ordering Information

4.1 Orders for material under this specification shall include the following information:
4.1.1 Quantity (length of strand or weight of quantity ordered, or both; see 14.1 and Table 1),
4.1.2 Name of material (stainless steel),
4.1.3 Form (wire strand in coils or on reels),
4.1.4 Applicable dimensions (for nominal strand diameter, see Table 1),
4.1.5 Number of wires per strand (Table 1),
4.1.6 Minimum breaking strength (medium or high strength),
4.1.7 Type designation (see Section 7),
4.1.8 ASTM designation, and date of issue, and
4.1.9 Special requirements, if any.

NOTE 1—A typical ordering description is as follows: 1000 ft, stainless steel 7-wire strand, $7/16$-in. diameter, medium strength, on reel, Type 302, ASTM A 368 dated _____.

5. Stranding

5.1 Three-wire strand shall have a left lay with a uniform pitch of not less than 10 nor more than 16 times the nominal diameter of the strand. Seven-wire strand and the outer layer of 19-wire strand, shall have a left lay with a uniform pitch of not less than 12 nor more than 16 times the nominal diameter of the strand. A left lay is defined as a counter-clockwise twist away from the observer. All wires shall be stranded with uniform tension. Stranding shall be sufficiently close to ensure no appreciable reduction in diameter when stressed to 10 % of the specified strength.

5.2 All wires in the strand shall lie naturally in their true positions in the completed strand, and when the strand is cut, the ends shall remain in position or be readily replaced by hand and then remain in position. This may be accomplished by any means or process, such as preforming, post forming, or form setting.

6. Joints

6.1 There shall be no strand joints or strand splices in any length of the completed strand.

---

[1] This specification is under the jurisdiction of ASTM Committee A-1 on Steel, Stainless Steel and Related Alloys, and is the direct responsibility of Subcommittee A01.08 on Wrought Stainless Steel Products.
Current edition approved July 30, 1982. Published December 1982. Originally published as A 368 – 53 T. Last previous edition A 368 – 81.

[2] *Annual Book of ASTM Standards*, Vol 01.03.

6.2 In 3-wire strand, there shall be no joints in the individual wires.

6.3 In 7-wire strand, joints in individual wires shall be acceptable provided there is not more than one joint in any 150-ft (46-m) section of the completed strand and the location of each wire joint is marked on the strand with paint or some other distinguishing mark.

6.4 In 19-wire strand, joints in the individual wires of the outer layer of 12 wires shall be acceptable provided there is not more than one joint in any 150-ft (46-m) section and the location of each wire joint is marked on the strand with paint or some other distinguishing mark. Joints in the 7-wire inner layer of 19-wire strand shall be acceptable provided there is not more than one joint in any 150-ft section.

6.5 Joints in the individual wires shall be flash or upset butt-welded. Care shall be taken to prevent injury to the wire during welding.

7. Chemical Requirements

7.1 The steel shall be Type 302, 304, 305, 316, 316Cb, or 316Ti and shall conform to the requirements as to chemical composition specified in Table 2.

7.2 Methods and practices relating to chemical analysis required by this specification shall be in accordance with Methods, Practices, and Definitions A 751.

8. Mechanical Requirements

8.1 The tensile strength requirements, based upon the nominal strand diameter and the number of wires in each strand, shall conform to the requirements specified in Table 1. All tension tests shall be made upon lengths of strand that do not contain welds in the individual wires.

8.2 The individual wires of the completed strand shall not fracture when wrapped in a close helix of at least two turns upon itself as a mandrel at a rate not exceeding 15 turns/min.

9. Sampling

9.1 Sampling for determination of compliance to this specification shall be performed on each lot of completed strand. A lot shall consist of all strand of one size (nominal strand diameter, nominal wire diameter, and number of wires per strand) and grade in each shipment.

10. Number of Tests

10.1 The number of samples tested shall be as follows:

Lot Size	No. of Tests
Up to 500 ft (152 m), incl	1
Over 500 to 5000 ft (152 to 1524 m), incl	2
Over 5000 to 10 000 ft (1524 to 3048 m), incl[A]	3
Over 10 000 ft (3048 m)[A]	4

[A] When a lot consists of only one reel, tests must be of necessity be limited to two in number (one from each end).

10.1.1 Each strand sample shall be subjected to the tension test as specified in 8.1.

10.1.2 In addition to the strand testing specified in 10.1.1, the individual wires shall be subjected to the wrapping test specified in 8.2. The number of individual wires to be tested from each strand shall be as follows:

3-wire strand: test all 3 wires
7-wire strand: test any 4 wires
19-wire strand: test 3 wires from each layer (inner and outer layer), for a total of 6 tests

11. Retests

11.1 In case of reasonable doubt in the first tests as to the failure of the wire or strand to meet any requirement of this specification, two additional tests shall be made on samples of wire or strand from the same coil (or reel). If failure occurs in either of these tests, the lot shall be rejected. However, the producer may, at his option, requalify individual coils (or reels) by performing the required tests on an individual coil (or reel) basis.

12. Permissible Variations in Dimensions

12.1 The nominal diameter of the finished strand and of the individual wires, the number of wires per strand, and the approximate weight per 1000 ft (or per 304.80 m) of strand are shown in Table 1.

12.2 The diameter of the individual wires forming the strand shall not vary from the nominal wire diameters by more than ± 0.001 in. (± 0.025 mm).

13. Workmanship and Finish

13.1 The finished strand shall be tight, smooth, and free of imperfections not consistent with good commercial practice.

13.2 The diameter of the finished strand shall be uniform, except for a minor increase in diameter due to a wire joint.

A 368

14. Packaging and Marking

14.1 Wire strand shall be furnished in standard lengths (see 14.1.1) and in compact coils or on reels (see 14.1.2) as specified by the purchaser; otherwise lengths shall be as agreed upon at the time of purchase.

14.1.1 Standard lengths of strand are as follows: 100 ft (30.48 m), 250 ft (76.2 m), 500 ft (152.40 m), 1000 ft (304.80 m), 2500 ft (762.00 m), and 5000 ft (1524.00 m).

14.1.2 Standard practice is to furnish all strand of 7/16 in. (11.11 mm) and over in diameter on reels in lengths of 1000 ft (304.80 m) and over. Strand lengths of less than 1000 ft are regularly furnished in coils.

14.2 The strand shall be protected against damage in ordinary handling and shipping as agreed upon at the time of purchase. Each coil (or reel) shall have a strong weather-proof tag securely fastened to it showing the minimum breaking strength, the nominal diameter of the strand, the length of the strand, the steel type designation, ASTM Designation A 368 dated _____, and the name or mark of the manufacturer.

TABLE 1 Mechanical Properties of Stainless and Heat-Resisting Steel Wire Strand

Nominal Diameter of Strand, in. (mm)	Number of Wires in Strand	Nominal Diameter of Stainless Wires, in. (mm)	Approximate Weight of Strand per 1000 ft (304.80 m), lb (kg)	Minimum Breaking Strength of Strand, lbf (kN)	
				Medium Strength	High Strength
13/64 (5.16)	3	0.093 (2.36)	72 (32.66)	3 150 (14.01)	4 500 (20.02)
7/32 (5.56)	3	0.104 (2.64)	90 (40.82)	3 950 (17.57)	5 650 (25.13)
1/4 (6.35)	3	0.120 (3.05)	120 (54.43)	5 300 (23.58)	7 550 (33.58)
5/16 (7.94)	3	0.145 (3.68)	175 (79.38)	7 700 (34.25)	11 000 (48.93)
3/8 (9.52)	3	0.165 (4.19)	225 (102.06)	10 000 (44.48)	14 300 (63.61)
7/32 (5.56)	7	0.072 (1.83)	100 (45.36)	4 500 (20.02)	6 300 (28.02)
1/4 (6.35)	7	0.083 (2.11)	132 (59.87)	5 950 (26.47)	8 500 (37.81)
9/32 (7.14)	7	0.093 (2.36)	167 (75.75)	7 350 (32.69)	10 500 (46.71)
5/16 (7.94)	7	0.104 (2.64)	208 (94.35)	9 200 (40.92)	13 200 (58.72)
3/8 (9.52)	7	0.120 (3.05)	278 (126.10)	12 500 (55.60)	18 000 (80.07)
7/16 (11.11)	7	0.145 (3.68)	405 (183.71)	18 200 (80.96)	26 000 (115.65)
1/2 (12.70)	7	0.165 (4.19)	525 (238.14)	23 600 (104.98)	33 700 (149.90)
3/8 (9.52)	19	0.075 (1.90)	295 (133.81)	11 800 (52.49)	16 800 (74.73)
7/16 (11.11)	19	0.087 (2.21)	400 (181.44)	15 800 (70.28)	22 500 (100.08)
1/2 (12.70)	19	0.100 (2.54)	530 (240.40)	21 000 (93.41)	30 000 (133.45)
9/16 (14.29)	19	0.110 (2.79)	640 (290.30)	25 400 (112.98)	36 200 (161.02)
5/8 (15.88)	19	0.125 (3.18)	825 (374.21)	33 000 (146.79)	47 000 (209.07)
3/4 (19.05)	19	0.150 (3.81)	1,190 (539.78)	47 500 (211.29)	67 500 (300.25)
7/8 (22.22)	19	0.175 (4.44)	1,620 (734.82)	64 000 (284.69)	91 400 (406.57)

A 368

TABLE 2 Chemical Requirements

UNS Designation	Type	Composition, %									Other Elements
		Carbon, max	Manganese, max	Phosphorus, max	Sulfur, max	Silicon, max	Chromium	Nickel	Molybdenum	Nitrogen, max	
S30200	302	0.15	2.00	0.045	0.030	1.00	17.00–19.00	8.00–10.00	...	0.10	...
S30400	304	0.08	2.00	0.045	0.030	1.00	18.00–20.00	8.00–10.50	...	0.10	...
S30500	305	0.12	2.00	0.045	0.030	1.00	17.00–19.00	10.50–13.00	...	...	...
S31600	316	0.08	2.00	0.045	0.030	1.00	16.00–18.00	10.00–14.00	2.00–3.00	0.10	...
S31635	316Ti	0.08	2.00	0.045	0.030	1.00	16.00–18.00	10.00–14.00	2.00–3.00	0.10	Ti 5 X (C + N) min, 0.70 max
S31640	316Cb	0.08	2.00	0.045	0.030	1.00	16.00–18.00	10.00–14.00	2.00–3.00	0.10	Cb + Ta 10 X C min, 1.10 max

The American Society for Testing and Materials takes no position respecting the validity of any patent rights asserted in connection with any item mentioned in this standard. Users of this standard are expressly advised that determination of the validity of any such patent rights, and the risk of infringement of such rights, are entirely their own responsibility.

This standard is subject to revision at any time by the responsible technical committee and must be reviewed every five years and if not revised, either reapproved or withdrawn. Your comments are invited either for revision of this standard or for additional standards and should be addressed to ASTM Headquarters. Your comments will receive careful consideration at a meeting of the responsible technical committee, which you may attend. If you feel that your comments have not received a fair hearing you should make your views known to the ASTM Committee on Standards, 1916 Race St., Philadelphia, Pa. 19103.

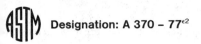

Designation: A 370 – 77[e2]

An American National Standard
American Association State Highway and
Transportation Officials Standard
AASHTO No.: T 244

Standard Methods and Definitions for
MECHANICAL TESTING OF STEEL PRODUCTS[1]

This standard is issued under the fixed designation A 370; the number immediately following the designation indicates the year of original adoption or, in the case of revision, the year of last revision. A number in parentheses indicates the year of last reapproval. A superscript epsilon (ϵ) indicates an editorial change since the last revision or reapproval.

These methods have been approved for use by agencies of the Department of Defense and for listing in the DoD Index of Specifications and Standards.

[e1] NOTE—Paragraph 18.2 was editorially changed in May 1979.
[e2] NOTE—Fig. 21 was editorially corrected in October 1980.

1. Scope

1.1 These methods[2] cover procedures and definitions for the mechanical testing of wrought and cast steel products. The various mechanical tests herein described are used to determine properties required in the product specifications. Variations in testing methods are to be avoided and standard methods of testing are to be followed to obtain reproducible and comparable results. In those cases where the testing requirements for certain products are unique or at variance with these general procedures, the product specification testing requirements shall control.

1.2 The following mechanical tests are described:

	Sections
Tension	5 to 13
Bend	14
Hardness:	15
Brinell	16 and 17
Rockwell	18
Impact	19 to 23

1.3 Supplements covering details peculiar to certain products are appended to these methods as follows:

	Sections
Bar Products (Supplement I)	S 1 to S 4
Tubular Products (Supplement II)	S 5 to S 9
Fasteners (Supplement III)	S 10 to S 15
Round Wire Products (Supplement IV)	S 16 to S 22
Significance of Notched Bar Impact Testing (Supplement V)	S 23 to S 28
Converting Percentage Elongation of Round Specimens to Equivalents for Flat Specimens (Supplement VI)	S 29 to S 31
Testing Seven Wire Stress-Relieved Strand (Supplement VII)	S 32 to S 36
Rounding Test Data (Supplement VIII)	

1.4 The values stated in inch-pound units are to be regarded as the standard.

2. Applicable Documents

2.1 *ASTM Standards:*

A 416 Specification for Uncoated Seven-Wire Stress-Relieved Strand for Prestressed Concrete[3]

E 4 Methods of Load Verification of Testing Machines[4]

E 6 Definitions of Terms Relating to Methods of Mechanical Testing[4]

E 8 Methods of Tension Testing of Metallic Materials[4]

E 10 Test Method for Brinell Hardness of Metallic Materials[4]

E 18 Test Methods for Rockwell hardness and Rockwell Superficial Hardness of Metallic Materials[4]

E 23 Methods for Notched Bar Impact Testing of Metallic Materials[4]

E 83 Method of Verification and Classification of Extensometers[4]

E 110 Test Method for Indentation Hardness of Metallic Materials by Portable Hardness Testers[4]

E 208 Method for Conducting Drop-Weight Test to Determine Nil-Ductility Transition Temperature of Ferritic Steels[4]

---

[1] These methods are under the jurisdiction of ASTM Committee A-1 on Steel, Stainless Steel and Related Alloys and are the direct responsibility of Subcommittee A01.13 on Mechanical Testing.
Current edition approved June 24, 1977. Published August 1977. Originally published as A 370 – 53 T. Last previous edition A 370 – 76.
[2] For ASME Boiler and Pressure Vessel Code applications see related Specification SA-370 in Section II of that Code.
[3] *Annual Book of ASTM Standards*, Vol 01.04.
[4] *Annual Book of ASTM Standards*, Vol 03.01.

3. General Precautions

3.1 Certain methods of fabrication such as bending, forming, and welding, or operations involving heating, may affect the properties of the material under test. Therefore, the product specifications cover the stage of manufacture at which mechanical testing is to be performed. The properties shown by testing prior to fabrication may not necessarily be representative of the product after it has been completely fabricated.

3.2 Improper machining or preparation of test specimens may give erroneous results. Care should be exercised to assure good workmanship in machining. Improperly machined specimens should be discarded and other specimens substituted.

3.3 Flaws in the specimen may also affect results. If any test specimen develops flaws, the retest provision of the applicable product specification shall govern.

3.4 If any test specimen fails because of mechanical reasons such as failure of testing equipment or improper specimen preparation, it may be discarded and another specimen taken.

4. Orientation of Test Specimens

4.1 The terms "longitudinal test" and "transverse test" are used only in material specifications for wrought products and are not applicable to castings. When such reference is made to a test coupon or test specimen, the following definitions apply:

4.1.1 *Longitudinal Test*, unless specifically defined otherwise, signifies that the lengthwise axis of the specimen is parallel to the direction of the greatest extension of the steel during rolling or forging. The stress applied to a longitudinal tension test specimen is in the direction of the greatest extension, and the axis of the fold of a longitudinal bend test specimen is at right angles to the direction of greatest extension (Figs. 1, 2(a), and 2(b)).

4.1.2 *Transverse Test*, unless specifically defined otherwise, signifies that the lengthwise axis of the specimen is at right angles to the direction of the greatest extension of the steel during rolling or forging. The stress applied to a transverse tension test specimen is at right angles to the greatest extension, and the axis of the fold of a transverse bend test specimen is parallel to the greatest extension (Fig. 1).

4.2 The terms "radial test" and "tangential test" are used in material specifications for some wrought circular products and are not applicable to castings. When such reference is made to a test coupon or test specimen, the following definitions apply:

4.2.1 *Radial Test*, unless specifically defined otherwise, signifies that the lengthwise axis of the specimen is perpendicular to the axis of the product and coincident with one of the radii of a circle drawn with a point on the axis of the product as a center (Fig. 2(a)).

4.2.2 *Tangential Test*, unless specifically defined otherwise, signifies that the lengthwise axis of the specimen is perpendicular to a plane containing the axis of the product and tangent to a circle drawn with a point on the axis of the product as a center (Figs. 2(a), 2(b), 2(c), and 2(d)).

TENSION TEST

5. Description

5.1 The tension test related to the mechanical testing of steel products subjects a machined or full-section specimen of the material under examination to a measured load sufficient to cause rupture. The resulting properties sought are defined in Definitions E 6.

5.2 In general the testing equipment and methods are given in Methods E 8. However, there are certain exceptions to Methods E 8 practices in the testing of steel, and these are covered in these methods.

6. Test Specimen Parameters

6.1 *Selection*—Test coupons shall be selected in accordance with the applicable product specifications.

6.1.1 *Wrought Steels*—Wrought steel products are usually tested in the longitudinal direction, but in some cases, where size permits and the service justifies it, testing is in the transverse, radial, or tangential directions (see Figs. 1 and 2).

6.1.2 *Forged Steels*—For open die forgings, the metal for tension testing is usually provided by allowing extensions or prolongations on one or both ends of the forgings, either on all or a representative number as provided by the applicable product specifications. Test

A 370

specimens are normally taken at mid-radius. Certain product specifications permit the use of a representative bar or the destruction of a production part for test purposes. For ring or disk-like forgings test metal is provided by increasing the diameter, thickness, or length of the forging. Upset disk or ring forgings, which are worked or extended by forging in a direction perpendicular to the axis of the forging, usually have their principal extension along concentric circles and for such forgings tangential tension specimens are obtained from extra metal on the periphery or end of the forging. For some forgings, such as rotors, radial tension tests are required. In such cases the specimens are cut or trepanned from specified locations.

6.1.3 *Cast Steels*—Test coupons for castings from which tension test specimens are prepared shall be attached to the castings where practicable. If the design of the casting is such that test coupons should not be attached thereon, test coupons shall be cast attached to separate cast blocks (Fig. 3 and Table 1).

6.2 *Size and Tolerances*—Test specimens shall be the full thickness or section of material as-rolled, or may be machined to the form and dimensions shown in Figs. 4 to 7, inclusive. The selection of size and type of specimen is prescribed by the applicable product specification. Full section specimens shall be tested in 8-in. (200-mm) gage length unless otherwise specified in the product specification.

6.3 *Procurement of Test Specimens*—Specimens shall be sheared, blanked, sawed, trepanned, or oxygen-cut from portions of the material. They are usually machined so as to have a reduced cross section at mid-length in order to obtain uniform distribution of the stress over the cross section and to localize the zone of fracture. When test coupons are sheared, blanked, sawed, or oxygen-cut, care shall be taken to remove by machining all distorted, cold-worked, or heat-affected areas from the edges of the section used in evaluating the test.

6.4 *Aging of Test Specimens*—Unless otherwise specified, it shall be permissible to age tension test specimens. The time-temperature cycle employed must be such that the effects of previous processing will not be materially changed. It may be accomplished by aging at room temperature 24 to 48 h, or in shorter time at moderately elevated temperatures by boiling in water, heating in oil or in an oven.

6.5 *Measurement of Dimensions of Test Specimens:*

6.5.1 *Standard Rectangular Tension Test Specimens*—These forms of specimens are shown in Fig. 4. To determine the cross-sectional area, the center width dimension shall be measured to the nearest 0.005 in. (0.13 mm) for the 8-in. (200-mm) gage length specimen and 0.001 in. (0.025 mm) for the 2-in. (50-mm) gage length specimen in Fig. 4. The center thickness dimension shall be measured to the nearest 0.001 in. for both specimens.

6.5.2 *Standard Round Tension Test Specimens*—These forms of specimens are shown in Figs. 5 and 6. To determine the cross-sectional area, the diameter shall be measured at the center of the gage length to the nearest 0.001 in.

6.6 *General*—Test specimens shall be either substantially full size or machined, as prescribed in the product specifications for the material being tested.

6.6.1 Improperly prepared test specimens often cause unsatisfactory test results. It is important, therefore, that care be exercised in the preparation of specimens, particularly in the machining, to assure good workmanship.

6.6.2 It is desirable to have the cross-sectional area of the specimen smallest at the center of the gage length to ensure fracture within the gage length. This is provided for by the taper in the gage length permitted for each of the specimens described in the following sections.

6.6.3 For brittle materials it is desirable to have fillets of large radius at the ends of the gage length.

7. Plate-Type Specimen

7.1 The standard plate-type test specimen is shown in Fig. 4. This specimen is used for testing metallic materials in the form of plate, structural and bar-size shapes, and flat material having a nominal thickness of $3/16$ in. (5 mm) or over. When product specifications so permit, other types of specimens may be used.

NOTE 1—When called for in the product specification, the 8-in. gage length specimen of Fig. 4 may be used for sheet and strip material.

8. Sheet-Type Specimen

8.1 The standard sheet-type test specimen is shown in Fig. 4. This specimen is used for testing metallic materials in the form of sheet, plate, flat wire, strip, band, and hoop ranging in nominal thickness from 0.005 to ¾ in. (0.13 to 19 mm). When product specifications so permit, other types of specimens may be used, as provided in Section 7.

9. Round Specimens

9.1 The standard 0.500-in. (12.5-mm) diameter round test specimen shown in Fig. 5 is used quite generally for testing metallic materials, both cast and wrought.

9.2 Figure 5 also shows small size specimens proportional to the standard specimen. These may be used when it is necessary to test material from which the standard specimen or specimens shown in Fig. 4 cannot be prepared. Other sizes of small round specimens may be used. In any such small size specimen it is important that the gage length for measurement of elongation be four times the diameter of the specimen (see Note 4, Fig. 5).

9.3 The shape of the ends of the specimens outside of the gage length shall be suitable to the material and of a shape to fit the holders or grips of the testing machine so that the loads are applied axially. Figure 6 shows specimens with various types of ends that have given satisfactory results.

10. Gage Marks

10.1 The specimens shown in Figs. 4, 5, and 7 shall be gage marked with a center punch, scribe marks, multiple device, or drawn with ink. The purpose of these gage marks is to determine the percent elongation. Punch marks shall be light, sharp, and accurately spaced. The localization of stress at the marks makes a hard specimen susceptible to starting fracture at the punch marks. The gage marks for measuring elongation after fracture shall be made on the flat or on the edge of the flat tension test specimen and within the parallel section; for the 8-in. gage length specimen, Fig. 4, one or more sets of 8-in. gage marks may be used, intermediate marks within the gage length being optional. Rectangular 2-in. gage length specimens, Fig. 4, and round specimens, Fig. 5, are gage marked with a double-pointed center punch or scribe marks. In both cases the gage points shall be approximately equidistant from the center of the length of the reduced section. These same precautions shall be observed when the test specimen is full section.

11. Testing Apparatus and Operations

11.1 *Loading Systems*—There are two general types of loading systems, mechanical (screw power) and hydraulic. These differ chiefly in the variability of the rate of load application. The older screw power machines are limited to a small number of fixed free running crosshead speeds. Some modern screw power machines and all hydraulic machines permit stepless variation throughout the range of speeds.

11.2 The tension testing machine shall be maintained in good operating condition, used only in the proper loading range, and calibrated periodically in accordance with the latest revision of Methods E 4.

NOTE 2—Many machines are equipped with stress-strain recorders for autographic plotting of stress-strain curves. It should be noted that some recorders have a load measuring component entirely separate from the load indicator of the testing machine. Such recorders are calibrated separately.

11.3 *Loading*—It is the function of the gripping or holding device of the testing machine to transmit the load from the heads of the machine to the specimen under test. The essential requirement is that the load shall be transmitted axially. This implies that the centers of the action of the grips shall be in alignment, insofar as practicable, with the axis of the specimen at the beginning and during the test, and that bending or twisting be held to a minimum. Gripping of the specimen shall be restricted to the section outside the gage length. In the case of certain sections tested in full size, nonaxial loading is unavoidable and in such cases shall be permissible.

11.4 *Speed of Testing*—The speed of testing shall not be greater than that at which

A 370

load and strain readings can be made accurately. In production testing, speed of testing is commonly expressed (*1*) in terms of free running crosshead speed (rate of movement of the crosshead of the testing machine when not under load), or (*2*) in terms of rate of separation of the two heads of the testing machine under load, or (*3*) in terms of rate of stressing the specimen. Speed of testing may also be expressed in terms of rate of straining the specimen. However, it is not practicable to control the rate of straining on machines currently used in production testing. The following limitations on the speed of testing are recommended as adequate for most steel products:

11.4.1 Any convenient speed of testing may be used up to one half the specified yield point or yield strength. When this point is reached, the rate of separation of the crossheads under load shall be adjusted so as not to exceed $\frac{1}{16}$ in. per min per inch of gage length, or the distance between the grips for test specimens not having reduced sections. This speed shall be maintained through the yield point or yield strength. In determining the tensile strength, the rate of separation of the heads under load shall not exceed $\frac{1}{2}$ in. per min per inch of gage length. In any event the minimum speed of testing shall not be less than $\frac{1}{10}$ of the specified maximum rates for determining yield point or yield strength and tensile strength.

11.4.2 It shall be permissible to set the speed of the testing machine by adjusting the free running crosshead speed to the above specified values, inasmuch as the rate of separation of heads under load at these machine settings is less than the specified values of free running crosshead speed.

11.4.3 As an alternative, if the machine is equipped with a device to indicate the rate of loading, the speed of the machine from half the specified yield point or yield strength through the yield point or yield strength may be adjusted so that the rate of stressing does not exceed 100,000 psi (690 MPa)/min. However, the minimum rate of stressing shall not be less than 10,000 psi (70 MPa)/min.

12. Definitions

12.1 For definitions of terms pertaining to tension testing, including tensile strength, yield point, yield strength, elongation, and reduction of area, reference should be made to Definitions E 6.

13. Determination of Tensile Properties

13.1 *Yield Point*—Yield point is the first stress in a material, less than the maximum obtainable stress, at which an increase in strain occurs without an increase in stress. Yield point is intended for application only for materials that may exhibit the unique characteristic of showing an increase in strain without an increase in stress. The stress-strain diagram is characterized by a sharp knee or discontinuity. Determine yield point by one of the following methods:

13.1.1 *Drop of the Beam or Halt of the Pointer Method*—In this method apply an increasing load to the specimen at a uniform rate. When a lever and poise machine is used, keep the beam in balance by running out the poise at approximately a steady rate. When the yield point of the material is reached, the increase of the load will stop, but run the poise a trifle beyond the balance position, and the beam of the machine will drop for a brief but appreciable interval of time. When a machine equipped with a load-indicating dial is used there is a halt or hesitation of the load-indicating pointer corresponding to the drop of the beam. Note the load at the "drop of the beam" or the "halt of the pointer" and record the corresponding stress as the yield point.

13.1.2 *Autographic Diagram Method*—When a sharp-kneed stress-strain diagram is obtained by an autographic recording device, take the stress corresponding to the top of the knee (Fig. 8), or the stress at which the curve drops as the yield point (Fig. 8).

13.1.3 *Total Extension Under Load Method*—When testing material for yield point and the test specimens may not exhibit a well-defined disproportionate deformation that characterizes a yield point as measured by the drop of the beam, halt of the pointer, or autographic diagram methods described in 13.1.1 and 13.1.2, a value equivalent to the yield point in its practical significance may be determined by the following method and may be recorded as yield point: Attach a Class C or better extensometer (Notes 3 and 4) to the specimen. When the load producing a specified extension (Note 5) is reached record the stress corresponding to the load as the yield point, and remove the extensometer (Fig. 9).

NOTE 3—Automatic devices are available that determine the load at the specified total extension without plotting a stress-strain curve. Such devices may be used if their accuracy has been demonstrated. Multiplying calipers and other such devices are acceptable for use provided their accuracy has been demonstrated as equivalent to a Class C extensometer.

NOTE 4—Reference should be made to Method E 83.

NOTE 5—For steel with a yield point specified not over 80 000 psi (550 MPa), an appropriate value is 0.005 in./in. of gage length. For values above 80 000 psi, this method is not valid unless the limiting total extension is increased.

13.2 *Yield Strength*—Yield strength is the stress at which a material exhibits a specified limiting deviation from the proportionality of stress to strain. The deviation is expressed in terms of strain, percent offset, total extension under load, etc. Determine yield strength by one of the following methods:

13.2.1 *Offset Method*—To determine the yield strength by the "offset method," it is necessary to secure data (autographic or numerical) from which a stress-strain diagram may be drawn. Then on the stress-strain diagram (Fig. 10) lay off *Om* equal to the specified value of the offset, draw *mn* parallel to *OA*, and thus locate *r*, the intersection of *mn* with the stress-strain curve corresponding to load *R* which is the yield strength load. In reporting values of yield strength obtained by this method, the specified value of "offset" used should be stated in parentheses after the term yield strength, thus:

Yield strength (0.2% offset)
= 52 000 psi (360 MPa)

In using this method, a minimum extensometer magnification of 250 to 1 is required. A Class B1 extensometer meets this requirement (see Note 5). See also Note 7 for automatic devices.

13.2.2 *Extension Under Load Method*—For tests to determine the acceptance or rejection of material whose stress-strain characteristics are well known from previous tests of similar material in which stress-strain diagrams were plotted, the total strain corresponding to the stress at which the specified offset (see Note 7) occurs will be known within satisfactory limits. The stress on the specimen, when this total strain is reached, is the value of the yield strength. The total strain can be obtained satisfactorily by use of a Class B1 extensometer (Notes 3 and 4).

NOTE 6—Automatic devices are available that determine offset yield strength without plotting a stress-strain curve. Such devices may be used if their accuracy has been demonstrated.

NOTE 7—The appropriate magnitude of the extension under load will obviously vary with the strength range of the particular steel under test. In general, the value of extension under load applicable to steel at any strength level may be determined from the sum of the proportional strain and the plastic strain expected at the specified yield strength. The following equation is used:

Extension under load, in./in. of gage length
$= (YS/E) + r$

where:

YS = specified yield strength, psi or MPa,
E = modulus of elasticity, psi or MPa, and
r = limiting plastic strain, in./in.

13.3 *Tensile Strength*—Calculate the tensile strength by dividing the maximum load the specimen sustains during a tension test by the original cross-sectional area of the specimen.

13.4 *Elongation:*

13.4.1 Fit the ends of the fractured specimen together carefully and measure the distance between the gage marks to the nearest 0.01 in. (0.25 mm) for gage lengths of 2 in. and under, and to the nearest 0.5 percent of the gage length for gage lengths over 2 in. A percentage scale reading to 0.5 percent of the gage length may be used. The elongation is the increase in length of the gage length, expressed as a percentage of the original gage length. In reporting elongation values, give both the percentage increase and the original gage length.

13.4.2 If any part of the fracture takes place outside of the middle half of the gage length or in a punched or scribed mark within the reduced section, the elongation value obtained may not be representative of the material. If the elongation so measured meets the minimum requirements specified, no further testing is indicated, but if the elongation is less than the minimum requirements, discard the test and retest.

13.5 *Reduction of Area*—Fit the ends of the fractured specimen together and measure the mean diameter or the width and thickness at the smallest cross section to the same accuracy as the original dimensions. The difference between the area thus found and the area of the original cross section expressed as a percentage of the original area, is the reduction of area.

BEND TEST

14. Description

14.1 The bend test is one method for evaluating ductility, but it cannot be considered as a quantitative means of predicting service performance in bending operations. The severity of the bend test is primarily a function of the angle of bend and inside diameter to which the specimen is bent, and of the cross section of the specimen. These conditions are varied according to location and orientation of the test specimen and the chemical composition, tensile properties, hardness, type, and quality of the steel specified.

14.2 Unless otherwise specified, it shall be permissible to age bend test specimens. The time-temperature cycle employed must be such that the effects of previous processing will not be materially changed. It may be accomplished by aging at room temperature 24 to 48 h, or in shorter time at moderately elevated temperatures by boiling in water, heating in oil, or in an oven.

14.3 Bend the test specimen at room temperature to an inside diameter, as designated by the applicable product specifications, to the extent specified without major cracking on the outside of the bent portion. The speed of bending is ordinarily not an important factor.

HARDNESS TEST

15. General

15.1 A hardness test is a means of determining resistance to penetration and is occasionally employed to obtain a quick approximation of tensile strength. Tables 3A, 3B, 3C, and 3D are for the conversion of hardness measurements from one scale to another or to approximate tensile strength. These conversion values have been obtained from computer-generated curves and are presented to the nearest 0.1 point to permit accurate reproduction of those curves. Since all converted hardness values must be considered approximate, however, all converted Rockwell hardness numbers shall be rounded to the nearest whole number.

16. Brinell Test

16.1 *Description:*

16.1.1 A specified load is applied to a flat surface of the specimen to be tested, through a hard ball of specified diameter. The average diameter of the indentation is used as a basis for calculation of the Brinell hardness number. The quotient of the applied load divided by the area of the surface of the indentation, which is assumed to be spherical, is termed the Brinell hardness number (HB) in accordance with the following equation:

$$HB = P/[(\pi D/2)(D - \sqrt{D^2 - d^2}\,)]$$

where:
HB = Brinell hardness number,
P = applied load, kgf,
D = diameter of the steel ball, mm, and
d = average diameter of the indentation, mm.

NOTE 8—The Brinell hardness number is more conveniently secured from standard tables which show numbers corresponding to the various indentation diameters, usually in increments of 0.05 mm.

16.1.2 The standard Brinell test using a 10-mm ball employs a 3000-kgf load for hard materials and a 1500 or 500-kgf load for thin sections or soft materials (see Supplement II on Steel Tubular Products, Section S 8). Other loads and different size indentors may be used when specified. In reporting hardness values, the diameter of the ball and the load must be stated except when a 10-mm ball and 3000-kgf load are used.

16.1.3 A range of hardness can properly be specified only for quenched and tempered or normalized and tempered material. For annealed material a maximum figure only should be specified. For normalized material a minimum or a maximum hardness may be specified by agreement. In general, no hardness requirements should be applied to untreated material.

16.1.4 Brinell hardness may be required when tensile properties are not specified. When agreed upon, hardness tests can be substituted for tension tests in order to expedite testing of a large number of duplicate pieces from the same lot.

16.2 *Apparatus*—Equipment shall meet the following requirements:

16.2.1 *Testing Machine*—A Brinell hard-

A 370

ness testing machine is acceptable for use over a loading range within which its load measuring device is accurate within 3 percent.

16.2.2 *Micrometer Microscope*—The micrometer microscope or equivalent device for measuring diameter or depth of indentation is adjusted so that throughout the range covered the error of reading does not exceed 0.02 mm.

16.2.3 *Standard Ball*—The standard ball for Brinell hardness testing is 10 mm (0.3937 in.) in diameter with a deviation from this value of not more than 0.01 mm (0.0004 in.) in any diameter. A ball suitable for use must not show a permanent change in diameter greater than 0.01 mm (0.0004 in.) when pressed with a force of 3000 kgf against the test specimen.

16.3 *Test Specimen*—Brinell hardness tests are made on prepared areas and sufficient metal must be removed from the surface to eliminate decarburized metal and other surface irregularities. The thickness of the piece tested must be such that no bulge or other marking showing the effect of the load appears on the side of the piece opposite the indentation.

16.4 *Procedure:*

16.4.1 It is essential that the applicable product specifications state clearly the position at which Brinell hardness indentations are to be made and the number of such indentations required. The distance of the center of the indentation from the edge of the specimen or edge of another indentation must be at least three times the diameter of the indentation.

16.4.2 Apply the load for a minimum of 10 s.

16.4.3 Measure two diameters of the indentation at right angles to the nearest 0.1 mm, estimate to the nearest 0.05 mm, and average to the nearest 0.05 mm. If the two diameters differ by more than 0.1 mm, discard the readings and make a new indentation.

16.4.4 Do not use a steel ball on steels having a hardness over 444 HB nor a carbide ball over 627 HB. The Brinell test is not recommended for materials having a HB over 627.

16.5 *Detailed Procedure*—For detailed requirements of this test, reference shall be made to the latest revision of Method E 10.

17. Portable Hardness Test

17.1 *Portable Testers*—Under certain circumstances, it may be desirable to substitute a portable Brinell testing instrument, which is calibrated to give equivalent results to those of a standard Brinell machine on a comparison test bar of approximately the same hardness as the material to be tested.

17.2 *Detailed Procedure*—For detailed requirements of the portable test, reference shall be made to the latest revision of Method E 110.

18. Rockwell Test

18.1 *Description:*

18.1.1 In this test a hardness value is obtained by using a direct-reading testing machine which measures hardness by determining the depth of penetration of a diamond point or a steel ball into the specimen under certain arbitrarily fixed conditions. A minor load of 10 kgf is first applied which causes an initial penetration, sets the penetrator on the material and holds it in position. A major load which depends on the scale being used is applied increasing the depth of indentation. The major load is removed and, with the minor load still acting, the Rockwell number, which is proportional to the difference in penetration between the major and minor loads, is read directly on the dial gage. This is an arbitrary number which increases with increasing hardness. The scales most frequently used are as follows:

Scale Symbol	Penetrator	Major Load, kgf	Minor Load, kgf
B	$\frac{1}{16}$-in. steel ball	100	10
C	Diamond brale	150	10

18.1.2 Rockwell superficial hardness machines are used for the testing of very thin steel or thin surface layers. Loads of 15, 30, or 45 kgf are applied on a hardened steel ball or diamond penetrator, to cover the same range of hardness values as for the heavier loads. The superficial hardness scales are as follows:

A 370

Scale Symbol	Penetrator	Major Load, kgf	Minor Load, kgf
15T	1/16-in. steel ball	15	3
30T	1/16-in. steel ball	30	3
45T	1/16-in. steel ball	45	3
15N	Diamond brale	15	3
30N	Diamond brale	30	3
45N	Diamond brale	45	3

18.2 *Reporting Hardness*—In reporting hardness values, the hardness number should always precede the scale symbol, 96 HRB, 40 HRC, 75 HR15N, or 77 HR30T.

18.3 *Test Blocks*—Machines should be checked to make certain they are in good order by means of standardized Rockwell test blocks.

18.4 *Detailed Procedure*—For detailed requirements of this test, reference shall be made to the latest revision of Methods E 18.

CHARPY IMPACT TESTING

18. Description

19.1 A Charpy impact test is a dynamic test in which a selected specimen, machined or surface ground and notched, is struck and broken by a single blow in a specially designed testing machine and the energy absorbed in breaking the specimen is measured. The energy values determined are qualitative comparisons on a selected specimen and although frequently specified as an acceptance criterion, they cannot be converted into energy figures that would serve for engineering calculations. Percentage shear fracture and mils of lateral expansion opposite the notch are other frequently used criteria of acceptance for Charpy V-notch impact test specimens.

19.2 Testing temperatures other than ambient temperature are often specified in the individual product specifications. Although the testing temperature is sometimes governed by the service temperature, the two may not be identical.

19.3 Further information on the significance of impact testing appears in Supplement V.

20. Test Specimens

20.1 *Selection and Number of Tests:*
20.1.1 Unless otherwise specified, longitudinal test specimens shall be used with the notch perpendicular to the surface of the object being tested.

20.1.2 An impact test shall consist of three specimens taken from a single test coupon or test location.

20.2 *Size and Type:*
20.2.1 The type of specimen desired, Charpy V-notch Type A or Charpy keyhole notch Type B, shown in Fig. 11, should be specified.

20.2.2 For material less than $7/16$ in. (11 mm) thick, subsize test specimens shall be used. They shall be made to the following dimensions and to the tolerances shown in Fig. 11:

10 by 7.5 mm
10 by 6.7 mm
10 by 5 mm
10 by 3.3 mm
10 by 2.5 mm

The base of the notch shall be perpendicular to the 10-mm-wide face.

20.2.3 When subsize specimens are required, the specified energy level or test temperature, or both, shall be reduced as agreed upon by purchaser and supplier.

NOTE 9—The Charpy U-notch specimen may be substituted for the keyhold specimen. A sketch of the U-notch specimen may be found as Fig. 4 (Specimen Type C) in Methods E 23.

20.3 *Notch Preparation:*
20.3.1 Particular attention must be paid to the machining of V-notches as it has been demonstrated that extremely minor variations in notch radius may result in very erratic test data. Tool marks at the bottom of the notch must be carefully avoided.

20.3.2 Keyhole notches shall be made by drilling the round hole and then cutting the slot by any feasible means. The drilling must be done carefully with a slow feed. Care must also be exercised in cutting the slot to see that the surface of the drilled hole is not damaged.

21. Testing Apparatus and Conditions

21.1 *General Characteristics:*
21.1.1 A Charpy impact machine is one in which a notched specimen is broken by a single blow of a freely swinging pendulum. The pendulum is released from a fixed height, so that the energy of the blow is fixed and known. The height to which the pendulum

rises in its swing after breaking the specimen is measured and used to determine the residual energy of the pendulum. The specimen is supported horizontally as a simple beam with the axis of the notch vertical. It is struck in the middle of the face opposite the notch.

21.1.2 Charpy machines used for testing steel generally strike the specimen with an energy of from 220 to 265 ft·lbf (298 to 359 J) and a linear velocity at the point of impact of 16 to 19 ft (4.88 to 5.80 m)/s. Sometimes machines of lighter capacity are used.

21.2 *Calibration (Accuracy and Sensitivity):*

21.2.1 Charpy impact machines shall be calibrated and adjusted in accordance with the requirements of the latest revision of Methods E 23.

21.2.2 The indicator should have an error not greater than 1 ft·lbf (1.4 J) as calibrated by the prescribed procedure.

21.2.3 The dimensions of the pendulum should be such that the center of percussion is at the point of impact with an error not greater than 1 percent of the distance from the axis of rotation to the point of impact.

21.2.4 The dimensions of the specimen supports and striking edge shall conform to Fig. 12.

21.3 *Temperature:*

21.3.1 The effect of variations in temperature on Charpy test results is sometimes very great and this variable shall be closely controlled. The actual temperature at which each specimen is broken shall be reported.

21.3.2 Tests are often specified to be run at low temperatures. These low temperatures can be obtained readily in the laboratory by the use of chilled liquids such as: water, ice plus water, dry ice plus organic solvents, liquid nitrogen, or chilled gases. Specimens to be tested at low temperatures shall be held at the specified temperature for at least 5 min in liquid coolants and 60 min in gaseous environments.

21.3.3 For elevated-temperature tests, the specimens shall preferably be immersed in an agitated oil, or other suitable liquid bath and held at temperature for at least 10 min; if samples are heated in an oven they must be held in the oven for at least 60 min.

21.3.4 When tested at temperatures other than ambient, specimens shall be inserted in the machine and broken within 5 s so as to minimize the change of temperature prior to breaking.

21.3.5 Tongs for handling the test specimens, and centering devices used to ensure proper location of the test on the anvil of the impact tester, shall be at the same relative temperature as the test specimen prior to each test so as not to affect the temperature of the test specimen at the notch.

22. Test Results

22.1 The result of an impact test shall be the average (arithmetic mean) of the results of the three specimens.

22.2 When the acceptance criteria are based on absorbed energy, not more than one specimen may exhibit a value below the specified minimum average, and in no case shall an individual value be below either two thirds of the specified minimum average or 5 ft·lbf (6.8 J), whichever is greater, subject to the retest provisions of 22.2.1.

22.2.1 If more than one specimen is below the specified minimum average, or if one value is below two thirds of the specified minimum average, a retest of three additional specimens shall be made, each of which shall have a value equal to or exceeding the specified minimum average value.

22.3 When the acceptance criteria are based on lateral expansion, the value for each of the specimens must equal or exceed the specified minimum value subject to the retest provision of 22.3.1.

22.3.1 If the value on one specimen falls below the specified minimum value, and not below two thirds of the specified minimum value, and if the average of the three specimens equals or exceeds the specified minimum value, a retest of three additional specimens shall be made. The value for each of the three retest specimens must equal or exceed the specified minimum value.

23. Acceptance Criteria

23.1 *Impact Strength*—In some applications, impact tests are specified to determine the behavior of the metal when subjected to a single application of a load that produces multiaxial stresses associated with a notch with high rates of loading, in some cases at high or low temperature. Data are reported in terms of foot-pounds of absorbed energy at the test temperature.

23.2 *Ductile-to-Brittle Transition Temperature*—Body-centered-cubic or ferritic alloys exhibit a significant change in behavior when impact tested over a range of temperatures. At elevated temperatures, impact specimens fracture by a shear mechanism absorbing large amounts of energy; at low temperatures they fracture brittlely by a cleavage mechanism absorbing little energy. The transition from one type of behavior to the other has been defined in various ways for specification purposes: (*1*) the temperature corresponding to a specific energy level; (*2*) the temperature at which Charpy V-notch specimens exhibit some specific value of cleavage (shiny, facetted appearance, often termed brittle or crystalline) and shear (often termed ductile or fibrous) fractures. This temperature is commonly called the fracture appearance transition temperature or $FATT_n$ where "n" is the percentage of shear fracture. $FATT_{50}$ is most frequently specified; (*3*) the temperature at which the lateral expansion (increase in specimen width on the compression side, opposite the notch, of the fractured Charpy V-notch specimen, Fig. 13) is some specified amount measured in thousandths of an inch (mils).

23.2.1 *Energy Level*—Energy level as determined on the Charpy V-notch impact test has been shown to have fairly good correlation with service failures and also with the nil-ductility transition temperature determined by the drop-weight test (Method E 208). Specific requirements should be based on material capability and either service experience or correlations with the drop weight test or other valid tests for fracture toughness. The test temperature must be specified.

23.2.2 *Fracture Appearance Transition Temperature, $FATT_n$:*

23.2.2.1 *Determination of Percent Shear Fracture*—The percentage of shear fracture may be determined by any of the following methods: (*1*) Measure the length and width of the cleavage portion of the fracture surface, as shown in Fig. 14, and determine the percent shear from either Table 4 or Table 5 depending on the units of measurement; (*2*) compare the appearance of the fracture of the specimen with a fracture appearance chart such as that shown in Fig. 15; (*3*) magnify the fracture surface and compare it to a precalibrated overlay chart or measure the percent shear fracture by means of a planimeter; or (*4*) photograph the fracture surface at a suitable magnification and measure the percent shear fracture by means of a planimeter.

23.2.2.2 *Determination of Transition Temperature*—For determining the transition temperature, break at least four specimens that have been taken from comparable locations. Break each specimen at a different temperature, but in a range of temperature that will produce fractures within the range of ±25 percent of the specified value, *n*, of shear. Plot the percent shear fracture against the test temperature and determine the transition by graphic interpolation (extrapolation is not permitted).

23.2.3 *Mils of Lateral Expansion:*

23.2.3.1 *Determination of Lateral Expansion*—The method for measuring lateral expansion must take into account the fact that the fracture path seldom bisects the point of maximum expansion on both sides of a specimen. One half of a broken specimen may include the maximum expansion for both sides, one side only, or neither. The technique used must therefore provide an expansion value equal to the sum of the higher of the two values obtained for each side by measuring the two halves separately. The amount of expansion on each side of each half must be measured relative to the plane defined by the undeformed portion of the side of the specimen. Expansion may be measured by using a gage similar to that shown in Figs. 16 and 17. Measure the two broken halves individually. First, though, check the sides perpendicular to the notch to ensure that no burrs were formed on these sides during impact testing; if such burrs exist, they must be removed, for example, by rubbing on emery cloth, making sure that the protrusions being measured are not rubbed during the removal of the burr. Next, place the halves together so that the compression sides are facing one another. Take one half and press it firmly against the reference supports, with the protrusion against the gage anvil. Note the reading, then repeat this step with the other broken half, ensuring that the same side of the specimen is measured. The larger of the two values is the expansion of that side of the specimen. Next, repeat this procedure to measure the protrusions on the opposite side, then add the larger values obtained for each side. Measure each specimen.

NOTE 10—Examine each fracture surface to ascertain that the protrusions have not been damaged by contacting the anvil, machine mounting surface, etc. Such samples should be discarded since this may cause erroneous readings.

23.2.3.2 *Determination of Transition Temperature*—For determining the transition temperature, break a sufficient number of specimens over a range of temperatures such that the temperature producing the specified lateral expansion may be determined by graphic interpolation (extrapolation is not permitted).

23.3 *Report*—Test reports shall include the test temperature and energy value (foot-pounds) for each test specimen broken. When specified in the product specification the percent shear fracture or mils of lateral expansion, or both, shall also be reported for each test specimen broken.

SUPPLEMENTS

I. STEEL BAR PRODUCTS

S1. Scope

S1.1 This supplement delineates only those details which are peculiar to hot-rolled and cold-finished steel bars and are not covered in the general section of these methods.

S2. Orientation of Test Specimens

S2.1 Carbon steel bars and bar-size shapes, due to their relatively small cross-sectional dimensions, are customarily tested in the longitudinal direction.

S2.2 Alloy steel bars and bar-size shapes are usually tested in the longitudinal direction. In special cases where size permits and the fabrication or service of a part justifies testing in a transverse direction, the selection and location of test or tests are a matter of agreement between the manufacturer and the purchaser.

S3. Tension Test

S3.1 *Carbon Steel Bars*—Carbon steel bars are not commonly specified to tensile requirements in the as-rolled condition for sizes of rounds, squares, hexagons, and octagons under $1/2$ in. (13 mm) in diameter or distance between parallel faces nor for other bar-size sections, other than flats, less than 1 in.^{2} (645. mm^2) in cross-sectional area.

S3.2 *Alloy Steel Bars*—Alloy steel bars are usually not tested in the as-rolled condition.

S3.3 When tension tests are specified, the recommended practice for selecting test specimens for hot-rolled and cold-finished steel bars of various sizes shall be in accordance with Table 7, unless otherwise specified.

S4. Bend Test

S4.1 When bend tests are specified, the recommended practice for hot-rolled and cold-finished steel bars shall be in accordance with Table 6.

II. STEEL TUBULAR PRODUCTS

S5. Scope

S5.1 This supplement covers definitions and methods of testing peculiar to tubular products which are not covered in the general section of these methods.

S6. Tension Test

S6.1 *Longitudinal Test Specimens:*

S6.1.1 It is standard practice to use tension test specimens of full-size tubular sections within the limit of the testing equipment (Fig. 20 (*d*)). Snug-fitting metal plugs should be inserted far enough in the end of such tubular specimens to permit the testing machine jaws to grip the specimens properly without crushing. A design that may be used for such plugs is shown in Fig. 18. The plugs shall not extend into that part of the specimen on which the elongation is measured (Fig. 18). Care should be exercised to see that insofar as practicable, the load in such cases is applied axially. The length of the full-section specimen depends on the gage length prescribed for measuring the elongation.

S6.1.2 Unless otherwise required by the individual product specification, the gage

A 370

length for furnace-welded pipe is normally 8 in. (200 mm), except that for nominal sizes ¾ in. and smaller, the gage length shall be as follows:

Nominal Size, in.	Gage Length, in. (mm)
¼ and ½	6 (150)
⅛ and ¼	4 (100)
⅛	2 (50)

S6.1.3 For seamless and electric-welded pipe and tubes the gage length is 2 in. However, for tubing having an outside diameter of ⅜ in. (10 mm) or less, it is customary to use a gage length equal to four times the outside diameter when elongation values comparable to larger specimens are required.

S6.1.4 To determine the cross-sectional area of the full-section specimen, measurements shall be recorded as the average or mean between the greatest and least measurements of the outside diameter and the average or mean wall thickness, to the nearest 0.001 in. (0.025 mm) and the cross-sectional area is determined by the following equation:

$$A = 3.1416t(D - t)$$

where:
A = sectional area, in.²
D = outside diameter, in., and
t = thickness of tube wall, in.

NOTE 11—There exist other methods of cross-sectional area determination, such as by weighing of the specimens, which are equally accurate or appropriate for the purpose.

S6.2 *Longitudinal Strip Test Specimens:*

S6.2.1 For larger sizes of tubular products which cannot be tested in full-section, longitudinal test specimens are obtained from strips cut from the tube or pipe as indicated in Fig. 19. For furnace-welded tubes or pipe the 8-in. gage length specimen as shown in Fig. 20 (b), or with both edges parallel as in Fig. 20 (a) is standard, the specimen being located at approximately 90 deg from the weld. For seamless and electric-welded tubes or pipe, the 2-in. gage length specimen as shown in Fig. 20 (c) is standard, the specimen being located approximately 90 deg from the weld in the case of electric-welded tubes. The specimen shown in Fig. 20 (a) may be used as an alternate for seamless and electric-welded tubes or pipe. Specimens of the type shown in Fig. 20 (a), (b), (c), may be tested with grips having a surface contour corresponding to the curvature of the tubes. When grips with curved faces are not available, the ends of the specimens may be flattened without heating. Standard tension test specimens, as shown in specimen No. 4 of Fig. 21, are nominally 1½ in. (38 mm) wide in the gage length section. When sub-size specimens are necessary due to the dimensions and character of the material to be tested, specimens 1, 2, or 3 shown in Fig. 21 where applicable, are considered standard. For tubes ¾ in. (19 mm) and over in wall thickness, the test specimen shown in Fig. 5 (Note 12) may be used.

NOTE 12—Standard round tension test specimen with 2-in. gage length.

S6.2.2 The width should be measured at each end of the gage length to determine parallelism and also at the center. The thickness should be measured at the center and used with the center measurement of the width to determine the cross-sectional area. The center width dimension should be recorded to the nearest 0.005 in. (0.127 mm), and the thickness measurement to the nearest 0.001 in. When the specimen shown in Fig. 5 (Note 12) is used, the diameter is measured at the center of the specimen to the nearest 0.001 in. (0.025 mm).

S6.3 *Transverse Test Specimens.*

S6.3.1 In general, transverse tension tests are not recommended for tubular products, in sizes smaller than 8 in. in nominal diameter. When required, transverse tension test specimens may be taken from rings cut from ends of tubes or pipe as shown in Fig. 22. Flattening of the specimen may be done either after separating it from the tube as in Fig. 22 (a), or before separating it as in Fig. 22 (b), and may be done hot or cold; but if the flattening is done cold, the specimen may subsequently be normalized. Specimens from tubes or pipe for which heat treatment is specified, after being flattened either hot or cold, shall be given the same treatment as the tubes or pipe. For tubes or pipe having a wall thickness of less than ¾ in. (19 mm), the transverse test specimen shall be of the form and dimensions shown in Fig. 23 and either or both surfaces may be machined to secure uniform thickness. For tubes having a sufficiently heavy wall thickness the test specimen shown in Fig. 5 (Note 12) may be used. The elongation requirements for the 2-in. gage length in the product specification shall apply to the gage length as specified in Fig. 5. Specimens for transverse tension tests on welded steel tubes or pipe to determine strength of welds, shall be located perpendicular to the welded seams with the weld at about the middle of their length.

S6.3.2 The width should be measured at each end of the gage length to determine parallelism and also at the center. The thickness should be measured at the center and used with the center measurement of the width to determine the cross-sectional area. The center width dimension should be recorded to the nearest 0.005 in. (0.127 mm), and the thickness measurement to the nearest 0.001 in. (0.025 mm). When the specimen shown in Fig. 5 (Note 12) is used, the diameter is measured at the center of the specimen to the nearest 0.001 in.

S7. Determination of Transverse Yield Strength, Hydraulic Ring-Expansion Method

S7.1 Until recently, the transverse yield strength, when required on tubular products, has been determined, as described in the general section of these methods, from standard tension test coupons cut transversely from the tubular sections. Due to the curvature on such coupons it is necessary to cold straighten them. It has long been recognized that the cold work introduced by straightening changes the mechanical properties so that the yield strength obtained is not truly representative of the yield strength in the original tubular section. The transverse yield strength is highly important on some classes of tubular products, such as line pipe, and a method for determining the true yield strength has been desirable for some time.

S7.2 A testing machine and method for determining the transverse yield strength from an annular ring specimen, have been developed and described in S7.3 through S7.5.

S7.3 A diagrammatic vertical cross-sectional sketch of the testing machine is shown in Fig. 24.

S7.4 In determining the transverse yield strength on this machine, a short ring (commonly 3 in. (76 mm) in length) test specimen is used. After the large circular nut is removed from the machine, the wall thickness of the ring specimen is determined and the specimen is telescoped over the oil resistant rubber gasket. The nut is then replaced, but is not turned down tight against the specimen. A slight clearance is left between the nut and specimen for the purpose of permitting free radial movement of the specimen as it is being tested. Oil under pressure is then admitted to the interior of the rubber gasket through the pressure line under the control of a suitable valve. An accurately calibrated pressure gage serves to measure oil pressure. Any air in the system is removed through the bleeder line. As the oil pressure is increased, the rubber gasket expands which in turn stresses the specimen circumferentially. As the pressure builds up, the lips of the rubber gasket act as a seal to prevent oil leakage. With continued increase in pressure, the ring specimen is subjected to a tension stress and elongates accordingly. The entire outside circumference of the ring specimen is considered as the gage length and the strain is measured with a suitable extensometer which will be described later. When the desired total strain or extension under load is reached on the extensometer, the oil pressure in pounds per square inch is read and by employing Barlow's formula, the unit yield strength is calculated. The yield strength, thus determined, is a true result since the test specimen has not been cold worked by flattening and closely approximates the same condition as the tubular section from which it is cut. Further, the test closely simulates service conditions in pipe lines. One testing machine unit may be used for several different sizes of pipe by the use of suitable rubber gaskets and adapters.

NOTE 13—Barlow's formula may be stated two ways:

(1) $P = 2St/D$

(2) $S = PD/2t$

where:
P = internal hydrostatic pressure, psi,
S = unit circumferential stress in the wall of the tube produced by the internal hydrostatic pressure, psi,
t = thickness of the tube wall, in., and
D = outside diameter of the tube, in.

S7.5 A roller chain type extensometer which has been found satisfactory for measuring the elongation of the ring specimen is shown in Figs. 25 and 26. Figure 25 shows the extensometer in position, but unclamped, on a ring specimen. A small pin, through which the strain is transmitted to and measured by

A 370

the dial gage, extends through the hollow threaded stud. When the extensometer is clamped, as shown in Fig. 26, the desired tension which is necessary to hold the instrument in place and to remove any slack, is exerted on the roller chain by the spring. Tension on the spring may be regulated as desired by the knurled thumb screw. By removing or adding rollers, the roller chain may be adapted for different sizes of tubular sections.

S8. Hardness Tests

S8.1 Hardness tests are made either on the outside or the inside surfaces on the end of the tube as appropriate.

S8.2 The standard 3000-kgf Brinell load may cause too much deformation in a thin-walled tubular specimen. In this case the 500-kgf load shall be applied, or inside stiffening by means of an internal anvil should be used. Brinell testing shall not be applicable to tubular products less than 2 in. (51 mm) in outside diameter, or less than 0.200 in. (5.1 mm) in wall thickness.

S8.3 The Rockwell hardness tests are normally made on the inside surface, a flat on the outside surface, or on the wall cross-section depending upon the product limitation. Rockwell hardness tests are not performed on tubes smaller than $5/16$ in. (7.9 mm) in outside diameter, nor are they performed on the inside surface of tubes with less than $1/4$ in. (6.4 mm) inside diameter. Rockwell hardness tests are not performed on annealed tubes with walls less than 0.065 in. (1.65 mm) thick or cold worked or heat treated tubes with walls less than 0.049 in. (1.24 mm) thick. For tubes with wall thicknesses less than those permitting the regular Rockwell hardness test, the Superficial Rockwell test is sometimes substituted. Transverse Rockwell hardness readings can be made on tubes with a wall thickness of 0.187 in. (4.75 mm) or greater. The curvature and the wall thickness of the specimen impose limitations on the Rockwell hardness test. When a comparison is made between Rockwell determinations made on the outside surface and determinations made on the inside surface, adjustment of the readings will be required to compensate for the effect of curvature. The Rockwell B scale is used on all materials having an expected hardness range of B 0 to B 100. The Rockwell C scale is used on material having an expected hardness range of C 20 to C 68.

S8.4 Superficial Rockwell hardness tests are normally performed on the outside surface whenever possible and whenever excessive spring back is not encountered. Otherwise, the tests may be performed on the inside. Superficial Rockwell hardness tests shall not be performed on tubes with an inside diameter of less than $1/4$ in. (6.4 mm). The wall thickness limitations for the Superficial Rockwell hardness test are given in Tables 8 and 9.

S8.5 When the outside diameter, inside diameter, or wall thickness precludes the obtaining of accurate hardness values, tubular products shall be specified to tensile properties and so tested.

S9. Manipulating Tests

S9.1 The following tests are made to prove ductility of certain tubular products:

S9.1.1 *Flattening Test*—The flattening test as commonly made on specimens cut from tubular products is conducted by subjecting rings from the tube or pipe to a prescribed degree of flattening between parallel plates (Fig. 22). The severity of the flattening test is measured by the distance between the parallel plates and is varied according to the dimensions of the tube or pipe. The flattening test specimen should not be less than $2\frac{1}{2}$ in. (63.5 mm) in length and should be flattened cold to the extent required by the applicable material specifications.

S9.1.2 *Reverse Flattening Test*—The reverse flattening test is designed primarily for application to electric-welded tubing for the detection of lack of penetration or overlaps resulting from flash removal in the weld. The specimen consists of a length of tubing approximately 4 in. (102 mm) long which is split longitudinally 90 deg on each side of the weld. The sample is then opened and flattened with the weld at the point of maximum bend (Fig. 27).

S9.1.3 *Crush Test*—The crush test, sometimes referred to as an upsetting test, is usually made on boiler and other pressure tubes, for evaluating ductility (Fig. 28). The specimen is a ring cut from the tube, usually about $2\frac{1}{2}$ in. (63.5 mm) long. It is placed on end and crushed endwise by hammer or press to the distance prescribed by the applicable material specifications.

S9.1.4 *Flange Test*—The flange test is intended to determine the ductility of boiler tubes and their ability to withstand the operation of bending into a tube sheet. The test is made on a ring cut from a tube, usually not less than 4 in. (100 mm) long and consists of having a flange turned over at right angles to the body of the tube to the width required by the applicable material specifications. The flaring tool and die block shown in Fig. 29 are recommended for use in making this test.

S9.1.5 *Flaring Test*—For certain types of pressure tubes, an alternate to the flange test is made. This test consists of driving a tapered mandrel having a slope of 1 in 10 as shown in Fig. 30 (*a*) or a 60 deg included angle as shown in Fig. 30 (*b*) into a section cut from the tube, approximately 4 in. (100 mm) in length, and thus expanding the specimen until the inside diameter has been increased to the extent required by the applicable material specifications.

S9.1.6 *Bend Test*—For pipe used for coiling in sizes 2 in. and under a bend test is made to determine its ductility and the soundness of weld. In this test a sufficient length of full-size pipe is bent cold through 90 deg around a cylindrical mandrel having a diameter 12 times the nominal diameter of the pipe. For close coiling, the pipe is bent cold through 180 deg around a mandrel having a diameter 8 times the nominal diameter of the pipe.

S9.1.7 *Transverse Guided Bend Test of Welds*—This bend test is used to determine the ductility of fusion welds. The specimens used are approximately $1\frac{1}{2}$ in. (38 mm) wide, at least 6 in. (152 mm) in length with the weld at the center, and are machined in accordance with Fig. 31(*a*) for face and root bend tests and in accordance with Fig. 31(*b*) for side bend tests. The dimensions of the plunger shall be as shown in Fig. 32 and the other dimensions of the bending jig shall be substantially as given in this same figure. A test shall consist of a face bend specimen and a root bend specimen or two side bend specimens. A face bend test requires bending with the inside surface of the pipe against the plunger; a root bend test requires bending with the outside surface of the pipe against the plunger; and a side bend test requires bending so that one of the side surfaces becomes the convex surface of the bend specimen.

S9.1.7.1 Failure of the bend test depends upon the appearance of cracks in the area of the bend, of the nature and extent described in the product specifications.

III. STEEL FASTENERS

S10. Scope

S10.1 This supplement covers definitions and methods of testing peculiar to steel fasteners which are not covered in the general section of Methods A 370. Standard tests required by the individual product specifications are to be performed as outlined in the general section of these methods.

S10.2 These tests are set up to facilitate production control testing and acceptance testing with certain more precise tests to be used for arbitration in case of disagreement over test results.

S11. Tension Tests

S11.1 It is preferred that bolts be tested full size, and it is customary, when so testing bolts to specify a minimum ultimate load in pounds, rather than a minimum ultimate strength in pounds per square inch. Three times the bolt nominal diameter has been established as the minimum bolt length subject to the tests described in the remainder of this section. Sections S11.1.1 through S11.1.3 apply when testing bolts full size. Section S11.1.4 shall apply where the individual product specifications permit the use of machined specimens.

S11.1.1 *Proof Load*—Due to particular uses of certain classes of bolts it is desirable to be able to stress them, while in use, to a specified value without obtaining any permanent set. To be certain of obtaining this quality the proof load is specified. The proof load test consists of stressing the bolt with a specified load which the bolt must withstand without permanent set. An alternate test which determines yield strength of a full size bolt is also allowed. Either of the following Methods, 1 or 2, may be used but Method 1 shall be the arbitration method in case of any dispute as to acceptance of the bolts.

S11.1.2 *Proof Load Testing Long Bolts*—When full size tests are required, proof load Method 1 is to be limited in application to bolts whose length does not exceed 8 in. (203 mm) or 8 times the nominal diameter, whichever is greater. For bolts longer than 8 in. or 8 times the nominal diameter, whichever is greater, proof load Method 2 shall be used.

S11.1.2.1 *Method 1, Length Measurement*—The overall length of a straight bolt shall be measured at its true center line with an instrument capable of measuring changes in length of 0.0001 in. (0.0025 mm) with an accuracy of 0.0001 in. in any 0.001-in. (0.025-mm) range. The preferred method of measuring the length shall be between conical centers machined on the center line of the bolt, with mating centers on the measuring anvils. The head or body of the bolt shall be marked so that it can be placed in the same position for all measurements. The bolt shall be assembled in the testing equipment as outlined in S11.1.4, and the proof load specified in the product specification shall be applied. Upon release of this load the length of the bolt shall be again measured and shall show no permanent elongation. A tolerance of ±0.0005 in. (0.0127 mm) shall be allowed between the measurement made before loading and that made after loading. Variables, such as straightness and thread alignment (plus measurement error), may result in apparent elongation of the fasteners when the proof load is initially applied. In such cases, the fastener may be retested using a 3 percent greater load, and may be considered satisfactory if the length after this loading is the same as before this loading (within the 0.0005-in. tolerance for measurement error).

S11.1.3 *Proof Load-Time of Loading*—The proof load is to be maintained for a period of 10 s before release of load, when using Method 1.

S11.1.3.1 *Method 2, Yield Strength*—The bolt shall be assembled in the testing equipment as outlined in S11.1.4. As the load is applied, the total elongation of the bolt or any part of the bolt which includes the exposed six threads shall be measured and recorded to produce a load-strain or a stress-strain diagram. The load or stress at an offset equal to 0.2 percent of the length of bolt occupied by 6 full threads shall be determined by the method described in 13.2.1 of these methods, A 370. This load or stress shall not be less than that prescribed in the product specification.

S11.1.4 *Axial Tension Testing of Full Size Bolts*—Bolts are to be tested in a holder with the load axially applied between the head and a nut or suitable fixture (Fig. 33), either of which shall have sufficient thread engagement to develop the full strength of the bolt. The nut or fixture shall be assembled on the bolt leaving six complete bolt threads unengaged between the grips, except for heavy hexagon structural bolts which shall have four complete threads unengaged between the grips. To meet the requirements of this test there shall be a tensile failure in the body or threaded section with no failure at the junction of the body and head. If it is necessary to record or report the tensile strength of bolts as psi values the stress area shall be calculated from the mean of the mean root and pitch diameters of Class 3 external threads as follows:

$$A_s = 0.7854 (D - (0.9743)/n)^2$$

where:
A_s = stress area, in.^{2},
D = nominal diameter, in., and
n = number of threads per inch.

S11.1.5 *Tension Testing of Full-Size Bolts with a Wedge*—The purpose of this test is to obtain the tensile strength and demonstrate the "head quality" and ductility of a bolt with a standard head by subjecting it to eccentric loading. The ultimate load on the bolt shall be determined as described in S11.1.4, except that a 10-deg wedge shall be placed under the same bolt previously tested for the proof load (see S11.1.1). The bolt head shall be so placed that no corner of the hexagon or square takes a bearing load, that is, a flat of the head shall be aligned with the direction of uniform thickness of the wedge (Fig. 34). The wedge shall have an included angle of 10 deg between its faces and shall have a thickness of one-half of the nominal bolt diameter at the short side of the hole. The hole in the wedge shall have the following clearance over the nominal size of the bolt, and its edges, top and bottom, shall be rounded to the following radius:

Nominal Bolt Size, in.	Clearance in Hole, in. (mm)	Radius on Corners of Hole, in. (mm)
¼ to ½	0.030 (0.76)	0.030 (0.76)
⁹/₁₆ to ¾	0.050 (1.3)	0.060 (1.5)
⅞ to 1	0.063 (1.5)	0.060 (1.5)
1⅛ to 1¼	0.063 (1.5)	0.125 (3.2)
1⅜ to 1½	0.094 (2.4)	0.125 (3.2)

A 370

S11.1.6 *Wedge Testing of HT Bolts Threaded to Head*—For heat-treated bolts over 100 000 psi (690 MPa) minimum tensile strength and that are threaded 1 diameter and closer to the underside of the head, the wedge angle shall be 6 deg for sizes ¼ through ¾ in. (6.35 to 19.0 mm) and 4 deg for sizes over ¾ in.

S11.1.7 *Tension Testing of Bolts Machined to Round Test Specimens:*

S11.1.7.1 Bolts under 1½ in. (38 mm) in diameter which require machined tests shall use a standard ½-in., (13 mm) round 2-in. (51-mm) gage length test specimen, turned concentric with the axis of the bolt, leaving the head and threaded section intact as in Fig. 35. Bolts of small cross-section which will not permit taking this standard test specimen shall have a turned section as large as feasible and concentric with the axis of the bolt. The gage length for measuring the elongation shall be four times the diameter of the specimen. Figure 36 illustrates examples of these small size specimens.

S11.1.7.2 For bolts 1½ in. and over in diameter, a standard ½-in. round 2-in. gage length test specimen shall be turned from the bolt, having its axis midway between the center and outside surface of the body of the bolt as shown in Fig. 37.

S11.1.7.3 Machined specimens are to be tested in tension to determine the properties prescribed by the product specifications. The methods of testing and determination of properties shall be in accordance with Section 13 of these methods, A 370.

S12. Speed of Testing

S12.1 Speed of testing shall be as prescribed in the individual product specifications.

S13. Hardness Tests for Bolts

S13.1 When specified, the bolts shall meet a hardness test. The Brinell or Rockwell hardness test is usually taken on the side or top of the bolt head. For final arbitration the hardness shall be taken on a transverse section through the threaded section of the bolt at a point one-quarter of the nominal diameter from the axis of the bolt. This section shall be taken at a distance from the end of the bolt which is equivalent to the diameter of the bolt. Due to possible distortion from the Brinell load, care shall be taken to see that this test meets all the provisions of 17.2 of the general section of these methods. Where the Brinell hardness test is impractical, the Rockwell hardness test shall be substituted. Rockwell hardness test procedures shall conform to Section 18 of these methods.

S14. Testing of Nuts

S14.1 *Proof Load*—A sample nut shall be assembled on a hardened threaded mandrel or on a bolt conforming to the particular specification. A load axial with the mandrel or bolt and equal to the specified proof load of the nut shall be applied. The nut shall resist this load without stripping or rupture. If the threads of the mandrel are damaged during the test the individual test shall be discarded. The mandrel shall be threaded to American National Standard Class 3 tolerance, except that the major diameter shall be the minimum major diameter with a tolerance of $+0.002$ in. (0.051 mm).

S14.2 *Hardness Test*—Rockwell hardness of nuts shall be determined on the top or bottom face of the nut. Brinell hardness shall be determined on the side of the nuts. Either method may be used at the option of the manufacturer, taking into account the size and grade of the nuts under test. When the standard Brinell hardness test results in deforming the nut it will be necessary to use a minor load or substitute a Rockwell hardness test.

S15. Bars Heat Treated or Cold Drawn for Use in the Manufacture of Studs, Nuts or Other Bolting Material

S15.1 When the bars as received by the manufacturer have been processed and proved to meet certain specified properties, it is not necessary to test the finished product when these properties have not been changed by the process of manufacture employed for the finished product.

IV. ROUND WIRE PRODUCTS

S16. Scope

S16.1 This supplement covers the apparatus, specimens and methods of testing peculiar to steel wire products which are not covered in the general section of Methods A 370.

A 370

S17. Apparatus

S17.1 *Gripping Devices*—Grips of either the wedge or snubbing types as shown in Figs. 38 and 39 shall be used (Note 14). When using grips of either type, care shall be taken that the axis of the test specimen is located approximately at the center line of the head of the testing machine (Note 15). When using wedge grips the liners used behind the grips shall be of the proper thickness.

NOTE 14—Testing machines usually are equipped with wedge grips. These wedge grips, irrespective of the type of testing machine, may be referred to as the "usual type" of wedge grips. The usual type of wedge grips generally furnish a satisfactory means of gripping wire. For tests of specimens of wire which are liable to be cut at the edges by the "usual type" of wedge grips, the snubbing type gripping device has proved satisfactory.

For testing round wire, the use of cylindrical seat in the wedge gripping device is optional.

NOTE 15—Any defect in a testing machine which may cause nonaxial application of load should be corrected.

S17.2 *Pointed Micrometer*—A micrometer with a pointed spindle and anvil suitable for reading the dimensions of the wire specimen at the fractured ends to the nearest 0.001 in. (0.025 mm) after breaking the specimen in the testing machine shall be used.

S18. Test Specimens

S18.1 Test specimens having the full cross-sectional area of the wire they represent shall be used. The standard gage length of the specimens shall be 10 in. (254 mm). However, if the determination of elongation values is not required, any convenient gage length is permissible. The total length of the specimens shall be at least equal to the gage length (10 in.) plus twice the length of wire required for the full use of the grip employed. For example, depending upon the type of testing machine and grips used, the minimum total length of specimen may vary from 14 to 24 in. (360 to 610 mm) for a 10-in. gage length specimen.

S18.2 Any specimen breaking in the grips shall be discarded and a new specimen tested.

S19. Elongation

S19.1 In determining permanent elongation, the ends of the fractured specimen shall be carefully fitted together and the distance between the gage marks measured to the nearest 0.01 in. (0.25 mm) with dividers and scale or other suitable device. The elongation is the increase in length of the gage length, expressed as a percentage of the original gage length. In reporting elongation values, both the percentage increase and the original gage length shall be given.

S19.2 In determining total elongation (elastic plus plastic extension) autographic or extensometer methods may be employed.

S19.3 If fracture takes place outside of the middle third of the gage length, the elongation value obtained may not be representative of the material.

S20. Reduction of Area

S20.1 The ends of the fractured specimen shall be carefully fitted together and the dimensions of the smallest cross section measured to the nearest 0.001 in. (0.025 mm) with a pointed micrometer. The difference between the area thus found and the area of the original cross section, expressed as a percentage of the original area, is the reduction of area.

S20.2 The reduction of area test is not recommended in wire diameters less than 0.092 in. (2.34 mm) due to the difficulties of measuring the reduced cross sections.

S21. Rockwell Hardness Test

S21.1 With the exception of heat treated wire of diameter 0.100 in. (2.54 mm) and larger, the Rockwell hardness test is not recommended for round wire. On such heat-treated wire the specimen shall be flattened on two parallel sides by grinding. For round wire the tensile strength test is greatly to be preferred to the Rockwell hardness test.

S22. Wrapping Test

S22.1 This test, also referred to as a coiling test or as a wrap-around bend test, is sometimes used as a means for testing the ductility of certain kinds of wire. The wrapping may be done by any hand or power device that will coil the wire closely about a mandrel of the specified diameter for a required number of turns without damage to the wire surface. The sample shall be considered to have failed if any cracks occur in the wire after the first complete turn. The test shall be repeated if a crack occurs in the first turn since the wire may have been bent locally to a radius less than that specified.

S22.2 When the wrapping test is used to determine the adherence of coating for coated wires, the mandrel diameter is commonly larger than that used in the test when used as a measure of ductility.

V. NOTES ON SIGNIFICANCE OF NOTCHED-BAR IMPACT TESTING

S23. Notch Behavior

S23.1 The Charpy and Izod type tests bring out notch behavior (brittleness versus ductility) by applying a single overload of stress. The energy values determined are quantitative comparisons on a selected specimen but cannot be converted into energy values that would serve for engineering design calculations. The notch behavior indicated in an individual test applies only to the specimen size, notch geometry, and testing conditions involved and cannot be generalized to other sizes of specimens and conditions.

S23.2 The notch behavior of the face-centered cubic metals and alloys, a large group of nonferrous materials and the austenitic steels can be judged from their common tensile properties. If they are brittle in tension they will be brittle when notched, while if they are ductile in tension, they will be ductile when notched, except for unusually sharp or deep notches (much more severe than the standard Charpy or Izod specimens). Even low temperatures do not alter this characteristic of these materials. In contrast, the behavior of the ferritic steels under notch conditions cannot be predicted from their properties as revealed by the tension test. For the study of these materials the Charpy and Izod type tests are accordingly very useful. Some metals that display normal ductility in the tension test may nevertheless break in brittle fashion when tested or when used in the notched condition. Notched conditions include restraints to deformation in directions perpendicular to the major stress, or multiaxial stresses, and stress concentrations. It is in this field that the Charpy and Izod tests prove useful for determining the susceptibility of a steel to notch-brittle behavior though they cannot be directly used to appraise the serviceability of a structure.

S23.3 The testing machine itself must be sufficiently rigid or tests on high-strength low-energy materials will result in excessive elastic energy losses either upward through the pendulum shaft or downward through the base of the machine. If the anvil supports, the pendulum striking edge, or the machine foundation bolts are not securely fastened, tests on ductile materials in the range of 80 ft·lbf (108 J) may actually indicate values in excess of 90 to 100 ft·lbf (122 to 136 J).

S24. Notch Effect

S24.1 The notch results in a combination of multiaxial stresses associated with restraints to deformation in directions perpendicular to the major stress, and a stress concentration at the base of the notch. A severely notched condition is generally not desirable, and it becomes of real concern in those cases in which it initiates a sudden and complete failure of the brittle type. Some metals can be deformed in a ductile manner even down to the low temperatures of liquid air, while others may crack. This difference in behavior can be best understood by considering the cohesive strength of a material (or the property that holds it together) and its relation to the yield point. In cases of brittle fracture, the cohesive strength is exceeded before significant plastic deformation occurs and the fracture appears crystalline. In cases of the ductile or shear type of failure, considerable deformation precedes the final fracture and the broken surface appears fibrous instead of crystalline. In intermediate cases the fracture comes after a moderate amount of deformation and is part crystalline and part fibrous in appearance.

S24.2 When a notched bar is loaded, there is a normal stress across the base of the notch which tends to initiate fracture. The property that keeps it from cleaving, or holds it together, is the "cohesive strength." The bar fractures when the normal stress exceeds the cohesive strength. When this occurs without the bar deforming it is the condition for brittle fracture.

S24.3 In testing, though not in service because of side effects, it happens more commonly that plastic deformation precedes fracture. In addition to the normal stress, the applied load also sets up shear stresses which are about 45 deg to the normal stress. The elastic behavior terminates as soon as the shear stress exceeds the shear strength of the material and deformation or plastic yielding sets in. This is the condition for ductile failure.

S24.4 This behavior, whether brittle or ductile, depends on whether the normal stress exceeds the cohesive strength before the shear stress exceeds the shear strength. Several important facts of notch behavior follow from this. If the notch is made sharper or more drastic, the normal stress at the root of the notch will be increased in relation to the shear stress and the bar will be more prone to brittle fracture (see Table 10). Also, as the speed of deformation increases, the shear strength increases and the likelihood of brittle fracture increases. On the other hand, by raising the temperature, leaving the notch and the speed of deformation the same, the shear strength is lowered and ductile behavior is promoted, leading to shear failure.

S24.5 Variations in notch dimensions will seriously affect the results of the tests. Tests on E4340 steel specimens[9] have shown the effect of dimensional variations on Charpy results (see Table 10).

S25. Size Effect

S25.1 Increasing either the width or the depth of the specimen tends to increase the volume of metal subject to distortion, and by this factor tends to increase the energy absorption when breaking the specimen. However, any increase in size, particularly in width, also tends to increase the degree of restraint and by tending to induce brittle fracture, may decrease the amount of energy absorbed. Where a standard-size specimen is on the verge of brittle fracture, this is particularly true, and a double-width specimen may actually require less energy for rupture than one of standard width.

S25.2 In studies of such effects where the size of the material precludes the use of the standard specimen, as for example when the material is $1/4$-in. plate, subsize specimens are necessarily used. Such specimens (see Fig. 6 of Method E 23) are based on the Type A specimen of Fig. 4 of Method E 23.

S25.3 General correlation between the energy values obtained with specimens of different size or shape is not feasible, but limited correlations may be established for specification purposes on the basis of special studies of particular materials and particular specimens. On the other hand, in a study of the relative effect of process variations, evaluation by use of some arbitrarily selected specimen with some chosen notch will in most instances place the methods in their proper order.

S26. Effects of Testing Conditions

S26.1 The testing conditions also affect the notch behavior. So pronounced is the effect of temperature on the behavior of steel when notched that comparisons are frequently made by examining specimen fractures and by plotting energy value and fracture appearance versus temperature from tests of notched bars at a series of temperatures. When the test temperature has been carried low enough to start cleavage fracture, there may be an extremely sharp drop in impact value or there may be a relatively gradual falling off toward the lower temperatures. This drop in energy value starts when a specimen begins to exhibit some crystalline appearance in the fracture. The transition temperature at which this embrittling effect takes place varies considerably with the size of the part or test specimen and with the notch geometry.

S26.2 Some of the many definitions of transition temperature currently being used are: (*1*) the lowest temperature at which the specimen exhibits 100 percent fibrous fracture, (*2*) the temperature where the fracture shows a 50 percent crystalline and a 50 percent fibrous appearance, (*3*) the temperature corresponding to the energy value 50 percent of the difference between values obtained at 100 percent and 0 percent fibrous fracture,

---
[9] Fahey, N. H., "Effects of Variables in Charpy Impact Testing," *Materials Research & Standards*, MTRSA Vol 1, No. 11, Nov., 1961, p. 872.

and (4) the temperature corresponding to a specific energy value.

S26.3 A problem peculiar to Charpy-type tests occurs when high-strength, low-energy specimens are tested at low temperatures. These specimens may not leave the machine in the direction of the pendulum swing but rather in a sidewise direction. To ensure that the broken halves of the specimens do not rebound off some component of the machine and contact the pendulum before it completes its swing, modifications may be necessary in older model machines. These modifications differ with machine design. Nevertheless the basic problem is the same in that provisions must be made to prevent rebounding of the fractured specimens into any part of the swinging pendulum. Where design permits, the broken specimens may be deflected out of the sides of the machine and yet in other designs it may be necessary to contain the broken specimens within a certain area until the pendulum passes through the anvils. Some low-energy high-strength steel specimens leave impact machines at speeds in excess of 50 ft (15.3 m)/s although they were struck by a pendulum traveling at speeds approximately 17 ft (5.2 m)/s. If the force exerted on the pendulum by the broken specimens is sufficient, the pendulum will slow down and erroneously high energy values will be recorded. This problem accounts for many of the inconsistencies in Charpy results reported by various investigators within the 10 to 25-ft·lbf (14 to 34 J) range. Section 5.5 of Methods E 23 discusses the two basic machine designs and a modification found to be satisfactory in minimizing jamming.

S27. Velocity of Straining

S27.1 Velocity of straining is likewise a variable that affects the notch behavior of steel. The impact test shows somewhat higher energy absorption values than the static tests above the transition temperature and yet, in some instances, the reverse is true below the transition temperature.

S28. Correlation with Service

S28.1 While Charpy or Izod tests may not directly predict the ductile or brittle behavior of steel as commonly used in large masses or as components of large structures, these tests can be used as acceptance tests of identity for different lots of the same steel or in choosing between different steels, when correlation with reliable service behavior has been established. It may be necessary to make the tests at properly chosen temperatures other than room temperature. In this, the service temperature or the transition temperature of full-scale specimens does not give the desired transition temperatures for Charpy or Izod tests since the size and notch geometry may be so different. Chemical analysis, tension, and hardness tests may not indicate the influence of some of the important processing factors that affect susceptibility to brittle fracture nor do they comprehend the effect of low temperatures in inducing brittle behavior.

VI. PROCEDURE FOR CONVERTING PERCENTAGE ELONGATION OF A STANDARD ROUND TENSION TEST SPECIMEN TO EQUIVALENT PERCENTAGE ELONGATION OF A STANDARD FLAT SPECIMEN

S29. Scope

S29.1 This method specifies a procedure for converting percentage elongation after fracture obtained in a standard 0.500-in. (12.7 mm) diameter by 2-in. (51-mm) gage length test specimen to standard flat test specimens $1/2$ in. by 2 in. and $1 1/2$ in. by 8 in. (38.1 by 203 mm).

S30. Basic Equation

S30.1 The conversion data in this method are based on an equation by Bertella,[10] and used by Oliver[11] and others. The relationship between elongations in the standard 0.500-in. diameter by 2.0-in. test specimen and other standard specimens can be calculated as follows:

$$e = e_o (4.47 \sqrt{A}/L)^a$$

---

[10] Bertella, C. A., *Giornale del Genio Civile*, Vol 60, 1922, p. 343.
[11] Oliver, D. A., *Proceedings of Institute of Mechanical Engineers*, Vol 11, 1928, p. 827.

A 370

where:

e_o = percentage elongation after fracture on a standard test specimen having a 2-in. gage length and 0.500-in. diameter,

e = percentage elongation after fracture on a standard test specimen having a gage length L and a cross-sectional area A, and

a = constant characteristic of the test material.

S31. Application

S31.1 In applying the above equation the constant a is characteristic of the test material. The value $a = 0.4$ has been found to give satisfactory conversions for carbon, carbon-manganese, molybdenum, and chromium-molybdenum steels within the tensile strength range of 40,000 to 85,000 psi (275 to 585 MPa) and in the hot-rolled, in the hot-rolled and normalized, or in the annealed condition, with or without tempering. Note that the cold reduced and quenched and tempered states are excluded. For annealed austenitic stainless steels, the value $a = 0.127$ has been found to give satisfactory conversions.

S31.2 Table 11 has been calculated taking $a = 0.4$, with the standard 0.500-in. (12.7 mm) diameter by 2-in. (51 mm) gage length test specimen as the reference specimen. In the case of the subsize specimens 0.350 in. (8.89 mm) in diameter by 1.4 in. (35.6 mm) gage length, and 0.250 (6.35 mm) diameter by 1.0 in. (25.4 mm) gage length the factor in the equation is 4.51 instead of 4.37. The small error introduced by using Table 11 for the subsized specimens may be neglected. Table 12 for annealed austenitic steels has been calculated taking $a = 0.127$, with the standard 0.500-in. diameter by 2-in. gage length test specimen as the reference specimen.

S31.3 Elongation given for a standard 0.500-in. diameter by 2-in. gage length specimen may be converted to elongation for $1/2$ in. by 2 in. or $1\,1/2$ in. by 8 in. (38.1 by 203 mm) flat specimens by multiplying by the indicated factor in Tables 11 and 12.

S31.4 These elongation conversions shall not be used where the width to thickness ratio of the test piece exceeds 20, as in sheet specimens under 0.025 in. (0.635 mm) in thickness.

S31.5 While the conversions are considered to be reliable within the stated limitations and may generally be used in specification writing where it is desirable to show equivalent elongation requirements for the several standard ASTM tension specimens covered in Methods A 370, consideration must be given to the metallurgical effects dependent on the thickness of the material as processed.

VII. METHOD OF TESTING UNCOATED SEVEN-WIRE STRESS-RELIEVED STRAND FOR PRESTRESSED CONCRETE

S32. Scope

S32.1 This method provides procedures for the tension testing of uncoated seven-wire stress-relieved strand for prestressed concrete. This method is intended for use in evaluating the strand for the properties prescribed in Specification A 416.

S33. General Precautions

S33.1 Premature failure of the test specimens may result if there is any appreciable notching, cutting, or bending of the specimen by the gripping devices of the testing machine.

S33.2 Errors in testing may result if the seven wires constituting the strand are not loaded uniformly.

S33.3 The mechanical properties of the strand may be materially affected by excessive heating during specimen preparation.

S33.4 These difficulties may be minimized by following the suggested methods of gripping described in Section S35.

S34. Gripping Devices

S34.1 The true mechanical properties of the strand are determined by a test in which fracture of the specimen occurs in the free span between the jaws of the testing machine. Therefore, it is desirable to establish a test procedure with suitable apparatus which will consistently produce such results. Due to inherent physical characteristics of individual machines, it is not practical to recommend a universal gripping procedure that is suitable for all testing machines. Therefore, it is necessary to determine which of the methods of gripping described in S34.2 to S34.8 is most suitable for the testing equipment available.

S34.2 *Standard V-Grips with Serrated Teeth (Note 16)*.

S34.3 *Standard V-Grips with Serrated Teeth (Note 16), Using Cushioning Material*—In this method, some material is placed between the grips and the specimen to minimize the notching effect of the teeth. Among the materials which have been used are lead foil, aluminum foil, carborundum cloth, bra shims, etc. The type and thickness of material required is dependent on the shape, condition, and coarseness of the teeth.

S34.4 *Standard V-Grips with Serrated Teeth (Note 16), Using Special Preparation of the Gripped Portions of the Specimen*—One of the methods used is tinning, in which the gripped portions are cleaned, fluxed, and coated by multiple dips in molten tin alloy held just above the melting point. Another method of preparation is encasing the gripped portions in metal tubing or flexible conduit, using epoxy resin as the bonding agent. The encased portion should be approximately twice the length of lay of the strand.

S34.5 *Special Grips with Smooth, Semi-Cylindrical Grooves (Note 17)*—The grooves and the gripped portions of the specimen are coated with an abrasive slurry which holds the specimen in the smooth grooves, preventing slippage. The slurry consists of abrasive such as Grade 3-F aluminum oxide and a carrier such as water or glycerin.

S34.6 *Standard Sockets of the Type Used for Wire Rope*—The gripped portions of the specimen are anchored in the sockets with zinc. The special procedures for socketing usually employed in the wire rope industry must be followed.

S34.7 *Dead-End Eye Splices*—These devices are available in sizes designed to fit each size of strand to be tested.

S34.8 *Chucking Devices*—Use of chucking devices of the type generally employed for applying tension to strands in casting beds is not recommended for testing purposes.

NOTE 16—The number of teeth should be approximately 15 to 30 per in., and the minimum effective gripping length should be approximately 4 in. (102 mm).

NOTE 17—The radius of curvature of the grooves is approximately the same as the radius of the strand being tested, and is located ½₂ in. (0.79 mm) above the flat face of the grip. This prevents the two grips from closing tightly when the specimen is in place.

S35. Specimen Preparation

S35.1 Nonuniform loading of the seven wires in the strand may result if slippage of the individual wires of the strand, either the outside wire or the center wire, occur during the tension test. Wire slippage may be minimized by fusing together the cut ends of the specimen. This fusing can be concurrent with torch cutting of the specimens.

S35.2 If the molten-metal temperatures employed during hot-dip tinning or socketing with metallic material are too high, over approximately 700 F (370 C), the specimen may be heat affected with a subsequent loss of strength and ductility. Careful temperature controls should be maintained if such methods of specimen preparation are used.

S36. Procedure

S36.1 *Yield Strength*—For determining the yield strength use a Class B-1 extensometer (Note 18) as described in Method E 83. Apply an initial load of 10 percent of the expected minimum breaking strength to the specimen, then attach the extensometer and adjust it to a reading of 0.001 in./in. of gage length. Then increase the load until the extensometer indicates an extension of 1 percent. Record the load for this extension as the yield strength. The extensometer may be removed from the specimen after the yield strength has been determined.

S36.2 *Elongation*—For determining the elongation use a Class D extensometer (Note 18), as described in Method E 83, having a gage length of not less than 24 in. (610 mm) (Note 19). Apply an initial load of 10 percent of the required minimum breaking strength to the specimen, then attach the extensometer (Note 18) and adjust it to a zero reading. The extensometer may be removed from the specimen prior to rupture after the specified minimum elongation has been exceeded. It is not necessary to determine the final elongation value.

S36.3 *Breaking Strength*—Determine the maximum load at which one or more wires of the strand are fractured. Record this load as the breaking strength of the strand.

Note 18—The yield-strength extensometer and the elongation extensometer may be the same instrument or two separate instruments. Two separate instruments are advisable since the more sensitive yield-strength extensometer, which could be damaged when the strand fractures, may be removed following the determination of yield strength. The elongation extensometer may be constructed with less sensitive parts or be constructed in such a way that little damage would result if fracture occurs while the extensometer is attached to the specimen.

Note 19—Specimens that break outside the extensometer or in the jaws and yet meet the minimum specified values are considered as meeting the mechanical property requirements of the product Specification A 416, regardless of what procedure of gripping has been used. Specimens that break outside of the extensometer or in the jaws and do not meet the minimum specified values are subject to retest in accordance with Specification A 416. Specimens that break between the jaws of the extensometer and do not meet the minimum specified values are subject to retest as provided in Section 14 of Specification A 416.

VIII. ROUNDING OF TEST DATA

S37. Rounding

S37.1 Recommended levels for rounding reported values of test data are given in Table 13. These values are designed to provide uniformity in reporting and data storage, and should be used in all cases except where they conflict with specific requirements of a product specification.

TABLE 1 Details of Test Coupon Design for Casting (See Fig. 3)

Note 1—*Test Coupons for Large and Heavy Steel Castings:* The test coupons in Fig. 3 are to be used for large and heavy steel castings. However, at the option of the foundry the cross-sectional area and length of the standard coupon may be increased as desired. This provision does not apply to ASTM Specification A 356, for Heavy-Walled Carbon and Low Alloy Steel Castings for Steam Turbines (*1983 Annual Book of ASTM Standards,* Vol 01.02).

Note 2—*Bend Bar:* If a bend bar is required, an alternate design (as shown by dotted lines in Fig. 3) is indicated.

Leg Design (125mm)		Riser Design	
1. *L* (length)	A 5 in. (125 mm) minimum length will be used. This length may be increased at the option of the foundry to accommodate additional test bars (see Note 1).	1. *L* (length)	The length of the riser at the base will be the same as the top length of the leg. The length of the riser at the top therefore depends on the amount of taper added to the riser.
2. End taper	Use of and size of end taper is at the option of the foundry.	2. Width	The width of the riser at the base of a multiple-leg coupon shall be $n (2\frac{1}{4})(57\ mm) - \frac{5}{8}$ (16 mm) where n equals the number of legs attached to the coupon. The width of the riser at the top is therefore dependent on the amount of taper added to the riser.
3. Height	1¼ in. (32 mm)		
4. Width (at top)	1¼ in. (32 mm) (see Note 1).		
5. Radius (at bottom)	½ in. (13 mm), max		
6. Spacing between legs	A ½-in. (13-mm) radius will be used between the legs.		
7. Location of test bars	The tensile, bend, and impact bars will be taken from the lower portion of the leg (see Note 2).		
8. Number of legs	The number of legs attached to the coupon is at the option of the foundry providing they are equispaced according to Item 6.	3. *T* (riser taper) Height	Use of and size is at the option of the foundry. The minimum height of the riser shall be 2 in. (51 mm). The maximum height is at the option of the foundry for the following reasons: (*a*) Many risers are cast open, (*b*) different compositions may require variation in risering for soundness, (*c*) different pouring temperatures may require variation in risering for soundness.
9. R_s	Radius from 0 to approximately 1/16 in. (2 mm).		

ASTM A 370

TABLE 2 Multiplying Factors to Be Used for Various Diameters of Round Test Specimens

Standard Specimen			Small Size Specimens Proportional to Standard					
0.500 in. Round			0.350 in. Round			0.250 in. Round		
Actual Diameter, in.	Area, in.^{2}	Multiplying Factor	Actual Diameter, in.	Area, in.^{2}	Multiplying Factor	Actual Diameter, in.	Area, in.^{2}	Multiplying Factor
0.490	0.1886	5.30	0.343	0.0924	10.82	0.245	0.0471	21.21
0.491	0.1893	5.28	0.344	0.0929	10.76	0.246	0.0475	21.04
0.492	0.1901	5.26	0.345	0.0935	10.70	0.247	0.0479	20.87
0.493	0.1909	5.24	0.346	0.0940	10.64	0.248	0.0483	20.70
0.494	0.1917	5.22	0.347	0.0946	10.57	0.249	0.0487	20.54
0.495	0.1924	5.20	0.348	0.0951	10.51	0.250	0.0491	20.37
0.496	0.1932	5.18	0.349	0.0957	10.45	0.251	0.0495 (0.05)[a]	20.21 (20.0)[a]
0.497	0.1940	5.15	0.350	0.0962	10.39	0.252	0.0499 (0.05)[a]	20.05 (20.0)[a]
0.498	0.1948	5.13	0.351	0.0968	10.33	0.253	0.0503 (0.05)[a]	19.89 (20.0)[a]
0.499	0.1956	5.11	0.352	0.0973	10.28	0.254	0.0507	19.74
0.500	0.1963	5.09	0.353	0.0979	10.22	0.255	0.0511	19.58
0.501	0.1971	5.07	0.354	0.0984	10.16	...	...	...
0.502	0.1979	5.05	0.355	0.0990	10.10	...	...	...
0.503	0.1987	5.03	0.356	0.0995 (0.1)[a]	10.05 (10.0)[a]	...	...	...
0.504	0.1995 (0.2)[a]	5.01 (5.0)[a]	0.357	0.1001 (0.1)[a]	9.99 (10.0)[a]	...	...	...
0.505	0.2003 (0.2)[a]	4.99 (5.0)[a]	...	...	...	...	...	...
0.506	0.2011 (0.2)[a]	4.97 (5.0)[a]	...	...	...	...	...	...
0.507	0.2019	4.95	...	...	...	...	...	...
0.508	0.2027	4.93	...	...	...	...	...	...
0.509	0.2035	4.91	...	...	...	...	...	...
0.510	0.2043	4.90	...	...	...	...	...	...

[a] The values in parentheses may be used for ease in calculation of stresses, in pounds per square inch, as permitted in Note 5 of Fig. 5.

TABLE 3A Approximate Hardness Conversion Numbers for Nonaustenitic Steels[A] (Rockwell C to other Hardness Numbers)

Rockwell C Scale, 150-kgf Load, Diamond Penetrator	Vickers Hardness Number	Brinell Indentation Diameter, mm	Brinell Hardness, 3000-kgf Load, 10-mm Ball	Knoop Hardness, 500-gf Load and Over	Rockwell A Scale, 60-kgf Load, Diamond Penetrator	Rockwell Superficial Hardness			Approximate Tensile Strength, ksi (MPa)
						15N Scale, 15-kgf Load, Diamond Penetrator	30N Scale, 30-kgf Load, Diamond Penetrator	45N Scale, 45-kgf Load, Diamond Penetrator	
68	940	...	...	920	85.6	93.2	84.4	75.4	...
67	900	...	...	895	85.0	92.9	83.6	74.2	...
66	865	...	...	870	84.5	92.5	82.8	73.3	...
65	832	2.26	739	846	83.9	92.2	81.9	72.0	...
64	800	2.28	722	822	83.4	91.8	81.1	71.0	...
63	772	2.31	706	799	82.8	91.4	80.1	69.9	...
62	746	2.34	688	776	82.3	91.1	79.3	68.8	...
61	720	2.37	670	754	81.8	90.7	78.4	67.7	...
60	697	2.40	654	732	81.2	90.2	77.5	66.6	...
59	674	2.44	634	710	80.7	89.8	76.6	65.5	351 (2420)
58	653	2.47	615	690	80.1	89.3	75.7	64.3	338 (2330)
57	633	2.51	595	670	79.6	88.9	74.8	63.2	325 (2240)
56	613	2.55	577	650	79.0	88.3	73.9	62.0	313 (2160)
55	595	2.59	560	630	78.5	87.9	73.0	60.9	301 (2070)
54	577	2.63	543	612	78.0	87.4	72.0	59.8	292 (2010)
53	560	2.67	525	594	77.4	86.9	71.2	58.6	283 (1950)
52	544	2.70	512	576	76.8	86.4	70.2	57.4	273 (1880)
51	528	2.75	496	558	76.3	85.9	69.4	56.1	264 (1820)
50	513	2.79	482	542	75.9	85.5	68.5	55.0	255 (1760)
49	498	2.83	468	526	75.2	85.0	67.6	53.8	246 (1700)
48	484	2.87	455	510	74.7	84.5	66.7	52.5	238 (1640)
47	471	2.91	442	495	74.1	83.9	65.8	51.4	229 (1580)
46	458	2.94	432	480	73.6	83.5	64.8	50.3	221 (1520)
45	446	2.98	421	466	73.1	83.0	64.0	49.0	215 (1480)
44	434	3.02	409	452	72.5	82.5	63.1	47.8	208 (1430)
43	423	3.05	400	438	72.0	82.0	62.2	46.7	201 (1390)
42	412	3.09	390	426	71.5	81.5	61.3	45.5	194 (1340)
41	402	3.13	381	414	70.9	80.9	60.4	44.3	188 (1300)
40	392	3.17	371	402	70.4	80.4	59.5	43.1	182 (1250)
39	382	3.21	362	391	69.9	79.9	58.6	41.9	177 (1220)
38	372	3.24	353	380	69.4	79.4	57.7	40.8	171 (1180)
37	363	3.28	344	370	68.9	78.8	56.8	39.6	166 (1140)
36	354	3.32	336	360	68.4	78.3	55.9	38.4	161 (1110)
35	345	3.36	327	351	67.9	77.7	55.0	37.2	156 (1080)
34	336	3.41	319	342	67.4	77.2	54.2	36.1	152 (1050)
33	327	3.45	311	334	66.8	76.6	53.3	34.9	149 (1030)
32	318	3.50	301	326	66.3	76.1	52.1	33.7	146 (1010)
31	310	3.54	294	318	65.8	75.6	51.3	32.5	141 (970)

TABLE 3A—Continued

Rockwell C Scale, 150-kgf Load, Diamond Penetrator	Vickers Hardness Number	Brinell Indentation Diameter, mm	Brinell Hardness, 3000-kgf Load, 10-mm Ball	Knoop Hardness, 500-gf Load and Over	Rockwell A Scale, 60-kgf Load, Diamond Penetrator	Rockwell Superficial Hardness			Approximate Tensile Strength, ksi (MPa)
						15N Scale, 15-kgf Load, Diamond Penetrator	30N Scale, 30-kgf Load, Diamond Penetrator	45N Scale, 45-kgf Load, Diamond Penetrator	
30	302	3.59	286	311	65.3	75.0	50.4	31.3	138 (950)
29	294	3.64	279	304	64.6	74.5	49.5	30.1	135 (930)
28	286	3.69	271	297	64.3	73.9	48.6	28.9	131 (900)
27	279	3.73	264	290	63.8	73.3	47.7	27.8	128 (880)
26	272	3.77	258	284	63.3	72.8	46.8	26.7	125 (860)
25	266	3.81	253	278	62.8	72.2	45.9	25.5	123 (850)
24	260	3.86	247	272	62.4	71.6	45.0	24.3	119 (820)
23	254	3.89	243	266	62.0	71.0	44.0	23.1	117 (810)
22	248	3.93	237	261	61.5	70.5	43.2	22.0	115 (790)
21	243	3.98	231	256	61.0	69.9	42.3	20.7	112 (770)
20	238	4.02	226	251	60.5	69.4	41.5	19.6	110 (760)

[A] This table gives the approximate interrelationships of hardness values and approximate tensile strength of steels. It is possible that steels of various compositions and processing histories will deviate in hardness-tensile strength relationship from the data presented in this table. The data in this table should not be used for austenitic stainless steels, but have been shown to be applicable for ferritic and martensitic stainless steels. Where more precise conversions are required, they should be developed specially for each steel composition, heat treatment, and part.

ASTM A 370

TABLE 3B Approximate Hardness Conversion Numbers for Nonaustenitic Steels (Rockwell B to other Hardness Numbers)

Rockwell B Scale, 100-kgf Load 1/16-in. (1.588-mm) Ball	Vickers Hardness Number	Brinell Indentation Diameter, mm	Brinell Hardness, 3000-kgf Load, 10-mm Ball	Knoop Hardness, 500-gf Load and Over	Rockwell A Scale, 60-kgf Load, Diamond Penetrator	Rockwell F Scale, 60-kgf Load, 1/16-in. (1.588-mm) Ball	Rockwell Superficial Hardness			Approximate Tensile Strength ksi (MPa)
							15T Scale, 15-kgf Load, 1/16-in. (1.588-mm) Ball	30T Scale, 30-kgf Load, 1/16-in. (1.588-mm) Ball	45T Scale, 45-kgf Load, 1/16-in. (1.588-mm) Ball	
100	240	3.91	240	251	61.5	...	93.1	83.1	72.9	116 (800)
99	234	3.96	234	246	60.9	...	92.8	82.5	71.9	114 (785)
98	228	4.01	228	241	60.2	...	92.5	81.8	70.9	109 (750)
97	222	4.06	222	236	59.5	...	92.1	81.1	69.9	104 (715)
96	216	4.11	216	231	58.9	...	91.8	80.4	68.9	102 (705)
95	210	4.17	210	226	58.3	...	91.5	79.8	67.9	100 (690)
94	205	4.21	205	221	57.6	...	91.2	79.1	66.9	98 (675)
93	200	4.26	200	216	57.0	...	90.8	78.4	65.9	94 (650)
92	195	4.32	195	211	56.4	...	90.5	77.8	64.8	92 (635)
91	190	4.37	190	206	55.8	...	90.2	77.1	63.8	90 (620)
90	185	4.43	185	201	55.2	...	89.9	76.4	62.8	89 (615)
89	180	4.48	180	196	54.6	...	89.5	75.8	61.8	88 (605)
88	176	4.53	176	192	54.0	...	89.2	75.1	60.8	86 (590)
87	172	4.58	172	188	53.4	...	88.9	74.4	59.8	84 (580)
86	169	4.62	169	184	52.8	...	88.6	73.8	58.8	83 (570)
85	165	4.67	165	180	52.3	...	88.2	73.1	57.8	82 (565)
84	162	4.71	162	176	51.7	...	87.9	72.4	56.8	81 (560)
83	159	4.75	159	173	51.1	...	87.6	71.8	55.8	80 (550)
82	156	4.79	156	170	50.6	...	87.3	71.1	54.8	77 (530)
81	153	4.84	153	167	50.0	...	86.9	70.4	53.8	73 (505)
80	150	4.88	150	164	49.5	...	86.6	69.7	52.8	72 (495)
79	147	4.93	147	161	48.9	...	86.3	69.1	51.8	70 (485)
78	144	4.98	144	158	48.4	...	86.0	68.4	50.8	69 (475)
77	141	5.02	141	155	47.9	...	85.6	67.7	49.8	68 (470)
76	139	5.06	139	152	47.3	...	85.3	67.1	48.8	67 (460)
75	137	5.10	137	150	46.8	99.6	85.0	66.4	47.8	66 (455)
74	135	5.13	135	147	46.3	99.1	84.7	65.7	46.8	65 (450)
73	132	5.18	132	145	45.8	98.5	84.3	65.1	45.8	64 (440)
72	130	5.22	130	143	45.3	98.0	84.0	64.4	44.8	63 (435)
71	127	5.27	127	141	44.8	97.4	83.7	63.7	43.8	62 (425)
70	125	5.32	125	139	44.3	96.8	83.4	63.1	42.8	61 (420)
69	123	5.36	123	137	43.8	96.2	83.0	62.4	41.8	60 (415)
68	121	5.40	121	135	43.3	95.6	82.7	61.7	40.8	59 (405)
67	119	5.44	119	133	42.8	95.1	82.4	61.0	39.8	58 (400)
66	117	5.48	117	131	42.3	94.5	82.1	60.4	38.7	57 (395)
65	116	5.51	116	129	41.8	93.9	81.8	59.7	37.7	56 (385)
64	114	5.54	114	127	41.4	93.4	81.4	59.0	36.7	...
63	112	5.58	112	125	40.9	92.8	81.1	58.4	35.7	...

TABLE 3B - Continued

Rockwell B Scale, 100-kgf Load 1/16-in. (1.588-mm) Ball	Vickers Hardness Number	Brinell Indentation Diameter, mm	Brinell Hardness, 3000-kgf Load, 10-mm Ball	Knoop Hardness, 500-gf Load and Over	Rockwell A Scale, 60-kgf Load, Diamond Penetrator	Rockwell F Scale, 60-kgf Load, 1/16-in. (1.588-mm) Ball	Rockwell Superficial Hardness			Approximate Tensile Strength ksi (MPa)
							15T Scale, 15-kgf Load, 1/16-in. (1.588-mm) Ball	30T Scale, 30-kgf Load, 1/16-in. (1.588-mm) Ball	45T Scale, 45-kgf Load, 1/16-in. (1.588-mm) Ball	
62	110	5.63	110	124	40.4	92.2	80.8	57.7	34.7	...
61	108	5.68	108	122	40.0	91.7	80.5	57.0	33.7	...
60	107	5.70	107	120	39.5	91.1	80.1	56.4	32.7	...
59	106	5.73	106	118	39.0	90.5	79.8	55.7	31.7	...
58	104	5.77	104	117	38.6	90.0	79.5	55.0	30.7	...
57	103	5.81	103	115	38.1	89.4	79.2	54.4	29.7	...
56	101	5.85	101	114	37.7	88.8	78.8	53.7	28.7	...
55	100	5.87	100	112	37.2	88.2	78.5	53.0	27.7	...
54	...	...	...	111	36.8	87.7	78.2	52.4	26.7	...
53	...	...	...	110	36.3	86.5	77.9	51.7	25.7	...
52	...	...	...	109	35.9	86.0	77.5	51.0	24.7	...
51	...	...	...	108	35.5	85.4	77.2	50.3	23.7	...
50	...	...	...	107	35.0	84.8	76.9	49.7	22.7	...
49	...	...	...	106	34.6	84.3	76.6	49.0	21.7	...
48	...	...	...	105	34.1	83.7	76.2	48.3	20.7	...
47	...	...	...	104	33.7	83.1	75.9	47.7	19.7	...
46	...	...	...	103	33.3	...	75.6	47.0	18.7	...
45	...	...	...	102	32.9	82.6	75.3	46.3	17.7	...
44	...	...	...	101	32.4	82.0	74.9	45.7	16.7	...
43	...	...	...	100	32.0	81.4	74.6	45.0	15.7	...
42	...	...	...	99	31.6	80.8	74.3	44.3	14.7	...
41	...	...	...	98	31.2	80.3	74.0	43.7	13.6	...
40	...	...	...	97	30.7	79.7	73.6	43.0	12.6	...
39	...	...	...	96	30.3	79.1	73.3	42.3	11.6	...
38	...	...	...	95	29.9	78.6	73.0	41.6	10.6	...
37	...	...	...	94	29.5	78.0	72.7	41.0	9.6	...
36	...	...	...	93	29.1	77.4	72.3	40.3	8.6	...
35	...	...	...	92	28.7	76.9	72.0	39.6	7.6	...
34	...	...	...	91	28.2	76.3	71.7	39.0	6.6	...
33	...	...	...	90	27.8	75.7	71.4	38.3	5.6	...
32	...	...	...	89	27.4	75.2	71.0	37.6	4.6	...
31	...	...	...	88	27.0	74.6	70.7	37.0	3.6	...
30	...	...	...	87	26.6	74.0	70.4	36.3	2.6	...

[A] This table gives the approximate interrelationships of hardness values and approximate tensile strength of steels. It is possible that steels of various compositions and processing histories will deviate in hardness-tensile strength relationship from the data presented in this table. The data in this table should not be used for austenitic stainless steels, but have been shown to be applicable for ferritic and martensitic stainless steels. Where more precise conversions are required, they should be developed specially for each steel composition, heat treatment, and part.

A 370

TABLE 3C Approximate Hardness Conversion Numbers for Austenitic Steels (Rockwell C to other Hardness Numbers)

Rockwell C Scale, 150-kgf Load, Diamond Penetrator	Rockwell A Scale, 60-kgf Load, Diamond Penetrator	Rockwell Superficial Hardness		
		15N Scale, 15-kgf Load, Diamond Penetrator	30N Scale, 30-kgf Load, Diamond Penetrator	45N Scale, 45-kgf Load, Diamond Penetrator
48	74.4	84.1	66.2	52.1
47	73.9	83.6	65.3	50.9
46	73.4	83.1	64.5	49.8
45	72.9	82.6	63.6	48.7
44	72.4	82.1	62.7	47.5
43	71.9	81.6	61.8	46.4
42	71.4	81.0	61.0	45.2
41	70.9	80.5	60.1	44.1
40	70.4	80.0	59.2	43.0
39	69.9	79.5	58.4	41.8
38	69.3	79.0	57.5	40.7
37	68.8	78.5	56.6	39.6
36	68.3	78.0	55.7	38.4
35	67.8	77.5	54.9	37.3
34	67.3	77.0	54.0	36.1
33	66.8	76.5	53.1	35.0
32	66.3	75.9	52.3	33.9
31	65.8	75.4	51.4	32.7
30	65.3	74.9	50.5	31.6
29	64.8	74.4	49.6	30.4
28	64.3	73.9	48.8	29.3
27	63.8	73.4	47.9	28.2
26	63.3	72.9	47.0	27.0
25	62.8	72.4	46.2	25.9
24	62.3	71.9	45.3	24.8
23	61.8	71.3	44.4	23.6
22	61.3	70.8	43.5	22.5
21	60.8	70.3	42.7	21.3
20	60.3	69.8	41.8	20.2

TABLE 3D Approximate Hardness Conversion Numbers for Austenitic Steels (Rockwell B to other Hardness Numbers)

Rockwell B Scale, 100-kgf Load, 1/16-in. (1.588-mm) Ball	Brinell Indentation Diameter, mm	Brinell Hardness, 3000-kgf Load, 10-mm Ball	Rockwell A Scale, 60-kgf Load, Diamond Penetrator	Rockwell Superficial Hardness		
				15T Scale, 15-kgf Load, 1/16-in. (1.588-mm) Ball	30T Scale, 30-kgf Load, 1/16-in. (1.588-mm) Ball	45T Scale, 45-kgf Load, 1/16-in. (1.588-mm) Ball
100	3.79	256	61.5	91.5	80.4	70.2
99	3.85	248	60.9	91.2	79.7	69.2
98	3.91	240	60.3	90.8	79.0	68.2
97	3.96	233	59.7	90.4	78.3	67.2
96	4.02	226	59.1	90.1	77.7	66.1
95	4.08	219	58.5	89.7	77.0	65.1
94	4.14	213	58.0	89.3	76.3	64.1
93	4.20	207	57.4	88.9	75.6	63.1
92	4.24	202	56.8	88.6	74.9	62.1
91	4.30	197	56.2	88.2	74.2	61.1
90	4.35	192	55.6	87.8	73.5	60.1
89	4.40	187	55.0	87.5	72.8	59.0
88	4.45	183	54.5	87.1	72.1	58.0
87	4.51	178	53.9	86.7	71.4	57.0
86	4.55	174	53.3	86.4	70.7	56.0
85	4.60	170	52.7	86.0	70.0	55.0
84	4.65	167	52.1	85.6	69.3	54.0
83	4.70	163	51.5	85.2	68.6	52.9
82	4.74	160	50.9	84.9	67.9	51.9
81	4.79	156	50.4	84.5	67.2	50.9
80	4.84	153	49.8	84.1	66.5	49.9

TABLE 4 Percent Shear for Measurements Made in Inches

NOTE—Since Table 4 is set up for finite measurements or dimensions A and B, 100 percent shear is to be reported when either A or B is zero.

Dimension B, in.	Dimension A, in.																	
	0.05	0.10	0.12	0.14	0.16	0.18	0.20	0.22	0.24	0.26	0.28	0.30	0.32	0.34	0.36	0.38	0.40
0.05	98	96	95	94	94	93	92	91	90	90	89	88	87	86	85	85	84
0.10	96	92	90	89	87	85	84	82	81	79	77	76	74	73	71	69	68
0.12	95	90	88	86	85	83	81	79	77	75	73	71	69	67	65	63	61
0.14	94	89	86	84	82	80	77	75	73	71	68	66	64	62	59	57	55
0.16	94	87	85	82	79	77	74	72	69	67	64	61	59	56	53	51	48
0.18	93	85	83	80	77	74	72	68	65	62	59	56	54	51	48	45	42
0.20	92	84	81	77	74	72	68	65	61	58	55	52	48	45	42	39	36
0.22	91	82	79	75	72	68	65	61	57	54	50	47	43	40	36	33	29
0.24	90	81	77	73	69	65	61	57	54	50	46	42	38	34	30	27	23
0.26	90	79	75	71	67	62	58	54	50	46	41	37	33	29	25	20	16
0.28	89	77	73	68	64	59	55	50	46	41	37	32	28	23	18	14	10
0.30	88	76	71	66	61	56	52	47	42	37	32	27	23	18	13	9	3
0.31	88	75	70	65	60	55	50	45	40	35	30	25	20	18	10	5	0

TABLE 5 Percent Shear for Measurements Made in Millimeters

NOTE—Since Table 5 is set up for finite measurements or dimensions A and B, 100 percent shear is to be reported when either A or B is zero.

Dimension B, mm	Dimension A, mm																		
	1.0	1.5	2.0	2.5	3.0	3.5	4.0	4.5	5.0	5.5	6.0	6.5	7.0	7.5	8.0	8.5	9.0	9.5	10
1.0	99	98	98	97	96	96	95	94	94	93	92	92	91	91	90	89	89	88	88
1.5	98	97	96	95	94	93	92	92	91	90	89	88	87	86	85	84	83	82	81
2.0	98	96	95	94	92	91	90	89	88	86	85	84	82	81	80	79	77	76	75
2.5	97	95	94	92	91	89	88	86	84	83	81	80	78	77	75	73	72	70	69
3.0	96	94	92	91	89	87	85	83	81	79	77	76	74	72	70	68	66	64	62
3.5	96	93	91	89	87	85	82	80	78	76	74	72	69	67	65	63	61	58	56
4.0	95	92	90	88	85	82	80	77	75	72	70	67	65	62	60	57	55	52	50
4.5	94	92	89	86	83	80	77	75	72	69	66	63	61	58	55	52	49	46	44
5.0	94	91	88	85	81	78	75	72	69	66	62	59	56	53	50	47	44	41	37
5.5	93	90	96	83	79	76	72	69	66	62	59	55	52	48	45	42	38	35	31
6.0	92	89	85	81	77	74	70	66	62	59	55	51	47	44	40	36	33	29	25
6.5	92	88	84	80	76	72	67	63	59	55	51	47	43	39	35	31	27	23	19
7.0	91	87	82	78	74	69	65	61	56	52	47	43	39	34	30	26	21	17	12
7.5	91	86	81	77	72	67	62	58	53	48	44	39	34	30	25	20	16	11	6
8.0	90	85	80	75	70	65	60	55	50	45	40	35	30	25	20	15	10	5	0

TABLE 6 Recommended Practice for Selecting Bend Test Specimens

NOTE 1—The length of all specimens is to be not less than 6 in. (150 mm).

NOTE 2—The edges of the specimen may be rounded to a radius not exceeding 1/16 in. (1.6 mm).

Flats		
Thickness, in. (mm)	Width, in. (mm)	Recommended Size
Up to 1/2 (13), incl	Up to 3/4 (19), incl	Full section.
	Over 3/4 (19)	Full section or machine to not less than 3/4 in. (19 mm) in width by thickness of specimen.
Over 1/2 (13)	All	Full section or machine to 1 by 1/2 in. (25 by 13 mm) specimen from midway between center and surface.

Rounds, Squares, Hexagons, and Octagons	
Diameter or Distance Between Parallel Faces, in. (mm)	Recommended Size
Up to 1 1/2 (38), incl	Full section.
Over 1 1/2 (38)	Machine to 1 by 1/2-in. (25 by 13-mm) specimen from midway between center and surface.

TABLE 7 Recommendations for Selecting Tension Test Specimens

NOTE 1—For bar sections where it is difficult to determine the cross-sectional area by simple measurement, the area in square inches may be calculated by dividing the weight per linear inch of specimen in pounds by 0.2833 (weight of 1 in.³ of steel) or by dividing the weight per linear foot of specimen by 3.4 (weight of steel 1 in. square and 1 ft long).

Thickness, in. (mm)	Width, in. (mm)	Hot-Rolled Bars	Cold-Finished Bars
		Flats	
Under ⅝ (16)	Up to 1½ (38), incl	Full section by 8-in. (203-mm) gage length (Fig. 4).	Mill reduced section to 2-in. (51-mm) gage length and approximately 25 percent less than test specimen width.
	Over 1½ (38)	Full section, or mill to 1½ in. (38 mm) wide by 8-in. (203-mm) gage length (Fig. 4).	Mill reduced section to 2-in. gage length and 1½ in. wide.
⅝ to 1½ (16 to 38), excl	Up to 1½ (38), incl	Full section by 8-in. gage length or machine standard ½ by 2-in. (13 by 51-mm) gage length specimen from center of section (Fig. 5).	Mill reduced section to 2-in. (51-mm) gage length and approximately 25 percent less than test specimen width or machine standard ½ by 2-in. (13 by 51-mm) gage length specimen from center of section (Fig. 5).
	Over 1½ (38)	Full section, or mill 1½ in. (38 mm) width by 8-in. (203-mm) gage length (Fig. 4) or machine standard ½ by 2-in. gage (13 by 51-mm) gage length specimen from midway between edge and center of section (Fig. 5).	Mill reduced section to 2-in. gage length and 1½ in. wide or machine standard ½ by 2-in. gage length specimen from midway between edge and center of section (Fig. 5)
1½ (38) and over		Full section by 8-in. (203-mm) gage length, or machine standard ½ by 2-in. (13 by 51-mm) gage length specimen from midway between surface and center (Fig. 5).	Machine standard ½ by 2-in. (13 by 51-mm) gage length specimen from midway between surface and center (Fig. 5).
		Rounds, Squares, Hexagons, and Octagons	
Diameter or Distance Between Parallel Faces, in. (mm)		Hot-Rolled Bars	Cold-Finished Bars
Under ⅝		Full section by 8-in. (203-mm) gage length or machine to sub-size specimen (Fig. 5).	Machine to sub-size specimen (Fig. 5).
⅝ to 1½ (16 to 38), excl		Full section by 8-in. (203-mm) gage length or machine standard ½ in. by 2-in. (13 by 51-mm) gage length specimen from center of section (Fig. 5).	Machine standard ½ in. by 2-in. gage length specimen from center of section (Fig. 5).
1½ (38) and over		Full section by 8-in. (203-mm) gage length or machine standard ½ in. by 2-in. (13 by 51-mm) gage length specimen from midway between surface and center of section (Fig. 5).	Machine standard ½ in. by 2-in. (13 by 51-mm gage length specimen from midway between surface and center of section (Fig. 5).
		Other Bar-Size Sections	
All sizes		Full section by 8-in. (203-mm) gage length or prepare test specimen 1½ in. (38 mm) wide (if possible) by 8-in. (203-mm) gage length.	Mill reduced section to 2-in. (51-mm) gage length and approximately 25 percent less than test specimen width.

A 370

TABLE 8 Wall Thickness Limitations of Superficial Hardness Test on Annealed or Ductile Materials[a]
("T" Scale (1/16-in. Ball))

Wall Thickness, in. (mm)	Load, kgf
Over 0.050 (1.27)	45
Over 0.035 (0.89)	30
0.020 and over (0.51)	15

[a] The heaviest load recommended for a given wall thickness is generally used.

TABLE 9 Wall Thickness Limitations of Superficial Hardness Test on Cold Worked or Heat Treated Material[a]
("N" Scale (Diamond Penetrator))

Wall Thickness, in. (mm)	Load, kgf
Over 0.035 (0.89)	45
Over 0.025 (0.51)	30
0.015 and over (0.38)	15

[a] The heaviest load recommended for a given wall thickness is generally used.

TABLE 10 Effect of Varying Notch Dimensions on Standard Specimens

	High-Energy Specimens, ft·lbf (J)	High-Energy Specimens, ft·lbf (J)	Low-Energy Specimens, ft·lbf (J)
Specimen with standard dimensions	76.0 ± 3.8 (103.0 ± 5.2)	44.5 ± 2.2 (60.3 ± 3.0)	12.5 ± 1.0 (16.9 ± 1.4)
Depth of notch, 0.084 in. (2.13 mm)[a]	72.2 (97.9)	41.3 (56.0)	11.4 (15.5)
Depth of notch, 0.0805 in. (2.04 mm)[a]	75.1 (101.8)	42.2 (57.2)	12.4 (16.8)
Depth of notch, 0.0775 in. (1.77 mm)[a]	76.8 (104.1)	45.3 (61.4)	12.7 (17.2)
Depth of notch, 0.074 in. (1.57 mm)[a]	79.6 (107.9)	46.0 (62.4)	12.8 (17.3)
Radius at base of notch, 0.005 in. (0.127 mm)[b]	72.3 (98.0)	41.7 (56.5)	10.8 (14.6)
Radius at base of notch, 0.015 in. (0.381 mm)[b]	80.0 (108.5)	47.4 (64.3)	15.8 (21.4)

[a] Standard 0.079 ± 0.002 in. (2.00 ± 0.05 mm).
[b] Standard 0.010 ± 0.001 in. (0.25 ± 0.025 mm).

A 370

TABLE 11 Carbon and Alloy Steels—Material Constant $a = 0.4$. Multiplication Factors for Converting Percent Elongation from ½-in. Diameter by 2-in. Gage Length Standard Tension Test Specimen to Standard ½ by 2-in. and 1½ by 8-in. Flat Specimens

Thickness, in.	½ by 2-in. Specimen	1½ by 8-in. Specimen	Thickness, in.	1½ by 8-in. Specimen
0.025	0.574	...	0.800	0.822
0.030	0.596	...	0.850	0.832
0.035	0.614	...	0.900	0.841
0.040	0.631	...	0.950	0.850
0.045	0.646	...	1.000	0.859
0.050	0.660	...	1.125	0.880
0.055	0.672	...	1.250	0.898
0.060	0.684	...	1.375	0.916
0.065	0.695	...	1.500	0.932
0.070	0.706	...	1.625	0.947
0.075	0.715	...	1.750	0.961
0.080	0.725	...	1.875	0.974
0.085	0.733	...	2.000	0.987
0.090	0.742	0.531	2.125	0.999
0.100	0.758	0.542	2.250	1.010
0.110	0.772	0.553	2.375	1.021
0.120	0.786	0.562	2.500	1.032
0.130	0.799	0.571	2.625	1.042
0.140	0.810	0.580	2.750	1.052
0.150	0.821	0.588	2.875	1.061
0.160	0.832	0.596	3.000	1.070
0.170	0.843	0.603	3.125	1.079
0.180	0.852	0.610	3.250	1.088
0.190	0.862	0.616	3.375	1.096
0.200	0.870	0.623	3.500	1.104
0.225	0.891	0.638	3.625	1.112
0.250	0.910	0.651	3.750	1.119
0.275	0.928	0.664	3.875	1.127
0.300	0.944	0.675	4.000	1.134
0.325	0.959	0.686	...	...
0.350	0.973	0.696	...	...
0.375	0.987	0.706	...	...
0.400	1.000	0.715	...	...
0.425	1.012	0.724	...	...
0.450	1.024	0.732	...	...
0.475	1.035	0.740	...	...
0.500	1.045	0.748	...	...
0.525	1.056	0.755	...	...
0.550	1.066	0.762	...	...
0.575	1.075	0.770	...	...
0.600	1.084	0.776	...	...
0.625	1.093	0.782	...	...
0.650	1.101	0.788	...	...
0.675	1.110	...	...	...
0.700	1.118	0.800	...	...
0.725	1.126	...	...	...
0.750	1.134	0.811	...	...

TABLE 12 Annealed Austenitic Stainless Steels—Material Constant $a = 0.127$. Multiplication Factors for Converting Percent Elongation from ½-in. Diameter by 2-in. Gage Length Standard Tension Test Specimen to Standard ½ by 2-in. and 1½ by 8-in. Flat Specimens

Thickness, in.	½ by 2-in. Specimen	1½ by 8-in. Specimen	Thickness, in.	1½ by 8-in. Specimen
0.025	0.839	...	0.800	0.940
0.030	0.848	...	0.850	0.943
0.035	0.857	...	0.900	0.947
0.040	0.864	...	0.950	0.950
0.045	0.870	...	1.000	0.953
0.050	0.876	...	1.125	0.960
0.055	0.882	...	1.250	0.966
0.060	0.886	...	1.375	0.972
0.065	0.891	...	1.500	0.978
0.070	0.895	...	1.625	0.983
0.075	0.899	...	1.750	0.987
0.080	0.903	...	1.875	0.992
0.085	0.906	...	2.000	0.996
0.090	0.909	0.818	2.125	1.000
0.095	0.913	0.821	2.250	1.003
0.100	0.916	0.823	2.375	1.007
0.110	0.921	0.828	2.500	1.010
0.120	0.926	0.833	2.625	1.013
0.130	0.931	0.837	2.750	1.016
0.140	0.935	0.841	2.875	1.019
0.150	0.940	0.845	3.000	1.022
0.160	0.943	0.848	3.125	1.024
0.170	0.947	0.852	3.250	1.027
0.180	0.950	0.855	3.375	1.029
0.190	0.954	0.858	3.500	1.032
0.200	0.957	0.860	3.625	1.034
0.225	0.964	0.867	3.750	1.036
0.250	0.970	0.873	3.875	1.038
0.275	0.976	0.878	4.000	1.041
0.300	0.982	0.883	...	...
0.325	0.987	0.887	...	...
0.350	0.991	0.892	...	...
0.375	0.996	0.895	...	...
0.400	1.000	0.899	...	...
0.425	1.004	0.903	...	...
0.450	1.007	0.906	...	...
0.475	1.011	0.909	...	...
0.500	1.014	0.912	...	...
0.525	1.017	0.915	...	...
0.550	1.020	0.917	...	...
0.575	1.023	0.920	...	...
0.600	1.026	0.922	...	...
0.625	1.029	0.925	...	...
0.650	1.031	0.927	...	...
0.675	1.034	...	...	...
0.700	1.036	0.932	...	...
0.725	1.038	...	...	...
0.750	1.041	0.936	...	...

TABLE 13 Recommended Values for Rounding Test Data

Test Quantity	Test Data Range	Rounded Value[A]
Yield Point, Yield Strength, Tensile Strength	up to 50 000 psi, excl	100 psi
	50 000 to 100 000 psi, excl	500 psi
	100 000 psi and above	1000 psi
	up to 500 MPa, excl	1 MPa
	500 to 1000 MPa, excl	5 MPa
	1000 MPa and above	10 MPa
Elongation	0 to 10 %, excl	0.5 %
	10 % and above	1 %
Reduction of Area	0 to 10 %, excl	0.5 %
	10 % and above	1 %
Impact Energy	0 to 240 ft·lbf (or 0 to 325 J)	1 ft·lbf (or 1 J)[B]
Brinell Hardness	all values	tabular value[C]
Rockwell Hardness	all scales	1 Rockwell Number

[A] Round test data to the nearest integral multiple of the values in this column. If the data value is exactly midway between two rounded values, round to the higher value.

[B] These units are not equivalent but the rounding occurs in the same numerical ranges for each. (1 ft·lbf = 1.356 J.)

[C] Round the mean diameter of the Brinell impression to the nearest 0.05 mm and report the corresponding Brinell hardness number read from the table without further rounding.

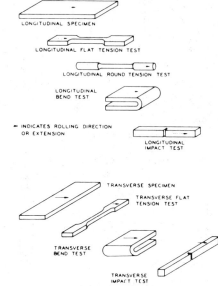

FIG. 1 The Relation of Test Coupons and Test Specimens to Rolling Direction or Extension (Applicable to General Wrought Products).

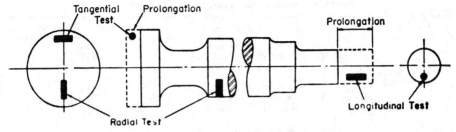

(a) Shafts and Rotors

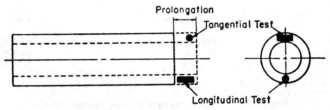

(b) Hollow Forgings.

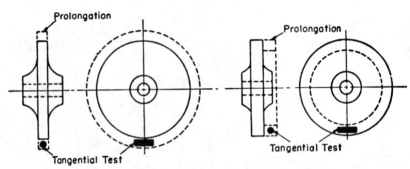

(c) Disk Forgings

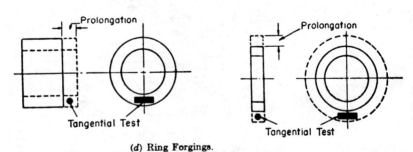

(d) Ring Forgings.

FIG. 2 Locations of Test Specimens for Various Types of Forgings.

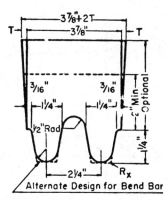
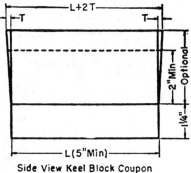

Side View Keel Block Coupon

(*a*) Design for Double Keel Block Coupon.

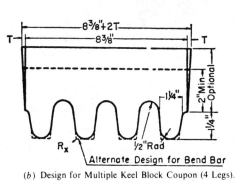
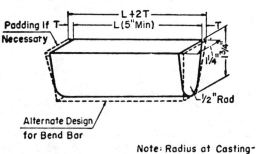

Note: Radius at Casting-Coupon Interface at Option of Foundry

(*b*) Design for Multiple Keel Block Coupon (4 Legs).

(*c*) Design for "Attached" Coupon.

Metric Equivalents

in.	3/16	1/2	1 1/4	1 3/4	2	2 1/4	3 7/8	5	8 1/8
mm	4.8	13	32	45	51	57	98	127	213

FIG. 3 Test Coupons for Castings (see Table 1 for Details of Design).

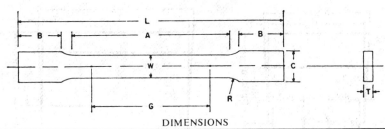

DIMENSIONS

	Standard Specimens				Subsize Specimen	
	Plate-Type, 1½-in. Wide		Sheet-Type, ½-in. Wide		¼-in. Wide	
	in.	mm	in.	mm	in.	mm
G—Gage length (Notes 1 and 2)	8.00 ± 0.01	200 ± 0.25	2.000 ± 0.005	50.0 ± 0.10	1.000 ± 0.003	25.0 ± 0.08
W—Width (Notes 3, 4, and 5)	1½ + ⅛ − ¼	40 + 3 − 6	0.500 ± 0.010	12.5 ± 0.25	0.250 ± 0.002	6.25 ± 0.05
T—Thickness (Note 6)			thickness of material			
R—Radius of fillet, min	½	13	½	13	¼	6
L—Over-all length, min (Notes 2 and 7)	18	450	8	200	4	100
A—Length of reduced section, min	9	225	2¼	60	1¼	32
B—Length of grip section, min (Note 8)	3	75	2	50	1¼	32
C—Width of grip section, approximate (Notes 4, 9, and 10)	2	50	¾	20	⅜	10

NOTE 1—For the 1½-in. (40-mm) wide specimen, punch marks for measuring elongation after fracture shall be made on the flat or on the edge of the specimen and within the reduced section. Either a set of nine or more punch marks 1 in. (25 mm) apart, or one or more pairs of punch marks 8 in. (200 mm) apart may be used.

NOTE 2—When elongation measurements of 1½-in. (40-mm) wide specimens are not required, a gage length (G) of 2.000 in. ± 0.005 in. (50.0 mm ± 0.10 mm) with all other dimensions similar to the plate-type specimen may be used.

NOTE 3—For the three sizes of specimens, the ends of the reduced section shall not differ in width by more than 0.004, 0.002 or 0.001 in. (0.10, 0.05 or 0.025 mm), respectively. Also, there may be a gradual decrease in width from the ends to the center, but the width at either end shall not be more than 0.015 in., 0.005 in., or 0.003 in. (0.40, 0.10 or 0.08 mm), respectively, larger than the width at the center.

NOTE 4—For each of the three sizes of specimens, narrower widths (W and C) may be used when necessary. In such cases the width of the reduced section should be as large as the width of the material being tested permits; however, unless stated specifically, the requirements for elongation in a product specification shall not apply when these narrower specimens are used. If the width of the material is less than W, the sides may be parallel throughout the length of the specimen.

NOTE 5—The specimen may be modified by making the sides parallel throughout the length of the specimen, the width and tolerances being the same as those specified above. When necessary a narrower specimen may be used, in which case the width should be as great as the width of the material being tested permits. If the width is 1½ in. (38 mm) or less, the sides may be parallel throughout the length of the specimen.

NOTE 6—The dimension T is the thickness of the test specimen as provided for in the applicable material specifications. Minimum nominal thickness of 1½-in. (40-mm) wide specimens shall be ³⁄₁₆ in. (5 mm), except as permitted by the product specification. Maximum nominal thickness of ½-in. (12.5-mm) and ¼-in. (6-mm) wide specimens shall be ³⁄₄ in. (19 mm) and ¼ in. (6 mm), respectively.

NOTE 7—To aid in obtaining axial loading during testing of ¼-in. (6-mm) wide specimens, the over-all length should be as the material will permit.

NOTE 8—It is desirable, if possible, to make the length of the grip section large enough to allow the specimen to extend into the grips a distance equal to two thirds or more of the length of the grips. If the thickness of ½-in. (13-mm) wide specimens is over ⅜ in. (10 mm), longer grips and correspondingly longer grip sections of the specimen may be necessary to prevent failure in the grip section.

NOTE 9—For standard sheet-type specimens and subsize specimens the ends of the specimen shall be symmetrical with the center line of the reduced section within 0.01 and 0.005 in. (0.25 and 0.13 mm), respectively. However, for steel if the ends of the ½-in. (12.5-mm) wide specimen are symmetrical within 0.05 in. (1.0 mm) a specimen may be considered satisfactory for all but referee testing.

NOTE 10—For standard plate-type specimens the ends of the specimen shall be symmetrical with the center line of the reduced section within 0.25 in. (6.35 mm) except for referee testing in which case the ends of the specimen shall be symmetrical with the center line of the reduced section within 0.10 in. (2.5 mm).

FIG. 4 Rectangular Tension Test Specimens.

A 370

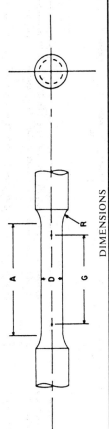

DIMENSIONS

Nominal Diameter	Standard Specimen		Small-Size Specimens Proportional to Standard					
	in.	mm	in.	mm	in.	mm	in.	mm
	0.500	12.5	0.350	8.75	0.250	6.25	0.160	4.00	0.113	2.50
G—Gage length	2.000 ± 0.005	50.0 ± 0.10	1.400 ± 0.005	35.0 ± 0.10	1.000 ± 0.005	25.0 ± 0.10	0.640 ± 0.005	16.0 ± 0.10	0.450 ± 0.005	10.0 ± 0.10
—Diameter (Note 1)	0.500 ± 0.010	12.5 ± 0.25	0.350 ± 0.007	8.75 ± 0.18	0.250 ± 0.005	6.25 ± 0.12	0.160 ± 0.003	4.00 ± 0.08	0.113 ± 0.002	2.50 ± 0.05
R—Radius of fillet, min	3/8	10	1/4	6	3/16	5	5/32	4	3/32	2
A—Length of reduced section, min (Note 2)	2 1/4	60	1 3/4	45	1 1/4	32	3/4	20	5/8	16

NOTE 1—The reduced section may have a gradual taper from the ends toward the center, with the ends not more than 1 percent larger in diameter than the center (controlling dimension).

NOTE 2—If desired, the length of the reduced section may be increased to accommodate an extensometer of any convenient gage length. Reference marks for the measurement of elongation should, nevertheless, be spaced at the indicated gage length.

NOTE 3—The gage length and fillets shall be as shown, but the ends may be of any form to fit the holders of the testing machine in such a way that the load shall be axial (see Fig. 9). If the ends are to be held in wedge grips it is desirable, if possible, to make the length of the grip section great enough to allow the specimen to extend into the grips a distance equal to two thirds or more of the length of the grips.

NOTE 4—On the round specimens in Figs. 5 and 6, the gage lengths are equal to four times the nominal diameter. In some product specifications other specimens may be provided for, but unless the 4-to-1 ratio is maintained within dimensional tolerances, the elongation values may not be comparable with those obtained from the standard test specimen.

NOTE 5—The use of specimens smaller than 0.250-in. (6.25-mm) diameter shall be restricted to cases when the material to be tested is of insufficient size to obtain larger specimens or when all parties agree to their use for acceptance testing. Smaller specimens, require suitable equipment and greater skill in both machining and testing.

NOTE 6—Five sizes of specimens often used have diameters of approximately 0.505, 0.357, 0.252, 0.160, and 0.113 in., the reason being to permit easy calculations of stress from loads, since the corresponding cross sectional areas are equal or close to 0.200, 0.100, 0.0500, 0.0200, and 0.0100 in.^{2}, respectively. Thus, when the actual diameters agree with these values, the stresses (or strengths) may be computed using the simple multiplying factors 5, 10, 20, 50, and 100, respectively. (The metric equivalents of these fixed diameters do not result in correspondingly convenient cross sectional areas and multiplying factors.)

FIG. 5 Standard 0.500-in. (12.5-mm) Round Tension Test Specimen with 2-in. (50-mm) Gage Length and Examples of Small-Size Specimens Proportional to the Standard Specimen.

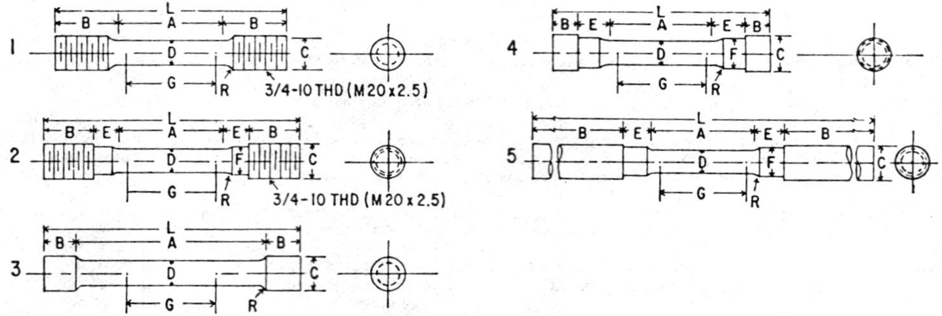

DIMENSIONS

	Specimen 1		Specimen 2		Specimen 3		Specimen 4		Specimen 5	
	in.	mm	in.	mm	in.	mm	in.	mm	in.	mm
G—Gage length	2.000 ± 0.005	50.0 ± 0.10	2.000 ± 0.005	50.0 ± 0.10	2.000 ± 0.005	50.0 ± 0.10	2.000 ± 0.005	50.0 ± 0.10	2.000 ± 0.005	50.0 ± 0.10
D—Diameter (Note 1)	0.500 ± 0.010	12.5 ± 0.25	0.500 ± 0.010	12.5 ± 0.25	0.500 ± 0.010	12.5 ± 0.25	0.500 ± 0.010	12.5 ± 0.25	0.500 ± 0.010	12.5 ± 0.25
R—Radius of fillet, min	⅜	10	⅜	10	1⁄16	2	⅜	10	⅜	10
A—Length of reduced section	2¼, min	60, min	2¼, min	60, min	4, approximately	100, approximately	2¼, min	60, min	2¼, min	60, min
L—Over-all length, approximate	5	125	5½	140	5½	140	4¾	120	9½	240
B—Length of end section (Note 2)	1⅜, approximately	35, approximately	1, approximately	25, approximately	¾, approximately	20, approximately	½, approximately	13, approximately	3, min	75, min
C—Diameter of end section	¾	20	¾	20	23⁄32	18	⅞	22	¾	20
E—Length of shoulder and fillet section, approximate	. . .	. . .	⅝	16	. . .	. . .	¾	20	⅝	16
F—Diameter of shoulder	. . .	. . .	⅝	16	. . .	. . .	⅝	16	19⁄32	15

NOTE 1—The reduced section may have a gradual taper from the ends toward the center with the ends not more than 0.005 in. (0.10 mm) larger in diameter than the center.

NOTE 2—On Specimen 5 it is desirable, if possible, to make the length of the grip section great enough to allow the specimen to extend into the grips a distance equal to two thirds or more of the length of the grips.

NOTE 3—The use of UNF series of threads (¾ by 16, ½ by 20, ⅜ by 24, and ¼ by 28) is recommended for high-strength, brittle materials to avoid fracture in the thread portion.

FIG. 6 Various Types of Ends for Standard Round Tension Test Specimen.

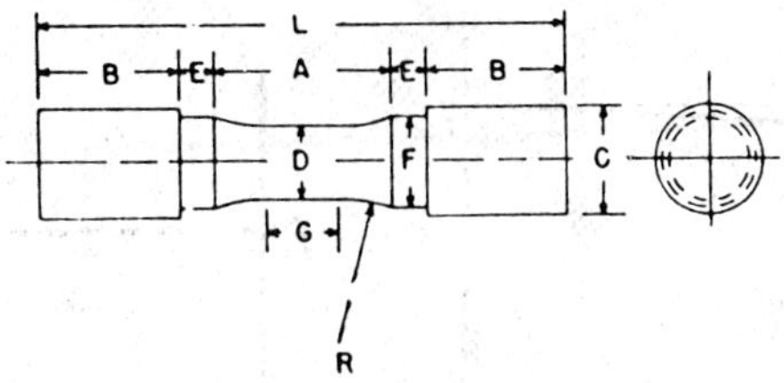

ASTM A 370

DIMENSIONS

	Specimen 1		Specimen 2		Specimen 3	
	in.	mm	in.	mm	in.	mm
G—Length of parallel	Shall be equal to or greater than diameter D					
D—Diameter	0.500 ± 0.010	12.5 ± 0.25	0.750 ± 0.015	20.0 ± 0.40	1.25 ± 0.025	30.0 ± 0.60
R—Radius of fillet, min	1	25	1	25	2	50
A—Length of reduced section, min	1¼	32	1½	38	2¼	60
L—Over-all length, min	3¾	95	4	100	6⅜	160
B—Length of end section, approximate	1	25	1	25	1¾	45
C—Diameter of end section, approximate	¾	20	1⅛	30	1⅞	48
E—Length of shoulder, min	¼	6	¼	6	5/16	8
F—Diameter of shoulder	⅝ ± 1/64	16.0 ± 0.40	15/16 ± 1/64	24.0 ± 0.40	17/16 ± 1/64	36.5 ± 0.40

NOTE—The reduced section and shoulders (dimensions A, D, E, F, G, and R) shall be shown, but the ends may be of any form to fit the holders of the testing machine in such a way that the load shall be axial. Commonly the ends are threaded and have the dimensions B and C given above.

FIG. 7   Standard Tension Test Specimen for Cast Iron.

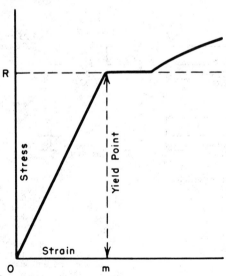

FIG. 8   Stress-Strain Diagram Showing Yield Point Corresponding with Top of Knee.

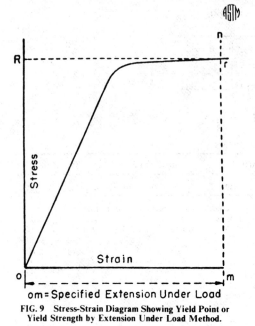

FIG. 9 Stress-Strain Diagram Showing Yield Point or Yield Strength by Extension Under Load Method.

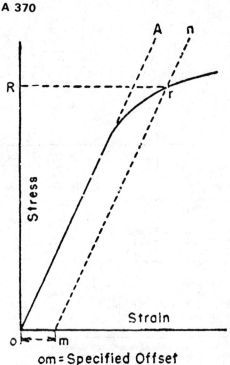

FIG. 10 Stress-Strain Diagram for Determination of Yield Strength by the Offset Method.

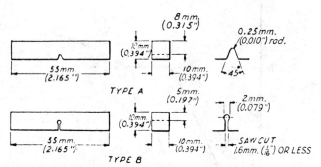

NOTE—Permissible variations shall be as follows:
Adjacent sides shall be at   90 deg ± 10 min
Cross section dimensions   ±0.025 mm (0.001 in.)
Length of specimen   +0, −2.5 mm (0.100 in.)
Angle of notch   ±1 deg
Radius of notch   ±0.025 mm (0.001 in.)
Dimensions to bottom of notch:
    Specimen, Type A   8 ± 0.025 mm (0.315 ± 0.001 in.)
    Specimen, Type B   5 ± 0.05 mm (0.197 ± 0.002 in.)
Finish   63 µin. (1.6 µm) max on notched surface and opposite face; 125 µin. (3.2 µm) max on other two surfaces

FIG. 11 Simple Beam Impact Test Specimens, Types A and B.

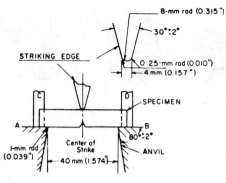

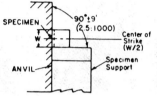

All dimensional tolerances shall be ± 0.05 mm (0.002 in.) unless otherwise specified.

NOTE 1—A shall be parallel to B within 2:1000 and coplanar with B within 0.05 mm (0.002 in.).
NOTE 2—C shall be parallel to D within 20:1000 and coplanar with D within 0.125 mm (0.005 in.).
NOTE 3—Finish on unmarked parts shall be 4 μm (125 μin.

FIG. 12 Charpy (Simple-Beam) Impact Test.

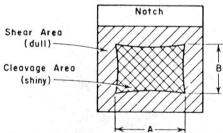

NOTE 1—Measure average dimensions A and B to the nearest 0.02 in. or 0.5 mm.
NOTE 2—Determine the percent shear fracture using Table 4 or Table 5.

FIG. 14 Determination of percent Shear Fracture.

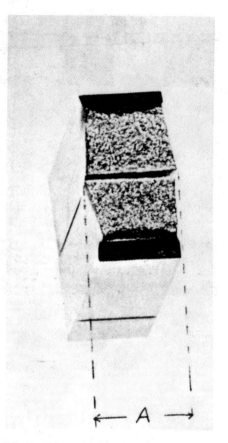

FIG. 13 Halves of Broken Charpy V-Notch Impact Specimen Joined for the Measurement of Lateral Expansion, Dimension A.

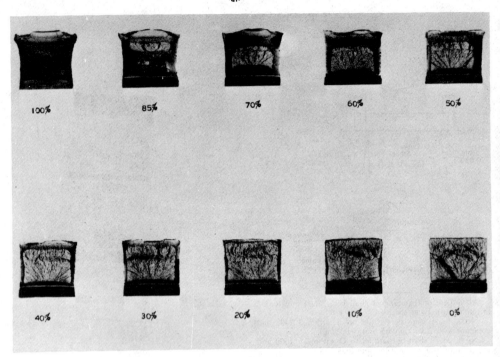

FIG. 15 Fracture Appearance Charts and percent Shear Fracture Comparator.

FIG. 16 Lateral Expansion Gage for Charpy Impact Specimens.

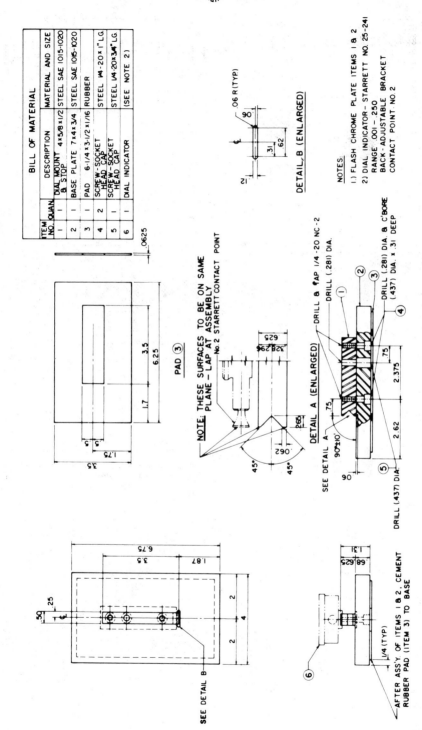

FIG. 17 Assembly and Details for Lateral Expansion Gage.

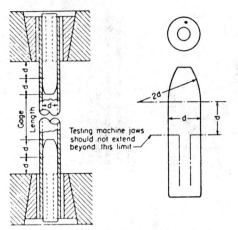

FIG. 18 Metal Plugs for Testing Tubular Specimens, Proper Location of Plugs in Specimen and of Specimen in Heads of Testing Machine.

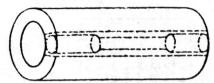

FIG. 19 Location of Longitudinal Tension Test Specimens in Large Diameter Tubing.

(*a*) Specimen for 8-in. Gage Length Test (Welded Only).

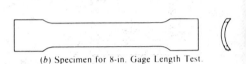

(*b*) Specimen for 8-in. Gage Length Test.

(*c*) Specimen for 2-in. Gage Length Test.

(*d*) Specimen for Full-Section Test.

FIG. 20 Longitudinal Tension Test Specimens for Large Diameter Tubing.

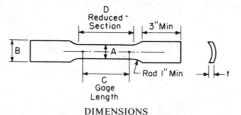

DIMENSIONS

Specimen No.	Dimensions, in.			
	A	B	C	D
1	½ ± 0.015	†¹¹⁄₁₆ approximately	2 ± 0.005	2¼ min
2	¾ ± 0.031	1 approximately	2 ± 0.005	2¼ min
			4 ± 0.005	4½ min
3	1 ± 0.062	1½ approximately	2 ± 0.005	2¼ min
			4 ± 0.005	4½ min
4	1½ ± ⅛	2 approximately	2 ± 0.010	2¼ min
			4 ± 0.015	4½ min
			8 ± 0.020	9 min

† Editorially corrected.
NOTE 1—Cross-sectional area may be calculated by multiplying A and t.
NOTE 2—The dimension t is the thickness of the test specimen as provided for in the applicable material specifications.
NOTE 3—The reduced section shall be parallel within 0.010 in. and may have a gradual taper in width from the ends toward the center, with the ends not more than 0.010 in. wider than the center.
NOTE 4—The ends of the specimen shall be symmetrical with the center line of the reduced section within 0.10 in.
NOTE 5—Metric equivalent: 1 in. = 25.4 mm.

FIG. 21 Dimensions and Tolerances for Longitudinal Tension Test Specimens for Large Diameter Tubing.

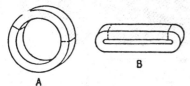

FIG. 22 Location of Transverse Tension Test Specimens in Ring Cut from Tubular Products.

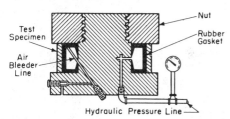

FIG. 24 Testing Machine for Determination of Transverse Yield Strength from Annular Ring Specimens.

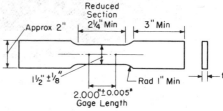

NOTE 1—The dimension t is the thickness of the test specimen as provided for in the applicable material specifications.
NOTE 2—The reduced section shall be parallel within 0.010 in. and may have a gradual taper in width from the ends toward the center, with the ends not more than 0.010 in. wider than the center.
NOTE 3—The ends of the specimen shall be symmetrical with the center line of the reduced section within 0.10 in.
NOTE 4—Metric equivalent: 1 in. = 25.4 mm.

FIG. 23 Transverse Tension Test Specimen Machined from Ring Cut from Tubular Products.

FIG. 25 Roller Chain Type Extensometer, Unclamped.

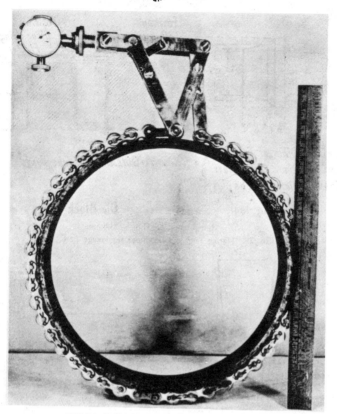

FIG. 26 Roller Chain Type Extensometer, Clamped.

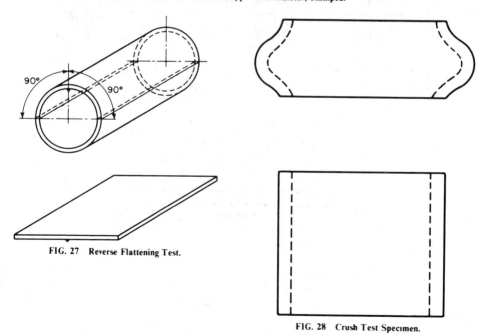

FIG. 27 Reverse Flattening Test.

FIG. 28 Crush Test Specimen.

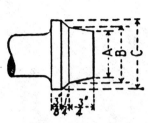 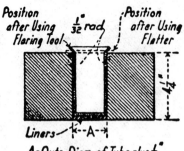

A = Outs. Diam. of Tube less $\frac{5}{8}"$
B = Outs. Diam. of Tube less $\frac{3}{8}"$
C = Outs. Diam. of Tube plus $\frac{1}{16}"$

A = Outs. Diam. of Tube plus $\frac{1}{32}"$

Flaring Tool   **Die Block**

NOTE—Metric equivalent: 1 in. = 25.4 mm.
FIG. 29 Flaring Tool and Die Block for Flange Test.

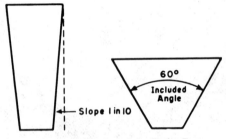

FIG. 30 Tapered Mandrels for Flaring Test.

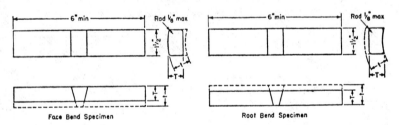

NOTE—Metric equivalent: 1 in. = 25.4 mm.

Pipe Wall Thickness (t), in.	Test Specimen Thickness, in.
Up to $\frac{3}{8}$, incl	t
Over $\frac{3}{8}$	$\frac{3}{8}$

FIG. 31(a) Transverse Face- and Root-Bend Test Specimens

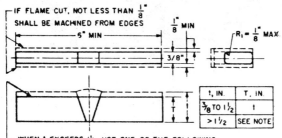

FIG. 31(b) Side-Bend Specimen for Ferrous Materials

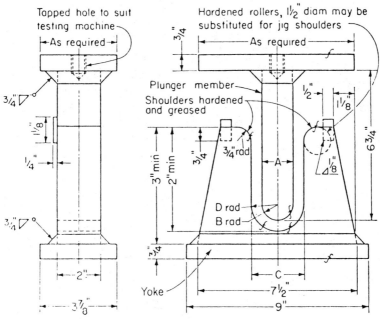

NOTE: Metric equivalent: 1 in. = 25.4 mm.

Test Specimen Thickness, in.	A	B	C	D
3/8	1½	¾	2⅜	1 3/16
t	4t	2t	6t + ⅛	3t + 1/16

FIG. 32 Guided-Bend Test Jig.

FIG. 33 Tension Testing Full-Size Bolt.

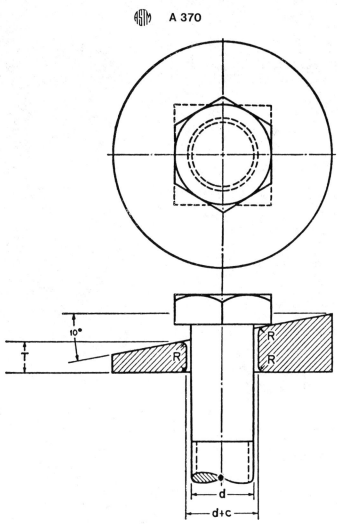

c = Clearance of wedge hole.
d = Diameter of bolt.
R = Radius.
T = Thickness of wedge at short side of hole equal to one-half diameter of bolt.

FIG. 34 Wedge Test Details.

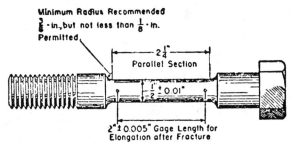

NOTE—Metric equivalent: 1 in. = 25.4 mm.

FIG. 35 Tension Test Specimen for Bolt with Turned-Down Shank.

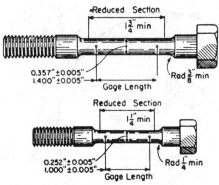

NOTE— Metric equivalent: 1 in. = 25.4 mm.
FIG. 36 Examples of Small Size Specimens Proportional to Standard 2-in. Gage Length Specimen.

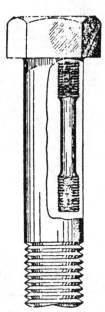

FIG. 37 Location of Standard Round 2-in. Gage Length Tension Test Specimen When Turned from Large Size Bolt.

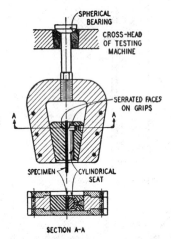

FIG. 38 Wedge-Type Gripping Device.

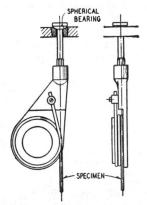

FIG. 39 Snubbing-Type Gripping Device.

The American Society for Testing and Materials takes no position respecting the validity of any patent rights asserted in connection with any item mentioned in this standard. Users of this standard are expressly advised that determination of the validity of any such patent rights, and the risk of infringement of such rights, are entirely their own responsibility.

This standard is subject to revision at any time by the responsible technical committee and must be reviewed every five years and if not revised, either reapproved or withdrawn. Your comments are invited either for revision of this standard or for additional standards and should be addressed to ASTM Headquarters. Your comments will receive careful consideration at a meeting of the responsible technical committee, which you may attend. If you feel that your comments have not received a fair hearing you should make your views known to the ASTM Committee on Standards, 1916 Race St., Philadelphia, Pa. 19103.

Designation: A 380 – 78

An American National Standard

Standard Recommended Practice for
CLEANING AND DESCALING STAINLESS STEEL PARTS, EQUIPMENT, AND SYSTEMS[1]

This standard is issued under the fixed designation A 380; the number immediately following the designation indicates the year of original adoption or, in the case of revision, the year of last revision. A number in parentheses indicates the year of last reapproval. A superscript epsilon (ϵ) indicates an editorial change since the last revision or reapproval.

1. Scope

1.1 This recommended practice covers recommendations and precautions for cleaning and descaling of new stainless steel parts, assemblies, equipment, and installed systems. These recommendations are presented as procedures for guidance when it is recognized that for a particular service it is desired to remove surface contaminants that may impair the normal corrosion resistance, or result in the later contamination of the particular stainless steel grade, or cause product contamination. For certain exceptional applications, additional requirements which are not covered by this standard may be specified upon agreement between the purchaser and the contractor. Although they apply primarily to materials in the composition ranges of the austenitic, ferritic, and martensitic stainless steels, the practices described may also be useful for cleaning other metals if due consideration is given to corrosion and possible metallurgical effects.

1.2 The standard does not cover decontamination or cleaning of equipment or systems that have been in service, nor does it cover descaling and cleaning of materials at the mill. On the other hand, some of the practices may be applicable for these purposes. While the standard provides recommendations and information concerning the use of acids and other cleaning and descaling agents, it cannot encompass detailed cleaning procedures for specific types of equipment or installations. It therefore in no way precludes the necessity for careful planning and judgment in the selection and implementation of such procedures.

1.3 These practices may be applied when free iron, oxide scale, rust, grease, oil, carbonaceous or other residual chemical films, soil, particles, metal chips, dirt, or other nonvolatile deposits might adversely affect the metallurgical or sanitary condition or stability of a surface, the mechanical operation of a part, component, or system, or contaminate a process fluid. The degree of cleanness required on a surface depends on the application. In some cases, no more than degreasing or removal of gross contamination is necessary. Others, such as food-handling, pharmaceutical, aerospace, and certain nuclear applications, may require extremely high levels of cleanness, including removal of all detectable residual chemical films and contaminants that are invisible to ordinary inspection methods.

NOTE 1 — The term "iron," when hereinafter referred to as a surface contaminant, shall denote free iron.

1.4 Attainment of surfaces that are free of iron, metallic deposits, and other contamination depends on a combination of proper design, fabrication methods, cleaning and descaling, and protection to prevent recontamination of cleaned surfaces. Meaningful tests to establish the degree of cleanness of a surface are few, and those are often difficult to administer and to evaluate objectively. Visual inspection is suitable for the detection

---

[1] This recommended practice is under the jurisdiction of ASTM Committee A-1 on Steel, Stainless Steel and Related Alloys, and is the direct responsibility of Subcommittee A01.14 on Methods of Corrosion Testing.
Current edition approved July 28, 1978. Published September 1978. Originally published as A 380 – 54 T. Last previous edition A 380 – 72.

A 380

of gross contamination, scale, rust, and particulates, but may not reveal the presence of thin films of oil or residual chemical films. In addition, visual inspection of internal surfaces is often impossible because of the configuration of the item. Methods are described for the detection of free iron and transparent chemical and oily deposits.

2. Applicable Documents

2.1 *ASTM Standards:*
F 21 Test Method for Hydrophobic Surface Films by the Atomizer Test[2]
F 22 Test Method for Hydrophobic Surface Films by the Water-Break Test[2]

2.2 *Other Documents:*
209a Federal Standard for Clean Room and Work Station Requiring Controlled Environments[3]

3. Design

3.1 Consideration should be given in the design of parts, equipment, and systems that will require cleaning to minimize the presence of crevices, pockets, blind holes, undrainable cavities, and other areas in which dirt, cleaning solutions, or sludge might lodge or become trapped, and to provide for effective circulation and removal of cleaning solutions. In equipment and systems that will be cleaned in place or that cannot be immersed in the cleaning solution, it is advisable to slope lines for drainage: to provide vents at high points and drains at low points of the item or system; to arrange for removal or isolation of parts that might be damaged by the cleaning solution of fumes from the cleaning solutions; to provide means for attaching temporary fill and circulation lines; and to provide for inspection of cleaned surfaces.

3.2 In a complex piping system it may be difficult to determine how effective a cleaning operation has been. One method of designing inspectability into the system is to provide a short flanged length of pipe (that is, a spool piece) at a location where the cleaning is, likely to be least effective; the spool piece can then be removed for inspection upon completion of cleaning.

4. Precleaning

4.1 Precleaning is the removal of grease, oil, paint, soil, grit, and other gross contamination preparatory to a fabrication process or final cleaning. Precleaning is not as critical and is generally not as thorough as subsequent cleaning operations. Materials should be precleaned before hot-forming, annealing, or other high-temperature operation, before any descaling operation, and before any finish-cleaning operation where the parts will be immersed or where the cleaning solutions will be reused. Items that are subject to several redraws or a series of hot-forming operations, with intermediate anneals, must be cleaned after each forming operation, prior to annealing. Precleaning may be accomplished by vapor degreasing; immersion in, spraying, or swabbing with alkaline or emulsion cleaners, steam, or high-pressure water-jet (see 6.2).

5. Descaling

5.1 *General*—Descaling is the removal of heavy, tightly adherent oxide films resulting from hot-forming, heat-treatment, welding, and other high-temperature operations. Because mill products are usually supplied in the descaled condition, descaling (except removal of localized scale resulting from welding) is generally not necessary during fabrication of equipment or erection of systems (see 6.3). When necessary, scale may be removed by one of the chemical methods listed below, by mechanical methods (for example, abrasive blasting, sanding, grinding, power brushing), or by a combination of these.

5.2 *Chemical Descaling (Pickling)*—Chemical descaling agents include aqueous solutions of sulfuric, nitric, and hydrofluoric acid as described in Annex A1, molten alkali or salt baths, and various proprietary formulations.

5.2.1 *Acid Pickling*—Nitric-hydrofluoric acid solution is most widely used by fabricators of stainless steel equipment and removes both metallic contamination, and welding and heat-treating scales. Its use should be carefully controlled and is not recommended for descaling sensitized austenitic stainless steels or hardened martensitic stainless steels or where it can come into contact with carbon steel parts, assemblies, equipment, and systems. See also A1.3. Solutions of nitric acid

[2] *Annual Book of ASTM Standards,* Vol 10.05.
[3] Available from Naval Publications and Forms Center, 5801 Tabor Ave., Philadelphia, Pa. 19120.

alone are usually not effective for removing heavy oxide scale.

5.2.2 Surfaces to be descaled are usually precleaned prior to chemical treatment. When size and shape of product permit, total immersion in the pickling solution is preferred. Where immersion is impractical, descaling may be accomplished by (*1*) wetting the surfaces by swabbing or spraying; or (*2*) by partially filling the item with pickling solution and rotating or rocking to slosh the solution so that all surfaces receive the required chemical treatment. The surface should be kept in contact with agitated solution for about 15 to 30 min or until inspection shows that complete scale removal has been accomplished. Without agitation, additional exposure time may be required. If rocking or rotation are impracticable, pickling solution may be circulated through the item or system until inspection shows that descaling has been accomplished.

5.2.3 Over-pickling must be avoided. Uniform removal of scale with acid pickling depends on the acid used, acid concentration, solution temperature, and contact time (see Annex A1). Continuous exposure to pickling solutions for more than 30 min is not recommended. The item should be drained and rinsed after 30 min and examined to check the effectiveness of the treatment. Additional treatment may be required. Most pickling solutions will loosen weld and heat-treating scale but may not remove them completely. Intermittent scrubbing with a stainless steel brush or fiber-bristle brush, in conjunction with pickling or the initial rinse, may facilitate the removal of scale particles and products of chemical reaction (that is, pickling *smut*).

5.2.4 After chemical descaling, surfaces must be thoroughly rinsed to remove residual chemicals; a neutralization step is sometimes necessary before final rinsing. To minimize staining, surfaces must not be permitted to dry between successive steps of the acid descaling and rinsing procedure, and thorough drying should follow the final water rinse. Chemical descaling methods, factors in their selection, and precautions in their use are described in the *Metals Handbook*.[4] When chemical descaling is necessary, it should be done while the part is in its simplest possible geometry, before subsequent fabrication or installation steps create internal crevices or undrainable spaces that may trap descaling agents, sludge, particles, or contaminated rinse water that might either result in eventual corrosion or adversely affect operation of the item after it is placed in service.

5.3 *Mechanical Descaling*—Mechanical descaling methods include abrasive blasting, power brushing, sanding, grinding, and chipping. Procedural requirements and precautions for some of these methods are given in the *Metals Handbook*.[4] Mechanical descaling methods have the advantage that they do not produce such physical or chemical conditions as intergranular attack, pitting, hydrogen embrittlement, cracks, or smut deposits. For some materials, in particular the austenitic stainless steels when in the sensitized condition and the martensitic stainless steels when in the hardened condition, mechanical descaling may be the only suitable method. Grinding is usually the most effective means of removing localized scale such as that which results from welding. Disadvantages of mechanical descaling are cost, as compared to chemical descaling, and the fact that surface defects (for example, laps, pits, slivers) may be obscured, making them difficult to detect.

5.3.1. Surfaces to be descaled may have to be precleaned. Particular care must be taken to avoid damage by mechanical methods when descaling thin sections, polished surfaces, and close-tolerance parts. After mechanical descaling, surfaces should be cleaned by scrubbing with hot water and fiber brushes, followed by rinsing with clean, hot water.

5.3.2 Grinding wheels and sanding materials should not contain iron, iron oxide, zinc, or other undersirable materials that may cause contamination of the metal surface. Grinding wheels, sanding materials, and wire brushes previously used on other metals should not be used on stainless steel. Wire brushes should be of a stainless steel which is equal in corrosion resistance to the material being worked on.

5.3.3 Clean, previously unused glass beads or ion-free silica or alumina sand are recom-

---

[4] "Heat Treating, Cleaning, and Finishing," *Metals Handbook*, Am. Soc. Metals, 8th ed., Vol 2, 1964.

A 380

mended for abrasive blasting. Steel shot or grit is generally not recommended because of the possibility of embedding iron particles. The use of stainless steel shot or grit reduces the danger of rusting and iron contamination, but cannot completely eliminate the possibility of embedding residues of iron-oxide scale.

5.3.4 If a totally iron and scale free surface is required, most abrasive blasting may be followed by a brief acid dip (see Annex A2).

6. Cleaning

6.1 *General*—Cleaning includes all operations necessary for the removal of surface contaminants from metals to ensure (*1*) maximum corrosion resistance of the metal; (*2*) prevention of product contamination; and (*3*) achievement of desired appearance. Cleanness is a perishable condition. Careful planning is necessary to achieve and maintain clean surfaces, especially where a high degree of cleanness is required. Selection of cleaning processes is influenced mainly by the type of contaminant to be removed, the required degree of cleanness, and cost. If careful control of fabrication processes, sequencing of cleaning and fabrication operations, and measures to prevent recontamination of cleaned surfaces are exercised, very little special cleaning of the finished item or system may be necessary to attain the desired level of cleanness. If there is a question concerning the effectiveness of cleaning agents or procedures, or the possible adverse effects of some cleaning agents or procedures on the materials to be cleaned, trial runs, using test specimens and sensitive inspection techniques may be desirable. Descriptions, processes, and precautions to be observed in cleaning are given in the *Metals Handbook*.[4] Proprietary cleaners may contain harmful ingredients, such as chlorides or sulfur, which could adversely affect the performance of a part, equipment, or system under service conditions. It is recommended that the manufacturer of the cleaner be consulted if there is reason for concern.

NOTE 2—Instances are known where stainless steel vessels have stress cracked before start-up due to steaming out or boiling out with a chloride-containing detergent.

6.2 *Cleaning Methods*—Degreasing and general cleaning may be accomplished by immersion in, swabbing with, or spraying with alkaline, emulsion, solvent, or detergent cleaners or a combination of these; by vapor degreasing; by ultrasonics using various cleaners; by steam, with or without a cleaner; or by high-pressure water-jetting. The cleaning method available at any given time during the fabrication or installation of a component or system is a function of the geometric complexity of the item, the type of contamination present, the degree of cleanliness required, and cost. Methods commonly used for removing deposited contaminants (as opposed to scale) are described briefly below and in greater detail (including factors to be considered in their selection and use) in the *Metals Handbook*[4] and the *SSPC Steel Structures Painting Handbook*.[5] The safety precautions of 8.6 must be observed in the use of these methods. Particular care must be exercised when cleaning closed systems and items with crevices or internal voids to prevent retention of cleaning solutions and residues.

6.2.1 *Alkaline Cleaning* is used for the removal of oily, semisolid, and solid contaminants from metals. To a great extent the solutions used depend on their detergent qualities for cleaning action and effectiveness. Agitation and temperature of the solution are important.

6.2.2 *Emulsion Cleaning* is a process for removing oily deposits and other common contaminants from metals by the use of common organic solvents dispersed in an aqueous solution with the aid of a soap or other emulsifying agent (an emulsifying agent is one which increases the stability of a dispersion of one liquid in another). It is effective for removing a wide variety of contaminants including pigmented and unpigmented drawing compounds and lubricants, cutting fluids, and residues resulting from liquid penetrant inspection. Emulsion cleaning is used when rapid, superficial cleaning is required and when a light residual film of oil is not objectionable.

6.2.3 *Solvent Cleaning* is a process for removing contaminants from metal surfaces by immersion or by spraying or swabbing with common organic solvents such as the aliphatic

---

[5] *Good Painting Practices*, Steel Structures Painting Council, Vol 1, 1954, Chapters 2 and 3.

A 380

petroleums, chlorinated hydrocarbons, or blends of these two classes of solvents. Cleaning is usually performed at or slightly above room temperature. Except for parts with extremely heavy contamination or with hard-to-reach areas, or both, good agitation will usually eliminate the need for prolonged soaking. Virtually all metal can be cleaned with the commonly used solvents unless the solvent has become contaminated with acid, alkali, oil, or other foreign material. Chlorinated solvents are not recommended for degreasing of closed systems or items with crevices or internal voids.

6.2.4 *Vapor Degreasing* is a generic term applied to a cleaning process that employs hot vapors of a volatile chlorinated solvent to remove contaminants, and is particularly effective against oils, waxes, and greases. The cleanness and chemical stability of the degreasing solvent are critical factors in the efficiency of the vapor and possible chemical attack of the metal. Water in the degreasing tank or on the item being cleaned may react with the solvent to form hydrochloric acid, which may be harmful to the metal. No water should be present in the degreasing tank or on the item being cleaned. Acids, oxidizing agents, and cyanides must be prevented from contaminating the solvent. Materials such as silicones cause foaming at the liquid-vapor interface and may result in recontamination of the workpiece as it is removed from the degreaser. Vapor degreasing with chlorinated solvents is not recommended for closed systems or items with internal voids or crevices.

6.2.5 *Ultrasonic Cleaning* is often used in conjunction with certain solvent and detergent cleaners to loosen and remove contaminants from deep recesses and other difficult to reach areas, particularly in small workpieces. Cavitation in the liquid produced by the high frequency sound causes micro agitation of the solvent in even tiny recesses of the workpiece, making the method especially desirable for cleaning parts or assemblies having an intricate configuration. For extremely high levels of surface cleanness, high-purity solvents (1 ppm total nonvolatile residue) are required.

6.2.6 *Synthetic Detergents* are extensively used as surface-active agents because they are freer rinsing than soaps, aid in soils dispersion, and prevent recontamination. They are effective for softening hard water and in lowering the surface and interfacial tensions of the solutions. Synthetic detergents, in particular, should be checked for the presence of harmful ingredients as noted in 6.1

6.2.7 *Chelate Cleaning* — Chelates are chemicals that form soluble, complex molecules with certain metal ions, inactivating the ions in solution so they cannot normally react with another element or ions to produce precipitates or scale. They enhance the solubility of scales and certain other contaminants, do not precipitate different scales when the cleaning solution becomes spent, and can be used on some scales and contaminants that even mineral acids will not attack. When properly used (chelating agents must be coninuously circulated and must be maintained within carefully controlled temperature limits), intergranular attack, pitting, and other harmful effects are minimal. Chelating agents are particularly useful for cleaning installed equipment and systems.

6.2.8 *Mechanical Cleaning* (also see 5.3) — Abrasive blasting, vapor blasting using a fine abrasive suspended in water, grinding, or wire brushing are often desirable for removing surface contaminants and rust. Cleanliness of abrasives and cleaning equipment is extremely important to prevent recontamination of the surfaces being cleaned. Although surfaces may appear visually clean following such procedures, residual films which could prevent the formation of an optimum passive condition may still be present. Subsequent treatment such as additional iron-free abrasive cleaning methods, acid cleaning, passivation, or combinations of these is, therefore, required for stainless steel parts, equipment, and systems to be used where corrosion resistance is a prime factor to satisfy performance and service requirements, or where product contamination must be avoided.

6.2.9 *Steam Cleaning* is used mostly for cleaning bulky objects that are too large for soak tanks or spray-washing equipment. It may be used with cleaning agents such as emulsions, solvents, alkalis, and detergents. Steam lances are frequently used for cleaning piping assemblies. Steam pressures from 50 to

A 380

75 psi (345 to 515 kPa) are usually adequate (see 6.1).

6.2.10 *Water-Jetting* at water pressures of up to 10 000 psi (70 mPa) is effective for removing grease, oils, chemical deposits (except adsorbed chemicals), dirt, loose and moderately adherent scale, and other contaminants that are not actually bonded to the metal. The method is particularly applicable for cleaning piping assemblies which can withstand the high pressures involved; self-propelled nozzles or "moles" are generally used for this purpose.

6.2.11 *Acid Cleaning* is a process in which a solution of a mineral or organic acid in water, sometimes in combination with a wetting agent or detergent or both, is employed to remove iron and other metallic contamination, light oxide films, shop soil, and similar contaminants. Suggested solutions, contact times, and solution temperatures for various alloys are given in Annex A2. Acid cleaning is not generally effective for removal of oils, greases, and waxes. Surfaces should be precleaned to remove oils and greases before acid cleaning. Common techniques for acid cleaning are immersion, swabbing, and spraying. Maximum surface quality is best achieved by using a minimum cleaning time at a given acid concentration and temperature. After acid cleaning the surfaces must be thoroughly rinsed with clean water to remove all traces of the acid and thoroughly dried after the final water rinse. To minimize staining, surfaces must not be permitted to dry between successive steps of the acid cleaning and rinsing procedure. A neutralizing treatment may be required under some conditions; if used, neutralization must be followed by repeated water rinsing to remove all trace of the neutralizing agent and then thorough drying after the final water rinse. Acid cleaning is not recommended where mechanical cleaning or other chemical methods will suffice on the basis of intended use and, as may be necessary, on inspection tests (see 7.2 and 7.3). Requirements for superfluous cleaning and inspection testing can result in excessive costs. Acid cleaning, if not carefully controlled, may damage the surface and may result in further contamination of the surface.

6.3 *Cleaning of Welds and Weld-Joint Areas* — The joint area and surrounding metal for several inches back from the joint preparation, on both faces of the weld, should be cleaned immediately before starting to weld. Cleaning may be accomplished by brushing with a clean, stainless steel brush or scrubbing with a clean, lint-free cloth moistened with solvent, or both. When the joint has cooled after welding, remove all accessible weld spatter, welding flux, scale, arc strikes, etc., by grinding. According to the application, some scale or heat temper may be permissible on the nonprocess side of a weld, but should be removed from the process side if possible. If chemical cleaning of the process side of the weld is deemed necessary, the precautions of this standard must be observed. Austenitic stainless steels in the sensitized condition should generally not be descaled with nitrichydrofluoric acid solutions. Welds may also be cleaned as described in Table A2.1, Part III, Treatment *P* and *Q* (also see 5.2.3 and 5.2.4).

6.4 *Final Cleaning or Passivation, or Both* — If proper care has been taken in earlier fabrication and cleaning, final cleaning may consist of little more than scrubbing with hot water or hot water and detergent (such as trisodium phosphate, TSP), using fiber brushes. Detergent washing must be followed by a hot-water rinse to remove residual chemicals. Spot cleaning to remove localized contamination may be accomplished by wiping with a clean, solvent-moistened cloth. If the purchaser specifies passivation, the final cleaning shall be in accordance with the requirements of Table A2.1, Part II. When the stainless steel parts are to be used for applications where corrosion resistance is a prime factor to achieve satisfactory performance and service requirements, or where product contamination must be avoided, passivation followed by thorough rinsing several times with hot water and drying thoroughly after the final water rinse is recommended, whenever practical.

NOTE 3 — The term passivation is used to indicate a chemically inactive surface condition of stainless steels. It was at one time considered that an oxidizing treatment such as a nitric acid dip was essential to establish a passive film. However, it has more recently been found that mere contact with air or other oxygen-containing environment is usu-

A 380

ally sufficient to establish a *passive* film. A passivation treatment following acid or mechanical cleaning or descaling generally is not necessary provided that thorough cleaning has been accomplished and there is subsequent exposure to air or other oxygen-containing environment.

6.5 *Precision Cleaning* — Certain nuclear, space, and other especially critical applications may require that only very high-purity alcohols, acetone, ketones, trichlorotrifluoroethane, or other *precision cleaning agents* be used for final cleaning or recleaning of critical surfaces after fabrication advances to the point that internal crevices, undrainable spaces, blind holes, or surfaces that are not accessible for thorough scrubbing, rinsing, and inspection are formed. Such items are often assembled under clean-room conditions (see 8.5.5) and require approval, by the purchaser, of carefully prepared cleaning procedures before the start of fabrication.

6.6 *Cleaning of Installed Systems* — There are two approaches to cleaning installed systems. In the first, which is probably adequate for most applications, cleaning solutions are circulated through the completed system after erection, taking care to remove or protect items that could be damaged during the cleaning operation. In the second approach, which may be required for gaseous or liquid oxygen, liquid metal, or other reactive-process solutions, piping and components are installed in a manner to avoid or minimize contamination of process-solution surfaces during erection so that little additional cleaning is necessary after erection; post-erection flushing, if necessary, is done with the process fluid. If process surfaces are coated with an appreciable amount of iron oxide, a chelating treatment or high-pressure water-jetting treatment should be considered in place of acid treatment (see 6.2.7 and 6.2.10).

6.6.1 *Post-Erection Cleaning* — Circulate hot water to which a detergent has been added, for at least 4 to 8 h. A water temperature of at least 140 to 160°F (60 to 71°C) is recommended (see 6.1). Rinse by circulating clean hot water until the effluent is clear. If excessive particulate matter is present, the cleaning cycle may be preceded with a high-pressure steam blow, repeating as necessary until a polished-aluminum target on the outlet of the system is no longer dulled and scratched by particulates loosened by the high-velocity steam. Valves and similar items must be protected from damage during a steam blow.

6.6.2 If metallic iron is indicated by one of the methods suggested in Section 7, it can be removed by circulating one of the acid cleaning solutions suggested in Annex A2 at room temperature until laboratory determination for iron, made on samples of the solution taken hourly, indicate no further increase in iron content, after which circulation may be stopped and the system drained. After this treatment, circulate clean hot water (that is, without detergent) through the system for 4 h to remove all traces of acid and corrosion product resulting from the acid treatment, or until the pH of the rinse water returns to neutral.

6.6.3 In critical systems where post-erection cleaning is not desirable (for example, liquid oxygen or nuclear reactor primary coolant systems), on-site erection may be conducted under clean-room conditions. Erection instructions may require that wrapping and seals of incoming materials and equipment be kept intact until the item is inside the clean area, and that careful surveillance be exercised to prevent foreign materials (for example, cleaning swabs or tools) from being dropped or left in the system. Where contamination does occur, the cleaning procedure usually is developed through consultation between the erector and the purchaser (or his site representative). Frequently, post-erection flushing is accomplished by circulating the process fluid through the system until contamination is reduced to tolerable levels.

6.6.4 When cleaning critical installed systems, do not permit the process surfaces to dry between successive cleaning and rinsing steps, or between the final rinse and filling with the layup solution.

7. Inspection After Cleaning

7.1 *General* — Inspection techniques should represent careful, considered review of end use requirements of parts, equipment, and systems. There is no substitute for good, uniform, cleaning practices which yield a metallurgically sound and smooth surface, followed by adequate protection to preserve that condition. Establishment of the most reliable tests and test standards for cleanness are helpful in attaining the desired performance of parts, equipment, and systems. Testing should be sufficiently extensive to ensure the cleanness of all surfaces exposed to process fluids when in service. The following represent some tests that have been successfully applied to stainless steels. The purchaser shall have the option of specifying in his purchase documents that any of these quality assurance tests be used as the basis for acceptability of the cleanness or state of passivity of the stainless steel item.

7.2 *Gross Inspection:*

7.2.1 *Visual* — Items cleaned in accordance with this standard should be free of paint, oil, grease, welding flux, slag, heat-treating and hot-forming scale (tightly adherent scale resulting from welding may be permissible on some surfaces), dirt, trash, metal and abrasive particles and chips, and other gross contamination. Some deposited atmospheric dust will normally be present on exterior surfaces but should not be present on interior surfaces. Visual inspection should be carried out under a lighting level, including both general and supplementary lighting, of at least 100 footcandles (1076 lx), and preferably 250 footcandles (2690 lx) on the surfaces being inspected. Visual inspection should be supplemented with borescopes, mirrors, and other aids, as necessary, to properly examine inaccessible or difficult-to-see surfaces. Lights should be positioned to prevent glare on the surfaces being inspected.

7.2.2 *Wipe Tests* — Rubbing of a surface with a clean, lint-free, white cotton cloth or filter paper moistened (but not saturated) with high-purity solvent (see 6.5), may be used for evaluating the cleanness of surfaces not accessible for direct visual inspection. Wipe tests of small diameter tubing are made by blowing a clean white felt plug, slightly larger in diameter than the inside diameter of the tube, through the tube with clean, dry, filtered compressed air. Cleanness in wipe tests is evaluated by the type of contamination rubbed off on the swab or plug. The presence of a smudge on the cloth is evidence of contamination. In cases of dispute concerning the harmful nature of the contamination, a sample of the smudge may be transferred to a clean quartz microscope slide for infrared analysis. The wipe test is sometimes supplemented by repeating the test with a black cloth to disclose contaminants that would be invisible on a white cloth.

7.2.3 *Residual Pattern* — Dry the cleaned surface after finish-cleaning at 120°F (49°C) for 20 min. The presence of stains or water spots on the dried surfaces indicates the presence of residual soil and incomplete cleaning. The test is rapid but not very sensitive.

7.2.4 *Water-Break Test* is a test for the presence of hydrophobic contaminants on a cleaned surface. It is applicable only for items that can be dipped in water and should be made with high-purity water. The test procedure and interpretation of results are described in Test Method F 22. The test is moderately sensitive.

7.2.5 *Tests for Free Iron: Gross Indications* — When required before testing, items should be cleaned in accordance with this standard.

7.2.5.1 *Water-Wetting and Drying* — Formation of rust stains may be accelerated by wetting the surface once with preferably distilled or deionized water or clean, fresh, potable tap water and allowing it to remain dry for 8 h during a 24-h period. After completion of this test, the surface should show no evidence of rust stains or other corrosion products.

7.2.5.2 *High-Humidity Test* — Subject the surface to a 95 to 100 % humidity at 100 to 115°F (38 to 46°C) in a suitable humidity cabinet for 24 to 26 h. After completion of this test, the surface should show no evidence of rust stains or other corrosion products.

7.2.5.3 *Copper Sulfate Test* — This method is recommended for the detection of metallic iron or iron oxide on the surface of austenitic 200 and 300 Series, the precipitation hardening alloys, and the ferritic 400 Series stainless steels containing 16% chromium or more. It is not recommended for the martensitic and lower chromium ferritic stainless steels of the 400 Series since the test will show a positive reaction on these materials. This test is hypersensitive and should be used and interpreted only by personnel familiar with its limitations. **Caution:** This test must not be applied to surfaces of items to be used in food processing. The test solution is prepared by first adding sulfuric acid to distilled water (**Caution:** Always add acid to cold water) and then dissolving copper sulfate in the following proportions:

	250-cm³ Batch
Distilled water	250 cm³
Sulfuric acid (H_2SO_4, sp gr 1.84)	1 cm³
Copper sulfate ($CuSO_4 \cdot 5H_2O$)	4 g

Swab the surface to be inspected with test solution, applying additional solution if needed to keep the surface wet for a period of 6 min. The specimen shall be rinsed and dried in a manner not to remove any deposited copper. Copper deposit will indicate the presence of free iron.

NOTE 4—If the copper sulfate test solution is more than 2 weeks old, it shall not be used.

NOTE 5—The copper sulfate test as set forth above is not applicable to surgical and dental instruments made of hardened martensitic stainless steels. Instead, a specialized copper sulfate test is extensively used for the purpose of detecting free iron and determining overall good manufacturing practice. Copper deposits at the surface of such instruments are wiped with moderate vigor to determine if the copper is adherent or nonadherent. Instruments with nonadherent copper are considered acceptable. The specialized test solution is prepared by first adding 5.4 cm³ of sulfuric acid (H_2SO_4, sp gr 1.84) to 90 cm³ of distilled water and then dissolving 4 g of copper sulfate ($CuSO_4 \cdot 5H_2O$).

7.3 Precision Inspection:

7.3.1 *Solvent-Ring Test* is a test to reveal the presence of tightly adherent transparent films that may not be revealed by visual inspection or wipe tests. A comparison standard is prepared by placing on a clean quartz microscope slide a single drop of high-purity solvent and allowing it to evaporate. Next place another drop on the surface to be evaluated, stir briefly, and transfer, using a clean capillary or glass rod, to a clean quartz microscope slide and allow the drop to evaporate. Make as many test slides as necessary to give a reasonable sample of the surface being examined. If foreign material has been dissolved by the solvent, a distinct ring will be formed on the outer edge of the drop as it evaporates. The nature of the contaminant can be determined by infrared analysis, comparing the infrared analysis with that of the standard.

7.3.2 *Black Light Inspection* is a test suitable for the detection of certain oil films and other transparent films that are not detectable under white light. In an area that is blacked out to white light, inspect all visible accessible surfaces with the aid of a new, flood-type, ultraviolet lamp. For inaccessible areas, use a wipe test as described in 7.2.2 and subject the used cloth or plug to ultraviolet lamp inspection in a blacked-out area. Fluorescence of the surface, cloth, or plug indicates the presence of contaminants. The nature of the contamination can be determined by subjecting a sample of the contaminant, that has been transferred to a clean quartz microscope slide, to infrared analysis. The test will not detect straight-chain hydrocarbons such as mineral oils.

7.3.3 *Atomizer Test* is a test for the presence of hydrophobic films. It is applicable to both small and large surfaces that are accessible for direct visual examination, and is about 100 times more sensitive than the water-break test. The test procedure and interpretation of results are described in Test Method F 21. High-purity water should be used for the test.

7.3.4 *Ferroxyl Test for Free Iron* is a highly sensitive test and should be used only when even traces of free iron or iron oxide might be objectionable. It should be made only by personnel familiar with its limitations. The test can be used on stainless steel to detect iron contamination, including iron-tool marks, residual-iron salts from pickling solutions, iron dust, iron deposits in welds, embedded iron or iron oxide, etc. The test solution is prepared by first adding nitric acid to distilled water and then adding potassium ferricyanide, in the following proportions:

Distilled water	94 weight %	1000 cm³	1 gal
Nitric acid (60–67 %)	3 weight %	20 cm³	⅕ pt
Potassium ferricyanide	3 weight %	30 g	4 oz

Apply solution with an aluminum, plastic, glass, or rubber atomizer having no iron or steel parts, or be swabbing (atomizer spray is preferred).

7.3.4.1 The appearance of a blue stain (within 15 s of application) is evidence of surface iron contamination (several minutes may be required for detection of oxide scale). The solution should be removed from the surface as quickly as possible after testing using water or, if necessary, white vinegar or a solution of 5 to 20 weight % acetic acid and scrubbing with a fiber brush. Flush the surface with water several times after use of vinegar or acetic acid.[6]

NOTE 6 — Potassium ferricyanide is not a dangerous poison as are the simple cyanides. However, when heated to decomposition or in contact with concentrated acid, it emits higher toxic cyanide fumes.

NOTE 7 — Rubber gloves, clothing, and face shields should be worn when applying the test solution, and inhalation of the atomized spray should be avoided.

NOTE 8 — The test is not recomended for process-surfaces of equipment that will be used for processing food, beverages, pharmaceuticals, or other products for human consumption unless all traces of the test solution can be thoroughly removed.

NOTE 9 — The test solution will change color on standing and must be mixed fresh prior to each use.

8. Precautions

8.1 *Minimizing Iron Contamination* — Iron contamination on stainless steel parts, components, and systems is almost always confined to the surface. It reasonable care is taken in fabrication, simple inexpensive cleaning procedures may suffice for its removal, and very little special cleaning should be required. Fabrication should be confined to an area where only the one grade of material is being worked. Powder cutting should be minimized or prohibited. Handling equipment such as slings, hooks, and lift-truck forks should be protected with clean wood, cloth, or plastic buffers to reduce contact with the iron surfaces. Walking on corrosion-resistant alloy surfaces should be avoided; where unavoidable, personnel should wear clean shoe covers each time they enter. Kraft paper, blotting paper, paperboard, flannel, vinyl-backed adhesive tape or paper, or other protective material should be laid over areas where personnel are required to walk. Shearing tables, press breaks, layout stands, and other carbon-steel work surfaces should be covered with clean kraft paper, cardboard, or blotting paper to reduce the amount of contact with the carbon steel. Hand tools, brushes, molding tools, and other tools and supplies required for fabrication should be segregated from similar items used in the fabrication of carbon steel equipment, and should be restricted to use on the one material; tools and supplies used with other materials should not be brought into the fabrication area. Tools and fixtures should be made of hardened tool steel or chrome-plated steel. Wire brushes should be stainless steel, or of an alloy composition similar to the steel being cleaned, and should not have been previously used on other materials. Only new, washed sand, free of iron particles, and stainless steel chills and chaplets should be used for casting.

8.2 *Reuse of Cleaning and Pickling Solutions* — Cleaning and pickling agents are weakened and contaminated by materials and soil being removed from surfaces as they are cleaned. Solutions may become spent or depleted in concentration after extended use, and it is necessary to check concentrations and to replace or replenish solutions when cleaning or pickling action slows. It may be impractical or uneconomical to discard solutions after a single use, even in precision cleaning operations (that is, finish-cleaning using very high-purity solvents and carried out under clean-room and rigidly controlled environmental conditions). When solutions are re-used, care must be taken to prevent the accumulation of sludge in the bottom of cleaning tanks; the formation of oil, scums, and undissolved matter on liquid surfaces; and high concentrations of emulsified oil, metal or chemical ions, and suspended solids in the liquids. Periodic cleaning of vats and degreasing tanks, decanting, periodic bottom-drain, agitation of solutions, and similar provisions are essential to maintain the effectiveness of solutions. Care must be taken to prevent water contamination of trichloroethylene and other halogenated solvents, both while in storage and in use. Redistillation and filtering of solvents and vapor-degreasing agents are necessary before reuse. Makeup is often required to maintain concentrations and pH of cleaning solutions at effective levels. Do not overuse chemical cleaners, particularly acids and vapor-degreasing solvents; if light films or oily residues remain on the metal surfaces after use of such agents, additional scrubbing with hot water and detergent, followed by

---

[6] For further information see *Journal of Materials*, Am. Soc. Testing Mats., Vol 3, No. 4, December 1968, pp. 983–995.

repeated rinsing with large quantities of hot water, may be necessary.

8.3 *Rinse Water*—Ordinary industrial or potable waters are usually suitable for most metal-cleaning applications. Biologically tested potable water should be used for final rinsing of food-handling, pharmaceutical, dairy, potable-water, and other sanitary equipment and systems. Rinsing and flushing of critical components and systems after finish-cleaning often requires high-purity deionized water, having strict controls on halide content, pH, resistivity, turbidity, and nonvolatile residues. Analytical methods that may be used for establishing the purity of rinse water should be demonstrated to have the sensitivity necessary to detect specified impurity levels; the analytical methods given in the *Annual of ASTM Standards,* Vols 11.01 and 11.02 are recommended for referee purposes in case of dispute. To minimize the use of costly high-purity water, preliminary rinses can often be made with somewhat lesser quality water, followed by final rinsing with the high-purity water. It is also possible in many cases to use effluent or overflow from the final rinse operation for preliminary rinsing of other items.

8.4 *Circulation of Cleaning Solutions and Rinse Water* — For restricted internal surfaces (for example, small diameter piping systems or the shell or tube side of a heat exchanger), high-velocity, turbulent flow of cleaning solutions and rinse water may be necessary to provide the scrubbing action needed for effective cleaning and rinsing. The velocity required is a function of the degree of cleanness required and the size of particles that are permissible in the system after the start of operation. For example, if particles between 500 and 1000 μm are acceptable to remain, a mean flushing velocity of 1 to 2 ft/s (0.3 to 0.6 m/s) may be sufficient for pipe diameters of 2 in. and smaller; to remove 100 to 200-μm particles, a mean flushing velocity of 3 to 4 ft/s (0.9 to 1.2 m/s) may be required.

8.5 *Protection of Cleaned Surfaces* — Measures to protect cleaned surfaces should be taken as soon as final cleaning is completed, and should be maintained during all subsequent fabrication, shipping, inspection, storage, and installation.

8.5.1 Do not remove wrappings and seals from incoming materials and components until they are at the use site, ready to be used or installed. If wrappings and seals must be disturbed for receiving inspection, do not damage them, remove no more than necessary to carry out the inspection, and rewrap and reseal as soon as the inspection is complete. For critical items that were cleaned by the supplier, and that will not be given further cleaning at the use site or after installation, the condition of seals and wrappings should be inspected regularly and at fairly short intervals while the item is in storage.

8.5.2 Finish-cleaned materials and components should not be stored directly on the ground or floor, and should not be permitted, insofar as practicable, to come in contact with asphalt, galvanized or carbon steel, mercury, zinc, lead, brass, low-melting point metals, or alloys or compounds of such materials. Acid cleaning of surfaces that have been in contact with such materials may be necessary to prevent failure of the item when subsequently heated. The use of carbon or galvanized steel wire for bundling and galvanized steel identification tags should be avoided.

8.5.3 Store materials and equipment, when in process, on wood skids or pallets or on metal surfaces that have been protected to prevent direct contact with stainless steel surfaces. Keep openings of hollow items (pipe, tubing, valves, tanks, pumps, pressure vessels, etc.,) capped or sealed at all times except when they must be open to do work on the item, using polyethylene, nylon, TFE-fluorocarbon plastic, or stainless steel caps, plugs, or seals. Where cleanness of exterior surfaces is important, keep the item wrapped with clear polyethylene or TFE-fluorocarbon plastic sheet at all times except when it is actually being worked on. Avoid asphalt-containing materials. Canvas, adhesive paper or plastics such as poly(vinyl chloride) may decompose in time to form corrosive substances; for example, when exposed to sunlight or ultraviolet light. The reuse of caps, plugs, or packaging materials should be avoided unless they have been cleaned prior to reuse.

8.5.4 Clean stainless steel wire brushes and hand tools before reuse on corrosion-resistant materials; if they have not been cleaned and if they could have been used on electrolytically different materials, the surfaces contacted by the tools should be acid-cleaned. The use of soft-face hammers or terne (lead)-coated, galvanized, or unprotected carbon steel tables, jigs, racks, slings, or fixtures should be avoided (see 8.5.2).

8.5.5 If close control of particulate contamination is required, particularly of internal surfaces, the latter stages of assembly and fabrication may have to be carried out in a clean room. For most large items an air cleanliness class (see Federal Standard 209a) at the work surface of Class 50,000 to 100,000 (that is, a maximum of from 50,000 to 100,000 particles 0.5 μm or larger suspended in the air) is probably sufficient.

NOTE 10 — Clean room is a specially constructed enclosure in which intake air is filtered so that the air at a work station contains no more than a specified number of particles of a specified size; special personnel and housekeeping procedures are required to maintain cleanness levels in a clean room (see Federal Standard 209a).

8.5.6 Workmen handling finish-cleaned surfaces of critical items should wear clean cotton or synthetic-fiber gloves. Rubber or plastic gloves are suitable during precleaning operations or cleaning of non-critical surfaces.

8.5.7 Installed piping systems are often *laid up wet;* that is, they are filled with water (or process fluid) after in-lace cleaning until ready to be placed in service. Storage water should be of the same quality as the makeup water for the system, and should be introduced in a manner that it directly replaces the final flush water without permitting the internal surfaces of the system to dry.

8.5.8 Equipment and assemblies for critical applications may be stored and shipped with pressurized, dry, filtered, oil-free nitrogen to prevent corrosion until they are ready to be installed. Means must be provided for maintaining and monitoring the gas pressure during shipping and storage. If the item is to be shipped to or through mountains or other areas where the altitude varies greatly from that where it was pressurized, consideration must be given to the effect of that change in altitude on the pressure inside the item, and possible rupture or loss of seals.

8.5.9 Pressure-sensitive tape is often used for sealing or protective covers, seals, caps, plugs, and wrappings. It possible, the gummed surface of the tape should not come in contact with stainless steel surfaces. If tape has come in contact with the metal, clean it with solvent or hot water, and vigorous scrubbing.

8.5.10 Protective adhesive papers or plastics are often used to protect the finish of sheet stock and parts. These materials may harden or deteriorate when subjected to pressure or sunlight, and damage the surface.

8.6 *Safety* — Cleaning operations often present numerous hazards to both personnel and facilities. Data sheets of the Manufacturing Chemists Association should be consulted to determine the hazards of handling specific chemicals.

8.6.1 Precautions must be taken to protect personnel, equipment, and facilities. This includes provisions for venting of explosive or toxic reaction-product gases, safe disposal of used solutions, provision of barriers and warning signs, provisions for safe transfer of dangerous chemicals, and maintenance of constant vigilance for hazards and leaks during the cleaning operation.

8.6.2 The physical capacity of the item or system to be cleaned, together with its foundations, to withstand the loads produced by the additional weight of fluids used in the cleaning operation, must be established before the start of cleaning operations.

8.6.3 Insofar as possible, chemicals having explosive, toxic, or obnoxious fumes should be handled out of doors.

8.6.4 The area in which the cleaning operation is being conducted should be kept clean and free of debris at all times, and should be cleaned upon completion of the operation.

8.7 *Disposal of Used Solutions and Water* — Federal, state, and local safety and water pollution control regulations should be consulted, particularly when large volumes of chemical solutions must be disposed of. Controlled release of large volumes of rinse water may be necessary to avoid damaging sewers or stream beds.

A 380

ANNEXES

A1. RECOMMENDATIONS AND PRECAUTIONS FOR ACID DESCALING (PICKLING) OF STAINLESS STEEL

A1.1 Where size and shape permit, immersion in the acid solution is preferred; when immersion is not practicable, one of the following room-temperature methods may be used:

A1.1.1 For interior surfaces, partially fill item with solution and rock, rotate, or circulate so that all inside surfaces are thoroughly wetted. Keep surfaces in contact with acid solution until inspection shows that soale is completely removed. Additional exposure without agitation may be needed. Treat exterior surfaces in accordance with A1.1.2.

A1.1.2 Surfaces that cannot be pickled by filling the item may be descaled by swabbing or spraying with acid solution for about 30 min, or until inspection shows that scale is completely removed.

A1.2 Severe pitting may result from prolonged exposure to certain acid solutions if the solution becomes depleted or if the concentration of metallic salts becomes too high as a result of prolonged use of the solution; the concentration of iron should not exceed 5 weight %; take care to prevent overpickling.

A1.3. Nitric-hydrofluoric acid solutions may intergranularly corrode certain alloys if they have been sensitized by improprer heat treatment or by welding. Crevices resulting from intergranular attack can collect and concentrate halogens under service conditions or during cleaning or processing with certain chemicals; these halogens can cause stress-corrosion cracking. These alloys should generally not be acid-pickled while in the sensitized condition. Consideration should be given to stabilized or low-carbon grades if acid pickling after welding is unavoidable.

A1.4 Some latitude is permissible in adjusting acid concentrations, temperatures, and contact times. In general, lower values in this table apply to lower alloys, and higher values to higher alloys. Close control over these variables is necessary once proper values are established in order to preserve desired finishes or close dimensional tolerances, or both.

A1.5 Materials must be degreased before acid pickling and must be vigorously brushed with hot water and a bristle brush or with high-pressure water jet on completion of pickling; pH of final rinse water should be between 6 and 8 for most applications, or 6.5 to 7.5 for critical applications. To minimize staining, surfaces must not be permitted to dry between successive steps of the acid descaling and rinsing procedure. Thorough drying should follow the final water rinse.

A1.6 Hardenable 400 Series alloys, maraging alloys, and precipitation-hardening alloys in the hardened condition are subject to hydrogen embrittlement or intergranular attack by acids. Descaling by mechanical methods is recommended where possible. If acid pickling is unavoidable, parts should be heated at 250 to 300°F (121 to 149°C) for 24 h immediately following acid treatment to drive off the hydrogen and reduce the susceptibility to embrittlement.

A1.7 Proper personnel protection, including face shields, rubber gloves, and rubber protective clothing, must be provided when handling acids and other corrosive chemicals. Adequate ventilation and strict personnel-access controls must be maintained in areas where such chemicals are being used.

TABLE A1.1 Acid Descaling (Pickling) of Stainless Steel

Alloy	Condition	Treatment			
		Code	Solution, Volume, %[A]	Temperature °F (°C)	Time, Minutes
200, 300, and 400 Series, precipitation hardening, and maraging alloys (except free-machining alloys).	fully annealed only	A	H_2SO_4, 8–11 %[B] Follow by treatment D or F, Annex A2, as appropriate	150–180 (66–82)	5–45 max[C]
200 and 300 Series; 400 Series containing Cr 16 % or more; precipitation-hardening alloys (except free-machining alloys)	fully annealed only	B	HNO_3, 15–25 % plus HF, 1–8 %[D,E]	70–140 max (21–60)	5–30[C]
All free-machining alloys and 400 Series containing less than Cr 16 %	fully annealed only	C	HNO_3, 10–15 % plus HF, $1/2$–$1\,1/2$ %[D,E]	70 (up to 140 with caution) (21–60)	5–30[C]

[A] Solution prepared from reagents of following weight %; H_2SO_4, 98; HNO_3, 67; HF, 70.

[B] Tight scale may be removed by a dip in this solution for a few minutes followed by water rinse and nitric-hydrofluoric acid treatment as noted.

[C] Minimum contact times necessary to obtain the desired surface should be used in order to prevent over-picking. Tests should be made to establish correct procedures for specific applications.

[D] For reasons of convenience and handling safety, commercial formulations containing fluoride salts may be found useful in place of HF for preparing nitric-hydrofluoric acid solutions.

[E] After pickling and water rinsing, an aqueous caustic permanganate solution containing NaOH, 10 weight % and $KMnO_4$, 4 weight %, 160 to 180°F, (71 to 82°C) 5 to 60 min, may be used as a final dip for removal of smut, followed by thorough water rinsing and drying.

A2. RECOMMENDATIONS AND PRECAUTIONS FOR ACID CLEANING OF STAINLESS STEEL

A2.1 Treatments shown are generally adequate for removal of contamination without seriously changing surface appearance of parts. Passivated parts should exhibit a clean surface and should show no etching, pitting, or frosting. The purchaser shall specify whether a slight discoloration is acceptable. Passivated parts should not exhibit staining attributable to the presence of free iron particles imbedded in the surface when subjected to the test described in 7.2.5.1. For specific requriements for items to be used in corrosive service or where surface appearance is critical, trials should be conducted to establish satisfactory procedures.

A2.2 The high-carbon and free-machining alloys may be subject to etching or discoloration in nitric acid. This tendency can be minimized by the use of high acid concentrations with inhibitors such as $Na_2Cr_2O_7 \cdot 2H_2O$ and $CuSO_4 \cdot 5H_2O$. Oxidizing action increases with increasing concentration of nitric acid; additional oxidizing action is provided by $Na_2Cr_2O_7 \cdot 2H_2O$. Avoid acid cleaning when possible; use mechanical cleaning followed by scrubbing with hot water and detergent, final thorough water rinsing and drying.

A2.3 Inhibitors may not always be required to maintain bright finishes on 200 and 300 Series, maraging, and precipitation-hardening alloys.

A2.4 Hardenable 400 Series, maraging, and precipitation-hardening alloys in the hardened condition are subject to hydrogen embrittlement or intergranular attack when exposed to acids. Cleaning by mechanical methods or other chemical methods is recommended. If acid treatment is unavoidable, parts should be heated at 250 and 300°F (121 to 149°C) for 24 h immediately following acid cleaning to drive off hydrogen and reduce susceptibility to embrittlement.

A2.5 Nitric-hydrofluoric acid solutions may intergranularly corrode certain alloys if they have been sensitized by improper heat treatment or by welding. Crevices resulting from intergranular attack can collect and concentrate halogens under service conditions or during cleaning or subsequent processing; these halogens can cause stress-corrosion cracking. Such alloys should not be cleaned with nitric-hydrofluoric acid solutions while in the sensitized condition. Consideration should be given to use of stabilized or low-carbon alloys if this kind of cleaning after welding is unavoidable.

A2.6 Severe pitting may result from prolonged exposure to certain acids if the solution becomes depleted or if the concentration of metallic salts becomes too high as a result of prolonged use of the solution; the concentration of iron should not exceed 2 weight %; take care to avoid overexposure.

A2.7 Nitric acid solutions are effective for removing free iron and other metallic contamination, but are not effective against scale, heavy deposits of corrosion products, temper films, or greasy or oily contaminants. Refer to Annex A1 for recommended practices where scale, heavy deposits of corrosion products, or heat-temper discoloration must be removed. Use conventional degreasing methods for removal of greasy or oil contaminants before any acid treatment.

A2.8 The citric acid-sodium nitrate treatment is

the least hazardous for removal of free iron and other metallic contamination and light surface contamination. Spraying of the solution, as compared to immersion, tends to reduce cleaning time.

A2.9 Some latitude is permissible in adjusting acid concentrations, temperatures, and contact times; close control over these variables is essential once proper values have been established. Care must be taken to prevent acid depletion and buildup of metallic salt concentrations with prolonged use of solutions. In general, increasing the treatment temperature may accelerate or improve the overall cleaning action but it may also increase the risk of surface staining or damage.

A2.10 Materials must be degreased before acid treatment, and must be vigorously scrubbed with hot water and bristle brushes or with high-pressure water-jet immediately after completion of acid treatment; pH of final rinse water should be between 6 and 8 for most applications, or 6.5 to 7.5 for critical applications. To minimize staining, surfaces must not be permitted to dry between successive steps of the acid cleaning or passivation and rinsing procedure. Thorough drying should follow the final water rinse.

A2.11 Proper personnel protection, including face shields, rubber gloves, and rubber protective clothing, must be provided when handling acids and other corrosive chemicals. Adequate ventilation and strict personnel access controls must be maintained where such chemicals are being used.

A2.12 Pickling and cleaning or passivating solutions containing nitric acid will severely attack carbon steel items including the carbon steel in stainless steel-clad assemblies.

A 380

TABLE A2.1 Acid Cleaning of Stainless Steel

Alloy	Condition	Code	Solution, Volume, %[A]	Temperature, °F (°C)	Time, minutes
PART I — Cleaning with Nitric-Hydrofluoric Acid					

Purpose — For use after descaling by mechanical or other chemical methods as a further treatment to remove residual particles of scale or products of chemical action (that is, smut), and to produce a uniform "white pickled" finish.

Alloy	Condition	Code	Solution	Temperature	Time
200 and 300 Series, 400 Series containing Cr 16 % or more, and precipitation-hardening alloys (except free-machining alloys).	fully annealed only	D	HNO_3, 6–25 % plus HF, $1/2$ to 8 %[B,C]	70–140 (21–60)	as necessary
Free-machining alloys, maraging alloys, and 400 Series containing less than Cr 16 %.	fully annealed only	E	HNO_3, 10 % plus HF, $1/2$ to $1 1/2$ %[B,C]	70 (up to 140 with caution) (21–60)	1–2

PART II — Cleaning-Passivation with Nitric Acid Solution

Purpose — For removal of soluble salts, corrosion products, and free ion and other metallic contamination resulting from handling, fabrication, or exposure to contaminated atmospheres (see 6.2.11).

Alloy	Condition	Code	Solution	Temperature	Time
200 and 300 Series, 400 Series, precipitation hardening and maraging alloys containing Cr 16 % or more (except free-machining alloys).[D]	annealed, cold-rolled, or work-hardened, with dull or nonreflective surfaces	F	HNO_3 20–50 %	120–160 (49–71) 70–100 (21–38)	10–30 30–60[C]
Same[D]	annealed, cold-rolled, or work-hardened with bright-machined or polished surfaces	G	HNO_3, 20–40 % plus $Na_2CrO_7 \cdot 2H_2O$, 2–6 weight %	120–155 (49–69) 70–100 (21–38)	10–30 30–60[C]
400 Series, maraging and precipitation-hardening alloys containing less than Cr 16 % high-carbon-straight Cr alloys (except free-machining alloys).[D]	annealed or hardened with dull or nonreflective surfaces	H	HNO_3, 20–50 %	110–130 (43–54) 70–100 (21–38)	20–30 60
Same[D]	annealed or hardened with bright machined or polished surfaces	I[F]	HNO_3, 20–25 % plus $Na_2Cr_2O_7 \cdot 2H_2O$, 2–6 weight %	120–130 (49–54) 70–100 (21–38)	15–30 30–60
200, 300, and 400 Series free-machining alloys.[D]	annealed or hardened, with bright-machined or polished surfaces	J[F]	HNO_3, 20–50 % plus $Na_2Cr_2O_7 \cdot 2H_2O$, 2–6 weight %[G]	70–120 (21–49)	25–40
Same[D]	same	K[E]	HNO_3, 1–2 % plus $Na_2Cr_2O_7 \cdot 2H_2O$, 1–5, weight %	120–140 (49–60)	10
Same[D]	same	L[F]	HNO_3, 12 % plus $CuSO_4 \cdot 5H_2O$, 4 weight %	120–140 (49–60)	10
Special free-machining 400 Series alloys with more than Mn 1.25 % or more than S 0.40 %[D]	annealed or hardened with bright-machined or polished surfaces	M[F]	HNO_3, 40–60 % plus $Na_2Cr_2O_7 \cdot 2H_2O$, 2–6 weight %	120–160 (49–71)	20–30

PART III — Cleaning with Other Chemical Solutions

Purpose — General cleaning.

Alloy	Condition	Code	Solution	Temperature	Time
200, 300, and 400 Series (except free-machining alloys), precipitation hardening and maraging alloys	fully annealed only	N	citric acid, 1 weight % plus ($NaNO_3$), 1 weight %	70 (21)	60
Same	same	O	ammonium citrate, 5–10 weight %	120–160 (49–71)	10–60

TABLE A2.1 *Continued*

PART III — Cleaning with Other Chemical Solutions *Continued*

Alloy	Condition	Code	Solution, Volume, %	Temperature, °F (°C)	Time, minutes
Assemblies of stainless and carbon steel (for example, heat exchanger with stainless steel tubes and carbon steel shell)	sensitized	P	inhibited solution of hydroxyacetic acid, 2 weight % and formic acid, 1 weight %	200 (93)	6 h
Same	same	Q	inhibited ammonia-neutralized solution of EDTA (ethylene-diamene-tetraacetic acid) followed by hot-water rinse and dip in solution of 10 ppm ammonium hydroxide plus 100 ppm hydrazine	up to 250 (121)	6 h

[A] Solution prepared from reagents of following weight %; HNO_3, 67; HF, 70.

[B] For reasons of convenience and handling safety, commercial formulations containing fluoride salts may be found useful in place of HF for preparing nitric-hydrofluoric acid solutions.

[C] After acid cleaning and water rising, a caustic permanganate solution containing NaOH, 10 weight %, and $KMnO_4$, 4 weight %, 160 to 180°F (71 to 82°C), 5 to 60 min, may be used as a final dip for removal of smut, followed by thorough water rinsing and drying.

[D] The purchaser shall have the option of specifying in his purchase documents that all 400 Series ferritic or martensitic parts receive additional treatment as follows: Within 1 h after the water rinse following the specified passivation treatment, all parts shall be immersed in an aqueous solution containing 4 to 6 weight % $Na_2Cr_2O_7 \cdot 2H_2O$, at 140 to 160°F (60 to 71°C), 30 min. This immersion shall be followed by thorough rinsing with clean water. The parts then shall be thoroughly dried.

[E] Shorter times may be acceptable where established by test and agreed upon by the purchaser.

[F] See A2.2.

[G] If flash attack (clouding of stainless steel surface) occurs, a fresh (clean) passivating solution or a higher HNO_3 concentration will usually eliminate it.

The American Society for Testing and Materials takes no position respecting the validity of any patent rights asserted in connection with any item mentioned in this standard. Users of this standard are expressly advised that determination of the validity of any such patent rights, and the risk of infringement of such rights, are entirely their own responsibility.

This standard is subject to revision at any time by the responsible technical committee and must be reviewed every five years and if not revised, either reapproved or withdrawn. Your comments are invited either for revision of this standard or for additional standards and should be addressed to ASTM Headquarters. Your comments will receive careful consideration at a meeting of the responsible technical committee, which you may attend. If you feel that your comments have not received a fair hearing you should make your views known to the ASTM Committee on Standards, 1916 Race St., Philadelphia, Pa. 19103.

Designation: A 401 – 77 (Reapproved 1983)

Standard Specification for
CHROMIUM-SILICON ALLOY STEEL SPRING WIRE[1]

This standard is issued under the fixed designation A 401; the number immediately following the designation indicates the year of original adoption or, in the case of revision, the year of last revision. A number in parentheses indicates the year of last reapproval. A superscript epsilon (ϵ) indicates an editorial change since the last revision or reapproval.

1. Scope

1.1 This specification covers round chromium-silicon alloy steel spring wire, having properties and quality intended for the manufacture of springs resistant to set when used at moderately elevated temperatures. The wire shall be either in the annealed and cold-drawn or oil-tempered condition as specified by the purchaser.

1.2 The values stated in inch-pound units are to be regarded as the standard.

2. Applicable Documents

2.1 *ASTM Standards:*
A 370 Methods and Definitions for Mechanical Testing of Steel Products[2]
A 700 Recommended Practices for Packaging, Marking, and Loading Methods for Steel Products for Domestic Shipment[3]
E 29 Recommended Practice for Indicating Which Places of Figures Are to Be Considered Significant in Specified Limiting Values[4]
E 380 Standard for Metric Practice[4]

3. Ordering Information

3.1 Orders for material under this specification shall include the following information for each ordered item:

3.1.1 Quantity (weight),
3.1.2 Name of material (chromium silicon alloy steel spring wire),
3.1.3 Packaging (Section 11),
3.1.4 Condition (Sections 1 and 6),
3.1.5 Dimensions (Section 9),
3.1.6 Cast or heat analysis report (if desired) (5.3), and
3.1.7 ASTM designation and date of issue.

NOTE 1—A typical description is as follows: 40 000 lb oil-tempered chromium silicon alloy steel spring wire, size 0.250 in. in 350-lb coils to ASTM A 401 dated ———.

4. Materials and Manufacture

4.1 The steel shall be made by the open-hearth, basic-oxygen, or electric-furnace process.

4.2 A sufficient discard shall be made to ensure freedom from injurious piping and undue segregation.

5. Chemical Requirements

5.1 The steel shall conform to the requirements as to chemical composition specified in Table 1.

5.2 An analysis may be made by the purchaser from finished wire representing each heat of steel. The average of all the separate determinations made shall be within the limits specified in the analysis column. Individual determinations may vary to the extent shown in the product analysis tolerance column, except that the several determinations of a single element in any one heat shall not vary both above and below the specified range.

5.3 An analysis of each heat of steel shall be furnished by the manufacturer, when requested, showing the percentages of the elements specified in Table 1.

---

[1] This specification is under the jurisdiction of ASTM Committee A-1 on Steel, Stainless Steel and Related Alloys, and is the direct responsibility of Subcommittee A01.03 on Steel Rod and Wire.
Current edition approved June 24, 1977. Published August 1977. Originally published as A 401 – 56 T. Last previous edition A 401 – 75.
[2] *Annual Book of ASTM Standards*, Vol 01.04.
[3] *Annual Book of ASTM Standards*, Vols 01.01, 01.03, 01.04, and 01.05.
[4] *Annual Book of ASTM Standards*, Vol 14.02.

6. Mechanical Requirements

6.1 *Annealed and Cold Drawn*—When purchased in the annealed and cold-drawn condition, the wire shall have been given a sufficient amount of cold working to meet the purchaser's coiling requirements and shall be in a suitable condition to respond properly to heat treatment. In special cases the hardness, if desired, shall be stated in the purchase order.

6.2 *Oil Tempered*—When purchased in the oil-tempered condition, the tensile strength and minimum percent reduction of area (sizes 0.092 in. (2.34 mm) and coarser) of the wire shall conform to the requirements prescribed in Table 2.

6.2.1 *Number of Tests*—One test specimen shall be taken for each ten coils, or fraction thereof, in a lot. Each cast or heat in a given lot shall be tested.

6.2.2 *Location of Tests*—Test specimens shall be taken from either end of the coil.

6.2.3 *Test Method*—The tension test shall be made in accordance with Methods and Definitions A 370.

6.3 *Wrap Test:*

6.3.1 Wire, oil tempered or cold drawn, 0.162 in. (4.11 mm) and smaller in diameter, shall wind on itself as an arbor without breakage. Larger diameter wire up to and including 0.312 in. (7.92 mm) in diameter shall wrap without breakage on a mandrel twice the wire diameter. The wrap test is not applicable to wire over 0.312 in. in diameter.

6.3.2 *Number of Tests*—One test specimen shall be taken for each ten coils or fraction thereof, in a lot. Each cast or heat in a given lot shall be tested.

6.3.3 *Location of Test*—Test specimens shall be taken from either end of the coil.

6.3.4 *Test Method*—The wrap test shall be made in accordance with Supplement IV of Methods and Definitions A 370.

7. Metallurgical Requirements

7.1 *Surface Condition:*

7.1.1 The surface of the wire as received shall be free of rust and excessive scale. No serious die marks, scratches, or seams may be present. Based upon examination of etched end specimen, seams shall not exceed 3.5 % of the wire diameter, or 0.010 in. (0.25 mm), whichever is the smaller as measured on a transverse section.

7.1.2 *Number of Tests*—The number of tests shall be as agreed upon by the supplier and purchaser.

7.1.3 *Location of Test*—Test specimens shall be taken from either or both ends of the coil.

7.1.4 *Test Method*—The surface shall be examined after etching in a solution of equal parts of hydrochloric acid and water that has been heated to approximately 175°F (80°C). Test ends shall be examined using 10× magnification. Any specimens showing the presence of a seam shall have a transverse section taken from the unetched area, properly mounted and polished and examined to measure the depth of the seam.

8. Retests

8.1 If any test specimen exhibits obvious defects or shows the presence of a weld, it may be discarded and another specimen substituted.

9. Permissible Variations in Dimension

9.1 The permissible variations in the diameter of the wire shall be as specified in Table 3.

10. Workmanship

10.1 *Annealed and Cold Drawn*—The wire shall not be kinked or improperly cast. To test for cast, a few convolutions of wire shall be cut loose from the coil and placed on a flat surface. The wire shall lie flat on itself and not spring up nor show a wavy condition.

10.2 *Oil Tempered*—The wire shall be uniform in quality and temper and shall not be wavy or crooked.

10.3 Each coil shall be one continuous length of wire properly coiled. Welds made prior to cold drawing are permitted. If unmarked welds are unacceptable to the purchaser, special arrangements should be made with the manufacturer at the time of purchase.

11. Packaging, Marking, and Loading

11.1 The coil weight, coil dimensions, and method of packaging shall be as agreed upon between the manufacturer and the purchaser.

11.2 The size of the wire, purchaser's order number, the ASTM specification number and condition (Section 1 and 6.1), heat number, and name or mark of the manufacturer shall be marked on a tag securely attached to each coil of wire.

11.3 When specified in the purchaser's order, packaging, marking, and loading for shipments shall be in accordance with those procedures recommended by Recommended Practices A 700.

12. Inspection

12.1 The manufacturer shall afford the inspector representing the purchaser all reasonable facilities to satisfy him that the material is being furnished in accordance with this specification. All tests (except product analysis) and inspections may be made at the place of manufacture prior to shipment, and shall be so conducted as not to interfere unnecessarily with the operation of the works.

13. Rejection and Rehearing

13.1 Unless otherwise specified, any rejection based on tests made in accordance with this specification shall be reported to the manufacturer within a reasonable length of time.

13.2 The material must be adequately protected and correctly identified in order that the manufacturer may make a proper investigation.

TABLE 1 Chemical Requirements

	Analysis, %	Product Analysis Tolerance, %
Carbon	0.51 to 0.59	±0.02
Manganese	0.60 to 0.80	±0.03
Phosphorus	0.035 max	+0.005
Sulfur	0.040 max	+0.005
Silicon	1.20 to 1.60	±0.05
Chromium	0.60 to 0.80	±0.03

TABLE 2 Tensile Requirements

Diameter, in. (mm)[a]	Tensile Strength, ksi (MPa) min	Tensile Strength, ksi (MPa) max	Reduction of Area, min, %
0.032 (0.81)	300 (2070)	325 (2240)	[b]
0.041 (1.04)	298 (2050)	323 (2230)	[b]
0.054 (1.37)	292 (2010)	317 (2190)	[b]
0.062 (1.57)	290 (2000)	315 (2170)	[b]
0.080 (2.03)	285 (1960)	310 (2140)	[b]
0.092 (2.34)	280 (1930)	305 (2100)	45
0.120 (3.05)	275 (1900)	300 (2070)	45
0.135 (3.43)	270 (1860)	295 (2030)	40
0.162 (4.11)	265 (1830)	290 (2000)	40
0.177 (4.50)	260 (1790)	285 (1960)	40
0.192 (4.88)	260 (1790)	283 (1950)	40
0.219 (5.56)	255 (1760)	278 (1920)	40
0.250 (6.35)	250 (1720)	275 (1900)	40
0.312 (7.92)	245 (1690)	270 (1860)	40
0.375 (9.52)	240 (1660)	265 (1830)	40
0.438 (11.12)	235 (1620)	260 (1790)	40

[a] Tensile strength values for intermediate sizes may be interpolated.
[b] The reduction of area test is not applicable to wire under 0.092 in. (2.34 mm) in diameter.

TABLE 3 Permissible Variations in Wire Diameter

NOTE – For purposes of determining conformance with this specification all specified limits are absolute as defined in Recommended Practice E 29.

Diameter, in. (mm)	Permissible Variations, plus and minus, in. (mm)	Permissible Out-of-Round, in. (mm)
0.032 to 0.072 (0.81 to 1.83), incl	0.001 (0.03)	0.001 (0.03)
Over 0.072 to 0.438 (1.83 to 11.12), incl	0.002 (0.05)	0.002 (0.05)

The American Society for Testing and Materials takes no position respecting the validity of any patent rights asserted in connection with any item mentioned in this standard. Users of this standard are expressly advised that determination of the validity of any such patent rights, and the risk of infringement of such rights, are entirely their own responsibility.

This standard is subject to revision at any time by the responsible technical committee and must be reviewed every five years and if not revised, either reapproved or withdrawn. Your comments are invited either for revision of this standard or for additional standards and should be addressed to ASTM Headquarters. Your comments will receive careful consideration at a meeting of the responsible technical committee, which you may attend. If you feel that your comments have not received a fair hearing you should make your views known to the ASTM Committee on Standards, 1916 Race St., Philadelphia, Pa. 19103.

Standard Specification for
STEEL WIRE, COLD-DRAWN, FOR COILED-TYPE SPRINGS[1]

This standard is issued under the fixed designation A 407; the number immediately following the designation indicates the year of original adoption or, in the case of revision, the year of last revision. A number in parentheses indicates the year of last reapproval. A superscript epsilon (ϵ) indicates an editorial change since the last revision or reapproval.

1. Scope

1.1 This specification covers round, cold-drawn, steel spring wire having properties and quality intended for the manufacture of the following types of upholstery springs:

1.1.1 *Type A*—Coiled (Marshall pack),
1.1.2 *Type B*—Coiled and knotted,
1.1.3 *Type C*—Coiled and knotted (offset style),
1.1.4 *Type D*—Coiled and hooked (single and cross helicals),
1.1.5 *Type E*—Coiled and hooked (short tension—regular tensile strength),
1.1.6 *Type F*—Coiled and hooked (short tension—high tensile strength),
1.1.7 *Type G*—Regular lacing, and
1.1.8 *Type H*—Automatic lacing.

1.2 These types of upholstery springs are used in the manufacture of bed spring units, mattresses, furniture cushions, and automobile seats. This wire is not intended for the manufacture of mechanical springs.

Note 1—A complete metric companion to Specification A 407 has been developed—A 407M; therefore, no metric equivalents are presented in this specification.

2. Applicable Documents

2.1 *ASTM Standards*:
A 370 Methods and Definitions for Mechanical Testing of Steel Products[2]
A 510 Specification for General Requirements for Wire Rods and Coarse Round Wire, Carbon Steel[3]
A 700 Recommended Practices for Packaging, Marking, and Loading Methods for Steel Products for Domestic Shipment[4]
E 29 Recommended Practice for Indicating Which Places of Figures Are to Be Considered Significant in Specified Limiting Values[5]
E 30 Methods for Chemical Analysis of Steel, Cast Iron, Open-Hearth Iron, and Wrought Iron[6]

3. Ordering Information

3.1 Orders for material under this specification shall include the following information for each ordered item:
3.1.1 Quantity (weight),
3.1.2 Name of material (name of specific type required) (Section 1 and Table 1),
3.1.3 Diameter (Table 2),
3.1.4 Packaging, marking, and loading (Section 10),
3.1.5 ASTM designation and date of issue, and
3.1.6 Heat (cast) analysis (if desired).

Note 2—A typical ordering description is as follows: 50 000 lb, cold-drawn upholstery spring wire Type B for coiling and knotting, size 0.080 in., 1500-lb coils on tubular carriers to ASTM A 407 dated _____.

---

[1] This specification is under the jurisdiction of ASTM Committee A-1 on Steel, Stainless Steel and Related Alloys, and is the direct responsibility of Subcommittee A01.03 on Steel Rod and Wire.
Current edition approved June 24, 1977. Published August 1977. Originally issued as A 407 – 57 T. Last previous edition A 407 – 74.
[2] *Annual Book of ASTM Standards*, Vol 01.04.
[3] *Annual Book of ASTM Standards*, Vol 01.03.
[4] *Annual Book of ASTM Standards*, Vols. 01.01, 01.03, 01.04, and 01.05.
[5] *Annual Book of ASTM Standards*, Vol 14.02.
[6] *Annual Book of ASTM Standards*, Vol 03.05.

A 407

4. Manufacture

4.1 The steel shall be made by any of the following processes: open-hearth, basic-oxygen, or electric-furnace.

4.2 A sufficient discard shall be made to ensure freedom from injurious piping and undue segregation.

4.3 The wire shall be cold-drawn to produce the desired mechanical properties.

5. Chemical Requirements

5.1 Upholstery spring wire for coiled-type springs is customarily produced within the chemical ranges shown below. Chemical composition and processing may vary depending on the gage of wire and specific use.

Carbon, %	0.45 to 0.70[a]
Manganese, %	0.60 to 1.20[a]

[a] In any lot in which all the wire is of the same size and type, and submitted for inspection at the same time, the carbon content shall not vary more than 0.20 %, and the manganese content shall not vary more than 0.30 %.

5.2 An analysis of each heat (cast) shall be made by the producer to determine the percentage of elements specified above. The analysis shall be made from a test sample preferably taken during the pouring of the heat (cast). The chemical composition thus determined shall be reported to the purchaser or his representative upon request.

6. Mechanical Requirements

6.1 *Tension Test:*

6.1.1 *Requirements*—The material as represented by tension test specimens shall conform to the requirements prescribed in Table 1 for the various sizes and specified types.

6.1.2 *Number of Tests*—One test specimen shall be taken for each ten coils, or fraction thereof, in a lot. Each heat in a given lot shall be tested.

6.1.3 *Location of Tests*—Test specimens shall be taken from either end of the coil.

6.1.4 *Test Method*—The tension test shall be made in accordance with Methods and Definitions A 370.

6.2 *Wrap Test:*

6.2.1 *Requirements*—The wire, except that for Type A (Marshall pack), shall wrap on itself as an arbor without breakage.

6.2.2 *Number of Tests*—One test specimen shall be taken for each ten coils, or fraction thereof, in a lot. Each heat in a given lot shall be tested.

6.2.3 *Location of Tests*—Test specimens shall be taken from either end of the coil.

6.2.4 *Test Method*—The wrap test shall be made in accordance with Supplement IV of Methods and Definitions A 370.

7. Retests

7.1 If any test specimen exhibits obvious defects or shows the presence of a weld, it may be discarded and another specimen substituted.

8. Permissible Variations in Dimensions

8.1 The diameter of the wire shall not vary from that specified by more than the tolerances specified in Table 2.

9. Workmanship

9.1 *Surface Condition*—The surface of the wire as received shall be smooth and have a uniform finish suitable for coiling the various types of springs. No serious die marks, scratches, or seams may be present.

9.2 The wire shall be properly cast. To test for cast, a few convolutions of wire shall be cut from the coil and allowed to fall on a flat surface. The wire shall lie substantially flat on itself and shall not spring up and show a wavy condition.

9.3 Each coil shall be one continuous length of wire, properly coiled. Welds made prior to cold drawing are permitted. Weld areas need not meet the mechanical requirements of this specification.

10. Packaging, Marking, and Loading

10.1 Packaging of the coils of wire shall be by agreement between the producer and the purchaser. This agreement shall include coil dimensions and weights.

10.2 When specified, the packaging, marking, and loading shall be in accordance with Recommended Practices A 700.

10.3 Marking shall be by a tag securely attached to each coil of wire and shall show the identity of the producer, size of the wire, type, and ASTM specification number.

11. Inspection

11.1 The manufacturer shall afford the purchaser's inspector all reasonable facilities necessary to satisfy him that the material is being produced and furnished in accordance with this specification. Mill inspection by the purchaser shall not interfere unnecessarily with the manufacturer's operations. All tests and inspections shall be made at the place of manufacture, unless otherwise agreed to.

12. Rejection and Rehearing

12.1 Unless otherwise specified, any rejection based on tests made in accordance with this specification shall be reported to the manufacturer within a reasonable length of time.

12.2 Failure of any of the test specimens to comply with the requirements of this specification shall constitute grounds for rejection of the lot represented by the specimen. The lot may be resubmitted for inspection by testing every coil for the characteristic in which the specimen failed and sorting out the defective coils.

12.3 The material must be adequately protected and correctly identified in order that the producer may make proper investigation.

TABLE 1 Tensile Strength Requirements[A]

Diameter, in.	Wire Gage	Tensile Strength, ksi	
		min	max
Type A, Marshall Pack			
0.048	18	255	295
0.054	17	250	290
0.062	16	250	290
0.072	15	240	280
0.080	14	230	270
0.092	13	225	265
0.106	12	220	260
Type B, Coiled and Knotted			
0.062	16	235	270
0.072	15	230	265
0.080	14	225	260
0.092	13	215	250
0.106	12	205	235
0.120	11	195	225
0.135	10	190	220
0.148	9	185	215
0.162	8	180	210
Type C, Coiled and Knotted (Offset Type)			
0.072	15	215	245
0.080	14	210	240
0.092	13	200	230
0.106	12	195	225
Type D, Coiled and Hooked (Cross Helicals)			
0.048	18	215	255
0.054	17	210	250
0.062	16	210	250
Type E, Coiled and Hooked (Short Tension, Regular Tensile Strength)			
0.080	14	200	240
0.092	13	200	240
0.106	12	195	235
Type F, Coiled and Hooked (Short Tension, High Tensile Strength)			
0.080	14	225	260
0.092	13	220	255
0.106	12	215	250
Type G, Regular Lacing Wire			
0.041	19	235	275
0.048	18	230	270
0.054	17	225	265
0.062	16	225	265
Type H, Automatic Lacing Wire			
0.041	19	250	290
0.048	18	245	285
0.054	17	240	280
0.062	16	235	275

[A] Tensile strength values for diameters not shown in this table shall conform to that shown for the next larger diameter (for example, for diameter 0.128 in. the value shall be the same as for 0.135 in.).

TABLE 2 Permissible Variations in Wire Diameter

NOTE—For purposes of determining conformance with this specification, all specified limits are absolute as defined in Recommended Practice E 29.

Diameter, in.	Variations, plus and minus, in.	Permissible Out-of-Round, in.
Sizes finer than 0.076	0.001	0.001
Sizes 0.076 to 0.162, incl	0.002	0.002

The American Society for Testing and Materials takes no position respecting the validity of any patent rights asserted in connection with any item mentioned in this standard. Users of this standard are expressly advised that determination of the validity of any such patent rights, and the risk of infringement of such rights, are entirely their own responsibility.

This standard is subject to revision at any time by the responsible technical committee and must be reviewed every five years and if not revised, either reapproved or withdrawn. Your comments are invited either for revision of this standard or for additional standards and should be addressed to ASTM Headquarters. Your comments will receive careful consideration at a meeting of the responsible technical committee, which you may attend. If you feel that your comments have not received a fair hearing you should make your views known to the ASTM Committee on Standards, 1916 Race St., Philadelphia, Pa. 19103.

Designation: A 407M – 80

Metric

Standard Specification for
STEEL-WIRE, COLD-DRAWN, FOR COILED-TYPE SPRINGS [METRIC][1]

This standard is issued under the fixed designation A 407M; the number immediately following the designation indicates the year of original adoption or, in the case of revision, the year of last revision. A number in parentheses indicates the year of last reapproval. A superscript epsilon (ϵ) indicates an editorial change since the last revision or reapproval.

1. Scope

1.1 This specification covers round, cold-drawn, steel spring wire having properties and quality intended for the manufacture of upholstery-type springs.

1.2 These coiled upholstery springs are used in the manufacture of bed units, mattresses, furniture cushions, and automobile seats. This wire is not intended for the manufacture of mechanical springs.

NOTE 1—This specification is the metric counterpart of Specification A 407.

2. Applicable Documents

2.1 *ASTM Standards*:
A 370 Methods and Definitions for Mechanical Testing of Steel Products[2]
A 510M Specification for General Requirements for Wire Rods and Coarse Round Wire, Carbon Steel[3]
A 700 Recommended Practices for Packaging, Marking, and Loading Methods for Steel Products for Domestic Shipment[4]
A 751 Methods, Practices, and Definitions for Chemical Analysis of Steel Products[5]
E 29 Recommended Practice for Indicating Which Places of Figures Are to Be Considered Significant in Specified Limiting Values[6]

3. Classification

3.1 The wire can be furnished in eight types:
3.1.1 *Type I*—Coiled (Marshall pack),
3.1.2 *Type II*—Coiled and knotted,
3.1.3 *Type III*—Coiled and knotted (offset style),
3.1.4 *Type IV*—Coiled and hooked (single and cross helicals),
3.1.5 *Type V*—Coiled and hooked (short tension—regular tensile strength),
3.1.6 *Type VI*—Coiled and hooked (short tension—high tensile strength),
3.1.7 *Type VII*—Regular lacing, and
3.1.8 *Type VIII*—Automatic lacing.

4. Ordering Information

4.1 Orders for material under this specification shall include the following information for each ordered item:
4.1.1 Quantity (mass),
4.1.2 Name of material (name of specific type required) (Section 3 and Table 1),
4.1.3 Diameter (mm),
4.1.4 Certification or test report, or both, if specified (Section 13),
4.1.5 Packaging, marking, and loading (Section 14),
4.1.6 ASTM designation and date of issue,
4.1.7 Heat (cast) analysis, if specified,

NOTE 2—A typical ordering description is as follows: 15 000 kg, cold-drawn upholstery spring wire, Type II for coiling and knotting size 2.0 mm, 700-kg coils shipped tubular carriers to ASTM A 407M dated _____.

---

[1] This specification is under the jurisdiction of ASTM Committee A-1 on Steel, Stainless Steel, and Related Alloys and is the direct responsibility of Subcommittee A01.03 on Steel Rod and Wire.
Current edition approved Sept. 26, 1980. Published November 1980.
[2] *Annual Book of ASTM Standards*, Vol 01.04.
[3] *Annual Book of ASTM Standards*, Vol. 01.03.
[4] *Annual Book of ASTM Standards*, Vol 01.01, 01.03, 01.04, and 01.05.
[5] *Annual Book of ASTM Standards*, Vol 01.01, 01.02, 01.03, 01.04, 01.05, and 03.05.
[6] *Annual Book of ASTM Standards*, Vol 14.02.

A 407M

5. Materials and Manufacture

5.1 The steel shall be of such quality that, when processed, the finished wire shall be free of detrimental pipe and undue segregation.

5.2 The wire shall be cold-drawn to produce the desired mechanical properties.

6. Chemical Requirements

6.1 Upholstery spring wire for coiled-type springs is commonly produced within the chemical ranges shown below. Chemical composition and processing may vary depending on the wire size and specified use.

Carbon, %	0.45 to 0.75[A]
Manganese, %	0.60 to 1.20[A]

[A] In any lot in which all the wires are of the same size and type, and submitted for inspection at the same time, the carbon content shall not vary more than 0.20 %, and the manganese content shall not vary more than 0.30 %.

6.2 An analysis of each heat (cast) shall be made by the producer to determine the percentage of the elements specified above. The analysis shall be made from a test sample preferably taken during the pouring of the heat (cast). The chemical composition thus determined shall be reported to the purchaser or his representative upon request.

6.3 Chemical analysis shall be made in accordance with Methods A 751.

7. Mechanical Requirements

7.1 *Tensile Strength Test*—The material as represented by test specimens shall conform to the requirements prescribed in Table 1 for the specified size and type.

7.2 *Wrap Test*—The material as represented by test specimens, except Type I (Marshall pack), shall wrap on itself as an arbor without fracture.

7.3 The tensile strength test and the wrap test shall be made in accordance with Methods A 370.

8. Permissible Variation in Diameter

8.1 The diameter of the wire shall not vary from that specified by more than the tolerances shown in Table 2.

9. Workmanship, Finish, and Appearance

9.1 The surface of the wire as received shall be smooth and have a uniform finish suitable for coiling the various types of springs. No serious die marks, scratches, or seams may be present.

9.2 The wire shall be properly cast. To test for cast, a few convolutions of wire shall be cut from the coil and allowed to fall on a flat surface. The wire shall lie substantially flat on itself and shall not spring up and show a wavy condition.

9.3 Each coil of wire shall be one continuous length and properly coiled. Welds made prior to cold drawing are permitted. Weld areas need not meet the mechanical requirements of this specification.

10. Sampling and Number of Tests

10.1 A lot shall consist of all of the coils of wire of the same size and type and offered for inspection at one time.

10.2 One test specimen shall be taken for each ten coils, or fraction thereof, in a lot. Each heat in a given lot shall be tested.

10.3 Test specimens may be taken from either end of the coil.

10.4 If any test specimen shows the presence of a weld or an obvious defect, it may be discarded and another specimen substituted.

10.5 Test specimens shall be tested for tensile strength (7.1) and wrap test (7.2).

11. Inspection

11.1 Unless otherwise specified in the contract or purchase order, the manufacturer is responsible for the performance of all inspection and test requirements specified in this specification. Except as otherwise specified in the contract or purchase order, the manufacturer may use his own or any other suitable facilities for the performance of the inspection and test requirements unless disapproved by the purchaser at the time the order is placed. The purchaser shall have the right to perform any of the inspections and tests set forth in this specification when such inspections and tests are deemed necessary to assure that the material conforms to prescribed requirements.

12. Rejection and Rehearing

12.1 Material that fails to conform to the requirements of this specification may be rejected. Rejections should be reported to the producer or supplier promptly and in writing.

A 407M

In case of dissatisfaction with the results of the test, the producer or supplier may make a claim for a rehearing.

12.2 The material must be adequately protected and correctly identified in order that the producer may make proper investigation.

13. Certification

13.1 When specified in the purchase order or contract, a producer or supplier's certification shall be furnished to the purchaser that the material was manufactured, sampled, tested, and inspected in accordance with this specification and has been found to meet the requirements. When specified in the purchase order or contract, a report of the chemical composition and test results shall be furnished.

14. Marking and Packaging

14.1 Marking shall be by a tag securely attached to each coil of wire and shall show the identity of the producer, size of wire, type, heat number, and ASTM designation.

14.2 Packaging of the coils of wire shall be by agreement between the producer and the purchaser. This agreement shall include coil dimensions and weight.

14.3 Unless otherwise specified, the packaging, marking, and loading shall be in accordance with Method A 700.

A 407M

TABLE 1 Tensile Strength Requirements

Diameter, mm	Tensile Strength, MPa	
	min	max
Type I, Marshall Pack		
1.2	1800	2060
1.4	1760	2010
1.6	1710	1950
1.8	1680	1920
2.0	1650	1890
2.4	1610	1830
2.8	1560	1780
Type II, Coiled and Knotted		
1.6	1550	1790
1.8	1530	1770
2.0	1500	1740
2.4	1460	1680
2.6	1420	1640
3.0	1380	1600
3.5	1350	1560
3.8	1330	1530
4.2	1300	1500
Type III, Coiled and Knotted (Offset Type)		
1.8	1450	1690
2.0	1430	1670
2.4	1390	1610
2.6	1360	1580
Type IV, Coiled and Hooked (Cross Helicals)		
1.2	1490	1750
1.4	1440	1700
1.6	1410	1650
Type V, Coiled and Hooked (Short Tension, Regular Tensile Strength)		
2.0	1410	1650
2.4	1360	1580
2.6	1330	1550
Type VI, Coiled and Hooked (Short Tension, High Tensile Strength)		
2.0	1550	1790
2.4	1520	1740
2.6	1490	1710
Type VII, Regular Lacing Wire		
1.0	1640	1910
1.2	1600	1860
1.4	1560	1810
1.6	1520	1760
Type VIII, Automatic Lacing Wire		
1.0	1740	2020
1.2	1700	1960
1.4	1660	1910
1.6	1620	1860

TABLE 2 Permissible Variations in Wire Diameter

NOTE—For purposes of determining conformance with this specification, all specified limits are absolute as defined in Recommended Practice E 29.

Diameter, mm	Variations, plus and minus, mm	Permissible Out-of-Round, mm
Sizes finer than 2.0	0.02	0.02
Sizes 2.0 to 4.2	0.05	0.05

The American Society for Testing and Materials takes no position respecting the validity of any patent rights asserted in connection with any item mentioned in this standard. Users of this standard are expressly advised that determination of the validity of any such patent rights, and the risk of infringement of such rights, are entirely their own responsibility.

This standard is subject to revision at any time by the responsible technical committee and must be reviewed every five years and if not revised, either reapproved or withdrawn. Your comments are invited either for revision of this standard or for additional standards and should be addressed to ASTM Headquarters. Your comments will receive careful consideration at a meeting of the responsible technical committee, which you may attend. If you feel that your comments have not received a fair hearing you should make your views known to the ASTM Committee on Standards, 1916 Race St., Philadelphia, Pa. 19103.

Designation: A 412 – 82

Standard Specification for
STAINLESS AND HEAT-RESISTING CHROMIUM-NICKEL-MANGANESE STEEL PLATE, SHEET, AND STRIP[1]

This standard is issued under the fixed designation A 412; the number immediately following the designation indicates the year of original adoption or, in the case of revision, the year of last revision. A number in parentheses indicates the year of last reapproval. A superscript epsilon (ε) indicates an editorial change since the last revision or reapproval.

This specification has been approved for use by agencies of the Department of Defense and for listing in the DoD Index of Specifications and Standards.

1. Scope

1.1 This specification[2] covers annealed, and annealed and cold-rolled, chromium-nickel-manganese austenitic stainless and heat-resisting steel plate, sheet, and strip.

1.2 The values stated in inch-pound units are to be regarded as the standard.

2. Applicable Documents

2.1 *ASTM Standards:*
A 370 Methods and Definitions for Mechanical Testing of Steel Products[3]
A 480/A 480M Specification for General Requirements for Flat-Rolled Stainless and Heat-Resisting Steel Plate, Sheet, and Strip[3]
A 751 Methods, Practices, and Definitions for Chemical Analysis of Steel Products[3]
E 527 Practice for Numbering Metals and Alloys (UNS)[3]

2.2 *Society of Automotive Engineers Standards:*
J1086 Unified Numbering Systems for Metals and Alloys[4]

3. General Requirements for Delivery

3.1 In addition to the requirements of this specification, all requirements of the current edition of Specification A 480/A 480M shall apply.

4. Chemical Requirements

4.1 An analysis of each cast or heat shall be made to determine the percentages of the elements specified in Table 1.

4.2 Methods and practices relating to chemical analysis required by this specification shall be in accordance with Methods, Practices, and Definitions A 751.

5. Mechanical Requirements

5.1 The material shall conform to the requirements for tensile properties specified in Table 2.

5.1.1 The yield strength shall be determined by the offset method as described in Methods and Definitions A 370. The limiting permanent offset shall be 0.2 % of the gage length of the specimen. An alternative method of determining yield strength may be used, based on the following total extensions under load:

Yield Strength, min, ksi (MPa)	Total Extension Under Load in 2-in. or 50-mm Gage Length, in. (mm)
45 (310)	0.0071 (0.180)
75 (515)	0.0098 (0.249)
110 (760)	0.0125 (0.318)
135 (930)	0.0144 (0.366)
140 (965)	0.0148 (0.376)

5.1.2 The requirements of this specification for yield strength will be considered as having been fulfilled if the extension under load for

---

[1] This specification is under the jurisdiction of ASTM Committee A-1 on Steel, Stainless Steel, and Related Alloys and is the direct responsibility of Subcommittee A01.17 on Flat Stainless Steel Products.
Current edition approved Oct. 29, 1982. Published December 1982. Originally published as A 412 – 57 T. Last previous edition A 412 – 81.
[2] For ASME Boiler and Pressure Vessel Code applications, see related Specification SA-412 in Section II of that code.
[3] *Annual Book of ASTM Standards,* Vol 01.03.
[4] Available from the Society of Automotive Engineers, 400 Commonwealth Drive, Warrendale, Pa. 15096.

the specified yield strength does not exceed the specified values. The value obtained in this manner, should not, however, be taken as the actual yield strength of 0.2 % permanent set. In case of dispute, the offset method of determining yield strength shall be used.

5.1.3 Between the yield strength and fracture of the specimens, the crosshead speed of the testing machine shall be between ⅛ and ½ in./in. (0.125 mm and 0.500 mm/mm) of gage length/min.

5.2 The sheet and strip product shall conform to the bend test requirements specified in Tables 3 or 4, respectively.

5.2.1 The bend test specimens shall withstand cold bending without cracking, when subjected to either the free-bend method or the controlled-bend (V-block) method. The choice of method of test for materials in physical conditions other than annealed shall be at the option of the seller.

5.2.2 The axis of the bend test specimen shall be parallel to the direction of rolling.

5.2.3 The bend test specimen shall be bent around a diameter equal to the product of the bend factor times the nominal thickness of the test specimen.

5.2.4 Free-bend test specimens shall be bent cold, either by pressure or by blows. However, in case of dispute, tests shall be made by pressure.

5.2.5 Controlled-bend (V-block) test specimens shall be bent cold by means of V-blocks or a mating punch and die having an included angle of 45° and with proper curvature of surface at the bend areas to impart the desired shape and diameter of bend to the specimen.

TABLE 1 Chemical Requirements

UNS Designation[A]	Type	Composition, %							
		Carbon, max	Manganese	Phosphorus, max	Sulfur, max	Silicon, max	Chromium	Nickel	Other
S20100	201	0.15	5.50–7.50	0.060	0.030	1.00	16.00–18.00	3.50–5.50	N 0.25 max
S20200	202	0.15	7.50–10.00	0.060	0.030	1.00	17.00–19.00	4.00–6.00	N 0.25 max
S21900	XM-10	0.08	8.00–10.00	0.060	0.030	1.00	19.00–21.50	5.50–7.50	N 0.15–0.40
S21904	XM-11	0.04	8.00–10.00	0.060	0.030	1.00	19.00–21.50	5.50–7.50	N 0.15–0.40
S21460	XM-14	0.12	14.00–16.00	0.060	0.030	1.00	17.00–19.00	5.00–6.00	N 0.35–0.50
S20910	XM-19	0.06	4.00–6.00	0.040	0.030	1.00	20.50–23.50	11.50–13.50	N 0.20–0.40 Cb 0.10–0.30 V 0.10–0.30 Mo 1.50–3.00
S24000	XM-29	0.08	11.50–14.50	0.060	0.030	1.00	17.00–19.00	2.25–3.75	N 0.20–0.40

[A] New designation established in accordance with ASTM E 527 and SAE J1086, Recommended Practice for Numbering Metals and Alloys (UNS).

TABLE 2 Tensile Requirements

Condition	Tensile Strength, min, ksi (MPa)	Yield Strength, min, ksi (MPa)	Elongation in 2 in. or 50 mm, min, %		
			Up to 0.015 in. (0.38 mm) in Thickness	0.016 (0.40 mm) to 0.030 in. (0.76 mm) in Thickness	0.031 in. (0.79 mm) and over in Thickness
Type 201					
Annealed	95 (655)	38 (260)	40	40	40
¼ hard	125 (860)	75 (515)	20	20	20
½ hard	150 (1025)	110 (760)	9	10	10
¾ hard	175 (1205)	135 (930)	3	5	7
Full hard	185 (1275)	140 (965)	3	4	5
Type 202					
Annealed	90 (620)	38 (260)	40	40	40
¼ hard	125 (860)	75 (515)	12	12	12
Type XM-10 and XM-11					
Annealed[A]	100 (690)	60 (415)	40	40	40
10 % cold rolled[A]	130 (895)	115 (795)	15	15	15
Annealed plate	90 (620)	50 (345)	...	...	45
Type XM-14					
Annealed	105 (725)	55 (380)	40	40	40
Type XM-19					
Annealed[A]	120 (825)	75 (515)	30	30	30
Annealed plate	100 (690)	55 (380)	...	...	35
Type XM-29					
Annealed[A]	100 (690)	60 (415)	40	40	40
Annealed plates	100 (690)	55 (380)	...	...	40

[A] Sheet and strip.

TABLE 3 Free-Bend Requirements

Condition	0.030 in. (0.76 mm) and Less in Thickness[A]		0.031 (0.79 mm) to 0.050 in. (1.27 mm) in Thickness[A]		Over 0.050 in. (1.27 mm) in Thickness[A]	
	Angle of Bend		Angle of Bend		Angle of Bend	
	Deg	Factor[B]	Deg	Factor[B]	Deg	Factor[B]
Type 201						
Annealed	180	1	180	1	180	1
¼ hard	180	2	180	2	90	2
½ hard	180	4	180	8	90	2
¾ hard	180	8	90	2½	...	...
Full hard	180	12	90	3½	...	...
Type 202						
Annealed	180	1	180	1	180	1
¼ hard	180	2	180	2	90	2
Type XM-10 and XM-11						
Annealed	180	1	180	1	180	1
10 % cold rolled	90	2	90	2	90	2
Types XM-14, XM-19, and XM-29						
Annealed	180	1	180	1	180	

[A] Ordered thickness.
[B] See 5.2.3.

TABLE 4 Controlled-Bend (V-Block) Test Requirements

Condition	0.050 in. (1.27 mm) and Less in Thickness[A]		0.051 (1.30 mm) to 0.187 in. (4.76 mm) in Thickness[A]		Over 0.187 in. (4.76 mm) in Thickness[A]	
	Angle of Bend		Angle of Bend		Angle of Bend	
	Deg	Factor[B]	Deg	Factor[B]	Deg	Factor[B]
Type 201						
¼ hard	135	2	135	3	...	...
½ hard	135	4	135	4	...	...
¾ hard	135	6	...	...	...	...
Full hard	135	6	...	...	...	...
Type 202						
¼ hard	135	4	135	4	...	...

[A] Ordered thickness.
[B] See 5.2.3.

The American Society for Testing and Materials takes no position respecting the validity of any patent rights asserted in connection with any item mentioned in this standard. Users of this standard are expressly advised that determination of the validity of any such patent rights, and the risk of infringement of such rights, are entirely their own responsibility.

This standard is subject to revision at any time by the responsible technical committee and must be reviewed every five years and if not revised, either reapproved or withdrawn. Your comments are invited either for revision of this standard or for additional standards and should be addressed to ASTM Headquarters. Your comments will receive careful consideration at a meeting of the responsible technical committee, which you may attend. If you feel that your comments have not received a fair hearing you should make your views known to the ASTM Committee on Standards, 1916 Race St., Philadelphia, Pa. 19103.

Designation: A 414/A 414M – 83

Standard Specification for
CARBON STEEL SHEET FOR PRESSURE VESSELS[1]

This standard is issued under the fixed designation A 414/A 414M; the number immediately following the designation indicates the year of original adoption or, in the case of revision, the year of last revision. A number in parentheses indicates the year of last reapproval. A superscript epsilon (ε) indicates an editorial change since the last revision or reapproval.

1. Scope

1.1 This specification[2] covers hot-rolled carbon steel sheet for pressure vessels involving fusion welding or brazing. Welding and brazing technique is of fundamental importance and shall be in accordance with commercial practices.

1.2 The following grades are included in this specification:

Mechanical Requirements

Grade	Yield Strength, min ksi	Yield Strength, min MPa	Tensile Strength, min ksi	Tensile Strength, min MPa
A	25	170	45	310
B	30	205	50	345
C	33	230	55	380
D	35	240	60	415
E	38	260	65	450
F	42	290	70	485
G	45	310	75	515

1.3 Hot-rolled carbon steel sheet is generally furnished in cut lengths and to decimal thickness only. Coils may be furnished provided tension test specimens are taken to represent the middle of the slab as required by 6.1.4. The purchaser should recognize this requires cutting the coils to obtain test samples and results in half-size coils. The sheet is furnished to the following size limits:

Thickness, in. [mm]	Width, in. [mm] Over 12 to 48 [Over 300 to 1200]	Width, in. [mm] Over 48 [Over 1200]
0.230 to 0.180 [6.0 to 4.5]	sheet	Sheet (coils only)
Under 0.180 to 0.057 under [4.5 to 1.5]	sheet	sheet

1.4 The values stated in either U.S. inch-pound units or SI (metric) units are to be regarded separately as standard. Within the text the SI units are shown in brackets. The values stated in each system are not exact equivalents, therefore each system must be used independently of the other. Combining values of the two systems may result in nonconformance with the specification.

2. Applicable Documents

2.1 *ASTM Standards:*
A 568 Specification for General Requirements for Steel, Carbon and High-Strength Low-Alloy Hot-Rolled Sheet and Cold-Rolled Sheet[3]
A 568M Specification for General Requirements for Steel, Carbon and High-Strength Low-Alloy Hot-Rolled Sheet and Cold-Rolled Sheet [Metric][3]

3. General Requirements for Delivery

3.1 Material furnished under this specification shall conform to the applicable requirements of the current editions of Specifications A 568 or A 568M unless otherwise provided herein.

4. Ordering Information

4.1 Orders for material under this specification shall include the following information, as required, to describe the material adequately:
4.1.1 Designation or specification number, date of issue, and grade,
4.1.2 Copper bearing steel, when required,
4.1.3 Special requirements, if required,
4.1.4 Condition—pickled (or blast cleaned), if

---

[1] This specification is under the jurisdiction of ASTM Committee A-1 on Steel, Stainless Steel, and Related Alloys and is the direct responsibility of Subcommittee A01.19 on Sheet Steel and Strip.
Current edition approved May 27, 1983. Published November 1983. Originally published as A 414 – 71. Last previous edition A 414 – 79.
[2] For ASME Boiler and Pressure Vessel Code applications see related Specification SA-414 in Section 11 of that Code.
[3] *Annual Book of ASTM Standards*, Vol 01.03.

A 414/A 414M

required. (Material so ordered will be oiled unless ordered dry.),

4.1.5 Dimensions, including type of edges, and

4.1.6 Cast analysis or test report request or both, if required.

NOTE 1—A typical ordering description is as follows: "ASTM A 414, Grade A, Hot-Rolled Sheet, 0.100 in. [2.54 mm] by 36 in. [914.4 mm] by 96 in. [2438 mm], cut edges."

5. Chemical Requirements

5.1 *Cast or Heat Analysis*—The analysis of the steel shall conform to the requirements prescribed in Table 1.

5.2 *Product, Check, or Verification Analysis*—Analyses may be made by the purchaser from finished material representing each heat.

6. Mechanical Requirements

6.1 *Tensile Strength:*

6.1.1 *Requirements*—The material as represented by the tension test specimens shall conform to the tensile properties prescribed in Table 2.

6.1.2 *Number of Tests*—Two tension tests shall be made from the product of each slab as rolled.

6.1.3 *Test Specimen Condition and Size*—Test specimens shall be prepared for testing from sheets or coils in their finished condition and shall be the full thickness of the sheet.

6.1.4 *Test Specimen Location and Orientation*—Tension test specimens shall be taken at locations representing the middle and the back end of the slab as rolled. They shall be located approximately midway between the center and edge of the sheet and shall be transverse to the longitudinal axis (see Fig. 1).

7. Finish and Appearance

7.1 *Surface Finish:*

7.1.1 Unless otherwise specified, the material shall be furnished without removing the hot-rolled oxide or scale.

7.1.2 When required, the material may be specified to be pickled or blast cleaned.

7.2 *Oiling:*

7.2.1 Unless otherwise specified, the material shall be furnished not oiled.

7.2.2 When specified to be pickled or blast cleaned, the material shall be furnished oiled. When required, pickled or blast-cleaned material may be specified to be furnished dry.

7.3 *Edges*—Unless otherwise specified, mill edges shall be furnished on material which has not had the hot-rolled oxide or scale removed and cut edges shall be furnished on material which has had the hot-rolled oxide or scale removed.

8. Workmanship

8.1 The material shall be free from injurious defects (see Specifications A 568/A 568M).

9. Marking

9.1 The name or brand of the manufacturer, heat and slab number, specification designation number, and grade shall be legibly and durably marked on each cut length sheet in two places not less than 12 in. [300 mm] from the edges. Cut length sheets, the maximum lengthwise and crosswise, dimensions of which do not exceed 72 in. [1800 mm], shall be legibly and durably marked in one place approximately midway between the center and a side edge. The manufacturer's test identification number shall be legibly and durably marked on each test specimen. Steel-die marking of sheets is prohibited.

9.2 For coil product, the information required in 9.1 shall be legibly and durably marked both on each coil and on a tag affixed to each coil.

10. Certification and Reports

10.1 When requested, the manufacturer shall furnish copies of a test report showing the results of the cast analysis and mechanical property tests made to determine compliance with this specification. The report shall include the purchase order number, specification number, and test identification number correlating the test results with material represented.

A 414/A 414M

TABLE 1 Chemical Requirements

Element	Composition, %						
	Grade A Ladle	Grade B Ladle	Grade C Ladle	Grade D Ladle	Grade E Ladle	Grade F Ladle	Grade G Ladle
Carbon, max	0.15	0.22	0.25	0.25	0.27	0.31	0.31
Manganese, max	0.90	0.90	0.90	1.20	1.20	1.20	1.35
Phosphorus, max	0.035	0.035	0.035	0.035	0.035	0.035	0.035
Sulfur, max	0.040	0.040	0.040	0.040	0.040	0.040	0.040
Silicon[A]	...	...	...	..	...	...	..
Copper, when copper steel is specified, min	0.20	0.20	0.20	0.20	0.20	0.20	0.20

[A] Killed steel may be supplied upon request to the manufacturer. When silicon killed steel is specified, a range from 0.15 to 0.30 shall be supplied.

TABLE 2 Tensile Requirements

	Grade A	Grade B	Grade C	Grade D	Grade E	Grade F	Grade G
Tensile strength:							
min, ksi [MPa]	45 [310]	50 [345]	55 [380]	60 [415]	65 [450]	70 [485]	75 [515]
max, ksi [MPa]	60 [415]	65 [450]	70 [485]	75 [515]	85 [585]	90 [620]	95 [655]
Yield strength, min, ksi (MPa)[A]	25 [170]	30 [205]	33 [230]	35 [240]	38 [260]	42 [290]	45 [310]
Elongation in 2 in. (50 mm), min, percent:							
Under 0.230 to 0.145 [Under 6.0 to 3.8]	26	24	22	20	18	16	16
Under 0.145 to 0.089 [Under 3.8 to 2.2]	24	22	20	18	16	14	14
Under 0.089 to 0.057 [Under 2.2 to 1.5]	23	21	19	17	15	13	13
Elongation in 8 in. (200 mm), min, percent:							
Under 0.230 to 0.145 [Under 6.0 to 3.8]	20	18	16	14	12	10	10

[A] Yield strength determined by the 0.2 % offset or 0.5 % extension under load methods.

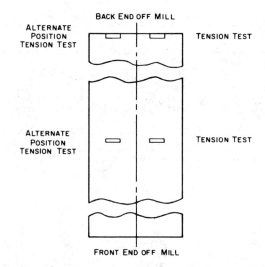

FIG. 1 Location of Test Specimens

The American Society for Testing and Materials takes no position respecting the validity of any patent rights asserted in connection with any item mentioned in this standard. Users of this standard are expressly advised that determination of the validity of any such patent rights, and the risk of infringement of such rights, are entirely their own responsibility.

This standard is subject to revision at any time by the responsible technical committee and must be reviewed every five years and if not revised, either reapproved or withdrawn. Your comments are invited either for revision of this standard or for additional standards and should be addressed to ASTM Headquarters. Your comments will receive careful consideration at a meeting of the responsible technical committee, which you may attend. If you feel that your comments have not received a fair hearing you should make your views known to the ASTM Committee on Standards, 1916 Race St., Philadelphia, Pa. 19103.

Designation: A 417 – 74 (Reapproved 1980)[ϵ1]

An American National Standard

Standard Specification for
STEEL WIRE, COLD-DRAWN, FOR ZIG-ZAG, SQUARE-FORMED, AND SINUOUS-TYPE UPHOLSTERY SPRING UNITS[1]

This standard is issued under the fixed designation A 417; the number immediately following the designation indicates the year of original adoption or, in the case of revision, the year of last revision. A number in parentheses indicates the year of last reapproval. A superscript epsilon (ϵ) indicates an editorial change since the last revision or reapproval.

[ϵ1] NOTE—The title was editorially changed and metric equivalents were editorially deleted from Tables 1 and 2 in September 1980.

1. Scope

1.1 This specification covers round, uncoated, cold-drawn spring wire in coils having properties and quality intended for the manufacture of the following upholstery springs:

1.1.1 *Type A*—Zig-zag (U-formed),
1.1.2 *Type B*—Square-formed, and
1.1.3 *Type C*—Sinuous for furniture spring units.

1.2 These types of upholstery springs are used in the manufacture of automotive seat springs and furniture springs. The wire is not intended for the manufacture of mechanical springs.

NOTE 1—A complete metric companion to Specification A 417 has been developed—A 417M; therefore, no metric equivalents are presented in this specification.

2. Applicable Documents

2.1 *ASTM Standards:*
A 370 Methods and Definitions for Mechanical Testing of Steel Products[2]
A 510 Specification for General Requirements for Wire Rods and Coarse Round Wire, Carbon Steel[3]
A 700 Recommended Practices for Packaging, Marking, and Loading Methods for Steel Products for Domestic Shipment[4]
E 29 Recommended Practice for Indicating Which Places of Figures Are to Be Considered Significant in Specified Limiting Values[5]
E 30 Methods for Chemical Analysis of Steel, Cast Iron, Open-Hearth Iron, and Wrought Iron[6]
E 380 Standard for Metric Practice[7]

3. Ordering Information

3.1 Orders for material under this specification shall include the following information for each ordered item:
3.1.1 Quantity (weight),
3.1.2 Name of material (name of specific type required) (Section 1 and Table 2),
3.1.3 Diameter (Table 2),
3.1.4 Packaging, marking, and loading (Section 10),
3.1.5 ASTM designation and date of issue, and
3.1.6 Heat (cast) analysis (if desired).

NOTE 2—A typical ordering description is as follows: 50 000 lb cold-drawn, upholstery, spring wire Type A for zig-zag-type springs, Class 1, 200 to 300 ksi, size 0.148 in. in 2000-lb coils to ASTM A 417, dated ——.

---

[1] This specification is under the jurisdiction of ASTM Committee A-1 on Steel, Stainless Steel and Related Alloys, and is the direct responsibility of Subcommittee A01.03 on Steel Rod and Wire.
Current edition approved July 29, 1974. Published September 1974. Originally published as A 417 – 57 T. Last previous edition A 417 – 68.
[2] *Annual Book of ASTM Standards*, Vol 01.04.
[3] *Annual Book of ASTM Standards*, Vol 01.03.
[4] *Annual Book of ASTM Standards*, Vols 01.01, 01.03, 01.04, and 01.05.
[5] *Annual Book of ASTM Standards*, Vol 14.02.
[6] *Annual Book of ASTM Standards*, Vol 03.05.
[7] *Annual Book of ASTM Standards*, Vol 14.02.

A 417

4. Manufacture

4.1 The steel shall be made by any of the following processes: open-hearth, basic-oxygen, or electric-furnace.

4.2 A sufficient discard shall be made to ensure freedom from injurious piping and undue segregation.

4.3 The wire shall be cold-drawn to produce the desired mechanical properties.

5. Chemical Requirements

5.1 Upholstery spring wire for these types of springs is customarily produced within the chemical ranges shown below. Chemical composition and processing may vary depending on the gage of wire and specific use.

Carbon, %	0.50 to 0.75[a]
Manganese, %	0.60 to 1.20[a]
Phosphorus, max, %	0.040
Sulfur, max, %	0.050

[a] In any lot in which all the wire is of the same size and type, and submitted for inspection at the same time, the carbon content shall not vary more than 0.20 %, and the manganese content shall not vary more than 0.30 %.

5.2 An analysis of each heat (cast) shall be made by the manufacturer to determine the percentage of elements specified above. The analysis shall be made from a test sample preferably taken during the pouring of the heat (cast). The chemical composition thus determined shall be reported to the purchaser or his representative upon request.

6. Mechanical Requirements

6.1 *Tension Test*:

6.1.1 *Requirements*—The material as represented by tension test specimens shall conform to the requirements prescribed in Table 1 for the various sizes and specified types.

6.1.2 *Number of Tests*—One test specimen shall be taken for each ten coils, or fraction thereof, in a lot. Each heat in a given lot shall be tested.

6.1.3 *Location of Tests*—The test specimen shall be taken from either end of the coil.

6.1.4 *Test Method*—The tension test shall be made in accordance with Methods and Definitions A 370.

6.2 *Wrap Test*:

6.2.1 *Requirements*—The wire for zig-zag-type and for square-formed-type springs for automobile seat and back spring units shall wrap on itself as an arbor without breakage. The wire for sinuous-type furniture spring units shall wrap on a mandrel twice the diameter of the wire without breakage.

6.2.2 *Number of Tests*—One test specimen shall be taken for each ten coils, or fraction thereof in a lot. Each heat in a given lot shall be tested.

6.2.3 *Location of Tests*—The test specimen shall be taken from either end of the coil.

6.2.4 *Test Method*—The wrap test shall be made in accordance with Supplement IV of Methods and Definitions A 370.

7. Retests

7.1 If any test specimen exhibits obvious defects or shows the presence of a weld, it may be discarded and another specimen substituted.

8. Permissible Variations in Dimensions

8.1 The diameter of the wire shall not vary from that specified by more than the tolerances specified in Table 2.

9. Workmanship

9.1 *Surface Condition*—The surface of the wire as received shall be smooth and shall have a uniform finish suitable for shaping the various type springs. No serious die marks, scratches, or seams may be present.

9.2 The wire shall be properly cast. To test for cast, a few convolutions of wire shall be cut from the coil and allowed to fall on a flat surface. The wire shall lie substantially flat on itself and shall not spring up and show a wavy condition.

9.3 Each coil shall be one continuous length of wire, properly coiled. Welds made prior to cold drawing are permitted. Weld areas need not meet the mechanical requirements of this specification.

10. Packaging, Marking, and Loading

10.1 Packaging of the coils of wire shall be by agreement between the manufacturer and the purchaser. This agreement shall include coil dimensions and weights.

10.2 When specified, the packaging, marking, and loading shall be in accordance with Recommended Practices A 700.

10.3 Marking shall be by a tag securely attached to each coil of wire and shall show the identity of the producer, size of the wire, type, and ASTM specification number.

11. Inspection

11.1 The manufacturer shall afford the purchaser's inspector all reasonable facilities necessary to satisfy him that the material is being produced and furnished in accordance with this specification. Mill inspection by the purchaser shall not interfere unnecessarily with the manufacturer's operations. All tests and inspections shall be made at the place of manufacture, unless otherwise agreed to.

12. Rejection and Rehearing

12.1 Unless otherwise specified, any rejection based on tests made in accordance with this specification shall be reported to the manufacturer within a reasonable length of time.

12.2 Failure of any of the test specimens to comply with the requirements of this specification shall constitute grounds for rejection of the lot represented by the specimen. The lot may be resubmitted for inspection by testing every coil for the characteristic in which the specimen failed and sorting out the defective coils.

12.3 The material must be adequately protected and correctly identified in order that the manufacturer may make proper investigation.

TABLE 1 Tensile Strength Requirements[A]

Diameter[B] Decimal Size, in.	Wire Gage	Tensile Strength, ksi min	Tensile Strength, ksi max	Tensile Strength, ksi min	Tensile Strength, ksi max
Type A, Zig-Zag Type Automobile Seat and Back Spring Units					
		Class I		Class II	
0.092	13	220	250	230	260
0.106	12	215	245	225	255
0.120	11	210	240	215	245
0.135	10	205	235	210	240
0.148	9	200	230	210	240
0.162	8	190	220	200	230
Type B, Square-Formed Type Automobile Seat and Back Spring Units					
0.092	13	215	245	225	255
0.106	12	210	240	220	250
0.120	11	205	235	215	245
0.135	10	200	230	210	240
0.148	9	190	220	200	230
0.162	8	180	210	190	220
Type C, Sinuous Type Furniture Spring Units					
0.092	13	235	265		
0.106	12	235	265		
0.120	11	230	260		
0.135	10	225	255		
0.148	9	220	250		
0.162	8	215	245		
0.177	7	210	240		
0.192	6	207	237		

[A] Tensile strength values for diameters not shown in this table shall conform to that shown for the next larger diameter. (For example, for diameter 0.128 in., the value shall be the same as for 0.135 in.)

[B] Decimal size is rounded to three significant places in accordance with Recommended Practice E 29.

A 417

TABLE 2 Permissible Variations in Wire Diameter

NOTE—For purposes of determining conformance with this specification, all limits are considered absolute as defined in Recommended Practice E 29.

Diameter, in.	Variations, plus and minus, in.	Permissible Out-of-Round, in.
Sizes 0.092 to 0.192	0.002	0.002

The American Society for Testing and Materials takes no position respecting the validity of any patent rights asserted in connection with any item mentioned in this standard. Users of this standard are expressly advised that determination of the validity of any such patent rights, and the risk of infringement of such rights, are entirely their own responsibility.

This standard is subject to revision at any time by the responsible technical committee and must be reviewed every five years and if not revised, either reapproved or withdrawn. Your comments are invited either for revision of this standard or for additional standards and should be addressed to ASTM Headquarters. Your comments will receive careful consideration at a meeting of the responsible technical committee, which you may attend. If you feel that your comments have not received a fair hearing you should make your views known to the ASTM Committee on Standards, 1916 Race St., Philadelphia, Pa. 19103.

Designation: A 417M – 80
Metric

Standard Specification for
STEEL WIRE, COLD-DRAWN FOR ZIG-ZAG SQUARE-FORMED, AND SINUOUS-TYPE UPHOLSTERY SPRING UNITS [METRIC][1]

This standard is issued under the fixed designation A 417M; the number immediately following the designation indicates the year of original adoption or, in the case of revision, the year of last revision. A number in parentheses indicates the year of last reapproval. A superscript epsilon (ϵ) indicates an editorial change since the last revision or reapproval.

1. Scope

1.1 This specification covers round, uncoated, cold-drawn spring wire in coils having properties and quality intended for the manufacture of upholstery springs.

1.2 These upholstery springs are used in the manufacture of automobile seat springs and furniture springs. This wire is not intended for the manufacture of mechanical springs.

NOTE 1—This specification is the metric counterpart of Specification A 417.

2. Applicable Documents

2.1 *ASTM Standards*:
A 370 Methods and Definitions for Mechanical Testing of Steel Products[2]
A 510M Specification for General Requirements for Wire Rods and Coarse Round Wire, Carbon Steel[3]
A 700 Recommended Practices for Packaging, Marking, and Loading Methods for Steel Products for Domestic Shipment[4]
A 751 Methods, Practices, and Definitions for Chemical Analysis of Steel Products[5]
E 29 Recommended Practice for Indicating Which Places of Figures Are to Be Considered Significant in Specified Limiting Values[6]

3. Classification

3.1 The upholstery spring wire can be furnished in two types:
3.1.1 *Type I*—Zig-zag or square formed for automobile seat and back spring units and
3.1.2 *Type II*—Sinuous for furniture spring units.

4. Ordering Information

4.1 Orders for material under this specification shall include the following information for each ordered item:
4.1.1 Quantity (mass),
4.1.2 Name of material (name of specific type required) (Section 3 and Table 1),
4.1.3 Diameter (mm)
4.1.4 Certification or test reports, or both, if specified (Section 13),
4.1.5 Packaging, marking, and loading (Section 14),
4.1.6 ASTM designation and date of issue, and
4.1.7 Heat (cast) analysis (if specified).

NOTE 2—A typical ordering description is as follows: 15 000 kg cold-drawn upholstery spring wire Type I, for Zig-Zag type springs, size 3.0 mm in 700-kg coils on tubular carriers to ASTM A 417M dated

---

[1] This specification is under the jurisdiction of ASTM Committee A-1 on Steel, Stainless Steel, and Related Alloys and is the direct responsibility of Subcommittee A01.03 on Steel Rod and Wire.
Current edition approved Sept. 26, 1980. Published November 1980.
[2] *Annual Book of ASTM Standards*, Vol 01.04.
[3] *Annual Book of ASTM Standards*, Vol 01.03.
[4] *Annual Book of ASTM Standards*, Vols 01.01, 01.03, 01.04, and 01.05.
[5] *Annual Book of ASTM Standards*, Vols 01.01, 01.02, 01.03, 01.04 and 01.05.
[6] *Annual Book of ASTM Standards*, Vol 14.02.

A 417M

5. Materials and Manufacture

5.1 The steel shall be of such quality that when processed, the finished wire shall be free of detrimental pipe and undue segregation.

5.2 The wire shall be cold-drawn to produce the desired mechanical properties.

6. Chemical Requirements

6.1 Upholstery spring wire for these types of springs is customarily produced within the chemical ranges shown below. Chemical composition and processing may vary depending on the size of wire and specific use.

Carbon, %	0.50 to 0.75[A]
Manganese, %	0.60 to 1.20[A]

[A] In any lot in which all the wire is of the same size and type and submitted for inspection at the same time, the carbon content shall not vary more than 0.20 %, and the manganese content shall not vary more than 0.30 %.

6.2 An analysis of each heat (cast) shall be made by the manufacturer to determine the percentage of elements specified above. The analysis shall be made from a test sample preferably taken during the pouring of the heat (cast). The chemical analysis thus determined shall be reported to the purchaser or his representative upon request.

6.3 Chemical analysis shall be made in accordance with Method A 751.

7. Mechanical Requirements

7.1 *Tensile Strength Test*—The material as represented by test specimens shall conform to the requirements prescribed in Table 1 for the specified size and type.

7.2 *Wrap Test:*

7.2.1 Type I material for zig-zag and square formed springs for automobile seat and back spring units shall wrap on itself as an arbor without fracture.

7.2.2 Type II material for sinuous furniture spring units shall wrap on a mandrel twice the diameter of the wire without fracture.

7.3 The tensile strength test and wrap test shall be made in accordance with Method A 370.

8. Permissible Variations in Diameter

8.1 The diameter of the wire shall not vary from that specified by more than the tolerances shown in Table 2.

9. Workmanship, Finish, and Appearance

9.1 the surface of the wire as received shall be smooth and shall have a uniform finish suitable for shaping the various type springs. No serious die marks, scratches, or seams may be present.

9.2 The wire shall be properly cast. To test for cast, a few convolutions of wire shall be cut from the coil and allowed to fall on a flat surface. The wire shall lie substantially flat on itself and shall not spring up and show a wavy condition.

9.3 Each coil shall be one continuous length of wire, properly coiled. Welds made prior to cold drawing are permitted. Weld areas need not meet the mechanical requirements of this specification.

10. Sampling and Number of Tests

10.1 A lot shall consist of all of the coils of wire of the same size, and type offered for inspection at one time.

10.2 One test specimen shall be taken for each ten coils or fraction there of, in a lot. Each heat in a given lot shall be tested.

10.3 Test specimens may be taken from either end of the coil.

10.4 If any test specimen shows the presence of a weld or an obvious detect, it may be discarded and another specimen substituted.

10.5 Test specimens shall be tested for tensile strength (7.1) and wrap test (7.2).

11. Inspection

11.1 Unless otherwise specified in the contract or purchase order, the manufacturer is responsible for the performance of all inspection and test requirements specified in this specification. Except as otherwise specified in the contract or purchase order, the manufacturer may use his own or any other suitable facilities for the performance of the inspection and test requirements unless disapproved by the purchaser at the time the order is placed. The purchaser shall have the right to perform any of the inspections and tests set forth in this specification when such inspections and tests are deemed necessary to assure that the material conforms to prescribed requirements.

12. Rejection and Rehearing

12.1 Material that fails to conform to the

A 417M

requirements of this specification may be rejected. Rejections should be reported to the producer or supplier promptly and in writing. In case of dissatisfaction with the results of the test, the producer or supplier may make a claim for a rehearing.

12.2 The material must be adequately protected and correctly identified in order that the producer may make proper investigation.

13. Certification

13.1 When specified in the purchase order or contract, a producer or supplier certification shall be furnished to the purchaser that the material was manufactured, sampled, tested, and inspected in accordance with this specification and has been found to meet the requirements. When specified in the purchase order or contract, a report of the test results shall be furnished.

14. Marking and Packaging

14.1 Marking shall be by a tag securely attached to each coil of wire and shall show the identity of the producer, size of wire, type, heat number, and ASTM designation.

14.2 Packaging of the coils of wire shall be by agreement between producer and the purchaser. This agreement shall include coil dimensions and weight.

14.3 Unless otherwise specified, the packaging, marking, and loading shall be in accordance with Method A 700.

TABLE 1 Tensile Strength Requirements

Diameter, mm	Tensile Strength, MPa	
	min	max
Type I, Zig-Zag or Square Formed for Automobile Seat and Back Spring Units		
2.4	1500	1730
2.6	1480	1700
3.0	1440	1660
3.5	1400	1610
3.8	1380	1580
4.2	1350	1550
Type II, Sinuous Type for Furniture Spring Units		
2.4	1640	1870
2.6	1620	1840
3.0	1580	1790
3.5	1530	1740
3.8	1510	1720
4.2	1490	1690
4.5	1460	1670
5.0	1430	1630

TABLE 2 Permissible Variations in Wire Diameter

NOTE—For purposes of determining conformance with this specification, all limits are considered absolute as defined in Recommended Practice E 29.

Diameter, mm	Variation, plus and minus, mm	Permissible Out-Of-Round, mm
Sizes 2.2 to 5.2	0.05	0.05

The American Society for Testing and Materials takes no position respecting the validity of any patent rights asserted in connection with any item mentioned in this standard. Users of this standard are expressly advised that determination of the validity of any such patent rights, and the risk of infringement of such rights, are entirely their own responsibility.

This standard is subject to revision at any time by the responsible technical committee and must be reviewed every five years and if not revised, either reapproved or withdrawn. Your comments are invited either for revision of this standard or for additional standards and should be addressed to ASTM Headquarters. Your comments will receive careful consideration at a meeting of the responsible technical committee, which you may attend. If you feel that your comments have not received a fair hearing you should make your views known to the ASTM Committee on Standards, 1916 Race St., Philadelphia, Pa. 19103.

 Designation: A 424 – 80

An American National Standard

Standard Specification for
STEEL SHEET FOR PORCELAIN ENAMELING[1]

This standard is issued under the fixed designation A 424; the number immediately following the designation indicates the year of original adoption or, in the case of revision, the year of last revision. A number in parentheses indicates the year of last reapproval. A superscript epsilon (ϵ) indicates an editorial change since the last revision or reapproval.

This specification has been approved for use by agencies of the Department of Defense and for listing in the DoD Index of Specifications and Standards.

1. Scope

1.1 This specification[2] covers sheet steel in coils and cut lengths for porcelain enameling. This material is chemically constituted and processed to make it suitable for the fabricating and enameling requirements of articles for porcelain enameling under proper conditions. It is furnished as Type I, Type II Composition A, Type II Composition B, and in three qualities, Commercial Quality, Drawing Quality, and Drawing Quality Special Killed.

1.2 The values stated in inch-pound units are to be regarded as the standard.

2. Applicable Document

2.1 *ASTM Standard*:
A 568 Specification for General Requirements for Steel, Carbon and High-Strength Low-Alloy Hot-Rolled Sheet and Cold-Rolled Sheet[3]

3. Descriptions of Terms

3.1 *Types:*

3.1.1 *Type I* has an extremely low carbon level commonly produced by decarburizing in an open-coil process, in which the coil laps are separated for easy flow of annealing gases. This material is suitable for direct cover coat enameling practice, but this requirement must be indicated by the purchaser in accordance with 5.1.6. This material is also suitable for ground and cover coat enameling practice. It has good sag resistance and good formability.

3.1.2 *Type II* has moderately low carbon and manganese levels as produced in the melting operation. This material is suitable for ground and cover coat enameling practice. Type II, Composition A, is intended for use where resistance to sag is of prime importance, and Type II, Composition B, is intended for use where formability is of prime importance.

3.2 *Quality Designations:*

3.2.1 *Commercial Quality* is intended for parts where bending, moderate forming, or moderate drawing may be involved.

3.2.2 *Drawing Quality* is intended for fabricating identified parts where drawing or severe forming may be involved.

3.2.3 *Drawing Quality Special Killed* is generally available in Type I only and is intended for fabricating identified parts where the draw is particularly severe or where the material shall be essentially free of changes in mechanical properties over a period of time. Drawing Quality Special Killed should be specified where the formed material shall be essentially free of such surface disturbances as stretcher strains or fluting without the need of prior roller leveling.

4. General Requirements for Delivery

4.1 Material furnished under this specification shall conform to the applicable requirements of the current edition of Specification A 568.

4.2 Products covered by this specification

---

[1] This specification is under the jurisdiction of ASTM Committee A-1 on Steel, Stainless Steel and Related Alloys and is the direct responsibility of Subcommittee A01.19 on Steel Sheet and Strip.
Current edition approved Aug. 1, 1980. Published January 1981. Originally published as A 424 – 58 T. Last previous edition A 424 – 73.
[2] Type II has been divided into two chemical compositions, A and B, and the format has been updated to reference Specification A 568.
[3] *Annual Book of ASTM Standards*, Vol 01.03.

are produced to decimal thickness only and decimal thickness tolerances apply.

5. Ordering Information

5.1 Orders for material under this specification shall include the following information, as required, to describe adequately the desired material:
  5.1.1 Quantity,
  5.1.2 ASTM specification number and date of issue,
  5.1.3 Name of material (Porcelain Enameling Sheet, Type I, Drawing Quality),
  5.1.4 Condition (not oiled unless otherwise specified),
  5.1.5 Finish (matte (dull) finish will be supplied unless otherwise specified),
  5.1.6 Enameling practice (for Type I, indicate if direct cover coat practice will be used),
  5.1.7 Dimensions (thickness, width, and length for cut lengths, or thickness and width if coils),
  5.1.8 Coil size and weight requirements (must include inside and outside diameters and maximum weight),
  5.1.9 Application (show part identification and description),
  5.1.10 Special requirements (if required), and
  5.1.11 Cast or heat analysis (if required) (see 9.1).

NOTE 2—A typical ordering description is as follows: 40 000 lb, ASTM A 424 (latest issue) Porcelain Enameling Sheet, Type I, Drawing Quality, direct cover coat, 0.048 in. by 35 in. by 96 in., Part 2587, Range Top.

6. Chemical Requirements

6.1 The cast or heat analysis (formerly ladle) of Type I and Type II shall conform to the requirements prescribed in Table 1 except for carbon in Type I as noted.

7. Mechanical Requirements

7.1 *Bend Test*—Commercial Quality sheet shall be capable of being bent at room temperature in any direction through 180 deg flat on itself without cracking on the outside of the bent portion.

7.2 *Formability*—Drawing Quality and Drawing Quality Special Killed sheet shall produce an identified part within a properly established breakage allowance.

8. Finish and Appearance

8.1 *Surface Finish*—Unless otherwise specified, the sheet shall have a matte (dull) finish suitable for porcelain enameling.

8.2 *Oiling*—Unless otherwise specified, the sheet shall not be oiled.

9. Certification and Results

9.1 When requested, the manufacturer shall furnish the results of the cast analysis made to determine compliance with this specification and shall include the purchase order number, ASTM designation number, and cast number correlating the results with the material represented.

TABLE 1 Chemical Requirements

Element	Composition, max, percent		
	Type I	Type II	
		A	B
Carbon	0.008[a]	0.04	0.08
Manganese	0.60	0.12	0.20
Phosphorus	...	0.015	0.015
Sulfur[b]	0.040	0.040	0.040

[a] Cast analysis of carbon is not appropriate for Type I. Sheet product analysis is appropriate for checking proper type of material. Extremely low carbon levels can be checked accurately with carbon-combustion chromatographic type of analyzers.

[b] Drawing Quality, 0.035 max.

The American Society for Testing and Materials takes no position respecting the validity of any patent rights asserted in connection with any item mentioned in this standard. Users of this standard are expressly advised that determination of the validity of any such patent rights, and the risk of infringement of such rights, are entirely their own responsibility.

This standard is subject to revision at any time by the responsible technical committee and must be reviewed every five years and if not revised, either reapproved or withdrawn. Your comments are invited either for revision of this standard or for additional standards and should be addressed to ASTM Headquarters. Your comments will receive careful consideration at a meeting of the responsible technical committee, which you may attend. If you feel that your comments have not received a fair hearing you should make your views known to the ASTM Committee on Standards, 1916 Race St., Philadelphia, Pa. 19103.

Designation: A 457 – 82

Standard Specification for
HOT-WORKED, HOT-COLD-WORKED, AND COLD-WORKED ALLOY STEEL PLATE, SHEET, AND STRIP FOR HIGH STRENGTH AT ELEVATED TEMPERATURES[1]

This standard is issued under the fixed designation A 457; the number immediately following the designation indicates the year of original adoption or, in the case of revision, the year of last revision. A number in parentheses indicates the year of last reapproval. A superscript epsilon (ϵ) indicates an editorial change since the last revision or reapproval.

1. Scope

1.1 This specification covers Grade 651 high-strength alloy-steel plate, and strip for elevated temperature service. The mechanical properties of the plate, sheet, and strip are developed by suitable hot, cold, or hot-cold working.

1.2 The values stated in inch-pound units are to be regarded as the standard.

2. Applicable Documents

2.1 *ASTM Standards:*
A 751 Methods, Practices, and Definitions for Chemical Analysis of Steel Products[2]
E 8 Methods of Tension Testing of Metallic Materials[3]
E 10 Test Method for Brinell Hardness of Metallic Materials[3]
E 18 Test Methods for Rockwell Hardness and Rockwell Superficial Hardness for Metallic Materials[3]

3. Ordering Information

3.1 Orders for material under this specification shall include the following information:
3.1.1 Quantity (weight or number of pieces),
3.1.2 Name of material (hot-cold rolled alloy steel for high strength at elevated temperatures),
3.1.3 Form (plate, sheet, or strip),
3.1.4 Dimensions,
3.1.5 Grade (651),
3.1.6 Condition (Table 2),
3.1.7 Finish (as agreed upon),
3.1.8 ASTM designation and date of issue,
3.1.9 Additions to the specification or special requirements, and
3.1.10 End use (especially when such information will enable the manufacturer to produce more satisfactory material for the purchaser's process and product).

NOTE 1—A typical description is as follows: Hot-cold rolled alloy steel for high strength at elevated temperatures, Grade 651, hot-cold rolled No. 2D finish, ASTM A 457 dated ____, 100 pieces, 0.060 by 48 by 120 in.

4. Process

4.1 The alloy shall be made by the electric-furnace process, or other process approved by the purchaser.

4.2 Plate, sheet, or strip shall be either hot-rolled or hot-cold rolled under suitably controlled finishing temperatures, or cold rolled to develop, after the heat treatment specified in Table 2, the mechanical properties specified.

5. Discard

5.1 A sufficient discard shall be made from each ingot to ensure freedom from injurious piping and undue segregation.

6. Chemical Requirements

6.1 The material shall conform to the requirements as to chemical composition prescribed in Table 1.

---

[1] This specification is under the jurisdiction of ASTM Committee A-1 on Steel, Stainless Steel and Related Alloys, and is the direct responsibility of Subcommittee A01.17 on Flat Stainless Steel Products.
Current edition approved Oct. 29, 1982. Published December 1982. Originally published as A 457 – 61 T. Last previous edition A 457 – 71 (1979).
[2] *Annual Book of ASTM Standards*, Vol 01.03.
[3] *Annual Book of ASTM Standards*, Vol 03.01.

6.2 Methods and practices relating to chemical analysis required by this specification shall be in accordance with Methods, Practices, and Definitions A 751.

7. Ladle Analysis

7.1 An analysis of each melt shall be made by the manufacturer to determine the percentages of the elements specified in Table 1. The analysis shall be made from a test sample taken during pouring of the heat. The chemical composition thus determined shall be reported to the purchaser, or his representative, and shall conform to the requirements prescribed in Table 1.

8. Product Analysis

8.1 A product analysis may be made by the purchaser. The chemical composition thus determined shall conform to the requirements prescribed in Table 1.

9. Tensile Requirements

9.1 The material shall conform to the requirements as to mechanical properties prescribed in Table 2.

9.2 The tensile properties shall be determined in accordance with Methods E 8. Tension test specimens shall be taken from finished material in either the longitudinal or transverse direction for material in the annealed condition; for stress relieved material the specimens shall be taken in the transverse direction whenever possible.

9.3 The yield strength shall be determined by the offset method as described in Methods E 8. The limiting permanent offset shall be 0.2 % of the gage length of the specimen. Between yield and fracture of the specimen the test shall be conducted at a constant strain rate between ⅛ in./in·min (0.125 mm/mm·min) and ½ in./in·min (0.5 mm/mm·min) incl, or at a crosshead speed that will give a strain rate within this range. For the purpose of this specification, the rate of strain may be determined by a strain rate pacer, indicator, or controller, or by dividing the unit elongation by the elapsed time for yield strength to fracture.

10. Hardness Requirements

10.1 The material shall conform to requirements as to hardness, prescribed in Table 2, the values being obtained on the finished material.

10.2 Hardness tests shall be made in accordance with Test Method E 10, or Methods E 18.

11. Bending Requirements

11.1 Bend tests shall be conducted by the free-bend method, that is by applying forces to the ends of a specimen without the application of force at the point of maximum bending and shall meet the requirements prescribed in Table 3. Sheet and strip bend specimens and annealed plate bend specimens shall be tested in the full thickness of the material and shall be approximately 1 in. (25 mm) in width except that plate specimens may be a maximum of 2 in. (51 mm) in width. Stress-relieved plate specimens shall be machined to a thickness in the range from 0.125 to 0.250 in. (3.18 to 6.35 mm) located centrally between the plate faces, and this thickness shall be considered the nominal thickness. The edges of test specimens may be rounded to a radius equal to one-half the thickness.

12. Reheat Treatment

12.1 If any specimen selected to represent any heat fails to meet any of the test requirements, the material represented by such specimens may be reheat treated and resubmitted for test.

13. Variations in Dimensions and Weights

13.1 *Sheet*—The material referred to as sheet shall conform to the permissible variations in dimensions and weight prescribed in Tables 4 to 10.

13.2 *Strip*—The material referred to as strip shall conform to the permissible variations in dimensions and weight prescribed in Tables 11 to 16.

13.3 *Plate*—The material referred to as plate shall conform to the permissible variations in dimensions and weight prescribed in Tables 17 to 21.

14. Workmanship and Finish

14.1 The material shall be free of injurious defects, shall have a workmanlike appearance, and shall correspond to the finish as agreed upon between the manufacturer and the purchaser.

15. Number of Tests

15.1 Two tension tests, two hardness tests, and one bend test shall be taken from plate,

A 457

sheet, or strip pieces from each heat to determine whether the material properly conforms to the requirements prescribed in Table 2.

16. Retests

16.1 If any test specimen shows defective machining or develops flaws, it may be discarded and another specimen substituted.

17. Inspection

17.1 The manufacturer shall afford the purchaser's inspector all reasonable facilities necessary to satisfy him that the material is being produced and furnished in accordance with this specification. Mill inspection by the purchaser shall not interfere unnecessarily with the manufacturer's operations. All tests (except product analysis) and inspections shall be made at the place of manufacture unless agreed upon otherwise.

18. Rejection

18.1 Unless otherwise specified, any rejection based on tests made in accordance with this specification shall be reported to the manufacturer within 10 days from the date of test.

18.2 Material that shows injurious defects subsequent to its acceptance at the purchaser's plant will be rejected and the manufacturer shall be notified.

TABLE 1 Chemical Requirements

	651 (was Grade 2)	
	%	Product Variation, Over or Under, %
Carbon	0.28–0.35	0.02
Manganese	0.75–1.50	0.04
Silicon	0.30–0.80	0.05
Sulfur	0.030 max	0.005 over
Phosphorus	0.040 max	0.005 over
Chromium	18.00–20.00	0.25
Nickel	8.00–11.00	0.15
Molybdenum	1.00–1.75	0.05
Tungsten	1.00–1.75	0.05
Columbium plus tantalum	0.25–0.60	0.05
Titanium	0.10–0.35	0.05
Copper	0.50 max	0.03 over
Iron	balance	...

TABLE 2 Mechanical Property Requirements for Hot-Rolled Plate and Sheet and Cold-Finished Strip

	Condition, Stress Relieved		Condition, Annealed
	Under 0.040 in. (1.02 mm) Thick	0.040 to 0.250 in. (1.02 to 6.35 mm) Thick	All Thicknesses
Tensile strength, min, psi (MPa)	120 000 (827)	125 000 (862)	95 000 (655)
Yield strength (0.2% offset), min, psi (MPa)	80 000 (552)	90 000 (620)	45 000 (310)
Elongation in 2 in. (or 50 mm) min, %	12	12	12
Brinell hardness, max	321	321	200
Rockwell C, max	35	35	25

TABLE 3 Bend Test Requirements[A]

Nominal Thickness, in. (mm)	Bend Angle, min, °	Bend Factor
Stress Relieved:		
0.125 (3.18) and under	90	1
0.126 to 0.250 (3.20 to 6.35), incl	90	2
Annealed:		
Under 0.050 (1.27)	180	1
0.050 (1.27) and over	90	2

[A] Material shall withstand bending at room temperature without cracking, through the angle indicated above, around a diameter equal to the bend factor times the nominal thickness of the material, with axes of bends both perpendicular and parallel to the direction of rolling.

TABLE 4 Permissible Variations in Thickness of Hot-Rolled and Cold-Rolled Sheets

NOTE—Thickness measurements shall be taken at least ⅜ in. (9.52 mm) from the edge of the sheet.

Specified Thickness, in. (mm)	Permissible Variations from Specified Thickness, Over and Under, in. (mm)
Over 0.145 to ³⁄₁₆ (3.68 to 4.76), excl	0.014 (0.36)
Over 0.130 to 0.145 (3.30 to 3.68), incl	0.012 (0.30)
Over 0.114 to 0.130 (2.89 to 3.30), incl	0.010 (0.25)
Over 0.098 to 0.114 (2.49 to 2.89), incl	0.009 (0.23)
Over 0.083 to 0.098 (2.11 to 2.49), incl	0.008 (0.20)
Over 0.072 to 0.083 (1.83 to 2.11), incl	0.007 (0.18)
Over 0.058 to 0.072 (1.47 to 1.83), incl	0.006 (0.15)
Over 0.040 to 0.058 (1.01 to 1.47), incl	0.005 (0.13)
Over 0.026 to 0.040 (0.66 to 1.01), incl	0.004 (0.10)
Over 0.016 to 0.026 (0.41 to 0.66), incl	0.003 (0.08)
Over 0.007 to 0.016 (0.18 to 0.41), incl	0.002 (0.05)
Over 0.005 to 0.007 (0.13 to 0.18), incl	0.0015 (0.04)
0.005 (0.13)	0.001 (0.03)

A 457

TABLE 5 Permissible Variations in Width and Length of Hot-Rolled and Cold-Rolled Resquared Sheets (Stretcher Leveled Standard of Flatness)

NOTE—Polished sheets with finishes No. 4 and higher shall be produced to the tolerances in this table.

Specified Dimensions	Permissible Variations in Width and Length, in. (mm)	
	Over	Under
For thicknesses under 0.131 in. (3.33 mm):		
Widths up to 48 in. (1219 mm), excl	1/16 (1.59)	0
Widths 48 in. (1219 mm) and over	1/8 (3.18)	0
Lengths up to 120 in. (3048 mm), excl	1/16 (1.59)	0
Lengths 120 in. (3048 mm) and over	1/8 (3.18)	0
For thicknesses 0.131 in. (3.33 mm) and over:		
All widths and lengths	1/4 (6.35)	0

TABLE 6 Permissible Variations in Width and Length of Hot-Rolled and Cold-Rolled Sheets not Resquared

Specified Thickness, in. (mm)	Permissible Variations in Width for Widths Given, in. (mm)	
	24 to 48 (610 to 1220), excl	48 (1220) and Over
Under 3/16 (4.76)	1/16 (1.59) over, 0 under	1/8 (3.18) over, 0 under

Specified Length, ft (m)	Permissible Variations from Specified Length, in. (mm)	
	Over	Under
10 (3.0) and under	1/4 (6.35)	0
Over 10 to 20 (3.0 to 6.1), incl	1/2 (12.7)	0

TABLE 7 Permissible Camber of Hot-Rolled and Cold-Rolled Sheets

NOTE—Camber is the greatest deviation of a side edge from a straight line, the measurement being taken on the concave side with a straight edge.

Specified Width, in. (mm)	Permissible Camber; per Unit Length of any 8 ft (2.4 m), in. (mm)
24 to 36 (610 to 914), incl	1/8 (3.18)
Over 36 (914)	3/32 (2.38)

TABLE 8 Permissible Variations in Flatness of Hot-Rolled and Cold-Rolled Sheets

Sheets not Specified to Stretcher Leveled Standard of Flatness[A]		
Specified Thickness, in. (mm)	Width, in. (mm)	Permissible Variations in Flatness (Maximum Deviation from a Horizontal Flat Surface), in. (mm)
0.062 (1.57) and over	To 60 (1524), incl	1/2 (12.7)
	Over 60 to 72 (1524 to 1829), incl	3/4 (19.0)
	Over 72 (1829)	1 (25)
Under 0.062 (1.57)	To 36 (914), incl	1/2 (12.7)
	Over 36 to 60 (914 to 1524), incl	3/4 (19.0)
	Over 60 (1524)	1 (25)

Sheets Specified to Stretcher Leveled Standard of Flatness			
Specified Thickness, in. (mm)	Width, in. (mm)	Length, in. (mm)	Permissible Variations in Flatness (Maximum Deviation from a Horizontal Flat Surface), in. (mm)
Under 3/16 (4.76)	To 48 (1220), incl	To 96 (2438), incl	1/8 (3.18)
		Over 96 (2438)	1/4 (6.35)
	Over 48 (1220)	To 96 (2438), incl	1/4 (6.35)
		Over 96 (2438)	1/4 (6.35)

[A] Exclusive of quarter-hard, half-hard, three-quarter-hard, and full-hard tempers of the chromium-nickel grades and dead soft and deep drawing sheets.

A 457

TABLE 9 Permissible Variations in Diameter of Sheared Circles of Hot-Rolled and Cold-Rolled Sheets

Specified Thickness, in. (mm)	Permissible Variations over Specified Diameter for Diameters Given, in. (mm)[A]		
	Under 30 (762)	30 to 48 (762 to 1219), incl	Over 48 (1219)
0.0972 (2.469) and over	⅛ (3.18)	³⁄₁₆ (4.76)	¼ (6.35)
0.0971 to 0.0568 (2.466 to 1.443), incl	³⁄₃₂ (2.38)	⁵⁄₃₂ (3.97)	⁷⁄₃₂ (5.56)
0.567 (1.440) and under	¹⁄₁₆ (1.59)	⅛ (3.18)	³⁄₁₆ (4.76)

[A] No permissible variations are permitted under the specified diameter.

TABLE 10 Permissible Variations in Weight of Hot-Rolled and Cold-Rolled Sheets

It is not practicable to produce hot-rolled and cold-rolled sheets to exact theoretical weight. Sheets of any one item of a specific thickness and size in any finish may be over-weight to the following extent:

(1) Any item of 5 sheets or less, or any item estimated to weigh 200 lb (91 kg) or less, may actually weigh as much as 10% over the theoretical weight.

(2) Any item of more than 5 sheets and estimated to weigh more than 200 lb, may actually weigh as much as 7½% over the theoretical weight.

The underweight variations for stainless steel sheets are limited by the under thickness tolerances shown in Table 4.

For determining estimated weights the following factors shall be used:

Stainless and heat-resisting steel sheets:

	lb/ft²·in. of thickness
Chromium-nickel grades	42.0
Chromium grades	41.2

TABLE 11 Permissible Variations in Thickness of Cold-Rolled Strip

NOTE 1—For thicknesses under 0.010 to 0.005 in., incl, in widths up to and including 16 in., a tolerance of ±10% of the thickness shall apply. For thicknesses under 0.010 to 0.005 in., incl, in widths over 16 to 23¹⁵⁄₁₆ in., incl, a tolerance of ±15% of the thickness shall apply. For thickness tolerances on thicknesses under 0.005 in., in widths up to 23¹⁵⁄₁₆ in., incl, the manufacturer shall be consulted.

NOTE 2—Thickness measurements shall be taken ⅜ in. in from the edge of the strip, except that on widths less than 1 in. the tolerances are applicable for measurements at all locations.

NOTE 3—The tolerances in this table do not include crown tolerances (see Table 13).

Specified Thickness, in.[A]	Permissible Variations in Thickness for Thicknesses and Widths Given, Over and Under, in.[A]							
	³⁄₁₆, incl to 1 excl	1, incl to 3, excl	3 to 6, incl	Over 6 to 9, incl	Over 9 to 12, incl	Over 12 to 16, incl	Over 16 to 20, incl	Over 20 to 23¹⁵⁄₁₆, incl
Under ³⁄₁₆ to 0.161, incl	0.002	0.003	0.004	0.004	0.004	0.005	0.006	0.006
0.160 to 0.100, incl	0.002	0.002	0.003	0.004	0.004	0.004	0.005	0.005
0.099 to 0.069, incl	0.002	0.002	0.003	0.003	0.003	0.004	0.004	0.004
0.068 to 0.050, incl	0.002	0.002	0.003	0.003	0.003	0.003	0.004	0.004
0.049 to 0.040, incl	0.002	0.002	0.0025	0.003	0.003	0.003	0.004	0.004
0.039 to 0.035, incl	0.002	0.002	0.0025	0.003	0.003	0.003	0.003	0.003
0.034 to 0.029, incl	0.0015	0.0015	0.002	0.0025	0.0025	0.0025	0.003	0.003
0.028 to 0.026, incl	0.001	0.0015	0.0015	0.002	0.002	0.002	0.0025	0.003
0.025 to 0.020, incl	0.001	0.001	0.0015	0.002	0.002	0.002	0.0025	0.0025
0.019 to 0.017, incl	0.001	0.001	0.001	0.0015	0.0015	0.002	0.002	0.002
0.016 to 0.013, incl	0.001	0.001	0.001	0.0015	0.0015	0.0015	0.002	0.002
0.012	0.001	0.001	0.001	0.001	0.0015	0.0015	0.0015	0.0015
0.011	0.001	0.001	0.001	0.001	0.001	0.0015	0.0015	0.0015
0.010	0.001	0.001	0.001	0.001	0.001	0.001	0.0015	0.0015

[A] 1 in. = 25.4 mm.

TABLE 12 Permissible Crown of Cold-Rolled Strip

NOTE—Cold-rolled strip may be thicker at the middle than at the edges by the amounts specified in this table.

Specified Thickness, in. (mm)	Additional Thickness at Middle of Strip Over that Shown in Table 12 for Edge Measurement for Widths and Thicknesses Given, in. (mm)		
	To 5 (127), incl	Over 5 to 12 (127 to 305), incl	Over 12 to 24 (305 to 610), excl
0.005 to 0.010 (0.13 to 0.25), incl	0.00075 (0.019)	0.001 (0.25)	0.0015 (0.38)
Over 0.010 to 0.025 (0.25 to 0.64), incl	0.001 (0.25)	0.0015 (0.38)	0.002 (0.51)
Over 0.025 to 0.065 (0.64 to 1.65), incl	0.0015 (0.38)	0.002 (0.51)	0.0025 (0.64)
Over 0.065 to ³⁄₁₆ (1.65 to 4.76), excl	0.002 (0.51)	0.0025 (0.64)	0.003 (0.76)

TABLE 13 Permissible Variations in Width for Cold-Rolled Strip of Edge Numbers 1 and 5

Specified Edge No.	Width, in. (mm)	Thickness, in. (mm)	Permissible Variations in Width for Thicknesses and Widths Given, in. (mm)	
			Over	Under
1 and 5	9/32 (7.14) and under	1/16 (1.59) and under	0.005 (0.13)	0.005 (0.13)
	Over 9/32 to 3/4 (7.14 to 19.05), incl	3/32 (2.38) and under	0.005 (0.13)	0.005 (0.13)
	Over 3/4 to 5 (19.05 to 127), incl	1/8 (3.18) and under	0.005 (0.13)	0.005 (0.13)
5	Over 5 to 9 (127 to 229), incl	1/8 to 0.008 (3.18 to 0.203), incl	0.010 (0.25)	0.010 (0.25)
	Over 9 to 20 (229 to 508), incl	0.005 to 0.015 (0.127 to 0.38)	0.010 (0.25)	0.010 (0.25)
	Over 20 to 23 15/16 (508 to 608.0), incl	0.080 to 0.023 (2.03 to 0.58)	0.015 (0.38)	0.015 (0.38)

TABLE 14 Permissible Variations in Width for Cold-Rolled Strip of Edge Number 3

Specified Thickness, in.[A]	Permissible Variations in Width for Thicknesses and Widths Given, Over and Under, in.[A]					
	3/16, incl, to 1/2, excl	1/2 to 6, incl	Over 6 to 9, incl	Over 9 to 12, incl	Over 12 to 20, incl	Over 20 to 23 15/16, incl
Under 3/16 to 0.161, incl	...	0.016	0.020	0.020	0.031	0.031
0.160 to 0.100, incl	0.010	0.010	0.016	0.016	0.020	0.020
0.099 to 0.069, incl	0.008	0.008	0.010	0.010	0.016	0.020
0.068 and under	0.005	0.005	0.005	0.010	0.016	0.020

[A] 1 in. = 25.4 mm.

TABLE 15 Permissible Variations in Length of Cold-Rolled Strip

Specified Length, ft (m)	Permissible Variations over Specified Length, in. (mm)[A]
5 (1.52) and under	3/8 (9.52)
Over 5 to 10 (1.52 to 3.05), incl	1/2 (12.7)
Over 10 to 20 (3.05 to 6.10), incl	5/8 (15.88)

[A] No permissible variations are permitted under the specified length.

TABLE 16 Permissible Camber of Cold-Rolled Strip

NOTE—Camber is the deviation of a side edge from a straight line, and the measurement is taken by placing an 8-ft (243-cm) straightedge on the concave side of the strip and measuring the greatest distance between the strip edge and the straightedge.

Specified Width, in. (mm)	Permissible Camber per Unit Length of any 8 ft or 243 cm, in. (mm)
1½ (38.1) and under	½ (12.7)
Over 1½ to 24 (38.1 to 610), excl	¼ (6.35)

TABLE 17 Permissible Variations in Thickness of Plates

Specified Thickness, in. (mm)[A]	Permissible Variations Over Specified Thickness for Thicknesses and Widths Given, in. (mm)[B]	
	To 84 (2134), incl	Over 84 to 129 (2134 to 3277), incl
3/16 to 3/8 (4.76 to 9.52), excl	0.046 (1.17)	0.050 (1.27)
3/8 to 3/4 (9.52 to 19.0), excl	0.054 (1.37)	0.058 (1.47)
3/4 to 1 (19.0 to 25), excl	0.060 (1.52)	0.064 (1.63)
1 to 2 (25 to 51), incl[C]	0.070 (1.78)	0.074 (1.88)

[A] For plates up to 2 in. (51 mm), incl, in thickness the tolerance under the specified thickness is 0.01 in. (2.5 mm).

[B] For circles the over-thickness tolerances in this table apply to the diameter of the circle corresponding to the width ranges shown. For plates of irregular shape the over-thickness tolerances apply to the greatest width corresponding to the width ranges shown.

[C] For thickness tolerances for plates over 2 in. in thickness the manufacturer shall be consulted.

A 457

TABLE 18 Permissible Variations in Width and Length of Rectangular Sheared Mill Plates and Universal Mill Plates

Width, in.[A]	Length, in.[A]	Permissible Variations Over Specified Width and Length, in.[A,B]					
		Under 3/8 in Thickness		3/8 to 1/2, incl, in Thickness		Over 1/2 in Thickness[C]	
		Width	Length	Width	Length	Width	Length
48 and under	144 and under	1/8	3/16	3/16	1/4	5/16	3/8
Over 48 to 60, incl		3/16	1/4	1/4	5/16	3/8	7/16
Over 60 to 84, incl		1/4	5/16	5/16	3/8	7/16	1/2
Over 84 to 108, incl		5/16	3/8	3/8	7/16	1/2	9/16
Over 108		3/8	7/16	7/16	1/2	5/8	11/16
48 and under	Over 144 to 240	3/16	3/8	1/4	1/2	5/16	5/8
Over 48 to 60, incl		1/4	7/16	5/16	5/8	3/8	3/4
Over 60 to 84, incl		3/8	1/2	7/16	11/16	1/2	3/4
Over 84 to 108, incl		7/16	9/16	1/2	3/4	5/8	7/8
Over 108		1/2	5/8	5/8	7/8	11/16	1
48 and under	Over 240 to 360	1/4	1/2	5/16	5/8	3/8	3/4
Over 48 to 60, incl		5/16	5/8	3/8	3/4	1/2	3/4
Over 60 to 84, incl		7/16	11/16	1/2	3/4	5/8	7/8
Over 84 to 108, incl		9/16	3/4	5/8	7/8	3/4	1
Over 108		5/8	7/8	11/16	1	7/8	1
60 and under	Over 360 to 480	7/16	1 1/8	1/2	1 1/4	5/8	1 3/8
Over 60 to 84, incl		1/2	1 1/4	5/8	1 3/8	3/4	1 1/2
Over 84 to 108, incl		9/16	1 1/4	3/4	1 3/8	7/8	1 1/2
Over 108		3/4	1 3/8	7/8	1 1/2	1	1 5/8
60 and under	Over 480 to 600	7/16	1 1/4	1/2	1 1/2	5/8	1 5/8
Over 60 to 84, incl		1/2	1 3/8	5/8	1 1/2	3/4	1 5/8
Over 84 to 108, incl		5/8	1 3/8	3/4	1 1/2	7/8	1 5/8
Over 108		3/4	1 1/2	7/8	1 5/8	1	1 3/4
60 and under	Over 600	1/2	1 3/4	5/8	1 7/8	3/4	1 7/8
Over 60 to 84, incl		5/8	1 3/4	3/4	1 7/8	7/8	1 7/8
Over 84 to 108, incl		5/8	1 3/4	3/4	1 7/8	7/8	1 7/8
Over 108		7/8	1 3/4	1	2	1 1/8	2 1/4

[A] 1 in. = 25.4 mm.
[B] The tolerance under specified width and length is 1/4 in.
[C] Rectangular plates over 1 in. in thickness are not commonly sheared and are machined or otherwise cut to length and width or produced in the size as rolled, uncropped.

TABLE 19 Permissible Variations in Diameter of Circular Plates

Specified Diameter, in. (mm)	Permissible Variations Over the Specified Diameter for Diameters and Thicknesses Given, in. (mm)[A]		
	To 3/8 (9.52), excl	3/8 to 5/8 (9.52 to 15.88), excl	5/8 (15.88) and Over[B]
Under 60 (1524), excl	1/4 (6.35)	3/8 (9.52)	1/2 (12.7)
60 to 84 (1524 to 2134), excl	5/16 (7.94)	7/16 (11.11)	9/16 (14.29)
84 to 108 (2134 to 2743), excl	3/8 (9.52)	1/2 (12.7)	5/8 (15.88)
108 to 130 (2743 to 3302), excl	7/16 (11.11)	9/16 (14.29)	11/16 (17.46)

[A] No permissible variations are permitted under the specified diameter.
[B] Circular and sketch plates over 5/8 in. (15.88 mm) in thickness are not commonly sheared and are machined or otherwise cut.

TABLE 20 Permissible Variations in Flatness of Annealed Plates

Note 1—Tolerances in this table apply to plates up to 15 ft in length, or to any 15 ft of longer plates.
Note 2—If the longer dimension is under 36 in., the tolerance is not greater than ¼ in.
Note 3—The shorter dimension specified is considered the width, and the flatness deviation across the width does not exceed the tabular amount for that dimension.
Note 4—The maximum deviation from a horizontal flat surface does not customarily exceed the tolerance for the longer dimension specified.

Specified Thickness, in.[A]	Permissible Variations in Flatness (Deviation from a Horizontal Flat Surface), for Thicknesses and Widths Given, in.[A]								
	48 and under	Over 48 to 60, excl	60 to 72, excl	72 to 84, excl	84 to 96, excl	96 to 108, excl	108 to 120, excl	120 to 144, excl	144 and over
3/16 to ¼, excl	¾	1 1/16	1¼	1 3/8	1 5/8	1 5/8	1 7/8	2	...
¼ to 3/8, excl	11/16	¾	15/16	1 1/8	1 3/8	1 7/16	1 9/16	1 7/8	...
3/8 to ½, excl	½	9/16	11/16	¾	15/16	1 1/8	1¼	1 7/16	1¾
½ to ¾, excl	½	9/16	5/8	5/8	13/16	1 1/8	1 1/8	1 1/8	1 3/8
¾ to 1, excl	½	9/16	5/8	5/8	¾	13/16	15/16	1	1 1/8
1 to 1½, excl	½	9/16	9/16	9/16	11/16	11/16	11/16	¾	1
1½ to 4, excl	3/16	5/16	3/8	7/16	½	9/16	5/8	¾	7/8
4 to 6, excl	¼	3/8	½	9/16	5/8	¾	7/8	1	1 1/8

[A] 1 in. = 25.4 mm.

TABLE 21 Permissible Camber of Sheared Mill and Universal Mill Plates

Permissible camber = 1/8 in. × number of feet of length/5

The American Society for Testing and Materials takes no position respecting the validity of any patent rights asserted in connection with any item mentioned in this standard. Users of this standard are expressly advised that determination of the validity of any such patent rights, and the risk of infringement of such rights, are entirely their own responsibility.

This standard is subject to revision at any time by the responsible technical committee and must be reviewed every five years and if not revised, either reapproved or withdrawn. Your comments are invited either for revision of this standard or for additional standards and should be addressed to ASTM Headquarters. Your comments will receive careful consideration at a meeting of the responsible technical committee, which you may attend. If you feel that your comments have not received a fair hearing you should make your views known to the ASTM Committee on Standards, 1916 Race St., Philadelphia, Pa. 19103.

Designation: A 478 – 82

Standard Specification for
CHROMIUM-NICKEL STAINLESS AND HEAT-RESISTING STEEL WEAVING WIRE[1]

This standard is issued under the fixed designation A 478; the number immediately following the designation indicates the year of original adoption or, in the case of revision, the year of last revision. A number in parentheses indicates the year of last reapproval. A superscript epsilon (ϵ) indicates an editorial change since the last revision or reapproval.

NOTE—The deletion of improperly added 309Cb and 310Cb in Table 1 was editorially corrected. Table 1 was rearranged by UNS number.

1. Scope

1.1 This specification covers the more commonly used types of round stainless and heat-resisting steel wire intended especially for weaving.

1.2 The values stated in inch-pound units are to be regarded as the standard.

2. Applicable Documents

2.1 *ASTM Standards:*
A 555 Specification for General Requirements for Stainless and Heat-Resisting Steel Wire[2]
A 751 Methods, Practices, and Definitions for Chemical Analysis of Steel Products[2]

3. General Requirements for Delivery

3.1 In addition to the requirements of this specification, all requirements of the current edition of Specification A 555 shall apply.

4. Ordering Information

4.1 Orders for material under this specification shall include the following information:
4.1.1 Quantity (weight),
4.1.2 Name of material (stainless steel),
4.1.3 Condition (see Section 5),
4.1.4 Finish (see Section 6),
4.1.5 Cross section (round),
4.1.6 Form (wire),
4.1.7 Applicable dimensions (diameter),
4.1.8 Type designation (see Table 1),
4.1.9 ASTM designation and date of issue, and
4.1.10 Special requirements.
4.2 If possible, the intended end use of the item should be given on the purchase order especially when the item is ordered for a specific end use or uses.

NOTE 1—A typical ordering description is as follows: 1000 lb; stainless steel, dead soft, bright annealed wire, 0.015 in. diameter, spools, Type 304, ASTM Specification A 478 dated _____, End Use Wire Screen.

5. Conditions

5.1 Wire may be furnished in one of the following conditions:
5.1.1 Annealed,
5.1.2 Bright annealed, or
5.1.3 Cold drawn.

6. Finish

6.1 The types of finish procurable are as follows:
6.1.1 Pickled finish, and
6.1.2 Bright finish.

7. Chemical Requirements

7.1 The steel shall conform to the requirements as to chemical composition specified in Table 1.

7.2 Methods and practices relating to chemical analysis required by this specification shall be in accordance with Methods, Practices, and Definitions A 751.

---

[1] This specification is under the jurisdiction of ASTM Committee A-1 on Steel, Stainless Steel and Related Alloys, and is the direct responsibility of Subcommittee A01.08 on Wrought Stainless Steel Products.
Current edition approved July 30, 1982. Published December 1982. Originally published as A 478 – 62 T. Last previous edition A 478 – 81.

[2] *Annual Book of ASTM Standards*, Vol 01.03.

8. Mechanical Requirements

8.1 The material shall conform to the requirements as to mechanical properties specified in Table 2.

9. Packaging

9.1 Each coil or spool shall be one continuous length of wire. Each coil shall be firmly tied and each spool shall be tightly wound. Unless otherwise specified, coils shall be placed in drums or shall be paper wrapped, and spools shall be boxed in such a manner as to assure safe delivery to their destination when properly transported by any common carrier.

TABLE 1 Chemical Requirements

UNS Designation[A]	Type	Composition, %									
		Carbon, max	Manganese, max	Phosphorus, max	Sulfur, max	Silicon, max	Chromium	Nickel	Molybdenum	Nitrogen, max	Other Elements
S 30200	302	0.15	2.00	0.045	0.030	1.00	17.00–19.00	8.00–10.00	...	0.10	...
S 30400	304	0.08	2.00	0.045	0.030	1.00	18.00–20.00	8.00–10.50	...	0.10	...
S 30403	304L	0.03	2.00	0.045	0.030	1.00	18.00–20.00	8.00–12.00	...	0.10	...
S 30500	305	0.12	2.00	0.045	0.030	1.00	17.00–19.00	10.50–30.00	...	0.10	...
S 30940	309Cb	0.08	2.00	0.045	0.030	1.00	22.00–24.00	12.00–16.00	...	0.10	Cb + Ta 10 × C min, 1.10 max
S 31040	310Cb	0.08	2.00	0.045	0.030	1.50	24.00–26.00	19.00–22.00	...	0.10	Cb + Ta 10 × C min, 1.10 max
S 31600	316	0.08	2.00	0.045	0.030	1.00	16.00–18.00	10.00–14.00	2.00–3.00	0.10	...
S 31640	316Cb	0.08	2.00	0.045	0.030	1.00	16.00–18.00	10.00–14.00	2.00–3.00	0.10	Cb + Ta 10 × C min, 1.10 max
S31603	316L	0.03	2.00	0.045	0.030	1.00	16.00–18.00	10.00–14.00	2.00–3.00	0.10	...
S 31635	316Ti	0.08	2.00	0.045	0.030	1.00	16.00–18.00	10.00–14.00	2.00–3.00	0.10	Ti 5 × (C + N) min, 0.70 max
S 3170	317	0.08	2.00	0.045	0.030	1.00	18.00–20.00	11.00–15.00	3.00–4.00	0.10	...

TABLE 2 Mechanical Requirements

Condition	Diameter, in. (mm)	Tensile Strength, psi (MPa)	Elongation in 10 in. or 254 mm, min, %
Annealed or bright annealed[A]	0.002 (0.05) to 0.005 (0.13), incl	145 000 (1000), max	30
	Over 0.005 (0.13) to 0.009 (0.23), incl	135 000 (930), max	30
	Over 0.009 (0.23) to 0.015 (0.38), incl	130 000 (900), max	35
	Over 0.015 (0.38) to 0.020 (0.51), incl	125 000 (860), max	40
	Over 0.020 (0.51) to 0.025 (0.64), incl	120 000 (830), max	40
	Over 0.025 (0.64) to 0.035 (0.89), incl	115 000 (790), max	40
	Over 0.035 (0.89) to 0.043 (1.09), incl	110 000 (760), max	45
	Over 0.043 (1.09)	105 000 (720), max	45
Cold drawn[B]	0.030 (0.76) to 0.125 (3.18), incl	120 000 (830) to 150 000 (1030)	15
	Over 0.125 (3.18)	110 000 (760) to 140 000 (970)	15

[A] In the annealed or bright annealed condition, for Type 302 and Type 304, tensile strength maximum is 10 000 psi (70 MPa) higher.

[B] Wire ordered in the cold-drawn condition can be supplied to higher tensile strength levels as specified by the purchaser.

The American Society for Testing and Materials takes no position respecting the validity of any patent rights asserted in connection with any item mentioned in this standard. Users of this standard are expressly advised that determination of the validity of any such patent rights, and the risk of infringement of such rights, are entirely their own responsibility.

This standard is subject to revision at any time by the responsible technical committee and must be reviewed every five years and if not revised, either reapproved or withdrawn. Your comments are invited either for revision of this standard or for additional standards and should be addressed to ASTM Headquarters. Your comments will receive careful consideration at a meeting of the responsible technical committee, which you may attend. If you feel that your comments have not received a fair hearing you should make your views known to the ASTM Committee on Standards, 1916 Race St., Philadelphia, Pa. 19103.

Designation: A 480/A 480M – 83a

Standard Specification for
GENERAL REQUIREMENTS FOR FLAT-ROLLED STAINLESS AND HEAT-RESISTING STEEL PLATE, SHEET, AND STRIP[1]

This standard is issued under the fixed designation A 480/A 480M – 83a; the number immediately following the designation indicates the year of original adoption or, in the case of revision, the year of last revision. A number in parentheses indicates the year of last reapproval. A superscript epsilon (ϵ) indicates an editorial change since the last revision or reapproval.

This specification has been approved for use by agencies of the Department of Defense and for listing in the DoD Index of Specifications and Standards.

1. Scope

1.1 This specification[2] covers a group of general requirements which, unless otherwise specified in the purchase order or in an individual specification, shall apply to rolled steel plate, sheet, and strip, under each of the following specifications issued by ASTM: Specifications A 167, A 176, A 177, A 240, and A 412.

1.2 In case of any conflicting requirements, the requirements of the purchase order, the individual material specification, and this general specification shall prevail in the sequence named.

1.3 The values stated in either inch-pound units or SI units are to be regarded separately as standard. Within the text, the SI units are shown in brackets. The values stated in each system are not exact equivalents; therefore, each system must be used independently of the other. Combining values from the two systems may result in nonconformance with the specification.

1.4 This specification and the applicable material specifications are expressed in both inch-pound units and SI units. However, unless the order specifies the applicable "M" specification designation (SI units), the material shall be furnished in inch-pound units.

2. Applicable Documents

2.1 *ASTM Standards:*
A 167 Specification for Stainless and Heat-Resisting Chromium-Nickel Steel Plate, Sheet, and Strip[3]
A 176 Specification for Stainless and Heat-Resisting Chromium Steel Plate, Sheet, and Strip[3]
A 177 Specification for High-Strength Stainless and Heat-Resisting Chromium-Nickel Steel Sheet and Strip[3]
A 240 Specification for Heat-Resisting Chromium and Chromium-Nickel Stainless Steel Plate, Sheet, and Strip for Fusion-Welded Unfired Pressure Vessels[3]
A 262 Practices for Detecting Susceptibility to Intergranular Attack in Austenitic Stainless Steels[3]
A 370 Methods and Definitions for Mechanical Testing of Steel Products[4]
A 412 Specification for Stainless and Heat-Resisting Chromium-Nickel-Manganese Steel Plate, Sheet, and Strip[3]
A 700 Practices for Packaging, Marking, and Loading Methods for Steel Products for Domestic Shipment[4]
A 751 Methods, Practices, and Definitions for Chemical Analysis of Steel Products[4]
E 140 Hardness Conversion Tables for Metals (Relationship Between Brinell Hardness, Vickers Hardness, Rockwell Hardness, Rockwell Superficial Hardness, and Knoop Hardness)[5]

---

[1] This specification is under the jurisdiction of ASTM Committee A-1 on Steel, Stainless Steel and Related Alloys, and is the direct responsibility of Subcommittee A01.17 on Flat Stainless Steel Product.
Current edition approved Oct. 28, 1983. Published December 1983. Originally published as A 480 – 62 T. Last previous edition A 480 – 83.
[2] For ASME Boiler and Pressure Vessel Code applications see related Specification SA – 48 in Section II of that Code.
[3] *Annual Book of ASTM Standards,* Vol 01.05.
[4] *Annual Book of ASTM Standards,* Vol 01.04.
[5] *Annual Book of ASTM Standards,* Vol 03.01.

2.2 *Federal Standard:*
Fed. Std. No. 123 Marking for Shipment (Civil Agencies)[6]

2.3 *Military Standards:*
MIL-STD-129 Marking for Shipment and Storage[6]
MIL-STD-163 Steel Mill Products, Preparation for Shipment and Storage[6]

3. Definitions

3.1 Plate, sheet, and strip as used in this specification apply to the following:

3.1.1 *plate*—material $3/16$ in. [5.00 mm] and over in thickness and over 10 in. [250 mm] in width. Finishes are shown in Section 9 for Plate.

3.1.2 *sheet*—material under $3/16$ in. [5.00 mm] in thickness and 24 in. [600 mm] and over in width. Finishes are shown in Section 7 for Sheet.

3.1.3 *strip*—cold-rolled material under $3/16$ in. [5.00 mm] in thickness and under 24 in. [600 mm] in width. Finishes are detailed in Section 8 for Strip, and strip edges in Section 10 for Cold-Rolled Strip.

4. Ordering Information

4.1 Orders for material under this specification shall include the following information:

4.1.1 Quantity (weight and number of pieces),

4.1.2 Name of material (stainless steel),

4.1.3 Condition (hot-rolled, cold-rolled, annealed, heat-treated),

4.1.4 Finish (see Section 7 for Sheet, Section 8 for Strip, and Section 9 for Plates). In the case of polished finishes, specify whether one or both sides are to be polished,

4.1.5 Form (plate, sheet, or strip),

4.1.6 Dimensions (thickness, width, length),

4.1.7 Edge, strip only (see Section 10 for Cold-Rolled Strip),

4.1.8 Type,

4.1.9 Specification designation and date of issue,

4.1.10 Additions to specification or special requirements,

4.1.11 Restrictions (if desired) on methods for determining yield strength (see appropriate footnote to mechanical properties table of the basic material specification),

4.1.12 Marking requirements (see Section 18), and

4.1.13 Preparation for delivery (see Section 18).

5. Heat Analysis

5.1 An analysis of each heat shall be made by the steel producer to determine the percentages of the elements specified in the applicable material specification. This analysis shall be made from a test sample taken during the pouring of the heat. The chemical composition thus determined shall conform to the applicable material specification.

5.2 Methods and practices relating to chemical analysis shall be in accordance with Methods, Practices, and Definitions A 751.

6. Product Analysis

6.1 Product analysis (formerly check analysis) may be made by the purchaser to verify the identity of the finished material representing each heat or lot. Such analysis may be made by any of the commonly accepted methods which will positively identify the material.

6.2 The chemical composition determined in accordance with 6.1 shall conform to the limits of the material specification within the tolerances of Table 1, unless otherwise specified in the applicable material specification or the purchase order. The allowable variation of a particular element in a single sample for product analysis may be either above or below the specified range. However, percentages must exhibit the same tendencies in all samples; that is, the several determinations of any individual element in a heat may not vary both above and below the specified range.

7. Finish for Sheet

7.1 The types of finish available on sheet products are:

7.1.1 *No. 1 Finish*—Hot-rolled, annealed, and descaled.

7.1.2 *No. 2D Finish*—Cold-rolled, dull finish.

7.1.3 *No. 2B Finish*—Cold-rolled, bright finish.

7.1.3.1 *Bright Annealed Finish*—A bright cold-rolled finish retained by final annealing in a controlled atmosphere furnace.

7.1.4 *No. 3 Finish*—Intermediate polished finish, one or both sides.

7.1.5 *No. 4 Finish*—General purpose polished finish, one or both sides.

---

[6] Available from Naval Publications and Forms Center, 5801 Tabor Ave., Philadelphia, Pa. 19120.

7.1.6 *No. 6 Finish*—Dull satin finish, Tampico brushed, one or both sides.

7.1.7 *No. 7 Finish*—High luster finish.

7.1.8 *No. 8 Finish*—Mirror finish.

NOTE 2—*Explanation of Finish:*
No. 1—Produced on hand sheet mills by hot rolling to specified thicknesses followed by annealing and descaling. Generally used in industrial applications, such as for heat and corrosion resistance, where smoothness of finish is not of particular importance.

No. 2D—Produced on either hand sheet mills or continuous mills by cold rolling to the specified thickness, annealing, and descaling. The dull finish may result from the descaling or pickling operation or may be developed by a final light cold-rolled pass on dull rolls. The dull finish is favorable for retention of lubricants on the surface in deep drawing operations. This finish is generally used in forming deep-drawn articles which may be polished after fabrication.

No. 2B—Commonly produced the same as 2D, except that the annealed and descaled sheet receives a final light cold-rolled pass on polished rolls. This is a general purpose cold-rolled finish. It is commonly used for all but exceptionally difficult deep drawing applications. This finish is more readily polished than No. 1 or No. 2D Finish.

Bright Annealed Finish is a bright cold-rolled highly reflective finish retained by final annealing in a controlled atmosphere furnace. The purpose of the atmosphere is to prevent scaling or oxidation during annealing. The atmosphere is usually comprised of either dry hydrogen or a mixture of dry hydrogen and dry nitrogen (sometimes known as dissociated ammonia).

No. 3—For use as a finish-polished surface or as a semifinished-polished surface when it is required to receive subsequent finishing operations following fabrication. Where sheet or articles made from it will not be subjected to additional finishing or polishing operations, No. 4 finish is recommended.

No. 4—Widely used for restaurant equipment, kitchen equipment, store fronts, dairy equipment, etc. Following initial grinding with coarser abrasives, sheets are generally finished last with abrasives approximately 120 to 150 mesh.

No. 6—Has a lower reflectivity than No. 4 finish. It is produced by Tampico brushing No. 4 finish sheets in a medium of abrasive and oil. It is used for architectural applications and ornamentation where high luster is undesirable; it is also used effectively to contrast with brighter finishes.

No. 7—Has a high degree of reflectivity. It is produced by buffing a finely ground surface, but the grit lines are not removed. It is chiefly used for architectural or ornamental purposes.

No. 8—The most reflective finish that is commonly produced. It is obtained by polishing with successively finer abrasives and buffing extensively with very fine buffing rouges. The surface is essentially free of grit lines from preliminary grinding operations. This finish is most widely used for press plates, as well as for small mirrors and reflectors.

7.1.9 Sheets can be produced with one or two sides polished. When polished on one side only, the other side may be rough ground in order to obtain the necessary flatness.

8. Finish for Strip

8.1 The various types of finish procurable on cold-rolled strip products are:

8.1.1 *No. 1 Finish*—Cold rolled to specified thickness, annealed, and descaled.

8.1.2 *No. 2 Finish*—Same as No. 1 Finish, followed by a final light cold-roll pass, generally on highly polished rolls.

8.1.3 *Bright Annealed Finish*—A bright cold-rolled finish retained by final annealing in a controlled atmosphere furnace.

8.1.3.1 *Polished Finish*—Stainless steel strip is also available in polished finishes such as No. 3 and No. 4, which are explained in Note 2.

NOTE 3—*Explanation of Finish:*
No. 1—Appearance of this finish varies from dull gray matte finish to a fairly reflective surface, depending largely upon composition. This finish is used for severely drawn or formed parts, as well as for applications where the brighter No. 2 Finish is not required, such as parts for heat resistance.

No. 2—This finish has a smoother and more reflective surface, the appearance of which varies with composition. This is a general purpose finish, widely used for household and automotive trim, tableware, utensils, trays, etc.

Bright Annealed Finish is a bright cold-rolled highly reflective finish retained by final annealing in a controlled atmosphere furnace. The purpose of the atmosphere is to prevent scaling or oxidation during annealing. The atmosphere is usually comprised of either dry hydrogen or a mixture of dry hydrogen and dry nitrogen (sometimes known as dissociated ammonia).

9. Finish for Plates

9.1 The types of finish available on plates are:

9.1.1 *Hot-Rolled, or Cold-Rolled and Annealed, or Heat Treated*—Scale not removed, an intermediate finish. Use of plates in this condition is generally confined to heat-resisting applications, scale impairs corrosion resistance.

9.1.2 *Hot-Rolled, or Cold-Rolled and Annealed, or Heat Treated and Blast Cleaned, or Pickled*—Condition and finish commonly preferred for corrosion-resisting and most heat-resisting applications, essentially a No. 1 Finish.

9.1.3 *Hot-Rolled, or Cold-Rolled and Annealed or Heat Treated, and Surface Cleaned and Polished*—Polish finish is generally No. 4 Finish.

9.1.4 *Hot-Rolled or Cold-Rolled, and Annealed or Heat Treated, and Descaled and Temper Passed—Smoother finish for specialized applications.

9.1.5 *Hot-Rolled, or Cold-Rolled and Annealed, or Heat Treated and Descaled and Cold-Rolled and Annealed, or Heat Treated and Descaled and Optionally Temper Passed*—Smooth finish with greater freedom from surface imperfections than the above.

10. Edges for Cold-Rolled Strip

10.1 The types of edges available on strip products are:

10.1.1 *No. 1 Edge*—A rolled edge, either round or square as specified.

10.1.2 *No. 3 Edge*—An edge produced by slitting.

10.1.3 *No. 5 Edge*—An approximately square edge produced by rolling or filing after slitting.

11. Test Specimens

11.1 Tension test specimens shall be taken from finished material and shall be selected in either or both longitudinal and transverse direction. The tension test specimen shall conform to the appropriate Sections 7, 8, or 9 of Methods A 370.

11.1.1 *Specimens for Specific Products (Sheet, Strip, and Plate)*—In testing sheet, strip, and plate, one of the following types of specimens shall be used:

11.1.1.1 For material ranging in nominal thickness from 0.005 [0.15 mm] to ⅝ in. [15.00 mm] (Note 4), the sheet type specimen described in Section 8 of Methods A 370.

11.1.1.2 For material having a nominal thickness of ³⁄₁₆ in. [5.00 mm] or over (Note 5), the plate type specimen described in Section 7 of Methods A 370. For plate material up to and including ¾ in. [20.00 mm] thick, the sheet-type tension specimen described in Section 8 (Fig. 4) of Methods A 370 is also permitted.

11.1.1.3 For material having a nominal thickness of ½ in. [15 mm] or over (Note 4), the largest practical size of specimen described in Section 9 of Methods A 370.

NOTE 4—Either of the flat specimens described in Sections 7 and 8 of Methods A 370 may be used for material from ³⁄₁₆ in. [5.00 mm] to ¾ in. [20.00 mm] in thickness, and one of the round specimens described in Section 9 of Methods A 370 may also be used for material ½ in. [15.00 mm] or more in thickness.

11.1.1.4 Transverse bend test specimens from sheet and strip shall be the full thickness of the material and approximately 1 in. [25 mm] in width. The edges of the test specimens may be rounded to a radius equal to one half the thickness.

11.1.1.5 The width of strip for which bend tests can be made is subject to practical limitations on the length of the bend test specimen. For narrow strip, the following widths can be tested:

Strip thickness, in. (mm)	Minimum Strip Width and Minimum Specimen Length for Transverse Bend Tests, in. (mm)
0.100 [2.50] and under	½ [15.00]
Over 0.100 [2.50] to 0.140 [3.50], excl	1 [25.00]
0.140 [3.50] and over	1½ [40.00]

11.1.1.6 Bend test specimens taken from plates shall be in full thickness of the material up to and including ½ in. [15.00 mm] in thickness, of suitable length, and between 1 and 2 in. [25.00 and 50.00 mm] in width. The sheared edges may be removed to a depth of at least ⅛ in. [3.00 mm] and the sides may be smoothed with a file. The corners of the cross section of the specimen may be broken with a file, but no appreciable rounding of corners may be permitted.

11.1.1.7 In the case of plates over ½ in. [15.00 mm] in thickness, bend test specimens machined to 1 by ½ in. [25.00 by 15.00 mm] in cross section and at least 6 in. [150.00 mm] in length may be used. The edges may be rounded to a ¹⁄₁₆-in. [1.50-mm] radius. When permitted upon agreement between the manufacturer and the purchaser, the cross section may be modified to ½ in. [15.00 mm] square. The center of cross section of the test specimen shall be halfway between center and the outer surface of the plate, unless otherwise specified.

11.1.1.8 Hardness tests may be made on the grip ends of the tension specimens before they are subjected to the tension test.

12. Number of Tests

12.1 In the case of sheet, strip, or plate produced in coil form, two or more hardness tests (one from each end of the coil); one bend test, when required; and one or more tension tests shall be made on specimens taken from each coil. If the hardness difference between the two ends of the coil exceeds 5 HRB, or equivalent, tensile properties must be determined on both ends.

12.2 One intergranular corrosion test, when required, shall be selected from each heat and thickness subjected to the same heat treatment practice. Such specimens may be obtained from specimens selected for mechanical testing.

13. Special Tests

13.1 If other tests which are thought to be pertinent to the intended application of the material covered by the purchase specification are required, the methods and limits shall be agreed upon between the seller and the purchaser and specified in the purchase order.

14. Test Methods

14.1 The properties enumerated in applicable specifications shall be determined in accordance with the following ASTM methods.

14.1.1 *Tension Tests*—Methods A 370.
14.1.2 *Brinell Hardness*—Methods A 370.
14.1.3 *Rockwell Hardness*—Methods A 370.
14.1.3.1 *Hardness Equivalents*—Standard E 140.
14.1.4 *Intergranular Corrosion (When Required)*—Practice E of Practices A 262.

15. Retests

15.1 If any test specimen shows defective machining or develops flaws, it may be discarded and another specimen substituted.

15.1.1 If the percentage of elongation of any tension test specimen is less than that specified and any part of the fracture is more than ¾ in. [20.00 mm] from the center of the gage length of a 2-in. [50.00-mm] specimen or is outside the middle half of the gage length of an 8-in. [200.00-mm] specimen, as indicated by scribe marks placed on the specimen before testing, a retest shall be allowed.

15.1.2 If a bend test specimen fails, due to conditions of bending more severe than required by the specification, a retest shall be permitted, either on a duplicate specimen or on a remaining portion of the failed specimen.

15.1.3 If the results of any test lot are not in conformance with the requirements of the applicable material specification, such lots may be retreated at the option of the producer. The material shall be acceptable if the results of retests on retreated material are within the specified requirements.

15.1.4 If the product analysis fails to conform to the specified limits, analysis shall be made on a new sample. The results of this retest shall be within the specified requirements.

16. Permissible Variations in Dimensions and Weight

16.1 *Sheet*—Sheets shall conform to the permissible variations in dimensions specified in Tables A1.2 [A2.1] to A1.11 [A2.10], incl.

16.2 *Cold-Rolled Strip*—Cold-rolled strip shall conform to the permissible variations in dimensions specified in Tables A1.12 [A2.11] to A1.16 [A2.15], incl.

16.3 *Plates*—Plates shall conform to the permissible variations in dimensions specified in Tables A1.17 [A2.16] to A1.23 [A2.21], incl.

16.4 *Sheet, Strip, and Plate*—Material with No. 1 finish may be ground to remove surface defects, provided such grinding does not reduce the thickness or width at any point beyond the permissible variations in dimensions.

17. Workmanship

17.1 The material shall be of uniform quality consistent with good manufacturing and inspection practices. The steel quality shall be satisfactory for the production or fabrication of finished parts.

18. Packaging, Marking, and Loading

18.1 *For Commercial Procurement*—Unless otherwise specified in the individual specification, packaging, marking, and loading shall be in accordance with the procedures recommended by Practices A 700.

18.2 *For U.S. Government Procurement:*

18.2.1 When specified in the contract or order, and for direct procurement by or direct shipment to the Government, marking for shipment shall be in accordance with Fed. Std. No. 123 for civil agencies and MIL-STD-129 for military agencies.

18.2.2 When specified in the contract or order, material shall be preserved, packaged, and packed in accordance with the requirements of MIL-STD-163. The applicable levels shall be as specified in the contract or order.

19. Inspection

19.1 Inspection of the material by the purchaser's representative at the producing plant shall be made as agreed upon between the purchaser and the seller as part of the purchase order.

19.2 Unless otherwise specified in the contract or purchase order: (*1*) the seller is responsible for the performance of all inspection and test requirements in this specification, (*2*) the seller may use his own or other suitable facilities for the performance of the inspection and testing, and (*3*) the purchaser shall have the right to perform any of the inspection and tests set forth in this specification. The manufacturer shall afford the purchaser's inspector all reasonable facilities necessary to satisfy him that the material is being furnished in accordance with the specification. Inspection by the purchaser shall not interfere unnecessarily with the manufacturer.

20. Rejection

20.1 Unless otherwise specified, any rejection based on tests made in accordance with this specification shall be reported to the seller within 60 working days from the receipt of the material by the purchaser.

20.2 Material that shows injurious defects subsequent to its acceptance at the seller's works will be rejected and the seller shall be notified.

21. Rehearing

21.1 Samples tested in accordance with this specification that represent rejected material shall be retained for 3 weeks from the date of the notification to the seller of the rejection. In case of dissatisfaction with the results of the test, the seller may make claim for a rehearing within that time.

22. Material Test Report and Certification

22.1 A report of the results of all tests required by the applicable material specification shall be supplied to the purchaser.

22.2 Upon request of the purchaser, the test report shall include the manufacturer's certification that the material was manufactured and tested in accordance with the applicable specification.

22.3 Upon request of the purchaser in the contract or order the test report and certification shall be furnished at the time of shipment.

ANNEXES

(Mandatory Information)

A1. PERMISSIBLE VARIATIONS IN DIMENSIONS, ETC.—INCH-POUND UNITS

A1.1 Listed in Annex A1 are tables showing the permissible variations in dimensions expressed in inch-pound units of measurement.

TABLE A1.1 Chemical Requirements (Product Analysis Tolerances)[A]

Elements	Limit or Maximum of Specified Range, %	Tolerance Over the Maximum Limit or Under the Minimum Limit	Elements	Limit or Maximum of Specified Range, %	Tolerance Over the Maximum Limit or Under the Minimum Limit
Carbon	to 0.010, incl	0.002	Titanium	to 1.00, incl	0.05
	over 0.010 to 0.030, incl	0.005		over 1.00 to 3.00, incl	0.07
	over 0.030 to 0.20, incl	0.01			
	over 0.20 to 0.60, incl	0.02	Cobalt	over 0.05 to 0.50, incl	0.01[B]
	over 0.60 to 1.20, incl	0.03		over 0.50 to 2.00, incl	0.02
				over 2.00 to 5.00, incl	0.05
Manganese	to 1.00, incl	0.03			
	over 1.00 to 3.00, incl	0.04	Columbium plus tantalum	to 1.50, incl	0.05
	over 3.00 to 6.00, incl	0.05			
	over 6.00 to 10.00, incl	0.06			
	over 10.00 to 15.00, incl	0.10			
	over 15.00 to 20.00, incl	0.15	Tantalum	to 0.10, incl	0.02
Phosphorus	to 0.040, incl	0.005	Copper	to 0.50, incl	0.03
	over 0.040 to 0.20, incl	0.010		over 0.50 to 1.00, incl	0.05
				over 1.00 to 3.00, incl	0.10
Sulfur	to 0.040, incl	0.005		over 3.00 to 5.00, incl	0.15
	over 0.040 to 0.20, incl	0.010		over 5.00 to 10.00, incl	0.20
	over 0.20 to 0.50, incl	0.020			
			Aluminum	to 0.15, incl	−0.005, +0.01
Silicon	to 1.00, incl	0.05		over 0.15 to 0.50, incl	0.05
	over 1.00 to 3.00, incl	0.10		over 0.50 to 2.00, incl	0.10
Chromium	over 4.00 to 10.00, incl	0.10			
	over 10.00 to 15.00, incl	0.15	Nitrogen	to 0.02, incl	0.005
	over 15.00 to 20.00, incl	0.20		over 0.02 to 0.19, incl	0.01
	over 20.00 to 30.00, incl	0.25		over 0.19 to 0.25, incl	0.02
				over 0.25 to 0.35, incl	0.03
Nickel	to 1.00, incl	0.03		over 0.35 to 0.45, incl	0.04
	over 1.00 to 5.00, incl	0.07			
	over 5.00 to 10.00, incl	0.10	Tungsten	to 1.00, incl	0.03
	over 10.00 to 20.00, incl	0.15		over 1.00 to 2.00, incl	0.05
	over 20.00 to 30.00, incl	0.20			
			Vanadium	to 0.50, incl	0.03
Molybdenum	over 0.20 to 0.60, incl	0.03		over 0.50 to 1.50, incl	0.05
	over 0.60 to 2.00, incl	0.05			
	over 2.00 to 7.00, incl	0.10	Selenium	all	0.03

[A] This table does not apply to heat analysis.
[B] Product analysis limits for cobalt under 0.05 % have not been established, and the manufacturer should be consulted for those limits.

TABLE A1.2 Permissible Variations in Thickness for Hot-Rolled Sheets in Cut Lengths, Cold-Rolled Sheet in Cut Lengths and Coils

Specified Thickness,[A] in. [mm]	Permissible Variations, Over and Under[B]	
	in.	mm
Over 0.145 [3.68] to less than 3/16 [4.76]	0.014	0.36
Over 0.130 [3.30] to 0.145 [3.68], incl	0.012	0.30
Over 0.114 [2.90] to 0.130 [3.30], incl	0.010	0.25
Over 0.098 [2.49] to 0.114 [2.90], incl	0.009	0.23
Over 0.083 [2.11] to 0.098 [2.49], incl	0.008	0.20
Over 0.072 [1.83] to 0.083 [2.11], incl	0.007	0.18
Over 0.058 [1.47] to 0.072 [1.83], incl	0.006	0.15
Over 0.040 [1.02] to 0.058 [1.47], incl	0.005	0.13
Over 0.026 [0.66] to 0.040 [1.02], incl	0.004	0.10
Over 0.016 [0.41] to 0.026 [0.66], incl	0.003	0.08
Over 0.007 [0.18] to 0.016 [0.41], incl	0.002	0.05
Over 0.005 [0.13] to 0.007 [0.18], incl	0.0015	0.04
0.005 [0.13]	0.001	0.03

[A] Thickness measurements are taken at least 3/8 in. [9.52 mm] from the edge of the sheet.
[B] Cold-rolled sheets in cut lengths and coils are produced in some type numbers and some widths and thickness to tolerances less than those shown in the table.

TABLE A1.3 Permissible Variations in Width and Length for Hot-Rolled and Cold-Rolled Resquared Sheets (Stretcher Leveled Standard of Flatness)

NOTE—Polished sheets with Finishes No. 4 and higher are produced to tolerances given in this table.

Specified Dimensions, in. [mm]	Tolerances		
	Over		Under
	in.	mm	
For thicknesses under 0.131 [3.33]:			
Widths up to 48 [1219] excl	1/16	1.59	0
Widths 48 [1219] and over	1/8	3.18	0
Lengths up to 120 [3048] excl	1/16	1.59	0
Lengths 120 [3048] and over	1/8	3.18	0
For thicknesses 0.131 [3.33] and over:			
All widths and lengths	1/4	6.35	0

TABLE A1.4 Permissible Variations in Weight for Hot-Rolled and Cold-Rolled Sheets

Any item of five sheets or less, or any item estimated to weigh 200 lb [90.72 kg] or less, may actually weigh as much as 10 % over the theoretical weight	weigh 200 lb [90.72 kg] or less
Any item of more than five sheets and estimated to weigh more than 200 lb [90.72 kg], may actually weigh as much as 7½ % over the theoretical weight	weigh more than 200 lb [90.72 kg]
Chromium-manganese-nickel	40.7 lb/ft²·in. thickness [7.82 kg/m²·mm thick]
Chromium-nickel	42.0 lb/ft²·in. thickness [8.07 kg/m²·mm thick]
Chromium	41.2 lb/ft²·in. thickness [7.92 kg/m²·mm thick]

TABLE A1.5 Permissible Variations in Width for Hot-Rolled and Cold-Rolled Sheets not Resquared and Cold-Rolled Coils

Specified Thickness, in. [mm]	Tolerances for Specified Width, in. [mm]	
	24 [610] to 48 [1219], excl	48 [1219] and Over
Less than 3/16 [4.76]	1/16 [1.59] over, 0 under	1/8 [3.18] over, 0 under

TABLE A1.6 Permissible Variations in Length for Hot-Rolled and Cold-Rolled Sheets Not Resquared

Length, ft [mm]	Tolerances, in. [mm]
Up to 10 [3048], incl	1/4 [6.35] over, 0
Over 10 [3048] to 20 [6096], incl	1/2 [12.70] over, 0 under

TABLE A1.7 Permissible Variations in Camber for Hot-Rolled and Cold-Rolled Sheets Not Resquared and Cold-Rolled Coils[A]

Specified Width, in. [mm]	Tolerance per Unit Length of Any 8 ft [2438 mm], in. [mm]
24 [610] to 36 [914], incl	1/8 [3.18]
Over 36 [914]	3/32 [2.38]

[A] Camber is the greatest deviation of a side edge from a straight line and measurement is taken by placing an 8-ft [2438-mm] straightedge on the concave side and measuring the greatest distance between the sheet edge and the straightedge.

 A 480/A 480M

TABLE A1.8 Permissible Variations in Flatness for Hot-Rolled and Cold-Rolled Sheets Specified to Stretcher-Leveled Standard of Flatness (Not Including Hard Tempers of 2XX and 3XX Series)

Specified Thickness, in. [mm]	Width, in. [mm]	Length, in. [mm]	Flatness Tolerance,[A] in. [mm]
Under 3/16 [4.76]	to 48 [1219], incl	to 96 [2438], incl	1/8 [3.18]
Under 3/16 [4.76]	to 48 [1219], incl	over 96 [2438]	1/4 [6.35]
Under 3/16 [4.76]	over 48 [1219]	to 96 [2438], incl	1/4 [6.35]
Under 3/16 [4.76]	over 48 [1219]	over 96 [2438]	1/4 [6.35]

[A] Maximum deviation from a horizontal flat surface.

TABLE A1.9 Permissible Variations in Flatness for Hot-Rolled and Cold-Rolled Sheets Not Specified to Stretcher-Leveled Standard of Flatness (Not Including Hard Tempers of 2XX and 3XX Series, Dead-Soft Sheets and Deep-Drawing Sheets)

Specified Thickness, in. [mm]	Width, in. [mm]	Flatness Tolerance,[A] in. [mm]
0.062 [1.57] and over	to 60 [1524], incl	1/2 [12.70]
	over 60 [1524] to 72 [1829], incl	3/4 [19.05]
	over 72 [1829]	1 [25.40]
Under 0.062 [1.57]	to 36 [914], incl	1/2 [12.70]
	over 36 [914] to 60 [1524], incl	3/4 [19.05]
	over 60 [1524]	1 [25.40]

[A] Maximum deviation from a horizontal flat surface.

TABLE A1.10 Permissible Variations in Flatness for Cold-Rolled Sheets of 2XX and 3XX Series Specified to 1/4 and 1/2 Hard Tempers

Specified Thickness, in. [mm]	Width, in. [mm]	Flatness Tolerance,[A] in. [mm]	
		1/4 Hard	1/2 Hard
0.016 [0.41] and under	24 [610] to 36 [914], excl	1/2 [12.70]	3/4 [19.05]
Over 0.016 [0.41] to 0.030 [0.76], incl		5/8 [15.88]	7/8 [22.22]
Over 0.030 [0.76]		3/4 [19.05]	7/8 [22.22]
0.016 [0.41] and under	36 [914] to 48 [1219], incl	5/8 [15.88]	1 [25.40]
Over 0.016 [0.41] to 0.030 [0.76], incl		3/4 [19.05]	1 1/8 [28.58]
Over 0.030 [0.76]		1 [25.40]	1 1/8 [28.58]

[A] Maximum deviation from a horizontal flat surface.

TABLE A1.11 Permissible Variations in Diameter for Hot-Rolled and Cold-Rolled Sheets, Sheared Circles

Specified Thickness, in. [mm]	Tolerance Over Specified Diameter (No Tolerance Under), in. [mm]		
	Diameters Under 30 in. [762]	Diameters 30 [762] to 48 in. [1219]	Diameters Over 48 in. [1219]
0.0972 [2.46] and thicker	1/8 [3.18]	3/16 [4.76]	1/4 [6.35]
0.0971 [2.46] to 0.0568 [1.45], incl	3/32 [2.38]	5/32 [3.97]	7/32 [5.56]
0.0567 [1.45] and thinner	1/16 [1.59]	1/8 [3.18]	3/16 [4.76]

A 480/A 480M

TABLE A1.12 Permissible Variations in Thickness for Cold-Rolled Strip in Coils and Cut Lengths

NOTE 1—Thickness measurements are taken at least ⅜ in. (9.52 mm) in from the edge of the strip, except on widths less than 1 in. (25.4 mm) the measurements should be taken at least ⅛ in. (3.18 mm) from the strip edge.
NOTE 2—The tolerance in this table include crown tolerances.

Specified Thickness, in. [mm]	Thickness Tolerances, for the Thickness and Widths Given, Over and Under, in. [mm]		
	Width, in. [mm]		
	3/16 [4.76] to 6 [152], incl	Over 6 [152] to 12 [305], incl	Over 12 [305] to 24 [610], excl
	Thickness Tolerances[A]		
0.005 [0.13] to 0.010 [0.25], incl	10 %	10 %	10 %
Over 0.010 [0.25] to 0.011 [0.28], incl	0.0015 [0.04]	0.0015 [0.04]	0.0015 [0.04]
Over 0.011 [0.28] to 0.013 [0.33], incl	0.0015 [0.04]	0.0015 [0.04]	0.002 [0.05]
Over 0.013 [0.33] to 0.017 [0.43], incl	0.0015 [0.04]	0.002 [0.05]	0.002 [0.05]
Over 0.017 [0.43] to 0.020 [0.51], incl	0.0015 [0.04]	0.002 [0.05]	0.0025 [0.06]
Over 0.020 [0.51] to 0.029 [0.74], incl	0.002 [0.05]	0.0025 [0.06]	0.0025 [0.06]
Over 0.029 [0.74] to 0.035 [0.89], incl	0.002 [0.05]	0.003 [0.08]	0.003 [0.08]
Over 0.035 [0.89] to 0.050 [1.27], incl	0.0025 [0.06]	0.0035 [0.09]	0.0035 [0.09]
Over 0.050 [1.27] to 0.069 [1.75], incl	0.003 [0.08]	0.0035 [0.09]	0.0035 [0.09]
Over 0.069 [1.75] to 0.100 [2.54], incl	0.003 [0.08]	0.004 [0.10]	0.005 [0.13]
Over 0.100 [2.54] to 0.125 [2.98], incl	0.004 [0.10]	0.0045 [0.11]	0.005 [0.13]
Over 0.125 [2.98] to 0.161 [4.09], incl	0.0045 [0.11]	0.0045 [0.11]	0.005 [0.13]
Over 0.161 [4.09] to under 3/16 [4.76]	0.005 [0.13]	0.005 [0.13]	0.006 [0.15]

[A] Thickness tolerances given in in. [mm] unless otherwise indicated.

TABLE A1.13 Permissible Variations in Width for Cold-Rolled Strip in Coils and Cut Lengths for Edge Nos. 1 and 5

Specified Edge No.	Width, in. [mm]	Thickness, in. [mm]	Width Tolerance for Thickness and Width Given, in. [mm]	
			Over	Under
1 and 5	9/32 [7.14] and under	1/16 [1.59] and under	0.005 [0.13]	0.005 [0.13]
1 and 5	over 9/32 [7.14] to 3/4 [19.05], incl	3/32 [2.38] and under	0.005 [0.13]	0.005 [0.13]
1 and 5	over 3/4 [19.05] to 5 [127], incl	1/8 [3.18] and under	0.005 [0.13]	0.005 [0.13]
5	over 5 [127.00] to 9 [228.60], incl	1/8 [3.18] to 0.008 [0.20], incl	0.010 [0.25]	0.010 [0.25]
5	over 9 [228.60] to 20 [508.00], incl	0.105 [2.67] to 0.015 [0.38]	0.010 [0.25]	0.010 [0.25]
5	over 20 [508.00]	0.080 [2.03] to 0.023 [0.58]	0.015 [0.38]	0.015 [0.38]

TABLE A1.14 Permissible Variations in Width for Cold-Rolled Strip in Coils and Cut Lengths for Edge No. 3

Specified Thickness, in. [mm]	Width Tolerance, Over and Under, for Thickness and Width Given, in. [mm]					
	Under ½ [12.70] to 3/16 [4.76], incl	½ [12.70] to 6 [152.40], incl	Over 6 [152.40] to 9 [228.60], incl	Over 9 [228.60] to 12 [304.80], incl	Over 12 [304.80] to 20 [508.00], incl	Over 20 [508.00] to 24 [609.60], incl
Under 3/16 [4.76] to 0.161 [4.09], incl	...	0.016 [0.41]	0.020 [0.51]	0.020 [0.51]	0.031 [0.79]	0.031 [0.79]
0.160 [4.06] to 0.100 [2.54], incl	0.010 [0.25]	0.010 [0.25]	0.016 [0.41]	0.016 [0.41]	0.020 [0.51]	0.020 [0.51]
0.099 [2.51] to 0.069 [1.75], incl	0.008 [0.20]	0.008 [0.20]	0.010 [0.25]	0.010 [0.25]	0.016 [0.41]	0.020 [0.51]
0.068 [1.73] and under	0.005 [0.13]	0.005 [0.13]	0.005 [0.13]	0.010 [0.25]	0.016 [0.41]	0.020 [0.51]

TABLE A1.15 Permissible Variations in Length for Cold-Rolled Strip in Cut Lengths

Specified Length, ft [mm]	Tolerance Over Specified Length (No Tolerance Under), in. [mm]
To 5 [1524], incl	⅜ [9.52]
Over 5 [1524] to 10 [3048], incl	½ [12.70]
Over 10 [3048] to 20 [6096], incl	⅝ [15.88]

TABLE A1.16 Permissible Variations in Camber for Cold-Rolled Strip in Coils and Cut Lengths[A]

Specified Width, in. [mm]	Tolerance per Unit Length of Any 8 ft [2438 mm], in. [mm]
To 1½ [38.10], incl	½ [12.70]
Over 1½ [38.10] to 24 [609.60], excl	¼ [6.35]

[A] Camber is the deviation of a side edge from a straight line and measurement is taken by placing an 8-ft [2438-mm] straightedge on the concave side and measuring the greatest distance between the strip edge and the straightedge.

TABLE A1.17 Permissible Variations in Thickness for Plates[A]

Specified Thickness, in. [mm]	Width, in. [mm]			
	To 84 [2134], incl	Over 84 [2134] to 120 [3048], incl	Over 120 [3048] to 144 [3658], incl	Over 144 [3658]
	Tolerance Over Specified Thickness,[B] in. [mm]			
3/16 [4.76] to ⅜ [9.52], excl	0.045 [1.14]	0.050 [1.27]	...	...
⅜ [9.52] to ¾ [19.05], excl	0.055 [1.40]	0.060 [1.52]	0.075 [1.90]	0.090 [2.29]
¾ [19.05] to 1 [25.40], excl	0.060 [1.52]	0.065 [1.65]	0.085 [2.16]	0.100 [2.54]
1 [25.40] to 2 [50.80], excl	0.070 [1.78]	0.075 [1.90]	0.095 [2.41]	0.115 [2.92]
2 [50.80] to 3 [76.20], excl	0.125 [3.18]	0.150 [3.81]	0.175 [4.44]	0.200 [5.08]
3 [76.20] to 4 [101.6], excl	0.175 [4.44]	0.210 [5.33]	0.245 [6.22]	0.280 [7.11]
4 [101.6] to 6 [152.4], excl	0.250 [6.35]	0.300 [7.62]	0.350 [8.89]	0.400 [10.16]
6 [152.4] to 8 [203.2], excl	0.350 [8.89]	0.420 [10.67]	0.490 [12.45]	0.560 [14.22]
8 [203.2] to 10 [254.0], excl	0.450 [11.43]	0.540 [13.72]	0.630 [16.00]	...

[A] Thickness is measured along the longitudinal edges of the plate at least ⅜ in. [9.52 mm], but not more than 3 in. [76.20 mm], from the edge.

[B] For circles, the over thickness tolerances in this table apply to the diameter of the circle corresponding to the width ranges shown. For plates of irregular shape, the over thickness tolerances apply to the greatest width corresponding to the width ranges shown. For plates up to 10 in. [254.0 mm], incl, in thickness, the tolerance under the specified thickness is 0.010 in. [0.25 mm].

A 480/A 480M

TABLE A1.18 Permissible Variations in Width and Length for Rectangular Sheared Mill Plates and Universal Mill Plates

Width, in. [mm]	Length, in. [mm]	Tolerances Over Specified Width and Length for Given Width, Length, and Thickness,[A] in. [mm]							
		Under 3/8 in. [9.52 mm] in Thickness		3/8 [9.52] to 1/2 [12.70] in., incl, in Thickness		1/2 [12.70] to 1 in. [25.40 mm], in Thickness		Over 1/2 [12.70 mm] to 1 in. [25.40 mm], in Thickness	
		Width	Length	Width	Length	Width	Length	Width	Length
48 [1219] and under	144 [3658] and under	1/8 [3.18]	3/16 [4.76]	3/16 [4.76]	1/4 [6.35]	5/16 [7.94]	3/8 [9.52]		
Over 48 [1219] to 60 [1524], incl		3/16 [4.76]	1/4 [6.35]	1/4 [6.35]	5/16 [7.94]	3/8 [9.52]	7/16 [11.11]		
Over 60 [1524] to 84 [2134], incl		1/4 [6.35]	5/16 [7.94]	5/16 [7.94]	3/8 [9.52]	7/16 [11.11]	1/2 [12.70]		
Over 84 [2134] to 108 [2743], incl		5/16 [7.94]	3/8 [9.52]	3/8 [9.52]	7/16 [11.11]	1/2 [12.70]	9/16 [14.29]		
Over 108 [2743]		3/8 [9.52]	7/16 [11.11]	7/16 [11.11]	1/2 [12.70]	5/8 [15.88]	11/16 [17.46]		
48 [1219] and under	over 144 [3658] to 240 [6096]	3/16 [4.76]	3/8 [9.52]	1/4 [6.35]	1/2 [12.70]	5/16 [7.94]	5/8 [15.88]		
Over 48 [1219] to 60 [1524], incl		1/4 [6.35]	7/16 [11.11]	5/16 [7.94]	5/8 [15.88]	3/8 [9.52]	3/4 [19.05]		
Over 60 [1524] to 84 [2134], incl		3/8 [9.52]	1/2 [12.70]	7/16 [11.11]	11/16 [17.46]	1/2 [12.70]	3/4 [19.05]		
Over 84 [2134] to 108 [2743], incl		7/16 [11.11]	9/16 [14.29]	1/2 [12.70]	3/4 [19.05]	5/8 [15.88]	7/8 [22.22]		
Over 108 [2743]		1/2 [12.70]	5/8 [15.88]	5/8 [15.88]	7/8 [22.22]	11/16 [17.46]	1 [25.40]		
48 [1219] and under	over 240 [6096] to 360 [9144]	1/4 [6.35]	1/2 [12.70]	5/16 [7.94]	5/8 [15.88]	3/8 [9.52]	3/4 [19.05]		
Over 48 [1219] to 60 [1524], incl		5/16 [7.94]	5/8 [15.88]	3/8 [9.52]	3/4 [19.05]	1/2 [12.70]	3/4 [19.05]		
Over 60 [1524] to 84 [2134], incl		7/16 [11.11]	11/16 [17.46]	1/2 [12.70]	3/4 [19.05]	5/8 [15.88]	7/8 [22.22]		
Over 84 [2134] to 108 [2743], incl		9/16 [14.29]	3/4 [19.05]	5/8 [15.88]	7/8 [22.22]	3/4 [19.05]	1 [25.40]		
Over 108 [2743]		5/8 [15.88]	7/8 [22.22]	11/16 [17.46]	1 [25.40]	7/8 [22.22]	1 [25.40]		
60 [1524] and under	over 360 [9144] to 480 [12192]	7/16 [11.11]	11/8 [28.58]	1/2 [12.70]	11/4 [31.75]	5/8 [15.88]	13/8 [34.92]		
Over 60 [1524] to 84 [2134], incl		1/2 [12.70]	11/4 [31.75]	5/8 [15.88]	13/8 [34.92]	3/4 [19.05]	11/2 [38.10]		
Over 84 [2134] to 108 [2743], incl		9/16 [14.29]	11/4 [31.75]	3/4 [19.05]	13/8 [34.92]	7/8 [22.22]	11/2 [38.10]		
Over 108 [2743]		3/4 [19.05]	13/8 [34.92]	7/8 [22.22]	11/2 [38.10]	1 [25.40]	15/8 [41.28]		
60 [1524] and under	over 480 [12192] to 600 [15240]	7/16 [11.11]	11/4 [31.75]	1/2 [12.70]	13/8 [38.10]	5/8 [15.88]	15/8 [41.28]		
Over 60 [1524] to 84 [2134], incl		1/2 [12.70]	13/8 [34.92]	5/8 [15.88]	11/2 [38.10]	3/4 [19.05]	15/8 [41.28]		
Over 84 [2134] to 108 [2743], incl		5/8 [15.88]	13/8 [34.92]	3/4 [19.05]	11/2 [38.10]	7/8 [22.22]	15/8 [41.28]		
Over 108 [2743]		3/4 [19.05]	11/2 [38.10]	7/8 [22.22]	15/8 [41.28]	1 [25.40]	13/4 [44.45]		
60 [1524] and under	over 600 [15240]	1/2 [12.70]	13/4 [44.45]	5/8 [15.88]	17/8 [47.62]	3/4 [19.05]	17/8 [47.62]		
Over 60 [1524] to 84 [2134], incl		5/8 [15.88]	13/4 [44.45]	3/4 [19.05]	17/8 [47.62]	7/8 [22.22]	17/8 [47.62]		
Over 84 [2134] to 108 [2743], incl		5/8 [15.88]	13/4 [44.45]	3/4 [19.05]	17/8 [47.62]	7/8 [22.22]	17/8 [47.62]		
Over 108 [2743]		7/8 [22.22]	13/4 [44.45]	1 [25.40]	2 [50.80]	11/8 [28.58]	21/4 [57.15]		

[A] The tolerance under specified width and length is 1/4 in. [6.35 mm].

267

TABLE A1.19 Permissible Variations in Annealed Plates

NOTE 1—Tolerances in this table apply to plates up to 15 ft (4572 mm) in length, or to any 15 ft (4572 mm) of longer plates.
NOTE 2—If the longer dimension is under 36 in. (914 mm), the tolerance is not greater than 1/4 in. (6.35 mm).
NOTE 3—For plates with specified minimum yield strengths of 35 ksi (240 MPa) or more, the permissible variations are increased to 1 1/2 times the amounts shown below.

Specified Thickness, in. [mm]	Flatness Tolerance (Deviation from a Horizontal Flat Surface) for Thicknesses and Widths Given, in. [mm] — Width, in. [mm]							
	48 [1219] or Under	Over 48 [1219] to 60 [1524], excl	60 [1524] to 72 [1829], excl	72 [1829] to 84 [2134], excl	84 [2134] to 96 [2438], excl	96 [2438] to 108 [2743], excl	108 [2743] to 120 [3048], excl	120 [3048] to 144 [3658], excl	144 [3658] and Over
3/16 [4.76] to 1/4 [6.35], excl	3/4 [19.05]	1 1/16 [26.99]	1 1/4 [31.75]	1 3/8 [34.92]	1 5/8 [41.28]	1 5/8 [41.28]	1 7/8 [47.62]	2 [50.80]	. . .
1/4 [6.35] to 3/8 [9.52], excl	11/16 [17.46]	3/4 [19.05]	15/16 [23.81]	1 1/8 [28.58]	1 3/8 [34.92]	1 7/16 [36.51]	1 9/16 [39.69]	1 7/8 [47.62]	. . .
3/8 [9.52] to 1/2 [12.70], excl	1/2 [12.70]	9/16 [14.29]	11/16 [17.46]	3/4 [19.05]	15/16 [23.81]	1 1/8 [28.58]	1 1/4 [31.75]	1 7/16 [36.51]	1 3/4 [44.45]
1/2 [12.70] to 3/4 [19.05], excl	1/2 [12.70]	9/16 [14.29]	5/8 [15.88]	5/8 [15.88]	13/16 [20.64]	1 1/8 [28.58]	1 1/8 [28.58]	1 1/8 [28.58]	1 3/8 [34.92]
3/4 [19.05] to 1 [25.40], excl	1/2 [12.70]	9/16 [14.29]	5/8 [15.88]	5/8 [15.88]	3/4 [19.05]	13/16 [20.64]	15/16 [23.81]	1 [25.40]	1 1/8 [28.58]
1 [25.40] to 1 1/2 [38.10], excl	1/2 [12.70]	9/16 [14.29]	9/16 [14.29]	9/16 [14.29]	11/16 [17.46]	11/16 [17.46]	11/16 [17.46]	3/4 [19.05]	1 [25.40]
1 1/2 [38.10] to 4 [101.60], excl	3/16 [4.76]	5/16 [7.94]	3/8 [9.52]	7/16 [11.11]	1/2 [12.70]	9/16 [14.29]	5/8 [15.88]	3/4 [19.05]	7/8 [22.22]
4 [101.60] to 6 [152.40], excl	1/4 [6.35]	3/8 [9.52]	1/2 [12.70]	9/16 [14.29]	5/8 [15.88]	3/4 [19.05]	7/8 [22.22]	1 [25.40]	1 1/8 [28.58]

A 480/A 480M

TABLE A1.20 Waviness Tolerances for Plates

NOTE—Waviness denotes the deviation of the top or bottom surface from a horizontal line contacting the surface, when the plate is resting on a flat surface, as measured in an increment of less than 15 ft [4.6 m] of length.

	Number of Waves in 15 ft (4.6 m)							
	1	2	3	4	5	6	7	8
Tolerance, expressed as % of Table 19 Tolerance	100	100	80	70	60	50	40	40

TABLE A1.21 Permissible Variations in Camber for Sheared Mill and Universal Mill Plates[A]

Maximum camber	= 1/8 in. in any 5 ft
	= 3.18 mm in any 1.524 m

[A] Camber is the deviation of a side edge from a straight line, and measurement is taken by placing a 5-ft straightedge on the concave side and measuring the greatest distance between the plate and the straightedge.

TABLE A1.22 Permissible Variations in Diameter for Circular Plates

Specified Diameter, in. [mm]	Tolerance Over Specified Diameter for Given Diameter and Thickness,[A] in. [mm]			
	To 3/8 [9.52] in., excl in Thickness	3/8 [9.52] to 5/8 [15.88] in., excl in Thickness	5/8 in. [15.88] and Over in Thickness[B]
To 60 [1524], excl	1/4 [6.35]	3/8 [9.52]	1/2 [12.70]
60 [1524 mm] to 84 [2134 mm], excl	5/16 [7.94]	7/16 [11.11]	9/16 [14.29]
84 [2134 mm] to 108 [2743 mm], excl	3/8 [9.52]	1/2 [12.70]	5/8 [15.88]
108 [2743 mm] to 180 [4572 mm], excl	7/16 [11.11]	9/16 [14.29]	11/16 [17.46]

[A] No tolerance under.
[B] Circular and sketch plates over 5/8 in. [15.88 mm] in thickness are not commonly sheared but are machined or flame cut.

TABLE A1.23 Recommended Flame Cutting Allowances to Clean up in Machining Plates, Circles, Rings, and Sketches

Specified Thickness, in. [mm]	Machining Allowance per Edge, in. [mm]
2 [50.80] and under	1/4 [6.35]
Over 2 [50.80] to 3 [76.20], incl	3/8 [9.52]
Over 3 [76.20] to 6 [152.40], incl	1/2 [12.70]

[A] Supplier assumes the appropriate clean-up allowances have been included in ordered dimension.

TABLE A1.24 Permissible Variations in Abrasive Cutting Width and Length for Plates

Specified Thickness, in. [mm]	Tolerance over Specified Width and Length,[A]	
	Width	Length
Up to 1 1/4 [31.75], incl	1/8 [3.18]	1/8 [3.18]
Over 1 1/4 [31.75] to 2 3/4 [69.85], incl[B]	3/16 [4.76]	3/16 [4.76]

[A] The tolerances under specified width and length is 1/8 in. [3.18 mm].
[B] Width and length tolerances for abrasive cut plates over 2 3/4 in. [69.85 mm] thick are not included in the table.

A 480/A 480M

A2. PERMISSIBLE VARIATIONS IN DIMENSIONS, ETC.—SI UNITS

A2.1 Listed in Annex A2 are tables showing the permissible variations in dimensions expressed in SI units of measurement.

TABLE A2.1 Permissible Variations in Thickness for Hot-Rolled Sheets in Cut Lengths, Cold-Rolled Sheets in Cut Lengths and Coils

NOTE 1—Thickness measurements are taken at least 10 mm from the edge of the sheet.
NOTE 2—Cold rolled sheets in cut lengths and coils are produced in some type numbers and some widths and thicknesses to tolerances less than those shown in the table.
NOTE 3—For specified thicknesses other than those shown, the tolerances for the next higher thickness shall apply.

Specified Thickness, mm	Permissible Variations, mm	
	Over	Under
4.99	0.36	0.36
4.00	0.36	0.36
3.75	0.36	0.36
3.50	0.30	0.30
3.25	0.30	0.30
3.00	0.25	0.25
2.75	0.25	0.25
2.50	0.23	0.23
2.25	0.20	0.20
2.00	0.18	0.18
1.75	0.15	0.15
1.50	0.15	0.15
1.25	0.13	0.13
1.00	0.13	0.13
0.75	0.10	0.10
0.50	0.08	0.08
0.25	0.05	0.05
0.20	0.05	0.05
0.15	0.04	0.04
0.10	0.03	0.03

TABLE A2.2 Permissible Variations in Width and Length for Hot-Rolled and Cold-Rolled Resquared Sheets (Stretcher-Leveled Standard of Flatness)

NOTE—Polished sheets with Finishes No. 4 and higher are produced to tolerances given in this table.

Specified Dimensions, mm			Width and Length Tolerance, mm	
Thickness	Width	Length	Over	Under
3.30 and over Under 3.30	All Up to 1200 1200 and over	All Up to 3000 3000 and over	7 2 3	0 0 0

TABLE A2.3 Permissible Variations in Weight for Hot-Rolled and Cold-Rolled Sheets

Any item of five sheets or less, and estimated to weigh 100 kg or less, may actually weigh 10 % over the theoretical weight	weigh 100 kg or less
Any item of more than five sheets and estimated to weigh more than 100 kg, may actually weigh 7½ % over the theoretical weight	weigh more than 100 kg
Chromium-manganese-nickel	7.82 kg/m²/mm thick
Chromium-nickel	8.07 kg/m²/mm thick
Chromium	7.92 kg/m²/mm thick

TABLE A2.4 Permissible Variations in Width for Hot-Rolled and Cold-Rolled Sheets not Resquared and Cold-Rolled Coils

Specified Dimension, mm		Tolerance, mm	
Thickness	Width	Over	Under
4.99 and under	1200 and over	4	0
	600 to 1200, excl	2	0

TABLE A2.5 Permissible Variations in Length for Hot-Rolled and Cold-Rolled Sheets Not Resquared

Specified Length, mm	Tolerance, mm	
	Over	Under
3000 to 6000 incl	13	0
Under 3000	7	0

TABLE A2.6 Permissible Variations in Camber for Hot-Rolled and Cold-Rolled Sheets Not Resquared and Cold-Rolled Coils[A]

Specified Width, mm	Tolerance per Unit Length of Any 2400 mm, mm
900 and over	3
600 to 900, excl	4

[A] Camber is the greatest deviation of a side edge from a straight line and measurement is taken by placing an 8-ft [2438-mm] straightedge on the concave side and measuring the greatest distance between the sheet edge and the straightedge.

TABLE A2.7 Permissible Variations in Flatness for Hot-Rolled and Cold-Rolled Sheets Specified to Stretcher-Leveled Standard of Flatness (Not Including Hard Tempers of 2XX and 3XX Series)

Specified Dimensions, mm			Flatness Tolerance[A], mm
Thickness	Width	Length	
4.99 and under	1200 and over	2400 and over	7
		to 2400 excl	7
	to 1200, excl	2400 and over	7
		to 2400 excl	4

[A] Maximum deviation from a horizontal flat surface.

TABLE A2.8 Permissible Variations in Flatness for Hot-Rolled and Cold-Rolled Sheets Not Specified to Stretcher-Leveled Standard of Flatness (Not Including Hard Tempers of 2XX and 3XX Series, Dead-Soft Sheets and Deep-Drawing Sheets)

Specified Dimensions, mm		Flatness Tolerance[A], mm
Thickness	Width	
1.50 and over	1800 and over	26
	1500 to 1800, excl	19
	to 1500, excl	13
Under 1.50	1500 and over	26
	900 to 1500 excl	19
	to 900, excl	13

[A] Maximum deviation from a horizontal flat surface.

TABLE A2.9 Permissible Variations in Flatness for Cold-Rolled Sheets of 2XX and 3XX Series Specified to ¼ and ½ Hard Tempers

Specified Dimensions, mm		Flatness Tolerance[A], mm	
Thickness	Width	¼ Hard	½ Hard
Over 0.80	600 to 900, excl	13	19
0.04 to 0.80 incl		16	23
0.04 and under		19	23
Over 0.80	900 to 1200, incl	16	26
0.04 to 0.80 incl		19	29
0.04 and under		26	29

[A] Maximum deviation from a horizontal flat surface.

TABLE A2.10 Permissible Variations in Diameter for Hot-Rolled and Cold-Rolled Sheets, Sheared Circles

Specified Thickness, mm	Tolerance Over Specified Diameter No Tolerance Under		
	Diameters Under 600	Diameters 600 to 1200 incl	Diameters Over 1200
2.50 and thicker	4	5	7
1.50 to 2.50 excl	3	4	6
Under 1.50	2	3	5

TABLE A2.11 Permissible Variations in Thickness for Cold-Rolled Strip in Coils and Cut Lengths

NOTE 1—Thickness measurements are taken at least 10 mm in from the edge of the strip, except that on widths less than 26 mm, the tolerances are applicable for measurements at all locations.

NOTE 2—The tolerances in this table include crown tolerances.

NOTE 3—For specified thicknesses other than those shown, the tolerances for the next higher thickness shall apply

Specified Thickness, mm	Thickness Tolerances[A], for the Thickness and Widths Given, Over and Under, mm		
	Width, mm		
	50 to 150, incl	Over 150 to 300, incl	Over 300 to 600, excl
4.99	0.13	0.13	0.15
4.00	0.13	0.13	0.15
3.00	0.11	0.11	0.13
2.50	0.10	0.11	0.13
2.00	0.08	0.10	0.13
1.75	0.08	0.10	0.13
1.50	0.08	0.09	0.09
1.25	0.08	0.09	0.09
1.00	0.06	0.09	0.09
0.75	0.06	0.09	0.09
0.50	0.05	0.06	0.06
0.25	0.04	0.04	0.05
0.15	10 %	10 %	10 %

[A] Thickness Tolerances given in mm unless otherwise indicated.

TABLE A2.12 Permissible Variations in Width for Cold-Rolled Strip in Coils and Cut Lengths for Edge Nos. 1 and 5

Specified Edge No.	Specified Dimension, mm		Width Tolerance, mm	
	Width	Thickness	Over	Under
5	600 and over	2.00 to 0.60 incl	0.40	0.40
5	300 to 600 excl	2.60 to 0.40 incl	0.25	0.25
5	100 to 300 excl	3.00 to 0.20 incl	0.25	0.25
1 and 5	20 to 100 excl	3.00 and under	0.13	0.13
1 and 5	10 to 20 excl	2.50 and under	0.13	0.13
1 and 5	under 10	1.50 and under	0.13	0.13

TABLE A2.13 Permissible Variations in Width for Cold-Rolled Strip in Coils and Cut Lengths for Edge No. 3

NOTE— For specified thickness other than those shown, the tolerances for the next higher thickness shall apply.

Specified Thickness, mm	Width Tolerance, Over and Under, for Thickness and Width Given, mm					
	5 to 12 excl	12 to 150 excl	150 to 200 excl	200 to 300 excl	300 to 500 excl	500 to 600 excl
4.99	...	0.40	0.50	0.50	0.80	0.80
4.00	0.25	0.25	0.40	0.40	0.50	0.50
2.50	0.20	0.20	0.25	0.25	0.40	0.50
1.75 and under	0.13	0.13	0.13	0.25	0.40	0.50

A 480/A 480M

TABLE A2.14 Permissible Variations in Length for Cold-Rolled Strip in Cut Lengths

Specified Length, mm	Tolerance Over Specified Length (No Tolerance Under), mm
Over 3000 to 6000, incl	10
Over 1500 to 3000, incl	13
To 1500 incl	16

TABLE A2.15 Permissible Variations in Camber for Cold-Rolled Strip in Coils and Cut Lengths[A]

Specified Width, mm	Tolerance Per Unit Length Of Any 2400 mm
Over 40 to 600, incl	7
to 40, incl	13

[A] Camber is the deviation of a side edge from a straight line and measurement is taken by placing a 2400-mm straightedge on the concave side and measuring the greatest distance between the strip edge and the straightedge.

TABLE A2.16 Permissible Variations in Thickness for Plates[A]

NOTE—For specified thicknesses other than those shown, the tolerances for the next highest thickness shall apply.

Specified Thickness, mm	Width, mm			
	To 2100, excl	2100 to 3000, excl	3000 to 3600, excl	Over 3600
	Tolerance over Specified Thickness,[B] mm			
250	11.45	13.75	16.00	...
200	8.96	10.70	12.45	14.25
150	6.35	7.65	8.90	10.15
100	4.45	5.35	6.25	7.10
75	3.20	3.80	4.45	5.10
50	1.80	1.90	2.40	2.95
25	1.55	1.65	2.15	2.55
20	1.40	1.55	1.90	2.30
10	1.15	1.30	...	...
5	1.15	1.30	...	...

[A] Thickness is measured along the longitudinal edges of the plate at least ⅜ in. [9.52 mm], but not more than 3 in. [76.20 mm], from the edge.

[B] For circles, the over thickness tolerances in this table apply to the diameter of the circle corresponding to the width ranges shown. For plates of irregular shape, the over thickness tolerances apply to the greatest width corresponding to the width ranges shown. For plates up to 10 in. [254.0 mm], incl, in thickness, the tolerance under the specified thickness is 0.010 in. [0.25 mm].

TABLE A2.17 Permissible Variations in Width and Length for Rectangular Sheared Mill Plates Tolerances Over Specified Width and Length for Given Width, Length, and Thickness[A]

Width, mm	Length, mm	Under 10 mm in Thickness		10 to 13 mm, incl in Thickness		Over 13 to 26 mm, incl in Thickness, mm	
		Width	Length	Width	Length	Width	Length
2700 and over	15 000 and over	23	45	26	51	29	51
2100 to 2700, excl		16	45	19	48	23	48
1500 to 2100, excl		16	45	19	48	23	48
250 to 1500, excl		13	45	16	48	19	48
2700 and over	12 000 to 5 000, excl	19	38	23	42	26	45
2100 to 2700, excl		16	35	19	38	23	43
1500 to 2100, excl		13	35	16	38	19	42
2500 to 1500, excl		11	32	13	38	16	42
2700 and over	9000 to 1200, excl	19	35	23	38	26	42
2100 to 2700, excl		15	32	19	35	23	38
1500 to 2100, excl		13	32	16	35	19	38
250 to 1500, excl		11	29	13	32	16	35
2700 and over	6000 to 9000, excl	16	23	18	26	23	26
2100 to 2700, excl		15	19	16	23	19	26
1500 to 2100, excl		11	18	13	19	16	23
1200 to 1500, excl		8	16	10	19	13	19
250 to 1200, excl		7	13	8	16	10	19
2700 and over	3600 to 6000, excl	13	16	16	23	18	26
2100 to 2700, excl		11	15	13	19	16	23
1500 to 2100, excl		10	13	11	18	13	19
1200 to 1500, excl		7	11	8	16	10	19
250 to 1200, excl		5	10	7	13	8	16
2700 and over	under 3600	10	11	11	13	16	18
2100 to 2700, excl		8	10	10	11	13	15
1500 to 2100, excl		7	8	8	10	11	13
1200 to 1500, excl		5	7	7	8	10	11
250 to 1200, excl		3	5	5	7	8	10

[A] The tolerance under specified width and length is 7 mm.

A 480/A 480M

TABLE A2.18 Permissible Flatness Variations in Annealed Plates

NOTE 1—Tolerances in this table apply to plates up to 4500 mm in length, or to any 4500 mm of longer plates.
NOTE 2—If the longer dimension is under 900 mm the tolerance is not greater than 7 mm.
NOTE 3—For plates with specified minimum yield strengths of 240 MPa or more, the permissible variations are increased to 1½ times the amounts shown below.
NOTE 4—For specified thicknesses other than those shown, the tolerances for the next higher thickness shall apply.

Specified Thickness, mm	Flatness Tolerance (Deviation from a Horizontal Flat Surface) for Thickness and Widths Given, mm								
	Width, mm								
	1200 or under	Over 1200 to 1500, excl	1500 to 1800, excl	1800 to 2100, excl	2100 to 2400, excl	2400 to 2700, excl	2700 to 3000, excl	3000 to 3600, excl	3600 and over
150	7	10	13	15	16	19	22	26	29
100	5	8	10	11	13	15	16	19	23
50	13	15	15	15	18	18	18	19	26
25	13	15	16	16	19	21	24	26	29
20	13	15	16	16	21	29	29	29	35
15	13	15	18	19	24	29	32	37	45
10	16	17	24	29	35	37	40	48	...
5	19	19	32	35	42	42	48	51	...

TABLE A2.19 Permissible Variations in Camber for Sheared Mill and Universal Mill Plates[A]

Maximum camber = 3 mm in any 1500 mm

[A] Camber is the deviation of a side edge from a straight line, and measurement is taken by placing a 1500-mm straightedge on the concave side and measuring the greatest distance between the plate and the straightedge.

TABLE A2.20 Permissible Variations in Diameter for Circular Plates

NOTE—For specific diameters other than those shown the tolerance for the next higher diameter shall apply.

Specified Diameter, mm	Tolerance Over Specified Diameter for Given Diameter and Thickness,[A] mm		
	Thickness of Plate		
	To 10 mm, excl	10 mm To 15 mm, excl	15 mm and over
4500	11	15	18
2700	10	13	16
2100	8	11	15
1500 and under	7	10	13

[A] No tolerance under.

TABLE A2.21 Recommended Flame Cutting Allowances to Clean Up in Machining Plates, Circles, Rings, and Sketches[A]

Specified Thickness, mm	Machining Allowance per Edge, mm
150	13
75	10
50 and under	7

[A] Supplier assumes the appropriate clean-up allowances have been included in the ordered dimension.

The American Society for Testing and Materials takes no position respecting the validity of any patent rights asserted in connection with any item mentioned in this standard. Users of this standard are expressly advised that determination of the validity of any such patent rights, and the risk of infringement of such rights, are entirely their own responsibility.

This standard is subject to revision at any time by the responsible technical committee and must be reviewed every five years and if not revised, either reapproved or withdrawn. Your comments are invited either for revision of this standard or for additional standards and should be addressed to ASTM Headquarters. Your comments will receive careful consideration at a meeting of the responsible technical committee, which you may attend. If you feel that your comments have not received a fair hearing you should make your views known to the ASTM Committee on Standards, 1916 Race St., Philadelphia, Pa. 19103.

ASTM Designation: A 492 – 82

Standard Specification for
STAINLESS AND HEAT-RESISTING STEEL ROPE WIRE[1]

This standard is issued under the fixed designation A 492; the number immediately following the designation indicates the year of original adoption or, in the case of revision, the year of last revision. A number in parentheses indicates the year of last reapproval. A superscript epsilon (ϵ) indicates an editorial change since the last revision or reapproval.

This specification has been approved for use by agencies of the Department of Defense and for listing in the DoD Index of Specifications and Standards.

1. Scope

1.1 This specification covers the more commonly used types of round stainless and heat-resisting steel wire intended especially for stranding into wire rope.

1.2 The values stated in inch-pound units are to be regarded as the standard.

2. Applicable Documents

2.1 *ASTM Standards:*
A 555 Specification for General Requirements for Stainless and Heat-Resisting Steel Wire[2]
A 751 Methods, Practices, and Definitions for Chemical Analysis of Steel Products[2]
E 527 Practice for Numbering Metals and Alloys (UNS)[2]

3. General Requirements for Delivery

3.1 In addition to the requirements of this specification, all requirements of the current edition of Specification A 555, shall apply.

4. Ordering Information

4.1 Orders for material under this specification shall include the following information:
 4.1.1 Quantity (weight),
 4.1.2 Name of material (stainless steel),
 4.1.3 Condition (see Section 5),
 4.1.4 Finish (see Section 6),
 4.1.5 Cross section (round),
 4.1.6 Form (wire),
 4.1.7 Applicable dimensions (diameter),
 4.1.8 Type designation (Table 1),
 4.1.9 ASTM designation and date of issue, and
 4.1.10 Special requirements.

4.2 If possible, the intended end use of the item should be given on the purchase order, especially when the item is ordered for a specific end use or uses.

NOTE 1—A typical ordering description is as follows: 1000 lb stainless steel, cold-drawn, bright-finish, wire, 0.009-in. diameter, spools. Type 304, ASTM Specification A 492 dated _____, End Use_____.

5. Condition

5.1 The wire is cold drawn and furnished in the finish (see Section 6) specified on the purchase order.

6. Finish

6.1 The types of finish procurable are pickled finish and bright finish (available only in sizes that can be diamond drawn which are generally 0.045 in. (1.14 mm) and smaller).

7. Chemical Requirements

7.1 The steel shall conform to the requirements as to chemical composition specified in Table 1.

7.2 Methods and practices relating to chemical analysis required by this specification shall be in accordance with Methods, Practices, and Definitions A 751.

---

[1] This specification is under the jurisdiction of ASTM Committee A-1 on Steel, Stainless Steel and Related Alloys, and is the direct responsibility of Subcommittee A01.08 on Wrought Stainless Steel Products.
Current edition approved July 30, 1982. Published September 1982. Originally published as A 492 – 63. Last previous edition A 492 – 80.
[2] *Annual Book of ASTM Standards*, Vol 01.03.

8. Mechanical Requirements

8.1 The material shall conform to the requirements as to mechanical properties specified in Table 2.

9. Packaging and Marking

9.1 Each coil or spool shall be one continuous length of wire. Each coil shall be firmly tied and each spool shall be tightly wound. Unless otherwise specified coils shall be placed in drums or paper wrapped and spools shall be boxed in such a manner as to assure safe delivery to their designation when properly transported by any common carrier.

9.2 Each coil or spool of wire shall be properly tagged showing heat number, grade, condition, ASTM specification number, and size.

TABLE 1 Chemical Requirements

UNS Designation	Type	Composition, %								
		Carbon, max	Manganese[B]	Phosphorus, max	Sulfur, max	Silicon, max	Chromium	Nickel	Molybdenum	Nitrogen[A]
S 30200	302	0.15	2.00	0.045	0.030	1.00	17.00–19.00	8.00–10.00	...	0.10
S 30400	304	0.08	2.00	0.045	0.030	1.00	18.00–20.00	8.00–10.50	...	0.10
S 30500	305	0.12	2.00	0.045	0.030	1.00	17.00–19.00	10.50–13.00	...	...
S 31600	316	0.08	2.00	0.045	0.030	1.00	16.00–18.00	10.00–14.00	2.00–3.00	0.10
S 21600	XM-17	0.08	7.50–9.00	0.045	0.030	1.00	17.50–22.00	5.00–7.00	2.00–3.00	0.25–0.50
S 21603	XM-18	0.03	7.50–9.00	0.045	0.030	1.00	17.50–22.00	5.00–7.00	2.00–3.00	0.25–0.50

[A] Maximum unless otherwise indicated.

TABLE 2 Tensile Strength Requirements

Diameter, in. (mm)	For Types XM-17, XM-18, 302, and 304				For Types 305 and 316	
	min		max		min	
	psi	MPa	psi	MPa	psi	MPa
0.007 (0.18) and smaller	320 000	2210	355 000	2450	245 000	1690
Over 0.007 (0.18) to 0.010 (0.25), incl	312 000	2150	350 000	2410	245 000	1690
Over 0.010 (0.25) to 0.015 (0.38), incl	310 000	2140	345 000	2380	240 000	1650
Over 0.015 (0.38) to 0.019 (0.48), incl	305 000	2100	340 000	2340	240 000	1650
Over 0.019 (0.48) to 0.025 (0.64), incl	295 000	2030	330 000	2280	235 000	1620
Over 0.025 (0.64) to 0.030 (0.76), incl	285 000	1960	315 000	2170	235 000	1620
Over 0.030 (0.76) to 0.035 (0.89), incl	275 000	1900	310 000	2140	235 000	1620
Over 0.035 (0.89) to 0.040 (1.02), incl	260 000	1790	300 000	2070	235 000	1620
Over 0.040 (1.02) to 0.050 (1.27), incl	255 000	1760	285 000	1970	230 000	1590
Over 0.050 (1.27) to 0.060 (1.52), incl	250 000	1720	280 000	1930	225 000	1550
Over 0.060 (1.52) to 0.070 (1.78), incl	245 000	1690	275 000	1900	220 000	1520
Over 0.070 (1.78) to 0.080 (2.03), incl	240 000	1650	270 000	1860	210 000	1450
Over 0.080 (2.03) to 0.090 (2.29), incl	240 000	1650	270 000	1860	210 000	1450
Over 0.090 (2.29) to 0.100 (2.54), incl	235 000	1620	265 000	1830	205 000	1410

The American Society for Testing and Materials takes no position respecting the validity of any patent rights asserted in connection with any item mentioned in this standard. Users of this standard are expressly advised that determination of the validity of any such patent rights, and the risk of infringement of such rights, are entirely their own responsibility.

This standard is subject to revision at any time by the responsible technical committee and must be reviewed every five years and if not revised, either reapproved or withdrawn. Your comments are invited either for revision of this standard or for additional standards and should be addressed to ASTM Headquarters. Your comments will receive careful consideration at a meeting of the responsible technical committee, which you may attend. If you feel that your comments have not received a fair hearing you should make your views known to the ASTM Committee on Standards, 1916 Race St., Philadelphia, Pa. 19103.

Designation: A 493 – 82a

Standard Specification for
STAINLESS AND HEAT-RESISTING STEEL FOR COLD HEADING AND COLD FORGING—BAR AND WIRE[1]

This standard is issued under the fixed designation A 493; the number immediately following the designation indicates the year of original adoption or, in the case of revision, the year of last revision. A number in parentheses indicates the year of last reapproval. A superscript epsilon (ϵ) indicates an editorial change since the last revision or reapproval.

This specification has been approved for use by agencies of the Department of Defense and for listing in the DoD Index of Specifications and Standards.

1. Scope

1.1 This specification covers cold-finished stainless and heat-resisting steel bar and wire for cold heading or cold forging for applications where corrosion resistance is a factor.

1.2 The values stated in inch-pound units are to be regarded as the standard.

2. Applicable Documents

2.1 *ASTM Standards*:
A 484 Specification for General Requirements for Stainless and Heat-Resisting Wrought Steel Products (Except Wire)[2]
A 555 Specification for General Requirements for Stainless and Heat-Resisting Steel Wire[3]
A 751 Methods, Practices, and Definitions for Chemical Analysis of Steel Products[4]

3. General Requirements for Delivery

3.1 In addition to the requirements of this specification, all requirements of the current editions of the following specifications shall apply:
3.1.1 *Bars*—Specification A 484.
3.1.2 *Wire*—Specification A 555.

4. Ordering Information

4.1 Orders for material under this specification shall include the following information:
4.1.1 Quantity (weight),
4.1.2 Name of material (stainless steel),
4.1.3 Form (bars, wire),
4.1.4 Condition (annealed, lightly drafted, etc.) (see Section 5),
4.1.5 Finish,
4.1.6 Applicable dimensions including wire diameter and coil size (inside and outside coil diameters),
4.1.7 Cross section (round),
4.1.8 Type designation (see Table 1),
4.1.9 ASTM designation and date of issue, and
4.1.10 Special requirements.

NOTE 1—A typical ordering description is as follows: 5000 lb (2268 kg) stainless steel wire, copper coated, lightly drafted, ¼ in. (6.35 mm) round, 32 in. (813 mm) max OD—22 in. (559 mm) min ID, coils, Type 305, ASTM Specification A 493 dated ——. End use: hex head machine bolts.

5. Condition

5.1 Bar and wire shall be furnished in one of the following conditions:
5.1.1 Lightly drafted (normal condition and need not be specified if this is condition desired),
5.1.2 Annealed, or
5.1.3 Drafted to a specified tensile strength range (as agreed upon between purchaser and producer).

6. Chemical Requirements

6.1 The steel shall conform to the requirements as to chemical composition specified in

---

[1] This specification is under the jurisdiction of ASTM Committee A-1 on Steel, Stainless Steel and Related Alloys, and is the direct responsibility of Subcommittee A01.08 on Wrought Stainless Steel Products.
Current edition approved July 30, 1982. Published September 1982. Originally published as A 493 – 63. Last previous edition A 493 – 82a.
[2] *Annual Book of ASTM Standards*, Vol 01.05.
[3] *Annual Book of ASTM Standards*, Vol 01.03.
[4] *Annual Book of ASTM Standards*, Vol 01.04.

Table 1.

6.2 Methods and practices relating to chemical analysis required by this specification shall be in accordance with Methods A 751.

7. Mechanical Requirements

7.1 The material shall conform to the requirements as to mechanical properties specified in Tables 2 and 3. Material can be manufactured to tensile strength ranges other than those shown in Table 2. The producer should be consulted when material is required to tensile strength ranges other than those specified.

8. Special Tests

8.1 If any special tests are required which are pertinent to the intended application of the material ordered, they shall be as agreed upon between the purchaser and the producer.

9. Coating and Lubricant

9.1 Coatings are necessary for most cold-heading or forming operations. An electroplated copper coating is generally used. The following coatings may be specified: copper, lime, or special (as agreed upon between purchaser and producer).

9.2 Lubricants are applied over the coating during the final drafting operation performed by the producer. Soap is generally used. The following lubricants may be specified: soap, grease, or special (as agreed upon between purchaser and producer).

10. Packaging and Marking

10.1 *Packaging*—Coils shall be bundled or boxed in such a manner as to assure safe delivery to their destination when properly transported by any common carrier.

10.2 *Markings*—Each lift, bundle, or box shall be properly tagged with durable tags (metal, plastic, or equivalent), showing heat number, type, condition, specification number (ASTM A 493), and size, in order to assure proper identification.

ASTM A 493

TABLE 1 Chemical Requirements

UNS Designation	AISI Type	Composition, %							
		Carbon	Manganese, max	Phosphorus, max	Sulfur, max	Silicon, max	Chromium	Nickel	Other Elements
S 28200	...	0.15	17.00–19.00	0.045	0.030	1.00	17.00–19.00	...	molybdenum 0.75–1.25; copper 0.75–1.25; nitrogen 0.40–0.60
S 30200	302	0.15 max	2.00	0.045	0.030	1.00	17.00–19.00	8.00 10.00	nitrogen 0.10 max
S 30400	304	0.08 max	2.00	0.045	0.030	1.00	18.00–20.00	8.00 10.50	nitrogen 0.10 max
S 30500	305	0.12 max	2.00	0.045	0.030	1.00	17.00–19.00	10.50 13.00	
S 31600	316	0.08 max	2.00	0.045	0.030	1.00	16.00–18.00	10.00 14.00	molybdenum 2.00–3.00; nitrogen 0.10 max
S 38400	384	0.08 max	2.00	0.045	0.030	1.00	15.00–17.00	17.00 19.00	
S 41000	410	0.15 max	1.00	0.040	0.030	1.00	11.50–13.50		
S 42900	429	0.12 max	1.00	0.040	0.030	1.00	14.00–16.00		
S 43000	430	0.12 max	1.00	0.040	0.030	1.00	16.00–18.00		
S 43100	431	0.20 max	1.00	0.040	0.030	1.00	15.00–17.00	1.25 2.50	
S 44004	440C	0.95–1.20	1.00	0.040	0.030	1.00	16.00–18.00		molybdenum 0.75 max
S 44401	...	0.025	1.00	0.040	0.030	...	17.5–19.5	1.00	molybdenum 1.75–2.50; nitrogen 0.035 max; (Ti + Nb) 0.20 + 4 (C + N) min to 0.80 max
S 30430	XM-7	0.10 max	2.00	0.045	0.030	1.00	17.00–19.00	8.00 10.00	copper 3.00–4.00
S 44625	XM-27[A]	0.010 max	0.40	0.020	0.020	0.40	25.0–27.5	0.50 max	molybdenum 0.75–1.50; copper 0.2 max; nitrogen 0.0150; max; nickel + copper 0.5 max
S 44700	...	0.010	0.30	0.025	0.020	0.20	28.00–30.00	0.15 max	Mo 3.50–4.20; N 0.020 max; C + N 0.025 max; Cu 0.15 max
S 44800	...	0.010	0.30	0.025	0.020	0.20	28.00–30.00	2.00 2.50	Mo 3.50–4.20; N 0.020 max; C + N 0.025 max; Cu 0.15 max

[A] Product analysis tolerance over the maximum limit for carbon and nitrogen to be 0.002 %.

TABLE 2 Mechanical Property Requirements for Lightly Drafted and Annealed Wire

Type	Tensile Strength, ksi(MPa)			
	Sizes 0.156 in. (3.96 mm) in Diameter and Over		Sizes Under 0.156 in. (3.96 mm) in Diameter	
	Lightly Drafted	Annealed	Lightly Drafted	Annealed
302	85–105 (585–725)	80–100 (550–690)	85–115 (585–795)	80–105 (550–725)
304	80–100 (550–690)	75–95 (520–655)	80–110 (550–760)	75–105 (520–725)
305	80–100 (550–690)	75–95 (520–655)	80–110 (550–760)	75–105 (520–725)
316	80–100 (550–690)	75–95 (520–655)	80–110 (550–760)	75–105 (520–725)
384	70–90 (485–620)	65–85 (450–585)	70–100 (485–690)	65–95 (450–655)
410	70–90 (485–620)	65–85 (450–585)	70–90 (485–620)	65–85 (450–585)
429	70–90 (485–620)	65–85 (450–585)	70–90 (485–620)	65–85 (450–585)
430	70–90 (485–620)	65–85 (450–585)	70–90 (485–620)	65–85 (450–585)
431	100–120 (690–830)	90–110 (620–760)	100–125 (690–860)	90–110 (620–760)
440C	100–120 (690–830)	90–110 (620–760)	100–125 (690–860)	90–115 (620–795)
XM-7	70–90 (485–620)	70–90 (485–620)	75–105 (520–725)	70–100 (485–690)
XM-27	75–95 (520–655)	60–80 (415–550)	75–95 (520–655)	60–80 (415–550)
S 28200	110–135 (760–935)	105–130 (725–900)	110–135 (760–935)	105–130 (725–900)
S 44401	70–90 (485–620)	60–80 (415–550)	70–90 (485–620)	60–80 (145–550)
S 44700	80–105 (550–725)	80–100 (550–690)	85–115 (585–795)	85–105 (585–725)
S 44800	80–105 (550–725)	80–100 (550–690)	85–115 (585–795)	85–105 (585–725)

TABLE 3 Mechanical Property Requirements for Lightly Drafted and Annealed Bars

Type	Tensile Strength, ksi(MPa)	
	Lightly Drafted	Annealed
302	80–100 (550–690)	75–95 (520–655)
304	75–95 (520–655)	70–90 (485–620)
305	75–95 (520–655)	70–90 (485–620)
316	75–95 (520–655)	70–90 (485–620)
384	65–85 (450–585)	60–80 (415–550)
410	70–90 (485–620)	65–85 (450–585)
429	70–90 (485–620)	65–85 (450–585)
430	70–90 (485–620)	65–85 (450–585)
431	100–120 (690–830)	90–110 (620–760)
440C	100–120 (690–830)	90–110 (620–760)
XM-7	70–90 (485–620)	65–85 (450–585)
XM-27	70–90 (485–620)	60–80 (415–550)
S 28200	110–135 (760–935)	105–130 (725–900)
S 44401	70–90 (485–620)	60–80 (415–550)
S 44700	80–105 (550–725)	80–100 (550–690)
S 44800	80–105 (550–725)	80–100 (550–690)

The American Society for Testing and Materials takes no position respecting the validity of any patent rights asserted in connection with any item mentioned in this standard. Users of this standard are expressly advised that determination of the validity of any such patent rights, and the risk of infringement of such rights, are entirely their own responsibility.

This standard is subject to revision at any time by the responsible technical committee and must be reviewed every five years and if not revised, either reapproved or withdrawn. Your comments are invited either for revision of this standard or for additional standards and should be addressed to ASTM Headquarters. Your comments will receive careful consideration at a meeting of the responsible technical committee, which you may attend. If you feel that your comments have not received a fair hearing you should make your views known to the ASTM Committee on Standards, 1916 Race St., Philadelphia, Pa. 19103.

Designation: A 505 – 78

An American National Standard

Standard Specification for
GENERAL REQUIREMENTS FOR STEEL AND STRIP, ALLOY, HOT-ROLLED AND COLD-ROLLED[1]

This standard is issued under the fixed designation A 505; the number immediately following the designation indicates the year of original adoption or, in the case of revision, the year of last revision. A number in parentheses indicates the year of last reapproval. A superscript epsilon (ε) indicates an editorial change since the last revision or reapproval.

This specification has been approved for use by agencies of the Department of Defense and for listing in the DoD Index of Specifications and Standards.

1. Scope

1.1 This specification covers the general requirements for hot-rolled and cold-rolled alloy steel sheet and strip. Alloy sheet and strip are used where specific mechanical properties are required over that which can be obtained from carbon sheet and strip. They should be heat treated to obtain optimum mechanical properties.

1.2 The values stated in inch-pound units are to be regarded as the standard.

2. Applicable Documents

2.1 *ASTM Standards:*
A 370 Methods and Definitions for Mechanical Testing of Steel Products[2]
A 506 Specification for Steel Sheet and Strip, Alloy, Hot-Rolled and Cold-Rolled, Regular Quality[2]
A 507 Specification for Steel Sheet and Strip, Alloy, Hot-Rolled and Cold-Rolled, Drawing Quality[2]
A 700 Practices for Packaging, Marking, and Loading Methods for Steel Products for Domestic Shipment[2]
E 8 Methods of Tension Testing of Metallic Materials[3]
E 18 Test Methods for Rockwell Hardness and Rockwell Superficial Hardness of Metallic Materials[3]
E 30 Methods for Chemical Analysis of Steel, Cast Iron, Open-Hearth Iron, and Wrought Iron[4]
E 59 Method of Sampling Steel and Iron for Determination of Chemical Composition[4]
E 112 Methods for Determining Average Grain Size[5]

3. Description of Terms

3.1 Hot-rolled and cold-rolled alloy steel sheet and strip shall be classified as prescribed in Table 1.

4. Chemical Requirements

4.1 *Standard Steels*—Hot-rolled and cold-rolled alloy steel sheet and strip shall be furnished in accordance with the chemical compositions prescribed in Table 2.

NOTE—Table 2 is a list of chemical compositions that are supplied as sheet and strip products. The inclusion of an analysis in this list should not be construed to mean that all analyses shown are equally available in sheet and strip. Those compositions indicated by footnote *b* are commonly produced in sheet and strip and are, in general, more readily available than other compositions. They should be specified whenever possible.

---

[1] This specification is under the jurisdiction of ASTM Committee A-1 on Steel, Stainless Steel and Related Alloys, and is the direct responsibility of Subcommittee A01.19 on Steel Sheet and Strip.
Current edition approved July 28, 1978. Published September 1978. Originally published as A 505 – 64. Last previous edition A 505 – 75.
[2] *Annual Book of ASTM Standards*, Vol 01.03.
[3] *Annual Book of ASTM Standards*, Vol 03.01.
[4] *Annual Book of ASTM Standards*, Vol 03.05.
[5] *Annual Book of ASTM Standards*, Vol 03.03.

4.2 *Steels Other Than Standard*—When compositions other than those shown in Table 2 are required, the composition limits shall be prepared using the ranges and limits shown in Table 3.

4.3 *Cast or Heat (formerly Ladle) Analysis*—An analysis of each cast or heat of steel shall be made by the manufacturer to determine the percentage of elements specified in the grade shown on the purchase order. The chemical composition thus determined shall be reported to the purchaser or his representative upon request, and shall conform to the requirements specified in Table 2 for the grade ordered or to such other analysis as was agreed upon.

4.4 *Product, Check, or Verification Analysis*—A product analysis may be made by the purchaser from finished material representing each cast or heat. The composition thus determined shall conform to the requirements specified in Table 2 for the grade ordered or to such other composition as was agreed upon subject to the product analysis tolerances specified in Table 4.

4.5 *Referee Analysis*—In case a referee analysis is required to resolve a dispute concerning the results of a chemical analysis, the procedure for performing the referee analysis shall be in accordance with the latest issue of Methods E 30.

5. Metallurgical Structure

5.1 *Grain Size:*

5.1.1 The austenitic grain size shall be predominantly 5 or finer with an occasional grain as large as 3 permissible.

5.1.2 *Number of Tests*—One test shall be taken from each cast or heat.

5.1.3 *Test Location*—The sample shall be representative of the cast or heat.

5.1.4 *Test Method*—The sample shall be evaluated in accordance with Methods E 112.

6. Mechanical Requirements

6.1 *Requirements*—This specification is applicable to material that may be furnished to numerous qualities, chemical compositions, and physical conditions. With the same chemistry, different mechanical properties may be expected for each of the qualities and physical conditions specified. For this reason, mechanical properties are frequently negotiated and when required, shall be specified in accordance with the quality, chemical compositions, and physical conditions ordered.

6.2 *Number of Tests*—The number of specimens, test location, and specimen orientation shall be in accordance with the applicable product specification.

6.3 *Test Specimens*—Test specimens shall be the full thickness of the material and shall be taken from the material in its finished condition.

6.3.1 Tension test specimens shall be taken approximately midway between the center and the edge of the sheet parallel to the direction of rolling. Tension test specimens shall conform to the form and dimensions shown in Method E 8.

6.3.2 Bend test specimens may be taken from any location parallel with or transverse to the direction of rolling, as required to perform the bend requirements of the product specification. When required, transverse specimens shall have the axis of the specimen 90 deg to the final rolling (see Methods and Definitions A 370).

6.3.2.1 Bend test specimens shall be at least ¾ in. (19 mm) wide or the same as the width of the material if it is less than ¾ in. wide. The length shall be sufficient to permit bending to the angle specified.

6.3.2.2 The edges of bend test specimens shall be practically free of burrs. Filing or machining to remove burrs is permissible.

6.4 *Test Methods*—When the tests shown below are required by the product specification, the method of test shall be in accordance with the following applicable ASTM methods:

Type of Test	ASTM Designation
Product Analysis	E 59
Tension	E 8
Bend	A 370
Hardness	E 18
Grain Size	E 112

7. Retests

7.1 If any test specimen shows defective machining or develops flaws, it may be discarded and another specimen substituted.

7.2 If the percentage elongation of any tension test specimen is less than that specified and any part of the fracture is more than ¾ in. (19 mm) from the center of a 2-in. (50-mm) specimen, as indicated by scribe scratches marked on the specimen before testing, a retest will be allowed.

7.3 If the results of any tests do not conform to the specified requirements, retests may be made on double the original number of specimens from the same lot, each of which shall conform to the requirements specified. If the results of the retests do not conform to the specified requirements, the lot shall be rejected.

8. Rework and Retreatment

8.1 If test specimens selected to represent the material fail to conform to the requirements specified, the material may be retreated. The producer may retreat the material on metallurgical evidence that the cause of failure is curable and that the quality of the material shall conform to the applicable specification. Retests shall be made in accordance with the product specification.

9. Dimensions

9.1 The dimensional tolerances (see Table 5) for hot-rolled and cold-rolled alloy steel sheet and strip shall be in accordance with Tables 6 to 24.

10. Finish and Appearance

10.1 *Surface Finish*—The degree or amount of surface imperfections of material in cut lengths shall be such that only a reasonable amount of metal finishing is required. Slight surface imperfections that are completely removable without reducing the section thickness below the minimum permissible dimensional tolerance limits shall not be considered injurious defects. Coils may contain some abnormal defects such as welds, holes, etc., which render a portion of the coil unusable since the inspection of coils does not afford the same opportunity to reject portions containing defects as is the case with cut lengths. However, an excessive number of abnormal imperfections is cause for rejection.

10.2 *Edges*—The types of edges procurable in hot-rolled and cold-rolled alloy steel sheet and strip are as follows:

10.2.1 *Hot-Rolled Sheet and Strip:*

10.2.1.1 *Mill Edge*—Normal edge produced in hot rolling which does not conform to any definite contour.

10.2.1.2 *Cut Edge*—Approximate square edge resulting from the cutting of flat-rolled alloy sheet into one or more desired widths by means of rotary knives (slit edge) or blade shears (sheared edge).

10.2.1.3 *Square Edge*—Type of mill edge produced by hot-edge rolling. Furnished on strip only.

10.2.2 *Cold Rolled Sheet:*

10.2.2.1 *Cut Edge*—Same description as shown for hot-rolled sheet and strip.

10.2.3 *Cold Rolled Strip:*

10.2.3.1 *No. 1 Edge*—Prepared edge of a specified round or square contour that is produced when a very accurate width is required.

10.2.3.2 *No. 2 Edge*—Natural mill edge carried through the cold rolling from the hot-rolled strip without additional processing of the edge.

10.2.3.3 *No. 3 Edge*—Approximately square edge produced by slitting.

10.2.3.4 *No. 4 Edge*—Rounded edge produced by edge rolling either the natural edge of hot-rolled strip or slit-edge strip. Used when an approximately round edge is desired and when the finish of the edge is not important.

10.2.3.5 *No. 5 Edge*—Square edge produced by edge rolling or filing for the purpose of eliminating burr.

10.2.3.6 *No. 6 Edge*—Square edge produced by edge rolling the natural edge of hot-rolled strip or slit-edge strip, when the width tolerances and finish required are not as exacting as for the No. 1 edge.

11. Workmanship

11.1 The material shall be clean, sound, of uniform quality and conditions, and free of foreign material, and internal or external defects which would make the material unsuitable for the intended application. Care shall be taken to avoid cracks, seams, slivers, grooves, laminations, pits, blisters, buckles, coil breaks, creases, holes, pickling stains and patches, pipe, ragged and torn edges, and "rolled-in" dirt and scale.

12. Packaging

12.1 Unless otherwise specified, the material shall be packaged and protected in accordance

A 505

with the methods shown in Recommended Practices A 700.

13. Marking

13.1 As a minimum requirement, the material shall be identified by having the producer's name, ASTM designation, weight, purchaser's order number, and material identification legibly stenciled on top of each lift or shown on a tag attached to each coil or shipping unit. If each sheet is to be marked the purchaser must specify.

14. Inspection

14.1 When purchaser's order stipulates that inspection and tests (except product analyses) for acceptance on the steel be made prior to shipment from the mill, the producer shall afford the purchaser's inspector all reasonable facilities to satisfy him that the steel is being produced and furnished in accordance with the specification. Mill inspection by the purchaser shall not interfere unnecessarily with the manufacturer's operation.

15. Certification and Reports

15.1 Upon request of the purchaser in the contract or order, a producer's certification that the material was produced and tested in accordance with this specification shall be furnished. A report of the test results may be included if required.

TABLE 1 Classification of Hot-Rolled and Cold-Rolled Sheet and Strip

	HOT-ROLLED SHEET AND STRIP			
	Width, in.[a]			
Thickness, in.[a]	Up to 6, incl	Over 6 to $23^{15}/_{16}$, incl	24 to 48, incl	Over 48
0.2299 to 0.2031, incl	...	strip	sheet	...
0.2030 to 0.1800, incl	strip	strip	sheet	...
0.1799 and thinner	strip	strip	sheet	sheet
	COLD-ROLLED SHEET AND STRIP			
	Width, in.[a]			
Thickness, in.[a]	Up to $23^{15}/_{16}$, incl		24 to 48, incl	Over 48
0.2499 to 0.2300, incl	strip		...	...
0.2299 to 0.1800, incl	strip		sheet	...
0.1799 and thinner	strip		sheet	sheet

[a] 1 in. = 25.4 mm.

TABLE 2 Chemical Composition Limits[a]

NOTE 1—Grades shown in this table with prefix letter E generally are manufactured by the basic electric-furnace process. All others are normally manufactured by the basic open-hearth process but may be manufactured by the basic electric-furnace process with adjustments in phosphorus and sulfur.

NOTE 2—The phosphorus and sulfur limitations for each process are as follows:

| Basic electric-furnace | 0.025 max, % | Acid electric-furnace | 0.050 max, % |
| Basic open-hearth or basic-oxygen | 0.040 max, % | Acid open-hearth | 0.050 max, % |

NOTE 3—Minimum silicon limit for acid open-hearth or acid electric-furnace alloy steel is 0.15%.

NOTE 4—Small quantities of certain elements are present in alloy steels which are not specified or required. These elements are considered as incidental and may be present to the following maximum amounts: copper, 0.35%; nickel, 0.25%; chromium, 0.20%, and molybdenum, 0.06%.

NOTE 5—Where minimum and maximum sulfur content is shown, it is indicative of resulfurized steels.

NOTE 6—The chemical ranges and limits shown in Table 2 are subject to the product analysis tolerances shown in Table 4.

UNS Designation[c]	AISI Number	Chemical Composition Ranges and Limits, %								Corresponding SAE Number
		Carbon	Manganese	Phosphorus, max	Sulfur, max	Silicon	Nickel	Chromium	Molybdenum	
...	E3310	0.08–0.13	0.45–0.60	0.025	0.025	0.15–0.30	3.25–3.75	1.40–1.75	...	3310
G 40120	4012	0.09–0.14	0.75–1.00	0.040	0.040	0.15–0.30	...	...	0.15–0.25	4012
G 41180	4118	0.18–0.23	0.70–0.90	0.040	0.040	0.15–0.30	...	0.40–0.60	0.08–0.15	4118
G 41300	4130[b]	0.28–0.33	0.40–0.60	0.040	0.040	0.15–0.30	...	0.80–1.10	0.15–0.25	4130
G 41350	4135[b]	0.33–0.38	0.70–0.90	0.040	0.040	0.15–0.30	...	0.80–1.10	0.15–0.25	4135
G 41370	4137	0.35–0.40	0.70–0.90	0.040	0.040	0.15–0.30	...	0.80–1.10	0.15–0.25	4137
G 41400	4140[b]	0.38–0.43	0.75–1.00	0.040	0.040	0.15–0.30	...	0.80–1.10	0.15–0.25	4140
G 41420	4142[b]	0.40–0.45	0.75–1.00	0.040	0.040	0.15–0.30	...	0.80–1.10	0.15–0.25	4142
G 41450	4145[b]	0.43–0.48	0.75–1.00	0.040	0.040	0.15–0.30	...	0.80–1.10	0.15–0.25	4145
G 41470	4147	0.45–0.50	0.75–1.00	0.040	0.040	0.15–0.30	...	0.80–1.10	0.15–0.25	4147
G 41500	4150[b]	0.48–0.53	0.75–1.00	0.040	0.040	0.15–0.30	...	0.80–1.10	0.15–0.25	4150
G 43200	4320	0.17–0.22	0.45–0.65	0.040	0.040	0.15–0.30	1.65–2.00	0.40–0.60	0.20–0.30	4320
G 43400	4340[b]	0.38–0.43	0.60–0.80	0.040	0.040	0.15–0.30	1.65–2.00	0.70–0.90	0.20–0.30	4340
G 43406	E4340[b]	0.38–0.43	0.65–0.85	0.025	0.025	0.15–0.30	1.65–2.00	0.70–0.90	0.20–0.30	E4340
G 45200	4520	0.18–0.23	0.45–0.65	0.040	0.040	0.15–0.30	...	...	0.45–0.60	4520
G 46150	4615	0.13–0.18	0.45–0.65	0.040	0.040	0.15–0.30	1.65–2.00	...	0.20–0.30	4615
G 46200	4620	0.17–0.22	0.45–0.65	0.040	0.040	0.15–0.30	1.65–2.00	...	0.20–0.30	4620
G 47180	4718	0.16–0.21	0.70–0.90	0.040	0.040	0.15–0.30	0.90–1.20	0.35–0.55	0.30–0.40	4718
G 48150	4815	0.13–0.18	0.40–0.60	0.040	0.040	0.15–0.30	3.25–3.75	...	0.20–0.30	4815
G 48200	4820	0.18–0.23	0.50–0.70	0.040	0.040	0.15–0.30	3.25–3.75	...	0.20–0.30	4820
G 50150	5015	0.12–0.17	0.30–0.50	0.040	0.040	0.15–0.30	...	0.30–0.50	...	5015
G 50460	5046	0.43–0.50	0.75–1.00	0.040	0.040	0.15–0.30	...	0.20–0.35	...	5046
G 51150	5115	0.13–0.18	0.70–0.90	0.040	0.040	0.15–0.30	...	0.70–0.90	...	5115
G 51300	5130	0.28–0.33	0.70–0.90	0.040	0.040	0.15–0.30	...	0.80–1.10	...	5130
G 51320	5132	0.30–0.35	0.60–0.80	0.040	0.040	0.15–0.30	...	0.75–1.00	...	5132
G 51400	5140[b]	0.38–0.43	0.70–0.90	0.040	0.040	0.15–0.30	...	0.70–0.90	...	5140
G 51500	5150[b]	0.48–0.53	0.70–0.90	0.040	0.040	0.15–0.30	...	0.70–0.90	...	5150
G 51600	5160[b]	0.55–0.65	0.75–1.00	0.040	0.040	0.15–0.30	...	0.70–0.90	...	5160
G 15116	E51100	0.95–1.10	0.25–0.45	0.025	0.025	0.15–0.30	...	0.90–1.15	...	51100
G 15216	E52100	0.95–1.10	0.25–0.45	0.025	0.025	0.15–0.30	...	1.30–1.60	...	52100
G 61500	6150[b]	0.48–0.53	0.70–0.90	0.040	0.040	0.15–0.30	...	0.80–1.10	(0.15 min V)†	6150
G 86150	8615[b]	0.13–0.18	0.70–0.90	0.040	0.040	0.15–0.30	0.40–0.70	0.40–0.60	0.15–0.25	8615
G 86170	8617[b]	0.15–0.20	0.70–0.90	0.040	0.040	0.15–0.30	0.40–0.70	0.40–0.60	0.15–0.25	8617
G 86200	8620[b]	0.18–0.23	0.70–0.90	0.040	0.040	0.15–0.30	0.40–0.70	0.40–0.60	0.15–0.25	8620
G 86300	8630[b]	0.28–0.33	0.70–0.90	0.040	0.040	0.15–0.30	0.40–0.70	0.40–0.60	0.15–0.25	8630
G 86400	8640[b]	0.38–0.43	0.75–1.00	0.040	0.040	0.15–0.30	0.40–0.70	0.40–0.60	0.15–0.25	8640
G 86420	8642	0.40–0.45	0.75–1.00	0.040	0.040	0.15–0.30	0.40–0.70	0.40–0.60	0.15–0.25	8642
G 86450	8645	0.43–0.48	0.75–1.00	0.040	0.040	0.15–0.30	0.40–0.70	0.40–0.60	0.15–0.25	8645
G 86500	8650	0.48–0.53	0.75–1.00	0.040	0.040	0.15–0.30	0.40–0.70	0.40–0.60	0.15–0.25	8650
G 86550	8655	0.50–0.60	0.75–1.00	0.040	0.040	0.15–0.30	0.40–0.70	0.40–0.60	0.15–0.25	8655
G 86600	8660	0.55–0.65	0.75–1.00	0.040	0.040	0.15–0.30	0.40–0.70	0.40–0.60	0.15–0.25	8660
G 87200	8720	0.18–0.23	0.70–0.90	0.040	0.040	0.15–0.30	0.40–0.70	0.40–0.60	0.20–0.30	8720
G 87350	8735[b]	0.33–0.38	0.75–1.00	0.040	0.040	0.15–0.30	0.40–0.70	0.40–0.60	0.20–0.30	...
G 87400	8740	0.38–0.43	0.75–1.00	0.040	0.040	0.15–0.30	0.40–0.70	0.40–0.60	0.20–0.30	8740
G 92600	9260	0.55–0.65	0.70–1.00	0.040	0.040	1.80–2.20	...	...	...	9260
G 92620	9262[b]	0.55–0.65	0.75–1.00	0.040	0.040	1.80–2.20	...	0.25–0.40	...	9262
...	E9310	0.08–0.13	0.45–0.65	0.025	0.025	0.20–0.35	3.00–3.50	1.00–1.40	0.08–0.15	9310

A 505

TABLE 2 *Continued*

[a] The ranges and limits in this table apply to steel not exceeding 200 in.² (12 900 mm²) in cross-sectional area.
[b] These compositions are commonly produced as sheet and strip and are more readily available than other analysis.
[c] New designation established in accordance with ASTM E 527 and SAE J1086. Recommended Practice for Numbering Metals and Alloys (UNS).
† Editorially revised.

TABLE 3 Cast or Heat (formerly Ladle) Chemical Ranges and Limits for Open-Hearth, Basic-Oxygen, and Electric-Furnace Alloy Steels

NOTE 1—Boron steels can be expected to have 0.0005%, min, boron content.
NOTE 2—The chemical ranges and limits of alloy steels are subject to the product analysis tolerances shown in Table 4.

Element	When Maximum is of Specified Range, %	Range, %		
		Open-Hearth Steels	Electric-Furnace Steels	Maximum Limit, %[a]
Carbon	To 0.55, incl	0.05	0.05	...
	Over 0.55–0.70, incl	0.08	0.07	...
	Over 0.70–0.80, incl	0.10	0.09	...
	Over 0.80–0.95, incl	0.12	0.11	...
	Over 0.95–1.35, incl	0.13	0.12	...
Manganese	To 0.60, incl	0.20	0.15	...
	Over 0.60–0.90, incl	0.20	0.20	...
	Over 0.90–1.05, incl	0.25	0.25	...
	Over 1.05–1.90, incl	0.30	0.30	...
	Over 1.90–2.10, incl	0.40	0.35	...
Phosphorus	Basic open-hearth steel	...	...	0.035
	Basic oxygen steel	...	...	0.035
	Basic electric-furnace steel	...	...	0.025
Sulfur	To 0.050	0.015	0.015	...
	Over 0.050–0.07, incl	0.020	0.02	...
	Over 0.07–0.10, incl	0.040	0.04	...
	Over 0.10–0.14, incl	0.050	0.05	...
	Basic open-hearth steel	...	...	0.040
	Basic oxygen steel	...	...	0.040
	Basic electric-furnace steel	...	...	0.025
Silicon	To 0.15, incl	0.08	0.08	...
	Over 0.15–0.20, incl	0.10	0.10	...
	Over 0.20–0.40, incl	0.15	0.15	...
	Over 0.40–0.60, incl	0.20	0.20	...
	Over 0.60–1.00, incl	0.30	0.30	...
	Over 1.00–2.20, incl	0.40	0.35	...
Copper	To 0.60, incl	0.20	0.20	...
	Over 0.60–1.50, incl	0.30	0.30	...
	Over 1.50–2.00, incl	0.35	0.35	...
Nickel	To 0.50, incl	0.20	0.20	...
	Over 0.50–1.50, incl	0.30	0.30	...
	Over 1.50–2.00, incl	0.35	0.35	...
	Over 2.00–3.00, incl	0.40	0.40	...
	Over 3.00–5.30, incl	0.50	0.50	...
	Over 5.30–10.00, incl	1.00	1.00	...

TABLE 3 *Continued*

Element	When Maximum is of Specified Range, %	Range, %		Maximum Limit, %[a]
		Open-Hearth Steels	Electric-Furnace Steels	
Chromium	To 0.40, incl	0.15	0.15	...
	Over 0.40–0.90, incl	0.20	0.20	...
	Over 0.90–1.05, incl	0.25	0.25	...
	Over 1.05–1.60, incl	0.30	0.30	...
	Over 1.60–1.75, incl	[b]	0.35	...
	Over 1.75–2.10, incl	[b]	0.40	...
	Over 2.10–3.99, incl	[b]	0.50	...
Molybdenum	To 0.10, incl	0.05	0.05	...
	Over 0.10–0.20, incl	0.07	0.07	...
	Over 0.20–0.50, incl	0.10	0.10	...
	Over 0.50–0.80, incl	0.15	0.15	...
	Over 0.80–1.15, incl	0.20	0.20	...
Tungsten	To 0.50, incl	0.20	0.20	...
	Over 0.50–1.00, incl	0.30	0.30	...
	Over 1.00–2.00, incl	0.50	0.50	...
	Over 2.00–4.00, incl	0.60	0.60	...
Vanadium	To 0.25, incl	0.05	0.05	...
	Over 0.25–0.50, incl	0.10	0.10	...
Aluminum	To 0.10, incl	0.05	0.05	...
	Over 0.10–0.20, incl	0.10	0.10	...
	Over 0.20–0.30, incl	0.15	0.15	...
	Over 0.30–0.80, incl	0.25	0.25	...
	Over 0.80–1.30, incl	0.35	0.35	...
	Over 1.30–1.80, incl	0.45	0.45	...

[a] Applies to nonrephosphorized and nonresulfurized steels.
[b] Not normally produced as open-hearth or basic-oxygen.

TABLE 4 Product Analysis Tolerances Over or Under Specified Range or Limit

Element	Limit or Maximum of Specified Element, %	Tolerance Over Maximum Limit or Under Minimum Limit
Carbon	To 0.30, incl	0.01
	Over 0.30–0.75, incl	0.02
	Over 0.75	0.03
Manganese	To 0.90, incl	0.03
	Over 0.90–2.10, incl	0.04
Phosphorus	Over max only	0.005
Sulfur	Over max only	0.005
Silicon	To 0.35, incl	0.02
	Over 0.35–2.20, incl	0.05
Copper	To 1.00, incl	0.03
	Over 1.00–2.00, incl	0.05
Nickel	To 1.00, incl	0.03
	Over 1.00–2.00, incl	0.05
	Over 2.00–5.30, incl	0.07
	Over 5.30–10.00, incl	0.10
Chromium	To 0.90, incl	0.03
	Over 0.90–2.10, incl	0.05
	Over 2.10–3.99, incl	0.10
Molybdenum	To 0.20, incl	0.01
	Over 0.20–0.40, incl	0.02
	Over 0.40–1.15, incl	0.03
Vanadium	To 0.10, incl	0.01
	Over 0.10–0.25, incl	0.02
	Over 0.25–0.50, incl	0.03
	Min value specified, check under minimum limit	0.01
Tungsten	To 1.00, incl	0.04
	Over 1.00–4.00, incl	0.08
Aluminum	To 0.10, incl	0.03
	Over 0.10–0.20, incl	0.04
	Over 0.20–0.30, incl	0.05
	Over 0.30–0.80, incl	0.07
	Over 0.80–1.80, incl	0.10

A 505

TABLE 5 Dimension Tolerance Tables

NOTE—The following table shows the dimensional tolerance table number applicable to the different product classifications and rolling methods. Continuous mill tolerances apply unless otherwise stated.

	Hot-Rolled					Cold-Rolled							
	Sheet				Strip	Sheet		Strip					
	Hand Mill		Continuous Mill			Continuous Mill		Continuous Mill					
											Edges		
	Mill Edge	Cut Edge	Mill Edge	Cut Edge		Mill Edge	Cut Edge	1	2	3	4	5	6
									Mill	Slit			
Thickness	7	7	6	6	15	8	8	20	20	20	20	20	20
Width	9	10	9	10	17	9	10	23	21	22	23	23	23
Length	11	11	11	11	18	11	11	24	24	24	24	24	24
Flatness for Cut Lengths	12	12	12	12	19	12	12						
Camber	13	13	13	13	13	13	13	13	13	13	13	13	13
Out-of-Square	...	14	...	14	...	...	...	...	...	...	...	...	...
Crown	...	...	...	...	16	...	...	...	...	...	...	...	...

TABLE 6 Thickness[a] Tolerances for Hot-Rolled Sheet (Continuous Mill Product), Coils, or Cut Lengths

Width, in. (mm)	Thickness, in. (mm)				
	0.2299 to 0.1800 (5.84 to 4.57), incl	0.1799 to 0.0972 (4.56 to 2.47), incl	0.0971 to 0.0822 (2.46 to 2.09), incl	0.0821 to 0.0710 (2.08 to 1.80), incl	0.0709 to 0.0568 (1.79 to 1.44), incl
	Thickness Tolerances, Over and Under, in. (mm), for Specified Widths and Thicknesses				
24 to 32 (610 to 810), incl	0.009 (0.23)	0.008 (0.20)	0.007 (0.18)	0.007 (0.18)	0.006 (0.15)
Over 32 to 40 (810 to 1020), incl	0.009 (0.23)	0.009 (0.23)	0.008 (0.20)	0.007 (0.18)	0.006 (0.15)
Over 40 to 48 (1020 to 1220), incl	0.010 (0.25)	0.010 (0.25)	0.008 (0.20)	0.007 (0.18)	0.006 (0.15)
Over 48 to 60 (1220 to 1520), incl	...	0.010 (0.25)	0.008 (0.20)	0.007 (0.18)	0.007 (0.18)
Over 60 to 70 (1520 to 1780), incl	...	0.011 (0.28)	0.009 (0.23)	0.008 (0.20)	0.007 (0.18)
Over 70 to 80 (1780 to 2030), incl	...	0.012 (0.30)	0.009 (0.23)	0.008 (0.20)	...
Over 80 to 90 (2030 to 2290), incl	...	0.012 (0.30)	0.010 (0.25)	...	...
Over 90 (2290)	...	0.012 (0.30)	...	...	...

[a] Thickness is measured at any point on the sheet not less than ⅜ in. (9.5 mm) from a cut edge and not less than ¾ in. (19 mm) from a mill edge.

A 505

TABLE 7 Thickness[a] Tolerances, Hot-Rolled Sheets, Hand Mill Product

Thickness, in.[b]	Tolerance, in.[b], Over and Under
0.229 to 0.188, incl	0.015
Under 0.188 to 0.146, incl	0.014
Under 0.146 to 0.131, incl	0.012
Under 0.131 to 0.115, incl	0.010
Under 0.115 to 0.099, incl	0.009
Under 0.099 to 0.084, incl	0.008
Under 0.084 to 0.073, incl	0.007
Under 0.073 to 0.059, incl	0.006
Under 0.059 to 0.041, incl	0.005
Under 0.041 to 0.027, incl	0.004
Under 0.027 to 0.019, incl	0.003

[a] Thickness is measured at any point on the sheet not less than 3/8 in. from a cut edge and not less than 3/4 in. from a mill edge.
[b] 1 in. = 25.4 mm.

TABLE 8 Thickness[a] Tolerances for Cold-Rolled Sheets, Coils, or Cut Lengths

Specified Width, in.[b]	Thickness, in.[b]								
	0.2299 to 0.1800, incl	0.1799 to 0.1420, incl	0.1419 to 0.0972, incl	0.0971 to 0.0822, incl	0.0821 to 0.0710, incl	0.0709 to 0.0568, incl	0.0567 to 0.0509, incl	0.0508 to 0.0314, incl	0.0313 to 0.0195, incl
	Thickness Tolerances, Over and Under, in.[b], for Specified Widths and Thicknesses								
24 to 32, incl	0.008	0.008	0.007	0.006	0.005	0.005	0.005	0.004	0.003
Over 32 to 40, incl	0.009	0.009	0.008	0.007	0.006	0.005	0.005	0.004	0.003
Over 40 to 48, incl	0.010	0.010	0.009	0.007	0.006	0.005	0.005	0.004	0.003
Over 48 to 60, incl	...	0.010	0.010	0.008	0.006	0.006	0.005	0.004	0.003
Over 60 to 70, incl	...	0.011	0.010	0.009	0.007	0.006	0.006	0.005	...
Over 70 to 80, incl	...	0.012	0.011	0.009	0.007	...	...	...	...
Over 80 to 90, incl	...	0.012	0.012	...	...	...	...	...	...
Over 90	...	0.012	0.012	...	...	...	...	...	...

[a] Thickness is measured at any point on the sheet not less than 3/8 in. from a cut edge and not less than 3/4 in. from a mill edge.
[b] 1 in. = 25.4 mm.

TABLE 9 Width Tolerances for Mill Edge of Hot or Cold-Rolled Sheets, Coils, or Cut Lengths

Specified Widths, in.[a]	Tolerances Over Specified Width, in.[a], No Tolerance Under
24 to 26, excl	13/16
26 to 28, excl	15/16
28 to 35, excl	1 1/8
35 to 50, excl	1 1/4
50 to 60, excl	1 1/2
60 to 65, excl	1 5/8
65 to 70, excl	1 3/4
70 to 80, excl	1 7/8
80 and over	2

[a] 1 in. = 25.4 mm.

A 505

TABLE 10 Width Tolerances for Cut Edge of Hot or Cold-Rolled Sheets, Coils, or Cut Lengths

Specified Width, in.[a]	Tolerance Over Specified Width, in.[a], No Tolerance Under
24 to 30, incl	3/16
Over 30 to 50, incl	1/4
Over 50 to 80, incl	5/16
Over 80	3/8

[a] 1 in. = 25.4 mm.

TABLE 11 Length Tolerances for Hot or Cold-Rolled Sheet

Specified Length, in. (mm)	Tolerance Over Specified Length, in. (mm), No Tolerance Under
24 to 30 (610 to 760), incl	1/4 (6.4)
Over 30 to 60 (760 to 1520), incl	1/2 (12.7)
Over 60 to 120 (1520 to 3050), incl	3/4 (19)
Over 120 to 156 (3050 to 3960), incl	1 (25)
Over 156 to 192 (3960 to 4880), incl	1 1/4 (31)
Over 192 to 240 (4880 to 6100), incl	1 1/2 (38)
Over 240 (6100)	1 3/4 (44)

TABLE 12 Flatness Tolerances for Cut Lengths of Hot or Cold-Rolled Sheets[a]

Specified Thickness, in.[b]	Specified Width, in.[b]	Flatness Tolerance, in.[b,c]
From 0.0195 to 0.0567, incl	24 to 36, incl	1/2
	Over 36 to 60, incl	3/4
	Over 60	1
From 0.0568 to 0.2299, incl	24 to 60, incl	1/2
	Over 60 to 72, incl	3/4
	Over 72	1

[a] Rolled or thermally treated and flattened.
[b] 1 in. = 25.4 mm.
[c] Deviations from flatness are measured by laying the sheet on a horizontal flat surface and measuring the maximum elevation above that surface to any point on the bottom surface of the sheet.

TABLE 13 Camber[a] Tolerances for Hot or Cold-Rolled Sheets and Strip

For sheet	1/4 in. (6.35 mm) in any 8 ft (2440 mm)
For strip to 1 1/2 in. (38 mm) in width, incl	1/2 in. (12.7 mm) in any 8 ft (2440 mm)
For strip over 1 1/2 to 23 15/16 in. (38.1 to 608.0 mm) in width, incl	1/4 in. (6.35 mm) in any 8 ft (2440 mm)

[a] Camber is determined by placing an 8-ft straightedge on the concave edge of the sheet or strip and measuring the greatest distance between the sheet edge or strip edge and the straightedge. When the camber shown in Table 13 is not suitable for a particular purpose, hot-rolled strip is sometimes machine straightened to a specified camber. For that requirement, the producer should be consulted.

TABLE 14 Out-Of-Square Tolerance for Hot-Rolled Sheets (Cut Edge Not Resquared)

Out-of-square is the greatest deviation of an end edge from a straight line at a right angle to a side and touching one corner. The tolerance for sheets of all gages and all sizes is 1/16 in./6 in., 10.5 mm/m or fraction thereof, of width.

It is also obtained by measuring the difference between the diagonals of the sheet. The out-of-square deviation is one half of that difference.

A 505

TABLE 15 Thickness Tolerances for Hot-Rolled Strip Coils, or Cut Lengths[a,b]

Specified Width, in.[c]	Thickness, in.[c]							
	0.2299 to 0.2031, incl	0.2030 to 0.1875, incl	0.1874 to 0.1719, incl	0.1718 to 0.1420, incl	0.1419 to 0.1121, incl	0.1120 to 0.0972, incl	0.0971 to 0.0710, incl	0.0709 to 0.0568, incl
	Thickness Tolerances, in.[c], Over and Under for Specified Widths and Thicknesses							
To 6, incl	...	0.006	0.006	0.005	0.005	0.005	0.005	0.005
Over 6 to 12, incl	0.007	0.006	0.006	0.006	0.006	0.005	0.005	0.005
Over 12 to 15, incl	0.008	0.008	0.007	0.007	0.007	0.007	0.006	0.006
Over 15 to 20, incl	0.009	0.009	0.009	0.008	0.008	0.008	0.007	0.006
Over 20 to 23 15/16, incl	0.009	0.009	0.009	0.009	0.008	0.008	0.007	0.006

[a] Thickness measurements are taken 3/8 in. from edge of strip on 1 in. or wider; and at any place on the strip when narrower than 1 in.
[b] See Table 16 for applicable crown tolerances.
[c] 1 in. = 25.4 mm.

TABLE 16 Crown Tolerances for Hot-Rolled Strip, Coils, or Cut Lengths[a]

Specified Widths, in.[b]	Additional Thickness at Center, in.[b]
Over 1 to 3½, incl	0.002
Over 3½ to 6, incl	0.003
Over 6 to 12, incl	0.004
Over 12 to 15, incl	0.005
Over 15 to 23 15/16, incl	0.006

[a] Hot-rolled alloy strip may be thicker at the center than at a point 3/8 in. in from the edge as shown in this table.
[b] 1 in. = 25.4 mm.

TABLE 17 Width Tolerances For Hot-Rolled Strip, Coils, or Cut Lengths

Specified Width, in. (mm)	Mill Edge	Cut Edge[a]		Sheared Edge
		Slit Edge		
		Thickness, in. (mm)		
		To 0.109 (2.77), incl	Over 0.109 (2.77)	
	Width Tolerance, in. (mm), Over and Under	Width Tolerance, in. (mm), Over and Under		Width Tolerance, in (mm), Over Width No Tolerance Under
To 2 (50), incl	1/32 (0.79)	0.010 (0.25)	0.016 (0.41)	1/8 (3.2)
Over 2 to 5 (50 to 130), incl	1/16 (1.59)	0.010 (0.25)	0.020 (0.51)	1/8 (3.2)
Over 5 to 9 (130 to 230), incl	3/32 (2.38)	0.016 (0.41)	0.024 (0.61)	1/8 (3.2)
Over 9 to 12 (230 to 300), incl	1/8 (3.20)	0.016 (0.41)	0.031 (0.79)	1/8 (3.2)
Over 12 to 20 (300 to 510), incl	1/4 (6.35)	0.020 (0.51)	0.031 (0.79)	1/8 (3.2)
Over 20 to 23 15/16 (510 to 610), incl	3/8 (9.50)	0.031 (0.79)	0.031 (0.79)	1/8 (3.2)

[a] Some alloy grades require annealing prior to cutting due to their high hardenability.

A 505

TABLE 18 Length Tolerances for Hot-Rolled Strip

Specified Widths, in.[a]	Length, in.[a]					
	To 60, incl	Over 60 to 120, incl	Over 120 to 240, incl	Over 240 to 360, incl	Over 360 to 480, incl	Over 480
	Length Tolerances, in.[a], Over Specified Length, No Tolerance Under					
To 3, incl	¼	⅜	½	¾	1	1½
Over 3 to 6, incl	⅜	½	⅝	¾	1	1½
Over 6 to 23¹⁵⁄₁₆, incl	½	¾	1	1¼	1½	1¾

[a] 1 in. = 25.4 mm.

TABLE 19 Flatness Tolerances for Cut Lengths of Hot-Rolled Strip As-Rolled or Thermally Treated and Flattened

½ in., max, in 8 ft (5.2 mm/m) of length

ASTM A 505

TABLE 20 Thickness Tolerances for Cold-Rolled Strip[a]

Specified Thickness, in. (mm)	Width, in. (mm)							
	3/16 to 1 (4.8 to 25), excl	1 to 3 (25 to 75), excl	3 to 6 (75 to 150), incl	Over 6 to 9 (150 to 230), incl	Over 9 to 12 (230 to 300), incl	Over 12 to 16 (300 to 410), incl	Over 16 to 20 (410 to 510), incl	Over 20 to 23 15/16 (510 to 610), incl
	Thickness Tolerances, in. (mm) Over and Under for Specified Thicknesses and Widths							
Under 0.010 (0.25)	...	...	...	...	...	...	...	...
0.010 (0.25)	0.001 (0.03)	0.001 (0.03)	0.001 (0.03)	0.001 (0.03)	0.001 (0.03)	0.001 (0.03)	0.0015 (0.04)	0.0015 (0.04)
0.011 (0.28)	0.001 (0.03)	0.001 (0.03)	0.001 (0.03)	0.001 (0.03)	0.001 (0.03)	0.0015 (0.04)	0.0015 (0.04)	0.0015 (0.04)
0.012 (0.30)	0.001 (0.03)	0.001 (0.03)	0.001 (0.03)	0.0015 (0.04)	0.0015 (0.04)	0.0015 (0.04)	0.0015 (0.04)	0.0015 (0.04)
Over 0.012 to 0.016 (0.30 to 0.41), incl	0.001 (0.03)	0.001 (0.03)	0.001 (0.03)	0.0015 (0.04)	0.0015 (0.04)	0.002 (0.05)	0.002 (0.05)	0.002 (0.05)
Over 0.016 to 0.019 (0.41 to 0.48), incl	0.001 (0.03)	0.001 (0.03)	0.0015 (0.04)	0.002 (0.05)	0.002 (0.05)	0.002 (0.05)	0.002 (0.05)	0.002 (0.05)
Over 0.019 to 0.025 (0.48 to 0.64), incl	0.001 (0.03)	0.0015 (0.04)	0.0015 (0.04)	0.002 (0.05)	0.0025 (0.06)	0.002 (0.05)	0.0025 (0.06)	0.0025 (0.06)
Over 0.025 to 0.028 (0.64 to 0.71), incl	0.0015 (0.04)	0.0015 (0.04)	0.002 (0.05)	0.0025 (0.06)	0.0025 (0.06)	0.002 (0.05)	0.0025 (0.06)	0.003 (0.08)
Over 0.028 to 0.034 (0.71 to 0.86), incl	0.002 (0.05)	0.002 (0.05)	0.0025 (0.06)	0.003 (0.08)	0.003 (0.08)	0.0025 (0.06)	0.003 (0.08)	0.003 (0.08)
Over 0.034 to 0.039 (0.86 to 0.99), incl	0.002 (0.05)	0.002 (0.05)	0.0025 (0.06)	0.003 (0.08)	0.003 (0.08)	0.003 (0.08)	0.003 (0.08)	0.003 (0.08)
Over 0.039 to 0.049 (0.99 to 1.25), incl	0.002 (0.05)	0.002 (0.05)	0.003 (0.08)	0.003 (0.08)	0.003 (0.08)	0.003 (0.08)	0.004 (0.10)	0.004 (0.10)
Over 0.049 to 0.068 (1.25 to 1.73), incl	0.002 (0.05)	0.002 (0.05)	0.003 (0.08)	0.003 (0.08)	0.003 (0.08)	0.003 (0.08)	0.004 (0.10)	0.004 (0.10)
Over 0.068 to 0.099 (1.73 to 2.52), incl	0.002 (0.05)	0.002 (0.05)	0.003 (0.08)	0.004 (0.10)	0.004 (0.10)	0.004 (0.10)	0.004 (0.10)	0.004 (0.10)
Over 0.099 to 0.160 (2.52 to 4.06), incl	0.002 (0.05)	0.003 (0.08)	0.003 (0.08)	0.004 (0.10)	0.004 (0.10)	0.004 (0.10)	0.004 (0.10)	0.005 (0.13)
Over 0.160 to 0.187 (4.06 to 4.75), incl	0.002 (0.05)	0.0035 (0.09)	0.004 (0.10)	0.004 (0.10)	0.004 (0.10)	0.005 (0.13)	0.005 (0.13)	0.006 (0.15)
Over 0.187 to 0.2499 (4.75 to 6.35), incl	...	...	...	...	...	...	...	...

[a] Thickness measurements are taken 3/8 in. (9.5 mm) in from edge of strip, except that on widths less than 1 in. (25 mm), the tolerances are applicable for measurements at all locations.

TABLE 21 Width Tolerances for No. 2 Edge (Mill Edge) Cold-Rolled Strip

Specified Width, in.[a]	Tolerances for Specified Width, Plus and Minus, in.[a]
To 2, incl	1/32
Over 2 to 5, incl	1/16
Over 5 to 9, incl	3/32
Over 9 to 12, incl	1/8
Over 12 to 20, incl	1/4
Over 20 to 23 15/16, incl	3/8

[a] 1 in. = 25.4 mm.

TABLE 22 Width Tolerances for No. 3 Edge (Slit Edge) of Cold-Rolled Strip

Specified Thickness, in.[a]		Width, in.[a]					
		3/16 to 1/2, excl	1/2 to 6, incl	Over 6 to 9, incl	Over 9 to 12, incl	Over 12 to 20, incl	Over 20 to 23 15/16, incl
Over	To and incl	Width Tolerances, in.,[a] Over and Under for Indicated Thicknesses and Widths					
0.187	0.2499	...	...	...	...	...	...
0.160	0.187	...	0.016	0.020	0.020	0.031	0.031
0.099	0.160	0.010	0.010	0.016	0.016	0.020	0.020
0.068	0.099	0.008	0.008	0.010	0.010	0.016	0.020
0.0099	0.068	0.005	0.005	0.005	0.010	0.016	0.020
0.0099 and under		...	...	...	...	...	...

[a] 1 in. = 25.4 mm.

TABLE 23 Width Tolerances for Edge Nos. 1, 4, 5, and 6 for Cold-Rolled Strip

Edge No.	Width, in. (mm)	Thickness, in. (mm)	Tolerances for Specified Width, Plus and Minus, in. (mm)
1	To 3/4 (19)	0.0938 (2.38) and thinner	0.005 (0.13)
1	Over 3/4 to 5 (19 to 130), incl	0.125 (3.2) and thinner	0.005 (0.13)
4	To 1 (25), incl	0.1875 (4.8) to 0.025 (0.64), incl	0.015 (0.38)
4	Over 1 to 2 (25 to 50), incl	0.2499 to 0.025 (6.35 to 0.64), incl	0.025 (0.64)
4	Over 2 to 4 (50 to 100), incl	0.2499 to 0.035 (6.35 to 0.89), incl	0.047 (1.19)
4	Over 4 to 6 (100 to 150), incl	0.2499 to 0.035 (6.35 to 0.89), incl	0.047 (1.19)
5	To 3/4 (19), incl	0.0938 (2.38) and thinner	0.005 (0.13)
5	Over 3/4 to 5 (19 to 130), incl	0.125 (3.2) and thinner	0.005 (0.13)
5	Over 5 to 9 (130 to 230), incl	0.125 to 0.008 (3.2 to 0.203), incl	0.010 (0.25)
5	Over 9 to 20 (230 to 510), incl	0.105 to 0.015 (2.67 to 0.38), incl	0.010 (0.25)
5	Over 20 to 23 15/16 (510 to 610), incl	0.080 to 0.023 (2.03 to 0.58), incl	0.015 (0.38)
6	To 1 (25), incl	0.1875 (4.8) to 0.025 (0.64), incl	0.015 (0.38)
6	Over 1 to 2 (25 to 50), incl	0.2499 to 0.025 (6.35 to 0.64), incl	0.025 (0.64)
6	Over 2 to 4 (50 to 100), incl	0.2499 to 0.035 (6.35 to 0.89), incl	0.047 (1.19)
6	Over 4 to 6 (100 to 150), incl	0.2499 to 0.047 (6.35 to 1.19), incl	0.047 (1.19)

A 505

TABLE 24 Length Tolerances for Cold-Rolled Strip

Specified Width, in. (mm)	Length, in. (mm)		
	24 to 60 (610 to 1520), incl	Over 60 to 120 (1520 to 3050), incl	Over 120 to 240 (3050 to 6100), incl
	Length Tolerance, in. (mm), Over Specified Length, No Tolerance Under		
To 12 (300), incl	¼ (6.4)	½ (12.7)	¾ (19)
Over 12 to 23¹⁵⁄₁₆ (300 to 610), incl	½ (12.7)	¾ (19)	1 (25)

The American Society for Testing and Materials takes no position respecting the validity of any patent rights asserted in connection with any item mentioned in this standard. Users of this standard are expressly advised that determination of the validity of any such patent rights, and the risk of infringement of such rights, are entirely their own responsibility.

This standard is subject to revision at any time by the responsible technical committee and must be reviewed every five years and if not revised, either reapproved or withdrawn. Your comments are invited either for revision of this standard or for additional standards and should be addressed to ASTM Headquarters. Your comments will receive careful consideration at a meeting of the responsible technical committee, which you may attend. If you feel that your comments have not received a fair hearing you should make your views known to the ASTM Committee on Standards, 1916 Race St., Philadelphia, Pa. 19103.

Designation: A 506 – 73 (Reapproved 1980)

An American National Standard

Standard Specification for
STEEL SHEET AND STRIP, ALLOY, HOT-ROLLED AND COLD-ROLLED, REGULAR QUALITY[1]

This standard is issued under the fixed designation A 506; the number immediately following the designation indicates the year of original adoption or, in the case of revision, the year of last revision. A number in parentheses indicates the year of last reapproval. A superscript epsilon (ϵ) indicates an editorial change since the last revision or reapproval.

This specification has been approved for use by agencies of the Department of Defense and for listing in the DoD Index of Specifications and Standards.

1. Scope

1.1 This specification covers hot-rolled and cold-rolled regular quality alloy-steel sheet and strip. This quality is intended primarily for general or miscellaneous use where bending and moderate forming is a requirement. If material of a higher degree of uniformity of internal soundness or freedom from surface imperfections is required, reference should be made to Specification A 507.

1.2 Regular quality is furnished in the conditions, heat treatments, surface finishes, and edges specified herein.

1.3 The values stated in inch-pound units are to be regarded as the standard.

2. Applicable Documents

2.1 *ASTM Standards:*
A 505 Specification for General Requirements for Steel Sheet and Strip, Alloy, Hot-Rolled and Cold-Rolled[2]
A 507 Specification for Steel Sheet and Strip, Alloy, Hot-Rolled and Cold-Rolled, Drawing Quality[2]

3. Definition

3.1 *regular quality*—alloy steel sheet and strip intended for general or miscellaneous applications where normal surface defects are not objectionable and a good finish is not the prime requirement.

4. General Requirements for Delivery

4.1 Material furnished under this specification shall conform to the applicable requirements of the current edition of Specification A 505, unless otherwise provided herein.

5. Ordering Information

5.1 Orders for material under this specification shall include the following information, as required, to describe the required material adequately:

5.1.1 ASTM specification number and date of issue,
5.1.2 Chemical composition (for example 41XX, 43XX, 86XX, etc.),
5.1.3 Condition (hot rolled or cold rolled),
5.1.4 Heat treatment (as rolled, annealed, normalized, or normalized and tempered),
5.1.5 Surface condition (as rolled or descaled, oiled or not oiled),
5.1.6 Edges (mill, cut, etc.),
5.1.7 Dimensions, including decimal thickness, width, and length (continuous mill thickness tolerances shall apply unless otherwise agreed),
5.1.8 Quantity (weight),
5.1.9 Mechanical properties if required. (Show properties desired, which must be consistent with quality and grade),
5.1.10 Application, and
5.1.11 Report of test results (when required).

NOTE—A typical ordering description is as follows: ASTM A 506-XX, Steel 4130, Hot Rolled, Annealed, Descaled, Oiled, Cut Edges, 0.190 by 36 by 72 in., 10 000 lb, Bumper brackets.

---

[1] This specification is under the jurisdiction of ASTM Committee A-1 on Steel, Stainless Steel and Related Alloys, and is the direct responsibility of Subcommittee A01.19 on Steel Sheet and Strip.
Current edition approved March 29, 1973. Published May 1973. Originally published as A 506 – 64. Last previous edition A 506 – 64.
[2] *Annual Book of ASTM Standards*, Vol 01.03.

6. Manufacture

6.1 *Melting Practice*—This steel shall be made by the open-hearth, basic-oxygen, or electric-furnace process.

6.2 *Hot or Cold Working*—The sheet and strip shall be furnished hot rolled or cold rolled as specified.

6.3 *Heat Treatment*:

6.3.1 *Cold Rolled*—Cold-rolled material shall be furnished annealed.

6.3.2 *Hot Rolled*—Hot-rolled material shall be furnished as rolled, annealed, normalized, or normalized and tempered as specified.

7. Chemical Requirements

7.1 *Cast or Heat (Formerly Ladle) Analysis*—The cast analysis shall conform to that specified in Specification A 505 for the steel number ordered; or, to such other limits as may be specified using the standard ranges in Specification A 505.

8. Metallurgical Structure

8.1 *Decarburization*:

8.1.1 *Requirements*—The maximum depth of total decarburization (complete plus partial) on each surface shall be such that the average hardness at the surface of a hardened specimen as determined by Rockwell A hardness tests shall be within two Rockwell A hardness numbers of the average subsurface hardness of the same specimen. This test is not applicable to material less than 0.025 in. (0.64 mm) thick.

8.1.2 *Number of Tests*—Two decarburization tests shall be made to represent approximately each 200 sheets or strips or from each 10 coils of material from the same cast, thickness, and condition and submitted for acceptance at one time.

8.1.3 *Test Specimen Location*—Test specimens shall be of the full thickness of the material and may be taken from any convenient location. The average hardness (surface or subsurface) shall be the average of three tests made adjacent to each other on the same specimen. Surface hardness tests shall be made on a clean but unground or unpolished surface. Subsurface hardness tests shall be made in a depression ground to a depth approximately 0.020 in. (0.51 mm) or ⅓ the thickness of the specimen, whichever is less.

9. Mechanical Requirements

9.1 *Tension and Hardness Tests*:

9.1.1 *Requirements*—The mechanical properties will vary depending on the chemical composition, condition, and heat treatment specified. These properties are altered by inplant treatment which may be required to meet certain acceptance tests important in fabrication. Producers are frequently consulted as to grade, resultant mechanical properties, recommended heat treatment, and other information for steels to meet the end use requirements. The following mechanical properties may be specified: tensile strength, yield strength, elongation, and Rockwell hardness tests.

9.1.2 *Number of Tests*—When mechanical properties are specified, not less than two longitudinal tension tests and two hardness tests shall be made to represent material from each cast and heat treatment charge.

9.2 *Bend Test*:

9.2.1 *Requirements*—When ordered in the annealed, normalized, or normalized and tempered condition, the material shall be capable of meeting the bend requirements in Table 1 without cracking on the outside of the bent portion when tested at room temperature.

9.2.2 *Number of Tests*—Not less than one bend test shall be made to represent material from each lot. A lot shall consist of material from the same cast, of the same condition and finish, the same thickness, subjected to the same heat treatment, and submitted for inspection at one time.

10. Finish and Appearance

10.1 *Surface Finish*:

10.1.1 *Hot Rolled*—Unless otherwise specified, hot-rolled material shall be furnished without removing the hot-rolled oxide or scale (that is, as rolled). When required, the material may be specified to be descaled.

10.1.2 *Cold Rolled*—Unless otherwise specified, cold-rolled material shall be furnished with a commercial dull matte finish.

10.2 *Oiling*:

10.2.1 *Hot Rolled*—Unless otherwise specified, as-rolled material shall be furnished not oiled, and descaled material shall be oiled. Descaled material may be specified to be furnished dry.

10.2.2 *Cold Rolled*—Unless otherwise specified, cold-rolled material shall be furnished oiled. When required, cold-rolled material may be specified to be furnished dry.

10.3 *Edges*:

10.3.1 *Hot Rolled*—Unless otherwise specified, hot-rolled sheet and strip shall be furnished with mill edges. When required, sheet may be specified to have cut edges, and strip may be specified to have cut or square edges.

10.3.2 *Cold Rolled*—Unless otherwise specified, cold-rolled sheet shall have cut edges and strip shall have No. 3 slit edges. When required, strip may be specified to have No. 1, No. 2, No. 4, or No. 6 edges.

TABLE 1 Bend Requirements

Thickness, in. (mm)	Carbon Content, %	Degree of Bend	Ratio of Bend Radius to Thickness of Specimen	Relation of Bend to Rolling Direction
All	up to 0.30, incl	180	½t	perpendicular
0.1250 (3.175) and less	over 0.30	180	½t	perpendicular
Over 0.1250 (3.175) to 0.2499 (6.347), incl	over 0.30	180	t	perpendicular

The American Society for Testing and Materials takes no position respecting the validity of any patent rights asserted in connection with any item mentioned in this standard. Users of this standard are expressly advised that determination of the validity of any such patent rights, and the risk of infringement of such rights, are entirely their own responsibility.

This standard is subject to revision at any time by the responsible technical committee and must be reviewed every five years and if not revised, either reapproved or withdrawn. Your comments are invited either for revision of this standard or for additional standards and should be addressed to ASTM Headquarters. Your comments will receive careful consideration at a meeting of the responsible technical committee, which you may attend. If you feel that your comments have not received a fair hearing you should make your views known to the ASTM Committee on Standards, 1916 Race St., Philadelphia, Pa. 19103.

ASTM Designation: A 507 – 73 (Reapproved 1980)

An American National Standard

Standard Specification for
STEEL SHEET AND STRIP, ALLOY, HOT-ROLLED AND COLD-ROLLED, DRAWING QUALITY[1]

This standard is issued under the fixed designation A 507; the number immediately following the designation indicates the year of original adoption or, in the case of revision, the year of last revision. A number in parentheses indicates the year of last reapproval. A superscript epsilon (ϵ) indicates an editorial change since the last revision or reapproval.

This specification has been approved for use by agencies of the Department of Defense and for listing in the DoD Index of Specifications and Standards.

1. Scope

1.1 This specification covers hot-rolled and cold-rolled drawing quality alloy-steel sheet and strip. Sheet and strip of this quality are produced principally for applications involving severe cold plastic deformation such as deep drawn or severely formed parts. This quality is produced by closely controlled steelmaking practice designed to assure internal soundness, relative uniformity of chemical composition, and freedom from injurious imperfections.

1.2 Drawing quality is furnished in the conditions, heat treatments, surface finishes, and edges specified herein.

1.3 The values stated in inch-pound units are to be regarded as the standard.

2. Applicable Document

2.1 *ASTM Standard*:
A 505 Specification for General Requirements for Steel Sheet and Strip, Alloy, Hot-Rolled and Cold-Rolled[2]

3. Definition

3.1 *drawing quality*—alloy steel sheet and strip intended for applications involving severe cold plastic deformation such as deep drawn or severely formed parts.

4. General Requirements for Delivery

4.1 Material furnished under this specification shall conform to the applicable requirements of the current edition of Specification A 505, unless otherwise provided herein.

5. Ordering Information

5.1 Orders for material under this specification shall include the following information, as required, to describe the required material adequately:

5.1.1 ASTM specification number and date of issue,

5.1.2 Chemical composition (for example 41XX, 43XX, 86XX, etc.),

5.1.3 Condition (hot rolled or cold rolled),

5.1.4 Surface condition (descaled, oiled, or not oiled),

5.1.5 Edges (mill, cut, etc.),

5.1.6 Dimensions, including decimal thickness, width and length (continuous mill thickness tolerances shall apply unless otherwise agreed),

5.1.7 Quantity (weight),

5.1.8 Mechanical properties if required (Show properties desired, which must be consistent with quality and grade),

5.1.9 Application, and

5.1.10 Report of test results (if required).

NOTE—A typical ordering description is as follows: ASTM A 507-XX, Steel 4130, Hot Rolled, Descaled, Oiled, Cut Edges, 0.190 by 36 by 72 in., 10 000 lb, Air cylinders.

6. Manufacture

6.1 *Melting Practice*—The steel shall be made by the open-hearth, basic-oxygen, or electric-furnace process.

6.2 *Hot Rolling or Cold Working*—The sheet

---

[1] This specification is under the jurisdiction of ASTM Committee A-1 on Steel, Stainless Steel, and Related Alloys, and is the direct responsibility of Subcommittee A01.19 on Steel Sheet and Strip.
Current edition approved March 29, 1973. Published May 1973. Originally published as A 507 – 64. Last previous edition A 507 – 64.

[2] *Annual Book of ASTM Standards*, Vol 01.03.

and strip shall be furnished hot rolled or cold rolled as specified.

6.3 *Heat Treatment*—Unless otherwise specified, the material shall be furnished spheroidize annealed. If material is to be spheroidize annealed by other than the producer, it may be ordered as rolled.

7. Chemical Requirements

7.1 *Cast or Heat (Formerly Ladle) Analysis*:

7.1.1 The cast analysis shall conform to that specified in Specification A 505 for the steel number ordered; or, to such other limits as may be specified using the standard ranges in Specification A 505.

7.1.2 Grade 4130 is the most commonly supplied drawing quality grade.

8. Metallurgical Structure

8.1 *Microstructure*:

8.1.1 *Requirements*—The carbide microstructure shall be at least 75% of the globular type.

8.1.2 *Number of Tests*—The number of tests shall be in accordance with the manufacturer's standard quality control procedures. A specific number of tests is not required but the material shall be produced by manufacturing practices and subjected to mill tests and inspection procedures to assure compliance with the specified requirements. Material tested by the purchaser that fails to meet the specified requirements shall be subject to rejection.

8.2 *Decarburization*:

8.2.1 *Requirements*—The maximum depth of total decarburization (complete plus partial) on each surface shall be such that the average hardness at the surface of a hardened specimen as determined by Rockwell A hardness tests shall be within two Rockwell A hardness numbers of the average subsurface hardness of the same specimen. This test is not applicable to material less than 0.025 in. (0.64 mm) thick.

8.2.2 *Number of Tests*—Two decarburization tests shall be made to represent approximately each 200 sheets or strips or from each 10 coils of material from the same cast, thickness, and condition and submitted for acceptance at one time.

8.2.3 *Test Specimen Location*—Test specimens shall be of the full thickness of the material and may be taken from any convenient location. The average hardness (surface or subsurface) shall be the average of three tests made adjacent to each other on the same specimen. Surface hardness tests shall be made on a clean but unground or unpolished surface. Subsurface hardness tests shall be made in a depression ground to a depth approximately 0.020 in. (0.51 mm) or ⅓ the thickness of the specimen, whichever is less.

9. Mechanical Requirements

9.1 *Tension and Hardness Tests*:

9.1.1 *Requirements*—The mechanical properties will vary depending on the chemical composition specified. In any case, the producer supplies a material to meet the end use requirements if these are known. Producers are frequently consulted as to grade, resultant mechanical properties, recommended heat treatment, and other information for steels to meet the end use requirements. The following mechanical properties may be specified: tensile strength, yield strength, elongation, and Rockwell hardness tests.

9.1.2 *Number of Tests*—When mechanical properties are specified, not less than two longitudinal tension tests and sufficient Rockwell hardness tests, or both of these, shall be taken at random to represent material from each cast and annealing charge.

10. Other Tests and Requirements

10.1 *End Use Suitability*:

10.1.1 When furnished spheroidize annealed, the material shall be capable of producing an identified part.

10.1.2 Alloy-steel sheet and strip in the spheroidize annealed condition generally work harden and may require stress relief annealing between drawing operations.

11. Finish and Appearance

11.1 *Surface Finish*:

11.1.1 *Hot Rolled*—Unless otherwise specified, hot-rolled material shall be furnished descaled and oiled. When required, hot-rolled material may be specified to be furnished not oiled.

11.1.2 *Cold Rolled*—Unless otherwise specified, cold-rolled material shall be furnished with a commerical dull matte finish and shall be oiled. When required, cold-rolled material may be specified to be furnished not oiled.

11.2 *Edges*:

11.2.1 *Hot Rolled*—Unless otherwise specified, hot-rolled sheet and strip shall be furnished with mill edges. When required, sheet may be specified to have cut edges, and strip may be specified to have cut or square edges.

11.2.2 *Cold Rolled*—Unless otherwise specified, cold-rolled sheet shall have cut edges and strip shall have No. 3 slit edges. When required, strip may be specified to have No. 1, No. 2, No. 4, or No. 6 edges.

The American Society for Testing and Materials takes no position respecting the validity of any patent rights asserted in connection with any item mentioned in this standard. Users of this standard are expressly advised that determination of the validity of any such patent rights, and the risk of infringement of such rights, are entirely their own responsibility.

This standard is subject to revision at any time by the responsible technical committee and must be reviewed every five years and if not revised, either reapproved or withdrawn. Your comments are invited either for revision of this standard or for additional standards and should be addressed to ASTM Headquarters. Your comments will receive careful consideration at a meeting of the responsible technical committee, which you may attend. If you feel that your comments have not received a fair hearing you should make your views known to the ASTM Committee on Standards, 1916 Race St., Philadelphia, Pa. 19103.

Designation: A 510 – 82

Standard Specification for
GENERAL REQUIREMENTS FOR WIRE RODS AND COARSE ROUND WIRE, CARBON STEEL[1]

This standard is issued under the fixed designation A 510; the number immediately following the designation indicates the year of original adoption or, in the case of revision, the year of last revision. A number in parentheses indicates the year of last reapproval. A superscript epsilon (ϵ) indicates an editorial change since the last revision or reapproval.

This specification has been approved for use by agencies of the Department of Defense and for listing in the DoD Index of Specifications and Standards.

1. Scope

1.1 This specification covers general requirements for carbon steel wire rods and uncoated coarse round wire in coils or straightened and cut lengths.

1.2 In case of conflict, the requirements in the purchase order, on the drawing, in the individual specification, and in this general specification shall prevail in the sequence named.

NOTE 1—A complete metric companion to Specification A 510 has been developed—Specification A 510M; therefore, no metric equivalents are presented in this specification.

2. Applicable Documents

2.1 *ASTM Standards*:
A 370 Methods and Definitions for Mechanical Testing of Steel Products[2]
A 700 Practices for Packaging, Marking, and Loading Methods for Steel Products for Domestic Shipment[2]
E 29 Recommended Practice for Indicating Which Places of Figures Are to Be Considered Significant in Specified Limiting Values[3]
E 30 Methods for Chemical Analysis of Steel, Cast Iron, Open-Hearth Iron, and Wrought Iron[4]
E 112 Methods for Determining Average Grain Size[5]

3. Description of Terms

3.1 *Carbon Steel*—Steel is considered to be carbon steel when no minimum content is specified or required for aluminum, chromium, cobalt, columbium, molybdenum, nickel, titanium, tungsten, vanadium, or zirconium, or any other element added to obtain a desired alloying effect; when the specified minimum for copper does not exceed 0.40 %; or when the maximum content specified for any of the following elements does not exceed these percentages: manganese 1.65, silicon 0.60, or copper 0.60.

3.1.1 In all carbon steels small quantities of certain residual elements unavoidably retained from raw materials are sometimes found which are not specified or required, such as copper, nickel, molybdenum, chromium, etc. These elements are considered as incidental and are not formally determined or reported.

3.1.2 Elements may be specified to improve machinability of carbon steels such as sulfur and lead.

3.2 Wire rods are hot rolled from billets to an approximate round cross section into coils of one continuous length. Rods are not comparable to hot-rolled bars in accuracy of cross section or surface finish and as a semifinished product are intended primarily for the manufacture of wire.

---

[1] This specification is under the jurisdiction of ASTM Committee A-1 on Steel, Stainless Steel and Related Alloys, and is the direct responsibility of Subcommittee A01.03 on Steel Rod and Wire.
Current edition approved July 30, 1982. Published September 1982. Originally published as A 510 – 64. Last previous edition A 510 – 77.
[2] *Annual Book of ASTM Standards*, Vol 01.03.
[3] *Annual Book of ASTM Standards*, Vol 14.02.
[4] *Annual Book of ASTM Standards*, Vol 03.05.
[5] *Annual Book of ASTM Standards*, Vol 03.03.

3.2.1 Rod sizes from 7/32 to 47/64 in. in diameter, inclusive are designated by fractions or decimal parts of an inch as shown in Table 1.

3.3 Coarse round wire from 0.035 to 0.999 in. in diameter, inclusive is produced from hot-rolled wire rods or hot-rolled coiled bars by one or more cold reductions primarily for the purpose of obtaining a desired size with dimensional accuracy, surface finish, and mechanical properties. By varying the amount of cold reduction and other wire mill practices, including thermal treatment, a wide diversity of mechanical properties and finishes are made available.

3.3.1 Coarse round wire is designated by Steel Wire Gage numbers, common fractions, or decimal parts of an inch. The Steel Wire Gage system is shown in Table 11.

3.4 Straightened and cut wire is produced from coils of wire by means of special machinery which straightens the wire and cuts it to a specified length.

3.4.1 The straightening operation may alter the mechanical properties of the wire especially the tensile strength. The straightening operation may also induce changes in the diameter of the wire. The extent of the changes in the properties of the wire after cold straightening depends upon the kind of wire and also on the normal variations in the adjustments of the straightening equipment. It is therefore not possible to forecast the properties of straightened and cut wire and each kind of wire needs individual consideration. In most cases, the end use of straightened and cut wire is not seriously influenced by these changes.

4. Ordering Information

4.1 Orders for hot-rolled wire rods under this specification shall include the following information:

4.1.1 Quantity (pounds),

4.1.2 Name of material (wire rods),

4.1.3 Diameter (Table 1),

4.1.4 Chemical composition grade no. (Tables 6, 7, 8, and 9),

4.1.5 Packaging,

4.1.6 ASTM designation and date of issue, and

4.1.7 Special requirements, if any.

NOTE 2—A typical ordering description is as follows: 100,000 lb Wire Rods, 7/32 in., Grade 1010 in approximately 1000 lb Coils to ASTM A 510 dated———.

4.2 Orders for coarse round wire under this specification shall include the following information:

4.2.1 Quantity (pounds or pieces),

4.2.2 Name of material (uncoated carbon steel wire),

4.2.3 Diameter (see 3.3.1),

4.2.4 Length (straightened and cut only),

4.2.5 Chemical composition (Tables 6, 7, 8, and 9),

4.2.6 Packaging,

4.2.7 ASTM designation and date of issue, and

4.2.8 Special requirements, if any.

NOTE 3—A typical ordering description is as follows: 40,000 lb Uncoated Carbon Steel Wire, 0.148 in. diameter, Grade 1008 in 500 lb Coils on Tubular Carriers to ASTM A 510 dated———, or
2500 Pieces, Carbon Steel Wire, 0.375 in. diameter, Straightened and Cut 29½ in., Grade 1015, in 25 Piece Bundles on Pallets to ASTM A 510 dated———.

5. Manufacture

5.1 The steel shall be made by the open-hearth, electric-furnace, or basic-oxygen process. The steel may be either ingot cast or strand cast.

6. Chemical Requirements

6.1 The chemical composition for steel under this specification shall conform to the requirements set forth in the purchase order. Chemical compositions are specified by ranges or limits for carbon and other elements. The grades commonly specified for carbon steel wire rods and coarse round wire are shown in Tables 6, 7, 8, and 9.

6.2 *Cast or Heat Analysis (Formerly Ladle Analysis)*—An analysis of each cast or heat shall be made by the producer to determine the percentage of the elements specified. The analysis shall be made from a test sample, preferably taken during the pouring of the cast or heat. The chemical composition thus determined shall be reported, if required, to the purchaser, or his representative.

6.3 *Product Analysis (Formerly Check Analysis)*—A product analysis may be made by the purchaser. The analysis is not used for a duplicate analysis to confirm a previous result. The purpose of the product analysis is to verify that the chemical composition is within specified limits for each element, including applicable

A 510

permissible variations in product analysis. The results of analyses taken from different pieces of a heat may differ within permissible limits from each other and from the heat or cast analysis. Table 10 shows the permissible variations for product analysis of carbon steel. The results of the product analysis obtained, except lead, shall not vary both above and below the permissible limits.

6.3.1 Rimmed or capped steels are characterized by a lack of uniformity in their chemical composition, especially for the elements carbon, phosphorus, and sulfur, and for this reason product analysis is not technologically appropriate for these elements unless misapplication is clearly indicated.

6.3.2 Because of the degree to which phosphorus and sulfur segregate, product analysis for these elements is not technologically appropriate for rephosphorized or resulfurized steels or both unless misapplication is clearly indicated.

6.3.3 The location at which chips for product analysis are obtained from the sample is important because of segregation. For rods and wire, chips are taken by milling or machining the full cross section of the sample.

6.3.3.1 Steel subjected to certain thermal treatment operations by the purchaser may not give chemical analysis results that properly represent its original composition. Therefore, purchasers should analyze chips taken from the steel in the condition in which it is received from the producer.

6.3.3.2 When samples are returned to the producer for product analysis, the samples should consist of pieces of the full cross section.

6.3.4 For referee purposes, Methods E 30 shall be used.

7. Metallurgical Structure

7.1 Grain size when specified shall be determined in accordance with the requirements of Methods E 112.

8. Mechanical Requirements

8.1 The properties enumerated in individual specifications shall be determined in accordance with Methods and Definitions A 370.

8.2 Because of the great variety in the kinds of wire and the extensive diversity of end uses, a number of formal mechanical test procedures have been developed. These tests are used as control tests by producers during the intermediate stages of wire processing, as well as for final testing of the finished product, and apply particularly to specification wire and wires for specified end uses. A number of these tests are further described in Supplement IV, Round Wire Products, of Methods and Definitions A 370.

8.3 Since the general utility of rods and wire requires continuity of length, in the case of rods, tests are commonly made on samples taken from the ends of coils after removing two to three rings. In the case of wire, tests are commonly made on samples taken from the ends of coils, thereby not impairing the usefulness of the whole coil.

9. Dimensions and Permissible Variations

9.1 The diameter and out-of-roundness of the wire rod shall not vary from that specified by more than that prescribed in Table 2.

9.2 The diameter and out-of-roundness of the coarse round wire and straightened and cut wire shall not vary from that specified by more than that prescribed in Table 3.

9.3 The length of straightened and cut wire shall not vary from that specified by more than that prescribed in Table 4.

9.4 The burrs formed in cutting straightened and cut wire shall not exceed the diameter specified by more than that prescribed in Table 5.

10. Workmanship, Finish, and Appearance

10.1 The wire rod shall be free of detrimental surface imperfections, tangles, and sharp kinks.

10.1.1 Two or more rod coils may be welded together to produce a larger coil. The weld zone may not be as sound as the original material. The mechanical properties existing in the weld metal may differ from those in the unaffected base metal. The weld may exceed the standard dimensional permissible variations on the minus side and must be within the permissible variations on the plus side.

10.2 The wire as received shall be smooth and substantially free from rust, shall not be kinked or improperly cast. No detrimental die marks or scratches may be present. Each coil shall be one continuous length of wire. Welds made during cold drawing are permitted.

10.3 The straightened and cut wire shall be

A 510

substantially straight and not be kinked or show excessive spiral marking.

11. Retests

11.1 The difficulties in obtaining truly representative samples of wire rod and coarse round wire without destroying the usefulness of the coil of wire account for the generally accepted practice of allowing retests for mechanical tests and surface examination. Two additional test pieces are cut from each end of the coil from which the original sample was taken. A portion of the coil may be discarded prior to cutting the sample for retest. If any of the retests fails to comply with the requirements, the coil of wire may be rejected. Before final rejection, however, it is frequently advisable to base final decision on an actual trial of the material to determine whether or not it will do the job for which it is intended.

12. Inspection

12.1 The manufacturer shall afford the purchaser's inspector all reasonable facilities necessary to satisfy him that the material is being produced and furnished in accordance with this specification. Mill inspection by the purchaser shall not interfere unnecessarily with the manufacturer's operations. All tests and inspections shall be made at the place of manufacture, unless otherwise agreed to.

13. Rejection

13.1 Any rejection based on tests made in accordance with this specification shall be reported to the producer within a reasonable length of time. The material must be adequately protected and correctly identified in order that the producer may make a proper investigation.

14. Certification

14.1 When specified in the purchase order or contract, a producer's or supplier's certification shall be furnished to the purchaser that the material was manufactured, sampled, tested, and inspected in accordance with this specification and has been found to meet the requirements. When specified in the purchase order or contract, a report of the test results shall be furnished.

15. Packaging, Marking, and Loading

15.1 A tag shall be securely attached to each coil or bundle and shall be marked with the size, ASTM specification number, heat or cast number, grade number, and name or mark of the manufacturer.

15.2 When specified in the purchase order, packaging, marking, and loading for shipments shall be in accordance with those procedures recommended by Practices A 700.

A 510

TABLE 1 Sizes of Wire Rods[A]

Inch Fraction	Decimal Equivalent, in.	Inch Fraction	Decimal Equivalent, in.
7/32	0.219	31/64	0.484
15/64	0.234	1/2	0.500
1/4	0.250	33/64	0.516
17/64	0.266	17/32	0.531
9/32	0.281	35/64	0.547
19/64	0.297	9/16	0.562
5/16	0.312	37/64	0.578
21/64	0.328	19/32	0.594
11/32	0.344	39/64	0.609
23/64	0.359	5/8	0.625
3/8	0.375	41/64	0.641
25/64	0.391	21/32	0.656
13/32	0.406	43/64	0.672
27/64	0.422	11/16	0.688
7/16	0.438	45/64	0.703
29/64	0.453	23/32	0.719
15/32	0.469	47/64	0.734

[A] Rounded off to 3 decimal places in decimal equivalents according to procedures outlined in Recommended Practice E 29.

TABLE 2 Permissible Variations in Diameter for Wire Rod in Coils

NOTE—For purposes of determining conformance with this specification, all specified limits are absolute as defined in Recommended Practice E 29.

Diameter of Rod		Permissible Variation, Plus and Minus, in.	Permissible Out-of-Round, in.
Fractions	Decimal		
7/32 to 47/64 in., incl	0.219 to 0.734 in., incl	0.016	0.025

TABLE 3 Permissible Variations in Diameter for Uncoated Coarse Round Wire

NOTE—For purposes of determining conformance with this specification, all specified limits are absolute as defined in Recommended Practice E 29.

In Coils		
Diameter of Wire, in.	Permissible Variations, Plus and Minus, in.	Permissible Out-Of-Round, in.
0.035 to under 0.076	0.001	0.001
0.076 to under 0.500	0.002	0.002
0.500 and over	0.003	0.003
Straightened and Cut		
Diameter of Wire, in.	Permissible Variations, Plus and Minus, in.	Permissible Out-of-Round, in.
0.035 to under 0.076	0.001	0.001
0.076 to 0.148, incl	0.002	0.002
Over 0.148 to under 0.500	0.003	0.003
0.500 and over	0.004	0.004

TABLE 4 Permissible Variations in Length for Straightened and Cut Wire

NOTE—For purposes of determining conformance with this specification, all specified limits are absolute as defined in Recommended Practice E 29.

Cut Length, ft	Permissible Variations, Plus and Minus, in.
Under 3	1/16
3 to 12, incl	3/32
Over 12	1/8

TABLE 5 Permissible Variations for Burrs for Straightened and Cut Wire

NOTE—For purposes of determining conformance with this specification, all specified limits are absolute as defined in Recommended Practice E 29.

Diameter of Wire, in.	Permissible Variation over Measured Diameter, in.
Up to 0.125, incl	0.004
Over 0.125 to 0.250, incl	0.006
Over 0.250 to 0.500, incl	0.008
Over 0.500	0.010

A 510

TABLE 6 Nonresulfurized Carbon Steel Cast or Heat Chemical Ranges and Limits

NOTE 1: *Silicon*—When silicon is required the following ranges and limits are commonly used for nonresulfurized carbon steels: 0.10 max, %, 0.07 to 0.15 %, 0.10 to 0.20 %, 0.15 to 0.35 %, 0.20 to 0.40 %, or 0.30 to 0.60 %.
NOTE 2: *Copper*—When required, copper is specified as an added element.
NOTE 3: *Lead*—When lead is required as an added element, a range from 0.15 to 0.35 % is specified. Such a steel is identified by inserting the letter "L" between the second and third numerals of the grade number, for example 10L18.

UNS Designation[A]	Grade No.	Chemical Composition Limits, %				SAE No.
		Carbon	Manganese	Phosphorus, max	Sulfur, max	
G 10050	1005	0.06 max	0.35 max	0.040	0.050	1005
G 10060	1006	0.08 max	0.25 to 0.40	0.040	0.050	1006
G 10080	1008	0.10 max	0.30 to 0.50	0.040	0.050	1008
G 10100	1010	0.08 to 0.13	0.30 to 0.60	0.040	0.050	1010
G 10110	1011	0.08 to 0.13	0.60 to 0.90	0.040	0.050	1011
G 10120	1012	0.10 to 0.15	0.30 to 0.60	0.040	0.050	1012
G 10130	1013	0.11 to 0.16	0.50 to 0.80	0.040	0.050	1013
G 10150	1015	0.13 to 0.18	0.30 to 0.60	0.040	0.050	1015
G 10160	1016	0.13 to 0.18	0.60 to 0.90	0.040	0.050	1016
G 10170	1017	0.15 to 0.20	0.30 to 0.60	0.040	0.050	1017
G 10180	1018	0.15 to 0.20	0.60 to 0.90	0.040	0.050	1018
G 10190	1019	0.15 to 0.20	0.70 to 1.00	0.040	0.050	1019
G 10200	1020	0.18 to 0.23	0.30 to 0.60	0.040	0.050	1020
G 10210	1021	0.18 to 0.23	0.60 to 0.90	0.040	0.050	1021
G 10220	1022	0.18 to 0.23	0.70 to 1.00	0.040	0.050	1022
G 10230	1023	0.20 to 0.25	0.30 to 0.60	0.040	0.050	1023
G 10250	1025	0.22 to 0.28	0.30 to 0.60	0.040	0.050	1025
G 10260	1026	0.22 to 0.28	0.60 to 0.90	0.040	0.050	1026
G 10290	1029	0.25 to 0.31	0.60 to 0.90	0.040	0.050	1029
G 10300	1030	0.28 to 0.34	0.60 to 0.90	0.040	0.050	1030
G 10340	1034	0.32 to 0.38	0.50 to 0.80	0.040	0.050	...
G 10350	1035	0.32 to 0.38	0.60 to 0.90	0.040	0.050	1035
G 10370	1037	0.32 to 0.38	0.70 to 1.00	0.040	0.050	1037
G 10380	1038	0.35 to 0.42	0.60 to 0.90	0.040	0.050	1038
G 10390	1039	0.37 to 0.44	0.70 to 1.00	0.040	0.050	1039
G 10400	1040	0.37 to 0.44	0.60 to 0.90	0.040	0.050	1040
G 10420	1042	0.40 to 0.47	0.60 to 0.90	0.040	0.050	1042
G 10430	1043	0.40 to 0.47	0.70 to 1.00	0.040	0.050	1043
G 10440	1044	0.43 to 0.50	0.30 to 0.60	0.040	0.050	1044
G 10450	1045	0.43 to 0.50	0.60 to 0.90	0.040	0.050	1045
G 10460	1046	0.43 to 0.50	0.70 to 1.00	0.040	0.050	1046
G 10490	1049	0.46 to 0.53	0.60 to 0.90	0.040	0.050	1049
G 10500	1050	0.48 to 0.55	0.60 to 0.90	0.040	0.050	1050
G 10530	1053	0.48 to 0.55	0.70 to 1.00	0.040	0.050	1053
G 10550	1055	0.50 to 0.60	0.60 to 0.90	0.040	0.050	1055
G 10590	1059	0.55 to 0.65	0.50 to 0.80	0.040	0.050	1059
G 10600	1060	0.55 to 0.65	0.60 to 0.90	0.040	0.050	1060
G 10640	1064	0.60 to 0.70	0.50 to 0.80	0.040	0.050	1064
G 10650	1065	0.60 to 0.70	0.60 to 0.90	0.040	0.050	1065
G 10690	1069	0.65 to 0.75	0.40 to 0.70	0.040	0.050	1069
G 10700	1070	0.65 to 0.75	0.60 to 0.90	0.040	0.050	1070
G 10740	1074	0.70 to 0.80	0.50 to 0.80	0.040	0.050	1074
G 10750	1075	0.70 to 0.80	0.40 to 0.70	0.040	0.050	1075
G 10780	1078	0.72 to 0.85	0.30 to 0.60	0.040	0.050	1078
G 10800	1080	0.75 to 0.88	0.60 to 0.90	0.040	0.050	1080
G 10840	1084	0.80 to 0.93	0.60 to 0.90	0.040	0.050	1084
G 10850	1085	0.80 to 0.93	0.70 to 1.00	0.040	0.050	1085
G 10860	1086	0.80 to 0.93	0.30 to 0.50	0.040	0.050	1086
G 10900	1090	0.85 to 0.98	0.60 to 0.90	0.040	0.050	1090
G 10950	1095	0.90 to 1.03	0.30 to 0.50	0.040	0.050	1095

[A] New designation established in accordance with ASTM E 527 and SAE J 1086, Recommended Practice for Numbering Metals and Alloys (UNS).

TABLE 7 Nonresulfurized Carbon Steel, High Manganese, Cast or Heat Chemical Ranges and Limits

NOTE 1: *Silicon*—When silicon is required the following ranges and limits are commonly used for nonresulfurized carbon steels: 0.10 max, %, 0.07 to 0.15 %, 0.10 to 0.20 %, 0.15 to 0.35 %, 0.20 to 0.40 %, or 0.30 to 0.60 %.
NOTE 2: *Copper*—When required, copper is specified as an added element.
NOTE 3: *Lead*—When lead is required as an added element a range from 0.15 to 0.35 % is specified. Such a steel is identified by inserting the letter "L" between the second and third numerals of the grade number, for example 15L18.

UNS Designation[A]	Grade No.	Chemical Composition Limits, %				SAE No.
		Carbon	Manganese	Phosphorus, max	Sulfur, max	
G 15130	1513	0.10 to 0.16	1.10 to 1.40	0.040	0.050	1513
G 15180	1518	0.15 to 0.21	1.10 to 1.40	0.040	0.050	1518
G 15220	1522	0.18 to 0.24	1.10 to 1.40	0.040	0.050	1522
G 15240	1524[B]	0.19 to 0.25	1.35 to 1.65	0.040	0.050	1524
G 15250	1525	0.23 to 0.29	0.80 to 1.10	0.040	0.050	1525
G 15260	1526	0.22 to 0.29	1.10 to 1.40	0.040	0.050	1526
G 15270	1527[B]	0.22 to 0.29	1.20 to 1.50	0.040	0.050	1527
G 15360	1536[B]	0.30 to 0.37	1.20 to 1.50	0.040	0.050	1536
G 15410	1541[B]	0.36 to 0.44	1.35 to 1.65	0.040	0.050	1541
G 15470	1547	0.43 to 0.51	1.35 to 1.65	0.040	0.050	1547
G 15480	1548[B]	0.44 to 0.52	1.10 to 1.40	0.040	0.050	1548
G 15510	1551[B]	0.45 to 0.56	0.85 to 1.15	0.040	0.050	1551
G 15520	1552[B]	0.47 to 0.55	1.20 to 1.50	0.040	0.050	1552
G 15610	1561[B]	0.55 to 0.65	0.75 to 1.05	0.040	0.050	1561
G 15660	1566[B]	0.60 to 0.71	0.85 to 1.15	0.040	0.050	1566
G 15720	1572[B]	0.65 to 0.76	1.00 to 1.30	0.040	0.050	1572

[A] New designation established in accordance with ASTM E 527 and SAE J1086, Recommended Practice for Numbering Metals and Alloys (UNS).
[B] These grades were formerly designated as 10XX steels.

TABLE 8 Resulfurized Carbon Steels, Cast or Heat Chemical Ranges and Limits

NOTE 1: *Silicon*—When silicon is required, the following ranges and limits are commonly used: Up to 1110, incl. 0.10 max, %; 1116 and over, 0.10 max, %, 0.10 to 0.20 %, or 0.15 to 0.35 %.
NOTE 2—Because of the degree to which sulfur segregates, product analysis for sulfur in resulfurized carbon steel is not technologically appropriate unless misapplication is clearly indicated.

UNS Designation[A]	Grade No.	Chemical Composition Limits, %				SAE No.
		Carbon	Manganese	Phosphorus, max	Sulfur	
G 11080	1108	0.08 to 0.13	0.50 to 0.80	0.040	0.08 to 0.13	1108
G 11090	1109	0.08 to 0.13	0.60 to 0.90	0.040	0.08 to 0.13	1109
G 11100	1110	0.08 to 0.13	0.30 to 0.60	0.040	0.08 to 0.13	1110
G 11160	1116	0.14 to 0.20	1.10 to 1.40	0.040	0.16 to 0.23	1116
G 11170	1117	0.14 to 0.20	1.00 to 1.30	0.040	0.08 to 0.13	1117
G 11180	1118	0.14 to 0.20	1.30 to 1.60	0.040	0.08 to 0.13	1118
G 11190	1119	0.14 to 0.20	1.00 to 1.30	0.040	0.24 to 0.33	1119
G 11320	1132	0.27 to 0.34	1.35 to 1.65	0.040	0.08 to 0.13	1132
G 11370	1137	0.32 to 0.39	1.35 to 1.65	0.040	0.08 to 0.13	1137
G 11390	1139	0.35 to 0.43	1.35 to 1.65	0.040	0.13 to 0.20	1139
G 11400	1140	0.37 to 0.44	0.70 to 1.10	0.040	0.08 to 0.13	1140
G 11410	1141	0.37 to 0.45	1.35 to 1.65	0.040	0.08 to 0.13	1141
G 11440	1144	0.40 to 0.48	1.35 to 1.65	0.040	0.24 to 0.33	1144
G 11450	1145	0.42 to 0.49	0.70 to 1.00	0.040	0.04 to 0.07	1145
G 11460	1146	0.42 to 0.49	0.70 to 1.00	0.040	0.08 to 0.13	1146
G 11510	1151	0.48 to 0.55	0.70 to 1.00	0.040	0.08 to 0.13	1151

[A] New designation established in accordance with ASTM E 527 and SAE J1086, Recommended Practice for Numbering Metals and Alloys (UNS).

TABLE 9 Rephosphorized and Resulfurized Carbon Steel Cast or Heat Chemical Ranges and Limits

NOTE 1—It is not common practice to produce the 12XX series of steel to specified limits for silicon. Silicon impairs machineability.

NOTE 2—Because of the degree to which phosphorus and sulfur segregate, product analysis for phosphorus and sulfur in the 12XX series steel is not technologically appropriate unless misapplication is clearly indicated.

UNS Designation[A]	Grade No.	Chemical Composition Limits, %					SAE No.
		Carbon, max	Manganese	Phosphorus	Sulfur	Lead	
G 12110	1211	0.13	0.60 to 0.90	0.07 to 0.12	0.10 to 0.15	...	1211
G 12120	1212	0.13	0.70 to 1.00	0.07 to 0.12	0.16 to 0.23	...	1212
G 12130	1213	0.13	0.70 to 1.00	0.07 to 0.12	0.24 to 0.33	...	1213
G 12150	1215	0.09	0.75 to 1.05	0.04 to 0.09	0.26 to 0.35	...	1215
...	12L13	0.13	0.70 to 1.00	0.07 to 0.12	0.24 to 0.33	0.15 to 0.35	12L13
...	12L14	0.15	0.85 to 1.15	0.04 to 0.09	0.26 to 0.35	0.15 to 0.35	12L14
...	12L15	0.09	0.75 to 1.05	0.04 to 0.09	0.26 to 0.35	0.15 to 0.35	12L15

[A] New designation established in accordance with ASTM E 527 and SAE J1086, Recommended Practice for Numbering Metals and Alloys (UNS).

TABLE 10 Permissible Variations for Product Analysis of Carbon Steel

Element	Limit, or Max of Specified Range, %	Over Max Limit, %	Under Min Limit, %
Carbon	0.25 and under	0.02	0.02
	over 0.25 to 0.55, incl	0.03	0.03
	over 0.55	0.04	0.04
Manganese	0.90 and under	0.03	0.03
	over 0.90 to 1.65, incl	0.06	0.06
Phosphorus	to 0.040, incl	0.008	...
Sulfur	to 0.060, incl	0.008	...
Silicon	0.35 and under	0.02	0.02
	over 0.35 to 0.60, incl	0.05	0.05
Copper	under minimum only	...	0.02
Lead[A]	0.15 to 0.35, incl	0.03	0.03

[A] Product analysis permissible variations for lead applies to both over and under the specified range.

TABLE 11 Steel Wire Gage[A]

Gage No.	Decimal Equivalent, in.	Gage No.	Decimal Equivalent, in.
7/0	0.490	9	0.148*
6/0	0.462*	9½	0.142
5/0	0.430*	10	0.135
4/0	0.394*	10½	0.128
3/0	0.362*	11	0.120*
2/0	0.331	11½	0.113
1/0	0.306	12	0.106*
1	0.283	12½	0.099
1½	0.272	13	0.092*
2	0.262*	13½	0.086
2½	0.253	14	0.080
3	0.244*	14½	0.076
3½	0.234	15	0.072
4	0.225*	15½	0.067
4½	0.216	16	0.062*
5	0.207	16½	0.058
5½	0.200	17	0.054
6	0.192	17½	0.051
6½	0.184	18	0.048*
7	0.177	18½	0.044
7½	0.170	19	0.041
8	0.162	19½	0.038
8½	0.155	20	0.035*

[A] The steel wire gage outlined in this table has been taken from the original Washburn and Moen Gage chart. In 20 gage and coarser, sizes originally quoted to 4 decimal equivalent places have been rounded to 3 decimal places in accordance with rounding procedures of Recommended Practice E 29. All rounded U.S. customary values are indicated by an asterisk.

The American Society for Testing and Materials takes no position respecting the validity of any patent rights asserted in connection with any item mentioned in this standard. Users of this standard are expressly advised that determination of the validity of any such patent rights, and the risk of infringement of such rights, are entirely their own responsibility.

This standard is subject to revision at any time by the responsible technical committee and must be reviewed every five years and if not revised, either reapproved or withdrawn. Your comments are invited either for revision of this standard or for additional standards and should be addressed to ASTM Headquarters. Your comments will receive careful consideration at a meeting of the responsible technical committee, which you may attend. If you feel that your comments have not received a fair hearing you should make your views known to the ASTM Committee on Standards, 1916 Race St., Philadelphia, Pa. 19103.

Designation: A 510M – 82

Metric

Standard Specification for
GENERAL REQUIREMENTS FOR WIRE RODS AND COARSE ROUND WIRE, CARBON STEEL [METRIC][1]

This standard is issued under the fixed designation A 510M; the number immediately following the designation indicates the year of original adoption or, in the case of revision, the year of last revision. A number in parentheses indicates the year of last reapproval. A superscript epsilon (ϵ) indicates an editorial change since the last revision or reapproval.

This specification has been approved for use by agencies of the Department of Defense and for listing in the DoD Index of Specifications and Standards.

1. Scope

1.1 This specification covers general requirements for carbon steel wire rods and uncoated coarse round wire in coils or straightened and cut lengths.

1.2 In case of conflict, the requirements in the purchase order, on the drawing, in the individual specification, and in this general specification shall prevail in the sequence named.

NOTE 1—This metric specification is equivalent to Specification A 510, and is compatible in technical content.

2. Applicable Documents

2.1 *ASTM Standards:*

A 370 Methods and Definitions for Mechanical Testing of Steel Products[2]

A 700 Practices for Packaging, Marking, and Loading Methods for Steel Products for Domestic Shipment[2]

E 29 Recommended Practice for Indicating Which Places of Figures Are to Be Considered Significant in Specified Limiting Values[3]

E 30 Methods for Chemical Analysis of Steel, Cast Iron, Open-Hearth Iron, and Wrought Iron[4]

E 112 Methods for Determining Average Grain Size[5]

3. Description of Terms

3.1 *Carbon Steel*—Steel is considered to be carbon steel when no minimum content is specified or required for aluminum, chromium, cobalt, columbium, molybdenum, nickel, titanium, tungsten, vanadium, or zirconium, or any other element added to obtain a desired alloying effect; when the specified minimum for copper does not exceed 0.40 %; or when the maximum content specified for any of the following elements does not exceed these percentages: manganese 1.65, silicon 0.60, or copper 0.60.

3.1.1 In all carbon steels small quantities of certain unspecified and unrequired residual elements (such as copper, nickel, molybdenum, chromium, etc.) unavoidably retained from raw materials are sometimes found. These elements are considered as incidental and are not normally determined or reported.

3.1.2 Elements (such as sulfur and lead) may be specified to improve machinability of carbon steels.

3.2 Wire rods are hot rolled from billets to an approximately round cross section and into coils of one continuous length. Rods are not comparable to hot-rolled bars in accuracy of cross section or surface finish and as a semifinished product are intended primarily for the manufacture of wire.

3.2.1 Table 1 shows the nominal diameter for hot-rolled wire rods. Sizes are shown in 0.5-

---

[1] This specification is under the jurisdiction of ASTM Committee A-1 on Steel, Stainless Steel and Related Alloys, and is the direct responsibility of Subcommittee A01.03 on Steel Rod and Wire.

Current edition approved July 30, 1982. Published September 1982. Originally published as A 510M – 77. Last previous edition A 510M – 77.

[2] *Annual Book of ASTM Standards*, Vol 01.03.
[3] *Annual Book of ASTM Standards*, Vol 14.02.
[4] *Annual Book of ASTM Standards*, Vol 03.05.
[5] *Annual Book of ASTM Standards*, Vol 03.03.

mm increments from 5.5 to 19 mm.

3.3 Coarse round wire from 0.90 to 25 mm in diameter, inclusive, is produced from hot-rolled wire rods or hot-rolled coiled rounds by one or more cold reductions primarily for the purpose of obtaining a desired size with dimensional accuracy, surface finish, and mechanical properties. By varying the amount of cold reduction and other wire mill practices, including thermal treatment, a wide diversity of mechanical properties and finishes are made available. Suggested wire diameters are shown in Table 11.

3.4 Straightened and cut wire is produced from coils of wire by means of special machinery that straightens the wire and cuts it to a specified length.

3.4.1 The straightening operation may alter the mechanical properties of the wire, especially the tensile strength. The straightening operation may also induce changes in the diameter of the wire. The extent of the changes in the properties of the wire after cold straightening depends upon the kind of wire and also on the normal variation in the adjustments of the straightening equipment. It is therefore not possible to forecast the properties of straightened and cut wire. Each kind of wire needs individual consideration. In most cases, the application of straightened and cut wire is not seriously influenced by these changes.

4. Ordering Information

4.1 Orders for hot-rolled wire rods under this specification shall include the following information:

4.1.1 Quantity (kilograms or megagrams),

4.1.2 Name of material (wire rods),

4.1.3 Diameter (Table 1),

4.1.4 Chemical composition grade number (Tables 6, 7, 8, and 9),

4.1.5 Packaging,

4.1.6 ASTM designation and date of issue, and

4.1.7 Special requirements, if any.

NOTE 2—A typical ordering description is as follows: 50 000 kg Steel Wire Rods, 5.5 mm, Grade G10100 in approximately 600-kg Coils to ASTM A 510M dated ____.

4.2 Orders for coarse round wire under this specification shall include the following information:

4.2.1 Quantity (kilograms or pieces),

4.2.2 Name of material (uncoated carbon steel wire),

4.2.3 Diameter (Table 11),

4.2.4 Length (straightened and cut only),

4.2.5 Chemical composition (Tables 6, 7, 8, and 9),

4.2.6 Packaging,

4.2.7 ASTM designation and date of issue, and

4.2.8 Special requirements, if any.

NOTE 3—A typical ordering description is as follows: 15 000 kg Uncoated Carbon Steel Wire, 3.8 mm diameter, Grade G10080 in 1000-kg Coils on Tubular Carriers, to ASTM A 510M dated ____, or 2500 Pieces, Carbon Steel Wire, 9.5 mm diameter, Straightened and Cut, 0.76 m, Grade G10150, in 25-Piece Bundles on Pallets to ASTM A 510M dated

5. Manufacture

5.1 The steel may be made by any commercially accepted steel making process. The steel may be either ingot cast or strand cast.

6. Chemical Requirements

6.1 The chemical composition for steel under this specification shall conform to the requirements set forth in the purchase order. Chemical compositions are specified by ranges or limits for carbon and other elements. The grades commonly specified for carbon steel wire rods and coarse round wire are shown in Tables 6, 7, 8, and 9.

6.2 *Cast or Heat Analysis*—An analysis of each heat shall be made by the producer to determine the percentage of the elements specified. The analysis shall be made from a test sample, preferably taken during the pouring of the heat. The chemical composition thus determined shall be reported, if required, to the purchaser or his representative.

6.3 *Product Analysis*—A product analysis may be made by the purchaser. The analysis is not used for a duplicate analysis to confirm a previous result. The purpose of the product analysis is to verify that the chemical composition is within specified limits for each element, including applicable permissible variations in product analysis. The results of analyses taken from different pieces of a heat may differ within permissible limits from each other and from the heat analysis. Table 10 shows the permissible variations for product analysis of carbon steel. The results of the product analysis, except lead, shall not vary both above and below the permissible limits.

6.3.1 Rimmed or capped steels are characterized by a lack of uniformity in their chemical composition, especially for the elements carbon, phosphorus, and sulfur, and for this reason product analysis is not technologically appropriate for these elements unless misapplication is clearly indicated.

6.3.2 Because of the degree to which phosphorus and sulfur segregate, product analysis for these elements is not technologically appropriate for rephosphorized or resulfurized steels or both unless misapplication is clearly indicated.

6.3.3 The location at which chips for product analysis are obtained from the sample is important because of segregation. For rods and wire, chips must be taken by milling or machining the full cross section of the sample.

6.3.3.1 Steel subjected to certain heat treating operations by the purchaser may not give chemical analysis results that properly represent its original composition. Therefore, purchasers should analyze chips taken from the steel in the condition in which it is received from the producer.

6.3.3.2 When samples are returned to the producer for product analysis, the samples should consist of pieces of the full cross section.

6.3.4 For referee purposes, Methods E 30 shall be used.

7. Metallurgical Structure

7.1 Grain size when specified shall be determined in accordance with the requirements of Methods E 112.

8. Mechanical Requirements

8.1 The properties enumerated in individual specifications shall be determined in accordance with Methods and Definitions A 370.

8.2 Because of the great variety in the kinds of wire and the extensive diversity of end uses, a number of formal mechanical test procedures have been developed. These tests are used as control tests by producers during the intermediate stages of wire processing, as well as for final testing of the finished product, and apply particularly to specification wire and wires for specific applications. A number of these tests are further described in Supplement IV, Round Wire Products, of Methods and Definitions A 370.

8.3 Since the general utility of rods and wire requires continuity of length, in the case of rods, tests are commonly made on samples taken from the ends of coils after removing two to three rings. In the case of wire, tests are commonly made on samples taken from the ends of coils, thereby not impairing the usefulness of the whole coil.

9. Dimensions and Permissible Variations

9.1 The diameter and out-of-roundness of the wire rod shall not vary from that specified by more than that prescribed in Table 2.

9.2 The diameter and out-of-roundness of the coarse round wire and straightened and cut wire shall not vary from that specified by more than that prescribed in Table 3.

9.3 The length of straightened and cut wire shall not vary from that specified by more than that prescribed in Table 4.

9.4 The burrs formed in cutting straightened and cut wire shall not exceed the diameter specified by more than that prescribed in Table 5.

10. Workmanship, Finish, and Appearance

10.1 The wire rod shall be free of detrimental surface imperfections, tangles, and sharp kinks.

10.1.1 Two or more rod coils may be welded together to produce a larger coil. The weld zone may not be as sound as the original material. The mechanical properties existing in the weld material may differ from those in the unaffected base metal. The weld may exceed the standard dimensional permissible variations on the minus side and must be within the permissible variations on the plus side.

10.2 The wire as received shall be smooth and substantially free from rust, shall not be kinked or improperly cast. No detrimental die marks or scratches may be present. Each coil shall be one continuous length of wire. Welds made during cold drawing are permitted.

10.3 The straightened and cut wire shall be substantially straight and not be kinked or show excessive spiral marking.

11. Retests

11.1 The difficulties in obtaining truly representative samples of wire rod and coarse round wire without destroying the usefulness of the coil of wire account for the generally

A 510M

accepted practice of allowing retests for mechanical tests and surface examination. An additional test piece is cut from each end of the coil from which the original sample was taken. A portion of the coil may be discarded prior to cutting the sample for retest. If any of the retests fail to comply with the requirements, the coil of wire may be rejected. Before final rejection, however, it is frequently advisable to base final decision on an actual trial of the material to determine whether or not it will do the job for which it is intended.

12. Inspection

12.1 The manufacturer shall afford the purchaser's inspector all reasonable facilities necessary to satisfy him that the material being produced and furnished is in accordance with this specification. Mill inspection by the purchaser shall not interfere unnecessarily with the manufacturer's operations. All tests and inspections shall be made at the place of manufacture, unless otherwise agreed upon.

13. Rejection

13.1 Any rejection based on tests made in accordance with this specification shall be reported to the producer within a reasonable length of time. The material must be adequately protected and correctly identified in order that the producer may make a proper investigation.

14. Certification

14.1 Upon request of the purchaser in the contract or order, a manufacturer's certification that the material was manufactured and tested in accordance with this specification together with a report of the test results shall be furnished at the time of shipment.

15. Packaging, Marking, and Loading

15.1 A tag shall be securely attached to each coil or bundle and shall be marked with the size, ASTM specification number, heat or cast number, grade number, and name or mark of the manufacturer.

15.2 When specified in the purchase order packaging, marking, and loading for shipments shall be in accordance with those procedures recommended by Practices A 700.

TABLE 1 Sizes of Wire Rods, mm

5.5	12.5
6	13
6.5	13.5
7	14
7.5	14.5
8	15
8.5	15.5
9	16
9.5	16.5
10	17
10.5	17.5
11	18
11.5	18.5
12	19

TABLE 2 Permissible Variation in Diameter for Wire Rod in Coils

NOTE—For purposes of determining conformance with this specification, all specified limits in this table are absolute limits as defined in Recommended Practice E 29.

Diameter of Rod, mm	Permissible Variation, Plus and Minus, mm	Permissible Out-of-Round, mm
5.5 to 19	0.40	0.60

TABLE 3 Permissible Variation in Diameter for Uncoated Coarse Round Wire

NOTE—For purposes of determining conformance with this specification, all specified limits in this table are absolute limits as defined in Recommended Practice E 29.

	In Coils	
Diameter of Wire, mm	Permissible Variation, Plus and Minus, mm	Permissible Out-of-Round, mm
0.90 to under 1.90	0.03	0.03
1.90 to under 12.5	0.05	0.05
12.5 and over	0.08	0.08
Straightened and Cut		
0.90 to under 1.90	0.03	0.03
1.90 to under 3.80	0.05	0.05
3.80 to under 12.5	0.08	0.08
12.5 and over	0.10	0.10

TABLE 4 Permissible Variation in Length for Straightened and Cut Wire

NOTE—For purposes of determining conformance with this specification, all specified limits in this table are absolute limits as defined in Recommended Practice E 29.

Cut Length, m	Permissible Variations, Plus and Minus, mm
Under 1.0	1.6
1.0 to 4.0	2.4
Over 4.0	3.0

A 510M

TABLE 5 Permissible Variation for Burrs for Straightened and Cut Wire

NOTE—For purposes of determining conformance with this specification, all specified limits in this table are absolute limits as defined in Recommended Practice E 29.

Diameter of Wire, mm	Permissible Variations, over Measured Diameter, mm
Up to 3.0, incl	0.10
Over 3.0 to 6.5, incl	0.15
Over 6.5 to 12.5, incl	0.20
Over 12.5	0.25

ASTM A 510M

TABLE 6 Nonresulfurized Carbon Steel Cast or Heat Chemical Ranges and Limits

NOTE 1: *Silicon*—When silicon is required the following ranges and limits are commonly used for nonresulfurized carbon steels: 0.10 max %, 0.07 to 0.15 %, 0.10 to 0.20 %, 0.15 to 0.35 %, 0.20 to 0.40 %, or 0.30 to 0.60 %.
NOTE 2: *Copper*—When required, copper is specified as an added element.
NOTE 3: *Lead*—When lead is required as an added element, a range from 0.15 to 0.35 % is specified. Such a steel is identified by inserting the letter "L" between the second and third numerals of the grade number, for example 10L18.

UNS Designation[A]	Grade No.	Chemical Composition Limits, %				SAE No.
		Carbon	Manganese	Phosphorus, max	Sulfur, max	
G 10050	1005	0.06 max	0.35 max	0.040	0.050	1005
G 10060	1006	0.08 max	0.25 to 0.40	0.040	0.050	1006
G 10080	1008	0.10 max	0.30 to 0.50	0.040	0.050	1008
G 10100	1010	0.08 to 0.13	0.30 to 0.60	0.040	0.050	1010
G 10110	1011	0.08 to 0.13	0.60 to 0.90	0.040	0.050	1011
G 10120	1012	0.10 to 0.15	0.30 to 0.60	0.040	0.050	1012
G 10130	1013	0.11 to 0.16	0.50 to 0.80	0.040	0.050	1013
G 10150	1015	0.13 to 0.18	0.30 to 0.60	0.040	0.050	1015
G 10160	1016	0.13 to 0.18	0.60 to 0.90	0.040	0.050	1016
G 10170	1017	0.15 to 0.20	0.30 to 0.60	0.040	0.050	1017
G 10180	1018	0.15 to 0.20	0.60 to 0.90	0.040	0.050	1018
G 10190	1019	0.15 to 0.20	0.70 to 1.00	0.040	0.050	1019
G 10200	1020	0.18 to 0.23	0.30 to 0.60	0.040	0.050	1020
G 10210	1021	0.18 to 0.23	0.60 to 0.90	0.040	0.050	1021
G 10220	1022	0.18 to 0.23	0.70 to 1.00	0.040	0.050	1022
G 10230	1023	0.20 to 0.25	0.30 to 0.60	0.040	0.050	1023
G 10250	1025	0.22 to 0.28	0.30 to 0.60	0.040	0.050	1025
G 10260	1026	0.22 to 0.28	0.60 to 0.90	0.040	0.050	1026
G 10290	1029	0.25 to 0.31	0.60 to 0.90	0.040	0.050	1029
G 10300	1030	0.28 to 0.34	0.60 to 0.90	0.040	0.050	1030
G 10340	1034	0.32 to 0.38	0.50 to 0.80	0.040	0.050	...
G 10350	1035	0.32 to 0.38	0.60 to 0.90	0.040	0.050	1035
G 10370	1037	0.32 to 0.38	0.70 to 1.00	0.040	0.050	1037
G 10380	1038	0.35 to 0.42	0.60 to 0.90	0.040	0.050	1038
G 10390	1039	0.37 to 0.44	0.70 to 1.00	0.040	0.050	1039
G 10400	1040	0.37 to 0.44	0.60 to 0.90	0.040	0.050	1040
G 10420	1042	0.40 to 0.47	0.60 to 0.90	0.040	0.050	1042
G 10430	1043	0.40 to 0.47	0.70 to 1.00	0.040	0.050	1043
G 10440	1044	0.43 to 0.50	0.30 to 0.60	0.040	0.050	1044
G 10450	1045	0.43 to 0.50	0.60 to 0.90	0.040	0.050	1045
G 10460	1046	0.43 to 0.50	0.70 to 1.00	0.040	0.050	1046
G 10490	1049	0.46 to 0.53	0.60 to 0.90	0.040	0.050	1049
G 10500	1050	0.48 to 0.55	0.60 to 0.90	0.040	0.050	1050
G 10530	1053	0.48 to 0.55	0.70 to 1.00	0.040	0.050	1053
G 10550	1055	0.50 to 0.60	0.60 to 0.90	0.040	0.050	1055
G 10590	1059	0.55 to 0.65	0.50 to 0.80	0.040	0.050	1059
G 10600	1060	0.55 to 0.65	0.60 to 0.90	0.040	0.050	1060
G 10640	1064	0.60 to 0.70	0.50 to 0.80	0.040	0.050	1064
G 10650	1065	0.60 to 0.70	0.60 to 0.90	0.040	0.050	1065
G 10690	1069	0.65 to 0.75	0.40 to 0.70	0.040	0.050	1069
G 10700	1070	0.65 to 0.75	0.60 to 0.90	0.040	0.050	1070
G 10740	1074	0.70 to 0.80	0.50 to 0.80	0.040	0.050	1074
G 10750	1075	0.70 to 0.80	0.40 to 0.70	0.040	0.050	1075
G 10780	1078	0.72 to 0.85	0.30 to 0.60	0.040	0.050	1078
G 10800	1080	0.75 to 0.88	0.60 to 0.90	0.040	0.050	1080
G 10840	1084	0.80 to 0.93	0.60 to 0.90	0.040	0.050	1084
G 10850	1085	0.80 to 0.93	0.70 to 1.00	0.040	0.050	1085
G 10860	1086	0.80 to 0.93	0.30 to 0.50	0.040	0.050	1086
G 10900	1090	0.85 to 0.98	0.60 to 0.90	0.040	0.050	1090
G 10950	1095	0.90 to 1.03	0.30 to 0.50	0.040	0.050	1095

[A] New designation established in accordance with ASTM E 527 and SAE J 1086, Recommended Practice for Numbering Metals and Alloys (UNS).

A 510M

TABLE 7 Nonresulfurized Carbon Steel, High Manganese, Cast or Heat Chemical Ranges and Limits

NOTE 1: *Silicon*—When silicon is required the following ranges and limits are commonly used for nonresulfurized carbon steels: 0.10 max %, 0.07 to 0.15 %, 0.10 to 0.20 %, 0.15 to 0.35 %, 0.20 to 0.40 %, or 0.30 to 0.60 %.
NOTE 2: *Copper*—When required, copper is specified as an added element.
NOTE 3: *Lead*—When lead is required as an added element a range from 0.15 to 0.35 % is specified. Such a steel is identified by inserting the letter "L" between the second and third numerals of the grade number, for example 15L18.

UNS Designation[A]	Grade No.	Chemical Composition Limits, %				SAE No.
		Carbon	Manganese	Phosphorus, max	Sulfur, max	
G 15130	1513	0.10 to 0.16	1.10 to 1.40	0.040	0.050	1513
G 15180	1518	0.15 to 0.21	1.10 to 1.40	0.040	0.050	1518
G 15220	1522	0.18 to 0.24	1.10 to 1.40	0.040	0.050	1522
G 15240	1524[B]	0.19 to 0.25	1.35 to 1.65	0.040	0.050	1524
G 15250	1525	0.23 to 0.29	0.80 to 1.10	0.040	0.050	1525
G 15260	1526	0.22 to 0.29	1.10 to 1.40	0.040	0.050	1526
G 15270	1527[B]	0.22 to 0.29	1.20 to 1.50	0.040	0.050	1527
G 15360	1536[B]	0.30 to 0.37	1.20 to 1.50	0.040	0.050	1536
G 15410	1541[B]	0.36 to 0.44	1.35 to 1.65	0.040	0.050	1541
G 15470	1547	0.43 to 0.51	1.35 to 1.65	0.040	0.050	1547
G 15480	1548[B]	0.44 to 0.52	1.10 to 1.40	0.040	0.050	1548
G 15510	1551[B]	0.45 to 0.56	0.85 to 1.15	0.040	0.050	1551
G 15520	1552[B]	0.47 to 0.55	1.20 to 1.50	0.040	0.050	1552
G 15610	1561[B]	0.55 to 0.65	0.75 to 1.05	0.040	0.050	1561
G 15660	1566[B]	0.60 to 0.71	0.85 to 1.15	0.040	0.050	1566
G 15720	1572[B]	0.65 to 0.76	1.00 to 1.30	0.040	0.050	1572

[A] New designation established in accordance with ASTM E 527 and SAE J1086, Recommended Practice for Numbering Metals and Alloys (UNS).
[B] These grades were formerly designated as 10XX steels.

TABLE 8 Resulfurized Carbon Steels, Cast or Heat Chemical Ranges and Limits

NOTE 1: *Silicon*—When silicon is required, the following ranges and limits are commonly used: Up to 1110, incl, 0.10 max %, 1116 and over, 0.10 max %, 0.10 to 0.20 %, or 0.15 to 0.35 %.
NOTE 2—Because of the degree to which sulfur segregates, products analysis for sulfur in resulfurized carbon steel is not technologically appropriate unless misapplication is clearly indicated.

UNS Designation[A]	Grade No.	Chemical Composition Limits, %				SAE No.
		Carbon	Manganese	Phosphorus, max	Sulfur	
G 11080	1108	0.08 to 0.13	0.50 to 0.80	0.040	0.08 to 0.13	1108
G 11090	1109	0.08 to 0.13	0.60 to 0.90	0.040	0.08 to 0.13	1109
G 11100	1110	0.08 to 0.13	0.30 to 0.60	0.040	0.08 to 0.13	1110
G 11160	1116	0.14 to 0.20	1.10 to 1.40	0.040	0.16 to 0.23	1116
G 11170	1117	0.14 to 0.20	1.00 to 1.30	0.040	0.08 to 0.13	1117
G 11180	1118	0.14 to 0.20	1.30 to 1.60	0.040	0.08 to 0.13	1118
G 11190	1119	0.14 to 0.20	1.00 to 1.30	0.040	0.24 to 0.33	1119
G 11320	1132	0.27 to 0.34	1.35 to 1.65	0.040	0.08 to 0.13	1132
G 11370	1137	0.32 to 0.39	1.35 to 1.65	0.040	0.08 to 0.13	1137
G 11390	1139	0.35 to 0.43	1.35 to 1.65	0.040	0.13 to 0.20	1139
G 11400	1140	0.37 to 0.44	0.70 to 1.00	0.040	0.08 to 0.13	1140
G 11410	1141	0.37 to 0.45	1.35 to 1.65	0.040	0.08 to 0.13	1141
G 11440	1144	0.40 to 0.48	1.35 to 1.65	0.040	0.24 to 0.33	1144
G 11450	1145	0.42 to 0.49	0.70 to 1.00	0.040	0.04 to 0.07	1145
G 11460	1146	0.42 to 0.49	0.70 to 1.00	0.040	0.08 to 0.13	1146
G 11510	1151	0.48 to 0.55	0.70 to 1.00	0.040	0.08 to 0.13	1151

[A] New designation established in accordance with ASTM E 527 and SAE J1086, Recommended Practice for Numbering Metals and Alloys (UNS).

TABLE 9 Rephosphorized and Resulfurized Carbon Steel Cast or Heat Chemical Ranges and Limits

NOTE 1—It is not common practice to produce the 12XX series of steel to specified limits for silicon. Silicon impairs machineability.

NOTE 2—Because of the degree to which phosphorus and sulfur segregate, product analysis for phosphorus and sulfur in the 12XX series steel is not technologically appropriate unless misapplication is clearly indicated.

UNS Designation[A]	Grade No.	Chemical Composition Limits, %					SAE No.
		Carbon, max	Manganese	Phosphorus	Sulfur	Lead	
G 12110	1211	0.13	0.60 to 0.90	0.07 to 0.12	0.10 to 0.15	...	1211
G 12120	1212	0.13	0.70 to 1.00	0.07 to 0.12	0.16 to 0.23	...	1212
G 12130	1213	0.13	0.70 to 1.00	0.07 to 0.12	0.24 to 0.33	...	1213
G 12150	1215	0.09	0.75 to 1.05	0.04 to 0.09	0.26 to 0.35	...	1215
...	12L13	0.13	0.70 to 1.00	0.07 to 0.12	0.24 to 0.33	0.15 to 0.35	12L13
...	12L14	0.15	0.85 to 1.15	0.04 to 0.09	0.26 to 0.35	0.15 to 0.35	12L14
...	12L15	0.09	0.75 to 1.05	0.04 to 0.09	0.26 to 0.35	0.15 to 0.35	12L15

[A] New designation established in accordance with ASTM E 527 and SAE J1086, Recommended Practice for Numbering Metals and Alloys (UNS).

TABLE 10 Permissible Variations for Product Analysis of Carbon Steel

Element	Limit, or Max of Specified Range, %	Over Max Limit, %	Under Min Limit, %
Carbon	0.25 and under	0.02	0.02
	over 0.25 to 0.55, incl	0.03	0.03
	over 0.55	0.04	0.04
Manganese	0.90 and under	0.03	0.03
	over 0.90 to 1.65, incl	0.06	0.06
Phosphorus	to 0.040, incl	0.008	...
Sulfur	to 0.060, incl	0.008	...
Silicon	0.35 and under	0.02	0.02
	over 0.35 to 0.60, incl	0.05	0.05
Copper	under minimum only	...	0.02
Lead[A]	0.15 to 0.35, incl	0.03	0.03

[A] Product analysis permissible variations for lead applies to both over and under the specified range.

TABLE 11 Suggested Diameters for Steel Wire, mm

0.90	6.0
1.00	6.5
1.10	7.0
1.20	7.5
1.30	8.0
1.40	8.5
1.60	9.0
1.80	9.5
2.0	10.0
2.1	11.0
2.2	12.0
2.4	13.0
2.5	14.0
2.6	15.0
2.8	16.0
3.0	17.0
3.2	18.0
3.5	19.0
3.8	20.0
4.0	21.0
4.2	22.0
4.5	23.0
4.8	24.0
5.0	25.0
5.5	

The American Society for Testing and Materials takes no position respecting the validity of any patent rights asserted in connection with any item mentioned in this standard. Users of this standard are expressly advised that determination of the validity of any such patent rights, and the risk of infringement of such rights, are entirely their own responsibility.

This standard is subject to revision at any time by the responsible technical committee and must be reviewed every five years and if not revised, either reapproved or withdrawn. Your comments are invited either for revision of this standard or for additional standards and should be addressed to ASTM Headquarters. Your comments will receive careful consideration at a meeting of the responsible technical committee, which you may attend. If you feel that your comments have not received a fair hearing you should make your views known to the ASTM Committee on Standards, 1916 Race St., Philadelphia, Pa. 19103.

Designation: A 544 – 82

Standard Specification for
STEEL WIRE, CARBON, SCRAPLESS NUT QUALITY[1]

This standard is issued under the fixed designation A 544; the number immediately following the designation indicates the year of original adoption or, in the case of revision, the year of last revision. A number in parentheses indicates the year of last reapproval. A superscript epsilon (ε) indicates an editorial change since the last revision or reapproval.

1. Scope

1.1 This specification covers carbon steel, round, scrapless nut quality wire intended for the manufacture of scrapless nuts.

1.2 The values stated in inch-pound units are to be regarded as the standard.

2. Applicable Documents

2.1 *ASTM Standards*:

A 510 Specification for General Requirements for Wire Rods and Coarse Round Wire, Carbon Steel[2]

A 510M Specification for General Requirements for Wire Rods and Coarse Round Wire, Carbon Steel [Metric][2]

A 700 Recommended Practices for Packaging, Marking, and Loading Methods for Steel Products for Domestic Shipment[3]

A 751 Methods, Practices, and Definitions for Chemical Analysis of Steel Products[3,5]

E 29 Recommended Practice for Indicating Which Places of Figures Are to Be Considered Significant in Specified Limiting Values[4]

E 30 Chemical Analysis of Steel, Cast Iron, Open-Hearth Iron, and Wrought Iron[5]

3. General Requirements for Delivery

3.1 Material furnished to this specification shall conform to the applicable requirements of the current edition of Specification A 510.

4. Ordering Information

4.1 Orders for material to this specification shall include the following for each item:

4.1.1 Quantity,

4.1.2 Name of material (carbon steel wire for scrapless nuts),

4.1.3 Dimensions (see 9.1),

4.1.4 Thermal treatment (see 5.3),

4.1.5 Steel grade (see Table 1),

4.1.6 End use (type of nut to be formed: conventional, flange nut, keps nut, clinch nut or other special types),

4.1.7 Packaging (see 11.1),

4.1.8 Cast or heat analysis, if desired (see 6.2), and

4.1.9 ASTM designation and date of issue.

NOTE—A typical ordering description is as follows: 30,000 lb round, carbon steel wire for scrapless nuts, 0.625 in. diameter, annealed at finish size, 1035, for flange nuts, paper wrapped in 1000-lb coils 30 in. nominal diameter to ASTM A 544 dated ___.

5. Materials and Manufacture

5.1 *Process*—The steel shall be made by the open-hearth, basic-oxygen, or electric-furnace process.

5.2 *Internal Soundness*—Sufficient discard shall be taken to ensure freedom from injurious pipe and segregation.

5.3 *Thermal Treatment*—Wire to this specification may be ordered without thermal treatment, but when required and depending on end use, wire may be ordered as follows:

5.3.1 Drawn from normalized rod,

5.3.2 Drawn from annealed or spheroidize annealed rod,

5.3.3 Annealed or spheroidize annealed in process, or

---

[1] This specification is under the jurisdiction of ASTM Committee A-1 on Steel, Stainless Steel and Related Alloys, and is the direct responsibility of Subcommittee A01.03 on Steel Rod and Wire.
Current edition approved July 30, 1982. Published September 1982. Originally published as A 544 – 65. Last previous edition A 544 – 77.
[2] *Annual Book of ASTM Standards*, Vol 01.03.
[3] *Annual Book of ASTM Standards*, Vols 01.03 and 01.05.
[4] *Annual Book of ASTM Standards*, Vol 14.02.
[5] *Annual Book of ASTM Standards*, Vol 03.05.

A 544

5.3.4 Annealed or spheroidize annealed at finished size.

6. Chemical Requirements

6.1 The steel shall conform to the requirements as to chemical composition specified in Table 1, for the grade ordered.

6.2 *Cast or Heat Analysis (Formerly Ladle Analysis)*—Each cast or heat of steel shall be analyzed by the manufacturer to determine the percentage of elements prescribed in Table 1. This analysis shall be made from a specimen preferably taken during the pouring of the cast or heat. When requested, this analysis shall be reported to the purchaser.

6.3 *Product Analysis (Formerly Check Analysis)*—An analysis may be made by the purchaser from finished wire representing each cast or heat of steel. The chemical composition of samples taken for product analysis shall conform to the permissible limits of Table 10 of Specification A 510.

6.4 For referee purposes, Methods E 30 shall be applied.

7. Decarburization Limits

7.1 *Decarburization of Killed Steels*—The entire periphery of a properly prepared sample of the wire shall be examined for decarburization at a magnification of 100 diameters and the limits of Table 3 shall apply.

8. Mechanical Requirements

8.1 Mechanical properties shall be compatible with the steel grade, size, thermal treatment, and end use. Grades 1030, 1035, and 1038 should be used in a spheroidize annealed in process or spheroidized at finished size condition.

9. Dimensions and Permissible Variations

9.1 The diameter of the wire shall not vary from the specified diameter by more than that prescribed in Table 2.

10. Workmanship and Finish

10.1 Each coil shall be one continuous length of wire, properly coiled, and firmly tied, or banded.

10.2 Welded coils are subject to agreement between the manufacturer and the purchaser.

10.3 The surface of the wire shall be free of defects of a nature or degree that will be detrimental to the forming, machining, or fabrication of the scrapless nuts.

10.4 The wire, as received, shall have a finish suitable for production of scrapless nuts or as agreed upon between the manufacturer and purchaser.

11. Packaging, Marking, and Loading

11.1 *Packaging*—The coil weight, dimensions, and method of packaging shall be agreed upon by the manufacturer and purchaser.

11.2 *Marking*—The size, steel grade, purchase order number, cast or heat number, thermal treatment (when required), ASTM designation number and name, brand, or trademark of the manufacturer shall be shown on a tag securely attached to each coil of wire.

11.3 When specified on the purchaser's order, packaging, marking, and loading for shipments shall be in accordance with those procedures recommended by Recommended Practices A 700.

12. Inspection

12.1 The manufacturer shall afford the inspector, representing the purchaser, all reasonable facilities to satisfy him that the material is being furnished in accordance with this specification. All tests and inspection may be made at the place of manufacture prior to shipment and shall be so conducted as not to interfere unnecessarily with the operations of the works.

13. Rejection and Rehearing

13.1 Unless otherwise specified, any rejection based on tests made in accordance with this specification shall be reported to the manufacturer within a reasonable length of time.

13.2 The wire rejected under 13.1 shall be protected from damage and loss of identity by the purchaser for a reasonable time to permit the manufacturer to evaluate the rejection.

TABLE 1 Chemical Requirements[A]

UNS Designation[C]	Grade	Carbon, %	Manganese, %	Phosphorus,[B] max, %	Sulfur, %
...	1106	0.08 max	0.30–0.60	0.040	0.08–0.13
...	1108	0.08–0.13	0.50–0.80	0.040	0.08–0.13
...	1109	0.08–0.13	0.60–0.90	0.040	0.08–0.13
...	1110	0.08–0.13	0.30–0.60	0.040	0.08–0.13
...	Modified 1106	0.08 max	0.30–0.60	0.040	0.04–0.09
...	Modified 1108	0.08–0.13	0.50–0.80	0.040	0.04–0.09
...	Modified 1109	0.08–0.13	0.60–0.90	0.040	0.04–0.09
...	Modified 1110	0.08–0.13	0.30–0.60	0.040	0.04–0.09
G 10170	1017	0.15–0.20	0.30–0.60	0.040	0.050 max[B]
G 10180	1018	0.15–0.20	0.60–0.90	0.040	0.050 max[B]
G 10200	1020	0.18–0.23	0.30–0.60	0.040	0.050 max[B]
G 10220	1022	0.18–0.23	0.70–1.00	0.040	0.050 max[B]
G 10300	1030	0.28–0.34	0.60–0.90	0.040	0.050 max[B]
G 10350	1035	0.32–0.38	0.60–0.90	0.040	0.050 max[B]
G 10380	1038	0.35–0.42	0.60–0.90	0.040	0.050 max[B]

[A] In non-resulfurized grades, silicon may be specified as 0.10 max %, 0.10 to 0.20 %, or 0.15 to 0.35 %.
[B] Ordinarily furnished phosphorus 0.035 max %, and sulfur 0.045 max %.
[C] New designation in accordance with ASTM E 527 and SAE J1086, Recommended Practice for Numbering Metals and Alloys (UNS).

TABLE 2 Permissible Variations in Wire Diameter

NOTE—For purposes of determining conformance with this specification all specified limits are absolute as defined in Recommended Practice E 29.

Diameter, in. (mm)	Permissible Variations, plus and minus, in. (mm)	Permissible Out-of-Round, in. (mm)
0.076 to 0.500 (1.93 to 12.70), excl	0.0015 (0.04)	0.0015 (0.04)
0.500 (12.70) and over	0.002 (0.05)	0.002 (0.05)

TABLE 3 Decarburization Limits[A]

NOTE—For purposes of determining conformance with this specification all specified limits are absolute as defined in Recommended Practice E 29.

Diameter, in. (mm)	Average Depth Free Ferrite, in. (mm)	Average Total Affected Depth (Free Ferrite Plus Partial Decarburization), in. (mm)
Up to 0.375 (9.52), incl	0.003 (0.08)	0.010 (0.25)
Over 0.375 to 0.500 (9.52 to 12.70), incl	0.004 (0.10)	0.012 (0.30)
Over 0.500 to 0.703 (12.70 to 17.86), incl	0.005 (0.13)	0.014 (0.36)
Over 0.703 to 0.999 (17.86 to 25.37), incl	0.006 (0.15)	0.016 (0.41)

[A] Applicable to killed steels only. Grade 1017 and higher.

The American Society for Testing and Materials takes no position respecting the validity of any patent rights asserted in connection with any item mentioned in this standard. Users of this standard are expressly advised that determination of the validity of any such patent rights, and the risk of infringement of such rights, are entirely their own responsibility.

This standard is subject to revision at any time by the responsible technical committee and must be reviewed every five years and if not revised, either reapproved or withdrawn. Your comments are invited either for revision of this standard or for additional standards and should be addressed to ASTM Headquarters. Your comments will receive careful consideration at a meeting of the responsible technical committee, which you may attend. If you feel that your comments have not received a fair hearing you should make your views known to the ASTM Committee on Standards, 1916 Race St., Philadelphia, Pa. 19103.

Designation: A 545 – 82

Standard Specification for
STEEL WIRE, CARBON, COLD-HEADING QUALITY, FOR MACHINE SCREWS[1]

This standard is issued under the fixed designation A 545; the number immediately following the designation indicates the year of original adoption or, in the case of revision, the year of last revision. A number in parentheses indicates the year of last reapproval. A superscript epsilon (ϵ) indicates an editorial change since the last revision or reapproval.

This specification has been approved for use by agencies of the Department of Defense and for listing in the DoD Index of Specifications and Standards.

1. Scope

1.1 This specification covers carbon steel, cold-heading quality, round wire intended for the manufacture of machine screws.

1.2 The values stated in inch-pound units are to be regarded as standard.

2. Applicable Documents

2.1 *ASTM Standards*:
A 510 Specification for General Requirements for Wire Rods and Coarse Round Wire, Carbon Steel[2]
A 510M Specification for General Requirements for Wire Rods and Coarse Round Wire, Carbon Steel [Metric][2]
A 700 Recommended Practices for Packaging, Marking, and Loading Methods for Steel Products for Domestic Shipment[3]
A 751 Methods, Practices, and Definitions for Chemical Analysis of Steel Products[3,5]
E 29 Recommended Practice for Indicating Which Places of Figures Are to Be Considered Significant in Specified Limiting Values[4]
E 30 Chemical Analysis of Steel, Cast Iron, Open-Hearth Iron, and Wrought Iron[5]
E 112 Determining Average Grain Size[6]

3. General Requirements for Delivery

3.1 Material furnished to this specification shall conform to the applicable requirements of the current edition of Specification A 510.

4. Ordering Information

4.1 Orders for wire to this specification shall include the following items to adequately describe the material:
4.1.1 Quantity (weight),
4.1.2 Name of material (cold-heading wire for machine screws),
4.1.3 Dimension (see 9.1),
4.1.4 Thermal treatment (see 5.3),
4.1.5 Steel grade (see Table 1),
4.1.6 End use (type of head to be formed—conventional slotted, flat, oval and round head, recessed heads, and other special head types),
4.1.7 Packaging (see 11.1),
4.1.8 Cast or heat analysis, if desired (see 6.2),
4.1.9 Grain size, if desired (see 7.1), and
4.1.10 ASTM designation and date of issue.

NOTE—A typical ordering description is as follows: 30,000 lb Round Carbon Steel Wire, Cold Heading Quality, 0.375 in. diameter, Annealed in Process, Grade 1008 for Slotted Round Head Machine Screws, Coils 500 lb, 26 in. normal diameter. Furnish analysis for each cast or heat, to ASTM A 545 dated ——.

5. Manufacture

5.1 *Process*—The steel shall be made by the open-hearth, basic-oxygen, or electric-furnace process.

5.2 *Internal Soundness*—Sufficient discard shall be taken to ensure freedom from injurious pipe and segregation:

---

[1] This specification is under the jurisdiction of ASTM Committee A-1 on Steel, Stainless Steel and Related Alloys, and is the direct responsibility of Subcommittee A01.03 on Steel Rod and Wire.
Current edition approved July 30, 1982. Published September 1982. Originally published as A 545 – 65. Last previous edition A 545 – 77.
[2] *Annual Book of ASTM Standards*, Vol 01.03.
[3] *Annual Book of ASTM Standards*, Vols 01.03 and 01.05.
[4] *Annual Book of ASTM Standards*, Vol 14.02.
[5] *Annual Book of ASTM Standards*, Vol 03.05.
[6] *Annual Book of ASTM Standards*, Vol 03.03.

5.3 *Thermal Treatment*—Wire to this specification may be ordered without thermal treatment, but when required, and depending on end use, wire may be ordered as follows:

5.3.1 Drawn from normalized rod,
5.3.2 Drawn from annealed or spheroid annealed rod,
5.3.3 Annealed or spheroidize annealed in process, or
5.3.4 Annealed or spheroidize annealed at finished size.

6. Chemical Requirements

6.1 The steel shall conform to the requirements as to chemical composition specified in Table 1 for the grade ordered.

6.2 *Cast or Heat Analysis (Formerly Ladle Analysis)*—Each cast or heat of steel shall be analyzed by the manufacturer to determine the percentage of elements prescribed in Table 1. This analysis shall be made from a specimen preferably taken during the pouring of cast or heat. When requested, this analysis shall be reported to the purchaser.

6.3 *Product Analysis (Formerly Check Analysis)*—An analysis may be made by the purchaser from finished wire representing each cast or heat of steel. The chemical composition of samples taken for product analysis shall conform to the permissible limits of Table 10 of Specification A 510.

6.4 For reference purposes, Methods E 30 shall be applied.

7. Metallurgical Structure

7.1 *Grain Size*, when specified, shall be determined in accordance with Methods E 112. The steel is designated as coarse grained when the grain size falls within the limits of 1 to 5 inclusive, and fine grained when the grain size falls within the limits of 5 to 8 inclusive.

7.2 *Decarburization of Killed Steels*—The entire periphery of a properly prepared sample of the wire shall be examined for decarburization at a magnification of 100 diameters and the limits of Table 3 shall apply.

8. Mechanical Requirements

8.1 Mechanical properties of the wire should be compatible with the steel grade, size, and thermal treatment.

9. Dimensions and Permissible Variations

9.1 The wire diameter shall not vary from the specified diameter by more than that prescribed in Table 2.

10. Workmanship, Finish, and Appearance

10.1 Each coil of wire shall be one continuous length of wire, properly coiled and firmly tied or banded.

10.2 Welded coils are subject to agreement between the manufacturer and the purchaser.

10.3 The surface of the wire shall be free of defects of a nature or degree that will be detrimental to the forming, machining, or fabrication of the machine screws.

10.4 The wire as received shall have a finish suitable for production of machine screws or as agreed upon by the manufacturer and the purchaser.

11. Packaging, Marking and Loading

11.1 *Packaging*—The coil weight, dimensions, and method of packaging shall be agreed upon by the manufacturer and purchaser.

11.2 *Marking*—The size, steel grade, purchase order number, cast or heat number, thermal treatment (when required), ASTM designation number and name, brand, or trademark of the manufacturer shall be shown on a tag securely attached to each coil of wire.

11.3 When specified on the purchaser's order, packaging, marking, and loading for shipment shall be in accordance with those procedures recommended by Recommended Practices A 700.

12. Inspection

12.1 The manufacturer shall afford the inspector representing the purchaser all reasonable facilities to satisfy him that the material is being furnished in accordance with this specification. All tests and inspection may be made at the place of manufacture prior to shipment and shall be so conducted as not to interfere unnecessarily with the operation of the works.

13. Rejection and Rehearing

13.1 Unless otherwise specified, any rejection based on tests made in accordance with this specification shall be reported to the manufacturer within a reasonable length of time.

13.2 Wire rejected under 13.1 shall be protected from damage and loss of identity by the purchaser for a reasonable time to permit the manufacturer to evaluate the rejection.

ASTM A 545

TABLE 1 Chemical Requirements[A]

UNS Designation[B]	Grade	Carbon, %	Manganese, %	Phosphorus,[C] max, %	Sulfur,[C] max, %
G 10060	1006	0.08 max	0.25–0.40	0.040	0.050
G 10080	1008	0.10 max	0.30–0.50	0.040	0.050
G 10100	1010	0.08–0.13	0.30–0.60	0.040	0.050
G 10120	1012	0.10–0.15	0.30–0.60	0.040	0.050
G 10150	1015	0.13–0.18	0.30–0.60	0.040	0.050
G 10160	1016	0.13–0.18	0.60–0.90	0.040	0.050
G 10180	1018	0.15–0.20	0.60–0.90	0.040	0.050
G 10190	1019	0.15–0.20	0.70–1.00	0.040	0.050
G 10210	1021	0.18–0.23	0.60–0.90	0.040	0.050
G 10220	1022	0.18–0.23	0.70–1.00	0.040	0.050
G 10260	1026	0.22–0.28	0.60–0.90	0.040	0.050
G 10300	1030	0.28–0.34	0.60–0.90	0.040	0.050
G 10350	1035	0.32–0.38	0.60–0.90	0.040	0.050
G 10380	1038	0.35–0.42	0.60–0.90	0.040	0.050
G 15240	1524	0.19–0.25	1.35–1.65	0.040	0.050
G 15410	1541	0.36–0.44	1.35–1.65	0.040	0.050

[A] Killed steels, when specified, may be ordered as aluminum killed or silicon killed. Silicon may be specified as 0.10 max, %, 0.10 to 0.20 %, or 0.15 to 0.35 %.
[B] New designation established in accordance with ASTM E 527 and SAE J1086, Recommended Practice for Numbering Metals and Alloys (UNS).
[C] Ordinarily furnished phosphorus 0.035 % max, and sulfur 0.045 % max.

TABLE 2 Permissible Variations in Wire Diameter

NOTE—For purposes of determining conformance with this specification all specified limits are absolute as defined in Recommended Practice E 29.

Diameter, in. (mm)	Permissible Variation, Plus and Minus, in. (mm)	Permissible Out-of-Round, in. (mm)
0.035 to under 0.076 (0.89 to under 1.93)	0.001 (0.03)	0.001 (0.03)
0.076 to under 0.500 (1.93 to under 12.70)	0.0015 (0.04)	0.0015 (0.04)

TABLE 3 Decarburization Limits[A]

NOTE—For purposes of determining conformance with this specification all specified limits are absolute as defined in Recommended Practice E 29.

Diameter, in. (mm)	Average Depth Free Ferrite, in. (mm)	Average Total Affected Depth (Free Ferrite plus Partial Decarburization), in. (mm)
Up to 0.375, incl (up to 9.52, incl)	0.003 (0.08)	0.010 (0.25)
Over 0.375 to 0.500 (over 9.52 to 12.70)	0.004 (0.10)	0.012 (0.30)

[A] Applicable to killed steels only, Grades 1015 and higher.

The American Society for Testing and Materials takes no position respecting the validity of any patent rights asserted in connection with any item mentioned in this standard. Users of this standard are expressly advised that determination of the validity of any such patent rights, and the risk of infringement of such rights, are entirely their own responsibility.

This standard is subject to revision at any time by the responsible technical committee and must be reviewed every five years and if not revised, either reapproved or withdrawn. Your comments are invited either for revision of this standard or for additional standards and should be addressed to ASTM Headquarters. Your comments will receive careful consideration at a meeting of the responsible technical committee, which you may attend. If you feel that your comments have not received a fair hearing you should make your views known to the ASTM Committee on Standards, 1916 Race St., Philadelphia, Pa. 19103.

Designation: A 546 – 82

Standard Specification for
STEEL WIRE, MEDIUM-HIGH-CARBON, COLD-HEADING QUALITY, FOR HEXAGON-HEAD BOLTS[1]

This standard is issued under the fixed designation A 546; the number immediately following the designation indicates the year of original adoption or, in the case of revision, the year of last revision. A number in parentheses indicates the year of last reapproval. A superscript epsilon (ϵ) indicates an editorial change since the last revision or reapproval.

This specification has been approved for use by agencies of the Department of Defense and for listing in the DoD Index of Specifications and Standards.

1. Scope

1.1 This specification covers round wire made from medium-high-carbon steels and some boron steels intended for the manufacture of hexagon-head bolts by the cold-heading process.

1.2 The values stated in inch-pound units are to be regarded as standard.

2. Applicable Documents

2.1 *ASTM Standards*:

A 510 Specification for General Requirements for Wire Rods and Coarse Round Wire, Carbon Steel[2]

A 510M Specification for General Requirements for Wire Rods and Coarse Round Wire, Carbon Steel [Metric][2]

A 700 Recommended Practices for Packaging, Marking, and Loading Methods for Steel Products for Domestic Shipment[3]

A 751 Methods, Practices, and Definitions for Chemical Analysis of Steel Products[3,5]

E 29 Recommended Practice for Indicating Which Places of Figures Are to Be Considered Significant in Specified Limiting Values[4]

E 30 Chemical Analysis of Steel, Cast Iron, Open-Hearth Iron, and Wrought Iron[5]

E 112 Determining Average Grain Size[6]

3. General Requirements for Delivery

3.1 Material furnished to this specification shall conform to the applicable requirements of the current edition of Specification A 510.

4. Ordering Information

4.1 Orders for material to this specification shall include the following information:

4.1.1 Quantity (weight),

4.1.2 Name of material (medium-high-carbon steel wire, cold heading quality),

4.1.3 Dimensions (see 9.1),

4.1.4 Thermal treatment (see 5.3),

4.1.5 Steel grade (Table 1),

4.1.6 Grain size (see 7.1) (if desired),

4.1.7 Packaging (see 11.1),

4.1.8 Cast or heat analysis, when required (see 6.2), and

4.1.9 ASTM designation and date of issue.

NOTE—A typical ordering description is as follows: 60,000 lb Round Medium High Carbon Wire, Cold-Heading-Quality for Hexagon-Head Bolts, 0.220 in. diameter, Annealed in Process, 1038 Coarse Grained, Coils 1,000 lb, 26 in. nominal diameter to ASTM A 546 dated ____.

5. Manufacture

5.1 *Process*—The steel shall be made by the open-hearth, basic-oxygen, or the electric-furnace process.

---

[1] This specification is under the jurisdiction of ASTM Committee A-1 on Steel, Stainless Steel and Related Alloys, and is the direct responsibility of Subcommittee A01.03 on Steel Rod and Wire.
Current edition approved July 30, 1982. Published September 1982. Originally published as A 546 – 65. Last previous edition A 546 – 77.
[2] *Annual Book of ASTM Standards*, Vol 01.03.
[3] *Annual Book of ASTM Standards*, Vols 01.03 and 01.05.
[4] *Annual Book of ASTM Standards*, Vol 14.02.
[5] *Annual Book of ASTM Standards*, Vol 03.05.
[6] *Annual Book of ASTM Standards*, Vol 03.03.

5.2 *Internal Soundness*—Sufficient discard shall be taken to ensure freedom from injurious pipe and segregation.

5.3 *Thermal Treatment*—When required or when indicated by the grade, the wire may be specified as:

5.3.1 Drawn from normalized rods (except Grade 1541).

5.3.2 Drawn from annealed or spheroidize annealed rods,

5.3.3 Annealed or spheroidize annealed in process, or

5.3.4 Annealed or spheroidize annealed at finished size.

6. Chemical Requirements

6.1 The steel shall conform to the requirements as to chemical composition specified in Table 1 for the grade ordered.

6.2 *Cast or Heat Analysis (Formerly Ladle Analysis)*—Each cast or heat of steel shall be analyzed by the manufacturer to determine the percentage of elements prescribed in Table 1. This analysis shall be made from a specimen preferably taken during the pouring of the cast or heat. When requested, this analysis shall be reported to the purchaser.

6.3 *Product Analysis (Formerly Check Analysis)*—An analysis may be made by the purchaser from finished wire representing each cast or heat of steel. The chemical composition of samples taken for product analysis shall conform to the permissible limits of Table 10 of Specification A 510.

6.4 For reference purposes, Methods E 30 shall be applied.

7. Metallurgical Structure

7.1 *Grain Size*, when specified shall be determined in accordance with Methods E 112. The steel is designated as coarse grained when the grain size falls within the limits of 1 to 5 inclusive, and fine grained when the grain size falls within the limits of 5 to 8 inclusive.

7.2 *Decarburization of Killed Steels*—The entire periphery of a properly prepared sample of the wire shall be examined for decarburization at a magnification of 100 diameters and the limits of Table 3 shall apply.

8. Mechanical Requirements

8.1 The mechanical properties of the wire should be compatible with the steel grade, size, and thermal treatment.

9. Dimensions and Permissible Variations

9.1 The diameter of the wire shall not vary from the specified diameter by more than that prescribed in Table 2.

10. Workmanship, Finish, and Appearance

10.1 Each coil of wire shall be one continuous length, properly coiled, and firmly tied or banded.

10.2 Welded coils are subject to agreement between the manufacturer and the purchaser.

10.3 The surface of the wire shall be free of defects of a nature or degree that will be detrimental to the forming, machining, or fabrication of the hexagon-head bolts.

10.4 The wire as received shall have a finish suitable for production of hexagon-head bolts or as agreed upon between the manufacturer and purchaser.

11. Packaging, Marking, and Loading

11.1 *Packaging and Loading Methods* shall be as agreed to by manufacturer and purchaser.

11.2 *Marking*—The size, steel grade, purchase order number, cast or heat number, thermal treatment (when required), ASTM designation number and name, brand, or trademark of the manufacturer shall be shown on a tag securely attached to each coil of wire.

11.3 When specified on the purchaser's order, packaging, marking, and loading shall be in accordance with the procedures of Recommended Practices A 700.

12. Inspection

12.1 The manufacturer shall afford the inspector representing the purchaser all reasonable facilities to satisfy him that the material is being furnished in accordance with this specification. All tests and inspections may be made at the place of manufacture prior to shipment.

13. Rejection and Rehearing

13.1 Unless otherwise specified, any rejection based on tests made in accordance with this specification shall be reported to the manufacturer within a reasonable length of time.

13.2 Wire rejected under 13.1 shall be protected from damage and loss of identity by the purchaser for a reasonable time to permit the manufacturer to evaluate the rejection.

A 546

TABLE 1 Chemical Requirements[A]

UNS Designation[B]	Grade	Carbon, %	Manganese, %	Phosphorus, max %[B]	Sulfur, max, %[C]	Boron, min, %
...	10B21	0.18–0.23	0.80–1.10	0.040	0.050	0.0005
...	10B22	0.17–0.23	1.00–1.30	0.040	0.050	0.0005
G 10300	1030	0.28–0.34	0.60–0.90	0.040	0.050	...
G 10350	1035	0.32–0.38	0.60–0.90	0.040	0.050	...
G 10380	1038	0.35–0.42	0.60–0.90	0.040	0.050	...
G 10390	1039	0.37–0.44	0.70–1.00	0.040	0.050	...
G 10400	1040	0.37–0.44	0.60–0.90	0.040	0.050	...
G 15410	1541	0.36–0.44	1.35–1.65	0.040	0.050	...

[A] Silicon may be specified as 0.10 max, %, 0.10 to 0.20 % or 0.15 to 0.35 %.
[B] New designation established in accordance with ASTM E 527 and SAE J1086, Recommended Practice for Numbering Metals and Alloys (UNS).
[C] Ordinarily furnished phosphorus 0.035 max, %, and sulfur 0.045 max, %.

TABLE 2 Permissible Variations in Wire Diameter

NOTE—For purposes of determining conformance with this specification all specified limits are absolute as defined in Recommended Practice E 29.

Diameter, in. (mm)	Permissible Variation, Plus and Minus, in. (mm)	Permissible Out-of-Round, in. (mm)
0.076 to 0.500 (1.93 to 12.70), excl	0.0015 (0.04)	0.0015 (0.04)
0.500 (12.70) and over	0.002 (0.05)	0.002 (0.05)

TABLE 3 Decarburization Limits

NOTE—For purposes of determining conformance with this specification all specified limits are absolute as defined in Recommended Practice E 29.

Diameter, in. (mm)	Average Depth Free Ferrite, in. (mm)	Average Total Affected Depth (Free Ferrite plus Partial Decarburization), in. (mm)
Up to 0.375 (9.52), incl	0.003 (0.08)	0.010 (0.25)
Over 0.375 to 0.500 (9.52 to 12.70), incl	0.004 (0.10)	0.012 (0.30)
Over 0.500 to 0.703 (12.70 to 17.86), incl	0.005 (0.13)	0.014 (0.36)
Over 0.703 to 0.999 (17.86 to 25.37), incl	0.006 (0.15)	0.016 (0.41)

The American Society for Testing and Materials takes no position respecting the validity of any patent rights asserted in connection with any item mentioned in this standard. Users of this standard are expressly advised that determination of the validity of any such patent rights, and the risk of infringement of such rights, are entirely their own responsibility.

This standard is subject to revision at any time by the responsible technical committee and must be reviewed every five years and if not revised, either reapproved or withdrawn. Your comments are invited either for revision of this standard or for additional standards and should be addressed to ASTM Headquarters. Your comments will receive careful consideration at a meeting of the responsible technical committee, which you may attend. If you feel that your comments have not received a fair hearing you should make your views known to the ASTM Committee on Standards, 1916 Race St., Philadelphia, Pa. 19103.

Designation: A 547 – 82

Standard Specification for
STEEL WIRE, ALLOY, COLD-HEADING QUALITY, FOR HEXAGON-HEAD BOLTS[1]

This standard is issued under the fixed designation A 547; the number immediately following the designation indicates the year of original adoption or, in the case of revision, the year of last revision. A number in parentheses indicates the year of last reapproval. A superscript epsilon (ϵ) indicates an editorial change since the last revision or reapproval.

This specification has been approved for use by agencies of the Department of Defense and for listing in the DoD Index of Specifications and Standards.

1. Scope

1.1 This specification covers round alloy steel wire cold-heading quality intended for the manufacture of hexagon-head bolts by the cold-heading process.

1.2 The values stated in inch-pound units are to be regarded as standard.

2. Applicable Documents

2.1 *ASTM Standards*:
A 700 Recommended Practices for Packaging, Marking, and Loading Methods for Steel Products for Domestic Shipment[2]
E 29 Recommended Practice for Indicating Which Places of Figures Are to Be Considered Significant in Specified Limiting Values[3]
E 30 Chemical Analysis of Steel, Cast Iron, Open-Hearth Iron, and Wrought Iron[4]
E 112 Determining Average Grain Size[5]

3. Ordering Information

3.1 Orders for material to this specification shall include the following for each item ordered.

3.1.1 Quantity (weight),
3.1.2 Name of material (alloy cold-heading wire for hexagon-head bolts),
3.1.3 Dimension (see 8.1),
3.1.4 Thermal treatment (see 4.3),
3.1.5 Steel grade (see Table 1),
3.1.6 Grain size (see 6.1) (fine or coarse),
3.1.7 Packaging (see 10.1),
3.1.8 Cast or heat analysis, if required (see 5.2),
3.1.9 ASTM designation and date of issue, and
3.1.10 Special requirements if any.

NOTE—A typical ordering description is as follows: 30,000 lb Round Alloy Cold-Heating Wire for Hexagon-Head Bolts, 0.625 in. diameter, Annealed in Process, Grade 4037, Fine Grained, Coils 500 lb, 30 in. nominal diameter, report heat (cast) analysis for each heat, to ASTM A 547 dated ____.

4. Manufacture

4.1 *Process*—The steel shall be made by the open-hearth, basic-oxygen, or electric-furnace process.

4.2 *Internal Soundness*—Sufficient discard shall be taken to ensure freedom from injurious pipe and segregation.

4.3 *Thermal Treatment*—Wire to this specification may be ordered as follows:

4.3.1 Drawn from normalized rod,
4.3.2 Drawn from annealed or spheriodize annealed rod,
4.3.3 Annealed or spheroidize annealed in process,
4.3.4 Annealed or spheroidize annealed at finish size,

---

[1] This specification is under the jurisdiction of ASTM Committee A-1 on Steel, Stainless Steel and Related Alloys, and is the direct responsibility of Subcommittee A01.03 on Steel Rod and Wire.
Current edition approved July 30, 1982. Published September 1982. Originally published as A 547 – 65. Last previous edition A 547 – 77.
[2] *Annual Book of ASTM Standards*, Vols 01.03 and 01.04.
[3] *Annual Book of ASTM Standards*, Vol 14.02.
[4] *Annual Book of ASTM Standards*, Vol 03.05.
[5] *Annual Book of ASTM Standards*, Vol 03.03.

A 547

4.3.5 Drawn from normalized rod and annealed or spheroidize annealed in process, or

4.3.6 Drawn from annealed or spheroidize annealed rod and annealed or spheroidize annealed at finish size.

5. Chemical Requirements

5.1 The steel shall conform to the requirements as to chemical composition specified in Table 1 for the grade ordered.

5.2 *Cast or Heat Analysis (Formerly Ladle Analysis)*—Each cast or heat of steel shall be analyzed by the manufacturer to determine the percentage of elements prescribed in Table 1. This analysis shall be made from a specimen preferably taken during the pouring of the cast or heat. When requested, this analysis shall be reported to the purchaser.

5.3 *Product Analysis (Formerly Check Analysis)*—An analysis may be made by the purchaser from finished wire representing each cast or heat of steel. The chemical composition shall conform to the permissible limits of Table 4.

5.4 For referee purposes, Methods E 30 shall be applied.

6. Metallurgical Structure

6.1 *Grain Size*, shall be determined in accordance with Methods E 112. The steel is designated as coarse grained when the grain size falls within the limits of 1 to 5 inclusive, and fine grained when the grain size falls within the limits of 5 to 8 inclusive.

6.2 *Decarburization*—The entire periphery of a properly prepared sample of the wire shall be examined for decarburization at a magnification of 100 diameters and the limits of Table 3 shall apply.

7. Mechanical Requirements

7.1 The mechanical properties of the wire should be compatible with the grade, size, and thermal treatment.

8. Dimensions and Permissible Variations

8.1 The wire shall conform to the permissible variations in diameter as prescribed in Table 2.

9. Workmanship, Finish, and Appearance

9.1 Each coil of wire shall be one continuous length, properly coiled, and securely tied or banded. Welded coils are subject to agreement between the manufacturer and the purchaser.

9.2 The surface of the wire shall be free of defects of a nature or degree that will be detrimental to the forming, machining, or fabrication of the hexagon-head bolts.

9.3 The wire as received by the purchaser shall have a finish suitable for production of hexagon-head bolts or as agreed upon between the manufacturer and purchaser.

10. Packaging, Marking, and Loading

10.1 *Packaging*—The coil size, coil weight, and method or type of package shall be as agreed between manufacturer and purchaser.

10.2 *Marking*—The size, steel grade, purchase order number, cast or heat number, thermal treatment (when required), ASTM designation number and name, brand, or trademark of the manufacturer shall be shown on a tag securely attached to each coil of wire.

10.3 When specified on the purchaser's order, packaging, marking, and loading shall be in accordance with those procedures recommended by Recommended Practices A 700.

11. Inspection

11.1 The manufacturer shall afford the inspector representing the purchaser all reasonable facilities necessary to satisfy him that the material is being produced and furnished in accordance with this specification. All tests and inspections may be made at the place of manufacture prior to shipment and shall be so conducted as not to interfere unnecessarily with the operation of the works.

12. Rejection and Rehearing

12.1 Unless otherwise specified, any rejection based on tests made in accordance with this specification shall be reported to the manufacturer within a reasonable length of time.

12.2 Wire rejected under 12.1 shall be protected from damage and loss of identity by the purchaser for a reasonable time to permit the manufacturer to evaluate the rejection.

TABLE 1 Chemical Requirements[A]

UNS Designation[B]	Grade	Carbon, %	Manganese, %	Nickel, %	Chromium, %	Molybdenum, %
G 13350	1335	0.33–0.38	1.60–1.90	...	...	...
...	1335H	0.32–0.38	1.45–2.05	...	...	...
G 13400	1340	0.38–0.43	1.60–1.90	...	...	...
...	1340H	0.37–0.44	1.45–2.05	...	...	...
...	4037H	0.34–0.41	0.60–1.00	...	...	0.20–0.30
G 40370	4037	0.35–0.40	0.70–0.90	...	...	0.20–0.30
...	4135H	0.32–0.38	0.60–1.00	...	0.75–1.20	0.15–0.25
G 41350	4135	0.33–0.38	0.70–0.90	...	0.80–1.10	0.15–0.25
...	4137H	0.34–0.41	0.60–1.00	...	0.75–1.20	0.15–0.25
G 41370	4137	0.35–0.40	0.70–0.90	...	0.80–1.10	0.15–0.25
...	4140H	0.37–0.44	0.65–1.10	...	0.75–1.20	0.15–0.25
G 41400	4140	0.38–0.43	0.75–1.00	...	0.80–1.10	0.15–0.25
...	4142H	0.39–0.46	0.65–1.10	...	0.75–1.20	0.15–0.25
G 41420	4142	0.40–0.45	0.75–1.00	...	0.80–1.10	0.15–0.25
...	4340H	0.37–0.44	0.55–0.90	1.55–2.00	0.65–0.95	0.20–0.30
G 43400	4340	0.38–0.43	0.60–0.80	1.65–2.00	0.70–0.90	0.20–0.30
...	8637H	0.34–0.41	0.70–1.05	0.35–0.75	0.35–0.65	0.15–0.25
G 86370	8637	0.35–0.40	0.75–1.00	0.40–0.70	0.40–0.60	0.15–0.25
...	8640H	0.37–0.44	0.70–1.05	0.35–0.75	0.35–0.65	0.15–0.25
G 86400	8640	0.38–0.43	0.75–1.00	0.40–0.70	0.40–0.60	0.15–0.25
...	8642H	0.39–0.46	0.70–1.05	0.35–0.75	0.35–0.65	0.15–0.25
G 86420	8642	0.40–0.45	0.75–1.00	0.40–0.70	0.40–0.60	0.15–0.25
...	8740H	0.37–0.44	0.70–1.05	0.35–0.75	0.35–0.65	0.20–0.30
G 87400	8740	0.38–0.43	0.75–1.00	0.40–0.70	0.40–0.60	0.20–0.30

	Basic Open Hearth or Basic Oxygen	Electric
Phosphorus content, max, %	0.035	0.025
Sulfur content, max, %	0.040	0.025

[A] Silicon content 0.15 to 0.35 %. When electric-furnace steels are specified, the prefix letter E shall be used.
[B] New designation established with ASTM E 527 and SAE J1086, Recommended Practice for Numbering Metals and Alloys (UNS).

TABLE 2 Permissible Variations in Wire Diameter

NOTE—For purposes of determining conformance with this specification all specified limits are absolute as defined in Recommended Practice E 29.

Diameter, in. (mm)	Permissible Variation, Plus and Minus, in. (mm)	Permissible Out-of-Round, in. (mm)
0.076 to 0.500 (1.93 to 12.70), excl	0.0015 (0.04)	0.0015 (0.04)
0.500 (12.70) and over	0.002 (0.05)	0.002 (0.05)

TABLE 3 Decarburization Limits

NOTE—For purposes of determining conformance with this specification all specified limits are absolute as defined in Recommended Practice E 29.

Diameter, in. (mm)	Average Depth Free Ferrite, in. (mm)	Average Total Affected Depth (Free Ferrite plus Partial Decarburization), in. (mm)
Up to 0.375 (9.52), incl	0.003 (0.08)	0.010 (0.25)
Over 0.375 to 0.500 (9.52 to 12.70), incl	0.004 (0.10)	0.012 (0.30)
Over 0.500 to 0.703 (12.70 to 17.86), incl	0.005 (0.13)	0.014 (0.36)
Over 0.703 to 0.999 (17.86 to 25.37), incl	0.006 (0.15)	0.016 (0.41)

TABLE 4 Permissible Variations for Product Analysis of Alloy Steels

Element	Limit, or Maximum of Specific Range, %	Tolerance, %, Over Max Limit or Under Min Limit
Carbon	To 0.30, incl	0.01
	Over 0.30 to 0.75, incl	0.02
	Over 0.75	0.03
Manganese	To 0.90, incl	0.03
	Over 0.90 to 2.10, incl	0.04
Phosphorus	Over max, only	0.005
Sulfur	To 0.060, incl[A]	0.005
Silicon	To 0.40, incl	0.02
	Over 0.40 to 2.20, incl	0.05
Nickel	To 1.00, incl	0.03
	Over 1.00 to 2.00, incl	0.05
	Over 2.00 to 5.30, incl	0.07
	Over 5.30 to 10.00, incl	0.10
Chromium	To 0.90, incl	0.03
	Over 0.90 to 2.10, incl	0.05
	Over 2.10 to 3.99, incl	0.10
Molybdenum	To 0.20, incl	0.01
	Over 0.20 to 0.40, incl	0.02
	Over 0.40 to 1.15, incl	0.03

[A] Sulfur over 0.60 % is not subject to product analysis.

The American Society for Testing and Materials takes no position respecting the validity of any patent rights asserted in connection with any item mentioned in this standard. Users of this standard are expressly advised that determination of the validity of any such patent rights, and the risk of infringement of such rights, are entirely their own responsibility.

This standard is subject to revision at any time by the responsible technical committee and must be reviewed every five years and if not revised, either reapproved or withdrawn. Your comments are invited either for revision of this standard or for additional standards and should be addressed to ASTM Headquarters. Your comments will receive careful consideration at a meeting of the responsible technical committee, which you may attend. If you feel that your comments have not received a fair hearing you should make your views known to the ASTM Committee on Standards, 1916 Race St., Philadelphia, Pa. 19103.

Designation: A 548 – 82

Standard Specification for
STEEL WIRE, CARBON, COLD-HEADING QUALITY, FOR TAPPING OR SHEET METAL SCREWS[1]

This standard is issued under the fixed designation A 548; the number immediately following the designation indicates the year of original adoption or, in the case of revision, the year of last revision. A number in parentheses indicates the year of last reapproval. A superscript epsilon (ϵ) indicates an editorial change since the last revision or reapproval.

This specification has been approved for use by agencies of the Department of Defense and for listing in the DoD Index of Specifications and Standards.

1. Scope

1.1 This specification covers carbon steel round wire, cold-heading quality, for the manufacture of tapping or sheet metal screws.

1.2 The values stated in inch-pound units are to be regarded as standard.

2. Applicable Documents

2.1 *ASTM Standards:*

A 510 Specification for General Requirements for Wire Rods and Coarse Round Wire, Carbon Steel[2]

A 510M Specification for General Requirements for Wire Rods and Coarse Round Wire, Carbon Steel [Metric][2]

A 700 Recommended Practices for Packaging, Marking, and Loading Methods for Steel Products for Domestic Shipment[3]

A 751 Methods, Practices, and Definitions for Chemical Analysis of Steel Products[3,6]

E 30 Chemical Analysis of Steel, Cast Iron, Open-Hearth Iron, and Wrought Iron[6]

E 29 Recommended Practice for Indicating Which Places of Figures Are to Be Considered Significant in Specified Limiting Values[4]

E 112 Determining Average Grain Size[5]

3. General Requirements for Delivery

3.1 Material furnished under this specification shall conform to the applicable requirements of the current edition of Specification A 510.

4. Ordering Information

4.1 Orders for wire under this specification shall include the following information:

4.1.1 Quantity (weight),

4.1.2 Name of material (cold-heading wire for tapping or sheet metal screws),

4.1.3 Dimensions (see 8.1),

4.1.4 Thermal treatment (see 5.3),

4.1.5 Steel grade (see Table 1),

4.1.6 End use (type of head to be formed—conventional slotted, flat, oval and round heads, recessed heads and other special head types),

4.1.7 Grain size, if desired (see 7.1),

4.1.8 Packaging (see 10.1),

4.1.9 Cast or heat analysis, if desired (see 6.2), and

4.1.10 ASTM designation and date of issue.

NOTE—A typical ordering description is as follows: 30,000 lb Round Cold-Heading Quality wire for Pan Head Tapping Screws, 0.225 in. diameter, Annealed in Process, Grade 1022, in 1,000 lb coils, 26 in. nominal diameter to ASTM A 548 dated ___.

5. Manufacture

5.1 The steel shall be made by the open-hearth, basic-oxygen, or electric-furnace process.

5.2 Sufficient discard shall be taken to en-

---

[1] This specification is under the jurisdiction of ASTM Committee A-1 on Steel, Stainless Steel and Related Alloys, and is the direct responsibility of Subcommittee A01.03 on Steel Rod and Wire.
Current edition approved July 30, 1982. Published September 1982. Originally published as A 548 – 65. Last previous edition A 548 – 77.
[2] *Annual Book of ASTM Standards*, Vol 01.03.
[3] *Annual Book of ASTM Standards*, Vol 01.04.
[4] *Annual Book of ASTM Standards*, Vol 14.02.
[5] *Annual Book of ASTM Standards*, Vol 03.05.
[6] *Annual Book of ASTM Standards*, Vol 03.03.

sure freedom from injurious pipe and segregation.

5.3 Wire to this specification may be ordered without thermal treatment, but when required and depending on end use, wire may be ordered as follows:

5.3.1 Drawn from normalized rod,
5.3.2 Drawn from annealed or spheroidize annealed rod.
5.3.3 Annealed or spheroidize annealed in process, or
5.3.4 Annealed or spheroidize annealed at finished size.

6. Chemical Requirements

6.1 The material shall conform to the requirements as to chemical composition of the grades specified in Table 1, for the grade ordered.

6.2 *Cast or Heat Analysis (Formerly Ladle Analysis)*—Each cast or heat of steel shall be analyzed by the manufacturer to determine the percentages of the elements prescribed in Table 1. This analysis shall be made from a specimen preferably taken during the pouring of the cast or heat. When requested, this analysis shall be reported to the purchaser.

6.3 *Product Analysis (Formerly Check Analysis)*—An analysis may be made by the purchaser from finished wire representing each cast or heat of steel. The chemical composition of samples taken for product analysis shall conform to the permissible limits of Table 10 of Specification A 510.

6.4 For referee purposes, Methods E 30 shall be applied.

7. Metallurgical Structure

7.1 *Grain Size,* when specified, shall be determined in accordance with Methods E 112. The steel is designated as coarse grained when the grain structure falls within the limits of 1 to 5 inclusive, and fine grained when the grain structure falls within the limits of 5 to 8 inclusive.

7.2 *Decarburization of Killed Steels*—The entire periphery of a properly prepared sample of the wire shall be examined for decarburization at a magnification of 100 diameters and the limits of Table 3 shall apply.

8. Dimensions and Permissible Variations

8.1 The diameter of the wire shall not vary from the specified diameter by more than that prescribed in Table 2.

9. Workmanship, Finish, and Appearance

9.1 Each coil of wire shall be one continuous length properly coiled and securely tied or banded.

9.2 Welded coils are subject to agreement between the manufacturer and the purchaser.

9.3 The surface of the wire shall be free of defects of a nature or degree that will be detrimental to the forming, machining, or fabrication of the tapping or sheet metal screws.

9.4 The wire, as received, shall have a finish suitable for production of tapping or sheet metal screws or as agreed upon between the manufacturer and purchaser.

10. Packaging, Marking, and Loading

10.1 *Packaging*—The coil weight, dimensions, and method of packaging shall be agreed upon by the manufacturer and purchaser.

10.2 *Marking*—The size, steel grade, purchase order number, cast or heat number, thermal treatment (when required), ASTM designation number and name, brand, or trademark of the manufacturer shall be shown on a tag securely attached to each coil of wire.

10.3 When specified on the purchaser's order, packaging, marking, and loading shall be in accordance with the requirements of Recommended Practices A 700.

11. Inspection

11.1 The manufacturer shall afford the inspector representing the purchaser all reasonable facilities necessary to satisfy him that the material is being produced and furnished in accordance with this specification. All tests and inspections may be made at the place of manufacture prior to shipment and shall be so conducted as not to interfere unnecessarily with the operation of the works.

12. Rejection and Rehearing

12.1 Unless otherwise specified, any rejection based on tests made in accordance with this specification shall be reported to the manufacturer within a reasonable length of time.

12.2 The wire rejected under 12.1 shall be protected from damage or loss of identity by the purchaser for a reasonable time to permit the manufacturer to evaluate the rejection.

A 548

TABLE 1 Chemical Requirements[A]

UNS Designation[B]	Grade	Carbon, %	Manganese, %	Phosphorus,[C] max, %	Sulfur,[C] max, %
G 10130	1013	0.11–0.16	0.50–0.80	0.040	0.050
G 10160	1016	0.13–0.18	0.60–0.90	0.040	0.050
G 10180	1018	0.15–0.20	0.60–0.90	0.040	0.050
G 10190	1019	0.15–0.20	0.70–1.00	0.040	0.050
G 10210	1021	0.18–0.23	0.60–0.90	0.040	0.050
G 10220	1022	0.18–0.23	0.70–1.00	0.040	0.050
...	1019 Modified	0.15–0.20	0.80–1.10	0.040	0.050
...	1022 Modified	0.18–0.23	0.80–1.10	0.040	0.050

[A] Silicon, for all grades above, may be specified as 0.10 % max, 0.10 to 0.20 %, or 0.15 to 0.35 %.
[B] New designation established in accordance with ASTM E 527 and SAE J1086. Recommended Practice for Numbering Metals and Alloys (UNS).
[C] Ordinarily furnished phosphorus 0.035 % max, and sulfur 0.045 % max.

TABLE 2 Permissible Variation in Wire Diameter

NOTE—For purposes of determining conformance with this specification all specified limits are absolute as defined in Recommended Practice E 29.

Diameter, in. (mm)	Permissible Variation, Plus and Minus, in. (mm)	Permissible Out-of-Round, in. (mm)
0.035 to 0.076 (0.89 to 1.93), excl	0.001 (0.03)	0.001 (0.03)
0.076 to 0.500 (1.93 to 12.70), excl	0.0015 (0.04)	0.0015 (0.04)
0.500 (12.70) and over	0.002 (0.05)	0.002 (0.05)

TABLE 3 Decarburization Limits[A]

NOTE—For purposes of determining conformance with this specification all specified limits are absolute as defined in Recommended Practice E 29.

Diameter, in. (mm)	Average Depth Free Ferrite, in. (mm)	Average Total Affected Depth (Free Ferrite plus Partial Decarburization), in. (mm)
Up to 0.375 (9.52), incl	0.003 (0.08)	0.010 (0.25)
Over 0.375 to 0.500 (9.52 to 12.70), incl	0.004 (0.10)	0.012 (0.30)
Over 0.500 to 0.703 (12.70 to 17.86), incl	0.005 (0.13)	0.014 (0.36)

[A] Applicable to killed steels only, Grades 1016 and higher.

The American Society for Testing and Materials takes no position respecting the validity of any patent rights asserted in connection with any item mentioned in this standard. Users of this standard are expressly advised that determination of the validity of any such patent rights, and the risk of infringement of such rights, are entirely their own responsibility.

This standard is subject to revision at any time by the responsible technical committee and must be reviewed every five years and if not revised, either reapproved or withdrawn. Your comments are invited either for revision of this standard or for additional standards and should be addressed to ASTM Headquarters. Your comments will receive careful consideration at a meeting of the responsible technical committee, which you may attend. If you feel that your comments have not received a fair hearing you should make your views known to the ASTM Committee on Standards, 1916 Race St., Philadelphia, Pa. 19103.

Designation: A 549 – 82

Standard Specification for
STEEL WIRE, CARBON, COLD-HEADING QUALITY, FOR WOOD SCREWS[1]

This standard is issued under the fixed designation A 549; the number immediately following the designation indicates the year of original adoption or, in the case of revision, the year of last revision. A number in parentheses indicates the year of last reapproval. A superscript epsilon (ϵ) indicates an editorial change since the last revision or reapproval.

This specification has been approved for use by agencies of the Department of Defense and for listing in the DoD Index of Specifications and Standards.

1. Scope

1.1 This specification covers carbon steel round wire, cold-heading quality intended for wood screws.

1.2 The values stated in inch-pound units are to be regarded as standard.

2. Applicable Documents

2.1 *ASTM Standards:*
A 510 Specification for General Requirements for Wire Rods and Coarse Round Wire, Carbon Steel[2]
A 510M Specification for General Requirements for Wire Rods and Coarse Round Wire, Carbon Steel [Metric][2]
A 700 Recommended Practices for Packaging, Marking, and Loading Methods for Steel Products for Domestic Shipment[3]
A 751 Methods, Practices, and Definitions for Chemical Analysis of Steel Products[3,5]
E 29 Recommended Practice for Indicating Which Places of Figures Are to Be Considered Significant in Specified Limiting Values[4]
E 30 Chemical Analysis of Steel, Cast Iron, Open-Hearth Iron, and Wrought Iron[5]

3. General Requirements

3.1 Material furnished under this specification shall conform to the applicable requirements of the current edition of Specification A 510.

4. Ordering Information

4.1 Orders for material to this specification shall include the following for each item:

4.1.1 Quantity (weight),
4.1.2 Name of the material (cold-heading wire for wood screws),
4.1.3 Dimensions (see 7.1),
4.1.4 Thermal treatment, if required (see 5.3),
4.1.5 Steel grade (see Table 1),
4.1.6 End use (type of head to be formed (conventional slotted, flat, oval, and round heads, recessed head, and other special types)),
4.1.7 Packaging (see 9.1),
4.1.8 Cast or heat analysis, if required (see 6.2), and
4.1.9 ASTM designation and date of issue.

NOTE—A typical ordering description is as follows: 30,000 lb Round Carbon Cold-Heading Quality Wire, 0.250 in. diameter, drawn, Grade 1108, for Slotted Flat Head Wood Screws, in 500 lb Coils, 30 in. nominal diameter to ASTM A 549 dated ___.

5. Manufacture

5.1 The steel shall be made by the open-hearth, basic-oxygen, or electric-furnace process.

5.2 Sufficient discard shall be taken to ensure freedom from injurious pipe and segregation.

---

[1] This specification is under the jurisdiction of ASTM Committee A-1 on Steel, Stainless Steel and Related Alloys, and is the direct responsibility of Subcommittee A01.03 on Steel Rod and Wire.
Current edition approved July 30, 1982. Published September 1982. Originally published as A 549 – 65. Last previous edition A 549 – 77.
[2] *Annual Book of ASTM Standards*, Vol 01.03.
[3] *Annual Book of ASTM Standards*, Vol 01.04.
[4] *Annual Book of ASTM Standards*, Vol 14.02.
[5] *Annual Book of ASTM Standards*, Vol 03.05.

5.3 *Thermal Treatment*—Wire to this specification may be ordered without thermal treatment, but when required and depending on end use, wire may be ordered as follows:

5.3.1 Drawn from normalized rod,

5.3.2 Drawn from annealed or spheroidize annealed rod,

5.3.3 Annealed or spheroidize annealed in process, or

5.3.4 Annealed or spheroidize annealed at finished size.

6. Chemical Requirements

6.1 The steel shall conform to the requirements as to chemical composition specified in Table 1, for the grade ordered.

6.2 *Cast or Heat Analysis (Formerly Ladle Analysis)*—Each cast or heat of steel shall be analyzed by the manufacturer to determine the percentages of the elements prescribed in Table 1. This analysis shall be made from a specimen preferably taken during the pouring of the cast or heat. When requested, this analysis shall be reported to the purchaser.

6.3 *Product Analysis (Formerly Check Analysis)*—An analysis may be made by the purchaser from finished wire representing each cast or heat of steel. The chemical composition of samples taken for product analysis shall conform to the permissible limits of Table 10 of Specification A 510.

6.4 For referee purposes, Methods E 30 shall be applied.

7. Dimensions and Permissible Variations

7.1 The diameter of the wire shall not vary from the specified diameter by more than that prescribed in Table 2.

8. Workmanship, Finish, and Appearance

8.1 Each coil of wire shall be one continuous length of wire properly coiled and securely tied or banded.

8.2 Welded coils are subject to agreement between the manufacturer and the purchaser.

8.3 The surface of the wire shall be free of defects of a nature or degree that will be detrimental to the forming, machining, or fabrication of the wood screws.

8.4 The wire as received shall have a finish suitable for production of wood screws or as agreed upon by the manufacturer and the purchaser.

9. Packaging, Marking, and Loading

9.1 *Packaging*—The coil weight, dimensions, and method of packaging shall be agreed upon by the manufacturer and the purchaser.

9.2 *Marking*—The size, steel grade, purchase order number, cast or heat number, thermal treatment (when required), ASTM designation number and name, brand, or trademark of the manufacturer shall be shown on a tag securely attached to each coil of wire.

9.3 When specified on purchaser's order, packaging, marking, and loading shall be in accordance with the procedures recommended by Recommended Practices A 700.

10. Inspection

10.1 The manufacturer shall afford the inspector representing the purchaser all reasonable facilities necessary to satisfy him that the material is being produced and furnished in accordance with this specification. All tests and inspections may be made at the place of manufacture prior to shipment and shall be conducted as not to interfere unnecessarily with the operation of the works.

11. Rejection and Rehearing

11.1 Unless otherwise specified, any rejection based on tests made in accordance with this specification shall be reported to the manufacturer within a reasonable length of time.

11.2 The wire rejected under 11.1 shall be protected from damage or loss of identity by the purchaser for a reasonable length of time to permit the manufacturer to evaluate the rejection.

TABLE 1 Chemical Requirements[A]

UNS Designation[D]	Grade	Carbon, %	Manganese, %	Phosphorus, %	Sulfur, %
G 10080	1008[C]	0.10 max	0.30–0.50	0.040 max	0.050 max
G 10100	1010[C]	0.08–0.13	0.30–0.60	0.040 max	0.050 max
G 10120	1012[C]	0.10–0.15	0.30–0.60	0.040 max	0.050 max
G 10150	1015[C]	0.13–0.18	0.30–0.60	0.040 max	0.050 max
G 10160	1016[C]	0.13–0.18	0.60–0.90	0.040 max	0.050 max
G 10170	1017[C]	0.15–0.20	0.30–0.60	0.040 max	0.050 max
G 10180	1018[C]	0.15–0.20	0.60–0.90	0.040 max	0.050 max
G 11080	1108[B]	0.08–0.13	0.50–0.80	0.04 max	0.08–0.13
G 11090	1109[B]	0.08–0.13	0.60–0.90	0.04 max	0.08–0.13
G 11100	1110[B]	0.08–0.13	0.30–0.60	0.04 max	0.08–0.13
...	B1010[B]	0.13 max	0.30–0.60	0.07–0.12	0.060 max

[A] In non-resulfurized grades, silicon may be specified as 0.10 % max, 0.10 to 0.20 %, or 0.15 to 0.35 %.
[B] Not recommended for recessed-head or struck-slot applications.
[C] Ordinarily furnished phosphorus 0.035 % max, and sulfur 0.045 % max.
[D] New designation established in accordance with ASTM E 527 and SAE J1086. Recommended Practice for Numbering Metals and Alloys (UNS).

TABLE 2 Permissible Variations in Wire Diameter

NOTE—For purposes of determining conformance with this specification all specified limits are absolute as defined in Recommended Practice E 29.

Diameter, in. (mm)	Permissible Variation, Plus and Minus, in. (mm)	Permissible Out-of-Round, in. (mm)
0.035 to 0.076 (0.89 to 1.93), excl	0.001 (0.03)	0.001 (0.03)
0.076 to 0.500 (1.93 to 12.70), excl	0.0015 (0.04)	0.0015 (0.04)
0.500 (12.70) and over	0.002 (0.05)	0.002 (0.05)

The American Society for Testing and Materials takes no position respecting the validity of any patent rights asserted in connection with any item mentioned in this standard. Users of this standard are expressly advised that determination of the validity of any such patent rights, and the risk of infringement of such rights, are entirely their own responsibility.

This standard is subject to revision at any time by the responsible technical committee and must be reviewed every five years and if not revised, either reapproved or withdrawn. Your comments are invited either for revision of this standard or for additional standards and should be addressed to ASTM Headquarters. Your comments will receive careful consideration at a meeting of the responsible technical committee, which you may attend. If you feel that your comments have not received a fair hearing you should make your views known to the ASTM Committee on Standards, 1916 Race St., Philadelphia, Pa. 19103.

Designation: A 555 – 80

An American National Standard

Standard Specification for
GENERAL REQUIREMENTS FOR STAINLESS AND HEAT-RESISTING STEEL WIRE[1]

This standard is issued under the fixed designation A 555; the number immediately following the designation indicates the year of original adoption or, in the case of revision, the year of last revision. A number in parentheses indicates the year of last reapproval. A superscript epsilon (ϵ) indicates an editorial change since the last revision or reapproval.

This specification has been approved for use by agencies of the Department of Defense and for listing in the DoD Index of Specifications and Standards.

1. Scope

1.1 This specification covers general requirements that shall apply to stainless and heat-resisting wire for cold forming, including coiling, stranding, weaving, heading, and forging, under the latest revision of each of the following specifications issued by ASTM: Specifications A 313, A 368, A 478, A 492, A 493, A 580, and A 581.

1.2 In case of conflicting requirements, the individual material specification and this general requirement specification shall prevail in the order named.

1.3 General requirements for flat products other than wire are covered in Specification A 480.

1.4 General requirements for other than flat and wire products are covered in Specification A 484.

1.5 The values stated in inch-pound units are to be regarded as the standard.

2. Applicable Documents

2.1 *ASTM Standards*:

A 262 Practices for Detecting Susceptibility to Intergranular Attack in Stainless Steels[2]

A 313 Specification for Chromium-Nickel Stainless and Heat-Resisting Steel Spring Wire[2]

A 368 Specification for Stainless and Heat-Resisting Steel Wire Strand[2]

A 370 Methods and Definitions for Mechanical Testing of Steel Products[3]

A 478 Specification for Chromium-Nickel Stainless and Heat-Resisting Steel Weaving Wire[2]

A 480/A 480M Specification for General Requirements for Flat-Rolled Stainless and Heat-Resisting Steel Plate, Sheet, and Strip[3]

A 484 Specification for General Requirements for Stainless and Heat-Resisting Wrought Steel Products (Except Wire)[2]

A 492 Specification for Stainless and Heat-Resisting Steel Rope Wire[2]

A 493 Specification for Stainless and Heat-Resisting Steel for Cold Heading and Cold Forging—Bar and Wire[2]

A 571 Specification for Austenitic Ductile Iron Castings for Pressure-Containing Parts Suitable for Low-Temperature Service[4]

A 580 Specification for Stainless and Heat-Resisting Steel Wire[2]

A 581 Specification for Free-Machining Stainless and Heat-Resisting Steel Wire[2]

A 700 Practices for Packaging, Marking, and Loading Methods for Steel Products for Domestic Shipment[3]

A 751 Methods, Practices, and Definitions for Chemical Analysis of Steel Products[3]

E 112 Methods for Determining Average Grain Size[5]

---

[1] This specification is under the jurisdiction of ASTM Committee A-1 on Steel, Stainless Steel and Related Alloys and is the direct responsibility of Subcommittee A01.08 on Wrought Stainless Steel Products.
Current edition approved Aug. 1, 1980. Published October 1980. Originally published as A 555 – 65. Last previous edition A 555 – 79.
[2] *Annual Book of ASTM Standards*, Vol 01.05.
[3] *Annual Book of ASTM Standards*, Vol 01.03.
[4] *Annual Book of ASTM Standards*, Vol 01.02.
[5] *Annual Book of ASTM Standards*, Vol 03.03.

A 555

2.2 *Federal Standard:*
Fed. Std. No. 123 Marking for Shipment (Civil Agencies)[6]

2.3 *Military Standards:*
MIL-STD-129 Marking for Shipment and Storage[6]
MIL-STD-163 Preservation of Steel Products for Domestic Shipment[6]

3. Description of Terms

3.1 Wire as covered by this specification and the specifications itemized in 1.1 is any cold-finished or cold-drawn round or shape within the following limitations:

3.1.1 Round, square, octagon, hexagon, and shape wire ½ in. (12.70 mm) and under in diameter or size.

3.1.2 Flat wire 1/16 in. (1.59 mm) to under 3/8 in. (9.52 mm) in width and 0.010 in. (0.25 mm) to under 3/16 in. (4.76 mm) in thickness.

4. Manufacture

4.1 The material may be furnished in one of the conditions detailed in the applicable material specification, that is, annealed, bright annealed, cold drawn, or as otherwise specified on the purchase order.

4.2 A variety of finishes, coatings, and lubricants are available. The particular type used is dependent upon the specific end use. Unless otherwise specified, the finish, coating, and lubricant will be furnished as required by the individual material specification.

5. Cast or Heat Analysis

5.1 The chemical analysis of each heat shall be determined in accordance with the applicable material specification and Methods, Practices, and Definitions A 751.

6. Product Analysis

6.1 When a product analysis (formerly check or verification analysis) is performed, one sample shall be taken from each lot. Such analysis may be made by any of the commonly accepted methods that will positively identify the material. The range of the specified chemical composition is expanded to take into account deviations associated with analytical reproducibility and heterogeneity of the steel. The chemical composition thus determined shall conform to the tolerances shown in Table 1. If several determinations of any element in a heat are made, they may not vary both above and below the specified range.

7. Lot Size

7.1 A lot for product analysis shall consist of all wire made from the same heat.

7.2 For other tests required by the product specification for wire, a lot shall consist of all product of the same size, heat, and heat treatment charge in a batch-type furnace or under the same conditions in a continuous heat-treating furnace.

8. Number of Tests

8.1 Unless otherwise specified in the product specification, one sample per heat shall be selected for chemical analysis and one mechanical test sample shall be selected from each lot of wire. All tests shall conform to the chemical and mechanical requirements of the material specification.

8.2 One intergranular corrosion test, when required, and one grain size test, when required, shall be made from each lot. It is often convenient to obtain test material from the specimen selected for mechanical testing.

9. Test Methods

9.1 The properties enumerated in applicable specifications shall be determined in accordance with the following ASTM methods:

9.1.1 *Chemical Analysis*—Methods A 571.

9.1.2 *Tension Test*—Methods and Definitions A 370.

9.1.3 *Intergranular Corrosion* (when required)—Practice E of Practices A 262.

9.1.4 *Grain Size* (when required)—Methods E 112.

10. Retests and Retreatment

10.1 If any test specimen shows imperfections that may affect the test results, it may be discarded and another specimen substituted.

10.2 If the results of any test lot are not in conformance with the requirements of this specification and the applicable product specification, a retest sample of two specimens may be tested to replace each failed specimen of the original sample. If one of the retest specimens fails, the lot shall be rejected.

---

[6] Available from Naval Publications and Forms Center, 5801 Tabor Ave., Philadelphia, Pa. 19120.

A 555

10.3 Where failure of any lot is due to inadequate heat treatment, the material may be reheat treated and resubmitted for test.

11. Permissible Variations in Dimensions

11.1 Unless otherwise specified in the purchase order, the wire shall conform to the permissible variations in dimensions as specified in the following tables:

11.1.1 *Round Wire (Drawn; Polished; Centerless Ground; Centerless Ground and Polished)*—See Table 2.

11.1.2 *Drawn Wire in Hexagons, Octagons, and Squares*—See Table 3.

11.1.3 *Round and Shape, Straightened and Cut Wire, Exact Length Resheared Wire*—See Table 4.

11.1.4 *Cold-Finished Flat Wire*—See Table 5.

12. Workmanship

12.1 The material shall be of uniform quality consistent with good manufacturing and inspection practices. Imperfections that may be present shall be of such a nature or degree, for the type and quality ordered, that they will not adversely affect the forming, machining, or fabrication of finished parts.

13. Preparation for Delivery

13.1 Unless otherwise specified, the wire shall be packaged and loaded in accordance with Practices A 700.

13.2 When specified in the contract or order, and for direct procurement by or direct shipment to the Government, when Level A is specified, preservation, packaging, and packing shall be in accordance with the Level A requirements of MIL-STD-163.

14. Marking

14.1 *For Civilian Procurement*—Each lift, bundle, or box shall be marked using durable tags (metal, plastic, or equivalent) showing heat number, type, condition, product specification number, and size.

14.2 *For U. S. Government Procurement*—In addition to any requirements specified in the contract or order, marking shall be in accordance with MIL-STD-129 for military agencies and in accordance with Fed. Std. No. 123 for civil agencies.

15. Inspection

15.1 *For Civilian Procurement*—Inspection of the material shall be as agreed upon between the purchaser and the supplier as part of the purchase contract.

15.2 *For Government Procurement*—Unless otherwise specified in the contract or purchase order: (*1*) the seller is responsible for the performance of all inspection and test requirements in this specification, (*2*) the seller may use his own or other suitable facilities for the performance of the inspection and testing, and (*3*) the purchaser shall have the right to perform any of the inspection and tests set forth in this specification. The manufacturer shall afford the purchaser's inspector all reasonable facilities necessary to satisfy him that the material is being furnished in accordance with the inspection. Inspection by the purchaser shall not interfere unnecessarily with the manufacturer.

16. Rejection and Rehearing

16.1 Material that fails to conform to the requirements of this specification may be rejected. Rejection should be reported to the producer or supplier promptly, preferably in writing. In case of dissatisfaction with the results of the test, the producer or supplier may make claim for a rehearing.

17. Certification

17.1 A certified report of the test results shall be furnished at the time of shipment.

A 555

TABLE 1 Product Analysis Tolerances

Note—This table specifies tolerances over the maximum limits or under the minimum limits of the chemical requirements of the applicable material specification (see 1.1); it does not apply to heat analysis.

Element	Upper Limit or Maximum of Specified Range, %	Tolerances over the Maximum (Upper Limit) or Under the Minimum (Lower Limit)	Element	Upper Limit or Maximum of Specified Range, %	Tolerances over the Maximum (Upper Limit) or Under the Minimum (Lower Limit)
Carbon	to 0.010, incl	0.002	Cobalt	over 0.05 to 0.50, incl	0.01
	over 0.010 to 0.030, incl	0.005		over 0.50 to 2.00, incl	0.02
	over 0.030 to 0.20, incl	0.01		over 2.00 to 5.00, incl	0.05
	over 0.20 to 0.60, incl	0.02		over 5.00 to 10.00, incl	0.10
	over 0.60 to 1.20, incl	0.03		over 10.00 to 15.00, incl	0.15
				over 15.00 to 22.00, incl	0.20
Manganese	to 1.00, incl	0.03		over 22.00 to 30.00, incl	0.25
	over 1.00 to 3.00, incl	0.04			
	over 3.00 to 6.00, incl	0.05	Columbium	to 1.50, incl	0.05
	over 6.00 to 10.00, incl	0.06	+	over 1.50 to 5.00, incl	0.10
	over 10.00 to 15.00, incl	0.10	tantalum	over 5.00	0.15
	over 15.00 to 20.00, incl	0.15			
			Tantalum	to 0.10, incl	0.02
Phosphorus	to 0.040, incl	0.005			
	over 0.040 to 0.20, incl	0.010	Copper	to 0.50, incl	0.03
				over 0.50 to 1.00, incl	0.05
Sulfur	to 0.040, incl	0.005		over 1.00 to 3.00, incl	0.10
	over 0.040 to 0.20, incl	0.010		over 3.00 to 5.00, incl	0.15
	over 0.20 to 0.50, incl	0.020		over 5.00 to 10.00, incl	0.20
Silicon	to 1.00, incl	0.05	Aluminum	to 0.15, incl	0.005 + 0.01
	over 1.00 to 3.00, incl	0.10			
				over 0.15 to 0.50, incl	0.05
Chromium	over 4.00 to 10.00, incl	0.10		over 0.50 to 2.00, incl	0.10
	over 10.00 to 15.00, incl	0.15		over 2.00 to 5.00, incl	0.20
	over 15.00 to 20.00, incl	0.20		over 5.00 to 10.00, incl	0.35
	over 20.00 to 30.00, incl	0.25			
			Nitrogen	to 0.02, incl	0.005
Nickel	to 1.00, incl	0.03		over 0.02 to 0.19, incl	0.01
	over 1.00 to 5.00, incl	0.07		over 0.19 to 0.25, incl	0.02
	over 5.00 to 10.00, incl	0.10		over 0.25 to 0.35, incl	0.03
	over 10.00 to 20.00, incl	0.15		over 0.35 to 0.45, incl	0.04
	over 20.00 to 30.00, incl	0.20			
	over 30.00 to 40.00, incl	0.25	Tungsten	to 1.00, incl	0.03
	over 40.00	0.30		over 1.00 to 2.00, incl	0.05
				over 2.00 to 5.00, incl	0.07
Molybdenum	over 0.20 to 0.60, incl	0.03		over 5.00 to 10.00, incl	0.10
	over 0.60 to 2.00, incl	0.05		over 10.00 to 20.00, incl	0.15
	over 2.00 to 7.00, incl	0.10			
	over 7.00 to 15.00, incl	0.15	Vanadium	to 0.50, incl	0.03
	over 15.00 to 30.00, incl	0.20		over 0.50 to 1.50, incl	0.05
Titanium	to 1.00, incl	0.05	Selenium	all	0.03
	over 1.00 to 3.00, incl	0.07			
	over 3.00	0.10			

A 555

TABLE 2 Diameter and Out-of-Round Tolerances for Round Wire (Drawn; Polished; Centerless Ground; Centerless Ground and Polished)[a,b]

Specified Diameter, in. (mm)	Diameter tolerance, in. (mm)	
	Over	Under
0.5000 (12.70)	0.002 (0.05)	0.002 (0.05)
Under 0.5000 (12.70) to 0.3125 (7.94), incl	0.0015 (0.04)	0.0015 (0.04)
Under 0.3125 (7.94) to 0.0440 (1.12), incl	0.001 (0.03)	0.001 (0.03)
Under 0.0440 (1.12) to 0.0330 (0.84), incl	0.0008 (0.02)	0.0008 (0.02)
Under 0.0330 (0.84) to 0.0240 (0.61), incl	0.0005 (0.013)	0.0005 (0.013)
Under 0.0240 (0.61) to 0.0120 (0.30), incl	0.0004 (0.010)	0.0004 (0.010)
Under 0.0120 (0.30) to 0.0080 (0.20), incl	0.003 (0.008)	0.0003 (0.008)
Under 0.0080 (0.20) to 0.0048 (0.12), incl	0.0002 (0.005)	0.0002 (0.005)
Under 0.0048 (0.12) to 0.0030 (0.08), incl	0.0001 (0.003)	0.0001 (0.003)

[a] The maximum out-of-round for round wire is one half of the total size tolerance given in this table.
[b] When it is necessary to heat treat or heat treat and pickle after cold finishing, size tolerances are double those shown above for sizes 0.024 in. (0.61 mm) and over.

TABLE 3 Size Tolerances for Drawn Wire in Hexagons, Octagons, and Squares[a]

Specified Size[b] in. (mm)	Size tolerance, in. (mm)	
	Over	Under
½ (12.70)	0	0.004 (0.10)
Under ½ (12.70) to 5/16 (7.94), incl	0	0.003 (0.08)
Under 5/16 (7.94) to ⅛ (3.18), incl	0	0.002 (0.05)

[a] When it is necessary to heat treat or heat treat and pickle after cold finishing, size tolerances are double those shown above.
[b] Distance across flats.

TABLE 4 Length Tolerances for Round and Shape, Straightened and Cut Wire, Exact Length Resheared Wire

Diameter, in. (mm)	Length, ft (mm)	Tolerance, in. (mm)	
		Over	Under
0.125 (3.18) and under	Up to 12 (3.658), incl	1/16 (1.59)	0
0.125 (3.18) and under	Over 12 (3.658)	⅛ (3.18)	0
Over 0.125 (3.18) to 0.500 (12.70), incl	Under 3 (914)	1/32 (0.79)	0
Over 0.125 (3.18) to 0.500 (12.70), incl	3 (914) to 12 (3.658), incl	1/16 (1.59)	0
Over 0.125 (3.18) to 0.500 (12.70), incl	Over 12 (3.658)	⅛ (3.18)	0

TABLE 5 Thickness and Width Tolerances for Cold-Finished Flat Wire

Specified Width, in. (mm)	Thickness Tolerance, in., Over or Under, for Given Thicknesses, in. (mm)			Width tolerance, in. (mm)	
	Under 0.029 (0.74)	0.029 (0.74) to 0.035 (0.89), excl	0.035 (0.89) to 3/16 (4.76), excl	Over	Under
Under ⅜ (9.52) to 1/16 (1.59), incl	0.001 (0.03)	0.0015 (0.04)	0.002 (0.05)	0.005 (0.13)	0.005 (0.13)

The American Society for Testing and Materials takes no position respecting the validity of any patent rights asserted in connection with any item mentioned in this standard. Users of this standard are expressly advised that determination of the validity of any such patent rights, and the risk of infringement of such rights, are entirely their own responsibility.

This standard is subject to revision at any time by the responsible technical committee and must be reviewed every five years and if not revised, either reapproved or withdrawn. Your comments are invited either for revision of this standard or for additional standards and should be addressed to ASTM Headquarters. Your comments will receive careful consideration at a meeting of the responsible technical committee, which you may attend. If you feel that your comments have not received a fair hearing you should make your views known to the ASTM Committee on Standards, 1916 Race St., Philadelphia, Pa. 19103.

Designation: A 568 – 83

Standard Specification for
GENERAL REQUIREMENTS FOR STEEL, CARBON AND HIGH-STRENGTH LOW-ALLOY HOT-ROLLED SHEET AND COLD-ROLLED SHEET[1]

This standard is issued under the fixed designation A 568; the number immediately following the designation indicates the year of original adoption or, in the case of revision, the year of last revision. A number in parentheses indicates the year of last reapproval. A superscript epsilon (ϵ) indicates an editorial change since the last revision or reapproval.

This specification has been approved for use by agencies of the Department of Defense and for listing in the DoD Index of Specifications and Standards.

1. Scope

1.1 This specification covers the general requirements for steel sheet in coils and cut lengths. It applies to carbon steel and high-strength, low-alloy steel (HSLA) furnished as hot-rolled sheet and cold-rolled sheet.

1.2 This specification is not applicable to hot-rolled heavy-thickness carbon sheet coils (ASTM Specification A 635).

1.3 For the purposes of determining conformance with this and the appropriate product specification referenced under 2.1, values shall be rounded to the nearest unit in the right hand place of figures used in expressing the limiting values in accordance with the rounding method of Recommended Practice E 29.

NOTE 1—A complete metric companion to Specification A 568 has been developed—A 568M; therefore no metric equivalents are presented in this specification.

2. Applicable Documents

2.1 *ASTM Standards:*

A 366 Specification for Steel, Carbon, Cold-Rolled Sheet, Commercial Quality[2]

A 370 Methods and Definitions for Mechanical Testing of Steel Products[3]

A 414/A 414M Specification for Carbon Steel Sheet for Pressure Vessels[2]

A 424 Specification for Steel Sheet for Porcelain Enameling[2]

A 569 Specification for Steel, Carbon (0.15 Maximum, Percent) Hot-Rolled Sheet and Strip, Commercial Quality[2]

A 570 Specification for Hot-Rolled Carbon Steel Sheet and Strip, Structural Quality[2]

A 606 Specification for Steel Sheet and Strip, Hot-Rolled and Cold-Rolled, High-Strength, Low-Alloy, with Improved Atmospheric Corrosion Resistance[2]

A 607 Specification for Steel Sheet and Strip, Hot-Rolled and Cold-Rolled, High-Strength, Low-Alloy Columbium and/or Vanadium[2]

A 611 Specification for Steel, Cold-Rolled Sheet, Carbon, Structural[2]

A 619/A 619M Specification for Steel Sheet, Carbon, Cold-Rolled, Drawing Quality[2]

A 620/A 620M Specification for Steel Sheet, Carbon, Cold-Rolled, Drawing Quality, Special Killed[2]

A 621/A 621M Specification for Steel Sheet and Strip, Carbon, Hot-Rolled, Drawing Quality[2]

A 622/A 622M Specification for Steel Sheet and Strip, Carbon, Hot-Rolled, Drawing Quality, Special Killed[2]

A 659 Specification for Steel, Carbon (0.16 Maximum to 0.25 Maximum, Percent), Hot-Rolled Sheet and Strip, Commercial Quality[2]

---

[1] This specification is under the jurisdiction of ASTM Committee A-1 on Steel, Stainless Steel and Related Alloys, and is the direct responsibility of Subcommittee A01.19 on Steel Sheet and Strip.
Current edition approved July 29, 1983. Published October 1983. Originally published as A 568 – 66 T. Last previous edition A 568 – 83.

[2] *Annual Book of ASTM Standards*, Vol 01.03.

[3] *Annual Book of ASTM Standards*, Vols 01.03 and 01.04.

A 700 Practices for Packaging, Marking, and Loading Methods for Steel Products for Domestic Shipment[3]

A 715 Specification for Steel Sheet and Strip, Hot-Rolled, High-Strength, Low-Alloy, with Improved Formability[2]

E 11 Specification for Wire-Cloth Sieves for Testing Purposes[4]

E 29 Recommended Practice for Indicating Which Places of Figures are to be Considered Significant in Specified Limiting Values[4]

E 30 Methods for Chemical Analysis of Steel, Cast Iron, Open-Hearth Iron, and Wrought Iron[5]

E 350 Methods for Chemical Analysis of Carbon Steel, Low-Alloy Steel, Silicon Electrical Steel, Ingot Iron, and Wrought Iron[5]

2.2 *Military Standards:*
MIL-STD-129 Marking for Shipment and Storage[6]
MIL-STD-163 Steel Mill Products Preparation for Shipment and Storage[6]

2.3 *Federal Standard:*
Fed. Std. No. 123 Marking for Shipments (Civil Agencies)[6]

3. Description of Terms

3.1 *Steel Types:*

3.1.1 *Carbon Steel* is the designation for steel when no minimum content is specified or required for aluminum, chromium, cobalt, columbium, molybdenum, nickel, titanium, tungsten, vanadium, zirconium, or any element added to obtain a desired alloying effect; when the specified minimum for copper does not exceed 0.40 %; or when the maximum content specified for any of the following elements does not exceed the percentages noted: manganese 1.65, silicon 0.60, or copper 0.60.

3.1.1.1 In all carbon steels small quantities of certain residual elements unavoidably retained from raw materials are sometimes found which are not specified or required, such as copper, nickel, molybdenum, chromium, etc. These elements are considered as incidental and are not normally determined or reported.

3.1.2 *High-Strength Low-Alloy Steel* is a specific group of steels in which higher strength, and in some cases additional resistance to atmospheric corrosion or improved formability, are obtained by moderate amounts of one or more alloying elements.

3.2 *Product Types:*

3.2.1 *Hot-Rolled Sheet* is manufactured by hot rolling slabs in a continuous mill to the required thickness and can be supplied in coils or cut lengths as specified.

3.2.1.1 Hot-rolled carbon and high-strength low-alloy (HSLA) steel sheet is commonly classified by size as follows:

Coils and Cut Lengths

Width, in.	Thickness, in.
12 to 48, incl	0.044[A] to 0.230, incl
Over 48	0.044[A] to 0.180, incl

[A] 0.071 in. minimum for HSLA.

3.2.2 *Cold-Rolled Sheet* is manufactured from hot-rolled descaled coils by cold reducing to the desired thickness, generally followed by annealing to recrystallize the grain structure. If the sheet is not annealed after cold reduction it is known as full hard with a hardness of 84 HRB minimum and can be used for certain applications where ductility and flatness are not required.

3.2.2.1 Cold-rolled carbon sheet is commonly classified by size as follows:

Width, in.	Thickness, in.
2 through 12[A]	0.014 through 0.082
Over 12[B] through 23 15/16	0.014 and over
Over 23 15/16	0.014 and over

[A] Cold-rolled sheet coils and cut lengths, slit from wider coils with cut edge (only) and in thicknesses 0.014 in. through 0.082 in. carbon 0.25 % maximum by cast analysis.

[B] When no special edge or finish (other than matte, commercial bright, or luster finish) or single strand rolling of widths, or both under 24 in. is not specified or required.

3.2.2.2 Cold-rolled high-strength low-alloy sheet is commonly classified by size as follows:

Width, in.	Thickness, in.
2 through 12[A]	0.019 through 0.082
Over 12[B]	0.020 and over

[A] Cold-rolled sheet coils and cut lengths, slit from wider coils with cut edge (only) and in thicknesses 0.019 in. through 0.082 in. carbon 0.25 % maximum by cast analysis.

[B] When no special edge or finish (other than matte, commercial bright, or luster finish) or single strand rolling of widths, or both under 24 in. is not specified or required.

4. General Requirements for Delivery

4.1 Products covered by this specification are

---

[4] *Annual Book of ASTM Standards*, Vol 14.02.
[5] *Annual Book of ASTM Standards*, Vol 03.05.
[6] Available from Naval Publications and Forms Center, 5801 Tabor Ave., Philadelphia, Pa. 19103.

produced to decimal thickness only and customary thickness tolerances apply.

5. Manufacture

5.1 Unless otherwise specified, hot-rolled material shall be furnished hot-rolled, not annealed, not pickled.

5.2 Coil breaks, stretcher strains, and fluting can occur during the user's processing of hot-rolled or hot-rolled pickled sheet. When any of these features are detrimental to the application, the manufacturer shall be notified at time of ordering in order to properly process the sheet.

5.3 Cold-rolled carbon steel sheet is available as discussed in 10.2, 10.3, and in Table 1.

5.4 Unless specified as a full-hard product, cold-rolled sheet is annealed after being cold reduced to thickness. The annealed, cold-rolled sheet can be used as annealed last (dead soft) for unexposed end-use applications. When cold-rolled sheet is used for unexposed applications and coil breaks are a hazard in uncoiling, it may be necessary to further process the material. In this case the manufacturer should be consulted. After annealing, cold-rolled sheet is generally given a light skin pass to impart shape or may be given a heavier skin pass or temper pass to prevent the phenomenon known as stretcher straining or fluting, when formed. Temper passing also provides a required surface texture.

5.5 *Temper Rolling*:

5.5.1 Unless otherwise specified, cold-rolled sheet for exposed applications shall be temper rolled and is usually specified and furnished in the strain free condition as shipped. See Appendix X2, Effect of Aging of Cold-Rolled Carbon Steel Sheet on Drawing and Forming.

5.5.2 Cold-rolled sheet for unexposed applications may be specified and furnished "annealed last" or "temper rolled." "Annealed last" is normally produced without temper rolling, but may be lightly temper rolled during oiling or rewinding. Unexposed temper-rolled material may be specified strain-free or nonfluting. Where specific hardness range or limit, or a specified surface texture is required, the application is considered as exposed.

NOTE 2—Skin-passed sheet is subject to an aging phenomenon (see Appendix X2). Unless special killed (nonaging) steel is specified, it is to the user's interest to fabricate the sheet as soon as possible, for optimum performance.

6. Chemical Requirements

6.1 *Limits*:

6.1.1 The chemical composition shall be in accordance with the applicable product specification. However, if other compositions are required for carbon steel, they shall be prepared in accordance with Appendix X1.

6.1.2 Where the material is used for fabrication by welding, care must be exercised in selection of chemical composition or mechanical properties to assure compatibility with the welding process and its effect on altering the properties.

6.2 *Cast or Heat Analysis*:

6.2.1 An analysis of each cast or heat of steel shall be made by the manufacturer to determine the percentage of elements specified or restricted by the applicable specification.

6.2.2 When requested, cast or heat analysis for elements listed or required shall be reported to purchaser or to his representative.

6.3 *Product, Check, or Verification Analysis*:

6.3.1 Non-killed steels such as capped or rimmed steels are not technologically suited to product analysis due to the nonuniform character of their chemical composition and therefore, the tolerances in Table 2 do not apply. Product analysis is appropriate on these types of steel only when misapplication is apparent or for copper when copper steel is specified.

6.3.2 For steels other than non-killed (capped or rimmed), product analysis may be made by the purchaser. The chemical analysis shall not vary from the limits specified by more than the amounts in Table 2. The several determinations of any element in a cast shall not vary both above and below the specified range.

6.4 *Sampling for Product Analysis*:

6.4.1 To indicate adequately the representative composition of a cast by product analysis, it is general practice to select samples to represent the steel, as fairly as possible, from a minimum number of pieces as follows: 3 pieces for lots up to 15 tons incl, and 6 pieces for lots over 15 tons.

6.4.2 When the steel is subject to tension test requirements, samples for product analysis may be taken either by drilling entirely through the used tension test specimens themselves, or as covered in the following paragraph.

6.4.3 When the steel is not subject to tension test requirements, the samples for analysis must be taken by milling or drilling entirely through

A 568

the sheet in a sufficient number of places so that the samples are representative of the entire sheet or strip. The sampling may be facilitated by folding the sheet both ways, so that several samples may be taken at one drilling. Steel subjected to certain heating operations by the purchaser may not give chemical analysis results that properly represent its original composition. Therefore users must analyze chips taken from the steel in the condition in which it is received from the steel manufacturer.

6.5 *Specimen Preparation*—Drillings or chips must be taken without the application of water, oil, or other lubricant, and must be free of scale, grease, dirt, or other foreign substances. They must not be overheated during cutting to the extent of causing decarburization. Chips must be well mixed and those too coarse to pass a No. 10 sieve or too fine to remain on a No. 30 sieve are not suitable for proper analysis. Sieve size numbers are in accordance with Specification E 11.

6.6 *Test Methods*—In case a referee analysis is required and agreed upon to resolve a dispute concerning the results of a chemical analysis, the procedure of performing the referee analysis must be in accordance with the latest issue of Methods E 350, unless otherwise agreed upon between the manufacturer and the purchaser.

7. Mechanical Requirements

7.1 The mechanical property requirements, number of specimens, and test locations and specimen orientation shall be in accordance with the applicable product specification.

7.2 Unless otherwise specified in the applicable product specification, test specimens must be prepared in accordance with Methods and Definitions A 370.

7.3 Mechanical tests shall be conducted in accordance with Methods and Definitions A 370.

7.4 To determine conformance with the product specification, a calculated value should be rounded to the nearest 1 ksi tensile strength and yield point or yield strength, and to the nearest unit in the right hand place of figures used in expressing the limiting value for other values in accordance with the rounding off method given in Recommended Practice E 29.

8. Retests

8.1 If any test specimen shows defective machining or develops flaws, it must be discarded and another specimen substituted.

8.2 If the percent elongation of any test specimen is less than that specified and any part of the fracture is more than ¾ in. from the center of the gage length of a 2-in. specimen or is outside the middle half of the gage length of an 8-in. specimen, as indicated by scribe scratches marked on the specimen before testing, a retest is allowed.

8.3 If a bend specimen fails, due to conditions of bending more severe than required by the specification, a retest is permitted either on a duplicate specimen or on a remaining portion of the failed specimen.

9. Dimensions, Tolerances, and Allowances

9.1 Dimensions, tolerances, and allowances applicable to products covered by this specification are contained in Tables 4 through 23. The appropriate tolerance tables shall be identified in each individual specification.

9.2 Flatness tolerances are not applicable to "annealed last" cold-rolled sheet, but that product will normally be within two times standard flatness when shipped in cut lengths and after removal of coil set when shipped in coils.

10. Finish and Condition

10.1 Hot-rolled sheet has a surface with an oxide or scale resulting from the hot-rolling operation. The oxide or scale can be removed by pickling or blast cleaning when required for press-work operations or welding. Hot-rolled and hot-rolled descaled sheet is not generally used for exposed parts where surface is of prime importance.

10.1.1 Hot-rolled sheet can be supplied with mill edges or cut edges as specified. Mill edges are the natural edges resulting from the hot-rolling operation. They do not conform to any particular contour. They may also contain some edge imperfections, the more common types of which are cracked edges, thin edges (feather), and damaged edges due to handling or processing and which should not extend in beyond the ordered width. These edge conditions are detrimental where joining of the mill edges by welding is practiced. When the purchaser intends to shear or to blank, a sufficient width allowance should be made when purchasing to assure obtaining the desired contour and size of the pattern sheet. The manufacturer may be consulted for guidance. Cut edges are the normal edges which result from the shearing, slitting, or trimming of mill-edge sheet.

10.1.1.1 The ends of plain hot-rolled mill-edge coils are irregular in shape and are referred to as uncropped ends. Where such ends are not acceptable, the purchaser's order should so specify. Processed coils such as pickled or blast cleaned are supplied with square-cut ends.

10.2 Cold-rolled carbon sheet (exposed) is intended for those applications where surface appearance is of primary importance. This class will meet requirements for controlled surface texture, surface quality, and flatness. It is normally processed by the manufacturer to be free of stretcher strain and fluting. Subsequent user roller leveling immediately before fabrication will minimize strain resulting from aging.

10.2.1 Cold-rolled carbon sheet, when ordered for exposed applications, can be supplied in the following finishes:

10.2.1.1 Matte finish is a dull finish, without luster, produced by rolling on rolls that have been roughened by mechanical or chemical means to various degrees of surface texture depending upon application. With some surface preparation matte finish is suitable for decorative painting. It is not generally recommended for bright plating.

10.2.1.2 Commercial bright finish is a relatively bright finish having a surface texture intermediate between that of matte and luster finish. With some surface preparation commercial bright finish is suitable for decorative painting or certain plating applications. If sheet is deformed in fabrication the surface may roughen to some degree and areas so affected will require surface preparation to restore surface texture to that of the undeformed areas.

10.2.1.3 Luster finish is a smooth bright finish produced by rolling on ground rolls and is suitable for decorative painting or plating with additional special surface preparation by the user. The luster may not be retained after fabrication; therefore, the formed parts will require surface preparation to make them suitable for bright plating.

10.3 Cold-rolled carbon sheet, when intended for unexposed applications, is not subject to limitations on degree and frequency of surface imperfections, and restrictions on texture and mechanical properties are not applicable. When ordered as "annealed last," the product will have coil breaks and a tendency toward fluting and stretcher straining. Unexposed cold-rolled sheet may contain more surface imperfections than exposed cold-rolled sheet because steel applications, processing procedures, and inspection standards are less stringent.

10.4 Cold-rolled high-strength low-alloy sheet is supplied with a matte finish, unless otherwise specified.

10.5 The cold-rolled products covered by this specification are furnished with cut edges and square cut ends, unless otherwise specified.

10.6 *Oiling*:

10.6.1 Plain hot-rolled sheet is customarily furnished not oiled. Oiling must be specified, when required.

10.6.2 Hot-rolled pickled or descaled sheet is customarily furnished oiled. If the product is not to be oiled, it must be so specified since the cleaned surface is prone to rusting.

10.6.3 Cold-rolled products covered by this specification can be furnished oiled or not oiled as specified.

10.7 Sheet steel in coils or cut lengths may contain surface imperfections that can be removed with a reasonable amount of metal finishing by the purchaser.

11. Workmanship

11.1 Cut lengths shall have a workmanlike appearance and shall not have imperfections of a nature or degree for the product, the grade, class, and the quality ordered that will be detrimental to the fabrication of the finished part.

11.2 Coils may contain some abnormal imperfections that render a portion of the coil unusable since the inspection of coils does not afford the producer the same opportunity to remove portions containing imperfections as in the case with cut lengths.

11.3 *Surface Conditions*:

11.3.1 Exposed cold-rolled sheet is intended for applications where surface appearance is of primary importance, that is, exposed applications. Unexposed or annealed cold-rolled sheet is intended for applications where surface appearance is not of primary importance, that is, unexposed applications.

11.3.2 Cut lengths for exposed applications shall not include individual sheets having major surface imperfections (holes, loose slivers, and pipe) and repetitive minor surface imperfections. Cut lengths may contain random minor surface imperfections that can be removed with a reasonable amount of metal finishing by the purchaser. These imperfections shall be acceptable to the purchaser within the manufacturer's published standards.

11.3.3 For coils for exposed applications, it is not possible to remove the surface imperfections listed in 11.3.2. Coils will contain such imperfections which shall be acceptable to the purchaser within the manufacturer's published standards. Coils contain more surface imperfections than cut lengths because the producer does not have the same opportunity to sort portions containing such imperfections as is possible with cut lengths.

11.3.4 Cut lengths for unexposed applications shall not include individual sheets having major surface imperfections such as holes, loose slivers, and pipe. In addition, unexposed cut lengths can be expected to contain more minor imperfections such as pits, scratches, sticker breaks, edge breaks, pinchers, cross breaks, roll marks, and other surface imperfections than exposed. These imperfections shall be acceptable to the purchaser without limitation.

11.3.5 For coils for unexposed applications, it is not possible to remove the surface imperfections listed in 11.3.4. Coils will contain surface imperfections that are normally not repairable. Minor imperfections shall be acceptable to the purchaser within the manufacturer's published standards. Unexposed coils contain more surface imperfections than exposed coils.

12. Packaging

12.1 Unless otherwise specified, the sheet shall be packaged and loaded in accordance with Practices A 700.

12.2 When specified in the contract or order, and for direct procurement by or direct shipment to the government, when Level A is specified, preservation, packaging, and packing shall be in accordance with the Level A requirements of MIL-STD-163.

12.3 When coils are ordered, it is customary to specify a minimum or range of inside diameter, maximum outside diameter, and a maximum coil weight, if required. The ability of manufacturers to meet the maximum coil weights depends upon individual mill equipment. When required, minimum coil weights are subject to negotiation.

13. Marking

13.1 As a minimum requirement, the material shall be identified by having the manufacturer's name, ASTM designation, weight, purchaser's order number, and material identification legibly stenciled on top of each lift or shown on a tag attached to each coil or shipping unit.

13.2 When specified in the contract or order, and for direct procurement by or direct shipment to the government, marking for shipment in addition to requirements specified in the contract or order, shall be in accordance with MIL-STD-129 for military agencies and in accordance with Fed. Std. No. 123 for civil agencies.

14. Inspection

14.1 When purchaser's order stipulates that inspection and tests (except product analyses) for acceptance on the steel be made prior to shipment from the mill, the manufacturer shall afford the purchaser's inspector all reasonable facilities to satisfy him that the steel is being produced and furnished in accordance with the specification. Mill inspection by the purchaser shall not interfere unnecessarily with the manufacturer's operation.

15. Rejection and Rehearing

15.1 Unless otherwise specified, any rejection shall be reported to the manufacturer within a reasonable time after receipt of material by the purchaser.

15.2 Material that is reported to be defective subsequent to the acceptance at the purchaser's works shall be set aside, adequately protected, and correctly identified. The manufacturer shall be notified as soon as possible so that an investigation may be initiated.

15.3 Samples that are representative of the rejected material shall be made available to the manufacturer. In the event that the manufacturer is dissatisfied with the rejection, he may request a rehearing.

A 568

TABLE 1 Cold-Rolled Sheet Steel Class Comparison

	Exposed	Unexposed
Major imperfections:		
Cut lengths	Mill rejects	Mill rejects
Coils	Purchaser accepts within the manufacturer's published standards (policy)	Purchaser accepts within the manufacturer's published standards (policy)
Minor imperfections:		
Cut lengths	Mill rejections repetitive imperfections. May contain random imperfections which the purchaser accepts within the manufacturer's published standards (policy)	Purchaser accepts all minor imperfections
Coils	Purchaser accepts within the manufacturer's published standards (policy)	Purchaser accepts all minor imperfections
Finish	Matte unless otherwise specified	Purchaser accepts all finishes
Special oils	May be specified	May not be specified
Thickness, width and length tolerance:		
Standard	Will be met	Will be met
Restricted	May be specified	May not be specified
Flatness tolerance:		
Standard	Will be met	Will be met (temper rolled) Not guaranteed—normally within twice standard (annealed last)
Stretcher leveled	May be specified	May not be specified
Resquaring	May be specified	May not be specified
Coil wraps	Purchaser accepts within the manufacturer's published standards (policy)	Purchaser accepts all
Coil welds	Purchaser accepts within the manufacturer's published standards (policy)	Purchaser accepts within the manufacturer's published standards (policy)
Outside inspection	May be specified	May not be specified
Special testing	May be specified	May not be specified

TABLE 2 Tolerances for Product Analysis

Element	Limit, or Maximum of Specified Element, %	Tolerance Under Minimum Limit	Tolerance Over Maximum Limit
Carbon	to 0.15 incl	0.02	0.03
	over 0.15 to 0.40 incl	0.03	0.04
	over 0.40 to 0.80 incl	0.03	0.05
	over 0.80	0.03	0.06
Manganese	to 0.60 incl	0.03	0.03
	over 0.60 to 1.15 incl	0.04	0.04
	over 1.15 to 1.65 incl	0.05	0.05
Phosphorus		...	0.01
Sulfur		...	0.01
Silicon	to 0.30 incl	0.02	0.03
	over 0.30 to 0.60 incl	0.05	0.05
Copper		0.02	...

TABLE 3 List of Tables for Dimensions, Tolerances, and Allowances

Dimension	Table No.	
Carbon[A] and High-Strength Low-Alloy Steel		
	Hot-Rolled Sheet	Cold-Rolled Sheet
Allowances in width and length, stretcher leveled	15[B]	15
Camber tolerances	10	10, 21
Diameter tolerances of sheared circles	9	9
Flatness tolerances, not stretcher leveled	13	22
Flatness tolerances, stretcher leveled	14	23
Length tolerances	8	18, 19
Out-of-square tolerances	11	11
Resquared tolerances	12	12
Thickness tolerances	4, 5	16, 17
Width tolerances of cut edge	7	7, 20
Width tolerances of mill edge	6	...

[A] Tolerances for hot-rolled carbon sheet steel with 0.25 % maximum carbon, cast or heat analysis.
[B] Carbon steel only.

A 568

TABLE 4 Thickness Tolerances of Hot-Rolled Sheet (Carbon Steel)
(Coils and Cut Lengths, Including Pickled)

NOTE 1—Thickness is measured at any point across the width not less than ⅜ in. from a cut edge and not less than ¾ in. from a mill edge. This table does not apply to the uncropped ends of mill edge coils.

NOTE 2—The specified thickness range captions also apply when sheet is specified to a nominal thickness, and the tolerances are divided equally, over and under.

Specified Width, in.	Thickness Tolerances Over, in. No Tolerance Under					
	Specified Minimum Thickness, in.					
	Over 0.180 to 0.230 incl	Over 0.098 to 0.180 incl	Over 0.071 to 0.098 incl	Over 0.057 to 0.071 incl	Over 0.051 to 0.057 incl	0.044 to 0.051 incl
12 to 20 incl	0.014	0.014	0.012	0.012	0.010	0.010
Over 20 to 40 incl	0.016	0.014	0.014	0.012	0.010	0.010
Over 40 to 48 incl	0.018	0.016	0.014	0.012	0.012	0.010
Over 48 to 60 incl	...	0.016	0.014	0.014	0.012	...
Over 60 to 72 incl	...	0.016	0.016	0.014	0.014	...
Over 72	...	0.016	0.016	...	...	...

TABLE 5 Thickness Tolerances of Hot-Rolled Sheet (High-Strength Low-Alloy Steel)
(Coils and Cut Lengths, Including Pickled)

NOTE 1—Thickness is measured at any point across the width not less than ⅜ in. from a cut edge and not less than ¾ in. from a mill edge. This table does not apply to the uncropped ends of mill edge coils.

NOTE 2—The specified thickness range captions also apply when sheet is specified to a nominal thickness, and the tolerances are divided equally, over and under.

Specified Width, in.	Thickness Tolerances Over, in. No Tolerance Under			
	Specified Minimum Thickness, in.			
	Over 0.180 to 0.230	Over 0.098 to 0.180	Over 0.082 to 0.098	0.071 to 0.082
12 to 15 incl	0.014	0.014	0.012	0.012
Over 15 to 20 incl	0.016	0.016	0.014	0.014
Over 20 to 32 incl	0.018	0.016	0.014	0.014
Over 32 to 40 incl	0.018	0.018	0.016	0.014
Over 40 to 48 incl	0.020	0.020	0.016	0.014
Over 48 to 60 incl	...	0.020	0.016	0.014
Over 60 to 72 incl	...	0.022	0.018	0.016
Over 72 to 80 incl	...	0.024	0.018	0.016
Over 80	...	0.024	0.020	...

A 568

TABLE 6 Width Tolerances[A] of Hot-Rolled Mill Edge Sheet (Carbon and High-Strength Low-Alloy Steel) (Coils and Cut Lengths, Including Pickled)

Carbon	
Specified Width, in.	Tolerances Over Specified Width, in. No Tolerance Under
12 to 14 incl	7/16
Over 14 to 17 incl	1/2
Over 17 to 19 incl	9/16
Over 19 to 21 incl	5/8
Over 21 to 24 incl	11/16
Over 24 to 26 incl	13/16
Over 26 to 30 incl	15/16
Over 30 to 50 incl	1 1/8
Over 50 to 78 incl	1 1/2
Over 78	1 7/8
High-Strength Low-Alloy	
12 to 14 incl	7/16
Over 14 to 17 incl	1/2
Over 17 to 19 incl	9/16
Over 19 to 21 incl	5/8
Over 21 to 24 incl	11/16
Over 24 to 26 incl	13/16
Over 26 to 28 incl	15/16
Over 28 to 35 incl	1 1/8
Over 35 to 50 incl	1 1/4
Over 50 to 60 incl	1 1/2
Over 60 to 65 incl	1 5/8
Over 65 to 70 incl	1 3/4
Over 70 to 80 incl	1 7/8
Over 80	2

[A] The above tolerances do not apply to the uncropped ends of mill edge coils (10.1.1.1).

TABLE 7 Width Tolerances of Hot-Rolled Cut Edge Sheet and Cold-Rolled Sheet (Carbon and High-Strength Low-Alloy Steel) (Not Resquared, Coils and Cut Lengths, Including Pickled)

Specified Width, in.	Tolerances Over Specified Width, in. No Tolerance Under
Over 12 to 30 incl	1/8
Over 30 to 48 incl	3/16
Over 48 to 60 incl	1/4
Over 60 to 80 incl	5/16
Over 80	3/8

TABLE 8 Length Tolerances of Hot-Rolled Sheet (Carbon and High-Strength Low-Alloy Steel) (Cut Lengths Not Resquared, Including Pickled)

Specified Length, in.	Tolerances Over Specified Length, in. No Tolerance Under
To 15 incl	1/8
Over 15 to 30 incl	1/4
Over 30 to 60 incl	1/2
Over 60 to 120 incl	3/4
Over 120 to 156 incl	1
Over 156 to 192 incl	1 1/4
Over 192 to 240 incl	1 1/2
Over 240	1 3/4

TABLE 9 Diameter Tolerances of Circles Sheared from Hot-Rolled (Including Pickled) and Cold-Rolled Sheet (Over 12 in. Width) (Carbon and High-Strength Low-Alloy Steel)

Specified Thickness,[A] in.	Tolerances Over Specified Diameter, in. (No Tolerances Under)		
	Under 30	Over 30 to 48 incl	Over 48
0.044 to 0.057 incl	1/16	1/8	3/16
Over 0.057 to 0.098 incl	3/32	5/32	7/32
Over 0.098	1/8	3/16	1/4

[A] 0.071 in. minimum thickness for hot-rolled high-strength low-alloy steel sheet.

TABLE 10 Camber Tolerances[A] for Hot-Rolled (Including Pickled) and Cold-Rolled Sheet (Over 12 in. Width) (Carbon and High-Strength Low-Alloy Steel)

NOTE 1—Camber is the greatest deviation of a side edge from a straight line, the measurement being taken on the concave side with a straightedge.

NOTE 2—Camber tolerances for cut lengths, not resquared, as shown in the table.

Cut Length, ft	Camber Tolerances, in.
To 4 incl	1/8
Over 4 to 6 incl	3/16
Over 6 to 8 incl	1/4
Over 8 to 10 incl	5/16
Over 10 to 12 incl	3/8
Over 12 to 14 incl	1/2
Over 14 to 16 incl	5/8
Over 16 to 18 incl	3/4
Over 18 to 20 incl	7/8
Over 20 to 30 incl	1 1/4
Over 30 to 40 incl	1 1/2

[A] The camber tolerance for coils is 1 in. in any 20 ft.

A 568

TABLE 11 Out-of-Square Tolerances of Hot-Rolled Cut-Edge (Including Pickled) and Cold-Rolled Sheet (Over 12 in. Width) (Carbon and High-Strength Low-Alloy Steel) (Cut Lengths Not Resquared)

Out-of-square is the greatest deviation of an end edge from a straight line at right angle to a side and touching one corner. It is also obtained by measuring the difference between the diagonals of the cut length. The out-of-square deviation is one half of that difference. The tolerance for all thicknesses and all sizes is 1/16 in./6 in. of width or fraction thereof.

TABLE 12 Resquare Tolerances of Hot-Rolled (Including Pickled) and Cold-Rolled Sheet (Over 12 in. Width) (Carbon and High-Strength Low-Alloy Steel) (Cut Lengths)

When cut lengths are specified resquared, the width and the length are not less than the dimensions specified. The individual tolerance for over-width, over-length, camber, or out-of-square should not exceed 1/16 in. up to and including 48 in. in width and up to and including 120 in. in length. For cut lengths wider or longer, the applicable tolerance is 1/8 in.

TABLE 13 Flatness Tolerances[A] of Hot-Rolled Sheet Including Pickled Cut Lengths Not Specified to Stretcher-Leveled Standard of Flatness (Carbon and High-Strength Low-Alloy Steel)

Specified Minimum Thickness, in.	Specified Width, in.	Flatness Tolerances,[B] in.	
		Specified Yield Point, min, ksi	
		To 42 incl	45 to 50[C,D]
0.044 to 0.057 incl	over 12 to 36 incl	½	...
	over 36 to 60 incl	¾	...
	over 60	1	...
Over 0.057 to 0.180 incl	over 12 to 60 incl	½	¾
	over 60 to 72 incl	¾	1⅛
	over 72	1	1½
Over 0.180 to 0.230 incl	over 12 to 48 incl	½	¾

[A] The above table also applies to lengths cut from coils by the consumer when adequate flattening operations are performed.
[B] Maximum deviation from a horizontal flat surface.
[C] Tolerances for high-strength low-alloy steels with specified minimum yield point in excess of 50 ksi are subject to negotiation.
[D] 0.0710 minimum thickness of HSLA.

TABLE 14 Flatness Tolerances of Hot-Rolled Sheet Including Pickled Cut Lengths Specified to Stretcher-Leveled Standard of Flatness (Carbon and High-Strength Low-Alloy Steel)

Specified Thickness, in.	Specified Width, in.	Specified Length, in.	Flatness Tolerance,[A] in.	
			Specified Yield Point, min, ksi	
			To 45 incl	45 to 50
0.044 to 0.180 incl	To 48 incl	To 96 incl	⅛	
	wider or longer		¼	B
Over 0.180 to 0.230 incl	To 48 incl	To 96 incl	⅛	
	longer		¼	B

[A] Maximum deviation from a horizontal flat surface.
[B] Tolerances for high-strength, low-alloy steel are subject to negotiation.

A 568

TABLE 15 Allowances in Width and Length of Hot-Rolled Sheet (Carbon Steel) and Cold-Rolled Sheet (Carbon and High-Strength Low-Alloy Steel)
(Cut Length Sheets Specified to Stretcher-Leveled Standard of Flatness, Not Resquared, Including Pickled)

NOTE 1—When cut lengths are specified to stretcher-leveled standard of flatness and not resquared, the allowances over specified dimensions in width and length given in the following table apply. Under these conditions the allowances for width and length are added by the manufacturer to the specified width and length and the tolerances given in Tables 6, 7, 8, 18, 19, and 20 apply to the new size established. The camber tolerances in Table 10 do not apply.

NOTE 2—When cut lengths are not to have grip or entry marks within the specified length, the purchaser should specify "grip entry marks outside specified length." When cut lengths may have grip or entry marks within the specified length, the purchaser should specify "grip or entry marks inside specified length."

		Allowances Over Specified Dimensions	
		Length, in.	
Specified Length in.	Width, in.	Specified "Grip or entry marks outside specified length"	Specified "Grip or entry marks inside specified length"
To 120 incl	¾	4	3
Over 120 to 156 incl	1	4	3
Over 156	1¼	5	4

TABLE 16 Thickness Tolerances of Cold-Rolled Sheet (Carbon and High-Strength Low-Alloy Steel)[A]
(Coils and Cut Lengths Over 12 in. in Width)

NOTE 1—Thickness is measured at any point across the width not less than ⅜ in. from a side edge.
NOTE 2—The specified thickness range captions noted below also apply when sheet is specified to a nominal thickness, and the above tolerances are divided equally, over and under.

	Thickness Tolerances Over, in. No Tolerance Under					
Specified Width, in.	Specified Minimum Thickness, in.					
	Over 0.098 to 0.142 incl	Over 0.071 to 0.098 incl	Over 0.057 to 0.071 incl	Over 0.039 to 0.057 incl	Over 0.019[A] to 0.039 incl	0.014 to 0.019 incl
12 to 15 incl	0.010	0.010	0.010	0.008	0.006	0.004
Over 15 to 72 incl	0.012	0.010	0.010	0.008	0.006	0.004
Over 72	0.014	0.012	0.010	0.008	0.006	...

[A] 0.020 in. minimum thickness for high-strength low-alloy.

TABLE 17 Thickness Tolerances of Cold-Rolled Sheet (Carbon and High-Strength Low-Alloy Steel)[A]
Coils and Cut Lengths 2 in. to 12 in. in Width

NOTE 1—Thickness is measured at any point across the width not less than ⅜ in. from a side edge.
NOTE 2—The specified thickness range captions noted below also apply when sheet is specified to a nominal thickness, and the below tolerances are divided equally, over and under.

	Thickness Tolerances Over, in. No Tolerance Under			
Specified Width, in.	Specified Minimum Thickness, in.			
	Over 0.057 to 0.082 incl	Over 0.039 to 0.057 incl	Over 0.019 to 0.039 incl	0.014 to 0.019 incl
2 to 12 incl	0.010	0.008	0.006	0.004

[A] 0.020 in. minimum thickness for high-strength low-alloy.

TABLE 18 Length Tolerances of Cold-Rolled Sheet (Carbon and High-Strength Low-Alloy Steel)
(Cut Lengths Over 12 inches in Width, Not Resquared)

Specified Length, in.	Tolerances Over Specified Length, in. No Tolerances Under
Over 12 to 30 incl	⅛
Over 30 to 60 incl	¼
Over 60 to 96 incl	½
Over 96 to 120 incl	¾
Over 120 to 156 incl	1
Over 156 to 192 incl	1¼
Over 192 to 240 incl	1½
Over 240	1¾

A 568

TABLE 19 Length Tolerances of Cold-Rolled Sheet (Carbon and High-Strength Low-Alloy Steel) (Cut Length Sheets, 2 in. to 12 in. in Width and 0.014 in. to 0.082 in. in Thickness, Not Resquared)

NOTE—This table applies to widths produced by slitting from wider sheet.

Specified Length, in.	Tolerances Over Specified Length, in. No Tolerance Under
24 to 60 incl	½
Over 60 to 120 incl	¾
Over 120 to 240 incl	1

TABLE 20 Width Tolerances for Cold-Rolled Sheet (Carbon and High-Strength Low-Alloy Steel) (Coils and Cut Lengths Over 12 in. in Width, Not Resquared)

Specified Width, in.	Tolerances Over Specified Width, in. No Tolerance Under
12 to 30 incl	⅛
Over 30 to 48 incl	3/16
Over 48 to 60 incl	¼
Over 60 to 80 incl	5/16
Over 80	⅜

TABLE 20a Width Tolerances for Cold-Rolled Sheet (Carbon and High-Strength Low-Alloy Steel) (Coils and Cut Lengths 2 in. to 12 in. in Width, Not Resquared, and 0.014 in.^{A} to 0.082 in. in Thickness)

Specified Width, in.	Width Tolerance, Plus and Minus, in.
2 to 6 incl	0.012
Over 6 to 9 incl	0.016
Over 9 to 12 incl	0.032

^{A} 0.020 in. minimum thickness for high-strength low-alloy.

TABLE 21 Camber Tolerances for Cold-Rolled Sheet (Carbon and High-Strength Low-Alloy Steel) (Cut Lengths, Not Resquared, Over 12 in. Wide)

NOTE—Camber is the greatest deviation of a side edge from a straight line, the measurement being taken on the concave side with a straightedge.

Cut Length, ft	Camber Tolerance,^{A} in.
To 4 incl	⅛
Over 4 to 6 incl	3/16
Over 6 to 8 incl	¼
Over 8 to 10 incl	5/16
Over 10 to 12 incl	⅜
Over 12 to 14 incl	½
Over 14 to 16 incl	⅝
Over 16 to 18 incl	¾
Over 18 to 20 incl	⅞
Over 20 to 30 incl	1¼
Over 30 to 40 incl	1½

^{A} The camber tolerance for coils is 1 in. in any 20 ft.

TABLE 21a Camber Tolerances of Cold-Rolled Sheet in Coils (Carbon and High-Strength Low-Alloy Steel) (Coils 2 in. to 12 in. in Width 0.014 in. to 0.082 in. in Thickness)

NOTE 1—Camber is the greatest deviation of a side edge from a straight line, the measurement being taken on the concave side with a straightedge.

NOTE 2—This table applies to widths produced by slitting from wider sheet.

Width, in.	Camber Tolerance
2 to 12 incl	¼ in. in any 8 ft

^{A} 0.020 in. minimum thickness for high-strength low-alloy.

TABLE 22 Flatness Tolerances of Cold-Rolled Sheet (Carbon and High-Strength Low-Alloy Steel) (Cut Lengths Over 12 in. in Width, Not Specified to Stretcher-Leveled Standard or Flatness)

NOTE 1—This table does not apply when product is ordered full hard, to a hardness range, "annealed last" (dead soft), or as Class 2.

NOTE 2—This table also applies to lengths cut from coils by the consumer when adequate flattening measures are performed.

Specified Thickness, in.	Specified Width, in.	Flatness Tolerance,^{A} in.	
		Specified Yield Point, min, ksi	
		To 40 incl	45 to 50^B
To 0.044 incl	to 36 incl	⅜	¾
	over 36 to 60 incl	⅝	1⅛
	over 60	⅞	1½
Over 0.044	to 36 incl	¼	¾
	over 36 to 60 incl	⅜	¾
	over 60 to 72 incl	⅝	1⅛
	over 72	⅞	1½

^{A} Maximum deviation from a horizontal flat surface.
^{B} Tolerances for high-strength, low-alloy steel with specified minimum yield point in excess of 50 ksi are subject to negotiation.

TABLE 23 Flatness Tolerances of Cold-Rolled Sheet (Carbon and High-Strength Low-Alloy Steel)
(Cut Lengths Specified to Stretcher-Leveled Standard of Flatness)

Specified Thickness, in.	Specified Width, in.	Specified Length, in.	Flatness Tolerance,[A] in.	
			Specified Yield Point, min, ksi	
			To 40 incl	40 to 50[B]
0.015 to 0.028 incl	to 36 incl	to 120 incl	¼	⅜
	Wider or Longer		⅜	9/16
Over 0.028	to 48 incl	to 120 incl	⅛	3/16
	Wider or Longer		¼	⅜

[A] Maximum deviation from a flat surface.
[B] Tolerances for high-strength, low-alloy steel with specified minimum yield point in excess of 50 ksi are subject to negotiation.

APPENDIXES

(Nonmandatory Information)

X1. STANDARD CHEMICAL RANGES AND LIMITS

X1.1 Standard chemical ranges and limits are prescribed for carbon steels in Tables X1.1 and X1.2.

A 568

TABLE X1.1 Standard Steels—Carbon Sheet Steel Compositions

Steel Designation SAE or AISI No.	C	Mn	P max	S max
1006	0.08 max	0.45 max	0.040	0.050
1008	0.10 max	0.50 max	0.040	0.050
1009	0.15 max	0.60 max	0.040	0.050
1010	0.08–0.13	0.30–0.60	0.040	0.050
1012	0.10–0.15	0.30–0.60	0.040	0.050
1015	0.12–0.18	0.30–0.60	0.040	0.050
1016	0.12–0.18	0.60–0.90	0.040	0.050
1017	0.14–0.20	0.30–0.60	0.040	0.050
1018	0.14–0.20	0.60–0.90	0.040	0.050
1019	0.14–0.20	0.70–1.00	0.040	0.050
1020	0.17–0.23	0.30–0.60	0.040	0.050
1021	0.17–0.23	0.60–0.90	0.040	0.050
1022	0.17–0.23	0.70–1.00	0.040	0.050
1023	0.19–0.25	0.30–0.60	0.040	0.050
1025	0.22–0.28	0.30–0.60	0.040	0.050
1026	0.22–0.28	0.60–0.90	0.040	0.050
1030	0.27–0.34	0.60–0.90	0.040	0.050
1033	0.29–0.36	0.70–1.00	0.040	0.050
1035	0.31–0.38	0.60–0.90	0.040	0.050
1037	0.31–0.38	0.70–1.00	0.040	0.050
1038	0.34–0.42	0.60–0.90	0.040	0.050
1039	0.36–0.44	0.70–1.00	0.040	0.050
1040	0.36–0.44	0.60–0.90	0.040	0.050
1042	0.39–0.47	0.60–0.90	0.040	0.050
1043	0.39–0.47	0.70–1.00	0.040	0.050
1045	0.42–0.50	0.60–0.90	0.040	0.050
1046	0.42–0.50	0.70–1.00	0.040	0.050
1049	0.45–0.53	0.60–0.90	0.040	0.050
1050	0.47–0.55	0.60–0.90	0.040	0.050
1055	0.52–0.60	0.60–0.90	0.040	0.050
1060	0.55–0.66	0.60–0.90	0.040	0.050
1064	0.59–0.70	0.50–0.80	0.040	0.050
1065	0.59–0.70	0.60–0.90	0.040	0.050
1070	0.65–0.76	0.60–0.90	0.040	0.050
1074	0.69–0.80	0.50–0.80	0.040	0.050
1078	0.72–0.86	0.30–0.60	0.040	0.050
1080	0.74–0.88	0.60–0.90	0.040	0.050
1084	0.80–0.94	0.60–0.90	0.040	0.050
1085	0.80–0.94	0.70–1.00	0.040	0.050
1086	0.80–0.94	0.30–0.50	0.040	0.050
1090	0.84–0.98	0.60–0.90	0.040	0.050
1095	0.90–1.04	0.30–0.50	0.040	0.050
1524	0.18–0.25	1.30–1.65	0.040	0.050
1527	0.22–0.29	1.20–1.55	0.040	0.050
1536	0.30–0.38	1.20–1.55	0.040	0.050
1541	0.36–0.45	1.30–1.65	0.040	0.050
1548	0.43–0.52	1.05–1.40	0.040	0.050
1552	0.46–0.55	1.20–1.55	0.040	0.050

Note—When silicon is required, the following ranges and limits are commonly used:

To 1015, excl	0.10 max
1015 to 1025, incl	0.10 max, 0.10–0.25, or 0.15–0.30
Over 1025	0.10–0.25 or 0.15–0.30

TABLE X1.2 Standard Chemical Ranges and Limits

Note—The carbon ranges shown in the column headed "Range" apply when the specified maximum limit for manganese does not exceed 1.00 %. When the maximum manganese limit exceeds 1.00 %, add 0.01 to the carbon ranges shown below.

Element	Carbon Steels Only, Cast or Heat Analysis		
	Minimum of Specified Element, %	Range	Lowest max
Carbon (see Note)	to 0.15 incl	0.05	0.08
	over 0.15 to 0.30 incl	0.06	
	over 0.30 to 0.40 incl	0.07	
	over 0.40 to 0.60 incl	0.08	
	over 0.60 to 0.80 incl	0.11	
	over 0.80 to 1.35 incl	0.14	
Manganese	to 0.50 incl	0.20	0.40
	over 0.50 to 1.15 incl	0.30	
	over 1.15 to 1.65 incl	0.35	
Phosphorus	to 0.08 incl	0.03	0.04[A]
	over 0.08 to 0.15 incl	0.05	
Sulfur	to 0.08 incl	0.03	0.05[A]
	over 0.08 to 0.15 incl	0.05	
	over 0.15 to 0.23 incl	0.07	
	over 0.23 to 0.33 incl	0.10	
Silicon	to 0.15 incl	0.08	0.10
	over 0.15 to 0.30 incl	0.15	
	over 0.30 to 0.60 incl	0.30	
Copper	When copper is required 0.20 min is commonly specified.		

[A] Certain individual specifications provide for lower standard limits for phosphorus and sulfur.

X2. EFFECT OF AGING OF COLD-ROLLED CARBON STEEL SHEET ON DRAWING AND FORMING

X2.1 Commercial Quality and Drawing Quality

X2.1.1 Cold-rolled sheet is usually temper rolled after annealing. Maximum ductility and minimum hardness exist in cold-rolled sheet in the annealed or dead soft conditions (not temper rolled); however, such sheet is not suitable for most formations due to the fact that it will flute or stretcher strain. A certain amount of cold work (temper rolling) will prevent these conditions from occurring, but the effect of temper rolling, is only temporary due to the phenomenon commonly known as aging.

X2.1.2 Effective roller leveling of temper-passed steel immediately prior to fabrication will minimize the tendency of the sheet to stretcher strain and to flute, but such leveling will not restore softness and ductility. In fact, roller leveling further work hardens the sheet and hence further reduces ductility. Rotation of stock by fabricating the oldest material first is important. Stocking material for extended periods of time should be avoided.

X2.2 Special Killed, Drawing Quality

X2.2.1 Special killed steel is essentially nonaging. It will not stretcher strain or flute or suffer loss of ductility with elapsed time when suitably temper rolled after annealing. This quality should be ordered when stretcher strains or fluting during fabrication are objectionable and the user does not have roller leveling equipment. This quality also should be ordered when the user plans to store sheet over an extended period of time without experiencing adverse changes due to aging.

X2.2.2 Drawing quality, special killed steel has its maximum ductility in the annealed or dead-soft conditions, but in this state it is subject to stretcher straining or fluting during fabrication.

X2.2.3 The superiority of special killed steel is not limited to the fact that it is essentially nonaging. It also has the exceptional ductility required for draws that cannot be made of drawing quality.

X3. PROCEDURE FOR DETERMINING BREAKAGE ALLOWANCE LEVELS (APPLICABLE TO CARBON STEEL SHEET ONLY)

X3.1 In spite of the many extra precautions exercised in making sheet for drawing purposes, certain manufacturing variables may be encountered, all beyond the manufacturer's reasonable control, which may contribute to breakage in fabrication and must be considered as part of the normal hazard of the purchaser's use. The manufacturer will undertake to establish with the purchaser's concurrence a breakage allowance level.

X3.2 Breakage, for the purpose of this proposal, is defined as unrepairable parts, broken during drawing and classed as scrap. Parts showing laminations, resulting from pipe, may be excluded provided they are separately identified. Broken parts that can be salvaged are not covered in this procedure.

X3.3 This procedure is intended to establish a breakage allowance without the need for reinspection of each broken stamping. It will apply to overall breakage on a given part (as calculated by the method outlined below) in excess of 1 % up to and including 8 %. Inherent variations in steel sheet and normal variables in the stamping operation preclude 100 % satisfactory performance. Therefore, it is accepted that practical perfection is attained when 99 % of the stampings are produced without breakage. When the overall breakage is in excess of 8 %, it is considered to be the result of abnormal stamping conditions, and this method does not apply.

X3.4 When there are two or more suppliers, the recommended procedure for determining a breakage allowance for an identified part is based on the average percentage of breakage of at least 75 % of the blanks run on that part, on one set of dies, during at least one month (3000 piece minimum). The total production of all suppliers used to obtain this 75 % minimum is to be included in the calculation starting with the best performance. The average breakage thus determined shall be considered the allowance for the part.

X3.4.1 *Example*:

Vendor	Parts Produced	Parts Scrap	% Scrap
A	32 466	630	1.94
B	27 856	579	2.08
C	67 120	1477	2.20
D	56 200	1349	2.40
E	40 900	1125	2.75
F	850	60	7.05
All	225 392 total	5220 total	2.32 avg

X3.4.2 Seventy-five percent of 225,392 equals to 169,044; therefore, it is necessary to include the total production of vendors A, B, C, and D (A + B + C + D = total production of 183,642 parts) since the total of A, B, and C is only 127,442, which is less than 75 % of the total. Total production of 183,642 parts (A + B + C + D) with 4035 parts being rejected, results in a percentage allowance of 2.20 %. On this basis, vendors D, E, and F exceed the allowance.

X4. PROCEDURES FOR DETERMINING THE EXTENT OF PLASTIC DEFORMATION ENCOUNTERED IN FORMING OR DRAWING

X4.1 Introduction

X4.1.1 The preferred method for determining plastic strain is the circle grid and forming limit curve. The scribed square and change in thickness methods may also be used to evaluate deformation during the forming of a flat sheet into the desired shape.

X4.2 Circle Grid Method

X4.2.1 The test system employs photographic or electrochemically etched circle patterns on the surface of a sheet metal blank of known "quality" and a forming limit curve for the evaluation of strains developed by forming in press operations. It is useful in the laboratory and in the press room. Selection from the various steels that are commercially available can be done effectively by employing this technique. In addition, corrective action in die or part design to improve performance is indicated.

X4.2.2 The forming limit curve in Fig. X4.1 has been developed from actual measurements of the major (e_1) and associated minor (e_2) strains found in critical areas of production type stampings. Strain combinations that locate below this curve are safe, while those that fail above the curve are critical. For analysis of metal strain on production stampings, one must recognize that day-to-day variations of material, lubrication, and die settings will affect the strain level. To ensure trouble-free press performance a zone below the forming limit curve bounded by the dashed and solid lines is designated as the "safety band." Therefore, strain combinations falling below the dashed lines should not exceed the forming limit curve in normal production operations. The left of zero portion of the curve defines the limiting biaxial tension-compression strain combination while the right side defines the forming limit curve. Because the production stampings used to develop for forming limit curve represented all qualities of low-carbon light-gage sheet steel, this single forming limit curve can be used successfully for these products.

X4.2.3 The circle grid method can also be used for other low-carbon sheet categories if the following adjustments to the forming limit curve are made:

X4.2.3.1 *Material Thickness*—As the metal thickness increases the forming limit curve shifts upwards in a parallel manner, 0.2 % (e_1) strain for each 0.025-mm increase in metal thickness above 0.75 mm.

X4.2.3.2 *Material Properties*—When material properties are considerably different from that of conventional low-carbon sheet steel (for example, higher strength-low ductility), the forming limit curve is lower. The magnitude of the downgrade displacement is specific to each material; therefore, current material information should be consulted to determine placement of the forming limit curve.

X4.3 Procedure

X4.3.1 Obtain a sheet sample of "known quality," the sheet quality being established by either supplier designation, consumer purchase order, or most preferred tensile data obtained from a companion sheet sample.

X4.3.2 Obtain or prepare a negative on stencil with selected circles in a uniform pattern. The circles may be 2.5 to 25.0 mm in diameter; the most convenient diameter is 5.0 mm because it is easy to read and the gage spacing is short enough to show the maximum strain in a specific location on the part.

X4.3.3 The sheet metal blanks should be cleaned to remove excess oil and dirt; however, some precoated sheets can be etched without removing the coating. The area(s) to be etched should be determined from observation of panels previously formed; generally, the area that has a split problem is selected for etching. Normally, the convex side of the radius is gridded. If sufficient time is available, the entire blank may be etched, since valuable information can be obtained about the movement of metal in stamping a part when strains can be evaluated in what may appear to be noncritical areas. Additionally, for complex shapes it may be desirable to etch both surfaces of blanks so that the strains that occur in reverse draws can be determined.

X4.3.4 The sheet metal blanks may be etched by a photographic or electrochemical method. In the former method a photosensitive solution, for example, 50 % Kodak Photo Resist (KPR) emulsion and 50 % KPR-thinner, is sprayed onto the sheet. The emulsion is dried by baking the sheet at 65°C for 15 min or by just standing it for several hours at room temperature in a dark room. The latter should be employed in materials that age and, hence, become stronger when baked at 65°C. The negative is placed on the emulsion, held intimately in contact with the sheet, and exposed to a strong ultraviolet light source for 1 to 1½ min. The sheet is developed for 30 to 45 s in KPR developer, rinsed with water, and sprayed with alcohol to set the resist. It is again rinsed with water and then sprayed with KPR black dye to reveal the etched circles.

X4.3.5 In the electrochemical method, the etch pad is saturated with an appropriate electrolyte. Various electrolytes are available from suppliers of the etching equipment. Some electrolytes are more effective than others for etching certain surfaces, such as terne plate and other metallic coated steels. A rust-inhibiting solution is preferred for steel sheets.

X4.3.6 A ground clamp from the transformer of suitable amperage (10 to 50 A is usually used) is fastened to the blank and the second lead is attached to the etch pad. Although the current may be turned on at this time, caution should be taken not to lay the pad on the sheet blank as it will arc. It is advisable to refrain from touching the metal of the etch pad and the grounded sheet blank.

X4.3.7 The stencil is placed with the plastic coating against the sheet surface in the area to be etched. Wetting the stencil with a minimum amount of electrolyte will assist in smoothing out the wrinkles and gives a more uniform etch. The etch pad is now positioned on the stencil and the current turned on, if it is not already on. Apply suitable pressure to the pad. Only the minimum time necessary to produce a clear etched pattern should be used. The etching time will vary with the amperage available from the power source and the stencil area, as well as the pad area in contact with the stencil. Rocker-type etch pads give good prints and require less amperage than flat-surfaced pads. Excessive current causes stencil damage.

X4.3.8 The etching solution activates the surface of the metal and may cause rusting unless it is inhibited. After the desired area has been etched, the blank should be wiped or rinsed, dried, and neutralized.

X4.3.9 The etched blank is now ready for forming. The lubricants and press conditions should simulate production situations. If a sequence of operations is used in forming a part, it is desirable to etch sufficient blanks so that each operation can be studied.

X4.4 Measurement of Strain After Forming

X4.4.1 After forming, the circles are generally distorted into elliptical shapes (Fig. X4.2). These ellipses have major and minor strain axes. The major strain (e_1) is always defined to be the direction in which the greatest positive strain has occurred without regard to original blank edges or the sheet rolling direction. The minor strain (e_2) is defined to be 90° to the major strain direction.

X4.4.2 There are several methods for determining the major and minor strains of the formed panel. Typical tools are a pair of dividers and a scale ruled in 0.5 mm. For sharp radii, a thin plastic scale that can follow the contour of the stamping can be used to determine the dimensions of the ellipses. (Scales are available to read the percent strain directly.)

X4.5 Evaluation of Strain Measurements

X4.5.1 The e_1 strain is always positive while the e_2 strain may be zero, positive, or negative, as indicated on the forming limit curve chart (Fig. X4.1). The maximum e_1 and associated e_2 values measured in critical areas on the formed part are plotted on the graph paper containing the forming limit curve by locating the point of intersection of the e_1, e_2 strains.

X4.5.2 If this point is on or below the "safety band" of the forming limit curve, the strain should not cause breakage. Points further below the curve indicate that a less ductile material of a lower grade may be applied. Points above the "safety band" show that fabrication has induced strains that could result in breakage. Therefore, in evaluation on stampings exhibiting these strains, efforts should be made to provide an e_1, e_2 strain combination that would lie on or below the "safety band" of the forming limit curve. A different e_1, e_2 strain combination can be obtained through changes of one or more of the forming variables such as die conditions, lubricants, blank size, thickness, or material grade. It should be noted at this time that these conclusions are derived from a reference base being the steel "quality" used to fabricate the grid stamping.

X4.5.3 When attempting to change the relationship of e_1 and e_2 strains, it should be noted that on the forming limit curve the most severe condition for a given e_1 strain is at 0 % e_2 strain. This means the metal works best when it is allowed to deform in two dimensions, e_1 and e_2, rather than being restricted in one dimension. A change in e_2 to decrease the severity can be made by changing one of the previously mentioned forming variables of the die design, for example, improving lubrication on the tension-tension side will increase e_2 and decrease the severity.

X4.5.4 In addition to the forming limit curve, the e_1, e_2 strain measurements may be used to evaluate the material requirements on the basis of strain gradients, as illustrated in Fig. X4.3, or by plotting contours of equivalent strain levels on the surface of the formed part. Even when the level of strain is relatively low, parts in which the e_1 strain is changing rapidly either in magnitude or direction over a short span on the surface may require more ductile grades of sheet metal, change in lubrication, or change in part design.

X4.6 Example of Major and Minor Strain Distribution

X4.6.1 A formed panel (Fig. X4.4) with a cross section as shown in Fig. X4.3 is used to illustrate major and minor strain combinations. A plot of the major strain distribution should be made by finding the ellipse with the largest major strain (circle 7) and measuring both the major and minor strains in the row of ellipses running in the direction of the major strain. The solid dots (Fig. X4.3) are the measured major strains for each ellipse. The Xs are the critical major strains as determined from the forming limit curve at the corresponding minor strain (intersection of the measured minor strain and the severity curve).

X4.6.2 Usually a single row of ellipses will suffice to determine the most severe strain distribution. The resulting strain distribution plot (Fig. X4.3) illustrates both severity of the strain compared to the critical strain limits and the concentration of strain in the stamping. Steep strain gradients should be avoided because they are inherent to fracture sites.

X4.7 Example for Reducing Splitting Tendency

X4.7.1 In an area such as that represented in Fig. X4.3, the splitting tendency can be reduced as follows:

X4.7.1.1 If the radius of the part in the region of circle 1 is increased, some strain can be induced to take place in this area which will allow the major strain in circle 7 to be reduced sufficiently to bring the strain combination below the critical limit. This course of action requires no binding nor reshaping of the punch, only grinding in the radius.

X4.7.1.2 The total average major strain required to make this formation is only 17.5 %; yet in a 5.0-mm circle the strain is as high as 40 %. The strain distribution curve puts forth graphically the need to distribute the strain over the length of the line by some means as described above.

X4.7.1.3 Change in lubrication can also improve the strain distribution of a stamping. If the strain over the punch is critical, the amount of stretch (strain) required to make the shape can be reduced by allowing metal to flow in over the punch by decreasing the friction through the use of a more effective lubricant in the hold-down area.

X4.7.1.4 If the part is critical, a change in material may help. That is, a material having a better uniform elongation will distribute the strain more uniformly or a material having a higher "r" value will make it possible to "draw" in more metal from the hold-down area so that less stretch is necessary to form the part.

X4.8 Scribed Square Method

X4.8.1 The basic technique is to draw a panel from a blank that has been scribed both longitudinally and transversely with a series of parallel lines spaced at 25.0-mm intervals. The lines on the panel are measured after drawing and the stretch or draw calculated as the percent increase in area of a 25.0-mm square. This is a fairly simple procedure for panels having generous radii and fairly even stretch or draw. Many major panels fall in this category, and in these instances it is quite easy to pick out the square area exhibiting the greatest increase.

X4.8.2 If the square or line to be measured is no longer a flat surface, place a narrow strip of masking (or other suitable tape) on the formed surface and mark the points which are to be measured. Remove the tape, place on a plane surface, and determine the distance between the points with a steel scale.

X4.8.3 There will be cases of minor increase in area with major elongation in the one direction. In these instances, the percent elongation should be recorded.

X4.9 Thickness Method

X4.9.1 There are instances when the maximum stretch is continued to an area smaller than 645 mm^2 or the shape of the square has been distorted irregularly, making measurement difficult and calculation inaccurate. When either of these conditions exists, an electronic thickness gage may be used at the area in question or this area may be sectioned and the decrease in metal thickness measured with a ball-point micrometer. The increase in unit area can be calculated by dividing the original thickness by the final thickness.

X4.9.2 *Example*—Assuming the blank thickness to be 0.80 mm and the final thickness to be 0.60 mm, the increase in unit area would be a [(0.80 − 0.60)/ 0.80] × 100 = 25 % increase.

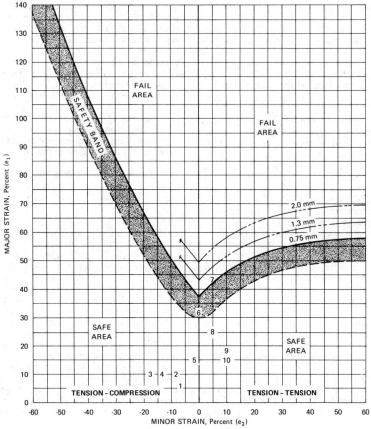

FIG. X4.1 Forming Limit Curve

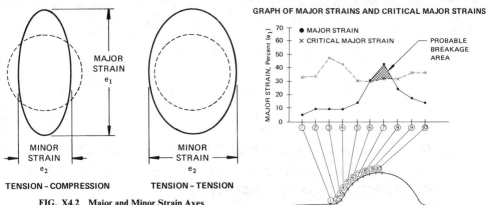

FIG. X4.2 Major and Minor Strain Axes

FIG. X4.3 Graph of Major Strains and Critical Major Strains and Cross Section of Etched Panel

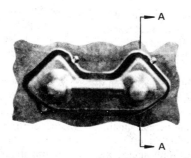

TOP VIEW OF PANEL BEFORE DIE MODIFICATIONS

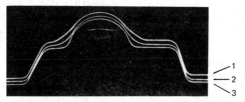

1 = Final Contour   2 = First Revision   3 = Original Contour

SECTION A - A

FIG. X4.4 Formed Panel and Cross Section

The American Society for Testing and Materials takes no position respecting the validity of any patent rights asserted in connection with any item mentioned in this standard. Users of this standard are expressly advised that determination of the validity of any such patent rights, and the risk of infringement of such rights, are entirely their own responsibility.

This standard is subject to revision at any time by the responsible technical committee and must be reviewed every five years and if not revised, either reapproved or withdrawn. Your comments are invited either for revision of this standard or for additional standards and should be addressed to ASTM Headquarters. Your comments will receive careful consideration at a meeting of the responsible technical committee, which you may attend. If you feel that your comments have not received a fair hearing you should make your views known to the ASTM Committee on Standards, 1916 Race St., Philadelphia, Pa. 19103.

Designation: A 568M – 77

An American National Standard

Metric

Standard Specification for
GENERAL REQUIREMENTS FOR STEEL, CARBON AND HIGH-STRENGTH LOW-ALLOY HOT-ROLLED SHEET AND COLD-ROLLED SHEET [METRIC][1]

This standard is issued under the fixed designation A 568M; the number immediately following the designation indicates the year of original adoption or, in the case of revision, the year of last revision. A number in parentheses indicates the year of last reapproval. A superscript epsilon (ϵ) indicates an editorial change since the last revision or reapproval.

This specification has been approved for use by agencies of the Department of Defense and for listing in the DoD Index of Specifications and Standards.

1. Scope

1.1 This specification covers the general requirements for steel sheet in coils and cut lengths. It applies to carbon steel and high-strength, low-alloy steel furnished as hot-rolled sheet and cold-rolled sheet.

1.2 This specification is not applicable to hot-rolled heavy-thickness carbon sheet coils, Specification A 635M for Steel Sheet and Strip, (0.15 % max) Hot-Rolled Commercial Quality, Heavy-Thickness Coils (Formerly Plate)).[2]

1.3 This specification covers only metric (SI) units and is not to be used or confused with inch-pound units.

2. Applicable Documents

2.1 *ASTM Standards:*
A 370 Methods and Definitions for Mechanical Testing of Steel Products[2]
A 700 Practices for Packaging, Marking, and Loading Methods for Steel Products for Domestic Shipment[2]
E 11 Specification for Wire-Cloth Sieves for Testing Purposes[3]
E 350 Methods for Chemical Analysis of Carbon Steel, Low-Alloy Steel, Silicon Electrical Steel, Ingot Iron, and Wrought Iron[4]

2.2 *Military Standards:*
MIL-STD-129 Marking for Shipment and Storage[5]
MIL-STD-163 Steel Mill Products, Preparation for Shipment and Storage[5]

2.3 *Federal Standards:*
Fed. Std. No. 123 Marking for Shipment (Civil Agencies)[5]
Fed. Std. No. 183 Continuous Identification Marking of Iron and Steel Products[5]

3. Descriptions of Terms

3.1 *Steel Types:*

3.1.1 *Carbon Steel* is the designation for steel when no minimum content is specified or required for aluminum, chromium, cobalt, columbium, molybdenum, nickel, titanium, tungsten, vanadium, zirconium, or any element added to obtain a desired alloying effect; when the specified minimum for copper does not exceed 0.40 %; or when the maximum content specified for any of the following elements does not exceed the percentages noted: manganese 1.65, silicon 0.60, or copper 0.60.

3.1.1.1 In all carbon steels small quantities of certain residual elements unavoidably retained from raw materials are sometimes

---

[1] This specification is under the jurisdiction of ASTM Committee A-1 on Steel, Stainless Steel and Related Alloys, and is the direct responsibility of Subcommittee A01.19 on Steel Sheet and Strip.
Current edition approved Nov. 3, 1977. Published January 1978.
[2] *Annual Book of ASTM Standards*, Vol 01.03.
[3] *Annual Book of ASTM Standards*, Vol 14.02.
[4] *Annual Book of ASTM Standards*, Vol 03.05.
[5] Available from Naval Publications and Forms Center, 5801 Tabor Ave., Philadelphia, Pa. 19120.

found which are not specified or required, such as copper, nickel, molybdenum, chromium, etc. These elements are considered as incidental and are not normally determined or reported.

3.1.2 *High-Strength Low-Alloy Steel* is a specific group of steels in which higher strength, and in some cases additional resistance to atmospheric corrosion are obtained by moderate amounts of one or more alloying elements.

3.2 *Product Types:*

3.2.1 *Hot-Rolled Sheet* is manufactured by hot rolling slabs in a continuous mill to the required thickness and can be supplied in coils or cut lengths as specified.

3.2.1.1 Hot-rolled carbon steel sheet is commonly available by size as follows:

Coils and Cut Lengths

Width, mm	Thickness, mm
Over 300 to 1200, incl	1.2 to 6.0, incl
Over 1200	1.2 to 4.5, incl

Coils Only[A]

Width, mm	Thickness, mm
Over 300 to 1200, incl	Over 6.0 to 12.5, incl
Over 1200	Over 4.5 to 12.5, incl

[A] One type of this product is covered by Specification A 635M.

3.2.1.2 Hot-rolled high-strength, low-alloy steel sheet is commonly available by size as follows:

Width, mm	Thickness, mm
Over 300 to 1200, incl	1.8 to 6.0, incl
Over 1200	1.8 to 4.5, incl

3.2.2 *Cold-Rolled Sheet* is manufactured from hot-rolled descaled coils by cold reducing to the desired thickness, generally followed by annealing to recrystallize the grain structure. If the sheet is not annealed after cold reduction it is known as full hard with a hardness of 84 HRB minimum and can be used for certain applications where ductility and flatness are not required.

3.2.2.1 Cold-rolled carbon sheet is commonly available by size as follows:

Width, mm	Thickness, mm
50 to 300, incl[A]	0.35 to 2.0, incl
Over 300[B]	0.35 and Over

[A] Cold-rolled sheet coils and cut lengths, slit from wider coils with cut edge (only) and in thicknesses 0.35 to 2.0 mm, incl, carbon 0.25 % maximum by heat analysis.

[B] When no special edge or finish (other than matte, commercial bright, or luster finish) or single strand rolling of widths, or both, up to and including 600 mm is not specified or required.

3.2.2.2 Cold-rolled high-strength, low-alloy sheet is commonly available by size as follows:

Width, mm	Thickness, mm
50 to 300, incl[A]	0.5 to 2.0, incl
Over 300[B]	0.7 and Over

[A] Cold-rolled sheet coils and cut lengths, slit from wider coils with cut edge (only) and in thicknesses 0.50 to 2.0 mm, incl, carbon 0.25 % maximum by heat analysis.

[B] When no special edge or finish (other than matte, commercial bright, or luster finish) or single strand rolling of widths, or both up to and including 600 mm is not specified or required.

3.2.2.3 Class 1 and Class 2 requirements are described in 9.2, 9.3 and 10.3 and prescribed in Appendix X3.

4. General Requirements for Delivery

4.1 Products covered by this specification are produced to metric decimal thicknesses only and metric thickness tolerances apply.

5. Manufacture

5.1 Unless otherwise specified, hot-rolled material shall be furnished hot-rolled, not annealed, not pickled.

5.2 Coil breaks, stretcher strains, and fluting can occur during the user's processing of hot-rolled or hot-rolled pickled sheet. When any of these features are detrimental to the application, the manufacturer shall be notified by ordering the proper class of cold-rolled sheet (see 5.3).

5.3 Cold-rolled sheet is available in two classes:

5.3.1 *Class 1* (see 9.2).

5.3.2 *Class 2* (see 9.3).

5.4 Unless specified as a full-hard product, cold-rolled sheet is annealed after being cold reduced to thickness. The annealed, cold-rolled sheet can be used as annealed last (dead soft) for unexposed end-use applications. When cold-rolled sheet is used for unexposed applications and coil breaks are a hazard in uncoiling, it may be necessary to process the material further. In this case the manufacturer should be consulted. After annealing, cold-rolled sheet is generally given a light skin pass to impart shape or may be given a heavier skin pass or temper pass to prevent the phenomenon known as stretcher straining or fluting, when formed. Temper passing also provides a required surface texture.

A 568M

5.5 *Temper Rolling:*

5.5.1 Unless otherwise specified, Class 1 sheet shall be temper rolled. Class 1 is normally specified and furnished temper rolled to be in the strain-free condition as shipped. See Appendix X4.

5.5.2 Class 2 sheet may be furnished annealed last or temper rolled at the manufacturer's option. Class 2 is normally produced without temper rolling but may be lightly temper rolled during oiling or rewinding. Temper rolled cannot be specified on Class 2. Material that is specified or required strain free or nonfluting at the time of shipment must be ordered as Class 1.

Note 1—Skin-passed sheet is subject to an aging phenomenon (see Appendix X4). Unless special killed (nonaging) steel is specified, it is to the user's interest to fabricate the sheet as soon as possible for optimum performance.

6. Chemical Requirements

6.1 *Limits:*

6.1.1 The chemical composition shall be in accordance with the applicable product specification. However, if other compositions are required for carbon steel, they shall be prepared in accordance with Appendix X2.

6.1.2 Where the material is used for fabrication by welding, care must be exercised in selection of chemical composition or mechanical properties to ensure compatibility with the welding process and its effect on altering the properties.

6.2 *Cast or Heat (Formerly Ladle) Analysis:*

6.2.1 An analysis of each cast or heat of steel shall be made by the manufacturer to determine the percentage of elements specified or restricted by the applicable specification.

6.2.2 When requested, cast or heat analysis for elements listed or required shall be reported to the purchaser or to his representative.

6.3 *Product, Check, or Verification (Formerly Check) Analysis:*

6.3.1 Non-killed steels (such as capped or rimmed) are not technologically suited to product analysis due to the nonuniform character of their chemical composition and therefore, the tolerances in Table 1 do not apply. Product analysis is appropriate on these types of steel only when misapplication is apparent or for copper when copper steel is specified.

6.3.2 For steels other than non-killed (capped or rimmed), product analysis may be made by the purchaser. The chemical analysis shall not vary from the limits specified by more than the amounts in Table 1. The several determinations of any element in a cast shall not vary both above and below the specified range.

6.4 *Sampling for Product Analysis:*

6.4.1 To indicate adequately the representative composition of a cast by product analysis, it is general practice to select samples to represent the steel, as fairly as possible, from a minimum number of pieces as follows: 3 pieces for lots up to 15.0 Mg, incl, and 6 pieces for lots over 15.0 Mg.

6.4.2 When the steel is subject to tension test requirements, samples for product analysis may be taken either by drilling entirely through the used tension test specimens themselves, or as covered in 6.4.3.

6.4.3 When the steel is not subject to tension test requirements, the samples for analysis must be taken by milling or drilling entirely through the sheet or strip in a sufficient number of places so that the samples are representative of the entire sheet or strip. The sampling may be facilitated by folding the sheet both ways, so that several samples may be taken at one drilling. Steel subjected to certain heating operations by the purchaser may not give chemical analysis results which properly represent its original composition. Therefore, users must analyze chips taken from the steel in the condition in which it is received from the steel manufacturer.

6.5 *Specimen Preparation*—Drillings or chips must be taken without the application of water, oil, or other lubricant, and must be free of scale, grease, dirt, or other foreign substances. They must not be overheated during cutting to the extent of causing decarburization. Chips must be well mixed and those too coarse to pass a No. 10 (2.00-mm) sieve or too fine to remain on a No. 30 (600-μm) sieve are not suitable for proper analysis. Sieve size numbers are in accordance with Specification E 11.

6.6 *Test Methods*—In case of referee analysis is required and agreed upon to resolve a dispute concerning the results of a chemical analysis, the procedure for performing the referee analysis must be in accordance with the latest issue of Methods E 350, unless otherwise agreed upon between the manufacturer and the purchaser.

7. Mechanical Requirements

7.1 The mechanical property requirements, number of specimens, and test locations and specimen orientation shall be in accordance with the applicable product specification.

7.2 Unless otherwise specified in the applicable product specification, test specimens must be prepared in accordance with Methods and Definitions A 370.

7.3 Mechanical tests shall be conducted in accordance with Methods and Definitions A 370.

8. Dimensions, Tolerances and Allowances

8.1 Dimensions, tolerances, and allowances applicable to products covered by this specification are contained in Tables 3 through 22. The appropriate tolerance tables shall be identified in each individual specification.

8.2 Flatness tolerances are not applicable to Class 2 sheet. Class 2 will normally be within two times standard flatness when shipped in cut lengths and after removal of coil set when shipped in coils.

9. Finish and Appearance

9.1 Hot-rolled sheet has a surface with an oxide or scale resulting from the hot-rolling operation. The oxide or scale can be removed by pickling or blast cleaning when required for press-work operations or welding. Hot-rolled and hot-rolled descaled sheet is not generally used for exposed parts where surface is of prime importance.

9.1.1 Hot-rolled sheet can be supplied with mill edges or cut edges as specified. Mill edges are the natural edges resulting from the hot-rolling operation. They do not conform to any particular contour. They may also contain some edge imperfections, the more common types of which are cracked edges, thin edges (feather), and damaged edges due to handling or processing and which should not extend in beyond the ordered width.

These edge conditions are detrimental where joining of the mill edges by welding is practiced. When the purchaser intends to shear or to blank, a sufficient width allowance should be made when purchasing to ensure obtaining the desired contour and size of the pattern sheet. The manufacturer may be consulted for guidance. Cut edges are the normal edges that result from the shearing, slitting, or trimming of mill-edge sheet.

9.1.1.1 The ends of plain hot-rolled mill-edge coils are irregular in shape and are referred to as uncropped ends. Where such ends are not acceptable, the purchaser's order should so specify. Processed coils such as pickled or blast cleaned are supplied with square-cut ends.

9.2 Cold-rolled carbon sheet Class 1 is intended for exposed applications where surface appearance is of primary importance. This class will meet requirements for controlled surface texture, surface quality, and flatness. It is normally processed by the manufacturer to be free of stretcher strain and fluting. Subsequent user roller levelling immediately before fabrication will minimize strain resulting from aging.

9.2.1 Cold-rolled carbon sheet Class 1 can be supplied in the following finishes:

9.2.1.1 Matte finish is a dull finish, without luster, produced by rolling on rolls that have been roughened by mechanical or chemical means to various degrees of surface texture depending upon application. With some surface preparation matte finish is suitable for decorative painting. It is not generally recommended for bright plating.

9.2.1.2 Commercial bright finish is a relatively bright finish having a surface texture intermediate between that of matte and luster finish. With some surface preparation commercial bright finish is suitable for decorative painting or certain plating applications. If sheet is deformed in fabrication the surface may roughen to some degree and areas so affected will require surface preparation to restore surface texture to that of the undeformed areas.

9.2.1.3 Luster finish is a smooth bright finish produced by rolling on ground rolls and is suitable for decorative painting or plating with additional special surface preparation by the user. The luster may not be retained after fabrication; therefore, the formed parts will require surface preparation to make them suitable for bright plating.

9.3 Cold-rolled carbon sheet Class 2 is intended for unexposed application. Limitations on degree and frequency of surface imperfections, as well as restrictions on texture and flatness, are not applicable. Normal processing by the manufacturer results in this class having coil breaks and a tendency toward fluting and stretcher straining. This class may contain more surface imperfections than Class 1 because steel applications, processing procedures, and inspection standards are less stringent.

9.4 Cold-rolled high-strength low-alloy sheet is supplied with a matte finish, unless otherwise specified.

9.5 The cold-rolled products covered by this specification are furnished with cut edges and square cut ends, unless otherwise specified.

9.6 *Oiling:*

9.6.1 Plain hot-rolled sheet is customarily furnished not oiled. Oiling must be specified when required.

9.6.2 Hot-rolled pickled or descaled sheet is customarily furnished oiled. If the product is not to be oiled, it must be so specified.

9.6.3 Cold-rolled products covered by this specification can be furnished oiled or not oiled as specified.

9.7 Sheet steel in coils or cut lengths may contain surface imperfections that can be removed with a reasonable amount of metal finishing by the purchaser.

10. Workmanship

10.1 Cut lengths shall have a workmanlike appearance and shall not have imperfections of a nature or degree for the product, the grade, class, and the quality ordered that will be detrimental to the fabrication of the finished part.

10.2 Coils may contain some abnormal imperfections that render a portion of the coil unusable since the inspection of coils does not afford the producer the same opportunity to remove portions containing imperfections as in the case with cut lengths.

10.3 *Surface Conditions:*

10.3.1 Class 1 is intended for applications where surface appearance is of primary importance, that is, exposed applications. Class 2 is intended for applications where surface appearance is not of primary importance, that is, unexposed applications.

10.3.2 Class 1 cut lengths shall not include individual sheets having major surface imperfections (holes, loose slivers, and pipe) and repetitive minor surface imperfections. Cut lengths may contain random minor surface imperfections which can be removed with a reasonable amount of metal finishing by the purchaser, which shall be acceptable to the purchaser within the manufacturer's published standards.

10.3.3 For Class 1 coils it is not possible to remove the surface imperfections listed in 10.3.2. Coils will contain such imperfections that shall be acceptable to the purchaser within the manufacturer's published standards. Class 1 coils contain more surface imperfections than cut lengths because the producer does not have the same opportunity to sort portions containing such imperfections as is possible with cut lengths.

10.3.4 Class 2 cut lengths shall not include individual sheets having major surface imperfections such as holes, loose slivers, and pipe. In addition, Class 2 cut lengths can be expected to contain more minor imperfections such as pits, scratches, sticker breaks, edge breaks, pinchers, cross breaks, roll marks, and other surface imperfections than Class 1 which shall be acceptable to the purchaser without limitation.

10.3.5 For Class 2 coils it is not possible to remove the surface imperfections listed in 10.3.4. Coils will contain surface imperfections that are normally not repairable. Minor imperfections shall be acceptable to the purchaser within the manufacturer's published standards. Class 2 coils contain more surface imperfections than Class 1 coils.

11. Retests

11.1 If any test specimen shows defective machining or develops flaws, it must be discarded and another specimen substituted.

11.2 If the percentage of elongation of any test specimen is less than that specified and any part of the fracture is more than 20 mm from the center of the gage length of a 50-mm specimen or is outside the middle half of the gage length of a 200-mm specimen, as indicated by scribe scratches marked on the specimen before testing, a retest is allowed.

11.3 If a bend specimen fails, due to conditions of bending more severe than required by the specification, a retest is permitted either on a duplicate specimen or on a remaining portion of the failed specimen.

12. Inspection

12.1 When the purchaser's order stipulates that inspection and tests (except product analyses) for acceptance on the steel be made prior to shipment from the mill, the manufacturer shall afford the purchaser's inspector all reasonable facilities to satisfy him that the steel is being produced and furnished in accordance with the specification. Mill inspection by the purchaser shall not interfere unnecessarily with the manufacturer's operation.

13. Rejection and Rehearing

13.1 Unless otherwise specified, any rejection shall be reported to the manufacturer within a reasonable time after receipt of material by the purchaser.

13.2 Material that is reported to be defective subsequent to the acceptance at the purchaser's works shall be set aside, adequately protected, and correctly identified. The manufacturer shall be notified as soon as possible so that an investigation may be initiated.

13.3 Samples that are representative of the rejected material shall be made available to the manufacturer. In the event that the manufacturer is dissatisfied with the rejection, he may request a rehearing.

14. Marking

14.1 As a minimum requirement, the material shall be identified by having the manufacturer's name, ASTM designation, weight, purchaser's order number, and material identification (grade, type, class, or other classification used to define the material) legibly stenciled on top of each lift or shown on a tag attached to each coil or shipping unit.

14.2 When specified in the contract or order, and for direct procurement by or direct shipment to the government, marking for shipment in addition to requirements specified in the contract or order, shall be in accordance with MIL-STD-129 for military agencies and in accordance with Fed. Std. No. 123 or 183 for civil agencies.

15. Packaging

15.1 Unless otherwise specified, the sheet shall be packaged and loaded in accordance with Practices A 700.

15.2 When specified in the contract or order, and for direct procurement by or direct shipment to the government, when Level A is specified, preservation, packaging, and packing shall be in accordance with the Level A requirements of MIL-STD-163.

15.3 When coils are ordered it is customary to specify a minimum or range of inside diameter, maximum outside diameter, and a maximum coil weight, if required. The ability of manufacturers to meet the maximum coil weights depends upon individual mill equipment. When required, minimum coil weights are subject to negotiation.

A 568M

TABLE 1 Tolerances for Product Analysis

Element	Limit, or Maximum of Specified Element, %	Tolerance Under Minimum Limit	Tolerance Over Maximum Limit
Carbon	to 0.15 incl	0.02	0.03
	over 0.15 to 0.40 incl	0.03	0.04
	over 0.40 to 0.80 incl	0.03	0.05
	over 0.80	0.03	0.06
Manganese	to 0.60 incl	0.03	0.03
	over 0.60 to 1.15 incl	0.04	0.04
	over 1.15 to 1.65 incl	0.05	0.05
Phosphorus		–	0.01
Sulfur		–	0.01
Silicon	to 0.30 incl	0.02	0.03
	over 0.30 to 0.60 incl	0.05	0.05
Copper		0.02	–

TABLE 2 List of Tables for Dimensions, Tolerances, and Allowances

Dimension	Table No.	
Carbon[A] and High-Strength Low-Alloy Steel		
	Hot-Rolled Sheet	Cold-Rolled Sheet
Allowances in width and length	16[B]	16[B]
Camber tolerances	11	11, 22
Diameter tolerances of sheared circles	10	10
Flatness tolerances, not stretcher leveled	14	23
Flatness tolerances, stretcher leveled	15	24
Length tolerances	9	19, 20
Out-of-square tolerances	12	12
Resquared tolerances	13	13
Thickness tolerances	5, 6	17, 18
Width tolerances of cut edge	8	8, 21
Width tolerances of mill edge	7	–

[A] Tolerances for hot-rolled carbon sheet steel with 0.25 % maximum carbon, cast or heat analysis.
[B] Carbon steel only.

TABLE 3 Thickness Tolerances of Hot-Rolled Sheet Ordered to Nominal Thickness (Carbon Steel)
(Coils and Cut Lengths, Including Pickled)

NOTE – Thickness is measured at any point across the width not less than 10.0 mm from a cut edge and not less than 20 mm from a mill edge. This table does not apply to the uncropped ends of mill edge coils.

Specified Width, mm		Thickness Tolerances, Over and Under, mm, for Specified Nominal Thickness, mm				
Over	Through	Through 2.0	Over 2.0 to 2.5, incl	Over 2.5 to 4.0, incl	Over 4.0 to 5.5, incl	Over 5.5 to 6.0, incl
300	600	0.15	0.15	0.18	0.20	0.28
600	1200	0.15	0.18	0.20	0.22	0.30
1200	1500	0.18	0.18	0.20	0.25	0.30
1500	1800	0.18	0.20	0.20	0.28	0.32
1800		0.18	0.20	0.20	0.30	0.38

TABLE 4 Thickness Tolerances of Hot-Rolled Sheet Ordered as Minimum Thickness (Carbon Steel)
(Coils and Cut Lengths, Including Pickled)

NOTE – Thickness is measured at any point across the width not less than 10.0 mm from a cut edge and not less than 20 mm from a mill edge. This table does not apply to the uncropped ends of mill edge coils.

Specified Width, mm		Thickness Tolerance, Over Only, mm, for Specified Minimum Thickness, mm				
Over	Through	Through 2.0	Over 2.0 to 2.5, incl	Over 2.5 to 4.0, incl	Over 4.0 to 5.5, incl	Over 5.5 to 6.0, incl
300	600	0.30	0.30	0.35	0.40	0.55
600	1200	0.30	0.35	0.40	0.45	0.60
1200	1500	0.35	0.35	0.40	0.50	0.60
1500	1800	0.35	0.40	0.40	0.55	0.65
1800		0.35	0.40	0.40	0.60	0.75

TABLE 5 Thickness Tolerances of Hot-Rolled Sheet Ordered to Nominal Thickness (High-Strength Low-Alloy Steel)

(Coils and Cut Lengths, Including Pickled)

Note—Thickness is measured at any point across the width not less than 10.0 mm from a cut edge and not less than 20 mm from a mill edge. This table does not apply to the uncropped ends of mill edge coils.

Specified Width, mm		Thickness Tolerances, Over and Under, mm, for Specified Nominal Thickness, mm			
Over	Through	Through 2.0	Over 2.0 to 2.5, incl	Over 2.5 to 4.5, incl	Over 4.5 to 6.0, incl
300	600	0.15	0.18	0.20	0.30
600	1200	0.18	0.20	0.23	0.32
1200	1500	0.18	0.20	0.25	
1500	1800	0.20	0.23	0.28	
1800	2000	0.20	0.23	0.30	
2000			0.25	0.30	

TABLE 6 Thickness Tolerances of Hot-Rolled Sheet Ordered as Minimum Thickness (High-Strength Low-Alloy Steel)

(Coils and Cut Lengths, Including Pickled)

Note—Thickness is measured at any point across the width not less than 10.0 mm from a cut edge and not less than 20 mm from a mill edge. This table does not apply to the uncropped ends of mill edge coils.

Specified Width, mm		Thickness Tolerances, Over Only, mm, for Specified Minimum Thickness, mm			
Over	Through	Through 2.0	Over 2.0 to 2.5, incl	Over 2.5 to 4.5, incl	Over 4.5 to 6.0, incl
300	600	0.30	0.35	0.40	0.60
600	1200	0.35	0.40	0.45	0.65
1200	1500	0.35	0.40	0.50	
1500	1800	0.40	0.45	0.55	
1800	2000	0.40	0.45	0.60	
2000			0.50	0.60	

TABLE 7 Width Tolerances[A] of Hot-Rolled Mill Edge Sheet (Carbon and High-Strength Low-Alloy Steel)

(Coils and Cut Lengths, Including Pickled)

Specified Width, mm		Width Tolerance, Over Only, mm	
Over	Through	Carbon	HSLA
300	600	16	16
600	1200	26	28
1200	1500	32	38
1500	1800	35	45
1800	–	48	50

[A] The above tolerances do not apply to the uncropped ends of mill edge coils (see 9.1.1).

TABLE 8 Width Tolerances of Hot-Rolled Cut Edge Sheet and Cold-Rolled Sheet (Carbon and High-Strength Low-Alloy Steel)

(Not Resquared, Coils and Cut Lengths, Including Pickled)

Specified Width, mm		Width Tolerance, Over Only, mm
Over	Through	
300	600	3
600	1200	5
1200	1500	6
1500	1800	8
1800	–	10

TABLE 9 Length Tolerances of Hot-Rolled Sheet (Carbon and High-Strength Low-Alloy Steel)

(Cut Lengths Not Resquared, Including Pickled)

Specified Length, mm		Length Tolerance, Over Only, mm
Over	Through	
300	600	6
600	900	8
900	1500	12
1500	3000	20
3000	4000	25
4000	5000	35
5000	6000	40
6000	–	45

TABLE 10 Diameter Tolerances of Circles from Hot-Rolled (Including Pickled) and Cold-Rolled Sheet (Over 300 mm Width)

(Carbon and High-Strength Low-Alloy Steel)

Specified Thickness[A], mm		Tolerances Over Specified Diameter, mm (No Tolerances Under)		
		Diameters, mm		
Over	Through	Through 600	Over 600 to 1200, incl	Over 1200
–	1.5	1.5	3.0	5.0
1.5	2.5	2.5	4.0	5.5
2.5	–	3.0	5.0	6.5

[A] 1.80 mm minimum thickness for hot-rolled high-strength low-alloy steel sheet.

TABLE 11 Camber Tolerances for Hot-Rolled (Including Pickled) and Cold-Rolled Sheet (Over 300 mm Width)

(Carbon and High-Strength Low-Alloy Steel)

NOTE 1 – Camber is the greatest deviation of a side edge from a straight line, the measurement being taken on the concave side with a straightedge.

NOTE 2 – Camber tolerances for cut lengths, not resquared, as shown in the table.

Cut Length, mm		Camber Tolerances[A], mm
Over	Through	
–	1200	4
1200	1800	5
1800	2400	6
2400	3000	8
3000	3700	10
3700	4300	13
4300	4900	16
4900	5500	19
5500	6000	22
6000	9000	32
9000	12200	38

[A] The camber tolerance for coils is 25.0 mm in any 6000 mm.

TABLE 12 Out-of-Square Tolerances of Hot-Rolled Cut-Edge (Including Pickled) and Cold-Rolled Sheet (Over 300 mm Width)

(Cut Lengths Not Resquared)

Out-of-square is the greatest deviation of an end edge from a straight line at right angle to a side and touching one corner. It is also obtained by measuring the difference between the diagonals of the cut length. The out-of-square deviation is one half of that difference. The tolerance for all thicknesses and all sizes is 1.0 mm/100 mm of width or fraction thereof.

TABLE 13 Resquare Tolerances of Hot-Rolled (Including Pickled) and Cold-Rolled Sheet (Over 300 mm Width) (Carbon and High-Strength Low-Alloy Steel)

(Cut Lengths)

When cut lengths are specified resquared, the width and the length are not less than the dimensions specified. The individual tolerance for over-width, over-length, camber, or out-of-square should not exceed 1.6 mm up to and including 1200 mm in width and up to and including 3000 mm in length. For cut lengths wider or longer, the applicable tolerance is 3.2 mm.

TABLE 14 Flatness Tolerances[A] of Hot-Rolled Sheet Including Pickled Cut Lengths Not Specified to Stretcher-Leveled Standard of Flatness (Carbon and High-Strength Low-Alloy Steel)

Specified Thickness, mm		Specified Width, mm	Flatness Tolerance[B], mm	
Over	Through		Carbon Steel	High-Strength[C,D] Low-Alloy Steel
1.2	1.5	to 900, incl	15	—
		over 900 to 1500, incl	20	—
		over 1500	25	—
1.5	4.5	to 1500, incl	15	20
		over 1500 to 1800, incl	20	30
		over 1800	25	40
4.5	6.0	to 1200, incl	15	20

[A] The above tolerances also apply to lengths cut from coils by the consumer when adequate flattening operations are performed.
[B] Maximum deviation from a horizontal flat surface.
[C] Tolerances for high-strength, low-alloy steels with specified minimum yield point in excess of 345 MPa are subject to negotiation.
[D] 1.8 minimum thickness.

TABLE 15 Flatness Tolerances of Hot-Rolled Sheet Including Pickled Cut Lengths Specified to Stretcher-Leveled Standard of Flatness (Carbon and High-Strength Low-Alloy Steel)

Specified Thickness, mm		Specified Width, mm	Specified Length, mm	Flatness Tolerance[A]	
Over	Through			Carbon Steel	High-Strength Low-Alloy Steel
1.2	4.5	Through 1200	Through 2400	3	
		Wider or longer		6	B
4.5	6.0	Through 1200	Through 2400	3	
		Longer		6	B

[A] Maximum deviation from a horizontal flat surface.
[B] Tolerances for high-strength, low-alloy steel are subject to negotiation.

TABLE 16 Allowances in Width and Length of Hot-Rolled Sheet (Carbon-Steel) and Cold-Rolled Sheet (Carbon and High-Strength Low-Alloy Steel)

(Cut Length Sheets Specified to Stretcher-Leveled Standard of Flatness, Not Resquared, Including Pickled)

NOTE 1—When cut lengths are specified to stretcher-leveled standard of flatness and not resquared, the allowances over specified dimensions in width and length given in the following table apply. Under these conditions the allowances for width and length are added by the manufacturer to the specified width and length and the tolerances given in Tables 5, 6, 7, 17, 18, and 19 apply to the new size established. The camber tolerances in Table 9 do not apply.

NOTE 2—When cut lengths are not to have grip or entry marks within the specified length, the purchaser should specify "gap entry marks outside specified length." When cut lengths may have grip or entry marks within the specified length, the purchaser should specify "grip or entry marks inside specified length."

Specified Length, mm		Width, mm	Allowances Over Specified Dimensions	
			Length, mm	
Over	Through		Specified "grip or entry marks outside specified length"	Specified "grip or entry marks inside specified length"
—	3000	20	100	75
3000	4000	25	100	75
4000	—	30	125	100

A 568M

TABLE 17 Thickness Tolerances of Cold-Rolled Sheet (Carbon and High-Strength Low-Alloy Steel)[A]
(Coils and Cut Lengths)

NOTE 1—Thickness is measured at any point across the width not less than 10.0 mm from a side edge.
NOTE 2—Widths up to and including 300 mm in this table apply to widths produced by slitting from wider sheet.

Specified Width, mm		Thickness Tolerances, Over and Under, mm, for Specified Nominal Thickness, mm				
Over	Through	Through 0.4	Over 0.4 to 1.0, incl	Over 1.0 to 1.2, incl	Over 1.2 to 2.5, incl	Over 2.5 to 4.0, incl
50	1800	0.05	0.08	0.10	0.12	0.15[B]
1800	2000	–	0.08	0.10	0.15	0.18
2000	–	–	0.15	0.15	0.18	0.20

[A] 0.50 mm minimum thickness for high-strength low-alloy.
[B] Not applicable to widths under 300 mm.

TABLE 18 Thickness Tolerances of Cold-Rolled Sheet (Carbon and High-Strength Low-Alloy Steel)[A]

NOTE 1—Thickness is measured at any point across the width not less than 10.0 mm from a side edge.
NOTE 2—Widths up to and including 300 mm in this table apply to widths produced by slitting from wider sheet.

Specified Width, mm		Thickness Tolerances, Over Only, mm, for Specified Minimum Thickness, mm				
Over	Through	Through 0.4	Over 0.4 to 1.0, incl	Over 1.0 to 1.2, incl	Over 1.2 to 2.5, incl	Over 2.5 to 4.0, incl
50	1800	0.10	0.15	0.20	0.25	0.30[B]
1800	2000	–	0.15	0.20	0.30	0.35
2000	–	–	0.30	0.30	0.35	0.40

[A] 0.50 mm minimum thickness for high-strength low-alloy.
[B] Not applicable to widths under 300 mm.

A 568M

TABLE 19 Length Tolerances of Cold-Rolled Sheet (Carbon and High-Strength Low-Alloy Steel)

(Cut Lengths Over 300 mm in Width, Not Resquared)

Specified Length, mm		Tolerance Over Specified Length (No Tolerance Under), mm
Over	Through	
300	1500	6
1500	3000	20
3000	6000	35
6000	–	45

TABLE 20 Length Tolerances of Cold-Rolled Sheet (Carbon and High-Strength Low-Alloy Steel)

(Cut Length Sheets, 50 to 300 mm in Width and 0.35 to 2.0 mm in Thickness, Not Resquared)

NOTE – This table applies to widths produced by slitting from wider sheet.

Specified Length, mm		Tolerances Over Specified Length (No Tolerance Under), mm
Over	Through	
600	1500	15
1500	3000	20
3000	6000	25

TABLE 21 Width Tolerances for Cold-Rolled Sheet (Carbon and High-Strength Low-Alloy Steel)

(Coils and Cut Lengths 50 to 300 mm in Width, Not Resquared, and 0.35[A] to 2.0 mm in Thickness)

NOTE – This table applies to widths produced by slitting from wider sheet.

Specified Width, mm		Width Tolerance, Over and Under, mm
Over	Through	
50	100	0.3
100	200	0.4
200	300	0.8

[A] 0.50 mm minimum thickness for high-strength low-alloy.

TABLE 22 Camber Tolerances of Cold-Rolled Sheet in Coils (Carbon and High-Strength Low-Alloy Steel 0.35[A] to 2.0 mm in Thickness)

NOTE – This table applies to widths produced by slitting from wider sheet.

Width, mm	Camber Tolerances
50 through 300, incl	5.0 mm in any 2000 mm

[A] 0.50 mm minimum thickness for high-strength low-alloy.

TABLE 23 Flatness Tolerances of Cold-Rolled Sheet (Carbon and High-Strength Low-Alloy Steel)

(Cut Lengths Over 300 mm in Width, Not Specified to Stretcher-Leveled Standard or Flatness)

NOTE 1 – This table does not apply when product is ordered full hard, to a hardness range, "annealed last" (dead soft), or as Class 2.

NOTE 2 – This table also applies to lengths cut from coils by the consumer when adequate flattening measures are performed.

Specified Thickness, mm	Specified Width, mm		Flatness Tolerance[A], mm	
	Over	Through	Carbon Steel	High-Strength, Low-Alloy[B] Steel
Through 1.0	–	900	10	20
	900	1500	15	30
	1500		20	40
Over 1.0	–	900	8	20
	900	1500	10	20
	1500	1800	15	30
	1800	–	20	40

[A] Maximum deviation from a horizontal flat surface.
[B] Tolerances for high-strength, low-alloy steel with specified minimum yield point in excess of 345 MPa are subject to negotiation.

TABLE 24 Flatness Tolerances of Cold-Rolled Sheet (Carbon and High-Strength Low-Alloy Steel)

(Cut Lengths Specified to Stretcher-Leveled Standard of Flatness)

Specified Thickness, mm	Specified Width, mm	Specified Length, mm	Flatness Tolerance[A], mm	
			Carbon Steel	High-Strenth, Low-Alloy Steel[B]
0.35 to 0.8, incl	Through 900	Through 3000	8	10
	Wider or Longer		10	15
Over 0.8	Through 1200	Through 3000	5	5
	Wider or Longer		8	10

[A] Maximum deviation from a flat surface.
[B] Tolerances for high-strength, low-alloy steel with specified minimum yield point in excess of 345 MPa are subject to negotiation.

APPENDIXES

X1. HOT-ROLLED AND COLD-ROLLED RECOMMENDED THICKNESSES

X1.1 Table X1.1 is based on ANSI B32.3, Preferred Metric Sizes for Flat Metal Products.[7]

[7] Available from American National Standards Institute, Inc., 1430 Broadway, New York, N.Y. 10018.

TABLE X1.1 Recommended Dimensions — Steel Sheet

Recommended Thicknesses, mm		Recommended Widths, mm	Recommended Lengths, mm
0.35	2.0	300	2000
0.40	2.2	400	2500
0.45	2.4	500	3000
0.50	2.5	600	3500
0.55	2.6	800	4000
0.60	2.8	1000	4500
0.65	3.0	1200	5000
0.7	3.2	1500	6000
0.75	3.4	2000	8000
0.8	3.5	2500	10 000
0.85	3.6	3000	12 000
0.9	3.8	3500	14 000
0.95	4.0	4000	16 000
1.0	4.2	5000	18 000
1.05	4.5		
1.1	4.8		
1.2	5.0		
1.4	5.5		
1.6	6.0		
1.8			

A 568M

X2. STANDARD CHEMICAL RANGES AND LIMITS

X2.1 Standard chemical ranges and limits are prescribed for carbon steels in Table X2.1.

TABLE X2.1 Standard Chemical Ranges and Limits

NOTE—The carbon ranges shown in the column headed "Range" apply when the specified maximum limit for manganese does not exceed 1.00 %. When the maximum manganese limit exceeds 1.00 % add 0.01 to the carbon ranges shown below.

Element	Carbon Steels, Only, Cast or Heat Analysis		
	Minimum of Specified Element, %	Range	Lowest max
Carbon (see Note)	to 0.15, incl	0.05	0.08
	over 0.15 to 0.30, incl	0.06	
	over 0.30 to 0.40, incl	0.07	
	over 0.40 to 0.60, incl	0.08	
	over 0.60 to 0.80, incl	0.11	
	over 0.80 to 1.35, incl	0.14	
Manganese	to 0.50, incl	0.20	0.40
	over 0.50 to 1.15, incl	0.30	
	over 1.15 to 1.65, incl	0.35	
Phosphorus	to 0.08, incl	0.03	0.04[A]
	over 0.08 to 0.15, incl	0.05	
Sulfur	to 0.08, incl	0.03	0.05[A]
	over 0.08 to 0.15, incl	0.05	
	over 0.15 to 0.23, incl	0.07	
	over 0.23 to 0.33, incl	0.10	
Silicon	to 0.15, incl	0.08	0.10
	over 0.15 to 0.30, incl	0.15	
	over 0.30 to 0.60, incl	0.30	
Copper	When copper is required 0.20 min is commonly specified		

[A] Certain individual specifications provide for lower standard limits for phosphorus and sulfur.

A 568M

X3. CLASS 1 AND CLASS 2 REQUIREMENTS

X3.1 Comparative requirements of Class 1 and Class 2 are as prescribed in Table X3.1.

TABLE X3.1 Cold-Rolled Sheet Steel Class Comparison

	Class 1	Class 2
Major imperfections:		
Cut lengths	Mill rejects	Mill rejects
Coils	Purchaser accepts within the manufacturer's published standards (policy)	Purchaser accepts within the manufacturer's published standards (policy)
Minor imperfections:		
Cut lengths	Mill rejects repetitive imperfections. May contain random imperfections which the purchaser accepts within the manufacturer's published standards (policy)	Purchaser accepts all minor imperfections
Coils	Purchaser accepts within the manufacturer's published standards (policy)	Purchaser accepts all minor imperfections
Finish	Matte unless otherwise specified	Purchaser accepts all finishes
Special oils	May be specified	May not be specified
Thickness, width and length tolerances:		
Standard	Will be met	Will be met
Restricted	May be specified	May not be specified
Flatness tolerances:		
Standard	Will be met	Not guaranteed
Stretcher leveled	May be specified	May not be specified
Resquaring	May be specified	May not be specified
Coil wraps	Purchaser accepts within the manufacturer's published standards (policy)	Purchaser accepts all
Coil welds	Purchaser accepts within the manufacturer's published standards (policy)	Purchaser accepts within the manufacturer's published standards (policy)
Outside inspection	May be specified	May not be specified
Special testing	May be specified	May not be specified

X4. EFFECT OF AGING OF COLD-ROLLED CARBON STEEL SHEET ON DRAWING AND FORMING

X4.1 Commercial Quality and Drawing Quality

X4.1.1 Cold-rolled sheet is usually temper rolled after annealing. Maximum ductility and minimum hardness exist in cold-rolled sheet in the annealed or dead-soft conditions (not temper rolled, Class 2); however, such sheet is not suitable for most formations due to the fact that they will flute or stretcher strain. A certain amount of cold work (temper rolling) will prevent these conditions from occurring, but the effect of temper rolling is only temporary due to the phenomenon commonly known as aging.

X4.1.2 Effective roller leveling of temper-passed steel immediately prior to fabrication will minimize the tendency of the sheet to stretcher strain and to flute but such leveling will not restore softness and ductility. In fact, roller leveling further work hardens the sheet and hence further reduces ductility. Rotation of stock by fabricating the oldest material first is important. Stocking material for extended periods of time should be avoided.

X4.2 Special Killed, Drawing Quality

X4.2.1 Special killed steel is essentially nonaging. It will not stretcher strain or flute or suffer loss of ductility with elapsed time when suitably temper rolled (Class 1) after annealing. This quality should be ordered when stretcher strains or fluting during fabrication are objectionable and the user does not have roller leveling equipment. This quality also should be ordered when the user plans to store sheet over an extended period of time without experiencing adverse changes due to aging.

X4.2.2 Drawing quality, special killed steel has its maximum ductility in the annealed or dead-soft conditions (Class 2) but in this state it is subject to stretcher straining or fluting during fabrication.

X4.2.3 The superiority of special killed steel is not limited to the fact that it is essentially nonaging. It also has the exceptional ductility required for draws which cannot be made of drawing quality.

X5. PROCEDURE FOR DETERMINING BREAKAGE ALLOWANCE LEVELS (APPLICABLE TO CARBON STEEL SHEET ONLY)

X5.1 In spite of the many extra precautions exercised in making sheet for drawing purposes, certain manufacturing variables may be encountered, all beyond the manufacturer's reasonable control, which may contribute to breakage in fabrication and must be considered as part of the normal hazard of the purchaser's use. The manufacturer will undertake to establish with the purchaser's concurrence a breakage allowance level.

X5.2 Breakage, for the purpose of this procedure, is defined as unrepairable parts, broken during drawing and classed as scrap. Parts showing laminations, resulting from pipe, may be excluded provided they are separately identified. Broken parts that can be salvaged are not covered in this procedure.

X5.3 this procedure is intended to establish a breakage allowance without the need for reinspection of each broken stamping. It will apply to overall breakage on a given part (as calculated by the method outlined below) in excess of 1 % up to and including 8 %. Inherent variations in steel sheet and normal variables in the stamping operation preclude 100 % satisfactory performance. Therefore, it is accepted that practical perfection is attained when 99 % of the stampings are produced without breakage. When the overall breakage is in excess of 8 %, it is considered to be the result of abnormal stamping conditions, and this method does not apply.

X5.4 When there are two or more suppliers, the recommended procedure for determining a breakage allowance for an identified part is based on the average percentage of breakage of at least 75 % of the blanks run on that part, on one set of dies, during at least one month (3000 piece minimum). The total production of all suppliers used to obtain this 75 % minimum is to be included in the calculation starting with the best performance. The average breakage thus determined shall be considered the allowance for the part.

X5.4.1 *Example:*

Vendor	Parts Produced	Parts Scrap	% Scrap
A	32 466	630	1.94
B	27 856	579	2.08
C	67 120	1477	2.20
D	56 200	1349	2.40
E	40 900	1125	2.75
F	850	60	7.05
All	225 392 total	5220 total	2.32 avg

X5.4.2 Seventy-five percent of 225 392 equals 169 044; therefore, it is necessary to include the total production of vendors A, B, C, and D (A + B + C + D = total production of 183 642 parts) since the total of A, B, and C is only 127 442, which is less than 75 % of the total. Total production of 183 642 parts (A + B + C + D) with 4035 parts being rejected, results in a percentage allowance of 2.20 %. On this basis, vendors D, E, and F exceed the allowance.

X5.5 When there is only one supplier in any one month, the recommended procedure for determining a breakage allowance for an identified part is based on the average percentage of breakage on that part, one set of dies, during at least two consecutive months (5000 piece minimum). This applies whether the supplier in the two consecutive months is the same or a different one. The average breakage thus determined shall be considered the allowance for the part.

X5.6 Individual lifts or coils that exhibit unreasonably high or unusually variable breakage will not be considered in determining the allowance. Such material should be set aside and the supplier notified.

X6. PROCEDURES FOR DETERMINING THE EXTENT OF PLASTIC DEFORMATION ENCOUNTERED IN FORMING OR DRAWING

X6.1 Introduction

X6.1.1 The preferred method for determining plastic strain is the circle grid and forming limit curve. The scribed square and change in thickness methods may also be used to evaluate deformation during the forming of a flat sheet into the desired shape.

X6.2 Circle Grid Method

X6.2.1 The test system employs photographic or electrochemically etched circle patterns on the surface of a sheet metal blank of known "quality" and a forming limit curve for the evaluation of strains developed by forming in press operations. It is useful in the laboratory and in the press room. Selection from the various steels that are commercially available can be done effectively by employing this technique. In addition, corrective action in die or part design to improve performance is indicated.

X6.2.2 The forming limit curve in Fig. X6.1 has been developed from actual measurements of the major (e_1) and associated minor (e_2) strains found in critical areas of production type stampings. Strain combinations that locate below this curve are safe, while those that fall above the curve are critical. For analysis of metal strain on production stampings, one must recognize that day to day variations of material, lubrication, and die settings will effect the strain level. To ensure trouble free press performance a zone below the forming limit curve bounded by the dashed and solid lines is designated as the "safety band." Therefore, strain combinations falling below the dashed lines should not exceed the forming limit curve in normal production operations. The left of zero portion of the curve defines the limiting biaxial tension-compression strain combination while the right side defines the forming limit curve represented all qualities of Because the production stampings used to develop the forming limit curve represented all qualities of low carbon light gage sheet steel, this single forming limit curve can be used successfully for these products.

X6.2.3 The circle grid method can also be used for other low carbon sheet categories if the following adjustments to the forming limit curve are made.

X6.2.3.1 *Material Gage* — As the metal gage increases the forming limit curve shifts upwards in a parallel manner, 0.2 % (e_1) strain for each 0.025-mm increase in metal thickness above 0.75 mm.

X6.2.3.2 *Material Properties* — When material properties are considerably different from that of conventional low-carbon sheet steel (for example, higher strength-low ductility), the forming limit curve is lower. The magnitude of the downgrade displacement is specific to each material; therefore, current material information should be consulted to determine placement of the forming limit curve.

X6.3 Procedure

X6.3.1 Obtain a sheet sample of "known quality," the sheet quality being established by either supplier designation, consumer purchase order, or most preferred tensile data obtained from a companion sheet sample.

X6.3.2 Obtain or prepare a negative on stencil with selected circles in a uniform pattern. The circles may be 2.5 to 25.0 mm in diameter; the most convenient diameter is 5.0 mm because it is easy to read and the gage spacing is short enough to show the maximum strain in a specific location on the part.

X6.3.3 The sheet metal blanks should be cleaned to remove excess oil and dirt; however, some precoated sheets can be etched without removing the coating. The area(s) to be etched should be determined from observation of panels previously formed; generally, the area that has a split problem is selected for etching. Normally, the convex side of the radius is gridded. If sufficient time is available, the entire blank may be etched, since valuable information can be obtained about the movement of metal in stamping a part when strains can be evaluated in what may appear to be noncritical areas. Additionally, for complex shapes it may be desirable to etch both surfaces of blanks so that the strains that occur in reverse draws can be determined.

X6.3.4 The sheet metal blanks may be etched by a photographic or electrochemical method. In the former method a photosensitive solution, for example, 50 % Kodak Photo Resist (KPR) emulsion and 50 % KPR-thinner, is sprayed onto the sheet. The emulsion is dried by baking the sheet at 65°C for 15 min or by just standing it for several hours at room temperature in a dark room. The latter should be employed in materials that age and, hence, become stronger when baked at 65°C. The negative is placed on the emulsion, held intimately in contact with the sheet, and exposed to a strong ultraviolet light source for 1 to 1½ min. The sheet is developed for 30 to 45 s in KPR developer, rinsed with water and sprayed with alcohol to set the resist. It is again rinsed with water and then sprayed with KPR black dye to reveal the etched circles.

X6.3.5 In the electrochemical method, the etch pad is saturated with an appropriate electrolyte. Various electrolytes are available from suppliers of the etching equipment. Some electrolytes are more effective than others for etching certain surfaces, such as terne plate and other metallic coated steels. A rust-inhibiting solution is preferred for steel sheets.

X6.3.6 A ground clamp from the transformer of suitable amperage (10 to 50 A is usually used) is fastened to the blank and the second lead is attached to the etch pad. Although the current may be turned on at this time, caution should be taken not to lay the pad on the sheet blank as it will arc. It is advisable to refrain from touching the metal of the etch pad and the grounded sheet blank.

X6.3.7 The stencil is placed with the plastic coating against the sheet surface in the area to be etched. Wetting the stencil with a minimum amount of electrolyte will assist in smoothing out the wrinkles and gives a more uniform etch. The etch pad is now positioned on the stencil and the current turned on, if it is not already on. Apply suitable pressure to the pad. Only the minimum time necessary to produce a clear etched pattern should be used. The etching time will vary with the amperage available from the power source and the stencil area, as well as the pad area in contact with the stencil. Rocker-type etch pads give good prints and require less amperage than flat-surfaced pads. Excessive current causes stencil damage.

X6.3.8 The etching solution activates the surface of the metal and may cause rusting unless it is inhibited. After the desired area has been etched the blank should be wiped or rinsed, dried, and neutralized.

X6.3.9 The etched blank is now ready for forming. The lubricants and press conditions should simulate production situations. If a sequence of operations is used in forming a part, it is desirable to etch sufficient blanks so that each operation can be studied.

X6.4 Measurement of Strain After Forming

X6.4.1 After forming, the circles are generally distorted into elliptical shapes (Fig. X6.2). These ellipses have major and minor strain axes. The major strain (e_1) is always defined to be the direction in which the greatest positive strain has occurred without regard to original blank edges or the sheet rolling direction. The minor strain (e_2) is defined to be 90 deg to the major strain direction.

X6.4.2 There are several methods for determining the major and minor strains of the formed panel. Typical tools are a pair of dividers and a scale ruled in 0.5 mm. For sharp radii, a thin plastic scale that can follow the contour of the stamping can be used to determine the dimension of the ellipses. (Scales are available to read the percent strain directly.)

X6.5 Evaluation of Strain Measurements

X6.5.1 The e_1 strain is always positive while the e_2 strain may be zero, positive, or negative, as indicated on the forming limit curve chart (Fig X6.1). The maximum e_1 and associated e_2 values measured in critical areas on the formed part are plotted on the graph paper containing the forming limit curve by locating the point of intersection of the e_1, e_2 strains.

X6.5.2 If this point is on or below the "safety band" of the forming limit curve, the strain should not cause breakage. Points further below the curve indicate that a less ductile material of a lower grade may be applied. Points above the "safety band" show that fabrication has induced strains that could result in breakage. Therefore, in evaluation on stampings exhibiting these strains, efforts should be made to provide an e_1, e_2 strain combination that would lie on or below the "safety band" of the forming limit curve. A different e_1, e_2 strain combination can be obtained through changes of one or more of the forming variables such as die conditions, lubricants, blank size, thickness, or material grade. It should be noted at this time that these conclusions are derived from a reference base being the steel "quality" used to fabricate the grid stamping.

X6.5.3 When attempting to change the relationship of e_1 and e_2 strains, it should be noted that on the forming limit curve the most severe condition for a given e_1 strain is at 0 % e_2 strain. This means the metal works best when it is allowed to deform in two dimensions, e_1 and e_2, rather than being restricted in one dimension. A change in e_2 to decrease the severity can be made by changing one of the previously mentioned forming variables or the die design, for example, improving lubrication on the tension-tension side will increase e_2 and decrease the severity.

X6.5.4 In addition to the forming limit curve, the e_1, e_2 strain measurements may be used to evaluate the material requirements on the basis of strain gradients, as illustrated in Fig. X6.3, or by plotting contours of equivalent strain levels on the surface of the formed part. Even when the level of strain is relatively low, parts in which the e_1 strain is changing rapidly either in magnitude or direction over a short span on the surface may require more ductile grades of sheet metal, change in lubrication, or change in part design.

X6.6 Example of Major and Minor Strain Distribution

X6.6.1 A formed panel (Fig. X6.4) with a cross section as shown in Fig. X6.3 is used to illustrate major and minor strain combinations. A plot of the major strain distribution should be made by finding the ellipse with the largest major strain (circle 7) and measuring both the major and minor strains in the row of ellipses running in the direction of the major strain. The solid dots (Fig. X6.3) are the measured major strains for each ellipse. The Xs are the critical major strains as determined from the forming limit curve at the corresponding minor strain (intersection of the measured minor strain and the severity curve).

X6.6.2 Usually a single row of ellipses will suffice to determine the most severe strain distribution. The resulting strain distribution plot (Fig. 6.3) illustrates both severity of the strain compared to the critical strain limits and the concentration of strain in the stamping. Steep strain gradients should be avoided because they are inherent to fracture sites.

X6.7 Example for Reducing Splitting Tendency

X6.7.1 In an area such as that represented in Fig. X6.3, the splitting tendency can be reduced as follows:

X6.7.1.1 If the radius of the part in the region of circle 1 is increased, some strain can be induced to take place in this area which will allow the major strain in circle 7 to be reduced sufficiently to bring the strain combination below the critical limit. This course of action requires no building nor reshaping of the punch, only grinding in the radius.

X6.7.1.2 The total average major strain required to make this formation is only 17.5 %; yet in a 5.0-mm circle the strain is as high as 40 %. The strain distribution curve puts forth graphically the need to distribute the strain over the length of the line by some means as described above.

X6.7.1.3 Change in lubrication can also improve the strain distribution of a stamping. If the strain over the punch is critical, the amount of stretch (strain) required to make the shape can be reduced by allowing metal to flow in over the punch by decreasing the friction through the use of a more effective lubricant in the hold-down area.

X6.7.1.4 If the part is critical, a change in material may help. That is, a material having a better uniform elongation will distribute the strain more uniformly or a material having a higher "r" value will make it possible to "draw" in more metal from the hold-down area so that less stretch is necessary to form the part.

X6.8 Scribed Square Method

X6.8.1 The basic technique is to draw a panel from a blank which has been scribed both longitudinally and transversely with a series of parallel lines spaced at 25.0-mm intervals. The lines on the panel are measured after drawing and the stretch or draw calculated as the percent increase in area of a 25.0-mm square. This is a fairly simple procedure for panels having generous radii and fairly even stretch or draw. Many major panels fall in this category and in these instances it is quite easy to pick out the square area exhibiting the greatest increase.

X6.8.2 If the square or line to be measured is no longer a flat surface, place a narrow strip of masking (or other suitable tape) on the formed surface and mark the points which are to be measured. Remove the tape, place on a plane surface, and determine the distance between the points with a steel scale.

X6.8.3 There will be cases of minor increase in area with major elongation in the one direction. In these instances, the percent elongation should be recorded.

X6.9 Thickness Method

X6.9.1 There are instances when the maximum stretch is confined to an area smaller than 645 mm^2 or the shape of the square has been distorted irregularly, making measurement difficult and calculation inaccurate. When either of these conditions exists, an electronic thickness gage may be used at the area in question or this area may be sectioned and the decrease in metal thickness measured with a ball point micrometer. The increase in unit area can be calculated by dividing the original thickness by the final thickness.

X6.9.2 *Example* – Assuming the blank thickness to be 0.80 mm and the final thickness to be 0.60 mm, the increase in unit area would be a 25 % (0.80 − 0.60/0.80 × 100) increase.

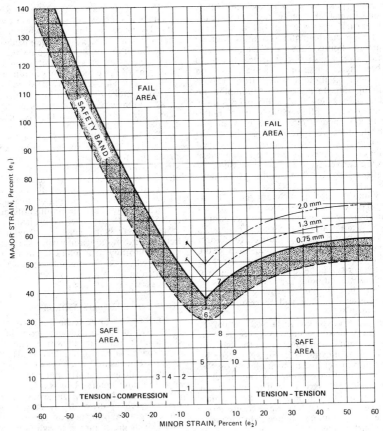

FIG. X6.1 Forming Limit Curve.

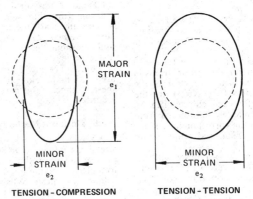

FIG. X6.2 Major and Minor Strain Axes.

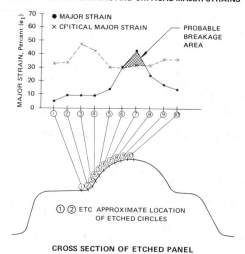

FIG. X6.3 Graph of Major Strains and Critical Major Strains and Cross Section of Etched Panel.

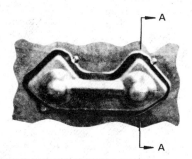

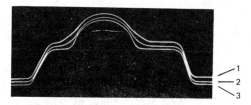

FIG. X6.4 Formed Panel and Cross Section.

The American Society for Testing and Materials takes no position respecting the validity of any patent rights asserted in connection with any item mentioned in this standard. Users of this standard are expressly advised that determination of the validity of any such patent rights, and the risk of infringement of such rights, are entirely their own responsibility.

This standard is subject to revision at any time by the responsible technical committee and must be reviewed every five years and if not revised, either reapproved or withdrawn. Your comments are invited either for revision of this standard or for additional standards and should be addressed to ASTM Headquarters. Your comments will receive careful consideration at a meeting of the responsible technical committee, which you may attend. If you feel that your comments have not received a fair hearing you should make your views known to the ASTM Committee on Standards, 1916 Race St., Philadelphia, Pa. 19103.

ASTM Designation: A 569 – 72 (Reapproved 1979)

Standard Specification for
STEEL, CARBON (0.15 MAXIMUM, PERCENT), HOT-ROLLED-SHEET AND STRIP, COMMERCIAL QUALITY[1]

This standard is issued under the fixed designation A 569; the number immediately following the designation indicates the year of original adoption or, in the case of revision, the year of last revision. A number in parentheses indicates the year of last reapproval. A superscript epsilon (ϵ) indicates an editorial change since the last revision or reapproval.

This specification has been approved for use by agencies of the Department of Defense and for listing in the DoD Index of Specifications and Standards.

1. Scope

1.1 This specification covers hot-rolled carbon steel sheet and strip of commercial quality, in coils and cut lengths, having a maximum carbon of 0.15 percent. This material is intended for parts where bending, moderate forming or drawing, and welding may be involved.

1.2 This specification is not applicable to the steel covered by Specification A 635.

1.3 The values stated in inch-pound units are to be regarded as the standard.

2. Applicable Documents

2.1 *ASTM Standards:*
A 370 Methods and Definitions for Mechanical Testing of Steel Products[2]
A 568 Specification for General Requirements for Steel, Carbon and High-Strength Low-Alloy Hot-Rolled Sheet and Cold-Rolled Sheet[2]
A 635 Specification for Hot-Rolled Carbon Steel Sheet and Strip, Commercial Quality, Heavy-Thickness Coils (Formerly Plate)[2]

3. General Requirements for Delivery

3.1 Material furnished under this specification shall conform to the applicable requirements of the current edition of Specification A 568, unless otherwise provided herein.

4. Ordering Information

4.1 Orders for material under this specification shall include the following information, as required, to describe the required material adequately:

4.1.1 ASTM designation number and date of issue,
4.1.2 Name of material (hot-rolled sheet or strip, commercial quality),
4.1.3 Copper-bearing steel (if required),
4.1.4 Condition:
4.1.4.1 Material to this specification is furnished in the hot-rolled condition. Pickled (or blast cleaned) should be specified, if required,
4.1.5 Type of edge (see 8.1),
4.1.6 Specify oiled or not oiled, as required (see 8.2),
4.1.7 Dimensions (thickness, width, and whether cut lengths or coils),
4.1.8 Coil size (must include inside diameter, outside diameter, and maximum weight),
4.1.9 Application (part identification and description),
4.1.10 Special requirements, if required, and
4.1.11 Cast or heat analysis report (request, if desired) (see 9.1).

NOTE—A typical ordering description is as follows: ASTM A 569 dated__, Hot-Rolled Sheet, Commercial Quality, Pickled and Oiled, Cut Edge, 0.075 by 36 by 96 in. for Part No. 6310, Shelf Bracket.

5. Manufacture

5.1 *Condition:*

---

[1] This specification is under the jurisdiction of ASTM Committee A-1 on Steel, Stainless Steel and Related Alloys, and is the direct responsibility of Subcommittee A01.19 on Steel Sheet and Strip.
Current edition approved May 1, 1972. Published June 1972. Originally published as A 569 – 66 T. Last previous edition A 569 – 66 T.
[2] *Annual Book of ASTM Standards*, Vol 01.03.

A 569

5.1.1 Unless otherwise specified, the material is furnished in the as-rolled condition (not annealed, not pickled).

6. Chemical Requirements

6.1 *The Cast or Heat (formerly Ladle) Analysis* of the steel shall conform to the requirements as to chemical composition shown in Table 1.

7. Bend Test

7.1 The material shall be capable of being bent at room temperature in any direction through 180 deg flat on itself without cracking on the outside of the bent portion (see Section 4 of Methods and Definitions A 370).

8. Finish and Appearance

8.1 *Edges:*

8.1.1 Sheet can be supplied with mill edge or cut edge.

8.1.2 Strip can be supplied with mill edge or slit (cut) edge.

8.2 *Oiling*—Hot-rolled, non-pickled material is commonly furnished not oiled while hot-rolled pickled (or blast cleaned) material is commonly furnished oiled. When required, pickled (or blast cleaned) material may be specified to be furnished not oiled, and nonpickled material may be specified to be furnished oiled.

9. Certification and Reports

9.1 When requested, the producer shall furnish copies of a report showing test results of the cast or heat analysis. The report shall include the purchase order number, the ASTM designation number, and the cast or heat number representing the material.

TABLE 1 Chemical Requirements

Element	Composition, percent
Carbon, max	0.15
Manganese, max	0.60
Phosphorus, max	0.035
Sulfur, max	0.040
Copper, when specified, min	0.20

The American Society for Testing and Materials takes no position respecting the validity of any patent rights asserted in connection with any item mentioned in this standard. Users of this standard are expressly advised that determination of the validity of any such patent rights, and the risk of infringement of such rights, are entirely their own responsibility.

This standard is subject to revision at any time by the responsible technical committee and must be reviewed every five years and if not revised, either reapproved or withdrawn. Your comments are invited either for revision of this standard or for additional standards and should be addressed to ASTM Headquarters. Your comments will receive careful consideration at a meeting of the responsible technical committee, which you may attend. If you feel that your comments have not received a fair hearing you should make your views known to the ASTM Committee on Standards, 1916 Race St., Philadelphia, Pa. 19103.

Designation: A 570 – 79

An American National Standard

Standard Specification for
HOT-ROLLED CARBON STEEL SHEET AND STRIP, STRUCTURAL QUALITY[1]

This standard is issued under the fixed designation A 570; the number immediately following the designation indicates the year of original adoption or, in the case of revision, the year of last revision. A number in parentheses indicates the year of last reapproval. A superscript epsilon (ε) indicates an editorial change since the last revision or reapproval.

This specification has been approved for use by agencies of the Department of Defense and for listing in the DoD Index of Specifications and Standards.

1. Scope

1.1 This specification covers hot-rolled carbon steel sheet and strip of structural quality in cut lengths or coils. This material is intended for structural purposes where mechanical test values are required, and is available in a maximum thickness of 0.2299 in. (5.8 mm) except as limited by Specification A 568.

1.1.2 The following grades are covered in this specification:

Mechanical Properties

Grade	Yield Point, min, ksi (MPa)	Tensile Strength, min, ksi (MPa)
30	30 (210)	49 (340)
33	33 (230)	52 (360)
36	36 (250)	53 (365)
40	40 (280)	55 (380)
45	45 (310)	60 (410)
50	50 (340)	65 (450)

1.2 The values stated in inch-pound units are to be regarded as the standard.

2. Applicable Documents

2.1 *ASTM Standard:*
A 568 Specification for General Requirements for Steel, Carbon and High-Strength Low-Alloy Hot-Rolled Sheet and Cold-Rolled Sheet[2]

3. General Requirements for Delivery

3.1 Material furnished under this specification shall conform to the applicable requirements of the current edition of Specification A 568, unless otherwise provided herein.

4. Ordering Information

4.1 Orders for material under this specification shall include the following information, as required, to describe the required material adequately:

4.1.1 ASTM specification number and date of issue, and grade,

4.1.2 Copper-bearing steel (if required),

4.1.3 Special requirements (if required),

4.1.4 Name of material (hot-rolled sheets or strip),

4.1.5 Condition (Material to this specification is furnished in the hot-rolled condition. Pickled (or blast cleaned) should be specified if required. Material so ordered will be oiled unless ordered dry),

4.1.6 Dimensions, including type of edges,

4.1.7 Coil size requirements, and

4.1.8 Cast or heat (formerly ladle) analysis or test report (request, if required).

NOTE—A typical ordering description is as follows: ASTM A 570, Grade 36, Hot-Rolled Sheets, 0.075 by 36 cut edge by 96 in.

5. Chemical Requirements

5.1 The cast or heat analysis of the steel shall conform to the requirements prescribed in Table 1.

6. Physical Requirements

6.1 *Tensile Properties*—The material a

---

[1] This specification is under the jurisdiction of ASTM Committee A-1 on Steel, Stainless Steel and Related Alloys, and is the direct responsibility of Subcommittee A01.19 on Sheet Steel and Steel Sheets.
Current edition approved Aug. 31, 1979. Published October 1979. Originally published as A 570 – 66 T. Last previous edition A 570 – 78.
[2] *Annual Book of ASTM Standards*, Vol 01.03.

A 570

represented by the test specimens shall conform to the requirements as to tensile properties prescribed in Table 2.

6.2 *Bending Properties* — The bend test specimens shall stand being bent at room temperature in any direction through 180 deg without cracking on the outside of the bent portion to an inside diameter which shall have a relation to the thickness of the specimen as prescribed in Table 3.

7. Test Specimens

7.1 Tension test specimens shall be taken longitudinally.

8. Number of Tests

8.1 Two tension tests and two bend tests shall be made from each heat or from each lot of 50 tons (45 Mg). When the amount of finished material from a heat or lot is less than 50 tons, only one tension test and one bend test shall be made. When material rolled from one heat differs 0.050 in. (1.27 mm) or more in thickness, one tension test and one bend test shall be made from both the thickest and thinnest material rolled regardless of the weight represented.

8.2 *Retests* — If one test fails, two more tests shall be run from the same lot, in which case both tests shall conform to the requirements prescribed in this specification; otherwise, the lot under test shall stand rejected.

9. Packaging

9.1 *Coil Size* — Small coils result from the cutting of full-size coils for center test purposes. These small coils are acceptable under this specification.

TABLE 1 Chemical Requirements

Element	Composition, %	
	Grades 30, 33, 36, and 40	Grades 45 and 50
Carbon, max	0.25	0.25
Manganese, max	0.90	1.35
Phosphorus, max	0.04	0.04
Sulfur, max	0.05	0.05
Copper, when copper is specified, min	0.20	0.20

TABLE 2 Tensile Requirements

	Grade 30	Grade 33	Grade 36	Grade 40	Grade 45	Grade 50
Tensile strength, min, ksi (MPa)	49 (340)	52 (360)	53 (365)	55 (380)	60 (410)	65 (450)
Yield point, min, ksi (MPa)	30 (210)	33 (230)	36 (250)	40 (280)	45 (310)	50 (340)
Elongation in 2 in. (50 mm), min, %, for thicknesses:						
0.2299 to 0.0972 in. (5.84 to 2.46 mm), incl	25.0	23.0	22.0	21.0	19.0	17.0
0.0971 to 0.0636 in. (2.45 to 1.62 mm), incl	24.0	22.0	21.0	20.0	18.0	16.0
0.0635 to 0.0255 in. (1.61 to 0.65 mm), incl	21.0	18.0	17.0	15.0	13.0	11.0
Elongation in 8 in. (200 mm), min, %, for thicknesses:						
0.2299 to 0.0972 in. (5.84 to 2.46 mm), incl	19.0	18.0	17.0	16.0	14.0	12.0
0.0971 to 0.0892 in. (2.45 to 2.26 mm), incl	17.0	16.0	15.0	14.0	12.0	10.0

TABLE 3 Bend Test Requirements

Grade	Ratio of Bend Diameter to Thickness of Specimen
30	1
33	1½
36	1½
40	2
45	2½
50	3

The American Society for Testing and Materials takes no position respecting the validity of any patent rights asserted in connection with any item mentioned in this standard. Users of this standard are expressly advised that determination of the validity of any such patent rights, and the risk of infringement of such rights, are entirely their own responsibility.

This standard is subject to revision at any time by the responsible technical committee and must be reviewed every five years and if not revised, either reapproved or withdrawn. Your comments are invited either for revision of this standard or for additional standards and should be addressed to ASTM Headquarters. Your comments will receive careful consideration at a meeting of the responsible technical committee, which you may attend. If you feel that your comments have not received a fair hearing you should make your views known to the ASTM Committee on Standards, 1916 Race St., Philadelphia, Pa. 19103.

ASTM Designation: A 580 – 83

Standard Specification for
STAINLESS AND HEAT-RESISTING STEEL WIRE[1]

This standard is issued under the fixed designation A 580; the number immediately following the designation indicates the year of original adoption or, in the case of revision, the year of last revision. A number in parentheses indicates the year of last reapproval. A superscript epsilon (ϵ) indicates an editorial change since the last revision or reapproval.

This specification has been approved for use by agencies of the Department of Defense and for listing in the DoD Index of Specifications and Standards.

1. Scope

1.1 This specification covers stainless and heat-resisting steel wire, except the free-machining types.[2] It includes round, square, octagon, hexagon, and shape wire in coils or in straightened-and-cut lengths in the more commonly used types of stainless steels for general corrosion resistance and high-temperature service.

1.2 The values stated in inch-pound units are to be regarded as the standard.

2. Applicable Documents

2.1 *ASTM Standards:*
A 370 Methods and Definitions for Mechanical Testing of Steel Products[3]
A 555 Specification for General Requirements for Stainless and Heat-Resisting Steel Wire[3]
A 751 Methods, Practices, and Definitions for Chemical Analysis of Steel Products[3]
E 527 Practice for Numbering Metals and Alloys (UNS)[3]

3. General Requirements for Delivery

3.1 In addition to the requirements of this specification, all requirements of the current edition of Specification A 555 shall apply.

4. Ordering Information

4.1 Orders for material to this specification shall include the following information:
  4.1.1 Quantity (weight),
  4.1.2 Name of material (stainless steel),
  4.1.3 Form (wire in straight lengths or coils),
  4.1.4 Condition (5.1),
  4.1.5 Finish (5.2),
  4.1.6 Applicable dimensions including size, thickness, width, and length or coil diameter (inside or outside diameter) and coil weights.
  4.1.7 Cross section (round, square, etc.),
  4.1.8 Type designation (Table 1),
  4.1.9 ASTM designation and date of issue, and
  4.1.10 Special requirements.

NOTE—A typical ordering description is as follows: 5000 lb (2268 kg) stainless steel wire, annealed and cold drawn, ½ in. (12.7 mm) round, 12-ft (3.66-m) lengths, Type 304, ASTM Specification A 580 dated _____. End use: machined hydraulic coupling parts.

5. Manufacture

5.1 *Condition*—Cold finished wire may be furnished in one of the conditions listed in Table 2.

5.2 *Finish:*
  5.2.1 Wire, cold finished to size, may be furnished with one of the following finishes:
    5.2.1.1 Cold drawn, (bright or matte finish—with or without drawing lubricant left on, as specified),

---

[1] This specification is under the jurisdiction of ASTM Committee A-1 on Steel, Stainless Steel and Related Alloys and is the direct responsibility of Subcommittee A01.08 on Wrought Stainless Steel Products.
Current edition approved July 29, 1983. Published October 1983. Originally published as A 580 – 67. Last previous edition A 580 – 82.
[2] For free-machining stainless wire designed especially for optimum machinability, see Specification A 581, for Free-Machining Stainless and Heat-Resisting Steel Wire, which appears in the *Annual Book of ASTM Standards*, Vol 01.03.
[3] *Annual Book of ASTM Standards*, Vol 01.03.

A 580

5.2.1.2 Centerless ground (round wire in straight lengths only), or

5.2.1.3 Centerless ground and polished (round wire in straight lengths only).

5.2.2 Wire annealed or heat treated and pickled as a final operation shall be furnished to the tolerances shown in Table 5 of Specification A 555.

6. Chemical Requirements

6.1 The steel shall conform to the requirements as to chemical composition specified in Table 1.

6.2 Methods and practices relating to chemical analysis required by this specification shall be in accordance with Methods, Practices, and Definitions A 751.

7. Mechanical Requirements

7.1 The material shall conform to the mechanical test requirements specified in Table 2.

8. Special Tests

8.1 If any special tests are required that are pertinent to the intended application of the material ordered, they shall be as agreed upon between the seller and the purchaser.

9. Permissible Variations in Dimensions

9.1 Total length tolerance is ¾ in. (19.05 mm) for straightened and cut wire unless subject to further operations or unless ordered as exact length resheared wire where tolerances in Table 4 of Specification A 555 apply.

The American Society for Testing and Materials takes no position respecting the validity of any patent rights asserted in connection with any item mentioned in this standard. Users of this standard are expressly advised that determination of the validity of any such patent rights, and the risk of infringement of such rights, are entirely their own responsibility.

This standard is subject to revision at any time by the responsible technical committee and must be reviewed every five years and if not revised, either reapproved or withdrawn. Your comments are invited either for revision of this standard or for additional standards and should be addressed to ASTM Headquarters. Your comments will receive careful consideration at a meeting of the responsible technical committee, which you may attend. If you feel that your comments have not received a fair hearing you should make your views known to the ASTM Committee on Standards, 1916 Race St., Philadelphia, Pa. 19103.

A 580

TABLE 1 Chemical Requirements

UNS Designation[A]	Type	Composition, %							
		Carbon, max[B]	Manganese, max[B]	Phosphorus, max	Sulfur, max	Silicon, max[B]	Chromium	Nickel	Other Elements
Austenitic Grades									
S20910	XM-19	0.06	4.00–6.00	0.040	0.030	1.00	20.50–23.50	11.50–13.70	Mo 1.50–3.00; N 0.20–0.40; Cb 0.10–0.30; V 0.10–0.30
S21400	XM-31	0.12	14.00–16.00	0.045	0.030	0.30–1.00	17.00–18.50	100 max	N 0.35 max
S21800	...	0.10	7.00–9.00	0.060	0.030	3.50–4.50	16.00–18.00	8.00–9.00	N 0.08–0.18
S21900	XM-10	0.08	8.00–10.00	0.060	0.030	1.00	19.00–21.50	5.50–7.50	N 0.15–0.40
S21904	XM-11	0.04	8.00–10.00	0.060	0.030	1.00	19.00–21.50	5.50–7.50	N 0.15–0.40
S24000	XM-29	0.08	11.50–14.50	0.060	0.030	1.00	17.00–19.00	2.25–3.75	N 0.20–0.40
S24100	XM-28	0.15	11.00–14.00	0.040	0.030	1.00	16.50–19.00	0.50–2.50	N 0.20–0.45
S28200	...	0.15	17.00–19.00	0.045	0.030	1.00	17.00–19.00	...	Mo 0.75–1.25; Cu 0.75–1.25; N 0.40–0.60
S30200	302	0.15	2.00	0.045	0.030	1.00	17.00–19.00	8.00–10.00	N 0.10 max
S30215	302B	0.15	2.00	0.045	0.030	2.00–3.00	17.00–19.00	8.00–10.00	
S30400	304	0.08	2.00	0.045	0.030	1.00	18.00–20.00	8.00–10.50	N 0.10 max
S30403	304L[C]	0.03	2.00	0.045	0.030	1.00	18.00–20.00	8.00–12.00	N 0.10 max
S30500	305	0.12	2.00	0.045	0.030	1.00	17.00–19.00	10.50–13.00	
S30800	308	0.08	2.00	0.045	0.030	1.00	19.00–21.00	10.00–12.00	
S30900	309	0.20	2.00	0.045	0.030	1.00	22.00–24.00	12.00–15.00	
S30908	309S	0.08	2.00	0.045	0.030	1.00	22.00–24.00	12.00–15.00	
S30940	309Cb	0.08	2.00	0.045	0.030	1.00	22.00–24.00	12.00–16.00	N 0.10 max; Cb + Ta 10 × C min, 1.10 max
S31000	310	0.25	2.00	0.045	0.030	1.50	24.00–26.00	19.00–22.00	
S31008	310S	0.08	2.00	0.045	0.030	1.50	24.00–26.00	19.00–22.00	
S31400	314	0.25	2.00	0.045	0.030	1.50–3.00	23.00–26.00	19.00–22.00	
S31600	316	0.08	2.00	0.045	0.030	1.00	16.00–18.00	10.00–14.00	Mo 2.00–3.00; N 0.10 max
S31603	316L[C]	0.03	2.00	0.045	0.030	1.00	16.00–18.00	10.00–14.00	Mo 2.00–3.00; N 0.10 max
S31700	317	0.08	2.00	0.045	0.030	1.00	18.00–20.00	11.00–15.00	Mo 3.00–4.00; N 0.10 max
S32100	321	0.08	2.00	0.045	0.030	1.00	17.00–19.00	9.00–12.00	Ti 5 × C min
S34700	347	0.08	2.00	0.045	0.030	1.00	17.00–19.00	9.00–13.00	Cb + Ta 10 × C min
S34800	348	0.08	2.00	0.045	0.030	1.00	17.00–19.00	9.00–13.00	Cb + Ta 10 × C min, Ta 0.10 max, Co 0.20 max
Ferritic Grades									
S40500	405	0.08	1.00	0.040	0.030	1.00	11.50–14.50	...	Al 0.10–0.30
S43000	430	0.12	1.00	0.040	0.030	1.00	16.00–18.00	...	
S44401	...	0.25	1.00	0.040	0.030	...	17.5–19.5	1.00	Mo 1.75–2.50; N 0.035 max (Ti + Nb) 0.20 + 4 (C + N) min 0.80 max
S44600	446	0.20	1.50	0.040	0.030	1.00	23.00–27.00	...	N 0.25 max
S44700	...	0.010	0.30	0.025	0.020	0.20	28.00–30.00	0.15 max	Mo 3.50–4.20 N 0.020 max; C + N 0.025 max; Cu 0.15 max
S44800	...	0.010	0.30	0.025	0.020	0.20	28.00–30.00	2.00–2.50	Mo 3.50–4.20; N 0.020 max; C + N 0.025 max; Cu 0.15 max
Martensitic Grades									
S40300	403	0.15	1.00	0.040	0.030	0.50	11.50–13.00	...	
S41000	410	0.15	1.00	0.040	0.030	1.00	11.50–13.50	...	
S41400	414	0.15	1.00	0.040	0.030	1.00	11.50–13.50	1.25–2.50	
S42000	420	over 0.15	1.00	0.040	0.030	1.00	12.00–14.00	...	
S43100	431	0.20	1.00	0.040	0.030	1.00	15.00–17.00	1.25–2.50	
S44002	440A	0.60–0.75	1.00	0.040	0.030	1.00	16.00–18.00	...	Mo 0.75 max
S44003	440B	0.75–0.95	1.00	0.040	0.030	1.00	16.00–18.00	...	Mo 0.75 max
S44004	440C	0.95–1.20	1.00	0.040	0.030	1.00	16.00–18.00	...	Mo 0.75 max

ASTM A 580

TABLE 1 *Continued*

[A] New designation established in accordance with ASTM E 527 and SAE J1086.
[B] Maximum, unless otherwise indicated.
[C] For some applications, the substitution of Type 304L for Type 304, or Type 316L for Type 316 may be undesirable because of design, fabrication, or service requirements. In such cases, the purchaser should so indicate on the order.

TABLE 2 Mechanical Test Requirements

Type	Condition (see 6.1)	Final Operation (see 6.2)	Tensile Strength,[A] min ksi	Tensile Strength,[A] min MPa	Yield Strength,[B] min ksi	Yield Strength,[B] min MPa	Elongation in Length 4 × Gage Diameter of Test Specimens,[C] min, %	Reduction of Area, min, %
302, 302B, 304, 304L, 305, 308, 309, 309Cb, 309S, 310, 310Cb, 310S, 314, 316, 316Cb, 316Ti, 316L, 317, 321, 347, and 348	A	cold finished annealed	90 75[D]	620 520[D]	45 30[D]	310 210[D]	30[E] 35[E]	40[E] 50[E]
302, 304, 316, and 317	B	cold finished	125	860	100	690	12	35
XM-10 and XM-11	A	cold finished annealed	90	620	50	340	45	60
XM-19	A	cold finished annealed	100	690	55	380	35	55
S21800	A	cold finished annealed	95	655	50	345	35	55
S28200	A	cold finished annealed	110	760	60	410	35	55
	B	cold finished	175	1370	150	1030	15	50
XM-28	A	cold finished annealed	100	690	55	380	30	50
XM-29	A	cold finished annealed	100	690	55	380	30	50
XM-31	A	cold finished annealed	130 100	900 690	85 50	590 340	24 40	60 65
XM-31	B	cold finished	220	1520	190	1310	5	50
403, 405,[F] 410, 430, 446, and S44401	A	cold finished annealed	70 70	480 480	40 40	280 280	16 20	45 45
403 and 410	T	cold finished	100	690	80	550	12	40
	H	cold finished	120	830	90	620	12	40
414[G]	A	cold finished	150 max	1030 max	...	...	...	...
420[G]	A	cold finished	125 max	860 max	...	...	...	...
431,[G] 440A,[G] 440B,[G] and 440C[G]	A	cold finished	140 max	970 max	...	...	...	...
S44700	A	cold finished annealed	75 70	520 480	60 55	415 380	15 20	30 40
S44800	A	cold finished annealed	75 70	520 480	60 55	415 380	15 20	30 40

[A] Minimum unless otherwise noted.
[B] Yield strength shall be determined by the 0.2 % offset method in accordance with ASTM Methods and Definitions A 370. An alternative method of determining yield strength, based on a total extension under load of 0.5 %, may be used.
[C] For wire products, it is generally necessary to use sub-size test specimens in accordance with Methods A 370.
[D] For Types 304L and 316L, the tensile strength shall be 70 000 psi (480 MPa) min and yield strength 25 000 psi (170 MPa) min.
[E] For material $\frac{5}{32}$ in. (3.96 mm) and under in size, the elongation and reduction in area shall be 25 % and 40 % respectively.
[F] Material shall be capable of being heat treated to a maximum hardness of HRC 25 when oil quenched from 1750°F (953°C).
[G] A section not exceeding ⅜ in. (9.52 mm) in thickness shall be capable of being heat treated by oil quenching from 1850 to 1950°F (1008 to 1063°C) to a minimum Rockwell C Hardness as follows: Type 414—45, Type 420—50, Type 431—45, Type 440A—54, Type 440B—56, Type 440C—59.

Designation: A 581 – 80

An American National Standard

Standard Specification for
FREE-MACHINING STAINLESS AND HEAT-RESISTING STEEL WIRE[1]

This standard is issued under the fixed designation A 581; the number immediately following the designation indicates the year of original adoption or, in the case of revision, the year of last revision. A number in parentheses indicates the year of last reapproval. A superscript epsilon (ε) indicates an editorial change since the last revision or reapproval.

This specification has been approved for use by agencies of the Department of Defense and for listing in the DoD Index of Specifications and Standards.

1. Scope

1.1 This specification covers cold-finished wire in coils or in straightened and cut lengths. It includes rounds, squares, and hexagons in the more commonly used types of stainless and heat-resisting free-machining steels designed especially for optimum machinability and for general corrosion and high temperature service. Stainless and heat-resisting steel wire other than the free-machining types are covered in a separate specification.[2]

1.2 The values stated in inch-pound units are to be regarded as the standard.

2. Applicable Documents

2.1 *ASTM Standards:*
A 555 Specification for General Requirements for Stainless and Heat-Resisting Steel Wire[3]
A 751 Methods, Practices, and Definitions for Chemical Analysis of Steel Products[3]

3. General Requirements for Delivery

3.1 In addition to the requirements of this specification, all requirements of the current edition of Specification A 555 shall apply.

4. Ordering Information

4.1 Orders for material under this specification shall include the following information:
4.1.1 Quantity (weight),
4.1.2 Name of material (stainless steel),
4.1.3 Form (wire in straight lengths or coils),
4.1.4 Condition (see Section 5),
4.1.5 Finish (see Section 15),
4.1.6 Applicable dimensions including size, thickness, width, and length or coil diameter (inside, or outside diameter), and coil weights,
4.1.7 Cross section (round, square, etc.),
4.1.8 Type designation (Table 1),
4.1.9 ASTM designation and date of issue, and
4.1.10 Exceptions to the specification or special requirements.

NOTE 1—A typical ordering description is as follows: 5000 lb (2268 kg) stainless steel wire, annealed and centerless ground, ¼ in. (6.35 mm) round, 10 to 12 ft (3.05 to 3.66 m) in length, Type 303, ASTM Specification A 581 dated ____. End use: machined valve parts.

5. Condition

5.1 Cold-finished wire may be furnished in one of the conditions listed in Table 2.

6. Chemical Requirements

6.1 The steel shall conform to the requirements as to chemical composition specified in Table 1.

6.2 Methods and practices relating to chem-

---

[1] This specification is under the jurisdiction of ASTM Committee A-1 on Steel, Stainless Steel and Related Alloys and is the direct responsibility of Subcommittee A01.08 on Wrought Stainless Steel Products.
Current edition approved Sept. 26, 1980. Published November 1980. Originally published as A 581 – 67. Last previous edition A 581 – 79.
[2] For wire other than those of the free-machining types, see ASTM Specification A 580, for Stainless and Heat-Resisting Steel Wire, which appears in the *Annual Book of ASTM Standards*, Vol 01.03.
[3] *Annual Book of ASTM Standards*, Vol 01.03.

A 581

ical analysis required by this specification shall be in accordance with Methods, Practices, and Definitions A 751.

7. Mechanical Requirements

7.1 The material shall conform to the mechanical test requirements specified in Table 3.

8. Lead Exudation Test

8.1 The lead exudation test applies to Type XM-3 only. The steel shall not show an exudation of lead greater than Standard 3 in Fig. 1. Samples shall be taken from semi-finished billets.

9. Test Method

9.1 The lead exudation specimens (Type XM-3 only) shall be lightly coated with a light oil over the fresh cut surface and placed in a furnace at 1290°F (679°C) and held for a period of 10 min/in. (25.4 mm) of thickness at this temperature. The specimens shall be removed and visually compared with Fig. 1.

10. Finish

10.1 Wire in the cold-finished condition, may be furnished with one of the following finishes:

10.1.1 Cold drawn,

10.1.2 Centerless ground (round wire in straight lengths only), or

10.1.3 Centerless ground and polished (round wire in straight lengths only).

10.2 Wire heat treated and pickled as a final operation shall be furnished to the tolerances shown in Table 5 of Specification A 555.

TABLE 1 Chemical Requirements

UNS Designation[A]	Type	Composition, %							
		Carbon, max	Manganese[B]	Phosphorus, max	Sulfur[B]	Silicon, max	Chromium	Nickel	Other Elements
S30300	303	0.15	2.00	0.20	0.15 min	1.00	17.00–19.00	8.00–10.00	...
S30323	303Se	0.15	2.00	0.20	0.06	1.00	17.00–19.00	8.00–10.00	Se 0.15 min
S41600	416	0.15	1.25	0.06	0.15 min	1.00	12.00–14.00	...	...
S41623	416Se	0.15	1.25	0.06	0.06	1.00	12.00–14.00	...	Se 0.15 min
S43020	430F	0.12	1.25	0.06	0.15 min	1.00	16.00–18.00	...	...
S43023	430FSe	0.12	1.25	0.06	0.06	1.00	16.00–18.00	...	Se 0.15 min
S20300	XM-1	0.08	5.00–6.50	0.04	0.18–0.35	1.00	16.00–18.00	5.00–6.50	Cu 1.75–2.25
S30345	XM-2	0.15	2.00	0.05	0.11–0.16	1.00	17.00–19.00	8.00–10.00	Mo 0.40–0.60 Al 0.60–1.00
S30360	XM-3	0.15	2.00	0.04	0.12–0.25	1.00	17.00–19.00	8.00–10.00	Pb 0.12–0.30
S30310	XM-5	0.15	2.50–4.50	0.20	0.25 min	1.00	17.00–19.00	7.00–10.00	...
S41610	XM-6	0.15	1.50–2.50	0.06	0.15 min	1.00	12.00–14.00	...	...
S18200	XM-34	0.08	2.50	0.04	0.15 min	1.00	17.50–19.50	...	Mo 1.50–2.50

[A] New designation established in accordance with ASTM E 527 and SAE J1086, Recommended Practice for Numbering Metals and Alloys (UNS).
[B] Maximum unless otherwise noted.

TABLE 2 Condition

Type	Condition A (Annealed)	Condition B[A] (Cold Worked)	Condition T (Intermediate Temper)	Condition H (Hard Temper)
303	A	B	...	...
303Se	A	B	...	...
416	A	...	T	H
416Se	A	...	T	H
430F	A	...	...	...
430FSe	A	...	...	...
XM-1	A	B	...	...
XM-2	A	B	...	...
XM-3	A	B	...	...
XM-5	A	B	...	...
XM-6	A	...	T	H
XM-34	A	...	...	...

[A] Condition B applies only to wire annealed and cold worked to produce high strength in chromium-nickel types not hardenable by heat treatment.

TABLE 3 Mechanical Test Requirements

Type	Condition (see Section 5)	Tensile Strength	
		ksi	MPa
All	A	85 to 125	590 to 860
303, 303Se, XM-1, XM-2, XM-3, and XM-5	B[A]	115 to 145	790 to 1000
416, 416Se, and XM-6	T	115 to 145	790 to 1000
416, 416Se, and XM-6	H	140 to 175	1000 to 1210

[A] Condition B applies only to wire annealed and cold worked to produce high strength in chromium-nickel types not hardenable by heat treatment.

FIG. 1 Lead Exudation, 303 Pb Basis.

APPENDIX

X1. CROSS REFERENCE

This table is intended to assist the user when Specification A 581 is referenced in a Government procurement. It shows the types of steels in Specification A 581 replacing the steels formerly specified in MIL-W-52263C(MR).

TABLE X1 Cross Reference

UNS Designation[A]	MIL-W-52263C(MR)	Specification A 581
S20300	203 EZ	XM-1
S30300	303	303
S30323	303 Se	303 Se
S30345	303 Ma	XM-2
S30360	303 Pb	XM-3
...	303 Cu[B]	...
S30310	303 plus X	XM-5
S41600	416	416
S41623	416 Se	416 Se
S41610	416 plus X	XM-6
S43020	430 F	430 F
S43023	430 Se	430 F Se

[A] New designation established in accordance with ASTM E 527 and SAE J1086, Recommended Practice for Numbering Metals and Alloys (UNS).
[B] Material no longer produced.

The American Society for Testing and Materials takes no position respecting the validity of any patent rights asserted in connection with any item mentioned in this standard. Users of this standard are expressly advised that determination of the validity of any such patent rights, and the risk of infringement of such rights, are entirely their own responsibility.

This standard is subject to revision at any time by the responsible technical committee and must be reviewed every five years and if not revised, either reapproved or withdrawn. Your comments are invited either for revision of this standard or for additional standards and should be addressed to ASTM Headquarters. Your comments will receive careful consideration at a meeting of the responsible technical committee, which you may attend. If you feel that your comments have not received a fair hearing you should make your views known to the ASTM Committee on Standards, 1916 Race St., Philadelphia, Pa. 19103.

Designation: A 604 – 77 (Reapproved 1982)

Standard Method for
MACROETCH TESTING OF CONSUMABLE ELECTRODE REMELTED STEEL BARS AND BILLETS[1]

This standard is issued under the fixed designation A 604; the number immediately following the designation indicates the year of original adoption or, in the case of revision, the year of last revision. A number in parentheses indicates the year of last reapproval. A superscript epsilon (ϵ) indicates an editorial change since the last revision or reapproval.

1. Scope

1.1 This method[2] of testing and inspection is applicable to bars, billets, and blooms of carbon, alloy, and stainless steel which have been consumable electrode remelted.

1.2 For the purpose of this specification, the consumable electrode remelting process is defined as a steel refining method wherein single or multiple electrodes are remelted into a crucible producing an ingot which is superior to the original electrode by virtue of improved cleanliness or lower gas content or reduced chemical or nonmetallic segregation. See Appendixes X1 and X2 for descriptions of applicable remelting processes.

1.3 This method and the accompanying comparison macrographs[3] are generally applicable to steel bar and billet sizes up to 225 in.^{2} in transverse cross section.

2. Description of Macroetch Testing

2.1 Macroetch testing, as described herein, is a method for examining and rating transverse sections of bars and billets to describe certain conditions of macro segregation which are often characteristic of consumable electrode remelted materials. This method is not intended to define major defects such as those described by ASTM Method E 381, Macroetch Testing, Inspection, and Rating Steel Products, Comprising Bars, Billets, Blooms, and Forgings[4].

2.2 This method employs the action of an acid or other corrosive agent to develop the characteristics of a suitably prepared specimen. After etching, the sections are compared visually, or at a very low magnification, if necessary for clarification of conditions, to standard plates describing the various conditions which may be found. Materials react differently to etching reagents because of variations in chemical composition, method of manufacture, heat treatment, and many other variables.

3. Application

3.1 When material is furnished subject to macroetch testing and inspection under this standard, the manufacturer and purchaser should be in agreement concerning the following:

3.1.1 The stage of manufacture at which the test shall be conducted,

3.1.2 The number and location of the sections to be tested,

3.1.3 The condition and preparation of the surface to be macroetched,

3.1.4 The etching reagent, temperature and time of etching, or degree of etching including any special techniques which must be used, and

3.1.5 The type and degree of conditions or combinations thereof that shall be considered acceptable or subject to metallurgical review.

---

[1] This method is under the jurisdiction of ASTM Committee A-1 on Steel, Stainless Steel and Related Alloys, and is the direct responsibility of Subcommittee A01.06 on Steel Forgings and Billets.
Current edition approved June 24, 1977. Published August 1977. Originally published as A 604 – 70. Last previous edition A 604 – 76.

[2] ASTM Committee A-1 gratefully acknowledges the help of the AISI Committee on General Metallurgy in preparing the appendix, assembling the macroetch photographs, and assisting with the text of this method.

[3] A complete set of the 20 macrographs on glossy paper is available at nominal cost from ASTM Headquarters. Request Adjunct No. 12-106040-00.

[4] *Annual Book of ASTM Standards*, Vol 03.03.

4. Sample Preparation

4.1 Unless otherwise specified, the test shall be performed on specimens, usually $1/2$ to 1 in. thick, cut to reveal a transverse surface.

4.2 Disks for macroetch inspection may be removed from billets by a variety of methods including torch cutting, sawing, machining, or high-speed abrasive wheels. Adequate preparation of the surface for macroetching must completely remove the effects of torch cutting or high-speed abrasive wheels.

4.3 Due to the nature of the conditions to be detected, further surface preparation is usually required.

4.4 When such further surface preparation is performed, grinding, machining, or sanding should be carried out in such a manner as not to mask the structure.

4.5 The surface of the disk to be etched must be free of dirt, grease, or other foreign material which might impair the result of the test.

5. Etching Reagents

5.1 The etching response and appearance is dependent upon the type and temperature of the etching reagent and the time of immersion. These details must be established by agreement between manufacturer and purchaser.

5.2 For illustrative purposes some of the commonly used etching reagents are as follows:

5.2.1 *Hydrochloric Acid*—A solution of 1 part commercial concentrated hydrochloric acid (HCl, sp gr 1.19) and 1 part water is more generally used than any other macroetching reagent. This solution may be heated without significant change in concentration, and may be reused if it has not become excessively contaminated or weakened. Etching is generally done with the solution at a temperature of approximately 160°F.

5.2.2 *Hydrochloric Acid-Sulfuric Acid Mixture*—A mixture containing 50 percent water, 38 percent commercial concentrated HCl, and 12 percent commercial concentrated sulfuric acid (H_2SO_4, sp gr 1.84) is sometimes used in place of the previously mentioned 50 percent HCl solution. The statements in the previous paragraph regarding reuse and temperature of etchant are applicable to this reagent.

5.2.3 *Aqua Regia*—A solution consisting of 1 part concentrated nitric acid (HNO_3, sp gr 1.42) and 2 parts concentrated HCl is used on corrosion and heat-resistant materials of the 18 percent chromium, 8 percent nickel type and higher alloy types. This reagent is used at room temperature.

NOTE—The reagents in 5.2.1, 5.2.2, and 5.2.3 should be used under ventilating hoods or with some provision to remove the corrosive fumes.

5.2.4 *Nitric Acid*—This etchant consists of 5 percent HNO_3 solution in alcohol or water, and is generally used at room temperature. When this reagent is used, the etch disk must have a smooth surface.

6. Etching Containers

6.1 Macroetching must be done in containers that are resistant to attack from the etching reagents. Caution must be exerted to prevent the occurrence of electrolytic couples which can cause uneven attacks and misleading results.

7. Preparation of Etched Surface and Examination

7.1 Upon completion of etching, surfaces of disks should be cleaned by either chemical or mechanical methods that do not affect the macroetch quality. Care should be taken to prevent rusting of the etched surface.

8. Interpretation of Conditions Found by Macroetching

8.1 Four distinct classes of conditions are defined and described under this standard:

8.1.1 *Class 1: Freckles*—Circular or near-circular dark etching areas generally enriched with carbides and carbide-forming elements.

8.1.2 *Class 2: White Spots*—Light etching areas, having no definitive configuration or orientation which are generally reduced in carbide or carbide-forming elements.

8.1.3 *Class 3: Radial Segregation*—Radially or spirally oriented dark etching elongated areas occurring most frequently at mid-radius which are generally carbide enriched. This condition may be easily confused with freckles in some materials.

8.1.4 *Class 4: Ring Pattern*—One or more

concentric rings evidenced by a differential in etch texture associated with minor composition gradients and ingot solidification.

8.2 Macroetch photographs show examples of each of the conditions revealed by macroetch testing, with five degrees of severity, identified as A, B, C, D, and E for each condition. Degree A exhibits the minimum occurrence of each condition detectable by visual examination of the etched surface, while degrees B, C, D, and E represent increasing severity of occurrence.

8.3 For each condition, or combination of conditions, ratings shall be obtained by comparing each macroetched section with the standard photographs. Bar or billet sections to 225 in.^{2} cross-sectional area may be rated against these standards. Larger sizes may be rated by agreement between manufacturer and purchaser, but caution must be exercised in interpretation of such results.

8.4 If the appearance of a given condition does not exactly match one of the five standard photographs, it shall be assigned the rating of the standard that it most nearly matches.

8.5 No standards for acceptance are stated or implied in these illustrations. The extent to which each condition may be permissible varies with the intended application, and such standards should be stated in the applicable product specification, or may be the subject of negotiation between manufacturer and purchaser.

FIG. 1 Class 1—Freckles—Severity A

FIG. 2 Class 1—Freckles—Severity B

FIG. 3 Class 1—Freckles—Severity C

FIG. 4 Class 1—Freckles—Severity D

FIG. 5 Class 1—Freckles—Severity E

FIG. 6 Class 2—White Spots—Severity A

FIG. 7 Class 2—White Spots—Severity B

FIG. 8 Class 2—White Spots—Severity C

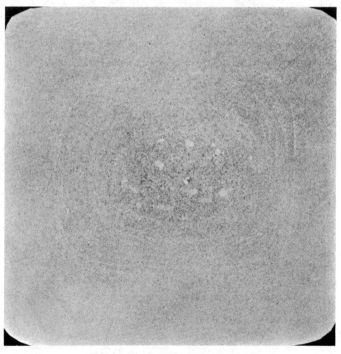

FIG. 9 Class 2—White Spots—Severity D

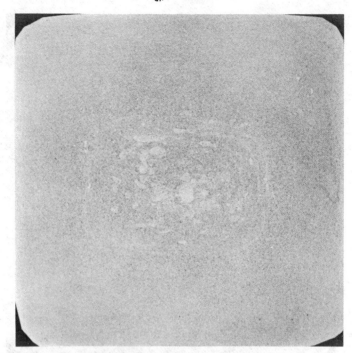

FIG. 10 Class 2—White Spots—Severity E

FIG. 11 Class 3—Radial Segregation—Severity A

FIG. 12 Class 3—Radial Segregation—Severity B

FIG. 13 Class 3—Radial Segregation—Severity C

FIG. 14 Class 3—Radial Segregation—Severity D

FIG. 15 Class 3—Radial Segregation—Severity E

FIG. 16 Class 4—Ring Pattern—Severity A

FIG. 17 Class 4—Ring Pattern—Severity B

FIG. 18 Class 4—Ring Pattern—Severity C

FIG. 19 Class 4—Ring Pattern—Severity D

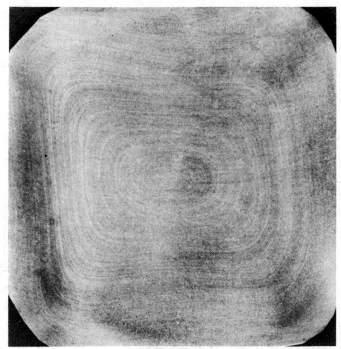

FIG. 20 Class 4—Ring Pattern—Severity E

APPENDIXES

X1. CONSUMABLE ELECTRODE VACUUM MELTING

X1.1 Process Description

X1.1.1 Within the last decade, consumable electrode vacuum melting (CEVM) of steel has grown from a laboratory process to a major production operation capable of producing ingots in certain grades up to 60 in. in diameter, weighing 50 tons. The available ingot sizes and weights vary from grade to grade, depending upon their complexity and alloy content. The total CEVM capacity in the country has multiplied several times during this period. Currently, a significant proportion of the ultra-high-strength steels for aircraft and missiles, bearing steels for aircraft engines, and other speciality alloys are being consumable electrode vacuum melted.

X1.1.2 The consumable electrode vacuum melting process is shown diagrammatically in Fig. X1.1. To start the melting operation, an electrode produced from conventional air-melted or vacuum-processed steel is suspended in the consumable electrode vacuum melting furnace. The system is evacuated and an arc is struck to a bottom starting pad. Molten metal is transferred across the arc from the electrode to the solidifying ingot contained within the water-cooled copper crucible. As melting proceeds and the ingot solidifies progressively upward, the electrode is fed downward to maintain the proper arc length. As the metal droplets pass through the arc, they are exposed to this vacuum at extremely high arc temperatures, producing extensive degassification, as well as some breakdown and dispersion of inclusions. Due to the rapid cooling provided by the copper crucible, only a portion of the ingot is molten at a time and solidification proceeds in a continuously progressive manner.

X1.2 Product Characteristics

X1.2.1 Essentially, the CEVM operation changes the properties of steel in three ways:
X1.2.1.1 By reducing gas content.
X1.2.1.2 By improving microcleanliness. The nonmetallic inclusion content is rated in a manner similar to that used for air melt except that the level is generally lower and a different chart is used.
X1.2.1.3 By changing the mode of solidification from that of the traditional static-cast ingot to a progressive solidification process, involving high heat input from an arc and rapid heat extraction by the water-cooled copper crucible.

X1.2.2 Depending upon the grade of steel and the application under consideration, consumable electrode vacuum melting is reported to significantly improve one or more of the following properties: transverse ductility in aircraft forging billets, fatigue strength or endurance limit, notched tensile strength or fracture toughness, Charpy V-notch impact strength, stress rupture, and creep strength. Furthermore, hot workability and yield of some grades are significantly improved. The CEVM process has also made possible the development of new alloys for extremely high-strength or high-temperature applications that did not exhibit satisfactory properties when melted by other methods.

X1.3 Macroetch Characteristics

X1.3.1 Consumable electrode vacuum-melted steels and alloys may contain discontinuities peculiar to this process which are disclosed upon macroetch examination.

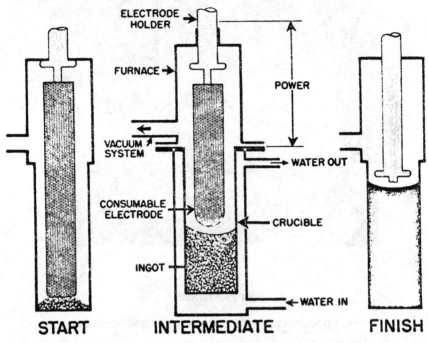

FIG. X1.1 Consumable Electrode Vacuum Melting Furnace.

X2. ELECTROSLAG REMELTING

X2.1 Process Description

X2.1.1 Electroslag remelting (ESR) was first introduced in an American patent by Hopkins, but most of the published work has been done by Russian engineers. The process has been shown to reduce inclusions, similar to vacuum-arc remelting, with the additional benefit of reducing sulfur content in critical alloys for aerospace and nuclear applications. Many variations of processing parameters, equipment design, ingot sizes and shapes are used.

X2.1.2 The ESR process consists of remelting a consumable electrode through a bath of molten slag using the electrical resistance of the slag to provide the required heat input. The slag composition will vary with the type of alloy and the processor's objectives. Single or three-phase a-c or d-c current may be applied. Water-cooled ingot molds may be square, round, or designed to produce rough tube rounds. Stationary molds or molds that can be raised as the ingot solidifies are used. Starting the process may be done with cold slag and starter chips or molten slag prepared in a small arc furnace. A diagram of this process is shown in Fig. X2.1.

X2.2 Product Characteristics

X2.2.1 The ingot surface is protected by a film of slag that solidifies on the ingot as it cools, providing an improved surface.

X2.2.2 Sulfur content may be reduced substantially to improve workability.

X2.2.3 A significant drop in oxide inclusions may be obtained.

X2.2.4 Improved uniformity occurs due to the solidification process.

X2.2.5 Little, if any, loss in alloying elements occurs with appropriate processing parameters.

X2.3 Macroetch Characteristics

X2.3.1 ESR steels and alloys may contain discontinuities peculiar to this process which are disclosed upon macroetch examination.

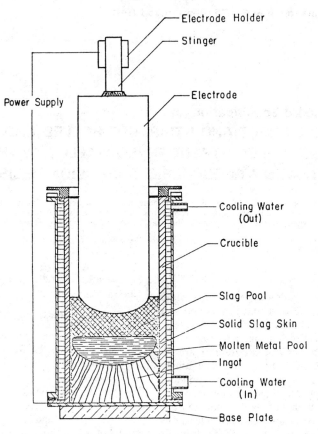

FIG. X2.1 Schematic of ESR Melting Process.

The American Society for Testing and Materials takes no position respecting the validity of any patent rights asserted in connection with any item mentioned in this standard. Users of this standard are expressly advised that determination of the validity of any such patent rights, and the risk of infringement of such rights, are entirely their own responsibility.

This standard is subject to revision at any time by the responsible technical committee and must be reviewed every five years and if not revised, either reapproved or withdrawn. Your comments are invited either for revision of this standard or for additional standards and should be addressed to ASTM Headquarters. Your comments will receive careful consideration at a meeting of the responsible technical committee, which you may attend. If you feel that your comments have not received a fair hearing you should make your views known to the ASTM Committee on Standards, 1916 Race St., Philadelphia, Pa. 19103.

Designation: A 606 – 75 (Reapproved 1981)

Standard Specification for
STEEL SHEET AND STRIP, HOT-ROLLED AND COLD-ROLLED, HIGH-STRENGTH, LOW-ALLOY, WITH IMPROVED ATMOSPHERIC CORROSION RESISTANCE[1]

This standard is issued under the fixed designation A 606; the number immediately following the designation indicates the year of original adoption or, in the case of revision, the year of last revision. A number in parentheses indicates the year of last reapproval. A superscript epsilon (ε) indicates an editorial change since the last revision or reapproval.

This specification has been approved for use by agencies of the Department of Defense and for listing in the DoD Index of Specifications and Standards.

1. Scope

1.1 This specification covers high-strength, low-alloy, hot- and cold-rolled sheet and strip in cut lengths or coils, intended for use in structural and miscellaneous purposes, where savings in weight or added durability are important. These steels have enhanced atmospheric corrosion resistance and are supplied in two types: Type 2 having corrosion resistance at least two times that of plain carbon steel and Type 4 having corrosion resistance at least four times that of plain carbon steel.[2] The degree of corrosion resistance is based on data acceptable to the consumer.

1.2 The values stated in inch-pound units are to be regarded as the standard.

2. Applicable Documents

2.1 *ASTM Standards:*
A 109 Specification for Steel, Carbon, Cold-Rolled Strip[3] (Dimensional Tolerance Tables 5, 6, 7, 8, and 9)
A 370 Methods and Definitions for Mechanical Testing of Steel Products[3]
A 568 Specification for General Requirements for Steel, Carbon and High-Strength Low-Alloy Hot-Rolled Sheet, and Cold-Rolled Sheet[3]

3. General Requirements for Delivery

3.1 Material furnished under this specification shall conform to the applicable requirements of the current edition of Specification A 568 and the dimensional tolerance tables of Specification A 109, unless otherwise provided herein.

4. Ordering Information

4.1 Orders for material under this specification shall include the following information, as required, to describe adequately the desired material:

4.1.1 ASTM specification number and date of issue, and type,

4.1.2 Name of material (high-strength low-alloy hot-rolled sheet or strip or high-strength low-alloy cold-rolled sheet or strip),

4.1.3 Condition (specify oiled or dry, as required),

4.1.4 Edges (must be specified for hot-rolled sheet or strip) (see 9.1),

4.1.5 Finish—Cold-rolled only (indicate exposed (E) or unexposed (U). Matte (dull) finish will be supplied unless otherwise specified),

---

[1] This specification is under the jurisdiction of ASTM Committee A-1 on Steel, Stainless Steel and Related Alloys, and is the direct responsibility of Subcommittee A01.19 on Steel Sheet and Strip.
Current edition approved July 25, 1975. Published October 1975. Originally published as A 606 – 70. Last previous edition A 606 – 71.
[2] Type 2 cold-rolled material is intended to replace ASTM Specification A 374, for High-Strength Low-Alloy Cold-Rolled Steel Sheets and Strip, and Type 2 hot-rolled material is intended to replace ASTM Specification A 375, for High-Strength Low-Alloy Hot-Rolled Steel Sheets and Strip, which appear in the 1971 *Annual Book of ASTM Standards*, Part 3.
[3] *Annual Book of ASTM Standards*, Vol 01.03.

4.1.6 Dimensions (thickness, width, and whether cut lengths or coils),

4.1.7 Coil size (must include inside diameter, outside diameter, and maximum weight),

4.1.8 Application (show part identification and description),

4.1.9 Special requirements (if required), and

4.1.10 Cast or heat (formerly ladle) analysis and mechanical properties report (if required) (see 10.1).

NOTE—A typical ordering description is as follows: "ASTM A 606 dated _____, Type 4 high-strength low-alloy hot-rolled sheet, dry, mill edge 0.106 by 48 by 96 in. for truck frame side members."

5. Manufacture

5.1 *Condition*—The material shall be furnished hot-rolled or cold-rolled as specified on the purchase order.

5.2 *Heat Treatment*—Unless otherwise specified, hot-rolled shall be furnished as rolled. When hot-rolled annealed or hot-rolled normalized material is required, it shall be specified on the purchase order.

6. Chemical Requirements

6.1 The maximum limits of carbon, manganese, and sulfur shall be as prescribed in Table 1, unless otherwise agreed upon between the manufacturer and the purchaser.

6.2 The manufacturer shall use such alloying elements, combined with the carbon, manganese, and sulfur within the limits prescribed in Table 1 to satisfy the mechanical properties prescribed in Tables 2 or 3. Such elements shall be included and reported in the specified heat or cast analysis. These steels have enhanced atmospheric corrosion resistance and are supplied in two types: Type 2 having at least two times that of carbon steel and Type 4 having at least four times that of carbon steel. When requested, the producer of the steel shall supply acceptable data of corrosion resistance to the purchaser.

6.3 When the steel is used in welded applications, welding procedure shall be suitable for the steel chemistry as described in 6.2 and the intended service.

7. Mechanical Requirements

7.1 *Tension Tests:*

7.1.1 *Hot-Rolled*—Hot-rolled material as represented by the test specimens shall conform to the mechanical properties specified in Table 2.

7.1.2 *Cold-Rolled*—Cold-rolled material as represented by the test specimen shall conform to the mechanical properties specified in Table 3.

7.1.3 *Number of Specimens*—Two tension tests shall be made from each heat, unless the finished material from a heat is less than 50 tons (45 metric tons), when one tension test will be sufficient. When material rolled from one heat differs 0.050 in. (1.27 mm) or more in thickness, one tension test shall be made from both the thickest and thinnest material rolled regardless of the weight represented.

7.1.4 *Test Specimen Orientation*—Tension test specimens should be taken longitudinally.

7.2 *Bend Test:*

7.2.1 *Requirements*—Bend test specimens taken in the direction specified in 7.2.3 shall stand being bent at room temperature through 180 deg to an inside diameter equivalent to the thickness of the material multiplied by the bend factor specified in Table 4 without cracking on the outside of the bent portion (see Section 14 of Methods and Definitions A 370).

7.2.2 *Number of Specimens*—Two bend tests (preferably transverse) shall be made from each heat, unless the finished material from a heat is less than 50 tons (45 metric tons), when one bend test will be sufficient. When material rolled from one heat differs 0.050 in. (1.27 mm) or more in thickness, one bend test shall be made from both the thickest and thinnest material rolled regardless of the weight represented.

7.2.3 *Test Specimen Orientation*—Bend test specimens shall be taken transversely, except for strip rolled single strand that is too narrow (less than 6.0 in. (150 mm) wide) for transverse specimens, in which case longitudinal specimens may be used.

8. Retests

8.1 If the results on an original tensile specimen are within 2 ksi (14 MPa) of the required tensile strength, within 1 ksi (7 MPa) of the required yield point, or within 2 percent of the required elongation, a retest shall be permitted for which one random specimen from the heat or test lot shall be used. If the results on this retest specimen meet the specified requirements,

9. Finish and Appearance

9.1 *Edges:*

9.1.1 *Hot-Rolled*—In the as-rolled condition the material has mill edges. Pickled or blast-cleaned material has cut edges. When required, as-rolled material may be specified to have cut edges. If mill edge material is required it must be specified.

9.1.2 *Cold-Rolled*—Cold-rolled material shall have cut edges only.

9.2 *Oiling:*

9.2.1 *Hot-Rolled*—Unless otherwise specified, hot-rolled as-rolled material shall be furnished dry, and hot-rolled pickled or blast-cleaned material shall be furnished oiled. When required, pickled or blast-cleaned material may be specified to be furnished dry, and as-rolled material may be specified to be furnished oiled.

9.2.2 *Cold-Rolled*—Unless otherwise specified, cold-rolled material shall be oiled. When required, cold-rolled material may be specified to be furnished dry, but is not recommended due to the increased possibility of rusting.

9.3 *Surface Finish:*

9.3.1 *Hot-Rolled*—Unless otherwise specified, hot-rolled material shall have an as-rolled, not pickled surface finish. When required, material may be specified to be pickled or blast-cleaned.

9.3.2 *Cold-Rolled*—Unless otherwise specified, cold-rolled material shall have a matte (dull) finish.

10. Certification and Reports

10.1 When requested, the manufacturer shall furnish copies of a test report showing the results of the heat or cast analysis and mechanical property tests made to determine compliance with this specification. The report shall include the purchase order number, the ASTM designation number, and the heat or lot number correlating the test results with the material represented.

TABLE 1 Chemical Requirements

	Composition, max, percent	
	Cast or Heat (Formerly Ladle) Analysis	Product Check, or Verification Analysis
Carbon[A]	0.22	0.26
Manganese	1.25	1.30
Sulfur	0.05	0.06

[A] For compositions with a maximum carbon content of 0.15 percent on heat or cast analysis, the maximum limit for manganese on heat or cast analysis may be increased to 1.40 percent (with product analysis limits of 0.19 percent carbon and 1.45 percent manganese).

TABLE 2 Tensile Requirements for Hot-Rolled Material

	As-Rolled		Annealed or Normalized Cut Lengths and Coils
	Cut Lengths	Coils[A]	
Tensile strength, min ksi (MPa)	70 (480)	65 (450)	65 (450)
Yield point, min, ksi (MPa)	50 (340)	45 (310)	45 (310)
Elongation in 2 in. or 50 mm, min, percent	22	22	22

[A] Coiled sheet and strip shall be produced and released to the same strength level as cut length product. Due to the producer's inability to test within the body of the coil, the strength levels are shown as being reduced by 5 ksi (35 MPa) to reflect the possibility of the inclusions of some lower strength material.

A 606

TABLE 3 Tensile Requirements for Cold-Rolled Material

	Cut Lengths and Coils
Tensile strength, min, ksi (MPa)	65 (450)
Yield point, min, ksi (MPa)	45 (310)
Elongation in 2 in. or 50 mm, min, percent	22[a]

[a] 0.0448 in. (1.14 mm) and under in thickness—20 percent.

TABLE 4 Bend Test Requirements

Thickness, in.(mm)	Ratio of Bend Diameter to Thickness of Specimen[a]	
	Transverse Specimen	Longitudinal Specimen[b]
0.090 (2.3) and less	2	1
Over 0.090 to 0.2299 (2.3 to 6), incl	3	1

[a] These bend tests apply to the bending performance of test specimens only. Where material is to be bent in fabricating operations, more liberal bend radii may be required and should be based on prior experience or consultation, or both, with the steel producer.

[b] Testing longitudinal specimens is not required by this specification except when the material is too narrow for transverse tests. However, experience has shown when transverse specimens meet the specified bend requirements, longitudinal specimens will normally pass the bend requirements in Table 4. See footnote a.

The American Society for Testing and Materials takes no position respecting the validity of any patent rights asserted in connection with any item mentioned in this standard. Users of this standard are expressly advised that determination of the validity of any such patent rights, and the risk of infringement of such rights, are entirely their own responsibility.

This standard is subject to revision at any time by the responsible technical committee and must be reviewed every five years and if not revised, either reapproved or withdrawn. Your comments are invited either for revision of this standard or for additional standards and should be addressed to ASTM Headquarters. Your comments will receive careful consideration at a meeting of the responsible technical committee, which you may attend. If you feel that your comments have not received a fair hearing you should make your views known to the ASTM Committee on Standards, 1916 Race St., Philadelphia, Pa. 19103.

Designation: A 607 – 83

Standard Specification for
SHEET AND STRIP, HOT ROLLED AND COLD ROLLED, HIGH-STRENGTH LOW-ALLOY, COLUMBIUM AND/OR VANADIUM[1]

This standard is issued under the fixed designation A 607; the number immediately following the designation indicates the year of original adoption or, in the case of revision, the year of last revision. A number in parentheses indicates the year of last reapproval. A superscript epsilon (ϵ) indicates an editorial change since the last revision or reapproval.

1. Scope

1.1 This specification covers high strength, low-alloy columbium and/or vanadium hot-rolled sheet and strip, and/or cold-rolled sheet in either cut lengths or coils, intended for applications where greater strength and savings in weight are important. The material is available as two classes. They are similar in strength level except that Class 2 offers improved weldability and more formability than Class 1. Atmospheric corrosion resistance of these steels is equivalent to plain carbon steels. With copper specified, the atmospheric corrosion resistance is twice that of plain carbon steel.

1.2 The value stated in inch-pound units are to be regarded as the standard.

1.3 Class 1 material was previously A 607 without a class designation.

2. Applicable Documents

2.1 *ASTM Standards:*
A 370 Methods and Definitions for Mechanical Testing of Steel Products[2]
A 568 Specification for General Requirements for Steel, Carbon and High-Strength Low-Alloy Hot-Rolled Sheet and Cold-Rolled Sheet[2]
A 749 Specification for General Requirements for Steel, Carbon, and High-Strength, Low-Alloy, Hot Rolled Strip[2]

3. General Requirements for Delivery

3.1 Material furnished under this specification shall conform to the applicable requirements of the current editions of Specifications A 568 or A 749 unless otherwise provided herein.

4. Ordering Information

4.1 Orders for material under this specification shall include the following information, as required, to describe adequately the desired material:

4.1.1 ASTM specification number and year of issue, grade, type, and class. When a class is not specified Class 1 will be furnished,

4.1.2 Name of material (high-strength low-alloy hot-rolled steel sheet or strip and cold-rolled steel sheet),

4.1.3 Copper bearing steel (when required),

4.1.4 Finish (cold-rolled)—indicate exposed (E) or unexposed (U). Matte (dull) finish will be supplied unless otherwise specified) (see 9.3),

4.1.5 Condition (specify oiled or dry, as required) (see 9.2),

4.1.6 Edges (must be specified for hot-rolled sheet and strip),

4.1.7 Dimensions (thickness, width, and whether cut lengths or coils),

4.1.8 Coil size and weight requirements (must include inside diameter, outside diameter, and maximum weight),

4.1.9 Application (show part identification and description),

4.1.10 Special requirements (if required) or supplementary requirements of S1, and

---

[1] This specification is under the jurisdiction of ASTM Committee A-1 on Steel, Stainless Steel, and Related Alloys and is the direct responsibility of Subcommittee A01.19 on Steel Sheet and Strip.
Current edition approved July 29, 1983. Published November 1983. Originally published as A 607 – 70. Last previous edition A 607 – 75 (1981).
[2] *Annual Book of ASTM Standards*, Vol 01.03.

4.1.11 Heat or cast (formerly ladle) analysis and mechanical property report (if required).

NOTE 1—A typical ordering description is as follows: "ASTM A 607 dated ____, Grade 45 Type I Class 2 hot-rolled high-strength low-alloy steel sheet, dry, mill edge, 0.075 by 36 by 96 in. for tote box frame members."

5. Manufacture

5.1 High-strength low-alloy columbium and/or vanadium hot-rolled sheet and strip, and cold-rolled sheet are ordinarily produced from capped or semi-killed steel. Should fully killed steel be required, the order should so indicate.

5.2 The material shall be furnished hot-rolled or cold-rolled as specified on the purchase order.

6. Chemical Requirements

6.1 The steel shall conform to the requirements as to chemical composition prescribed in Table 1 for the class specified.

6.2 When the steel is used in welded applications, welding procedure shall be suitable for the steel chemistry and the intended service.

6.3 When a class is not specified, Class 1 will be furnished.

7. Mechanical Requirements

7.1 *Tension Test:*

7.1.1 The material, as represented by the test specimen, shall conform to the mechanical requirements as outlined in Table 2.

7.1.2 *Number of Specimens*—Two tension tests shall be made from each heat, unless the finished material from a heat is less than 50 tons (45 metric tons), when one tension test will be sufficient. When material rolled from one heat differs 0.050 in. (1.27 mm) or more in thickness, one tension test shall be made from both the thickest and thinnest material rolled regardless of the weight represented.

7.1.3 *Test Specimen Orientation*—Tension test specimens should be taken longitudinally.

7.2 *Bend Test:*

7.2.1 *Requirements*—Bend test specimens taken in the direction specified in 7.2.3 shall stand being bent at room temperature through 180° to an inside diameter equivalent to the thickness of the material multiplied by the bend factor specified in Table 3, without cracking on the outside of the bent portion (see Section 14 of Methods A 370).

7.2.2 *Number of Specimens*—Two bend tests shall be made from each heat, unless the finished material from a heat is less than 50 tons (45 metric tons), when one bend test will be sufficient. When material rolled from one heat differs 0.050 in. (1.27 mm) or more in thickness, one bend test shall be made from both the thickest and thinnest material rolled regardless of the weight represented.

7.2.3 *Test Specimen Orientation*—Bend test specimens shall be taken transversely, except for strip rolled single strand that is too narrow (less than 6 in. (150 mm) wide) for transverse specimens in which case longitudinal specimens may be used.

8. Retests

8.1 If the results on an original tensile specimen are within 2000 psi (14 MPa) of the required tensile strength, within 1000 psi (7 MPa) of the required yield point, or within 2 % of the required elongation, a retest shall be permitted for which one random specimen from the heat or test lot shall be used. If the results on this retest specimen meet the specified mechanical requirements, the heat or lot will be accepted.

9. Finish and Appearance

9.1 *Edges:*

9.1.1 *Hot-Rolled*—In the as-rolled condition the material has mill edges. Pickled or blast-cleaned material has cut edges. When required, as-rolled material may be specified to have cut edges. If mill edge material is required it must be specified.

9.1.2 *Cold-Rolled*—Cold-rolled material shall have cut edges only.

9.2 *Oiling:*

9.2.1 *Hot-Rolled*—Unless otherwise specified, hot-rolled as-rolled material shall be furnished dry, and hot-rolled pickled or blast-cleaned material shall be furnished oiled. When required, pickled or blast-cleaned material may be specified to be furnished dry, and as-rolled material may be specified to be furnished oiled.

9.2.2 *Cold-Rolled*—Unless otherwise specified, cold-rolled material shall be oiled. When required, cold-rolled material may be specified to be furnished dry, but is not recommended due to the increased possibility of rusting.

9.3 *Surface Finish:*

9.3.1 *Hot-Rolled*—Unless otherwise specified, hot-rolled material shall have an as-rolled, not pickled surface finish. When required, material may be specified to be pickled or blast-cleaned.

9.3.2 *Cold-Rolled*—Unless otherwise specified cold-rolled material shall have a matte (dull) finish.

10. Certification and Reports

10.1 When requested, the manufacturer shall furnish copies of a test report showing the results of the heat or cast analysis and mechanical property tests made to determine compliance with this specification. The report shall include the purchase order number, ASTM designation number, and the heat or lot number correlating the test results with the material represented.

SUPPLEMENTARY REQUIREMENTS

The following supplementary requirements shall apply when specified in the order or contract:

S1. Types

S1.1 When a purchaser prefers to designate the specific elements (columbium, vanadium, nitrogen, or combinations thereof), one of the types listed below shall be specified. The type in addition to the grade must be shown on the order (see 4.1.1).

Type 1—Columbium
Type 2—Vanadium
Type 3—Columbium and vanadium
Type 4—Vanadium and nitrogen

S1.2 The composition limits of Section 6 shall apply for any of these types.

TABLE 1 Chemical Requirements[A]
Composition, %

NOTE—Class 2 requirements for manganese, phosphorus, sulfur, columbium, and vanadium are the same as shown for Class 1.

Element	Grade 45		Grade 50		Grade 55		Grade 60		Grade 65		Grade 70	
	Heat or Cast Analysis	Product Analysis	Heat or Cast Analysis	Product Analysis	Heat or Cast Analysis	Product Analysis	Heat or Cast Analysis	Product Analysis	Heat or Cast Analysis	Product Analysis	Heat or Cast Analysis	Product Analysis
Class 1												
Carbon, max	0.22	0.26	0.23	0.27	0.25	0.29	0.26	0.30	0.26	0.30	0.26	0.30
Manganese, max	1.35	1.40	1.35	1.40	1.35	1.40	1.50	1.55	1.50	1.55	1.65	1.70
Phosphorus, max	0.04	0.05	0.04	0.05	0.04	0.05	0.04	0.05	0.04	0.05	0.04	0.05
Sulfur, max	0.05	0.06	0.05	0.06	0.05	0.06	0.05	0.06	0.05	0.06	0.05	0.06
Columbium or vanadium, min	Cb 0.005	0.004	Cb 0.005	0.004	Cb 0.005	0.004	Cb 0.005	0.004	Cb 0.005	0.004	Cb 0.005	0.004
	V 0.01	0.005	V 0.01	0.005	V 0.01	0.005	V 0.01	0.005	V 0.01	0.005	V 0.01	0.005
Nitrogen, max	...	...	...	...	...	...	...	...	0.012	0.015	0.012	0.015
Class 2												
Carbon, max	0.15	0.18	0.15	0.18	0.15	0.18	0.15	0.18	0.15	0.18	0.15	0.18
Nitrogen	...	...	...	...	...	...	0.020	0.024	0.020	0.024	0.020	0.024

[A] Copper, when specified, shall have a minimum content of 0.20 % by heat or cast analysis (0.18 % by product analysis).

TABLE 2 Tensile Requirements[A]

ksi (MPa)	Grade 45	Grade 50	Grade 55	Grade 60	Grade 65	Grade 70
Class 1						
Tensile strength, min	60 (410)	65 (450)	70 (480)	75 (520)	80 (550)	85 (590)
Yield point, min	45 (310)	50 (340)	55 (380)	60 (410)	65 (450)	70 (480)
Elongation in 2 in., or 50 mm, min, %:						
Hot-Rolled over 0.097 (2.46 mm)	25.0	22.0	20.0	18.0	16.0	14.0
Up to 0.097 (2.46 mm) incl	23.0	20.0	18.0	16.0	14.0	12.0
Cold-Rolled	22.0	20.0	18.0	16.0	15.0	14.0
Class 2						
Tensile strength, min	55 (380)	60 (410)	65 (450)	70 (480)	75 (520)	80 (550)
Yield strength, min	45 (310)	50 (340)	55 (380)	60 (410)	65 (450)	70 (480)
Elongation in 2 in., % min:						
Hot-Rolled over 0.097 (2.46 mm)	25.0	22.0	20.0	18.0	16.0	14.0
Up to 0.097 (2.46 mm) incl	23.0	20.0	18.0	16.0	14.0	12.0
Cold-Rolled	22.0	20.0	18.0	16.0	15.0	14.0

[A] All product specified to be furnished in coils will be produced to the same strength level mill practices as cut length product. Because testing within the body of the coil cannot be performed by the producer, recognition must be given to the fact that some portions of coils could fall below the specified minimum.

TABLE 3 Bend Test Requirements[A]

Grade	Ratio of Bend Diameter to Thickness of Specimen	
	For Thicknesses up to 0.2299 in. (6 mm) incl[B]	
	Transverse Specimens[A]	Longitudinal Specimens[C]
45	1½	1
50	1½	1
55	2	1½
60	3	2
65	3½	2½
70	4	3

[A] Bend requirements for Class 2 product are reduced ½t.

[B] These bend tests apply to the bending performance of test specimens only. Where material is to be bent in fabricating operations, more liberal bend radii may be required and should be based on prior experience or consultation, or both, with the steel producers.

[C] Testing longitudinal specimens is not required by this specification except when material is too narrow for transverse tests. However, experience has shown when transverse specimens meet the specified bend requirements, longitudinal specimens will normally pass the bend requirements in Table 3.[B]

The American Society for Testing and Materials takes no position respecting the validity of any patent rights asserted in connection with any item mentioned in this standard. Users of this standard are expressly advised that determination of the validity of any such patent rights, and the risk of infringement of such rights, are entirely their own responsibility.

This standard is subject to revision at any time by the responsible technical committee and must be reviewed every five years and if not revised, either reapproved or withdrawn. Your comments are invited either for revision of this standard or for additional standards and should be addressed to ASTM Headquarters. Your comments will receive careful consideration at a meeting of the responsible technical committee, which you may attend. If you feel that your comments have not received a fair hearing you should make your views known to the ASTM Committee on Standards, 1916 Race St., Philadelphia, Pa. 19103.

Designation: A 611 – 82

Standard Specification for
STEEL, COLD-ROLLED SHEET, CARBON, STRUCTURAL[1]

This standard is issued under the fixed designation A 611; the number immediately following the designation indicates the year of original adoption or, in the case of revision, the year of last revision. A number in parentheses indicates the year of last reapproval. A superscript epsilon (ϵ) indicates an editorial change since the last revision or reapproval.

This specification has been approved for use by agencies of the Department of Defense and for listing in the DoD Index of Specifications and Standards.

1. Scope

1.1 This specification covers cold-rolled carbon structural steel sheet, in cut lengths or coils. It includes five strength levels designated as Grade A with yield point 25 ksi (170 MPa) minimum; Grade B with 30 ksi (205 MPa) minimum; Grade C with 33 ksi (230 MPa) minimum; Grade D Types 1 and 2 with 40 ksi (275 MPa) minimum; and Grade E with 80 ksi (550 MPa) minimum.

1.2 Grades A, B, C, and D have moderate ductility whereas Grade E is a full-hard product with no specified minimum elongation.

1.3 The values stated in inch-pound units are to be regarded as the standard.

2. Applicable Document

2.1 *ASTM Standard*:
A 568 Specification for General Requirements for Steel, Carbon and High-Strength Low-Alloy Hot-Rolled Sheet, and Cold-Rolled Sheet.[2]

3. Definitions

3.1 *structural steel sheet*—sheet produced to tensile property values as specified or required.

4. General Requirements for Delivery

4.1 Material furnished under this specification shall conform to the applicable requirements of the current edition of Specification A 568.

5. Ordering Information

5.1 Orders for material under this specification shall include the following information, as required, to describe the material adequately.

5.1.1 ASTM specification number, date of issue, and grade (if Grade D, indicate Type 1 or Type 2),

5.1.2 Copper-bearing steel (if required),

5.1.3 Special requirements (if required),

5.1.4 Name of material (cold-rolled sheet), structural quality,

5.1.5 Finish; matte (dull) finish will be supplied unless otherwise ordered,

5.1.6 Condition (oiled or dry),

5.1.7 Dimensions,

5.1.8 Coil size requirements, and

5.1.9 Cast or heat (formerly ladle) analysis and test report (request, if required).

NOTE 1—A typical ordering description is as follows: ASTM A 611, date, Grade C, Cold-Rolled Oiled Sheet, Structural Quality, 0.035 by 36 by 96 in. (0.89 by 914 by 2438 mm) for Roof Deck.

6. Chemical Requirements

6.1 The cast or heat analysis of the steel shall conform to the requirements prescribed in Table 1.

7. Mechanical Requirements

7.1 *Tension Tests:*

7.1.1 *Requirements*—The material as represented by the test specimens shall conform to the mechanical requirements prescribed in Table 2.

7.1.2 *Number of Tests*—Two tension tests shall be made from each heat or from each lot

---

[1] This specification is under the jurisdiction of ASTM Committee A-1 on Steel, Stainless Steel and Related Alloys, and is the direct responsibility of Subcommittee A01.19 on Steel Sheet and Strip.
Current edition approved July 30, 1982. Published September 1982. Originally published as A 611 – 70. Last previous edition A 611 – 72 (1979).

[2] *Annual Book of ASTM Standards*, Vol 01.03.

of 50 tons (45 Mg). When the amount of finished material from a heat or lot is less than 50 tons, only one tension test shall be made. When material rolled from one heat differs 0.050 in. (1.27 mm) or more in thickness, one tension test shall be made from both the thickest and thinnest material rolled regardless of the weight represented.

7.1.3 *Test Specimen Orientation*—Test specimens shall be taken longitudinally.

7.2 *Bend Test:*

7.2.1 *Requirements*—The bend test specimens shall stand being bent at room temperature in any direction through 180° without cracking on the outside of the bent portion to an inside diameter which shall have a relation to the thickness of the specimen as prescribed in Table 3.

7.2.2 *Number of Tests*—Two bend tests shall be made from each heat or from each lot of 50 tons (45 Mg). When the amount of finished material from a heat or lot is less than 50 tons, only one bend test shall be made. When material rolled from one heat differs 0.050 in. (1.27 mm) or more in thickness, one bend test shall be made from both the thickest and thinnest material rolled regardless of the weight represented.

7.2.3 *Retests*—If one test fails, two more tests shall be run from the same lot, in which case both tests shall conform to the requirements prescribed in this specification; otherwise, the lot under test shall stand rejected.

8. Finish and Condition

8.1 *Surface Finish*—Unless otherwise specified the sheet shall have a matte (dull) finish.

8.2 *Oiling*—The sheet shall be furnished oiled or dry, as specified.

9. Certification and Reports

9.1 When requested, the manufacturer shall furnish copies of a test report showing the results of the ladle or cast analysis and mechanical property tests made to determine compliance with this specification. The report shall include the purchase order number; ASTM designation number; and heat or lot number correlating the test results with the material represented.

10. Packaging

10.1 *Coil Size*—Small coils result from the cutting of full-size coils for center test purposes. These small coils are acceptable under this specification.

TABLE 1 Chemical Requirements

Element	Composition, %		
	Grades A, B, C, E	Grade D Type 1	Grade D Type 2
Carbon, max	0.20	0.20	0.15
Manganese, max	0.60	0.90	0.60
Phosphorus, max	0.04	0.04	0.20
Sulfur, max	0.04	0.04	0.04
Copper, when copper steel is specified, min	0.20	0.20	0.20

TABLE 2 Mechanical Requirements

Grade	Yield Point, min		Tensile Strength, min		Elongation in 2 in. or 50 mm, min, %
	ksi	MPa	ksi	MPa	
A	25	170	42	290	26
B	30	205	45	310	24
C	33	230	48	330	22
D, Types 1 and 2	40	275	52	360	20
E	80[A]	550	82	565	...

[A] On this full-hard product, the yield point approaches the tensile strength and since there is no halt in the gage or drop in the beam, the yield point shall be taken as the stress at 0.5 % elongation, under load.

TABLE 3 Bend Test Requirements

Grade	Ratio of the Bend Diameter to Thickness of the Specimen
A	0
B	1
C	1½
D, Types 1 and 2	2
E	bend test not applicable

The American Society for Testing and Materials takes no position respecting the validity of any patent rights asserted in connection with any item mentioned in this standard. Users of this standard are expressly advised that determination of the validity of any such patent rights, and the risk of infringement of such rights, are entirely their own responsibility.

This standard is subject to revision at any time by the responsible technical committee and must be reviewed every five years and if not revised, either reapproved or withdrawn. Your comments are invited either for revision of this standard or for additional standards and should be addressed to ASTM Headquarters. Your comments will receive careful consideration at a meeting of the responsible technical committee, which you may attend. If you feel that your comments have not received a fair hearing you should make your views known to the ASTM Committee on Standards, 1916 Race St., Philadelphia, Pa. 19103.

Designation: A 619/A 619M – 82

Standard Specification for
STEEL SHEET, CARBON, COLD-ROLLED, DRAWING QUALITY[1]

This standard is issued under the fixed designation A 619/A 619M; the number immediately following the designation indicates the year of original adoption or, in the case of revision, the year of last revision. A number in parentheses indicates the year of last reapproval. A superscript epsilon (ϵ) indicates an editorial change since the last revision or reapproval.

This specification has been approved for use by agencies of the Department of Defense and for listing in the DoD Index of Specifications and Standards.

1. Scope

1.1 This specification covers cold-rolled carbon steel sheet of drawing quality in coils or cut lengths. The material is intended for fabricating identified parts where drawing or severe forming may be involved.

1.2 This specification is applicable for orders in either inch-pound units (as A 619) or SI units (A 619M).

2. Applicable Documents

2.1 *ASTM Standards:*
A 568 Specification for General Requirements for Steel, Carbon and High-Strength Low-Alloy Hot-Rolled Sheet and Cold-Rolled Sheet[2]
A 568M Specification for General Requirements for Steel, Carbon and High-Strength Low-Alloy Hot-Rolled Sheet and Cold-Rolled Sheet [Metric][2]
A 620/A 620M Specification for Steel Sheet, Carbon, Cold-Rolled, Drawing Quality, Special Killed[2]

3. Classes

3.1 Cold-rolled sheet is supplied for either exposed or unexposed applications. Within the latter category, cold-rolled sheet is specified either "temper rolled" or "annealed last." For details on processing, attributes and limitations, and inspection standards, refer to Specifications A 568 or A 568M.

4. Definitions

4.1 *drawing quality*—sheet manufactured from specially produced or selected steels specially processed to have good uniform drawing properties for use in fabricating an identified part having severe deformations.

4.2 *aging*—loss of ductility with an increase in hardness, yield point, and tensile strength that occur when steel, which has been slightly cold worked (such as by temper rolling), is stored for some time. Aging also increases the tendency toward stretcher strains and fluting.

5. General Requirements

5.1 Material furnished under this specification shall conform to the applicable requirements of the current edition of Specifications A 568 or A 568M unless otherwise provided herein.

6. Ordering Information

6.1 Orders for material under this specification shall include the following information, as necessary, to describe adequately the required material:

6.1.1 ASTM specification number and date of issue,

6.1.2 Name of material (cold-rolled sheet, drawing quality),

6.1.3 Class (either exposed, unexposed temper rolled, or annealed last (see 3.1),

---

[1] This specification is under the jurisdiction of ASTM Committee A-1 on Steel, Stainless Steel and Related Alloys, and is the direct responsibility of Subcommittee A01.19 on Steel Sheet and Strip.
Current edition approved March 26, 1982. Published May 1982. Originally published as A 619 – 68. Replaces portions of A 365. Last previous edition A 619 – 75.
[2] *Annual Book of ASTM Standards*, Vol 01.03.

6.1.4 Finish (matte (dull) finish will be supplied on exposed unless otherwise specified and on unexposed) (see 10.1),

6.1.5 Oiling (material will be oiled unless ordered not oiled) (see 10.2),

6.1.6 Dimensions (actual thickness, width, and length where applicable),

6.1.7 Coil size and weight (include inside diameter, outside diameter, and maximum weight),

6.1.8 Quantity,

6.1.9 Application (show part identification and description),

6.1.10 Cast or heat (formerly ladle) analysis (if required) (see 12.1), and

6.1.11 Special requirements (if required).

NOTE—A typical ordering description is as follows: "ASTM A 619 dated _____, cold-rolled, drawing quality sheet; exposed; oiled; 0.035 by 53 in. by coil ID 24 in., OD 48 in., max; weight 15 000 lb, max; 100 000 lb; for deck lid (outer)" or

"ASTM A 619M dated _____, cold rolled, drawing quality sheet; exposed; oiled; 0.8 by 1340 mm by coil ID 600 mm, OD 1500 mm, max; weight 10 000 kg, max; 45 000 kg; for deck lid (outer)."

7. Manufacture

7.1 *Melting Practice*—The sheet is normally produced from selected rimmed, or equivalent steels.

8. Chemical Requirements

8.1 The cast or heat analysis of the steel shall conform to the requirements prescribed in Table 1.

9. Mechanical Requirements

9.1 The sheet shall be suitable for the production of identified deep drawn parts. The manufacturer shall assume responsibility for selection of steel, control of processing, and ability of the material to form identified parts within properly established breakage limits. (Refer to Appendix X4 of Specifications A 568 or A 568M.)

9.2 Deformations on identified parts are assessed by the use of the scribed-square test as described in Appendix X5 of Specifications A 568 or A 568M. Experience has shown that if the maximum percent increase in area of any drawn portion of a satisfactory untrimmed part is more than 25 %, drawing quality is required.

9.3 When these sheets are required to be free of stretcher strains or fluting during fabrication, exposed is required, and effective roller levelling is necessary immediately prior to use. Material so specified is subject to aging (refer to Appendix X3 of Specifications A 568 or A 568M).

9.4 Where the draw is particularly severe, or essential freedom from aging is required, drawing quality, special killed steel (in accordance with Specification A 620/A 620M) is recommended.

10. Finish and Appearance

10.1 *Surface Finish:*

10.1.1 Unless otherwise specified, exposed sheet shall have a matte (dull) finish. When required, a controlled surface texture and condition may be specified.

10.1.2 Unexposed sheet shall have a matte (dull) finish. Surface texture or condition may not be specified.

10.2 *Oiling:*

10.2.1 Unless otherwise specified, the sheet shall be oiled.

10.2.2 When required, the sheet may be specified to be furnished not oiled (dry).

10.3 *Edges*—The sheet shall have cut edges.

11. Marking

11.1 In addition to the requirements of Specifications A 568 or A 568M, each lift or coil shall be marked with the designation, "DQ."

12. Certification and Reports

12.1 While not normally required for this specification, upon request of the purchaser in the contract or order, a manufacturer's certification that the material was produced in accordance with this specification shall be furnished. The cast or heat analysis may be reported, if required.

TABLE 1 Chemical Requirements

Element	Composition, max, %
Carbon	0.10
Manganese	0.50
Phosphorus	0.025
Sulfur	0.035

The American Society for Testing and Materials takes no position respecting the validity of any patent rights asserted in connection with any item mentioned in this standard. Users of this standard are expressly advised that determination of the validity of any such patent rights, and the risk of infringement of such rights, are entirely their own responsibility.

This standard is subject to revision at any time by the responsible technical committee and must be reviewed every five years and if not revised, either reapproved or withdrawn. Your comments are invited either for revision of this standard or for additional standards and should be addressed to ASTM Headquarters. Your comments will receive careful consideration at a meeting of the responsible technical committee, which you may attend. If you feel that your comments have not received a fair hearing you should make your views known to the ASTM Committee on Standards, 1916 Race St., Philadelphia, Pa. 19103.

Designation: A 620/A 620M – 82

Standard Specification for
STEEL SHEET, CARBON, COLD-ROLLED, DRAWING QUALITY, SPECIAL KILLED[1]

This standard is issued under the fixed designation A 620/A 620M; the number immediately following the designation indicates the year of original adoption or, in the case of revision, the year of last revision. A number in parentheses indicates the year of last reapproval. A superscript epsilon (ϵ) indicates an editorial change since the last revision or reapproval.

This specification has been approved for use by agencies of the Department of Defense and for listing in the DoD Index of Specifications and Standards.

1. Scope

1.1 This specification covers cold-rolled carbon steel sheet of drawing quality, special killed, in coils or cut lengths. This material is intended for fabricating identified parts where particularly severe drawing or forming may be involved or essential freedom from aging is required.

1.2 This specification is applicable for orders in either inch-pound units (as A 620) or SI units (as A 620M).

2. Applicable Documents

2.1 *ASTM Standards:*
A 568 Specification for General Requirements for Steel, Carbon and High-Strength Low-Alloy Hot-Rolled Sheet and Cold-Rolled Sheet[2]
A 568M Specification for General Requirements for Steel, Carbon and High-Strength Low-Alloy Hot-Rolled Sheet and Cold-Rolled Sheet, [Metric][2]
A 619/A 619M Specification for Steel Sheet, Carbon, Cold-Rolled, Drawing Quality[2]

3. Classes

3.1 Cold-rolled sheet is supplied for either exposed or unexposed applications. Within the latter category, cold-rolled sheet is specified either "temper rolled" or "annealed last." For details on processing, attributes and limitations, and inspection standards, refer to Specifications A 568 or A 568M.

4. Definitions

4.1 *drawing quality special killed*—sheet manufactured from specially produced or selected killed steels (normally aluminum killed) specially processed to have good uniform drawing properties for use in fabricating an identified part having extremely severe deformations and to be essentially free from aging.

4.2 *aging*—loss of ductility with an increase in hardness, yield point, and tensile strength that occur when steel, which has been slightly cold worked (such as by temper rolling), is stored for some time. Aging also increases the tendency toward stretcher strains and fluting.

5. General Requirements

5.1 Material furnished under this specification shall conform to the applicable requirements of the current edition of Specifications A 568 or A 568M unless otherwise provided herein.

6. Ordering Information

6.1 Orders for material under this specification shall include the following information, as necessary, to describe adequately the required material:

6.1.1 ASTM specification number and date of issue,

6.1.2 Name of material (cold-rolled sheet, drawing quality, special killed),

---

[1] This specification is under the jurisdiction of ASTM Committee A-1 on Steel, Stainless Steel and Related Alloys, and is the direct responsibility of Subcommittee A01.19 on Steel Sheet and Strip.
Current edition approved March 26, 1982. Published May 1982. Originally published as A 620 – 68. Last previous edition A 620 – 75.
[2] *Annual Book of ASTM Standards*, Vol 01.03.

6.1.3 Class (either exposed, unexposed temper rolled, or annealed last) (see 3.1),

6.1.4 Finish (matte (dull) finish will be supplied on exposed, unless otherwise specified, and on unexposed) (see 10.1),

6.1.5 Oiling (material will be oiled unless ordered not oiled) (see 10.2),

6.1.6 Dimensions (decimal thickness, width, and length where applicable),

6.1.7 Coil size and weight (include inside diameter, outside diameter, and maximum weight),

6.1.8 Quantity,

6.1.9 Application (show part identification and description),

6.1.10 Cast or heat (formerly ladle) analysis (if required) (see 12.1), and

6.1.11 Special requirements (if required).

NOTE—A typical ordering description is as follows: "ASTM A 620 dated _____, exposed; cold-rolled, drawing quality, special-killed sheet; oiled; 0.035 by 53 in. by coil; ID 24 in.; OD 48 in., max; weight 15 000 lb, max; 100 000 lb; for instrument panel" or

"ASTM A 620M dated _____, cold-rolled, drawing quality, special-killed sheet, exposed, oiled; 0.8 by 1340 mm by coil; ID 600 mm; OD 1500 mm, max; weight 10 000 kg, max; 45 000 kg; for instrument panel."

7. Manufacture

7.1 *Melting Practice*—The sheet shall be produced from special killed steel. It is normally produced from aluminum killed steel but may be otherwise deoxidized provided the material is capable of meeting the specified requirements.

8. Chemical Requirements

8.1 The cast or heat analysis of the steel shall conform to the requirements prescribed in Table 1.

9. Mechanical Requirements

9.1 The sheet shall be suitable for the production of identified deep drawn parts. The manufacturer shall assume responsibility for selection of steel, control of processing, and ability of the material to form identified parts within properly established breakage limits. (Refer to Appendix X4 of Specifications A 568 or A 568M.)

9.2 Drawing quality, special killed steel is required when drawing quality (in accordance with Specification A 619/A 619M) will not provide a sufficient degree of ductility for the fabrication of parts having stringent drawing requirements.

9.3 Refer to X3.2 of Specifications A 568 or A 568M for additional information.

10. Finish and Appearance

10.1 *Surface Finish:*

10.1.1 Unless otherwise specified, exposed sheet shall have a matte (dull) finish. When required, a controlled surface texture and condition may be specified.

10.1.2 Unexposed sheet shall have a matte (dull) finish. Surface texture or condition may not be specified.

10.2 *Oiling:*

10.2.1 Unless otherwise specified, the sheet shall be oiled.

10.2.2 When required, the sheet may be specified to be furnished not oiled (dry).

10.3 *Edges*—The sheet shall have cut edges.

11. Marking

11.1 In addition to the requirements of Specifications A 568 or A 568M, each lift or coil shall be marked with the designation, "DQSK."

12. Certification and Reports

12.1 While not normally required for this specification, upon request of the purchaser in the contract or order, a manufacturer's certification that the material was produced in accordance with this specification shall be furnished. The cast or heat analysis may be reported, if required.

A 620/A 620M

TABLE 1 Chemical Requirements[A]

Element	Composition, max, %
Carbon	0.10
Manganese	0.50
Phosphorus	0.025
Sulfur	0.035

[A] If aluminum is used as the deoxidizing agent, the aluminum content total by product analysis is usually in excess of 0.010 %.

The American Society for Testing and Materials takes no position respecting the validity of any patent rights asserted in connection with any item mentioned in this standard. Users of this standard are expressly advised that determination of the validity of any such patent rights, and the risk of infringement of such rights, are entirely their own responsibility.

This standard is subject to revision at any time by the responsible technical committee and must be reviewed every five years and if not revised, either reapproved or withdrawn. Your comments are invited either for revision of this standard or for additional standards and should be addressed to ASTM Headquarters. Your comments will receive careful consideration at a meeting of the responsible technical committee, which you may attend. If you feel that your comments have not received a fair hearing you should make your views known to the ASTM Committee on Standards, 1916 Race St., Philadelphia, Pa. 19103.

Designation: A 621/A 621M – 82

Standard Specification for
STEEL SHEET AND STRIP, CARBON, HOT-ROLLED, DRAWING QUALITY[1]

This standard is issued under the fixed designation A 621/A 621M; the number immediately following the designation indicates the year of original adoption or, in the case of revision, the year of last revision. A number in parentheses indicates the year of last reapproval. A superscript epsilon (ϵ) indicates an editorial change since the last revision or reapproval.

This specification has been approved for use by agencies of the Department of Defense and for listing in the DoD Index of Specifications and Standards.

1. Scope

1.1 This specification covers hot-rolled carbon steel sheet and strip of drawing quality, in coils or cut lengths. This material is intended for use in fabricating identified parts where drawing or severe forming may be involved and surface appearance is not of primary importance.

1.2 This specification is applicable for orders in either inch-pound units (as A 621) or SI units (as A 621M).

2. Applicable Documents

2.1 *ASTM Standards:*
A 568 Specification for General Requirements for Steel, Carbon and High-Strength Low-Alloy Hot-Rolled Sheet and Cold-Rolled Sheet[2]
A 568M Specification for General Requirements for Steel, Carbon and High-Strength Low-Alloy Hot-Rolled Sheet and Cold-Rolled Sheet [Metric][2]
A 622/A 622M Specification for Steel Sheet and Strip, Carbon, Hot-Rolled, Drawing Quality, Special Killed[2]
A 749 Specification for General Requirements for Steel, Carbon and High-Strength, Low-Alloy Hot-Rolled Strip[2]
A 749M Specification for General Requirements for Steel, Carbon and High-Strength, Low-Alloy Hot-Rolled Strip, [Metric][2]

3. Definition

3.1 *drawing quality*—sheet manufactured from specially produced or selected steels specially processed to have good uniform drawing properties for use in fabricating an identified part having severe deformations.

4. General Requirements

4.1 Material furnished under this specification shall conform to the applicable requirements of the current edition of Specifications A 568, A 568M, A 749, or A 749M, unless otherwise provided herein.

5. Ordering Information

5.1 Orders for material under this specification shall include the following information, as necessary, to describe adequately the required material:
5.1.1 ASTM specification number and date of issue,
5.1.2 Name of material (hot-rolled sheet or hot-rolled strip, drawing quality),
5.1.3 Finish (indicate descaled, if required) (9.1),
5.1.4 Oiling (9.2),
5.1.5 Edges (9.3),
5.1.6 Dimensions (decimal thickness and width),
5.1.7 Coil size and weight (include inside diameter, outside diameter, and maximum weight),

---

[1] This specification is under the jurisdiction of ASTM Committee A-1 on Steel, Stainless Steel and Related Alloys, and is the direct responsibility of Subcommittee A01.19 on Steel Sheet and Strip.
Current edition approved March 26, 1982. Published May 1982. Originally published as A 621 – 68. Last previous edition A 621 - 75.
[2] *Annual Book of ASTM Standards*, Vol 01.03.

5.1.8 Quantity,

5.1.9 Application (show part identification and description),

5.1.10 Cast or heat (formerly ladle) analysis (if required) (11.1), and

5.1.11 Special requirements (if required).

NOTE—A typical ordering description is as follows: "ASTM A 621 dated _____, hot-rolled, drawing quality, pickled and oiled sheet, 0.147 by 46 by 96 in., 10 000 lb, for upper control arm" or

"ASTM A 621M dated _____, hot-rolled, drawing quality, pickled and oiled sheet, 3.7 mm by 117 mm by coil, ID 600 mm, OD 1500 mm max, weight 10 000 kg max, 50 000 kg, for upper control arm."

6. Manufacture

6.1 *Melting Practice*—The sheet is normally produced from selected rimmed, or equivalent steels.

7. Chemical Requirements

7.1 The cast or heat analysis of the steel shall conform to the requirements prescribed in Table 1.

8. Mechanical Requirements

8.1 The sheet and strip shall be suitable for the production of identified deep drawn parts. The manufacturer shall assume responsibility for selection of steel, control of processing, and ability of the material to form identified parts within properly established breakage limits. (Refer to Appendix X4 of Specifications A 568 or A 568M.)

8.2 Deformations on identified parts are assessed by the use of the scribed-square test as described in Appendix X5 of Specifications A 568 or A 568M. Experience has shown that if the percent increase in area of the most deformed 1-in. (25-mm) square of a satisfactory untrimmed part is more than 25 % for material 0.0822 to 0.187 in. (2.09 to 4.75 mm) thick or more than 30 % for material over 0.187 in., then drawing quality is required. This requirement is subject to negotiation when the thickness is less than 0.0822 in.

8.3 Where the draw is particularly severe or essential freedom from aging is required, drawing quality, special killed steel (Specification A 622/A 622M) is recommended.

9. Finish and Appearance

9.1 *Surface Finish:*

9.1.1 Unless otherwise specified, the material shall be furnished as rolled, that is, without removing the hot-rolled oxide or scale.

9.1.2 When required, the material may be specified to be pickled or blast cleaned (descaled).

9.2 *Oiling:*

9.2.1 Unless otherwise specified, hot-rolled as-rolled material shall be furnished not oiled (that is, dry) and hot-rolled pickled or blast-cleaned material shall be furnished oiled.

9.2.2 When required, as-rolled material may be specified to be furnished oiled and pickled or blast-cleaned material may be specified to be furnished dry.

9.3 *Edges*—Unless otherwise specified, sheet and strip shall have the "standard" edge designated below. When required, the "optional" edge may be specified.

	Edge	
Product	Standard	Optional (Must Be Specified)
Sheet (cut lengths and coils)	cut	mill
Strip	cut	mill
Widths 2.00 in. (50 mm) and less	mill or cut	(must be specified)
Widths over 2.00 in. (50 mm)	cut	mill

10. Marking

10.1 In addition to the requirements of Specifications A 568, A 568M, A 749, or A 749M, each lift or coil shall be marked with the designation "DQ."

11. Certification and Reports

11.1 While not normally required for this specification, upon request of the purchaser in the contract or order, a manufacturer's certification that the material was produced in accordance with this specification shall be furnished. The cast or heat analysis may be reported, if required.

A 621/A 621M

TABLE 1 Chemical Requirements

Element	Composition, max, %
Carbon	0.10
Manganese	0.50
Phosphorus	0.025
Sulfur	0.035

The American Society for Testing and Materials takes no position respecting the validity of any patent rights asserted in connection with any item mentioned in this standard. Users of this standard are expressly advised that determination of the validity of any such patent rights, and the risk of infringement of such rights, are entirely their own responsibility.

This standard is subject to revision at any time by the responsible technical committee and must be reviewed every five years and if not revised, either reapproved or withdrawn. Your comments are invited either for revision of this standard or for additional standards and should be addressed to ASTM Headquarters. Your comments will receive careful consideration at a meeting of the responsible technical committee, which you may attend. If you feel that your comments have not received a fair hearing you should make your views known to the ASTM Committee on Standards, 1916 Race St., Philadelphia, Pa. 19103.

Designation: A 622/A 622M – 82

Standard Specification for
STEEL SHEET AND STRIP, CARBON, HOT-ROLLED, DRAWING QUALITY, SPECIAL KILLED[1]

This standard is issued under the fixed designation A 622/A 622M; the number immediately following the designation indicates the year of original adoption or, in the case of revision, the year of last revision. A number in parentheses indicates the year of last reapproval. A superscript epsilon (ϵ) indicates an editorial change since the last revision or reapproval.

This specification has been approved for use by agencies of the Department of Defense and for listing in the DoD Index of Specifications and Standards.

1. Scope

1.1 This specification covers hot-rolled carbon steel sheet and strip of drawing quality, special killed, in coils or cut lengths. This material is intended for use in fabricating identified parts where particularly severe drawing or forming may be involved and surface appearance is usually not of primary importance.

1.2 This specification is applicable for orders in either inch-pound units (as A 622) of SI units (as A 622M).

2. Applicable Documents

2.1 *ASTM Standards:*
A 568 Specification for General Requirements for Steel, Carbon and High-Strength Low-Alloy Hot-Rolled Sheet and Cold-Rolled Sheet[2]
A 568M Specification for General Requirements for Steel, Carbon and High-Strength Low-Alloy Hot-Rolled Sheet and Cold-Rolled Sheet, [Metric][2]
A 621/A 621M Specification for Steel Sheet and Strip, Carbon, Hot-Rolled, Drawing Quality[2]
A 749 Specification for General Requirements for Steel, Carbon and High-Strength, Low-Alloy Hot-Rolled Strip[2]
A 749M Specification for General Requirements for Steel, Carbon and High-Strength, Low-Alloy Hot-Rolled Strip [Metric][2]

3. Definition

3.1 *drawing quality special killed*—sheet manufactured from specially produced or selected killed steels (normally aluminum) specially processed to have good uniform drawing properties for use in fabricating an identified part having extremely severe deformations.

4. General Requirements

4.1 Material furnished under this specification shall conform to the applicable requirements of the current edition of Specifications A 568, A 568M, A 749, or A 749M unless otherwise provided herein.

5. Ordering Information

5.1 Orders for material under this specification shall include the following information, as necessary, to adequately describe the required material:

5.1.1 ASTM specification number and date of issue,

5.1.2 Name of material (hot-rolled sheet or strip, drawing quality, special killed),

5.1.3 Finish (indicate descaled, if required) (9.1),

5.1.4 Oiling (9.2),

5.1.5 Edges (9.3),

5.1.6 Dimensions (decimal thickness and width),

5.1.7 Coil size and weight (include inside

---

[1] This specification is under the jurisdiction of ASTM Committee A-1 on Steel, Stainless Steel and Related Alloys, and is the direct responsibility of Subcommittee A01.19 on Steel Sheet and Strip.
Current edition approved March 26, 1982. Published May 1982. Originally published as A 622 - 68. Last previous edition A 622 - 75.
[2] *Annual Book of ASTM Standards*, Vol 01.03.

A 622/A 622M

diameter, outside diameter, and maximum weight),

5.1.8 Quantity,

5.1.9 Application (show part identification and description),

5.1.10 Special requirements (if required), and

5.1.11 Cast or heat (formerly ladle) analysis (if required) (11.1).

NOTE—A typical ordering description is as follows: "ASTM A 622 dated _____; hot-rolled, drawing quality, special-killed sheet; 0.089 by 48 in. by coil, ID 24 in., OD 48 in., max; weight 15 000 lb; 100 000 lb; for compressor housing" or

"ASTM A 622M dated _____; hot-rolled, drawing quality, special-killed sheet; 2.2 by 1180 mm by coil, ID 600 mm, OD 1500 mm, max; coil weight 10 000 kg, max; 50 000 kg, for compressor housing."

6. Manufacture

6.1 *Melting Practice*—The sheet shall be produced from special-killed steel. It is normally produced from aluminum-killed steel but may be otherwise deoxidized provided the material is capable of meeting the specified requirements.

7. Chemical Requirements

7.1 The cast or heat analysis of the steel shall conform to the requirements as to chemical composition prescribed in Table 1.

8. Mechanical Requirements

8.1 The sheet and strip shall be suitable for the production of identified deep drawn parts. The manufacturer shall assume responsibility for selection of steel, control of processing, and ability of the material to form identified parts within properly established breakage limits. (Refer to Specifications A 568 or A 568M, Appendix X4.)

8.2 Drawing quality, special killed is required when drawing quality in accordance with Specification A 621/A 621M will not provide a sufficient degree of ductility for the fabrication of parts having stringent drawing requirements.

8.3 Refer to Appendix X3 of Specification A 568M for additional information.

9. Finish and Appearance

9.1 *Surface Finish:*

9.1.1 Unless otherwise specified, the material shall be furnished as rolled, that is, without removing the hot-rolled oxide or scale.

9.1.2 When required, the material may be specified to be pickled or blast cleaned (descaled).

9.2 *Oiling:*

9.2.1 Unless otherwise specified, hot-rolled as-rolled material shall be furnished not oiled (that is, dry), and hot-rolled pickled or blast-cleaned material shall be furnished oiled.

9.2.2 When required, as-rolled material may be specified to be furnished oiled and pickled, or blast-cleaned material may be specified to be furnished dry.

9.3 *Edges*—Unless otherwise specified, sheet and strip shall have the "standard" edge designated below. When required, the "optional" edge may be specified.

	Edge	
Product	Standard	Optional (Must Be Specified)
Sheet (cut lengths and coils)	cut	mill
Strip Widths 2.00 in. (50 mm) and less	mill or cut	(must be specified)
Widths Over 2.00 in. (50 mm)	cut	mill

10. Marking

10.1 In addition to the requirements of Specifications A 568, A 568M, A 749, or A 749M, each lift or coil shall be marked with the designation "DQ SK."

11. Certification and Reports

11.1 While not normally required for this specification, upon request of the purchaser in the contract or order, a manufacturer's certification that the material was produced in accordance with this specification shall be furnished. The cast or heat analysis may be reported, if required.

TABLE 1 Chemical Requirements[A]

Element	Composition, max, %
Carbon	0.10
Manganese	0.50
Phosphorus	0.025
Sulfur	0.035

[A] If aluminum is used as the deoxidizing agent, the aluminum content total by product analysis is usually in excess of 0.010 %.

The American Society for Testing and Materials takes no position respecting the validity of any patent rights asserted in connection with any item mentioned in this standard. Users of this standard are expressly advised that determination of the validity of any such patent rights, and the risk of infringement of such rights, are entirely their own responsibility.

This standard is subject to revision at any time by the responsible technical committee and must be reviewed every five years and if not revised, either reapproved or withdrawn. Your comments are invited either for revision of this standard or for additional standards and should be addressed to ASTM Headquarters. Your comments will receive careful consideration at a meeting of the responsible technical committee, which you may attend. If you feel that your comments have not received a fair hearing you should make your views known to the ASTM Committee on Standards, 1916 Race St., Philadelphia, Pa. 19103.

Last ASTM Designation: A 628 – 74

Standard Specification for
TOOL-RESISTING COMPOSITE STEEL PLATES FOR SECURITY APPLICATIONS

This specification covers requirements for performance characteristics including simulated service tests and testing equipment for determining the characteristics of composite tool-resisting plates.

Formerly under the jurisdiction of Committee A-1 on Steel, Stainless Steel and Related Alloys. This specification was discontinued in 1983, without replacement.

Designation: A 635 – 81

Standard Specification for
HOT-ROLLED CARBON STEEL SHEET AND STRIP, COMMERCIAL QUALITY, HEAVY-THICKNESS COILS (FORMERLY PLATE)[1]

This standard is issued under the fixed designation A 635; the number immediately following the designation indicates the year of original adoption or, in the case of revision, the year of last revision. A number in parentheses indicates the year of last reapproval. A superscript epsilon (ϵ) indicates an editorial change since the last revision or reapproval.

This specification has been approved for use by agencies of the Department of Defense and for listing in the DoD Index of Specifications and Standards.

1. Scope

1.1 This specification covers hot-rolled, heavy-thickness, carbon-steel sheet and strip of commercial quality. The material is available in coil form only and was formerly designated "plate coils."

NOTE 1—A complete metric companion to Specification A 635 has been developed—A 635M; therefore, no metric equivalents are presented in this specification.

2. Applicable Documents

2.1 *ASTM Standards:*
A 700 Practices for Packaging, Marking, and Loading Methods for Steel Products for Domestic Shipment[2]
E 112 Methods for Determining Average Grain Size[3]
E 350 Methods for Chemical Analysis of Carbon Steel, Low-Alloy Steel, Silicon Electrical Steel, Ingot Iron, and Wrought Iron[4]

2.2 *Military Standards:*
MIL-STD-129 Marking for Shipment and Storage[5]
MIL-STD-163 Steel Mill Products, Preparation for Shipment and Storage[5]

2.3 *Federal Standards:*
Fed. Std. No. 123 Marking for Shipment (Civil Agencies)[5]
Fed. Std. No. 183 Continuous Identification Marking of Iron and Steel Products[5]

3. Description of Terms

3.1 *Heavy Thickness Sheet and Strip Coils*—This material is available as hot-rolled sheet and strip in coil form only, furnished in the following size classifications:

Product	Size Limits, Coils Only	
	Width, in.	Thickness, in.
Strip	over 8 to 12, incl	0.2300 to 0.5000, incl
Sheet	over 12 to 48, incl	0.2300 to 0.5000, incl
	over 48 to 72, incl	0.1800 to 0.5000, incl

4. Ordering Information

4.1 Orders for material under this specification shall include the following information, as required, to adequately describe the desired material:

4.1.1 ASTM specification number and date of issue, and grade designation or chemical composition or both.

4.1.2 Name of material (hot-rolled sheet coils or hot-rolled strip coils).

4.1.3 Copper bearing steel (when required).

4.1.4 Condition (material to this specification is furnished in the hot-rolled condition) (pickled, or blast cleaned, must be specified if required. Material so ordered will be oiled unless ordered not oiled) (see 8.1 and 8.2).

4.1.5 Edges (must be specified for hot-rolled sheet coils and strip coils) (see 8.3).

---

[1] This specification is under the jurisdiction of ASTM Committee A-1 on Steel, Stainless Steel and Related Alloys, and is the direct responsibility of Subcommittee A01.19 on Steel Sheet and Strip.
Current edition approved Sept. 25, 1981. Published December 1981. Originally published as A 635 – 70. Last previous edition A 635 – 80.
[2] *Annual Book of ASTM Standards*, Vol 01.03.
[3] *Annual Book of ASTM Standards*, Vol 03.03.
[4] *Annual Book of ASTM Standards*, Vol 03.05.
[5] Available from Naval Publications and Forms Center, 5801 Tabor Ave., Philadelphia, Pa. 19120.

4.1.6 Dimensions (decimal thickness and width of material).

4.1.7 Coil size and weight requirements (must include inside and outside diameters and maximum weight).

4.1.8 Quantity (weight).

4.1.9 Application (show part identification and description).

4.1.10 Special requirements, if required.

4.1.11 Cast or heat analysis if required.

4.1.12 Test reports, if required.

NOTE 2—A typical ordering description is as follows: ASTM A 635 dated_____; Grade 1023, Hot-Rolled Sheet Coils Pickled and Oiled, Mill Edge, 0.250 by 36 in. by coil; ID 24 in. OD 48 in., max; coil weight 15,000 lb, max; 100,000 lb for roll forming shapes.

5. Manufacture

5.1 *Melting Practice*—Hot-rolled commercial quality sheet and strip coils are normally produced from rimmed, capped, or semikilled steel. If either fine or coarse austenitic grain size practice is specified, killed steel will be furnished (see Methods E 112).

5.2 The steel shall be in the hot-rolled condition.

6. Chemical Requirements

6.1 *Cast or Heat Analysis*—An analysis of each heat or cast shall be made by the manufacturer to determine the percentages of the elements specified in Table 1. The analysis shall be from a test sample preferably taken during the pouring of the heat or cast and shall conform to the requirements in Table 1, or chemical compositions can be specified from carbon 0.08 maximum to 0.25 maximum, inclusive, % and manganese 1.65 maximum, inclusive, % that conform to the ranges and limits in Appendix A2 of Specification A 568.

6.2 *Product, Check, or Verification Analysis:*

6.2.1 A product analysis may be made by the purchaser for copper when copper-bearing steel is specified. The chemical composition thus determined shall conform to the requirements in Table 1 subject to the product analysis tolerances in Table 2.

6.2.2 Carbon, manganese, phosphorus, and sulfur are not subject to product analysis as rimmed or capped steels are characterized by a lack of uniformity in their chemical composition, and for this reason, product analysis is not technologically appropriate unless misapplication is indicated.

6.2.3 For steels other than capped or rimmed, product analysis may be made by the purchaser. The chemical composition thus determined shall conform to the requirements in Table 1 subject to the product analysis tolerances in Table 2.

6.3 *Test Methods*—In case a referee analysis is required and agreed upon to resolve a dispute concerning the results of a chemical analysis. The procedure for performing the referee analysis must be in accordance with the latest issue of Methods E 350, unless otherwise agreed upon between the manufacturer and the purchaser.

7. Dimensions and Tolerances

7.1 The permissible tolerances for dimensions shall not exceed the applicable limits specified in Tables 3 to 7 for hot-rolled and hot-rolled, pickled sheet coils and Tables 8 to 12 for hot-rolled and hot-rolled, pickled strip coils.

8. Finish and Appearance

8.1 *Surface Finish:*

8.1.1 Unless otherwise specified, the material shall be furnished without removing the hot-rolled oxide or scale.

8.1.2 When required, the material may be specified to be pickled or blast cleaned.

8.2 *Oiling:*

8.2.1 Unless otherwise specified, hot-rolled, as-rolled material shall be furnished not oiled and hot-rolled, pickled or blast-cleaned, material shall be furnished oiled.

8.2.2 When required, as-rolled material may be specified to be furnished oiled and pickled or blast-cleaned material may be specified to be furnished not oiled.

8.3 *Edges:*

8.3.1 As-rolled material has mill edges. Pickled or blast-cleaned material has cut edges; if mill-edge material is required, it must be specified.

8.3.2 When required, as-rolled material may be specified to have cut edges.

9. Workmanship

9.1 The steel shall have a workmanlike appearance and shall not have defects of a nature or degree that will be detrimental to the stamping or fabrication of finished parts.

9.2 Coils may contain some abnormal im-

A 635

perfections which render a portion of the coil unusable since the inspection of coils does not afford opportunity to remove portions containing imperfections.

10. Inspection

10.1 When purchaser's order stipulates that inspection and chemical tests (except product analyses) for acceptance on the steel be made prior to shipment from the mill, the manufacturer shall afford the purchaser's inspector all reasonable facilities to determine that the steel is being furnished in accordance with the specification. Steel sheet and strip products subject to purchaser's inspection and sampling are customarily inspected and sampled in conjunction with the manufacturer's inspection and sampling operations.

11. Rejection and Rehearing

11.1 Material that is reported to be defective subsequent to the acceptance at the purchaser's works shall be set aside, adequately protected, and correctly identified. The manufacturer shall be notified as soon as possible so that an investigation may be initiated.

11.2 Samples that are representative of the rejected material shall be made available to the manufacturer. In the event that the manufacturer is dissatisfied with the rejection, he may request a rehearing.

12. Certification and Reports

12.1 While not normally required for this specification, upon request of the purchaser in the contract or order, a manufacturer's certification that the material was produced and tested in accordance with this specification shall be furnished. A report of the cast or heat analysis may also be included if required.

13. Marking

13.1 As a minimum requirement, the material shall be identified by having the manufacturer's name, ASTM designation and grade, weight, purchaser's order number, and material identification legibly marked on a tag attached to each coil or shipping unit.

13.2 When specified in the contract or order, and for direct procurement by or direct shipment to the government, marking for shipment, in addition to requirements specified in the contract or order, shall be in accordance with MIL-STD-129 for military agencies and in accordance with Fed. Std. No. 123 for civil agencies.

13.3 For Government procurement by the Defense Supply Agency, strip material shall be continuously marked for identification in accordance with Fed. Std. No. 183.

14. Packaging

14.1 Unless otherwise specified, the sheet and strip shall be packaged and loaded in accordance with Recommended Practices A 700.

14.2 When specified in the contract or order, and for direct procurement by or direct shipment to the government, when Level A is specified, preservations, packaging, and packing shall be in accordance with the Level A requirements of MIL-STD-163.

14.3 When coils are ordered it is customary to specify a minimum or range of inside diameter, maximum outside diameter, and a maximum coil weight, if required. The ability of manufacturers to meet the maximum coil weights depends upon individual mill equipment. When required, minimum coil weights are subject to negotiation.

A 635

TABLE 1 Typical Grade Designations and Chemical Compositions (Heat Analysis)

Grade Designation	Composition Limits, %				
	C	Mn	P, max	S, max	Cu, min[A]
1006	0.08 max	0.45 max	0.040	0.050	0.20
1008	0.10 max	0.50 max	0.040	0.050	0.20
1009	0.15 max	0.60 max	0.040	0.050	0.20
1010	0.08–0.13	0.30–0.60	0.040	0.050	0.20
1012	0.10–0.15	0.30–0.60	0.040	0.050	0.20
1015	0.12–0.18	0.30–0.60	0.040	0.050	0.20
1016	0.12–0.18	0.60–0.90	0.040	0.050	0.20
1017	0.14–0.20	0.30–0.60	0.040	0.050	0.20
1018	0.14–0.20	0.60–0.90	0.040	0.050	0.20
1019	0.14–0.20	0.70–1.00	0.040	0.050	0.20
1020	0.17–0.23	0.30–0.60	0.040	0.050	0.20
1021	0.17–0.23	0.60–0.90	0.040	0.050	0.20
1022	0.17–0.23	0.70–1.00	0.040	0.050	0.20
1023	0.19–0.25	0.30–0.60	0.040	0.050	0.20
1524	0.18–0.25	1.30–1.65	0.040	0.050	0.20

[A] When specified.

TABLE 2 Tolerances for Product Analysis

Element	Limit, or Maximum of Specified Element, %	Tolerance	
		Under Minimum Limit	Over Maximum Limit
Carbon	to 0.15 incl	0.02	0.03
	over 0.15 to 0.25 incl	0.03	0.04
Manganese	to 0.60 incl	0.03	0.03
	over 0.60 to 1.15 incl	0.04	0.04
	over 1.15 to 1.65 incl	0.05	0.05
Phosphorus		...	0.01
Sulfur		...	0.01
Silicon	to 0.30 incl	0.02	0.03
	over 0.30 to 0.60 incl	0.05	0.05
Copper		0.02	...

TABLE 3 Thickness Tolerances for Heavy-Thickness Hot-Rolled Sheet Ordered to Nominal Thickness (Coils Only)

NOTE—Thickness is measured at any point across the width not less than 3/8 in. from a cut edge and not less than 3/4 in. from a mill edge. This table does not apply to the uncropped ends of mill edge coils.

Specified Width, in.	Thickness Tolerances for Widths and Thicknesses, Over and Under, in.			
	Specified Thickness, in.			
	Over 0.375 to 0.500, incl	Over 0.313 to 0.375, incl	Over 0.230 to 0.313, incl	Over 0.180 to 0.230, incl
Over 12 to 20, incl	0.014	0.012	0.010	
Over 20 to 40, incl	0.014	0.012	0.011	
Over 40 to 48, incl	0.014	0.013	0.012	
Over 48 to 60, incl	0.015	0.014	0.012	0.010
Over 60 to 72, incl	0.016	0.015	0.013	0.011
Over 72	0.018	0.016	0.015	0.012

TABLE 4 Thickness Tolerances for Heavy-Thickness Hot-Rolled Sheet Ordered to Minimum Thickness (Coils Only)

NOTE—Thickness is measured at any point across the width not less than ⅜ in. from a cut edge and not less than ¾ in. from a mill edge. This table does not apply to the uncropped ends of mill edge coils.

	Thickness Tolerances Plus Only in.			
Specified Width, in. (mm)	Specified Minimum Thickness, in.			
	Over 0.375 to 0.500, incl	Over 0.313 to 0.375, incl	Over 0.230 to 0.313, incl	Over. 0.180 to 0.230, incl
Over 12 to 20, incl	0.028	0.024	0.020	
Over 20 to 40, incl	0.028	0.024	0.022	
Over 40 to 48, incl	0.028	0.026	0.024	
Over 48 to 60, incl	0.030	0.028	0.024	0.020
Over 60 to 72, incl	0.032	0.030	0.026	0.022
Over 72	0.036	0.032	0.030	0.024

TABLE 5 Width Tolerances for Heavy-Thickness Mill Edge Sheet
(Coils Only)

NOTE—This table does not apply to the uncropped end of mill-edge coils.

Specified Width, in.	Tolerance Over Specified Width, in. (No Tolerance Under)
Over 12 to 14, incl	⁷⁄₁₆
Over 14 to 17, incl	½
Over 17 to 19, incl	⁹⁄₁₆
Over 19 to 21, incl	⅝
Over 21 to 24, incl	¹¹⁄₁₆
Over 24 to 26, incl	¹³⁄₁₆
Over 26 to 28, incl	¹⁵⁄₁₆
Over 28 to 35, incl	1⅛
Over 35 to 50, incl	1¼
Over 50 to 60, incl	1½
Over 60 to 65, incl	1⅝
Over 65 to 70, incl	1¾
Over 70 to 72, incl	1⅞

TABLE 6 Width Tolerances for Heavy-Thickness Cut-Edge Sheet
(Coils Only)

Specified Width, in.	Tolerance Over Specified Width, in. (No Tolerance Under)
Over 12 to 30, incl	⅛
Over 30 to 48, incl	³⁄₁₆
Over 48 to 60, incl	¼
Over 60 to 72, incl	⁵⁄₁₆

TABLE 7 Camber Tolerances for Heavy-Thickness Sheet
(Coils Only)

NOTE—Camber is the deviation of a side edge from a straight line. Such a deviation is measured by placing a straightedge on the concave side and measuring the greatest distance between the sheet edge and the straightedge.

Camber should not exceed 1 in. in any 20 ft of length.

TABLE 8 Thickness Tolerances for Hot-Rolled Heavy-Thickness Strip Ordered to Minimim Thickness
(Coils Only)

NOTE—Thickness measurements are taken ⅜ in. from edge of strip. These tolerances do not include crown and therefore the tolerances given in Table 10 are in addition to this table.

	Tolerances for Specified Thickness for Widths Given—Over and Under, in.		
Specified Widths, in.	0.5000 to 0.3751	0.3750 to 0.3125	0.3124 to 0.2300
Over 8 to 12, incl	0.020	0.018	0.016

TABLE 9 Thickness Tolerances for Hot-Rolled Heavy-Thickness Strip Ordered to Nominal Thickness
(Coils Only)

NOTE—Thickness measurements are taken ⅜ in. from edge of strip. These tolerances do not include crown and therefore the tolerances given in Table 10 are in addition to this table.

	Tolerances for Specified Thickness for Widths Given—Over and Under, in.		
Specified Widths, in.	0.5000 to 0.3751	0.3750 to 0.3125	0.3124 to 0.2300
Over 8 to 12, incl	0.010	0.009	0.008

TABLE 10 Crown Tolerances for Heavy-Thickness Strip
(Coils Only)

NOTE—Strip may be thicker at the center than at a point ⅜ in. from the edge the amounts given in the table.

Specified Widths, in.	Crown Tolerances for Specified Thickness and Width Given, in.
	0.2300 to 0.500 incl
Over 8 to 12, incl	0.002

A 635

TABLE 11 Width Tolerances for Heavy-Thickness Strip
(Coils Only)

Specified Width, in.	Tolerances for Specified Width for Thickness Given, Over and Under, in.	
	Mill Edge and Square Edge All Thicknesses	Slit or Cut Edge
Over 8 to 12	3/16	A

[A] Manufacturer must be consulted.

TABLE 12 Camber Tolerances for Heavy-Thickness Strip
(Coils Only)

NOTE—Camber is the deviation of a side edge from a straight line. Such a deviation is obtained by placing an 8-ft straightedge on the concave side and measuring the greatest distance between the strip edge and the straightedge.

For strip over 8 in. to 12 in., incl	¼ in. in any 8 ft

The American Society for Testing and Materials takes no position respecting the validity of any patent rights asserted in connection with any item mentioned in this standard. Users of this standard are expressly advised that determination of the validity of any such patent rights, and the risk of infringement of such rights, are entirely their own responsibility.

This standard is subject to revision at any time by the responsible technical committee and must be reviewed every five years and if not revised, either reapproved or withdrawn. Your comments are invited either for revision of this standard or for additional standards and should be addressed to ASTM Headquarters. Your comments will receive careful consideration at a meeting of the responsible technical committee, which you may attend. If you feel that your comments have not received a fair hearing you should make your views known to the ASTM Committee on Standards, 1916 Race St., Philadelphia, Pa. 19103.

ASTM Designation: A 635M – 81
Metric

Standard Specification for
HOT-ROLLED CARBON STEEL SHEET AND STRIP, COMMERCIAL QUALITY, HEAVY-THICKNESS COILS (FORMERLY PLATE) [METRIC][1]

This standard is issued under the fixed designation A 635M; the number immediately following the designation indicates the year of original adoption or, in the case of revision, the year of last revision. A number in parentheses indicates the year of last reapproval. A superscript epsilon (ϵ) indicates an editorial change since the last revision or reapproval.

1. Scope

1.1 This specification covers hot-rolled, heavy-thickness, carbon steel sheet and strip of commercial quality. The material is available in coil form only and was formerly designated "plate coils."

NOTE 1—This metric specification is equivalent to A 635, and is compatible in technical content.

	Size Limits, Coils Only			
	Width, mm		Thickness, mm	
Product	Over	Through	Over	Through
Strip	200	300	6.0	12.5
Sheet	300	1200	6.0	12.5
	1200	1800	4.5	12.5

2. Applicable Documents

2.1 *ASTM Standards:*
A 700 Practices for Packaging, Marking, and Loading Methods for Steel Products for Domestic Shipment[2]
E 112 Methods for Determining Average Grain Size[3]
E 350 Methods for Chemical Analysis of Carbon Steel, Low-Alloy Steel, Silicon Electrical Steel, Ingot Iron, and Wrought Iron[4]

2.2 *Military Standards:*
MIL-STD-129 Marking for Shipment and Storage[5]
MIL-STD-163 Steel Mill Products, Preparation for Shipment and Storage[5]

2.3 *Federal Standards:*
Fed. Std. No. 123 Marking for Shipment (Civil Agencies)[5]
Fed. Std. No. 183 Continuous Identification Marking of Iron and Steel Products[5]

3. Description of Terms

3.1 *Heavy Thickness Sheet and Strip Coils*—This material is available as hot-rolled sheet and strip in coil form only furnished in the following classifications:

4. Ordering Information

4.1 Orders for material under this specification shall include the following information, as required, to adequately describe the desired material:

4.1.1 ASTM specification number and date of issue, and grade designation or chemical composition or both.

4.1.2 Name of material (hot-rolled sheet coils or hot-rolled strip coils).

4.1.3 Copper-bearing steel (when required).

4.1.4 Condition (material to this specification is furnished in the hot-rolled condition) (pickled, or blast-cleaned, must be specified if required. Material so ordered will be oiled unless ordered not oiled) (see 8.1 and 8.2).

4.1.5 Edges (must be specified for hot-rolled sheet coils and strip coils) (see 8.3).

---

[1] This specification is under the jurisdiction of ASTM Committee A-1 on Steel, Stainless Steel and Related Alloys and is the direct responsibility of Subcommittee A01.19 on Steel Sheet and Strip.
Current edition approved Sept. 25, 1981. Published December 1981. Originally published as A 635M – 77. Last previous edition A 635M – 80.
[2] *Annual Book of ASTM Standards*, Vol 01.03.
[3] *Annual Book of ASTM Standards*, Vol 03.03.
[4] *Annual Book of ASTM Standards*, Vol 03.05.
[5] Available from Naval Publications and Forms Center, 5801 Tabor Ave., Philadelphia, Pa. 19120.

4.1.6 Dimensions (decimal thickness and width of material).

4.1.7 Coil size and weight requirements (must include inside and outside diameters and maximum weight).

4.1.8 Quantity (weight).

4.1.9 Application (show part identification and description).

4.1.10 Special requirements if required.

4.1.11 Cast or heat analysis, if required.

4.1.12 Test report, if required.

NOTE 2—A typical ordering description is as follows: ASTM A 635M dated —. Grade 1021, Hot-Rolled Sheet Coils, Pickled and Oiled, Mill Edge, 8.00 by 900 mm by coil; ID 600 mm, OD 1200 mm, max; coil weight 6000 kg, max; 50 000 kg for roll forming shapes.

5. Manufacture

5.1 *Melting Practice*—Hot-rolled commercial quality sheet and strip coils are normally produced from rimmed, capped, or semikilled steel. If either fine or coarse austenitic grain size practice is specified, killed steel will be furnished.

5.2 The steel shall be in the hot-rolled condition.

6. Chemical Requirements

6.1 *Cast or Heat Analysis*—An analysis of each heat or cast shall be made by the manufacturer to determine the percentages of the elements specified in Table 1. The analysis shall be from a test sample preferably taken during the pouring of the heat or cast, and shall conform to the requirements in Table 1, or chemical compositions can be specified from carbon 0.08 maximum to 0.25 maximum, inclusive, % and manganese 1.65 maximum, inclusive, % that conform to the ranges and limits in Appendix A2 of Specification A 568M.

6.2 *Product, Check, or Verification Analysis:*

6.2.1 A product analysis may be made by the purchaser for copper when copper-bearing steel is specified. The chemical composition thus determined shall conform to the requirements in Table 1 subject to the product analysis tolerances in Table 2.

6.2.2 Carbon, manganese, phosphorus, and sulfur are not subject to product analysis as rimmed or capped steels are characterized by a lack of uniformity in their chemical composition, and for this reason, product analysis is not technologically appropriate unless misapplication is indicated.

6.2.3 For steels other than capped or rimmed, product analysis may be made by the purchaser. The chemical composition thus determined shall conform to the requirements in Table 1 subject to the product analysis tolerances in Table 2.

6.3 *Test Methods*—In case a referee analysis is required and agreed upon to resolve a dispute concerning the results of a chemical analysis, the procedure for performing the referee analysis must be in accordance with the latest issue of Methods E 350, unless otherwise agreed upon between the manufacturer and the purchaser.

7. Dimensions and Tolerances

7.1 The permissible tolerances for dimensions shall not exceed the applicable limits specified in Tables 3 to 7 for hot-rolled and hot-rolled, pickled sheet coils and Tables 8 to 12 for hot-rolled and hot-rolled, pickled strip coils.

8. Finish and Appearance

8.1 *Surface Finish:*

8.1.1 Unless otherwise specified, the material shall be furnished without removing the hot-rolled oxide or scale.

8.1.2 When required, the material may be specified to be pickled or blast-cleaned.

8.2 *Oiling:*

8.2.1 Unless otherwise specified, hot-rolled, as-rolled material shall be furnished not oiled, and hot-rolled, pickled or blast-cleaned material shall be furnished oiled.

8.2.2 When required, as-rolled material may be specified to be furnished oiled and pickled or blast-cleaned material may be specified to be furnished not oiled.

8.3 *Edges:*

8.3.1 As-rolled material has mill edges. Pickled or blast-cleaned material has cut edges: if mill-edge material is required, it must be specified.

8.3.2 When required, as-rolled material may be specified to have cut edges.

9. Workmanship

9.1 The steel shall have a workmanlike appearance and shall not have imperfections of a nature or degree that will be detrimental to the stamping or fabrication of finished parts.

9.2 Coils may contain some abnormal imperfections that render a portion of the coil unusable since the inspection of coils does not afford opportunity to remove portions containing imperfections.

10. Inspection

10.1 When the purchaser's order stipulates that inspection and chemical tests (except product analyses) for acceptance on the steel be made prior to shipment from the mill, the manufacturer shall afford the purchaser's inspector all reasonable facilities to determine that the steel is being furnished in accordance with the specifications. Steel sheet and strip products subject to purchaser's inspection and sampling are customarily inspected and sampled in conjunction with the manufacturer's inspection and sampling operations.

11. Rejection and Rehearing

11.1 Material that is reported to be defective subsequent to the acceptance at the purchaser's works shall be set aside, adequately protected, and correctly identified. The manufacturer shall be notified as soon as possible so that an investigation may be initiated.

11.2 Samples that are representative of the rejected material shall be made available to the manufacturer. In the event that the manufacturer is dissatisfied with the rejection, he may request a rehearing.

12. Certification and Reports

12.1 While not normally required for this specification, upon request of the purchaser in the contract or order, a manufacturer's certification that the material was produced and tested in accordance with this specification shall be furnished. A report of the cast or heat analysis may also be included if required.

13. Marking

13.1 As a minimum requirement, the material shall be identified by having the manufacturer's name, ASTM designation and grade, weight, purchaser's order number, and material identification legibly stenciled on the outside wrap of each coil or shown on a tag attached to each coil or shipping unit.

13.2 When specified in the contract or order, and for direct procurement by or direct shipment to the government, marking for shipment, in addition to requirements specified in the contract or order, shall be in accordance with MIL-STD-129 for military agencies and in accordance with Fed. Std. No. 123 for civil agencies.

13.3 For Government procurement by the Defense Supply Agency, material shall be continuously marked for identification in accordance with Fed. Std. No. 183.

14. Packaging

14.1 Unless otherwise specified, the sheet and strip shall be packaged and loaded in accordance with Practices A 700.

14.2 When specified in the contract or order, and for direct procurement by or direct shipment to the government, when Level A is specified, preservations, packaging, and packing shall be in accordance with the Level A requirements of MIL-STD-163.

14.3 When coils are ordered it is customary to specify a minimum or range of inside diameter, maximum outside diameter, and a maximum coil weight, if required. The ability of manufacturers to meet the maximum coil weights depends upon individual mill equipment. When required, minimum coil weights are subject to negotiation.

A 635M

TABLE 1 Typical Grade Designations and Chemical Compositions (Heat Analysis)

Grade Designation	Composition Limits, %				
	C	Mn	P, max	S, max	Cu, min[A]
1006	0.08 max	0.45 max	0.040	0.050	0.20
1008	0.10 max	0.50 max	0.040	0.050	0.20
1009	0.15 max	0.60 max	0.040	0.050	0.20
1010	0.08–0.13	0.30–0.60	0.040	0.050	0.20
1012	0.10–0.15	0.30–0.60	0.040	0.050	0.20
1015	0.12–0.18	0.30–0.60	0.040	0.050	0.20
1016	0.12–0.18	0.60–0.90	0.040	0.050	0.20
1017	0.14–0.20	0.30–0.60	0.040	0.050	0.20
1018	0.14–0.20	0.60–0.90	0.040	0.050	0.20
1019	0.14–0.20	0.70–1.00	0.040	0.050	0.20
1020	0.17–0.23	0.30–0.60	0.040	0.050	0.20
1021	0.17–0.23	0.60–0.90	0.040	0.050	0.20
1022	0.17–0.23	0.70–1.00	0.040	0.050	0.20
1023	0.19–0.25	0.30–0.60	0.040	0.050	0.20
1524	0.18–0.25	1.30–1.65	0.040	0.050	0.20

[A] When specified.

TABLE 2 Tolerances for Product Analysis

Element	Limit or Maximum of Specified Element, %	Tolerance	
		Under Minimum Limit	Over Maximum Limit
Carbon	to 0.15 incl	0.02	0.03
	over 0.15 to 0.25 incl	0.03	0.04
Manganese	to 0.60 incl	0.03	0.03
	over 0.60 to 1.15 incl	0.04	0.04
	over 1.15 to 1.65 incl	0.05	0.05
Phosphorus		...	0.01
Sulfur		...	0.01
Silicon	to 0.30 incl	0.02	0.03
	over 0.30 to 0.60 incl	0.05	0.05
Copper		0.02	...

TABLE 3 Thickness Tolerances for Heavy-Thickness Hot-Rolled Sheet Steel Ordered to Nominal Thickness (Coils Only)

NOTE—Thickness is measured at any point across the width not less than 10 mm from a cut edge and not less than 20 mm from a mill edge. This table does not apply to the uncropped ends of mill edge coils.

Specified Width, mm		Thickness Tolerance, Plus and Minus, mm, for Specified Nominal Thickness, mm			
Over	Through	Over 4.5 to 6.0, incl	Over 6.0 to 8.0, incl	Over 8.0 to 10.0, incl	Over 10.0 to 12.5, incl
300	600	...	0.28	0.30	0.32
600	1200	...	0.30	0.32	0.35
1200	1500	0.25	0.30	0.35	0.38
1500	1800	0.28	0.32	0.38	0.40

TABLE 4 Thickness Tolerances for Heavy-Thickness Hot-Rolled Sheet Steel Ordered to Minimum Thickness (Coils Only)

NOTE—Thickness is measured at any point across the width not less than 10 mm from a cut edge and not less than 20 mm from a mill edge. This table does not apply to the uncropped ends of mill edge coils.

Specified Width, mm		Thickness Tolerance, Plus Only, mm, for Specified Minimum Thickness, mm			
Over	Through	Over 4.5 to 5.5, incl	Over 5.5 to 8.0, incl	Over 8.0 to 10.0, incl	Over 10.0 to 12.5, incl
300	600	...	0.55	0.60	0.65
600	1200	...	0.60	0.65	0.70
1200	1500	0.50	0.60	0.70	0.75
1500	1800	0.55	0.65	0.75	0.80

A 635M

TABLE 5 Width Tolerances for Heavy-Thickness Mill-Edge Sheet Steel (Coils Only)

NOTE—This table does not apply to the uncropped end of mill edge coils.

Specified Width, mm		Width Tolerance, Plus Only, mm
Over	Through	
300	600	16
600	1200	28
1200	1500	38
1500	1800	45

TABLE 6 Width Tolerances for Heavy-Thickness Cut-Edge Sheet Steel (Coils Only)

Specified Width, mm		Width Tolerance, Plus Only, mm
Over	Through	
300	600	3
600	1200	5
1200	1500	6
1500	1800	8

TABLE 7 Camber Tolerances for Heavy-Thickness Sheet Steel (Coils Only)

NOTE—Camber is the deviation of a side edge from a straight line. Such a deviation is measured by placing a straightedge on the concave side and measuring the greatest distance between the sheet edge and the straightedge.

Camber should not exceed 25 mm in any 6000 mm of length

TABLE 8 Thickness Tolerances for Hot-Rolled Heavy-Thickness Strip Ordered to Minimum Thickness (Coils Only)

NOTE—Thickness measurements for Table 8 are taken 10.0 mm from the edge of the strip. These tolerances do not include crown and therefore the tolerances given in Table 10 are in addition to Table 8.

Specified Widths, mm		Thickness Tolerances, Plus Only, mm, for Specified Minimum Thickness, mm	
Over	Through	Over 6.0 to 8.0, incl	Over 8.0 to 12.5, incl
200	300	0.40	0.50

TABLE 9 Thickness Tolerances for Hot-Rolled Heavy-Thickness Strip Steel Ordered to Nominal Thickness (Coils Only)

NOTE—Thickness measurements for Table 9 are taken 10.0 mm from the edge of the strip. These tolerances do not include crown and therefore the tolerances given in Table 10 are in addition to Table 9.

Specified Width, mm		Thickness Tolerances, Plus and Minus, mm, for Specified Nominal Thickness, mm	
Over	Through	Over 6.0 to 8.0, incl	Over 8.0 to 12.5, incl
200	300	0.20	0.25

A 635M

TABLE 10 Crown Tolerances for Heavy-Thickness Strip Steel (Coils Only)

Note—Strip may be thicker at the center than at a point 10.0 mm from the edge of the amounts given in the table.

Specified Widths, mm		Crown Tolerances for Specified Thickness and Width Given, mm
Over	Through	Over 6.0 to 12.5, incl
200	300	0.05

TABLE 11 Width Tolerances for Heavy-Thickness Strip Steel (Coils Only)

Specified Widths, mm		Tolerances for Specified Width for Thickness Given, Plus and Minus, mm	
		Mill Edge and Square Edge, All Thicknesses	Slit or Cut Edge
Over	Through		
200	300	5	A

[A] Manufacturer must be consulted.

TABLE 12 Camber Tolerances for Heavy-Thickness Strip Steel (Coils Only)

Note—Camber is the deviation of a side edge from a straight line. Such a deviation is obtained by placing a 2000-mm straightedge on the concave side and measuring the greatest distance between the strip edge and the straightedge.

For strip over 200 mm to 300 mm, incl	5 mm in any 2000 mm

The American Society for Testing and Materials takes no position respecting the validity of any patent rights asserted in connection with any item mentioned in this standard. Users of this standard are expressly advised that determination of the validity of any such patent rights, and the risk of infringement of such rights, are entirely their own responsibility.

This standard is subject to revision at any time by the responsible technical committee and must be reviewed every five years and if not revised, either reapproved or withdrawn. Your comments are invited either for revision of this standard or for additional standards and should be addressed to ASTM Headquarters. Your comments will receive careful consideration at a meeting of the responsible technical committee, which you may attend. If you feel that your comments have not received a fair hearing you should make your views known to the ASTM Committee on Standards, 1916 Race St., Philadelphia, Pa. 19103.

Designation: A 659 – 72 (Reapproved 1979)

An American National Standard

Standard Specification for
STEEL, CARBON (0.16 MAXIMUM TO 0.25 MAXIMUM, PERCENT), HOT-ROLLED SHEET AND STRIP, COMMERCIAL QUALITY[1]

This standard is issued under the fixed designation A 659; the number immediately following the designation indicates the year of original adoption or, in the case of revision, the year of last revision. A number in parentheses indicates the year of last reapproval. A superscript epsilon (ϵ) indicates an editorial change since the last revision or reapproval.

This specification has been approved for use by agencies of the Department of Defense and for listing in the DoD Index of Specifications and Standards.

1. Scope

1.1 This specification covers hot-rolled carbon steel sheet and strip of commercial quality, in coils and cut lengths, in which the maximum of the specified carbon range is over 0.15 and not over 0.25 percent and the maximum of the specified manganese range is not over 0.90 percent. This material is ordered to chemical composition.

1.2 This specification is not applicable to the steels covered by Specifications A 414/A 414M, A 569, A 570, and A 635.

2. Applicable Documents

2.1 *ASTM Standards:*
A 370 Methods and Definitions for Mechanical Testing of Steel Products[2]
A 414/A 414M Specification for Carbon Steel Sheet for Pressure Vessels[2]
A 568 Specification for General Requirements for Steel, Carbon and High-Strength Low-Alloy, Hot-Rolled Sheet and Cold-Rolled Sheet[2]
A 569 Specification for Steel, Carbon (0.15 Maximum, Percent), Hot-Rolled-Sheet and Strip, Commercial Quality[2]
A 570 Specification for Hot-Rolled Carbon Steel Sheet and Strip, Structural Quality[2]
A 635 Specification for Hot-Rolled Carbon Steel Sheet and Strip, Commercial Quality, Heavy-Thickness Coils (Formerly Plate)[2]

3. General Requirements for Delivery

3.1 Material furnished under this specification shall conform to the applicable requirements of the current edition of Specification A 568, unless otherwise provided herein.

4. Ordering Information

4.1 Orders for material under this specification shall include the following information, as required, to describe the required material adequately:

4.1.1 ASTM designation number and date of issue,

4.1.2 Name of material (hot-rolled sheet or strip, commercial quality),

4.1.3 Grade designation or chemical composition, or both,

4.1.4 Copper-bearing steel (if required),

4.1.5 Condition (as-rolled, pickled, or blast cleaned),

4.1.5.1 Material to this specification is furnished in the hot-rolled condition. Pickled (or blast cleaned) should be specified if required,

4.1.6 Type of edge (see 8.1),

4.1.7 Specify oiled or not oiled, as required (see 8.2),

4.1.8 Dimensions (thickness, width, and whether cut lengths or coils),

4.1.9 Coil size (must include inside diameter, outside diameter, and maximum weight),

4.1.10 Application (part identification and description),

4.1.11 Special requirements, if required, and

---

[1] This specification is under the jurisdiction of ASTM Committee A-1 on Steel and is the direct responsibility of Subcommittee A01.19 on Steel Sheet and Strip.
Current edition approved May 1, 1972, Published June 1972.
[2] *Annual Book of ASTM Standards*, Vol 01.03.

A 659

4.1.12 Cast or heat (formerly ladle) analysis report (request, if required) (See 9.1).

NOTE—A typical ordering description is as follows: ASTM A 659 dated ____. Hot-Rolled Sheet, Commercial Quality, Grade 1017, Cut Edge, Pickled, Oiled, 0.075 by 36 by 96 in. for Part No. 6509, Shelf Leg.

5. Manufacture

5.1 *Condition*:

5.1.1 Unless otherwise specified, the material is furnished in the as-rolled condition (not annealed, not pickled).

6. Chemical Requirements

6.1 *The Cast or Heat (formerly Ladle) Analysis* of the steel shall conform to the chemical requirements shown in Table 1, or chemical compositions can be specified from carbon 0.16 maximum to 0.25 maximum, inclusive, percent and manganese 0.90 maximum, inclusive, percent which conform to the ranges and limits in Appendix A2 of Specification A 568.

6.2 Where material is used for fabrication by welding, care must be exercised in selection of the chemical composition to assure compatibility with the welding process and its effects on altering the properties of the steel.

7. Bend Test

7.1 The material shall be capable of being bent at room temperature through 180 deg in any direction to inside diameters as shown in Table 2 without cracking on the outside of the bent portion (see Section 17 of Methods and Definitions A 370). When steel is subject to bending in a fabricating operation, more liberal bend radii should be used.

8. Finish and Appearance

8.1 *Edge*:

8.1.1 Sheet can be supplied with mill edge or cut edge.

8.1.2 Strip can be supplied with mill edge or slit (cut) edge.

8.2 *Oiling*:

8.2.1 Hot-rolled, non-pickled material is commonly furnished not oiled while hot-rolled pickled (or blast cleaned) material is commonly furnished oiled. When required, pickled (or blast cleaned) material may be specified to be furnished not oiled, and non-pickled material may be specified to be furnished oiled.

9. Certification and Reports

9.1 When requested, the producer shall furnish copies of a report showing test results of the cast or heat analysis. The report shall include the purchase order number, the ASTM designation number, and the cast or heat number representing the material.

TABLE 1 Typical Grade Designations and Chemical Compositions[A]

UNS Designation[B]	Grade Designation	Carbon, percent	Manganese, percent	Phosphorus, max, percent	Sulfur, max, percent
G10150	1015	0.12–0.18	0.30–0.60	0.040	0.050
G10160	1016	0.12–0.18	0.60–0.90	0.040	0.050
G10170	1017	0.14–0.20	0.30–0.60	0.040	0.050
G10180	1018	0.14–0.20	0.60–0.90	0.040	0.050
G10200	1020	0.17–0.23	0.30–0.60	0.040	0.050
G10210	1021	0.17–0.23	0.60–0.90	0.040	0.050
G10230	1023	0.19–0.25	0.30–0.60	0.040	0.050

[A] Copper, when specified, shall have a minimum content of 0.20 percent by cast or heat analysis.
[B] New designation established in accordance with ASTM E 527 and SAE J1086, Recommended Practice for Numbering Metals and Alloys (UNS).

TABLE 2 Bend test Requirements

Maximum of the Specified Manganese Range, percent	Ratio of Inside Bend Diameter to Thickness of Specimen
To 0.60 incl	1
Over 0.60 to 0.90 incl	1½

The American Society for Testing and Materials takes no position respecting the validity of any patent rights asserted in connection with any item mentioned in this standard. Users of this standard are expressly advised that determination of the validity of any such patent rights, and the risk of infringement of such rights, are entirely their own responsibility.

This standard is subject to revision at any time by the responsible technical committee and must be reviewed every five years and if not revised, either reapproved or withdrawn. Your comments are invited either for revision of this standard or for additional standards and should be addressed to ASTM Headquarters. Your comments will receive careful consideration at a meeting of the responsible technical committee, which you may attend. If you feel that your comments have not received a fair hearing you should make your views known to the ASTM Committee on Standards, 1916 Race St., Philadelphia, Pa. 19103.

ASTM Designation: A 666 – 82

Standard Specification for
AUSTENITIC STAINLESS STEEL, SHEET, STRIP, PLATE, AND FLAT BAR FOR STRUCTURAL APPLICATIONS[1]

This standard is issued under the fixed designation A 666; the number immediately following the designation indicates the year of original adoption or, in the case of revision, the year of last revision. A number in parentheses indicates the year of last reapproval. A superscript epsilon (ϵ) indicates an editorial change since the last revision or reapproval.

This specification has been approved for use by agencies of the Department of Defense and for listing in the DoD Index of Specifications Standards.

1. Scope

1.1 This specification covers six types of austenitic stainless steels in four strength levels primarily for use in architectural structural applications. Sheet, strip, plate, and flat bar forms are included.

1.2 The values stated in inch-pound units are to be regarded as the standard.

2. Applicable Documents

2.1 *ASTM Standards*:
A 370 Methods and Definitions for Mechanical Testing of Steel Products[2]
A 480 Specification for General Requirements for Flat-Rolled Stainless and Heat-Resisting Steel Plate, Sheet, and Strip[2]
A 484 Specification for General Requirements for Stainless and Heat-Resisting Wrought Steel Products (Except Wire)[3]
A 751 Methods, Practices, and Definitions for Chemical Analysis of Steel Products[4]
E 527 Practice for Numbering Metals and Alloys (UNS)[4]

3. Description of Terms

3.1 Plate, sheet, and strip as used in this specification are described as follows:

3.1.1 *Plate*—Material ³⁄₁₆ in. (4.76 mm) and over in thickness and over 10 in. (254 mm) in width.

3.1.2 *Sheet*—Material under ³⁄₁₆ in. in thickness and 24 in. (609.6 mm) and over in width.

3.1.3 *Strip*—Material under ³⁄₁₆ in. in thickness and under 24 in. in width.

3.1.4 *Flat Bar*—Hot-finished flats: ¼ to 10 in. (6.35 to 254 mm) inclusive, in width and ⅛ in. (3.17 mm) and over in thickness. Cold-finished flats: ⅜ in. (9.53 mm) and over in width, ⅛ in. and over in thickness.

4. General Requirements for Delivery

4.1 Sheet, strip, and plate material furnished under this specification shall conform to applicable requirements of the current edition of Specification A 480.

4.2 Flat bar material furnished under this specification shall conform to the applicable requirements of the current edition of Specification A 484.

5. Ordering Information

5.1 Orders for material under this specification shall include the following information:
5.1.1 Quantity (weight or number of pieces),
5.1.2 Name of material (stainless steel),
5.1.3 Form (plate, sheet, strip, or flat bar),
5.1.4 Dimensions,
5.1.5 Type (Table 1),
5.1.6 Grade: A (30 000 psi (205 MPa) minimum yield strength), B (40 000 or 45 000 psi (275 or 310 MPa) minimum yield strength), C

---

[1] This specification is under the jurisdiction of ASTM Committee A-1 on Steel, Stainless Steel and Related Alloys, and is the direct responsibility of Subcommittee A01.17 on Flat Stainless Steel Products.
Current edition approved Oct. 29, 1982. Published December 1982. Originally published as A 666 – 72. Last previous edition A 666 – 80.
[2] *Annual Book of ASTM Standards*, Vol 01.03.
[3] *Annual Book of ASTM Standards*, Vol 01.05.
[4] *Annual Book of ASTM Standards*, Vol 01.04.

(75 000 psi (515 MPa) minimum yield strength), or D (100 000 or 110 000 psi (690 or 760 MPa) minimum yield strength),

NOTE 1—All plate thicknesses and sizes of all types are not available in Grades B, C, and D. Producers should be consulted for information on availability of specific sizes and thicknesses.

5.1.7 Edge, strip only (see Specification A 480),

5.1.8 Finish (see Specification A 480; specify whether one or both sides are to be polished if polished finish is specified,

5.1.9 ASTM designation, and

5.1.10 Additions to the specification or special requirements.

NOTE 2—A typical ordering description is as follows: 200 pieces, stainless steel sheets, 0.060 by 48 by 120 in., Type 201, Grade C, No. 2B Finish, ASTM A 666, latest revision.

6. Process

6.1 If a specific type of melting is required by the purchaser, it shall be stated on the purchase order.

6.2 When specified on the purchase order, or when a specific type of melting has been specified on the purchase order, the material manufacturer shall indicate on the test report the type of melting used to produce the material.

7. Chemical Requirements

7.1 The steel shall conform to the chemical requirements specified in Table 1.

7.2 Methods and practices relating to chemical analysis shall be in accordance with Methods, Practices, and Definitions A 751.

8. Finish

8.1 The finish of products ordered to this specification shall be as specified and as described in Table 3.

9. Mechanical Requirements

9.1 *Tensile Strength Requirements*—The material shall conform to the requirements specified in Table 2.

9.2 *Bending Requirements*—The bend test specimens of material up to and including 0.187 in. (4.75 mm) thick shall withstand cold bending through the angle specified in Table 4 or Table 5, without cracking on the outside of the bent portion.

10. Special Tests

10.1 If any special tests are required which are pertinent to the intended applications of the material ordered, they shall be as agreed upon between the seller and the purchaser.

11. Test Specimens

11.1 Tension and bend test specimens shall be taken from finished material and shall be selected in the transverse direction, except in the case of strip under 9 in. (228.6 mm) in width, in which case tension test specimens shall be selected in the longitudinal direction.

11.2 The axis of the bend shall be parallel to the direction of rolling.

11.3 Flat bar tension test and bend test specimens taken from the finished material shall be selected in the longitudinal direction.

11.4 Test specimens shall conform to the appropriate sections of Methods and Definitions A 370.

12. Number of Tests

12.1 In case of sheet, strip, and plate produced in coil form, two or more tension tests (one from each end of the coil), and one or more bend tests shall be made on samples taken from each coil. When material is produced in sheet or strip form in cut lengths one tension test and one bend test shall be made, and for plate one tension test shall be made, from each 100 or less sheets, strips, or plates of the same gage and heat produced under the same processing conditions.

12.2 For flat bar products at least one tension test and one bend test shall be made on each size from each heat in a lot annealed in a single charge in a batch-type furnace or under the same conditions in a continuous furnace.

13. Test Methods

13.1 The properties enumerated in this specification shall be determined in accordance with the applicable sections of Methods and Definitions A 370.

14. Permissible Variations in Dimensions

14.1 Unless otherwise specified in the purchase order, sheet, strip, and plate material shall conform to the permissible tolerances shown in Specification A 480.

14.2 Unless otherwise specified in the order, all flat bars shall conform to the permissible

A 666

variations in dimensions shown in Tables 4, 7, 8, 9, and 10 of Specification A 484.

14.3 Sheet and strip material with a No. 1 finish, hot-rolled annealed or hot-rolled annealed and pickled plate material, and hot-rolled annealed and pickled or cold-rolled annealed and pickled flat bar material may be ground to remove surface defects, provided such grinding does not reduce the thickness or width at any point beyond the permissible variations in dimensions.

15. Workmanship

15.1 The material shall be of uniform quality consistent with good manufacturing and inspection practices. The steel shall have no defects of a nature or degree, for the grade and type ordered, that will be detrimental to the stamping, forming, machining, or fabrication of finished parts.

16. Inspection

16.1 Inspection of the material by the purchaser's representative at the producing plant shall be made as agreed upon by the purchaser and the seller as part of the purchase order.

17. Material Test Report

17.1 A report of the results of all tests required by this specification and the type of melting used when required by the purchase order shall be supplied to the purchaser.

18. Certification

18.1 Upon request of the purchaser in the contract or order, the producer's certification that the material was manufactured and tested in accordance with this specification together with a report of the test results shall be furnished at the time of shipment.

19. Rejection

19.1 Unless otherwise specified, any rejection based upon tests made in accordance with this specification shall be reported to the seller within 60 working days from the receipt of the material by the purchaser.

20. Rehearing

20.1 Samples tested by the purchaser in accordance with this specification that represent rejected material shall be retained for 3 weeks from the date of the notification of the rejection to the seller. In case of dissatisfaction with the results of the tests, the seller may make claim for a rehearing within that time.

TABLE 1 Chemical Requirements

UNS Designation[A]	Type	Composition, %								
		Carbon, max	Manganese	Phosphorus, max	Sulfur, max	Silicon, max	Chromium	Nickel	Molybdenum	Nitrogen, max
S20100	201	0.15	5.50–7.50	0.060	0.030	1.00	16.00–18.00	3.50–5.50		0.25
S20200	202	0.15	7.50–10.00	0.060	0.030	1.00	17.00–19.00	4.00–6.00		0.25
S30100	301	0.15	2.00 max	0.045	0.030	1.00	16.00–18.00	6.00–8.00		
S30200	302	0.15	2.00 max	0.045	0.030	1.00	17.00–19.00	8.00–10.00		
S30400	304	0.08	2.00 max	0.045	0.030	1.00	18.00–20.00	8.00–10.50		
S31600	316	0.08	2.00 max	0.045	0.030	1.00	16.00–18.00	10.00–14.00	2.00–3.00	

[A] New designations established in accordance with ASTM E 527 and SAE J1086, Recommended Practice for Numbering Metals and Alloys (UNS).

TABLE 2 Tensile Strength Requirements

Type	Product	Tensile Strength, min, psi (MPa)	Yield Strength, (0.2 % offset) min, psi (MPa)	Elongation in 2 in. or 50 mm, min, % Thickness, in. (mm)		
				Up to 0.015 (0.38)	Over 0.015 to 0.030 (0.38 to 0.76)	Over 0.030 (0.76)
		Grade A				
301	Plate, Sheet, Strip	75 000 (515)	30 000 (205)	40	40	40
302	Plate, Sheet, Strip	75 000 (515)	30 000 (205)	40	40	40
	Flat Bar	75 000 (515)	30 000 (205)	...	...	40
304	Plate, Sheet, Strip	75 000 (515)	30 000 (205)	40	40	40
	Flat Bar	75 000 (515)	30 000 (205)	...	...	40
316	Plate, Sheet, Strip	75 000 (515)	30 000 (205)	40	40	40
	Flat Bar	75 000 (515)	30 000 (205)	...	...	40
		Grade B				
201	Plate, Sheet, Strip	95 000 (655)	45 000 (310)	40	40	40
	Flat Bar	75 000 (515)	40 000 (275)	...	...	40
202	Plate, Sheet, Strip	90 000 (620)	45 000 (310)	40	40	40
	Flat Bar	75 000 (515)	40 000 (275)	...	...	40
301	Plate, Sheet, Strip	90 000 (620)	45 000 (310)	40	40	40
302	Plate, Sheet, Strip	85 000 (585)	45 000 (310)	40	40	40
	Flat Bar	90 000 (620)	45 000 (310)	...	...	40
304	Plate, Sheet, Strip	80 000 (550)	45 000 (310)	35	35	35
	Flat Bar	90 000 (620)	45 000 (310)	...	...	40
316	Plate, Sheet, Strip	85 000 (585)	45 000 (310)	35	35	35
	Flat Bar	90 000 (620)	45 000 (310)	...	...	40
		Grade C				
201	Plate, Sheet, Strip	125 000 (860)	75 000 (515)	20	20	20
202	Plate, Sheet, Strip	125 000 (860)	75 000 (515)	12	12	12
	Flat Bar (1 in. (25.4 mm) max)	115 000 (790)	75 000 (515)	...	...	15
301	Plate, Sheet, Strip	125 000 (860)	75 000 (515)	25	25	25
302	Plate, Sheet, Strip	125 000 (860)	75 000 (515)	10	10	12
	Flat Bar (1 in. (25.4 mm) max)	115 000 (790)	75 000 (515)	...	...	15
304	Plate, Sheet, Strip	125 000 (860)	75 000 (515)	10	10	12
	Flat Bar	115 000 (790)	75 000 (515)	...	...	15
316	Plate, Sheet, Strip	125 000 (860)	75 000 (515)	10	10	10
	Flat Bar	115 000 (790)	75 000 (515)	...	...	15
		Grade D				
201	Plate, Sheet, Strip	150 000 (1035)	110 000 (760)	9	10	10
202	Plate, Sheet, Strip	150 000 (1035)	110 000 (760)	9	10	10
	Flat Bar (¾ in. (19 mm) max)	125 000 (860)	100 000 (690)	...	...	12
301	Plate, Sheet, Strip	150 000 (1035)	110 000 (760)	15	18	18
302	Plate, Sheet, Strip	150 000 (1035)	110 000 (760)	9	10	10
	Flat Bar (¾ in. (19 mm) max)	125 000 (860)	100 000 (690)	...	...	12
304	Plate, Sheet, Strip	150 000 (1035)	110 000 (760)	6	7	7
	Flat Bar	125 000 (860)	100 000 (690)	...	...	7
316	Plate, Sheet, Strip	150 000 (1035)	110 000 (760)	6	7	7
	Flat Bar	125 000 (860)	100 000 (690)	...	...	7

ASTM A 666

TABLE 3 Types, Finishes, and Grades

Product	Finish[A]	Type					
		201	202	301	302	304	316
Plate	HRA	B	B	A	A	A	A
	HRAP	B	B	A	A	A	A
	HRAP and P	B	B	A	A	A	A
	CR	CD	C	BCD	BCD	BCD	BCD
	CR and P	CD	C	BCD	BCD	BCD	BCD
Sheet	No. 1	B	B	A	A	A	A
	No. 2D	B	B	A	A	A	A
	No. 2B	BCD	BCD	ABCD	ABCD	ABCD	ABCD
	BA	B	B	A	A	A	A
	No. 3	B	B	AB	AB	AB	AB
	No. 4	B	B	AB	AB	AB	AB
	No. 6	B	B	AB	AB	AB	AB
	No. 7	B	B	AB	AB	AB	AB
	No. 8	B	B	AB	AB	AB	AB
Strip	No. 1	B	B	A	A	A	A
	No. 2	BCD	BCD	ABCD	ABCD	ABCD	ABCD
	BA	B	B	A	A	A	A
	P	B	B	AB	AB	AB	AB
Flat bar	HRAP/CRAP	B	B	A	A	A	A
	CD/CR	...	CD	...	BCD	BCD	BCD

[A] Finish:
BA   Bright annealed
CD   Cold drawn
CR   Cold rolled
CRAP   Cold rolled, annealed, and pickled
CR and P   Cold rolled and polished
HRA   Hot rolled and annealed
HRAP   Hot rolled, annealed, and pickled
HRAP and P   Hot rolled, annealed, and polished
P   Polished

TABLE 4 Free Bend

Type and Grade	Thickness, in. (mm)			
	Up to 0.050 (1.27) incl, Angle	Bend Factor (×T)[A]	Over 0.050 (1.27) to 0.187 (4.75) incl, Angle	Bend Factor (×T)[A]
201:				
B	180	1	180	1
C[B]	180	2	180	4
D[B]	180	4	180	8
202:				
B	180	1	90	2
C[B]	180	1	90	2
D[B]	180	2	90	2
301:				
A	180	1	180	1
B[B]	180	1	90	2
C[B]	180	1	90	2
D[B]	180	2	90	2
302:				
A	180	1	180	1
B[B]	180	1	90	2
C[B]	180	1	90	2
D[B]	180	2	90	2
304:				
A	180	1	180	1
B[B]	180	1	90	2
C[B]	180	1	90	2
D[B]	180	2	90	2
316:				
A	180	1	180	1
B[B]	180	2	90	2
C[B]	180	2	90	2
D[B]	...	...	...	...

[A] Specimens shall be bent around a *diameter* equal to the product of the bend factor (×T) times the nominal thickness of the test specimen.
[B] Bend requirements for these strength levels in the polished finishes are not established and are subject to negotiation between purchaser and producer.

TABLE 5 V-Block Bend

Type and Grade	Thickness, in. (mm)			
	Up to 0.050 (1.27) incl, Angle	Bend Factor (×T)[A]	Over 0.050 (1.27) to 0.187 (4.75) incl, Angle	Bend Factor (×T)[A]
201:				
B	135	2	135	3
C[B]	135	2	135	3
D[B]	135	4	135	4
202:				
B	135	4	135	4
C[B]	135	4	135	4
D[B]	135	4	135	4
301:				
A	135	1	135	1
B[B]	135	2	135	2
C[B]	135	2	135	3
D[B]	135	4	135	4
302:				
A	135	1	135	1
B[B]	135	1	135	2
C[B]	135	2	135	3
D[B]	135	4	135	4
304:				
A	135	1	135	1
B[B]	135	2	135	2
C[B]	135	2	135	3
D[B]	135	4	135	4
316:				
A	135	1	135	1
B[B]	135	5	135	6
C[B]	135	5	135	6
D[B]	...	...	...	...

[A] Specimens shall be bent around a *diameter* equal to the product of the bend factor (×T) times the nominal thickness of the test specimen.
[B] Bend requirements for these strength levels in the polished finishes are not established and are subject to negotiation between purchaser and producer.

The American Society for Testing and Materials takes no position respecting the validity of any patent rights asserted in connection with any item mentioned in this standard. Users of this standard are expressly advised that determination of the validity of any such patent rights, and the risk of infringement of such rights, are entirely their own responsibility.

This standard is subject to revision at any time by the responsible technical committee and must be reviewed every five years and if not revised, either reapproved or withdrawn. Your comments are invited either for revision of this standard or for additional standards and should be addressed to ASTM Headquarters. Your comments will receive careful consideration at a meeting of the responsible technical committee, which you may attend. If you feel that your comments have not received a fair hearing you should make your views known to the ASTM Committee on Standards, 1916 Race St., Philadelphia, Pa. 19103.

Designation: A 679 – 77 (Reapproved 1983)

Standard Specification for
STEEL WIRE, HIGH TENSILE STRENGTH, HARD DRAWN, FOR MECHANICAL SPRINGS[1]

This standard is issued under the fixed designation A 679; the number immediately following the designation indicates the year of original adoption or, in the case of revision, the year of last revision. A number in parentheses indicates the year of last reapproval. A superscript epsilon (ε) indicates an editorial change since the last revision or reapproval.

1. Scope

1.1 This specification covers a round, uncoated, high tensile strength, hard drawn, steel spring wire in coils having properties and quality suitable for the manufacture of mechanical springs and wire forms subject to high static stresses or infrequent dynamic loading, or both.

1.2 The values stated in inch-pound units are to be regarded as the standard.

2. Applicable Documents

2.1 *ASTM Standards:*
A 370 Methods and Definitions for Mechanical Testing of Steel Products[2]
A 510 Specification for General Requirements for Wire Rods and Coarse Round Wire, Carbon Steel[3]
A 700 Recommended Practices for Packaging, Marking, and Loading Methods for Steel Products for Domestic Shipment[4]
E 4 Methods of Load Verification of Testing Machines[5]
E 29 Recommended Practice for Indicating Which Places of Figures Are to Be Considered Significant in Specified Limiting Values[6]
E 30 Methods for Chemical Analysis of Steel, Cast Iron, Open-Hearth Iron, and Wrought Iron[7]
E 380 Standard for Metric Practice[6]

3. Ordering Information

3.1 Orders for material under this specification shall include the following for each item:
3.1.1 Quantity (weight),
3.1.2 Name of material (high tensile strength hard drawn steel mechanical spring wire),
3.1.3 Dimension (Section 8),
3.1.4 Packaging (Section 10),
3.1.5 Cast or heat analysis report, if desired (Section 5), and
3.1.6 ASTM designation and date of issue.

NOTE 1—A typical ordering description is as follows: 40 000 lb, high tensile strength, hard drawn mechanical spring wire, size 0.192 in. in 1000-lb coils to ASTM A 679 dated ___.

4. Manufacture

4.1 The steel shall be made by any of the following processes: open-hearth, basic-oxygen, or electric-furnace.

4.2 A sufficient discard shall be made to ensure freedom from injurious piping and undue segregation.

4.3 The wire shall be cold drawn to produce the desired mechanical properties.

4.4 The wire finish shall be suitable for forming or coiling. It is not intended that this material be furnished with a metallic coating.

5. Chemical Requirements

5.1 The steel shall conform to the chemical composition specified in Table 1.

---

[1] This specification is under the jurisdiction of ASTM Committee A-1 on Steel, Stainless Steel and Related Alloys, and is the direct responsibility of Subcommittee A01.03 on Steel Rod and Wire.
Current edition approved June 24, 1977. Published August 1977. Originally published as A 679 – 73. Last previous edition A 679 – 75.
[2] *Annual Book of ASTM Standards*, Vol 01.04.
[3] *Annual Book of ASTM Standards*, Vol 01.03.
[4] *Annual Book of ASTM Standards*, Vols 01.01, 01.03, 01.04, and 01.05.
[5] *Annual Book of ASTM Standards*, Vol 03.01.
[6] *Annual Book of ASTM Standards*, Vol 14.02.
[7] *Annual Book of ASTM Standards*, Vol 03.05.

5.2 *Cast or Heat Analysis*—Each cast or heat of steel shall be analyzed by the manufacturer to determine the percentage of elements prescribed in Table 1. This analysis shall be made from a test specimen preferably taken during the pouring of the cast or heat. When requested, this analysis shall be reported to the purchaser.

5.3 *Product Analysis*—An analysis may be made by the purchaser from finished wire representing each cast or heat of steel. The chemical composition of samples taken for product analysis shall conform to the permissible limits of Table 10 in Specification A 510.

5.4 For referee purposes, Method E 30 shall be used.

6. Mechanical Requirements

6.1 *Tension Test:*

6.1.1 *Requirements*—The material as represented by tension test specimens shall conform to the requirements prescribed in Table 2 for the sizes ordered.

6.1.2 *Number of Tests*—One test specimen shall be taken for each ten coils, or fraction thereof, in a lot. Each heat in a given lot shall be tested.

6.1.3 *Location of Tests*—Test specimens shall be taken from either end of the coil.

6.1.4 *Test Method*—The tension test shall be made in accordance with Methods and Definitions A 370, Supplement IV.

6.2 *Wrap Test:*

6.2.1 *Requirements*—The material as represented by the wrap test specimens shall conform to the requirements specified in Table 3.

6.2.2 *Number of Tests*—One test specimen shall be taken for each ten coils, or fraction thereof, in a lot. Each heat in a given lot shall be tested.

6.2.3 *Location of Test*—Test specimens shall be taken from either end of the coil.

6.2.4 *Test Method*—The wrap test shall be made in accordance with Methods and Definitions A 370, Supplement IV.

7. Retests

7.1 If any test specimen exhibits obvious defects or shows the presence of a weld, it may be discarded and another specimen substituted.

8. Permissible Variations in Dimensions

8.1 The diameter of the wire shall not vary from that specified by more than the tolerances specified in Table 4.

9. Workmanship

9.1 The surface of the wire as received shall be smooth and free of rust. No serious die marks, scratches, or seams may be present.

9.2 The wire shall not be kinked or improperly cast. To test for cast, a convolution of wire shall be cut from the coil and allowed to fall on a flat surface. The wire shall lie substantially flat on itself and shall not spring up or show a wavy condition.

9.3 Each coil shall be one continuous length of wire, properly coiled. Welds made prior to cold drawing are permitted.

10. Packaging, Marking, and Loading

10.1 The coil weight, dimensions, and method of packaging shall be agreed upon between the manufacturer and purchaser.

10.2 The size of the wire, purchaser's order number, ASTM specification, cast (heat) number, and name or mark of the manufacturer shall be marked on a tag securely attached to each coil of wire.

10.3 When specified in the purchaser's order, packaging, marking, and loading for shipments shall be in accordance with those procedures recommended in Recommended Practices A 700.

11. Inspection

11.1 The manufacturer shall afford the inspector representing the purchaser all reasonable facilities to satisfy him that the material is being furnished in accordance with this specification. All tests (except product analysis) and inspections may be made at the place of manufacture prior to shipment, and shall be so conducted as not to interfere unnecessarily with the operation of the works.

12. Rejection and Rehearing

12.1 Unless otherwise specified, any rejection based on tests made in accordance with this specification shall be reported to the manufacturer within a reasonable length of time.

12.2 Failure of any of the test specimens to comply with the requirements of this specification shall constitute grounds for rejection of the lot represented by the specimen. The lot may be resubmitted for inspection after testing every coil for the characteristic in which the specimen failed and sorting out the defective coils.

12.3 The material must be adequately protected and correctly identified in order that the manufacturer may make a proper investigation.

TABLE 1 Chemical Requirements

Element	Composition, %
Carbon	0.65–1.00[A]
Manganese	0.20–1.30[B]
Phosphorus, max	0.040
Sulfur, max	0.050
Silicon	0.10–0.40

[A] Carbon in any one lot shall not vary more than 0.13 %.
[B] Manganese in any one lot shall not vary more than 0.30 %.

TABLE 2 Tensile Requirements

NOTE 1—Any retesting performed by the purchaser to confirm conformance to the tensile requirements of this specification shall be considered meeting the requirements if the tensile strength is within the limit of 2 % below the minimum required or 2 % above the maximum requirement (see Note 2 below). For purposes of determining conformance with this specification, the calculated tensile strength test value and the limit values shall be rounded to the nearest 1.0 ksi used in expressing the limiting value, in accordance with the rounding method of Recommended Practice E 29.

NOTE 2—The reason for tolerance limits in tension testing is recognized in Methods E 4, Section 16, and 16.1 stating, "The percentage of error for loads within the loading range of the testing machine shall not exceed ±1.0 %."

Diameter,[A] in. (mm)	Tensile Strength, ksi (MPa)	
	min	max
0.020 (0.51)	350 (2410)	387 (2670)
0.023 (0.58)	343 (2360)	380 (2620)
0.026 (0.66)	337 (2320)	373 (2570)
0.029 (0.74)	331 (2280)	366 (2520)
0.032 (0.81)	327 (2250)	361 (2490)
0.035 (0.89)	322 (2220)	356 (2450)
0.041 (1.04)	314 (2160)	347 (2390)
0.048 (1.22)	306 (2110)	339 (2340)
0.054 (1.37)	300 (2070)	331 (2280)
0.062 (1.57)	293 (2020)	324 (2230)
0.072 (1.83)	287 (1980)	317 (2190)
0.080 (2.03)	282 (1940)	312 (2150)
0.092 (2.34)	275 (1900)	304 (2100)
0.106 (2.69)	268 (1850)	296 (2040)
0.120 (3.05)	263 (1810)	290 (2000)
0.135 (3.43)	258 (1780)	285 (1970)
0.148 (3.76)	253 (1740)	279 (1920)
0.162 (4.11)	249 (1720)	275 (1900)
0.177 (4.50)	245 (1690)	270 (1860)
0.192 (4.88)	241 (1660)	267 (1840)
0.207 (5.26)	238 (1640)	264 (1820)

[A] Tensile strength values for intermediate sizes may be interpolated.

TABLE 3 Wrap Test Requirements

Diameter, in. (mm)	Mandrel Size
0.020 (0.51) to 0.162 (3.11), incl	2X[A]
Over 0.162 (3.11) to 0.207 (5.26), incl	4X

[A] The symbol "X" represents the diameter of the wire tested.

TABLE 4 Permissible Variations in Wire Diameter

NOTE—For purposes of determining conformance with this specification, all specified limits are considered absolute as defined in Recommended Practice E 29.

Diameter, in. (mm)	Permissible Variations, Plus and Minus, in. (mm)	Permissible Out-of-Round, in. (mm)
0.020 to 0.028 (0.51 to 0.71), incl	0.0008 (0.02)	0.0008 (0.02)
Over 0.028 to 0.075 (0.71 to 1.90), incl	0.001 (0.03)	0.001 (0.03)
Over 0.075 to 0.207 (1.90 to 5.26), incl	0.002 (0.05)	0.002 (0.05)

The American Society for Testing and Materials takes no position respecting the validity of any patent rights asserted in connection with any item mentioned in this standard. Users of this standard are expressly advised that determination of the validity of any such patent rights, and the risk of infringement of such rights, are entirely their own responsibility.

This standard is subject to revision at any time by the responsible technical committee and must be reviewed every five years and if not revised, either reapproved or withdrawn. Your comments are invited either for revision of this standard or for additional standards and should be addressed to ASTM Headquarters. Your comments will receive careful consideration at a meeting of the responsible technical committee, which you may attend. If you feel that your comments have not received a fair hearing you should make your views known to the ASTM Committee on Standards, 1916 Race St., Philadelphia, Pa. 19103.

Designation: A 680/A 680M – 81

Standard Specification for
STEEL, HIGH CARBON, STRIP, COLD-ROLLED HARD, UNTEMPERED QUALITY[1]

This standard is issued under the fixed designation A 680/A 680M; the number immediately following the designation indicates the year of original adoption or, in the case of revision, the year of last revision. A number in parentheses indicates the year of last reapproval. A superscript epsilon (ϵ) indicates an editorial change since the last revision or reapproval.

This specification has been approved for use by agencies of the Department of Defense and for listing in the DoD Index of Specifications and Standards.

1. Scope

1.1 This specification covers cold rolled high-carbon steel strip hard untempered. This type is a very stiff, springy product intended for flat work not requiring the ability to withstand cold forming.

1.2 This type is a cold rolled product produced with or without a preparatory thermal treatment to a minimum hardness of Rockwell B 98 or equivalent Rockwell superficial scale depending upon thickness.

1.3 The maximum of the specified carbon range is over 0.25 % to 1.35 %, incl.

1.4 This specification is applicable for orders in either inch-pound units (as A 680) or SI units (as A 680M).

2. Applicable Documents

2.1 *ASTM Standards:*

A 370 Methods and Definitions for Mechanical Testing of Steel Products[2]

A 682 Specification for General Requirements for Steel, High-Carbon, Strip, Cold-Rolled, Spring Quality[3]

A 682M Specification for General Requirements for Steel, High-Carbon, Strip, Cold-Rolled, Spring Quality [Metric][3]

E 18 Test Methods for Rockwell Hardness and Rockwell Superficial Hardness of Metallic Materials[4]

3. Definitions

3.1 *hard-type untempered carbon spring steel*—a stiff springy product produced with or without preparatory thermal treatment to arrive at full hardness after the final rolling. It is intended for flat work not requiring the ability to withstand cold forming and the parts may later be heat treated.

3.2 *Number 1 or matte (dull) finish*—without luster, produced by rolling on rolls roughened by mechanical or chemical means. This finish is especially suitable for lacquer or paint adhesion, and is beneficial in aiding drawing operations by reducing the contact friction between the die and the strip.

3.3 *No. 2 regular bright finish*—produced by rolling on rolls having a moderately smooth finish. It is suitable for many requirements but not generally applicable to plating.

3.4 *ground finish*—produced by mechanically roughing the surface of the strip. Numerous textures are available and the producer should be consulted for availability.

4. General Requirements for Delivery

4.1 Material furnished under this specification shall conform to the applicable requirements of the current edition of either Specification A 682 or A 682M.

5. Ordering Information

5.1 Orders for material under this specification shall include the following information:

---

[1] This specification is under the jurisdiction of ASTM Committee A-1 on Steel, Stainless Steel and Related Alloys, and is the direct responsibility of Subcommittee A01.19 on Steel Sheet and Strip.
Current edition approved Sept. 25, 1981. Published November 1981. Originally published as A 680 – 73. Last previous edition A 680 – 73.
[2] *Annual Book of ASTM Standards*, Vol 01.04.
[3] *Annual Book of ASTM Standards*, Vol 01.03.
[4] *Annual Book of ASTM Standards*, Vol 03.01.

5.1.1 ASTM designation and date of issue,
5.1.2 Name of material (cold-rolled high-carbon, strip, hard, untempered quality),
5.1.3 Chemical composition,
5.1.4 Application,
5.1.5 Dimensions,
5.1.6 Coil size requirements,
5.1.7 Edge (indicate No. 1 round, square, etc.),
5.1.8 Finish (indicate No. 2 regular bright, or No. 1 matte)–specify finish. Producers should be consulted for other types.
5.1.9 Conditions–specify oiled or dry,
5.1.10 Package (bare coils, skid, etc.),
5.1.11 Cast or heat analysis report (if required), and
5.1.12 Special requirements.

NOTE 2—A typical ordering description is as follows: "ASTM A 680 dated _____ Cold Rolled, High-Carbon, Hard, Untempered Quality Strip for Hand Saws, 0.062 in. by 7 in. by coil size (16 in. ID by 30 in. OD max), No. 3 Edge, No. 2 Finish, Dry, on Bare Skid" or
"ASTM A 680M dated _____ Cold Rolled, High-Carbon, Hard, Untempered Quality Strip for Hand Saws, 0.6 mm by 200 mm by coil size (400 mm ID by 7500 mm OD max), No. 3 Edge, No. 2 Finish, Oiled, on Bare Skid."

6. Manufacture

6.1 *Condition*—The strip shall be cold rolled with or without preparatory thermal treatment.

7. Chemical Requirements

7.1 *Cast or Heat Analysis*—The heat or cast analysis shall conform to that specified in Specification A 682 or A 682M for the steel grade ordered; or, to such other limits as may be specified using the standard ranges in Specification A 682 or A 682M.

8. Mechanical Requirements

8.1 *Hardness*:
8.1.1 The strip shall conform to the following hardness requirements:

Thickness		Rockwell Hardness, min	
in. (A 680)	mm (A 680M)		
Less than 0.015	Less than 0.4	15T	92.5
0.015 to 0.024	Over 0.4 to 0.6	30T	81
0.025 to 0.034	Over 0.6 to 0.9	45T	70
0.035 and over	Over 0.9	B	98

8.1.2 *Number of Tests*—At least one specimen shall be taken from each lot (see Specification A 682 or A 682M).
8.1.3 *Test Method*—Hardness tests shall be made in accordance with Methods and Definitions A 370 and Methods E 18. Alternative hardness scales to those provided in 8.1.1 may be used provided they are within the limits recommended by Methods E 18.

9. Finish and Edges

9.1 *Surface*—The strip shall be furnished with a No. 2 regular bright finish, No. 1 matte (dull) finish, or a ground finish, as specified. When ordering the ground finish, the surface texture must also be specified.
9.2 *Oiling*—The strip shall be furnished oiled or dry, as specified.
9.3 *Edges*:
9.3.1 Unless otherwise specified, the strip shall be furnished with a No. 3 slit edge.
9.3.2 When required, the strip may be specified to be furnished with a No. 1 round or square, approximate No. 2 natural mill edge, No. 4 rounded, No. 5 approximate square burr eliminated, or No. 6 square.

The American Society for Testing and Materials takes no position respecting the validity of any patent rights asserted in connection with any item mentioned in this standard. Users of this standard are expressly advised that determination of the validity of any such patent rights, and the risk of infringement of such rights, are entirely their own responsibility.

This standard is subject to revision at any time by the responsible technical committee and must be reviewed every five years and if not revised, either reapproved or withdrawn. Your comments are invited either for revision of this standard or for additional standards and should be addressed to ASTM Headquarters. Your comments will receive careful consideration at a meeting of the responsible technical committee, which you may attend. If you feel that your comments have not received a fair hearing you should make your views known to the ASTM Committee on Standards, 1916 Race St., Philadelphia, Pa. 19103.

Designation: A 682 – 77 (Reapproved 1983)

An American National Standard

Standard Specification for
STEEL, HIGH-CARBON, STRIP, COLD-ROLLED, SPRING QUALITY, GENERAL REQUIREMENTS[1]

This standard is issued under the fixed designation A 682; the number immediately following the designation indicates the year of original adoption or, in the case of revision, the year of last revision. A number in parentheses indicates the year of last reapproval. A superscript epsilon (ϵ) indicates an editorial change since the last revision or reapproval.

This specification has been approved for use by agencies of the Department of Defense and for listing in the DoD Index of Specifications and Standards.

1. Scope

1.1 This specification covers the general requirements for cold-rolled carbon spring steel strip in coils or cut lengths. Strip is classified by size as a product that is 0.2499 in. or less in thickness and over $1/2$ to $23^{15}/_{16}$ in. in width, inclusive.

1.2 The maximum of the specified carbon range is over 0.25 % to 1.35 %, inclusive.

1.3 The above shall apply to the cold-rolled carbon spring steel strip furnished under each of the following specifications issued by ASTM:

Title of Specification	ASTM Designation
Steel, Carbon, Strip, Cold-Rolled Hard, Untempered Spring Quality	A 680[2]
Steel, Carbon, Strip, Cold-Rolled Soft, Untempered Spring Quality	A 684[2]

NOTE — A complete metric companion to Specification A 682 has been developed — Specification A 682M; therefore, no metric equivalents are presented in this specification.

2. Applicable Documents

2.1 *ASTM Standards:*
A 370 Methods and Definitions for Mechanical Testing of Steel Products[3]
A 700 Recommended Practices for Packaging, Marking, and Loading Methods for Steel Products for Domestic Shipment[4]
E 3 Methods of Preparation of Metallographic Specimens[5]
E 18 Test Methods for Rockwell Hardness and Rockwell Superficial Hardness of Metallic Materials[6]
E 44 Definitions of Terms Relating to Heat Treatment of Metals[5]
E 112 Methods for Determining Average Grain Size[5]
E 350 Method for Chemical Analysis of Carbon Steel, Low-Alloy Steel, Silicon Electrical Steel, Ingot Iron, and Wrought Iron[7]

2.2 *Federal Standards*[8]*:*
Fed. Std. No. 123 Marking for Shipments (Civil Agencies)
Fed. Std. No. 183 Continuous Identification Marking of Iron and Steel Products

2.3 *Military Standards*[8]*:*
MIL-STD-129 Marking for Shipping and Storage
MIL-STD-163 Steel Mill Products Preparation for Shipment and Storage

3. Definitions

3.1 *burr* — metal displaced beyond the plane of the surface by slitting or shearing.

3.2 *decarburization* — refer to Definitions E 44.

3.3 *lot* — the quantity of material of the same type, size, and finish produced at one

---

[1] This specification is under the jurisdiction of ASTM Committee A-1 on Steel, Stainless Steel and Related Alloys, and is the direct responsibility of Subcommittee A01.19 on Sheet Steel and Steel Sheets.
Current edition approved Aug. 26, 1977. Published January 1978. Originally published as A 682–73. Last previous edition A 682–73.
[2] *Annual Book of ASTM Standards*, Vol 01.03.
[3] *Annual Book of ASTM Standards*, Vol 01.04.
[4] *Annual Book of ASTM Standards*, Vol 01.01, 01.03, 01.04, and 01.05.
[5] *Annual Book of ASTM Standards*, Vol 03.03.
[6] *Annual Book of ASTM Standards*, Vol 03.01.
[7] *Annual Book of ASTM Standards*, Vol 03.05.
[8] Available from Naval Publications and Forms Center, 5801 Tabor Ave., Phila., Pa. 19120.

A 682

time from the same cast or heat, and heat treated in the same heat-treatment cycle.

4. General Requirements for Delivery

4.1 The requirements of the purchase order, the individual material specification, and this general specification shall govern in the sequence stated.

4.2 Products covered by this specification are produced to decimal thickness only, and decimal thickness tolerances apply.

5. Manufacture

5.1 *Melting Practice*—The steel shall be made by either the open-hearth, basic-oxygen, or electric-furnace process. It is normally produced as a fully killed steel. Elements such as aluminum may be added in sufficient amounts to control the austenitic grain size.

5.2 *Cold Working Procedure:*

5.2.1 Prior to cold rolling, the hot-rolled strip shall be descaled by chemical or mechanical means.

5.2.2 The strip shall be cold rolled by reducing to thickness at room temperature (that is, below the recrystallization temperature).

6. Chemical Requirements

6.1 *Limits:*

6.1.1 When carbon steel strip is specified to chemical composition, the compositions are commonly prepared using the ranges and limits shown in Table 1. The elements comprising the desired chemical composition are specified in one of three ways:

6.1.1.1 By a maximum limit,

6.1.1.2 By a minimum limit, or

6.1.1.3 By minimum and maximum limits, termed the "range." By common usage, the range is the arithmetical difference between the two limits (for example, 0.60 to 0.71 is 0.11 range).

6.1.2 Steel grade numbers indicating chemical composition commonly produced to this specification are shown in Table 2 and may be used.

6.2 *Cast or Heat (Formerly Ladle) Analysis:*

6.2.1 An analysis of each cast or heat of steel shall be made by the manufacturer to determine the percentage of elements specified or restricted by the applicable specification.

6.2.2 When requested, cast or heat analysis for elements listed or required shall be reported to the purchaser or to his representative.

6.3 *Product Analysis (Formerly Check Analysis)*—Product analysis is the chemical analysis of the semi-finished product form. The strip may be subjected to product analysis by the purchaser either for the purpose of verifying that the chemical composition is within specified limits for each element, including applicable tolerance for product analysis, or to determine variations in compositions within a cast or heat. The results of analyses taken from different pieces within a case may differ from each other and from the cast analysis. The chemical composition thus determined shall not vary from the limits specified by more than the amounts shown in Table 3, but the several determinations of any element in any cast may not vary both above and below the specified range.

6.4 *Methods of Analysis*—Methods E 350 shall be used for referee purposes.

7. Metallurgical Structure

7.1 *Grain Size:*

7.1.1 The steel strip shall have an austenitic grain size of which a minimum of 70 % is 5 or finer.

7.1.2 One sample shall be taken from each lot.

7.1.3 The sample shall be evaluated in accordance with Method E 112.

7.2 *Decarburization:*

7.2.1 When specified, the steel strip shall have a maximum permissible depth of complete plus partial decarburization of 0.001 in. or 1.5 % of the thickness of the strip, whichever is greater, except that strip less than 0.011 in. thick shall show no complete decarburization.

7.2.2 At least one specimen from each lot shall be taken for microscopical examination.

7.2.3 The specimens shall be prepared for microscopical examination in accordance with Method E 3. The prepared specimen shall not be less than $1/2$ in. in length, representing the full thickness, and shall be perpendicular to the rolling direction. The examination of the specimen includes the periphery and therefore it must be polished in a single plane without edge rounding. The specimen shall be etched and shall be examined at 100× magnification. The depth of decarburization reported should be the average depth in both the amount of free ferrite (complete decarburization) and the affected depth to the point where carbon content appears to be the same as the carbon content of the strip (partial decarburization) under investigation. In some instances, it is necessary to resort to heat treatment of the specimens to reveal decarburized areas more accurately.

8. Mechanical Requirements

8.1 The mechanical property requirements, number of specimens, and test locations and specimen orientation shall be in accordance with the applicable product specification.

8.2 Unless otherwise specified in the applicable product specification, test specimens must be prepared in accordance with Methods A 370.

8.3 Mechanical tests shall be conducted in accordance with Methods A 370.

9. Dimensions, Weights, and Tolerances

9.1 The thickness, width, camber, and length tolerances shall conform to the requirements specified in Tables 4, 5, 6, 7, 8, and 9.

9.2 *Flatness* — It is not practical to formulate flatness tolerances for cold-rolled carbon spring steel strip to represent the range of widths and thicknesses in coils and cut lengths.

10. Finish and Edges

10.1 *Surface* — The surface requirements shall be as specified in the product specifications.

10.2 *Edges* — Cold-rolled carbon spring steel strip shall be supplied with one of the following edges as specified:

10.2.1 *No. 1* — A prepared edge of a specified contour (round or square) that is produced when a very accurate width is required or when an edge condition suitable for electroplating is required, or both.

10.2.2 *No. 2* — A natural mill edge carried through the cold rolling from the hot-rolled strip without additional processing of the edge.

10.2.3 *No. 3* — An approximately square edge produced by slitting on which the burr is not eliminated. This is produced when the edge condition is not a critical requirement for the finished part. Normal coiling or piling does not provide a definite positioning of the slitting burr.

10.2.4 *No. 4* — An approximately rounded edge. This edge is produced when the width tolerance and edge condition are not as exacting as for No. 1 edges.

10.2.5 *No. 5* — An approximately square edge produced from slit edge material on which the burr is eliminated.

10.2.6 *No. 6* — An approximately square edge. This edge is produced when the width tolerance and edge condition are not as exacting as for No. 1 edges.

11. Workmanship

11.1 The steel shall have a workmanlike appearance and shall not have defects of a nature or degree for the grade and quality ordered that will be detrimental to the fabrication of the finished part.

11.2 Coils may contain some abnormalities that render a portion of the coil unusable since the inspection of coils does not afford the same opportunity to remove portions containing imperfections as is the case with cut lengths.

12. Retests

12.1 The difficulties in obtaining truly representative samples of strip without destroying the usefulness of the coil account for the generally accepted practice of allowing retests for mechanical properties and surface examination. Two additional samples are secured from each end of the coil from which the original sample was taken. A portion of the coil may be discarded prior to cutting the samples for retest. If any of the retests fail to comply with the requirements, the coil shall be rejected.

13. Rework and Retreatment

13.1 Lots rejected for failure to meet the specified requirements may be resubmitted for test provided the manufacturer has reworked the lots as necessary to correct the deficiency or has removed the nonconforming material.

14. Inspection

14.1 The manufacturer shall afford the

A 682

purchaser's inspector all reasonable facilities necessary to satisfy him that the material is being produced and furnished in accordance with this specification. Mill inspection by the purchaser shall not interfere unnecessarily with the manufacturer's operations. Unless otherwise agreed to, all tests and inspections, except product analysis, shall be made at the place of production.

15. Rejection and Rehearing

15.1 Unless otherwise specified, any rejection based on tests made in accordance with this specification shall be reported to the purchaser within a reasonable time.

15.2 Material that shows injurious defects subsequent to its acceptance at the purchaser's works shall be rejected and the manufacturer shall be notified. The material must be adequately protected and correctly identified in order that the manufacturer may make a proper investigation. In case of dissatisfaction with the results of the test, the manufacturer may make claims for a rehearing.

16. Certification and Reports

16.1 Upon request of the purchaser in the contract or order, a manufacturer's certification that the material was produced and tested in accordance with this specification shall be furnished. A report of the test results may be included if required.

17. Marking

17.1 Unless otherwise specified, the material shall be identified by having the manufacturer's name or mark, ASTM designation, weight, purchase order number, and material identification legibly stenciled on top of each lift or shown on a tag attached to each coil or shipping unit.

17.2 When specified in the contract or order, and for direct procurement by or direct shipment to the Government, marking for shipment, in addition to requirements specified in the contract or order, shall be in accordance with MIL-STD-129 for military agencies and in accordance with Fed. Std. No. 123 for civil agencies.

17.3 For Government procurement by the Defense Supply Agency, strip material shall be continuously marked for identification in accordance with Fed. Std. No. 183.

18. Packaging

18.1 Unless otherwise specified, the strip shall be packaged and loaded in accordance with Recommended Practices A 700.

18.2 When Level A is specified in the contract or order, and for direct procurement by or direct shipment to the Government, preservation, packaging, and packing shall be in accordance with the Level A requirements of MIL-STD-163.

18.3 When coils are ordered it is customary to specify a minimum or range of inside diameter, maximum outside diameter, and a maximum coil weight, if required. The ability of manufacturers to meet the maximum coil weights depends upon individual mill equipment. When required, minimum coil weights are subject to negotiation.

TABLE 1 Cast or Heat (Formerly Ladle) Analysis Limits and Ranges

Element	Standard Chemical Limits and Ranges, Limit or Max of Specified Range	Range, %
Carbon[a]	over 0.25 to 0.30, incl	0.06
	over 0.30 to 0.40, incl	0.07
	over 0.40 to 0.60, incl	0.08
	over 0.60 to 0.80, incl	0.11
	over 0.80 to 1.35, incl	0.14
Manganese	to 0.50, incl	0.20
	over 0.50 to 1.15, incl	0.30
	over 1.15 to 1.65, incl	0.35
Phosphorus	to 0.08, incl	0.03
	over 0.08 to 0.15, incl	0.05
Sulfur	to 0.08, incl	0.03
	over 0.08 to 0.15, incl	0.05
	over 0.15 to 0.23, incl	0.07
	over 0.23 to 0.33, incl	0.10
Silicon	to 0.20, incl	0.10
	over 0.20 to 0.30, incl	0.15
	over 0.30 to 0.60, incl	0.30

[a] The carbon ranges shown in the column headed Range apply when the specified maximum limit for manganese does not exceed 1.00 %. When the maximum manganese limit exceeds 1.00 %, add 0.01 to the carbon ranges shown above.

TABLE 2 Cast or Heat (Formerly Ladle) Analysis Chemical Composition, %

UNS Designation[c]	Steel Grade	Carbon	Manganese	Phosphorus, max[a]	Sulfur, max[a]	Silicon[b]
G10300	1030	0.27 to 0.34	0.60 to 0.90	0.040	0.050	0.15 to 0.30
G10350	1035	0.31 to 0.38	0.60 to 0.90	0.040	0.050	0.15 to 0.30
G10400	1040	0.36 to 0.44	0.60 to 0.90	0.040	0.050	0.15 to 0.30
G10450	1045	0.42 to 0.50	0.60 to 0.90	0.040	0.050	0.15 to 0.30
G10500	1050	0.47 to 0.55	0.60 to 0.90	0.040	0.050	0.15 to 0.30
G10550	1055	0.52 to 0.60	0.60 to 0.90	0.040	0.050	0.15 to 0.30
G10600	1060	0.55 to 0.66	0.60 to 0.90	0.040	0.050	0.15 to 0.30
G10640	1064	0.59 to 0.70	0.50 to 0.80	0.040	0.050	0.15 to 0.30
G10650	1065	0.59 to 0.70	0.60 to 0.90	0.040	0.050	0.15 to 0.30
G10700	1070	0.65 to 0.76	0.60 to 0.90	0.040	0.050	0.15 to 0.30
G10740	1074	0.69 to 0.80	0.50 to 0.80	0.040	0.050	0.15 to 0.30
G10800	1080	0.74 to 0.88	0.60 to 0.90	0.040	0.050	0.15 to 0.30
G10850	1085	0.80 to 0.94	0.70 to 1.00	0.040	0.050	0.15 to 0.30
G10860	1086	0.80 to 0.94	0.30 to 0.50	0.040	0.050	0.15 to 0.30
G10950	1095	0.90 to 1.04	0.30 to 0.50	0.040	0.050	0.15 to 0.30

[a] Ordinarily produced to 0.025 % max phosphorus and 0.035 % max sulfur by cast or heat analysis.
[b] When specified, silicon may be ordered at 0.10 to 0.20 %.
[c] New designation established in accordance with ASTM E 527 and SAE J1086, Recommended Practice for Numbering Metals and Alloys (UNS).

A 682

TABLE 3 Permissible Variations from Specified Cast or Heat (Formerly Ladle) Analysis Ranges and Limits

Element	Limit or Max of Specification, %	Variations Over Max Limit or Under Min Limit	
		Under Min Limit	Over Max Limit
Carbon	over 0.25 to 0.40, incl	0.03	0.04
	over 0.40 to 0.80, incl	0.03	0.05
	over 0.80	0.03	0.06
Manganese	to 0.60, incl	0.03	0.03
	over 0.60 to 1.15, incl	0.04	0.04
	over 1.15 to 1.65, incl	0.05	0.05
Phosphorus	...	...	0.01
Sulfur	...	...	0.01
Silicon	to 0.30, incl	0.02	0.03
	over 0.30 to 0.60	0.05	0.05

TABLE 4 Thickness Tolerances[a] (Over and Under) for Cold-Rolled Carbon Spring Steel Strip

Specified Thickness, in.		Width, in.							
From	Up to and excl	Under 1 to $\frac{1}{2}$, excl	Under 3 to 1, incl	3 to 6 incl	Over 6 to 9, incl	Over 9 to 12, incl	Over 12 to 16, incl	Over 16 to 20, incl	Over 20 to $23\frac{15}{16}$, incl
...	0.007	0.0005	0.0005	0.0005	...	...	...	...	...
0.007	0.009	0.00075	0.00075	0.00075	...	...	...	...	...
0.009	0.013	0.00075	0.00075	0.00075	0.001	0.001	0.001	0.001	0.001
0.013	0.020	0.00075	0.00075	0.00075	0.001	0.001	0.0015	0.0015	0.0015
0.020	0.023	0.001	0.001	0.001	0.0015	0.0015	0.0015	0.0015	0.0015
0.023	0.026	0.001	0.001	0.001	0.0015	0.0015	0.002	0.002	0.002
0.026	0.029	0.001	0.0015	0.0015	0.002	0.002	0.002	0.002	0.002
0.029	0.032	0.0015	0.0015	0.0015	0.002	0.002	0.002	0.002	0.002
0.032	0.035	0.0015	0.0015	0.002	0.002	0.002	0.002	0.002	0.002
0.035	0.040	0.002	0.002	0.002	0.002	0.002	0.002	0.002	0.002
0.040	0.050	0.002	0.002	0.0025	0.0025	0.0025	0.0025	0.0025	0.0025
0.050	0.069	0.002	0.002	0.0025	0.0025	0.0025	0.003	0.003	0.003
0.069	0.100	0.002	0.002	0.0025	0.003	0.003	0.0035	0.0035	0.0035
0.100	0.161	0.002	0.002	0.003	0.003	0.003	0.0035	0.0045	0.005
0.161	0.200	0.0025	0.0035	0.004	0.004	0.0045	0.0045	0.005	0.005
0.200	0.250	0.003	0.004	0.0045	0.0045	0.005	0.0055	0.0055	0.0055

[a] Measured $\frac{3}{8}$ in. or more from the edge on 1.0 in. or wider; and at any place between the edges on narrower than 1.0 in.

A 682

TABLE 5 Width Tolerances for Edges Nos. 1, 4, 5, and 6, Cold-Rolled Carbon Spring Steel Strip

Width Tolerances, in.

Edge No.	Specified Width	Specified Thickness	Width Tolerance, Over and Under
1	Over ½ to ¾, incl	0.0938 and thinner	0.005
1	Over ¾ to 7, incl	0.125 and thinner	0.005
4	Over ½ to 1, incl	0.1875 to 0.025, incl	0.015
4	Over 1 to 2, incl	0.2499 to 0.025, incl	0.025
4	Over 2 to 4, incl	0.2499 to 0.035, incl	0.047
4	Over 4 to 6, incl	0.2499 to 0.047, incl	0.047
5	Over ½ to ¾, incl	0.0938 and thinner	0.005
5	Over ¾ to 5, incl	0.125 and thinner	0.005
5	Over 5 to 9, incl	0.125 to 0.008, incl	0.010
5	Over 9 to 20, incl	0.105 to 0.015, incl	0.010
5	Over 20 to 23 15/16, incl	0.080 to 0.023, incl	0.015
6	Over ½ to 1, incl	0.1875 to 0.025, incl	0.015
6	Over 1 to 2, incl	0.2499 to 0.025, incl	0.025
6	Over 2 to 4, incl	0.2499 to 0.035, incl	0.047
6	Over 4 to 6, incl	0.2499 to 0.047, incl	0.047

TABLE 6 Width Tolerances for Edge No. 2 (Mill) Cold-Rolled Carbon Spring Steel Strip

Width Tolerances, in.

Specified Width	Tolerances, Over and Under
Over ½ to 2, incl	1/32
Over 2 to 5, incl	3/64
Over 5 to 10, incl	5/64
Over 10 to 15, incl	3/32
Over 15 to 20, incl	⅛
Over 20 to 23 15/16	5/32

ASTM A 682

TABLE 7 Width Tolerances For Edge No. 3 (Slit) Cold-Rolled Carbon Spring Steel Strip

Width Tolerances, in.

Specified Width	Thickness	Tolerance for Specified Width, Over and Under
Over ½ to 6, incl	over 0.160 to 0.2499, incl	0.016
Over 6 to 12, incl	over 0.160 to 0.2499, incl	0.020
Over 12 to 23¹⁵⁄₁₆, incl	over 0.160 to 0.2499, incl	0.031
Over ½ to 6, incl	over 0.099 to 0.160, incl	0.010
Over 6 to 12, incl	over 0.099 to 0.160, incl	0.016
Over 12 to 23¹⁵⁄₁₆, incl	over 0.099 to 0.160, incl	0.020
Over ½ to 6, incl	over 0.068 to 0.099, incl	0.008
Over 6 to 12, incl	over 0.068 to 0.099, incl	0.010
Over 12 to 20, incl	over 0.068 to 0.099, incl	0.016
Over 20 to 23¹⁵⁄₁₆, incl	over 0.068 to 0.099, incl	0.020
Over ½ to 9, incl	over 0.016 to 0.068, incl	0.005
Over 9 to 12, incl	over 0.016 to 0.068, incl	0.010
Over 12 to 20, incl	over 0.016 to 0.068, incl	0.016
Over 20 to 23¹⁵⁄₁₆, incl	over 0.016 to 0.068, incl	0.020
Over ½ to 9, incl	up to 0.016, incl	0.005
Over 9 to 12, incl	up to 0.016, incl	0.010
Over 12 to 20, incl	up to 0.016, incl	0.016
Over 20 to 23¹⁵⁄₁₆, incl	up to 0.016, incl	0.020

TABLE 8 Camber Tolerances, Cold-Rolled Carbon Spring Steel Strip

NOTE—Camber is the deviation of a side edge from a straight line. The standard for measuring this deviation is based on any 8-ft length. It is obtained by placing an 8-ft straightedge on the concave side and measuring the maximum distance between the strip edge and the straightedge.

Width, in.	Camber Tolerances, max
Over ½ to 1½, incl	½ in. in any 8 ft
Over 1½ to 23¹⁵⁄₁₆, incl	¼ in. in any 8 ft

TABLE 9 Length Tolerances, Cold-Rolled Carbon Spring Steel Strip

Specified Width, in.	Specified Length, in.[a]		
	24 to 60, incl	Over 60 to 120, incl	Over 120 to 240, incl
Over ½ to 12, incl	¼	½	¾
Over 12 to 23¹⁵⁄₁₆, incl	½	¾	1

[a] Tolerance over specified length. No tolerances under.

The American Society for Testing and Materials takes no position respecting the validity of any patent rights asserted in connection with any item mentioned in this standard. Users of this standard are expressly advised that determination of the validity of any such patent rights, and the risk of infringement of such rights, are entirely their own responsibility.

This standard is subject to revision at any time by the responsible technical committee and must be reviewed every five years and if not revised, either reapproved or withdrawn. Your comments are invited either for revision of this standard or for additional standards and should be addressed to ASTM Headquarters. Your comments will receive careful consideration at a meeting of the responsible technical committee, which you may attend. If you feel that your comments have not received a fair hearing you should make your views known to the ASTM Committee on Standards, 1916 Race St., Philadelphia, Pa. 19103.

Designation: A 682M – 77 (Reapproved 1983)

An American National Standard

Metric

Standard Specification for
STEEL, HIGH-CARBON, STRIP, COLD-ROLLED, SPRING QUALITY, GENERAL REQUIREMENTS [METRIC][1]

This standard is issued under the fixed designation A 682M; the number immediately following the designation indicates the year of original adoption or, in the case of revision, the year of last revision. A number in parentheses indicates the year of last reapproval. A superscript epsilon (ϵ) indicates an editorial change since the last revision or reapproval.

This specification has been approved for use by agencies of the Department of Defense and for listing in the DoD Index of Specifications and Standards.

1. Scope

1.1 This specification covers the general requirements for cold-rolled carbon spring steel strip in coils or cut lengths. Strip is commonly available by size as a product that is 6 mm or less in thickness and 13 through 600 mm in width.

1.2 The maximum of the specified carbon range is over 0.25 to 1.35 %, inclusive.

1.3 The above shall apply to the cold-rolled carbon spring steel strip furnished under each of the following specifications issued by ASTM:

Title of Specification	ASTM Designation
Steel, Carbon, Strip, Cold-Rolled Hard, Untempered Spring Quality	A 680[2]
Steel, Carbon, Strip, Cold-Rolled Soft, Untempered Spring Quality	A 684[2]

1.4 This specification covers only metric (SI) units and is not to be used or confused with inch-pound units.

NOTE—This metric specification is the equivalent of A 682, and is compatible in technical content.

2. Applicable Documents

2.1 *ASTM Standards:*
A 370 Methods and Definitions for Mechanical Testing of Steel Products[3]
A 700 Recommended Practices for Packaging, Marking, and Loading Methods for Steel Products for Domestic Shipment[4]
E 3 Methods of Preparation of Metallographic Specimens[5]
E 44 Definitions of Terms Relating to Heat Treatment of Metals[5]
E 112 Methods for Determining Average Grain Size[5]
E 350 Method for Chemical Analysis of Carbon Steel, Low-Alloy Steel, Silicon Electrical Steel, Ingot Iron, and Wrought Iron[6]

2.2 *Military Standards:*[7]
MIL-STD-129 Marking for Shipment and Storage
MIL-STD-163 Steel Mill Products, Preparation for Shipment and Storage

2.3 *Federal Standards:*[7]
Fed. Std. No. 123 Marking for Shipments (Civil Agencies)
Fed. Std. No. 183 Continuous Identification Marking of Iron and Steel Products

3. Definitions

3.1 *burr* — metal displaced beyond the plane of the surface by slitting or shearing.

3.2 *decarburization* — Refer to Definitions E 44.

3.3 *lot* — the quantity of material of the same type, size, and finish produced at one time from the same cast or heat, and heat treated in the same heat-treatment cycle.

---

[1] This specification is under the jurisdiction of ASTM Committee A-1 on Steel, Stainless Steel and Related Alloys, and is the direct responsibility of Subcommittee A01.19 on Sheet Steel and Steel Sheets.
Current edition approved Nov. 3, 1977. Published January 1978.
[2] *Annual Book of ASTM Standards*, Vol 01.03.
[3] *Annual Book of ASTM Standards*, Vol 01.04.
[4] *Annual Book of ASTM Standards*, Vols 01.01, 01.03, 01.04, and 01.05.
[5] *Annual Book of ASTM Standards*, Vol 03.03.
[6] *Annual Book of ASTM Standards*, Vol 03.05.
[7] Available from Naval Publications and Forms Center, 5801 Tabor Ave., Philadelphia, Pa. 19120.

4. General Requirements for Delivery

4.1 Products covered by this specification are produced to decimal thickness only, and decimal thickness tolerances apply.

5. Manufacture

5.1 The steel is produced as a fully killed steel. Elements such as aluminum may be added in sufficient amounts to control the austenitic grain size.

5.2 *Cold Working Procedure:*

5.2.1 Prior to cold rolling, the hot-rolled strip shall be descaled by chemical or mechanical means.

5.2.2 The strip shall be cold rolled by reducing to thickness at room temperature (that is, below the recrystallization temperature).

6. Chemical Requirements

6.1 *Limits:*

6.1.1 When carbon steel strip is specified to chemical composition, the compositions are commonly prepared using the ranges and limits shown in Table 1. The elements comprising the desired chemical composition are specified in one of three ways:

6.1.1.1 By a maximum limit,

6.1.1.2 By a minimum limit, or

6.1.1.3 By minimum and maximum limits, termed the "range." By common usage, the range is the arithmetical difference between the two limits (for example, 0.60 to 0.71 is 0.11 range).

6.1.2 Steel grade numbers indicating chemical composition commonly produced to this specification are shown in Table 2 and may be used.

6.2 *Cast or Heat (Formerly Ladle) Analysis:*

6.2.1 An analysis of each cast or heat of steel shall be made by the manufacturer to determine the percentage of elements specified or restricted by the applicable specification.

6.2.2 When requested, cast or heat analysis for elements listed or required shall be reported to the purchaser or to his representative.

6.3 *Product, Check, or Verification (Formerly Check) Analysis*—Product analysis is the chemical analysis of the semi-finished product form. The strip may be subjected to product analysis by the purchaser either for the purpose of verifying that the chemical composition is within specified limits for each element, including applicable tolerance for product analysis, or to determine variations in compositions within a cast or heat. The results of analyses taken from different pieces within a case may differ from each other and from the cast analysis. The chemical composition thus determined shall not vary from the limits specified by more than the amounts shown in Table 3, but the several determinations of any element in any cast may not vary both above and below the specified range.

6.4 *Methods of Analysis*—Methods E 350 shall be used for referee purposes.

7. Metallurgical Structure

7.1 *Grain Size:*

7.1.1 The steel strip shall have an austenitic grain size of which a minimum of 70 % is 5 or finer.

7.1.2 One sample shall be taken from each lot.

7.1.3 The sample shall be evaluated in accordance with Methods E 112.

7.2 *Decarburization:*

7.2.1 When specified, the steel strip shall have a maximum permissible depth of complete plus partial decarburization of 0.025 mm or 1.5 % of the thickness of the strip, whichever is greater, except that strip less than 0.28 mm thick shall show no complete decarburization.

7.2.2 At least one specimen from each lot shall be taken for microscopical examination.

7.2.3 The specimens shall be prepared for microscopical examination in accordance with Method E 3. The prepared specimen shall not be less than 12.0 mm in length, representing the full thickness, and shall be perpendicular to the rolling direction. The examination of the specimen includes the periphery and therefore it must be polished in a single plane without edge rounding. The specimen shall be etched and shall be examined at $100\times$ magnification. The depth of decarburization reported should be the average depth in both the amount of free ferrite (complete decarburization) and the affected depth to the point where carbon content appears to be the same as the carbon content of the strip (partial decarburization) under investigation. In some instances, it is necessary to resort to heat treatment of the specimens to reveal decarburized areas more accurately.

8. Mechanical Requirements

8.1 The mechanical property requirements, number of specimens, and test locations and specimen orientation shall be in accordance with the applicable product specification.

8.2 Unless otherwise specified in the applicable product specification, test specimens must be prepared in accordance with Methods A 370.

8.3 Mechanical test shall be conducted in accordance with Methods A 370.

9. Dimensions, Weights, and Tolerances

9.1 The thickness, width, camber, and length tolerances shall conform to the requirements specified in Tables 4 through 10.

9.2 *Flatness* — It is not practical to formulate flatness tolerances for cold-rolled carbon spring steel strip to represent the range of widths and thicknesses in coils and cut lengths.

10. Finish and Edges

10.1 *Surface* — The surface requirements shall be as specified in the product specifications.

10.2 *Edges* — Cold-rolled carbon spring steel strip shall be supplied with one of the following edges as specified:

10.2.1 *No. 1* — A prepared edge of a specified contour (round or square) that is produced when a very accurate width is required or when an edge condition suitable for electroplating is required, or both.

10.2.2 *No. 2* — A natural mill edge carried through the cold rolling from the hot-rolled strip without additional processing of the edge.

10.2.3 *No. 3* — An approximately square edge produced by slitting on which the burr is not eliminated. This is produced when the edge condition is not a critical requirement for the finished part. Normal coiling or piling does not provide a definite positioning of the slitting burr.

10.2.4 *No. 4* — An approximately rounded edge. This edge is produced when the width tolerance and edge condition are not as exacting as for No. 1 edges.

10.2.5 *No. 5* — An approximately square edge produced from slit edge material on which the burr is eliminated.

10.2.6 *No. 6* — An approximately square edge. This edge is produced when the width tolerance and edge condition are not as exacting as for No. 1 edges.

11. Workmanship

11.1 The steel shall have a workmanlike appearance and shall not have imperfections of a nature or degree for the grade and quality ordered that will be detrimental to the fabrication of the finished part.

11.2 Coils may contain some abnormalities that render a portion of the coil unusable since the inspection of coils does not afford the same opportunity to remove portions containing imperfections as is the case with cut lengths.

12. Retests

12.1 The difficulties in obtaining truly representative samples of strip without destroying the usefulness of the coil account for the generally accepted practice of allowing retests for mechanical properties and surface examination. Two additional samples are secured from each end of the coil from which the original sample was taken. A portion of the coil may be discarded prior to cutting the samples for retest. If any of the retests fail to comply with the requirements, the coil shall be rejected.

13. Rework and Retreatment

13.1 Lots rejected for failure to meet the specified requirements may be resubmitted for test provided the manufacturer has reworked the lots as necessary to correct the deficiency or has removed the nonconforming material.

14. Inspection

14.1 The manufacturer shall afford the purchaser's inspector all reasonable facilities necessary to satisfy him that the material is being produced and furnished in accordance with this specification. Mill inspection by the purchaser shall not interfere unnecessarily with the manufacturer's operations. Unless otherwise agreed to, all tests and inspections, except product analysis, shall be made at the place of production.

15. Rejection and Rehearing

15.1 Unless otherwise specified, any rejection shall be reported to the manufacturer within a reasonable time after receipt of material by the purchaser.

15.2 Material that is reported to be defective subsequent to the acceptance at the purchaser's works shall be set aside, adequately protected, and correctly identified. The manufacturer shall be notified as soon as possible so that an investigation may be initiated.

15.3 Samples that are representative of the rejected material shall be made available to the manufacturer. In the event that the manufacturer is dissatisfied with the rejection, he may request a rehearing.

16. Certification and Reports

16.1 Upon request of the purchaser in the contract or order, a manufacturer's certification that the material was produced and tested in accordance with this specification shall be furnished. A report of the test results may be included if required.

17. Marking

17.1 As a minimum requirement, the material shall be identified by having the manufacturer's name, ASTM designation, weight, and the purchaser's order number and material identification legibly stenciled on top of each lift or shown on a tag attached to each coil or shipping unit.

17.2 When specified in the contract or order, and for direct procurement by or direct shipment to the government, marking for shipment, in addition to requirements specified in the contract or order, shall be in accordance with MIL-STD-129 for military agencies and in accordance with Fed. Std. No. 123 for civil agencies.

17.3 For Government procurement by the Defense Supply Agency, strip material shall be continuously marked for identification in accordance with Fed. Std. No. 183.

18. Packaging

18.1 Unless otherwise specified, the sheet and strip shall be packaged and loaded in accordance with Recommended Practices A 700.

18.2 When specified in the contract or order, and for direct procurement by or direct shipment to the government, when Level A is specified, preservation, packaging, and packing shall be in accordance with the Level A requirements of MIL-STD-163.

18.3 When coils are ordered it is customary to specify a minimum or range of inside diameter, maximum outside diameter, and a maximum coil weight, if required. The ability of manufacturers to meet the maximum coil weights depends upon individual mill equipment. When required, minimum coil weights are subject to negotiation.

A 682M

TABLE 1 Cast or Heat (Formerly Ladle) Analysis Limits and Ranges

Element	Standard Chemical Limits and Ranges, Limit or Maxium of Specified Range	Range, %
Carbon[A]	over 0.25 to 0.30, incl	0.06
	over 0.30 to 0.40, incl	0.07
	over 0.40 to 0.60, incl	0.08
	over 0.60 to 0.80, incl	0.11
	over 0.80 to 1.35, incl	0.14
Manganese	to 0.50, incl	0.20
	over 0.50 to 1.15, incl	0.30
	over 1.15 to 1.65, incl	0.35
Phosphorus	to 0.08, incl	0.03
	over 0.08 to 0.15, incl	0.05
Sulfur	to 0.08, incl	0.03
	over 0.08 to 0.15, incl	0.05
	over 0.15 to 0.23, incl	0.07
	over 0.23 to 0.33, incl	0.10
Silicon	to 0.20, incl	0.10
	over 0.20 to 0.30, incl	0.15
	over 0.30 to 0.60, incl	0.30

[A] The carbon ranges shown in the column headed "Range" apply when the specified maximum limit for manganese does not exceed 1.00 %. When the maximum manganese limit exceeds 1.00 %, add 0.01 to the carbon ranges shown above.

TABLE 2 Cast or Heat (Formerly Ladle) Analysis Chemical Composition, %

UNS No.	Former Steel Grade	Carbon	Manganese	Phosphorus, max[A]	Sulfur, max[A]	Silicon[B]
G10300	1030	0.27 to 0.34	0.60 to 0.90	0.040	0.050	0.15 to 0.30
G10350	1035	0.31 to 0.38	0.60 to 0.90	0.040	0.050	0.15 to 0.30
G10400	1040	0.36 to 0.44	0.60 to 0.90	0.040	0.050	0.15 to 0.30
G10450	1045	0.42 to 0.50	0.60 to 0.90	0.040	0.050	0.15 to 0.30
G10500	1050	0.47 to 0.55	0.60 to 0.90	0.040	0.050	0.15 to 0.30
G10550	1055	0.52 to 0.60	0.60 to 0.90	0.040	0.050	0.15 to 0.30
G10600	1060	0.55 to 0.66	0.60 to 0.90	0.040	0.050	0.15 to 0.30
G10640	1064	0.59 to 0.70	0.50 to 0.80	0.040	0.050	0.15 to 0.30
G10650	1065	0.59 to 0.70	0.60 to 0.90	0.040	0.050	0.15 to 0.30
G10700	1070	0.65 to 0.76	0.60 to 0.90	0.040	0.050	0.15 to 0.30
G10740	1074	0.69 to 0.80	0.50 to 0.80	0.040	0.050	0.15 to 0.30
G10800	1080	0.74 to 0.88	0.60 to 0.90	0.040	0.050	0.15 to 0.30
G10850	1085	0.80 to 0.94	0.70 to 1.00	0.040	0.050	0.15 to 0.30
G10860	1086	0.80 to 0.94	0.30 to 0.50	0.040	0.050	0.15 to 0.30
G10950	1095	0.90 to 1.04	0.30 to 0.50	0.040	0.050	0.15 to 0.30

[A] Ordinarily produced to 0.025 % max phosphorus and 0.035 % max sulfur by cast or heat analysis.
[B] When specified, silicon may be ordered at 0.10 to 0.20 %.

TABLE 3 Permissible Variations from Specified Cast or Heat (Formerly Ladle) Analysis Ranges and Limits

Element	Limit or Max of Specification, %	Variations Over Max Limit or Under Min Limit, mm	
		Under Min Limit	Over Max Limit
Carbon	over 0.25 to 0.40, incl	0.03	0.04
	over 0.40 to 0.80, incl	0.03	0.05
	over 0.80	0.03	0.06
Manganese	to 0.60, incl	0.03	0.03
	over 0.60 to 1.15, incl	0.04	0.04
	over 1.15 to 1.65, incl	0.05	0.05
Phosphorus	...	...	0.01
Sulfur	...	...	0.01
Silicon	to 0.30, incl	0.02	0.03
	over 0.30 to 0.60	0.05	0.05

TABLE 4 Thickness Tolerances[A] for Cold-Rolled Carbon Spring Steel Strip Ordered to Nominal Thickness

Specified Nominal Thickness, mm		Thickness Tolerance, Over and Under, mm, for Specified Width, mm				
Over	Through	Through 75	Over 75 to 150, incl	Over 150 to 300, incl	Over 300 to 450, incl	Over 450 to 600, incl
...	0.2	0.015	0.015	...	...	...
0.2	0.3	0.02	0.02	...	...	...
0.3	0.4	0.02	0.02	0.03	0.04	0.04
0.4	0.5	0.02	0.02	0.03	0.04	0.04
0.5	0.6	0.03	0.03	0.04	0.04	0.04
0.6	0.7	0.04	0.04	0.05	0.05	0.05
0.7	0.8	0.04	0.04	0.05	0.05	0.05
0.8	0.9	0.04	0.05	0.05	0.05	0.05
0.9	1.0	0.05	0.05	0.05	0.05	0.05
1.0	1.5	0.05	0.07	0.07	0.08	0.08
1.5	2.0	0.05	0.07	0.07	0.09	0.09
2.0	3.0	0.05	0.08	0.08	0.12	0.13
3.0	4.0	0.05	0.08	0.08	0.12	0.13
4.0	5.0	0.09	0.10	0.10	0.13	0.13
5.0	6.0	0.10	0.12	0.12	0.14	0.14

[A] Measured 10.0 mm or more in from the edge on 25 mm or wider, and at any place between the edges on narrower than 25.0 mm.

TABLE 5 Width Tolerances for Edge Numbers 1, 4, 5, and 6, Cold-Rolled Carbon Spring Steel Strip

Edge No.	Specified Width, mm		Specified Thickness, mm		Width Tolerance, mm
	Over	Through	min	max	Over and Under
1	...	200	...	3.0	0.13
4	...	25	0.6	5.0	0.38
4	25	50	0.6	6.0	0.65
4	50	150	1.0	6.0	1.20
5	...	100	...	3.0	0.13
5	100	500	0.4	3.0	0.25
5	500	600	0.6	2.0	0.38
6	...	25	0.6	5.0	0.38
6	25	50	0.6	6.0	0.65
6	50	150	1.0	6.0	1.20

TABLE 6 Width Tolerance for Edge Number 2 (Mill), Cold-Rolled Carbon Spring Steel Strip

Specified Width, mm		Width Tolerance, mm
Over	Through	Over and Under
...	50	0.8
50	100	1.2
100	200	1.6
200	400	2.5
400	500	3.0
500	600	4.0

TABLE 7 Width Tolerance for Edge Number 3 (Slit), Cold-Rolled Carbon Spring Steel Strip

Specified Width, mm		Width Tolerance, Over and Under, mm, for Specified Thickness, mm			
Over	Through	Through 1.5	Over 1.5 to 2.5, incl	Over 2.5 to 4.5, incl	Over 4.5 to 6.0, incl
...	100	0.13	0.20	0.25	0.40
100	200	0.13	0.25	0.40	0.50
200	300	0.25	0.25	0.40	0.50
300	450	0.40	0.40	0.50	0.80
450	600	0.50	0.50	0.50	0.80

TABLE 8 Length Tolerances, Cold-Rolled Carbon Spring Steel Strip

Specified Width, mm		Tolerance Over Specified Length, mm (No Tolerance Under)		
		Specified Length, mm		
Over	Through	600 to 1500, incl	Over 1500 to 3000, incl	Over 3000
...	300	10	15	25
300	600	15	20	25

Flatness Tolerances

It has not been practicable to formulate flatness tolerances for cold-rolled carbon steel strip to represent the wide range of widths and thicknesses and variety of tempers produced in coils and cut lengths.

A 682M

TABLE 9 Camber Tolerances, Cold-Rolled Carbon Spring Steel Strip

(Coils and Cut Lengths Applicable to All Types of Edges)

NOTE 1—Camber is the deviation of a side edge from a straight line. The standard for measuring this deviation is based on any 2000-mm length. It is obtained by placing a 2000-mm straightedge on the concave side and measuring the maximum distance between the strip edge and the straightedge.

NOTE 2—For strip less than 2000 mm, tolerances are to be established in each instance.

NOTE 3—When the camber tolerances shown in Table 9 are not suitable for a particular purpose, cold-rolled strip is sometimes machine straightened.

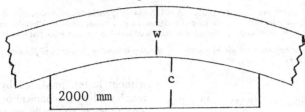

W = width of strip, mm
C = camber, mm

Width, mm		Standard Camber Tolerance, mm
Over	Through	
...	50	10
50	600	5

The American Society for Testing and Materials takes no position respecting the validity of any patent rights asserted in connection with any item mentioned in this standard. Users of this standard are expressly advised that determination of the validity of any such patent rights, and the risk of infringement of such rights, are entirely their own responsibility.

This standard is subject to revision at any time by the responsible technical committee and must be reviewed every five years and if not revised, either reapproved or withdrawn. Your comments are invited either for revision of this standard or for additional standards and should be addressed to ASTM Headquarters. Your comments will receive careful consideration at a meeting of the responsible technical committee, which you may attend. If you feel that your comments have not received a fair hearing you should make your views known to the ASTM Committee on Standards, 1916 Race St., Philadelphia, Pa. 19103.

Designation: A 684/A 684M – 81

Standard Specification for
STEEL, HIGH-CARBON, STRIP, COLD-ROLLED SOFT, UNTEMPERED QUALITY[1]

This standard is issued under the fixed designation A 684/A 684M; the number immediately following the designation indicates the year of original adoption or, in the case of revision, the year of last revision. A number in parentheses indicates the year of last reapproval. A superscript epsilon (ϵ) indicates an editorial change since the last revision or reapproval.

This specification has been approved for use by agencies of the Department of Defense and for listing in the DoD Index of Specifications and Standards.

1. Scope

1.1 This specification covers cold-rolled, high-carbon, soft untempered steel strip. It is furnished in the following types as specified:

1.1.1 Spheroidized high-carbon steel is intended for applications requiring maximum cold forming. It is produced to give the lowest maximum Rockwell hardness.

1.1.2 Annealed high-carbon steel is intended for applications requiring moderate cold forming. It is produced to a maximum Rockwell hardness.

1.1.3 Intermediate hardness high-carbon steel is intended for applications where cold forming is slight or a stiff, springy product is needed, or both. It is produced to specified Rockwell hardness ranges, the maximum being higher than obtained for the annealed type.

1.2 This specification is applicable for orders in either inch-pound units (as A 684) or SI units (A 684M).

2. Applicable Documents

2.1 *ASTM Standards*:
A 370 Methods and Definitions for Mechanical Testing of Steel Products[2]
A 682 Specification for General Requirements for Steel, High-Carbon, Strip, Cold-Rolled, Spring Quality[3]
A 682M Specification for General Requirements for Steel, High-Carbon, Strip, Cold-Rolled, Spring Quality [Metric][3]
E 3 Preparation of Metallographic Specimens[4]

3. Definitions

3.1 *Number 1 or matte (dull) finish*—finish without luster, produced by rolling on rolls roughened by mechanical or chemical means. This finish is especially suitable for lacquer or paint adhesion, and is beneficial in aiding drawing operations by reducing the contact friction between the die and the strip.

3.2 *No. 2 (regular bright finish)*—finish produced by rolling on rolls having a moderately smooth finish. It is suitable for many requirements, but not generally applicable to plating.

3.3 *spheroidizing*—the heating and cooling of the strip to produce a spheroidal or globular form of carbide microconstituent.

3.4 *stretcher strains*—elongated markings that appear on the surface of the strip when the material is deformed beyond its yield point.

4. General Requirements

4.1 Material furnished under this specification shall conform to the applicable requirements for the current edition of either Specification A 682 or A 682M.

5. Ordering Information

5.1 Orders for material under this specification shall include the following information:

5.1.1 ASTM designation and date of issue,
5.1.2 Name, type, and steel grade number,

---

[1] This specification is under the jurisdiction of ASTM Committee A-1 on Steel, Stainless Steel and Related Alloys, and is the direct responsibility of Subcommittee A01.19 on Steel Sheet and Strip.
Current edition approved Sept. 25, 1981. Published November 1981. Originally published as A 684 – 73. Last previous edition A 684 – 73.
[2] *Annual Book of ASTM Standards*, Vol 01.04.
[3] *Annual Book of ASTM Standards*, Vol 01.03.
[4] *Annual Book of ASTM Standards*, Vol 03.03.

A 684/A 684M

5.1.3 Hardness (if intermediate hardness is specified),
5.1.4 Decarburization (if required),
5.1.5 Application,
5.1.6 Dimensions,
5.1.7 Coil size requirements,
5.1.8 Edge (indicate No. 1 round, square, etc.),
5.1.9 Finish (indicate and specify),
5.1.10 Conditions (specify whether material is oiled or dry),
5.1.11 Package (bare coils, skid, etc.),
5.1.12 Cast or heat (formerly ladle) analysis report (if required), and
5.1.13 Special requirements (if required).

NOTE 1—A typical ordering description is as follows: ASTM A 684 dated ____ Cold Rolled, High-Carbon Soft, Untempered Quality Strip, Spheroidized 1064, 0.042 in. by 6 in. by coil (16 in. ID by 40 in. OD max), No. 5 Edge, No. 2 Finish, Oiled, Bare Skid or
"ASTM A 684M dated ____ Cold Rolled, High-Carbon, Soft, Untempered Quality Strip, Spheroidized 1064, 0.6 mm by 200 mm by coil (400 mm ID by 7500 mm OD max), No. 3 Edge, No. 2 Finish, Oiled, Bare Skid."

6. Manufacture

6.1 *Condition:*
6.1.1 The strip shall be furnished cold rolled spheroidized annealed, cold rolled annealed, or to intermediate hardness, as specified.
6.1.2 Intermediate hardness may be obtained by either rolling the strip after final annealing or by varying the annealing treatment, or both.
6.2 *Pinch Pass*—Spheroidized annealed and annealed material may be pinch rolled after the final anneal to improve flatness and minimize stretcher strains if required by the purchaser.

7. Chemical Requirements

7.1 *Cast or Heat Analysis*—The heat or cast analysis shall conform to that specified in Specification A 682 or A 682M for the steel grade ordered; or to such other limits as may be specified using the standard ranges in Specification A 682 or A 682M.

8. Metallurgical Structure

8.1 *Lamellar Pearlite:*
8.1.1 Cold rolled strip ordered as "spheroidized" shall be free of lamellar pearlite.
8.1.2 At least one specimen shall be taken from each lot (see Specification A 682 or A 682M) for microexamination.
8.1.3 The specimens shall be prepared for microscopical examination in accordance with Method E 3.

9. Mechanical Requirements

9.1 *Hardness:*
9.1.1 *Spheroidized Annealed and Annealed Types*—When furnished spheroidized annealed or annealed, the hardness of the strip shall not exceed the maximum values specified in Figs. 1 and 2 for the applicable carbon range and type.
9.1.2 *Intermediate Hardness Type*—When furnished as intermediate hardness, the hardness of the strip shall conform to the range specified on the purchase order. The maximum hardness limit and the corresponding minimum shall be specified by the purchaser in accordance with the limits and ranges in Table 1 for the applicable carbon range.
9.1.3 At least one specimen shall be taken from each lot (see Specification A 682 or A 682M).
9.1.4 The sample shall be tested in accordance with Methods and Definitions A 370.
9.2 *Bend Test:*
9.2.1 The steel strip produced as spheroidized, or the annealed type shall meet the cold bend requirement in Table 2. Any visible cracking on the tension side of the bend portion shall be cause for rejection.
9.2.2 At least one specimen shall be taken from each lot (see Specification A 682 or A 682M).
9.2.3 The specimen shall be the full thickness and shall be taken transverse to the rolling direction as described in Methods and Definitions A 370. The edges of the bend test specimens shall be rounded and free of burrs; filing or machining is permissible.

10. Finish and Edges

10.1 *Surface*—The strip shall be furnished with a No. 2 Regular Bright or No. 1 Matte (Dull) finish, as specified.
10.2 *Oiling*—The strip shall be furnished oiled or dry, as specified.
10.3 *Edges*—The strip shall be furnished with a No. 1 round or Square, No. 2 Mill, No. 3 Square Slit, No. 4 Approximately Round, No. 5 Approximately Square Burr Free, or No. 6 Approximately Square Edge, as specified.

A 684/A 684M

TABLE 1 Rockwell Hardness Ranges for Soft-Type Intermediate Hardness for Cold Rolled High-Carbon Steel Strip

Thickness	Hardness Scale	Hardness Range	Maximum of Range
0.030 and over	B	10	98
0.020 to 0.030	30T	6	81
Under 0.020	15T	4	92

TABLE 2 Cold Bending Requirements[A,B] for Spheroidized, Annealed, and Soft-Type Annealed Cold Rolled Carbon Steel Strip

Type	Degree of Bend	Inside Radius to Thickness	Relation of Bend Test Specimen to Rolling Direction
Annealed	180°	$3t$	transverse[C]
Spheroidized	180°	$2t$	transverse[C]

[A] Up to 0.100 in. or 2.5 mm, incl, thickness maximum. When bend radius for thickness is over 0.100 in. or 2.5 mm the producer should be consulted. These ratios apply to bending performance of the test specimen.

[B] These bend tests apply to the bending performance of test specimens only. Where material is to be bent in fabricating operations a more liberal bend radius may be required and should be based on prior experience or consultation with the steel producer, or both.

[C] If finished strip width prohibits taking a transverse bend test specimen, a longitudinal specimen may be substituted, except the bend radius shall be reduced by $1t$.

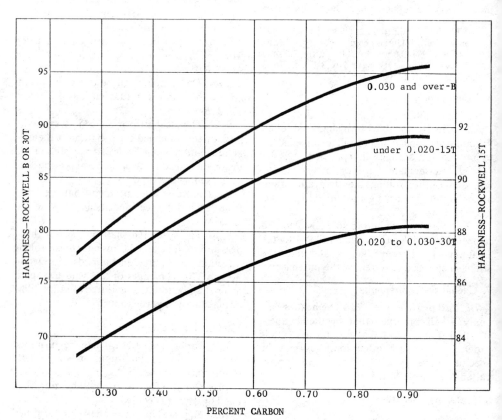

FIG. 1 Approximate Relationship Between Carbon Designations and the Maximum Hardness Limit of Soft Type Annealed Cold Rolled High Carbon Steel Strip

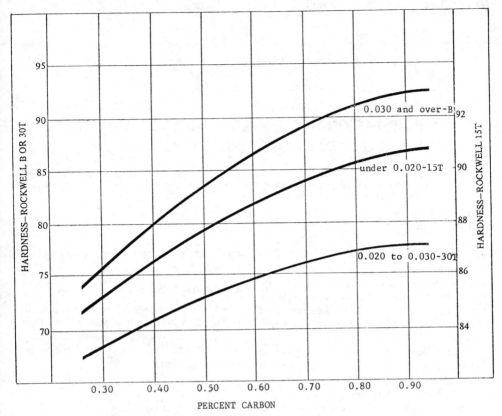

FIG. 2 Approximate Relationship Between Carbon Designations and the Maximum Hardness Limit of Spheroidized Annealed Cold Rolled High Carbon Steel Strip

The American Society for Testing and Materials takes no position respecting the validity of any patent rights asserted in connection with any item mentioned in this standard. Users of this standard are expressly advised that determination of the validity of any such patent rights, and the risk of infringement of such rights, are entirely their own responsibility.

This standard is subject to revision at any time by the responsible technical committee and must be reviewed every five years and if not revised, either reapproved or withdrawn. Your comments are invited either for revision of this standard or for additional standards and should be addressed to ASTM Headquarters. Your comments will receive careful consideration at a meeting of the responsible technical committee, which you may attend. If you feel that your comments have not received a fair hearing you should make your views known to the ASTM Committee on Standards, 1916 Race St., Philadelphia, Pa. 19103.

Designation: A 693 – 82

Standard Specification for
PRECIPITATION-HARDENING STAINLESS AND HEAT-RESISTING STEEL PLATE, SHEET, AND STRIP[1]

This standard is issued under the fixed designation A 693; the number immediately following the designation indicates the year of original adoption or, in the case of revision, the year of last revision. A number in parentheses indicates the year of last reapproval. A superscript epsilon (ϵ) indicates an editorial change since the last revision or reapproval.

This specification has been approved for use by agencies of the Department of Defense and for listing in the DoD Index of Specifications and Standards.

1. Scope

1.1 This specification covers precipitation-hardening stainless steel plate, sheet, and strip. The mechanical properties of these steels are developed by suitable low-temperature heat treatments generally referred to as precipitation hardening.

1.2 These steels are used for parts requiring corrosion resistance and high strength at room temperature or at temperatures up to 600°F (315°C). Some of these steels are particularly suitable for moderate to severe drawing and forming in the solution-treated condition. Others are capable of mild forming only. They are suitable for machining in the solution-annealed condition, after which they may be hardened to the mechanical properties specified in Table 3 without danger of cracking or distortion.

1.3 The values stated in inch-pound units are to be regarded as the standard.

2. Applicable Documents

2.1 *ASTM Standards:*
A 370 Methods and Definitions for Mechanical Testing of Steel Products[2]
A 480/A 480M Specification for General Requirements for Flat-Rolled Stainless and Heat-Resisting Steel Plate, Sheet, and Strip[2]
A 751 Methods, Practices, and Definitions for Chemical Analysis of Steel Products[2]
E 527 Practice for Numbering Metals and Alloys (UNS)[2]

3. General Requirements for Delivery

3.1 Material furnished under this specification shall conform to the applicable requirements of the current edition of Specification A 480/A 480M.

4. Ordering Information

4.1 Orders for material under this specification shall include the following information:
4.1.1 Quantity (weight or number of pieces),
4.1.2 Name of material (precipitation-hardening stainless steel),
4.1.3 Form (plate, sheet, or strip),
4.1.4 Dimensions (thickness, width, length),
4.1.5 Type (Table 1),
4.1.6 Heat treatment (Section 7),
4.1.7 Edge, strip only (see Specification A 480),
4.1.8 Finish (see Specification A 480/A 480M); specify whether one or both sides are to be polished if polished finish is specified,
4.1.9 ASTM designation and date of issue, and
4.1.10 Additions to the specification or special requirements.

NOTE 1—A typical ordering description is as follows: 200 pieces ASTM A 693 dated ———, Precipitation-Hardening Stainless Steel, Sheets, 0.060 by 48 by 120 in. (1.52 by 1219 by 3048 mm), Type 631, Solution-Annealed, No. 2B Finish.

---

[1] This specification is under the jurisdiction of ASTM Committee A-1 on Steel, Stainless Steel and Related Alloys, and is the direct responsibility of Subcommittee A01.17 on Flat Stainless Steel Products.
Current edition approved Oct. 29, 1982. Published December 1982. Originally published as A 693 – 74. Last previous edition A 693 – 81.
[2] *Annual Book of ASTM Standards*, Vol 01.03.

5. Manufacture

5.1 The steel shall be melted by one of the following processes:

5.1.1 Electric furnace (with separate degassing and refining optional),

5.1.2 Vacuum furnace, and

5.1.3 One of the former followed by:

5.1.3.1 Consumable remelting in vacuum, inert gas, or electroslag, or

5.1.3.2 Electron beam refining.

5.1.4 Other commercial melting methods as agreed upon by purchaser and seller are acceptable.

6. Chemical Requirements

6.1 The steel shall conform to the requirements as to chemical composition specified in Table 1.

6.2 Methods and practices relating to chemical analysis shall be in accordance with Methods, Practices, and Definitions A 751.

7. Heat Treatment of Product

7.1 Material of types other than XM-9 shall be furnished in the solution-annealed condition as noted in Tables 2 and 2a unless otherwise specified by the purchaser.

7.2 Type XM-9 shall be furnished in the annealed condition capable of hardening to the intermediate or high-strength condition as specified in Tables 2 and 2a.

8. Mechanical Requirements

8.1 The material, as represented by mechanical test specimens, shall conform to the mechanical property requirements specified in Table 3 and shall be capable of developing the properties in Table 5 when heat treated as specified in 10.1.

9. Bending Requirements

9.1 Samples cut from the solution-annealed plate, sheet, or strip shall withstand cold bending through the angle specified in Table 4 without cracking on the outside of the bent portion.

10. Heat Treatment of Test Specimens

10.1 Samples cut from the plate, sheet, or strip shall conform to the mechanical properties of Table 5 when precipitation hardened as specified in Tables 2 and 2a.

11. Test Methods

11.1 The properties enumerated in this specification shall be determined in accordance with methods as specified in Section 14 of Specification A 480/A 480M.

11.2 *Bend Test*—Materials shall be bent over pieces of flat stock as shown in Table 4, allowing the test material to form its natural curvature.

NOTE 2—The bend may be made over a diameter equal to the number of thicknesses of flat stock shown in Table 4, or over a single piece of flat stock equal to the number of thicknesses shown in Table 4.

TABLE 1 Chemical Requirements[A]

UNS Designation[G]	Type	Carbon	Manganese	Phosphorus	Sulfur	Silicon	Chromium	Nickel	Aluminum	Molybdenum	Titanium	Copper	Other Elements
S 17400	630	0.07	1.00	0.040	0.030	1.00	15.00–17.50	3.00–5.00	...	...	...	3.00–5.00	[B]
S 17700	631	0.09	1.00	0.040	0.030	1.00	16.00–18.00	6.50–7.75	0.75–1.50	...	...	...	...
S 15700	632	0.09	1.00	0.040	0.030	1.00	14.00–16.00	6.50–7.75	0.75–1.50	2.00–3.00	...	...	...
S 35000	633	0.07–0.11	0.50–1.25	0.040	0.030	0.50	16.00–17.00	4.00–5.00	...	2.50–3.25	...	...	[C]
S 35500	634	0.10–0.15	0.50–1.25	0.040	0.030	0.50	15.00–16.00	4.00–5.00	...	2.50–3.25	...	...	[C]
S 17600	635	0.08	1.00	0.040	0.030	1.00	16.00–17.50	6.00–7.50	...	...	0.40–1.20	...	[E]
S 36200	XM-9	0.05	0.50	0.030	0.030	0.30	14.00–14.50	6.25–7.00	0.40	...	0.60–0.90	...	...
S 15500	XM-12	0.07	1.00	0.040	0.030	1.00	14.00–15.50	3.50–5.50	0.10	0.30	...	2.50–4.50	[B]
S 13800	XM-13	0.05	0.20	0.010	0.008	0.10	12.25–13.25	7.50–8.50	0.90–1.35	2.00–2.50	...	...	[D]
S 45500	XM-16	0.05	0.50	0.040	0.030	0.50	11.00–12.50	7.50–9.50	...	0.50	0.80–1.40	1.50–2.50	[E]
S 45000	XM-25	0.05	1.00	0.030	0.030	1.00	14.00–16.00	5.00–7.00	...	0.50–1.00	...	1.25–1.75	[F]

[A] Limits are in percent maximum unless shown as a range or stated otherwise.
[B] Columbium plus tantalum 0.15–0.45.
[C] Nitrogen 0.07–0.13.
[D] Nitrogen 0.01.
[E] Columbium plus tantalum 0.10–0.50.
[F] Columbium 8 times carbon minimum.
[G] New designation established in accordance with ASTM E 527 and SAE J1086. Recommended Practice for Numbering Metals and Alloys (UNS).

ASTM A 693

TABLE 2 Heat Treatment, °F

Type	Solution Treatment	Precipitation Hardening Treatment[A]
630	1925 ± 50°F (cool as required)	900 ± 15°F, 1 h, air cool.
		925 ± 15°F, 4 h, air cool.
		1025 ± 15°F, 4 h, air cool.
		1075 ± 15°F, 4 h, air cool.
		1100 ± 15°F, 4 h, air cool.
		1150 ± 15°F, 4 h, air cool.
		(1400 ± 15°F, 2 h, air cool + 1150 ± 15°F, 4 h, air cool).
631	1950 ± 25°F (cool as required)	1750 ± 15°F, hold 10 min, cool rapidly to room temperature. Cool within 24 h, to −100 ± 10°F, hold not less than 8 h. Warm in air to room temperature. Heat to 950 ± 10°F, hold 1 h, air cool.
		Alternative Treatment:
		1400 ± 25°F, hold 90 min, cool to 55 ± 5°F within 1 h. Hold not less than 30 min, heat to 1050 ± 10°F, hold for 90 min, air cool.
632	1950 ± 25°F (cool as required)	Same as Type 631
633	1710 ± 25°F (water quench), hold not less than 3 h at −100°F or lower.	850 ± 15°F, 3 h, air cool.
		1000 ± 15°F, 3 h, air cool.
634[B]	1900 ± 25°F (quench), hold not less than 3 h at −100°F or lower.	1750 −10°F for not less than 10 min, but not more than 1 h, water quench. Cool to not higher than −100°F, hold for not less than 3 h. Temper at 1000 ± 25°F, holding for not less than 3 h.
635	1900 ± 25°F (air cool)	950 ± 15°F, 30 min, air cool.
		1000 ± 15°F, 30 min, air cool.
		1050 ± 15°F, 30 min, air cool.
XM-9	Supplied in fully heat treated condition by the producer.	
XM-12	1925 ± 50°F (cool as required)	Same as Type 630
XM-13	1700 ± 15°F (cool as required to below 60°F)	950 ± 10°F, 4 h, air cool.
		1000 ± 10°F, 4 h, air cool.
XM-16	1525 ± 25°F (water quench)	900 ± 10°F, 4 h, air cool. or 950 ± 10°F, 4 h, air cool.
XM-25	1900 ± 25°F (cool rapidly)	900 ± 15°F, 4 h, air cool.
		1000 ± 15°F, 4 h, air cool.
		1150 ± 15°F, 4 h, air cool.

[A] Times refer to time material is at temperature.
[B] Equalization and over-tempering treatment: 1425 ± 50°F for not less than 3 h, cool to room temperature, heat to 1075 ± 25°F for not less than 3 h.

TABLE 2A Heat Treatment, °C

Type	Solution Treatment	Precipitation Hardening Treatment[A]
630	1050 ± 25°C (cool as required)	482 ± 8°C, 1 h, air cool.
		496 ± 8°C, 4 h, air cool.
		552 ± 8°C, 4 h, air cool.
		579 ± 8°C, 4 h, air cool.
		593 ± 8°C, 4 h, air cool.
		621 ± 8°C, 4 h, air cool.
		(760 ± 8°C, 2 h, air cool + 621 ± 8°C, 4 h, air cool).
631	1065 ± 15°C (water quench)	954 ± 8°C, hold 10 min, cool rapidly to room temperature. Cool within 24 h to −73°C ± 6°C, hold not less than 8 h. Warm in air to room temperature. Heat to 510 ± 6°C, hold 1 h, air cool.
		Alternative Treatment
		760 ± 15°C, hold 90 min, cool to 15 ± 3°C within 1 h. Hold not less than 30 min, heat to 566 ± 6°C, hold for 90 min, air cool.
632	1038 ± 15°C (water quench)	Same as Type 631
633	930 ± 15°C (water quench), hold not less than 3 h at −75°C or lower.	455 ± 8°C, 3 h, air cool.
		540 ± 8°C, 3 h, air cool.
634[B]	1038 ± 15°C (quench), hold not less than 3 h at −73°C or lower.	954 ± 6°C for not less than 10 min, but not more than 1 h, water quench. Cool to not higher than −73°C, hold for not less than 3 h. Temper at 538 ± 15°C, holding for not less than 3 h.
635	1038 ± 15°C (air cool)	510 ± 8°C, 30 min, air cool.
		538 ± 8°C, 30 min, air cool.
		566 ± 8°C, 30 min, air cool.
XM-9	Supplied in fully heat treated condition by the producer	
XM-12	1050 ± 25°C (cool as required)	Same as Type 630
XM-13	927 ± 8°C (cool as required to below 16°C)	510 ± 6°C, 4 h, air cool.
		538 ± 6°C, 4 h, air cool.
XM-16	829 ± 14°C (water quench)	482 ± 6°C, 4 h, air cool. or 510 ± 6°C, 4 h, air cool.
XM-25	1038 ± 15°C (cool rapidly)	482 ± 8°C, 4 h, air cool.
		538 ± 8°C, 4 h, air cool.
		621 ± 8°C, 4 h, air cool.

[A] Times refer to time material is at temperature.

[B] Equalization and over-tempering treatment: 774 ± 28°C for not less than 3 h, cool to room temperature, heat to 579 ± 14°C for not less than 3 h.

TABLE 3 Mechanical Test Requirements in Solution-Treated Condition

Type	Size	Tensile Strength, max ksi	Tensile Strength, max MPa	Yield Strength, max ksi	Yield Strength, max MPa	Elongation in 2 in. or 50 mm, min,%	Hardness, max Rockwell	Hardness, max Brinell
630	0.015 to 4.0 in. (0.38 to 102 mm)	185	1255	160	1105	3	C38	364
631	0.010 in. (0.25 mm) and under	150	1035	65	450	...	...	...
	Over 0.010 to 4.0 in. (0.25 to 102 mm)	150	1035	55	380	20	B92	...
632	0.0015 to 4.0 in. (0.038 to 102 mm)	150	1035	65	450	25	B100	...
633	0.001 to 0.0015 in. (0.03 to 0.038 mm), excl	200	1380	90	620	8	C30	...
	0.0015 to 0.002 in. (0.03 to 0.05 mm), excl	200	1380	88	605	8	C30	...
	0.002 to 0.005 in. (0.05 to 0.13 mm), excl	200	1380	86	595	8	C30	...
	0.005 to 0.010 in. (0.13 to 0.25 mm), excl	200	1380	85	585	8	C30	...
	Over 0.010 in. (0.254 mm)	200	1380	85	585	12	C30	...
634[A]	Plate	...	...	...	...	...	C40	...
635	0.030 in. (0.76 mm) and under	120	825	75	515	3	C32	...
	Over 0.030 to 0.060 in. (0.76 to 1.52 mm)	120	825	75	515	4	C32	...
	Over 0.060 in. (1.52 mm)	120	825	75	515	5	C32	...
XM-9	Over 0.010 in. (0.25 mm)	150	1035	125	860	4	C28	...
XM-12	0.0015 to 4.00 in. (0.038 to 101.6 mm)	...	...	...	...	...	C38	363
XM-13	0.0015 to 4.00 in. (0.038 to 101.6 mm)	...	...	...	...	...	C38	363
XM-16	0.010 in. (0.25 mm) and greater	175	1205	160	1105	3	C36	331
XM-24	0.0015 to 4.00 in. (0.038 to 101.6 mm)	150	1035	65	450	20	B100	...
XM-25[B]	0.010 in. (0.25 mm) and greater	165	1205	150	1035	4	C33	311
		130	895	90	620	4	C25	255

[A] Solution-treated, equalized, and over-tempered plate only.
[B] XM-25 also furnished to the following minimum properties:

TABLE 4 Bend Test Requirements in Solution-Treated Condition

Type	Size, in. (mm)	Cold Bend Degrees	Bend Test Mandrel
630		none required	
631	0.187 (4.76) and under	180	1T[A]
	Over 0.187 to 0.275 (4.76 to 6.98)	180	3T
632	0.187 (4.76) and under	180	1T
	Over 0.187 to 0.275 (4.76 to 6.98)	180	3T
633	Under 0.1875 (4.762)	180	2T
634	0.187 to 0.249 (4.76 to 6.32)	130	3T
	Over 0.249 to 0.750 (6.32 to 19.08)	90	3T
635		none required	
XM-9	0.109 (2.77) and under	180	9T
XM-12		none required	
XM-13		none required	
XM-16	Under 0.1875 (4.762)	180	6T
XM-24	Under 0.1875 (4.762)	180	1T
XM-25	Under 0.1875 (4.762)	180	6T

[A] T = thickness of sheet being tested.

TABLE 5 Mechanical Test Requirements After Precipitation Hardening Treatment

Grade	Hardening or Precipitation Treatment or both, °F(°C)	Thickness, in. (mm)	Tensile Strength, min		Yield Strength, min		Elongation in 2 in. or 50 mm, min, %	Reduction of Area, min, %	Hardness, min			Impact Charpy V, min[A]	
			ksi	MPa	ksi	MPa			Rockwell, min/max	Brinell, min/max		ft·lbf	J
630 and XM-12	900 (482)	Under 0.1875 (4.762)	190	1310	170	1170	5	...	C40/C48	...		...	...
		0.1875 to 0.625 (4.762 to 15.88)	190	1310	170	1170	8	30	C40/C48	387/484		...	...
		0.626 to 4.0 (15.90 to 102)	190	1310	170	1170	10	35	C40/C48	387/484		...	...
	925 (496)	Under 0.1875 (4.762)	170	1170	155	1070	5	...	C38/C46	...		...	...
		0.1875 to 0.625 (4.762 to 15.88)	170	1170	155	1070	8	30	C38/C47	375/471		...	...
		0.626 to 4.0 (15.90 to 102)	170	1170	155	1070	10	35	C38/C47	375/471		...	...
	1025 (552)	Under 0.1875 (4.762)	155	1070	145	1000	5	...	C35/C43	...		...	...
		0.1875 to 0.625 (4.762 to 15.88)	155	1070	145	1000	8	35	C33/C42	322		10	14
		0.626 to 4.0 (15.90 to 102)	155	1070	145	1000	12	40	C33/C42	322		15	20
	1075 (579)	Under 0.1875 (4.762)	145	1000	125	860	5	...	C31/C40	...		...	...
		0.1875 to 0.625 (4.762 to 15.88)	145	1000	125	860	9	35	C29/C38	293/375		15	20
		0.626 to 4.0 (15.88 to 102)	145	1000	125	860	13	45	C29/C38	293/375		20	27
	1100 (593)	Under 0.1875 (4.762)	140	965	115	790	5	...	C31/C40	...		...	...
		0.1875 to 0.625 (4.762 to 15.88)	140	965	115	790	10	35	C29/C38	293/375		15	20
		0.626 to 4.0 (15.88 to 102)	140	965	115	790	14	45	C29/C38	293/375		20	27
	1150 (621)	Under 0.1875 (4.762)	135	930	105	725	8	...	C28/C38	...		...	...
		0.1875 to 0.625 (4.762 to 15.88)	135	930	105	725	10	40	C26/C36	270/351		25	34
		0.626 to 4.0 (15.88 to 102)	135	930	105	725	16	50	C26/C36	270/351		30	41
	1400 + 1150 (760 + 621)	Under 0.1875 (4.762)	115	790	75	515	9	...	C26/C36	255/331		...	...
		0.1875 to 0.625 (4.762 to 15.88)	115	790	75	515	11	45	C24/C34	248/321		55	75
		0.626 to 4.0 (15.88 to 102)	115	790	75	515	18	55	C24/C34	248/321		55	75
631	1400 (760) + plus 55 (15) + 1050 (566)	0.0015 to 0.0049 (0.038 to 0.124)	180	1240	150	1035	3	...	C38	...		...	...
		0.0050 to 0.0099 (0.127 to 0.251)	180	1240	150	1035	4	...	C38	...		...	...
		0.010 to 0.0199 (0.25 to 0.505)	180	1240	150	1035	5	...	C38	...		...	...
		0.020 to 0.1874 (0.51 to 4.760)	180	1240	150	1035	6	...	C38	...		...	...
		0.1875 to 0.625 (4.762 to 15.88)	170	1170	140	965	7	20	C38	352		...	...
		0.626 to 4.0 (15.90 to 102)	170	1170	140	965	6	20	C38	352		...	...
	1750 (954) + minus 100 (73) + 950 (510)	0.0015 to 0.0049 (0.038 to 0.124)	210	1450	190	1310	1	...	C44	...		...	...
		0.0050 to 0.0099 (0.127 to 0.251)	210	1450	190	1310	2	...	C44	...		...	...
		0.010 to 0.0199 (0.25 to 0.505)	210	1450	190	1310	3	...	C44	...		...	...

TABLE 5 Continued

Grade	Hardening or Precipitation Treatment or both, °F(°C)	Thickness, in. (mm)	Tensile Strength, min		Yield Strength, min		Elongation in 2 in. or 50 mm, min, %	Reduction of Area, min, %	Hardness, min			Impact Charpy V, min[A]	
			ksi	MPa	ksi	MPa			Rockwell, min/max	Brinell, min/max		ft·lbf	J
632	1400 (760) + plus 55 (15) + 1050 (566)	0.020 to 0.1874 (0.51 to 4.760)	210	1450	190	1310	4	...	C44	...	...	...
		0.1875 to 0.625 (4.762 to 15.88)	200	1380	180	1240	6	20	C43	404	...	...
		0.626 to 4.0 (15.90 to 102)	185	1255	150	1035	6	20	C41	382	...	...
	Cold rolled at mill	0.0015 to 0.050 (0.038 to 1.27)	200	1380	175	1205	1	...	C41	...	...	...
	Cold rolled at mill + 900 (482)	0.0015 to 0.050 (0.038 to 1.27)	240	1655	230	1580	:	...	C46	...	...	...
	1400 (760) + plus 55 (15) + 1050 (566)	0.0015 to 0.0049 (0.038 to 0.124)	190	1310	170	1170	2	...	C40	...	...	...
		0.0050 to 0.0099 (0.127 to 0.251)	190	1310	170	1170	3	...	C40	...	...	...
		0.010 to 0.0199 (0.25 to 0.505)	190	1310	170	1170	4	...	C40	...	...	...
		0.020 to 0.1874 (0.51 to 4.760)	190	1310	170	1170	5	...	C40	...	...	...
		0.1875 to 0.625 (4.762 to 15.88)	190	1310	170	1170	4	20	C40	372	...	...
		0.626 to 4.0 (15.90 to 102)	190	1310	170	1170	5	20	C38	352	...	...
	1750 (954) + minus 100 (73) + 950 (510)	0.0015 to 0.0049 (0.038 to 0.124)	225	1550	200	1380	1	...	C46	...	...	...
		0.0050 to 0.0099 (0.127 to 0.251)	225	1550	200	1380	2	...	C46	...	...	...
		0.010 to 0.0199 (0.25 to 0.505)	225	1550	200	1380	3	...	C46	...	...	...
		0.020 to 0.1874 (0.51 to 4.760)	225	1550	200	1380	4	...	C46	...	...	...
		0.1875 to 0.625 (4.762 to 15.88)	225	1550	200	1380	4	20	C45	426	...	...
		0.626 to 4.0 (15.90 to 102)	200	1380	175	1205	5	20	C43	404	...	...
	Cold rolled at mill	0.0015 to 0.050 (0.038 to 0.13)	200	1380	175	1205	1	...	C41	...	...	...
	Cold rolled at mill + 900 (482)	0.0015 to 0.050 (0.038 to 0.13)	240	1655	230	1585	1	...	C46	...	...	...
633	850 (455)	0.0005 to 0.0015 (0.022 to 0.038)	185	1275	150	1035	2	...	C42	...	...	...
		0.0015 to 0.0020 (0.038 to 0.041)	185	1275	150	1035	4	...	C42	...	...	...
		0.0020 to 0.0100 (0.041 to 0.254)	185	1275	150	1035	6	...	C42	...	...	...
		0.0100 to 0.1875 (0.254 to 4.762)	185	1275	150	1035	8	...	C42	...	...	...
	1000 (540)	0.0005 to 0.0015 (0.022 to 0.038)	165	1140	145	1000	2	...	C36	...	...	...
		0.0015 to 0.0020 (0.038 to 0.041)	165	1140	145	1000	4	...	C36	...	...	...
		0.0020 to 0.0100 (0.041 to 0.254)	165	1140	145	1000	6	...	C36	...	...	...
		0.0100 to 0.1875 (0.254 to 4.762)	165	1140	145	1000	8	...	C36	...	...	...

ASTM A 693

TABLE 5 *Continued*

Grade	Hardening or Precipitation Treatment or both, °F(°C)	Thickness, in. (mm)	Tensile Strength, min ksi	Tensile Strength, min MPa	Yield Strength, min ksi	Yield Strength, min MPa	Elongation in 2 in. or 50 mm, min, %	Reduction of Area, min, %	Hardness, min Rockwell, min/max	Hardness, min Brinell, min/max	Impact Charpy V, min[A] ft·lbf	Impact Charpy V, min[A] J
634	850 (455)		190	1310	165	1140	10	...	...	...	...	...
	1000 (540)		170	1170	150	1035	12	...	C37	...	...	...
635	950 (510)	0.030 (0.76) and under	190	1310	170	1170	3	...	C39	...	...	...
		0.030 to 0.060 (0.76 to 1.52)	190	1310	170	1170	4	...	C39	...	...	...
		Over 0.060 (1.52)	190	1310	170	1170	5	...	C39	363	...	...
		Plate	190	1310	170	1170	8	25	C39		...	...
	1000 (540)	0.030 (0.76) and under	180	1240	160	1105	3	...	C37	...	...	...
		0.030 to 0.060 (0.76 to 1.52)	180	1240	160	1105	4	...	C37	...	...	...
		Over 0.060 (1.52)	180	1240	160	1105	5	...	C37	352	...	...
		Plate	180	1240	160	1105	8	30	C38		...	...
	1050 (565)	0.030 (0.76) and under	170	1170	150	1035	3	...	C35	...	...	...
		0.030 to 0.060 (0.76 to 1.52)	170	1170	150	1035	4	...	C35	...	...	...
		Over 0.060 (1.52)	170	1170	150	1035	5	...	C35	331	...	...
		Plate	170	1170	150	1035	8	40	C36		...	...
XM-13	950 (510)	Under 0.020 (0.51)	220	1515	205	1410	6	...	C45	...	...	...
		0.020 to 0.1874 (0.51 to 4.760)	220	1515	205	1410	8	...	C45	...	...	...
		0.1875 to 0.625 (4.760 to 15.88)	220	1515	205	1410	10	...	C45	430	...	...
		0.626 to 4.0 (15.90 to 102)	220	1515	205	1410	10	...	C45		...	...
	1000 (538)	Under 0.020 (0.51)	200	1380	190	1310	6	...	C43	...	...	...
		0.020 to 0.1874 (0.51 to 4.760)	200	1380	190	1310	8	...	C43	...	...	...
		0.1875 to 0.625 (4.760 to 15.88)	200	1380	190	1310	10	...	C43	400	...	...
		0.626 to 4.0 (15.90 to 102)	200	1380	190	1310	10	...	C43		...	...
XM-16	950 (510)	Up to 0.020 (0.51)	222	1525	205	1410	...	...	C44	...	...	...
		Over 0.020 to 0.062 (0.51 to 1.57)	222	1525	205	1410	3	...	C44	...	...	...
		Over 0.062 (1.57)	222	1525	205	1410	4	...	C44	...	...	...
X-24	950 (510)	0.005 to 0.0099 (0.22 to 0.251)	220	1515	190	1310	2	...	...	...	...	...
		0.010 to 0.0199 (0.25 to 0.505)	220	1515	190	1310	3	...	...	...	...	...
		0.020 to 0.1874 (0.51 to 4.760)	220	1515	190	1310	4	...	...	...	...	...

490

TABLE 5 Continued

Grade	Hardening or Precipitation Treatment or both, °F(°C)		Tensile Strength, min		Yield Strength, min		Elongation in 2 in. or 50 mm, min, %	Reduction of Area, min, %	Hardness, min			Impact Charpy V, min[A]	
			ksi	MPa	ksi	MPa			Rockwell, min/max	Brinell, min/max		ft·lbf	J
	1050 (566)	0.005 to 0.0099 (0.22 to 0.251)	200	1380	180	1240	2	...	...	...		...	...
		0.010 to 0.199 (0.25 to 0.505)	200	1380	180	1240	3	...	...	...		...	...
		0.020 to 0.1874 (0.51 to 4.760)	200	1380	180	1240	4	...	...	...		...	...
XM-25	900 (482)	Up to 0.020 (0.51)	180	1240	170	1170	3	...	C40	...		...	...
		Over 0.020 to 0.062 (0.51 to 1.57)	180	1240	170	1170	4	...	C40	...		...	...
		Over 0.062 (1.57)	180	1240	170	1170	5	...	C40	...		...	...
	1000 (538)	Up to 0.020 (0.51)	160	1105	150	1035	5	...	C36	...		...	...
		Over 0.020 to 0.062 (0.51 to 1.57)	160	1105	150	1035	6	...	C36	...		...	...
		Over 0.062 (1.57)	160	1105	150	1035	7	...	C36	...		...	...
	1150 (621)	Up to 0.020 (0.51)	125	860	75	515	8	...	C26	...		...	...
		Over 0.020 to 0.062 (0.51 to 1.57)	125	860	75	515	9	...	C26	...		...	...
		Over 0.062 (1.57)	125	860	75	515	10	...	C26	...		...	...
XM-9	900 (482)	Over 0.010 (0.25)	180	1240	160	1105	3	...	C38	...		...	...

[A] Impact test is not required unless specified on the purchase order.

The American Society for Testing and Materials takes no position respecting the validity of any patent rights asserted in connection with any item mentioned in this standard. Users of this standard are expressly advised that determination of the validity of any such patent rights, and the risk of infringement of such rights, are entirely their own responsibility.

This standard is subject to revision at any time by the responsible technical committee and must be reviewed every five years and if not revised, either reapproved or withdrawn. Your comments are invited either for revision of this standard or for additional standards and should be addressed to ASTM Headquarters. Your comments will receive careful consideration at a meeting of the responsible technical committee, which you may attend. If you feel that your comments have not received a fair hearing you should make your views known to the ASTM Committee on Standards, 1916 Race St., Philadelphia, Pa. 19103.

ASTM Designation: A 700 – 81

Standard Practices for
PACKAGING, MARKING, AND LOADING METHODS FOR STEEL PRODUCTS FOR DOMESTIC SHIPMENT[1,2]

This standard is issued under the fixed designation A 700; the number immediately following the designation indicates the year of original adoption or, in the case of revision, the year of last revision. A number in parentheses indicates the year of last reapproval. A superscript epsilon (ϵ) indicates an editorial change since the last revision or reapproval.

These practices have been approved for use by agencies of the Department of Defense and for listing in the DoD Index of Specifications and Standards.

1. Scope

1.1 These practices cover the packaging, marking, and loading of steel products for domestic shipment. Assuming proper handling in transit, the practices are intended to deliver the products to their destination in good condition. It is also intended that these recommendations be used as guides for attaining uniformity, simplicity, adequacy, and economy in the domestic shipment of steel products.

1.2 These practices cover semi-finished steel products, bars, bar-size shapes and sheet piling, rods, wire and wire products, tubular products, plates, sheets, and strips, tin mill products, and castings. A glossary of packaging, marking, and loading terms is also included.

1.3 The practices are presented in the following sequence:

	Section
Applicable Documents	2
General Provisions	3
General	3.1
Railcar Loading	3.2
Truck Loading	3.3
Barge Loading	3.4
Packaging Materials	3.5
Package Identification	3.6
Weight and Count	3.7
Packaging Lists or Tally	3.8
Loss or Damage	3.9
Glossary of Terms	4
Semifinished Steel Products	5
Hot-Rolled Bars and Bar-Size Shapes	6
Cold-Finished Bars	7
Structural Shapes and Sheet Piling	8
Rods, Wire, and Wire Products	9
Tubular Products	10
Plates	11
Sheets and Strip	12
Tin Mill Products	13
Castings	14

2. Applicable Documents

2.1 *ASTM Standards:*

D 245 Methods for Establishing Structural Grades and Related Allowable Properties for Visually Graded Lumber[3]

D 689 Test Method for Internal Tearing Resistance of Paper[4]

D 774 Test Method for Bursting Strength of Paper[4]

D 781 Test Methods for Puncture and Stiffness of Paperboard and Corrugated and Solid Fiberboard[4]

D 828 Test Methods for Tensile Breaking Strength of Paper and Paperboard[4]

D 2555 Methods for Establishing Clear-Wood Strength Values[3]

2.2 *Other Documents:*

2.2.1 *Federal Specifications:*[5]

QQ-S-781 Steel Strapping, Flat

QQ-S-790 Steel Strapping, Round (Bare and Zinc-Coated)

2.2.2 *Association of American Railroads:*[6]

Rules Governing the Loading of Commodities on Open Top Cars

Pamphlet 23—The Rules Governing the Loading of Steel Products in Closed Cars

---

[1] These practices are under the jurisdiction of ASTM Committee A-1 on Steel, Stainless Steel and Related Alloys and are the direct responsibility of Subcommittee A01.94 on Government Specifications.
Current edition approved July 31, 1981. Published November 1981. Originally published as A 700 – 74. Last previous edition A 700 – 78.

[2] A revision of Simplified Practice Recommendation R 247-62, formerly published by the U. S. Department of Commerce.

[3] *Annual Book of ASTM Standards*, Vol 04.09.

[4] *Annual Book of ASTM Standards*, Vol 15.09.

[5] Available from Naval Publications and Forms Center, 5801 Tabor Ave., Philadelphia, Pa. 19120.

[6] Available from Association of American Railroads, American Railroads Bldg., 1920 L St., NW, Washington, D. C. 20036.

and Protection of Equipment

2.2.3 *American Society of Agricultural Engineers:*[7]

ASAE Standard S 229, Baling Wire for Automatic Balers

3. General Provisions

3.1 *General*—It is recommended that producers and users follow the packaging, marking, and loading methods for individual steel products so described and illustrated herein. It is the responsibility of the purchaser to provide the producer with his requirements concerning protective wrapping materials. When unusual or special conditions require packaging, marking, and loading methods not covered herein, the purchaser should consult with the supplier. Each load involves variables in lading and equipment which cannot be precisely covered by loading rules. Therefore, it is essential that the receiver supply the shipper with pertinent information on his unloading methods and equipment.

3.2 *Rail Car Loading*—All rail shipments of steel products are loaded in accordance with the latest rules governing the loading of either open top cars or closed cars as published by the Association of American Railroads. These publications are entitled "Rules Governing the Loading of Commodities on Open Top Cars" and "Pamphlet 23—The Rules Governing the Loading of Steel Products in Closed Cars and Protection of Equipment."

3.3 *Truck Loading*—The trucker is responsible for the arrangement and securement of the load for safe transit, the protection of the lading from damage by binders, and the prevention of damage to the lading from the elements. These loads shall be in accordance with applicable state and federal regulations.[8]

3.4 *Barge Loading*—There are no formal rules covering barge loading. Steel products are suitably packaged and the barge is loaded to provide ample clearance or blocking, or both, for subsequent handling and unloading. Covered or open-top barges may be used depending upon the nature of the product.

3.5 *Packaging Materials*:

3.5.1 *General*—Materials not covered by specifications or which are not specifically described herein shall be of a quality suitable for the intended purpose. Specifications described are intended as the minimum requirements for packaging of steel products. After the product has been delivered, purchasers are faced with the problems of disposal of the packaging materials. For this reason the simplest effective packaging is the most desirable. The packaging materials described are subject to change in accordance with the rapidly developing technology and the changing regulations affecting ecology.

3.5.2 *Lumber*—The proper selection of lumber for use in the packaging of steel products depends upon many factors, such as end use, compressive strength, beam strength, hardness, moisture content, nail-holding power, condition, etc. Detailed information is contained in the U. S. Department of Agriculture's *Wood Handbook* No. 72 and in ASTM Specifications D 245 and D 2555.

3.5.3 *Protective Wrapping Material*—Protective wrappings are used in packaging to (*1*) retard moisture penetration, (*2*) minimize loss of oil, and (*3*) provide protection from dirt.

3.5.3.1 *Paper*—The basis weight is determined by the number of pounds per 500 sheets of 24 by 36 in. For example, 50-lb kraft paper will equal 50 lb per 500 sheets of 24 by 36 in. The following tests may be used to determine the physical properties of paper:

Test	ASTM Method
Bursting strength	D 774
Beach puncture	D 781
Tearing resistance	D 689
Tensile strength	D 828

3.5.3.2 *Oil-Resistant Paper*—Paper treated, laminated, or constructed to resist absorption of oil from the packaged product.

3.5.3.3 *Waterproof Paper*—These papers are laminated, coated, or impregnated with a moisture-barrier material. The paper shall meet a minimum requirement of 12 h when tested by the dry indicator test, Method D 779.

3.5.4 *Protective Coatings*—In selecting corrosion-preventive materials to protect steel mill products during shipment and storage, consideration should be given to ease and method of application, coverage desired, severity of conditions expected, and ease of removal. The material and method of application determined

---

[7] Available from American Society of Agricultural Engineers, 2950 Niles Rd., St. Joseph, Mich. 49085.
[8] Code of Federal Regulations Title 49—Transportation, Chapter III-Federal Highway Administration, Department of Transportation, Subchapter B-Motor Carrier Safety Regulations, Part 393, Parts and Accessories Necessary for Safe Operation, Safe Loading of Motor Vehicles.

to be the best suited for protection of a product are based on experience. Therefore, selection of protective coatings should be left to the discretion of the steel supplier whenever possible. The protective coatings used on steel products are listed in Table 1.

3.5.5 *Package Ties*—Tying of packages shall be accomplished by tension-tying with bands or wire; or by hand tying and twisting heavy gage wire or rods. Either bands or wire may be used for package ties, regardless of which type of tie is shown in illustrations in the individual product sections of this practice.

3.5.5.1 *Breaking Strength Ties* used in packaging steel mill products shall have the minimum breaking strengths of Federal Specifications QQ-S-781 and QQ-S-790.

3.5.6 *Protectors*—Protectors are used with certain products to protect them from damage and to prevent shearing of the ties. Various materials, such as lumber, metal, plastic, fiber, or other suitable materials, are used under the package ties as required.

3.6 *Package Identification*:

3.6.1 All marking shall be legible and of a size consistent with the space available to be marked. All tags shall be securely affixed to the package to prevent loss in transit. Tags shall be of a size to show clearly all of the information required, and shall be able to withstand reasonable exposure to the elements.

3.6.2 *Marking Metal Surfaces*—Unless otherwise specified, metal surfaces shall be marked with either permanent ink or paint.

3.6.3 *Marking Containers*—All materials used for marking containers shall be resistant to the elements.

3.7 *Weight and Count*—When steel products are invoiced on mill scale weights and such weights are checked after shipment, variations from invoice weights up to 1 % are normally expected due to differences in the kind, type, and location of the scales. When invoiced on mill scale weights, where there are large quantities of one size or thickness, or where the number of pieces in a lift or bundle is required to be shown on the identification tags and shipping papers, the count is considered approximate and the weight is the more accurate. When steel products are invoiced on theoretical weights, the invoice weights are based on the number of pieces or lineal feet shipped.

3.8 *Packaging Lists or Tally*—Furnished as required. Such lists are compiled as accurately as practicable, subject to confirmation by the official shipping notice or invoice.

3.9 *Loss or Damage*—If upon delivery there is any evidence of loss or damage, exception should be taken by notation on the freight bill, and the carrier's representative should be called in to inspect the lading before unloading.

4. Glossary of Terms

4.1 The following glossary defines packaging, marking, and loading terms.

AAR—Association of American Railroads.

"A" end of car—arbitrary definition used to describe the end of a freight car opposite the end on which the manual brake control is located. In the event there is a manual brake control on both ends, the ends are designated by stenciling the letters "A" and "B" respectively on both sides near the ends.

air tool—tool operated by air pressure used for strap tensioning, sealing, nailing, etc.

anchor plate—a plate that is nailed to side or floor of car used to attach steel strapping for load securement.

anchor tie—a coil eye-tie that is applied in a special manner to resistant movement on bar or rod coils. A typical method is to wrap the tie around several strands, then around the complete coil.

anti-skid plate—a device with sharp projections placed under the package to retard shifting of the load in transit.

"A" rack—a rack built in the form of the letter "A" for storing steel bars.

asphalt-laminated paper—paper used for packaging or shrouding, or both, composed of two or more sheets of paper bonded by asphalt.

back-up cleat—wood strip nailed to floor or side of car to strengthen or prevent displacement of the primary blocking.

banding, band—See *strapping*.

band protector—material used under package or load ties to protect product from damage and to prevent shearing of the package ties.

bare—any product that has not been protectively wrapped or covered when packaged.

barrel, slack—wooden barrel, not watertight by construction, used for solid materials.

basis weight—standard weight accepted by trade customs, based upon standard size for the given class of material. The weights of all other standard sizes are proportionate to the size and weight established for the given class of material.

batten strips—strips of wood used to protect machined surface or projections on castings from damage by the securing tie or contact with other objects. Their location is optional but must be so located to afford maximum protection.

bearing pieces—supports beneath but not secured to lift, package, or load.

belt rails—perforated angle or channel, running lengthwise at various levels along wall of vehicle, used to affix load-securement devices such as cross members or bulkheads.

"B" end of car—the end of a freight car on which the manual brake control is located. In the event there is a manual brake control on both ends, the ends are designated by stenciling the letters "A" and "B" respectively, on both sides near the ends.

beveled—usually refers to a packaging or loading member with ends or edges cut at an angle other than 90 deg.

binder—a clamping device used to secure chains or cables.

blocking—material used to prevent or control movement of the unit or load or to facilitate handling.

box—a fully enclosed rigid container having length, width, and depth.

box car—a freight car completely enclosed by ends, sides, and roof equipped with doors to permit entry of loading equipment and lading.

bracing—material used to make the unit or load firm or rigid.

brand—producer's or consumer's identification marks.

bulkhead—fabricated and affixed barrier used to prevent lengthwise movements of a unit or load.

bulkhead, movable—bulkhead, part of railroad equipment, that is capable of being adjusted for load securement.

bumper block—material affixed to ends or sides of a unit or load to prevent damaging contact.

bundle—two or more pieces secured together.

cleat—a piece of material, such as wood or metal, attached to a structural body to strengthen, secure, or furnish a grip.

clinched tie—a coil eye-tie (round wire) that is tensioned after manual twisting. Normally done with special twisting tool or a bar.

coil—a continuous length of wire, bar, rod, strip, sheet, etc., cylindrically wound.

coil car—railroad car specially equipped for the transportation of sheet or strip coils.

coil carrier—a carrying and dispensing device primarily for wire coils.

coil group—two or more coils secured into a unit that can be handled as a single package.

coil skid—See (coil) *platform*.

core—a cylinder on which coiled products are wound and which remains in the inside diameter after winding.

corrosion inhibitor—any material used by the steel industry to inhibit corrosion. This includes chemicals, oils, treated packaging materials, etc.

corrugated box—shipping container made of corrugated fiber board.

covered—top, sides, and ends of package covered with paper under the ties.

crate—a container of open-frame construction.

cross member "DF"—a wood or metal support of rated strength that is attached to the belt rails of a vehicle and that may be used with or without a bulkhead to contain the load.

cushion underframe—a device affixed to the underframe of a railroad car to absorb longitudinal shocks caused by impacts.

damage-free box car—box car equipped with load securement.

deck—top surface of a platform or pallet.

desiccant—chemical used to absorb moisture.

double deck—two-level stacking.

double-door box car—box car equipped with two doors on each side. The doors may be staggered or directly opposite.

drums—fiber or metal cylindrical containers.

eye (of coil)—center opening of coil.

eye vertical—placement of coil with eye of coil vertical.

filler block—wood block used to fill voids when necessary for effective packaging or loading.

fixed bulkhead—immovable bulkhead permanently attached to car.

floating load—a rail load that is permitted to move in a longitudinal direction so that impact shocks are dissipated through movement of the load.

gondola—a freight car with sides and ends but without a top covering. May be equipped with high or low sides, drop or fixed ends, solid or drop bottoms, and is used for shipment of any commodity not requiring protection from the weather.

gondola, covered—a gondola with a movable or removable cover. Used for the shipment of any commodity that requires protection from the weather.

gondola, drop-end—a gondola with ends in the form of doors which can be lowered to facilitate loading and unloading, or for transporting long material that extends beyond the ends of the car.

gondola, fixed-end—a gondola with fixed ends and sides but without top covering.

gondola, low-side—a gondola with car sides under 45 in. (1.14 m).

greaseproof paper—paper treated to inhibit absorption of grease or oil.

gross weight—See definitions under *weights*.

guide strips—lumber secured to car floor to prevent lateral movement of lading.

hand bundle—a secured or unsecured unit that can be handled manually.

headerboard—bulkhead on the front end of a trailer to protect the cab from shifting of the load.

ID—inside diameter or inside dimension.

idler car—flat car or drop-end gondola placed adjacent to a car carrying an overhanging load.

insert—a support used in the inside diameter of a coil placed in position after the coil is formed to prevent collapse.

integral cover—a retractable permanently affixed cover on a gondola or flat car.

interleaving—placing paper between sheets in a lift or between coil wraps for protection against abrasion.

interlocking—procedure for stacking small channels and shapes.

joint strength—the tension measured in pounds that a tied joint can withstand before the joint slips or breaks.

keg—a small barrel.

knee brace—a triangular brace against the load consisting of a vertical and a diagonal member used to prevent shifting of the load. It is frequently supplemented with cleats.

kraft paper—wood pulp paper made by the sulfate process.

label—paper or other material affixed to the package containing identification of product, consignee, producer, etc.

lagging—narrow strips of protective material, usually wood, spaced at intervals around a cylindrical object as protection against mechanical damage.

laminant—the bonding agent used to combine two or more sheeted materials such as films, foils, paper, etc. Often selected to improve barrier qualities of the laminated product.

lift—a unit prepared for handling by mechanical equipment. It may be either secured or loose.

lift truck—a wheeled device used to lift and to transport material. May be a fork lift, ram lift, platform, or straddle truck.

light weight—See definition under *weights*.

load limit—the maximum load in pounds that the conveyance is designed to carry.

loose—often used to mean shipping unsecured.

LTL—less truck load; quantities shipped in amounts less than truck load.

marking—term applied to any of several methods of identifying steel products such as stenciling, stamping, free handwriting, or printing.

metal package—a paper-wrapped package enclosed with metal intended for overseas shipment.

multiple lift—usually refers to unsecured individual lifts of sheets combined one on top of another to make a package.

MVT—moisture vapor transmission.

nailable steel floor—steel floor designed with slots or perforations to permit nailing of lumber blocking.

nestable steel products—rolled or formed steel products or containers that can be fitted into each other when packaged or loaded.

net weight—See definition under *weights*.

OD—outside diameter or outside dimension.

oilproof—a term used to describe packaging materials that are oil resistant.

package—one or more articles or pieces contained or secured into a single unit.

pallet—a structure of wood, metal, or other materials having two faces separated by stringers. Either or both faces may be solid or skeleton construction.

piggy back—highway trailers transported on freight cars.

platform—a structure of wood, metal, or other materials consisting of a deck supported by runners used to facilitate mechanical handling. The deck may be solid or skeleton.

A 700

pneumatic tool—a tool operated by air pressure for purpose of tensioning, sealing, nailing, etc.

polyethylene—a synthetic material used as a free film or in combination with other materials (usually paper) as a protective wrap, cover, or shroud.

port mark—marking that identifies the port of discharge.

racks, storage—a structure on which material is stored.

reel—any device with a flange on each end of which material may be wound, having a flange diameter of 12 in. (305 mm) or over.

retarder plates—formed metal plates secured to the floor through which unit securement bands are threaded. They are used to retard movement of loads.

rub rail—(1) a rail extending around the perimeter of a flat-bed trailer.

(2) a buffer strip used in a conveyance between the side and the lading.

(3) a guide on flat cars used in TOFC service.

runner—member supporting platform deck.

rust inhibitor—a chemical agent used to retard oxidation.

seal—(1) means of effecting strapping joints.

(2) protective device used to provide evidence that closure has not been disturbed.

seal protector—a protector to prevent strapping seal indentation damage to the product.

secured lift—See *lift*.

separator—any material placed between units of the package or load to provide clearance.

shroud—a protective cover placed over the load, unit, or package, covering the top and four sides.

skeleton platform—See *platform*.

skid protector (stain protector)—any of various practices followed to prevent corrosion damage from packaging lumber.

skids—supporting members placed either lengthwise or crosswise beneath and secured to the material to facilitate handling.

solid platform—See *platform*.

spool—a device with a flange at each end on which material may be wound, having flange diameters up to 12 in. (305 mm).

stack—placement of materials or package in tiers.

stake pocket—a metal receptacle that is part of the vehicle and that is designed for the acceptance of stakes.

stakes—metal or lumber placed vertically along sides of vehicle to prevent movement of the lading beyond the side of the vehicle. Also used to provide clearance between the lading and the side of the vehicle.

stamp—to identify with either metal or rubber die.

stencil—to provide identification through the use of a precut stencil.

strapping—flxible material used as a medium to fasten, hold, or reinforce, for example, steel strapping; flat steel band designed for application with tensioning tools.

strapping joint—location or method of providing a strapping closure.

stringers—supporting members that separate the two faces of a pallet.

tag—material, such as paper, plastic, or metal, on which product or shipping data are furnished and which is fastened to a package or container by wires, staples, tacks, etc.

tally—a recapitulation of items comprising a load.

tare weight—weight of container or packaging materials.

tarpaulin—water-resistant material used to protect load or materials from the elements.

tension tie—strapping applied with mechanical tools.

theoretical weight—a calculated weight based on nominal dimensions and the density of material.

tier—one of two or more rows placed one above the other.

TOFC—trailer on flat car. See *piggy back*.

truck—a rubber-tired highway vehicle in the form of a straight truck, semi-trailer, full trailer, or any combination thereof.

(1) *flat bed*—a truck whose cargo-carrying area is a flat surface without sides, ends, or tops.

(2) *low side*—a truck whose cargo-carrying area is a flat surface equipped with side and ends and approximately 2 ft 6 in. to 4 ft (0.76 to 1.22 m) in height.

(3) *removable side*—a truck whose cargo-carrying area is a flat surface equipped with removable sides and rear door approximately 2 ft 6 in. to 8 ft (0.76 to 2.44 m) in height.

(4) *open top high side*—a truck whose

cargo-carrying area is a flat surface equipped with high sides and ends but no permanent top. The end at rear of vehicle opens to facilitate loading.

　　(5) *pole trailer*—highway trailer with a pole-like connection between the front and back wheels for transporting long material.

　　(6) *expandable trailer*—a flat trailer of more than one section which may be extended for long product.

　　(7) *van*—a truck or trailer with nonremovable top.

twist ties—round or oval ties in which the joint is made by twisting the two ends together.

unitized—segments of the load secured into one unit.

unsecured lifts—See *lift*.

VCI—volatile corrosion inhibitor. One type of rust inhibitor.

waster sheet—a secondary grade sheet, sometimes used in packaging to increase resistance to mechanical damage.

waterproof paper—paper constructed or treated to resist penetration of water in liquid form for specific lengths of time.

weights (package): (1) *gross weight*—total weight of commodity and all packaging.

　　(2) *lift weight*—the weight of the material in a lift.

　　(3) *net weight*—the weight of the commodity alone excluding the weight of all packaging material or containers.

　　(4) *tare weight*—weight of packaging components.

weights (transportation): (1) *gross weight*—total weight of lading and transporting vehicle.

　　(2) *light weight*—the weight of the empty transporting vehicle. On rail cars, the light weight is stenciled on car sides.

　　(3) *tare weight*—same as *light weight*.

wrapped—a package or shipping unit completely enclosed with protective material.

5. Semifinished Steel Products

5.1 Semifinished steel products are generally produced for further processing and, because of their nature, only the simple methods of packaging and loading described below are recommended.

5.2 *Product Grades*:

5.2.1 Carbon, alloy, and stainless steel ingots, blooms, billets, and slabs.

5.2.2 Carbon steel skelp in coils.

5.3 *Marking*:

5.3.1 It is normal practice to stamp or paint the heat number on each piece shipped loose and to show the heat number on a tag attached to each secured lift of smaller size billets. The ordered size and weight may be painted on at least one piece of each size when shipped loose or on at least one piece of each secured lift. Each skelp coil is tagged or marked with the heat number and the size.

5.3.2 *Color Marking*—There is no generally recognized color code for identification of steel grades. When specified, color marking to denote grade is applied. In such cases a dash of color on one end of loose pieces is sufficient. In the case of secured lifts of smaller sizes, the grade is shown on a tag attached to the lift or by a dash of one color on one end of the lift.

5.4 *Packaging*:

5.4.1 Semifinished steel products are usually shipped loose. When specified, lifts of billets 9 in.^{2} (58 cm^2) and under in cross section may be secured into lifts of 5 tons (4.5 Mg) or heavier. The securement of this type of package consists of ties of soft wire rod or tensioned flat bands. The number of ties to be used on any specific lift can best be determined by the shipper's experience.

5.4.2 Skelp in coils is secured with a minimum of two ties per coil.

5.4.3 Semifinished steel products are usually shipped in open-top equipment and require no further protection from the elements.

5.5 *Loading*—Semifinished steel products are usually shipped loose with different sizes or weights segregated. Unitizing requires additional labor and material.

6. Hot-Rolled Bars and Bar-Size Shapes

6.1 Hot-rolled bars and bar-size shapes are usually further processed by the purchaser. Simple methods of packaging and loading are recommended. The major consideration is the prevention of physical damage in transit, such as bending or twisting.

6.2 *Product Grades*:

6.2.1 Carbon, alloy, and stainless steel bars, and bar-size shapes.

6.2.2 Concrete reinforcing bars.

6.3 *Marking*:

6.3.1 *Carbon, Alloy, and Stainless Steel Bars, and Bar-Size Shapes*:

6.3.1.1 It is normal practice to identify each lift or coil with a tag containing the following information:

(1) Producer's name, brand, or trademark,
(2) Size,
(3) Specification number, grade (or type for stainless steel only),
(4) Heat number,
(5) Weight (except coils),
(6) Customer's name, and
(7) Customer's order number.

6.3.1.2 *Die Stamping of Carbon Steel Bars*—The ultimate uses of the products do not usually require die stamping. Therefore, this method of marking for other than mill identification requires additional labor and handling.

6.3.1.3 *Die Stamping of Alloy and Stainless Steel Bars*—When specified, heat numbers or symbols are stamped on one end or on the surface near the end of rounds, squares, hexagons, and octagons 2 in. (51 mm) and larger, and on flats 2 in. in width or 2 in. or over in thickness.

6.3.1.4 The above described marking is practicable on smaller sizes down to a minimum of 1 in. (25 mm) in thickness and 1 in. in width for flats, and not less than 1 in. in thickness or diameter for other bars, but because of its precise nature, such marking delays normal production.

6.3.1.5 Stamping of sizes under 1 in. is not practicable. These sizes are secured in lifts and tagged to show heat numbers or symbols.

6.3.1.6 *Color Marking*—There is no standard color code for identification of steel grades. When marking of bars with identification colors is required, the following practices are regularly employed:

(1) Sizes 2 in. (51 mm) and over are marked on one end with not more than two colors.

(2) Sizes 1½ in. (38 mm) up to 2 in. (51 mm) are marked on one end with not more than one color.

(3) Sizes smaller than 1½ in. (38 mm) are not marked individually; but the bundle, lift, or pile (any size bar or flats) is marked on one end with a dab of paint of one color or not more than two different colored stripes.

(4) Bars are regularly painted after assembly into lifts, and due to the nonuniformity of ends, it is not expected that paint will be on every bar in the lift. Any other paint marking slows normal production. Superimposed color marking requires additional labor and time for drying.

(5) When the back of the tag is color marked, one or two colors are used or the names of the colors are given.

6.3.2 *Concrete Reinforcing Bars*:

6.3.2.1 It is normal practice to identify each lift with a tag containing the following information:

(1) Producer's name, brand, or trademark,
(2) Size or bar designation number, and
(3) Grade and specification.

6.3.2.2 *Color Marking*—When specified, a dab of paint, one color only for each grade, is placed on one end of each lift to distinguish grades. Such marking augments but does not replace the marking requirements contained in the product specification.

6.4 *Packaging*:

6.4.1 *Carbon, Alloy, and Stainless Steel Bars, and Bar-Size Shapes*:

6.4.1.1 *Secured Lifts*—Bars are generally packaged into secured lifts (see Figs. 1 and 2). The recommended weight of hot-rolled bars in a secured lift is 10 000 lb (4.5 Mg). Lifts under 10 000 lb require additional material and handling. Producers recommend that purchasers specify the maximum possible weight for lifts because heavier units withstand transportation hazards better and result in greater economy to both the purchaser and the producer. The securement of this type of package consists of ties of soft wire rod or tensioned flat bands. The number of ties to be used on any specific lift can best be determined by the shipper's experience. This recommended securement is adequate for normal handling and transit requirements. Handling by means of the package ties or by magnet is considered an unsafe practice and is not recommended.

6.4.1.2 *Loose Bars*—The term "loose" means single pieces that can be handled individually. This method of loading is sometimes used when shipping to purchasers who unload by hand or magnet or for shipping large bars.

6.4.1.3 *Stack Piling*—This method of piling is regularly used for straightened flats and certain shapes and consists of arranging pieces in order and securing into lifts of 10 000 lb (4.5 Mg) minimum weight. Stack piling of bars under 1 in. (25 mm) in width is impractical.

When stack piling is specified for other than straightened flats or shapes, additional handling is generally required. Figure 3 illustrates a suitable lift of stack-piled straightened flats.

6.4.1.4 *Bar Coils*—Hot-rolled bar coils are regularly secured with two ties of soft wire or flat steel bands and loaded loose, unprotected, in open-top equipment. Bar coils that have had special treatment, such as cleaned and coated or cleaned and oiled, are loaded in closed or covered equipment and require additional labor and material. Securing two or more bar coils into a coil group requires additional labor and material.

6.4.1.5 *Protective Coatings*—The nature of hot-rolled bars or bar-size shapes is such that protective coatings are not regularly applied.

6.4.2 *Concrete Reinforcing Bars*—Concrete reinforcing bars are secured in lifts as illustrated in Fig. 1. The recommended weight of bars in the secured lift is 10 000 lb (4.5 Mg) or more. Lifts under 10 000 lb require additional labor and materials. The securement of this type of package consists of ties of soft wire rods or tensioned flat bands. The number of ties to be used on any specific lift can best be determined by the shipper's experience. Secured lifts in the smaller sizes may contain individually tied bundles within the lift. Bundling of the smaller sizes requires additional material and handling. Packaging of concrete reinforcing bars into units of specified count, weight, or dimensions requires additional handling and material.

6.5 *Loading*:

6.5.1 Carbon, alloy, and stainless steel bars, bar-size shapes, and concrete reinforcing bars are regularly shipped unprotected in open-top equipment. Loading of closed equipment and flatcars requires additional handling and materials.

6.5.2 When separation of lifts is required to allow sufficient clearances needed for unloading equipment, separators or bearing pieces are furnished up to a maximum of commercial 4-in. lumber.

6.5.3 *Weather Protection*—Hot-rolled bars, hot-rolled heat-treated bars, bar-size shapes, and concrete reinforcing bars generally require further processing or fabrication and, therefore, are regularly shipped in open-top equipment, unprotected. When the bars are scale-free or have been processed beyond the as-rolled or heat-treated condition, such as by pickling and oiling or by pickling and liming, producers usually recommend protection by shipment in covered equipment or by wrapping or shrouding when loaded in open-top equipment. In covered rail equipment, shrouding may be required. Figure 4 illustrates a suitable method of wrapping lifts for loading in open-top equipment. Figure 5 illustrates a suitable method of shrouding the carload. The material is a waterproof paper or plastic sheet placed over a number of lifts or over the entire carload and suitably secured.

7. Cold-Finished Bars

7.1 Cold-finished carbon, alloy, and stainless steel bars are among the most highly finished products of the steel industry. Because of their high finish and the exacting uses to which such products are put, packaging and loading methods are very important.

7.2 *Product Grades*—Carbon, alloy, and stainless steel bars.

7.3 *Marking*:

7.3.1 *Carbon, Alloy, and Stainless Steel Bars*:

7.3.1.1 It is normal practice to identify each lift with a tag containing the following information:

(1) Producer's name, brand, or trademark,

(2) Size,

(3) Specification or grade (or type for stainless steel only),

(4) Heat number,

(5) Weight,

(6) Customer's name, and

(7) Customer's order number.

7.3.1.2 *Die Stamping*—It is not regular practice to die-stamp cold-finished bars. Therefore, when specified, this method of marking retards the normal flow of materials.

7.3.1.3 *Color Marking*—When the marking of bars with identification colors is required, the following practices are employed:

(1) Sizes 1½ in. (38 mm) and over are marked on one end with not more than two colors.

(2) Sizes smaller than 1½ in. are not marked individually, but the bundle, lift, or pile is marked on one end with a dab of paint of one color or not more than two different colored stripes.

(3) Any other paint marking slows normal production.

(4) Superimposed color marking also re-

quires additional labor and time for drying.

(5) When the back of the tag is marked, one or two colors are used or the names of the colors are spelled out.

7.4 *Packaging*:

7.4.1 *Carbon, Alloy, and Stainless Steel Bars*:

7.4.1.1 *Secured Lifts* (Fig. 1)—The recommended minimum quantity of cold-finished bars in the secured lift is 6000 lb (2.7 Mg). Producers recommend that purchasers specify the maximum possible weight for lifts because heavier lifts withstand transportation hazards better and result in greater economy to both the purchaser and the producer. The packaging of bars into lifts for closed-car loading requires additional handling. The securement of this type of package consists of ties of soft wire or flat steel bands. Ties are regularly applied as follows:

Up to 15 ft (4.57 m), incl	3 ties
Over 15 ft to 22 ft (4.57 to 6.71 m), incl	4 ties
Over 22 ft to 33 ft (6.71 to 10.06 m), incl	5 ties
Over 33 ft (10.06 m)	6 ties

The recommended securement is adequate for normal handling and transportation requirements. Handling by means of the package ties or by magnet is considered an unsafe practice and is not recommended.

7.4.1.2 *Loose Bars*—The term "loose" means single pieces that can be handled individually. This method of loading is used by producers in the loading of large sizes.

7.4.1.3 *Stack Piling*—This method of piling is regularly used for straightened flats and certain shapes and consists of arranging pieces in order, in one or more piles, into secured lifts of 6000 lb (2.7 Mg) minimum weight. Stack piling of bars under 1 in. (25 mm) in width is impractical. When stack piling is specified for other than straightened flats, additional handling is generally required. The stacking or piling of all bars or bar-size shapes, including straightened flats, into lifts of specified count or dimensions involves additional handling. Figure 3 illustrates a suitable lift of stack-piled straightened flats.

7.4.1.4 *Bundling*—Cold-finished round, square, hexagon, or similar bar sections ⁵⁄₁₆ in. (7.9 mm) and under are put up in hand bundles because of the flexible nature of the material. Bundling of sizes over ⁵⁄₁₆ in. requires additional handling. Figure 6 illustrates a suitable hand bundle. Such bundles regularly contain not less than three pieces, the package weighs from 150 to 200 lb (68 to 91 kg), and is tied with No. 14 gage (1.63-mm) wire or its equivalent as follows:

Up to 8 ft (2.44 m), incl	2 ties
Over 8 ft to 16 ft (2.44 to 4.88 m), incl	3 ties
Over 16 ft to 20 ft (4.88 to 6.10 m), incl	4 ties
Over 20 ft to 24 ft (6.10 to 7.32 m), incl	5 ties

Figure 7 illustrates a bundle of bars banded to a board. Small quantity items unable to support their own weight without possible damage from bending or distortion are usually secured to boards or boxed.

7.4.1.5 *Containers*—Due to the special high finish and very close tolerances of some cold-finished bars, packaging in special containers for extra protection against damage is required. This type of packaging requires additional material and handling. Less than carload or less than truckload shipments of polished, turned ground and polished, cold-drawn ground and polished bars and shafting, or any bars produced to a high finish, are packaged in chipboard tubes, wood boxes, corrugated fiberboard boxes or other suitable containers. Figure 8 illustrates a suitable chipboard container. Such containers are made of heavy spirally wound chipboard with various end closures. Figure 9 illustrates a suitable wood box. Such boxes are made of seasoned lumber, lined with paper, and are reinforced with bands or wire at the ends and at intermediate points, as required.

7.4.1.6 *Protective Coatings*—Cold-finished bars are coated with corrosion preventatives or shipped without protective coating depending upon the use and the purchaser's specification.

7.5 *Loading*:

7.5.1 Cold-finished carbon, alloy, and stainless steel bars are normally shipped in closed or covered equipment. Loading in box cars requires additional handling.

7.5.2 When separation of lifts or piles in cars is required to allow sufficient clearances for unloading equipment, separators or bearing pieces are furnished up to a maximum of commercial 4-in. lumber. Loads are often shipped in bulkhead equipment or are rigidly braced for protection in transit.

7.5.3 Where additional protection is specified in covered gondolas, material may be wrapped or shrouded as illustrated in Figs. 8 or

9. Figure 4 illustrates a suitable method for wrapping lifts of cold-finished bars. Figure 5 illustrates a suitable method of shrouding the carload.

8. Structural Shapes and Steel Sheet Piling

8.1 *Product Grades*:
8.1.1 Carbon, high-strength low-alloy, and stainless steel structural shapes.
8.1.2 Steel sheet piling.
8.2 *Marking*:
8.2.1 *Carbon, High-Strength Low-Alloy, and Stainless Steel Structural Shapes*:
8.2.1.1 It is normal practice to mark each individual structural shape shipped loose or tag each secured lift with the following information:
(1) Producer's name, brand, or trademark,
(2) Section designation or size of section,
(3) Heat number,
(4) Length, and
(5) Grade or type (stainless steel).
8.2.1.2 *Die Stamping*—When specified, the heat number is die-stamped in one location. Die stamping or hot rolling the heat number into structural shapes is not universally practiced. The standard sizes of steel die-stamps are ¼ in., 5/16 in., and ⅛ in. (6.4 mm, 7.9 mm, and 9.5 mm). Any additional or different marking other than as indicated above or specifying stamping with steel die-stamps of sizes other than indicated is negotiated between purchaser and manufacturer.
8.2.1.3 *Color Marking*—On structural shapes made to certain ASTM specifications, color marking is required. Each structural shape shipped loose is marked with one or two color stripes. When shipped in secured lifts, the lift is marked with a vertical stripe for the full height of the lift. Each piece in the lift shall be marked by this stripe.
8.2.2 *Steel Sheet Piling*—It is normal practice to mark each steel sheet piling with the following:
(1) Producer's name, brand, or trademark,
(2) Heat number, and
(3) Length.
Additional or different marking requires additional handling and complicates the normal marking procedure.
8.3 *Packaging*:
8.3.1 *Carbon, High-Strength Low-Alloy and Stainless Steel Structural Shapes*—Structural shapes are normally shipped in unsecured lifts or units weighing approximately 10 000 to 20 000 lb (4.5 to 9.0 Mg). Various methods are used to maintain the unity of such lifts during transit. At manufacturer's option, small sizes may be secured to facilitate identification, handling, or transportation.
8.3.2 *Steel Sheet Piling*—Steel sheet piling is normally handled and loaded in lifts or units weighing approximately 10 000 to 20 000 lb (4.5 to 9.0 Mg), depending on the size of piling sections.
8.4 *Loading*:
8.4.1 *Carbon, High-Strength Low-Alloy, and Stainless Steel Structural Shapes*:
8.4.1.1 *Loading Practice*—Structural shapes are loaded unprotected in open-top equipment because of their nature and the universal use of mechanical unloading equipment. The method used to separate lifts in the car to facilitate unloading can best be determined at the time of loading. Wood blocking and endwise staggering are typical means of separating lifts. Segregation of sections by size, type, or item into separate cars requires additional handling.
8.4.1.2 *Weather Protection*—Structural shapes, due to their nature, are seldom protected from the weather in transit. Protection such as shrouding requires additional labor and material.
8.4.2 *Steel Sheet Piling*—Because of its nature and the universal use of mechanical unloading equipment, steel sheet piling is loaded unprotected in open-top equipment. The method used to separate lifts in the car and thus facilitate unloading can best be determined at the time of unloading. Wood blocking and endwise staggering are typical means of separating lifts.

9. Rods, Wire, and Wire Products

9.1 Hot-rolled wire rods are regularly produced for further processing, and because of their nature only simple methods of marking, packaging, and loading are required.
9.1.1 The major consideration is the prevention of physical damage in transit, such as bending and twisting.
9.1.2 Other wire and wire products however, are among the most highly finished products of the steel industry, and marking, packaging, and loading methods are very important.
9.1.3 Because of the many specific combi-

nations of size, grades, and types supplied in wire, no standard limits for types, diameters, weights, and coil sizes are established. Limitations for coil sizes are controlled by manufacturing practices and other factors.

9.1.4 The purchaser should give careful consideration to marking, packaging, and loading requirements when ordering, and if in question about a suitable method, should consult with the manufacturer. Consultation is usually essential to develop mutually satisfactory methods for packaging of specific products.

9.2 *Product Grades*:
9.2.1 Hot-rolled rods (all grades).
9.2.2 Merchant wire products.
9.2.3 Carbon, alloy, and stainless steel wire (in coils).
9.2.4 Carbon, alloy, and stainless steel wire (straightened and cut).

9.3 *Marking*:
9.3.1 *Hot-Rolled Rods in Coils*—It is normal practice to tag each coil with the following information:
9.3.1.1 Producer's name, brand, or trademark,
9.3.1.2 Grade, product identification or type (stainless steel only),
9.3.1.3 Size,
9.3.1.4 Heat number,
9.3.1.5 Customer's name, and
9.3.1.6 Customer's order number.

(1) When identification colors are specified, marking practice shall be limited to paint striping coil with one color.

9.3.2 *Merchant Wire Products*—It is normal practice to identify each package with the following information, as applicable:
9.3.2.1 Producer's name, brand, or trademark,
9.3.2.2 Product name:
(1) Design or construction
(2) Style
9.3.2.3 Size,
9.3.2.4 Type or class of coating,
9.3.2.5 Finish,
9.3.2.6 Length,
9.3.2.7 Width and mesh, and
9.3.2.8 Height.

9.3.3 *Carbon, Alloy, and Stainless Steel Wire*—It is normal practice to identify each coil or package with the following information:
9.3.3.1 Customer's name,
9.3.3.2 Customer's order number,
9.3.3.3 Producer's name, brand, or trademark,
9.3.3.4 Grade, product identification or type (stainless steel only),
9.3.3.5 Size,
9.3.3.6 Heat number,
9.3.3.7 Quality (when applicable),
9.3.3.8 Finish, and
9.3.3.9 Weight (except coil).

When identification colors are specified, marking practice shall be limited to paint striping coil, one end of bundle or lift with one color.

9.4 *Packaging*:
9.4.1 *Hot-Rolled Rods in Coils* are shipped as individual coils or in coil groups. Securement of individual coils is with a minimum of two twisted wire ties, or tensioned flat bands (Fig. 10). Coil groups are secured with a minimum of two tensioned flat bands (Fig. 11).

9.4.1.1 *Protective Coatings*—It is not standard practice to apply protective coatings to hot-rolled rods, as the product is generally intended for further processing.

9.4.2 *Merchant Wire Products* are finished products sold through distributors or merchandizers and are primarily intended for agricultural, building and home consumption. These products are packaged in various ways depending upon the end use as shown in Table 2.

9.4.3 *Carbon, Alloy, and Stainless Steel Wire in Coils*—Wire is among the most highly finished products of the steel industry. Packaging, marking, and preservation methods are very important and the purchaser should give careful consideration to these requirements when ordering. Wire is drawn from hot-rolled rods. The choice of the wire drawing block diameter for a given wire size varies from manufacturer-to-manufacturer and is dependent upon the equipment in the plants and the buyer's uncoiling equipment. Wire is commonly produced in catchweight coils of one single length and generally wound in a counterclockwise direction. For special requirements, wire may also be furnished in exact weight coils, exact length coils, or straightened and cut lengths. Carbon, alloy, and stainless steel wire in coils may be packaged as shown in Table 3. When protection is necessary it should be specified, depending on finish, end use, type of package, mode of transportation, etc. The following types of protection are available when specified:

Package	Protection
Single coil	Spiral wrap(s) up to approximate 600-lb (272-kg) maximum weight. Protection of heavier coils should be negotiated with manufacturer
Coil on carrier	Shroud
Reel-less coil	Shroud
Wood rack	Shroud
Reel	Wrap(s) between flanges
Container	Liner or shroud, depending on type of package

NOTE—If special finishes require additional protection, negotiate with manufacturer.

9.4.3.1 *Protective Coating*—Depending upon finish, end use, and shipping or storage conditions, oiling may be specified. The use of specified brands of oil involves special handling and interferes with normal processing. Spray oiling of packages may be helpful but affords inadequate protection under normal conditions. Shipment should be in closed equipment.

9.4.4 *Carbon, Alloy, and Stainless Steel Wire, Straightened and Cut Lengths*, is packed in containers, bundles, or lifts as shown in Table 4.

9.4.4.1 *Protective Coatings*—Oiling of straightened and cut length wire requires additional handling and material. Flat wire is generally oiled for protection in transit. The use of special brands of oil involves excessive inventory of oil and disrupts the normal manufacturing process. Spray oiling of packages may be helpful but affords inadequate protection under some conditions.

9.5 *Loading*—Hot-rolled wire rods are regularly shipped in open-top equipment except material that has had special treatment, such as cleaning and coating or oiling. Such material is generally loaded in closed equipment and may require additional handling and material. Due to the nature and high finish of steel wire and wire products, they are normally shipped in closed equipment. Special rail equipment, such as DF (damage free), compartment, and insulated cars, are suitable and can be used for wire products.

10. Tubular Products

10.1 Tubular products can be used in the as-shipped condition or further processed into a finished product. The end use directly affects the extent and types of packaging and marking required.

10.2 *Product Grades*:
10.2.1 Mechanical tubing.
10.2.2 Pressure tubing.
10.2.3 EMT conduit.
10.2.4 Rigid conduit.
10.2.5 Standard pipe.
10.2.6 Line pipe.
10.2.7 Oil country goods.
10.2.8 Couplings and fittings.
10.2.9 Stainless steel tubing and pipe.

10.3 *Marking*—It is normal practice to identify each piece of large diameter steel pipe or tubing shipped loose, or each secured lift or package of smaller sizes with the following information:

(1) Producer's name, brand, or trademark.

NOTE—The above practice is subject to modification as to standard specifications if applicable.

10.4 *Packaging*:

10.4.1 *Mechanical and Pressure Tubing*—This product is shipped loose or in packages (secured lifts) up to 10 000 lb (454 kg). The type of package normally depends on the length and surface quality of the tubing, the user handling facilities, and the method of storage. Thin-wall, polished, or bright finish tubing subject to possible damage during transit is furnished in wrapped packages, frame packages, or boxes. All packages are secured with tension ties. See Figs. 47 to 50 for types of packages. The number of ties are shown as follows:

Length, ft (m)	Minimum Number of Ties
Up to (3.05), incl	2
Over 10 to 15 (3.05 to 4.57), incl	3
Over 15 to 22 (4.57 to 6.71), incl	4
Over 22 to 33 (6.71 to 10.06), incl	5
Over 33 (10.06)	6

NOTE—Sub-bundles are used for EMT conduit (10.4.2), rigid conduit (10.4.3), and standard pipe (10.4.4).

10.4.2 *EMT Conduit*—This product is normally shipped in packages weighing 2000 lb (907 kg) or more. All EMT conduit of 2-in. nominal diameter and smaller is sub-bundled as listed in the following table. Before it is packaged, all sub-bundles are secured with either bands or tape. All packages are secured with tension ties. See 10.4.1 for number of ties.

Nominal Size, in.	Pieces	ft (m)	Weight, lb (kg)
½	10	100 (30.5)	32 (14.5)
¾	10	100 (30.5)	49 (22.2)
1	10	100 (30.5)	71 (32.2)
1¼	5	50 (15.2)	50 (22.7)
1½	5	50 (15.2)	59 (26.8)
2	3	30 (9.1)	45 (20.4)

A 700

10.4.3 *Rigid Conduit*—This product is normally shipped in packages weighing 2000 lb (907 kg) or more. All rigid conduit of 1½-in. nominal diameter and smaller is sub-bundled as in the following table. Before it is packaged, all sub-bundles are secured with either bands or tape. All packages are secured with tension ties. See 10.4.1 for number of ties.

Nominal Size, in.	Pieces	ft (m)	Weight, lb (kg)
½	10	100 (30.5)	79 (35.8)
¾	5	50 (15.2)	53 (24.0)
1	5	50 (15.2)	77 (34.9)
1¼	3	30 (9.1)	60 (27.2)
1½	3	30 (9.1)	75 (34.0)

10.4.4 *Standard Pipe, Line Pipe, and Oil Country Goods*—These products in sizes 1½ in. nominal diameter and smaller may be shipped in sub-bundles as shown in Table 5 or in larger lifts as requested. Sub-bundles are secured with soft annealed wire, tape or secured with tension ties. A minimum of two ties are used for lengths 22 ft (6.71 m) or less and a minimum of three ties for lengths over 22 ft. Sub-bundles may be shipped in packages (secured lifts) of up to 10 000 lb (4540 kg). Larger sizes are shipped loose. Thread protectors are used as indicated in Table 6.

10.4.4.1 *Protective Coatings*—Standard pipe, line pipe, and oil country goods are normally protected with a varnish-type coating (see 3.5.4). The purchaser may order the pipe shipped bare or with other coatings.

10.4.5 *Couplings and Fittings*:

10.4.5.1 *Conduit Couplings and Fittings*—These products are generally shipped on wires, in burlap sacks, or corrugated fiberboard cartons, dependent upon quantities. The weight of a carton generally does not exceed 200 lb (91 kg).

10.4.5.2 *Pipe Couplings*—These are generally shipped in either burlap sacks or wooden boxes, dependent upon quantities. The weight of a wooden box generally does not exceed 600 lb (272 kg).

10.4.5.3 *Pipe Fittings*—These are generally shipped loose, in burlap sacks, in wooden boxes, in corrugated fiberboard cartons, on pallets, and by other acceptable means at the option of the manufacturer.

10.4.6 *Stainless Steel Tubular Products*:

10.4.6.1 Stainless steel tubular products are variously packaged according to product, finish, size, and method of shipment. Stainless steel tubular products are pipe, pressure tubing, mechanical tubing, and structural tubing (including ornamental). Finishes are as-produced (welded or seamless), annealed and pickled, cold finished, ground and polished, and ornamental (including stainless clad). Due to the many sizes, grades, and finishes produced, the purchaser should give careful attention to the packaging, marking, and loading methods when ordering: if in doubt about a suitable method, the purchaser should consult with the supplier.

10.4.6.2 Stainless steel tubular products are packaged in bundles, boxes, or protective containers. Tubes over 6 in. in outside diameter may be shipped loose. Packages may be wrapped or bare. Length, outside diameter, wall, finish, and method of shipment will determine the most suitable packaging method. Polished tubing is always packed in boxes or containers of wood or other suitable material.

10.4.6.3 *Bundles*—If tubing is shipped in such quantities that a risk of its being bent, crushed, or distorted from handling exists, the bundle may require additional support. Bundles are normally secured with flat steel bands but other suitable materials may be used. The amount of securement required is dependent upon length and weight of bundle.

10.4.6.4 *Containers*—Special finishes, quantities ordered, methods of transportation, or other factors may require special containers such as fiberboard or clipboard tubes, fiberboard boxes, wooden boxes or crates, or similar containers.

10.5 *Weather Protection*—Wrapping, shrouding, or covering pipe involves additional labor and material. However, when outside diameter or inside diameter surfaces are critical, shrouding of rail shipments and tarping of trucks is normal practice. Some amount of dirt and oxidation may be expected on black or galvanized pipe and tubes noncoated, or, when coated with nondrying coating, regardless of the type of protection specified.

10.6 *Loading*—Certain steel tubular products are regularly shipped unprotected in open top-cars. It is common practice to load pipe nested without separators, except for external upset pipe and tubing. Securing or separating pipe into lifts, separating sizes and quantities, requires additional handling and material.

Loading tubular products in closed cars or closed trucks requires additional handling. Loading small outside diameter pipe on flat cars requires additional labor and material.

11. Plates

11.1 *Product Grades*:
11.1.1 Carbon, high-strength low-alloy, and alloy steel plates, cut length.
11.1.2 Carbon and alloy steel plate in coils.
11.1.3 Stainless steel plates.
11.1.4 Floor plates.

11.2 *Marking*—It is normal practice to identify each piece, lift, or coil with those requirements as specified in applicable specifications (ASTM, ASME, etc.).

11.3 *Packaging and Loading*:
11.3.1 It is regular practice to load carbon, high-strength low-alloy, and alloy steel plates unprotected in open-top equipment. When specified, loading in closed cars requires additional labor and handling. Carbon, high-strength low-alloy, and alloy steel plates are regularly loaded in unsecured lifts. Loading plates in lifts weighing less than 5 tons (4.5 Mg) involves additional labor and handling. The method used to maintain the unity of unsecured lifts is best determined by the shipper's experience. An example of a suitable method is the staggering of lifts. Segregation of sizes and items involves additional handling, often causes congestion in the manufacturer's plant, and may retard production. Such segregation is not considered feasible. The use of special or particular methods of loading or blocking and specifying the use of bands and wire ties to secure lifts disrupts the normal packaging and loading procedures. This requires additional labor and materials.

11.3.2 *Carbon and Alloy Steel Plates in Coils* are secured with a minimum of either one circumferential tie and one eye tie or with two eye ties.

11.3.3 *Stainless Steel Plates*—Packaging requirements of stainless steel plates are determined by the method of transportation, the finish specified, and the dimensions of the plates. Stainless steel hot-rolled and hot-rolled annealed plates are shipped loose, or when specified, in secured lifts and are loaded in open-top equipment. When processed beyond the as-rolled or annealed condition, such as by pickling or blast cleaning, the plates may also be shrouded or tarped if specified on the order or contract. Cold-rolled stainless steel plates may require greater protection such as wrapping or shrouding and the use of skids or platforms. Polished stainless steel plates are boxed when shipped in small quantities. Larger quantities are packaged on skids or platforms and are paper wrapped and may have additional protection when necessary.

11.3.4 *Floor Plates* are handled in the same manner as carbon and alloy steel plates.

12. Sheets and Strip

12.1 Sheets and strip, in cut lengths, coils, and circles, are among the most highly finished products of the steel industry. Because of their nature and the exacting uses to which such products are put, the marking, packaging, and loading methods are very important. The many sizes, grades, and finishes produced require various methods of packaging and loading, along with surface and weather protection. The methods exemplified in this section recognize these general requirements, the end use of the material, the quantity involved, and the methods of transportation. The purchaser should give careful consideration to the marking, packaging, and loading requirements when ordering and, if in question, about a suitable method, should consult with the manufacturer.

12.1.1 *Suitable Lifts*—In order to facilitate handling, the manufacturer generally prepares these products into lifts or packages so that various mechanical handling equipment can be utilized to advantage. The maximum acceptable package weight should be specified whenever possible, because the heavier packages withstand transportation hazards better and result in greater economy for both the manufacturer and the purchaser. The recommended minimum weight for single lift packages is 10 000 lb (4.5 Mg). Lifts lighter than 10 000 lb require additional labor, material, and handling.

12.1.2 *Skid Arrangements and Platforms*—Figures 53 to 60 show packages on skids only but platforms may be used when required. Suitable arrangements of skids and platform are covered in 12.5.

12.2 *Product Grades*:
12.2.1 Carbon steels.
12.2.2 Alloy steels.
12.2.3 Electrical steels.

12.2.4 Metallic coated (except in mill products).
12.2.5 Nonmetallic coated.
12.2.6 Painted.
12.2.7 Stainless steels.

12.3 *Marking*—It is normal practice to identify each coil, group of coils, or lift of cut lengths with the following information:

12.3.1 Producer's name, brand, or trademark,
12.3.2 Width and gage or thickness,
12.3.3 Product type,
12.3.4 Weight (except strip, coil),
12.3.5 Customer's name, and
12.3.6 Customer's order number.

Stainless steel coils and cut lengths are also identified with the following:

12.3.7 Finish, and
12.3.8 Heat number or coil number.

12.4 *Packaging*:

12.4.1 *Carbon Steel Sheets, Cut Lengths*:

12.4.1.1 *Bare*—Figures 51 and 52 illustrate suitable methods of packaging carbon steel sheets in unsecured lift and secured lift, bare. Hot-rolled carbon steel sheets, not pickled, in heavier gages can be shipped in unsecured lifts as illustrated by Fig. 51 or in secured lifts as illustrated by Fig. 52.

12.4.1.2 *Bare on Skids*—Figures 53 and 54 illustrate suitable methods of packaging bare unwrapped carbon steel sheet on lengthwise and crosswise skids for handling with mechanical equipment. When protection of surface is important, packaging bare as illustrated by these methods is not recommended and, when specified, the responsibility for damage due to inadequate protection rests with the purchaser. The customary weight of this package is 10 000 lb (4.5 Mg) or more. Lengthwise skids are not used on sheets over 192 in. (4.88 m) long or less than 22 in. (559 mm) wide.

12.4.1.3 *Covered*—Figure 55 illustrates a suitable method for covered sheet packages. The bottom of the package is not covered. Sheets lighter than 11 to 16 gage (2.30 to 1.29 mm), inclusive, for shipment in open-top equipment may be covered as illustrated by this figure, which requires additional labor and material. This type of packaging is not used for highly finished sheets for shipment in open-top equipment.

12.4.1.4 *Wrapped*—Figure 56 illustrates a suitable method of wrapping sheet packages. Hot-rolled pickled, and other more highly finished sheets, for shipment in open-top equipment may be wrapped as illustrated by this figure.

12.4.1.5 *Multiple Lift*—Figure 57 illustrates a method of packaging two or more smaller lifts into a secured lift. This method, because of the higher center of gravity of the unit and a tendency for the wood separators to mark the steel, is less desirable than the conventional single lift of 10 000 lb (4.5 Mg). Such packaging requires additional labor and materials. Separators are usually from 1 to 2 in. (25 to 51 mm) in thickness and from 2 to 4 in. (51 to 102 mm) wide, aligned with the skids, and extending the full dimension of the sheets. The individual lift weight in this type of package is generally not less than 2000 lb (907 kg), and the total weight of the multiple lift package is usually not less than 10 000 lb. The individual lifts are not tied, covered, or wrapped. Figure 57 shows lengthwise skids and separators, but crosswise skids and separators may be used.

12.4.1.6 *Short-Length or Narrow-Width Sheets, Crosswise*—Figure 58 illustrates a suitable method of packaging short length or narrow width sheets arranged crosswise, side by side, into secured lifts. Minimum weight of secured lift for such package is 10 000 lb (4.5 Mg). This package is recommended for short sheets 48 in. (1.22 m) or less in length or narrow sheets under 22 in. (559 mm) in width. Suitable vertical separators between piles are used when required. When handled as a unit lift, this package should be handled with a sheet lifter. Figure 58 shows material piled on skids. Platforms may also be used when required; however, such packaging requires additional labor and material. Figure 58 also shows a method of wrapping such a package when protection of steel in open-top equipment requires it to be fully wrapped.

12.4.1.7 *Short-Length Sheets, Lengthwise, End to End*—Figure 59 illustrates a suitable method of packaging short-length sheets arranged lengthwise end to end into secured lifts. This package is generally not used for sheets under 22 in. (559 mm) in width; otherwise, the method of packaging is similar to that shown for short length or narrow width sheets, crosswise, Fig. 58. When handled as a unit lift, this package should be handled with a sheet lifter. Figure 60 illustrates a suitable method of packaging narrow long sheets side by side.

12.4.1.8 *Protective Materials*—Hot-rolled pickled and better grades may require protection from contact with separator or skid lumber. Wrapping sheets with more than one layer of paper requires additional labor and materials. On highly finished sheets, protection against band seal damage is recommended. Protecting sheets with metal wrapping, or the use of metal protector sheets on top or bottom, or both, of lift or package involves additional labor and material.

12.4.1.9 *Protective Coatings*—Oiling to customers' specifications requires additional labor and material.

12.4.1.10 *Weather Protection*:

(*1*) *Open-Top Equipment, General*—Experience has shown that the amount of weather protection required for shipping sheets in open-top equipment depends upon the quality, size, and method of transportation. Hot-rolled sheets, due to their nature, are not generally protected from the weather when loaded in open-top equipment. Hot-rolled pickled and more highly finished sheets are regularly wrapped and shrouded when loaded on open-top equipment. Such wrapping and shrouding requires additional labor and material. Truck tarpaulins are considered to be the equivalent of waterproof paper shrouding.

(*2*) *Shrouded Package Open-Top Equipment*—Figures 61 and 62 illustrate suitable methods of shrouding lifts of cut-length sheets for shipment in open-top equipment.

(*3*) *Covered or Closed Equipment—General*—While this type of equipment is recommended for rail and affords better protection from the elements, covering, wrapping, or shrouding of sheets may be required for preservation of the surface. Such protection, when specified, requires additional labor and material.

12.4.2 *Carbon Steel Sheets, Coils*:

12.4.2.1 *General*—All coil weights are subject to mill manufacturing limits. When individual coil weights are required, narrow sheet coils are generally weighed in groups and the weight of the group averaged over the number of coils in the group. This average is not intended to be the actual weight of each individual coil of the group. Weighing such coils individually, recording, and marking the weight of each coil requires additional time and handling. Individual coils are usually secured with one to four flat steel bands. Hot-rolled coils are regularly shipped in the as-rolled condition, unprotected, in open-top equipment. It is not standard practice to ship hot-rolled coils on platforms. Hot-rolled pickled or other highly finished sheet coils may be packaged on platforms when required. However, the use of platforms requires additional labor and material. Supporting coils with special cores, or placing coils on spools, requires additional labor and material. Wrapping narrow coils individually requires additional labor and material.

12.4.2.2 *Bare, Unwrapped, Individual Coils*—Figure 63 illustrates a suitable method of packaging individual hot-rolled sheet coil in the as-rolled condition. Figure 64 illustrates a method of packaging often used on more highly finished coils.

12.4.2.3 *Bare, Unwrapped, Coil Group Package*—Figure 65 illustrates a suitable method of packaging two or more narrow sheet coils into a coil group package. Securing sheet coils into specified groups requires additional labor and material.

12.4.2.4 *Coils, Bare Unwrapped, on Platform*—Figure 66 illustrates a suitable method of packaging bare unwrapped sheet coils on skeleton platform with the eye of the coils vertical. The use of separators between coils requires additional labor and material.

12.4.2.5 *Wrapped Individual Coil*—Figure 67 illustrates a suitably wrapped individual sheet coil with eye of the coil horizontal. Wrapping coils requires additional labor and material.

12.4.2.6 *Wrapped Individual Coil on Cradle Platform*—Figure 68 illustrates a suitably wrapped individual coil on cradle platform with the eye of the coil horizontal.

12.4.2.7 *Wrapped Individual Coil on Platform*—Figure 69 illustrates a suitably wrapped individual coil on platform with the eye of the coil vertical.

12.4.2.8 *Surface Protection*:

(*1*) *Oiling*—Oiling coils to customer's specifications requires additional labor and material.

12.4.3 *Circles*:

12.4.3.1 *General*—Figure 70 illustrates suitably wrapped or covered sheet circles on skeleton platform. Circles 17 in. (432 mm) and over in diameter are packaged single pile on square or round platforms, or on crossed skids. To avoid top heaviness, the maximum height of the single pile package should not exceed the diameter of the circle. Circles under 17 in. in diameter may be packaged in several piles on square or rectangular platforms.

12.4.3.2 *Weather Protection*:

(*1*) Hot-rolled pickled and more highly finished sheet coils are regularly wrapped and shrouded when loaded in open-top equipment. Such wrapping and shrouding requires additional labor and material. Truck tarpaulins are considered to be the equivalent of waterproof paper shrouding.

(*2*) *Covered or Closed Equipment, General*— While this type of equipment affords better protection from the elements, covering, wrapping, or shrouding of sheets may be required for preservation of the surface. Such protection, when specified, requires additional labor and material.

12.4.3.3 *Loading*:

(*1*) *Open-Top Equipment, General*—Hot-rolled sheet coils, due to their nature, are not generally protected from the weather when loaded in open-top equipment.

12.4.4 *Stainless Steel Sheets, Cut Lengths*:

12.4.4.1 *General*—The minimum net weight for conventional single-lift packages of stainless sheets depends on the type of package specified. Small amounts regardless of finish are regularly packaged in boxes.

12.4.4.2 *Cut Lengths, Bare*—Figures 53 and 54 illustrate suitable methods of packaging bare, unwrapped, stainless steel sheets on lengthwise and crosswise skids. The figures show the package on skids only, but skeleton deck platforms are also used when required. The recommended weight for this type package is 5000 lb (2268 kg) or more. Finishes and gages generally confined to this type of package are:

(*1*) No. 1 Finish, 0.0418 in. (1.062 mm) and thicker, on skids.

(*2*) No. 1 Finish, under 0.0418 in. (1.062 mm), on skeleton platforms.

When protection of surface is important, packaging bare, as illustrated by these methods, is not recommended. Suitable arrangements of skids and platforms are shown by Figs. 53 to 60 and Tables 7 and 8.

12.4.4.3 *Cut Lengths, Wrapped*—Figures 71 and 72 illustrate suitably wrapped stainless steel sheets on lengthwise and crosswise skids. The illustrations show the package on skids only, but skeleton deck platforms are also used when required. Skeleton deck platforms may have either lengthwise or crosswise runners. The customary weight of this type package is 5000 lb (2268 kg) or more. Finishes and gages generally confined to this type of package are:

(*1*) No. 1 Finish, 0.0418 in. (1.062 mm) and thicker, on skids.

(*2*) No. 1 Finish, under 0.0418 in. (1.062 mm) on skeleton platforms.

When protection of surface is important, wrapped packages, as illustrated by these methods, are not recommended. Suitable arrangements of skids and platforms are shown in Figs. 53 to 60 and Tables 7 and 8.

12.4.4.4 *Cut Lengths, Fully Enclosed Packages 5000 lb (2268 kg) and Heavier*—Figure 73 illustrates a suitable method of packaging steel sheets in a fully enclosed package on a skeleton platform, using wood materials. Other materials used are hardboard, composition board, fiberboard, plywood, angles and channels, depending on the materials available, the type of package, and the discretion of the shipper. This package is designed for lifts 5000 lb and over, and is recommended for maximum protection of all domestic shipments of all gages and finishes. Sideboards are not usually needed if material is less than 1 in. (25 mm) piling height.

12.4.4.5 *Cut Lengths, Boxed*—Figure 74 illustrates a suitable method of packaging steel sheets in a wooden box of suitable solid protective material to provide an entirely closed flat container. This type container is designed for maximum protection of small quantities of all grades, gages, and finishes.

12.4.4.6 *Surface Protection*:

(*1*)*Protective Coverings*—The usual method of protecting surfaces is to interleave with nonabrasive antitarnish paper. Protection of surfaces by means of gluing or pasting paper or otherwise applying protective coverings requires additional labor and material. Protecting sheets with metal wrapping or the use of metal protector sheets on top or bottom, or both, of lift or package requires additional labor and material.

12.4.5 *Stainless Steel Sheets, Coils*:

12.4.5.1 *Bare Unwrapped Individual Coil*— Figure 63 illustrates a suitable method of packaging individual stainless steel hot-rolled sheet coil in the as-rolled condition. This type of packaging is confined to hot-rolled or hot-rolled annealed material.

12.4.5.2 *Wrapped Individual Coil*—Figure 67

A 700

illustrates a suitable method of packaging fully wrapped individual stainless steel sheet coil with eye of coil horizontal. This type of packaging is not recommended for light gage material or for any material when protection of the surface is important. Stainless steel coils, No. 1 Finish, 0.062 in. (1.57 mm) and thicker, are generally confined to this type of package. For thinner gages, platforms are recommended.

12.4.5.3 *Bare Individual Coil on Cradle Platform*—Figure 75 illustrates a suitable method of packaging bare, unwrapped, individual stainless steel sheet coil on cradle platform with the eye of the coil horizontal. This method of packaging provides adequate protection for most grades and gages, when surface protection is not important.

12.4.5.4 *Wrapped Individual Coil on Cradle Platform*—Figure 68 illustrates a suitable method of packaging a wrapped individual stainless steel coil on a cradle platform with the eye of the coil horizontal. This method of packaging is recommended for practically all domestic usage and for most finishes and gages. Gages and finishes requiring additional protection should be boxed.

12.4.5.5 *Bare Individual Coil on Platform*—Figure 76 illustrates a suitable method of packaging bare unwrapped individual sheet coil on platform with the eye of the coil vertical. This method of packaging provides adequate protection for most grades and gages, when surface protection is not important.

12.4.5.6 *Wrapped Individual Coil on Platform*—Figure 69 illustrates a suitably wrapped individual stainless steel coil on platform with the eye of the coil vertical. This method of packaging provides adequate protection for most grades and gages.

12.4.5.7 *Boxed on Platform with Eye of Coil Vertical*—Figure 77 illustrates a suitable method of packaging individual sheet coil or group of sheet coils in solid box, on platform, with the eye of the coil vertical. This type of package, an entirely enclosed container made of suitable solid material, is recommended for maximum protection of all finishes and gages.

12.4.5.8 *Surface Protection*:

(*1*) *Protective Coatings*—The usual method of protecting surfaces is to interleave with nonabrasive antitarnish paper. Protection of surfaces by means of gluing or pasting paper, or otherwise applying protective coverings, requires additional labor and material. Spiral wrapping is not applied to stainless steel sheet coils. The use of metal protective wrapping on coils requires additional labor and material.

12.4.6 *Stainless Steel Sheets, Circles*:

12.4.6.1 *Circles, Bare Unwrapped Single Pile on Platform*—Figure 78 illustrates a suitable method of packaging a single pile of bare stainless steel sheet circles on skeleton platform. This type of packaging is generally confined to No. 1 Finishes 0.062 in. (1.57 mm) and thicker. This package is not recommended for light gage material or for any material when protection of surface is important.

12.4.6.2 *Circles, Wrapped Single Pile on Platform*—Figure 70 illustrates a suitable method of packaging wrapped stainless steel sheet circles on skeleton platform. This type of package is generally recommended for practically all domestic usage, for all gages and finishes.

12.4.6.3 *Circles, Multiple Piles on Platform, Covered with Corrugated Fiberboard*—Figure 79 illustrates a typical method of packaging multiple piles of stainless steel circles on platforms, covered with corrugated fiberboard. This type of package is generally recommended for practically all domestic usage and for all gages and finishes in lots of 2000 lb (907 kg) or more. Quantities less than 2000 lb should be packaged in individual piles or in boxes.

12.4.6.4 *Boxes Wrapped Multiple Piles of Circles on Platform*—Figure 80 illustrates a suitable method of packaging multiple piles of stainless steel sheet circles in a box on a solid deck platform.

12.4.6.5 *Surface Protection*:

(*1*) *Protective Coverings*—The usual method of protecting surfaces is to interleave with nonabrasive antitarnish paper. Protection of surfaces by means of gluing or pasting paper, or otherwise applying protective coverings, requires additional labor and material.

12.4.6.6 *Loading*—Due to the nature of stainless products, shipment in covered or closed equipment is recommended. If shipped in open-top equipment, shrouding of the package or load is recommended.

12.4.7 *Carbon Steel Strip, Cut Lengths*:

12.4.7.1 *Bare Oval Lift*—Figure 2 illustrates a suitable method of packaging narrow hot-rolled steel strip, 5 in. (127 mm) or less in width into conventional oval lifts.

12.4.7.2 *Secured Lift, Strip Lengthwise*—Fig

ure 60 illustrates a suitable method of packaging narrow steel strip piled lengthwise on crosswide skids, in multiple rows, into secured lift. The illustration shows material piled on skids, but skeleton platforms are also used when required. Such packaging requires additonal labor and material. Light-gage reinforcing shields or channels are used to maintain alignment of strip in the piles. Figure 60 also shows the method of wrapping packages when protection of steel is required for shipping in open-top equipment.

12.4.7.3 *Surface Protection*:

(*1*) *Oiling*—Oiling to customer's specifications requires additional labor and material.

(*2*) *Protective Coverings*—Wrapping strip with more than one layer of paper has been found unnecessary for satisfactory delivery.

12.4.7.4 *Loading*:

(*1*) *Open-Top Equipment, General*—Hot-rolled strip, due to its nature, is not protected from the weather when loaded in open-top equipment. It is recommended that hot-rolled pickled and more highly finished strip be covered or shrouded when loaded in open-top equipment. When specified, such protection requires additional labor and material. Truck tarpaulins are considered to be the equivalent of waterproof paper shrouding.

(*2*) *Covered or Closed Equipment, General*— While this type of equipment affords better protection from the elements, covering, wrapping, or shrouding of strip may be required for preservation of the surface. Such protection, when specified, requires additional labor and material.

12.4.8 *Carbon Steel Strip, Coils*:

12.4.8.1 *General*—All coil weights are subject to mill manufacturing limits. When individual coil weights are required, coils are generally weighed in groups, and the weight of the group averaged over the number of coils in the group. This average is not intended to be the actual weight of each individual coil of the group. Weighing coils individually, recording, and marking the weight of each coil require additional time and handling. Individual coils are usually secured with one to four flat steel bands. Hot-rolled coils are regularly shipped in the as-rolled condition, unprotected, in open-top equipment. It is not regular practice to ship hot-rolled coils on platforms. When specified, the use of platforms requires additional labor

and material. Supporting coils with special cores or placing coils on spools requires additional labor and material.

12.4.8.2 *Individual Narrow Strip Coils*—Figures 81, 82, and 83 illustrate suitable methods of packaging individual narrow-strip coils.

12.4.8.3 *Bare Unwrapped Coil Group Package*—Figure 84 illustrates a suitable method of packaging narrow-strip coils into a coil group package. Banding coils into coil group package requires additional labor and material.

12.4.8.4 *Coils on Platforms*—Figure 85 illustrates a suitable method of packaging narrow-strip coils on skeleton platform with the eye of the coils vertical. The illustration shows the package on skeleton platform. Placing individual coils or stacking coils on platforms requires additional labor and material. Separators between coils decrease the security of the package, and requires additional labor and material.

12.4.8.5 *Coils Wrapped*—Figure 86 illustrates suitably wrapped individual strip coils or groups of coils. Wrapping individual coils or wrapping or shrouding coil group packages requires additional labor and material.

12.4.8.6 *Bare Coils in Container*—Figure 87 illustrates a suitable method of packaging narrow-strip coils in a container with the eye of the coils vertical. This type of package is an entirely enclosed container made of suitable solid materials, and is designed for maximum protection of all finishes and gages.

12.4.8.7 *Surface Protection*:

(*1*) *Oiling*—Oiling coils to customer's specifications requires additional labor and material.

(*2*) *Protective Coverings*—Wrapping coils requires additional labor and material. Wrapping individual coils or groups of coils with more than one layer of paper has been found unnecessary for satisfactory delivery. The use of metal protective wrapping on coils requires additional labor and material.

12.4.8.8 *Loading*:

(*1*) *Open-Top Equipment, General*—Hot-rolled strip coils, due to their nature, are not generally protected from the weather when loaded in open-top equipment. It is recommended that hot-rolled pickled and more highly finished strip coils be wrapped or shrouded when loaded in open-top equipment. Truck tarpaulins are considered to be the equivalent of waterproof paper shrouding.

(*2*) *Closed Equipment, General*—While this

A 700

type of equipment affords better protection from the elements, wrapping or shrouding of strip may be required for preservation of the surface. Such protection, when specified, requires additional labor and material.

12.4.9 *Stainless Steel Strip, Cut Lengths*:

12.4.9.1 *Cut Lengths, Bare*—Figures 88 and 89 illustrate suitable methods of packaging bare unwrapped stainless steel strip on crosswise skids or platforms. The recommended weight of this type package is 5000 lb (2268 kg) or more. Finishes and gages generally confined to this type of package are:

(1) No. 1 Finish, 0.0418 in. (1.062 mm) and thicker, on skids.

(2) No. 1 Finish, under 0.0418 in. (1.062 mm), on skeleton platforms.

When protection of surface is important, packaging bare, as illustrated by these methods, is not recommended. Suitable arrangement of skids and platforms is described in 12.5.

12.4.9.2 *Cut Lengths, Wrapped*—Figures 60 and 90 illustrate suitably wrapped stainless steel strip on crosswise skids or platforms. Platforms may have either lengthwise or crosswise runners. The recommended weight of this type package is 5000 lb (2268 kg) or more. Finishes and gages generally confined to this type of package are:

(1) No. 1 Finish, 0.0418 in. (1.062 mm) and thicker, on skids.

(2) No. 1 Finish, under 0.0418 in. (1.062 mm), on skeleton platforms.

When protection of surface is important, packages wrapped, as illustrated by these methods, are not recommended. Suitable arrangement of skids and platforms is shown in 12.5.

12.4.9.3 *Cut Lengths, Fully Enclosed Package 5000 lb (2268 kg) and Heavier*—Figure 91 illustrates a suitable method of packaging stainless steel strip in a fully enclosed package on a platform using wood materials. Other materials generally used are hardboard, composition board, fiberboard, plywood, angles and channels, depending on the materials available, the type of package, and the discretion of the shipper. This package is designed for lifts 5000 lb (2268 kg) and over, and is recommended for maximum protection of all domestic shipments of all gages and finishes. Sideboards are not usually needed if material is less than 1 in. (25 mm) piling height.

12.4.9.4 *Cut Lengths Boxed*—Figure 74 illustrates a suitable method of packaging stainless steel strip in a box of suitable solid protective material, lined or unlined, to provide an entirely closed flat container. This type container is designed for maximum protection of small quantities of all grades, gages, and finishes. Boxes are designed for packaging quantities of less than 5000 lb (2268 kg). Placing boxes on runners or platforms requires additional labor and material.

12.4.9.5 *Surface Protection*:

(1) *Protective Coverings*—The usual method of protecting surfaces is to interleave with nonabrasive antitarnish paper. Protection of surfaces by means of gluing or pasting paper, or otherwise applying protective coverings, requires additional labor and material. Protecting stainless steel strip with metal wrapping or using metal protector sheets on top or bottom, or both, of lift or package requires additional labor and material.

12.4.9.6 *Loading*—Due to the nature of stainless products, shipment in covered or closed equipment is recommended. If shipped in open-top equipment, shrouding of the package or load is recommended.

12.4.10 *Stainless Steel Strip, Coils:*

12.4.10.1 *Bare Unwrapped Individual Coils*—Figure 63 illustrates a suitable method of packaging individual hot-rolled stainless steel strip coil in the as-rolled condition. This type of packaging is generally confined to hot-rolled or hot-rolled annealed material.

12.4.10.2 *Coils on Platform*—Figure 92 illustrates a suitable method of packaging narrow stainless steel strip coils on skeleton platform with eye of the coils vertical. Placing individual coils or stacking coils on platforms requires additional labor and material. Separators between coils decreases the security of the package, and requires additional labor and material. This method of packaging is considered to be adequate for practically all domestic shipments of most gages and finishes. Material requiring maximum protection should be boxed.

12.4.10.3 *Coils, Fully Wrapped*—Figure 86 illustrates a suitably wrapped individual stainless steel strip coil or group of coils. This method of packaging is not recommended for light-gage material nor for any material when protection of surface is important. Wrapping individual coils or wrapping, covering, or shrouding coil group packages requires additional labor and material.

12.4.10.4 *Coils in a Container*—Figure 87 illustrates a suitable method of packaging narrow stainless steel strip coils in a container with the eye of the coils vertical. This type of package is an entirely enclosed container made of suitable solid material, and is recommended for maximum protection of all finishes and gages.

12.4.10.5 *Coils, Boxed on Platform with Eye of Coils Vertical*—Figure 77 illustrates a suitable method of packaging individual stainless steel strip coils or group of strip coils in solid box on platform with the eye of the coils vertical.

12.4.10.6 *Surface Protection:*

(*1*) *Protective Coverings*—The usual method of protecting surfaces is to interleave with non-abrasive antitarnish paper. Protection of surfaces by means of gluing or pasting paper, or otherwise applying protective coverings, requires additional labor and material. Protecting stainless steel coils with metal wrapping requires additional labor and material.

12.4.10.7 *Loading*—Due to the nature of stainless products, shipment in covered or closed equipment is recommended. If shipped in open-top equipment, shrouding of the package or load is recommended.

12.5 *Skid Arrangements and Platforms:*

12.5.1 *Skid Arrangements*—All skids shall be made of sound lumber of commercial sizes not less than 3 in. (76 mm) in width nor more than 4 in. (102 mm) in height. The overall length of skids shall be approximately equal to the full dimension of the package along the direction in which they are used. The number of skids required on packages using skids parallel to their lengthwise direction are shown in Table 7. The number of skids required on packages using skids parallel to their crosswise direction are shown in Table 8. Figures 53 to 60 illustrate typical packaging of cut length sheets on skids.

12.5.2 *Platforms for Cut Lengths*—Structures consisting of deckboards and runners. The arrangements shown in Fig. 93 are often used for packaging of wide sheets of light gage or for packaging long, narrow sheets and strip of any gage piled side by side on one platform. Deckboards shall be equal in length to full width or length of the unit and shall have a minimum thickness of 1 in. (25 mm). Deckboards may be nailed to the runners. The minimum number of deckboards shall be the same as the number of lengthwise or crosswise skids shown in Tables 7 and 8. Illustrations are general and actual construction may vary among producers.

13. Tin Mill Products

13.1 Tin mill products are among the most highly finished products of the steel industry; and marking, packaging, and loading methods are very important. The purchaser should give careful attention to these requirements when ordering and, if in question about a suitable method, should consult with the manufacturer.

13.2 *Product Grades:*

13.2.1 Tin plate.

13.2.2 Black plate.

13.2.3 Electrolytic chromium-coated steel (tin-free steel).

13.3 *Marking:*

13.3.1 *Cut Lengths*—Packages of cut length tin plate are identified with the following:

(*1*) Producer's name, brand, or trademark,

(*2*) Basis weight,

(*3*) Size,

(*4*) Type,

(*5*) Temper,

(*6*) Coating weight (when applicable),

(*7*) Product classification,

(*8*) Surface treatment (when applicable), and

(*9*) Differential markings (when applicable).

13.3.2 *Coils*—It is normal practice to identify each coil package with the following information:

(*1*) Producer's name, brand, or trademark,

(*2*) Width,

(*3*) Basis weight,

(*4*) Type,

(*5*) Temper,

(*6*) Coating weight (when applicable),

(*7*) Coil number,

(*8*) Lineal feet,

(*9*) Weight,

(*10*) Product classification,

(*11*) Differential markings (when applicable), and

(*12*) Surface treatment (when applicable).

13.4 *Packaging:*

13.4.1 *Cut Lengths*—Most tin mill products in cut lengths are shipped in multiple-package units secured to platforms. Such units consist of 10, 12, 15, or more packages, containing 112 sheets per package. The amount of protection and securement may vary with the method of transportation, the ultimate destination, and the experience of the shipper. The components of a typical package unit are as follows:

 (*1*) Standard platform with two or three runners.
 (*2*) Protection on top of platform when required.
 (*3*) Paper lining when specified or required.
 (*4*) Edge protectors under ties.
 (*5*) Wire or flat band ties.
 (*6*) Fiberboard covering.

Regardless of which type of ties are shown in the illustrations, either flat bands or wires may be used. Figure 94 illustrates a suitable method of packaging cut-length tin mill products in a multiple-package unit. This package has fiberboard covering. Edge protectors are used under ties. Ties may be bands or wire. Modifications may be made for units shipped to certain points. Standard platforms used for tin mill products are illustrated by Figs. 95 and 96. The illustrations show platforms with only two runners, but three runners are used when required. The two-runner platform is regularly used for sheets up to 30 in. (762 mm) maximum length. The third runner, when used, is placed midway between the two outside runners. Decks are usually of single thickness, made from lumber dressed not lighter than ⅜ in. (9.5 mm) nor more than 13/16 in. (20.6 mm) in thickness, depending on the size and weight of the package. Deck sizes should be the same or slightly smaller than plate size, never larger. Runners are regularly made from lumber dressed to 1¾ in. (44.4 mm) in width and not less than 2 in. (50.8 mm) or more than 4 in. (101.6 mm) in height with the ends beveled. Unless otherwise specified, they are placed parallel to the short dimension of the sheet.

13.4.2 *Coils*—It is regular practice to package tin mill coils on platforms. It is not recommended that coils be shipped eye horizontal either with or without cradle platforms on those products where transit abrasion might be detrimental. Coil packages are secured with tension-tied bands. The number of ties depends upon the size and weight of the coil, and the method of handling. Edge protectors are used under all ties. Platform runners not over 6 in. (152 mm) high have been found adequate in practically all instances. Supporting coils with special cores or spools is unnecessary and requires additional labor and material. It is regular practice to paper wrap coils. (See Figs. 68 and 97.) Fiberboard covers may be used for further protection when required. Protecting coils with metal wrapping requires additional labor and material.

13.5 *Loading*—Cut length tin mill products are shipped in closed cars or by truck. Coiled tin mill products are usually shipped in covered or closed cars or by truck.

14. Castings

14.1 All castings shall be separated by class type, and condition when packed for shipment.

14.1.1 When castings are packed into containers they shall be adequately blocked, braced, or otherwise secured to prevent their movement within the containers.

14.1.2 Finished or polished castings shall be adequately protected from mechanical damage. Where practical the castings shall be boxed. All polished or finished surfaces shall be protected with a suitable cover such as paper or plastic.

14.1.2.1 When boxing is not practical because of size or weight, the castings shall be secured on skids or pallets. Large polished or finished castings shall have the polished or finished surfaces protected with batten strips. The entire surfaces may be covered with protective cover such as paper or plastic.

14.1.3 Rough castings unless otherwise specified may be shipped unpacked or bundled unless by so doing the castings may be subject to damage.

14.1.3.1 Large castings weighing more than 250 lb (114 kg) may be secured on skids or pallets for convenience in handling.

 (*1*) When shipped on skids or pallets they may be secured by ties of soft wire or tensioned flat bands. The number of ties is at the shipper option but must be adequate to secure the load.

14.1.4 Castings having projections that may be damaged in handling or shipping may be boxed, crated, or secured on skids or pallets with the projections adequately protected with batten strips.

14.2 Containers when used shall afford maximum protection from the normal hazards of transportation and shall be so constructed as to ensure safe delivery by common carrier.

14.3 *Marking*—It is normal practice to have the heat number, alloy type, and pattern number cast or stamped on the surface of castings. The purchase order number may be shown on a tag attached to each box, skid, pallet or loose casting.

TABLE 1 Protective Coatings Used to Protect Steel Mill Products

Type	Method of Application	Purpose
Type A—Thin soft film preservative consisting of a rust inhibitor in petroleum oil	cold; spray, dip or brush	to provide protection against corrosion and staining of steel mill products for short-term preservation periods (up to 3 months indoor storage)
Type C—Hard drying varnish resinous or plastic coating	cold; spray, dip or brush	to provide protection against corrosion of steel mill products for intermediate-term preservation periods (up to 6 months outdoor storage)
Type D—Medium soft film preservative in a solvent	cold; spray, dip or brush	to provide protection for edges of coils or cut lengths

TABLE 2 Packaging Merchant Wire Products

Bale ties (3 to 20-ft (0.91 to 6.10-m) lengths)	Ends protected, secured with spiral tie wire the entire length of the bundle (Fig. 12).

Size, gage	Ties per Bundle
11	125
12, 13, and 14	250
14½, 15, 15½, 16, and 16½	500

Baling wire:	
6500-ft (1981-m) minimum length coil (100 lb (45.4 kg) approximate weight)	One coil in self-dispensing corrugated carton (Fig. 13).
3150-ft (960-m) minimum length coil (48.5 lb (22 kg) approximate weight)	Two coils in corrugated box. NOTE—Packaging must comply with ASAE Standard S229 (latest revision).
Barbed wire	80-rod spool, secured with wire ties (Fig. 14).
Fence and netting	In rolls secured with wire ties (Fig. 15).
Fence panels	Ten sheets per bundle, inverted; five bundles per lift (Fig. 16). Bundles secured at the four corners with wire ties. Lift secured in the four corners with rod ties.
Fence posts	Five posts per bundle, 40 or 50 bundles per lift (manufacturer's option), secured (Figs. 17 and 18). Bundle is secured with minimum of two flat bands. Lift is secured with minimum of two flat bands.
Fence wire	150-lb (68-kg) catchweight coil secured with four wire ties (Fig. 19).
Fence assemblies/accessories:	
End and corner posts	Secured into a set.
Brace, complete with bolts	Five braces per bundle.
Stretchers and tools	Single unit.
Stays	100 per bundle, secured with a minimum of three ties.
Fasteners (clamps)	25 or 50 fasteners in a bag; 1000 or 2500 fasteners in a shipping bag or container (manufacturer's option).
Gates, complete with screws, fittings, and latches	Single unit.
Lath-tie wire	One 25-lb (11-kg) bundle in corrugated box (Fig. 23).
Merchant quality wire	One or more pieces of wire in a 100-lb (45-kg) coil secured with a minimum of three wire ties or flat bands (Fig. 19). 100-lb coil group secured with a minimum of three wire ties or flat bands segregated in increments of 10 or 25 lb (4.5 or 11 kg), each secured with three wire ties or flat bands (Fig. 19). When specified, two or more 100-lb coils may be combined into coil groups secured with a minimum of three wire ties or flat bands (Fig. 19).
Nails, brads, staples, spikes:	
Bulk	50-lb (22-kg) corrugated box (Fig. 20).
Packaged	1 and 5-lb (0.5 and 2-kg) boxes, packed in 50-lb shipping containers (Figs. 21 and 22).
Reinforcing bar tie wire	Twenty, approximate 4-lb (1.8-kg) coils in corrugated box (Fig. 24).

The following items may be furnished on pallets: baling wire, barbed wire, lath-tie wire, netting, nails, brads, staples and spikes, and reinforcing bar tie wire (Fig. 25).

TABLE 3 Packaging Carbon, Alloy, and Stainless Steel Wire in Coils

Single coil	Secured with a minimum of two ties (Fig. 26).
Coil group (16 in. (406 mm) inside diameter and larger)	Individually tied coils secured into a unit with minimum of two tensioned flat bands (Fig. 27).
Coil carrier[b]	Single or multiple coils on carrier; normally not secured to carrier (Fig. 28).
Reel-less coil	Approximate 600 to 1000-lb (272 to 474-kg) coil wound on a fiber core and secured with minimum of three tensioned flat bands; pack eye vertical on wood pallet (Fig. 29 or Fig. 31).
Wood rack	Small single coils secured with minimum of two ties nested in rack. Approximate maximum weight 2000 lb (907 kg) (Fig. 30).
Fiber drum[A]	Small single coils secured with a minimum of two ties, nested in drum; or a single coil laid loose in drum. Maximum diameter of drum 23 in. (584 mm). Approximate maximum weight 550 lb (249 kg) (Fig. 32 or Fig. 33).
Pay-off drum[A]	Single coil laid in drum with a fiber core. Core diameters: 11, 11½, 13, or 16 in. (279, 292, 330, or 406 mm). Maximum diameter of drum 23 in. (584 mm). Approximate maximum weight 550 lb (249 kg) (Fig. 34).
Reel	Single or multiple lengths wound on a reel. Reel size and weight vary by product and manufacturer (Fig. 36).

[A] Available loose; or palletized on wood pallets, to improve handling (Fig. 35).
List of commonly used sizes of coil carriers:

Arbor	Base	Height	Tube Diameter and Gage	Identification
11	23	35	1 × 16	pink
*13	23	35	1 × 16	orange
13	32	46	1¼ × 13	purple
*15	32	46	1¼ × 13	green
*16¾	36	48	1¼ × 13	yellow
*18½	37	46	1¼ × 13	red
20½	34	46	1¼ × 13	white
22½	42	46	1¼ × 13	aluminum
*22½	42	46	1⅝ × 13	blue
*26	50	50	1⅝ × 13	brown
*30	50	50	1⅝ × 13	black

*Preferred sizes.

A 700

TABLE 4 Packaging Table for Carbon, Alloy, and Stainless Steel Wire, Straightened and Cut Lengths[A]

Length, in. (mm)	Package	Approximate Weight, lb (kg)	
		Bundle/Lift	Package
12 (305) and under	loose in corrugated box (Fig. 37)	...	125 (57) max
Over 12 to 36 (305 to 914), incl	loose in container (Fig. 46)	...	1500 to 1800 (680 to 816)
Over 12 to 18 (305 to 457), incl	hand bundles	25 to 50 (11 to 23)	...
Over 18 to 36 (457 to 914), incl	hand bundles	50 to 100 (23 to 45)	...
	in containers (Fig. 46)	...	1500 to 1800 (680 to 816)
	on skids	...	2000 (907)
	on platform (Fig. 43)	...	2000 (907)
Over 36 (914)	hand bundles (Fig. 38)		
	loose	100 to 200 (45 to 91)	...
	secured lifts (Fig. 42)	4000 to 6000 (1814 to 2722)	...
Over 36 to 96 (914 to 2438)	secured lift (Fig. 41)	2000 to 4000 (907 to 1814)	...
Over 96 (2438)	secured lift	4000 to 6000 (1814 to 2722)	...

Securement—The number of bands or wire ties depends upon the length and weight of the lift or bundle, or both, as follows:

Length, ft (m)	Number of Ties
8 (2.44) and under	2
Over 8 to 16 (2.44 to 4.88), incl	3
Over 16 to 20 (4.88 to 6.10), incl	4
Over 20 (6.10)	5

Protection[A]—Where protection is required, wire may be packaged as follows:

Package	Protection
Hand bundles	Ends wrapped (Fig. 39)
Hand bundles	Wrapped entire length (Fig. 40)
Lifts	Shrouded (Fig. 44)
Lifts	Wrapped (Fig. 45)
Loose or hand bundles	Special containers constructed of fiberboard, chipboard, wood, or other suitable material. Containers are to be lined when required (Fig. 46).

[A] Some manufacturers determine type of packaging and protection by gage and other factors, and these should be considered in ordering.

TABLE 5 Pieces, Feet, and Weight per Sub-bundle for Standard Pipe, Line Pipe, and Oil Country Goods[A]

Nominal Size, in.	Pieces	ft (m)	Weight, lb (kg)
1/8	30	630 (192)	151 (69)
1/4	24	504 (154)	212 (96)
3/8	18	378 (115)	215 (98)
1/2	12	252 (77)	214 (97)
3/4	7	147 (45)	166 (75)
1	5	105 (32)	176 (80)
1 1/4	3	63 (19)	144 (65)
1 1/2	3	63 (19)	172 (78)
Extra Strong Pipe:			
1/8	30	630 (192)	195 (89)
1/4	24	504 (154)	272 (123)
3/8	18	378 (115)	280 (127)
1/2	12	252 (77)	275 (125)
3/4	7	147 (45)	216 (98)
1	5	105 (32)	228 (104)
1 1/4	3	63 (19)	189 (86)
1 1/2	3	63 (19)	229 (104)
Double Extra Strong Pipe:			
1/2	7	147 (45)	251 (114)
3/4	5	105 (32)	256 (116)
1	3	63 (19)	230 (104)
1 1/4	3	63 (19)	328 (149)
1 1/2	3	63 (19)	404 (184)

[A] Other bundling practices may be available, subject to agreement between the purchaser and the manufacturer.

TABLE 6 Thread Protection for Standard Pipe, Line Pipe, and Oil Country Goods

Types of Pipe	Nominal Size		
	1 1/2 in. and Smaller	2 to 3 1/2 in., incl	4 in. and Over
Standard pipe	none	none	threads protected[A]
Standard pipe, reamed and drifted	none	threads protected[A]	threads protected[A]
Line pipe	threads protected[B]	threads protected[A]	threads protected[A]
Oil country pipe	threads protected[C]	threads protected[D]	threads protected[D]

[A] Thread protectors are used on pipe threads not protected by a coupling.
[B] Either burlap cloth or waterproof paper is used to wrap the ends of sub-bundles or lifts, or thread protectors are used to protect exposed threads.
[C] Burlap cloth or waterproof paper is used to wrap the end of sub-bundles or lifts to protect the exposed threads.
[D] Thread protectors are used on pipe threads not protected by a coupling. The exposed coupling threads are protected with either a protective coating or thread protectors.

TABLE 7 Number of Lengthwise Skids for Sheet Steel Packages[A]

Sheet Gage (in.) (mm)	Use 2 Skids, in. (mm)	Use 3 Skids, in. (mm)	Use 4 Skids, in. (mm)
28 to 24 (0.0149 to 0.0239) (0.378 to 0.607)	22 to 36 (559 to 914) wide	36 to 56 (1422) wide	56 to 75 (1905) wide
Under 24 to 20 (0.0239 to 0.0359) (0.607 to 0.912)	22 to 42 (559 to 1067) wide	42 to 68 (1727) wide	68 to 96 (2438) wide
Under 20 to 16 (0.0359 to 0.0598) (0.912 to 1.519)	22 to 50 (559 to 1270) wide	50 to 84 (2134) wide	Over 84 wide
Under 16 to 12 (0.0598 to 0.1046) (1.519 to 2.657)	all widths		

[A] Lengthwise skids are not used on sheets over 192 in. (4877 mm) long or less than 22 in. (559 mm) wide.

A 700

TABLE 8 Number of Crosswise Skids for Sheet Steel Packages[A]

Sheet Gage (in.) (mm)	Use 2 Skids, in. (mm)	Use 3 Skids, in. (mm)	Use 4 Skids, in. (mm)	Use 5 Skids, in. (mm)	Use 6 Skids, in. (mm)
24 and lighter (0.0239) (0.607)	22 to 36 (559 to 914) long	36 to 56 (1422) long	56 to 76 (1930) long	76 to 96 (2438) long	96 to 120 (3048) long
Under 24 to 20 (0.0239 to 0.0359) (0.607 to 0.912)	22 to 42 (559 to 1067) long	42 to 68 (1727) long	68 to 96 (2438) long	96 to 122 (3099) long	122 to 149 (3785) long
Under 20 to 16 (0.0359 to 0.0598) (0.912 to 1.519)	22 to 50 (599 to 1270) long	50 to 84 (2134) long	84 to 120 (3048) long	120 to 154 (3912) long	154 to 188 (4775) long
Under 16 to 12 (0.0598 to 0.1046) (1.519 to 2.657)	22 to 72 (559 to 1829) long	72 to 120 (3048) long	120 to 164 (4166) long	164 to 208 (5283) long	208 to 253 (6426) long
Heavier than 12 (0.1046) (2.657)	22 to 82 (559 to 2083) long	86 to 146 (3708) long	146 to 206 (5232) long	206 to 266 (6756) long	266 to 327 (8306) long

[A] The arrangements shown in Figs. 53 and 54 illustrate lengthwise and crosswise skid arrangements used for packaging cut length sheets.

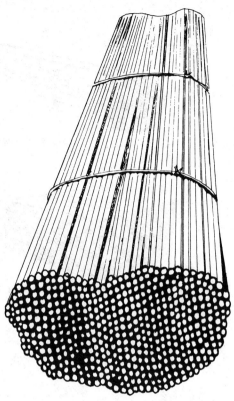

FIG. 1 Suitable Secured Lift—Hot-Rolled and Cold-Finished Bars and Bar-Size Shapes

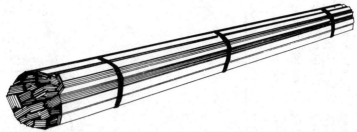

FIG. 2 Suitable Secured Lift—Flats

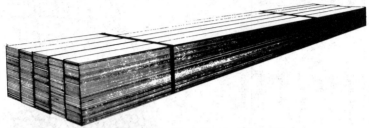

FIG. 3 Suitable Lift of Stack-Piled Straightened Flats

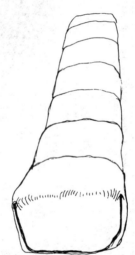

FIG. 4 Suitable Method of Wrapping Lifts for Loading in Open-Top Equipment

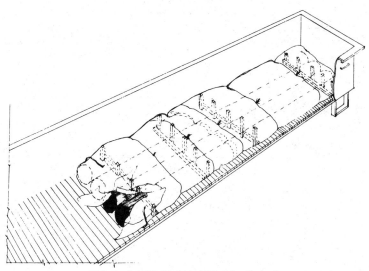

FIG. 5 Suitable Method of Shrouding Carload

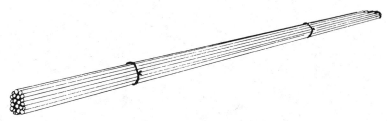

FIG. 6 Suitable Hand Bundle of Cold-Finished Bars

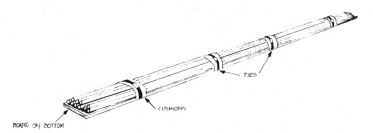

FIG. 7 Bundle of Cold-Finished Bars Secured to a Board

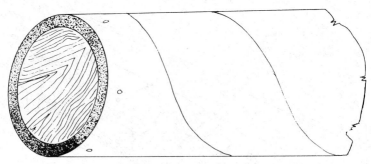

FIG. 8 Suitable Chipboard Container

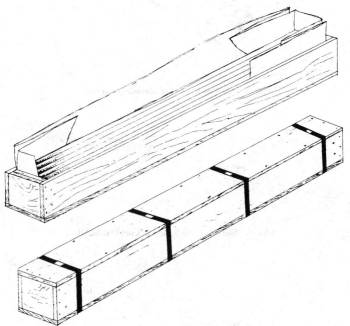

FIG. 9 Suitable Wood Box for Cold-Finished Bars

FIG. 10 Securement of Hot-Rolled Rods in Individual Coils

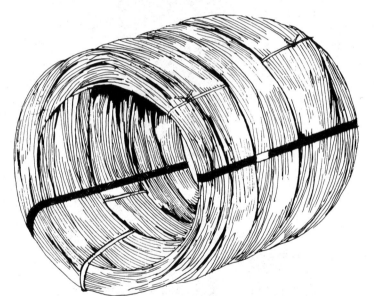

FIG. 11 Securement of Hot-Rolled Rods in Coil Group

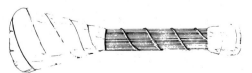

FIG. 12 Bale Ties

FIG. 14 Spool of Barbed Wire

FIG. 13 Coil of Baling Wire and Self-Dispensing Carton

FIG. 15 Roll of Fence/Netting

FIG. 16 Secured Lift of Fence Panels

FIG. 17 5-Post Bundle

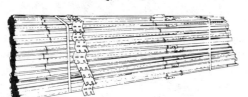

FIG. 18 Secured Lift of 5-Post Bundles

Single Coil   Coil Group

FIG. 19 Coils of Merchant Quality Wire

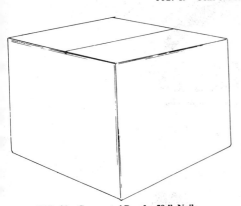

FIG. 20 Corrugated Box for 50-lb Nails

FIG. 21 Box for 1-lb and 5-lb Nails

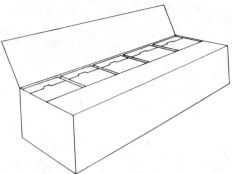

FIG. 22 Shipping Container for Packaged Nails

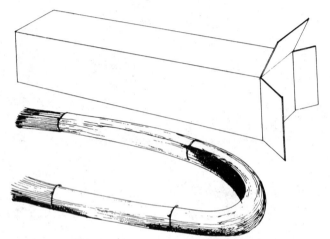

FIG. 23 Bundle of Wire in Corrugated Box

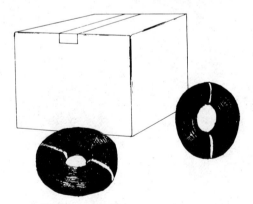

FIG. 24 Coils of Wire in Corrugated Box

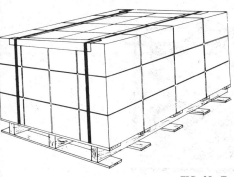

FIG. 25 Typical Palletizing

FIG. 26 Single Coil, Bare

FIG. 27 Coil Group, Bare

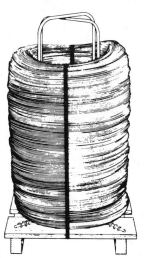

FIG. 28 Single Coil, Bare on Coil Carrier

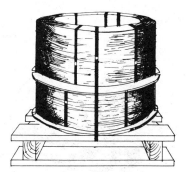

FIG. 29 Reel-less Coils

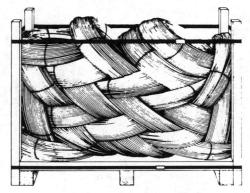

FIG. 30 Coils Nested in Wood Rack

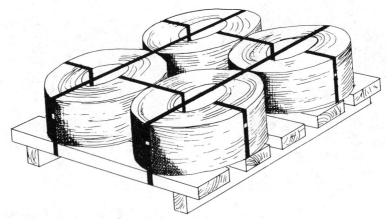

FIG. 31 Reel-less Coils

FIG. 32 Coils Nested in Fiber Drum

FIG. 33 Coil in Fiber Drum

FIG. 34 Single Length Coil in Pay-Off Drum

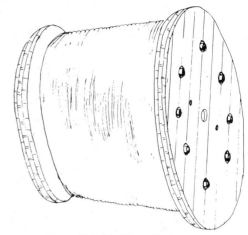

FIG. 36 Wire on Reel

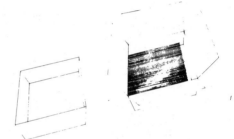

FIG. 37 Short Lengths of Straightened and Cut Wire in Corrugated Box

FIG. 35 Palletized Drums

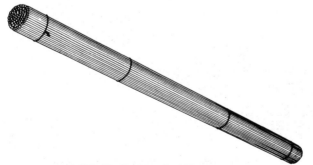

FIG. 38 Hand Bundle of Wire, Bare

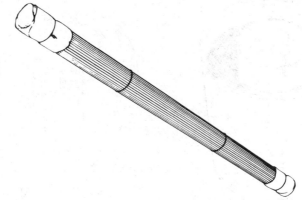

FIG. 39 Hand Bundle of Wire, Ends Wrapped

FIG. 40 Hand Bundle of Wire, Wrapped Entire Length

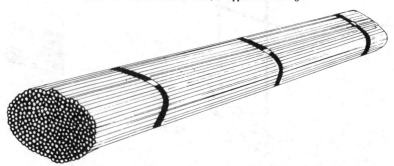

FIG. 41 Secured Lift of Wire, Bare

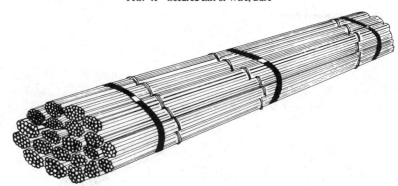

FIG. 42 Secured Lift of Hand Bundles of Wire, Bare

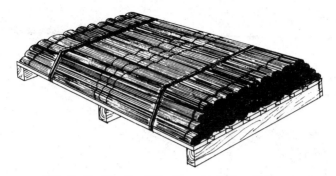

FIG. 43 Hand Bundles of Wire Secured to Skids or Platforms

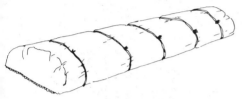

FIG. 44 Shrouded Lift of Wire

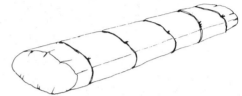

FIG. 45 Wrapped Lift of Wire

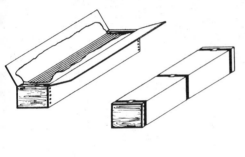

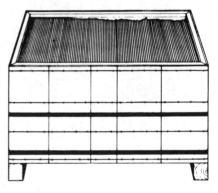

FIG. 46 Loose or Hand Bundles of Wire in Containers

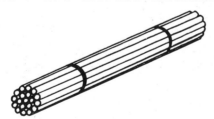

FIG. 47 Secured Lift

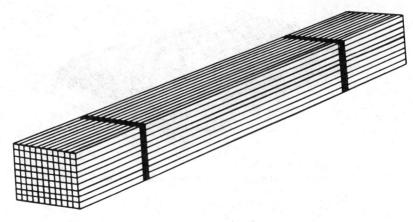

FIG. 48 Rectangular Package

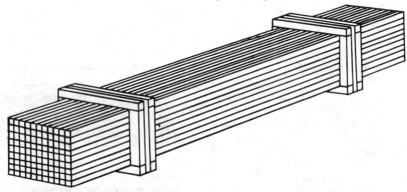

FIG. 49 Frame Package

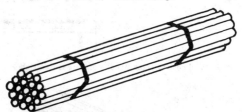

FIG. 50 Hexagonal Package

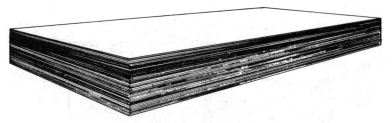

FIG. 51 Suitable Method of Packaging Carbon Steel Sheets in Unsecured Lift, Bare

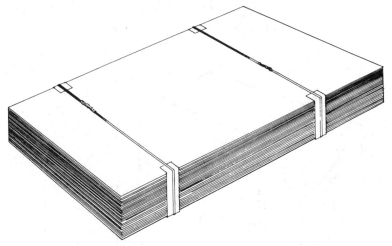

FIG. 52 Suitable Method of Packaging Carbon Steel Sheets in Secured Lift, Bare

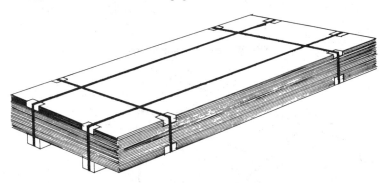

FIG. 53 Bare Package on Lengthwise Skids

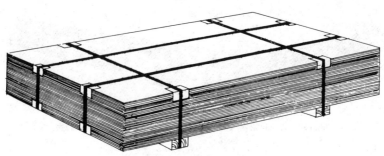

FIG. 54 Bare Package on Crosswise Skids

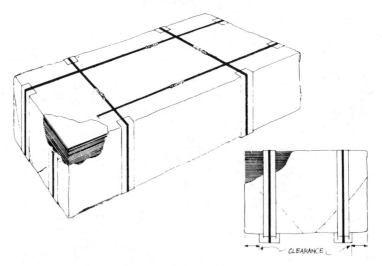

FIG. 55 Covered Package on Skids

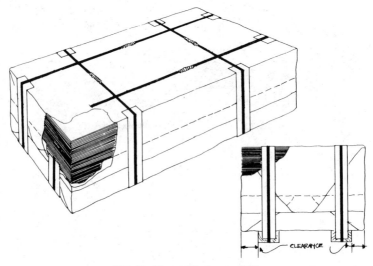

FIG. 56 Wrapped Package on Skids

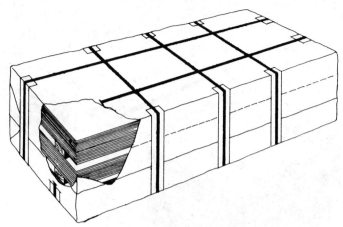

FIG. 57 Multiple-Lift Package on Skids

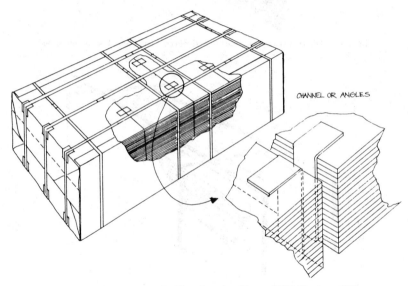

FIG. 58 Suitable Package for Short-Length or Narrow-Width Sheets on Skids

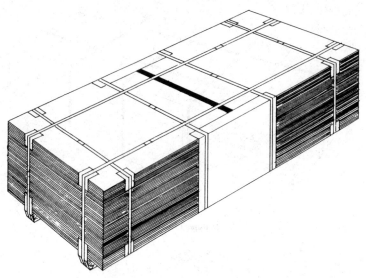

FIG. 59 Suitable Package for Short-Length Sheets Lengthwise, End to End, on Skids

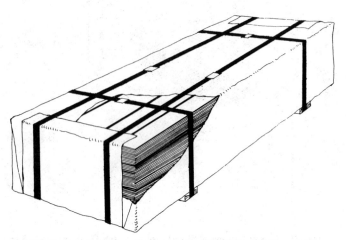

FIG. 60 Suitable Package for Narrow Long Sheets Side by Side, on Skids

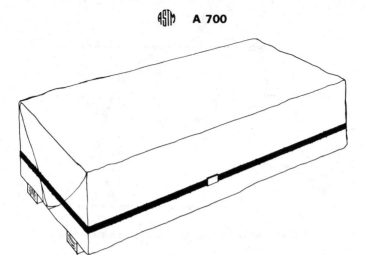

FIG. 61 Suitable Shrouded Package of Cut-Length Sheets, Banded

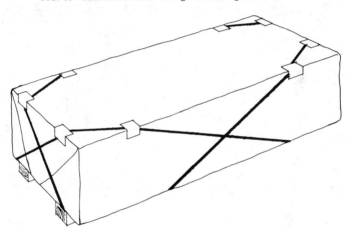

FIG. 62 Suitable Shrouded Package of Cut-Length Sheets, Wired

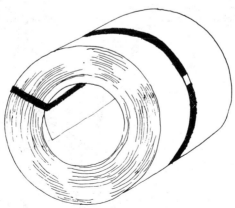

FIG. 63 Suitable Method of Packaging Individual Hot-Rolled Sheet or Strip Coil in the As-Rolled Condition

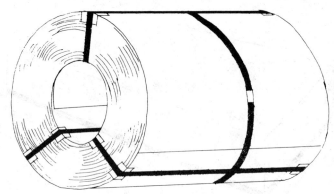

FIG. 64 Suitable Packaging of Highly Finished Individual Coil

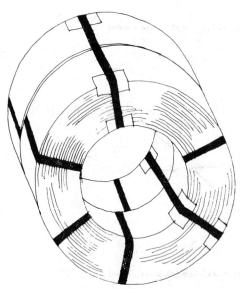

FIG. 65 Suitable Packaging of Two or More Narrow Sheet Coils into a Coil Group Package

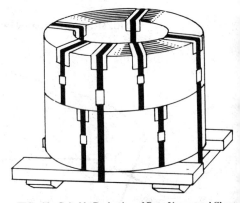

FIG. 66 Suitable Packaging of Bare Unwrapped Sheet Coils on Skeleton Platform with the Eye of the Coils Vertical

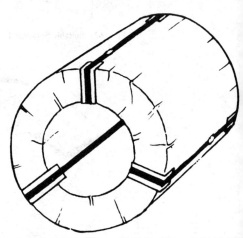

FIG. 67 Suitably Wrapped Individual Sheet Coil with Eye of the Coil Horizontal

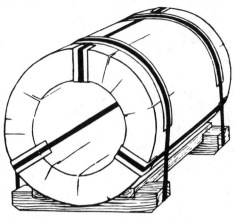

FIG. 68 Suitably Wrapped Individual Coil on Cradle Platform with the Eye of the Coil Horizontal

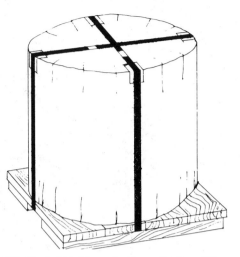

FIG. 70 Suitably Wrapped or Covered Package of Sheet Circles on Skeleton Platform

FIG. 69 Suitably Wrapped Individual Coil on Platform with the Eye of the Coil Vertical

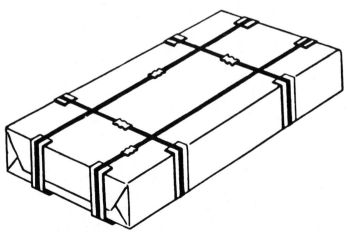

FIG. 71 Suitably Wrapped Package on Lengthwise Skids

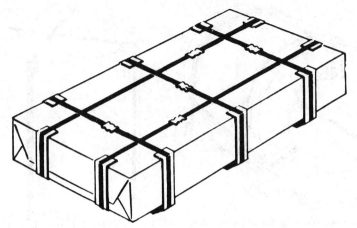

FIG. 72 Suitably Wrapped Package on Crosswise Skids

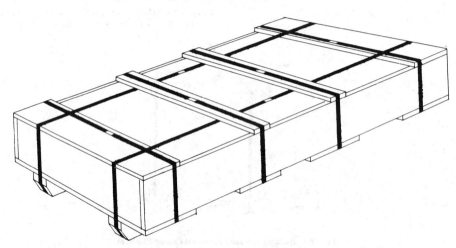

FIG. 73 Steel Sheets in a Fully Enclosed Package on a Skeleton Platform, Using Wood Materials

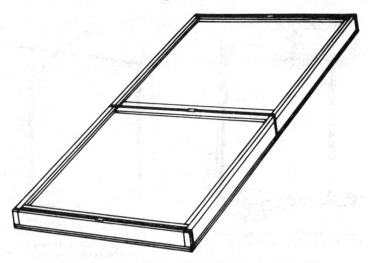

FIG. 74 Steel Sheets in a Box of Suitable Solid Protective Material

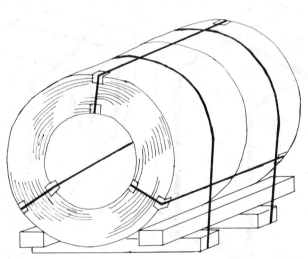

FIG. 75 Suitably Packaged Bare Unwrapped Individual Stainless Steel Sheet Coil on Cradle Platform, with the Eye of the Coil Horizontal

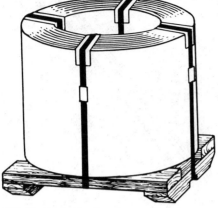

FIG. 76 Suitably Packaged Bare Unwrapped Individual Sheet Coil on Platform with the Eye of the Coil Vertical

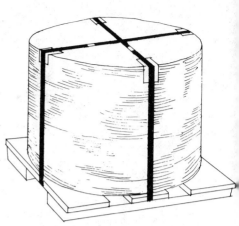

FIG. 78 Suitably Packaged Single Pile of Bare Stainless Steel Sheet Circles on Skeleton Platform

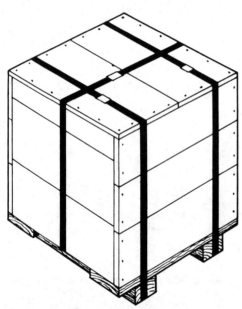

FIG. 77 Suitably Packaged Individual Sheet or Strip Coil or Group of Sheet or Strip Coils in Solid Box, on Platform with the Eye of the Coil Vertical

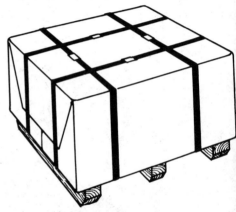

FIG. 79 Multiple Piles of Circles on Solid Platform, Covered with Corrugated Fiberboard

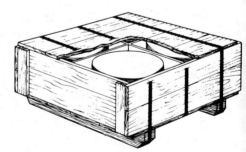

FIG. 80 Multiple Piles of Circles in Box on Solid Deck Platform

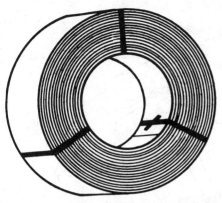

FIG. 81 Narrow Strip Coil with Flat Twist Bands

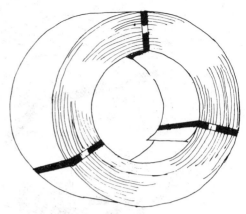

FIG. 82 Narrow Strip Coil with Machine Tension Bands

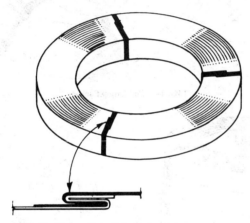

FIG. 83 Narrow Strip Coil with Knockdown or Buckle Bands

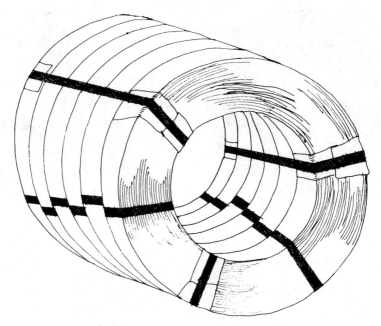

FIG. 84 Coil Group Package

FIG. 85 Narrow-Strip Coils on Skeleton Platform with the Eye of the Coils Vertical

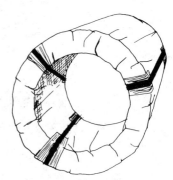

FIG. 86 Suitably Wrapped Individual Strip Coils or Groups of Coils

A 700

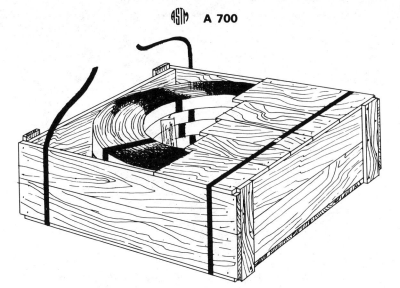

FIG. 87 Suitably Packaged Bare Narrow Strip Coils in Container

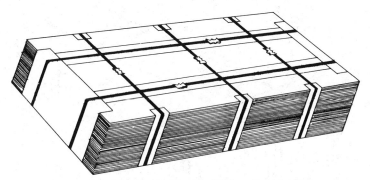

FIG. 88 Bare Package of Stainless Steel Strip on Crosswise Skids

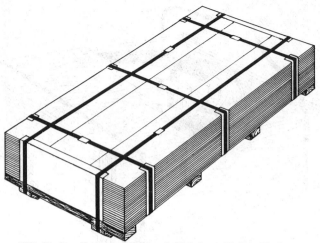

FIG. 89 Bare Package of Stainless Steel Strips on Skeleton Platform

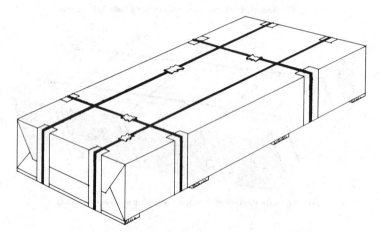

FIG. 90 Suitably Wrapped Stainless Steel Strip on Platform

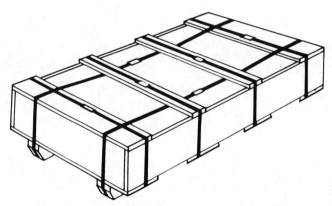

FIG. 91 Suitably Packaged Stainless Steel Strip in a Fully Enclosed Package on a Platform Using Wood Materials

FIG. 92 Suitably Packaged Narrow Stainless Steel Strip Coils on Skeleton Platform

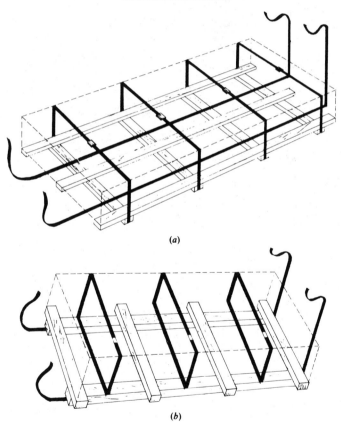

FIG. 93 Two Types of Skeleton Platform Systems

A 700

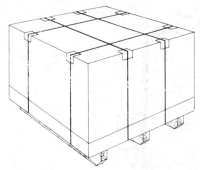

FIG. 94 Suitable Method of Packaging Cut-Length Tin Mill Products in Multiple Package Unit

FIG. 96 Standard Solid Deck Two-Runner Platform

FIG. 95 Standard Skeleton Deck Two-Runner Platform

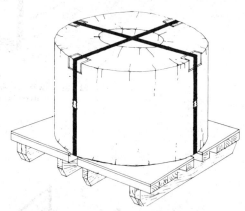

FIG. 97 Paper Wrapped Tin Plate Coil on a Platform

The American Society for Testing and Materials takes no position respecting the validity of any patent rights asserted in connection with any item mentioned in this standard. Users of this standard are expressly advised that determination of the validity of any such patent rights, and the risk of infringement of such rights, are entirely their own responsibility.

This standard is subject to revision at any time by the responsible technical committee and must be reviewed every five years and if not revised, either reapproved or withdrawn. Your comments are invited either for revision of this standard or for additional standards and should be addressed to ASTM Headquarters. Your comments will receive careful consideration at a meeting of the responsible technical committee, which you may attend. If you feel that your comments have not received a fair hearing you should make your views known to the ASTM Committee on Standards, 1916 Race St., Philadelphia, Pa. 19103.

Designation: A 702 – 81

Standard Specification for
STEEL FENCE POSTS AND ASSEMBLIES, HOT-WROUGHT[1]

This standard is issued under the fixed designation A 702; the number immediately following the designation indicates the year of original adoption or, in the case of revision, the year of last revision. A number in parentheses indicates the year of last reapproval. A superscript epsilon (ϵ) indicates an editorial change since the last revision or reapproval.

1. Scope

1.1 This specification covers steel fence posts and assemblies manufactured from hot-wrought sections and intended for use in field and line fencing.

1.2 The posts are available in tee, channel or U, and Y-bar shapes, and are either painted or galvanized.

1.3 The values stated in inch-pound units are to be regarded as the standard.

2. Applicable Documents

2.1 *ASTM Standards*:
A 36, Specification for Structural Steel[2]
A 123, Specification for Zinc (Hot-Galvanized) Coatings on Products Fabricated from Rolled, Pressed, and Forged Steel Shapes, Plates, Bars, and Strip[3]
A 499, Specification for Steel Bars and Shapes, Carbon Rolled from "T" Rails[4]
A 641, Specification for Zinc-Coated (Galvanized) Carbon Steel Wire[3]

3. Definitions

3.1 Items covered by this standard are defined as follows:

3.1.1 *line posts*—posts with anchor plates that support the straight-away body of the fence.

3.1.2 *assemblies*—components to provide for proper installation of gates, fence end or corners, and intermediate bracing.

4. Ordering Information

4.1 Orders for products under this standard shall include the following information:

4.1.1 Quantity (number of pieces) of line posts, end or gate assemblies, corner or intermediate brace assemblies. (It is customary to order line posts in multiples of five of the required length and the required number of end or corner assemblies.)

4.1.2 Type of section (if a specific section is required) (5.2 and Fig. 1).

4.1.3 Length or lengths required (6.2).

4.1.4 Finish: galvanized or painted.

4.1.5 ASTM designation and date of issue.

4.1.6 Anchor plates, when not required (5.4.3).

4.1.7 Wire fasteners (state weight of zinc coating) (5.6.2 and 5.6.3).

NOTE—A typical ordering description is as follows: 500 line posts; tee type; 8 ft long; galvanized A 123, C1 B-1; ASTM A 702 dated ————; omit anchor plates.

5. Materials and Manufacture

5.1 *Material*:

5.1.1 Line post shall be fabricated from Steels A or B and assemblies from Steels A, B, or C as specified in Table 1.

5.1.2 Except as provided in 5.1.3, the finished line post and assemblies shall conform to

---

[1] This specification is under the jurisdiction of ASTM Committee A-1 on Steel, Stainless Steel and Related Alloys, and is the direct responsibility of Subcommittee A01.15 on Bars. This standard is a revision of Commercial Standard CS 184-51, Steel Fence Posts—Field Line Type, formerly published by the United States Department of Commerce.
Current edition approved July 31, 1981. Published September 1981. Originally published as A 702 – 74. Last previous edition A 702 – 74.
[2] *Annual Book of ASTM Standards*, Vol 01.04.
[3] *Annual Book of ASTM Standards*, Vol 01.03.
[4] *Annual Book of ASTM Standards*, Vol 01.05.

the tensile properties specified in Table 1 for the applicable steel.

5.1.3 At the manufacturer's option, a Brinell or Rockwell B hardness test may be substituted for the tensile requirements in Table 1. In such cases the material shall conform to the Brinell or Rockwell B hardness specified in Table 2.

5.2 *Line Post Section Types*:

5.2.1 The posts shall be furnished as T, channel or U, or Y sections as illustrated in Fig. 1. The cross section of T posts shall approximate 1⅜ in. (34.9 mm) wide, 1⅜ in. deep, and ⅛ in. (3.2 mm) thick. Unless otherwise specified by the purchaser, the line post type is at the manufacturer's option.

5.2.2 Dimensions may vary slightly in individual design in maintaining the control weight per foot.

5.3 *Wire Attachments*—Line posts shall have corrugations, knobs, notches, holes, or studs so placed and formed as to engage a substantial number of fence line wires in proper positions.

5.4 *Anchor Plates*:

5.4.1 Unless otherwise specified, each line post shall be manufactured with an anchor plate. The anchor plate shall be made from carbon steel which shall be clamped, welded, swaged, or riveted to the section in such a manner as to prevent displacement when the posts are driven.

5.4.2 Anchor plates shall be tapered to facilitate driving, shall have a minimum area of 18 in.^{2} (116 cm^2), and shall weigh 0.67 lb (0.304 kg) ±5 %.

5.4.3 When specified, line posts may be furnished without anchor plates.

5.5 *Post Assemblies*:

5.5.1 Uprights shall consist of angles 2½ by 2½ by ¼ in. (64 by 64 by 6.4 mm) weighing 4.10 lb/ft (6.1 kg/m) prior to fabrication.

5.5.2 Braces shall consist of angles 2 by 2 by ¼ in. (51 by 51 by 6.4 mm) weighing 3.19 lb/ft (4.75 kg/m) prior to fabrication, or an alternative angle of equivalent weight.

5.5.3 Uprights and braces shall be furnished with the necessary holes and galvanized hardware for the required assembly.

5.5.4 All assemblies shall be furnished with one upright. End and gage assemblies shall be furnished with one brace, and corner and intermediate braces with two braces.

5.6 *Wire Fasteners*:

5.6.1 Unless otherwise specified by the purchaser, each line post shall be provided with not less than five suitable fasteners for attaching fence wire to the posts.

5.6.2 The fasteners shall be formed from zinc-coated steel wire not less than 0.120-in. (3.05-mm) diameter zinc coated in accordance with Specification A 641. The coating weight class (1, 2, or 3) shall be as specified by the purchaser.

5.6.3 When line posts are intended for range type western fencing using three line wires, it is satisfactory to provide only three fasteners for each post.

6. Permissible Variations in Dimensions and Weight

6.1 *Nominal Weights and Tolerances*:

6.1.1 *Nominal Weight*—Prior to fabrication by punching, drilling, attaching anchors, or finish coating, the line post sections shall have a nominal weight of 1.33 lb/ft (1.98 kg/m) of length.

6.1.2 The weight of anchor plates plus line posts prior to fabrication, drilling, or finish coating shall not vary from nominal weights specified in Table 3 by more than ±5 %. Weight shall be determined in lots of five line posts.

6.1.3 The weight of assembly components prior to fabrication, drilling, or finish coating shall not vary from the nominal weights specified in Table 4 by more than ±5 %. Single assembly components shall be used to determine weight.

6.2 *Standard Lengths and Tolerances*:

6.2.1 Line posts shall be furnished in standard lengths of 5 to 10 ft, inclusive, as specified by the purchaser. Standard length increments are shown in Table 3.

6.2.2 The length of line posts, uprights, and braces shall not vary from that specified more than −1.0 in. (−25.4 mm) or +2 in. (51 mm).

7. Workmanship, Finish, and Appearance

7.1 *Workmanship*—The posts shall be free of excessive bow, camber, twist, or other injurious imperfections in surface or coating. Such imperfections may be considered cause for rejection of individual posts.

7.2 Line posts, uprights, and braces shall be

A 702

furnished painted or galvanized as specified by the purchaser.

7.3 When specified to be painted, the posts shall be cleaned of all loose scale prior to finishing, and painted with one or more coats of weather resistant, air drying or baking, paint or enamel.

7.4 When specified to be galvanized, the posts shall be zinc coated by the hot-dip process in accordance with Specification A 123.

8. Packaging

8.1 Line posts shall be bound in bundles of five posts and by agreement with the purchaser may be supplied in master bundles containing up to 250 posts.

8.2 Orders are customarily on piece count rather than on weight of posts or assemblies. The nominal weights establish the basis for weight tolerance and transportation data (see Tables 3 and 4 for nominal weights).

TABLE 1 Materials for Line Posts and Assemblies

Steel	Steel Description	Line Posts		Assemblies	
		Yield Point, min, ksi (MPa)	Tensile Strength, min, ksi (MPa)	Yield Point, min, ksi (MPa)	Tensile Strength, min, ksi (MPa)
A	hot-wrought carbon steel 0.35 % carbon, min	40 (275)	70 (485)	40 (275)	70 (485)
B	hot-wrought carbon steel, or hot-wrought rail steel[a]	50 (345)	80 (550)	50 (345)	80 (550)
C	structural steel[b]	...	...	36 (250)	58 (400)

[a] In accordance with Specification A 499.
[b] In accordance with Specification A 36.

TABLE 2 Brinell Hardness

Steel	Steel Description	Brinell Hardness, min	Rockwell B Hardness, min
A	hot-wrought carbon steel, 0.35 % carbon, min	143	79
B	hot-wrought carbon steel or rail steel[a]	156	83
C	structural steel[b]	116	68

[a] In accordance with Specification A 499.
[b] In accordance with Specification A 36.

TABLE 3 Nominal Weights of (Raw) Line Posts

Post Length		Weight[a]	
ft	m	lb	kg
5	1.524	7.32	3.32
5½	1.675	7.99	3.61
6	1.83	8.65	3.92
6½	1.98	9.32	4.22
7	2.13	9.98	4.53
7½	2.28	10.64	4.83
8	2.44	11.31	5.13
9	2.74	12.64	5.74
10	3.05	13.97	6.34

[a] Includes weight of anchor plate.

TABLE 4 Nominal Weights of (Raw) Assemblies

	Length		Weight[a]	
	ft	m	lb	kg
Upright and one brace, each (end or gate)	7	2.13	51	23
Upright and two braces, each (corner of intermediate brace)	7	2.13	73	33
Upright and one brace, each	8	2.44	58	26
Upright and two braces, each	8	2.44	84	38
Upright and one brace, each	9	2.74	66	30
Upright and two braces, each	9	2.74	94	43

[a] Includes weight of bolts.

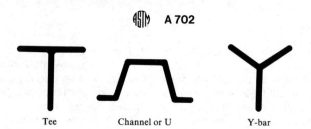

FIG. 1 Typical Cross Sections of Line Post Types.

The American Society for Testing and Materials takes no position respecting the validity of any patent rights asserted in connection with any item mentioned in this standard. Users of this standard are expressly advised that determination of the validity of any such patent rights, and the risk of infringement of such rights, are entirely their own responsibility.

This standard is subject to revision at any time by the responsible technical committee and must be reviewed every five years and if not revised, either reapproved or withdrawn. Your comments are invited either for revision of this standard or for additional standards and should be addressed to ASTM Headquarters. Your comments will receive careful consideration at a meeting of the responsible technical committee, which you may attend. If you feel that your comments have not received a fair hearing you should make your views known to the ASTM Committee on Standards, 1916 Race St., Philadelphia, Pa. 19103.

Designation: A 708 – 79

An American National Standard

Standard Recommended Practice for
DETECTION OF SUSCEPTIBILITY TO INTERGRANULAR CORROSION IN SEVERELY SENSITIZED AUSTENITIC STAINLESS STEEL[1]

This standard is issued under the fixed designation A 708; the number immediately following the designation indicates the year of original adoption or, in the case of revision, the year of last revision. A number in parentheses indicates the year of last reapproval. A superscript epsilon (ϵ) indicates an editorial change since the last revision or reapproval.

1. Scope

1.1 This recommended practice covers the procedure for conducting the acidified copper sulfate test for detection of susceptibility to intergranular corrosion of severely sensitized austenitic stainless steel (Note 1). The presence or absence of attack in this test is not necessarily a measure of the performance of the material in other corrosive environments.

NOTE 1—This method is less sensitive than those described in Practices A 262 and will detect only severe sensitization. This method shall not be used in place of Practices A 262 where Practices A 262 is specified in the basic material specification.

1.2 The test is to be used to detect severe sensitization resulting from the fabrication, welding, heat treatment, or combination of those applied to austenitic stainless steel.

1.3 This recommended practice may be applied to wrought products (including tubes), castings, and weldments of austenitic stainless steel.

2. Applicable Documents

2.1 *ASTM Standards:*
A 213 Specification for Seamless Ferritic and Austenitic Alloy-Steel Boiler, Superheater, and Heat Exchanger Tubes[2]
A 262 Practices for Detecting Susceptibility to Intergranular Attack in Stainless Steels[3]
2.2 *American Welding Society Standard:*
A 5.4 Specification for Corrosion-Resisting Chromium and Chromium-Nickel Steel Covered Welding Electrodes[4]

3. Apparatus

3.1 *Container*—A suitable glass or other inert container for the acidified copper sulfate solution, fitted with a condenser capable of avoiding loss of acid by evaporation. If a reflux condenser (see Fig. 8 in Practice A 262) is used, its connection to the flask should be made through a ground-glass joint. Rubber or cork stoppers are not suitable.

NOTE 2—Some samples, such as from castings or heavy bar stock, may require the use of a larger Erlenmeyer flask.

3.2 *Specimen Supports*—Glass hooks, stirrups, or cradles may be used for supporting the specimens in the flask. Their design should prevent specimens from coming in contact with each other when tested in the same container.

3.3 *Heater*—A means for heating the test solution and of keeping it boiling throughout the test period. A gas or electrically-heated hot plate is satisfactory for this purpose.[5]

4. Reagent

4.1 *Acidified Copper Sulfate Solution*—Dissolve 100 g of copper sulfate ($CuSO_4 \cdot 5H_2O$) in 700 ml of distilled water, add 100 ml of sulfuric acid (H_2SO_4, cp, sp gr 1.84), and make up to 1000 ml with distilled water.

NOTE 3—This solution will contain approxi-

---

[1] This recommended practice is under the jurisdiction of ASTM Committee A-1 on Steel, Stainless Steel and Related Alloys, and is the direct responsibility of Subcommittee A01.14 on Methods of Corrosion Testing.
Current edition approved July 27, 1979. Published October 1979. Originally published as A 708 – 74. Last previous edition A 708 – 74.
[2] *Annual Book of ASTM Standards*, Vol 01.01.
[3] *Annual Book of ASTM Standards*, Vol 01.05.
[4] Can be obtained from the American Welding Society, 2501 North West 7th St., Miami, Fla. 33125.
[5] A Glascol heater with a Variac has been found suitable.

mately 6 % of anhydrous $CuSO_4$ and 16 % of H_2SO_4 (by weight).

5. Test Specimens

5.1 *Size*:

5.1.1 The size of the test specimen and method of selection shall be that specified in the product specification or by agreement between the purchaser and seller as stated in the purchase contract. Consideration should be given to the volume of solution to be used, whether the product is wrought or cast, how the specimen is to be selected, and the method of bending the specimen for inspection. The size of the specimen should permit easy entrance and removal through the relatively narrow neck of the Erlenmeyer flask.

5.1.2 Weld metal specimens shall be prepared in accordance with AWS Specification A5.4 for Corrosion-Resisting Chromium and Chromium-Nickel Steel Covered Welding Electrodes.

5.1.3 The sizes given in Table 1 may be used as guides for referee purposes. The exact length will be governed by the equipment used for bending the specimen.

5.2 *Sheared Edges*—When specimens are cut by shearing, the sheared edges shall be refinished by machining or wet grinding prior to testing.

5.3 *Surface Finishing*:

5.3.1 Any scale on the as-received surface should be removed by mechanical finishing. Unless it is desired to test specimens with some particular surface finish all surfaces of the specimen, including edges, should be finished with 120-grit iron-free aluminum oxide abrasive, taking care to avoid overheating.

5.3.2 *Cleaning*—The specimen shall be degreased using a suitable solvent, such as acetone, a mixture of alcohol and ether, or a vapor degreaser, then dried.

5.4 *Sensitizing Treatment*—Heat treatment of test material will be specified in the product specifications or by agreement between the purchaser and seller as stated in the purchase contract.

6. Test Conditions

6.1 *Volume of Solution*—A sufficient quantity of the acidified copper sulfate test solution to cover the specimens and to provide a volume of at least 50 ml/in.^{2} (8 ml/cm^2) of specimen surface area shall be used.

6.2 *Number of Specimens per Flask*—The best practice is to use a separate container for each specimen. However, it is acceptable to treat as many as three specimens in the same container, provided they are all the same grade and that the specimen supports prevent the specimens from contacting each other, and provided the solution volume to sample area ratio is maintained. If any specimen tested in the same flask fails the test, new specimens representative of the failed material shall be retested separately.

6.3 *Boiling Time*:

6.3.1 After the test specimen has been placed in the test solution, it should be brought to a boil and kept boiling throughout the test period (Note 4).

6.3.2 The test should consist of one 72-h boiling period unless otherwise specified (Note 5).

NOTE 4—"Bumping" of the solution while boiling may be minimized by the use of glass beads, sections of porcelain crucibles, abrasive aluminum oxide rings, or boiling stones.

NOTE 5—When more than one 72-h boiling period is specified, fresh test solution should be used for each period.

7. Bend Test

7.1 *Bending*:

7.1.1 The test specimen shall be bent through 180 deg and over a diameter equal to the thickness of the specimen being bent (see Fig. 11 in Practices A 262). In no case shall the specimen be bent over a smaller radius or through a greater angle than that specified in the product specification. In cases of material having low ductility, such as severely cold-worked material, a 180-deg bend may prove impractical. Determine the maximum angle of bend without causing cracks in such material by bending an untested specimen of the same configuration as the specimen to be tested.

7.1.2 Sheet samples shall be bent to an "S" shape with the surface on both sides of the rolled samples tested in tension (Note 6).

7.1.3 Samples machined from round sections or cast material shall have the curved or original surface on the outside of the bend.

7.1.4 The specimens are generally bent by holding in a vise and starting the bend with a hammer. It is generally completed by bringing the two ends together in the vise. Heavy specimens may require bending in a fixture of suitable design.

7.1.5 Tubular products should be bent in accordance with the flattening test prescribed in the product specification (Note 7).

7.1.6 When agreed upon between the purchaser and the producer, the following shall apply to austenitic stainless steel plates 0.1875 in. (4.76 mm) and thicker:

7.1.6.1 Samples shall be prepared according to Table 1.

7.1.6.2 The radius of bend shall be two times the sample thickness, and the bend axis shall be perpendicular to the direction of rolling.

7.1.6.3 Welds on material 0.1875 in. and thicker shall have the above bend radius, and the weld-base metal interface shall be located approximately in the centerline of the bend.

7.1.6.4 Face, root, or side bend tests may be performed, and the type of bend test shall be agreed upon between the purchaser and the producer. The bend radius shall not be less than that required for mechanical testing in the appropriate material specification (for base metal) or in ASME Code Section IX (for welds).

NOTE 6—The "S" shape bend ensures detection of intergranular attack that may have resulted from carburization of one surface of sheet material during the final stages of rolling.

NOTE 7—A typical flattening test is described in Specification A 213.

7.2 *Evaluation*—The U-bend or fold shall be examined under low (5× to 15×) magnification. The appearance of fissures or cracks indicates the presence of intergranular attack.

NOTE 8—Cracking that originates at the edge of the specimen should be disregarded. The appearance of deformation lines, wrinkles, or "orange peel" on the surface, without accompanying cracks or fissures, should be disregarded also.

NOTE 9—For referee purposes, where the above evaluation is questioned, the presence or absence of grain boundary attack shall be determined by metallographic examination of a cross section of the specimen at a magnification of 100× to 250×.

TABLE 1 Sizes of Test Specimens

Type of Material	Size of Test Specimens
Wrought wire or rod:	
Up to ¼ in. (6.35 mm) in diameter, incl	full diameter by 3 to 5 in. (76 to 127 mm) long
Over ¼ in. (6.35 mm) in diameter	cylindrical segment ¼ in. (6.35 mm) thick by 1 in. (25.4 mm) (max) wide by 3 to 5 in. (76 to 127 mm) long[a]
Wrought sheet, strip, plates, or flat-rolled products:	
Up to ³⁄₁₆ in. (4.76 mm) thick, incl	full thickness by ⅜ to ½ in. (9.5 to 12.7 mm) wide by 3 to 5 in. (76 to 127 mm) long
Over ³⁄₁₆ in. (4.76 mm) thick	³⁄₁₆ in. (4.76 mm) thick by ⅜ to ½ in. (9.5 to 12.7 mm) wide by 3 to 5 in. (76 to 127 mm) long[b]
Tubing:	
Up to 1½ in. (38 mm) in diameter, incl	full ring, 1 in. (25.4 mm) wide[c]
Over 1½ in. (38 mm) in diameter	a circumferential segment 3 to 5 in. (76 to 127 mm) long cut from a 1-in. (25.4-mm) wide ring[d]
Castings[e]:	³⁄₁₆ in. (4.76 mm) thick by ⅜ to ½ in. (9.5 to 12.7 mm) wide by 3 to 5 in. (76 to 127 mm) long

[a] When bending such specimens, the curved surface shall be on the outside of the bend.
[b] One surface shall be an original surface of the material under test and it shall be on the outside of the bend. It shall be finished as prescribed in 5.3.1.
[c] Ring sections are not flattened or subjected to any mechanical work before they are subjected to the test solution.
[d] Specimens from welded tubes over 1½ in. (38 mm) in diameter shall be taken with the weld on the axis of the bend.
[e] Specimens to be machined from keel blocks or as near the outside surface of the casting as practical.

The American Society for Testing and Materials takes no position respecting the validity of any patent rights asserted in connection with any item mentioned in this standard. Users of this standard are expressly advised that determination of the validity of any such patent rights, and the risk of infringement of such rights, are entirely their own responsibility.

This standard is subject to revision at any time by the responsible technical committee and must be reviewed every five years and if not revised, either reapproved or withdrawn. Your comments are invited either for revision of this standard or for additional standards and should be addressed to ASTM Headquarters. Your comments will receive careful consideration at a meeting of the responsible technical committee, which you may attend. If you feel that your comments have not received a fair hearing you should make your views known to the ASTM Committee on Standards, 1916 Race St., Philadelphia, Pa. 19103.

Designation: A 713 – 77 (Reapproved 1983)

Standard Specification for
STEEL WIRE, HIGH-CARBON SPRING, FOR HEAT-TREATED COMPONENTS[1]

This standard is issued under the fixed designation A 713; the number immediately following the designation indicates the year of original adoption or, in the case of revision, the year of last revision. A number in parentheses indicates the year of last reapproval. A superscript epsilon (ϵ) indicates an editorial change since the last revision or reapproval.

1. Scope

1.1 This specification covers round carbon spring steel wire in coils intended for the manufacture of mechanical springs and wire forms that are heat treated (austenitized, quenched, and tempered) after fabrication.

1.2 The values stated in inch-pound units are to be regarded as the standard.

2. Applicable Documents

2.1 *ASTM Standards:*
A 370 Methods and Definitions for Mechanical Testing of Steel Products[2]
A 510 Specification for General Requirements for Wire Rods and Coarse Round Wire, Carbon Steel[3]
A 700 Recommended Practices for Packaging, Marking, and Loading Methods for Steel Products for Domestic Shipment[4]
E 29 Recommended Practice for Indicating Which Places of Figures Are to Be Considered Significant in Specified Limiting Values[5]
E 30 Methods for Chemical Analysis of Steel, Cast Iron, Open-Hearth Iron, and Wrought Iron[6]
E 44 Definitions of Terms Relating to Heat Treatment of Metals[7]
E 112 Methods for Determining Average Grain Size[8]
E 350 Methods for Chemical Analysis of Carbon Steel, Low-Alloy Steel, Silicon Electrical Steel, Ingot Iron, and Wrought Iron[6]
E 380 Standard for Metric Practice[5]

3. Definitions

3.1 *heat-treated components*—mechanical springs or wire forms that are austenitized, quenched, and tempered after fabrication.

3.2 Refer to Definitions E 44 for a more detailed description of heat-treating terms.

4. General Requirements for Delivery

4.1 Material furnished under this specification shall conform to the applicable requirements of the latest edition of Specification A 510 unless otherwise specified herein.

5. Ordering Information

5.1 Orders for material under this specification shall include the following information:
5.1.1 Quantity (weight),
5.1.2 Name of material (Sections 1 and 7),
5.1.3 Diameter (Table 3),
5.1.4 Packaging, marking, and loading (Section 12),
5.1.5 ASTM designation and date of issue,
5.1.6 Special requirements (Sections 8 and 9), and
5.1.7 End use.

---

[1] This specification is under the jurisdiction of ASTM Committee A-1 on Steel, Stainless Steel and Related Alloys, and is the direct responsibility of Subcommittee A01.03 on Steel Rod and Wire.
Current edition approved June 24, 1977. Published August 1977. Originally published as A 713 – 75. Last previous edition A 713 – 75.
[2] *Annual Book of ASTM Standards*, Vol 01.04.
[3] *Annual Book of ASTM Standards*, Vol 01.03.
[4] *Annual Book of ASTM Standards*, Vols 01.01, 01.03, 01.04, and 01.05.
[5] *Annual Book of ASTM Standards*, Vol 14.02.
[6] *Annual Book of ASTM Standards*, Vol 03.05.
[7] *Annual Book of ASTM Standards*, Vol 01.02.
[8] *Annual Book of ASTM Standards*, Vol 03.03.

A 713

NOTE 1—A typical ordering description is as follows: Steel Wire, High Carbon Spring, for Heat-Treated Components, Grade 1070, to ASTM A 713 dated ——, for Door Closer Springs, 30 000 lb, Size 0.250 in. in 500-lb Catch Weight Coils.

6. Manufacture

6.1 The steel shall be made by the open-hearth, basic-oxygen, or electric-furnace process.

6.2 The wire, prior to fabrication, shall be thermally treated or thermally treated and drawn.

6.3 The condition of wire (metallurgical and mechanical properties) to be used is at the discretion of the purchaser and is generally dependent on the severity of the component part to be formed.

7. Chemical Requirements

7.1 The steel shall conform to the requirements for chemical composition prescribed in Table 1 for the grade ordered.

7.2 A chemical composition other than those shown in Table 1 may be supplied when agreed upon by the manufacturer and purchaser.

7.3 An analysis of each cast or heat shall be made by the manufacturer to determine the percentage of elements specified in Table 1. The analysis shall be made from a test sample preferably taken during the pouring of the cast or heat. The chemical composition thus determined shall be reported to the purchaser or his representative upon request.

7.4 A product analysis may be made by the purchaser. The chemical composition thus determined, as to elements required or restricted, shall conform to permissible variations for product analysis as specified in Table 10 in Specification A 510. For referee purposes, Methods E 30 or Methods E 350 shall be used.

8. Metallurgical Structure

8.1 Austenitic grain size, when specified, shall be determined in accordance with the requirements of Methods E 112 or some other mutually agreeable method.

9. Mechanical Requirements

9.1 Tensile strength is not normally a requirement. Minimum or maximum values for tensile strength may be agreed upon between the purchaser and manufacturer and are dependent on the chemical composition, thermal treatment, and diameter of wire specified.

9.2 *Wrap Test:*

9.2.1 *Requirements*—Wire shall wind without fracture on a cylindrical mandrel of a diameter specified in Table 2. The wrap test is not applicable to wires over 0.312 in. (8 mm). Since the conventional methods will not accommodate wire sizes over 0.312 in., an alternative test procedure may be agreed upon by the purchaser and manufacturer.

9.2.2 *Number of Tests*—At least one test specimen shall be taken for each ten coils or fraction thereof in a lot.

9.2.3 *Location of Test*—The test specimen shall be taken from either end of the coil.

9.2.4 *Test Method*—The wrap test shall be made in accordance with Supplement IV of Methods and Definitions A 370.

10. Dimensions and Tolerances

10.1 The diameter of the wire shall not vary from the specified size by more than the tolerance shown in Table 3.

11. Workmanship

11.1 The surface of the wire as received shall be substantially free of rust and such other surface imperfections of a nature or degree, for the grade ordered, that will be detrimental to the fabrication of the parts.

11.2 Wire drawn as a final operation shall not be kinked or improperly cast. To test for a cast, a single convolution, or ring, of wire shall be cut from the bundle and placed on a flat surface. The wire shall lie substantially flat and not spring up. The wire shall not show a wavy condition.

11.3 Each coil of wire shall be one continuous length.

11.4 Wire may be processed with welds made prior to wire drawing. Weld areas need not meet the mechanical requirements of the specification. If unmarked welds are unacceptable to the purchaser, special arrangements should be made with the manufacturer at the time of purchase.

12. Packaging, Marking, and Loading

12.1 Packaging of the coils of wire shall be

A 713

by agreement between the manufacturer and the purchaser. This agreement shall include coil dimensions and weights.

12.2 When specified, the packaging, marking, and loading shall be in accordance with Recommended Practices A 700.

12.3 Marking shall be by tag securely attached to each coil of wire and shall show the identity of the manufacturer, size of the wire, grade, ASTM specification number, and cast or heat number.

13. Inspection

13.1 The manufacturer shall afford the inspector representing the purchaser all reasonable facilities to satisfy him that the material being furnished is in accordance with this specification. All tests (except product analysis) and inspections shall be made at the place of manufacture prior to shipment, and shall be so conducted as not to interfere unnecessarily with the operation of the works.

14. Certification

14.1 Upon request of the purchaser in the contract or purchase order, a manufacturer's certification that the material was manufactured and tested in accordance with this specification together with a report of the test results shall be furnished at the time of shipment.

15. Rejection and Rehearing

15.1 Unless otherwise specified, any rejection based on tests made in accordance with this specification shall be reported to the manufacturer within a reasonable length of time.

15.2 Failure of any of the test specimens to comply with the requirements of this specification shall constitute grounds for rejection of the lot represented by the specimen. The lot may be resubmitted for inspection by testing every coil for the characteristic in which the specimen failed and sorting out the nonconforming coils.

15.3 The material must be adequately protected and correctly identified in order that the manufacturer may make a proper investigation.

TABLE 1 Chemical Requirements

NOTE—The following ranges of silicon are commonly specified for high-carbon steels: 0.10 to 0.20 %; 0.15 to 0.30 %; 0.20 to 0.40 %; or 0.30 to 0.60 %.

UNS Designation[a]	Grade	Composition, %			
		Carbon	Manganese	Phosphorus, max	Sulfur, max
G 10550	1055	0.50–0.60	0.60–0.90	0.040	0.050
G 10590	1059	0.55–0.65	0.50–0.80	0.040	0.050
G 10600	1060	0.55–0.65	0.60–0.90	0.040	0.050
G 10640	1064	0.60–0.70	0.50–0.80	0.040	0.050
G 10650	1065	0.60–0.70	0.60–0.90	0.040	0.050
G 10690	1069	0.65–0.75	0.40–0.70	0.040	0.050
G 10700	1070	0.65–0.75	0.60–0.90	0.040	0.050
G 10740	1074	0.70–0.80	0.50–0.80	0.040	0.050
G 10750	1075	0.70–0.80	0.40–0.70	0.040	0.050
G 10780	1078	0.72–0.85	0.30–0.60	0.040	0.050
G 10800	1080	0.75–0.88	0.60–0.90	0.040	0.050
G 10840	1084	0.80–0.93	0.60–0.90	0.040	0.050
G 10860	1086	0.80–0.93	0.30–0.50	0.040	0.050
G 10900	1090	0.85–0.98	0.60–0.90	0.040	0.050
G 10950	1095	0.90–1.03	0.30–0.50	0.040	0.050
G 15610	1561	0.55–0.65	0.75–1.05	0.040	0.050
G 15660	1566	0.60–0.71	0.85–1.15	0.040	0.050
G 15720	1572	0.65–0.76	1.00–1.30	0.040	0.050

[a] New designation established in accordance with ASTM E 527 and SAE J1086, Recommended Practice for Numbering Metals and Alloys (UNS).

TABLE 2 Wrap Test Requirements

Wire Diameter, in. (mm)	Mandrel Sizes	
	Grades to 1090	Grades 1090 and Over
Up to 0.162 (4)	1×[a]	2×
Over 0.162 to 0.312 (4 to 8), incl	2×	3×

[a] The symbol × represents the diameter of the wire tested. For 1× mandrel, wire may be wrapped around itself.

TABLE 3 Permissible Variations in Wire Diameter

NOTE—For purposes of determining conformance with this specification, all specified limits are considered absolute as defined in Recommended Practice E 29.

Diameter, in. (mm)	Permissible Variations, Plus and Minus, in. (mm)	Permissible Out-of-Round, in. (mm)
0.035 to 0.075 (0.89 to 1.90), incl	0.001 (0.03)	0.001 (0.03)
Over 0.075 to 0.375 (1.90 to 9.52), incl	0.002 (0.05)	0.002 (0.05)
Over 0.375 to 0.625 (9.52 to 15.88), incl	0.003 (0.08)	0.003 (0.08)

The American Society for Testing and Materials takes no position respecting the validity of any patent rights asserted in connection with any item mentioned in this standard. Users of this standard are expressly advised that determination of the validity of any such patent rights, and the risk of infringement of such rights, are entirely their own responsibility.

This standard is subject to revision at any time by the responsible technical committee and must be reviewed every five years and if not revised, either reapproved or withdrawn. Your comments are invited either for revision of this standard or for additional standards and should be addressed to ASTM Headquarters. Your comments will receive careful consideration at a meeting of the responsible technical committee, which you may attend. If you feel that your comments have not received a fair hearing you should make your views known to the ASTM Committee on Standards, 1916 Race St., Philadelphia, Pa. 19103.

Designation: A 715 – 81[ϵ1]

Standard Specification for
STEEL SHEET AND STRIP, HOT-ROLLED, HIGH-STRENGTH, LOW-ALLOY, WITH IMPROVED FORMABILITY[1]

This standard is issued under the fixed designation A 715; the number immediately following the designation indicates the year of original adoption or, in the case of revision, the year of last revision. A number in parentheses indicates the year of last reapproval. A superscript epsilon (ϵ) indicates an editorial change since the last revision or reapproval.

[ϵ1] NOTE—Specification A 749 was included editorially in Section 2 and 3.1 in September 1982.

1. Scope

1.1 This specification covers high-strength, low-alloy, hot-rolled steel sheet and strip having improved formability when compared with steels covered by Specifications A 606 and A 607. The product is furnished as either cut lengths or coils and is available in four-strength levels, Grades 50, 60, 70, and 80 (corresponding to minimum yield point (see Table 2)), and in eight types (according to chemical composition (see Table 1)). Not all grades are available in all types. The steel is killed, made to a fine grain practice, and includes microalloying elements such as columbium, titanium, vanadium, zirconium, etc. The product is intended for structural and miscellaneous applications where higher strength, savings in weight, improved formability, and weldability are important.

1.2 The values stated in inch-pound units are to be regarded as the standard.

2. Applicable Documents

2.1 *ASTM Standards:*
A 370 Methods and Definitions for Mechanical Testing of Steel Products[2]
A 568 Specification for General Requirements for Steel, Carbon and High-Strength Low-Alloy Hot-Rolled Sheet and Cold-Rolled Sheet[2]
A 606 Specification for Steel Sheet and Strip, Hot-Rolled and Cold-Rolled, High-Strength, Low-Alloy, with Improved Atmospheric Corrosion Resistance[2]
A 607 Specification for Steel Sheet and Strip, Hot-Rolled and Cold-Rolled, High-Strength, Low-Alloy Columbium and/or Vanadium[2]
A 749 Specification for General Requirements for Steel, Carbon and High-Strength, Low-Alloy Hot-Rolled Strip[2]

3. General Requirements for Delivery

3.1 Material furnished under this specification shall conform to the applicable requirements of the current edition of Specifications A 568 and A 749.

4. Ordering Information

4.1 Orders for material under this specification shall include the following information, as required, to describe adequately the desired material:

4.1.1 ASTM specification number and date of issue.

4.1.2 Grade (see Table 2) and type designation, if required (see Table 1).

4.1.3 Name of material (hot-rolled high-strength low-alloy steel sheet or strip),

4.1.4 Condition (material to this specification is furnished in the hot-rolled condition. Pickled, or blast cleaned, must be specified if required. Material so ordered will be oiled unless ordered "not oiled" (see 9.1 and 9.2)),

---

[1] This specification is under the jurisdiction of ASTM Committee A-1 on Steel, Stainless Steel and Related Alloys, and is the direct responsibility of Subcommittee A01.19 on Steel Sheet and Strip.
Current edition approved July 31, 1981. Published September 1981. Originally published as A 715 – 75. Last previous edition A 715 – 79.

[2] *Annual Book of ASTM Standards*, Vol 01.03.

4.1.5 Edges (see 9.3),

4.1.6 Dimensions (thickness, width, and length for cut lengths, or thickness and width for coils),

4.1.7 Coil size and weight requirements (must include inside diameter, outside diameter, and maximum weight),

4.1.8 Application (part identification and description),

4.1.9 Special requirements (if required), and

4.1.10 If required, cast analysis or mechanical property report, or both (see 10.1).

NOTE—A typical ordering description is as follows: ASTM A 715 dated ———, Grade 80, hot-rolled high-strength low-alloy steel sheet, pickled and oiled, cut edge 0.100 by 48 by 96 in. for Part 83479, bumper reinforcement bracket.

5. Manufacture

5.1 Sheet or strip to this specification is produced from killed steel, made to a fine grain practice.

6. Chemical Requirements

6.1 The cast or heat analysis of the steel shall conform to the chemical requirements shown in Table 1. Depending on the thickness of the material and its intended application, variations in the chemical composition might be used with the limits shown in Table 1, for each grade and type (if specified). Where it is of particular importance and the producer should be consulted for specific chemical composition.

6.2 Steel to this specification contains microalloying elements such as columbium, titanium, vanadium, zirconium, etc. which should be considered when selecting a welding procedure to assure the procedure is compatible with the chemical composition for the grade and type to be welded.

6.3 Unless otherwise specified, the type (see Table 1) is at the discretion of the producer. If a type is required, one of the types listed in Table 1 shall be specified, and type, in addition to the grade, must be shown on the order (see 4.1.2).

7. Mechanical Requirements

7.1 *Tension Test:*

7.1.1 *Requirements*—The material as represented by the test specimens shall conform to the requirements as listed in Table 2.

7.1.2 *Number of Tests*—Two tension tests shall be made from each cast unless the finished material from a case is less than 50 tons (45.4 Mg), when one tension test will be sufficient. When material from one case differs 0.050 in. (1.27 mm) or more in thickness, one tension test shall be made from both the thickest and thinnest material rolled regardless of the weight represented.

7.1.3 *Tension Specimen Orientation*—Tension test specimens shall be taken longitudinally to the direction of rolling.

7.2 *Bend Test:*

7.2.1 *Requirements*—The bend test specimen shall stand being bent at room temperature through 180 deg to an inside diameter equivalent to thicknesses shown in Table 3, without cracking on the outside of the bent portion (see Section 14 of Methods and Definitions A 370).

7.2.2 *Number of Tests*—Two bend tests shall be made from each case, unless the finished material from a cast is less than 50 tons (45.4 Mg), when one bend test will be sufficient. When material rolled from one case differs 0.050 in. (1.27 mm) or more in thickness, one bend test shall be made from both the thickest and thinnest material rolled regardless of the weight represented.

7.2.3 *Test Specimen Orientation*—Bend test specimens shall be taken transversely to the direction of rolling.

8. Retests

8.1 If the results of an original tensile specimen are within 2 ksi (14 MPa) of the required tensile strength, within 1 ksi (7 MPa) of the required yeild point, or within 2 % of the required elongation, a retest shall be permitted for which one random specimen from the cast or test lot shall be used. If the results of this retest specimen meet the specified requirements, the cast or test lot will be accepted.

9. Finish and Appearance

9.1 *Surface Finish*—Unless otherwise specified, hot-rolled material shall have an as-rolled, not pickled surface finish. When required, material may be specified to be pickled or blast cleaned.

9.2 *Oiling*—Unless otherwise specified, hot-rolled (as-rolled) material shall be furnished not oiled, and hot-rolled pickled or blast-cleaned material shall be furnished oiled. When required, pickled or blast-cleaned material may be specified to be furnished not oiled, and as-rolled material may be specified to be furnished oiled.

9.3 *Edges:*

9.3.1 Hot-rolled sheet can be furnished as mill edge or cut edge, as specified.

9.3.2 Hot-rolled strip can be furnished as mill edge, square edge, or slit (cut) edge, as specified.

10. Certification and Reports

10.1 When requested, the producer shall furnish the test results of the cast analysis or mechanical properties or both to determine compliance with this specification. The information shall include the purchase order number, ASTM designation number and date of issue, and the cast or lot number correlating the test data with the material represented.

TABLE 1 Chemical Requirements

Element	Composition, max, % Cast or Heat (formerly Ladle) Analysis
Carbon	0.15
Manganese	1.65
Phosphorus	0.025
Sulfur	0.035
Types, by Added Elements, %	
Type 1:	
Titanium, min	0.05
Silicon, max	0.10
Type 2:	
Vanadium, min	0.02
Silicon,[a] max	0.60
Nitrogen,[a] min	0.005
Type 3:	
Columbium, min	0.005
Vanadium,[a] max	0.08
Silicon,[a] max	0.60
Nitrogen,[a] max	0.020
Type 4:	
Zirconium, min	0.05
Silicon, max	0.90
Chromium,[a] max	0.80
Titanium,[a] max	0.10
Boron,[a] max	0.0025
Columbium[b]	0.005–0.06
Type 5:	
Columbium,[c] min	0.03
Molybdenum,[c] min	0.20
Silicon, max	0.30
Type 6:	
Columbium	0.005–0.10
Silicon, max	0.90
Type 7:	
Columbium or vanadium or both, min	0.005
Silicon, max	0.60
Nitrogen, max	0.020
Type 8:	
Columbium	0.005–0.15
Zirconium, min	0.05
Type 9:	
Columbium, min	0.01
Vanadium,[a] min	0.05
Silicon,[a] max	0.60

[a] Not added to Grades 50 and 60.
[b] Might not be added to Grade 50.
[c] Available as Grade 80 only.

TABLE 2 Tensile Requirements

	Grade 50	Grade 60	Grade 70	Grade 80
Yield point, min, ksi (MPa)	50 (345)	60 (415)	70 (485)	80 (550)
Tensile strength, min, ksi (MPa)	60 (415)	70 (485)	80 (550)	90 (620)
Elongation in 2 in. or 50 mm, min, % for thickness:				
Over 0.097 in. (2.46 mm)	24.0	22.0	20.0	18.0
Up to 0.097 in. (2.46 mm), incl	22.0	20.0	18.0	16.0

[a] All product specified to be furnished in coils will be produced to the same strength level mill practices as cut length product. Because testing within the body of the coil cannot be performed by the producer, recognition must be given to the fact that some portions of coils could fall below the specified minimums.

TABLE 3 Bend Test Requirements

Grade	Ratio of Bend Diameter to Thickness of Specimen[a] for Thicknesses up to 0.2299 in. (6 mm) incl	
	Transverse Specimens	Longitudinal Specimens[b]
50	1	0
60	1	0
70	1½	1
80	1½	1

[a] These ratios apply to bending performance of the test specimen only. Where material is to be bent in fabricating operations a more liberal bend radius may be required and should be based on prior experience or consultation with the steel producer or both.

[b] Testing longitudinal specimens is not required by this specification except when material is too narrow for transverse tests. However, experience has shown that when transverse specimens meet the specified bend requirements, longitudinal specimens will normally pass the bend requirements in Table 3. See footnote a.

The American Society for Testing and Materials takes no position respecting the validity of any patent rights asserted in connection with any item mentioned in this standard. Users of this standard are expressly advised that determination of the validity of any such patent rights, and the risk of infringement of such rights, are entirely their own responsibility.

This standard is subject to revision at any time by the responsible technical committee and must be reviewed every five years and if not revised, either reapproved or withdrawn. Your comments are invited either for revision of this standard or for additional standards and should be addressed to ASTM Headquarters. Your comments will receive careful consideration at a meeting of the responsible technical committee, which you may attend. If you feel that your comments have not received a fair hearing you should make your views known to the ASTM Committee on Standards, 1916 Race St., Philadelphia, Pa. 19103.

Designation: A 749 – 83

Specification for
GENERAL REQUIREMENTS FOR STEEL, CARBON AND HIGH-STRENGTH, LOW-ALLOY HOT-ROLLED STRIP[1]

This standard is issued under the fixed designation A 749; the number immediately following the designation indicates the year of original adoption or, in the case of revision, the year of last revision. A number in parentheses indicates the year of last reapproval. A superscript epsilon (ϵ) indicates an editorial change since the last revision or reapproval.

1. Scope

1.1 This specification covers the general requirements for hot-rolled steel strip in coils and cut lengths. It applies to carbon steel and high-strength, low-alloy steel furnished as hot-rolled.

1.2 This specification is not applicable to hot-rolled heavy-thickness carbon sheet and strip coils (ASTM Specification A 635),[2] cold-rolled carbon steel strip (ASTM Specification A 109),[2] high-strength, low-alloy cold-rolled steel (ASTM Specifications A 606 and A 607)[2] or cold-rolled carbon spring steel (ASTM Specification A 682).[2]

1.3 For the purposes of determining conformance with this and the appropriate product specification referenced under 2.1, values shall be rounded to the nearest unit in the right hand place of figures used in expressing the limiting values in accordance with the rounding method of Recommended Practice E 29.

NOTE 1—A complete metric companion to Specification A 749 has been developed—A 749M; therefore no metric equivalents are presented in this specification.

2. Applicable Documents

2.1 *ASTM Standards:*
A 370 Methods and Definitions for Mechanical Testing of Steel Products[2]
A 700 Practices for Packaging, Marking, and Loading Methods for Steel Products for Domestic Shipment[2]
E 11 Specification for Wire-Cloth Sieves for Testing Purposes[3]
E 29 Recommended Practice for Indicating Which Places of Figures are to be Considered Significant in Specified Limiting Values[3]
E 350 Methods for Chemical Analysis of Carbon Steel, Low-Alloy Steel, Silicon Electrical Steel, Ingot Iron, and Wrought Iron[4]

2.2 *Military Standards:*
MIL-STD-129 Marking for Shipment and Storage[5]
MIL-STD-163 Steel Mill Products, Preparation for Shipment and Storage[5]

2.3 *Federal Standards:*
Fed. Std. No. 123 Marking for Shipments (Civil Agencies)[5]
Fed. Std. No. 183 Continuous Identification Marking of Iron and Steel Products[5]

3. Descriptions of Terms

3.1 *Steel Types:*

3.1.1 *Carbon Steel* is the designation for steel when no minimum content is specified or required for aluminum, chromium, cobalt, columbium, molybdenum, nickel, titanium, tungsten, vanadium, zirconium, or any element added to obtain a desired alloying effect; when the specified minimum for copper does not exceed 0.40 %; or when the maximum content specified for any of the following elements does not exceed the percentages noted; manganese 1.65, silicon 0.60, or copper 0.60.

3.1.1.1 In all carbon steels small quantities

---

[1] This specification is under the jurisdiction of ASTM Committee A-1 on Steel, Stainless Steel, and Related Alloys, and is the direct responsibility of Subcommittee A01.19 on Steel Sheet and Strip.
Current edition approved July 29, 1983. Published November 1983. Originally published as A 749 – 81. Last previous edition A 749 – 81.
[2] *Annual Book of ASTM Standards*, Vol 01.03.
[3] *Annual Book of ASTM Standards*, Vol 14.02.
[4] *Annual Book of ASTM Standards*, Vol 03.05.
[5] Available from Naval Publications and Forms Center, 5801 Tabor Ave., Philadelphia, Pa. 19120.

of certain residual elements unavoidably retained from raw materials are sometimes found which are not specified or required, such as copper, nickel, molybdenum, chromium, etc. These elements are considered as incidental and are not normally determined or reported.

3.1.2 *High-Strength, Low-Alloy Steel* is a specific group of steels in which higher strength, and in some cases additional resistance to atmospheric corrosion, are obtained by moderate amounts of one or more alloying elements.

3.2 *Product Types:*

3.2.1 *Hot-Rolled Strip* is manufactured by hot rolling billets or slabs to the required thickness. It may be produced single width or by rolling multiple width and slitting to the desired width. It can be supplied in coils or cut lengths as specified.

Width, in.		Thickness, in.	
Over	Through	Over	Through
...	3½	0.044	0.203
3½	6	0.044	0.203
6	12	0.044	0.230

3.2.2 Hot-rolled, high-strength, low-alloy strip is commonly available by size as follows:

Width, in		Thickness, in.	
Over	Through	From	Through
			Coils & Cut Lengths / Coils Only
...	6	0.071	0.203 / 0.230
6	12	0.071	0.230 / 0.230

4. General Requirements for Delivery

4.1 Products covered by this specification are produced to decimal thickness only and thickness tolerances apply.

5. Manufacture

5.1 Unless otherwise specified, hot-rolled material shall be furnished hot-rolled, not annealed or pickled.

6. Chemical Requirements

6.1 *Limits:*

6.1.1 The chemical composition shall be in accordance with the applicable product specification. However, if other compositions are required for carbon steel, they shall be prepared in accordance with Appendix X1.

6.1.2 Where the material is used for fabrication by welding, care must be exercised in the selection of chemical composition or mechanical properties to ensure compatibility with the welding process and its effect on altering the properties.

6.2 *Cast or Heat (Formerly Ladle) Analysis:*

6.2.1 An analysis of each cast or heat of steel shall be made by the manufacturer to determine the percentage of elements specified or restricted by the applicable specification.

6.2.2 When requested, cast or heat analysis for elements listed or required shall be reported to the purchaser or to his representative.

6.3 *Product, Check, or Verification Analysis:*

6.3.1 Non-killed steels (such as capped or rimmed) are not technologically suited to product analysis due to the nonuniform character of their chemical composition and therefore, the tolerances in Table 1 do not apply. Product analysis is appropriate on these types of steel only when misapplication is apparent or for copper when copper steel is specified.

6.3.2 For steels other than non-killed (capped or rimmed), product analysis may be made by the purchaser. The chemical analysis shall not vary from the limits specified by more than the amounts in Table 1. The several determinations of any element in a cast shall not vary both above and below the specified range.

6.4 *Sampling for Product Analysis:*

6.4.1 To indicate adequately the representative composition of a cast by product analysis, it is general practice to select samples to represent the steel, as fairly as possible, from a minimum number of pieces as follows: 3 pieces for lots up to 15 tons inclusive, and 6 pieces for lots over 15 tons.

6.4.2 When the steel is subject to tension test requirements, samples for product analysis may be taken either by drilling entirely through the used tension test specimens themselves or in accordance with 6.4.3.

6.4.3 When the steel is not subject to tension test requirements, the samples for analysis must be taken by milling or drilling entirely through the strip in a sufficient number of places so that the samples are representative of the entire strip. The sampling may be facilitated by folding the strip both ways, so that several samples may be taken at one drilling. Steel subjected to certain heating operations by the purchaser may not give chemical analysis results that properly represent its original composition. Therefore, users must analyze chips taken from the steel in the condition in which it is received from the steel manufacturer.

6.5 *Specimen Preparation*—Drillings or chips must be taken without the application of water, oil, or other lubricant, and must be free of scale, grease, dirt, or other foreign substances. They must not be overheated during cutting to the extent of causing decarburization. Chips must be well mixed, and those too coarse to pass a No. 10 (2.00-mm) sieve or too fine to remain on a No. 30 (600-μm) sieve are not suitable for proper analysis. Sieve size numbers are in accordance with Specification E 11.

6.6 *Test Methods*—In case a referee analysis is required and agreed upon to resolve a dispute concerning the results of a chemical analysis, the procedure for performing the referee analysis must be in accordance with the latest issue of Methods E 350, unless otherwise agreed upon between the manufacturer and the purchaser.

7. Mechanical Requirements

7.1 The mechanical property requirements, number of specimens, test locations, and specimen orientation shall be in accordance with the applicable product specification.

7.2 Unless otherwise specified in the applicable product specification, test specimens must be prepared in accordance with Methods and Definitions A 370.

7.3 Mechanical tests shall be conducted in accordance with Methods and Definitions A 370.

7.4 To determine conformance with the product specification, a calculated value should be rounded to the nearest 1 ksi tensile strength and yield point or yield strength, and to the nearest unit in the right hand place of figures used in expressing the limiting value for other values in accordance with the rounding off method given in Recommended Practice E 29.

8. Dimensions, Tolerances, and Allowances

8.1 Dimensions, tolerances, and allowances applicable to products covered by this specification are contained in Tables 3 through 9. The appropriate tolerance tables shall be identified in each individual specification.

9. Finish and Condition

9.1 Hot-rolled strip has a surface with an oxide or scale resulting from the hot-rolling operation. The oxide or scale can be removed by pickling or blast cleaning when required for press-work operations or welding. Hot-rolled and hot-rolled descaled strip are not generally used for exposed parts where surface is of prime importance. However, hot-rolled surface might be of importance, as in the case of weathering steels for exposed parts.

9.1.1 Hot-rolled strip can be supplied with mill edges, square edges, or cut (slit) edges as specified.

9.1.1.1 Mill edges are the natural edges resulting from the hot-rolling operation and are generally round and smooth without any definite contour.

9.1.1.2 Square edges are the edges resulting from rolling through vertical edging rolls during the hot-rolling operations. These edges are square and smooth, with the corners slightly rounded.

9.1.1.3 Cut (slit) edges are the normal edges that result from the shearing, slitting, or trimming of mill edges.

9.1.2 The ends of plain hot-rolled mill edge coils are irregular in shape and are referred to as uncropped ends. Where such ends are not acceptable, the purchaser's order should so specify. Processed coils such as pickled or blast cleaned are supplied with square-cut ends.

9.2 *Oiling*:

9.2.1 Plain hot-rolled strip is customarily furnished not oiled. Oiling must be specified when required.

9.2.2 Hot-rolled pickled or descaled strip is customarily furnished oiled. If the product is not to be oiled, it must be so specified since the cleaned surface is prone to rusting.

10. Workmanship

10.1 Cut lengths shall have a workmanlike appearance and shall not have imperfections of a nature or degree for the product, the grade, and the quality ordered that will be detrimental to the fabrication of the finished part.

10.2 Coils may contain some abnormal imperfections that render a portion of the coil unusable since the inspection of coils does not afford the producer the same opportunity to remove portions containing imperfections as in the case with cut lengths.

11. Retests

11.1 If any test specimen shows defective machining or develops flaws, it must be discarded and another specimen substituted.

11.2 If the percentage of elongation of any tension test specimen is less than that specified and any part of the fracture is more than ¾ in.

from the center of the gage length of a 2-in. specimen or is outside the middle half of the gage length of an 8-in. specimen, as indicated by scribe scratches marked on the specimen before testing, a retest is allowed.

11.3 If a bend specimen fails, due to conditions of bending more severe than required by the specification, a retest is permitted either on a duplicate specimen or on a remaining portion of the failed specimen.

12. Inspection

12.1 When the purchaser's order stipulates that inspection and test (except product analyses) for acceptance on the steel be made prior to shipment from the mill, the manufacturer shall afford the purchaser's inspector all reasonable facilities to satisfy him that the steel is being produced and furnished in accordance with the specification. Mill inspection by the purchaser shall not interfere unnecessarily with the manufacturer's operation.

13. Rejection and Rehearing

13.1 Unless otherwise specified, any rejection shall be reported to the manufacturer within a reasonable time after receipt of material by the purchaser.

13.2 Material that is reported to be defective subsequent to the acceptance at the purchaser's works shall be set aside, adequately protected, and correctly identified. The manufacturer shall be notified as soon as possible so that an investigation may be initiated.

13.3 Samples that are representative of the rejected material shall be made available to the manufacturer. In the event that the manufacturer is dissatisfied with the rejection, he may request a rehearing.

14. Marking

14.1 As a minimum requirement, the material shall be identified by having the manufacturer's name, ASTM designation, weight, purchaser's order number, and material identification legibly stenciled on top of each lift or shown on a tag attached to each coil or shipping unit.

14.2 When specified in the contract or order, and for direct procurement by or direct shipment to the government, marking for shipment, in addition to requirements specified in the contract or order, shall be in accordance with MIL-STD-129 for military agencies and in accordance with Fed. Std. No. 123 for civil agencies.

14.3 For Government procurement by the Defense Supply Agency, strip material shall be continuously marked for identification in accordance with Fed. Std. No. 183.

15. Packaging

15.1 Unless otherwise specified, the strip shall be packaged and loaded in accordance with Practices A 700.

15.2 When specified in the contract or order, and for direct procurement by or direct shipment to the government, when Level A is specified, preservation, packaging, and packing shall be in accordance with the Level A requirements of MIL-STD-163.

15.3 When coils are ordered it is customary to specify a minimum or range of inside diameter, maximum outside diameter, and a maximum coil weight, if required. The ability of manufacturers to meet the maximum coil weights depends upon individual mill equipment. When required, minimum coil weights are subject to negotiation.

A 749

TABLE 1 Tolerances for Product Analysis[A]

Element	Limit, or Maximum of Specified Element, %	Tolerances Under Minimum Limit	Tolerances Over Maximum Limit
Carbon	to 0.15, incl	0.02	0.03
	over 0.15 to 0.40, incl	0.03	0.04
	over 0.40 to 0.80, incl	0.03	0.05
	over 0.80	0.03	0.06
Manganese	to 0.60, incl	0.03	0.03
	over 0.60 to 1.15, incl	0.04	0.04
	over 1.15 to 1.65, incl	0.05	0.05
Phosphorus		...	0.01
Sulfur		...	0.01
Silicon	to 0.30, incl	0.02	0.03
	over 0.30 to 0.60, incl	0.05	0.05
Copper		0.02	...

[A] See 6.3.1.

TABLE 2 List of Tables for Dimensions, Tolerances, and Allowances

Dimensions	Table No.
Camber tolerances	8
Crown tolerances	5
Flatness tolerances	9
Length tolerances	7
Thickness tolerances	3, 4
Width tolerances	6

TABLE 3 Thickness Tolerances of Hot-Rolled Strip[A] **(Carbon and High-Strength Low-Alloy Steel)**[B] **Ordered to Nominal Thickness**

(Coils and Cut Lengths, Including Pickled)

Specified Width, in.		Thickness Tolerance, Over and Under, in., for Specified Nominal Thickness, in.				
Over	Through	Through 0.057	Over 0.057 to 0.118, incl	Over 0.118 to 0.187, incl	Over 0.187 to 0.203, incl	Over 0.203 to 0.230, incl
...	3½	0.003	0.004	0.005	0.006	...
3½	6	0.003	0.005	0.005	0.006	...
6	12	0.004	0.005	0.005	0.006	0.006

[A] Measurements for the above table are taken ⅜ in. from the edge of a strip on 1 in. or wider; and at any place on the strip when narrower than 1 in. The given tolerances do not include crown and therefore the tolerances for crown as shown in Table 5 are in addition to tolerances in Table 3.
[B] 0.071 in. minimum thickness for high-strength low alloy.

TABLE 4 Thickness Tolerances of Hot-Rolled Strip[A] **(Carbon and High-Strength Low-Alloy Steel)**[B] **Ordered to Minimum Thickness**

(Coils and Cut Lengths, Including Pickling)

Specified Width, in.		Thickness Tolerance, Over Only, for Specified Minimum Thickness, in.				
Over	Through	Through 0.057	Over 0.057 to 0.118, incl	Over 0.118 to 0.187, incl	Over 0.187 to 0.203, incl	Over 0.203 to 0.230, incl
...	3½	0.006	0.008	0.010	0.012	...
3½	6	0.006	0.010	0.010	0.012	...
6	12	0.008	0.010	0.010	0.012	0.012

[A] Measurements for the above table are taken ⅜ in. from the edge of a strip on 1 in. or wider; and at any place on the strip when narrower than 1 in. The given tolerances do not include crown and therefore the tolerances for crown as shown in Table 5 are in addition to tolerances in Table 4.
[B] 0.071 in. minimum thickness for high-strength low alloy.

TABLE 5 Crown Tolerances of Hot-Rolled Strip (Carbon and High-Strength Low-Alloy Steel)
(Coils and Cut Lengths, Including Pickled)
Strip may be thicker at the center than at a point ⅜ in. from the edge by the following amounts:

Specified Width, in.		Crown Tolerance, Over Only, for Specified Minimum Thickness, in.		
Over	Through	Through 0.118	Over 0.118 to 0.187, incl	Over 0.187 to 0.230, incl
...	3½	0.002	0.002	0.001
3½	6	0.003	0.002	0.002
6	12	0.004	0.003	0.003

TABLE 6 Width Tolerances of Hot-Rolled Strip (Carbon and High-Strength Low-Alloy Steel)
(Coils and Cut Lengths, Including Pickled)

Specified Width, in.		Width Tolerance, Over and Under, in.		
Over	Through	Mill Edge and Square Edge Strip	Cut Edge	
			Through 0.109 in.	Over 0.109 in. Through 0.230
...	2	¹⁄₃₂	0.008	0.016
2	5	³⁄₆₄	0.008	0.016
5	10	¹⁄₁₆	0.010	0.016
10	12	³⁄₃₂	0.016	0.016

TABLE 7 Length Tolerances of Hot-Rolled Strip (Carbon and High Strength Alloy)
(Cut Lengths, Including Pickled)

Specified Widths, in.	Length Tolerances over Specified Length, ft for Widths Given, in. No Tolerance Under						
	To 5 ft, incl	Over 5 to 10 ft, incl	Over 10 to 20 ft, incl	Over 20 to 30 ft, incl	Over 30 to 40 ft, incl	Over 40 ft, incl
To 3, incl	¼	⅜	½	¾	1	1½
Over 3 to 6, incl	⅜	½	⅝	¾	1	1½
Over 6 to 12, incl	½	¾	1	1¼	1½	1¾

TABLE 8 Camber Tolerances[A] of Hot-Rolled Strip
(Carbon and High-Strength Low-Alloy)
(Coils and Cut Lengths, Including Pickled, Applicable to Mill Edge, Square Edge, and Slit or Cut Edge)

NOTE—Camber is the deviation of a side edge from a straight line. The standard for measuring this deviation is based on any 8-ft length.[B] It is obtained by placing an 8-ft straightedge on the concave side and measuring the maximum distance between the strip edge and the straightedge.

For strip wider than 1½ in.—¼ in. in any 8 ft.
For strip 1½ in. and narrower—½ in. in any 8 ft.

[A] When the camber tolerances shown in the above table are not suitable for a particular purpose, hot-rolled strip is sometimes machine straightened.

[B] For strip less than 8 ft. tolerances are to be established in each instance.
A formula for calculating camber is as follows:

$$\frac{L^2 \times C_1}{64} = C_2 \text{ in } L$$

where:
C_1 = Camber in 8 ft and
C_2 = Camber in any given length L.

TABLE 9 Flatness Tolerances of Hot-Rolled Strip
(Carbon and High-Strength Low-Alloy)

It has not been practicable to formulate flatness tolerances for hot-rolled carbon strip steel because of the wide range of widths and thicknesses, and variety of chemical compositions, mechanical properties and types, produced in coils and cut lengths.

APPENDIX

(Nonmandatory Information)

X1. STANDARD CHEMICAL RANGES AND LIMITS

X1.1 Standard chemical ranges and limits are prescribed for carbon steels in Table X1.1

TABLE X1.1 Standard Chemical Ranges and Limits

NOTE—The carbon ranges shown in the column headed "Range" apply when the specified maximum limit for manganese does not exceed 1.00 %. When the maximum manganese limit exceeds 1.00 %, add 0.01 to the carbon ranges shown below.

Element	Carbon Steels Only, Cast or Heat Analysis		
	Minimum of Specified Element, %	Range	Lowest, max
Carbon (see Note)	to 0.15, incl	0.05	0.08
	over 0.15 to 0.30, incl	0.06	
	over 0.30 to 0.40, incl	0.07	
	over 0.40 to 0.60, incl	0.08	
	over 0.60 to 0.80, incl	0.11	
	over 0.80 to 1.35, incl	0.14	
Manganese	to 0.50, incl	0.20	0.40
	over 0.50 to 1.15, incl	0.30	
	over 1.15 to 1.65, incl	0.35	
Phosphorus	to 0.08, incl	0.03	0.04[A]
	over 0.08 to 0.15, incl	0.05	
Sulfur	to 0.08, incl	0.03	0.05[A]
	over 0.08 to 0.15, incl	0.05	
	over 0.15 to 0.23, incl	0.07	
	over 0.23 to 0.33, incl	0.10	
Silicon	to 0.15, incl	0.08	0.10
	over 0.15 to 0.30, incl	0.15	
	over 0.30 to 0.60, incl	0.30	
Copper	When copper is required 0.20 min is commonly specified.		

[A] Certain individual specifications provide for lower standard limits for phosphorus and sulfur.

The American Society for Testing and Materials takes no position respecting the validity of any patent rights asserted in connection with any item mentioned in this standard. Users of this standard are expressly advised that determination of the validity of any such patent rights, and the risk of infringement of such rights, are entirely their own responsibility.

This standard is subject to revision at any time by the responsible technical committee and must be reviewed every five years and if not revised, either reapproved or withdrawn. Your comments are invited either for revision of this standard or for additional standards and should be addressed to ASTM Headquarters. Your comments will receive careful consideration at a meeting of the responsible technical committee, which you may attend. If you feel that your comments have not received a fair hearing you should make your views known to the ASTM Committee on Standards, 1916 Race St., Philadelphia, Pa. 19103.

ASTM Designation: A 749M – 83

An American National Standard

Standard Specification for
GENERAL REQUIREMENTS FOR STEEL, CARBON AND HIGH-STRENGTH, LOW-ALLOY HOT-ROLLED STRIP [METRIC][1]

This standard is issued under the fixed designation A 749M; the number immediately following the designation indicates the year of original adoption or, in the case of revision, the year of last revision. A number in parentheses indicates the year of last reapproval. A superscript epsilon (ϵ) indicates an editorial change since the last revision or reapproval.

This specification has been approved for use by agencies of the Department of Defense for listing in the DoD Index of Specifications and Standards.

NOTE—1.2 and Section 2 were corrected editorially and the designation changed on Aug. 30, 1983.

1. Scope

1.1 This specification covers the general requirements for hot-rolled steel strip in coils and cut lengths. It applies to carbon steel and high-strength, low-alloy steel furnished as hot-rolled.

1.2 This specification is not applicable to hot-rolled heavy-thickness carbon sheet and strip coils (ASTM Specification A 635M),[2] cold-rolled carbon steel strip (ASTM Specification A 109M),[2] high-strength, low-alloy cold-rolled steel strip (ASTM Specifications A 606M and A 607M)[2] or cold-rolled carbon spring steel (ASTM Specification A 682M).[2]

NOTE—This specification is the metric counterpart of Specification A 749.

2. Applicable Documents

2.1 *ASTM Standards:*
A 370 Methods and Definitions for Mechanical Testing of Steel Products[2]
A 700 Practices for Packaging, Marking, and Loading Methods for Steel Products for Domestic Shipment[2]
E 11 Specification for Wire-Cloth Sieves for Testing Purposes[3]
E 350 Methods for Chemical Analysis of Carbon Steel, Low-Alloy Steel, Silicon Electrical Steel, Ingot Iron, and Wrought Iron[4]

2.2 *Military Standards:*
MIL-STD-129 Marking for Shipment and Storage[5]
MIL-STD-163 Steel Mill Products, Preparation for Shipment and Storage[5]

2.3 *Federal Standards:*
Fed. Std. No. 123 Marking for Shipment (Civil Agencies)[5]
Fed. Std. No. 183 Continuous Identification Marking of Iron and Steel Products[5]

3. Descriptions of Terms

3.1 *Steel Types:*

3.1.1 *Carbon Steel* is the designation for steel when no minimum content is specified or required for aluminum, chromium, cobalt, columbium, molybdenum, nickel, titanium, tungsten, vanadium, zirconium, or any element added to obtain a desired alloying effect; when the specified minimum for copper does not exceed 0.40 %; or when the maximum content specified for any of the following elements does not exceed the percentages noted; manganese 1.65, silicon 0.60, or copper 0.60.

3.1.1.1 In all carbon steels small quantities of certain residual elements unavoidably retained

---

[1] This specification is under the jurisdiction of ASTM Committee A-1 on Steel, Stainless Steel and Related Alloys, and is the direct responsibility of Subcommittee A01.19 on Steel Sheet and Strip.
Current edition approved Aug. 30, 1983. Published November 1983.
Originally published as A 749M - 77. Last previous edition A 749M - 77.
[2] *Annual Book of ASTM Standards*, Vol 01.03.
[3] *Annual Book of ASTM Standards*, Vol 14.02.
[4] *Annual Book of ASTM Standards*, Vol 03.05.
[5] Available from Naval Publications and Forms Center, 5801 Tabor Ave., Philadelphia, Pa. 19120.

from raw materials are sometimes found which are not specified or required, such as copper, nickel, molybdenum, chromium, etc. These elements are considered as incidental and are not normally determined or reported.

3.1.2 *High-Strength, Low-Alloy Steel* is a specific group of steels in which higher strength, and in some cases additional resistance to atmospheric corrosion, are obtained by moderate amounts of one or more alloying elements.

3.2 *Product Types:*

3.2.1 *Hot-Rolled Strip* is manufactured by hot rolling billets or slabs to the required thickness. It may be produced single width or by rolling multiple width and slitting to the desired width. It can be supplied in coils or cut lengths as specified.

3.2.1.1 Hot-rolled carbon strip is commonly available by size as follows:

Coils and Cut Lengths

Width, mm		Thickness, mm	
Over	Through	Over	Through
...	100	0.65	5.0
100	200	0.9	5.0
200	300	1.2	6.0

Coils Only[A]

200	300	6.0	12.5

[A] This product is covered by Specification A 635M.

3.2.2 Hot-rolled, high-strength, low-alloy strip is commonly available by size as follows:

Width, mm		Thickness, mm	
Over	Through	Over	Through
...	200	1.8	5.0
200	300	1.8	6.0

4. General Requirements for Delivery

4.1 Products covered by this specification are produced to decimal thickness only and thickness tolerances apply.

5. Manufacture

5.1 Unless otherwise specified, hot-rolled material shall be furnished hot-rolled, not annealed or pickled.

6. Chemical Requirements

6.1 *Limits:*

6.1.1 The chemical composition shall be in accordance with the applicable product specification. However, if other compositions are required for carbon steel, they shall be prepared in accordance with Appendix X2.

6.1.2 Where the material is used for fabrication by welding, care must be exercised in the selection of chemical composition or mechanical properties to ensure compatibility with the welding process and its effect on altering the properties.

6.2 *Cast or Heat (Formerly Ladle) Analysis:*

6.2.1 An analysis of each cast or heat of steel shall be made by the manufacturer to determine the percentage of elements specified or restricted by the applicable specification.

6.2.2 When requested, cast or heat analysis for elements listed or required shall be reported to the purchaser or to his representative.

6.3 *Product, Check, or Verification Analysis (Formerly Check Analysis):*

6.3.1 Non-killed steels (such as capped or rimmed) are not technologically suited to product analysis due to the nonuniform character of their chemical composition and therefore, the tolerances in Table 1 do not apply. Product analysis is appropriate on these types of steel only when misapplication is apparent or for copper when copper steel is specified.

6.3.2 For steels other than non-killed (capped or rimmed), product analysis may be made by the purchaser. The chemical analysis shall not vary from the limits specified by more than the amounts in Table 1. The several determinations of any element in a cast shall not vary both above and below the specified range.

6.4 *Sampling for Product Analysis:*

6.4.1 To indicate adequately the representative compositions of a cast by product analysis, it is general practice to select samples to represent the steel, as fairly as possible, from a minimum number of pieces as follows: 3 pieces for lots up to 15.0 Mg incl, and 6 pieces for lots over 15.0 Mg.

6.4.2 When the steel is subject to tension test requirements, samples for product analysis may be taken either by drilling entirely through the used tension test specimens themselves or in accordance with 6.4.3.

6.4.3 When the steel is not subject to tension test requirements, the samples for analysis must be taken by milling or drilling entirely through the sheet or strip in a sufficient number of places so that the samples are representative of the entire strip. The sampling may be facilitated by folding the strip both ways, so that several samples may be taken at one drilling. Steel subjected to certain heating operations by the purchaser may not give chemical analysis results that properly represent its original composition. Therefore, users must analyze chips taken from the steel in the condition in which it is received from the steel manufacturer.

6.5 *Specimen Preparation*—Drillings or chips must be taken without the application of water, oil, or other lubricant, and must be free from scale, grease, dirt, or other foreign substances. They must not be overheated during cutting to the extent of causing decarburization. Chips must be well mixed and those too coarse to pass a No. 10 (2.00-mm) sieve or too fine to remain on a No. 30 (600-μm) sieve are not suitable for proper analysis. Sieve size numbers are in accordance with Specification E 11.

6.6 *Test Methods*—In case a referee analysis is required and agreed upon to resolve a dispute concerning the results of a chemical analysis, the procedure for performing the referee analysis must be in accordance with the latest issue of Methods E 350, unless otherwise agreed upon between the manufacturer and the purchaser.

7. Mechanical Requirements

7.1 The mechanical property requirements, number of specimens, and test locations and specimen orientation shall be in accordance with the applicable product specification.

7.2 Unless otherwise specified in the applicable product specification, test specimens must be prepared in accordance with Methods and Definitions A 370.

7.3 Mechanical tests shall be conducted in accordance with Methods and Definitions A 370.

8. Dimensions, Tolerances, and Allowances

8.1 Dimensions, tolerances, and allowances applicable to products covered by this specification are contained in Tables 3 through 9. The appropriate tolerance tables shall be identified in each individual specification.

9. Finish and Condition

9.1 Hot-rolled strip has a surface with an oxide or scale resulting from the hot-rolling operation. The oxide or scale can be removed by pickling or blast cleaning when required for press-work operations or welding. Hot-rolled and hot-rolled descaled strip are not generally used for exposed parts where surface is of prime importance.

9.1.1 Hot-rolled strip can be supplied with mill edges, square edges, or cut (slit) edges as specified.

9.1.1.1 Mill edges are the natural edges resulting from the hot-rolling operation and are generally round and smooth without any definite contour.

9.1.1.2 Square edges are the edges resulting from rolling through vertical edging rolls during the hot-rolling operations. These edges are square and smooth, with the corners slightly rounded.

9.1.1.3 Cut (slit) edges are the normal edges that result from the shearing, slitting, or trimming of mill edges.

9.1.2 The ends of plain hot-rolled milledge coils are irregular in shape and are referred to as uncropped ends. Where such ends are not acceptable, the purchaser's order should so specify. Processed coils such as pickled or blast cleaned are supplied with square-cut ends.

9.2 *Oiling:*

9.2.1 Plain hot-rolled strip is customarily furnished not oiled. Oiling must be specified when required.

9.2.2 Hot-rolled pickled or descaled strip is customarily furnished oiled. If the product is not to be oiled, it must be so specified.

10. Workmanship

10.1 Cut lengths shall have a workmanlike appearance and shall not have imperfections of a nature or degree for the product, the grade, and the quality ordered that will be detrimental to the fabrication of the finished part.

10.2 Coils may contain some abnormal imperfections that render a portion of the coil unusable since the inspection of coils does not afford the producer the same opportunity to remove portions containing imperfections as in the case with cut lengths.

11. Retests

11.1 If any test specimen shows defective machining or develops flaws, it must be discarded and another specimen substituted.

11.2 If the percentage of elongation of any test specimen is less than that specified and any part of the fracture is more than 20 mm from the center of the gage length of a 50-mm specimen or is outside the middle half of the gage length of a 200-mm specimen, as indicated by scribe

A 749M

scratches marked on the specimen before testing, a retest is allowed.

11.3 If a bend specimen fails, due to conditions of bending more severe than required by the specification, a retest is permitted either on a duplicate specimen or on a remaining portion of the failed specimen.

12. Inspection

12.1 When the purchaser's order stipulates that inspection and tests (except product analyses) for acceptance on the steel be made prior to shipment from the mill, the manufacturer shall afford the purchaser's inspector all reasonable facilities to satisfy him that the steel is being produced and furnished in accordance with the specification. Mill inspection by the purchaser shall not interfere unnecessarily with the manufacturer's operation.

13. Rejection and Rehearing

13.1 Unless otherwise specified, any rejection shall be reported to the manufacturer within a reasonable time after receipt of material by the purchaser.

13.2 Material that is reported to be defective subsequent to the acceptance at the purchaser's works shall be set aside, adequately protected, and correctly identified. The manufacturer shall be notified as soon as possible so that an investigation may be initiated.

13.3 Samples that are representative of the rejected material shall be made available to the manufacturer. In the event that the manufacturer is dissatisfied with the rejection, he may request a rehearing.

14. Marking

14.1 As a minimum requirement, the material shall be identified by having the manufacturer's name, ASTM designation, weight, purchaser's order number, and material identification legibly stenciled on top of each lift or shown on a tag attached to each coil or shipping unit.

14.2 When specified in the contract or order, and for direct procurement by or direct shipment to the government, marking for shipment, in addition to requirements specified in the contract or order, shall be in accordance with MIL-STD-129 for military agencies and in accordance with Fed. Std. No. 123 for civil agencies.

14.3 For Government procurement by the Defense Supply Agency, strip material shall be continuously marked for identification in accordance with Fed. Std. No. 183.

15. Packaging

15.1 Unless otherwise specified, the strip shall be packaged and loaded in accordance with Practices A 700.

15.2 When specified in the contract or order, and for direct procurement by or direct shipment to the government, when Level A is specified, preservation, packaging, and packing shall be in accordance with the Level A requirements of MIL-STD-163.

15.3 When coils are ordered it is customary to specify a minimum or range of inside diameter, maximum outside diameter, and a maximum coil weight, if required. The ability of manufacturers to meet the maximum coil weights depends upon individual mill equipment. When required, minimum coil weights are subject to negotiation.

TABLE 1 Tolerances for Product Analysis[A]

Element	Limit, or Maximum of Specified Element, %	Tolerances Under Minimum Limit	Tolerances Over Maximum Limit
Carbon	to 0.15, incl	0.02	0.03
	over 0.15 to 0.40, incl	0.03	0.04
	over 0.40 to 0.80, incl	0.03	0.05
	over 0.80	0.03	0.06
Manganese	to 0.60, incl	0.03	0.03
	over 0.60 to 1.15, incl	0.04	0.04
	over 1.15 to 1.65, incl	0.05	0.05
Phosphorus		. . .	0.01
Sulfur		. . .	0.01
Silicon	to 0.30, incl	0.02	0.03
	over 0.30 to 0.60, incl	0.05	0.05
Copper		0.02	. . .

[A] See 6.3.1.

TABLE 2 List of Tables for Dimensions, Tolerances, and Allowances

Dimensions	Table No.
Camber tolerances	8
Crown tolerances	5
Flatness tolerances	9
Length tolerances	7
Thickness tolerances	3, 4
Width tolerances	6

TABLE 3 Thickness Tolerances of Hot-Rolled Strip[A] (Carbon and High-Strength Low-Alloy Steel)[B] Ordered to Nominal Thickness

(Coils and Cut Lengths, Including Pickled)

Specified Width, mm		Thickness Tolerance, Over and Under, mm, for Specified Nominal Thickness, mm					
Over	Through	Through 1.5	Over 1.5 to 3.0, incl	Over 3.0 to 4.5, incl	Over 4.5 to 5.0, incl	Over 5.0 to 6.0, incl	
. . .	100	0.08	0.10	0.13	0.15	. . .	
100	200	0.08	0.13	0.13	0.15	. . .	
200	300	0.10	0.13	0.13	0.15	0.15	

[A] Measurements for the above table are taken 10 mm from the edge of a strip on 25 mm or wider; and at any place on the strip when narrower than 25 mm. The given tolerances do not include crown and therefore the tolerances for crown as shown in Table 5 are in addition to tolerances in Table 3.

[B] 1.80 minimum thickness for high-strength low alloy.

TABLE 4 Thickness Tolerances of Hot-Rolled Strip[A] (Carbon and High-Strength Low-Alloy Steel)[B] Ordered to Minimum Thickness

(Coils and Cut Lengths, Including Pickling)

Specified Width, mm		Thickness Tolerance, Over Only, for Specified Minimum Thickness, mm				
Over	Through	Through 1.5	Over 1.5 to 3.0, incl	Over 3.0 to 4.5, incl	Over 4.5 to 5.0, incl	Over 5.0 to 6.0, incl
. . .	100	0.15	0.20	0.25	0.30	. . .
100	200	0.15	0.25	0.25	0.30	. . .
200	300	0.20	0.25	0.25	0.30	0.30

[A] Measurements for the above table are taken 10 mm from the edge of a strip on 25 mm or wider; and at any place on the strip when narrower than 25 mm. The given tolerances do not include crown and therefore the tolerances for crown as shown in Table 5 are in addition to tolerances in Table 4.

[B] 1.80 minimum thickness for high-strength low alloy.

TABLE 5 Crown Tolerances of Hot-Rolled Strip (Carbon and High-Strength Low-Alloy Steel)
(Coils and Cut Lengths, Including Pickled)

Specified Width, mm		Crown Tolerance, Over Only, for Specified Minimum Thickness, mm			
Over	Through	Through 3.0	Over 3.0 to 4.5, incl	Over 4.5 to 6.0, incl	Over 6.0 to 9.5, incl
...	100	0.05	0.05	0.03	...
100	200	0.10	0.08	0.05	...
200	300	0.10	0.08	0.08	0.05

TABLE 6 Width Tolerances of Hot-Rolled Strip (Carbon and High-Strength Low-Alloy Steel)
(Coils and Cut Lengths, Including Pickled)

Specified Width, mm		Width Tolerance, Over and Under, mm		
Over	Through	Mill Edge and Square Edge Strip	Cut Edge	
			Through 2.5 mm	Over 2.5 mm
...	50	0.8	0.2	0.4
50	100	1.2	0.2	0.4
100	200	1.6	0.3	0.4
200	300	2.4	0.4	0.4

TABLE 7 Length Tolerances of Hot-Rolled Strip (Carbon and High-Strength Alloy Steel)
(Cut Lengths, Including Pickled)

Specified Widths, mm		Length Tolerances Over Specified Length, for Widths Given, No Tolerance Under, mm						
Over	Through	Through 1500	Over 1500 to 3000, incl	Over 3000 to 6000, incl	Over 6000 to 9000, incl	Over 9000 to 12 000, incl	Over 12 000	
...	100	10	10	15	20	25	40	
100	200	10	15	15	20	25	40	
200	300	15	20	25	30	40	45	

TABLE 8 Camber Tolerances[A] of Hot-Rolled Strip (Carbon and High-Strength Low-Alloy)
(Coils and Cut Lengths, Including Pickled, Applicable to Mill Edge, Square Edge, and Slit or Cut Edge)

NOTE—Camber is the deviation of a side edge from a straight line. The standard for measuring this deviation is based on any 2000-mm length.[B] It is obtained by placing a 2000-mm straightedge on the concave side and measuring the maximum distance between the strip edge and the straightedge.

For strip wider than 50 mm—5.0 mm in any 2000 mm
For strip 50 mm and narrower—10.0 mm in any 2000 mm

[A] When the camber tolerances shown in the above table are not suitable for a particular purpose, hot-rolled strip is sometimes machine straightened.

[B] For strip less than 2000 mm tolerances are to be established in each instance.

TABLE 9 Flatness Tolerances of Hot-Rolled Strip (Carbon and High-Strength Low-Alloy)

It has not been practicable to formulate flatness tolerances for hot-rolled strip because of the wide range of widths and thicknesses, and variety of chemical compositions and qualities, produced in coils and cut lengths.

A 749M

APPENDIXES

X1. PREFERRED METRIC SIZES

X1.1 Table X1.1 is based on ANSI B32.3, Preferred Metric Sizes for Flat Metal Products.[6]

[6] Available from American National Standards Institute, Inc., 1430 Broadway, New York, N.Y. 10018.

TABLE X1.1 Preferred Metric Sizes for Flat Metal Products, mm

Recommended Thicknesses — Steel Strip	Recommended Widths — Steel Strip	Recommended Lengths — Steel Strip
0.60	10	2000
0.65	12	2500
0.7	16	3000
0.75	20	3500
0.8	25	4000
0.85	30	4500
0.9	35	5000
0.95	40	6000
1.0	45	8000
1.05	50	10 000
1.1	55	12 000
1.2	60	14 000
1.4	70	16 000
1.6	80	18 000
1.8	90	
2.0	100	
2.2	110	
2.4	120	
2.5	130	
2.6	140	
2.8	150	
3.0	160	
3.2	180	
3.4	200	
3.5	225	
3.6	250	
3.8	300	
4.0		
4.2		
4.5		
4.8		
5		
5.5		
6		
6.5		
7		
7.5		
8		
9		
10		
11		
12		

A 749M

X2. STANDARD CHEMICAL RANGES AND LIMITS

X2.1 Standard chemical ranges and limits are prescribed for carbon steels in Table X2.1.

TABLE X2.1 Standard Chemical Ranges and Limits

NOTE—The carbon ranges shown in the column headed "Range" apply when the specified maximum limit for manganese does not exceed 1.00 %. When the maximum manganese limit exceeds 1.00 %, add 0.01 to the carbon ranges shown.

Element	Minimum of Specified Element, %	Carbon Steels Only, Cast or Heat Analysis	
		Range	Lowest, max
Carbon (see Note)	to 0.15, incl	0.05	0.08
	over 0.15 to 0.30, incl	0.06	
	over 0.30 to 0.40, incl	0.07	
	over 0.40 to 0.60, incl	0.08	
	over 0.60 to 0.80, incl	0.11	
	over 0.80 to 1.35, incl	0.14	
Manganese	to 0.50, incl	0.20	0.40
	over 0.50 to 1.15, incl	0.30	
	over 1.15 to 1.65, incl	0.35	
Phosphorus	to 0.08, incl	0.03	0.04[A]
	over 0.08 to 0.15, incl	0.05	
Sulfur	to 0.08, incl	0.03	0.05[A]
	over 0.08 to 0.15, incl	0.05	
	over 0.15 to 0.23, incl	0.07	
	over 0.23 to 0.33, incl	0.10	
Silicon	to 0.15, incl	0.08	0.10
	over 0.15 to 0.30, incl	0.15	
	over 0.30 to 0.60, incl	0.30	
Copper	When copper is required 0.20 min is commonly specified.		

[A] Certain individual specifications provide for lower standard limits for phosphorus and sulfur.

The American Society for Testing and Materials takes no position respecting the validity of any patent rights asserted in connection with any item mentioned in this standard. Users of this standard are expressly advised that determination of the validity of any such patent rights, and the risk of infringement of such rights, are entirely their own responsibility.

This standard is subject to revision at any time by the responsible technical committee and must be reviewed every five years and if not revised, either reapproved or withdrawn. Your comments are invited either for revision of this standard or for additional standards and should be addressed to ASTM Headquarters. Your comments will receive careful consideration at a meeting of the responsible technical committee, which you may attend. If you feel that your comments have not received a fair hearing you should make your views known to the ASTM Committee on Standards, 1916 Race St., Philadelphia, Pa. 19103.

Designation: A 751 – 82

Standard Methods, Practices, and Definitions for
CHEMICAL ANALYSIS OF STEEL PRODUCTS[1]

This standard is issued under the fixed designation A 751; the number immediately following the designation indicates the year of original adoption or, in the case of revision, the year of last revision. A number in parentheses indicates the year of last reapproval. A superscript epsilon (ϵ) indicates an editorial change since the last revision or reapproval.

INTRODUCTION

This standard was prepared to answer the need for a single document that would include all aspects of obtaining and reporting the chemical analysis of steel, stainless steel, and related alloys. Such subjects as definitions of terms and product (check analysis variations (tolerances) required clarification. Requirements for sampling, meeting specified limits, and treatment of data usually were not clearly established in product specifications.

It is intended that this standard will contain all requirements for the determination of chemical composition of steel, stainless steel, or related alloys so that product specifications will need contain only special modifications and exceptions.

1. Scope

1.1 This standard covers definitions, reference methods, and practices relating to the chemical analysis of steel, stainless steel, and related alloys. It includes both wet chemical and instrumental techniques.

1.2 Directions are provided for handling chemical requirements, product analyses, residual elements, and reference standards, and for the treatment and reporting of chemical analysis data.

1.3 This standard applies only to those product standards which include this standard or parts thereof as a requirement.

1.4 In cases of conflict, the product specification requirements shall take precedence over the requirements of this standard.

2. Applicable Documents

2.1 *ASTM Standards*:

E 29 Recommended Practice for Indicating Which Places of Figures Are to Be Considered Significant in Specified Limiting Values[2]

E 30 Chemical Analysis of Steel, Cast Iron, Open-Hearth Iron, and Wrought Iron[3]

E 38 Chemical Analysis of Nickel-Chromium and Nickel-Chromium–Iron Alloys[3]

E 59 Sampling Steel and Iron for Determination of Chemical Composition[3]

E 212 Spectrographic Analysis of Carbon and Low-Alloy Steel by the Rod-to-Rod Technique[4]

E 281 Spectrographic Analysis of Carbon and Low-Alloy Steel by the Pellet Technique[4]

E 282 Spectrographic Analysis of Carbon and Low-Alloy Steel by the Point-to-Plane Technique[4]

E 311 Recommended Practice for Sampling and Sample Preparation Techniques in Spectrochemical Analysis[4]

E 322 X-Ray Emission Spectrometric Analysis of Low-Alloy Steels and Cast Irons[4]

E 327 Optical Emission Spectrometric Analysis of Stainless Type 18-8 Steels by the Point-to-Plane Technique[4]

---

[1] These methods, practices, and definitions are under the jurisdiction of ASTM Committee A-1 on Steel, Stainless Steel and Related Alloys, and are the direct responsibility of Subcommittee A01.13 on Methods of Mechanical Testing.
Current edition approved Oct. 29, 1982. Published December 1982. Originally published as A 751 – 77. Last previous edition A 751 – 77.
[2] *Annual Book of ASTM Standards*, Vols 03.05 and 14.02.
[3] *Annual Book of ASTM Standards*, Vol 03.05.
[4] *Annual Book of ASTM Standards*, Vol 03.06.

E 350 Chemical Analysis of Carbon Steel, Low-Alloy Steel, Silicon Electrical Steel, Ingot Iron, and Wrought Iron[3]

E 352 Chemical Analysis of Tool Steels and Other Similar Medium- and High-Alloy Steels[3]

E 353 Chemical Analysis of Stainless, Heat-Resisting, Maraging, and Other Similar Chromium-Nickel-Iron Alloys[3]

E 354 Chemical Analysis of High-Temperature, Electrical, Magnetic, and Other Similar Iron, Nickel, and Cobalt Alloys[3]

E 403 Optical Emission Spectrometric Analysis of Carbon and Low-Alloy Steel by the Point-to-Plane Technique[4]

E 404 Spectrographic Determination of Boron in Carbon and Low-Alloy Steel by the Point-to-Plane Technique[4]

E 415 Optical Emission Vacuum Spectrometric Analysis of Carbon and Low-Alloy Steel[4]

E 572 X-Ray Emission Spectrometric Analysis of Stainless Steel[4]

3. Definitions

3.1 *Pertaining to Analyses:*

3.1.1 *cast or heat (formerly ladle) analysis*—applies to chemical analyses representative of a heat of steel as reported to the purchaser and determined by analyzing a test sample, preferably obtained during the pouring of the steel, for the elements designated in a specification.

3.1.2 *product, check or verification analysis*—a chemical analysis of the semifinished or finished product, usually for the purpose of determining conformance to the specification requirements. The range of the specified composition applicable to product analysis is normally greater than that applicable to heat analysis in order to take into account deviations associated with analytical reproducibility (Note 1) and the heterogeneity of the steel.

NOTE 1—The chemical analysis procedures in Methods E 350, E 352, E 353, and E 354 include precision statements with regard to reproducibility data.

3.1.3 *product analysis tolerances* (Note 2)—a permissible variation over the maximum limit or under the minimum limit of a specified element and applicable only to product analyses, not cast or heat analyses.

NOTE 2—The term "analysis tolerance" is often misunderstood. It does not apply to cast or heat analyses determined to show conformance to specified chemical limits. It applies only to product analysis and becomes meaningful only when the heat analysis of an element falls close to one of the specified limits. For example, stainless steel UNS 30400 limits for chromium are 18.00 to 20.00 %. A heat that the producer reported as 18.01 % chromium may be found to show 17.80 % chromium by a user performing a product analysis. If the product analysis tolerance for such a chromium level is 0.20 %, the product analysis of 17.80 % chromium would be acceptable. A product analysis of 17.79 % would not be acceptable.

3.1.4 *proprietary analytical method*—a nonstandard analytical method, not published by ASTM, utilizing reference standards traceable to the National Bureau of Standards (NBS) (when available) or other sources referenced in Section 9.

3.1.5 *referee analysis*—performed using ASTM methods listed in 8.1.1 and NBS reference standards or methods and reference standards agreed upon between parties. The selection of a laboratory to perform the referee analysis shall be a matter of agreement between the supplier and the purchaser.

3.1.6 *standard reference material*—a specimen of material specially prepared, analyzed, and certified for chemical content under the jurisdiction of a recognized standardizing agency or group, such as the National Bureau of Standards, for use by analytical laboratories as an accurate basis for comparison. Reference samples should bear sufficient resemblance to the material to be analyzed so that no significant differences are required in procedures or corrections (for example, for interferences or inter-element effects).

3.1.7 *working reference materials*—reference materials used for routine analytical control and traceable to NBS standards and other recognized standards when appropriate standards are available.

3.2 *Pertaining to Analyzed Elements:*

3.2.1 *intentionally added unspecified element*—an element added in controlled amounts at the option of the producer to obtain desirable characteristics.

3.2.2 *residual element*—a specified or unspecified element, not intentionally added, originating in raw materials, refractories, or air.

3.2.3 *specified element*—an element controlled to a specified range, maximum or minimum, in accordance with the requirements of the product specification.

3.2.4 *trace element*—a residual element that may occur in very low concentrations, generally less than 0.01 %.

4. Cast or Heat Analysis

4.1 The producer shall perform analyses for those elements specified in the material specification. The results of such analyses shall conform to the requirements specified in the material specification.

4.1.1 For multiple heats, either individual heat or cast analysis or an average heat or cast analysis shall be reported. If significant variations in heat or cast size are involved, a weighted average heat or cast analysis, based on the relative quantity of metal in each heat or cast, shall be reported.

4.1.2 For consumable electrode remelted material, a heat is defined as all the ingots remelted from a primary heat. The heat analysis shall be obtained from one remelted ingot or the product of one remelted ingot of each primary melt. If the initial heat analysis does not meet the heat (ladle) analysis requirements of the specification, one sample from the product of each remelted ingot shall be analyzed and the analysis shall meet the heat (ladle) analysis requirements.

4.2 If the test samples taken for the heat analysis are lost, inadequate, or not representative of the heat, a product analysis of the semifinished or finished product may be used to establish the heat analysis.

4.2.1 If a product analysis is made to establish the heat analysis, the product analysis shall meet the specified limits for heat analysis and the product analysis tolerances described in Section 5 do not apply.

5. Product Analysis Requirements

5.1 For product analysis, the range of the specified chemical composition is normally greater (designated product analysis tolerances) than that applicable to heat analyses to take into account deviations associated with analytical reproducibility and the heterogeneity of the steel. If several determinations of any element in the heat are made, they may not vary both above and below the specified range.

5.2 Product analysis tolerances may not be used to determine conformance to the specified heat or cast analysis unless permitted by the individual material specification.

5.3 Product analysis tolerances, where available, are given in the individual material specifications or in the general requirement specifications.

6. Unspecified, Residual, and Trace Elements (Note 3)

6.1 Reporting analyses of unspecified, residual, and trace elements is permitted.

NOTE 3—All commercial metals contain small amounts of various elements in addition to those which are specified. It is neither practical nor necessary to specify limits for every residual element that might be present, despite the fact that the presence of many of these elements is often routinely determined by the producer.

6.2 Unspecified, residual, and trace elements are acceptable within concentrations that do not significantly and deleteriously affect mechanical properties, physical properties, metallurgical characteristics, or corrosion resistance.

6.3 Analysis limits shall be established for specific elements rather than groups of elements such as "all others," "rare earths," and "balance."

7. Sampling

7.1 *Cast or Heat Analyses:*

7.1.1 Samples shall be taken, insofar as possible, during the casting of a heat, at a time which, in the producer's judgment, best represents the composition of the cast.

7.1.2 In case the heat analysis samples or analyses are lost or inadequate, or when it is evident that the sample does not truly represent the heat, representative samples may be taken from the semifinished or finished product, in which case such samples may be analyzed to satisfy the specified requirements. The analysis shall meet the specified limits for heat analysis.

7.2 *Check, Product, or Verification Analyses:*

7.2.1 Unless otherwise specified, the latest revision of Method E 59 shall be used as a guide for sampling.

7.3 When employing spectrochemical analytical techniques, the latest revision of Method E 311 shall be used as a guide for sampling.

8. Test Methods

8.1 This section lists some methods that have been found acceptable for chemical analysis of

A 751

steels.

8.1.1 The following ASTM wet chemical methods have been found acceptable as referee test methods and as a base for standardizing instrumental analysis techniques:

Methods	General Description
E 30	— antecedent to Methods E 350 through E 354
E 38	— antecedent to Methods E 350 through E 354
E 350	— the basic wet chemical procedure for steels
E 352	— wet chemical procedure for tool steels
E 353	— wet chemical procedure for stainless steels
E 354	— wet chemical procedure for high nickel steels

8.1.2 The following ASTM instrumental methods may be employed for chemical analysis of steels or may be useful as a guide in the calibration and standardization of instrumental equipment for routine analysis of steels:

Methods	General Description
E 212	— spectrographic analysis of steels (rod-to-rod technique)
E 281	— spectrographic analysis of steels (pellet technique)
E 282	— spectrographic analysis of steels (photographic process)
E 322	— X-ray fluorescence for steels
E 327	— spectrometric analysis of stainless steels
E 403	— spectrometric analysis of steels
E 404	— spectrographic determination of steels for boron (point-to-plane technique)
E 415	— vacuum spectrometric analysis of steels
E 572	— X-ray emission spectrometric analysis of stainless steels

8.2 The following are some of the commonly accepted techniques employed for routine chemical analysis of steels. Proprietary methods are permissible provided the results are equivalent to those obtained from standard methods when applicable.

8.2.1 Analysis of stainless steels using an X-ray fluroescence spectrometer.

8.2.2 Analysis of stainless steels using a vacuum emission spectrometer.

8.2.3 Analysis of solutions using an atomic absorption spectrophotometer.

8.2.4 Determination of carbon or sulfur, or both, by combustion (in oxygen) and measurement of CO_2 by thermal conductivity or infrared detectors.

8.3 There are additional common techniques often used for chemical analysis of standards for instrument analysis such as: polarographic analysis, ion exchange separations, radioactivation, and mass spectrometry.

9. Reference Materials

9.1 For referee analyses, reference standards of a recognized standardizing agency shall be employed with preference given to NBS standard reference materials when applicable. (NBS does not produce reference standards suitable for all elements or all alloys.[5])

9.1.1 When standard reference materials for certain alloys are not available from NBS, reference materials may be produced by employing ASTM standard procedures and NBS standard reference materials to the extent that such procedures and reference standards are available.

9.2 Working reference materials may be used for routine analytical control.

10. Significant Numbers

10.1 Laboratories shall report each element to the same number of significant numbers as used in the pertinent material specifications.

10.2 When a chemical determination yields a greater number of significant numbers than is specified for an element, the result shall be rounded in accordance with Section 11.

11. Rounding Procedure

11.1 To determine conformance with the specification requirements, an observed value or calculated value shall be rounded in accordance with Recommended Practice E 29 to the nearest unit in the last right-hand place of values listed in the table of chemical requirements.

11.2 In the special case of rounding the number "5" when no additional numbers other than "0" follow the "5", rounding shall be done in the direction of the specification analysis limits if following Recommended Practice E 29 would cause rejection of material.

12. Report

12.1 The chemical analysis of a heat shall be reported only when specified by the user or the material specification.

12.2 Certification or identification to an ASTM material standard requires that the specific heat of steel meets the chemical requirements of the ASTM standard.

12.3 Any report that is legally binding on

---

[5] Some sources of reference materials are listed in ASTM Data Series Publication No. DS2, issued 1963.

the manufacturer is acceptable as a certified test report.

12.3.1 Notarization of Certificates of Test is neither required nor prohibited.

The American Society for Testing and Materials takes no position respecting the validity of any patent rights asserted in connection with any item mentioned in this standard. Users of this standard are expressly advised that determination of the validity of any such patent rights, and the risk of infringement of such rights, are entirely their own responsibility.

This standard is subject to revision at any time by the responsible technical committee and must be reviewed every five years and if not revised, either reapproved or withdrawn. Your comments are invited either for revision of this standard or for additional standards and should be addressed to ASTM Headquarters. Your comments will receive careful consideration at a meeting of the responsible technical committee, which you may attend. If you feel that your comments have not received a fair hearing you should make your views known to the ASTM Committee on Standards, 1916 Race St., Philadelphia, Pa. 19103.

Designation: A 752 – 83

Standard Specification for
GENERAL REQUIREMENTS FOR WIRE RODS AND COARSE ROUND WIRE, ALLOY STEEL[1]

This standard is issued under the fixed designation A 752; the number immediately following the designation indicates the year of original adoption or, in the case of revision, the year of last revision. A number in parentheses indicates the year of last reapproval. A superscript epsilon (ϵ) indicates an editorial change since the last revision or reapproval.

1. Scope

1.1 This specification covers general requirements for alloy steel rods and uncoated coarse round alloy wire in coils that are not required to meet hardenability band limits.

1.2 In case of conflict, the requirements in the purchase order, on the drawing, in the individual specification, and in this general specification shall prevail in the sequence named.

1.3 The values stated in inch-pound units are to be regarded as the standard.

2. Applicable Documents

2.1 *ASTM Standards:*
A 370 Methods and Definitions for Mechanical Testing of Steel Products[2]
A 700 Practices for Packaging, Marking, and Loading Methods for Steel Products for Domestic Shipment[2]
E 29 Recommended Practice for Indicating Which Places of Figures Are to Be Considered Significant in Specified Limiting Values[3]
E 30 Methods for Chemical Analysis of Steel, Cast Iron, Open-Hearth Iron, and Wrought Iron[4]
E 44 Definitions of Terms Relating to Heat Treatment of Metals[5]
E 112 Methods for Determining Average Grain Size[5]

3. Description of Terms

3.1 *Alloy Steel*—Steel is considered to be alloy steel when the maximum of the range given for the content of alloying elements exceeds one or more of the following limits: manganese 1.65 %, silicon 0.60 %, copper 0.60 %; or in which a definite range or a definite minimum quantity of any of the following elements is specified or required within the limits of the recognized field of constructional alloy steels: aluminum, chromium up to 3.99 %, cobalt, columbium, molybdenum, nickel, titanium, tungsten, vanadium, zirconium, or any other alloying elements added to obtain a desired alloying effect. Note that aluminum, columbium, and vanadium may also be used for grain refinement purposes.

3.1.1 Boron treatment of alloy steels, which are fine grain, may be specified to improve hardenability.

3.1.2 Other elements, such as lead, selenium, tellurium, or bismuth, may be specified to improve machinability.

3.2 Wire rods are hot rolled from billets into an approximate round cross section and into coils of one continuous length. Rods are not comparable to hot-rolled bars in accuracy of cross section or surface finish and as a semifinished product are primarily for the manufacture of wire.

3.2.1 Rod sizes from $7/32$ to $47/64$ in. (5.6 to 18.7 mm) in diameter, inclusive, are designated by fractions or decimal parts of an inch as shown in Table 1.

---

[1] This specification is under the jurisdiction of ASTM Committee A-1 on Steel, Stainless Steel and Related Alloys, and is the direct responsibility of Subcommittee A01.03 on Steel Rod and Wire.
Current edition approved July 29, 1983. Published November 1983. Originally published as A 752 – 77. Last previous edition A 752 – 77.
[2] *Annual Book of ASTM Standards*, Vol 01.03.
[3] *Annual Book of ASTM Standards*, Vol 14.02.
[4] *Annual Book of ASTM Standards*, Vol 03.05.
[5] *Annual Book of ASTM Standards*, Vol 03.03.

A 752

3.3 Coarse round wire from 0.035 to 0.999 in. (0.89 to 25.4 mm) in diameter, inclusive, is produced from hot-rolled wire rods or hot-rolled coiled bars by one or more cold reductions primarily for the purpose of obtaining a desired size with dimensional accuracy, surface finish, and mechanical properties. By varying the amount of cold reduction and other wire mill practices, including thermal treatment, a wide diversity of mechanical properties and finishes are made available.

3.3.1 Coarse round wire is designated by common fractions or decimal parts of an inch, or millimetres.

4. Ordering Information

4.1 Orders for hot-rolled wire rods under this specification shall include the following information:

4.1.1 Quantity (pounds or kilograms),
4.1.2 Name of material (wire rods),
4.1.3 Diameter (Table 1),
4.1.4 Chemical composition grade number (Table 4),
4.1.5 Thermal treatment, if required,
4.1.6 Packaging, and
4.1.7 ASTM designation and date of issue.

NOTE 1—A typical ordering description is as follows: 80 000 lb Hot-Rolled Alloy Steel Wire Rods, ¼ in., Grade 4135 in 2 000-lb maximum coils to ASTM A 752 dated ———.

4.2 Orders for coarse round wire under this specification shall include the following information:

4.2.1 Quantity (pounds or kilograms),
4.2.2 Name of material (alloy steel wire),
4.2.3 Diameter (see 3.3.1),
4.2.4 Chemical composition (Table 4 or Table 6),
4.2.5 Thermal treatment, if required,
4.2.6 Packaging,
4.2.7 ASTM designation A 752 and date of issue, and
4.2.8 Special requirements, if any.

NOTE 2—A typical ordering description is as follows: 40 000 lb, Alloy Steel Wire, 0.312 in. diameter, Grade 8620, annealed at finish size, in 500-lb Catch Weight Coils on Tubular Carriers to ASTM A 752 dated ———.

5. Manufacture

5.1 The product of the steel making processes is either cast into ingots that are hot rolled to blooms or billets, or strand cast directly into blooms or billets for subsequent processing into rods.

6. Chemical Requirements

6.1 The chemical composition for alloy steel under this specification shall conform to the requirements set forth in the purchase order. The grades commonly specified for alloy steel wire rods and alloy steel wire are shown in Table 4. For specified compositions not contained in Table 4 the ranges and limits expressed in Table 6 shall apply unless other such ranges and limits shall have been agreed upon between the purchaser and the manufacturer.

6.2 *Cast or Heat Analysis*—An analysis of each cast or heat shall be made by the producer to determine the percentage of the elements specified. The analysis shall be made from a test sample preferably taken during the pouring of the cast or heat. The chemical composition thus determined shall be reported, if required, to the purchaser or his representative.

6.3 *Product Analysis*—A product analysis may be made by the purchaser. The analysis is not used for a duplicate analysis to confirm a previous result. The purpose of the product analysis is to verify that the chemical composition is within specified limits for each element, including applicable permissible variations in product analysis. The results of analyses taken from different pieces of a heat may differ within permissible limits from each other and from the heat analysis. Table 5 shows the permissible variations for product analysis of alloy steel. The results of the product analysis, except lead, shall not vary both above and below the specified ranges.

6.3.1 The location from which chips for product analysis are obtained is important because of normal segregation. For rods and wire, chips must be taken by milling or machining the full cross section of the sample.

6.3.1.1 Steel subjected to certain thermal treatments by the purchaser may not give chemical analysis results that properly represent its original composition. Therefore, purchasers should analyze chips taken from the steel in the condition in which it is received from the producer.

6.3.1.2 When samples are returned to the producer for product analysis, the samples should consist of pieces of the full cross section.

6.3.2 For referee purposes, Methods E 30 shall be used.

7. Metallurgical Structure

7.1 Grain size when specified shall be determined in accordance with the requirements of Methods E 112.

7.2 Alloy steel wire rods may be specified as annealed, spheroidize annealed, or patented. Refer to Definitions E 44 for definitions.

7.3 Alloy steel wire may be specified as drawn from annealed, spheroidize annealed or patented wire rod or bars; also as annealed, spheroidize annealed, patented, or oil tempered at finish size. Refer to Definitions E 44 for definitions.

8. Mechanical Requirements

8.1 The properties enumerated in individual specifications shall be determined in accordance with Methods and Definitions A 370.

8.2 The maximum expected tensile strengths for the more common grades of alloy rods with regular mill annealing (non-spheroidized) are shown in Table 7.

8.3 Because of the great variety in the kinds and grades of rods and wire and the extensive diversity of application, a number of formal mechanical test procedures have been developed. These tests are used as control tests by producers during intermediate stages of wire processing, as well as for final testing of the finished product, and apply particularly to rods and wire for specific applications. A number of these tests are further described in Supplement IV, Round Wire Products, of Methods and Definitions A 370.

8.4 Since the general utility of rods and wire require continuity of length, in the case of rods, tests are commonly made on samples taken from the ends of coils after removing two to three rings; in the case of wire, tests are commonly taken from the ends of the coils, thereby not impairing the usefulness of the whole coil.

9. Dimensions and Permissible Variation

9.1 The diameter and out-of-roundness of the alloy wire rod shall not vary from that specified by more than that prescribed in Table 2.

9.2 The diameter and out-of-roundness of the alloy coarse round wire shall not vary from that specified by more than that prescribed in Table 3.

10. Workmanship, Finish, and Appearance

10.1 The alloy wire rod shall be free from detrimental surface imperfections, tangles, and sharp kinks.

10.1.1 Two or more rod coils may be welded together to produce a larger coil. The weld zone may not be as sound as the original material. The mechanical properties existing in the weld zone may differ from those in the unaffected base metal. The weld may exceed the permissible variations for diameter and out-of-roundness on the minus side of the permissible variation but not on the plus side.

10.2 The wire as received shall be smooth and substantially free of rust, shall not be kinked or improperly cast. No detrimental die marks or scratches may be present. Each coil shall be one continuous length of wire. Welds made during cold drawing are permitted.

11. Retests

11.1 The difficulties in obtaining truly representative samples of wire rod and coarse round wire without destroying the usefulness of the coil account for the generally accepted practice of allowing retests for mechanical tests and surface examination. An additional test piece is cut from each end of the coil from which the original sample was taken. A portion of the coil may be discarded prior to cutting the sample for retest. If any of the retests fail to comply with the requirements, the coil of rod or wire may be rejected. Before final rejection, however, it is frequently advisable to base the final decision on an actual trial of the material to determine whether or not it will perform the function for which it is intended.

12. Inspection

12.1 The manufacturer shall afford the purchaser's inspector all reasonable facilities necessary to satisfy him that the material is being produced and furnished in accordance with this specification. Mill inspection by the purchaser shall not interfere unnecessarily with the manufacturer's operations. All tests and inspections shall be made at the place of manufacture, unless otherwise agreed to.

13. Rejection

13.1 Any rejection based on tests made in accordance with this specification shall be reported to the manufacturer within a reasonable length of time. The material must be adequately

A 752

protected and correctly identified in order that the manufacturer may make a proper investigation.

14. Certification

14.1 Upon request of the purchaser in the contract or order, a manufacturer's certification that the material was manufactured and tested in accordance with this specification, together with a report of the test results, shall be furnished at the time of shipment.

15. Packaging, Marking, and Loading

15.1 A tag shall be securely attached to each coil and shall be marked with the size, heat number, grade number, ASTM specification number, and name or mark of the manufacturer.

15.2 When specified in the purchase order, packaging, marking, and loading for shipment shall be in accordance with those procedures recommended by Practices A 700.

TABLE 1 Sizes of Alloy Steel Wire Rods

Inch Fraction	Decimal Equivalent, in.	Metric Equivalent, mm	Inch Fraction	Decimal Equivalent, in.	Metric Equivalent, mm
7/32	0.219	5.6	31/64	0.484	12.3
15/64	0.234	6.0	1/2	0.500	12.7
1/4	0.250	6.4	33/64	0.516	13.1
17/64	0.266	6.7	17/32	0.531	13.5
9/32	0.281	7.1	35/64	0.547	13.9
19/64	0.297	7.5	9/16	0.562	14.3
5/16	0.312	7.9	37/64	0.578	14.7
21/64	0.328	8.3	19/32	0.594	15.1
11/32	0.344	8.7	39/64	0.609	15.5
23/64	0.359	9.1	5/8	0.625	15.9
3/8	0.375	9.5	41/64	0.641	16.3
25/64	0.391	9.9	21/32	0.656	16.7
13/32	0.406	10.3	43/64	0.672	17.1
27/64	0.422	10.7	11/16	0.688	17.5
7/16	0.438	11.1	45/64	0.703	17.9
29/64	0.453	11.5	23/32	0.719	18.3
15/32	0.469	11.9	47/64	0.734	18.7

TABLE 2 Permissible Variation in Diameter of Alloy Wire Rod in Coils

Note — For purposes of determining conformance with this specification, all specified limits in Table 2 are absolute limits as defined in Recommended Practice E 29.

Diameter of Rod		Permissible Variation, Plus and Minus, in. (mm)	Permissible Out-of-Round in. (mm)
Fractions, in.	Decimal, in. (mm)		
7/32 to 47/64, incl	0.219 to 0.734 (5.6 to 18.7), incl	0.016 (0.41)	0.025 (0.64)

TABLE 3 Permissible Variation in Diameter for Alloy Coarse Round Wire

Note — For purposes of determining conformance with this specification, all specified limits in Table 3 are absolute limits as defined in Recommended Practice E 29.

Diameter of Wire, in. (mm)	Permissible Variation, Plus and Minus, in. (mm)	Permissible Out-of-Round, in. (mm)
0.035 to 0.075 (0.89 to 1.90), incl	0.001 (0.03)	0.001 (0.03)
Over 0.075 to 0.148 (1.90 to 3.76), incl	0.0015 (0.04)	0.0015 (0.04)
Over 0.148 to 0.500 (3.76 to 12.70), incl	0.002 (0.05)	0.002 (0.05)
Over 0.500 to 0.625 (12.70 to 15.87), incl	0.0025 (0.06)	0.0025 (0.06)
Over 0.625 (15.87)	0.003 (0.08)	0.003 (0.08)

A 752

TABLE 4 Chemical Composition Ranges and Limits for Cast or Heat Analysis

NOTE 1—Grades shown in this table with prefix letter E are normally only made by the basic electric furnace process. All others are normally manufactured by the basic open hearth or basic oxygen processes but may be manufactured by a basic electric furnace process. If the electric furnace process is specified or required for grades other than those designated above, the limits for phosphorus and sulfur are respectively 0.025 % max.

NOTE 2—Small quantities of certain elements, which are not specified or required, are present in alloy steels. These elements are considered as incidental and may be present to the following maximum amounts: copper, 0.35 %, nickel, 0.25 %, chromium, 0.20 %, molybdenum, 0.06 %.

NOTE 3—Where minimum and maximum sulfur content is shown it is indicative of resulfurized steel.

NOTE 4—The chemical ranges and limits shown in Table 4 are produced to check, product, or verification analysis tolerances shown in Table 5.

NOTE 5—Standard alloy steels can be produced with a lead range of 0.15 to 0.35 %. Such steels are identified by inserting the letter "L" between the second and third numerals of the Grade number, for example, 41L40. Lead is reported only as a range of 0.15 to 0.35 % since it is added to the mold as the steel is poured.

UNS Designation	Grade No.	Chemical Composition, Ranges and Limits, %							
		Carbon	Manganese	Phosphorus, max	Sulfur, max	Silicon	Nickel	Chromium	Molybdenum
				STANDARD ALLOY STEELS					
G13300	1330	0.28 to 0.33	1.60 to 1.90	0.035	0.040	0.15 to 0.30	...	...	...
G13350	1335	0.33 to 0.38	1.60 to 1.90	0.035	0.040	0.15 to 0.30	...	...	...
G13400	1340	0.38 to 0.43	1.60 to 1.90	0.035	0.040	0.15 to 0.30	...	...	...
G13450	1345	0.43 to 0.48	1.60 to 1.90	0.035	0.040	0.15 to 0.30	...	...	...
G40120	4012	0.09 to 0.14	0.75 to 1.00	0.035	0.040	0.15 to 0.30	...	...	0.15 to 0.25
G40230	4023	0.20 to 0.25	0.70 to 0.90	0.035	0.040	0.15 to 0.30	...	...	0.20 to 0.30
G40240	4024	0.20 to 0.25	0.70 to 0.90	0.035	0.035 to 0.050	0.15 to 0.30	...	...	0.20 to 0.30
G40270	4027	0.25 to 0.30	0.70 to 0.90	0.035	0.040	0.15 to 0.30	...	...	0.20 to 0.30
G40280	4028	0.25 to 0.30	0.70 to 0.90	0.035	0.035 to 0.050	0.15 to 0.30	...	...	0.20 to 0.30
G40370	4037	0.35 to 0.40	0.70 to 0.90	0.035	0.040	0.15 to 0.30	...	...	0.20 to 0.30
G40470	4047	0.45 to 0.50	0.70 to 0.90	0.035	0.040	0.15 to 0.30	...	...	0.20 to 0.30
G41180	4118	0.18 to 0.23	0.70 to 0.90	0.035	0.040	0.15 to 0.30	...	0.40 to 0.60	0.08 to 0.15
G41300	4130	0.28 to 0.33	0.40 to 0.60	0.035	0.040	0.15 to 0.30	...	0.80 to 1.10	0.15 to 0.25
G41370	4137	0.35 to 0.40	0.70 to 0.90	0.035	0.040	0.15 to 0.30	...	0.80 to 1.10	0.15 to 0.25
G41400	4140	0.38 to 0.43	0.75 to 1.00	0.035	0.040	0.15 to 0.30	...	0.80 to 1.10	0.15 to 0.25
G41420	4142	0.40 to 0.45	0.75 to 1.00	0.035	0.040	0.15 to 0.30	...	0.80 to 1.10	0.15 to 0.25
G41450	4145	0.43 to 0.48	0.75 to 1.00	0.035	0.040	0.15 to 0.30	...	0.80 to 1.10	0.15 to 0.25
G41470	4147	0.45 to 0.50	0.75 to 1.00	0.035	0.040	0.15 to 0.30	...	0.80 to 1.10	0.15 to 0.25
G41500	4150	0.48 to 0.53	0.75 to 1.00	0.035	0.040	0.15 to 0.30	...	0.80 to 1.10	0.15 to 0.25
G41610	4161	0.56 to 0.64	0.75 to 1.00	0.035	0.040	0.15 to 0.30	...	0.70 to 0.90	0.25 to 0.35
G43200	4320	0.17 to 0.22	0.45 to 0.65	0.035	0.040	0.15 to 0.30	1.65 to 2.00	0.40 to 0.60	0.20 to 0.30
G43400	4340	0.38 to 0.43	0.60 to 0.80	0.035	0.040	0.15 to 0.30	1.65 to 2.00	0.70 to 0.90	0.20 to 0.30
G43406	E4340	0.38 to 0.43	0.65 to 0.85	0.025	0.025	0.15 to 0.30	1.65 to 2.00	0.70 to 0.90	0.20 to 0.30
G44190	4419	0.18 to 0.23	0.45 to 0.65	0.035	0.040	0.15 to 0.30	...	...	0.45 to 0.60

ASTM A 752

TABLE 4 *Continued*

UNS Designation	Grade No.	Chemical Composition, Ranges and Limits, %							
		Carbon	Manganese	Phosphorus, max	Sulfur, max	Silicon	Nickel	Chromium	Molybdenum
STANDARD ALLOY STEELS									
G46150	4615	0.13 to 0.18	0.45 to 0.65	0.035	0.040	0.15 to 0.30	1.65 to 2.00	...	0.20 to 0.30
G46200	4620	0.17 to 0.22	0.45 to 0.65	0.035	0.040	0.15 to 0.30	1.65 to 2.00	...	0.20 to 0.30
G46210	4621	0.18 to 0.23	0.70 to 0.90	0.035	0.040	0.15 to 0.30	1.65 to 2.00	...	0.20 to 0.30
G46260	4626	0.24 to 0.29	0.45 to 0.65	0.035	0.040	0.15 to 0.30	0.70 to 1.00	...	0.15 to 0.25
G47180	4718	0.16 to 0.21	0.70 to 0.90	0.035	0.040	0.15 to 0.30	0.90 to 1.20	0.35 to 0.55	0.30 to 0.40
G47200	4720	0.17 to 0.22	0.50 to 0.70	0.035	0.040	0.15 to 0.30	0.90 to 1.20	0.35 to 0.55	0.15 to 0.25
G48150	4815	0.13 to 0.18	0.40 to 0.60	0.035	0.040	0.15 to 0.30	3.25 to 3.75	...	0.20 to 0.30
G48170	4817	0.15 to 0.20	0.40 to 0.60	0.035	0.040	0.15 to 0.30	3.25 to 3.75	...	0.20 to 0.30
G48200	4820	0.18 to 0.23	0.50 to 0.70	0.035	0.040	0.15 to 0.30	3.25 to 3.75	...	0.20 to 0.30
G50150	5015	0.12 to 0.17	0.30 to 0.50	0.035	0.040	0.15 to 0.30	...	0.30 to 0.50	...
G51200	5120	0.17 to 0.22	0.70 to 0.90	0.035	0.040	0.15 to 0.30	...	0.70 to 0.90	...
G51300	5130	0.28 to 0.33	0.70 to 0.90	0.035	0.040	0.15 to 0.30	...	0.80 to 1.10	...
G51320	5132	0.30 to 0.35	0.60 to 0.80	0.035	0.040	0.15 to 0.30	...	0.75 to 1.00	...
G51350	5135	0.33 to 0.38	0.60 to 0.80	0.035	0.040	0.15 to 0.30	...	0.80 to 1.05	...
G51400	5140	0.38 to 0.43	0.70 to 0.90	0.035	0.040	0.15 to 0.30	...	0.70 to 0.90	...
G51450	5145	0.43 to 0.48	0.70 to 0.90	0.035	0.040	0.15 to 0.30	...	0.70 to 0.90	...
G51470	5147	0.46 to 0.51	0.70 to 0.95	0.035	0.040	0.15 to 0.30	...	0.85 to 1.15	...
G51500	5150	0.48 to 0.53	0.70 to 0.90	0.035	0.040	0.15 to 0.30	...	0.70 to 0.90	...
G51550	5155	0.51 to 0.59	0.70 to 0.90	0.035	0.040	0.15 to 0.30	...	0.70 to 0.90	...
G51600	5160	0.56 to 0.64	0.75 to 1.00	0.035	0.040	0.15 to 0.30	...	0.70 to 0.90	...
...	E51100	0.98 to 1.10	0.25 to 0.45	0.025	0.025	0.15 to 0.30	...	0.90 to 1.15	...
...	E52100	0.98 to 1.10	0.25 to 0.45	0.025	0.025	0.15 to 0.30	...	1.30 to 1.60	...

UNS Designation	Grade No.	Chemical Composition, Ranges and Limits, %							
		Carbon	Manganese	Phosphorus, max	Sulfur, max	Silicon	Nickel	Chromium	Other Elements
									Vanadium
G61180	6118	0.16 to 0.21	0.50 to 0.70	0.035	0.040	0.15 to 0.30	...	0.50 to 0.70	0.10 to 0.15
G61500	6150	0.48 to 0.53	0.70 to 0.90	0.035	0.040	0.15 to 0.30	...	0.80 to 1.10	0.15 min
									Molybdenum
G86150	8615	0.13 to 0.18	0.70 to 0.90	0.035	0.040	0.15 to 0.30	0.40 to 0.70	0.40 to 0.60	0.15 to 0.25
G86170	8617	0.15 to 0.20	0.70 to 0.90	0.035	0.040	0.15 to 0.30	0.40 to 0.70	0.40 to 0.60	0.15 to 0.25
G86200	8620	0.18 to 0.23	0.70 to 0.90	0.035	0.040	0.15 to 0.30	0.40 to 0.70	0.40 to 0.60	0.15 to 0.25
G86220	8622	0.20 to 0.25	0.70 to 0.90	0.035	0.040	0.15 to 0.30	0.40 to 0.70	0.40 to 0.60	0.15 to 0.25
G86250	8625	0.23 to 0.28	0.70 to 0.90	0.035	0.040	0.15 to 0.30	0.40 to 0.70	0.40 to 0.60	0.15 to 0.25
G86270	8627	0.25 to 0.30	0.70 to 0.90	0.035	0.040	0.15 to 0.30	0.40 to 0.70	0.40 to 0.60	0.15 to 0.25
G86300	8630	0.28 to 0.33	0.70 to 0.90	0.035	0.040	0.15 to 0.30	0.40 to 0.70	0.40 to 0.60	0.15 to 0.25

A 752

TABLE 4 *Continued*

UNS Designation	Grade No.	Chemical Composition, Ranges and Limits, %							
		Carbon	Manganese	Phosphorus, max	Sulfur, max	Silicon	Nickel	Chromium	Molybdenum
STANDARD ALLOY STEELS									
G86370	8637	0.35 to 0.40	0.75 to 1.00	0.035	0.040	0.15 to 0.30	0.40 to 0.70	0.40 to 0.60	0.15 to 0.25
G86400	8640	0.38 to 0.43	0.75 to 1.00	0.035	0.040	0.15 to 0.30	0.40 to 0.70	0.40 to 0.60	0.15 to 0.25
G86420	8642	0.40 to 0.45	0.75 to 1.00	0.035	0.040	0.15 to 0.30	0.40 to 0.70	0.40 to 0.60	0.15 to 0.25
G86450	8645	0.43 to 0.48	0.75 to 1.00	0.035	0.040	0.15 to 0.30	0.40 to 0.70	0.40 to 0.60	0.15 to 0.25
G86550	8655	0.51 to 0.59	0.75 to 1.00	0.035	0.040	0.15 to 0.30	0.40 to 0.70	0.40 to 0.60	0.15 to 0.25
G87200	8720	0.18 to 0.23	0.70 to 0.90	0.035	0.040	0.15 to 0.30	0.40 to 0.70	0.40 to 0.60	0.20 to 0.30
G87400	8740	0.38 to 0.43	0.75 to 1.00	0.035	0.040	0.15 to 0.30	0.40 to 0.70	0.40 to 0.60	0.20 to 0.30
G88220	8822	0.20 to 0.25	0.75 to 1.00	0.035	0.040	0.15 to 0.30	0.40 to 0.70	0.40 to 0.60	0.30 to 0.40
G92540	9254	0.51 to 0.59	0.60 to 0.80	0.035	0.040	1.20 to 1.60	...	0.60 to 0.80	...
G92550	9255	0.51 to 0.59	0.70 to 0.95	0.035	0.040	1.80 to 2.20	...	...	...
G92600	9260	0.56 to 0.64	0.75 to 1.00	0.035	0.040	1.80 to 2.20	...	...	...
STANDARD BORON ALLOY STEELS[A]									
G50441	50B44	0.43 to 0.48	0.75 to 1.00	0.035	0.040	0.15 to 0.30	...	0.40 to 0.60	...
G50461	50B46	0.44 to 0.49	0.75 to 1.00	0.035	0.040	0.15 to 0.30	...	0.20 to 0.35	...
G50501	50B50	0.48 to 0.53	0.75 to 1.00	0.035	0.040	0.15 to 0.30	...	0.40 to 0.60	...
G50601	50B60	0.56 to 0.64	0.75 to 1.00	0.035	0.040	0.15 to 0.30	...	0.40 to 0.60	...
G51601	51B60	0.56 to 0.64	0.75 to 1.00	0.035	0.040	0.15 to 0.30	...	0.70 to 0.90	...
G81451	81B45	0.43 to 0.48	0.75 to 1.00	0.035	0.040	0.15 to 0.30	0.20 to 0.40	0.35 to 0.55	0.08 to 0.15
G94171	94B17	0.15 to 0.20	0.75 to 1.00	0.035	0.040	0.15 to 0.30	0.30 to 0.60	0.30 to 0.50	0.08 to 0.15
G94301	94B30	0.28 to 0.33	0.75 to 1.00	0.035	0.040	0.15 to 0.30	0.30 to 0.60	0.30 to 0.50	0.08 to 0.15

[A] These steels can be expected to a minimum boron content of 0.0005 %.

TABLE 5 Product or Verification Analysis Tolerances – Alloy Steels

Element	Limit or Maximum of Specified Range, %	Tolerance Over Maximum Limit or Under Minimum Limit, %
Carbon	To 0.30, incl	0.01
	Over 0.30 to 0.75, incl	0.02
	Over 0.75	0.03
Manganese	To 0.90, incl	0.03
	Over 0.90 to 2.10, incl	0.04
Phosphorus	Over max only	0.005
Sulfur	To 0.060, incl[A]	0.005
Silicon	To 0.40, incl	0.02
	Over 0.40 to 2.20, incl	0.05
Nickel	To 1.00, incl	0.03
	Over 1.00 to 2.00, incl	0.05
	Over 2.00 to 5.30, incl	0.07
	Over 5.30 to 10.00, incl	0.10
Chromium	To 0.90, incl	0.03
	Over 0.90 to 2.10, incl	0.05
	Over 2.10 to 3.99, incl	0.10
Molybdenum	To 0.20, incl	0.01
	Over 0.20 to 0.40, incl	0.02
	Over 0.40 to 1.15, incl	0.03
Vanadium	To 0.10, incl	0.01
	Over 0.10 to 0.25, incl	0.02
	Over 0.25 to 0.50, incl	0.03
	Min value specified, check under min limit	0.01
Tungsten	To 1.00, incl	0.04
	Over 1.00 to 4.00, incl	0.08
Aluminum	Up to 0.10, incl	0.03
	Over 0.10 to 0.20, incl	0.04
	Over 0.20 to 0.30, incl	0.05
	Over 0.30 to 0.80, incl	0.07
	Over 0.80 to 1.80, incl	0.10
Lead	0.15 to 0.35, incl	0.03[B]
Copper	To 1.00, incl	0.03
	Over 1.00 to 2.00, incl	0.05

[A] Sulfur over 0.060 % is not subject to check, product, or verification analysis.
[B] Tolerance is over *and* under.

A 752

TABLE 6 Alloy Steels – Chemical Composition Ranges and Limits for Cast or Heat Analysis

NOTE 1 – Boron steels can be expected to have a 0.0005 % minimum boron content.
NOTE 2 – Alloy steels can be produced with a lead range of 0.15 to 0.35 %. Lead is reported only as a range of 0.15 to 0.35 % since it is added to the mold as the steel is poured.
NOTE 3 – The chemical ranges and limits of alloy steels are produced to the check, product, or verification analysis tolerances shown in Table 5.

Element	When Maximum of Specified Element is, %	Range, % Open-Hearth or Basic Oxygen Steel	Range, % Electric Furnace Steel	Maximum Limit, %[A]
Carbon	To 0.55, incl	0.05	0.05	
	Over 0.55 to 0.70, incl	0.08	0.07	
	Over 0.70 to 0.80, incl	0.10	0.09	
	Over 0.80 to 0.95, incl	0.12	0.11	
	Over 0.95 to 1.35, incl	0.13	0.12	
Manganese	To 0.60, incl	0.20	0.15	
	Over 0.60 to 0.90, incl	0.20	0.20	
	Over 0.90 to 1.05, incl	0.25	0.25	
	Over 1.05 to 1.90, incl	0.30	0.30	
	Over 1.90 to 2.10, incl	0.40	0.35	
Phosphorus	Basic open-hearth or basic oxygen steel			0.035
	Basic electric furnace steel			0.025
Sulfur	To 0.050, incl	0.015	0.015	
	Over 0.050 to 0.07, incl	0.02	0.02	
	Over 0.07 to 0.10, incl	0.04	0.04	
	Over 0.10 to 0.14, incl	0.05	0.05	
	Basic open hearth or basic oxygen steel			0.040
	Basic electric furnace steel			0.025
Silicon	To 0.15, incl	0.08	0.08	
	Over 0.15 to 0.20, incl	0.10	0.10	
	Over 0.20 to 0.40, incl	0.15	0.15	
	Over 0.40 to 0.60, incl	0.20	0.20	
	Over 0.60 to 1.00, incl	0.30	0.30	
	Over 1.00 to 2.20, incl	0.40	0.35	
Nickel	To 0.50, incl	0.20	0.20	
	Over 0.50 to 1.50, incl	0.30	0.30	
	Over 1.50 to 2.00, incl	0.35	0.35	
	Over 2.00 to 3.00, incl	0.40	0.40	
	Over 3.00 to 5.30, incl	0.50	0.50	
	Over 5.30 to 10.00, incl	1.00	1.00	
Chromium	To 0.40, incl	0.15	0.15	
	Over 0.40 to 0.90, incl	0.20	0.20	
	Over 0.90 to 1.05, incl	0.25	0.25	
	Over 1.05 to 1.60, incl	0.30	0.30	
	Over 1.60 to 1.75, incl	B	0.35	
	Over 1.75 to 2.10, incl	B	0.40	
	Over 2.10 to 3.99, incl	B	0.50	
Molybdenum	To 0.10, incl	0.05	0.05	
	Over 0.10 to 0.20, incl	0.07	0.07	
	Over 0.20 to 0.50, incl	0.10	0.10	
	Over 0.50 to 0.80, incl	0.15	0.15	
	Over 0.80 to 1.15, incl	0.20	0.20	
Tungsten	To 0.50, incl	0.20	0.20	
	Over 0.50 to 1.00, incl	0.30	0.30	
	Over 1.00 to 2.00, incl	0.50	0.50	
	Over 2.00 to 4.00, incl	0.60	0.60	
Vanadium	To 0.25, incl	0.05	0.05	
	Over 0.25 to 0.50, incl	0.10	0.10	
Aluminum	Up to 0.10, incl	0.05	0.05	
	Over 0.10 to 0.20, incl	0.10	0.10	
	Over 0.20 to 0.30, incl	0.15	0.15	
	Over 0.30 to 0.80, incl	0.25	0.25	
	Over 0.80 to 1.30, incl	0.35	0.35	
	Over 1.30 to 1.80, incl	0.45	0.45	

TABLE 6 *Continued*

Element	When Maximum of Specified Element is, %	Open-Hearth or Basic Oxygen Steel	Electric Furnace Steel	Maximum Limit, %[A]
Copper	To 0.60, incl	0.20	0.20	
	Over 0.60 to 1.50, incl	0.30	0.30	
	Over 1.50 to 2.00, incl	0.35	0.35	

[A] Applies to only nonrephosphorized and nonresulfurized steels.
[B] Not normally produced in open hearth.

TABLE 7 Maximum Expected Tensile Strengths for Annealed Alloy Steel Rods

NOTE — Specific microstructures such as spheroidize anneal or lamellar pearlite anneal may require modification of these tensile strength values.

UNS Designation	Grade No.	Tensile Strength, ksi (MPa)	UNS Designation	Grade No.	Tensile Strength, ksi (MPa)
G13300	1330	88 (610)	G50501	50B50	97 (670)
G13350	1335	90 (620)	G50601	50B60	103 (710)
G13400	1340	92 (630)	G51200	5120	82 (570)
G13450	1345	97 (670)	G51300	5130	84 (580)
G40120	4012	71 (490)	G51320	5132	84 (580)
G40230	4023	73 (500)	G51350	5135	86 (590)
G40240	4024	73 (500)	G51400	5140	90 (620)
G40270	4027	82 (570)	G51450	5145	94 (650)
G40280	4028	82 (570)	G51470	5147	99 (680)
G40370	4037	89 (610)	G51500	5150	97 (670)
G40470	4047	97 (670)	G51550	5155	103 (710)
G41180	4118	82 (570)	G51600	5160	104 (720)
G41300	4130	86 (590)	G51601	51B60	109 (750)
G41370	4137	92 (630)	G61180	6118	80 (550)
G41400	4140	94 (650)	G61500	6150	97 (670)
G41420	4142	97 (670)	G81451	81B45	92 (630)
G41450	4145	99 (680)	G86150	8615	80 (550)
G41470	4147	100 (690)	G86170	8617	80 (550)
G41500	4150	100 (690)	G86200	8620	82 (570)
G41610	4161	111 (770)	G86220	8622	84 (580)
G43200	4320	94 (650)	G86250	8625	84 (580)
G43400	4340	104 (720)	G86270	8627	84 (580)
G44190	4419	82 (570)	G86300	8630	88 (610)
G46150	4615	80 (550)	G86370	8637	92 (630)
G46200	4620	82 (570)	G86400	8640	94 (650)
G46210	4621	82 (570)	G86420	8642	97 (670)
G46260	4626	84 (580)	G86450	8645	99 (680)
G47180	4718	84 (580)	G86550	8655	104 (720)
G47200	4720	82 (570)	G87200	8720	82 (570)
G48150	4815	92 (630)	G87400	8740	97 (670)
G48170	4817	94 (650)	G88220	8822	88 (610)
G48200	4820	94 (650)	G92540	9254	111 (770)
G50150	5015	73 (500)	G92550	9255	111 (770)
G50441	50B44	94 (650)	G92600	9260	111 (770)
G50461	50B46	92 (630)	G94171	94B17	73 (500)
			G94301	94B30	84 (580)

The American Society for Testing and Materials takes no position respecting the validity of any patent rights asserted in connection with any item mentioned in this standard. Users of this standard are expressly advised that determination of the validity of any such patent rights, and the risk of infringement of such rights, are entirely their own responsibility.

This standard is subject to revision at any time by the responsible technical committee and must be reviewed every five years and if not revised, either reapproved or withdrawn. Your comments are invited either for revision of this standard or for additional standards and should be addressed to ASTM Headquarters. Your comments will receive careful consideration at a meeting of the responsible technical committee, which you may attend. If you feel that your comments have not received a fair hearing you should make your views known to the ASTM Committee on Standards, 1916 Race St., Philadelphia, Pa. 19103.

Designation: A 752M – 83

Standard Specification for
GENERAL REQUIREMENTS FOR WIRE RODS AND COARSE ROUND WIRE, ALLOY STEEL (METRIC)[1]

This standard is issued under the fixed designation A 752M; the number immediately following the designation indicates the year of original adoption or, in the case of revision, the year of last revision. A number in parentheses indicates the year of last reapproval. A superscript epsilon (ϵ) indicates an editorial change since the last revision or reapproval.

1. Scope

1.1 This specification covers general requirements for alloy steel rods and uncoated coarse round alloy wire in coils that are not required to meet hardenability band limits.

1.2 In case of conflict, the requirements in the purchase order, on the drawing, in the individual specification, and in this general specification shall prevail in the sequence named.

NOTE 1—This specification is the metric counterpart of Specification A752.

2. Applicable Documents

2.1 *ASTM Standards:*
A 370 Methods and Definitions for Mechanical Testing of Steel Products[2]
A 700 Practices for Packaging, Marking, and Loading Methods for Steel Products for Domestic Shipment[2]
A 751 Methods, Practices, and Definitions for Chemical Analysis of Steel Products[2]
E 29 Recommended Practice for Indicating Which Places of Figures Are to Be Considered Significant in Specified Limiting Values[3]
E 44 Definitions of Terms Relating to Heat Treatment of Metals[4]
E 112 Methods for Determining Average Grain Size[4]

2.2 *Military Standards:*
MIL-STD-163 Steel Mill Products, Preparation for Shipment and Storage[5]

2.3 *Federal Standard:*
Fed. Std. No. 123 Marking for Shipment (Civil Agencies)[5]

3. Description of Terms

3.1 *Alloy Steel*—Steel is considered to be alloy steel when the maximum of the range given for the content of alloying elements exceeds one or more of the following limits: manganese 1.65 %, silicon 0.60 %, copper 0.60 %; or in which a definite range or a definite minimum quantity of any of the following elements is specified or required within the limits of the recognized field of constructional alloy steels: chromium up to 3.99 %, cobalt, columbium, molybdenum, nickel, titanium, tungsten, vanadium, zirconium, or any other alloying elements added to obtain a desired alloying effect.

3.1.1 Boron treatment of alloy steels, which are fine grain, may be specified to improve hardenability.

3.1.2 Other elements, such as lead, selenium, tellurium, or bismuth, may be specified to improve machinability.

3.2 Wire rods are hot rolled from billets into an approximate round cross section and into coils of one continuous length. Rods are not comparable to hot-rolled bars in accuracy of cross section or surface finish and as a semifinished product are primarily for the manufacture of wire.

---

[1] This specification is under the jurisdiction of ASTM Committee A-1 on Steel, Stainless Steel and Related Alloys and is the direct responsibility of Subcommittee A01.03 on Steel Rod and Wire.
Current edition approved July 29, 1983. Published November 1983.
[2] *Annual Book of ASTM Standards,* Vol 01.03.
[3] *Annual Book of ASTM Standards,* Vol 14.02.
[4] *Annual Book of ASTM Standards,* Vol 03.02.
[5] Available from Naval Publications and Forms Center, 5801 Tabor Ave., Phila., Pa. 19120.

3.2.1 Table 1 shows the nominal diameter for hot-rolled wire rods. Sizes are shown in 0.5-mm increments from 5.5 to 19 mm.

3.3 Coarse round wire from 0.90 to 25 mm in diameter, inclusive, is produced from hot-rolled wire rods or hot-rolled coiled rounds by one or more cold reductions primarily for the purpose of obtaining a desired size with dimensional accuracy, surface finish, and mechanical properties. By varying the amount of cold reduction and other wire mill practices, including thermal treatment, a wide diversity of mechanical properties and finishes are made available. Suggested wire diameters are shown in Table 2.

3.4 Straightened and cut wire is produced from coils of wire by means of special machinery that straightens the wire and cuts it to a specified length.

3.4.1 The straightening operation may alter the mechanical properties of the wire, especially the tensile strength. The straightening operation may also induce changes in the diameter of the wire. The extent of the changes in the properties of the wire after cold straightening depends upon the kind of wire and also on the normal variation in the adjustments of the straightening equipment. It is therefore not possible to forecast the properties of straightened and cut wire. Each kind of wire needs individual consideration. In most cases, the application of straightened and cut wire is not seriously influenced by these changes.

4. Ordering Information

4.1 Orders for hot-rolled wire rods under this specification shall include the following information:

4.1.1 Quantity (kilograms or megagrams),
4.1.2 Name of material (wire rods),
4.1.3 Diameter (Table 1),
4.1.4 Chemical composition grade number (Table 3),
4.1.5 Packaging (Section 14),
4.1.6 Heat analysis report, if requested (Section 6),
4.1.7 Certification or test report, or both, if specified (Section 13),
4.1.8 ASTM designation and date of issue, and
4.1.9 Special requirements, if any.

NOTE 2—A typical ordering description is as follows: 50 000 kg Steel Wire Rods, 5.5 mm, Grade G 41400 W approximately 600 kg coils to ASTM A 752M dated _____.

4.2 Orders for coarse round wire under this specification shall include the following information:

4.2.1 Quantity (kilograms or pieces),
4.2.2 Name of material (alloy steel wire),
4.2.3 Diameter (Table 2),
4.2.4 Length (straightened and cut only),
4.2.5 Chemical composition (Table 3),
4.2.6 Packaging (Section 14),
4.2.7 Heat analysis report, if requested (Section 6),
4.2.8 Certification or test report, or both, if specified (Section 13),
4.2.9 ASTM designation and date of issue, and
4.2.10 Special requirements, if any.

NOTE 3—A typical ordering description is as follows: 15 000 kg Alloy Steel Wire, 3.8 mm diameter, Grade 4135 in 1000-kg Coils on Tubular Carriers, to ASTM A 752M dated _____, or 2500 Pieces, Alloy Steel Wire, 9.5 mm diameter, Straightened and Cut, 0.76 m, Grade G 41400 in 25-Piece Bundles on Pallets to ASTM A 752M dated _____.

5. Manufacture

5.1 The steel may be produced by any commercially accepted steel making process. The steel may be either ingot cast or strand cast.

6. Chemical Requirements

6.1 The chemical composition for alloy steel under this specification shall conform to the requirements set forth in the purchase order. The grades commonly specified for alloy steel wire rods and alloy steel wire are shown in Table 3. For specified compositions not contained in Table 3 the ranges and limits expressed in Table 4 shall apply unless other such ranges and limits shall have been agreed upon between the purchaser and the manufacturer.

6.2 *Heat Analysis*—An analysis of each heat shall be made by the producer to determine the percentage of the elements specified. The analysis shall be made from a test sample preferably taken during the pouring of the heat. The chemical composition thus determined shall be reported, if required, to the purchaser or his representative.

6.3 *Product Analysis*—A product analysis may be made by the purchaser. The analysis is not used for a duplicate analysis to confirm a previous result. The purpose of the product analysis is to verify that the chemical composition is within specified limits for each element, including applicable permissible variations in product analysis. The results of analyses taken from different pieces of a heat may differ within permissible limits from each other and from the heat analysis. Table 5 shows the permissible variations for product analysis of alloy steel. The results of the product analysis, except lead, shall not vary both above and below the specified ranges.

6.3.1 The location from which chips for product analysis are obtained is important because of normal segregation. For rods and wire, chips must be taken by milling or machining the full cross section of the sample.

6.3.1.1 Steel subjected to certain thermal treatments by the purchaser may not give chemical analysis results that properly represent its original composition. Therefore, purchasers should analyze chips taken from the steel in the condition in which it is received from the producer.

6.3.1.2 When samples are returned to the producer for product analysis, the samples should consist of pieces of the full cross section.

6.3.2 For referee purposes, Practice and Definitions A 751 shall be used.

7. Metallurgical Structure

7.1 Grain size when specified shall be determined in accordance with the requirements of Methods E 112.

7.2 Alloy steel wire rods may be specified as annealed, spheroidize annealed, or patented. Refer to Definitions E 44 for definitions.

7.3 Alloy steel wire may be specified as drawn from annealed or spheroidize annealed wire, rod, or bars, spheroidize annealed at finish size, patented, or oil tempered. Refer to Definitions E 44 for definitions.

8. Mechanical Requirements

8.1 The properties enumerated in individual specifications shall be determined in accordance with Methods A 370.

8.2 The maximum expected tensile strengths for the more common grades of alloy rods with regular mill annealing (nonspheroidized) are shown in Table 6.

8.3 Because of the great variety in the kinds and grades of rods and wire and the extensive diversity of application, a number of formal mechanical test procedures have been developed. These tests are used as control tests by producers during intermediate stages of wire processing, as well as for final testing of the finished product, and apply particularly to rods and wire for specific applications. A number of these tests are further described in Supplement IV, Round Wire Products, of Methods A 370.

8.4 Since the general utility of rods and wire require continuity of length, in the case of rods, tests are commonly made on samples taken from the ends of coils after removing two to three rings; in the case of wire, tests are commonly taken from the ends of the coils, thereby not impairing the usefulness of the whole coil.

9. Dimensions and Permissible Variation

9.1 The diameter and out-of-roundness of the alloy wire rod shall not vary from that specified by more than that prescribed in Table 7.

9.2 The diameter and out-of-roundness of the alloy coarse round wire shall not vary from that specified by more than that prescribed in Table 8.

9.3 The length of straightened and cut wire shall not vary from that specified by more than that prescribed in Table 9.

9.4 The burrs formed in cutting straightened and cut wire shall not exceed the diameter specified by more than that prescribed in Table 10.

10. Workmanship, Finish, and Appearance

10.1 The alloy wire rod shall be free from detrimental surface imperfections, tangles, and sharp kinks.

10.1.1 Two or more rod coils may be welded together to produce a larger coil. The weld zone may not be as sound as the original material. The mechanical properties existing in the weld zone may differ from those in the unaffected base metal. The weld may exceed the permissible variations for diameter and out-of-roundness on the minus side of the permissible variation but not on the plus side.

10.2 The wire as received shall be smooth and substantially free from rust, shall not be kinked or improperly cast. No detrimental die marks or scratches may be present. Each coil shall be one continuous length of wire. Welds made during cold drawing are permitted.

10.3 The straightened and cut wire shall be

A 752M

substantially straight and not be kinked or show excessive spiral marking.

11. Inspection

11.1 The manufacturer shall afford the purchaser's inspector all reasonable facilities necessary to satisfy him that the material is being produced and furnished in accordance with this specification. Mill inspection by the purchaser shall not interfere unnecessarily with the manufacturer's operations. All tests and inspections shall be made at the place of manufacture, unless otherwise agreed to.

12. Rejection

12.1 Any rejection based on tests made in accordance with this specification shall be reported to the manufacturer within a reasonable length of time. The material must be adequately protected and correctly identified in order that the manufacturer may make a proper investigation.

13. Certification

13.1 Upon request of the purchaser in the contract or order, a manufacturer's certification that the material was manufactured and tested in accordance with this specification, together with a report of the test results, shall be furnished at the time of shipment.

14. Packaging, Marking, and Loading for Shipment

14.1 The coil mass, dimensions, and the method of packaging shall be agreed upon between the manufacturer and purchaser.

14.2 The size of the wire, purchaser's order number, ASTM specification number, heat number, and name or mark of the manufacturer shall be marked on a tag securely attached to each coil of wire.

14.3 Unless otherwise specified in the purchaser's order, packaging, marking, and loading for shipments shall be in accordance with those procedures recommended by Practice A 700.

14.4 *For Government Procurement*—Packaging, packing, and marking of material for military procurement shall be in accordance with the requirements of MIL-STD-163, Level A, Level C, or commercial as specified in the contract or purchase order. Marking for shipment of material for civil agencies shall be in accordance with Fed. Std. No. 123.

TABLE 1 Sizes of Alloy Steel Wire Rods, mm

5.5	12.5
6	13
6.5	13.5
7	14
7.5	14.5
8	15
8.5	15.5
9	16
9.5	16.5
10	17
10.5	17.5
11	18
11.5	18.5
12	19

TABLE 2 Suggested Diameters for Alloy Steel Wire, mm

0.90	6.0
1.00	6.5
1.10	7.0
1.20	7.5
1.30	8.0
1.40	8.5
1.60	9.0
1.80	9.5
2.0	10.0
2.1	11.0
2.2	12.0
2.4	13.0
2.5	14.0
2.6	15.0
2.8	16.0
3.0	17.0
3.2	18.0
3.5	19.0
3.8	20.0
4.0	21.0
4.2	22.0
4.5	23.0
4.8	24.0
5.0	25.0
5.5	...

TABLE 3 Alloy Steels—Chemical Composition Ranges and Limits for Heat Analysis

NOTE 1—Grades shown in this table with prefix letter E are normally only made by the basic electric furnace process. All others are normally manufactured by the basic open hearth or basic oxygen processes but may be manufactured by a basic electric furnace process. If the electric furnace process is specified or required for grades other than those designated above, the limits for phosphorus and sulfur are respectively 0.025 % max.

NOTE 2—Small quantities of certain elements, which are not specified or required, are present in alloy steels. These elements are considered as incidental and may be present to the following maximum amounts: copper, 0.35 %, nickel, 0.25 %, chromium, 0.20 %, molybdenum, 0.06 %.

NOTE 3—Where minimum and maximum sulfur content is shown it is indicative of resulfurized steel.

NOTE 4—The chemical ranges and limits shown in Table 4 are produced to product analysis tolerances shown in Table 5.

NOTE 5—Standard alloy steels can be produced with a lead range of 0.15 to 0.35 %. Such steels are identified by inserting the letter "L" between the second and third numerals of the Grade number, for example, 41L40. Lead is reported only as a range of 0.15 to 0.35 % since it is added to the mold as the steel is poured.

UNS Designation	Grade No.	Chemical Composition, Ranges and Limits, %							
		Carbon	Manganese	Phosphorus, max	Sulfur, max	Silicon	Nickel	Chromium	Molybdenum
STANDARD ALLOY STEELS									
G13300	1330	0.28 to 0.33	1.60 to 1.90	0.035	0.040	0.15 to 0.35	...	...	...
G13350	1335	0.33 to 0.38	1.60 to 1.90	0.035	0.040	0.15 to 0.35	...	...	...
G13400	1340	0.38 to 0.43	1.60 to 1.90	0.035	0.040	0.15 to 0.35	...	...	...
G13450	1345	0.43 to 0.48	1.60 to 1.90	0.035	0.040	0.15 to 0.35	...	...	...
G40120	4012	0.09 to 0.14	0.75 to 1.00	0.035	0.040	0.15 to 0.35	...	...	0.15 to 0.25
G40230	4023	0.20 to 0.25	0.70 to 0.90	0.035	0.040	0.15 to 0.35	...	...	0.20 to 0.30
G40240	4024	0.20 to 0.25	0.70 to 0.90	0.035	0.035 to 0.050	0.15 to 0.35	...	...	0.20 to 0.30
G40270	4027	0.25 to 0.30	0.70 to 0.90	0.035	0.040	0.15 to 0.35	...	...	0.20 to 0.30
G40280	4028	0.25 to 0.30	0.70 to 0.90	0.035	0.035 to 0.050	0.15 to 0.35	...	...	0.20 to 0.30
G40370	4037	0.35 to 0.40	0.70 to 0.90	0.035	0.040	0.15 to 0.35	...	...	0.20 to 0.30
G40470	4047	0.45 to 0.50	0.70 to 0.90	0.035	0.040	0.15 to 0.35	...	...	0.20 to 0.30
G41180	4118	0.18 to 0.23	0.70 to 0.90	0.035	0.040	0.15 to 0.35	...	0.40 to 0.60	0.08 to 0.15
G41300	4130	0.28 to 0.33	0.40 to 0.60	0.035	0.040	0.15 to 0.35	...	0.80 to 1.10	0.15 to 0.25
G41370	4137	0.35 to 0.40	0.70 to 0.90	0.035	0.040	0.15 to 0.35	...	0.80 to 1.10	0.15 to 0.25
G41400	4140	0.38 to 0.43	0.75 to 1.00	0.035	0.040	0.15 to 0.35	...	0.80 to 1.10	0.15 to 0.25
G41420	4142	0.40 to 0.45	0.75 to 1.00	0.035	0.040	0.15 to 0.35	...	0.80 to 1.10	0.15 to 0.25
G41450	4145	0.43 to 0.48	0.75 to 1.00	0.035	0.040	0.15 to 0.35	...	0.80 to 1.10	0.15 to 0.25
G41470	4147	0.45 to 0.50	0.75 to 1.00	0.035	0.040	0.15 to 0.35	...	0.80 to 1.10	0.15 to 0.25
G41500	4150	0.48 to 0.53	0.75 to 1.00	0.035	0.040	0.15 to 0.35	...	0.80 to 1.10	0.15 to 0.25
G41610	4161	0.56 to 0.64	0.75 to 1.00	0.035	0.040	0.15 to 0.35	...	0.70 to 0.90	0.25 to 0.35
G43200	4320	0.17 to 0.22	0.45 to 0.65	0.035	0.040	0.15 to 0.35	1.65 to 2.00	0.40 to 0.60	0.20 to 0.30
G43400	4340	0.38 to 0.43	0.60 to 0.80	0.035	0.040	0.15 to 0.35	1.65 to 2.00	0.70 to 0.90	0.20 to 0.30
G43406	E4340	0.38 to 0.43	0.65 to 0.85	0.025	0.025	0.15 to 0.35	1.65 to 2.00	0.70 to 0.90	0.20 to 0.30
G44190	4419	0.18 to 0.33	0.45 to 0.65	0.035	0.040	0.15 to 0.35	...	...	0.45 to 0.60

TABLE 3 *Continued*

UNS Designation	Grade No.	Chemical Composition, Ranges and Limits, %							
		Carbon	Manganese	Phosphorus, max	Sulfur, max	Silicon	Nickel	Chromium	Molybdenum
	STANDARD ALLOY STEELS
G46150	4615	0.13 to 0.18	0.45 to 0.65	0.035	0.040	0.15 to 0.35	1.65 to 2.00	...	0.20 to 0.30
G46200	4620	0.17 to 0.22	0.45 to 0.65	0.035	0.040	0.15 to 0.35	1.65 to 2.00	...	0.20 to 0.30
G46210	4621	0.18 to 0.23	0.70 to 0.90	0.035	0.040	0.15 to 0.35	1.65 to 2.00	...	0.20 to 0.30
G46260	4626	0.24 to 0.29	0.45 to 0.65	0.035	0.040	0.15 to 0.35	0.70 to 1.00	...	0.15 to 0.25
G47180	4718	0.16 to 0.21	0.70 to 0.90	0.035	0.040	0.15 to 0.35	0.90 to 1.20	0.35 to 0.55	0.30 to 0.40
G47200	4720	0.17 to 0.22	0.50 to 0.70	0.035	0.040	0.15 to 0.35	0.90 to 1.20	0.35 to 0.55	0.15 to 0.25
G48150	4815	0.13 to 0.18	0.40 to 0.60	0.035	0.040	0.15 to 0.35	3.25 to 3.75	...	0.20 to 0.30
G48170	4817	0.15 to 0.20	0.40 to 0.60	0.035	0.040	0.15 to 0.35	3.25 to 3.75	...	0.20 to 0.30
G48200	4820	0.18 to 0.23	0.50 to 0.70	0.035	0.040	0.15 to 0.35	3.25 to 3.75	...	0.20 to 0.30
G50150	5015	0.12 to 0.17	0.30 to 0.50	0.035	0.040	0.15 to 0.35	...	0.30 to 0.50	...
G51200	5120	0.17 to 0.22	0.70 to 0.90	0.035	0.040	0.15 to 0.35	...	0.70 to 0.90	...
G51300	5130	0.28 to 0.33	0.70 to 0.90	0.035	0.040	0.15 to 0.35	...	0.80 to 1.10	...
G51320	5132	0.30 to 0.35	0.60 to 0.80	0.035	0.040	0.15 to 0.35	...	0.75 to 1.00	...
G51350	5135	0.33 to 0.38	0.60 to 0.80	0.035	0.040	0.15 to 0.35	...	0.80 to 1.05	...
G51400	5140	0.38 to 0.43	0.70 to 0.90	0.035	0.040	0.15 to 0.35	...	0.70 to 0.90	...
G51450	5145	0.43 to 0.48	0.70 to 0.90	0.035	0.040	0.15 to 0.35	...	0.70 to 0.90	...
G51470	5147	0.46 to 0.51	0.70 to 0.95	0.035	0.040	0.15 to 0.35	...	0.85 to 1.15	...
G51500	5150	0.48 to 0.53	0.70 to 0.90	0.035	0.040	0.15 to 0.35	...	0.70 to 0.90	...
G51550	5155	0.51 to 0.59	0.70 to 0.90	0.035	0.040	0.15 to 0.35	...	0.70 to 0.90	...
G51600	5160	0.56 to 0.64	0.75 to 1.00	0.035	0.040	0.15 to 0.35	...	0.70 to 0.90	...
...	E51100	0.98 to 1.10	0.25 to 0.45	0.025	0.025	0.15 to 0.35	...	0.90 to 1.15	...
...	E52100	0.98 to 1.10	0.25 to 0.45	0.025	0.025	0.15 to 0.35	...	1.30 to 1.60	...

UNS Designation	Grade No.	Chemical Composition, Ranges and Limits, %							
		Carbon	Manganese	Phosphorus, max	Sulfur, max	Silicon	Nickel	Chromium	Other Elements
									Vanadium
G61180	6118	0.16 to 0.21	0.50 to 0.70	0.035	0.040	0.15 to 0.35	...	0.50 to 0.70	0.10 to 0.15
G61500	6150	0.48 to 0.53	0.70 to 0.90	0.035	0.040	0.15 to 0.35	...	0.80 to 1.10	0.15 min
									Molybdenum
G86150	8615	0.13 to 0.18	0.70 to 0.90	0.035	0.040	0.15 to 0.35	0.040 to 0.70	0.040 to 0.60	0.15 to 0.25
G86170	8617	0.15 to 0.20	0.70 to 0.90	0.035	0.040	0.15 to 0.35	0.040 to 0.70	0.040 to 0.60	0.15 to 0.25
G86200	8620	0.18 to 0.23	0.70 to 0.90	0.035	0.040	0.15 to 0.35	0.040 to 0.70	0.040 to 0.60	0.15 to 0.25
G86220	8622	0.20 to 0.25	0.70 to 0.90	0.035	0.040	0.15 to 0.35	0.040 to 0.70	0.040 to 0.60	0.15 to 0.25
G86250	8625	0.23 to 0.28	0.70 to 0.90	0.035	0.040	0.15 to 0.35	0.040 to 0.70	0.040 to 0.60	0.15 to 0.25
G86270	8627	0.25 to 0.30	0.70 to 0.90	0.035	0.040	0.15 to 0.35	0.040 to 0.70	0.040 to 0.60	0.15 to 0.25
G86300	8630	0.28 to 0.33	0.70 to 0.90	0.035	0.040	0.15 to 0.35	0.040 to 0.70	0.040 to 0.60	0.15 to 0.25

A 752M

TABLE 3 *Continued*

UNS Designation	Grade No.	Chemical Composition, Ranges and Limits, %							
		Carbon	Manganese	Phosphorus, max	Sulfur, max	Silicon	Nickel	Chromium	Molybdenum
			STANDARD ALLOY STEELS						
G86370	8637	0.35 to 0.40	0.75 to 1.00	0.035	0.040	0.15 to 0.35	0.40 to 0.70	0.40 to 0.60	0.15 to 0.25
G86400	8640	0.38 to 0.43	0.75 to 1.00	0.035	0.040	0.15 to 0.35	0.40 to 0.70	0.40 to 0.60	0.15 to 0.25
G86420	8642	0.40 to 0.45	0.75 to 1.00	0.035	0.040	0.15 to 0.35	0.40 to 0.70	0.40 to 0.60	0.15 to 0.25
G86450	8645	0.43 to 0.48	0.75 to 1.00	0.035	0.040	0.15 to 0.35	0.40 to 0.70	0.40 to 0.60	0.15 to 0.25
G86550	8655	0.51 to 0.59	0.75 to 1.00	0.035	0.040	0.15 to 0.35	0.40 to 0.70	0.40 to 0.60	0.15 to 0.25
G87200	8720	0.18 to 0.23	0.70 to 0.90	0.035	0.040	0.15 to 0.35	0.40 to 0.70	0.40 to 0.60	0.20 to 0.30
G87400	8740	0.38 to 0.43	0.75 to 1.00	0.035	0.040	0.15 to 0.35	0.40 to 0.70	0.40 to 0.60	0.20 to 0.30
G88220	8822	0.20 to 0.25	0.75 to 1.00	0.035	0.040	0.15 to 0.35	0.40 to 0.70	0.40 to 0.60	0.30 to 0.40
G92540	9254	0.51 to 0.59	0.60 to 0.80	0.035	0.040	1.20 to 1.60	...	0.60 to 0.80	...
G92550	9255	0.51 to 0.59	0.70 to 0.95	0.035	0.040	1.80 to 2.20	...	...	...
G92600	9260	0.56 to 0.64	0.75 to 1.00	0.035	0.040	1.80 to 2.20	...	...	...
			STANDARD BORON ALLOY STEELS[A]						
G50441	50B44	0.43 to 0.48	0.75 to 1.00	0.035	0.040	0.15 to 0.35	...	0.40 to 0.60	...
G50461	50B46	0.44 to 0.49	0.75 to 1.00	0.035	0.040	0.15 to 0.35	...	0.20 to 0.35	...
G50501	50B50	0.48 to 0.53	0.75 to 1.00	0.035	0.040	0.15 to 0.35	...	0.40 to 0.60	...
G50601	50B60	0.56 to 0.64	0.75 to 1.00	0.035	0.040	0.15 to 0.35	...	0.40 to 0.60	...
G51601	51B60	0.56 to 0.64	0.75 to 1.00	0.035	0.040	0.15 to 0.35	...	0.70 to 0.90	...
G81451	81B45	0.43 to 0.48	0.75 to 1.00	0.035	0.040	0.15 to 0.35	0.20 to 0.40	0.35 to 0.55	0.08 to 0.15
G94171	94B17	0.15 to 0.20	0.75 to 1.00	0.035	0.040	0.15 to 0.35	0.30 to 0.60	0.30 to 0.50	0.08 to 0.15
G94301	94B30	0.28 to 0.33	0.75 to 1.00	0.035	0.040	0.15 to 0.35	0.30 to 0.60	0.30 to 0.50	0.08 to 0.15

[A] These steels can be expected to a minimum boron content of 0.0005 %.

TABLE 4 Alloy Steels—Chemical Composition Ranges and Limits for Cast or Heat Analysis

NOTE 1—Boron steels can be expected to have a 0.0005 % minimum boron content.

NOTE 2—Alloy steels can be produced with a lead range of 0.15 to 0.35 %. Lead is reported only as a range of 0.15 to 0.35 % since it is added to the mold as the steel is poured.

NOTE 3—The chemical ranges and limits of alloy steels are produced to the check, product, or verification analysis tolerances shown in Table 5.

Element	When Maximum of Specified Element is, %	Range, % Open-Hearth or Basic Oxygen Steel	Range, % Electric Furnace Steel	Maximum Limit, %[A]
Carbon	To 0.55, incl	0.05	0.05	
	Over 0.55 to 0.70, incl	0.08	0.07	
	Over 0.70 to 0.80, incl	0.10	0.09	
	Over 0.80 to 0.95, incl	0.12	0.11	
	Over 0.95 to 1.35, incl	0.13	0.12	
Manganese	To 0.60, incl	0.20	0.15	
	Over 0.60 to 0.90, incl	0.20	0.20	
	Over 0.90 to 1.05, incl	0.25	0.25	
	Over 1.05 to 1.90, incl	0.30	0.30	
	Over 1.90 to 2.10, incl	0.40	0.35	
Phosphorus	Basic open-hearth or basic oxygen steel			0.035
	Basic electric furnace steel			0.025
Sulfur	To 0.050, incl	0.015	0.015	
	Over 0.050 to 0.07, incl	0.02	0.02	
	Over 0.07 to 0.10, incl	0.04	0.04	
	Over 0.10 to 0.14, incl	0.05	0.05	
	Basic open hearth or basic oxygen steel			0.040
	Basic electric furnace			0.025
Silicon	To 0.15, incl	0.08	0.08	
	Over 0.15 to 0.20, incl	0.10	0.10	
	Over 0.20 to 0.40, incl	0.15	0.15	
	Over 0.40 to 0.60, incl	0.20	0.20	
	Over 0.60 to 1.00, incl	0.30	0.30	
	Over 1.00 to 2.20, incl	0.40	0.35	
Nickel	To 0.50, incl	0.20	0.20	
	Over 0.50 to 1.50, incl	0.30	0.30	
	Over 1.50 to 2.00, incl	0.35	0.35	
	Over 2.00 to 3.00, incl	0.40	0.40	
	Over 3.00 to 5.30, incl	0.50	0.50	
	Over 5.30 to 10.00, incl	1.00	1.00	
Chromium	To 0.40, incl	0.15	0.15	
	Over 0.40 to 0.90, incl	0.20	0.20	
	Over 0.90 to 1.05, incl	0.25	0.25	
	Over 1.05 to 1.60, incl	0.30	0.30	
	Over 1.60 to 1.75, incl	[B]	0.35	
	Over 1.75 to 2.10, incl	[B]	0.40	
	Over 2.10 to 3.99, incl	[B]	0.50	
Molybdenum	To 0.10, incl	0.05	0.05	
	Over 0.10 to 0.20, incl	0.07	0.07	
	Over 0.20 to 0.50, incl	0.10	0.10	
	Over 0.50 to 0.80, incl	0.15	0.15	
	Over 0.80 to 1.15, incl	0.20	0.20	
Tungsten	To 0.50, incl	0.20	0.20	
	Over 0.50 to 1.00, incl	0.30	0.30	
	Over 1.00 to 2.00, incl	0.50	0.50	
	Over 2.00 to 4.00, incl	0.60	0.60	
Vanadium	To 0.25, incl	0.05	0.05	
	Over 0.25 to 0.50, incl	0.10	0.10	
Aluminum	Up to 0.10, incl	0.05	0.05	
	Over 0.10 to 0.20, incl	0.10	0.10	
	Over 0.20 to 0.30, incl	0.15	0.15	
	Over 0.30 to 0.80, incl	0.25	0.25	
	Over 0.80 to 1.30, incl	0.35	0.35	
	Over 1.30 to 1.80, incl	0.45	0.45	
Copper	To 0.60, incl	0.20	0.20	
	Over 0.60 to 1.50, incl	0.30	0.30	
	Over 1.50 to 2.00, incl	0.35	0.35	

[A] Applies to only nonrephosphorized and nonresulfurized steels.
[B] Not normally produced in open hearth.

TABLE 5 Alloy Steels—Product or Verification Analysis Tolerances

Element	Limit or Maximum of Specified Range, %	Tolerance Over Maximum Limit or Under Minimum Limit, %
Carbon	To 0.30, incl	0.01
	Over 0.30 to 0.75, incl	0.02
	Over 0.75	0.03
Manganese	To 0.90, incl	0.03
	Over 0.90 to 2.10, incl	0.04
Phosphorus	Over max only	0.005
Sulfur	To 0.060, incl[A]	0.005
Silicon	To 0.40, incl	0.02
	Over 0.40 to 2.20, incl	0.05
Nickel	To 1.00, incl	0.03
	Over 1.00 to 2.00, incl	0.05
	Over 2.00 to 5.30, incl	0.07
	Over 5.30 to 10.00, incl	0.10
Chromium	To 0.90, incl	0.03
	Over 0.90 to 2.10, incl	0.05
	Over 2.10 to 3.99, incl	0.10
Molybdenum	To 0.20, incl	0.01
	Over 0.20 to 0.40, incl	0.02
	Over 0.40 to 1.15, incl	0.03
Vanadium	To 0.10, incl	0.01
	Over 0.10 to 0.25, incl	0.02
	Over 0.25 to 0.50, incl	0.03
	Min value specified, check under min limit	0.01
Tungsten	To 1.00, incl	0.04
	Over 1.00 to 4.00, incl	0.08
Aluminum	Up to 0.10, incl	0.03
	Over 0.10 to 0.20, incl	0.04
	Over 0.20 to 0.30, incl	0.05
	Over 0.30 to 0.80, incl	0.07
	Over 0.80 to 1.80, incl	0.10
Lead	0.15 to 0.35, incl	0.03[B]
Copper	To 1.00, incl	0.03
	Over 1.00 to 2.00, incl	0.05

[A] Sulfur over 0.060 % is not subject to product analysis.
[B] Tolerance is over *and* under.

TABLE 6 Maximum Expected Tensile Strengths for Annealed Alloy Steel Rods

NOTE—Specific microstructures such as spheroidize anneal or lamellar pearlite anneal may require modification of these tensile strength values.

UNS designation	Grade no.	Tensile Strength, MPa	UNS Designation	Grade No.	Tensile Strength, MPa
G13300	1330	610	G50601	50B60	710
G13350	1335	620	G51200	5120	570
G13400	1340	630	G51300	5130	580
G13450	1345	670	G51320	5132	580
G40120	4012	490	G51350	5135	590
G40230	4023	500	G51400	5140	620
G40240	4024	500	G51450	5145	650
G40270	4027	570	G51470	5147	680
G40280	4028	570	G51500	5150	670
G40370	4037	610	G51550	5155	710
G40470	4047	670	G51600	5160	720
G41180	4118	570	G51601	51B60	750
G41300	4130	590	G61180	6118	550
G41370	4137	630	G61500	6150	670
G41400	4140	650	G81451	81B45	630
G41420	4142	670	G86150	8615	550
G41450	4145	680	G86170	8617	550
G41470	4147	690	G86200	8620	570
G41500	4150	690	G86220	8622	580
G41610	4161	770	G86250	8625	580
G43200	4320	650	G86270	8627	580
G43400	4340	720	G86300	8630	610
G44190	4419	570	G86370	8637	630
G46150	4615	550	G86400	8640	650
G46200	4620	570	G86420	8642	670
G46210	4621	570	G86450	8645	680
G46260	4626	580	G86550	8655	720
G47180	4718	580	G87200	8720	570
G47200	4720	570	G87400	8740	670
G48150	4815	630	G88220	8822	610
G48170	4817	650	G92540	9254	770
G48200	4820	650	G92550	9255	770
G50150	5015	500	G92600	9260	770
G50441	50B44	650	G94171	94B17	500
G50461	50B46	630	G94301	94B30	580
G50501	50B50	670			

A 752M

TABLE 7 Permissible Variation in Diameter for Alloy Steel Wire Rod in Coils

NOTE—For purposes of determining conformance with this specification, all specified limits in this table are absolute limits as defined in Recommended Practice E 29.

Diameter of Rod, mm	Permissible Variation, Plus and Minus, mm	Permissible Out-of-Round, mm
3.5 to 19	0.40	0.60

TABLE 8 Permissible Variation in Diameter for Alloy Steel Coarse Round Wire

NOTE—For purposes of determining conformance with this specification, all specified limits in this table are absolute limits as defined in Recommended Practice E 29.

Diameter of Wire, mm	Permissible Variation, Plus and Minus, mm	Permissible Out-of-Round, mm
In Coils		
0.90 to under 1.90	0.03	0.03
1.90 to under 12.5	0.05	0.05
12.5 and over	0.08	0.08
Straightened and Cut		
0.90 to under 1.90	0.03	0.03
1.90 to under 3.80	0.05	0.05
3.80 to under 12.5	0.08	0.08
12.5 and over	0.10	0.10

TABLE 9 Permissible Variation in Length for Straightened and Cut Alloy Steel Wire

NOTE—For purposes of determining conformance with this specification, all specified limits in this table are absolute limits as defined in Recommended Practice E 29.

Cut Length, m	Permissible Variations, Plus and Minus, mm
Under 1.0	1.6
1.0 to 4.0	2.4
Over 4.0	3.0

TABLE 10 Permissible Variation for Burrs for Straightened and Cut Alloy Steel Wire

NOTE—For purposes of determining conformance with this specification, all specified limits in this table are absolute limits as defined in Recommended Practice E 29.

Diameter of Wire, mm	Permissible Variations, over Measured Diameter, mm
Up to 3.0, incl	0.10
Over 3.0 to 6.5, incl	0.15
Over 6.5 to 12.5, incl	0.20
Over 12.5	0.25

The American Society for Testing and Materials takes no position respecting the validity of any patent rights asserted in connection with any item mentioned in this standard. Users of this standard are expressly advised that determination of the validity of any such patent rights, and the risk of infringement of such rights, are entirely their own responsibility.

This standard is subject to revision at any time by the responsible technical committee and must be reviewed every five years and if not revised, either reapproved or withdrawn. Your comments are invited either for revision of this standard or for additional standards and should be addressed to ASTM Headquarters. Your comments will receive careful consideration at a meeting of the responsible technical committee, which you may attend. If you feel that your comments have not received a fair hearing you should make your views known to the ASTM Committee on Standards, 1916 Race St., Philadelphia, Pa. 19103.

Designation: A 763 – 83

Standard Practices for
DETECTING SUSCEPTIBILITY TO INTERGRANULAR ATTACK IN FERRITIC STAINLESS STEELS[1]

This standard is issued under the fixed designation A 763; the number immediately following the designation indicates the year of original adoption or, in the case of revision, the year of last revision. A number in parentheses indicates the year of last reapproval. A superscript epsilon (ϵ) indicates an editorial change since the last revision or reapproval.

1. Scope

1.1 These practices cover the following four tests:

1.1.1 *Practice W*—Oxalic acid etch test for detecting susceptibility to intergranular attack in stabilized ferritic stainless steels by classification of the etching structures (see Sections 3 through 10).

1.1.2 *Practice X*—Ferric sulfate-sulfuric acid test for detecting susceptibility to intergranular attack in ferritic stainless steels (Sections 11 to 16).

1.1.3 *Practice Y*—Copper-copper sulfate-50 % sulfuric acid test for detecting susceptibility to intergranular attack in ferritic stainless steels (Sections 17 to 22).

1.1.4 *Practice Z*—Copper-copper sulfate-16 % sulfuric acid test for detecting susceptibility to intergranular attack in ferritic stainless steels (Sections 23 to 29).

1.2 The following factors govern the application of these practices (1-5)[2]:

1.2.1 Practice W, oxalic acid test, is a rapid method of identifying, by simple, electrolytic etching, those specimens of certain ferritic alloys that are not susceptible to intergranular corrosion associated with chromium carbide precipitation. Practice W is used as a screening test to avoid the necessity, for acceptable specimens, of more extensive testing required by Practices X, Y, and Z. See Table 2 for a listing of alloys for which Practice W is appropriate.

1.2.2 Practices X, Y, and Z can be used to detect the susceptibility of certain ferritic alloys to intergranular attack associated with the precipitation of chromium carbides or nitrides.

1.2.3 Practices W, X, Y, and Z can also be used to evaluate the effect of heat treatment or of fusion welding on susceptibility to intergranular corrosion.

1.2.4 Table 1 lists the chemical composition of ferritic stainless steels for which data on the application of at least one of the standard practices is available.

1.2.5 Some stabilized ferritic stainless steels may show high rates when tested by Practice X because of metallurgical factors not associated with chromium carbide or nitride precipitation. This possibility must be considered in selecting the test method. Combinations of alloys and test methods for which successful experience is available are shown in Table 2. Application of these standard tests to the other ferritic stainless steels will be by specific agreement between producer and user.

1.3 Depending on the test and alloy, evaluations may be accomplished by weight loss determination, microscopical examination, or bend test (Sections 30 to 31). The choices are listed in Table 2.

2. Applicable Document

2.1 *ASTM Standard*:
A 370 Methods and Definitions for Mechanical Testing of Steel Products[3]

3. Apparatus

3.1 *Apparatus for Practice W, Oxalic Acid Etch Test*:

---

[1] These practices are under the jurisdiction of ASTM Committee A-1 on Steel, Stainless Steel, and Related Alloys, and are the direct responsibility of Subcommittee A01.14 on Methods of Corrosion Testing.
Current edition approved July 29, 1983. Published November 1983. Originally published as A 763 – 79. Last previous edition A 763 – 82.
[2] The boldface numbers in parentheses refer to the list of references appended to these practices.
[3] *Annual Book of ASTM Standards*, Vol 01.03.

3.1.1 *Source of DC*—Battery, generator, or rectifier capable of supplying 15 V and 20 A.

3.1.2 *Ammeter*, range 0 to 30 A.

3.1.3 *Variable Resistance*, for control of specimen current.

3.1.4 *Cathode*—One-litre stainless steel beaker or suitable piece of stainless steel.

3.1.5 *Electric Clamp*, to hold etched specimen.

3.1.6 *Metallurgical Microscope*, for examination of etched structures at 250 to 500×.

3.1.7 *Electrodes*—The specimen is made the anode and the beaker or other piece of stainless steel the cathode.

3.1.8 *Electrolyte*—Oxalic acid ($H_2C_2O_4 \cdot 2H_2O$) reagent grade, 10 weight % solution.

3.2 The apparatus common to Practices X, Y, and Z is listed below. Supplementary requirements are noted as required.

3.2.1 The apparatus used is shown in Fig. 1.

NOTE 1—No substitution for this equipment may be used. The cold-finger type of condenser with standard Erlenmeyer flasks may not be used.

3.2.2 *Allihn or Soxhlet Condenser*, four-bulb with a 45/50 ground-glass joint. Overall length shall be about 13 in. (330 mm) with condensing section, 9½ in. (241 mm).

3.2.3 *Erlenmeyer Flask*, 1-L with a 45/50 ground-glass joint. The ground-glass opening is somewhat over 1½ in. (38 mm) wide.

3.2.4 *Glass Cradles* (Note 2) can be supplied by a glass blowing shop. The size of the cradles should be such that they can pass through the ground-glass joint of the Erlenmeyer flask. They should have three or four holes in them to increase circulation of the test solution around the specimen.

NOTE 2—Other equivalent means of specimen support such as glass hooks or stirrups may also be used.

3.2.5 *Boiling Chips* must be used to prevent bumping.

3.2.6 *Silicone Grease* is recommended for the ground-glass joint.

3.2.7 *Electrically Heated Hot Plate* or other device to provide heat for continuous boiling of the solution.

4. Preparation of Test Specimens

4.1 The preparation of test specimens is common among Practices X, Y, and Z. Additional requirements are noted where necessary.

4.2 A specimen having a total surface area of 5 to 20 cm^2 is recommended for Practices X, Y, and Z. As-welded specimens should be cut so that no more than ½ in. (13 mm) width of unaffected base metal is included on either side of the weld and heat-affected zone.

4.3 The intent is to test a specimen representing as nearly as possible the surface of the material as used in service. Only such surface finishing should be performed as is required to remove foreign material and obtain a standard, uniform finish as specified. For very heavy sections, specimens should be prepared to represent the appropriate surface while maintaining reasonable specimen size for convenience in testing. Ordinarily, removal of more material than necessary will have little influence on the test results. However, in the special case of surface carburization (sometimes encountered, for instance, in tubing when carbonaceous lubricants are employed) it may be possible by heavy grinding or machining to remove the carburized layer completely. Such treatment of test specimens is not permissible, except in tests undertaken to demonstrate such surface effects.

4.4 *Sensitization of Test Specimens*:

4.4.1 Specimens from material that is going to be used in the as-received condition without additional welding or heat treatment may be tested in the as-received condition without any sensitizing treatment.

4.4.2 Specimens from material that is going to be welded or heat treated should be welded or heat treated in as nearly the same manner as the material will experience in service.

4.4.3 The specific sensitizing or welding treatment, or both, should be agreed upon between the supplier and the purchaser.

4.5 For Practice W, a cross section of the sample including material at both surfaces and a cross section of any weld and its heat affected zones should be prepared. If the sample is too thick, multiple specimens should be used. Grind the cross section on wet or dry 80 or 120-grit abrasive paper followed by successively finer papers until a number 400 or 3/0 finish is obtained. Avoid excessive heat when dry-grinding.

4.6 For Practices X, Y, and Z, all surfaces of the specimen including edges should be ground on wet or dry 80 or 120-grit abrasive paper. Avoid excessive heat when dry-grinding. Do not use sand- or grit-blasting. All traces of oxide scale formed during heat treatment must be removed. To avoid scale entrapment, stamp specimens for identification after heat treatment and grinding.

4.7 Degrease and dry the sample using suitable nonchlorinated agents.

PRACTICE W—OXALIC ACID ETCH TEST FOR DETECTING SUSCEPTIBILITY TO INTERGRANULAR ATTACK BY CLASSIFICATION OF MICROSTRUCTURE FOR SCREENING OF CERTAIN FERRITIC STAINLESS STEELS

5. Scope

5.1 The oxalic acid etch test is intended and may be used for screening of certain ferritic stainless steels to precede or preclude the need for corrosion testing as described in Practices X, Y, or Z. Specimens with unacceptable microstructures should be subjected to Practices X, Y, or Z to better determine their susceptibility to intergranular attack. See Table 2 for a listing of alloys for which Practice W is appropriate.

6. Etching Conditions

6.1 The polished specimens should be etched at 1 A/cm^2 for 1.5 min. This may be accomplished with the apparatus prescribed in 3.1 by adjusting the variable resistance until the ammeter reading in amperes equals the immersed specimen area in square centimetres. Immersion of the specimen-holding clamp in the etching solution should be avoided.

7. Etching Precautions

7.1 Etching should be carried out under a ventilating hood. Gas evolved at the electrodes with entrained oxalic acid is poisonous and irritating. The temperature of the etching solution, which increases during etching, should be kept below 50°C by using two beakers of acid, one of which may be cooled while the other is in use.

8. Rinsing Prior to Examination

8.1 Following etching, the specimen should be rinsed in hot water then acetone or alcohol to avoid oxalic acid crystallization on the etched surface during forced air-drying.

9. Examination

9.1 Examine etched specimens on a metallurgical microscope at 250 to 500× as appropriate for classification of etched microstructure type as defined in Section 10.

10. Classification of Etched Structures

10.1 Acceptable structures indicating resistance to chromium carbide-type intergranular attack:

10.1.1 *Step structure*—Steps only between grains—no ditches at grain boundaries (see Fig. 2).

10.1.2 *Dual structure*—Some ditches at grain boundaries in addition to steps, but no single grain completely surrounded by ditches (see Fig. 3).

10.2 Unacceptable structures requiring additional testing (Practices X, Y, or Z):

10.2.1 *Ditch structure*—One or more grains completely surrounded by ditches (see Fig. 4).

PRACTICE X—FERRIC SULFATE-SULFURIC ACID TEST FOR DETECTING SUSCEPTIBILITY TO INTERGRANULAR ATTACK IN FERRITIC STAINLESS STEELS

11. Scope

11.1 This practice describes the procedure for conducting the boiling ferric sulfate-sulfuric acid test which measures the susceptibility of ferritic stainless steels to intergranular attack. This test detects susceptibility to intergranular attack associated with the precipitation of chromium carbides and nitrides in stabilized and unstabilized ferric stainless steels. It may also detect the presence of chi or sigma phase in these steels. The test will not differentiate between intergranular attack resulting from carbides and that due to intermetallic phases. The ferric sulfate-sulfuric acid solution may also selectively attack titanium carbides and nitrides in stabilized steels. The alloys on which the test has been successfully applied are shown in Table 2.

11.2 This test may be used to evaluate the susceptibility of as-received material to intergranular corrosion caused by chromium carbide or nitride precipitation. It may be applied to wrought products and weld metal.

11.3 This procedure may be used on ferritic stainless steels after an appropriate sensitizing heat treatment or welding procedure as agreed upon between the supplier and the purchaser.

12. Apparatus

12.1 The basic apparatus is described in Section 3. Also needed are:

12.1.1 For weight loss determination, an analytical balance capable of weighing to at least the nearest 0.001 g.

12.1.2 For microscopical examination, a microscope with magnification to at least 40×.

13. Ferric Sulfate-Sulfuric Acid Test Solution

13.1 Prepare 600 mL of test solution as follows. **Caution**—Protect the eyes and use rubber gloves and apron for handling acid. Place the test flask under a hood.

13.1.1 First, measure 400.0 mL of distilled water in a 500-mL graduate and pour into the Erlenmeyer flask.

13.1.2 Then measure 236.0 mL of reagent grade sulfuric acid of a concentration that must be in the range from 95.0 to 98.0 weight % in at 250-mL graduate. Add the acid slowly to the water in the Erlenmeyer flask to avoid boiling by the heat evolved.

NOTE 3—Loss of vapor results in concentration of the acid.

13.1.3 Weigh 25 g of reagent grade ferric sulfate (contains about 75 % $Fe_2(SO_4)_3$) and add to the sulfuric acid solution. A trip balance may be used.

13.1.4 Drop boiling chips into the flask.

13.1.5 Lubricate the ground-glass joint with silicone grease.

13.1.6 Cover the flask with the condenser and circulate cooling water.

13.1.7 Boil the solution until all the ferric sulfate is dissolved.

14. Preparation of Test Specimens

14.1 Prepare test specimens as described in Section 4.

15. Procedure

15.1 When weight loss is to be determined, measure the sample prior to final cleaning and then weigh.

15.1.1 Measure the sample including the inner surfaces of any holes, and calculate the total exposed surface area.

15.1.2 Degrease and dry the sample using suitable nonchlorinated agents, and then weigh to the nearest 0.001 g.

15.2 Place the specimen in a glass cradle and immerse in boiling solution.

15.3 Mark the liquid level on the flask with wax crayon to provide a check on vapor loss which would result in concentration of acid. If there is an appreciable change in the level, repeat the test with fresh solution and a reground specimen.

15.4 Continue immersion of the specimen for the time shown in Table 2, then remove the specimen, rinse in water and acetone, and dry. Times for steels not listed in Table 2 are subject to agreement between the supplier and the purchaser.

15.5 For weight loss determination, weigh the specimen and subtract this weight from the original weight.

15.6 No intermediate weighings are usually necessary. The tests can be run without interruption for the time specified in Table 2. However, if preliminary results are desired, the specimen can be removed at any time for weighing.

15.7 No changes in solution are necessary during the test period.

15.8 Additional ferric sulfate inhibitor may have to be added during the test if the corrosion rate is extraordinarily high as evidenced by a change in the color of the solution. More ferric sulfate must be added if the total weight loss of all specimens exceeds 2 g. (During the test, ferric sulfate is consumed at a rate of 10 g for each 1 g of dissolved stainless steel.)

15.9 Testing of a single specimen in a flask is preferred. However, several specimens may be tested simultaneously. The number is limited only by the number of glass cradles that can be fitted into the flask (usually three or four). Each sample must be in a separate cradle so that the samples do not touch.

15.10 During testing, there is some deposition of iron oxides on the upper part of the Erlenmeyer flask. This can be readily removed, after test completion, by boiling a solution of 10 % hydrochloric acid in the flask.

16. Evaluation

16.1 Depending on the agreement between the supplier and the puchaser, the results of the test may be evaluated by weight loss or microscopical examination as indicated in Table 2. See Sections 30 and 31.

PRACTICE Y—COPPER-COPPER SULFATE- 50 % SULFURIC ACID TEST FOR DETERMINING SUSCEPTIBILITY TO INTERGRANULAR ATTACK IN FERRITIC STAINLESS STEELS

17. Scope

17.1 This practice describes the procedure

for conducting the boiling copper-copper sulfate-50 % sulfuric acid test which measures the susceptibility of stainless steels to intergranular attack. This test detects susceptibility to intergranular attack associated with the precipitation of chromium carbides or nitrides in unstabilized and stabilized ferritic stainless steels.

17.2 This test may be used to evaluate the susceptibility of as-received material to intergranular corrosion caused by chromium carbide or nitride precipitation. It may also be used to evaluate the resistance of high purity or stabilized grades to sensitization to intergranular attack caused by welding or heat treatments. It may be applied to wrought products.

17.3 This test should not be used to detect susceptibility to intergranular attack resulting from the formation or presence of chi phase, sigma phase, or titanium carbides or nitrides. For detecting susceptibility to environments known to cause intergranular attack due to these phases use Practice X.

18. Apparatus

18.1 The basic apparatus is described in Section 3. Also needed are:

18.1.1 For weight loss determination, an analytical balance capable of weighing to the nearest 0.001 g.

18.1.2 For microscopical examination, a microscope with magnification to at least 40×.

18.1.3 A piece of copper metal about ⅛ by ¾ by 1½ in. (3.2 by 19 by 38 mm) with a bright, clean finish. An equivalent area of copper shot or chips may be used. The copper should be washed and degreased before use. A rinse in 5 % H_2SO_4 will clean corrosion products from the copper.

19. Copper-Copper Sulfate-50 % Sulfuric Acid Test Solution

19.1 Prepare 600 mL of test solution as follows. **Caution**—Protect the eyes and face by face shield and use rubber gloves and apron when handling acid. Place flask under hood.

19.1.1 First, measure 400.0 mL of distilled water in a 500-mL graduate and pour into the Erlenmeyer flask.

19.1.2 Then measure 236.0 mL of reagent grade sulfuric acid of a concentration that must be in the range from 95.0 to 98.0 weight % in a 250-mL graduate. Add the acid slowly to the water in the Erlenmeyer flask to avoid boiling by the heat evolved.

19.1.3 Weigh 72 g of reagent grade cupric sulfate ($CuSO_4 \cdot 5H_2O$) and add to the sulfuric acid solution. A trip balance may be used.

19.1.4 Place the copper piece into one glass cradle and put it into the flask.

19.1.5 Drop boiling chips into the flask.

19.1.6 Lubricate the ground-glass joint with silicone grease.

19.1.7 Cover the flask with the condenser and circulate cooling water.

19.1.8 Boil the solution until all of the copper sulfate is dissolved.

20. Preparation of Test Specimens

20.1 Prepare test specimens as described in Section 4.

21. Procedure

21.1 When weight loss is to be determined, measure the sample prior to final cleaning and then weigh.

21.1.1 Measure the sample including the inner surfaces of any holes, and calculate the total area.

21.1.2 Degrease and dry the specimen using suitable nonchlorinated agents, such as soap and acetone, and then weigh to the nearest 0.001 g.

21.2 Place the specimen in another glass cradle and immerse in boiling solution.

21.3 Mark the liquid level on the flask with wax crayon to provide a check on vapor loss which would result in concentration of the acid. If there is an appreciable change in the level, repeat the test with fresh solution and a reground specimen.

21.4 Continue immersion of the specimen for the time shown in Table 2, then remove the specimen, rinse in water and acetone, and dry. Times for alloys not listed in Table 2 are subject to agreement between the supplier and the purchaser.

21.5 For weight loss determination, weigh the specimen and subtract this weight from the original weight.

21.6 No intermediate weighings are usually necessary. The tests can be run without interruption. However, if preliminary results are desired, the specimen can be removed at any time for weighing.

21.7 No changes in solution are necessary during the test period.

22. Evaluation

22.1 Depending on the agreement between the supplier and the purchaser, the results of the test may be evaluated by weight loss or microscopical examination as indicated in Table 2. See Sections 30 and 31.

PRACTICE Z—COPPER-COPPER SULFATE-16 % SULFURIC ACID TEST FOR DETECTING SUSCEPTIBILITY TO INTERGRANULAR ATTACK IN FERRITIC STAINLESS STEELS

23. Scope

23.1 This practice describes the procedure by which the copper-copper sulfate-16 % sulfuric acid test is conducted to determine the susceptibility of ferritic stainless steels to intergranular attack. This test detects susceptibility to intergranular attack associated with the precipitation of chromium carbides or nitrides in stabilized and unstabilized ferritic stainless steels.

23.2 This test may be used to evaluate the heat treatment accorded as-received material. It may also be used to evaluate the effectiveness of stabilizing element additions (Cb, Ti, etc.) and reductions in interstitial content to aid in resistance to intergranular attack. It may be applied to all wrought products and weld metal.

23.3 This test does not detect susceptibility associated with chi phase, sigma phase, or titanium carbides or nitrides. For detecting susceptibility in environments known to cause intergranular attack due to these phase, use Practice X.

24. Apparatus

24.1 The basic apparatus is described in Section 3.

25. Copper-Copper Sulfate-16 % Sulfuric Acid Test Solution

25.1 Dissolve 100 g of copper sulfate ($CuSO_4 \cdot 5H_2O$) in 700 mL of distilled water, add 100 mL of sulfuric acid (H_2SO_4, reagent grade, sp gr 1.84), and dilute to 1000 mL with distilled water.

NOTE 4—The solution will contain approximately 6 weight % of anhydrous $CuSO_4$, and 16 weight % of H_2SO_4.

26. Copper Addition

26.1 Electrolytic grade copper shot or grindings may be used. Shot is preferred for its ease of handling before and after the test.

26.2 A sufficient quantity of copper shot or grindings shall be used to cover all surfaces of the specimen whether it is in a vented cradle or embedded in a layer of copper shot on the bottom of the test flask.

26.3 The amount of copper used, assuming an excess of metallic copper is present, is not critical. The effect of galvanic coupling between copper and the test specimen may have importance **(6)**.

26.4 The copper shot or grindings may be reused if they are cleaned in warm tap water after each test.

27. Preparation of Test Specimens

27.1 Prepare test specimens as described in Section 4.

28. Procedure

28.1 The volume of acidified copper sulfate test solution used should be sufficient to completely immerse the specimens and provide a minimum of 50 mL/in.^{2} (8 mL/cm^2).

28.1.1 As many as three specimens can be tested in the same container. It is ideal to have all the specimens in one flask to be of the same grade, but it is not absolutely necessary. The solution volume-to-sample area ratio shall be maintained.

NOTE 5—It may be necessary to embed large specimens, such as from heavy bar stock, in copper shot on the bottom of the test flask. A copper cradle may also be used.

28.1.2 The test specimen(s) should be immersed in ambient test solution which is then brought to a boil and maintained boiling throughout the test period. Begin timing the test period when the solution reaches the boiling point.

NOTE 6—Measures should be taken to minimize bumping of the solution when glass cradles are used to support specimens. A small amount of copper shot (eight to ten pieces) on the bottom of the flask will conveniently serve this purpose.

28.1.3 The test shall consist of one 24-h boiling period unless a longer time is specified. See Table 2. Times longer than 24 h should be included in the test report. Fresh test solution would not be needed if the test were to run 48 or 72 h. (If any adherent copper remains on the specimen, it may be removed by a brief immersion in concentrated nitric acid at room temperature. The sample is then rinsed in water and dried.)

29. Evaluation

29.1 As shown in Table 2, the results of this test are evaluated by a bend test. See Section 32.

EVALUATION METHODS

30. Evaluation by Weight Loss

30.1 Measure the effect of the acid solution on the material by determining the loss of weight of the specimen. Report the corrosion rates as inches of penetration per month, calculated as follows:

$$\text{Inches per month} = 287 \times W/A \times t \times d$$

where:
t = time of exposure, h,
A = area, cm^2,
W = weight loss, g, and
d = density, g/cm^3. For steels 14-20Cr, d = 7.7 g/cm^3; for steels with more than 20Cr, d = 7.6 g/cm^3.

NOTE 7—Conversion factors to other commonly used units for corrosion rates are as follows:
Inches per month × 12 = inches per year
Inches per month × 1000 = mils per month
Inches per month × 12 000 = mils per year
Inches per month × 8350 × density = milligrams per square decimetre per day
Inches per month × 34.8 × density = grams per square metre per hour
1.00 in.^{2} = 6.45 cm^2

30.2 What corrosion rate is indicative of intergranular attack depends on the alloy and must be determined by agreement between the supplier and the purchaser. Some experience with corrosion rates of ferritic stainless steels in Practices X and Y is given in the literature (5).

31. Evaluation by Microscopical Examination

31.1 Examine the test specimens for Practices X and Y under a binocular microscope at 40× magnification. Grain dropping is usually an indication of intergranular attack, but the number of dropped grains per unit area that can be tolerated is subject to agreement between the supplier and the purchaser.

32. Evaluation by Bend Test

32.1 Bend the test specimen through 180° and over a radius equal to twice the thickness of the specimen being bent (see Fig. 5). In no case shall the specimen be bent over a smaller radius or through a greater angle than that specified in the product specification. In cases of material having low ductility, such as severely cold worked material, a 180° bend may prove impractical. Determine the maximum angle of bend without causing cracks in such material by bending an untested specimen of the same configuration as the specimen to be tested. Welded samples should be bent in such a manner that weld and the heat-affected zone are strained.

32.1.1 Obtain duplicate specimens from sheet material so that both sides of the rolled samples may be bent through a 180° bend. This will assure detection of intergranular attack resulting from carburizing of one surface of sheet material during the final stages of rolling.

NOTE 8—Identify the duplicate specimens in such a manner as to ensure both surfaces of sheet material being tested are subjected to the tension side of the 180° bends.

32.1.2 Samples machined from round sections shall have the curved or original surface on the outside of the bend.

32.1.3 The specimens are generally bent by holding in a vise and starting the bend with a hammer. It is generally completed by bringing the two ends together in the vise. Heavy specimens may require bending in a fixture of suitable design. An air or hydraulic press may also be used for bending the specimens.

32.1.4 Flatten tubular products in accordance with the flattening test prescribed in Methods and Definitions A 370.

32.2 Examine the bent specimen under low (5 to 20×) magnification (see Fig. 6). The appearance of fissures or cracks indicates the presence of intergranular attack (see Fig. 7).

32.2.1 When an evaluation is questionable, determine presence or absence of intergranular attack by metallographic examination of a longitudinal section of the specimen at a magnification of 100 to 250×.

NOTE 9—Cracking that originates at the edge of the specimen should be disregarded. The appearance of deformation lines, wrinkles, or "orange peel" on the surface, without accompanying cracks or fissures, should be disregarded also.

NOTE 10—Cracks suspected as arising through poor ductility may be investigated by bending a similar specimen that was not exposed to the boiling test solution. A visual comparison between these specimens should assist in interpretation.

REFERENCES

(1) Streicher, M. A., "Theory and Application of Evaluation Tests for Detecting Susceptibility to Intergranular Attack in Stainless Steels and Related Alloys—Problems and Opportunities," *Intergranular Corrosion of Stainless Alloys, ASTM STP 656*, R. F. Steigerwald, Ed., 1978, pp. 3–84.

(2) Dundas, H. J., and Bond, A. P., "Niobium and Titanium Requirements for Stabilization of Ferritic Stainless Steels," *Intergranular Corrosion of Stainless Alloys, ASTM STP 656*, R. F. Steigerwald, Ed., 1978, pp. 154–178.

(3) Nichol, T. J., and Davis, J. A., "Intergranular Corrosion Testing and Sensitization of Two High-Chromium Ferritic Stainless Steels," *Intergranular Corrosion of Stainless Alloys, ASTM STP 656*, R. F. Steigerwald, Ed., 1978, pp. 179–196.

(4) Sweet, A. J., "Detection of Susceptibility of Alloy 26-1S to Intergranular Attack," *Intergranular Corrosion of Stainless Alloys, ASTM STP 656*, R. F. Steigerwald, Ed., 1978, pp. 197–232.

(5) Streicher, M. A., "The Role of Carbon, Nitrogen, and Heat Treatment in the Dissolution of Iron Chromium Alloys in Acids," *Corrosion*, Vol 29, pp 337–360.

(6) Herbsleb, G., and Schwenk, W., "Untersuchungen zur Einstellung des Redoxpotentials der Strausschen Lösung mit Zusatz von Metalleischem Kufer," *Corrosion Science*, Vol 7, 1967, pp. 501–511.

TABLE 1 Nominal Compositions of Ferritic Stainless Steels

UNS Designation	Alloy	Weight Percent Maximum Unless Otherwise Specified													Practice(s)
		C	Mn	P	S	Si	Cr	Ni	Mo	Cu	N	Ti	Nb	Other	
S43000	430[A]	0.12	1.00	0.040	0.030	0.75	16.0–18.0	0.5	...	...	...	...	...	...	X, Z
S43400	434[A]	0.12	1.00	0.040	0.030	1.00	16.0–18.0	...	0.75–1.25	...	...	...	...	...	Z
S43600	436[A]	0.12	1.00	0.040	0.030	1.00	16.0–18.0	...	0.75–1.25	...	...	...	5 × C 0.70 max	...	Z
S43035	XM8	0.07	1.00	0.040	0.030	0.75	17.0–19.0	0.5	...	...	0.04	0.2 + 4(C + N) min, 1.10 max	...	0.15 Al	Z
S44400	18Cr-2Mo	0.025	1.00	0.040	0.030	1.00	17.5–19.5	1.00	1.75–2.50	...	0.025	Ti + Nb = 0.2 + 4(C + N) min, 0.80 max	...	...	W, Z
S44600	446[A]	0.20	1.50	0.040	0.030	0.75	23.0–27.0	0.50	...	...	0.10 0.25	...	...	...	X, Y
S44626	XM33	0.06	0.75	0.040	0.020	0.75	25.0–27.0	0.50	0.75–1.50	0.20	0.040	0.2–1.0, 7(C + N) min	...	...	W, Y
S44627	XM27	0.01	0.4	0.02	0.02	0.40	25.0–27.5	0.5[B]	0.75–1.50	0.2	0.015	...	0.05–0.2	...	W, X, Y
S44700	29-4Mo	0.010	0.3	0.025	0.02	0.2	28.0–30.0	0.15	3.5–4.2	0.15	0.020[C]	...	...	...	X, Y
S44735	29-4C	0.030	1.0	0.040	0.030	1.00	28.0–30.0	1.0	3.6–4.2	...	0.045	Ti + Cb = 0.2–1.0, 6(C + N) min	...	...	Y
S44800	29Cr-4Mo-2Ni	0.010	0.3	0.025	0.02	0.2	28.0–30.0	2.0–2.5	3.5–4.2	0.15	0.020[C]	...	...	...	X, Y

[A] Types 430, 434, 436, and 446 are nonstabilized grades that are generally not used in the as-welded or sensitized condition in other than mildly corrosive environments. In the annealed condition, they are not subject to intergranular corrosion. For any studies of IGA on Types 430, 434, 436, or 446, the indicated test methods are suggested.
[B] Nickel plus copper.
[C] Carbon plus nitrogen = 0.025 max.

ASTM A 763

TABLE 2 Methods for Evaluating Ferritic Stainless Steels for Susceptibility to Intergranular Corrosion

Alloy	Time of Test, h	Weight Loss	Microscopical Examination	Bend Test
\multicolumn{5}{c}{PRACTICE W—OXALIC ACID ETCH TEST}				
18Cr-2Mo	0.025	NA	A^E	NA
XM27	0.025	NA	A^E	NA
XM33	0.025	NA	A^E	NA
\multicolumn{5}{c}{PRACTICE X—FERRIC SULFATE - SULFURIC ACID TEST}				
430	24	$A^{A,B}$	A	NA
446	72	A^B	A	NA
XM27	120	A^C	A^B	NA
29Cr-4Mo	120	NA^D	A^B	NA
29Cr-4Mo-2Ni	120	NA	A^B	NA
\multicolumn{5}{c}{PRACTICE Y—COPPER-COPPER SULFATE - 50% SULFURIC ACID TEST}				
446	96	A^B	A	NA
XM27	120	A^C	A^B	NA
XM33	120	A^C	A^B	NA
29-4C	120	A^C	A^B	NA
29Cr-4Mo	120	NA	A^B	NA
29Cr-4Mo-2Ni	120	NA	A^B	NA
\multicolumn{5}{c}{PRACTICE Z—COPPER-COPPER SULFATE - 16% SULFURIC ACID TEST}				
430	24	NA	NA	no fissures
434	24	NA	NA	no fissures
436	24	NA	NA	no fissures
XM8	24	NA	NA	no fissures
18Cr-2Mo	24	NA	NA	no fissures

[A] A = Applicable.
[B] Preferred criterion, these criteria are the most sensitive for the particular combination of alloy and test.
[C] Weight loss measurements can be used to detect severely sensitized material, but they are not very sensitive for alloys noted with this superscript and may not detect slight or moderate sensitization.
[D] NA = Not applicable.
[E] Polished surface examined at 250 to 500× with a metallurgical microscope (see 3.1.6). All other microscopical examinations are of the corroded surface under 40× binocular examination (see Section 27).

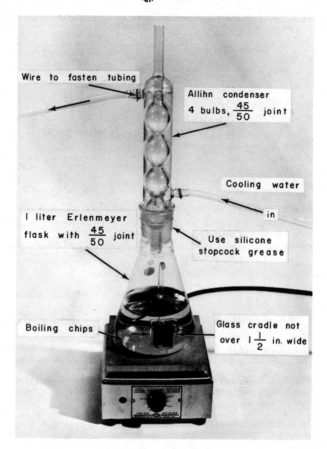

FIG. 1 Test Apparatus

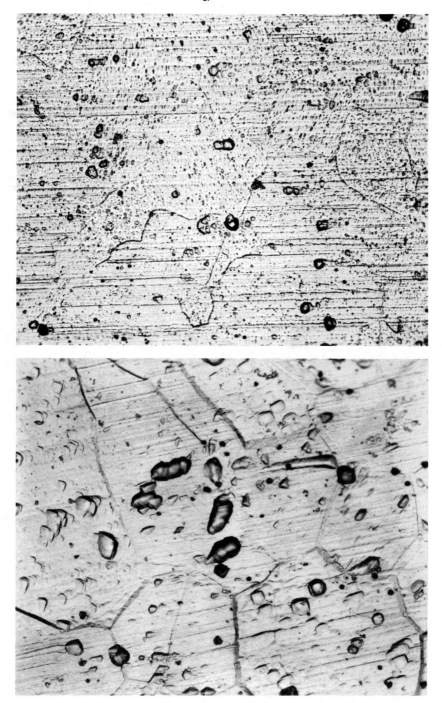

FIG. 2 Acceptable Structures Practice W—Oxalic-Acid Etch Test Steps Between Grains No Ditching

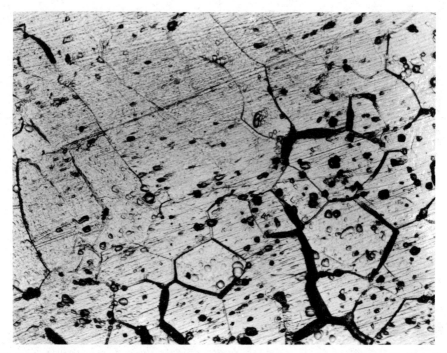

FIG. 3 Acceptable Structure Practice W—Oxalic Acid Etch Test Dual Structure Some Ditches But No Single Grain Completely Surrounded

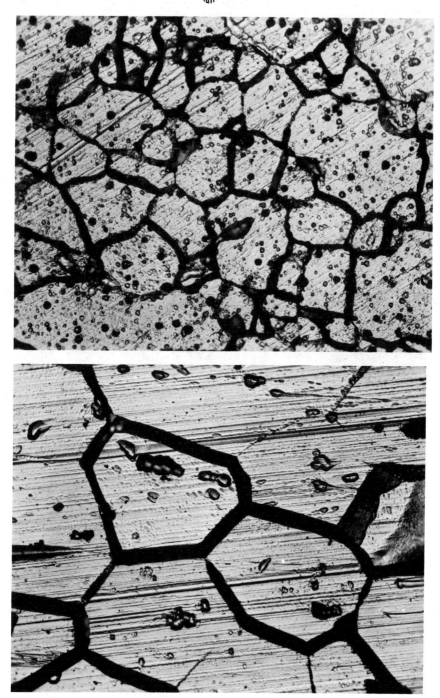

FIG. 4 Unacceptable Structures Practice W—Oxalic-Acid Etch Test Ditched Structure—One Or More Grains Completely Surrounded

FIG. 5 Bend Test Specimen

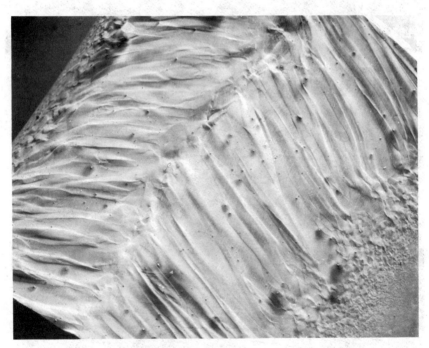

FIG. 6 Bend Test Specimen That Does Not Show Fissures

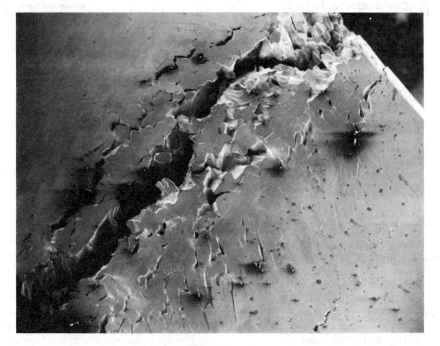

FIG. 7 Bend Test Specimen Showing Intergranular Fissures

The American Society for Testing and Materials takes no position respecting the validity of any patent rights asserted in connection with any item mentioned in this standard. Users of this standard are expressly advised that determination of the validity of any such patent rights, and the risk of infringement of such rights, are entirely their own responsibility.

This standard is subject to revision at any time by the responsible technical committee and must be reviewed every five years and if not revised, either reapproved or withdrawn. Your comments are invited either for revision of this standard or for additional standards and should be addressed to ASTM Headquarters. Your comments will receive careful consideration at a meeting of the responsible technical committee, which you may attend. If you feel that your comments have not received a fair hearing you should make your views known to the ASTM Committee on Standards, 1916 Race St., Philadelphia, Pa. 19103.

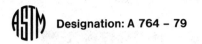

Designation: A 764 – 79

An American National Standard

Standard Specification for
STEEL WIRE, CARBON, DRAWN GALVANIZED AND GALVANIZED AT SIZE FOR MECHANICAL SPRINGS[1]

This standard is issued under the fixed designation A 764; the number immediately following the designation indicates the year of original adoption or, in the case of revision, the year of last revision. A number in parentheses indicates the year of last reapproval. A superscript epsilon (ϵ) indicates an editorial change since the last revision or reapproval.

1. Scope

1.1 This specification covers two finishes of round, zinc-coated, hard-drawn, carbon steel spring wire having properties and quality for the manufacture of mechanical springs and wire forms that are not subject to high stress or require high fatigue properties.

1.2 The values stated in inch-pound units are to be regarded as the standard.

2. Applicable Documents

2.1 *ASTM Standards:*
A 90 Test Method for Weight of Coating on Zinc-Coated (Galvanized) Iron or Steel Articles[2]
A 370 Methods and Definitions for Mechanical Testing of Steel Products[3]
A 510 Specification for General Requirements for Wire Rods and Coarse Round Wire, Carbon Steel[2]
A 700 Recommended Practices for Packaging, Marking, and Loading Methods for Steel Products for Domestic Shipment[4]
A 751 Methods, Practices, and Definitions for Chemical Analyses of Steel Products[5]
B 6 Specification for Zinc (Slab Zinc)[6]
E 29 Recommended Practice for Indicating Which Places of Figures Are to Be Considered Significant in Specified Limiting Values[7]

2.2 *U. S. Government Standards:*[8]
Fed. Std. No. 123 Marking for Shipment (Civil Agencies)
MIL-STD-129 Marking for Shipment and Storage
MIL-STD-163 Steel Mill Products, Preparation for Shipment and Storage

3. Classification

3.1 Wire covered by this specification may be either hot dipped or electro-galvanized in the following finishes:

Finish 1—"Drawn galvanized" (zinc coating is applied prior to the final wire drawing operation) with a Regular or Type 1 coating on the final wire, in two classes of tensile strength.

Finish 2—"Galvanized at size" (zinc coating is applied after the final wire drawing operation) with a Regular, Type 1, Type 2, or Type 3 coating, in two classes of tensile strength.

4. Ordering Information

4.1 Orders for material under this specification shall include the following information:
4.1.1 Quantity (weight),
4.1.2 Diameter, finish, type of coating, class of tensile strength, name of material,
4.1.3 ASTM designation and date of issue,
4.1.4 Packaging (Section 15),
4.1.5 Cast or heat analysis report, if desired (Section 6), and
4.1.6 Test report, if desired (Section 14).

---

[1] This specification is under the jurisdiction of ASTM Committee A-1 on Steel, Stainless Steel, and Related Alloys, and is the direct responsibility of Subcommittee A01.03 on Steel Rod and Wire.
Current edition approved July 27, 1979. Published November 1979.
[2] *Annual Book of ASTM Standards,* Vol 01.06.
[3] *Annual Book of ASTM Standards,* Vol 01.04.
[4] *Annual Book of ASTM Standards,* Vols 01.01, 01.03, 01.04, and 01.05.
[5] *Annual Book of ASTM Standards,* Vols 01.01, 01.02, 01.03, 01.04, and 01.05.
[6] *Annual Book of ASTM Standards,* Vol 02.04.
[7] *Annual Book of ASTM Standards,* Vol 14.02.
[8] Available from Naval Publications and Forms Center, 5801 Tabor Ave., Philadelphia, Pa. 19120.

A 764

NOTE 1—A typical ordering description is as follows: 30 000 lb, 0.120 in. Finish 1, Type 1 Coating Drawn Galvanized, Class I, Steel Mechanical Spring Wire in 400-lb, 22-in. coils to ASTM A 764, dated _____.

5. Materials and Manufacture

5.1 The steel shall be made by any of the following processes: open hearth, basic oxygen, or electric furnace.

5.2 A sufficient discard shall be made to ensure freedom from injurious piping and undue segregation.

5.3 The material shall be galvanized prior to cold drawing, or at an intermediate stage of cold drawing (Finish 1), or shall be galvanized after the final cold drawing (Finish 2).

5.4 The slab zinc when used shall be any grade of zinc conforming to Specification B 6.

6. Chemical Requirements

6.1 Since the primary criteria of spring wire are the mechanical properties developed, steel composition is usually at the producer's discretion, compatible with the processing method.

6.2 *Cast or Heat Analysis (formerly Ladle Analysis)*—Each cast or heat of steel shall be analyzed by the manufacturer to determine compliance with the percentage of elements prescribed in Table 1. This analysis shall be from a test specimen preferably taken during the pouring of the cast or heat. When required, this shall be reported to the purchaser.

6.3 *Product Analysis (formerly Check Analysis)*—An analysis may be made by the purchaser from finished wire representing each cast or heat of steel. The chemical composition thus determined, as to elements required or restricted, shall conform to the product analysis requirements specified in Table 10 in Specification A 510.

6.4 For referee purposes, Methods A 751 shall be used.

7. Mechanical Requirements

7.1 *Tension Test:*

7.1.1 *Requirements*—The material as represented by tension test specimens shall conform to the requirements prescribed in Table 2 or 3 for the various sizes and specified class.

7.1.2 *Test Method*—The tension test shall be made in accordance with Methods A 370, Supplement IV.

7.2 *Wrap Test:*

7.2.1 *Requirement*—Finish 1 (drawn galvanized) wire shall withstand wrapping on a mandrel as shown in Table 7 without the steel base fracturing or the zinc coating peeling or flaking to such an extent that zinc can be removed by rubbing with the bare fingers. Finish 2 (galvanized at size) wire shall withstand wrapping on a mandrel as shown in Table 8 without the steel base fracturing or the zinc coating peeling or flaking to such an extent that zinc can be removed by rubbing with the bare fingers.

NOTE 2—Loosening or detachment, during the wrap test, of superficial, small particles of zinc formed by mechanical polishing of the zinc-coated wire shall not be considered cause for rejection.

7.2.2 *Test Method*—The wrap test shall be made in accordance with Methods A 370, Supplement IV.

7.3 *Galvanized Coating:*

7.3.1 *Requirement*—The wire shall conform to the weight of zinc coating requirements prescribed in Table 6.

7.3.2 *Test Method*—The weight of zinc coating test shall be made in accordance with Method A 90.

8. Permissible Variations in Dimensions

8.1 The diameter of the wire shall not vary from that specified by more than the tolerances specified in Tables 4 or 5.

9. Workmanship

9.1 The surface of Finish 1, drawn galvanized wire, shall be smooth. No serious die marks, scratches, or seams may be present.

9.2 The surface of Finish 2, galvanized at size wire, shall be free of slivers, scale, and other imperfections not consistent with good commercial practice. The zinc coating shall be reasonably smooth and continuous. As footnoted in Table 5, it is recognized the surface of heavy zinc coatings, particularly those produced by hot dip galvanizing, are not perfectly smooth and devoid of irregularities.

9.3 The wire shall be smooth wound in its package so that it can be unwound in a trouble-free manner.

9.4 Each coil shall be of one continuous length. Only welds made prior to cold drawing are permitted, unless otherwise agreed upon at the time of the purchase or contract.

10. Sampling

10.1 A lot is defined as all of the wire of one size, tensile strength class, finish, that is produced from one heat or cast and is offered for inspection at one time.

10.2 Test specimens shall be taken from either end of the package.

11. Number of Tests and Retests

11.1 A minimum of one test specimen shall be taken for each ten packages, or fraction thereof, in a lot.

11.2 If any test specimen exhibits obvious imperfections or shows the presence of a weld, it may be discarded and another specimen substituted.

12. Inspection

12.1 The manufacturer shall afford the inspector representing the purchaser all reasonable facilities, to satisfy him that the material is being furnished in accordance with this specification. Unless otherwise agreed upon, all tests and inspections may be made at the place of manufacture, prior to shipment, and shall be so conducted as not to interfere unnecessarily with the operation of the works.

13. Rejection and Rehearing

13.1 Unless otherwise specified, any rejection based on tests made in accordance with this specification shall be reported to the manufacturer within a reasonable length of time after receipt of the material.

13.2 The material must be adequately protected and correctly identified in order that the producer may make a proper investigation.

14. Certification

14.1 When specified on the purchase order or contract, a producer's or supplier's certification shall be furnished to the purchaser that the material was manufactured, sampled, tested, and inspected in accordance with this specification and has been found to meet the requirements. When specified in the purchase order or contract, a report of the test results shall be furnished.

15. Packaging, Marking, and Loading

15.1 The coil weight, dimensions, and method of packaging shall be agreed upon between the manufacturer and the purchaser.

15.2 The size of wire, ASTM specification, finish, type, and class number, and name or mark of the manufacturer shall be shown on a tag securely attached to each package of wire.

15.3 Unless otherwise specified in the purchaser's order, packaging, marking, and loading for shipments shall be in accordance with those procedures outlined in Recommended Practices A 700.

15.4 When specified in the contract or order, and for direct procurement by or direct shipment to the U. S. Government, when Level A is specified, preservation, packaging, and packing shall be in accordance with the Level A requirement of MIL-STD-163.

15.5 When specified in the contract or order, and for direct procurement by or direct shipment to the U. S. Government, marking for shipment, in addition to requirements specified in the contract or order, shall be in accordance with MIL-STD-129 for U. S. Military Agencies and in accordance with Fed. Std. No. 123 for U. S. Government Civil Agencies.

TABLE 1 Chemical Requirements

Element	Composition, %
Carbon	0.45–0.85[A]
Manganese	0.30–1.30[B]
Phosphorus, max	0.040
Sulfur, max	0.050
Silicon	0.10–0.35

[A] Carbon in any one lot shall not vary more than 0.13 %.
[B] Manganese in any one lot shall not vary more than 0.30 %.

TABLE 2 Tensile Requirements[A]—Finish 1 (Drawn Galvanized) Regular or Type 1 Coating

Diameter[B] Decimal Size, in. (mm)	Class I Tensile Strength, ksi (MPa)		Class II Tensile Strength, ksi (MPa)	
	min	max	min	max
0.032 (0.81)	253 (1740)	306 (2110)	292 (2010)	347 (2390)
0.035 (0.89)	248 (1710)	301 (2080)	287 (1980)	342 (2360)
0.041 (1.04)	242 (1670)	293 (2020)	279 (1920)	332 (2290)
0.048 (1.22)	236 (1630)	286 (1970)	273 (1880)	325 (2240)
0.054 (1.37)	231 (1590)	279 (1920)	266 (1830)	316 (2180)
0.062 (1.57)	225 (1550)	272 (1880)	259 (1790)	308 (2120)
0.072 (1.83)	220 (1520)	266 (1830)	254 (1750)	301 (2080)
0.080 (2.03)	216 (1490)	261 (1800)	249 (1720)	296 (2040)
0.092 (2.34)	209 (1440)	253 (1740)	241 (1660)	287 (1980)
0.106 (2.69)	205 (1410)	248 (1710)	237 (1630)	281 (1940)
0.120 (3.05)	200 (1380)	241 (1660)	230 (1590)	273 (1880)
0.135 (3.43)	196 (1350)	237 (1630)	226 (1560)	269 (1850)
0.148 (3.76)	193 (1330)	234 (1610)	223 (1540)	266 (1830)

[A] Tensile strength values for intermediate diameters may be interpolated.
[B] Decimal size is rounded to three significant places in accordance with Recommended Practice E 29.

TABLE 3 Tensile Requirements[A]—Finish 2 (Galvanized at Size) Regular, Type 1, Type 2, or Type 3 Coating

Diameter[B] Decimal Size, in. (mm)	Class I Tensile Strength, ksi (MPa)		Class II Tensile Strength, ksi (MPa)	
	min	max	min	max
0.062 (1.57)	213 (1470)	272 (1880)	232 (1600)	291 (2010)
0.072 (1.83)	209 (1440)	266 (1830)	227 (1570)	284 (1960)
0.080 (2.03)	204 (1410)	261 (1800)	223 (1540)	280 (1930)
0.092 (2.34)	198 (1370)	253 (1740)	216 (1490)	271 (1870)
0.106 (2.69)	194 (1340)	248 (1710)	212 (1460)	266 (1830)
0.120 (3.05)	189 (1300)	241 (1660)	206 (1420)	258 (1780)
0.135 (3.43)	185 (1280)	237 (1630)	202 (1390)	254 (1750)
0.148 (3.76)	183 (1260)	234 (1610)	200 (1380)	251 (1730)
0.162 (4.11)	180 (1240)	230 (1590)	196 (1350)	246 (1700)
0.177 (4.50)	176 (1210)	225 (1550)	192 (1320)	241 (1660)
0.192 (4.88)	173 (1190)	221 (1520)	189 (1300)	237 (1630)
0.207 (5.26)	171 (1180)	218 (1500)	186 (1280)	233 (1610)
0.225 (5.72)	167 (1150)	214 (1480)	183 (1260)	230 (1590)
0.250 (6.35)	164 (1130)	210 (1450)	179 (1230)	225 (1550)

[A] Tensile strength values for intermediate diameters may be interpolated.
[B] Decimal size is rounded to three significant places in accordance with Recommended Practice E 29.

TABLE 4 Permissible Variations in Dimensions Finish 1, Drawn Galvanized

Wire Diameter, in. (mm)	Permissible Variations, Plus and Minus, in. (mm)	Permissible Out-of-Round, in. (mm)
0.032 (0.81) to under 0.076 (1.93)	0.001 (0.03)	0.001 (0.03)
0.076 (1.93) to 0.148 (3.76) incl	0.002 (0.05)	0.002 (0.05)

TABLE 5 Permissible Variations in Dimensions Finish 2, Galvanized at Size

Wire Diameter		Tolerance,[A] Plus and Minus, in. (mm)		
in.	mm	Regular and Type 1 Coating	Type 2 Coating	Type 3 Coating
0.062 to under 0.076	1.57 under 1.93	0.002 (0.05)	0.002 (0.05)	0.002 (0.05)
0.076 to under 0.250 incl	1.93 to under 6.35	0.003 (0.08)	0.003 (0.08)	0.004 (0.102)

[A] It is recognized that the surface of heavy zinc coatings, particularly those produced by hot galvanizing, are not perfectly smooth and devoid of irregularities. If the tolerances shown above are rigidly applied to such irregularities that are inherent to the product, unjustified rejections of wire that would actually be satisfactory for use could occur. Therefore, it is intended that these tolerances be used in gaging the uniform areas of the galvanized wire.

[B] For the purpose of determining conformance with this specification, an observed value shall be rounded to the nearest 0.001 in. in accordance with the rounding method of Recommended Practice E 29.

TABLE 6 Minimum Weight of Zinc per Unit Area of Uncoated Wire Surface

Wire Diameter,[A] in. (mm)	Regular Coating		Type 1 Coating		Type 2 Coating		Type 3 Coating	
	oz/ft^2	g/m^2	oz/ft^2	g/m^2	oz/ft^2	g/m^2	oz/ft^2	g/m^2
0.032 (0.81)			0.05	16	0.20	61	0.25	76
0.035 (0.89)			0.10	31	0.30	92	0.40	122
0.041 (1.04)			0.10	31	0.30	92	0.40	122
0.048 (1.22)			0.15	46	0.30	92	0.40	122
0.054 (1.37)			0.15	46	0.30	92	0.40	122
0.062 (1.57)			0.15	46	0.35	107	0.50	153
0.072 (1.83)			0.15	46	0.35	107	0.50	153
0.076 (1.93)			0.20	61	0.40	122	0.60	183
0.080 (2.03)			0.25	76	0.45	137	0.65	198
0.092 (2.34)			0.30	92	0.50	153	0.70	214
0.099 (2.51)	no minimum required		0.30	92	0.50	153	0.80	244
0.106 (2.69)			0.30	92	0.50	153	0.80	244
0.120 (3.05)			0.30	92	0.50	153	0.80	244
0.135 (3.43)			0.30	92	0.50	153	0.80	244
0.148 (3.76)			0.40	122	0.60	183	0.80	244
0.162 (4.11)			0.40	122	0.60	183	0.80	244
0.177 (4.50)			0.40	122	0.60	183	0.80	244
0.192 (4.88)			0.50	153	0.70	214	0.90	275
0.207 (5.26)			0.65	198	0.75	229	0.90	275
0.225 (5.72)			0.65	198	0.75	229	0.90	275
0.250 (6.35)			0.65	198	0.75	229	0.90	275

[A] Diameters, other than those shown above, are produced with zinc coating equivalent to those of the next smaller size.

TABLE 7 Mandrel Diameters for Steel Ductility and Adherence of Zinc Coating Test for Finish 1 Wire (Drawn Galvanized)

Wire Diameter		Mandrel Diameter	
in.	mm	Class I Tensile	Class II Tensile
0.032 to 0.148 incl	0.81 to 3.76 incl	1D[A]	2D

[A] D equals nominal wire diameter being tested. For 1D mandrel, wire may be wound on itself.

TABLE 8 Mandrel Diameters for Steel Ductility and Adherence of Zinc Coating Test for Finish 2 Wire (Galvanized at Size)

Wire Diameter		Mandrel Diameters for Various Coating and Tensile Strength Classes					
		Regular and Type I Coating		Type 2 Coating		Type 3 Coating	
in.	mm	Class I Tensile	Class II Tensile	Class I Tensile	Class II Tensile	Class I Tensile	Class II Tensile
0.062 to under 0.076	1.57 to under 1.93	$1D^A$	$2D$	$1D$	$2D$	$2D$	$3D$
0.076 to under 0.148	1.93 to under 3.76	$2D$	$3D$	$2D$	$3D$	$3D$	$4D$
0.148 to under 0.162	3.76 to 4.11 incl	$2D$	$4D$	$3D$	$4D$	$4D$	$5D$
0.162 to 0.250 incl	4.11 to 6.35 incl	$3D$	$5D$	$3D$	$5D$	$4D$	$5D$

[A] D equals nominal wire diameter being tested.

The American Society for Testing and Materials takes no position respecting the validity of any patent rights asserted in connection with any item mentioned in this standard. Users of this standard are expressly advised that determination of the validity of any such patent rights, and the risk of infringement of such rights, are entirely their own responsibility.

This standard is subject to revision at any time by the responsible technical committee and must be reviewed every five years and if not revised, either reapproved or withdrawn. Your comments are invited either for revision of this standard or for additional standards and should be addressed to ASTM Headquarters. Your comments will receive careful consideration at a meeting of the responsible technical committee, which you may attend. If you feel that your comments have not received a fair hearing you should make your views known to the ASTM Committee on Standards, 1916 Race St., Philadelphia, Pa. 19103.

Designation: A 793 – 81

Standard Specification for
ROLLED FLOOR PLATE, STAINLESS STEEL[1]

This standard is issued under the fixed designation A 793; the number immediately following the designation indicates the year of original adoption or, in the case of revision, the year of last revision. A number in parentheses indicates the year of last reapproval. A superscript epsilon (ε) indicates an editorial change since the last revision or reapproval.

1. Scope

1.1 This specification covers stainless steel floor plates for use in galley spaces, washrooms, engine rooms, and machinery spaces, and for ladder treads, gun platforms, and deck treads. For these uses, Patterns A and B are considered interchangeable (see Figs. 1 and 2).

2. Applicable Documents

2.1 *ASTM Standards:*
A 340 Definitions of Terms, Symbols, and Conversion Factors Relating to Magnetic Testing[2]
A 370 Methods and Definitions for Mechanical Testing of Steel Products[3]
A 480/A 480M Specification for General Requirements for Flat-Rolled Stainless and Heat-Resisting Steel Plate, Sheet, and Strip[3]
A 700 Practices for Packaging, Marking, and Loading Methods for Steel Products for Domestic Shipment[3]

2.2 *Military Standards:*
MIL-I-17214 Indicator, Permeability; Low Mu (Go-No-Go)[4]
MIL-STD-163 Preservation of Steel Products for Domestic Shipment (Storage and Overseas Shipment)[4]

3. General Requirements for Delivery

3.1 Material furnished under this specification shall conform to applicable requirements of the current edition of Specification A 480/A 480M.

4. Ordering Information

4.1 Orders for material under this specification shall include the following information:
4.1.1 Quantity—number of pieces or weight,
4.1.2 Dimensions (thickness, width, and length),
4.1.3 Name of material—stainless steel,
4.1.4 ASTM designation and date of issue,
4.1.5 Special requirements such as magnetic permeability test,
4.1.6 Preparation for delivery, and
4.1.7 Marking requirements.

NOTE 1—A typical ordering description is as follows: 100 pieces, stainless steel floor plates, 0.1875 by 60 by 120 in., Type 304 (S30400) ASTM A 793, dated _____.

5. Manufacture

5.1 All plates shall be hot-rolled, annealed, and pickled.
5.2 Annealing shall be the last heat treatment to which the material will be subjected by the manufacturer.
5.3 The stainless steel floor plates shall be of the following patterns, at the option of the manufacturer.
5.3.1 *Pattern A*—Angular (see Fig. 1).
5.3.2 *Pattern B*—Angular (see Fig. 2).

6. Chemical Composition

6.1 The heat chemical composition shall be reported to the purchaser, or his representative, and shall conform to the requirements specified in Table 1.
6.2 For the purpose of chemical analysis, a lot shall consist of all floor plates and sheets

---

[1] This specification is under the jurisdiction of ASTM Committee A-1 on Steel, Stainless Steel, and Related Alloys, and is the direct responsibility of Subcommittee A01.17 on Flat Stainless Steel Products.
Current edition approved Oct. 30, 1981. Published December 1981.
[2] *Annual Book of ASTM Standards*, Vol 03.04.
[3] *Annual Book of ASTM Standards*, Vol 01.03.
[4] Available from Naval Publications and Forms Center, 5801 Tabor Ave., Philadelphia, Pa. 19120.

made from the same heat. In case the material cannot be identified by melt or heat, a lot shall consist of not more than 25 tons (22.7 Mg) of floor plates and sheets offered for delivery at the same time.

6.3 One sample of suitable size shall be selected from each lot identified by heat. When the material cannot be identified by melt or heat, five separate samples from each lot shall be selected. Samples may be taken from bend test specimens representative of the lot. The chemical compositions thus determined shall conform to the expanded tolerances for product analysis shown in Table 1 of Specification A 480/A 480M. If any sample fails to conform, this shall be cause for rejection of the lot represented by the sample.

7. Mechanical Properties

7.1 *Lot Size*—A lot shall consist of all floor plates of the same thickness made from the same heat. In case the material cannot be identified by melt, a lot shall consist of not more than 25 tons (22.7 Mg) of floor plates of the same thickness.

7.2 *Bend Test:*

7.2.1 From each lot, two longitudinal cold bend test specimens, 18 in. long by 2 in. wide (457 by 51 mm), shall be selected. When a lot cannot be identified by melt, five longitudinal cold bend test specimens, 18 in. long by 2 in. wide, shall be selected. Each of the specimens shall be taken from a different plate.

7.2.2 The specimens shall be bent cold, with the figures in the inside of the bend test specimens, through an angle of 180° flat on itself. The raised figures shall not be removed and the material shall not fracture nor develop cracks or flaws when subjected to the cold bend test. In the case of failure of any of the bend test specimens to conform, this shall be cause for rejection of the lot represented by the sample.

7.3 *Magnetic Permeability Test:*

7.3.1 When magnetic permeability is specified, a lot shall consist of all floor plates of the same thickness made from the same melt. In case the material cannot be identified by melt, a lot shall consist of not more than 25 tons (22.7 Mg) of floor plates of the same thickness.

7.3.2 Specimen for magnetic permeability test shall be 1-3/16 in. wide by 2 in. by not more than 1/2 in. thick (30 by 51 by 12.7 mm). Samples representing annealed material may be annealed and pickled after machining and before testing for permeability.

7.3.3 The magnetic permeability shall not exceed 1.20 when the magnetic permeability indicator of MIL-I-17214 is used. In case of failure of the specimen to comply, this shall be cause for rejection of the lot represented by the sample.

8. Permissible Variations in Dimensions

8.1 Dimensions, weights, and special characteristics of Patterns A and B shall be as specified in Table 2. The stainless steel floor plates shall have raised figures on one surface of the plate. The reverse side shall be flat except that the portion below the raised figure on plates having a nominal weight below 3.50 lb/ft^2 (17 kg/m^2) may be hollow.

8.2 Plates shall not exceed the respective weight specified in Table II by more than 8 percent.

8.3 Variations over the specified width and length shall not exceed the amounts permitted in Table 3. Variations under the specified width and length shall not exceed 1/4 in. (6.3 mm).

8.4 The thickness of the floor plates shall conform to the requirements of Table 2 and shall be measured at least 3/8 in. (9.5 mm) from the edge of the plate exclusive of the raised figures.

8.5 *Camber Tolerances*—The camber tolerance for like raised figures of rolled floor plates shall be determined by the following formula:

Camber tolerance, in.

$$= \frac{3/8 \times \text{number of feet of length}}{5}$$

NOTE 2—Length shall be taken as the direction along which the camber is to be measured.

8.6 Floor plates of stainless steel shall have figures not less than 5/8 in. (15.9 mm) nor more than 1-1/2 in. (38.1 mm) long at the base. The figures shall have a uniform pitch of 7/8 in. (22.2 mm) apart from center to center. A variation of ±1/32 in. (0.8 mm) will be permitted. Angular pattern plates and sheets shall have the figures arranged so that they are in an angular position from a vertical or horizontal line. Plates shall have a raised figure and raised portion of the plates shall cover at least 30% of the total surface of one side of the plate. (Figures 1 and 2 approximate actual horizontal dimensions and show the plates acceptable as to pattern of raised figures.)

8.7 Floor plates shall be sheared on a line of 45° to the axis of the figures and located so far as practicable in such position as to cut through the minimum amount of raised figure.

9. Workmanship

9.1 Floor plates of stainless steel shall be uniform in quality and condition, free of injurious defects, that, due to their nature, or severity may detrimentally affect the suitability for the service intended.

10. Inspection

10.1 Inspection of the material by the purchaser's representative at the producing plant shall be made as agreed upon between the purchaser and the seller as part of the purchase order.

11. Packaging, Marking, and Loading

11.1 Unless otherwise specified, packaging, marking, and loading shall be in accordance with those procedures recommended by Practices A 700.

11.2 *For Government Procurement*—When specified in the contract or order, marking or preparation for shipment shall be in accordance with MIL-STD-163.

TABLE 1 Heat Chemical Composition, %

Type	UNS Designation	C, max	Mn, max	P, max	S, max	Si, max	N, max	Ni	Cr
304	S30400	0.08	2.00	0.045	0.030	1.00	0.10	8.00–10.50	18.00–20.00

TABLE 2 Dimensions, Weights, and Special Characteristics of Patterns A and B

Thickness of Plate at Base of Raised Figure (Nominal), in. (mm)	Thickness of Plate at Base of Raised Figure, min, in. (mm)	Weight (Approximate), lb/ft² (kg/m²)	Height of Raised Figures, min, in. (mm)	Styles
0.025 (0.64)	0.015 (0.38)	1.10 (5.37)	0.025 (0.64)	hollow back
0.03125 (0.79)	0.02125 (0.54)	1.40 (6.84)	0.025 (0.64)	hollow back
0.0375 (0.95)	0.027 (0.69)	1.165 (5.69)	0.025 (0.64)	hollow back
0.05 (1.27)	0.039 (0.99)	2.30 (11.23)	0.035 (0.89)	hollow back
0.0625 (1.59)	0.051 (1.30)	2.90 (14.16)	0.035 (0.89)	hollow back
0.078 (1.98)	0.063 (1.60)	3.50 (17.09)	0.035 (0.89)	hollow back
0.09375 (2.38)	0.079 (2.01)	4.875 (23.80)	0.035 (0.89)	flat back
0.109 (2.77)	0.096 (2.44)	5.50 (26.85)	0.045 (1.14)	flat back
0.125 (3.18)	0.110 (2.79)	6.125 (29.90)	0.055 (1.40)	flat back
0.140 (3.56)	0.126 (3.20)	6.750 (32.96)	0.055 (1.40)	flat back
0.156 (3.96)	0.141 (3.58)	7.375 (36.01)	0.055 (1.40)	flat back
0.172 (4.37)	0.157 (3.99)	8.000 (39.06)	0.055 (1.40)	flat back
0.1875 (4.76)	0.173 (4.39)	8.625 (42.11)	0.060 (1.52)	flat back
0.203 (5.16)	0.189 (4.80)	9.25 (45.16)	0.060 (1.52)	flat back
0.218 (5.54)	0.204 (5.18)	9.875 (48.21)	0.060 (1.52)	flat back
0.234 (5.94)	0.219 (5.56)	10.50 (51.27)	0.600 (1.52)	flat back
0.250 (6.35)	0.235 (5.97)	11.125 (54.32)	0.600 (1.52)	flat back

TABLE 3 Permissible Variations over Specified Width and Length of Regular Sheared Plates, in. (mm)

Specified Dimensions		Variations over Specified Width and Length for Given Width, Length, and Thickness			
		Thickness Under ⅜		Thickness ⅜ to ½, incl	
Width	Length	Width	Length	Width	Length
48 (1219) and under	240 (6096) and under	⅛ (3.2)	³⁄₁₆ (4.8)	³⁄₁₆ (4.8)	¼ (6.4)
Over 48 to 60 (1219 to 1524), incl	240 (6096) and under	³⁄₁₆ (4.8)	¼ (6.4)	¼ (6.4)	⁵⁄₁₆ (7.9)

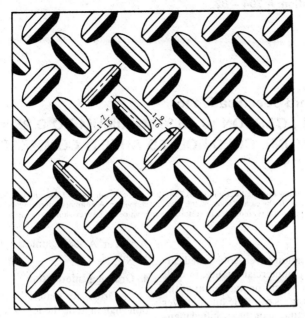

FIG. 1 Pattern A (percent of reduction, 61 %)

FIG. 2 Pattern B (original of raised figures approximately 1¼ in., percent of reduction 61 %)

The American Society for Testing and Materials takes no position respecting the validity of any patent rights asserted in connection with any item mentioned in this standard. Users of this standard are expressly advised that determination of the validity of any such patent rights, and the risk of infringement of such rights, are entirely their own responsibility.

This standard is subject to revision at any time by the responsible technical committee and must be reviewed every five years and if not revised, either reapproved or withdrawn. Your comments are invited either for revision of this standard or for additional standards and should be addressed to ASTM Headquarters. Your comments will receive careful consideration at a meeting of the responsible technical committee, which you may attend. If you feel that your comments have not received a fair hearing you should make your views known to the ASTM Committee on Standards, 1916 Race St., Philadelphia, Pa. 19103.

ASTM Designation: A 794 – 82

Standard Specification for
STEEL, CARBON (0.16 % MAXIMUM TO 0.25 % MAXIMUM), COLD-ROLLED SHEET, COMMERCIAL QUALITY[1]

This standard is issued under the fixed designation A 794; the number immediately following the designation indicates the year of original adoption or, in the case of revision, the year of last revision. A number in parentheses indicates the year of last reapproval. A superscript epsilon (ϵ) indicates an editorial change since the last revision or reapproval.

1. Scope

1.1 This specification covers cold-rolled carbon steel sheet of commercial quality, in coils and cut lengths, in which the maximum of the specified carbon range is over 0.15 and not over 0.25 %, and the maximum of the specified manganese range is not over 0.90 %. This material is ordered to chemical composition.

1.2 This specification is not applicable to the steels covered in Specifications A 366, A 611, A 109, and A 109M.

2. Applicable Documents

2.1 *ASTM Standards:*

A 109 Specification for Steel, Carbon, Cold-Rolled Strip[2]

A 109M Specification for Steel, Carbon, Cold-Rolled Strip [Metric][2]

A 366 Specification for Steel, Carbon, Cold-Rolled Sheet, Commercial Quality[2]

A 370 Methods and Definitions for Mechanical Testing of Steel Products[2]

A 568 Specification for General Requirements for Steel, Carbon and High-Strength Low-Alloy Hot-Rolled Sheet and Cold-Rolled Sheet[2]

A 568M Specification for General Requirements for Steel, Carbon and High-Strength Low-Alloy Hot-Rolled Sheet and Cold-Rolled Sheet, [Metric][2]

A 611 Specification for Steel, Cold-Rolled Sheet, Carbon, Structural[2]

3. General Requirements for Delivery

3.1 Material furnished under this specification shall conform to the applicable requirements of the current edition of Specifications A 568 or A 568M, unless otherwise provided herein.

4. Ordering Information

4.1 Orders for material under this specification shall include the following information, as required to describe the required material adequately:

4.1.1 ASTM specification number and date of issue,

4.1.2 Name of material (cold-rolled sheet, commercial quality),

4.1.3 Grade designation or chemical composition or both,

4.1.4 Copper-bearing steel (if required),

4.1.5 Finish; indicate unexposed with matte (dull) finish, or exposed with either matte (dull), commercial bright or luster finish as required,

4.1.6 Specify oiled or not oiled, as required,

4.1.7 Dimensions (thickness, width, and whether cut lengths or coils),

4.1.8 Quantity,

4.1.9 Coil size (must include inside diameter, outside diameter, and maximum mass),

4.1.10 Appliction (part identification and description),

4.1.11 Special requirements (if required), and

4.1.12 Cast or heat analysis report (request, if required).

---

[1] This specification is under the jurisdiction of ASTM Committee A-1 on Steel, Stainless Steel, and Related Alloys and is the direct responsibility of Subcommittee A01.19 on Steel Sheet and Strip.
Current edition approved Feb. 25, 1982. Published April 1982.

[2] *Annual Book of ASTM Standards*, Vol 01.03.

A 794

NOTE—A typical ordering description is as follows: ASTM A 794 dated _____, Cold-Rolled Sheet, Commercial Quality, Class 1, Matte Finish, Oiled, Grade 1017, 0.030 by 36 by 96 in. for part No. 5226 Steel Shelving.

5. Manufacture

5.1 *Condition*—The material shall be furnished in the annealed and temper-rolled condition but may be supplied full hard if specified.

6. Chemical Composition

6.1 The cast or heat (formerly ladle) analysis of the steel shall conform to the chemical requirements shown in Table 1, or chemical compositions can be specified from carbon 0.16 % maximum to 0.25 % maximum, inclusive, and manganese 0.90 % maximum, inclusive, which conforms to the ranges and limits in Appendix X2 of Specifications A 568 or A 568M.

6.2 Where material is used for fabrication by welding, care must be exercised in selection of the chemical composition to assure compatibility with the welding process and its effects on altering the properties of the steel.

7. Mechanical Requirements

7.1 *Bend Test*—The material shall be capable of being bent at room temperature through 180° in any direction to inside diameters as shown in Table 2 without cracking on the outside of the bent portion (see Section 17 of Methods and Definitions A 370). When steel is subject to bending in a fabricating operation, more liberal bend radii should be used.

8. Certification and Reports

8.1 When requested, the producer shall furnish copies of a report showing test results of the cast or heat analysis. The report shall include the purchase order number, ASTM designation number, and the cast or heat number representing the material.

TABLE 1 Typical Grade Designations and Chemical Compositions[A]

UNS Designation[B]	Grade Designation	Carbon, %	Mn, %	P, Max, %	S, Max, %
G10150	1015	0.12–0.18	0.30–0.60	0.040	0.050
G10160	1016	0.12–0.18	0.60–0.90	0.040	0.050
G10170	1017	0.14–0.20	0.30–0.60	0.040	0.050
G10180	1018	0.14–0.20	0.60–0.90	0.040	0.050
G10200	1020	0.17–0.23	0.30–0.60	0.040	0.050
G10210	1021	0.17–0.23	0.60–0.90	0.040	0.050
G10230	1023	0.19–0.25	0.30–0.60	0.040	0.050

[A] Copper, when specified, shall have a minimum content of 0.20 % by cast or heat analysis.
[B] New designation established in accordance with Specification E 527 and SAE J1086, Recommended Practice for Numbering Metals and Alloys (UNS).

TABLE 2 Bend Test Requirements[A]

Maximum of the Specified Manganese Range, %	Ratio of Inside Bend Diameter to Thickness of Specimen
To 0.60 incl	1
Over 0.60 to 0.90 incl	1½

[A] Applicable only to material furnished in the annealed and temper-rolled condition.

The American Society for Testing and Materials takes no position respecting the validity of any patent rights asserted in connection with any item mentioned in this standard. Users of this standard are expressly advised that determination of the validity of any such patent rights, and the risk of infringement of such rights, are entirely their own responsibility.

This standard is subject to revision at any time by the responsible technical committee and must be reviewed every five years and if not revised, either reapproved or withdrawn. Your comments are invited either for revision of this standard or for additional standards and should be addressed to ASTM Headquarters. Your comments will receive careful consideration at a meeting of the responsible technical committee, which you may attend. If you feel that your comments have not received a fair hearing you should make your views known to the ASTM Committee on Standards, 1916 Race St., Philadelphia, Pa. 19103.

Designation: A 805 – 82

Standard Specification for
COLD-ROLLED CARBON STEEL FLAT WIRE[1]

This standard is issued under the fixed designation A 805; the number immediately following the designation indicates the year of original adoption or, in the case of revision, the year of last revision. A number in parentheses indicates the year of last reapproval. A superscript epsilon (ϵ) indicates an editorial change since the last revision or reapproval.

1. Scope

1.1 This specification covers carbon steel flat wire in coils or cut lengths. Flat wire is classified as a cold-rolled section, rectangular in shape, 0.500 in. (12.7 mm) or less in width and under 0.250 in. (6.35 mm) in thickness.

1.2 Low-carbon steel flat wire is produced from steel compositions with a maximum carbon content of 0.25 % by cast or heat analysis.

1.3 Carbon spring steel flat wire is produced to a carbon range in which the specified or required maximum is over 0.25 % by cast or heat analysis.

1.3.1 Two types of carbon spring steel flat wire are produced:

1.3.1.1 Untempered cold-rolled carbon spring steel flat wire, produced to several desirable combinations of properties and

1.3.1.2 Hardened and tempered carbon spring steel wire.

1.4 Definite application flat wire is a product developed for a specific application and may be specified only by size and descriptive name.

2. Applicable Documents

2.1 *ASTM Standards:*
A 370 Methods and Definitions for Mechanical Testing of Steel Products[2]
A 510 Specification for General Requirements for Wire Rods and Coarse Round Wire, Carbon Steel[2]
E 30 Methods for Chemical Analysis of Steel, Cast Iron, Open-Hearth Iron, and Wrought Iron[3]
E 45 Practice for Determining the Inclusion Content of Steel[4]
E 112 Methods for Determining Average Grain Size[4]
E 140 Standard Hardness Conversion Tables for Metals (Relationship between Brinel Hardness, Vickers Hardness, Rockwell Hardness, Rockwell Superficial Hardness and Knoop Hardness)[5]

2.2 *Military Standards:*
MIL-STD-129 Marking for Shipment and Storage[6]
MIL-STD-163 Steel Mill Products, Preparation for Shipment and Storage[6]

2.3 *Federal Standard:*
Fed. Std. No. 123 Marking for Shipment (Civil Agencies)[6]

2.4 *Automotive Engineers Standard:*
Recommended Practice SAE J419 Methods of Measuring Decarburization[7]

3. Descriptions of Terms Specific to This Standard

3.1 *cold reduction*—the process of reducing the thickness of the strip at room temperature. The amount of reduction is greater than that used in skin-rolling (see 3.7).

3.2 *annealing*—the process of heating to and holding at a suitable temperature and then cooling at a suitable rate, for such purposes as reducing hardness, facilitating cold working, producing a desired microstructure, or obtain-

---

[1] This specification is under the jurisdiction of ASTM Committee A-1 on Steel, Stainless Steel, and Related Alloys and is the direct responsibility of Subcommittee A01.19 on Sheet Steel and Strip.
Current edition approved Oct. 29, 1982. Published December 1982.
[2] *Annual Book of ASTM Standards*, Vol 01.03.
[3] *Annual Book of ASTM Standards*, Vol 03.05.
[4] *Annual Book of ASTM Standards*, Vol 03.03.
[5] *Annual Book of ASTM Standards*, Vol 03.01.
[6] Available from Naval Publications and Forms Center, 5801 Tabor Ave., Philadelphia, Pa. 19120.
[7] Available from Society of Automotive Engineers, 400 Commonwealth Drive, Warrendale, Pa. 15096.

ing desired mechanical, physical, or other properties.

3.2.1 *batch annealing*—annealing that is generally performed in large cylindrical bell type or large rectangular box or car-type furnaces. The product is protected from scaling and decarburization by the use of a controlled atmosphere that envelops the charge in an inner chamber sealed to prevent the influx of air or products of combustion. The coils or bundles are heated to a temperature in the vicinity of the lower critical temperature for the grade of steel, and held at that temperature for a definite length of time; after which the steel is allowed to cool slowly to room temperature. The time of holding at the annealing temperature varies with the grade of the steel and the desired degree of softness.

3.2.2 *spheroidize annealing*—an operation consisting of prolonged heating and prolonged cooling cycles to produce a globular or spheroidal condition of the carbide for maximum softness.

3.2.3 *salt annealing*—annealing that is accomplished by immersing bundles or coils of flat wire in a molten salt bath at a desired temperature for a definite time. Following the annealing, the coils are permitted to cool slowly, after which they are immersed in hot water to remove any adhering salts.

3.2.4 *continuous or strand annealing*—annealing that consists of passing a number of individual strands of flat wire continuously through either a muffle furnace or a bath of molten lead or salt, thus heating the flat wire to the desired temperature for a definite time. The hardness obtained by this type of annealing, as measured by Rockwell hardness number, is normally somewhat higher than is secured by batch-type annealing. Other characteristics peculiar to strand-annealed steel require this type of annealing for some flat wire products.

3.3 *patenting*—a thermal treatment usually confined to steel over 0.25 % carbon. In this process individual strands of rods or wire are heated well above the upper critical temperature followed by comparatively rapid cooling in air, molten salt, or molten lead. This treatment is generally employed to prepare the material for subsequent processing.

3.4 *hardening and tempering*—a heat treatment for steel over 0.25 % carbon by cast or heat analysis involving continuous strand heating at finish size to an appropriate temperature above the critical temperature range, followed by quenching in oil and finally passing the strands through a tempering bath. This heat treatment is used in the production of such commodities as oil-tempered spring wire for use in certain types of mechanical springs that are not subjected to a final heat treatment after forming. Oil-tempered wire is intended primarily for the manufacture of products that are required to withstand high stresses. The mechanical properties and resiliency of oil-tempered wire provide resistance to permanent set under repeated and continuous stress applications.

3.5 *temper*—a designation by number to indicate the hardness as a minimum, as a maximum, or as a range. The tempers are obtained by the selection and control of chemical composition, by amounts of cold reduction, and by thermal treatment.

3.6 *skin-rolled*—a term denoting a relatively light cold-rolling operation following annealing. It serves to reduce the tendency of the steel to flute or stretcher strain during fabrication. It is also used to impart surface finish, or affect hardness or other mechanical properties.

3.7 *finish*—the degree of smoothness or lustre of the flat wire. The production of specific finishes requires special preparation and control of the roll surfaces employed.

4. Ordering Information

4.1 Orders for material to this specification shall include the following information, as necessary, to describe adequately the desired product:

4.1.1 Quantity,

4.1.2 Name of material (flat wire identified by type),

4.1.3 Analysis or grade, if required (Section 6),

4.1.4 Temper of low carbon or type of spring steel (Sections 9, 10, and 11),

4.1.5 Edge (Section 7),

4.1.6 Finish or coating (Sections 12 and 13),

4.1.7 Dimensions,

4.1.8 Coil type and size requirements (Section 15),

4.1.9 Packaging (15.1),

4.1.10 Condition (oiled or not oiled) (12.4),

4.1.11 ASTM designation and date of issue,

4.1.12 Copper-bearing steel, if required,

4.1.13 Application (part identification or description),

4.1.14 Case or heat analysis (request, if desired), and

4.1.15 Exceptions to the specification, if required.

NOTE 1—A typical ordering description is as follows: 18 000 lb Low-Carbon Cold-Rolled Carbon Steel Flat Wire, Temper 4, Edge 4, Finish 2, 0.125 by 0.450-in. vibrated coils, 2000 lb max, coil weight, 16 to 20 in. ID, 36 in. max OD, Face dimension 6 to 10 in., ASTM A 805 dated ___, for Stove Frames.

5. Manufacture

5.1 Low-carbon steel flat wire is normally produced from rimmed, capped, or semi-killed steel. When required, killed steel may be specified, with silicon or aluminum as the deoxidizer.

5.2 Untempered-carbon spring steel flat wire is commonly produced from killed steel, although semi-killed steel is sometimes used.

5.3 Hardened and tempered carbon spring steel flat wire customarily has a carbon content over 0.60 %.

5.4 Flat wire is generally produced from hot-rolled rods or round wire, by one or more cold-rolling operations, primarily for the purpose of obtaining the size and section desired and for improving surface finish, dimensional accuracy, and varying mechanical properties. Flat wire can also be produced from slitting hot- or cold-rolled flat steel to the desired width. The hot-rolled slit flat steel is subsequently cold reduced. The width to thickness ratio and the specified type of edge generally determine the process that is necessary to produce a specific flat-wire item.

5.5 The production of good surface quality flat wire is dependent upon scale-free and clean wire, rod, or hot-rolled steel prior to cold-rolling. Scale removal can be accomplished by chemical or mechanical cleaning.

5.6 Edge rolls, machined with contour grooves, may be used in conjunction with flat-rolling passes to produce the desired edge shape.

5.7 Straightness in flat wire may be controlled by the use of roll straighteners alone or in conjunction with cold-rolling passes.

5.8 Edges of flat wire produced by slitting wider flat-rolled steel can be dressed, depending upon requirements by:

5.8.1 *Deburring*—A process by which burrs are removed by rolling or filing to obtain an approximate square edge;

5.8.2 *Rolling*—A process by which the slit edge is dressed by edge rolling to the desired contour; and

5.8.3 *Filing*—A process by which the slit edge is filed to a specific contour and dimension by passing one or more times against a series of files mounted at various angles.

6. Chemical Composition

6.1 *Limits*:

6.1.1 When carbon steel flat wire is specified to chemical composition, the compositions are commonly prepared using the ranges and limits shown in Table 1. The elements comprising the desired chemical composition are specified in one of three ways:

6.1.1.1 By a maximum limit,

6.1.1.2 By a minimum limit, or

6.1.1.3 By minimum and maximum limits termed the "range." By common usage, the range is the arithmetical difference between the two limits (for example, 0.60 to 0.71 is 0.11 range).

6.1.2 When carbon steel flat wire is produced from round rods or wire it may be designated by grade number. In such cases the chemical ranges and limits of Tables 6, 7, 8 and 9 of Specification A 510 shall apply.

6.2 *Cast or Heat Analysis*:

6.2.1 An analysis of each cast or heat of steel shall be made by the manufacturer to determine the percentage of elements specified or restricted by the applicable specification.

6.2.2 When requested, cast or heat analysis for elements listed or required shall be reported to the purchaser or his representative.

6.3 *Product Analysis* may be made by the purchaser on the finished material.

6.3.1 Capped or rimmed steels are not technologically suited to product analysis due to the nonuniform character of their chemical composition and, therefore, the tolerances in Table 2 do not apply. Product analysis is appropriate on these types of steel only when misapplication is apparent, or for copper when copper steel is specified.

6.3.2 For steels other than rimmed or capped, when product analysis is made by the purchaser, the chemical analysis shall not vary from the limits specified by more than the amounts in Table 2. The several determinations of any element shall not vary both above and below the specified range.

6.3.3 When flat wire is produced from round rods or wire, and when a grade number is used to specify the chemical composition, the values obtained on a product analysis shall not vary from the limits specified by more than the amounts in Table 10 of Specification A 510.

6.4 For referee purposes, if required, Methods E 30 shall be used.

7. Edge

7.1 The desired edge shall be specified as follows:

7.1.1 *Number 1 Edge* is a prepared edge of a specified contour (round or square) which is produced when a very accurate width is required or when the finish of the edge suitable for electroplating is required, or both.

7.1.2 *Number 2 Edge* is not applicable to flat wire products.

7.1.3 *Number 3 Edge* is an approximately square edge produced by slitting.

7.1.4 *Number 4 Edge* is a rounded edge produced either by edge rolling or resulting from the flat rolling of a round section. Width tolerance and edge condition are not as exacting as for a No. 1 Edge.

7.1.5 *Number 5 Edge* is an approximately square edge produced from slit–edge material on which the burr is eliminated by rolling or filing.

7.1.6 *Number 6 Edge* is a square edge produced by edge rolling when the width tolerance and edge condition are not as exacting as for No. 1 Edge.

8. Dimensional Tolerances

8.1 The dimensional tolerances shall be in accordance with the following:

Tolerances	Table Number
Thickness	3
Width	4
Length	5

8.2 If restricted tolerances closer than those shown in Tables 3, 4, and 5 are required, the degree of restriction should be established between the purchaser and manufacturer.

8.3 Tolerances for camber should be established between the purchaser and manufacturer. Camber is the greatest deviation of a side edge from a straight line, the measurement being taken on the concave side with a straight edge.

9. Temper and Bend Test Requirement for Low-Carbon Steel Flat Wire

9.1 Low-carbon steel flat wire specified to temper numbers shall approximate the hardness or tensile strength values shown in Table 6.

9.2 Bend test specimens shall stand being bent at room temperatures as required in Table 7.

9.3 All mechanical tests are to be conducted in accordance with Methods and Definitions A 370.

10. Types of Untempered-Carbon Spring Steel Flat Wire

10.1 The following types are produced:

10.1.1 *Hard-Type Carbon Spring Steel Flat Wire* is a very stiff, springy product intended for flat work not requiring ability to withstand cold forming. It is cold reduced with or without preparatory treatment to a minimum Rockwell value of B98.

10.1.2 *Soft-Type Spring Steel Flat Wire* is intended for application where varying degrees of cold forming are encountered, that necessitates control of both carbon content and hardness. Maximum values for carbon vary from 0.25 to 1.35 %, inclusive. This type also involves one of the following hardness restrictions; a maximum only designated as "soft-type annealed" or a range only designated as "soft-type intermediate hardness."

10.1.2.1 *Soft-Type Annealed Carbon Spring Steel Flat Wire*, intended for moderately severe cold forming, is produced to a specific maximum hardness value. The final anneal is at the finish thickness. Lowest maximum expected hardness values or tensile strength for specific carbon maximums for steel to 0.90 % maximum manganese are shown in Table 8.

10.1.2.2 *Soft-Type Intermediate Carbon Spring Steel Flat Wire* is produced to a specified hardness range, somewhat higher than the category covered in 10.1.2.1. The product is produced by rolling after annealing or by varying the annealing treatment, or both.

10.1.2.3 The Rockwell hardness range which can be produced varies with the carbon content, the required hardness, and the thickness of the material. In Tables 9, 10, and 11 are shown the applicable hardness ranges for various carbon contents and several thickness ranges. If hardness values other than those shown in the tables are required, the applicable ranges should be agreed upon between the purchaser and the manufacturer. Rockwell hardness range is the arithmetical difference between two limits (for example B82 to B90 is an eight-point range).

10.1.3 *Spheroidize-Type Carbon Spring Steel Flat Wire* is best suited for the severest cold-forming application, where heat treatment after forming is employed. Spheroidize annealing treatment is employed in its production. Lowest maximum expected hardness values by carbon maximums for steel to 0.90 % maximum manganese are shown in Table 12. For thicknesses under 0.025 in. (0.64 mm) the values for the "Soft-Type Annealed" as contained in Table 8 shall apply.

11. Hardness and Tensile Properties of Hardened and Tempered Carbon Spring Steel Flat Wire

11.1 This product is commonly produced to meet a range of Rockwell hardness as shown in Table 13.

11.2 The hardness scale appropriate to each thickness range is shown in Table 14. Although conversion tables for hardness numbers are available, the recommended practice is to specify the same scale as that to be used in testing. A Rockwell hardness range is the arithmetic difference between two limits (for example C42 to C46 is a four-point range). Below a thickness of 0.008 in. (0.20 mm) the Rockwell 15N test becomes inaccurate, and the use of the tensile test is recommended. The values of ultimate tensile strength cited in Fig. 1 apply only to thicknesses less than 0.008 in. (0.20 mm). When necessary to specify tensile properties for thicknesses of 0.008 in. (0.20 mm) and greater, the manufacturer should be consulted.

11.3 Shown in Fig. 1 is the relationship of thickness and carbon content with Rockwell hardness or tensile strength for hardened and tempered spring steel flat wire appropriate for spring applications. When mechanical properties are specified, they should be compatible with the application.

12. Finish and Condition

12.1 The finish of low-carbon steel flat wire normally specified is one of the following:

12.1.1 *Number 2 or Regular Bright Finish* is produced by rolling on rolls having a moderately smooth finish. It is not generally applicable to plating.

12.1.2 *Number 3 or Best Bright Finish* is generally of high lustre produced by selective-rolling practices, including the use of specially prepared rolls. Number 3 finish is the highest quality finish produced and is particularly suited for electroplating. The production of this finish requires extreme care in processing and extensive inspection.

12.2 Untempered-carbon spring steel flat wire is commonly supplied in a Number 2 regular bright finish, as in 12.1.1. The manufacturer should be consulted if another finish is required.

12.3 Hardened and tempered spring steel flat wire is usually supplied in one of the following recognized finishes:

12.3.1 *Black-tempered,*
12.3.2 *Scaleless-tempered,*
12.3.3 *Bright-tempered,*
12.3.4 *Tempered and polished,*
12.3.5 *Tempered, polished, and colored* (*blue or straw*), and
12.4 *Oiled.*

12.4.1 Unless otherwise specified, flat wire is coated with oil to minimize scratching and to retard rusting in transit. If the product is not to be oiled, it must be so specified.

13. Coatings

13.1 Low-carbon steel flat wire can be produced with various coatings, such as liquor finish, white-liquor finish, lacquer, paint, copper, zinc (galvanized), cadmium, chromium, nickel, and tin. Metallic coatings can be applied by the hot-dip method or by electrodeposition. The flat steel can be coated prior to slitting to wire widths. In this case the slit edges will not be coated.

13.1.1 Copper or liquor coatings consist of thin deposits of either copper or bronze produced by immersion of the material in an acid solution of metallic salts. Because of the nature of liquor coatings no appreciable corrosion protection is afforded by them.

13.1.2 Hot-dipped coatings are produced by passing strands of cleaned flat wire continuously through a molten bath of metal or alloy. Zinc and tin are commonly applied in this manner.

13.1.3 Electrodeposited coatings are produced by passing strands of cleaned flat wire through an electroplating tank containing a solution of a metallic salt, wherein the metal is deposited on the flat wire. Zinc, tin, nickel, cadmium, and copper are applied in this manner.

13.2 Coatings applicable to untempered-carbon spring steel flat wire are the same as those covered in 13.1.

13.3 Metallic coatings are seldom applied to hardened and tempered carbon steel flat wire. If they are required the manufacturer should be consulted.

14. Workmanship

14.1 Cut lengths shall have a workmanlike appearance and shall not have defects of a nature or degree for the product, the grade, and the quality ordered that will be detrimental to the fabrication of the finished part.

14.2 Coils may contain more frequent imperfections that render a portion of the coil unusable since the inspection of coils does not afford the manufacturer the same opportunity to remove portions containing imperfections as in the case with cut lengths.

15. Packaging

15.1 Flat wire is prepared for shipment in a number of ways. The material may be bare, paper or burlap wrapped, boxed, skidded or palletized, skidded and shrouded, palletized and shrouded, barrelled, or a combination thereof. The purchaser should specify the method desired.

15.2 When specified in the contract or order, and for direct procurement by or direct shipment to the government, when Level A is specified, preservation, packaging, and packing shall be in accordance with Level A requirements of MIL-STD-163.

15.3 When coils are ordered it should be specified whether a ribbon or tape wound or a vibrated coil is desired. Since coil diameters and weights vary by the manufacturers, the manufacturer should be consulted for specific capability and limitations. When coil weight is specified for low-carbon steel flat wire or for untempered-carbon spring steel flat wire, it is common practice to ship not more than 10 % of the total weight of an item in short coils, which are those weighing between 25 and 75 % of the maximum coil weight.

15.4 For flat wire in cut lengths, when the specified length is over 36 in. (915 mm), it is permissible to ship up to 10 % of the item in short lengths, but not shorter than 36 in. (915 mm), unless otherwise agreed upon.

16. Marking

16.1 As a minimum requirement, the material shall be identified by having the manufacturer's name, ASTM designation, weight, purchaser's order number, and material identification legibly stenciled on top of each lift or shown on a tag attached to each coil or shipping unit.

16.2 When specified in the contract or order, and for direct procurement by or direct shipment to the government, marking for shipment, in addition to requirements specified in the contract or order, shall be in accordance with MIL-STD-129 for military agencies and in accordance with Fed. Std. No. 123 for civil agencies.

17. Inspection

17.1 When the purchaser's order stipulates that inspection and tests (except product analysis) for acceptance on the steel be made prior to shipment from the mill, the manufacturer shall afford the purchaser's inspector all reasonable facilities to satisfy him that the steel is being produced and furnished in accordance with the specification. Mill inspection by the purchaser shall not interfere unnecessarily with the manufacturer's operation.

18. Rejection and Rehearing

18.1 Unless otherwise specified, any rejection shall be reported to the manufacturer within a reasonable time after receipt of material by the purchaser.

18.2 Material that is reported to be defective subsequent to the acceptance at the manufacturer's works shall be set aside, adequately protected, and correctly identified. The manufacturer shall be notified as soon as possible so that an investigation may be initiated.

18.3 Samples that are representative of the rejected material shall be made available to the manufacturer. In the event that the manufacturer is dissatisfied with the rejection, he may request a rehearing.

A 805

TABLE 1 Cast or Heat Analysis

Element	When Maximum of Specified Element is	Range
Carbon[A]		
	to 0.15 incl	0.05
	over 0.15 to 0.30 incl	0.06
	over 0.30 to 0.40 incl	0.07
	over 0.40 to 0.60 incl	0.08
	over 0.60 to 0.80 incl	0.11
	over 0.80 to 1.35 incl	0.14
Manganese		
	to 0.50 incl	0.20
	over 0.50 to 1.15 incl	0.30
	over 1.15 to 1.65 incl	0.35
Phosphorus[B]		
	to 0.08 incl	0.03
	over 0.08 to 0.15 incl	0.05
Sulfur[B]		
	to 0.08 incl	0.03
	over 0.08 to 0.15 incl	0.05
	over 0.15 to 0.23 incl	0.07
	over 0.23 to 0.33 incl	0.10
Silicon[C]		
	to 0.15 incl	0.08
	over 0.15 to 0.30 incl	0.15
	over 0.30 to 0.60 incl	0.30
Copper		
	When copper is required 0.20 minimum is commonly specified.	...

[A] *Carbon*—The carbon ranges shown in the column headed "Range" apply when the specified maximum limit for manganese does not exceed 1.00%. When the maximum manganese limit exceeds 1.00%, add 0.01 to the carbon ranges shown above.
[B] *Phosphorus and Sulfur*—The standard lowest maximum limits for phosphorus and sulfur are 0.040% and 0.050% respectively. Certain qualities, descriptions, or specifications are furnished to lower standard maximum limits.
[C] *Silicon*—The standard lowest maximum for silicon is 0.10%.

TABLE 2 Tolerances for Product Analysis[A]

Element	Limit, or Maximum of Specified Element, %	Tolerance, % Under Minimum Limit	Tolerance, % Over Maximum Limit
Carbon	to 0.15 incl	0.02	0.03
	over 0.15 to 0.40 incl	0.03	0.04
	over 0.40 to 0.80 incl	0.03	0.05
	over 0.80	0.03	0.06
Manganese	to 0.60 incl	0.03	0.03
	over 0.60 to 1.15 incl	0.04	0.04
	over 1.15 to 1.65 incl	0.05	0.05
Phosphorus	...	...	0.01
Sulfur	...	...	0.01
Silicon	to 0.30 incl	0.02	0.03
	over 0.30 to 0.60 incl	0.05	0.05
Copper	...	0.02	...

[A] When produced from round wire or rod the producer may use the tolerances for product analysis that appear in Specification A 510 (see 6.3.3).

TABLE 3 Thickness Tolerances

Specified Thickness in. (mm)	Tolerances for Specified Thickness, Plus and Minus, in. (mm)
0.005 (0.13) to 0.010 (0.25), excl	0.0005 (0.013)
0.010 (0.25) to 0.029 (0.74), excl	0.001 (0.03)
0.029 (0.74) to 0.0625 (1.59), excl	0.0015 (0.04)
0.0625 (1.59) to 0.250 (6.35), excl	0.002 (0.05)

TABLE 4 Tolerances for Specified Width

Edge Number	Specified Width, in. (mm)	Specified Thickness, in. (mm)		
		Under 0.0625 (1.60)	0.0625 (1.59) to 0.126 (3.20) excl	0.126 (3.20) to 0.250 (6.35) excl
1	Under 0.0625 (1.60)	0.003 (0.08)	...	...
	0.0625 (1.60) to 0.126 (3.20) excl	0.004 (0.10)	0.004 (0.10)	...
	0.126 (3.20) to 0.500 (12.70) incl	0.005 (0.13)	0.005 (0.13)	0.005 (0.13)
4 and 6	Under 0.0625 (1.60)	0.006 (0.15)	...	...
	0.0625 (1.60) to 0.126 (3.20) excl	0.008 (0.20)	0.008 (0.20)	...
	0.126 (3.20) to 0.500 (12.70) incl	0.010 (0.25)	0.010 (0.25)	0.010 (0.25)
3 and 5	0.125 (3.18) to 0.500 (12.70) incl	0.005 (0.13)	0.008 (0.02)	...

TABLE 5 Length Tolerances

Specified Length, in. (mm)	Tolerances Over the Specified Length in. (mm)—No Tolerance Under
24 (600) to 60 (1500), incl	¼ (6.4)
Over 60 (1500) to 120 (3000), incl	½ (12.7)
Over 120 (3000) to 240 (6100), incl	¾ (19.1)

TABLE 6 Temper, Hardness and Tensile Strength Requirement for Low-Carbon Steel Flat Wire

Temper	Thickness, in. (mm)	Rockwell Hardness min	Rockwell Hardness max (approximate)	Approximate Tensile Strength, ksi (MPa) min	Approximate Tensile Strength, ksi (MPa) max
No. 1 (hard)	Under 0.010 (0.25)			85 (586)	...
	0.010 (0.25) to 0.025 (0.64) excl	15T90	...		
	0.025 (0.64) to 0.040 (1.02) excl	30T76	...		
	0.040 (1.02) to 0.070 (1.78) excl	B90	...		
	0.070 (1.78) and over	B84	...		
No. 2 (half-hard)	Under 0.010 (0.25)			65 (448)	90 (621)
	0.010 (0.25) to 0.025 (0.64) excl	15T83.5	15T88		
	0.025 (0.64) to 0.040 (1.02) excl	30T63.5	30T74		
	0.040 (1.02) and over	B70	B85		
No. 3 (quarter-hard)	Under 0.010 (0.25)			55 (379)	80 (552)
	0.010 (0.25) to 0.025 (0.64) excl	15T80	15T85		
	0.025 (0.64) to 0.040 (1.02) excl	30T56.5	30T67		
	0.040 (1.02) and over	B60	B75		
No. 4 (skin-rolled)	Under 0.010 (0.25)			...	65 (448)
	0.010 (0.25) to 0.025 (0.64) excl	...	15T82		
	0.025 (0.64) to 0.040 (1.02) excl	...	30T60		
	0.040 (1.02) and over	...	B65		
No. 5 (dead-soft)	Under 0.010 (0.25)			...	60 (414)
	0.010 (0.25) to 0.025 (0.64) excl	...	15T78.5		
	0.025 (0.64) to 0.040 (1.02) excl	...	30T53		
	0.040 (1.02) and over	...	B55		

TABLE 7 Temper and Bend Test Requirement for Low-Carbon Steel Flat Wire

Temper	Bend Test Requirement
No. 1 (hard)	Not required to make bends in any direction.
No. 2 (half-hard)	Bend 90° across[A] the direction of rolling around a radius equal to that of the thickness.
No. 3 (quarter-hard)	Bend 180° across[A] the direction of rolling over one thickness of the wire.
No. 4 (skin-rolled)	Bend flat upon itself in any direction.
No. 5 (dead-soft)	Bend flat upon itself in any direction.

[A] To bend "across the direction of rolling" means that the bend axis (crease of the bend) shall be at a right angle to the length of the wire.

TABLE 8 Soft-Type Annealed Carbon Spring Steel Flat Wire Lowest Expected Maximum Rockwell Hardness or Tensile Strength

Maximum of Carbon Range, %	Flat Wire Thickness, in. (mm)			
	Under 0.010 (0.25)	0.010 (0.25) to 0.025 (0.64) excl	0.025 (0.64) to 0.040 (1.02) excl	0.040 (1.02) and Over
	Tensile Strength ksi (MPa)	Rockwell Hardness, 15T Scale	Rockwell Hardness, 30T Scale	Rockwell Hardness, B Scale
0.30	66 (455)	84	67	74
0.35	68 (469)	84	68	76
0.40	70 (483)	85	70	78
0.45	72 (496)	85	71	80
0.50	74 (510)	86	72	82
0.55	76 (524)	87	73	84
0.60	78 (538)	87	74	85
0.65	80 (552)	88	75	87
0.70	82 (565)	88	76	88
0.75	83 (572)	88	76	89
0.80	85 (586)	89	77	90
0.85	87 (600)	89	77	91
0.90	88 (607)	89	78	92
0.95 and over	90 (621)	90	78	92

TABLE 9 Rockwell Hardness Ranges for Soft-Type Intermediate Hardness Carbon Spring Steel Flat Wire Thickness Under 0.025 in. (0.64 mm)

Maximum of Carbon Range, %	For Maximum of Specified Rockwell Hardness Range, 15T Scale												
	83.5	84.5	85	85.5	86	86.5	87	87.5	88	88.5	89	89.5	90/92	Over 92[A]
0.26–0.30[B]	...	5	5	5	5	5	4	4	4	4	4	4	3	...
0.31–0.35[B]	...	...	5	5	5	5	4	4	4	4	4	4	3	...
0.36–0.40[B]	...	...	...	5	5	5	4	4	4	4	4	4	3	...
0.41–0.45[B]	...	...	...	...	5	5	4	4	4	4	4	4	3	...
0.46–0.50[B]	...	...	...	...	...	5	4	4	4	4	4	4	3	3
0.51–0.55[B]	...	...	...	...	...	...	4	4	4	4	4	4	3	3
0.56–0.60[B]	...	...	...	...	...	...	...	4	4	4	4	4	3	3
0.61–0.65[B]	...	...	...	...	...	...	...	...	4	4	4	4	3	3
0.66–0.70[B]	...	...	...	...	...	...	...	...	...	4	4	4	3	3
0.71–0.75[B]	...	...	...	...	...	...	...	...	...	4	4	4	3	3
0.76–0.80[B]	...	...	...	...	...	...	...	...	...	...	4	4	3	3
0.81–0.90[B]	...	...	...	...	...	...	...	...	...	...	...	4	3	3
0.91–1.35	...	...	...	...	...	...	...	...	...	...	...	...	3	3

[A] Rockwell 15T Scale is not recommended for values over 15T93.
[B] Indicates soft-type annealed cold-rolled carbon spring steel flat wire which is furnished to a maximum (hardness) shown in Table 8.

A 805

TABLE 10 Rockwell Hardness Ranges for Soft-Type Intermediate Hardness Carbon Spring Steel Flat Wire Thickness 0.025 to 0.040 in. (0.64 to 1.02 mm) excl

Maximum of Carbon Range, %	For Maximum of Specified Rockwell Hardness Range, 30T Scale												
	66.5	68	69.5	70.5	71.5	72.5	73.5	74.5	75.5	76	76.5	77.5	78	78.5/80.5	Over 80.5[A]
	Rockwell Hardness Range														
0.26–0.30[B]	...	8	8	6	6	6	6	6	6	5	5	5	5	4	...
0.31–0.35[B]	...	...	8	6	6	6	6	6	6	5	5	5	5	4	...
0.36–0.40[B]	...	...	...	6	6	6	6	6	6	5	5	5	5	4	...
0.41–0.45[B]	...	...	...	...	6	6	6	6	6	5	5	5	5	4	...
0.46–0.50[B]	...	...	...	...	...	6	6	6	6	5	5	5	5	4	4
0.51–0.55[B]	...	...	...	...	...	...	6	6	6	5	5	5	5	4	4
0.56–0.60[B]	...	...	...	...	...	...	...	6	6	5	5	5	5	4	4
0.61–0.65[B]	...	...	...	...	...	...	...	...	6	5	5	5	5	4	4
0.66–0.70[B]	...	...	...	...	...	...	...	...	...	5	5	5	5	4	4
0.71–0.75[B]	...	...	...	...	...	...	...	...	...	...	5	5	5	4	4
0.76–0.80[B]	...	...	...	...	...	...	...	...	...	...	...	5	5	4	4
0.81–0.90[B]	...	...	...	...	...	...	...	...	...	...	...	...	5	4	4
0.91–1.35[B]	...	...	...	...	...	...	...	...	...	...	...	...	...	4	4

[A] Rockwell 30T Scale is not recommended for values over 30T83.
[B] Indicates soft-type annealed cold-rolled carbon spring steel flat wire which is furnished to a maximum hardness as shown in Table 8.

TABLE 11 Rockwell Hardness Ranges for Soft-Type Intermediate Hardness Carbon Spring Steel Flat Wire Thickness 0.040 in. (1.02 mm) and Over

Maximum of Carbon Range, %	For Maximum of Specified Rockwell Hardness Range, B Scale												
	74	76	78	80	82	83.5	85	86.5	88	89	90	91	92	93/97	Over 97[A]
0.26–0.30[B]	...	12	12	10	10	10	10	10	10	8	8	8	8	6	...
0.31–0.35[B]	...	...	12	10	10	10	10	10	10	8	8	8	8	6	...
0.36–0.40[B]	...	...	...	10	10	10	10	10	10	8	8	8	8	6	...
0.41–0.45[B]	...	...	...	...	10	10	10	10	10	8	8	8	8	6	...
0.46–0.50[B]	...	...	...	...	...	10	10	10	10	8	8	8	8	6	...
0.51–0.55[B]	...	...	...	...	...	...	10	10	10	8	8	8	8	6	...
0.56–0.60[B]	...	...	...	...	...	...	...	10	10	8	8	8	8	6	...
0.61–0.65[B]	...	...	...	...	...	...	...	...	10	8	8	8	8	6	5
0.66–0.70[B]	...	...	...	...	...	...	...	...	...	8	8	8	8	6	5
0.71–0.75[B]	...	...	...	...	...	...	...	...	...	...	8	8	8	6	5
0.76–0.80[B]	...	...	...	...	...	...	...	...	...	...	...	8	8	6	5
0.81–0.90[B]	...	...	...	...	...	...	...	...	...	...	...	...	8	6	5
0.91–1.35[B]	...	...	...	...	...	...	...	...	...	...	...	...	...	6	5

[A] Rockwell B Scale is not recommended for values over B100.
[B] Indicates soft-type annealed cold-rolled carbon spring steel flat wire which is furnished to a maximum hardness as shown in Table 8.

A 805

TABLE 12 Spheroidize Type Carbon Spring Steel Flat Wire Lowest Expected Maximum Rockwell Hardness

Maximum of Carbon Range, %	Flat Wire Thickness, in. (mm)	
	0.025 (0.64) to 0.040 (1.02) excl	0.040 (1.02) and Over
	Rockwell Hardness, 30T Scale	Rockwell Hardness, B Scale
0.30	63	68
0.35	65	70
0.40	66	72
0.45	67	74
0.50	68	77
0.55	69	78
0.60	70	80
0.65	71	82
0.70	72	83
0.75	73	84
0.80	73	86
0.85	74	87
0.90	75	87
0.95 and over	75	88

TABLE 13 Hardened- and Tempered-Carbon Spring Steel Flat Wire Rockwell Hardness Ranges

NOTE—A Rockwell hardness range is the arithmetic difference between two limits (for example, C42 to C46 is a four-point range). It is customary to specify Rockwell range requirements within the above ranges for each grade of hardened and tempered carbon spring steel flat wire in accordance with the following:

Rockwell Hardness Scale	Specified Range
C	Any 4 points
30N	Any 4 points
15N	Any 3 points

Thickness in. (mm)	Rockwell Scale	Maximum of Carbon Range, %						
		0.75	0.80	0.85	0.90	0.95	1.00	1.05
		Rockwell Hardness Ranges						
Over 0.005 (0.13) to 0.015 (0.38), incl	15N	78–84	80.5–84.5	81–85	81.5–85.5	82–86	82.5–86.5	83–87
Over 0.015 (0.38) to 0.035 (0.89), incl	30N	57–68	62–69	63–70	64–71	64.5–71.5	65–72	66–73
Over 0.035 (0.89) to 0.055 (1.40), incl	C	37–49	42–50	43–51	44–52	45–53	46–54	47–55
Over 0.055 (1.40) to 0.070 (1.78), incl	C	36–48	41–49	42–50	43–51	44–52	45–53	46–54
Over 0.070 (1.78) to 0.085 (2.16), incl	C	35–47	40–48	41–49	42–50	43–51	44–52	45–53
Over 0.085 (2.16) to 0.100 (2.54), incl	C	34–46	39–47	40–48	41–49	42–50	43–51	44–52
Over 0.100 (2.54) to 0.115 (2.92), incl	C	33–45	38–46	39–47	40–48	41–49	42–50	43–51
Over 0.115 (2.92) to 0.125 (3.17), incl	C	32–44	37–45	38–46	39–47	40–48	41–49	42–50

ASTM A 805

TABLE 14 Rockwell Hardness Scales for Various Thicknesses (A Guide for Selection of Scales Using the Diamond Penetrator Hardened and Tempered Cold-Rolled Carbon Spring Steel)

NOTE—For a given thickness, any hardness greater than that corresponding to that thickness can be tested. For a given hardness, material of any greater thickness than that corresponding to that hardness can be tested on the indicated scale.

Thickness, in. (mm)	Rockwell Scale								
	A		C	15N		30N		45N	
	Dial Reading	Approximate Hardness C-Scale[A]	Dial Reading	Dial Reading	Approximate Hardness C-Scale[A]	Dial Reading	Approximate Hardness C-Scale[A]	Dial Reading	Approximate Hardness C-Scale[A]
0.008[B] (0.20)	...	...	...	90	60	...	...	...	...
0.010 (0.25)	...	...	...	88	55	...	...	...	...
0.012 (0.30)	...	...	...	83	45	82	65	77	69.5
0.014 (0.36)	...	...	...	76	32	78.5	61	74	67
0.016 (0.41)	86	69	...	68	18	74	56	72	65
0.018 (0.46)	84	65	...	...	...	66	47	68	61
0.020 (0.51)	82	61.5	...	...	...	57	37	63	57
0.022 (0.56)	79	56	69	...	...	47	26	58	52.5
0.024 (0.71)	76	50	67	...	...	...	...	51	47
0.026 (0.66)	71	41	65	...	...	...	...	37	35
0028 (0.71)	67	32	64	...	...	...	...	20	20.5
0.030 (0.76)	60	19	57	...	...	...	...	...	...
0.032 (0.81)	...	...	52	...	...	...	...	...	...
0.034 (0.86)	...	...	45	...	...	...	...	...	...
0.036 (0.91)	...	...	37	...	...	...	...	...	...
0.038 (0.97)	...	...	28	...	...	...	...	...	...
0.040 (1.02)	...	...	20	...	...	...	...	...	...

[A] These approximate hardness numbers are for use in selecting a suitable scale, and should not be used as hardness conversions. If necessary to convert test readings to another scale, refer to the ASTM Standard Hardness Conversion Tables 140, for Metals (Relationship Between Brinell Hardness, Vickers Hardness, Rockwell Hardness, and Rockwell Superficial Hardness).

[B] For thickness less than 0.008 in. (0.20 mm) use of the tension test is recommended.

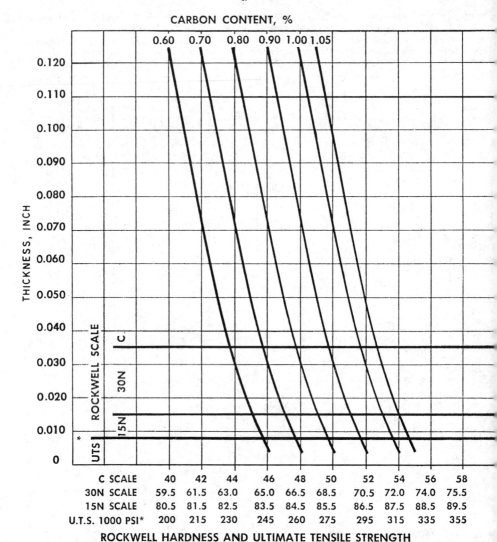

NOTE—For thicknesses less than 0.008 in. (0.20 mm) use of the tension test is recommended.

FIG. 1 Approximate Relationship Between Thickness, Carbon Content, Rockwell Hardness, and Tensile Strength for Hardened- and Tempered-Spring Steel Flat Wire Heat-Treated to Combinations of Mechanical Properties Appropriate for Spring Applications

APPENDIX
(Nonmandatory Information)

X1. GENERAL INFORMATION AND METALLURGICAL ASPECTS

X1.1 Aging Phenomenon

X1.1.1 Although the maximum ductility is obtained in low-carbon steel flat wire in its dead-soft (annealed last) condition, such flat wire is unsuited for some forming operations due to its tendency to stretcher strain or flute. A small amount of cold-rolling (skin-rolling) will prevent this tendency, but the effect is only temporary due to a phenomenon called aging. Aging is accompanied by a loss of ductility with an increase in hardness, yield point and tensile strength. For those uses in which stretcher straining, fluting, or breakage due to aging of the steel is likely to occur, the steel should be fabricated as promptly as possible after skin-rolling. When the above aging characteristics are undesirable, special killed (generally aluminum-killed) steel is used.

X1.2 Uncoiling Characteristics of Annealed or Spheroidized Flat Wire

X1.2.1 Carbon spring steel coiled flat wire annealed or spheroidized at finished thickness does not always possess optimum uncoiling characteristics during subsequent forming. If uncoiling characteristics are important, it may be necessary for the manufacturer to recoil such material with a concurrent very light skin pass.

X1.3 Definite Application Flat Wire

X1.3.1 Definite application carbon steel flat wire is a product developed for a specific application and is commonly specified only by size and descriptive name. Frequently, the characteristics that measure performance of the product cannot be described in terms of test limits. Satisfactory performance is primarily dependent upon the processing and control developed by the flat wire producer as a result of intensive intimate studies of the purchaser's problems in fabrication.

X1.3.2 Some examples of definite application flat wire are given below:

X1.3.2.1 *Low-Carbon Flat Wire*—Stitching wire, bookbinder's wire, shoe pattern wire, stapling wire.

X1.3.2.2 *Untempered-Carbon Spring Steel Flat Wire*—Umbrella rib wire, metal-band saw steel.

X1.3.2.3 *Hardened and Tempered Carbon Spring Steel Flat Wire*—Tape line, brush wire, heddle wire.

X1.4 Restrictive Requirements

X1.4.1 The requirements that are described below concern characteristics of carbon steel flat wire that are adapted to the particular conditions encountered in the fabrication or use for which the wire is produced. The practices used to meet such requirements necessitate appropriate control and close supervision. These requirements entail one or more of the practices in the manufacture of carbon steel flat wire as follows:

X1.4.1.1 Careful selection of raw materials for melting, which vary with each requirement;

X1.4.1.2 More exacting steelmaking practices;

X1.4.1.3 Selection of heats or portions of heats with consequent higher loss than normal;

X1.4.1.4 Additional discard specified or required;

X1.4.1.5 Special supervision and inspection;

X1.4.1.6 Extensive testing;

X1.4.1.7 Test methods not commonly used for production control; and

X1.4.1.8 Possible processing delays.

X1.4.2 As the application becomes more severe the steel producer is more limited in applying steel for the several requirements described below. The processing methods used to meet these requirements vary among producers because of differences in production facilities.

NOTE X1.1—It is customary to specify only one kind of a mechanical test requirement on any one item.

X1.4.3 Restricted temper requirements for low-carbon steel flat wire are sometimes specified or required and the special properties may include restricted Rockwell ranges or restricted tensile strength ranges.

X1.4.3.1 This type of low-carbon steel flat wire is sometimes required to produce identified parts, within properly established allowances, combined with requirements for Rockwell ranges: 15 points, the minimum of which is not less than B60; 10 points, the minimum of which is not less than 30T58; or 5 points, the minimum of which is not less than 15T81.

X1.4.3.2 For certain applications, this type of low-carbon steel flat wire is required to meet separate temper restrictions, such as Rockwell ranges of less than 15 points but not less than 10 points, when the minimum of the range is not less than B60; Rockwell ranges of less than 10 points but not less than 7 points, when the minimum of the range is not less than 30T58; Rockwell ranges of less than 5 points but not less than 3.5 points, when the minimum of the range is not less than 15T81; or tensile strength ranges restricted to less than 25 000 psi (170 MPa).

X1.4.4 *Restricted hardness requirements for carbon spring steel flat wire are sometimes specified or required*:

X1.4.4.1 For certain applications, lower Rockwell hardnesses than shown in Table 8 are required to meet severe forming operations in annealed-carbon

A 805

spring steel in thicknesses 0.025 in. (0.64 mm) and thicker. In this thickness range, carbon spring steel flat wire must be spheroidize-annealed to produce the lowest possible hardness. In producing spheroidize-annealed carbon spring steel flat wire in thicknesses 0.025 in. (0.64 mm) and thicker, the lowest expected maximum Rockwell hardness is shown in Table 12.

X1.4.4.2 For certain applications, in intermediate-hardness untempered-carbon spring steel flat wire, ranges other than the values shown in Tables 9, 10, and 11 may be required.

X1.4.4.3 For certain applications of hard-type untempered carbon spring steel flat wire minimum Rockwell hardness values over B98 are required.

X1.4.4.4 For certain applications of hardened and tempered carbon spring steel flat wire, Rockwell hardness ranges closer than or Rockwell values higher than those shown in Table 14 are required.

X1.4.5 *Heat-Treating Requirements*—When heat-treating requirements must be met in the purchaser's end product, all phases of heat treatment procedure and mechanical property requirements should be clearly specified. The specified mechanical properties should be compatible with the nature of the steel involved and the full range of the specified chemical composition when conventional hardening and tempering practices are employed.

X1.4.6 *Testing*—The following tests are not normally made or required except for some special applications:

X1.4.6.1 *Tension Test*—The measurement of tensile properties, such as tensile strength, yield point, and elongation, is not commonly used as a production control for untempered spring steel flat wire in thicknesses 0.010 in. (0.25 mm) and heavier. If, however, a purchaser finds it necessary to specify tensile strength a range of at least 20 000 psi (140 MPa) is commonly used.

X1.4.6.2 *Extensometer Test*—The measurement of elastic properties such as proportional limit, proof stress, yield strength by the offset method, etc., requires the use of special testing equipment and testing procedures such as the use of an extensometer or the plotting of a stress-strain diagram.

X1.4.6.3 *Specified Austenite Grain Size* is determined in accordance with Methods E 112. For an specified fine- or coarse-grain size it is customary that not more than 30 % of the grain structure be outside grain illustrations 5 to 8, inclusive, in the case of fine grain steel, and grain illustrations 1 to 5 inclusive, in the case of coarse grain steel. The foregoing testing procedures involve one or more of the following:

(*a*) Selection and preparation of special test specimens;

(*b*) Additional handling and identification of product;

(*c*) Special testing equipment, unusual testing procedures, or both; and

(*d*) Possible processing delays due to storage of product while awaiting results of such tests.

X1.4.6.4 *Decarburization*—The loss of carbon at the surface of carbon steel when heated for processing or to modify mechanical properties. Microscopical chemical and hardness test methods are used to determine the extent of decarburization. A definition and method for determination of decarburization is described in the Society of Automotive Engineers Recommended Practice SAE J419.

X1.4.6.5 *Macroetch Test*—This test consists of immersing a carefully prepared section of the steel in hot acid to evaluate the soundness and homogeneity of the products being tested. Because there are no recognized standards, the location and number of tests, details of testing technique, and interpretation of test results are established in each instance.

X1.4.6.6 *Nonmetallic Inclusion Examination (Macroscopical)*—The samples for the determination of the inclusion count are taken longitudinally. The rating is based upon Recommended Practice E 45. Because there are no recognized standards, the area to be examined and the interpretation of test results are established in each instance.

The American Society for Testing and Materials takes no position respecting the validity of any patent rights asserted in connection with any item mentioned in this standard. Users of this standard are expressly advised that determination of the validity of any such patent rights, and the risk of infringement of such rights, are entirely their own responsibility.

This standard is subject to revision at any time by the responsible technical committee and must be reviewed every five years and if not revised, either reapproved or withdrawn. Your comments are invited either for revision of this standard or for additional standards and should be addressed to ASTM Headquarters. Your comments will receive careful consideration at a meeting of the responsible technical committee, which you may attend. If you feel that your comments have not received a fair hearing you should make your views known to the ASTM Committee on Standards, 1916 Race St., Philadelphia, Pa. 19103.

Designation: A 812/A 812M – 83

Standard Specification for
STEEL SHEET, HOT-ROLLED, HIGH-STRENGTH, LOW-ALLOY, FOR WELDED LAYERED PRESSURE VESSELS[1]

This standard is issued under the fixed designation A 812/A 812M; the number immediately following the designation indicates the year of original adoption or, in the case of revision, the year of last revision. A number in parentheses indicates the year of last reapproval. A superscript epsilon (ε) indicates an editorial change since the last revision or reapproval.

1. Scope

1.1 This specification covers hot-rolled, high-strength, low-alloy steel sheet intended primarily for use in coil form in the construction of welded layered pressure vessels.

1.2 The following grades are included in this specification:

	Mechanical Requirements			
	Yield Strength, min		Tensile Strength, min	
Grade	ksi	MPa	ksi	MPa
65	65	450	85	585
80	80	550	100	690

1.3 This material is produced in coil form. However, testing is to represent the back end and the middle of the slab as rolled—see 6.1.4. The purchaser should therefore recognize this requires cutting the as-rolled coil to obtain test samples and results in half-size coils.

1.4 The values stated in either inch-pound units or SI (metric) units are to be regarded separately as standard. Within the text the SI units are shown in brackets. The values stated in each system are not exact equivalents, therefore each system must be used independently of the other. Combining values of the two systems may result in nonconformance with the specification.

2. Applicable Documents

2.1 *ASTM Standards:*
A 568 Specification for General Requirements for Steel, Carbon and High-Strength Low-Alloy Hot-Rolled Sheet and Cold-Rolled Sheet[2]
A 568M Specification for General Requirements for Steel, Carbon and High-Strength Low-Alloy Hot-Rolled Sheet and Cold-Rolled Sheet [Metric][2]

3. General Requirements for Delivery

3.1 Material furnished under this specification shall conform to the applicable requirements of the current edition of Specifications A 568 or A 568M unless otherwise provided herein.

4. Ordering Information

4.1 Orders for material under this specification shall include the following information, as required, to describe the material adequately:

4.1.1 Specification number, grade, and date of issue,
4.1.2 Name of material (hot-rolled sheet),
4.1.3 Special requirements, if required, and
4.1.4 Dimensions, including type of edges.

NOTE 1—A typical ordering description is as follows: ASTM Axxx, Grade 65, Hot-Rolled Sheet, 0.157 in. [4.00 mm] by 48 in. [1220 mm] by coil, cut edges.

5. Chemical Requirements

5.1 *Cast or Heat Analysis*—The analysis of the steel shall conform to the requirements in Table 1.

6. Mechanical Requirements

6.1 *Tensile Strength:*
6.1.1 *Requirements*—The material as represented by the tension test specimens shall con-

---
[1] This specification is under the jurisdiction of ASTM Committee A-1 on Steel, Stainless Steel, and Related Alloys and is the direct responsibility of Subcommittee A01.19 on Sheet Steel and Strip.
Current edition approved May 27, 1983. Published December 1983.
[2] *Annual Book of ASTM Standards*, Vol 01.03.

form to the tensile properties prescribed in Table 2.

6.1.2 *Number of Tests*—Two tension tests shall be made from the product of each slab as rolled.

6.1.3 *Test Specimen Condition and Size*—Test specimens shall be prepared for testing from coils in their finished condition and shall be the full thickness of the sheet.

6.1.4 *Test Specimen Location and Orientation*—Tension test specimens shall be taken at points representing the back end and the middle of the slab as rolled. These specimens shall be located approximately midway between the center and edge of the coil and shall be transverse to the longitudinal axis.

7. Finish and Appearance

7.1 *Surface Finish:*

7.1.1 Unless otherwise specified, the material shall be furnished without removing the hot-rolled oxide or scale.

7.2 *Edges*—Unless otherwise specified, mi edges shall be furnished.

8. Marking

8.1 The name or the brand of the manufac turer, heat number, slab or coil number, specif cation number and year of issue, and grade sha be legibly and durably marked on each finishe coil. Steel-die marking of coils is prohibited.

9. Certifications and Reports

9.1 The manufacturer shall furnish copies a test report showing the results of the cast heat analysis and mechanical property tests mac to determine compliance with this specificatio The report shall include the purchase order nun ber, specification designation, and test identif cation number correlating the test results wit material represented.

TABLE 1 Chemical Requirements

Element	Composition, %	
	Grade 65, Heat or Cast	Grade 80, Heat or Cast
Carbon, max	0.23	0.23
Manganese, max	1.40	1.50
Phosphorus, max	0.035	0.035
Sulfur, max	0.040	0.040
Silicon	[A]	0.15–0.50
Columbium plus vanadium	0.02–0.15[B]	0.02–0.15[B]
Chromium, max	...	[C]

[A] Killed steel may be supplied when specified. When silicon-killed steel is ordered, a range from 0.15 to 0.50 % shall be supplied.

[B] Columbium content not to exceed 0.05 % on heat analysis, 0.06 % on product analysis.

[C] Chromium may be present as an additional element but shall not exceed 0.35 %.

TABLE 2 Tensile Requirements

	Grade 65	Grade 80
Tensile strength:		
ksi	85–110	100–125
MPa	585–760	690–860
Yield strength, min:[A]		
ksi	65	80
MPa	450	550
Elongation in 2 in. [50 mm], min, %:		
Under 0.230 to 0.145 in. [under 6.0 to 3.8 mm], incl. in thickness	15	13
Under 0.145 to 0.089 in. [under 3.8 to 2.2 mm], incl. in thickness	14	12
Under 0.089 to 0.057 in. [under 2.2 to 1.4 mm], incl. in thickness	13	11
Elongation in 8 in. [200 mm], min, %:		
Under 0.230 to 0.145 in. [under 6.0 to 3.8 mm], incl. in thickness	9	7

[A] Yield strength determined by the 0.2 % offset or 0.5 extension under load methods.

The American Society for Testing and Materials takes no position respecting the validity of any patent rights asserted in connectio with any item mentioned in this standard. Users of this standard are expressly advised that determination of the validity of any suc patent rights, and the risk of infringement of such rights, are entirely their own responsibility.

This standard is subject to revision at any time by the responsible technical committee and must be reviewed every five years an if not revised, either reapproved or withdrawn. Your comments are invited either for revision of this standard or for additiono standards and should be addressed to ASTM Headquarters. Your comments will receive careful consideration at a meeting of th responsible technical committee, which you may attend. If you feel that your comments have not received a fair hearing you shoul make your views known to the ASTM Committee on Standards, 1916 Race St., Philadelphia, Pa. 19103.

Designation: B 670 – 80

An American National Standard

Standard Specification for
PRECIPITATION-HARDENING NICKEL ALLOY (UNS N07718) PLATE, SHEET, AND STRIP FOR HIGH-TEMPERATURE SERVICE[1]

This standard is issued under the fixed designation B 670; the number immediately following the designation indicates the year of original adoption or, in the case of revision, the year of last revision. A number in parentheses indicates the year of last reapproval. A superscript epsilon (ε) indicates an editorial change since the last revision or reapproval.

1. Scope

1.1 This specification covers rolled precipitation hardenable nickel alloy (N07718)* plate, sheet, and strip in the annealed condition (temper).

1.2 The values stated in inch-pound units are to be regarded as the standard.

2. Applicable Documents

2.1 *ASTM Standards:*

B 637 Specification for Precipitation-Hardening Nickel Alloy Bars, Forgings, and Forging Stock for High-Temperature Service[2]

E 8 Methods of Tension Testing of Metallic Materials[3]

E 29 Recommended Practice for Indicating Which Places of Figures Are to Be Considered Significant in Specified Limiting Values[4]

E 38 Methods for Chemical Analysis of Nickel-Chromium and Nickel-Chromium-Iron Alloys[5]

E 139 Recommended Practice for Conducting Creep, Creep-Rupture, and Stress-Rupture Tests of Metallic Materials[3]

E 354 Methods for Chemical Analysis of High-Temperature, Electrical, Magnetic, and Other Similar Iron, Nickel, and Cobalt Alloys[5]

3. Description of Terms

3.1 The terms given in Table 1 shall apply.

4. Ordering Information

4.1 Orders for material under this specification shall include the following information:

4.1.1 Alloy name or UNS number.

4.1.2 *Condition*—Table 3 and Appendix X1.

4.1.3 *Finish*—Appendix X1.

4.1.4 *Dimensions*—Thickness, width, and length.

4.1.5 *Optional Requirements:*

4.1.5.1 *Sheet and Strip*—Whether to be furnished in coil, in cut straight lengths, or in random straight lengths.

4.1.5.2 *Strip*—Whether to be furnished with commercial slit edge, square edge, or round edge.

4.1.5.3 *Plate*—Whether to be furnished specially flattened (7.7.1); also how plate is to be cut (Tables 7 and 10).

4.1.6 *Fabrication Details*—Not mandatory but helpful to the manufacturer:

4.1.6.1 *Welding or Brazing*—Process to be employed.

4.1.6.2 *Plate*—Whether material is to be hot-formed.

4.1.7 *Certification*—State if certification is required (Section 15).

4.1.8 *Samples for Product (Check) Analysis*—Whether samples should be furnished (5.2).

---

[1] This specification is under the jurisdiction of ASTM Committee B-2 on Nonferrous Metals and Alloys.
Current edition approved July 3, 1980. Published August 1980. Originally published as A 670 – 72. Redesignated B 670 in 1978. Last previous edition A 670 – 78.
* New designation established in accordance with ASTM E 527 and SAE J1086, Recommended Practice for Numbering Metals and Alloys (UNS).
[2] *Annual Book of ASTM Standards*, Vol 02.04.
[3] *Annual Book of ASTM Standards*, Vol 03.01.
[4] *Annual Book of ASTM Standards*, Vol 14.02.
[5] *Annual Book of ASTM Standards*, Vol 03.05.

4.1.9 *Purchaser Inspection*—If the purchaser wishes to witness the tests or inspection of material at the place of manufacture, the purchase order must so state indicating which tests or inspections are to be witnessed (Section 13).

5. Chemical Requirements

5.1 The material shall conform to the requirements as to chemical composition prescribed in Table 2.

5.2 If a product (check) analysis is performed by the purchaser, the material shall conform to the product (check) analysis variations prescribed in Table 2.

6. Mechanical and Other Requirements

6.1 *Tensile Properties*—The material after precipitation hardening shall conform to the tensile properties prescribed in Table 3.

6.2 *Stress-Rupture Properties*—The material after precipitation hardening shall conform to the stress-rupture properties prescribed in Table 4.

7. Dimensions and Permissible Variations

7.1 *Thickness and Weight:*

7.1.1 *Plate*—The permissible variation under the specified thickness and permissible excess in overweight shall not exceed the amounts prescribed in Table 5.

7.1.1.1 For use with Table 5, plate shall be assumed to weigh 0.296 lb/in^3 (8.17 g/cm^3).

7.1.2 *Sheet and Strip*—The permissible variations in thickness of sheet and strip shall be as prescribed in Table 6. The thickness of strip and sheet shall be measured with the micrometer spindle $\frac{3}{8}$ in. (9.52 mm) or more from either edge for material 1 in. (25.4 mm) or over in width and at any place on the strip under 1 in. in width.

7.2 *Width or Diameter:*

7.2.1 *Plate*—The permissible variations in width of rectangular plates and diameter of circular plates shall be as prescribed in Tables 7 and 8.

7.2.2 *Sheet and Strip*—The permissible variations in width for sheet and strip shall be as prescribed in Table 9.

7.3 *Length:*

7.3.1 Sheet and strip of all sizes may be ordered to cut lengths, in which case a variation of ⅛ in. (3.18 mm) over the specified length shall be permitted.

7.3.2 Permissible variations in length of rectangular plate shall be as prescribed in Table 10.

7.4 *Straightness:*

7.4.1 The edgewise curvature (depth of chord) of flat sheet, strip, and plate shall not exceed 0.05 in. multiplied by the length of the product in feet (0.04 mm multiplied by the length of the product in centimetres).

7.4.2 Straightness for coiled strip material subject to agreement between the manufacturer and the purchaser.

7.5 *Edges:*

7.5.1 When finished edges of strip are specified in the contract or purchase order, the following descriptions shall apply:

7.5.1.1 Square-edge strip shall be supplied with finished edges, with sharp, square corners and without bevel or rounding.

7.5.1.2 Round-edge strip shall be supplied with finished edges, semicircular in form, and the diameter of the circle forming the edge being equal to the strip thickness.

7.5.1.3 When no description of any required form of strip edge is given, it shall be understood that edges such as those resulting from slitting or shearing will be acceptable.

7.5.1.4 Sheet shall have sheared or slit edges.

7.5.1.5 Plate shall have sheared or cut (machined, abrasive-cut, powder-cut, or inert-arc cut) edges, as specified.

7.6 *Squareness (Sheet)*—For sheets of all thicknesses, the angle between adjacent sides shall be 90 ± 0.15 deg (¹⁄₁₆ in. in 24 in.) (1.5 mm in 610 mm).

7.7 *Flatness*—Standard flatness tolerance for plate shall conform to the requirements prescribed in Table 11. "Specially flattened" plate when so specified, shall have permissible variations in flatness as agreed upon between the manufacturer and the purchaser.

8. Workmanship, Finish, and Appearance

8.1 The material shall be uniform in quality and temper, smooth, commercially straight or flat, and free of injurious imperfections.

9. Sampling

9.1 *Lot*—Definition:

9.1.1 A lot for chemical analysis shall consist of one heat.

9.1.2 A lot for tension and stress-rupture testing shall consist of all material from the same heat, nominal thickness, and condition temper).

9.1.2.1 Where material cannot be identified by heat, a lot shall consist of not more than 500 lb (227 kg) of material in the same thickness and condition, except for plates weighing over 500 lb, in which case only one specimen shall be taken.

9.2 *Test Material Selection:*

9.2.1 *Chemical Analysis*—Representative samples shall be taken during pouring or subsequent processing.

9.2.1.1 *Product (Check) Analysis* shall be wholly the responsibility of the purchaser.

9.2.2 *Tension and Stress-Rupture Testing*—Samples of the material to provide test specimens for tension and stress-rupture testing shall be taken from such locations in each lot as to be representative of that lot.

10. Number of Tests

10.1 *Chemical Analysis*—One test per lot.
10.2 *Tension*—One test per lot.
10.3 *Stress Rupture*—One test per lot.

11. Specimen Preparation

11.1 Tension test specimens shall be taken from material in the annealed condition (temper). The specimen shall be transverse to the direction of rolling when width will permit. The test specimen shall be precipitation heat treated (see Table 3) prior to testing.

11.2 Tension test specimens shall be any of the standard or subsize specimens shown in Methods E 8.

11.3 In the event of disagreement, referee specimens shall be as follows:

11.3.1 Full thickness of the material machined to the form and dimensions shown for the sheet-type specimen in Methods E 8 for material under ½ in. (12.7 mm) in thickness.

11.3.2 The largest possible round specimen shown in Methods E 8 for material ½ in. (12.7 mm) and over.

11.4 Stress-rupture specimens shall be the same as tension specimens except modified as necessary for stress-rupture testing in accordance with Recommended Practice E 139.

12. Test Methods

12.1 The chemical composition, mechanical, and other properties of the material as enumerated in this specification shall be determined, in case of disagreement, in accordance with the following methods:

Test	ASTM Designation
Chemical analysis	E 38, E 354[1]
Tension	E 8
Rounding procedure	E 29
Stress rupture	E 139

[1] Methods E 38 is to be used only for elements not covered by Methods E 354.

12.2 For purposes of determining compliance with the specified limits for requirements of the properties listed in the following table, an observed value or a calculated value shall be rounded as indicated in accordance with the rounding method of Recommended Practice E 29.

Test	Rounded Unit for Observed or Calculated Value
Chemical composition and tolerances (when expressed in decimals)	nearest unit in the last right-hand place of figures of the specified limit If two choices are possible, as when the digits dropped are exactly a 5, or a 5 followed only by zeros, choose the one ending in an even digit, with zero defined as an even digit.
Tensile strength and yield strength	nearest 1000 psi (6.9 MPa)
Elongation	nearest 1 %
Rupture life	1 h

13. Inspection

13.1 Inspection of the material shall be agreed upon between the manufacturer and the purchaser as part of the purchase contract.

14. Rejection and Rehearing

14.1 Material that fails to conform to the requirements of this specification may be rejected. Rejection should be reported to the producer or supplier promptly and in writing. In case of dissatisfaction with the results of the test, the producer or supplier may make claim for a rehearing.

15. Certification

15.1 Upon request of the purchaser in the contract or purchase order, a manufacturer's certification that the material was manufactured and tested in accordance with this specification together with a report of the test results shall be furnished.

B 670

16. Marking

16.1 Each bundle or shipping container shall be marked with the name of the material; condition (temper); this specification number; the size; gross, tare, and net weight; consignor and consignee address; contract or order number; or such other information as may be defined in the contract or purchase order.

TABLE 1 Product Description

Product	Thickness, in. (mm)	Width
Hot-rolled plate[A]	$^3/_{16}$ to $2^{1}/_{4}$ (4.76 to 57.2) (Table 5)	Tables 7[C] and 8
Cold-rolled sheet[B]	0.010 to 0.250 (0.254 to 6.35), incl (Table 6)	Table 9
Cold-rolled strip[B]	0.005 to 0.250 (0.127 to 6.35), incl (Table 6)	Table 9

[A] Material $^3/_{16}$ to $^1/_4$ in. (4.76 to 6.35 mm), incl, in thickness may be furnished as sheet or plate provided the material meets the specification requirements for the condition ordered.

[B] Material under 48 in. (1219 mm) in width may be furnished as sheet or strip provided the material meets the specification requirements for the condition ordered.

[C] Hot-rolled plate, in widths 10 in. (250 mm) and under, may be furnished as hot-finished rectangles with sheared or cut edges in accordance with Specification B 637, UNS N07718, provided the mechanical property requirements of this specification are met.

TABLE 2 Chemical Composition

Element	Composition, %	Product (Check) Analysis Variations, under min or over max, of the Specified Limit of Element
Carbon	0.08 max	0.01
Manganese	0.35 max	0.03
Silicon	0.35 max	0.03
Phosphorus	0.015 max	0.005
Sulfur	0.015 max	0.003
Chromium	17.0 to 21.0	0.25
Cobalt[A]	1.0 max	0.03
Molybdenum	2.80 to 3.30	0.10
Columbium + tantalum	4.75 to 5.50	0.20
Titanium	0.65 to 1.15	0.05
Aluminum	0.20 to 0.80	0.10
Iron[B]	remainder	...
Copper	0.30 max	0.03
Nickel[C]	50.0 to 55.0	0.35
Boron	0.006 max	0.002

[A] If determined.
[B] Iron shall be determined arithmetically by difference.
[C] Nickel plus cobalt.

TABLE 3 Tensile Properties for Plate, Sheet, and Strip[A]

Nominal Thickness, in. (mm)	Tensile Strength min, ksi (MPa)	Yield Strength (0.2 % offset), min, ksi (MPa)	Elongation in 2 in. or 50 mm (or 4D), min, %
Up to 1.0 (25.4), incl	180 (1241)	150 (1034)	12
Over 1.0 to 2.25 (25.4 to 57.2), incl	180 (1241)	150 (1034)	10

[A] Material shall be supplied in the annealed condition (temper). The manufacturer shall demonstrate that annealed material is capable of meeting the properties prescribed in Table 3 after precipitation heat treatment. Precipitation heat treatment shall consist of heating to 1325 ± 25°F (718 ± 14°C), hold at temperature for 8 h, furnace cool to 1150 ± 25°F (62 ± 14°C), hold until total precipitation heat treatment time has reached 18 h, and then air cool.

TABLE 4 Stress-Rupture Test at 1200°F (649°C) for Plate, Sheet, and Strip[1]

Nominal Thickness, in. (mm)	Stress,[B] ksi (MPa)	Life, min, h	Elongation in 2 in. or 50 mm (or 4D), min, %
Up to 0.015 (0.38), incl	95 (655)	23	...
Over 0.015 to 0.025 (0.38 to 0.64), incl	95 (655)	23	4
Over 0.025 to 1.5 (0.64 to 38.1), incl	100 (690)	23	4

[A] Material shall be supplied in the annealed condition (temper). The manufacturer shall demonstrate that annealed material is capable of meeting the properties prescribed in Table 4 after precipitation heat treatment. Precipitation heat treatment is as specified in footnote A of Table 3.

[B] Testing may be conducted at a stress higher than that specified but stress shall not be changed while test is in process. Time to rupture and elongation requirements shall be as specified in Table 4.

Testing may also be conducted using incremental loading. In such case, the stress specified in Table 4 shall be maintained to rupture or for 48 h, whichever occurs first. After the 48 h and at intervals of 8 to 16 h, preferably 8 to 10 h, thereafter, the stress shall be increased in increments of 5000 psi (34.5 MPa). Time to rupture and elongation requirements shall be as specified in Table 4.

TABLE 5 Permissible Variations in Thickness and Overweight of Rectangular Plates

NOTE—All plates shall be ordered to thickness and not to weight per square foot (centimetre). No plates shall vary more than 0.01 in. (0.25 mm) under the thickness ordered, and the overweight of each lot[A] in each shipment shall not exceed the amount in the table. Spot grinding is permitted to remove surface imperfections, such spots not to exceed 0.01 in (0.25 mm) under the specified thickness.

Specified Thickness, in. (mm)	Permissible Excess in Average Weight[B,C] per Square Foot of Plates for Widths Given in Inches (Millimetres) Expressed in Percentage of Nominal Weights									
	Under 48 (1220)	48 to 60 (1220 to 1520), excl	60 to 72 (1520 to 1830), excl	72 to 84 (1830 to 2130), excl	84 to 96 (2130 to 2440), excl	96 to 108 (2440 to 2740), excl	108 to 120 (2740 to 3050), excl	120 to 132 (3050 to 3350), excl	132 to 144 (3350 to 3660), excl	144 to 160 (3660 to 4070), incl
3/16 to 5/16 (4.76 to 7.94), excl	9.0	10.5	12.0	13.5	15.0	16.5	18.0	...	...	...
5/16 to 3/8 (7.94 to 9.52), excl	7.5	9.0	10.5	12.0	13.5	15.0	16.5	18.0	...	...
3/8 to 7/16 (9.52 to 11.11), excl	7.0	7.5	9.0	10.5	12.0	13.5	15.0	16.5	18.0	19.5
7/16 to 1/2 (11.11 to 12.70), excl	6.0	7.0	7.5	9.0	10.5	12.0	13.5	15.0	16.5	18.0
1/2 to 5/8 (12.70 to 15.88), excl	5.0	6.0	7.0	7.5	9.0	10.5	12.0	13.5	15.0	16.5
5/8 to 3/4 (15.88 to 19.05), excl	4.5	5.5	6.0	7.0	7.5	9.0	10.5	12.0	13.5	15.0
3/4 to 1 (19.05 to 25.4), excl	4.0	4.5	5.5	6.0	7.0	7.5	9.0	10.5	12.0	13.5
1 to 2¼ (25.4 to 57.2), incl	5.0	5.0	5.5	6.5	7.0	8.0	8.5	10.0	11.5	13.0

[A] The term "lot" applied to this table means all of the plates of each group width and each group thickness.
[B] The permissible overweight for lots of circular and sketch plates shall be 25 % greater than the amounts given in this table.
[C] The weight of individual plates shall not exceed the nominal weight by more than 1¼ times the amount given in this table and Footnote B.

B 670

TABLE 6 Permissible Variations in Thickness of Sheet and Strip
(Permissible Variations, Plus and Minus, in Thickness, in. (mm), for Widths Given in in. (mm))

Specified Thickness, in. (mm)	Sheet			
	Hot-Rolled		Cold-Rolled	
	48 (1220) and Under[A]	Over 48 to 60 (1220 to 1520), incl[A]	48 (1220) and Under[A]	Over 48 to 60 (1220 to 1520), incl[A]
0.018 to 0.025 (0.457 to 0.635), incl	0.003 (0.076)	0.004 (0.100)	0.002 (0.051)	0.003 (0.076)
Over 0.025 to 0.034 (0.635 to 0.864), incl	0.004 (0.100)	0.005 (0.130)	0.003 (0.076)	0.004 (0.100)
Over 0.034 to 0.43 (0.864 to 1.09), incl	0.005 (0.130)	0.006 (0.150)	0.004 (0.100)	0.005 (0.130)
Over 0.043 to 0.056 (1.09 to 1.42), incl	0.005 (0.130)	0.006 (0.150)	0.004 (0.100)	0.005 (0.130)
Over 0.056 to 0.070 (1.42 to 1.78), incl	0.006 (0.150)	0.007 (0.180)	0.005 (0.130)	0.006 (0.150)
Over 0.070 to 0.078 (1.78 to 1.98), incl	0.007 (0.180)	0.008 (0.200)	0.006 (0.150)	0.007 (0.180)
Over 0.078 to 0.093 (1.98 to 2.36), incl	0.008 (0.200)	0.009 (0.230)	0.007 (0.180)	0.008 (0.200)
Over 0.093 to 0.109 (2.36 to 2.77), incl	0.009 (0.230)	0.010 (0.254)	0.007 (0.180)	0.009 (0.230)
Over 0.109 to 0.125 (2.77 to 3.18), incl	0.010 (0.254)	0.012 (0.305)	0.008 (0.200)	0.010 (0.254)
Over 0.125 to 0.140 (3.18 to 3.56), incl	0.012 (0.305)	0.014 (0.356)	0.008 (0.200)	0.010 (0.254)
Over 0.140 to 0.171 (3.56 to 4.34), incl	0.014 (0.356)	0.016 (0.406)	0.009 (0.230)	0.012 (0.305)
Over 0.171 to 0.187 (4.34 to 4.76), incl	0.015 (0.381)	0.017 (0.432)	0.010 (0.254)	0.013 (0.330)
Over 0.187 to 0.218 (4.76 to 5.54), incl	0.017 (0.432)	0.019 (0.483)	0.011 (0.279)	0.015 (0.381)
Over 0.218 to 0.234 (5.54 to 5.94), incl	0.018 (0.457)	0.020 (0.508)	0.012 (0.305)	0.016 (0.406)
Over 0.234 to 0.250 (5.94 to 6.35), incl	0.020 (0.508)	0.022 (0.559)	0.013 (0.330)	0.018 (0.457)

Cold-Rolled Strip	
Specified Thickness, in. (mm)	Widths 12 in. (305 mm) and under, plus and minus[A]
Up to 0.050 (1.27), incl	0.0015 (0.038)
Over 0.050 to 0.093 (1.27 to 2.39), incl	0.0025 (0.063)
Over 0.093 to 0.125 (2.39 to 3.18), incl[B]	0.004 (0.106)

[A] Measured ⅜ in. (9.52 mm) or more from either edge except for strip under 1 in. (25.4 mm) in width which is measured at any place.
[B] Standard sheet tolerances apply for thicknesses over 0.125 in. (3.18 mm) and for all thicknesses of strip over 12 in. (305 mm) wide.

TABLE 7 Permissible Variations in Width[A] of Sheared, Plasma-Torch-Cut, and Abrasive-Cut Rectangular Plate[B,C]

Specified Thickness	Permissible Variations in Widths for Widths Given, in. (mm)									
	Up to 30 (760), incl		Over 30 to 72 (760 to 1830), incl		Over 72 to 108 (1830 to 2740), incl		Over 108 to 144 (2740 to 3660), incl		Over 144 to 160 (3660 to 4070), incl	
	Plus	Minus	Plus	Minus	Plus	Minus	Plus	Minus	Plus	Minus
Inches										
Sheared:[D]										
3/16 to 5/16, excl	3/16	1/8	1/4	1/8	3/8	1/8	1/2	1/8	...	...
5/16 to 1/2, excl	1/4	1/8	3/8	1/8	3/8	1/8	1/2	1/8	5/8	1/8
1/2 to 3/4, excl	3/8	1/8	3/8	1/8	1/2	1/8	5/8	1/8	3/4	1/8
3/4 to 1, excl	1/2	1/8	1/2	1/8	5/8	1/8	3/4	1/8	7/8	1/8
1 to 1 1/4, incl	5/8	1/8	5/8	1/8	3/4	1/8	7/8	1/8	1	1/8
Abrasive-cut:[E,F]										
3/16 to 1 1/4, incl	1/8	1/8	1/8	1/8	1/8	1/8	1/8	1/8	1/8	1/8
over 1 1/4 to 2 1/4, incl	3/16	1/8	3/16	1/8	3/16	1/8	3/16	1/8	3/16	1/8
Plasma-torch-cut:[G]										
3/16 to 1 1/2, excl	3/4	0	3/4	0	3/4	0	3/4	0	3/4	0
1 1/2 to 2 1/4, incl	1	1/4	1	1/4	1	1/4	1	1/4	1	1/4
Millimetres										
Sheared:[D]										
4.8 to 7.9, excl	4.8	3.2	6.4	3.2	9.5	3.2	12.7	3.2	...	...
7.9 to 12.7, excl	6.4	3.2	9.5	3.2	9.5	3.2	12.7	3.2	15.9	3.2
12.7 to 19.0, excl	9.5	3.2	9.5	3.2	12.7	3.2	15.9	3.2	19.0	3.2
19.0 to 25.4, excl	12.7	3.2	12.7	3.2	15.9	3.2	19.0	3.2	22.2	3.2
25.4 to 31.8, incl	15.9	3.2	15.9	3.2	19.0	3.2	22.2	3.2	25.4	3.2
Abrasive-cut:[E,F]										
4.8 to 31.8, incl	3.2	3.2	3.2	3.2	3.2	3.2	3.2	3.2	3.2	3.2
over 31.8 to 57.2, incl	4.8	3.2	4.8	3.2	4.8	3.2	4.8	3.2	4.8	3.2
Plasma-torch-cut:[G]										
4.8 to 38.1, excl	19.0	0	19.0	0	19.0	0	19.0	0	19.0	0
38.1 to 57.2, incl	25.4	6.4	25.4	6.4	25.4	6.4	25.4	6.4	25.4	6.4

[A] Permissible variations in width for powder or inert-arc-cut plate shall be as agreed upon between the manufacturer and the purchaser.

[B] Permissible variations in machined, powder-, or inert-arc-cut circular plate shall be as agreed upon between the manufacturer and the purchaser.

[C] Permissible variations in plasma-torch-cut sketch plates shall be as agreed upon between the manufacturer and the purchaser.

[D] The minimum sheared width is 10 in. (254 mm) for material 3/4 in. (19.0 mm) and under in thickness and 20 in. (508 mm) for material over 3/4 in. in thickness.

[E] The minimum abrasive-cut width is 2 in. (51 mm) and increases to 4 in. (102 mm) for thicker plates.

[F] These tolerances are applicable to lengths of 240 in. (6100 mm), max. For lengths over 240 in., an additional 1/16 in. (1.6 mm) is permitted, both plus and minus.

[G] The tolerance spread shown for plasma-torch cutting may be obtained all on the minus side, or divided between the plus and minus side if so specified by the purchaser.

ASTM B 670

TABLE 8 Permissible Variations in Diameter for Circular Plates

Sheared Plate

Specified Diameter, in. (mm)	Permissible Variations Over Specified Diameter for Thickness Given, in. (mm)[A]
	To ⅜ (9.52), incl
20 to 32 (508 to 813), excl	¼ (6.35)
32 to 84 (813 to 2130), excl	⁵⁄₁₆ (7.94)
84 to 108 (2130 to 2740), excl	⅜ (9.52)
108 to 140 (2740 to 3560), incl	⁷⁄₁₆ (11.11)

Plasma-Torch-Cut Plate[B]

Specified Diameter, in. (mm)	Thickness, max, in. (mm)	Permissible Variations in Specified Diameter for Thickness Given, in. (mm)[C]			
		³⁄₁₆ to 1½ (4.76 to 38.1), excl		1½ to 2¼ (38.1 to 57.2), incl	
		Plus	Minus	Plus	Minus
19 to 20 (483 to 508), excl	2¼ (57.2)	¾ (19.05)	0	1 (25.4)	¼ (6.35)
20 to 22 (508 to 559), excl	2¼ (57.2)	¾ (19.05)	0	1 (25.4)	¼ (6.35)
22 to 24 (559 to 610), excl	2¼ (57.2)	¾ (19.05)	0	1 (25.4)	¼ (6.35)
24 to 28 (610 to 711), excl	2¼ (57.2)	¾ (19.05)	0	1 (25.4)	¼ (6.35)
28 to 32 (711 to 812), excl	2 (50.8)	¾ (19.05)	0	1 (25.4)	¼ (6.35)
32 to 34 (812 to 864), excl	1¾ (44.5)	¾ (19.05)	0	1 (25.4)	¼ (6.35)
34 to 38 (864 to 965), excl	1½ (38.1)	¾ (19.05)	0	1 (25.4)	¼ (6.35)
38 to 40 (965 to 1020), excl	1¼ (31.8)	¾ (19.05)	0	1 (25.4)	¼ (6.35)
40 to 140 (1020 to 3560), incl	2¼ (57.2)	¾ (19.05)	0	1 (25.4)	¼ (6.35)

[A] No permissible variations under.
[B] Permissible variations in plasma-torch-cut sketch plates shall be as agreed upon between the manufacturer and the purchaser.
[C] The tolerance spread shown may also be obtained all on the minus side or divided between the plus and minus sides if so specified by the purchaser.

TABLE 9 Permissible Variations in Width of Sheet and Strip

Specified Thickness, in. (mm)	Specified Width, in. (mm)	Permissible Variations in Specified Width, in. (mm)	
		Plus	Minus
	Sheet		
Up to 0.250 (6.35)	all	0.125 (3.18)	0
	Strip[A]		
Under 0.075 (1.9)	Up to 12 (305), incl	0.007 (0.18)	0.007 (0.18)
	Over 12 to 48 (305 to 1219), incl	0.062 (1.6)	0
0.075 to 0.100 (1.9 to 2.5), incl	Up to 12 (305), incl	0.009 (0.23)	0.009 (0.23)
	Over 12 to 48 (305 to 1219), incl	0.062 (1.6)	0
Over 0.100 to 0.125 (2.5 to 3.2), incl	Up to 12 (305), incl	0.012 (0.30)	0.012 (0.30)
	Over 12 to 48 (305 to 1219), incl	0.062 (1.6)	0
Over 0.125 to 0.160 (3.2 to 4.1), incl	Up to 12 (305) incl.	0.016 (0.41)	0.016 (0.41)
	Over 12 to 48 (305 to 1219), incl	0.062 (1.6)	0
Over 0.160 to 0.187 (4.1 to 4.7), incl	Up to 12 (305), incl	0.020 (0.51)	0.020 (0.51)
	Over 12 to 48 (305 to 1219), incl	0.062 (1.6)	0
Over 0.187 to 0.250 (4.7 to 6.4), incl	Up to 12 (305) incl	0.062 (1.6)	0.062 (1.6)
	Over 12 to 48 (305 to 1219), incl	0.062 (1.6)	0.062 (1.6)

[A] Rolled round or square-edge strip in thicknesses of 0.071 to 0.125 in. (1.80 to 3.18 mm), incl, in widths 3 in. (76.2 mm) and under, shall have permissible width variations of ±0.005 in. (±0.130 mm). Permissible variations for other sizes shall be as agreed upon between the manufacturer and the purchaser.

ASTM B 670

TABLE 10 Permissible Variations in Length[A] of Sheared, Plasma-Torch-Cut,[B] and Abrasive-Cut Rectangular Plate[C]

	Permissible Variation in Length for Lengths Given, in. (mm)																	
Specified Thickness	Up to 60 (1520), incl		Over 60 to 96 (1520 to 2440), incl		Over 96 to 120 (2440 to 3050), incl		Over 120 to 240 (3050 to 6096), incl		Over 240 to 360 (6096 to 9144), incl		Over 360 to 450 (9144 to 11 430), incl		Over 450 to 540 (11 430 to 13 716), incl		Over 540 (13 716)	
	Plus	Minus	Plus	Minus	Plus	Minus	Plus	Minus	Plus	Minus	Plus	Minus	Plus	Minus	Plus	Minus
	Inches															
Sheared:[D]																
3/16 to 5/16, excl	1/4	1/8	1/4	1/8	3/8	1/8	1/2	1/8	5/8	1/8	3/4	1/8	7/8	1/8	...	...
5/16 to 1/2, excl	3/8	1/8	1/2	1/8	1/2	1/8	1/2	1/8	5/8	1/8	3/4	1/8	7/8	1/8	1	1/8
1/2 to 3/4, excl	1/2	1/8	1/2	1/8	5/8	1/8	5/8	1/8	3/4	1/8	1 1/8	1/8	1 3/8	1/8	1 3/8	1/8
3/4 to 1, excl	5/8	1/8	5/8	1/8	3/4	1/8	3/4	1/8	7/8	1/8	1 1/8	1/8	1 3/8	1/8	1 3/8	1/8
1 to 1 1/4, incl	3/4	1/8	3/4	1/8	3/4	1/8	7/8	1/8	1 3/8	1/8						
Abrasive-cut:[E]																
3/16 to 1 1/4, incl	1/8	1/8	1/8	1/8	1/8	1/8	1/8	1/8	1/8	1/8						
over 1 1/4 to 2 1/4, incl	3/16	1/8	3/16	1/8	3/16	1/8	3/16	1/8	3/16	1/8						
Plasma-torch-cut:[E]																
3/16 to 1 1/2, excl	3/4	0	3/4	0	3/4	0	3/4	0	3/4	0	3/4	0	3/4	0	3/4	0
1 1/2 to 2 1/4, incl	1	1/4	1	1/4	1	1/4	1	1/4	1	1/4	1	1/4	1	1/4	1	1/4
	Millimetres															
Sheared:[D]																
4.8 to 7.9, excl	4.8	3.2	6.4	3.2	9.5	3.2	12.7	3.2	15.9	3.2	19.0	3.2	22.2	3.2	...	...
7.9 to 12.7, excl	9.5	3.2	12.7	3.2	12.7	3.2	12.7	3.2	15.9	3.2	19.0	3.2	22.2	3.2	25.4	3.2
12.7 to 19.0, excl	12.7	3.2	12.7	3.2	15.9	3.2	15.9	3.2	19.0	3.2	22.2	3.2	28.6	3.2	34.9	3.2
19.0 to 25.4, excl	15.9	3.2	15.9	3.2	15.9	3.2	19.0	3.2	22.2	3.2	28.6	3.2	34.9	3.2	41.2	3.2
25.4 to 31.8, incl	19.0	3.2	19.0	3.2	19.0	3.2	22.2	3.2	28.6	3.2	34.9	3.2	41.2	3.2	...	...
Abrasive-cut:[E]																
4.8 to 31.8, incl	3.2	3.2	3.2	3.2	3.2	3.2	3.2	3.2	3.2	3.2						
over 31.8 to 57.2, incl	4.8	3.2	4.8	3.2	4.8	3.2	4.8	3.2	4.8	3.2						
Plasma-torch-cut:[F]																
4.8 to 38.1, excl	19.0	0	19.0	0	19.0	0	19.0	0	19.0	0	19.0	0	19.0	0	19.0	0
38.1 to 57.2, incl	25.4	6.4	25.4	6.4	25.4	6.4	25.4	6.4	25.4	6.4	25.4	6.4	25.4	6.4	25.4	6

[A] Permissible variations in length for powder- or inert-arc-cut plate shall be as agreed upon between the manufacturer and the purchaser.
[B] The tolerance spread shown for plasma-torch cutting may be obtained all on the minus side or divided between the plus and minus sides if so specified by the purchaser.
[C] Permissible variations in machined, powder-, or inert-arc-cut circular plate shall be as agreed upon between the manufacturer and the purchaser.
[D] The minimum sheared length is 10 in. (254 mm).
[E] Abrasive cut applicable to a maximum length of 144 to 400 in. (3658 to 10 160 mm) depending on the thickness and width ordered.
[F] The tolerance spread shown for plasma-torch-cut sketch plates shall be as agreed upon between the manufacturer and the purchaser.

TABLE 11 Permissible Variations From Flatness of Rectangular, Circular, and Sketch Plates

NOTE 1—Permissible variations apply to plates up to 12 ft (3.66 m) in length, or to any 12 ft of longer plates.
NOTE 2—If the longer dimension is under 36 in. (914 mm) the permissible variation is not greater than ¼ in. (6.35 mm).
NOTE 3—The shorter dimension specified is considered the width, and the permissible variation in flatness across the width does not exceed the tabular amount for that dimension.
NOTE 4—The maximum deviation from a flat surface does not customarily exceed the tabular tolerance for the longer dimension specified.

Specified Thickness	Permissible Variations from a Flat Surface for Thickness and Widths Given, in. (mm)								
	To 48 (1220), excl	48 to 60 (1220 to 1520), excl	60 to 72 (1520 to 1830), excl	72 to 84 (1830 to 2130), excl	84 to 96 (2130 to 2440), excl	96 to 108 (2440 to 2740), excl	108 to 120 (2740 to 3050), excl	120 to 144 (3050 to 3660), excl	144 (3660) and over
			Inches						
³⁄₁₆ to ¼, excl	¾	1¹⁄₁₆	1¼	1⅜	1⅝	1⅝	...	...	...
¼ to ⅜, excl	11⁄16	¾	15⁄16	1⅛	1⅜	1⁷⁄₁₆	1⁹⁄₁₆	1⅞	...
⅜ to ½, excl	½	9⁄16	11⁄16	¾	15⁄16	1⅛	1¼	1⁷⁄₁₆	1¾
½ to ¾, excl	½	9⁄16	⅝	⅝	13⁄16	1⅛	1⅛	1⅛	1⅜
¾ to 1, excl	½	9⁄16	⅝	⅝	¾	13⁄16	15⁄16	1	1⅛
1 to 2, excl	½	9⁄16	9⁄16	9⁄16	11⁄16	11⁄16	11⁄16	¾	1
2 to 2¼, incl	¼	5⁄16	⅜	7⁄16	½	9⁄16	⅝	¾	⅞
			Millimetres						
4.76 to 6.35, excl	19.05	27.0	31.7	34.9	41.3	41.3	...	...	...
6.35 to 9.52, excl	17.46	19.05	23.81	28.6	35.0	36.5	39.7	47.6	...
9.52 to 12.70, excl	12.70	14.29	17.46	19.05	23.8	28.6	31.7	35.0	44.4
12.70 to 19.05, excl	12.70	14.29	15.88	15.88	20.64	28.6	28.6	28.6	34.9
19.05 to 25.4, excl	12.70	14.29	15.88	15.88	19.05	20.64	23.81	25.4	28.6
25.4 to 50.8, excl	12.70	14.29	14.29	14.29	17.46	17.46	17.46	19.05	25.4
50.8 to 57.2, incl	6.35	7.94	9.52	11.11	12.70	14.29	15.88	19.05	22.22

APPENDIX

X1. CONDITIONS AND FINISHES NORMALLY SUPPLIED

X1.1 This appendix lists the conditions and finishes in which plate, sheet, and strip are normally supplied.

X1.2 *Plate*—Hot-rolled, annealed, and descaled.

X1.3 *Sheet*—Cold-rolled, annealed, descaled, or bright annealed.

X1.4 *Strip*—Cold-rolled, annealed, descaled, or bright annealed.

The American Society for Testing and Materials takes no position respecting the validity of any patent rights asserted in connection with any item mentioned in this standard. Users of this standard are expressly advised that determination of the validity of any such patent rights, and the risk of infringement of such rights, are entirely their own responsibility.

This standard is subject to revision at any time by the responsible technical committee and must be reviewed every five years and if not revised, either reapproved or withdrawn. Your comments are invited either for revision of this standard or for additional standards and should be addressed to ASTM Headquarters. Your comments will receive careful consideration at a meeting of the responsible technical committee, which you may attend. If you feel that your comments have not received a fair hearing you should make your views known to the ASTM Committee on Standards, 1916 Race St., Philadelphia, Pa. 19103.

Designation: E 437 – 80

An American National Standard
Adopted by Industrial Wire Cloth Institute

Standard Specification for
INDUSTRIAL WIRE CLOTH AND SCREENS (SQUARE OPENING SERIES)[1]

This standard is issued under the fixed designation E 437; the number immediately following the designation indicates the year of original adoption or, in the case of revision, the year of last revision. A number in parentheses indicates the year of last reapproval. A superscript epsilon (ϵ) indicates an editorial change since the last revision or reapproval.

This specification has been approved for use by agencies of the Department of Defense and for listing in the DoD Index of Specifications and Standards.

INTRODUCTION

Industrial wire cloth can be produced in many thousands of combinations of size and shape of opening, wire diameter, type of weave, and metal. Such variety is most confusing and, to the vast majority of wire cloth users, unnecessary, since each usually requires only a very few specifications.

The purpose of this specification is to simplify this problem by a condensed table of recommended specifications covering the entire range of openings in which industrial wire cloth is made with several recommended wire diameters for each opening, for various grades of service.

By making selections from this standard, the user will be guided to specifications that are normally carried in stock or are being regularly produced, thus avoiding inadvertent selection of specifications that, because of little or no demand, are unobtainable, except on special order (usually quite expensive unless the quantity ordered is sufficient to justify the cost of the special manufacturing set-up).

If a user has a specific application for industrial wire cloth that cannot be solved by a selection from this standard, it is recommended that he consult his wire cloth supplier on the availability of an acceptable alternative specification.

1. Scope

1.1 This specification covers the sizes of square opening wire cloth and screens for general industrial uses, including the separating or grading of materials according to designated nominal particle size, and lists standards for openings from 5 in. (125 mm) and finer, woven with wire diameters for various grades of service. Methods of checking and calibrating industrial wire cloth and screens are included as information in the Appendixes.

1.2 This specification does not apply to wire cloth with rectangular openings or to any of the following special-purpose wire cloth:

Testing Sieve Cloth
Mill Screen Cloth
Light Wire Bolting Cloth
Fourdrinier and Cylinder Cloth
Dutch Weave Filter Cloth
Spiral Weave Wire Cloth
Welded Wire Screen

1.3 The values stated in inch-pound units are to be regarded as the standard.

2. Applicable Documents

2.1 *ASTM Standard*:

---

[1] This specification is under the jurisdiction of ASTM Committee E-29 on Particle Size Measurement and is the direct responsibility of Subcommittee E 29.01 on Sieves, Sieving Methods, and Screening Media.
Current edition approved Oct. 31, 1980. Published December 1980. Originally published as E 437 – 71. Last previous edition E 437 – 77.

E 437

E 11 Specification for Wire-Cloth Sieves for Testing Purposes[2]
2.2 *Other Documents*:
Fed. Std. 123 Marking for Shipment (Civil Agencies)[3]
MIL-STD-129 Marking for Shipment and Storage[3]

3. Standard Specifications

3.1 Standard specifications for industrial wire cloth and screens are listed in Tables 1 and 2. Table 1 lists standard specifications for wire cloth that is used primarily for the separation and grading of materials according to particle size. Table 2 lists the standard specifications for general industrial use for wire cloth that is not commonly used for grading of materials according to particle size and is commonly sold by mesh rather than opening size.

3.2 *Openings*—The series of standard openings listed in Table 1 correspond to those of the USA Standard Sieve Series, Specification E 11, and to the ISO Recommended Apertures for Sieves[4] with supplemental openings.

3.3 *Wire Diameters*—A choice of four wire diameters is shown for each standard opening from 5 in. (125 mm) to 0.0117 in. (300 μm) opening, inclusive. For practical reasons, the number of wire diameters or grades for openings finer than 0.0117 in. is progressively reduced.

3.4 *Relationship of Grades*—The purpose of the several grades is to provide combinations of opening and wire diameter for various types of service, from medium light to heavy. The entire standard series has been designed for logical relationship of wire diameter to opening in each grade and between the grades.

3.5 *Equivalent Metric Specification*—Table X1, in the Appendix, shows the equivalent metric specifications to the USA Standard, woven from standard ISO metric wire diameters, for grading of materials according to particle size.

4. Types of Weave

4.1 The following types of weave are those most generally used for the industrial wire cloth covered by this specification:

Plain weave (Fig. 1)
Lock crimp (Fig. 2)
Intercrimped (Fig. 3)
Flat top (Fig. 4)
Twilled weave (Fig. 5)

For definitions of these and other types of weave used for industrial wire cloth and screens, see X4.2 in the Appendix.

5. Metal Composition of Wire

5.1 Industrial wire cloth can be woven from a great variety of metals and alloys, but the following are the most commonly used:

Steel, low-carbon
Steel, high-carbon, spring-grade
Steel, high-carbon, oil- or lead-tempered
Steel, electroplated with zinc, nickel, or copper before weaving
Steel, galvanized before weaving
Steel, galvanized after weaving
Stainless steel, Type 304 (Cr 18 %, Ni 8 %)
Stainless steel, Type 316 (Cr 18 %, Ni 8 % Mo added)
Brass (Cu 80 %, Zn 20 %)
Commercial bronze (Sn 4 to 90 %, P ¼ %, balance Cu)
Monel (high nickel-copper alloy)
Aluminum

For a more complete list of metals and alloys, including their applicable characteristics for use in industrial wire cloth, see Appendix X2.

6. Tolerances

6.1 *Openings*—Permissible variations in average openings in USA Standard Specifications for Industrial Wire Cloth and Screens (Table 1) shall be in accordance with those listed in Table 3.

6.2 *Wire Diameters*—Tolerances for wire diameters used in weaving USA Standard Specifications for Industrial Wire Cloth and Screens (Table 1) shall be in accordance with those listed in Table 4.

---

[2] *Annual Book of ASTM Standards*, Vol 14.02.
[3] Available from Naval Publications and Forms Center, 5801 Tabor Ave., Philadelphia, Pa. 19120.
[4] ISO Recommended Apertures for Sieves, ISO/TC 24, R 565-1967, augmented to include R-40/3 Series, ISO/TC 24, Doc. 54/76, July 1969.

SUPPLEMENTARY REQUIREMENTS

The following sections shall be applicable when U.S. government contractual matters are involved.

S1. Responsibility for Inspection

S1.1 Unless otherwise specified in the contract or purchase order, the producer is responsible for the performance of all inspection and test requirements specified herein. Except as otherwise specified in the contract or order, the producer may use his own or any other suitable facilities for the performance of the inspection and test requirements specified herein, unless disapproved by the purchaser. The purchaser shall have the right to perform any of the inspections and tests set forth in this specification where such inspections are deemed necessary to ensure that material conforms to prescribed requirements.

S2. Government Procurement

S2.1 Unless otherwise specified in the contract, the material shall be packaged in accordance with the suppliers' standard practice which will be acceptable to the carrier at lowest rates. Containers and packing shall comply with the Uniform Freight Classification rules or National Motor Freight Classification rules. Marking for shipment of such material shall be in accordance with Fed. Std. No. 123 for civil agencies, and MIL STD 129 for military agencies.

TABLE 1 USA Standard Specifications for Industrial Wire Cloth and Screens (Square Opening Series)

Opening Designation		D Medium Light				C Medium				B Medium Heavy				A Heavy			
Standard (Metric)	USA Industrial Standard	Mesh	Wire Diameter, in.	Opening, in.	Open Area, %	Mesh	Wire Diameter, in.	Opening, in.	Open Area, %	Mesh	Wire Diameter in.	Opening, in.	Open Area, %	Mesh	Wire Diameter, in.	Opening, in.	Open Area, %
(1)	(2)	(3)				(4)				(5)				(6)			
125 mm	5.0 in.	...	0.500	5	82.8	...	0.625	5	79.1	...	0.750	5	75.5	...	1.00	5	69.5
106	4.24	...	0.500	4¼	80.0	...	0.625	4¼	76.0	...	0.750	4¼	72.2	...	1.00	4¼	65.1
100	4.0	...	0.4375	4	81.3	...	0.625	4	74.8	...	0.750	4	70.9	...	1.00	4	64.0
90	3.5	...	0.375	3½	81.6	...	0.500	3½	76.6	...	0.750	3½	67.8	...	0.875	3½	63.9
75	3.0	...	0.375	3	79.0	...	0.500	3	73.5	...	0.625	3	68.5	...	0.750	3	64.0
...	...	...	0.375	2¾	77.4	...	0.500	2¾	71.6	...	0.625	2¾	66.4	...	0.750	2¾	61.7
63	2.5	...	0.375	2½	75.6	...	0.4375	2½	72.4	...	0.625	2½	64.0	...	0.750	2½	59.2
...	...	...	0.3125	2¼	77.1	...	0.4375	2¼	70.1	...	0.500	2¼	66.9	...	0.625	2¼	61.2
53	2.12	...	0.3125	2⅛	75.9	...	0.375	2⅛	72.2	...	0.500	2⅛	65.5	...	0.625	2⅛	59.7
50	2.0	...	0.283	2	76.7	...	0.375	2	70.9	...	0.500	2	64.0	...	0.625	2	58.0
45	1.75	...	0.283	1¾	74.1	...	0.375	1¾	67.8	...	0.4375	1¾	64.0	...	0.500	1¾	60.5
37.5	1.5	...	0.250	1½	73.4	...	0.3125	1½	68.5	...	0.375	1½	64.0	...	0.500	1½	56.3
...	...	...	0.250	1⅜	71.6	...	0.3125	1⅜	66.4	...	0.375	1⅜	61.7	...	0.500	1⅜	53.8
31.5	1.25	...	0.225	1¼	71.8	...	0.283	1¼	66.5	...	0.375	1¼	59.2	...	0.4375	1¼	54.8
...	...	...	0.207	1⅛	71.3	...	0.283	1⅛	63.8	...	0.3125	1⅛	61.2	...	0.4375	1⅛	51.8
26.5	1.06	...	0.192	1 1/16	71.7	...	0.250	1 1/16	65.5	...	0.3125	1 1/16	59.7	...	0.375	1 1/16	54.6
25	1.0	...	0.177	1	72.2	...	0.250	1	64.0	...	0.283	1	60.8	...	0.375	1	52.9
22.4	0.875	...	0.162	⅞	71.2	...	0.225	⅞	63.3	...	0.283	⅞	57.1	...	0.375	⅞	49.0
19	0.750	...	0.135	¾	71.8	...	0.192	¾	63.4	...	0.250	¾	56.3	...	0.3125	¾	49.8
16	0.625	...	0.120	⅝	70.3	...	0.177	⅝	60.7	...	0.225	⅝	54.0	...	0.283	⅝	47.4
...	...	...	0.120	9/16	67.9	...	0.177	9/16	57.7	...	0.192	9/16	56.0	...	0.250	9/16	47.9
13.2	0.530	...	0.105	0.530	69.2	...	0.162	0.530	58.6	...	0.192	0.530	53.9	...	0.250	0.530	46.2
12.5	0.500	...	0.105	½	68.3	...	0.162	½	57.1	...	0.192	½	52.2	...	0.250	½	44.4
11.2	0.4375	...	0.092	7/16	68.3	...	0.135	7/16	58.4	...	0.177	7/16	50.7	...	0.225	7/16	43.6
9.5	0.375	...	0.092	⅜	64.5	...	0.120	⅜	57.4	...	0.162	⅜	48.7	...	0.192	⅜	43.8
8.0	0.312	...	0.080	5/16	63.4	...	0.105	5/16	56.0	...	0.135	5/16	48.8	...	0.177	5/16	40.8
6.7	0.265	3	0.072	0.261	61.3	2¾	0.105	0.259	50.7	2½	0.120	0.280	49.0	2¼	0.162	0.282	40.3
6.3	0.250	...	0.072	¼	60.3	...	0.092	¼	53.4	...	0.105	¼	49.6	...	0.135	¼	42.2
5.6	0.223	3½	0.063	0.223	60.9	3¼	0.080	0.228	54.9	3	0.105	0.228	46.8	2¾	0.135	0.229	39.7
4.75	0.187	4	0.063	0.187	56.0	3¾	0.072	0.195	53.5	3½	0.092	0.194	46.1	3¼	0.120	0.188	37.3
4.00	0.157	5	0.047	0.153	58.5	4¾	0.054	0.157	55.6	4	0.092	0.158	39.9	3¾	0.105	0.162	36.9
3.35	0.132	6	0.041	0.126	57.2	5½	0.047	0.135	55.1	5	0.072	0.128	41.0	4½	0.092	0.130	34.2
...	...	...	0.041	⅛	56.7	...	0.047	⅛	52.8	...	0.072	⅛	40.3	...	0.092	⅛	33.2
2.80	0.111	7	0.035	0.108	57.2	6½	0.041	0.113	53.0	6	0.063	0.104	38.9	5	0.092	0.108	29.2
...	...	...	0.032	3/32	55.6	...	0.041	3/32	48.4	...	0.054	3/32	40.3	...	0.080	3/32	29.2
2.36	0.0937	8	0.032	0.093	55.4	7½	0.041	0.092	47.6	7	0.054	0.089	38.8	6	0.072	0.095	32.5

TABLE 1 *Continued*

Opening Designation		D Medium Light				C Medium				B Medium Heavy				A Heavy			
Standard (Metric)	USA Industrial Standard	Mesh	Wire Diameter, in.	Opening, in.	Open Area, %	Mesh	Wire Diameter, in.	Opening, in.	Open Area, %	Mesh	Wire Diameter, in.	Opening, in.	Open Area, %	Mesh	Wire Diameter, in.	Opening, in.	Open Area, %
(1)	(2)			(3)				(4)				(5)				(6)	
2.00 mm	0.0787 in.	9½	0.028	0.077	53.5	8¾	0.035	0.079	48.1	8	0.047	0.078	38.9	7	0.063	0.080	31.4
1.70	0.0661	11	0.025	0.066	52.7	10	0.032	0.068	46.2	9	0.047	0.064	33.2	8½	0.054	0.064	29.6
...	0.0555	...	0.023	1/16	53.4	...	0.035	1/16	41.1	...	0.041	1/16	36.4	...	0.054	1/16	28.7
1.40	...	13	0.023	0.054	49.3	12	0.028	0.055	43.6	11	0.035	0.056	37.9	10	0.047	0.053	28.1
1.18	0.0469	15	0.018	0.049	54.0	14	0.025	0.046	41.5	13	0.032	0.045	34.2	11	0.041	0.050	30.3
1.00	0.0394	18	0.016	0.0396	50.8	16	0.023	0.0395	39.9	15	0.028	0.039	34.2	13	0.035	0.042	29.8
850 µm	0.0331	22	0.014	0.0315	48.0	19	0.020	0.0331	39.6	18	0.025	0.0306	30.3	15	0.032	0.035	27.6
710	0.0278	24	0.013	0.0287	47.4	22	0.018	0.0275	36.6	20	0.023	0.0270	29.2				
600	0.0234	30	0.010	0.0233	48.9	26	0.015	0.0235	37.3	24	0.018	0.0237	32.4				
500	0.0197	35	0.009	0.0196	47.1	30	0.0135	0.0198	35.3	28	0.016	0.0197	30.4				
425	0.0165	40	0.0085	0.0165	43.6	35	0.012	0.0166	33.8	32	0.014	0.0173	30.6				
355	0.0139	45	0.0085	0.0137	38.0	42	0.010	0.0138	33.6	38	0.013	0.0133	25.5				
300	0.0117	55	0.0065	0.0117	41.4	50	0.0085	0.0115	33.1	45	0.0105	0.0117	27.7				
250	0.0098	70	0.005	0.0093	42.4	60	0.007	0.0097	33.9								
212	0.0083	80	0.004	0.0085	46.2	70	0.006	0.0083	33.8								
180	0.0070	94	0.0035	0.0071	45.0	80	0.0055	0.0070	31.4								
150	0.0059	120	0.0026	0.0057	47.3	100	0.0040	0.0060	36.0								
125	0.0049	145	0.0022	0.0047	46.4	120	0.0035	0.0048	33.2								
106	0.0041	165	0.0019	0.0042	47.1	140	0.0030	0.0041	32.9								
90	0.0035	200	0.0016	0.0034	46.2	170	0.0024	0.0035	35.1								
75	0.0029	230	0.0014	0.0029	46.0	200	0.0021	0.0029	33.6								
63	0.0025					230	0.0018	0.0025	34.3								
53	0.0021					280	0.0015	0.0021	34.6								
45	0.0017					350	0.0012	0.0017	35.4								
38	0.0015					400	0.0010	0.0015	36.0								

TABLE 2 USA Standard Specifications for General Industrial Use Wire Cloth

D Light				C Medium Light				B Medium				A Medium Heavy			
Mesh	Wire Diameter, in.	Opening, in.	Open Area, %	Mesh	Wire Diameter, in.	Opening, in.	Open Area, %	Mesh	Wire Diameter, in.	Opening, in.	Open Area, %	Mesh	Wire Diameter, in.	Opening, in.	Open Area, %
(3)				(4)				(5)				(6)			
1	0.063	0.937	87.8	1	0.080	0.920	84.6	1	0.105	0.895	80.1	1	0.120	0.880	77.4
2	0.041	0.459	84.3	2	0.063	0.437	76.4	2	0.080	0.420	70.6	2	0.105	0.395	62.4
2½	0.035	0.365	83.3	2½	0.047	0.353	77.9	2½	0.063	0.337	71.0	2½	0.080	0.320	64.0
3	0.032	0.301	81.5	3	0.041	0.292	76.7	3	0.054	0.279	70.1	3	0.063	0.270	65.6
4	0.025	0.225	81.0	4	0.035	0.215	74.0	4	0.041	0.209	69.9	4	0.047	0.203	65.9
5	0.023	0.177	78.3	5	0.032	0.168	70.6	5	0.035	0.165	68.1	5	0.041	0.159	63.2
6	0.020	0.147	77.8	6	0.025	0.142	72.6	6	0.032	0.135	65.6	6	0.035	0.132	65.6
8	0.017	0.108	74.6	8	0.020	0.105	70.6	8	0.028	0.097	60.2	8	0.032	0.090	51.8
10	0.015	0.085	72.3	10	0.018	0.082	67.2	10	0.025	0.075	56.3	10	0.032	0.068	46.2
12	0.014	0.069	68.6	12	0.017	0.066	62.7	12	0.023	0.060	51.8	12	0.028	0.055	43.6
14	0.013	0.058	65.9	14	0.016	0.055	59.3	14	0.020	0.051	51.0	14	0.025	0.046	41.5
16	0.012	0.051	65.3	16	0.015	0.048	57.8	16	0.018	0.045	50.7	16	0.023	0.040	39.9
18	0.011	0.045	64.4	18	0.014	0.042	56.1	18	0.017	0.039	48.3	18	0.020	0.036	41.1
20	0.010	0.040	64.0	20	0.013	0.037	54.8	20	0.016	0.034	46.2	20	0.018	0.032	41.0
22	0.009	0.037	64.5	22	0.012	0.034	54.3	22	0.015	0.031	45.0	22	0.017	0.029	39.3
24	0.008	0.034	65.4	24	0.011	0.031	54.3	24	0.014	0.028	44.2	24	0.016	0.026	38.0
26	0.0075	0.031	65.0	26	0.010	0.029	54.9	26	0.013	0.026	44.0	26	0.015	0.024	37.3
28	0.0075	0.028	62.3	28	0.010	0.026	51.8	28	0.012	0.024	44.0	28	0.014	0.022	36.9
30	0.007	0.026	62.4	30	0.010	0.023	48.9	30	0.011	0.022	44.8	30	0.013	0.020	37.1
35	0.007	0.022	57.2	35	0.009	0.020	47.1	35	0.010	0.019	42.4	35	0.011	0.018	37.9
40	0.006	0.019	57.8	40	0.007	0.018	51.8	40	0.009	0.016	41.0	40	0.010	0.015	36.0
								50	0.0075	0.013	39.1	50	0.009	0.011	30.3
								60	0.0065	0.010	37.5	60	0.0075	0.009	30.5
												70	0.0065	0.008	29.8
												80	0.0055	0.007	31.4
												90	0.005	0.006	30.1
												100	0.0045	0.006	30.3

TABLE 3 Tolerances on Openings of USA Standard Specifications for Industrial Wire Cloth and Screens

Opening Designation, in.	Tolerance in Average Opening, in.
5.0	±0.150
Under 5.0 to 3.5	±0.138
Under 3.5 to 2.5	±0.100
Under 2.5 to 1.5	±0.075
Under 1.5 to 1.0	±0.045
Under 1.0 to 0.750	±0.030
Under 0.750 to 0.500	±0.020
Under 0.500 to 0.375	±0.017
Under 0.375 to 0.250	±0.015
Under 0.250 to 0.187	±0.012
Under 0.187 to 0.132	±0.010
Under 0.132 to 0.0937	±0.0071
Under 0.0937 to 0.0661	±0.0051
Under 0.0661 to 0.0394	±0.0035
Under 0.0394 to 0.0197	±0.0021
Under 0.0197 to 0.0098	±0.0011
Under 0.0098 to 0.0049	±0.0006
Under 0.0049 to 0.0029	±0.0003
Under 0.0029 to 0.0021	±0.0002
Under 0.0021 to 0.0015	±0.0002

TABLE 4 Tolerances on Wire Diameters of USA Standard Specifications for Industrial Wire Cloth and Screens

Wire Diameters, in.	Tolerance, in.
1.0 to 0.5	±0.004
Under 0.5 to 0.140	±0.003
Under 0.140 to 0.078	±0.002
Under 0.078 to 0.035	±0.001
Under 0.035 to 0.028	±0.0008
Under 0.028 to 0.020	±0.0006
Under 0.020 to 0.16	±0.0005
Under 0.016 to 0.011	±0.0004
Under 0.011 to 0.006	±0.0003
Under 0.006 to 0.0045	±0.0002
Under 0.0045 to 0.002	±0.00015
Under 0.002	±0.0001

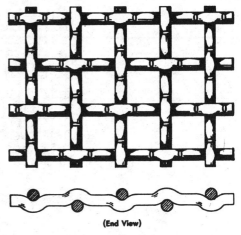

FIG. 2 Lock Crimp.

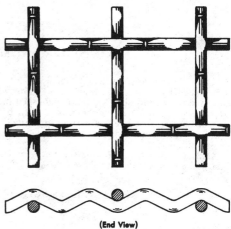

FIG. 3 Intercrimped.

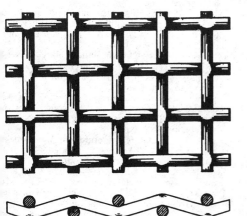

FIG. 1 Plain Weave.

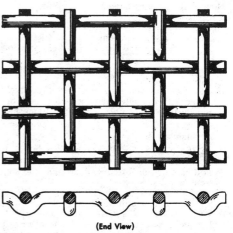

FIG. 4 Flat Top.

(End View)
FIG. 5 Twilled Weave.

APPENDIXES

X1. EQUIVALENT METRIC SPECIFICATIONS

X1.1 The metric equivalent to the USA Standard Specification for Industrial Wire Cloth and Screen is shown in Table X1 and is published for convenience of users in countries on full metric standard and to facilitate comparison of the USA Standard with national standards or locally available specifications.

X1.2 In the metric table, the standard opening designations are identical to those in Column 1 of the USA Standard (Table 1), but the wire diameters are the nearest Standard ISO metric (R20) wire diameter to the one used in the corresponding specification in the USA Standard.

X1.3 The tolerances on average openings, Table X2, and on wire diameters, Table X3, are the metric equivalents to those in the USA Standard Tables 2 and 3.

X2. METAL ALLOYS SUITABLE FOR INDUSTRIAL WIRE CLOTH

X2.1 Steels

Low carbon, plain finish.
High carbon, hard drawn (spring) for abrasion resistance.
High carbon, oil- or lead-tempered for extra resistance to abrasion and heavy screening loads.
Steel, electroplated with zinc, nickel, or copper before weaving.
Galvanized before woven—Steel wire coated with zinc before weaving.
Galvanized after weaving—Steel wire cloth, hot dip coated with zinc after weaving 1 through 8 mesh and electroplated with zinc through 40 mesh.

X2.2 Stainless Steels

Type 304 (Cr 18, Ni 8)—The basic stainless steel most extensively used for weaving industrial wire cloth and the most frequently available in manufacturer's stock.
Type 304L—Same as Type 304 except with extra low carbon to permit welding.
Type 316 (Cr 18, Ni 8 with molybdenum added)—Has increased resistance to chemical corrosion and can be used in bleach solutions where the hydrochloric acid content does not exceed 2 %
Type 316L—Same as Type 316 except with extra low carbon to permit welding.
Type 347 (Cr 18, Ni 8 with columbium added for stabilization of steel within the critical range from 800 to 1500°F (427 to 816°C). It is used where the cloth is to be welded.
Type 430—A straight chromium alloy (Cr 14 without nickel. Has a high degree of resistance to chemical and atmospheric corrosion and oxidation up to 1600°F (817°C).
Type 309—A chromium-nickel alloy (Cr 25, Ni 12 developed for increased heat resistance over the basic Type 304 analysis but not equal to the high nickel alloys.
Type 310—A high-temperature chromium-nickel alloy (Cr 25, Ni 20) similar to Type 309 but more stable due to its higher nickel content.
Type 317 (Cr 18, Ni 14, Mo 3-4)—The basic alloy for increased corrosion resistance over Type 316
Type 317L—Same as Type 317 except with extra low carbon content to permit welding.
Type 318—Combines the qualities of both Type 316 and 347. It is the basic 18-8 alloy with the addition of molybdenum and columbium for sta

bilization of the steel within the critical range from 800 to 1500°F (427 to 816°C).
Type 321 – The same as Type 347 except for the addition of titanium instead of columbium.
Type 330 – A nickel-chromium alloy (Cr 15, Ni 35) used mostly for heat treating baskets and fixtures for temperatures up to 1650°F (900°C).
Type 410 – A straight chromium alloy (Cr 11.5) without nickel, possessing excellent resistance to corrosion and oxidation, but not as generally used for wire cloth as Type 430.
Type 446 – A high-chromium steel (Cr 23) without nickel, possessing excellent resistance to chemical corrosion and oxidation up to 2000°F (1093°C).
Types 501 and 502 – Low-chromium alloys without nickel, possessing characteristics between carbon steel and regular stainless steel.

X2.3 Heat-Resisting Alloys

X2.3.1 A variety of nickel-chromium alloys (Note X1) are available (under various trade names) for varying degrees of temperature resistance, which can be classified as follows:

Approximate Analysis	Temperature °F	Resistance °C
Ni 32, Cr 20.5, Fe 46	1650	900
Ni 60, Cr 16, Fe 24	1700	916
Ni 76, Cr 15.8, Fe 7.2	1800	982
Ni 80, Cr 20	2000	1093

NOTE X1 – Used mostly for heat-treating baskets and fixtures.

X2.4 Copper Alloys

Copper – Because of its low tensile strength and high ductility, copper is not much used for wire cloth, except where desired for its electrical properties.
Brass (Cu 80, Zn 20) – Used for wire cloth where its nonrusting qualities are desired and where the strength and toughness of phosphor bronze is not required.
Commercial bronze (Cu 90, Zn 10) – Used where a slight increase in corrosion resistance over brass is desired.
Phosphor bronze (Sn 4-9, P $^1/_4$, balance Cu) – The ideal copper alloy for wire cloth, combining corrosion resistance with strength, toughness, and ability to withstand cold working. Used in preference to brass for wire cloth specifications 100 mesh (150 μm) and finer.

X2.5 Nickel Alloys

Nickel 200 – Used for wire cloth for food products to resist corrosion of chemicals such as caustics, some organic acids, and other corrosive products.
Monel – A high-nickel alloy (Ni 67, Cu 30) well suited for wire cloth since it has the strength of mild steel in addition to its excellent corrosion resistance and suitability for processing food products.

X2.6 Aluminum Alloys

1100 Aluminum – Pure aluminum is used where light weight and corrosion resistance is more important than strength.
5056 Aluminum – An aluminum alloy containing magnesium, manganese, and chromium for greater strength. The preferred aluminum alloy for industrial wire cloth.
Clad-aluminum 5056 – Has a core of 5056 aluminum encased or clad with pure aluminum. Combines added corrosion resistance with strength.

X3. CHECKING AND CALIBRATING INDUSTRIAL WIRE CLOTH AND SCREENS

X3.1 In checking specifications of industrial wire cloth it is necessary to determine two dimensions: (*a*) the diameter of the wire in both warp and shoot directions, see X3.2, and (*b*) the opening in both warp and shoot directions, see X3.3.

X3.2 *Wire Diameter* – The diameter of the warp wire and of the shoot wire shall be determined, separately, by the use of a micrometer (Fig. X1) or, for diameters greater than the capacity of the micrometer, by vernier outside caliper. For inch measurements, the micrometer or caliper should be able to determine the diameter to 0.0001 in. For metric measurements, the micrometer or caliper should be able to determine the diameter to 0.001 mm.

X3.3 *Openings, Inch:*

X3.3.1 For openings 5 to 0.250 in., measure the opening using a steel rule graduated to $^1/_{32}$ in. (Fig. X3), or by a vernier inside caliper (Fig. X4).

X3.3.2 For openings 0.250 in. to 20 mesh, count the number of meshes per linear inch, divide the result into 1.00 in., and subtract the wire diameter. The result will be the opening in decimals of an inch.

X3.3.3 To determine the opening in wire cloth 20 mesh to 60 mesh, use a magnifying counting glass (Fig. X2); count the number of meshes in $^1/_2$ in., divide the result into 0.500, and subtract the wire diameter. The result will be the opening in decimals of an inch.

X3.3.4 To determine the opening in wire cloth 60 mesh and finer, use the same procedure as X3.3.3, except count the number of meshes in $^1/_4$ in., divide the result into 0.250, and subtract the wire diameter. The result will be the opening in decimals of an inch.

X3.4 *Openings, Metric:*

X3.4.1 For openings 125 to 11.2 mm, measure the opening with a steel rule or an inside caliper graduated to 0.1 mm.

X3.4.2 For openings 11.2 to 1 mm, determine the space occupied by ten openings and ten wires, divide by 10 and from the result subtract the diameter of one wire. The result will be the opening in millimetres.

X3.4.3 For openings 1 mm to 250 μm, use a magnifying counting glass (Fig. X2) to count the number of meshes in 10 mm, divide the result by 10, and subtract the diameter of one wire.

X3.4.4 For openings 250 to 38 μm, use same procedure as X3.4.3, except count the number of meshes in 5 mm, divide the result by 5 and subtract the diameter of one wire.

X4. TERMS RELATING TO INDUSTRIAL WOVEN WIRE CLOTH AND SCREENS

X4.1 General
aperture or opening — dimensions defining an opening in a screen.
open area — ratio of the total area of the apertures to the total area of the screen.
screen — (*a*) a surface provided with apertures of uniform size. (*b*) a machine provided with one or more screening surfaces.
screening — the process of separating a mixture of different sizes by means of one or more screening surfaces.

X4.2 Woven Wire Cloth
mesh — (*a*) the number of apertures per unit of length. (*b*) in countries using English measure; the number of openings, and fraction thereof, per linear inch counting from the center of a wire.
square mesh — mesh with equal dimensions on each side of an aperture.
rectangular mesh or opening — mesh or opening with unequal dimensions of length and width of an aperture.
off-count mesh — same as **rectangular mesh**.
plain weave — wire cloth in which each warp and each shoot wire passes over one and under the next adjacent wire in both directions.
double crimp — wire cloth woven with approximately equal corrugations in both warp and shoot.
precrimped — wire cloth with both warp and shoot wires crimped before weaving.
lock crimp — precrimped wire cloth with deep crimps at points of intersection to lock the wires securely in place.
intercrimped — precrimped with cloth with extra crimps between the intersections.
flat top — wire cloth with deep crimps, as in "lock crimp," except that all crimps are on the under side leaving the top surface in one plane.
smooth top — same as **flat top**.
space cloth — wire cloth that is designated by the clear opening between the wires instead of by the mesh.
twilled weave — wire cloth in which each shoot wire passes successively over two and under two warp wires and each warp wire passes successively over two and under two shoot wires (Fig. 5).
shoot wires — the wires running crosswise of the cloth as woven (also called "shute wires" or "filler wires").
warp wires — the wires running the long way of the cloth as woven.
weft wires — same as **shoot wires**.

ASTM E 437

TABLE X1 Equivalent Metric Specifications of USA Standard Industrial Wire Cloth (Using ISO Standard Metric Wire Diameters)[A]

Opening Designation		D Medium Light		C Medium		B Medium Heavy		A Heavy	
Standard Metric	USA Industrial Standard, in.	Wire Diameter, mm	Open Area, %	Wire Diameter, mm	Open Area, %	Wire Diameter, mm	Open Area, %	Wire Diameter, mm	Open Area, %
125 mm	5.0	12.5	82.6	16.0	78.6	20.0	74.3	25.0	69.4
106	4.24	12.5	80.0	16.0	75.5	20.0	70.8	25.0	65.5
100	4.0	11.2	80.9	14.0	76.9	18.0	71.8	25.0	64.0
90	3.5	10.0	81.0	14.0	74.9	18.0	69.4	22.4	64.1
75	3.0	9.00	79.7	12.5	73.5	16.0	67.9	20.0	62.3
63	2.5	9.00	76.6	11.2	72.1	14.0	66.9	18.0	60.5
53	2.12	8.00	75.5	10.0	70.8	12.5	65.5	16.0	59.0
50	2.0	7.10	76.7	10.0	69.4	12.5	64.0	16.0	57.4
45	1.75	7.10	74.6	9.00	69.4	11.2	64.1	14.0	58.2
37.5	1.5	6.30	73.3	8.00	67.9	10.0	62.3	12.5	56.3
31.5	1.25	5.60	70.6	7.10	66.6	9.00	60.5	11.2	54.4
26.5	1.06	5.00	70.8	6.30	65.3	8.00	59.0	10.00	52.7
25	1.0	4.50	71.8	6.30	63.8	7.10	60.7	10.00	51.0
22.4	0.875	4.00	72.0	5.60	64.0	7.10	57.7	9.00	50.9
19.0	0.750	3.55	71.0	5.00	62.7	6.30	56.4	8.00	49.5
16.0	0.625	3.15	69.8	4.50	60.9	5.60	54.9	7.10	48.0
13.2	0.530	2.80	68.1	4.00	58.9	5.00	52.6	6.30	45.8
12.5	0.500	2.80	66.7	3.55	60.7	4.50	54.1	6.30	44.2
11.2	0.438	2.50	66.8	3.55	57.7	4.50	50.9	5.60	44.4
9.5	0.375	2.24	65.5	3.15	56.4	4.00	49.5	5.00	42.9
8.0	0.312	2.00	64.0	2.80	54.9	3.55	48.0	4.50	41.0
6.7	0.265	1.80	62.1	2.50	53.0	3.15	46.3	4.00	39.2
6.3	0.250	1.80	60.5	2.24	54.4	2.80	48.0	3.55	40.9
5.6	0.223	1.60	60.5	2.00	54.3	2.50	47.8	3.55	37.5
4.75	0.187	1.60	56.0	1.80	52.6	2.24	46.2	3.15	36.2
4.00	0.157	1.25	58.0	1.40	54.9	2.00	44.4	2.80	34.6
3.35	0.132	1.00	59.3	1.25	53.0	1.80	42.3	2.24	35.9
2.80	0.111	0.900	57.3	1.12	51.0	1.60	40.5	2.00	34.0
2.36	0.0937	0.800	55.8	1.00	49.3	1.40	39.4	1.80	32.2
2.00	0.0787	0.710	54.5	0.900	47.6	1.25	37.9	1.60	30.9
1.70	0.0661	0.630	53.2	0.800	46.2	1.12	36.3	1.40	30.1
1.40	0.0555	0.560	51.0	0.710	44.0	0.900	37.1	1.25	27.9
1.18	0.0469	0.450	52.4	0.630	42.5	0.800	35.5	1.00	29.3
1.00	0.0394	0.400	51.0	0.560	41.1	0.710	34.2	0.900	27.7
850 μm	0.0331	0.355	49.8	0.500	39.6	0.630	33.0	0.800	26.5
710	0.0278	0.315	48.0	0.450	37.5	0.560	31.3		
600	0.0234	0.280	46.5	0.400	36.0	0.450	32.7		
500	0.0197	0.224	47.7	0.355	34.2	0.400	30.9		
425	0.0165	0.200	46.2	0.280	36.3	0.355	29.7		
355	0.0139	0.180	44.0	0.250	34.4	0.315	28.1		
300	0.0117	0.160	42.5	0.224	32.8	0.250	29.8		
250	0.0098	0.125	44.4	0.180	33.8				
212	0.0083	0.100	46.2	0.160	32.5				
180	0.0070	0.090	44.4	0.125	34.8				
150	0.0059	0.080	42.5	0.112	32.8				
125	0.0049	0.063	44.2	0.090	33.8				
106	0.0041	0.056	42.8	0.080	32.5				
90	0.0035	0.045	44.4	0.063	34.6				
75	0.0029	0.040	42.5	0.056	32.8				
63	0.0025			0.045	34.0				
53	0.0021			0.036	35.5				
45	0.0017			0.032	34.2				
38	0.0015			0.025	36.4				

[A] ISO Recommendation R 388, 1964 (R 20 Series) for Wire Diameters.

E 437

TABLE X2. Tolerances on Openings of Equivalent Metric Specifications of USA Standard for Industrial Wire Cloth and Screens

Opening Designation, mm	Tolerance in Average Opening, mm
125	±3.8
Under 125 to 90	±3.5
Under 90 to 63	±2.54
Under 63 to 37.5	±1.9
Under 37.5 to 25	±1.14
Under 25 to 19	±0.76
Under 19 to 12.5	±0.51
Under 12.5 to 9.5	±0.43
Under 9.5 to 6.3	±0.38
Under 6.3 to 4.75	±0.30
Under 4.75 to 3.35	±0.25
Under 3.35 to 2.36	±0.18
Under 2.36 to 1.70	±0.13
Under 1.70 to 1.00	±0.088
Under 1.00 to 500 μm	±54 μm
Under 500 μm to 250	±27
Under 250 to 125	±14
Under 125 to 75	±8
Under 75 to 53	±6
Under 53 to 38	±5

TABLE X3 Tolerances for Wire Diameters for Equivalent Metric Specifications of USA Standard Industrial Wire Cloth and Screens

Wire Diameters, mm	Tolerance, mm
25.0 to 12.5	±0.102
Under 12.5 to 3.5	±0.076
Under 3.55 to 2.0	±0.050
Under 2.0 to 0.90	±0.025
Under 0.90 to 0.71	±0.020
Under 0.71 to 0.51	±0.015
Under 0.51 to 0.40	±0.0127
Under 0.40 to 0.28	±0.010
Under 0.28 to 0.152	±0.0076
Under 0.152 to 0.114	±0.005
Under 0.114 to 0.051	±0.0038
Under 0.051	±0.0025

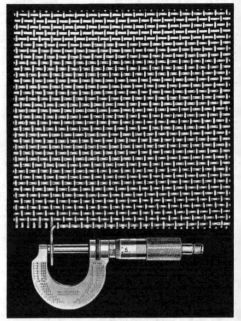

FIG. X1 Micrometer.

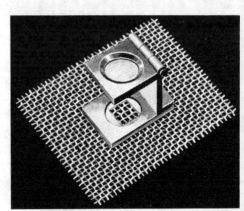

FIG. X2 Counting Glass for Fine Mesh Woven Wire Cloth.

⏻ E 437

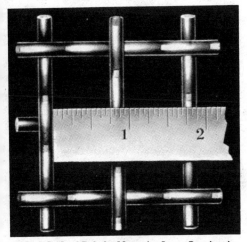

FIG. X3 Steel Rule for Measuring Large Openings in Wire Cloth.

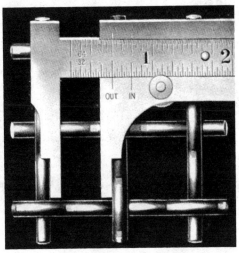

FIG. X4 Vernier Inside Caliper.

The American Society for Testing and Materials takes no position respecting the validity of any patent rights asserted in connection with any item mentioned in this standard. Users of this standard are expressly advised that determination of the validity of any such patent rights, and the risk of infringement of such rights, are entirely their own responsibility.

This standard is subject to revision at any time by the responsible technical committee and must be reviewed every five years and if not revised, either reapproved or withdrawn. Your comments are invited either for revision of this standard or for additional standards and should be addressed to ASTM Headquarters. Your comments will receive careful consideration at a meeting of the responsible technical committee, which you may attend. If you feel that your comments have not received a fair hearing you should make your views known to the ASTM Committee on Standards, 1916 Race St., Philadelphia, Pa. 19103.

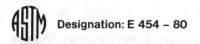

Designation: E 454 – 80

An American National Standard

Standard Specification for
INDUSTRIAL PERFORATED PLATE AND SCREENS (SQUARE OPENING SERIES)[1]

This standard is issued under the fixed designation E 454; the number immediately following the designation indicates the year of original adoption or, in the case of revision, the year of last revision. A number in parentheses indicates the year of last reapproval. A superscript epsilon (ε) indicates an editorial change since the last revision or reapproval.

This specification has been approved for use by agencies of the Department of Defense and for listing in the DoD Index of Specifications and Standards.

INTRODUCTION

Industrial perforated plate can be produced in many thousands of combinations of size and shape of opening, bar size, thickness of material, and type of metal. Such variety is often confusing and, to the vast majority of perforated plate users, unnecessary, since each usually requires only a very few specifications.

The purpose of this specification is to simplify this problem by a condensed table of recommended specifications covering a wide range of openings in which industrial perforated plate is made, with several recommended bar sizes and thicknesses of plate for each opening, for use in various grades of service.

By making selections from this standard, the user will be guided to specifications that are being regularly produced, thus avoiding inadvertent selection of specifications that, because of little or no demand, are unobtainable, except on special order (usually quite expensive unless the quantity ordered is sufficient to justify the cost of special tooling).

If a user has a specific application for industrial perforated plate that can not be solved by a selection from this standard, it is recommended that he consult his perforated plate supplier on the availability of an acceptable alternative specification.

1. Scope

1.1 This specification covers the sizes of square opening perforated plate and screens for general industrial uses, including the separating or grading of materials according to designated nominal particle size, and lists standards for openings from 5 in. (125 mm) to 0.127 (⅛) in. (3.35 mm) punched with bar sizes and thicknesses of plate for various grades of service. Methods of checking industrial perforated plate and screens are included as information in the Appendix.

1.2 This specification does not apply to perforated plate or screens with round, hexagon, slotted, or other shaped openings.

1.3 The values stated in inch-pound units are to be regarded as the standard.

2. Applicable Documents

2.1 *ASTM Standard*:
E 323 Specification for Perforated-Plate Sieves for Testing Purposes[2]
2.2 *Other Documents*:
Fed. Std. No. 123 Marking for Shipment (Civil Agencies)[3]
MIL-STD-129 Marking for Shipment and Storage[3]

---

[1] This specification is under the jurisdiction of ASTM Committee E-29 on Particle Size Measurement and is the direct responsibility of Subcommittee E 29.01 on Sieves, Sieving Methods, and Screening Media.
Current edition approved Oct. 31, 1980. Published December 1980. Originally published as E 454 – 72. Last previous edition E 454 – 72.
[2] *Annual Book of ASTM Standards*, Vol 14.02.
[3] Available from Naval Publications and Forms Center, 5801 Tabor Ave., Philadelphia, Pa. 19120.

3. Standard Specifications

3.1 Standard specifications for industrial perforated plate and screens are listed in Table 1.

3.2 *Openings*—The series of standard openings listed in Table 1 include those of the USA Standard Sieve Series, Specification E 323, and those of the ISO apertures for industrial plate screens,[4] with the addition of those openings in common usage.

3.3 *Relationship of Grades*—The purpose of the several grades is to provide combinations of opening and bar size for various types of service, from medium-light to heavy. Since it is possible to vary the bar size independently from the plate thickness, each of the service grades lists up to three combinations of bar and gage for each opening. The entire standard series has been designed for a logical relationship of bar size to opening in each grade and between grades with the capability of also being able to vary the plate thickness.

3.4 *Bar*—A choice of six bars is shown for each standard opening from 5-in. (125-mm) to 0.312-in. (8-mm) opening, inclusive. For practical reasons, the number of bars or grades available for openings finer than 0.312 in. is progressively reduced.

3.5 *Gage*—A choice of six gages is shown for each standard opening for 5 in. (125 mm) to 0.312 in. (8 mm). For practical reasons, the number of gages or grades available for openings finer than 0.312 in. is progressively reduced.

NOTE 1—The gages shown in Table 1 are practical for a low-carbon steel plate. For other materials, consult your perforated plate supplier.

3.6 *Equivalent Metric Specification*—Table A1, in the Appendix, shows the equivalent metric specifications to the USA Standard, punched in standard ISO recommended thickness of plate.[5]

4. Types of Perforated Pattern

4.1 This specification covers square openings arranged in a staggered pattern with their midpoints nominally at the vertices of isosceles triangles whose bases shall equal their heights, and also covers square openings arranged in line with their midpoints nominally at the vertices of squares (see Fig. 1).

NOTE 2—The percentage of open area for square apertures is identical for both staggered and straight-line patterns (see Fig. 2).

5. Metal Composition of Plate

5.1 Perforated plate can be punched from a great variety of metals and alloys, but the following are most commonly used:

Steel, low-carbon
Steel, high-carbon
Steel, heat-treated
Steel, galvanized
Stainless steel, Type 304
Stainless steel, Type 316
Stainless steel, Type 410
Brass (Cu 80, Zn 20)
Manganese bronze (Cu 61, Zn 37)
Monel (high nickel-copper alloy)
Aluminum (all grades)

6. Tolerances

6.1 *Openings*—Tolerances on openings in USA Standard Specifications for Industrial Perforated Plate and Screens (Tables 1 and A1) shall be in accordance with those listed in Table 2.

6.2 *Bars*—Tolerances on bars used in USA Standard Specification for Industrial Perforated Plate and Screens (Tables 1 and A1) shall be in accordance with those listed in Table 3.

6.3 *Gages*—Tolerances on gages used in USA Standard Specifications for Industrial Perforated Plate and Screens (Tables 1 and A1) shall be in accordance with those listed in Table 4.

NOTE 3—The tolerances expressed in inch-pound units are taken from the current AISI values.

---

[4] ISO 2194-1972, Wire Screens and Plate Screens for Industrial Purposes—Nominal Sizes of Apertures.
[5] ISO Recommendation R388-1964, Metric Series for Basic Thicknesses of Sheet and Diameters of Wire.

SUPPLEMENTARY REQUIREMENTS

The following sections shall be applicable when U.S. government contractual matters are involved.

S1. Responsibility for Inspection

S1.1 Unless otherwise specified in the contract or purchase order, the producer is responsible for the performance of all inspection and test requirements specified herein. Except as otherwise specified in the contract or order, the producer may use his own or any other suitable facilities for the performance of the inspection and test requirements specified herein, unless disapproved by the purchaser. The purchaser shall have the right to perform any of the inspections and tests set forth in this specification where such inspections are deemed necessary to ensure that material conforms to prescribed requirements.

S2. Government Procurement

S2.1 Unless otherwise specified in the contract, the material shall be packaged in accordance with the suppliers' standard practice which will be acceptable to the carrier at lowest rates. Containers and packing shall comply with the Uniform Freight Classification rules or National Motor Freight Classification rules. Marking for shipment of such material shall be in accordance with Fed. Std. No. 123 for civil agencies, and MIL STD 129 for military agencies.

TABLE 1 USA Standard Specifications for Industrial Perforated Plate and Screens (Square Opening Series)—(U.S. Customary Units)

Perforated Opening		Medium Light				Medium				Medium Heavy				Heavy			
Standard (metric), mm	USA Industrial Standard, in.	Opening, in.	Bar, in.	Gage-Steel, in.	Open Area, percent	Opening, in.	Bar, in.	Gage-Steel, in.	Open Area, percent	Opening, in.	Bar, in.	Gage-Steel, in.	Open Area, percent	Opening, in.	Bar, in.	Gage-Steel, in.	Open Area, percent
125	5	5	½	½	82.6	5	⅝	⅝	79.0	5	¾	¾	75.6	5	1	1	69.4
125	5	5	⅝	⅜	79.0	5	¾	½	75.6	5	⅞	⅝	72.4	5	1⅛	⅞	66.6
125	5	5	⅝	½	79.0	5	¾	⅝	75.6	5	⅞	¾	72.4	5	1⅛	1	66.6
...	...	4½	½	½	81.0	4½	⅝	⅝	77.1	4½	¾	¾	73.4	4½	1	1	66.9
...	...	4½	⅝	⅜	77.1	4½	¾	½	73.4	4½	⅞	⅝	70.1	4½	1⅛	⅞	64.0
...	...	4½	⅝	½	77.1	4½	¾	⅝	73.4	4½	⅞	¾	70.1	4½	1⅛	1	64.0
106	4¼	4¼	½	½	80.1	4¼	⅝	⅝	76.0	4¼	¾	¾	72.3	4¼	1	1	65.5
106	4¼	4¼	⅝	⅜	76.0	4¼	¾	½	72.3	4¼	⅞	⅝	68.8	4¼	1⅛	⅞	62.5
106	4¼	4¼	⅝	½	76.0	4¼	¾	⅝	72.3	4¼	⅞	¾	68.8	4¼	1⅛	1	62.5
100	4	4	½	½	79.0	4	⅝	⅝	74.8	4	¾	¾	70.9	4	1	1	64.0
100	4	4	⅝	⅜	74.8	4	¾	½	70.9	4	⅞	⅝	67.3	4	1⅛	⅞	60.9
100	4	4	⅝	½	74.8	4	¾	⅝	70.9	4	⅞	¾	67.3	4	1⅛	1	60.9
...	...	3¾	½	½	77.9	3¾	⅝	⅝	73.5	3¾	¾	¾	69.4	3¾	⅞	⅞	65.7
...	...	3¾	⅝	⅜	73.5	3¾	¾	½	69.4	3¾	⅞	⅝	65.7	3¾	1	¾	62.3
...	...	3¾	⅝	½	73.5	3¾	¾	⅝	69.4	3¾	⅞	¾	65.7	3¾	1	⅞	62.3
90	3½	3½	½	½	76.6	3½	⅝	⅝	72.0	3½	¾	¾	67.8	3½	⅞	⅞	64.0
90	3½	3½	⅝	⅜	72.0	3½	¾	½	67.8	3½	⅞	⅝	64.0	3½	1	¾	60.5
90	3½	3½	⅝	½	72.0	3½	¾	⅝	67.8	3½	⅞	¾	64.0	3½	1	⅞	60.5
...	...	3¼	⅜	⅜	80.4	3¼	½	½	75.1	3¼	⅝	⅝	70.3	3¼	¾	¾	66.0
...	...	3¼	½	5⁄16	75.1	3¼	⅝	⅜	70.3	3¼	¾	½	66.0	3¼	⅞	⅝	62.1
...	...	3¼	½	⅜	75.1	3¼	⅝	½	70.3	3¼	¾	⅝	66.0	3¼	⅞	¾	62.1
75	3	3	⅜	⅜	79.0	3	½	½	73.5	3	⅝	⅝	68.5	3	¾	¾	64.0
75	3	3	½	5⁄16	73.5	3	⅝	⅜	68.5	3	¾	½	64.0	3	⅞	⅝	59.9
75	3	3	½	⅜	73.5	3	⅝	½	68.5	3	¾	⅝	64.0	3	⅞	¾	59.9
...	...	2¾	⅜	⅜	77.4	2¾	½	½	71.6	2¾	⅝	⅝	66.4	2¾	¾	¾	61.7
...	...	2¾	½	5⁄16	71.6	2¾	⅝	⅜	66.4	2¾	¾	½	61.7	2¾	⅞	⅝	57.6
...	...	2¾	½	⅜	71.6	2¾	⅝	½	66.4	2¾	¾	⅝	61.7	2¾	⅞	¾	57.6
63	2½	2½	⅜	⅜	75.6	2½	½	½	69.4	2½	⅝	⅝	64.0	2½	¾	¾	59.2
63	2½	2½	½	5⁄16	69.4	2½	⅝	⅜	64.0	2½	¾	½	59.2	2½	⅞	⅝	54.9
63	2½	2½	½	⅜	69.4	2½	⅝	½	64.0	2½	¾	⅝	59.2	2½	⅞	¾	54.9

ASTM E 454

TABLE 1 Continued

Perforated Opening		Medium Light				Medium				Medium Heavy				Heavy			
Standard (metric), mm	USA Industrial Standard, in.	Opening, in.	Bar, in.	Gage-Steel, in.	Open Area, percent	Opening, in.	Bar, in.	Gage-Steel, in.	Open Area, percent	Opening, in.	Bar, in.	Gage-Steel, in.	Open Area, percent	Opening, in.	Bar, in.	Gage-Steel, in.	Open Area, percent
...	...	2¼	⅜	⅜	73.5	2¼	½	½	66.9	2¼	⅝	⅝	61.2	2¼	¾	¾	56.3
...	...	2¼	½	5⁄16	66.9	2¼	⅝	⅜	61.2	2¼	¾	½	56.3	2¼	⅞	⅝	51.8
...	...	2¼	½	⅜	66.9	2¼	⅝	½	61.2	2¼	¾	⅝	56.3	2¼	⅞	¾	51.8
53	2⅛	2⅛	5⁄16	5⁄16	76.0	2⅛	⅜	⅜	72.3	2⅛	½	½	65.5	2⅛	⅝	⅝	59.7
53	2⅛	2⅛	⅜	¼	72.3	2⅛	½	5⁄16	59.7	2⅛	⅝	⅜	59.7	2⅛	¾	½	54.6
53	2⅛	2⅛	⅜	5⁄16	72.3	2⅛	½	⅜	59.7	2⅛	⅝	½	59.7	2⅛	¾	⅝	54.6
50	2	2	5⁄16	5⁄16	74.8	2	⅜	⅜	70.9	2	½	½	64.0	2	⅝	⅝	58.0
50	2	2	⅜	¼	70.9	2	½	5⁄16	64.0	2	⅝	⅜	58.0	2	¾	½	52.9
50	2	2	⅜	5⁄16	70.9	2	½	⅜	64.0	2	⅝	½	58.0	2	¾	⅝	52.9
...	...	1⅞	5⁄16	5⁄16	73.5	1⅞	⅜	⅜	69.4	1⅞	½	½	62.3	1⅞	⅝	⅝	56.3
...	...	1⅞	⅜	¼	69.4	1⅞	½	5⁄16	62.3	1⅞	⅝	⅜	56.3	1⅞	¾	½	51.0
...	...	1⅞	⅜	5⁄16	69.4	1⅞	½	⅜	62.3	1⅞	⅝	½	56.3	1⅞	¾	⅝	51.0
45	1¾	1¾	5⁄16	5⁄16	72.0	1¾	⅜	⅜	67.8	1¾	½	½	60.5	1¾	⅝	⅝	54.3
45	1¾	1¾	⅜	¼	67.8	1¾	½	5⁄16	60.5	1¾	⅝	⅜	49.0	1¾	¾	½	49.0
45	1¾	1¾	⅜	5⁄16	67.8	1¾	½	⅜	60.5	1¾	⅝	½	49.0	1¾	¾	⅝	49.0
...	...	1⅝	¼	¼	75.1	1⅝	5⁄16	5⁄16	70.3	1⅝	⅜	⅜	66.0	1⅝	½	½	58.5
...	...	1⅝	5⁄16	3⁄16	70.3	1⅝	⅜	¼	66.0	1⅝	½	5⁄16	58.5	1⅝	⅝	⅜	52.1
...	...	1⅝	5⁄16	¼	70.3	1⅝	⅜	5⁄16	66.0	1⅝	½	⅜	58.5	1⅝	⅝	½	52.1
37.5	1½	1½	¼	¼	73.5	1½	5⁄16	5⁄16	68.5	1½	⅜	⅜	64.0	1½	½	½	56.3
37.5	1½	1½	5⁄16	3⁄16	68.5	1½	⅜	¼	64.0	1½	½	5⁄16	56.3	1½	⅝	⅜	49.8
37.5	1½	1½	5⁄16	¼	68.5	1½	⅜	5⁄16	64.0	1½	½	⅜	56.3	1½	⅝	½	49.8
...	...	1⅜	¼	¼	71.6	1⅜	5⁄16	5⁄16	66.4	1⅜	⅜	⅜	61.7	1⅜	½	½	53.8
...	...	1⅜	5⁄16	3⁄16	66.4	1⅜	⅜	¼	61.7	1⅜	½	5⁄16	53.8	1⅜	⅝	⅜	47.3
...	...	1⅜	5⁄16	¼	66.4	1⅜	⅜	5⁄16	61.7	1⅜	½	⅜	53.8	1⅜	⅝	½	47.3
31.5	1¼	1¼	¼	¼	69.4	1¼	5⁄16	5⁄16	64.0	1¼	⅜	⅜	59.2	1¼	½	½	51.0
31.5	1¼	1¼	5⁄16	3⁄16	64.0	1¼	⅜	¼	59.2	1¼	½	5⁄16	51.0	1¼	⅝	⅜	44.4
31.5	1¼	1¼	5⁄16	¼	64.0	1¼	⅜	5⁄16	59.2	1¼	½	⅜	51.0	1¼	⅝	½	44.4

TABLE 1 Continued

Perforated Opening		Medium Light				Medium				Medium Heavy				Heavy			
Standard (metric), mm	USA Industrial Standard, in.	Opening, in.	Bar, in.	Gage Steel, in.	Open Area, percent	Opening, in.	Bar, in.	Gage Steel, in.	Open Area, percent	Opening, in.	Bar, in.	Gage Steel, in.	Open Area, percent	Opening, in.	Bar, in.	Gage Steel, in.	Open Area, percent
...	...	13/16	3/16	3/16	74.6	13/16	1/4	1/4	68.2	13/16	5/16	5/16	62.7	13/16	3/8	3/8	57.8
...	...	13/16	1/4	8	68.2	13/16	5/16	3/16	62.7	13/16	3/8	1/4	57.8	13/16	1/2	5/16	49.5
...	...	13/16	1/4	3/16	68.2	13/16	5/16	1/4	62.7	13/16	3/8	5/16	57.8	13/16	1/2	3/8	49.5
...	...	1 1/8	3/16	3/16	73.5	1 1/8	1/4	1/4	66.9	1 1/8	5/16	5/16	61.2	1 1/8	3/8	3/8	56.3
...	...	1 1/8	1/4	8	66.9	1 1/8	5/16	3/16	61.2	1 1/8	3/8	1/4	56.3	1 1/8	1/2	5/16	47.9
...	...	1 1/8	1/4	3/16	66.9	1 1/8	5/16	1/4	61.2	1 1/8	3/8	5/16	56.3	1 1/8	1/2	3/8	47.9
26.5	1 1/16	1 1/16	3/16	3/16	72.2	1 1/16	1/4	1/4	65.5	1 1/16	5/16	5/16	59.7	1 1/16	3/8	3/8	54.6
26.5	1 1/16	1 1/16	1/4	8	65.5	1 1/16	5/16	3/16	59.7	1 1/16	3/8	1/4	54.6	1 1/16	1/2	5/16	46.2
26.5	1 1/16	1 1/16	1/4	3/16	65.5	1 1/16	5/16	1/4	59.7	1 1/16	3/8	5/16	54.6	1 1/16	1/2	3/8	46.2
25	1	1	3/16	3/16	70.9	1	1/4	1/4	64.0	1	5/16	5/16	58.0	1	3/8	3/8	52.9
25	1	1	1/4	8	64.0	1	5/16	3/16	58.0	1	3/8	1/4	52.9	1	1/2	5/16	44.4
25	1	1	1/4	3/16	64.0	1	5/16	1/4	58.0	1	3/8	5/16	52.9	1	1/2	3/8	44.4
...	...	15/16	3/16	3/16	69.4	15/16	1/4	1/4	62.3	15/16	5/16	5/16	56.2	15/16	3/8	3/8	51.0
...	...	15/16	1/4	8	62.3	15/16	5/16	3/16	56.2	15/16	3/8	1/4	51.0	15/16	1/2	5/16	42.5
...	...	15/16	1/4	3/16	62.3	15/16	5/16	1/4	56.2	15/16	3/8	5/16	51.0	15/16	1/2	3/8	42.5
22.4	7/8	7/8	3/16	3/16	67.8	7/8	1/4	1/4	60.5	7/8	5/16	5/16	54.3	7/8	3/8	3/8	49.0
22.4	7/8	7/8	1/4	8	60.5	7/8	5/16	3/16	54.3	7/8	3/8	1/4	49.0	7/8	1/2	5/16	40.5
22.4	7/8	7/8	1/4	3/16	60.5	7/8	5/16	1/4	54.3	7/8	3/8	5/16	49.0	7/8	1/2	3/8	40.5
...	...	13/16	3/16	3/16	66.0	13/16	1/4	1/4	58.5	13/16	5/16	5/16	52.2	13/16	3/8	3/8	46.8
...	...	13/16	1/4	8	58.5	13/16	5/16	3/16	52.2	13/16	3/8	1/4	46.8	13/16	1/2	5/16	38.3
...	...	13/16	1/4	3/16	58.5	13/16	5/16	1/4	52.2	13/16	3/8	5/16	46.8	13/16	1/2	3/8	38.3
19	3/4	3/4	3/16	3/16	64.0	3/4	1/4	1/4	56.3	3/4	5/16	5/16	49.8	3/4	3/8	3/8	44.4
19	3/4	3/4	1/4	8	56.3	3/4	5/16	3/16	49.8	3/4	3/8	1/4	44.4	3/4	1/2	5/16	36.0
19	3/4	3/4	1/4	3/16	56.3	3/4	5/16	1/4	49.8	3/4	3/8	5/16	44.4	3/4	1/2	3/8	36.0
...	...	11/16	3/16	3/16	61.7	11/16	1/4	1/4	53.8	11/16	5/16	5/16	47.2	11/16	3/8	3/8	41.9
...	...	11/16	1/4	8	53.8	11/16	5/16	3/16	47.2	11/16	3/8	1/4	41.9	11/16	1/2	5/16	33.5
...	...	11/16	1/4	3/16	53.8	11/16	5/16	1/4	47.2	11/16	3/8	5/16	41.9	11/16	1/2	3/8	33.5

TABLE 1 Continued

Perforated Opening		Medium Light				Medium				Medium Heavy				Heavy			
Standard (metric), mm	USA Industrial Standard, in.	Opening, in.	Bar, in.	Gage-Steel, in.	Open Area, percent	Opening, in.	Bar, in.	Gage-Steel, in.	Open Area, percent	Opening, in.	Bar, in.	Gage-Steel, in.	Open Area, percent	Opening, in.	Bar, in.	Gage-Steel, in.	Open Area, percent
16	5/8	5/8	5/32	8	64.0	5/8	3/16	3/16	59.2	5/8	1/4	1/4	51.0	5/8	5/16	5/16	44.4
16	5/8	5/8	3/16	10	59.2	5/8	1/4	8	51.0	5/8	5/16	3/16	44.4	5/8	3/8	1/4	39.1
16	5/8	5/8	3/16	8	59.2	5/8	1/4	3/16	51.0	5/8	5/16	1/4	44.4	5/8	3/8	1/4	39.1
...	...	9/16	5/32	8	61.2	9/16	3/16	3/16	56.2	9/16	1/4	1/4	47.9	9/16	5/16	5/16	41.3
...	...	9/16	3/16	10	56.2	9/16	1/4	8	47.9	9/16	5/16	1/4	41.3	9/16	3/8	1/4	36.0
...	...	9/16	3/16	8	56.2	9/16	1/4	3/16	47.9	9/16	5/16	1/4	41.3	9/16	3/8	5/16	36.0
13.2	17/32	17/32	1/8	10	65.5	17/32	5/32	8	59.7	17/32	3/16	3/16	54.6	17/32	1/4	1/4	46.2
13.2	17/32	17/32	3/16	11	59.7	17/32	3/16	10	54.6	17/32	1/4	8	46.2	17/32	5/16	1/4	39.6
13.2	17/32	17/32	5/32	10	59.7	17/32	3/16	8	54.6	17/32	1/4	3/16	46.2	17/32	5/16	1/4	39.6
12.5	1/2	1/2	1/8	10	64.0	1/2	5/32	8	58.0	1/2	3/16	3/16	52.9	1/2	1/4	1/4	44.4
12.5	1/2	1/2	3/16	11	58.0	1/2	3/16	10	52.9	1/2	1/4	8	44.4	1/2	5/16	1/4	37.9
12.5	1/2	1/2	5/32	10	58.0	1/2	3/16	8	52.9	1/2	1/4	3/16	44.4	1/2	5/16	1/4	37.9
...	...	15/32	1/8	10	62.3	15/32	5/32	8	56.2	15/32	3/16	3/16	51.0	15/32	1/4	1/4	42.5
...	...	15/32	3/16	11	56.2	15/32	3/16	10	51.0	15/32	1/4	8	42.5	15/32	5/16	1/4	36.0
...	...	15/32	5/32	10	56.2	15/32	3/16	8	51.0	15/32	1/4	3/16	42.5	15/32	5/16	1/4	36.0
11.2	7/16	7/16	1/8	10	60.5	7/16	5/32	8	54.3	7/16	3/16	3/16	49.0	7/16	1/4	1/4	40.5
11.2	7/16	7/16	3/16	11	54.3	7/16	3/16	10	49.0	7/16	1/4	8	40.5	7/16	5/16	1/4	34.0
11.2	7/16	7/16	5/32	10	54.3	7/16	3/16	8	49.0	7/16	1/4	3/16	40.5	7/16	5/16	1/4	34.0
9.5	3/8	3/8	3/32	11	64.0	3/8	1/8	8	56.3	3/8	5/32	8	49.8	3/8	3/16	3/16	44.4
9.5	3/8	3/8	1/8	12	56.3	3/8	5/32	11	49.8	3/8	3/16	10	44.4	3/8	1/4	8	36.0
9.5	3/8	3/8	1/8	11	56.3	3/8	5/32	10	49.8	3/8	3/16	8	44.4	3/8	1/4	3/16	36.0
8	5/16	5/16	3/32	11	59.2	5/16	1/8	10	51.0	5/32	5/32	7	44.4	5/16	3/16	3/16	39.0
8	5/16	5/16	1/8	12	51.0	5/16	5/32	11	44.4	5/16	3/16	10	39.0	5/16	1/4	8	30.9
8	5/16	5/16	1/8	11	51.0	5/16	5/32	10	44.4	5/16	3/16	8	39.0	5/16	1/4	3/16	30.9
6.7	17/64	17/64	...	...	...	17/64	3/32	11	54.6	17/64	1/8	10	46.2	17/64	5/32	8	39.6
6.7	17/64	17/64	3/32	14	54.6	17/64	1/8	12	46.2	17/64	5/32	11	39.6	17/64	3/16	10	34.4
6.7	17/64	17/64	3/32	12	54.6	17/64	1/8	11	46.2	17/64	5/32	10	39.6	17/64	3/16	8	34.4

TABLE 1 *Continued*

Perforated Opening		Medium Light				Medium				Medium Heavy				Heavy			
Standard (metric), mm	USA Industrial Standard, in.	Opening, in.	Bar, in.	Gage Steel, in.	Open Area, percent	Opening, in.	Bar, in.	Gage Steel, in.	Open Area, percent	Opening, in.	Bar, in.	Gage Steel, in.	Open Area, percent	Opening, in.	Bar, in.	Gage Steel, in.	Open Area, percent
6.3	1/4	...	...	...	...	1/4	3/32	11	52.9	1/4	1/8	10	44.4	1/4	5/32	8	37.9
6.3	1/4	1/4	3/32	14	52.9	1/4	1/8	12	44.4	1/4	5/32	11	37.9	1/4	3/16	10	32.7
6.3	1/4	1/4	3/32	12	52.9	1/4	1/8	11	44.4	1/4	5/32	10	37.9	1/4	3/16	8	32.7
5.6	7/32	...	...	...	...	...	...	...	...	7/32	3/32	11	49.0	7/32	1/8	10	40.5
5.6	7/32	...	...	...	...	7/32	3/32	14	49.0	7/32	1/8	12	40.5	7/32	5/32	11	34.0
5.6	7/32	...	...	...	...	7/32	3/32	12	49.0	7/32	1/8	11	40.5	7/32	5/32	10	34.0
4.75	3/16	...	...	...	...	...	...	...	...	3/16	3/32	11	44.4	3/16	1/8	10	36.0
4.75	3/16	...	...	...	...	3/16	3/32	14	44.4	3/16	1/8	12	36.0	3/16	5/32	11	29.8
4.75	3/16	...	...	...	...	3/16	3/32	12	44.4	3/16	1/8	11	36.0	3/16	5/32	10	29.8
4	5/32	...	...	...	...	...	...	...	...	...	...	...	...	5/32	3/32	11	39.1
4	5/32	...	...	...	...	...	...	...	...	5/32	3/32	14	39.1	5/32	1/8	12	30.9
4	5/32	...	...	...	...	...	...	...	...	5/32	3/32	12	39.1	5/32	1/8	11	30.9
3.35	1/8	...	...	...	...	...	...	...	...	...	...	...	...	...	...	...	...
3.35	1/8	...	...	...	...	...	...	...	...	...	...	...	...	1/8	3/32	14	32.7
3.35	1/8	...	...	...	...	...	...	...	...	...	...	...	...	1/8	3/32	12	32.7

TABLE 2 Tolerances on Openings of USA Standard Specifications for Industrial Perforated Plate and Screens

Perforated Opening			Tolerance on Openings	
Standard (metric), mm	USA Industrial Standard, in.	Additional Sizes, in.	Standard (metric), mm	USA Industrial Standard, in.
125.0	5	...	±2.5	±0.100
...	...	4½	...	±0.090
106.0	4¼	...	±2.1	±0.085
100.0	4	...	±2.0	±0.080
...	...	3¾	...	±0.075
90.0	3½	...	±1.8	±0.070
...	...	3¼	...	±0.065
75.0	3	...	±1.5	±0.060
...	...	2¾	...	±0.055
63.0	2½	...	±1.3	±0.050
...	...	2¼	...	±0.045
53.0	2⅛	...	±1.1	±0.043
50.0	2	...	±1.0	±0.040
...	...	1⅞	...	±0.038
45.0	1¾	...	±0.9	±0.035
...	...	1⅝	...	±0.033
37.5	1½	...	±0.8	±0.030
...	...	1⅜	...	±0.028
31.5	1¼	...	±0.6	±0.025
...	...	1³⁄₁₆	...	±0.024
...	...	1⅛	...	±0.023
26.5	1¹⁄₁₆	...	±0.5	±0.021
25.0	1	...	±0.5	±0.020
...	...	¹⁵⁄₁₆	...	±0.019
22.4	⅞	...	±0.46	±0.018
...	...	¹³⁄₁₆	...	±0.016
19.0	¾	...	±0.38	±0.015
...	...	¹¹⁄₁₆	...	±0.014
16.0	⅝	...	±0.32	±0.013
...	...	⁹⁄₁₆	...	±0.012
13.2	¹⁷⁄₃₂	...	±0.30	±0.012
12.5	½	...	±0.28	±0.011
...	...	¹⁵⁄₃₂	...	±0.011
11.2	⁷⁄₁₆	...	±0.28	±0.011
9.5	⅜	...	±0.28	±0.010
8.0	⁵⁄₁₆	...	±0.26	±0.010
6.7	¹⁷⁄₆₄	...	±0.25	±0.009
6.3	¼	...	±0.25	±0.009
5.6	⁷⁄₃₂	...	±0.24	±0.009
4.75	³⁄₁₆	...	±0.21	±0.008
4.00	⁵⁄₃₂	...	±0.19	±0.007
3.35	0.127 (⅛)	...	±0.17	±0.006

TABLE 3 Tolerances on Bars of USA Standard Specifications for Industrial Perforated Plate and Screens

Perforated Opening			Tolerance on Average Bar	
Standard (metric), mm	USA Industrial Standard, in.	Additional Sizes, in.	Standard (metric), mm	USA Industrial Standard, in.
125.0	5	...	±3.2	±0.125
...	...	4½	...	±0.122
106.0	4¼	...	±2.9	±0.113
100.0	4	...	±2.7	±0.107
...	...	3¾	...	±0.102
90.0	3½	...	±2.5	±0.097
...	...	3¼	...	±0.089
75.0	3	...	±2.1	±0.081
...	...	2¾	...	±0.076
63.0	2½	...	±1.8	±0.069
...	...	2¼	...	±0.063
53.0	2⅛	...	±1.5	±0.059
50.0	2	...	±1.4	±0.056
...	...	1⅞	...	±0.054
45.0	1¾	...	±1.3	±0.051
...	...	1⅝	...	±0.047
37.5	1½	...	±1.1	±0.043
...	...	1⅜	...	±0.040
31.5	1¼	...	±0.9	±0.037
...	...	1 3/16	...	±0.035
...	...	1⅛	...	±0.034
26.5	1 1/16	...	±0.8	±0.032
25.0	1	...	±0.8	±0.030
...	...	15/16	...	±0.029
22.4	⅞	...	±0.7	±0.028
...	...	13/16	...	±0.026
19.0	¾	...	±0.6	±0.024
...	...	11/16	...	±0.022
16.0	5/18	...	±0.5	±0.021
...	...	9/16	...	±0.019
13.2	17/32	...	±0.46	±0.018
12.5	½	...	±0.44	±0.017
...	...	15/32	...	±0.017
11.2	7/16	...	±0.41	±0.016
9.5	⅜	...	±0.36	±0.014
8.0	5/16	...	±0.32	±0.013
6.7	17/64	...	±0.29	±0.011
6.3	¼	...	±0.28	±0.011
5.6	7/32	...	±0.27	±0.011
4.75	3/16	...	±0.23	±0.009
4.00	5/32	...	±0.22	±0.009
3.5	0.127 (⅛)	...	±0.20	±0.008

TABLE 4 Tolerance on Thickness of USA Standard Specifications for Industrial Perforated Plate and Screens

Gage	Steel		Tolerance on Gage	
Standard (metric), mm	USA Industrial Standard, in.	USA Industrial Decimal Equivalent, in.	Standard (metric), mm	USA Industrial Standard, in.
25.4	1		+1.00 / −0.25	+0.040 / −0.010
22.4	⅞		+0.89 / −0.25	+0.035 / −0.010
19.0	¾		+0.84 / −0.25	+0.033 / −0.010
16.0	⅝		+0.79 / −0.25	+0.031 / −0.010
12.5	½		+0.76 / −0.25	+0.030 / −0.010
9.50	⅜		+0.66 / −0.25	+0.026 / −0.010
8.00	5/16		+0.64 / −0.25	+0.025 / −0.010
6.30	¼		+0.53 / −0.25	+0.021 / −0.010
4.75	3/16		+0.51 / −0.25	+0.020 / −0.010
4.25	No. 8 USS gage	0.1644	±0.25	±0.010
3.35	10	0.1345	±0.25	±0.010
3.00	11	0.1196	±0.25	±0.010
2.65	12	0.1046	±0.25	±0.010
1.90	14	0.0747	±0.18	±0.007

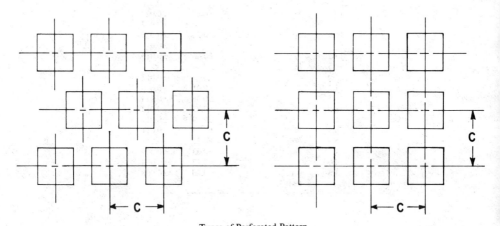

Types of Perforated Pattern

FIG. 1 Staggered Pattern   FIG. 2 Straight-Line Pattern

APPENDIXES

A1. EQUIVALENT METRIC SPECIFICATIONS

A1.1 The metric equivalent to the USA Standard Specification for Perforated Plate and Screens is shown in Table A1 and is published for the convenience of users in countries on full metric standard

E 454

and to facilitate comparison of the USA Standard with national standards on locally available specifications.

A1.2 In the metric table, the standard opening designations and bar sizes are identical to those in the USA Standard (Table 1), but the thicknesses of material are the nearest Standard ISO metric plate gages to the one used in the corresponding specification in the USA Standard.

A1.3 The tolerances on openings, average bar, and thickness of material are as shown in the appropriate columns of Tables 2, 3, and 4.

A2. MATERIALS USED FOR INDUSTRIAL PERFORATED PLATE AND SCREENS

A2.1 With a few exceptions, any material normally furnished in a flat sheet or plate can be successfully perforated, but the thickness of the material may have to be adjusted to keep the tonnage required for perforating within the limits of the perforating supplier's equipment.

A2.2 In selecting the material, it is recommended that the user consider the factors required in the end use of his product, such as abrasion resistance, corrosion resistance, impact resistance, weldability, formability, appearance, etc. When a suitable material has been selected, the perforating supplier should be contacted to determine the feasibility of supplying the required complete specification at reasonable cost.

A3. CHECKING AND CALIBRATING PERFORATED METAL AND SCREENS

A3.1 In checking specifications of industrial perforated plate it is necessary to determine three dimensions: (a) the dimension of the opening, (b) the dimensions of the bar, and (c) the thickness of the plate.

A3.2 *Dimensions of Openings:*

A3.2.1 For openings from 5 in. (125 mm) to ¼ in. (6.3 mm), measure the opening using a steel rule graduated to ¹⁄₆₄ in. or 0.5 mm or use a vernier inside caliper graduated to 0.001 in. or 0.02 mm. Since any perforated opening has an inherent taper, care should be taken to make all measurements from the surface of the plate that was uppermost during the punching operation (smoothside).

A3.2.2 For openings from ¼ in. (6.3 mm) to 0.127 in. (3.35 mm) measure the opening using a steel rule graduated to ¹⁄₆₄ in. or 0.5 mm.

A3.3 *Dimensions of Bars*—Bars may be checked with a steel rule graduated to ¹⁄₆₄ in. or 0.5 mm or with a vernier outside caliper graduated to 0.001 in. or 0.02 mm, if the nibs will fit into the adjacent openings.

A3.4 *Thickness of Plate*—Thickness of plate may be checked with a steel rule graduated to ¹⁄₆₄ in. or 0.5 mm or with a micrometer graduated to 0.001 in. or 0.02 mm.

A4. TERMS RELATING TO INDUSTRIAL PERFORATED PLATE AND SCREENS

A4.1 General

aperture or opening—dimensions defining an opening in a screen.
open area—ratio of the total area of the apertures to the total area of the screen, usually expressed in percentage.
screen—(a) a surface provided with openings of uniform size. (b) a machine provided with one or more screen surfaces.
screening—the process of separating a mixture of different sizes by means of one or more screen surfaces.

A4.2 Perforated Plate or Screens

perforation—an aperture or opening produced by punching.
bar—the metal between perforations measured at the point where perforations are the closest.
centers—the dimensional sum of one perforation and one bar, or the dimensional distance from the center of one perforation to the center of an adjacent perforation.
smooth side or punch side—the surface of the plate that was uppermost during the punching operation and through which the punch entered the plate.
break-out—a term applied to the action that occurs ahead of the punch in its going through the plate. The fracturing of the material results in a tapered hole with the small dimensions on the punch side.
die side—the surface of the plate that was against the die during the punching operation.
blank—an unperforated area located other than along the perimeter of a plate.
margin or border—an unperforated area located along the perimeter of a plate.
unfinished end pattern—the condition that occurs with some specifications of staggered pattern perforations as a result of tool design. On one end of the plate, the pattern will appear to be incomplete due to unperforated holes in the even numbered rows, while on the other end of the same plate, the pattern will appear to be incomplete due to unperforated holes in the odd numbered rows (Fig. A1).
finished end pattern—the condition that occurs with some specifications of staggered pattern perforations as a result of tool design where the pattern is completed on both ends of the plate (Fig. A2).

TABLE A1 USA Standard Specifications for Industrial Perforated Plate and Screens (Square Opening Series)—(Metric Equivalents)

Perforated Opening		Medium Light				Medium				Medium-Heavy				Heavy			
Standard (metric), mm	USA Industrial Standard, in.	Opening, mm	Bar, mm	Gage Steel, mm	Open Area, percent	Opening, mm	Bar, mm	Gage Steel, mm	Open Area, percent	Opening, mm	Bar, mm	Gage Steel, mm	Open Area, percent	Opening, mm	Bar, mm	Gage Steel, mm	Open Area, percent
125	5.0	125	12.50	12.50	82.6	125	16.00	16.00	78.6	125	19.00	19.00	75.4	125	25.40	25.40	69.1
125	5.0	125	16.00	9.50	78.6	125	19.00	12.50	75.4	125	22.40	16.00	71.9	125	28.50	22.40	66.3
125	5.0	125	16.00	12.50	78.6	125	19.00	16.00	75.4	125	22.40	19.00	71.9	125	28.50	25.40	66.3
106	4.25	106	12.50	12.50	80.0	106	16.00	16.00	75.5	106	19.00	19.00	71.9	106	25.40	25.40	65.1
106	4.25	106	16.00	9.50	75.5	106	19.00	12.50	71.9	106	22.40	16.00	68.2	106	28.50	22.40	62.1
106	4.25	106	16.00	12.50	75.5	106	19.00	16.00	71.9	106	22.40	19.00	68.2	106	28.50	25.40	62.1
100	4.0	100	12.50	12.50	79.0	100	16.00	16.00	74.3	100	19.00	19.00	70.6	100	25.40	25.40	63.6
100	4.0	100	16.00	9.50	74.3	100	19.00	12.50	70.6	100	22.40	16.00	66.7	100	28.50	22.40	60.6
100	4.0	100	16.00	12.50	74.3	100	19.00	16.00	70.6	100	22.40	19.00	66.7	100	28.50	25.40	60.6
90	3.5	90	12.50	12.50	77.1	90	16.00	16.00	72.1	90	19.00	19.00	68.2	90	25.40	22.40	.64.1
90	3.5	90	16.00	9.50	72.1	90	19.00	12.50	68.2	90	22.40	16.00	64.1	90	28.50	19.00	57.7
90	3.5	90	16.00	12.50	72.1	90	19.00	16.00	68.2	90	22.40	19.00	64.1	90	28.50	22.40	57.7
75	3.0	75	9.50	9.50	78.8	75	12.50	12.50	73.5	75	16.00	16.00	67.9	75	19.00	19.00	63.7
75	3.0	75	12.50	8.00	73.5	75	16.00	9.50	67.9	75	19.00	12.50	63.7	75	22.40	16.00	59.3
75	3.0	75	12.50	9.50	73.5	75	16.00	12.50	67.9	75	19.00	16.00	63.7	75	22.40	19.00	59.3
63	2.5	63	9.50	9.50	75.5	63	12.50	12.50	69.6	63	16.00	16.00	63.4	63	19.00	19.00	59.0
63	2.5	63	12.50	8.00	69.6	63	16.00	9.50	63.4	63	19.00	12.50	59.0	63	22.40	16.00	54.4
63	2.5	63	12.50	9.50	69.6	63	16.00	12.50	63.4	63	19.00	16.00	59.0	63	22.40	19.00	54.4
53	2.125	53	8.00	8.00	75.5	53	9.50	9.50	71.9	53	12.50	12.50	65.5	53	16.00	16.00	59.0
53	2.125	53	9.50	6.30	71.9	53	12.50	8.00	65.5	53	16.00	9.50	59.0	53	19.00	12.50	54.2
53	2.125	53	9.50	8.00	71.9	53	12.50	9.50	65.5	53	16.00	12.50	59.0	53	19.00	16.00	54.2
50	2.0	50	8.00	8.00	74.3	50	9.50	9.50	70.6	50	12.50	12.50	64.0	50	16.00	16.00	57.4
50	2.0	50	9.50	6.30	70.6	50	12.50	8.00	64.0	50	16.00	9.50	57.4	50	19.00	12.50	52.5
50	2.0	50	9.50	8.00	70.6	50	12.50	9.50	64.0	50	16.00	12.50	57.4	50	19.00	16.00	52.5
45	1.75	45	8.00	8.00	72.1	45	9.50	9.50	68.2	45	12.50	12.50	61.2	45	16.00	16.00	54.4
45	1.75	45	9.50	6.30	68.2	45	12.50	8.00	61.2	45	16.00	9.50	54.4	45	19.00	12.50	49.4
45	1.75	45	9.50	8.00	68.2	45	12.50	9.50	61.2	45	16.00	12.50	54.4	45	19.00	16.00	49.4

TABLE A1 Continued

Perforated Opening		Medium Light				Medium				Medium-Heavy				Heavy			
Standard (metric), mm	USA Industrial Standard, in.	Opening, mm	Bar, mm	Gage-Steel, mm	Open Area, percent	Opening, mm	Bar, mm	Gage-Steel, mm	Open Area, percent	Opening, mm	Bar, mm	Gage-Steel, mm	Open Area, percent	Opening, mm	Bar, mm	Gage-Steel, mm	Open Area, percent
37.5	1.5	37.5	6.30	6.30	73.3	37.5	8.00	8.00	67.9	37.5	9.50	9.50	63.7	37.5	12.50	12.50	56.3
37.5	1.5	37.5	8.00	4.75	67.9	37.5	9.50	6.30	63.7	37.5	12.50	8.00	56.3	37.5	16.00	9.50	49.1
37.5	1.5	37.5	8.00	6.30	67.9	37.5	9.50	8.00	63.7	37.5	12.50	9.50	56.3	37.5	16.00	12.50	49.1
31.5	1.25	31.5	6.30	6.30	69.4	31.5	8.00	8.00	63.6	31.5	9.50	9.50	59.0	31.5	12.50	12.50	51.3
31.5	1.25	31.5	8.00	4.75	63.6	31.5	9.50	6.30	59.0	31.5	12.50	8.00	51.3	31.5	16.00	9.50	44.0
31.5	1.25	31.5	8.00	6.30	63.6	31.5	9.50	8.00	59.0	31.5	12.50	9.50	51.3	31.5	16.00	12.50	44.0
26.5	1.06	26.5	4.75	4.75	71.9	26.5	6.30	6.30	65.3	26.5	8.00	8.00	59.0	26.5	9.50	9.50	54.2
26.5	1.06	26.5	6.30	4.25	65.3	26.5	8.00	4.75	59.0	26.5	9.50	6.30	54.2	26.5	12.50	8.00	46.2
26.5	1.06	26.5	6.30	4.75	65.3	26.5	8.00	6.30	59.0	26.5	9.50	8.00	54.2	26.5	12.50	9.50	46.2
25.0	1.0	25.0	4.75	4.75	70.6	25.0	6.30	6.30	63.8	25.0	8.00	8.00	57.4	25.0	9.50	9.50	52.5
25.0	1.0	25.0	6.30	4.25	63.8	25.0	8.00	4.75	57.4	25.0	9.50	6.30	52.5	25.0	12.50	8.00	44.4
25.0	1.0	25.0	6.30	4.75	63.8	25.0	8.00	6.30	57.4	25.0	9.50	8.00	52.5	25.0	12.50	9.50	44.4
22.4	0.875	22.4	4.75	4.75	68.1	22.4	6.30	6.30	60.9	22.4	8.00	8.00	54.3	22.4	9.50	9.50	49.3
22.4	0.875	22.4	6.30	4.25	60.9	22.4	8.00	4.75	54.3	22.4	9.50	6.30	49.3	22.4	12.50	8.00	41.2
22.4	0.875	22.4	6.30	4.75	60.9	22.4	8.00	6.30	54.3	22.4	9.50	8.00	49.3	22.4	12.50	9.50	41.2
19	0.750	19	4.75	4.75	64.0	19	6.30	6.30	56.4	19	8.00	8.00	49.5	19	9.50	9.50	44.4
19	0.750	19	6.30	4.25	56.4	19	8.00	4.75	49.5	19	9.50	6.30	44.4	19	12.50	8.00	36.4
19	0.750	19	6.30	4.75	56.4	19	8.00	6.30	49.5	19	9.50	8.00	44.4	19	12.50	9.50	36.4
16	0.625	16	4.25	4.25	64.0	16	6.30	4.75	59.5	16	6.30	6.30	51.5	16	8.00	8.00	44.4
16	0.625	16	4.75	3.35	59.5	16	4.75	4.25	51.5	16	8.00	4.75	44.4	16	9.50	6.30	39.4
16	0.625	16	4.75	4.25	59.5	16	6.30	4.75	51.5	16	8.00	6.30	44.4	16	9.50	8.00	39.4
13.2	0.531	13.2	3.15	3.35	65.2	13.2	4.00	4.25	58.9	13.2	4.75	4.75	54.1	13.2	6.30	6.30	45.8
13.2	0.531	13.2	4.00	3.00	58.9	13.2	4.75	3.35	54.1	13.2	4.25	4.25	45.8	13.2	8.00	4.75	38.8
13.2	0.531	13.2	4.00	3.35	58.9	13.2	4.75	4.25	54.1	13.2	4.75	4.75	45.8	13.2	8.00	6.30	38.8
12.5	0.500	12.5	3.15	3.35	63.8	12.5	4.00	4.25	57.4	12.5	4.75	4.75	52.5	12.5	6.30	6.30	44.2
12.5	0.500	12.5	4.00	3.00	57.4	12.5	4.75	3.35	52.5	12.5	6.30	4.25	44.2	12.5	8.00	4.75	37.2
12.5	0.500	12.5	4.00	3.35	57.4	12.5	4.75	4.25	52.5	12.5	6.30	4.75	44.2	12.5	8.00	6.30	37.2

TABLE A1 Continued

Perforated Opening		Medium Light				Medium				Medium-Heavy				Heavy			
Standard (metric), mm	USA Industrial Standard, in.	Opening, mm	Bar, mm	Gage-Steel, mm	Open Area, percent	Opening, mm	Bar, mm	Gage-Steel, mm	Open Area, percent	Opening, mm	Bar, mm	Gage-Steel, mm	Open Area, percent	Opening, mm	Bar, mm	Gage-Steel, mm	Open Area, percent
11.2	0.438	11.2	3.15	3.35	60.9	11.2	4.00	4.25	54.3	11.2	4.75	4.75	49.3	11.2	6.30	6.30	41.0
11.2	0.438	11.2	4.00	3.00	54.3	11.2	4.75	3.35	49.3	11.2	6.30	4.25	41.0	11.2	8.00	4.75	34.0
11.2	0.438	11.2	4.00	3.35	54.3	11.2	4.75	4.25	49.3	11.2	6.30	4.75	41.0	11.2	8.00	6.30	34.0
9.5	0.375	9.5	2.36	3.00	64.2	9.5	3.15	3.35	56.4	9.5	4.00	4.25	49.5	9.5	4.75	4.75	44.4
9.5	0.375	9.5	3.15	2.65	56.4	9.5	4.00	3.00	49.5	9.5	4.75	3.35	44.4	9.5	6.30	4.25	36.2
9.5	0.375	9.5	3.15	3.00	56.4	9.5	4.00	3.35	49.5	9.5	4.75	4.25	44.4	9.5	6.30	4.75	36.2
8	0.313	8	2.36	3.00	59.6	8	3.15	3.35	51.5	8	4.00	4.25	44.4	8	4.75	4.75	39.4
8	0.313	8	3.15	2.65	51.5	8	4.00	3.00	44.4	8	4.75	3.35	39.4	8	6.30	4.25	31.3
8	0.313	8	3.15	3.00	51.5	8	4.00	3.35	44.4	8	4.75	4.25	39.4	8	6.30	4.75	31.3
6.7	0.266	…	…	…	…	6.7	2.36	3.00	54.7	6.7	3.15	3.35	46.3	6.7	4.00	4.25	39.2
6.7	0.266	6.7	2.36	1.90	54.7	6.7	3.15	2.65	46.3	6.7	4.00	3.00	39.2	6.7	4.75	3.35	34.2
6.7	0.266	6.7	2.36	2.65	54.7	6.7	3.15	3.00	46.3	6.7	4.00	3.35	39.2	6.7	4.75	4.25	34.2
6.3	0.250	…	…	…	…	6.3	2.36	3.00	52.9	6.3	3.15	3.35	44.4	6.3	4.00	4.25	37.4
6.3	0.250	6.3	2.36	1.90	52.9	6.3	3.15	2.65	44.4	6.3	4.00	3.00	37.4	6.3	4.75	3.35	32.5
6.3	0.250	6.3	2.36	2.65	52.9	6.3	3.15	3.00	44.4	6.3	4.00	3.35	37.4	6.3	4.75	4.25	32.5
5.6	0.219	…	…	…	…	…	…	…	…	5.6	2.36	3.00	49.5	5.6	3.15	3.35	41.0
5.6	0.219	…	…	…	…	5.6	2.36	1.90	49.5	5.6	3.15	2.65	41.0	5.6	4.00	3.00	34.0
5.6	0.219	…	…	…	…	5.6	2.36	2.65	49.5	5.6	3.15	3.00	41.0	5.6	4.00	3.35	34.0
4.75	0.188	…	…	…	…	…	…	…	…	4.75	2.36	3.00	44.6	4.75	3.15	3.35	36.1
4.75	0.188	…	…	…	…	4.75	2.36	1.90	44.6	4.75	3.15	2.65	36.1	4.75	4.00	3.00	29.5
4.75	0.188	…	…	…	…	4.75	2.36	2.65	44.6	4.75	3.15	3.00	36.1	4.75	4.00	3.35	29.5
4	0.156	…	…	…	…	…	…	…	…	…	…	…	…	4	2.36	3.00	39.6
4	0.156	…	…	…	…	…	…	…	…	4	2.36	1.90	39.6	4	3.15	2.65	31.3
4	0.156	…	…	…	…	…	…	…	…	4	2.36	2.65	39.6	4	3.15	3.00	31.3
3.35	0.125	…	…	…	…	…	…	…	…	…	…	…	…	…	…	…	…
3.35	0.125	…	…	…	…	…	…	…	…	…	…	…	…	3.35	2.36	1.90	34.4
3.35	0.125	…	…	…	…	…	…	…	…	…	…	…	…	3.35	2.36	2.65	34.4

E 454

Arrangement of Staggered Pattern Openings

FIG. A1 Unfinished End Pattern   FIG. A2 Finished End Pattern

The American Society for Testing and Materials takes no position respecting the validity of any patent rights asserted in connection with any item mentioned in this standard. Users of this standard are expressly advised that determination of the validity of any such patent rights, and the risk of infringement of such rights, are entirely their own responsibility.

This standard is subject to revision at any time by the responsible technical committee and must be reviewed every five years and if not revised, either reapproved or withdrawn. Your comments are invited either for revision of this standard or for additional standards and should be addressed to ASTM Headquarters. Your comments will receive careful consideration at a meeting of the responsible technical committee, which you may attend. If you feel that your comments have not received a fair hearing you should make your views known to the ASTM Committee on Standards, 1916 Race St., Philadelphia, Pa. 19103.

Designation: E 527 – 83

Standard Practice for
NUMBERING METALS AND ALLOYS (UNS)[1]

This standard is issued under the fixed designation E 527; the number immediately following the designation indicates the year of original adoption or, in the case of revision, the year of last revision. A number in parentheses indicates the year of last reapproval. A superscript epsilon (ε) indicates an editorial change since the last revision or reapproval.

' Editorial changes were made throughout and the designation date was changed May 10, 1983.

1. Scope

1.1 This practice (Note 1) covers a unified numbering system (UNS) for metals and alloys that have a "commercial standing" (see Note 2), and covers the procedure by which such numbers are assigned. Section 2 describes the system of alphanumeric designations or "numbers" established for each family of metals and alloys. Section 3 outlines the organization established for administering the system. Section 4 describes the procedure for requesting number assignment to metals and alloys for which UNS numbers have not previously been assigned.

NOTE 1—UNS designations shall not be used for metals and alloys that are not registered under the system described herein, or for any metal or alloy whose composition differs from those registered.

NOTE 2—The terms "commercial standing," "production usage," and others are intended to portray a material in active industrial use, although the actual amount of such use will depend, among other things, upon the type of materials. (Obviously gold will not be used in the same "tonnages" as hot-rolled steel.)

Different standardizing groups use different criteria to define the status that a material has to attain before a standard number will be assigned to it. For instance, the American Iron and Steel Institute requires for stainless steels "two or more producers with combined production of 200 tons per year for at least two years"; the Copper Development Association requires that the material be "in commercial use (without tonnage limits)"; the Aluminum Association requires that the alloy be "offered for sale (not necessarily in commercial use)"; the SAE Aerospace Materials Division calls for "repetitive procurement by at least two users."

While it is apparent that no hard and fast usage definition can be set up for an all-encompassing system, the UNS numbers are intended to identify metals and alloys that are in more or less regular production and use. A UNS number will not ordinarily be issued for a material that has just been conceived or that is still in only experimental trial.

1.2 The UNS provides a means of correlating many nationally used numbering systems currently administered by societies, trade associations, and individual users and producers of metals and alloys, thereby avoiding confusion caused by use of more than one identification number for the same material; and by the opposite situation of having the same number assigned to two or more entirely different materials. It also provides the uniformity necessary for efficient indexing, record keeping, data storage and retrieval, and cross referencing.

1.3 A UNS number is not in itself a specification, since it establishes no requirements for form, condition, quality, etc. It is a unified identification of metals and alloys for which controlling limits have been established in specifications published elsewhere.

NOTE 3—Organizations that issue specifications should report to appropriate UNS number-assigning offices (3.1.2) any specification changes that affect descriptions shown in published UNS listings.

2. Description of Numbers (or Codes) Established for Metals and Alloys

2.1 The unified numbering system (UNS) establishes 18 series of numbers for metals and alloys, as shown in Table 1. Each UNS number consists of a single letter-prefix followed by five digits. In most cases the letter is suggestive of the family of metals identified; for example, A for aluminum, P for precious metals, and S for stainless steels.

---

[1] This practice is under the jurisdiction of ASTM Committee A-1 on Steel, Stainless Steel, and Related Alloys.
Current edition approved May 10, 1983. Published July 1983. Originally published as E 527 – 74. Last previous edition E 527 – 74 (1981).

2.2 Whereas some of the digits in certain UNS number groups have special assigned meaning, each series is independent of the others in such significance; this practice permits greater flexibility and avoids complicated and lengthy UNS numbers.

NOTE 4—This arrangement of alphanumeric six-character numbers is a compromise between the thinking that identification numbers should indicate many characteristics of the material, and the belief that numbers should be short and uncomplicated to be widely accepted and used.

2.3 Wherever feasible, identification "numbers" from existing systems are incorporated into the UNS numbers. For example: carbon steel, presently identified by AISI 1020 (American Iron and Steel Institute), is covered by "UNS G10200"; and free cutting brass, presently identified by CDA (Copper Development Association C36000), is covered by "UNS C36000." Table 2 shows the secondary division of some primary series of numbers.

2.4 Welding filler metals fall into two general categories: those whose compositions are determined by the filler metal analysis (e.g. solid bare wire or rods and cast rods) and those whose composition is determined by the weld deposit analysis (e.g. covered electrodes, flux-cored and other composite wire electrodes). The latter are assigned to a new primary series with the letter W as shown in Table 1. The solid bare wire and rods continue to be assigned in the established number series according to their composition.

NOTE 5—Readers are cautioned *not* to make their own assignments of numbers from such listings, as this can result in unintended and unexpected duplication and conflict.

2.5 ASTM and SAE periodically publish up-to-date listings of all UNS numbers assigned to specific metals and alloys, with appropriate reference information on each.[2] Many trade associations also publish similar listings related to materials of primary interest to their organizations.

3. Organization for Administering the UNS for Metals and Alloys

3.1 The organization for administering the UNS consists of the following:

3.1.1 *Advisory Board*—The Advisory Board has approximately 20 volunteer members who are affiliated with major producing and using industries, trade associations, government agencies, and standards societies, and who have extensive experience with identification, classification, and specification of materials. The Board is the administrative arm of SAE and ASTM on all matters pertaining to the UNS. It coordinates thinking on the format of each series of numbers and the administration of each by selected experts. It sets up ground rules for determining eligibility of any material for a UNS number, for requesting such numbers, and for appealing unfavorable rulings. It is the final referee on matters of disagreement between requesters and assigners.

3.1.2 *Several Number-Assigning Offices*—UNS number assigners for certain materials are set up at trade associations which have successfully administered their own numbering systems; for other materials, assigners are located at offices of SAE and ASTM. Each of these assigners has the responsibility for administering a specific series of numbers, as shown in Table 3. Each considers requests for assignment of new UNS numbers, and informs applicants of the action taken. Trade association UNS number assigners report immediately to both SAE and ASTM details of each number assignment. ASTM and SAE assigners collaborate with designated consultants when considering requests for assignment of new numbers.

3.1.3 *Corps of Volunteer Consultants*—Consultants are selected by the Advisory Board to provide expert knowledge of a specific field of materials. Since they are utilized primarily by the Board and the SAE and ASTM number assigners, they are not listed in this recommended practice. At the request of the ASTM (or SAE) number assigner, a consultant considers a request for a new number in the light of the ground rules established for the material involved, decides whether a new number is justified, and informs the ASTM or the SAE number assigner accordingly. This utilization of experts (consultants and number assigners) is intended to ensure prompt and fair consideration of all requests. It permits each decision to be based on current knowledge of the needs of a specific industry of producers and users.

3.1.4 *Staffs at ASTM and SAE*—Staff members at SAE and ASTM maintain duplicate mas-

---

[2] Request ASTM DS 56A and SAE Handbook Supplement HS 1086a, *Unified Numbering System for Metals and Alloys*, (a joint ASTM–SAE publication), PCN 05-056001-01.

ter listings of all UNS numbers assigned.

3.1.5 In addition, established SAE and ASTM committees which normally deal with standards and specifications for the materials covered by the UNS, and other knowledgeable persons, are called upon by the Advisory Board for advice when considering appeals from unfavorable rulings in the matter of UNS number assignments.

4. Procedure for Requesting Number Assignment to Metals and Alloys Not Already Covered by UNS Numbers (or Codes)

4.1 UNS numbers are assigned only to metals and alloys that have a commercial standing (as defined in Note 2).

4.2 The need for a new number should always be verified by determining from the latest complete listing of already assigned UNS numbers that a usable number is or is not available.

NOTE 6—In assigning UNS numbers, and consequently in searching complete listings of numbers, the predominant element of the metal or alloy usually determines the prefix letter of the series to which it is assigned. In certain instances where no one element predominates, arbitrary decisions are made as to what prefix letter to use, depending on the producing industry and other factors.

4.3 For a new UNS number to be assigned, the composition (or other properties, as applicable) must be significantly different from that of any metal or alloy which has already been assigned a UNS number.

4.3.1 In the case of metals or alloys that are normally identified or specified by chemical composition, the chemical composition limits must be reported.

4.3.2 In the case of metals or alloys that are normally identified or specified by mechanical (or other) properties, such properties and limits thereof must be reported. Only those chemical elements and limits, if any, which are significant in defining such materials need be reported.

4.4 Requests for new numbers shall be submitted on "Application for UNS Number Assignment" forms (see Fig. 1). Copies of these are available from any UNS number-assigning office (see Table 3) or facsimiles may be made of the one herein.

4.5 All instructions on the printed application form should be read carefully and all information provided as indicated.

NOTE 7—The application form is designed to serve also as a data input sheet to facilitate processing each request through to final print-out of the data on electronic data-processing equipment and to minimize transcription errors at number-assigning offices and data-processing centers.

4.6 To further assist in assigning UNS numbers, the requester is encouraged to suggest a possible UNS number in each request, giving appropriate consideration to any existing number presently used by a trade association, standards society, producer, or user.

4.7 Each completed application form shall be sent to the UNS number-assigning office having responsibility for the series of numbers that appears to most closely relate to the material described on the form (see Table 3).

TABLE 1 Primary Series of Numbers

Nonferrous Metals and Alloys
A00001–A99999	aluminum and aluminum alloys
C00001–C99999	copper and copper alloys
E00001–E99999	rare earth and rare earth-like metals and alloys (18 items; see Table 2)
L00001–L99999	low melting metals and alloys (15 items; see Table 2)
M00001–M99999	miscellaneous nonferrous metals and alloys (12 items; see Table 2)
N00001–N99999	nickel and nickel alloys
P00001–P99999	precious metals and alloys (8 items; see Table 2)
R00001–R99999	reactive and refractory metals and alloys (14 items; see Table 2)
Z00001–Z99999	zinc and zinc alloys

Ferrous Metals and Alloys
D00001–D99999	specified mechanical properties steels
F00001–F99999	cast irons and cast steels
G00001–G99999	AISI and SAE carbon and alloy steels
H00001–H99999	AISI H-steels
J00001–J99999	cast steels (except tool steels)
K00001–K99999	miscellaneous steels and ferrous alloys
S00001–S99999	heat and corrosion resistant (stainless) steels
T00001–T99999	tool steels

Specialized Metals and Alloys
W00001–W99999	welding filler metals, covered and tubular electrodes, classified by weld deposit composition (see Table 2)

TABLE 2 Secondary Division of Some Series of Numbers

E00001–E99999 Rare Earth and Rare Earth-Like Metals and Alloys		M05001–M05999	plutonium
		M06001–M06999	strontium
		M07001–M07999	tellurium
E00000–E00999	actinium	M08001–M08999	uranium
E01000–E20999	cerium	M10001–M19999	magnesium
E21000–E45999	mixed rare earths[A]	M20001–M29999	manganese
E46000–E47999	dysprosium	M30001–M39999	silicon
E48000–E49999	erbium		
E50000–E51999	europium	*P00001–P99999 Precious Metals and Alloys*	
E52000–E55999	gadolinium		
E56000–E57999	holmium	P00001–P00999	gold
E58000–E67999	lanthanum	P01001–P01999	iridium
E68000–E68999	lutetium	P02001–P02999	osmium
E69000–E73999	neodymium	P03001–P03999	palladium
E74000–E77999	praseodymium	P04001–P04999	platinum
E78000–E78999	promethium	P05001–P05999	rhodium
E79000–E82999	samarium	P06001–P06999	ruthenium
E83000–E84999	scandium	P07001–P07999	silver
E85000–E86999	terbium		
E87000–E87999	thulium	*R00001–R99999 Reactive and Refractory Metals and Alloys*	
E88000–E89999	ytterbium		
E90000–E99999	yttrium	R01001–R01999	boron
		R02001–R02999	hafnium
F00001–F9999 Cast Irons		R03001–R03999	molybdenum
		R04001–R04999	niubium (columbium)
K00001–K99999 Miscellaneous Steels and Ferrous Alloys		R05001–R05999	tantalum
		R06001–R06999	thorium
L00001–L99999 Low-Melting Metals and Alloys		R07001–R07999	tungsten
		R08001–R08999	vanadium
L00001–L00999	bismuth	R10001–R19999	beryllium
L01001–L01999	cadmium	R20001–R29999	chromium
L02001–L02999	cesium	R30001–R39999	cobalt
L03001–L03999	gallium	R40001–R49999	rhenium
L04001–L04999	indium	R50001–R59999	titanium
L05001–L05999	lead	R60001–R69999	zirconium
L06001–L06999	lithium		
L07001–L07999	mercury	*W00001–W99999 Welding Filler Metals Classified by Weld Deposit Composition*	
L08001–L08999	potassium		
L09001–L09999	rubidium		
L10001–L10999	selenium	W00001–W09999	carbon steel with no significant alloying elements
L11001–L11999	sodium	W10000–W19999	manganese-molybdenum low alloy steels
L12001–L12999	thallium		
L13001–L13999	tin	W20000–W29999	nickel low alloy steels
		W30000–W39999	austenitic stainless steels
M00001–M99999 Miscellaneous Nonferrous Metals and Alloys		W40000–W49999	ferritic stainless steels
		W50000–W59999	chromiun low alloy steels
M00001–M00999	antimony	W60000–W69999	copper base alloys
M01001–M01999	arsenic	W70000–W79999	surfacing alloys
M02001–M02999	barium	W80000–W89999	nickel base alloys
M03001–M03999	calcium		
M04001–M04999	germanium	*Z00001–Z99999 Zinc and Zinc Alloys*	

[A] Alloys in which the rare earths are used in the ratio of their natural occurrence (that is, unseparated rare earths). In this mixture, cerium is the most abundant of the rare earth elements.

ASTM E 527

TABLE 3 Number Assigners and Areas of Responsibility

The Aluminum Association 818 Connecticut Ave. N.W. Washington, D.C. 20006 Attention: Office for Unified Numbering System for Metals Telephone: (202)862-5100	Aluminum and Aluminum Alloys UNS Number Series: A00001–A99999
American Iron and Steel Institute 1000 16th St., N.W. Washington, D.C. 20036 Attention: Office for Unified Numbering System for Metals Telephone: (202)452-7236	Carbon and Alloy Steels UNS Number Series: G00001–G99999 H-Steels UNS Number Series: H00001–H99999 Tool Steels UNS Number Series: T00001–T99999
American Welding Society 550 N. W. LeJeune Road P.O. Box 351040 Miami, FL 33135 Attention: Office for Unified Numbering System for Metals Telephone: (305)642-7090	Welding Filler Metals UNS Number Series: W00001–W99999
Copper Development Association 405 Lexington Ave. New York, N. Y. 10017 Attention: Office for Unified Numbering System for Metals Telephone: (212)953-7321	Copper and Copper Alloys UNS Number Series: C00001–C99999
ASTM 1916 Race St. Philadelphia, Pa. 19103 Attention: Office for Unified Numbering System for Metals Telephone: (215)299-5521	Rare Earth and Rare Earth-Like Metals and Alloys UNS Number Series: E00001–E99999 Low Melting Metals and Alloys UNS Number Series: L00001–L99999 Miscellaneous Steels and Ferrous Alloys UNS Number Series: K00001–K99999 Miscellaneous Nonferrous Metals and Alloys UNS Number Series: M00001–M99999 Cast Steels UNS Number Series: J00001–J99999 Heat and Corrosion Resistant (Stainless) Steels UNS Number Series: S00001–S99999 Zinc and Zinc Alloys UNS Number Series: Z00001–Z99999 Precious Metals and Alloys UNS Number Series: P00001–P99999 Cast Irons and Cast Steels UNS Number Series: F00001–F99999
Society of Automotive Engineers 400 Commonwealth Drive Warrendale, Pa. 15096 Attention: Office for Unified Numbering System for Metals Telephone: (412)776-4841	Nickel and Nickel Alloys UNS Number Series: N00001–N99999 Steels Specified by Mechanical Properties UNS Number Series: D00001–D99999 Reactive and Refractory Metals and Alloys UNS Number Series: R00001–R99999

ASTM E 527

APPLICATION FOR UNS NUMBER ASSIGNMENT
and
Data Input Sheet for Entering a Specific Material in the
SAE-ASTM Unified Numbering System for Metals and Alloys
(See Reverse Side for Instructions for Completing This Form)

Material Description _____

_____

_____ Suggested UNS No. _____

*UNS Assigned Description _____

_____ *UNS Assigned No. _____

*Chemical Composition

Aluminum	Al	____	Indium	In	____	Selenium	Se	____
Antimony	Sb	____	Iridium	Ir	____	Silicon	Si	____
Arsenic	As	____	Iron	Fe	____	Silver	Ag	____
Beryllium	Be	____	Lead	Pb	____	Sulfur	S	____
Bismuth	Bi	____	Lithium	Li	____	Tantalum	Ta	____
Boron	B	____	Magnesium	Mg	____	Tellurium	Te	____
Cadmium	Cd	____	Manganese	Mn	____	Thorium	Th	____
Carbon	C	____	Mercury	Hg	____	Tin	Sn	____
Chromium	Cr	____	Molybdenum	Mo	____	Titanium	Ti	____
Cobalt	Co	____	Nickel	Ni	____	Tungsten	W	____
Columbium	Cb	____	Nitrogen	N	____	Uranium	U	____
Copper	Cu	____	Oxygen	O	____	Vanadium	V	____
Germanium	Ge	____	Phosphorus	P	____	Zinc	Zn	____
Gold	Au	____	Platinum	Pt	____	Zirconium	Zr	____
Hafnium	Hf	____	Rhenium	Re	____	Other		____
Hydrogen	H	____	Rhodium	Rh	____			____

*Cross References
AA _____
ACI _____
AISI _____
ANSI _____
AMS _____
ASME _____
ASTM _____
AWS _____
CDA _____
FED _____
MIL SPEC _____
SAE _____
OTHERS _____

Requesting Person and Organization (full address) _____

_____ Date of Request _____

*Assigning Org _____ *Date of UNS Assignment _____
Assigner's Name and Office _____

Applicant do not write in shaded areas.   *These items for Computer Operator.

NOTE—Reverse side of Fig. 1 is located on the next page.

FIG 1(a) Sample Application Form.

General:

Before attempting to complete this form, the applicant should be thoroughly familiar with the objectives of the UNS and the "ground rules" for assigning numbers, as stated in SAE J 1086 and ASTM E 527, Section 4.

Material Description:

Identify the base element; the single alloying element that constitutes 50 % or more of the total alloy content; other distinguishing predominant characteristics (such as "casting"); and common or generic names if any (such as "ounce metal" or "Waspalloy"). When no single element makes up 50 % or more of the total alloy content, list in decreasing order of abundance the two alloying elements that together constitute the largest portion of the total alloy contents; except that if no two elements make up at least 50 % of the total alloy content, list the three most abundant, and so on. Instead of "iron," use "steel" to identify the base element of those iron-low-carbon alloys commonly known as steels.

When mechanical properties or physical characteristics are the primary defining criteria and chemical composition is secondary or nonsignificant, enter such properties and characteristics with the appropriate values or limits for each.

Suggested UNS No.:

While applicant's suggestion may or may not be the one finally assigned, it will assist proper identification of the material by the UNS Number Assigner.

Chemical Composition:

Enter limits such as 0.13–0.18 (*not* .13–.18, or 0.13 to 0.18), 1.5 max, 0.040 min, and balance. In space designated "other" enter information such as "Each 0.05 max, Total 0.15 max" and "Sn plus Pb 2.0 min."

Cross References:

Letter-symbols listed indicate widely known trade associations and standards-issuing organizations. Enter after appropriate symbols any known specification numbers or identification numbers issued by such groups to cover material equivalent to, similar to, or closely resembling the subject material.

Examples: SAE J 404 (50B44), AISI 415, ASTM A 638 (660)

In space designated "other" enter any pertinent numbers issued by groups not listed above. In these instances, the full name and address of the issuing group shall be included.

SUBMIT COMPLETED FORM TO APPROPRIATE UNS NUMBER ASSIGNER, AS LISTED IN SAE J 1086 AND ASTM E 527.

FIG. 1(*b*)   Sample Application Form (Reverse Side).

The American Society for Testing and Materials takes no position respecting the validity of any patent rights asserted in connection with any item mentioned in this standard. Users of this standard are expressly advised that determination of the validity of any such patent rights, and the risk of infringement of such rights, are entirely their own responsibility.

This standard is subject to revision at any time by the responsible technical committee and must be reviewed every five years and if not revised, either reapproved or withdrawn. Your comments are invited either for revision of this standard or for additional standards and should be addressed to ASTM Headquarters. Your comments will receive careful consideration at a meeting of the responsible technical committee, which you may attend. If you feel that your comments have not received a fair hearing you should make your views known to the ASTM Committee on Standards, 1916 Race St., Philadelphia, Pa. 19103.

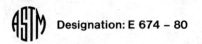 Designation: E 674 – 80

An American National Standard

Standard Specification for
INDUSTRIAL PERFORATED PLATE AND SCREENS (ROUND OPENING SERIES)[1]

This standard is issued under the fixed designation E 674; the number immediately following the designation indicates the year of original adoption or, in the case of revision, the year of last revision. A number in parentheses indicates the year of last reapproval. A superscript epsilon (ε) indicates an editorial change since the last revision or reapproval.

This specification has been approved for use by agencies of the Department of Defense and for listing in the DoD Index of Specifications and Standards.

INTRODUCTION

Industrial perforated plate can be produced in many thousands of combinations of size and shape of opening, bar size, thickness of material, and type of metal. Such variety is often confusing and, to the vast majority of perforated plate users, unnecessary, since each usually requires only a very few specifications.

The purpose of this specification is to simplify this problem by a condensed table of recommended specifications covering a wide range of openings in which industrial perforated plate is made, with several recommended bar sizes and thicknesses of plate for each opening, for use in various grades of service.

By making selections from this specification, the user will be guided to specifications that are being regularly produced, thus avoiding inadvertent selection of specifications that, because of little or no demand, are unobtainable, except on special order (usually quite expensive unless the quantity ordered is sufficient to justify the cost of special tooling).

If a user has a specific application for industrial perforated plate that can not be solved by a selection from this specification, it is recommended that he consult his perforated plate supplier on the availability of an acceptable alternative specification.

1. Scope

1.1 This specification covers the sizes of round opening perforated plate and screens for general industrial uses, including the separating or grading of materials according to designated nominal particle size, and lists standards for openings from 5 in. (125 mm) to 0.020 in. (500 μm) punched with bar sizes and thicknesses of plate for various grades of service. Methods of checking industrial perforated plate and screens are included as information in the Appendix.

1.2 This specification does not apply to perforated plate or screens with square, hexagon, slotted, or other shaped openings.

1.3 The values stated in inch-pound units are to be regarded as the standard.

2. Applicable Documents

2.1 *ASTM Standard:*

E 323 Specification for Perforated-Plate Sieves for Testing Purposes[2]

2.2 *Other Documents:*

Fed. Std. 123 Marking for Shipments (Civil Agencies)[3]

MIL-STD-129 Marking for Shipment and Storage[3]

3. Standard Specifications

3.1 Standard specifications for industrial perforated plate and screens are listed in Table 1.

---

[1] This specification is under the jurisdiction of ASTM Committee E-29 on Particle Size Measurement and is the direct responsibility of Subcommittee E29.01 on Sieves, Sieving Methods, and Screening Media.
Current edition approved Oct. 31, 1980. Published December 1980. Originally published as E 674 – 79. Last previous edition E 674 – 79.
[2] *Annual Book of ASTM Standards*, Vol 14.02.
[3] Available from Naval Publications and Forms Center, 5801 Tabor Ave., Philadelphia, Pa. 19120.

3.2 *Openings*—The series of standard openings listed in Table 1 include those of the USA Standard Sieve Series, Specification E 323, and those of the ISO apertures for industrial plate screens,[4] with the addition of those openings in common usage.

3.3 *Relationship of Grades*—The purpose of the several grades is to provide combinations of opening and bar size for various types of service, from medium-light to heavy. Since it is possible to vary the bar size independently from the plate thickness, each of the service grades lists up to three combinations of bar and gage for each opening. The entire standard series has been designed for a logical relationship of bar size to opening in each grade and between grades with the capability of also being able to vary the plate thickness.

3.4 *Bar*—A choice of several bars is shown for each standard opening from 5-in. (125-mm) to 0.078-in. (2-mm) opening, inclusive. For practical reasons, the number of bars or grades available for openings finer than 0.078 in. is progressively reduced.

3.5 *Gage*—A choice of several gages is shown for each standard opening for 5 in. (125 mm) to 0.078 in. (2 mm). For practical reasons, the number of gages or grades available for openings finer than 0.078 in. is progressively reduced.

NOTE 1—The gages shown in Table 1 are practical for a low-carbon steel plate. For other materials, consult your perforated plate supplier.

3.6 *Equivalent Metric Specification*—Table X1, in the Appendix, shows the equivalent metric specifications to the USA Standard, punched in standard ISO recommended thickness of plate.[5]

4. Types of Perforated Pattern

4.1 This specification covers round openings arranged in a staggered 60-deg pattern with their centers nominally at the vertices of equilateral triangles (See Fig. 1).

5. Metal Composition of Plate

5.1 Perforated plate can be punched from a great variety of metals and alloys, but the following are most commonly used:

Steel, low-carbon
Steel, high-carbon
Steel, heat-treated
Steel, galvanized
Stainless steel, Type 304
Stainless steel, Type 316
Stainless steel, Type 410
Brass (Cu 80, Zn 20)
Manganese bronze (Cu 61, Zn 37)
Monel (high nickel-copper alloy)
Aluminum (all grades)

6. Tolerances

6.1 *Openings*—Tolerances on openings in USA Standard Specifications for Industrial Perforated Plate and Screens (Tables 1 and X1.1) shall be in accordance with those listed in Table 2.

6.2 *Bars*—Tolerances on bars used in USA Standard Specification for Industrial Perforated Plate and Screens (Tables 1 and X1.1) shall be in accordance with those listed in Table 3.

6.3 *Gages*—Tolerances on gages used in USA Standard Specifications for Industrial Perforated Plate and Screens (Tables 1 and X1.1) shall be in accordance with those listed in Table 4.

NOTE 2—The tolerances expressed in inch-pound units are taken from the current AISI values.

---

[4] ISO 2194-1972, Wire Screens and Plate Screens for Industrial Purposes—Nominal Sizes of Apertures.
[5] ISO Recommendation R388-1964, Metric Series for Basic Thicknesses of Sheet and Diameters of Wire.

SUPPLEMENTARY REQUIREMENTS

The following sections shall be applicable when U.S. government contractual matters are involved.

S1. Responsibility for Inspection

S1.1 Unless otherwise specified in the contract or purchase order, the producer is responsible for the performance of all inspection and test requirements specified herein. Except as otherwise specified in the contract or order, the producer may use his own or any other suitable facilities for the performance of the inspection and test requirements specified herein, unless disapproved by the purchaser. The purchaser shall have the right to perform any of the inspections and tests set forth in this specification where such inspections are deemed necessary to ensure that material conforms to prescribed requirements.

S2. Government Procurement

S2.1 Unless otherwise specified in the contract, the material shall be packaged in accordance with the suppliers' standard practice which will be acceptable to the carrier at lowest rates. Containers and packing shall comply with the Uniform Freight Classification rules or National Motor Freight Classification rules. Marking for shipment of such material shall be in accordance with Fed. Std. No. 123 for civil agencies, and MIL STD 129 for military agencies.

TABLE 1 USA Standard Specifications for Industrial Perforated Plate and Screens (Round Opening Series)—(U.S. Customary Units)

Perforated Opening		Medium Light			Medium			Medium Heavy			Heavy		
Standard (metric), mm	USA Industrial Standard, in.	Opening, in.	Bar, in.	Gage Steel, in.	Open Area, %	Opening, in.	Bar, in.	Gage Steel, in.	Open Area, %	Opening, in.	Bar, in.	Gage Steel, in.	Open Area, %	Opening, in.	Bar, in.	Gage Steel, in.	Open Area, %
125	5	5	1/2	1/2	74.9	5	5/8	5/8	71.6	5	3/4	3/4	68.5	5	1	1	62.9
125	5	5	5/8	3/8	71.6	5	3/4	1/2	68.5	5	7/8	5/8	65.6	5	1 1/8	7/8	60.4
125	5	5	5/8	1/2	71.6	5	3/4	5/8	68.5	5	7/8	3/4	65.6	5	1 1/8	1	60.4
...	...	4 1/2	1/2	1/2	73.4	4 1/2	5/8	5/8	69.9	4 1/2	3/4	3/4	66.6	4 1/2	1	1	60.7
...	...	4 1/2	5/8	3/8	69.9	4 1/2	3/4	1/2	66.6	4 1/2	7/8	5/8	63.5	4 1/2	1 1/8	7/8	58
...	...	4 1/2	5/8	1/2	69.9	4 1/2	3/4	5/8	66.6	4 1/2	7/8	3/4	63.5	4 1/2	1 1/8	1	58
106	4 1/4	4 1/4	1/2	1/2	72.6	4 1/4	5/8	5/8	68.9	4 1/4	3/4	3/4	65.5	4 1/4	1	1	59.4
106	4 1/4	4 1/4	5/8	3/8	68.9	4 1/4	3/4	1/2	65.5	4 1/4	7/8	5/8	62.3	4 1/4	1 1/8	7/8	56.7
106	4 1/4	4 1/4	5/8	1/2	68.9	4 1/4	3/4	5/8	65.5	4 1/4	7/8	3/4	62.3	4 1/4	1 1/8	1	56.7
100	4	4	1/2	1/2	71.6	4	5/8	5/8	67.8	4	3/4	3/4	64.3	4	1	1	58
100	4	4	5/8	3/8	67.8	4	3/4	1/2	64.3	4	7/8	5/8	61	4	1 1/8	7/8	55.2
100	4	4	5/8	1/2	67.8	4	3/4	5/8	64.3	4	7/8	3/4	61	4	1 1/8	1	55.2
...	...	3 3/4	1/2	1/2	70.6	3 3/4	5/8	5/8	66.6	3 3/4	3/4	3/4	62.9	3 3/4	7/8	7/8	59.6
...	...	3 3/4	5/8	3/8	66.6	3 3/4	3/4	1/2	62.9	3 3/4	7/8	5/8	59.6	3 3/4	1	3/4	48.9
...	...	3 3/4	5/8	1/2	66.6	3 3/4	3/4	5/8	62.9	3 3/4	7/8	3/4	59.6	3 3/4	1	7/8	48.9
90	3 1/2	3 1/2	1/2	1/2	69.4	3 1/2	5/8	5/8	65.2	3 1/2	3/4	3/4	61.5	3 1/2	7/8	7/8	58
90	3 1/2	3 1/2	5/8	3/8	65.2	3 1/2	3/4	1/2	61.5	3 1/2	7/8	5/8	58	3 1/2	1	3/4	54.8
90	3 1/2	3 1/2	5/8	1/2	65.2	3 1/2	3/4	5/8	61.5	3 1/2	7/8	3/4	58	3 1/2	1	7/8	54.8
...	...	3 1/4	3/8	3/8	72.8	3 1/4	1/2	1/2	68.1	3 1/4	5/8	5/8	63.8	3 1/4	3/4	3/4	59.8
...	...	3 1/4	1/2	5/16	68.1	3 1/4	5/8	3/8	63.7	3 1/4	3/4	1/2	59.8	3 1/4	7/8	5/8	56.2
...	...	3 1/4	1/2	3/8	68.1	3 1/4	5/8	1/2	63.7	3 1/4	3/4	5/8	59.8	3 1/4	7/8	3/4	56.2
75	3	3	3/8	3/8	71.6	3	1/2	1/2	66.6	3	5/8	5/8	62.1	3	3/4	3/4	58
75	3	3	1/2	5/16	66.6	3	5/8	3/8	62.1	3	3/4	1/2	58	3	7/8	5/8	54.3
75	3	3	1/2	3/8	66.6	3	5/8	1/2	62.1	3	3/4	5/8	58	3	7/8	3/4	54.3
...	...	2 3/4	3/8	3/8	70.2	2 3/4	1/2	1/2	64.9	2 3/4	5/8	5/8	60.2	2 3/4	3/4	3/4	55.9
...	...	2 3/4	1/2	5/16	64.9	2 3/4	5/8	3/8	60.2	2 3/4	3/4	1/2	55.9	2 3/4	7/8	5/8	52.1
...	...	2 3/4	1/2	3/8	64.9	2 3/4	5/8	1/2	60.2	2 3/4	3/4	5/8	55.9	2 3/4	7/8	3/4	52.1

TABLE 1 *Continued*

Perforated Opening		Medium Light				Medium				Medium Heavy				Heavy			
Standard (metric), mm	USA Industrial Standard, in.	Opening, in.	Bar, in.	Gage-Steel, in.	Open Area, %	Opening, in.	Bar, in.	Gage-Steel, in.	Open Area, %	Opening, in.	Bar, in.	Gage-Steel, in.	Open Area, %	Opening, in.	Bar, in.	Gage-Steel, in.	Open Area, %
63	2 1/2	2 1/2	3/8	3/8	68.5	2 1/2	1/2	1/2	62.9	2 1/2	5/8	5/8	58	2 1/2	3/4	3/4	53.6
63	2 1/2	2 1/2	1/2	5/16	62.9	2 1/2	5/8	3/8	58	2 1/2	3/4	1/2	53.6	2 1/2	7/8	5/8	49.7
63	2 1/2	2 1/2	1/2	3/8	62.9	2 1/2	5/8	1/2	58	2 1/2	3/4	5/8	53.6	2 1/2	7/8	3/4	49.7
...	...	2 1/4	3/8	3/8	66.6	2 1/4	1/2	1/2	60.7	2 1/4	5/8	5/8	55.5	2 1/4	3/4	3/4	51
...	...	2 1/4	1/2	5/16	60.7	2 1/4	5/8	3/8	55.5	2 1/4	3/4	1/2	51	2 1/4	7/8	5/8	47
...	...	2 1/4	1/2	3/8	60.7	2 1/4	5/8	1/2	55.5	2 1/4	3/4	5/8	51	2 1/4	7/8	3/4	47
53	2 1/8	2 1/8	5/16	5/16	68.9	2 1/8	3/8	3/8	65.5	2 1/8	1/2	1/2	59.4	2 1/8	5/8	5/8	54.1
53	2 1/8	2 1/8	3/8	1/4	65.5	2 1/8	1/2	5/16	59.4	2 1/8	5/8	3/8	54.1	2 1/8	3/4	1/2	49.5
53	2 1/8	2 1/8	3/8	5/16	65.5	2 1/8	1/2	3/8	59.4	2 1/8	5/8	1/2	54.1	2 1/8	3/4	5/8	49.5
50	2	2	5/16	5/16	67.8	2	3/8	3/8	64.3	2	1/2	1/2	58	2	5/8	5/8	52.6
50	2	2	3/8	1/4	64.3	2	1/2	5/16	58	2	5/8	3/8	52.6	2	3/4	1/2	47.9
50	2	2	3/8	5/16	64.3	2	1/2	3/8	58	2	5/8	1/2	52.6	2	3/4	5/8	47.9
...	...	1 7/8	5/16	5/16	66.6	1 7/8	3/8	3/8	62.9	1 7/8	1/2	1/2	56.5	1 7/8	5/8	5/8	51
...	...	1 7/8	3/8	1/4	62.9	1 7/8	1/2	5/16	56.5	1 7/8	5/8	3/8	51	1 7/8	3/4	1/2	46.2
...	...	1 7/8	3/8	5/16	62.9	1 7/8	1/2	3/8	56.5	1 7/8	5/8	1/2	51	1 7/8	3/4	5/8	46.2
45	1 3/4	1 3/4	5/16	5/16	65.2	1 3/4	3/8	3/8	61.5	1 3/4	1/2	1/2	54.8	1 3/4	5/8	5/8	49.2
45	1 3/4	1 3/4	3/8	1/4	61.5	1 3/4	1/2	5/16	54.8	1 3/4	5/8	3/8	49.2	1 3/4	3/4	1/2	44.4
45	1 3/4	1 3/4	3/8	5/16	61.5	1 3/4	1/2	3/8	54.8	1 3/4	5/8	1/2	49.2	1 3/4	3/4	5/8	44.4
...	...	1 5/8	1/4	1/4	68.1	1 5/8	5/16	5/16	63.7	1 5/8	3/8	3/8	59.8	1 5/8	1/2	1/2	53
...	...	1 5/8	5/16	3/16	63.7	1 5/8	3/8	1/4	59.8	1 5/8	1/2	5/16	53	1 5/8	5/8	3/8	47.3
...	...	1 5/8	5/16	1/4	63.7	1 5/8	3/8	5/16	59.8	1 5/8	1/2	3/8	53	1 5/8	5/8	1/2	47.3
37.5	1 1/2	1 1/2	1/4	1/4	66.6	1 1/2	5/16	5/16	62.1	1 1/2	3/8	3/8	58	1 1/2	1/2	1/2	51
37.5	1 1/2	1 1/2	5/16	3/16	62.1	1 1/2	3/8	1/4	58	1 1/2	1/2	5/16	51	1 1/2	5/8	3/8	45.1
37.5	1 1/2	1 1/2	5/16	1/4	62.1	1 1/2	3/8	5/16	58	1 1/2	1/2	3/8	51	1 1/2	5/8	1/2	45.1
...	...	1 3/8	1/4	1/4	64.9	1 3/8	5/16	5/16	60.2	1 3/8	3/8	3/8	55.9	1 3/8	1/2	1/2	48.7
...	...	1 3/8	5/16	3/16	60.2	1 3/8	3/8	1/4	55.9	1 3/8	1/2	5/16	48.7	1 3/8	5/8	3/8	42.8
...	...	1 3/8	5/16	1/4	60.2	1 3/8	3/8	5/16	55.9	1 3/8	1/2	3/8	48.7	1 3/8	5/8	1/2	42.8

TABLE 1 *Continued*

Perforated Opening		Medium Light				Medium				Medium Heavy				Heavy			
Standard (metric), mm	USA Industrial Standard, in.	Opening, in.	Bar, in.	Gage-Steel, in.	Open Area, %	Opening, in.	Bar, in.	Gage-Steel, in.	Open Area, %	Opening, in.	Bar, in.	Gage-Steel, in.	Open Area, %	Opening, in.	Bar, in.	Gage-Steel, in.	Open Area, %
31.5	1 1/4	1 1/4	1/4	1/4	62.9	1 1/4	5/16	5/16	58	1 1/4	3/8	3/8	53.6	1 1/4	1/2	1/2	46.2
31.5	1 1/4	1 1/4	5/16	3/16	58	1 1/4	3/8	1/4	53.6	1 1/4	1/2	5/16	46.2	1 1/4	5/8	3/8	40.3
31.5	1 1/4	1 1/4	5/16	1/4	58	1 1/4	3/8	5/16	53.6	1 1/4	1/2	3/8	46.2	1 1/4	5/8	1/2	40.3
...	...	1 3/16	3/16	3/16	67.6	1 3/16	1/4	1/4	61.8	1 3/16	5/16	5/16	56.8	1 3/16	3/8	3/8	52.3
...	...	1 3/16	1/4	8	61.8	1 3/16	5/16	3/16	56.8	1 3/16	3/8	1/4	52.3	1 3/16	1/2	5/16	44.9
...	...	1 3/16	1/4	3/16	61.8	1 3/16	5/16	1/4	56.8	1 3/16	3/8	5/16	52.3	1 3/16	1/2	3/8	44.9
...	...	1 1/8	3/16	3/16	66.6	1 1/8	1/4	1/4	60.7	1 1/8	5/16	5/16	55.5	1 1/8	3/8	3/8	51
...	...	1 1/8	1/4	8	60.7	1 1/8	5/16	3/16	55.5	1 1/8	3/8	1/4	51	1 1/8	1/2	5/16	43.4
...	...	1 1/8	1/4	3/16	60.7	1 1/8	5/16	1/4	55.5	1 1/8	3/8	5/16	51	1 1/8	1/2	3/8	43.4
26.5	1 1/16	1 1/16	3/16	3/16	65.5	1 1/16	1/4	1/4	59.4	1 1/16	5/16	5/16	54.1	1 1/16	3/8	3/8	49.5
26.5	1 1/16	1 1/16	1/4	8	59.4	1 1/16	5/16	3/16	54.1	1 1/16	3/8	1/4	49.5	1 1/16	1/2	5/16	41.9
26.5	1 1/16	1 1/16	1/4	3/16	59.4	1 1/16	5/16	1/4	54.1	1 1/16	3/8	5/16	49.5	1 1/16	1/2	3/8	41.9
25	1	1	3/16	3/16	64.3	1	1/4	1/4	58	1	5/16	5/16	52.6	1	3/8	3/8	47.9
25	1	1	1/4	8	58	1	5/16	3/16	52.6	1	3/8	1/4	47.9	1	1/2	5/16	40.3
25	1	1	1/4	3/16	58	1	5/16	1/4	52.6	1	3/8	5/16	47.9	1	1/2	3/8	40.3
...	...	15/16	3/16	3/16	62.9	15/16	1/4	1/4	56.4	15/16	5/16	5/16	51	15/16	3/8	3/8	46.2
...	...	15/16	1/4	8	56.4	15/16	5/16	3/16	51	15/16	3/8	1/4	46.2	15/16	1/2	5/16	38.5
...	...	15/16	1/4	3/16	56.4	15/16	5/16	1/4	51	15/16	3/8	5/16	46.2	15/16	1/2	3/8	38.5
22.4	7/8	7/8	3/16	3/16	61.5	7/8	1/4	1/4	54.8	7/8	5/16	5/16	49.2	7/8	3/8	3/8	44.4
22.4	7/8	7/8	1/4	8	54.8	7/8	5/16	3/16	49.2	7/8	3/8	1/4	44.4	7/8	1/2	5/16	36.7
22.4	7/8	7/8	1/4	3/16	54.8	7/8	5/16	1/4	49.2	7/8	3/8	5/16	44.4	7/8	1/2	3/8	36.7
...	...	13/16	3/16	3/16	59.8	13/16	1/4	1/4	53	13/16	5/16	5/16	47.2	13/16	3/8	3/8	42.4
...	...	13/16	1/4	8	53	13/16	5/16	3/16	47.2	13/16	3/8	1/4	42.4	13/16	1/2	5/16	34.7
...	...	13/16	1/4	3/16	53	13/16	5/16	1/4	47.2	13/16	3/8	5/16	42.4	13/16	1/2	3/8	34.7
19	3/4	3/4	3/16	3/16	58	3/4	1/4	1/4	51	3/4	5/16	5/16	45.1	3/4	3/8	3/8	40.3
19	3/4	3/4	1/4	8	51	3/4	5/16	3/16	45.1	3/4	3/8	1/4	40.3	3/4	1/2	5/16	32.6
19	3/4	3/4	1/4	3/16	51	3/4	5/16	1/4	45.1	3/4	3/8	5/16	40.3	3/4	1/2	3/8	32.6

TABLE 1 Continued

Perforated Opening Standard (metric), mm	USA Industrial Standard, in	Medium Light Opening, in	Medium Light Bar, in	Medium Light Gage Steel, in	Medium Light Open Area, %	Medium Opening, in	Medium Bar, in	Medium Gage Steel, in	Medium Open Area, %	Medium Heavy Opening, in	Medium Heavy Bar, in	Medium Heavy Gage Steel, in	Medium Heavy Open Area, %	Heavy Opening, in	Heavy Bar, in	Heavy Gage Steel, in	Heavy Open Area, %
...	...	11/16	3/16	3/16	55.9	11/16	1/4	1/4	48.7	11/16	5/16	5/16	42.8	11/16	3/8	3/8	37.9
...	...	11/16	1/4	8	48.7	11/16	1/4	3/16	42.8	11/16	3/8	1/4	37.9	11/16	1/2	5/16	30.3
...	...	11/16	1/4	3/16	48.7	11/16	5/16	3/16	42.8	11/16	3/8	5/16	37.9	11/16	1/2	3/8	30.3
16	5/8	5/8	5/32	8	58	5/8	3/16	8	53.6	5/8	1/4	1/4	46.2	5/8	5/16	5/16	40.3
16	5/8	5/8	3/16	10	53.6	5/8	1/4	8	46.2	5/8	5/16	3/16	40.3	5/8	3/8	1/4	35.4
16	5/8	5/8	3/16	8	53.6	5/8	1/4	3/16	46.2	5/8	5/16	1/4	40.3	5/8	3/8	5/16	35.4
...	...	9/16	5/32	8	55.5	9/16	3/16	3/16	51	9/16	1/4	1/4	43.4	9/16	5/16	5/16	37.4
...	...	9/16	3/16	10	51	9/16	1/4	8	43.4	9/16	5/16	3/16	37.4	9/16	3/8	1/4	32.6
...	...	9/16	3/16	8	51	9/16	1/4	3/16	43.4	9/16	5/16	1/4	37.4	9/16	3/8	5/16	32.6
13.2	17/32	...	...	...	...	...	...	10	...	17/32	5/32	3/8	54.1	17/32	7/32	1/4	45.4
13.2	17/32	...	...	...	...	17/32	5/32	10	54.1	17/32	7/32	8	45.4	17/32	11/32	3/16	33.4
13.2	17/32	...	...	...	...	17/32	5/32	8	54.1	17/32	7/32	3/8	45.4	17/32	11/32	1/4	33.4
12.5	1/2	...	...	...	...	...	...	10	...	1/2	3/16	3/16	47.9	1/2	1/4	1/4	40.3
12.5	1/2	...	...	...	...	1/2	3/16	10	47.9	1/2	1/4	8	40.3	1/2	5/16	3/16	34.3
12.5	1/2	...	...	...	...	1/2	3/16	8	47.9	1/2	1/4	3/16	40.3	1/2	5/16	1/4	34.3
...	...	...	...	...	...	1/8	5/32	8	56.5	15/32	5/32	3/16	50.9	15/32	7/32	1/4	42.1
...	...	15/32	1/8	11	56.5	5/32	5/32	10	50.9	15/32	7/32	8	42.1	15/32	9/32	3/16	35.4
...	...	15/32	1/8	10	56.5	5/32	5/32	8	50.9	15/32	7/32	3/16	42.1	15/32	9/32	1/4	35.4
11.2	7/16	7/16	5/32	10	49.2	7/16	3/16	8	44.4	7/16	1/4	3/16	36.7	7/16	5/16	1/4	30.8
11.2	7/16	7/16	3/16	11	44.4	7/16	1/4	10	36.7	7/16	5/16	8	30.8	7/16	7/16	3/16	22.6
11.2	7/16	7/16	3/16	10	44.4	7/16	1/4	10	36.7	7/16	5/16	3/16	30.8	7/16	7/16	1/4	22.6
9.5	3/8	3/8	1/8	11	51	3/8	3/16	10	40.3	3/8	7/32	1/4	36.1	3/8	1/4	3/16	32.6
9.5	3/8	3/8	3/16	12	40.3	3/8	7/32	11	36.1	3/8	1/4	3/8	32.6	3/8	3/8	8	22.6
9.5	3/8	3/8	3/16	11	40.3	3/8	7/32	10	36.1	3/8	1/4	1/4	32.6	3/8	3/8	3/16	22.6
8	5/16	5/16	3/32	11	53.6	5/16	1/8	10	46.2	5/16	5/32	3/8	40.3	5/16	3/16	3/16	35.4
8	5/16	5/16	1/8	12	46.2	5/16	5/32	11	40.3	5/16	3/16	10	35.4	5/16	8	8	27.9
8	5/16	5/16	1/8	11	46.2	5/16	5/32	10	40.3	5/16	3/16	3/16	35.4	5/16	1/4	3/16	27.9

TABLE 1 *Continued*

Perforated Opening		Medium Light				Medium				Medium Heavy				Heavy			
Standard (metric), mm	USA Industrial Standard, in.	Opening, in.	Bar, in.	Gage-Steel, in.	Open Area, %	Opening, in.	Bar, in.	Gage-Steel, in.	Open Area, %	Opening, in.	Bar, in.	Gage-Steel, in.	Open Area, %	Opening, in.	Bar, in.	Gage-Steel, in.	Open Area, %
6.7	17/64	...	...	...	...	17/64	7/64	11	45.4	17/64	1/8	10	41.9	17/64	9/64	8	38.7
6.7	17/64	17/64	7/64	14	45.4	17/64	1/8	12	41.9	17/64	9/64	11	38.7	17/64	11/64	10	33.4
6.7	17/64	17/64	7/64	12	45.4	17/64	1/8	11	41.9	17/64	9/64	10	38.7	17/64	11/64	8	33.4
6.3	1/4	1/4	1/16	16	58	1/4	1/8	11	40.3	1/4	5/32	10	34.3	1/4	3/16	8	29.6
6.3	1/4	1/4	1/8	14	40.3	1/4	5/32	12	34.3	1/4	3/16	11	29.6	1/4	1/4	10	22.6
6.3	1/4	1/4	1/8	12	40.3	1/4	5/32	11	34.3	1/4	3/16	10	29.6	1/4	1/4	8	22.6
5.6	7/32	7/32	3/32	14	44.4	7/32	1/8	12	36.7	7/32	5/32	11	30.8	7/32	3/16	10	26.2
5.6	7/32	7/32	1/8	16	36.7	7/32	5/32	14	30.8	7/32	3/16	12	26.2	7/32	7/32	11	22.6
5.6	7/32	7/32	1/8	14	36.7	7/32	5/32	12	30.8	7/32	3/16	11	26.2	7/32	1/4	10	22.6
4.75	3/16	3/16	1/16	14	51	3/16	3/32	12	40.3	3/16	7/64	11	36.1	3/16	1/8	10	32.6
4.75	3/16	3/16	3/32	16	40.3	3/16	7/64	14	36.1	3/16	1/8	12	32.6	3/16	3/16	11	22.6
4.75	3/16	3/16	3/32	14	40.3	3/16	7/64	12	36.1	3/16	1/8	11	32.6	3/16	3/16	10	22.6
4	5/32	...	...	...	...	5/32	1/16	14	46.2	5/32	3/32	12	35.4	5/32	1/8	11	27.9
4	5/32	5/32	1/16	18	46.2	5/32	3/32	16	35.4	5/32	1/8	14	27.9	5/32	5/32	12	22.6
4	5/32	5/32	1/16	16	46.2	5/32	3/32	14	35.4	5/32	1/8	12	27.9	5/32	5/32	11	22.6
3.35	1/8	...	...	...	...	1/8	3/64	14	47.9	1/8	1/16	12	40.3	1/8	3/32	11	29.6
3.35	1/8	1/8	3/64	18	47.9	1/8	1/16	16	40.3	1/8	3/32	14	29.6	1/8	1/8	12	22.6
3.35	1/8	1/8	3/64	16	47.9	1/8	1/16	14	40.3	1/8	3/32	12	29.6	1/8	1/8	11	22.6
2.80	7/64	...	...	...	...	7/64	1/16	16	36.4	7/64	3/32	14	26.1	7/64	9/64	12	17.2
2.80	7/64	7/64	1/16	20	36.4	7/64	3/32	18	26.1	7/64	9/64	16	17.2	7/64	5/32	14	15.2
2.80	7/64	7/64	1/16	18	36.4	7/64	3/32	16	26.1	7/64	9/64	14	17.2	7/64	5/32	12	15.2
2.36	3/32	...	...	...	...	3/32	1/16	18	33.0	3/32	3/32	16	22.4	3/32	1/8	14	16.7
2.36	3/32	3/32	1/16	22	33.0	3/32	3/32	20	22.4	3/32	1/8	18	16.7	3/32	5/32	16	12.8
2.36	3/32	3/32	1/16	20	33.0	3/32	3/32	18	22.4	3/32	1/8	16	16.7	3/32	5/32	14	12.8
2.00	0.078	...	...	...	...	0.078	0.030	18	47.3	0.078	0.047	16	35.3	0.078	0.078	14	22.4
2.00	0.078	0.078	0.030	22	47.3	0.078	0.047	20	35.3	0.078	0.078	18	22.4	0.078	0.109	16	15.8
2.00	0.078	0.078	0.030	20	47.3	0.078	0.047	18	35.3	0.078	0.078	16	22.4	0.078	0.109	14	15.8

ASTM E 674

TABLE 1 Continued

Perforated Opening		Medium Light				Medium				Medium Heavy				Heavy			
Standard (metric), mm	USA Industrial Standard, in.	Opening, in.	Bar, in.	Gage-Steel, in.	Open Area, %	Opening, in.	Bar, in.	Gage-Steel, in.	Open Area, %	Opening, in.	Bar, in.	Gage-Steel, in.	Open Area, %	Opening, in.	Bar, in.	Gage-Steel, in.	Open Area, %
1.70	0.066	...	...	...	...	...	...	...	...	0.066	0.043	18	33.2	0.066	0.059	16	25.3
1.70	0.066	...	...	...	...	0.066	0.043	22	33.2	0.066	0.059	20	25.3	0.066	0.090	18	16.2
1.70	0.066	...	...	...	...	0.066	0.043	20	33.2	0.066	0.059	18	25.3	0.066	0.090	16	16.2
1.40	0.055	...	...	...	...	...	...	...	...	0.055	0.040	20	30.4	0.055	0.055	18	22.6
1.40	0.055	...	...	...	...	0.055	0.040	24	30.4	0.055	0.055	22	22.6	0.055	0.070	20	17.5
1.40	0.055	...	...	...	...	0.055	0.040	22	30.4	0.055	0.055	20	22.6	0.055	0.070	18	17.5
1.18	0.045	...	...	...	...	...	...	...	...	0.045	0.021	22	42.1	0.045	0.033	20	30.2
1.18	0.045	...	...	...	...	0.045	0.021	26	42.1	0.045	0.033	24	30.2	0.045	0.045	22	22.4
1.18	0.045	...	...	...	...	0.045	0.021	24	42.1	0.045	0.033	22	30.2	0.045	0.045	20	22.4
1.00	0.039	...	...	...	...	...	...	...	...	...	...	...	...	0.039	0.027	22	31.6
1.00	0.039	...	...	...	...	...	...	...	...	0.039	0.027	26	31.6	0.039	0.039	24	22.4
1.00	0.039	...	...	...	...	...	...	...	...	0.039	0.027	24	31.6	0.039	0.039	22	22.4
830	0.032	...	...	...	...	...	...	...	...	...	...	...	...	0.032	0.032	24	22.4
830	0.032	...	...	...	...	...	...	...	...	0.032	0.032	28	22.4	0.032	0.040	26	17.9
830	0.032	...	...	...	...	...	...	...	...	0.032	0.032	26	22.4	0.032	0.040	24	17.9
710	0.027	...	...	...	...	...	...	...	...	...	...	...	...	0.027	0.030	26	20.3
710	0.027	...	...	...	...	...	...	...	...	0.027	0.030	30	20.3	0.027	0.039	28	15.2
710	0.027	...	...	...	...	...	...	...	...	0.027	0.030	28	20.3	0.027	0.039	26	15.2
600	0.023	...	...	...	...	...	...	...	...	...	...	...	...	0.023	...	...	...
600	0.023	...	...	...	...	...	...	...	...	...	...	...	...	0.023	0.032	30	15.8
600	0.023	...	...	...	...	...	...	...	...	...	...	...	...	0.023	0.032	28	15.8
500	0.020	...	...	...	...	...	...	...	...	...	...	...	...	0.020	...	...	...
500	0.020	...	...	...	...	...	...	...	...	...	...	...	...	0.020	0.025	30	17.9
500	0.020	...	...	...	...	...	...	...	...	...	...	...	...	0.020	0.025	28	17.9

E 674

TABLE 2 Tolerances on Openings of USA Standard Specifications for Industrial Perforated Plate and Screens

Perforated Opening			Tolerance on Openings	
Standard (metric), mm	USA Industrial Standard, in.	Additional Sizes, in.	Standard (metric), mm	USA Industrial Standard, in.
125.0	5	...	±2.5	±0.100
...	...	4¹/₂	...	±0.090
106.0	4¹/₄	...	±2.1	±0.085
100.0	4	...	±2.0	±0.080
...	...	3³/₄	...	±0.075
90.0	3¹/₂	...	±1.8	±0.070
...	...	3¹/₄	...	±0.065
75.0	3	...	±1.5	±0.060
...	...	2³/₄	...	±0.055
63.0	2¹/₂	...	±1.3	±0.050
...	...	2¹/₄	...	±0.045
53.0	2¹/₈	...	±1.1	±0.043
50.0	2	...	±1.0	±0.040
...	...	1⁷/₈	...	±0.038
45.0	1³/₄	...	±0.9	±0.035
...	...	1⁵/₈	...	±0.033
37.5	1¹/₂	...	±0.8	±0.030
...	...	1³/₈	...	±0.028
31.5	1¹/₄	...	±0.6	±0.025
...	...	1³/₁₆	...	±0.024
...	...	1¹/₈	...	±0.023
26.5	1¹/₁₆	...	±0.5	±0.021
25.0	1	...	±0.5	±0.020
...	...	¹⁵/₁₆	...	±0.019
22.4	⁷/₈	...	±0.46	±0.018
...	...	¹³/₁₆	...	±0.016
19.0	³/₄	...	±0.38	±0.015
...	...	¹¹/₁₆	...	±0.014
16.0	⁵/₈	...	±0.32	±0.013
...	...	⁹/₁₆	...	±0.012
13.2	¹⁷/₃₂	...	±0.30	±0.012
12.5	¹/₂	...	±0.28	±0.011
...	...	¹⁵/₃₂	...	±0.011
11.2	⁷/₁₆	...	±0.28	±0.011
9.5	³/₈	...	±0.28	±0.010
8.0	⁵/₁₆	...	±0.26	±0.010
6.7	¹⁷/₆₄	...	±0.25	±0.009
6.3	¹/₄	...	±0.25	±0.009
5.6	⁷/₃₂	...	±0.24	±0.009
4.75	³/₁₆	...	±0.21	±0.008
4.00	⁵/₃₂	...	±0.19	±0.007
3.35	0.127 (¹/₈)	...	±0.17	±0.006
2.80	⁷/₆₄	...	±0.150	±0.006
2.36	³/₃₂	...	±0.135	±0.005
2.00	0.078	...	±0.125	±0.005
1.70	0.066	...	±0.110	±0.004
1.40	0.055	...	±0.100	±0.004
1.18	0.045	...	±0.085	±0.003
1.00	0.039	...	±0.070	±0.003
830 μm	0.032	...	±60 μm	±0.002
710	0.027	...	±50 μm	±0.002
600	0.023	...	±45 μm	±0.002
500	0.020	...	±40 μm	±0.002

TABLE 3 Tolerances on Bars of USA Standard Specifications for Industrial Perforated Plate and Screens

Perforated Opening			Tolerance on Average Bar	
Standard (metric), mm	USA Industrial Standard, in.	Additional Sizes, in.	Standard (metric), mm	USA Industrial Standard, in.
125.0	5	...	±3.2	±0.125
...	...	4¹/₂	...	±0.122
106.0	4¹/₄	...	±2.9	±0.113
100.0	4	...	±2.7	±0.107
...	...	3³/₄	...	±0.102
90.0	3¹/₂	...	±2.5	±0.097
...	...	3¹/₄	...	±0.089
75.0	3	...	±2.1	±0.081
...	...	2³/₄	...	±0.076
63.0	2¹/₂	...	±1.8	±0.069
...	...	2¹/₄	...	±0.063
53.0	2¹/₈	...	±1.5	±0.059
50.0	2	...	±1.4	±0.056
...	...	1⁷/₈	...	±0.054
45.0	1³/₄	...	±1.3	±0.051
...	...	1⁵/₈	...	±0.047
37.5	1¹/₂	...	±1.1	±0.043
...	...	1³/₈	...	±0.040
31.5	1¹/₄	...	±0.9	±0.037
...	...	1³/₁₆	...	±0.035
...	...	1¹/₈	...	±0.034
26.5	1¹/₁₆	...	±0.8	±0.032
25.0	1	...	±0.8	±0.030
...	...	¹⁵/₁₆	...	±0.029
22.4	⁷/₈	...	±0.7	±0.028
...	...	¹³/₁₆	...	±0.026
19.0	³/₄	...	±0.6	±0.024
...	...	¹¹/₁₆	...	±0.022
16.0	⁵/₈	...	±0.5	±0.021
...	...	⁹/₁₆	...	±0.019
13.2	¹⁷/₃₂	...	±0.46	±0.018
12.5	¹/₂	...	±0.44	±0.017
...	...	¹⁵/₃₂	...	±0.017
11.2	⁷/₁₆	...	±0.41	±0.016
9.5	³/₈	...	±0.36	±0.014
8.0	⁵/₁₆	...	±0.32	±0.013
6.7	¹⁷/₆₄	...	±0.29	±0.011
6.3	¹/₄	...	±0.28	±0.011
5.6	⁷/₃₂	...	±0.27	±0.011
4.75	³/₁₆	...	±0.23	±0.009
4.00	⁵/₃₂	...	±0.22	±0.009
3.35	0.127 (¹/₈)	...	±0.20	±0.008
2.80	⁷/₆₄	...	±0.18	±0.007
2.36	³/₃₂	...	±0.16	±0.006
2.00	0.078	...	±0.150	±0.006
1.70	0.066	...	±0.135	±0.005
1.40	0.055	...	±0.125	±0.005
1.18	0.045	...	±0.110	±0.004
1.00	0.039	...	±0.090	±0.004
830 μm	0.032	...	±80 μm	±0.003
710	0.027	...	±70 μm	±0.003
600	0.023	...	±65 μm	±0.003
500	0.020	...	±60 μm	±0.002

TABLE 4 Tolerance on Thickness of USA Standard Specifications for Industrial Perforated Plate and Screens

Gage			Tolerance on Gage	
Standard (metric), mm	USA Industrial Standard, in.	USA Industrial Decimal Equivalent, in.	Standard (metric), mm	USA Industrial Standard, in.
25.4	1		+1.00 −0.25	+0.040 −0.010
22.4	7/8		−0.25 +0.89	+0.035 −0.010
19.0	3/4		+0.84 −0.25	+0.033 −0.010
16.0	5/8		+0.79 −0.25	+0.031 −0.010
12.5	1/2		+0.76 −0.25	+0.030 −0.010
9.50	3/8		+0.66 −0.25	+0.026 −0.010
8.00	5/16		+0.64 −0.25	+0.025 −0.010
6.30	1/4		+0.53 −0.25	+0.021 −0.010
4.75	3/16		+0.51 −0.25	+0.020 −0.010
4.25	No. 8 USS gage	0.1644	±0.25	±0.010
3.35	10	0.1345	±0.25	±0.010
3.00	11	0.1196	±0.25	±0.010
2.65	12	0.1046	±0.25	±0.010
1.90	14	0.0747	±0.18	±0.007
1.52	16	0.0598	±0.13	±0.005
1.21	18	0.0478	±0.10	±0.004
0.91	20	0.0359	±0.08	±0.003
0.76	22	0.0299	±0.08	±0.003
0.61	24	0.0239	±0.08	±0.003
0.45	26	0.0179	±0.05	±0.002
0.38	28	0.0149	±0.05	±0.002
0.30	30	0.0120	±0.05	±0.002

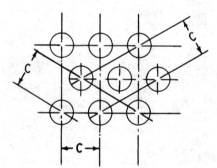

FIG. 1 Arrangement of Round Apertures.

APPENDIXES

X1. EQUIVALENT METRIC SPECIFICATIONS

X1.1 The metric equivalent to the USA Standard Specification for Perforated Plate and Screens is shown in Table X1.1 and is published for the convenience of users in countries on full metric standard and to facilitate comparison of the USA Standard with national standards on locally available specifications.

X1.2 In the metric table, the standard opening designations and bar sizes are identical to those in the USA Standard (Table 1), but the thicknesses of material are the nearest Standard ISO metric plate gages to the one used in the corresponding specification in the USA Standard.

X1.3 The tolerances on openings, average bar, and thickness of material are as shown in the appropriate columns of Tables 2, 3, and 4.

TABLE X1.1 USA Standard Specifications for Industrial Perforated Plate and Screens (Round Opening Series)—(Metric Equivalents)

Perforated Opening		Medium Light				Medium				Medium Heavy				Heavy			
Standard (metric), mm	USA Industrial Standard, in.	Opening, mm	Bar, mm	Gage Steel, mm	Open Area, %	Opening, mm	Bar, mm	Gage Steel, mm	Open Area, %	Opening, mm	Bar, mm	Gage Steel, mm	Open Area, %	Opening, mm	Bar, mm	Gage Steel, mm	Open Area, %
125	5.0	125	12.50	12.50	74.9	125	16.00	16.00	71.3	125	19.00	19.00	68.3	125	25.40	25.40	62.6
125	5.0	125	16.00	9.50	71.3	125	19.00	12.50	68.3	125	22.40	16.00	65.2	125	28.50	22.40	60.1
125	5.0	125	16.00	12.50	71.3	125	19.00	16.00	68.3	125	22.40	19.00	65.2	125	28.50	25.40	60.1
106	4.25	106	12.50	12.50	72.5	106	16.00	16.00	68.4	106	19.00	19.00	65.2	106	25.40	25.40	59.0
106	4.25	106	16.00	9.50	68.4	106	19.00	12.50	65.2	106	22.40	16.00	61.8	106	28.50	22.40	56.3
106	4.25	106	16.00	12.50	68.4	106	19.00	16.00	65.2	106	22.40	19.00	61.8	106	28.50	25.40	56.3
100	4.0	100	12.50	12.50	71.7	100	16.00	16.00	67.4	100	19.00	19.00	67.4	100	25.40	25.40	57.7
100	4.0	100	16.00	9.50	67.4	100	19.00	12.50	64.0	100	22.40	16.00	60.5	100	28.50	22.40	54.9
100	4.0	100	16.00	12.50	67.4	100	19.00	16.00	64.0	100	22.40	19.00	60.5	100	28.50	25.40	54.9
90	3.5	90	12.50	12.50	69.9	90	16.00	16.00	65.4	90	19.00	19.00	61.8	90	22.40	22.40	58.1
90	3.5	90	16.00	9.50	65.4	90	19.00	12.50	61.8	90	22.40	16.00	58.1	90	25.40	19.00	55.2
90	3.5	90	16.00	12.50	65.4	90	19.00	16.00	61.8	90	22.40	19.00	58.1	90	25.40	22.40	55.2
75	3.0	75	9.50	9.50	71.4	75	12.50	12.50	66.6	75	16.00	16.00	61.6	75	19.00	19.00	57.7
75	3.0	75	12.50	8.00	66.6	75	16.00	9.50	61.6	75	19.00	12.50	57.7	75	22.40	16.00	53.8
75	3.0	75	12.50	9.50	66.6	75	16.00	12.50	61.6	75	19.00	16.00	57.7	75	22.40	19.00	53.8
63	2.5	63	9.50	9.50	68.5	63	12.50	12.50	63.1	63	16.00	16.00	57.7	63	19.00	19.00	53.5
63	2.5	63	12.50	8.00	63.1	63	16.00	9.50	57.7	63	19.00	12.50	53.5	63	22.40	16.00	49.4
63	2.5	63	12.50	9.50	63.1	63	16.00	12.50	57.7	63	19.00	16.00	53.5	63	22.40	19.00	49.4
53	2.125	53	8.00	8.00	68.5	53	12.50	12.50	65.2	53	12.50	12.50	59.4	53	16.00	16.00	53.5
53	2.125	53	9.50	6.30	65.2	53	16.00	8.00	59.4	53	16.00	9.50	53.5	53	19.00	12.50	49.1
53	2.125	53	9.50	8.00	65.2	53	16.00	9.50	59.4	53	16.00	12.50	53.5	53	19.00	16.00	49.1
50	2.0	50	8.00	8.00	67.4	50	9.50	9.50	64.0	50	12.50	12.50	58.0	50	16.00	16.00	52.1
50	2.0	50	9.50	6.30	64.0	50	12.50	8.00	58.0	50	16.00	9.50	52.1	50	19.00	12.50	47.6
50	2.0	50	9.50	8.00	64.0	50	12.50	9.50	58.0	50	16.00	12.50	52.1	50	19.00	16.00	47.6
45	1.75	45	8.00	8.00	65.4	45	9.50	9.50	61.8	45	12.50	12.50	55.5	45	16.00	16.00	49.4
45	1.75	45	9.50	6.30	61.8	45	12.50	8.00	55.5	45	16.00	9.50	49.4	45	19.00	12.50	44.8
45	1.75	45	9.50	8.00	61.8	45	12.50	9.50	55.5	45	16.00	12.50	49.4	45	19.00	16.00	44.8

TABLE X1.1 *Continued*

Perforated Opening		Medium Light				Medium				Medium Heavy				Heavy			
Standard (metric), mm	USA Industrial Standard, in.	Opening, mm	Bar, mm	Gage-Steel, mm	Open Area, %	Opening, mm	Bar, mm	Gage-Steel, mm	Open Area, %	Opening, mm	Bar, mm	Gage-Steel, mm	Open Area, %	Opening, mm	Bar, mm	Gage-Steel, mm	Open Area, %
37.5	1.5	37.5	6.30	6.30	66.5	37.5	8.00	8.00	61.6	37.5	9.50	9.50	57.7	37.5	12.50	12.50	51.0
37.5	1.5	37.5	8.00	4.75	61.6	37.5	9.50	6.30	57.7	37.5	12.50	8.00	51.0	37.5	16.00	9.50	44.6
37.5	1.5	37.5	8.00	6.30	61.6	37.5	9.50	8.00	57.7	37.5	12.50	9.50	51.0	37.5	16.00	12.50	44.6
31.5	1.25	31.5	6.30	6.30	63.0	31.5	8.00	8.00	57.7	31.5	9.50	9.50	53.5	31.5	12.50	12.50	46.5
31.5	1.25	31.5	8.00	4.75	57.7	31.5	9.50	6.30	53.5	31.5	12.50	8.00	46.5	31.5	16.00	9.50	39.9
31.5	1.25	31.5	8.00	6.30	57.7	31.5	9.50	8.00	53.5	31.5	12.50	9.50	46.5	31.5	16.00	12.50	39.9
26.5	1.06	26.5	4.75	4.75	65.2	26.5	6.30	6.30	59.2	26.5	8.00	8.00	53.5	26.5	9.50	9.50	49.1
26.5	1.06	26.5	6.30	4.25	59.2	26.5	8.00	4.75	53.5	26.5	9.50	6.30	49.1	26.5	12.50	8.00	41.9
26.5	1.06	26.5	6.30	4.75	59.2	26.5	8.00	6.30	53.5	26.5	9.50	8.00	49.1	26.5	12.50	9.50	41.9
25	1.0	25	4.75	4.75	64.0	25	6.30	6.30	57.9	25	8.00	8.00	52.1	25	9.50	9.50	47.6
25	1.0	25	6.30	4.25	57.9	25	8.00	4.75	52.1	25	9.50	6.30	47.6	25	12.50	8.00	40.3
25	1.0	25	6.30	4.75	57.9	25	8.00	6.30	52.1	25	9.50	8.00	47.6	25	12.50	9.50	40.3
22.4	0.875	22.4	4.75	4.25	61.7	22.4	6.30	6.30	55.2	22.4	8.00	8.00	49.2	22.4	9.50	9.50	44.7
22.4	0.875	22.4	6.30	4.25	55.2	22.4	8.00	4.75	49.2	22.4	9.50	6.30	44.7	22.4	12.50	8.00	37.3
22.4	0.875	22.4	6.30	4.75	55.2	22.4	8.00	6.30	49.2	22.4	9.50	8.00	44.7	22.4	12.50	9.50	37.3
19	0.750	19	4.75	4.25	58.0	19	6.30	6.30	51.1	19	8.00	8.00	44.9	19	9.50	9.50	40.3
19	0.750	19	6.30	4.25	51.1	19	8.00	4.75	44.9	19	9.50	6.30	40.3	19	12.50	8.00	33.0
19	0.750	19	6.30	4.75	51.1	19	8.00	6.30	44.9	19	9.50	8.00	40.3	19	12.50	9.50	33.0
16	0.625	16	4.00	4.25	58.0	16	4.75	4.75	53.9	16	6.30	6.30	46.7	16	8.00	8.00	40.3
16	0.625	16	4.75	3.35	53.9	16	6.30	4.25	46.7	16	8.00	4.75	40.3	16	9.50	6.30	35.7
16	0.625	16	4.75	4.25	53.9	16	6.30	4.75	46.7	16	8.00	6.30	40.3	16	9.50	8.00	35.7
13.2	0.531	...	...	...	...	13.2	4.00	3.35	53.4	13.2	4.75	4.75	53.4	13.2	5.55	4.75	44.9
13.2	0.531	...	...	...	...	13.2	4.00	4.25	53.4	13.2	5.55	4.25	44.9	13.2	8.75	6.30	32.8
13.2	0.531	...	...	...	...	...	...	...	...	13.2	5.55	4.75	44.9	13.2	8.75	6.30	32.8
12.5	0.500	...	...	...	...	12.5	4.75	3.35	47.6	12.5	4.75	4.75	47.6	12.5	6.30	6.30	40.1
12.5	0.500	...	...	...	...	12.5	4.75	4.25	47.6	12.5	6.30	4.25	40.1	12.5	8.00	4.75	33.7
12.5	0.500	...	...	...	...	12.5	4.75	4.25	47.6	12.5	6.30	4.75	40.1	12.5	8.00	6.30	33.7

TABLE X1.1 Continued

Perforated Opening		Medium Light				Medium				Medium Heavy				Heavy			
Standard (metric), mm	USA Industrial Standard, in.	Opening, mm	Bar, mm	Gage-Steel, mm	Open Area, %	Opening, mm	Bar, mm	Gage-Steel, mm	Open Area, %	Opening, mm	Bar, mm	Gage-Steel, mm	Open Area, %	Opening, mm	Bar, mm	Gage-Steel, mm	Open Area, %
11.2	0.438	11.2	4.00	3.35	49.2	11.2	4.75	4.25	44.7	11.2	6.30	4.75	37.1	11.2	8.00	6.30	30.8
11.2	0.438	11.2	4.75	3.00	44.7	11.2	6.30	3.35	37.1	11.2	8.00	4.25	30.8	11.2	11.20	4.75	22.7
11.2	0.438	11.2	4.75	3.35	44.7	11.2	6.30	4.25	37.1	11.2	8.00	4.75	30.8	11.2	11.20	6.30	22.7
9.5	0.375	9.5	3.15	3.00	51.1	9.5	4.75	3.35	40.3	9.5	5.55	4.25	36.1	9.5	6.30	4.75	32.8
9.5	0.375	9.5	4.75	2.65	40.3	9.5	5.55	3.00	36.1	9.5	6.30	3.35	32.8	9.5	9.50	4.25	22.7
9.5	0.375	9.5	4.75	3.00	40.3	9.5	5.55	3.35	36.1	9.5	6.30	4.25	32.8	9.5	9.50	4.75	22.7
8	0.313	8	2.36	3.00	54.1	8	3.15	3.35	46.7	8	4.00	4.25	40.3	8	4.75	4.75	35.7
8	0.313	8	3.15	2.65	46.7	8	4.00	3.00	40.3	8	4.75	3.35	35.7	8	6.30	4.25	28.4
8	0.313	8	3.15	3.00	46.7	8	4.00	3.35	40.3	8	4.75	4.25	35.7	8	6.30	4.75	28.4
6.7	0.266	...	...	...	...	6.7	2.77	3.00	45.4	6.7	3.15	3.00	42.0	6.7	3.54	4.25	38.8
6.7	0.266	6.7	2.77	1.90	45.4	6.7	3.15	2.65	42.0	6.7	3.54	3.00	38.8	6.7	4.36	3.35	33.3
6.7	0.266	6.7	2.77	2.65	45.4	6.7	3.15	3.00	42.0	6.7	3.54	3.35	38.8	6.7	4.36	4.25	33.3
6.3	0.250	6.3	1.59	1.52	57.8	6.3	3.15	3.00	40.3	6.3	4.00	3.35	33.9	6.3	4.75	4.25	29.5
6.3	0.250	6.3	3.15	1.90	40.3	6.3	4.00	2.65	33.9	6.3	4.75	3.00	29.5	6.3	6.30	3.35	22.7
6.3	0.250	6.3	3.15	2.65	40.3	6.3	4.00	3.00	33.9	6.3	4.75	3.35	29.5	6.3	6.30	4.25	22.7
5.6	0.219	5.6	2.36	1.90	44.9	5.6	3.15	2.65	37.1	5.6	4.00	3.00	30.9	5.6	4.75	3.35	26.5
5.6	0.219	5.6	3.15	1.52	37.1	5.6	4.00	1.90	30.9	5.6	4.75	2.65	26.5	5.6	5.55	3.00	22.9
5.6	0.219	5.6	3.15	1.90	37.1	5.6	4.00	2.65	30.9	5.6	4.75	3.00	26.5	5.6	5.55	3.35	22.9
4.75	0.188	4.75	1.59	1.90	50.9	4.75	2.36	2.65	40.5	4.75	2.77	3.00	36.2	4.75	3.15	3.35	32.8
4.75	0.188	4.75	2.36	1.52	40.5	4.75	2.77	1.90	36.2	4.75	3.15	2.65	32.8	4.75	4.75	3.00	22.7
4.75	0.188	4.75	2.36	1.90	40.5	4.75	2.77	2.65	36.2	4.75	3.15	3.00	32.8	4.75	4.75	3.35	22.7
4	0.156	...	...	...	...	4	1.59	1.90	46.4	4	2.36	2.65	35.9	4	3.15	3.00	28.4
4	0.156	4	1.59	1.21	46.4	4	2.36	1.52	35.9	4	3.15	1.90	28.4	4	4.00	2.65	22.7
4	0.156	4	1.59	1.52	46.4	4	2.36	1.90	35.9	4	3.15	2.65	28.4	4	4.00	3.00	22.7
3.35	0.127	...	...	...	...	3.35	1.19	1.90	49.4	3.35	1.59	2.65	41.7	3.35	2.36	3.00	31.2
3.35	0.127	3.35	1.19	1.21	49.4	3.35	1.59	1.52	41.7	3.35	2.36	1.90	31.2	3.35	3.15	2.65	24.1
3.35	0.127	3.35	1.19	1.52	49.4	3.35	1.59	1.90	41.7	3.35	2.36	2.65	31.2	3.35	3.15	3.00	24.1

TABLE XI.1 *Continued*

Perforated Opening		Medium Light				Medium				Medium Heavy				Heavy			
Standard (metric), mm	USA Industrial Standard, in.	Opening, mm	Bar, mm	Gage-Steel, mm	Open Area, %	Opening, mm	Bar, mm	Gage-Steel, mm	Open Area, %	Opening, mm	Bar, mm	Gage-Steel, mm	Open Area, %	Opening, mm	Bar, mm	Gage-Steel, mm	Open Area, %
2.8	0.109	...	...	...	...	2.80	1.59	1.52	36.9	2.8	2.36	1.90	26.7	2.8	3.54	2.65	17.7
2.8	0.109	2.80	1.59	0.91	36.9	2.80	2.36	1.21	26.7	2.8	3.54	1.52	17.7	2.8	4.00	1.90	15.4
2.8	0.109	2.80	1.59	1.21	36.9	2.80	2.36	1.52	26.7	2.8	3.54	1.90	17.7	2.8	4.00	2.65	15.4
2.36	0.095	...	...	...	...	2.36	1.59	1.21	32.4	2.36	2.36	1.52	22.7	2.36	3.15	1.90	16.6
2.36	0.095	2.36	1.59	0.76	32.4	2.36	2.36	0.91	22.7	2.36	3.15	1.21	16.6	2.36	4.00	1.52	12.5
2.36	0.095	2.36	1.59	0.91	32.4	2.36	2.36	1.21	22.7	2.36	3.15	1.52	16.6	2.36	4.00	1.90	12.5
2	0.078	...	...	...	...	2	0.76	1.21	47.6	2	1.19	1.52	35.7	2	2.00	1.90	22.7
2	0.078	2	0.76	0.76	47.6	2	1.19	0.91	35.7	2	2.00	1.21	22.7	2	2.77	1.52	16.0
2	0.078	2	0.76	0.91	47.6	2	1.19	1.21	35.7	2	2.00	1.52	22.7	2	2.77	1.90	16.0
1.7	0.066	...	...	...	...	...	...	...	...	1.7	1.09	1.21	33.7	1.7	1.50	1.52	25.6
1.7	0.066	...	...	...	...	1.7	1.09	0.76	33.7	1.7	1.50	0.91	25.6	1.7	2.29	1.21	16.4
1.7	0.066	...	...	...	...	1.7	1.09	0.91	33.7	1.7	1.50	1.21	25.6	1.7	2.29	1.52	16.4
1.4	0.055	...	...	...	...	...	...	...	...	1.4	1.02	0.91	30.4	1.4	1.40	1.21	22.7
1.4	0.055	...	...	...	...	1.4	1.02	0.61	30.4	1.4	1.40	0.76	22.7	1.4	1.78	0.91	17.6
1.4	0.055	...	...	...	...	1.4	1.02	0.76	30.4	1.4	1.40	0.91	22.7	1.4	1.78	1.21	17.6
1.18	0.045	...	...	...	...	...	...	...	...	1.18	0.53	0.76	43.1	1.18	0.84	0.91	30.9
1.18	0.045	...	...	...	...	1.18	0.53	0.45	43.1	1.18	0.84	0.61	30.9	1.18	1.14	0.76	23.4
1.18	0.045	...	...	...	...	1.18	0.53	0.61	43.1	1.18	0.84	0.76	30.9	1.18	1.14	0.91	23.4
1	0.039	...	...	...	...	...	...	...	...	...	...	...	...	1	0.68	0.76	32.1
1	0.039	...	...	...	...	...	...	...	...	1	0.68	0.45	32.1	1	1.00	0.61	22.7
1	0.039	...	...	...	...	...	...	...	...	1	0.68	0.61	32.1	1	1.00	0.76	22.7
830 µm	0.033	...	...	...	...	...	...	...	...	830 µm	...	...	...	830 µm	0.81	0.61	23.2
830	0.033	...	...	...	...	...	...	...	...	830	0.81	0.38	23.2	830	1.02	0.45	18.3
830	0.033	...	...	...	...	...	...	...	...	830	0.81	0.45	23.2	830	1.02	0.61	18.3
710	0.027	...	...	...	...	...	...	...	...	710	...	...	...	710	0.76	0.45	21.2
710	0.027	...	...	...	...	...	...	...	...	710	0.76	0.30	21.2	710	1.00	0.38	15.6
710	0.027	...	...	...	...	...	...	...	...	710	0.76	0.38	21.2	710	1.00	0.45	15.6

TABLE X1.1 *Continued*

Perforated Opening		Medium Light				Medium				Medium Heavy				Heavy			
Standard (metric), mm	USA Industrial Standard, in.	Opening, mm	Bar, mm	Gage-Steel, mm	Open Area, %	Opening, mm	Bar, mm	Gage-Steel, mm	Open Area, %	Opening, mm	Bar, mm	Gage-Steel, mm	Open Area, %	Opening, mm	Bar, mm	Gage-Steel, mm	Open Area, %
600	0.023	...	...	...	...	...	...	...	...	...	...	...	...	...	...	...	...
600	0.023	...	...	...	...	...	...	...	...	...	...	...	...	600	0.81	0.30	16.4
600	0.023	...	...	...	...	...	...	...	...	...	...	...	...	600	0.81	0.38	16.4
500	0.020	...	...	...	...	...	...	...	...	...	...	...	...	...	...	...	...
500	0.020	...	...	...	...	...	...	...	...	...	...	...	...	500	0.63	0.30	17.7
500	0.020	...	...	...	...	...	...	...	...	...	...	...	...	500	0.63	0.38	17.7

X2. MATERIALS USED FOR INDUSTRIAL PERFORATED PLATE AND SCREENS

X2.1 With a few exceptions, any material normally furnished in a flat sheet or plate can be successfully perforated, but the thickness of the material may have to be adjusted to keep the tonnage required for perforating within the limits of the perforating supplier's equipment.

X2.2 In selecting the material, it is recommended that the user consider the factors required in the end use of his product, such as abrasion resistance, corrosion resistance, impact resistance, weldability, formability, appearance, etc. When a suitable material has been selected, the perforating supplier should be contacted to determine the feasibility of supplying the required complete specification at reasonable cost.

X3. CHECKING AND CALIBRATING PERFORATED METAL AND SCREENS

X3.1 In checking specifications of industrial perforated plate it is necessary to determine three dimensions: (a) the dimension of the opening, (b) the dimensions of the bar, and (c) the thickness of the plate.

X3.2 *Dimensions of Openings:*

X3.2.1 For openings from 5 in. (125 mm) to ¼ in. (6.3 mm), measure the opening using a steel rule graduated to ¹⁄₆₄ in. or 0.5 mm or use a vernier inside caliper graduated to 0.001 in. or 0.02 mm. Since any perforated opening has an inherent taper, care should be taken to make all measurements from the surface of the plate that was uppermost during the punching operation (smoothside).

X3.2.2 For openings from ¼ in. (6.3 mm) to 0.127 in. (3.35 mm) measure the opening using a steel rule graduated to ¹⁄₆₄ in. or 0.5 mm.

X3.2.3 For openings from ⁷⁄₆₄ in. (2.80 mm) to 0.020 in. (500 μm), measure the opening using a taper gage similar to that shown in Fig. X3.1.

X3.3 *Dimensions of Bars:*

X3.3.1 Bars from 1⅛ in. (28.5 mm) to ³⁄₃₂ in. (2.36 mm) may be checked with a steel rule graduated to ¹⁄₆₄ in. (0.4 mm) or with a vernier outside caliper graduated to 0.001 in. (0.02 mm) if the nibs will fit into the adjacent openings.

X3.3.2 Bars from 0.090 in. (2.29 mm) to 0.021 in. (0.53 mm) may be checked by use of a steel rule graduated in decimals of an inch or in metric dimensions. The procedure used is to line up the scale along a straight row of openings. Set the end of the scale exactly at the edge of a given opening. Follow along the scale until the same edge of another opening lines up exactly with another graduation on the scale. Count the number of openings in this dimension and divide this dimension by the number of included openings. From this figure, subtract the diameter of the opening and the resultant figure will be the dimension of the average bar along that straight row of openings. In order to check the consistency of the bar dimensions in all directions, it is necessary to rotate the scale 60 deg and repeat the above procedure along the new straight row of openings.

X3.4 *Thickness of Plate*—Thickness of plate may be checked with a steel rule graduated to ¹⁄₆₄ in. or 0.5 mm or with a micrometer graduated to 0.001 in. or 0.02 mm.

FIG. X3.1 Kwik-Chek Hole Gage.

X4. TERMS RELATING TO INDUSTRIAL PERFORATED PLATE AND SCREENS

X4.1 General

aperture or opening—dimensions defining an opening in a screen.

open area—ratio of the total area of the apertures to the total area of the screen, usually expressed in percentage.

screen—(a) a surface provided with openings of uniform size. (b) a machine provided with one or more screen surfaces.

screening—the process of separating a mixture of different sizes by means of one or more screen surfaces.

X4.2 Perforated Plate or Screens

perforation—an aperture or opening produced by punching.

bar—the metal between perforations measured at the point where perforations are the closest.

centers—the dimensional sum of one perforation and one bar, or the dimensional distance from the center of one perforation to the center of an adjacent perforation.

smooth side or punch side—the surface of the plate that was uppermost during the punching operation and through which the punch entered the plate.

break-out—a term applied to the action that occurs ahead of the punch in its going through the plate. The fracturing of the material results in a tapered hole with the small dimensions on the punch side.

die side—the surface of the plate that was against the die during the punching operation.

blank—an unperforated area located other than along the perimeter of a plate.

margin or border—an unperforated area located along the perimeter of a plate.

unfinished end pattern—the condition that occurs with some specifications of staggered pattern perforations as a result of tool design. On one end of the plate, the pattern will appear to be incomplete due to unperforated holes in the even numbered rows, while on the other end of the same plate, the pattern will appear to be incomplete due to unperforated holes in the odd numbered rows (Fig. X4.1).

finished end pattern—the condition that occurs with some specifications of staggered pattern perforations as a result of tool design where the pattern is completed on both ends of the plate (Fig. X4.2).

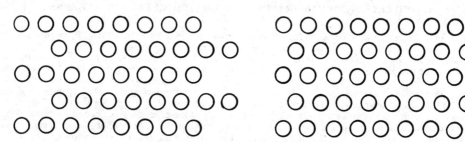

Arrangement of Staggered Pattern Openings

FIG. X4.1 Unfinished End Pattern.   FIG. X4.2 Finished End Pattern.

The American Society for Testing and Materials takes no position respecting the validity of any patent rights asserted in connection with any item mentioned in this standard. Users of this standard are expressly advised that determination of the validity of any such patent rights, and the risk of infringement of such rights, are entirely their own responsibility.

This standard is subject to revision at any time by the responsible technical committee and must be reviewed every five years and if not revised, either reapproved or withdrawn. Your comments are invited either for revision of this standard or for additional standards and should be addressed to ASTM Headquarters. Your comments will receive careful consideration at a meeting of the responsible technical committee, which you may attend. If you feel that your comments have not received a fair hearing you should make your views known to the ASTM Committee on Standards, 1916 Race St., Philadelphia, Pa. 19103.

RELATED MATERIAL

Any documents included in this section that are marked "proposed" are published for information only. They have received the approval of the sponsoring technical committee for publication as "proposed," but have not been officially accepted by the Society. Comments are solicited and should be addressed to ASTM, 1916 Race St., Philadelphia, Pa. 19103.

Excerpts from Standard for
METRIC PRACTICE[1]

Following are excerpts from Standard for Metric Practice E 380, which is available as a separate publication and which appears in its entirety in Part 41. Deleted are Appendixes X1, X2, X3, and X4. Added is a table of selected conversion factors from Appendix X3.

CONTENTS

	Section
Scope	1
SI Units and Symbols	2
Classes of Units	2.1
Base Units	2.2
Supplementary Units	2.3
Derived Units	2.4
SI Prefixes	2.5
Application of the Metric System	3
General	3.1
Application of SI Prefixes	3.2
Other Units	3.3
Other Recommendations Concerning Units	3.4
Style and Usage	3.5
Rules for Conversion and Rounding	4
General	4.1
Accuracy and Rounding	4.2
Significant Digits	4.3
Rounding Values	4.4
Conversion of Linear Dimensions of Interchangeable Parts	4.5
Other Units	4.6
Terminology	5
Appendixes	
Development of the International System of Units	X1
Organs of the Metre Convention: BIPM, CIPM, CGPM	X2
Conversion Factors	X3
Supplementary Metric Practice Guides	X4
Bibliography	
Index	

---

[1] This standard is under the jurisdiction of ASTM Committee E-43 on Metric Practice and is the direct responsibility of Subcommittee E43.10, Technical.
    Current edition approved Nov. 10, 1981, and March 1, 1982. Published March 1982. Originally published in 1964 without designation as ASTM *Metric Practice Guide*, revised 1966. Adopted as standard 1968. Last previous edition E 380 – 79.

1. Scope

1.1 This standard gives guidance for application of the modernized metric system in the United States. The International System of Units, developed and maintained by the General Conference on Weights and Measures (abbreviated CGPM from the official French name Conférence Générale des Poids et Mesures) is intended as a basis for worldwide standardization of measurement units. The name International System of Units and the international abbreviation SI[2] were adopted by the 11th CGPM in 1960. SI is a complete, coherent system that is being universally adopted.

1.2 Information is included on SI, a limited list of non-SI units recognized for use with SI units, and a list of conversion factors from non-SI to SI units, together with general guidance on proper style and usage.

1.3 It is hoped that an understanding of the system and its characteristics, and careful use according to this standard, will help to avoid the degradation that has occurred in all older measurement systems.

2. SI Units and Symbols

2.1 *Classes of Units*—SI units are divided into three classes:
- base units
- supplementary units
- derived units

2.2 *Base Units*—SI is based on seven well-defined units (see Table 1) which by convention are regarded as dimensionally independent.

2.3 *Supplementary Units*—This class contains two units, the radian and the steradian (see Table 2). At the time of the introduction of the International System, the 11th CGPM left open the question of the nature of these supplementary units. Considering that plane angle is generally expressed as the ratio between two lengths and solid angle as the ratio between an area and the square of length, the CIPM (1980) specified that in the International System the quantities plane angle and solid angle should be considered as dimensionless derived quantities. Therefore, the supplementary units radian and steradian are to be regarded as dimensionless derived units which may be used or omitted in the expressions for derived units.

2.4 *Derived Units*:

TABLE 1 Base SI Units

Quantity[3]	Unit	Symbol
length	metre	m
mass	kilogram	kg
time	second	s
electric current	ampere	A
thermodynamic temperature[A]	kelvin	K
amount of substance	mole	mol
luminous intensity	candela	cd

[A] For a discussion of Celsius temperature see 3.4.2.

TABLE 2 Supplementary SI Units

Quantity[3]	Unit	Symbol
plane angle	radian	rad
solid angle	steradian	sr

2.4.1 Derived units are formed by combining base units, supplementary units, and other derived units according to the algebraic relations linking the corresponding quantities. The symbols for derived units are obtained by means of the mathematical signs for multiplication, division, and use of exponents. For example, the SI unit for velocity is the metre per second (m/s or $m \cdot s^{-1}$), and that for angular velocity is the radian per second (rad/s or $rad \cdot s^{-1}$).

2.4.2 Those derived SI units which have special names and symbols approved by the CGPM are listed in Table 3.

2.4.3 It is frequently advantageous to express derived units in terms of other derived units with special names; for example, the SI unit for electric dipole moment is usually expressed as $C \cdot m$ instead of $A \cdot s \cdot m$.

2.4.4 Some common derived units are listed in Table 4.

2.5 *SI Prefixes* (see 3.2 for application):

2.5.1 The prefixes and symbols listed in Table 5 are used to form names and symbols of the decimal multiples and submultiples of the SI units except for kilogram.

2.5.2 *Unit of Mass*—Among the base and derived units of SI, the unit of mass (kilogram) is the only one whose name, for historical reasons, contains a prefix. Names of decimal multiples and submultiples of the unit of mass are formed by attaching prefixes to the word *gram* (g).

---

[2] From the French name, Le Système International d'Unités.

[3] "Quantity" as used in the headings of the tables of this standard means measurable attribute of phenomena or matter.

E 380

TABLE 3 Derived SI Units with Special Names

Quantity[3]	Unit	Symbol	Formula
frequency (of a periodic phenomenon)	hertz	Hz	1/s
force	newton	N	$kg \cdot m/s^2$
pressure, stress	pascal	Pa	N/m^2
energy, work, quantity of heat	joule	J	$N \cdot m$
power, radiant flux	watt	W	J/s
quantity of electricity, electric charge	coulomb	C	$A \cdot s$
electric potential, potential difference, electromotive force	volt	V	W/A
electric capacitance	farad	F	C/V
electric resistance	ohm	Ω	V/A
electric conductance	siemens	S	A/V
magnetic flux	weber	Wb	$V \cdot s$
magnetic flux density	tesla	T	Wb/m^2
inductance	henry	H	Wb/A
Celsius temperature	degree Celsius[A]	°C	K[see 3.4.2]
luminous flux	lumen	lm	$cd \cdot sr$
illuminance	lux	lx	lm/m^2
activity (of a radionuclide)	becquerel	Bq	1/s
absorbed dose[B]	gray	Gy	J/kg
dose equivalent	sievert	Sv	J/kg

[A] Inclusion in the table of derived SI units with special names approved by the CIPM in 1976.
[B] Related quantities using the same unit are: specific energy imparted, kerma, and absorbed dose index.

2.5.3 These prefixes or their symbols are directly attached to names or symbols of units, forming multiples and submultiples of the units. In strict terms these must be called "multiples and submultiples of SI units," particularly in discussing the coherence of the system (see Section 5). In common parlance, the base units and derived units, along with their multiples and submultiples, are all called SI units.

3. Application of the Metric System

3.1 *General*—SI is the form of the metric system that is preferred for all applications. It is important that this modernized form of the metric system be thoroughly understood and properly applied. Obsolete metric units and practices are widespread, particularly in those countries that long ago adopted the metric system, and much usage is improper. This section gives guidance concerning the limited number of cases in which units outside SI are appropriately used, and makes recommendations concerning usage and style.

3.2 *Application of SI Prefixes:*

3.2.1 *General*—In general the SI prefixes (2.5) should be used to indicate orders of magnitude, thus eliminating nonsignificant digits and leading zeros in decimal fractions, and providing a convenient alternative to the powers-of-ten notation preferred in computation. For example:

12 300 mm becomes 12.3 m
12.3×10^3 m becomes 12.3 km
0.00123 µA becomes 1.23 nA

3.2.2 *Selection*—When expressing a quantity by a numerical value and a unit, a prefix should preferably be chosen so that the numerical value lies between 0.1 and 1000. To minimize variety, it is recommended that prefixes representing 1000 raised to an integral power be used. However, three factors may justify deviation from the above:

3.2.2.1 In expressing area and volume, the prefixes hecto-, deka-, deci-, and centi- may be required, for example, square hectometre, cubic centimetre.

3.2.2.2 In tables of values of the same quantity, or in a discussion of such values within a given context, it is generally preferable to use the same unit multiple throughout.

3.2.2.3 For certain quantities in particular applications, one particular multiple is customarily used. For example, the millimetre is used for linear dimensions in mechanical engineering drawings even when the values lie far outside the range 0.1 to 1000 mm; the centimetre is often used for body measurements and clothing sizes.

3.2.3 *Prefixes in Compound Units*[4]—It is recommended that only one prefix be used in forming a multiple of a compound unit. Normally the prefix should be attached to a unit in the numerator. One exception to this is when the kilogram occurs in the denominator.
Examples:

V/m, *not* mV/mm, and MJ/kg, *not* kJ/g

3.2.4 *Compound Prefixes*—Compound prefixes, formed by the juxtaposition of two or more SI prefixes are not to be used. For example, use

1 nm, *not* 1 mµm
1 pF, *not* 1 µµF

[4] A compound unit is a derived unit that is expressed in terms of two or more units rather than by a single special name.

ASTM E 380

TABLE 4 Some Common Derived Units of SI

Quantity[3]	Unit	Symbol
absorbed dose rate	gray per second	Gy/s
acceleration	metre per second squared	m/s^2
angular acceleration	radian per second squared	rad/s^2
angular velocity	radian per second	rad/s
area	square metre	m^2
concentration (of amount of substance)	mole per cubic metre	mol/m^3
current density	ampere per square metre	A/m^2
density, mass	kilogram per cubic metre	kg/m^3
electric charge density	coulomb per cubic metre	C/m^3
electric field strength	volt per metre	V/m
electric flux density	coulomb per square metre	C/m^2
energy density	joule per cubic metre	J/m^3
entropy	joule per kelvin	J/K
exposure (X and gamma rays)	coulomb per kilogram	C/kg
heat capacity	joule per kelvin	J/K
heat flux density / irradiance	watt per square metre	W/m^2
luminance	candela per square metre	cd/m^2
magnetic field strength	ampere per metre	A/m
molar energy	joule per mole	J/mol
molar entropy	joule per mole kelvin	$J/(mol \cdot K)$
molar heat capacity	joule per mole kelvin	$J/(mol \cdot K)$
moment of force[A]	newton metre	$N \cdot m$
permeability (magnetic)	henry per metre	H/m
permittivity	farad per metre	F/m
power density	watt per square metre	W/m^2
radiance	watt per square metre steradian	$W/(m^2 \cdot sr)$
radiant intensity	watt per steradian	W/sr
specific heat capacity	joule per kilogram kelvin	$J/(kg \cdot K)$
specific energy	joule per kilogram	J/kg
specific entropy	joule per kilogram kelvin	$J/(kg \cdot K)$
specific volume	cubic metre per kilogram	m^3/kg
surface tension	newton per metre	N/m
thermal conductivity	watt per metre kelvin	$W/(m \cdot K)$
velocity	metre per second	m/s
viscosity, dynamic	pascal second	$Pa \cdot s$
viscosity, kinematic	square metre per second	m^2/s
volume	cubic metre	m^3
wave number	1 per metre	1/m

[A] See 3.4.4.

TABLE 5 SI Prefixes

Multiplication Factor	Prefix	Symbol
$1\ 000\ 000\ 000\ 000\ 000\ 000 = 10^{18}$	exa	E
$1\ 000\ 000\ 000\ 000\ 000 = 10^{15}$	peta	P
$1\ 000\ 000\ 000\ 000 = 10^{12}$	tera	T
$1\ 000\ 000\ 000 = 10^9$	giga	G
$1\ 000\ 000 = 10^6$	mega	M
$1\ 000 = 10^3$	kilo	k
$100 = 10^2$	hecto[A]	h
$10 = 10^1$	deka[A]	da
$0.1 = 10^{-1}$	deci[A]	d
$0.01 = 10^{-2}$	centi[A]	c
$0.001 = 10^{-3}$	milli	m
$0.000\ 001 = 10^{-6}$	micro	μ
$0.000\ 000\ 001 = 10^{-9}$	nano	n
$0.000\ 000\ 000\ 001 = 10^{-12}$	pico	p
$0.000\ 000\ 000\ 000\ 001 = 10^{-15}$	femto	f
$0.000\ 000\ 000\ 000\ 000\ 001 = 10^{-18}$	atto	a

[A] To be avoided where practical, except as noted in 3.2.2.

E 380

TABLE 6 Units in Use with SI

Quantity[3]	Unit	Symbol	Definition
time	minute	min	1 min = 60 s
	hour	h	1 h = 60 min = 3600 s
	day	d	1 d = 24 h = 86 400 s
	week, month, etc.	—	
plane angle	degree	°	1° = (π/180) rad
	minute[A]	′	1′ = (1/60)°
			= (π/10 800) rad
	second[A]	″	1″ = (1/60)′
			= (π/648 000) rad
volume	litre[B]	L	1 L = 1 dm^3 = 10^{-3} m^3
mass	metric ton	t	1 t = 10^3 kg
area	hectare	ha	1 ha = 1 hm^2 = 10^4 m^2

[A] Use discouraged except for special fields such as cartography.
[B] See 3.3.2.4.

If values are required outside the range covered by the prefixes, they should be expressed by using powers of ten applied to the base unit.

3.2.5 *Powers of Units*—An exponent attached to a symbol containing a prefix indicates that the multiple or submultiple of the unit (the unit with its prefix) is raised to the power expressed by the exponent. For example:

$$1 \text{ cm}^3 = (10^{-2} \text{ m})^3 = 10^{-6} \text{ m}^3$$
$$1 \text{ ns}^{-1} = (10^{-9} \text{ s})^{-1} = 10^9 \text{ s}^{-1}$$
$$1 \text{ mm}^2/\text{s} = (10^{-3} \text{ m})^2/\text{s} = 10^{-6} \text{ m}^2/\text{s}$$

3.2.6 *Calculations*—Errors in calculations can be minimized if the base and the coherent derived SI units are used and the resulting numerical values are expressed in powers-of-ten notation instead of using prefixes.

3.3 *Other Units*:

3.3.1 *Units from Different Systems*—To assist in preserving the advantage of SI as a coherent system, it is advisable to minimize the use with it of units from other systems. Such use should be limited to units listed in this section.

3.3.2 *Units in Use with SI (see Table 6)*:

3.3.2.1 *Time*—The SI unit of time is the second. This unit is preferred and should be used if practical, particularly when technical calculations are involved. In cases where time relates to life customs or calendar cycles, the minute, hour, day, and other calendar units may be necessary. For example, vehicle speed will normally be expressed in kilometres per hour.

3.3.2.2 *Plane Angle*—The SI unit for plane angle is the radian. Use of the degree and its decimal submultiples is permissible when the radian is not a convenient unit. Use of the minute and second is discouraged except for special fields such as cartography.

3.3.2.3 *Area*—The SI unit of area is the square metre (m^2). The hectare (ha) is a special name for square hectometre (hm^2). Large land or water areas are generally expressed in hectares or in square kilometres (km^2).

3.3.2.4 *Volume*—The SI unit of volume is the cubic metre. This unit, or one of the regularly formed multiples such as the cubic centimetre, is preferred. The special name *litre*[5] (L)[6] has been approved for the cubic decimetre, but use of this unit is restricted to volumetric capacity, dry measure, and measure of fluids (both gases and liquids). No prefix other than milli- or micro- should be used with litre.

3.3.2.5 *Mass*—The SI unit of mass is the kilogram. This unit, or one of the multiples formed by attaching an SI prefix to *gram* (g), is preferred for all applications. The megagram (Mg) is the appropriate unit for measuring large masses such as have been expressed in tons. However, the name *ton* has been given to several large mass units that are widely used in commerce and technology—the long ton of 2240 lb, the short ton of 2000 lb, and metric ton of 1000 kg (also called the *tonne*). None of these terms are SI. The term *metric ton* should be restricted to commercial usage, and no prefixes should be used with it. Use of the term *tonne* is deprecated.

3.3.3 *Units in Use with SI Temporarily* (see Table 7):

3.3.3.1 *Energy*—The SI unit of energy, the

---

[5] See Appendix X1.11.1.
[6] The CGPM in October 1979 approved L and l as alternative symbols for litre. Since the letter symbol l can easily be confused with the numeral 1, only the symbol L is recommended for USA use.

TABLE 7 Units in Use with SI Temporarily

Quantity[3]	Unit	Symbol	Definition
energy [see 3.3.3.1]	kilowatthour	kWh	1 kWh = 3.6 MJ
cross section	barn	b	1 b = 10^{-28} m^2
pressure [see 3.3.3.2]	bar	bar	1 bar = 10^5 Pa
activity (of a radionuclide)	curie	Ci	1 Ci = 3.7×10^{10} Bq
exposure (X and gamma rays)	roentgen	R	1 R = 2.58×10^{-4} C/kg
absorbed dose	rad	rd	1 rd = 0.01 Gy

joule, together with its multiples, is preferred for all applications. The kilowatthour is widely used, however, as a measure of electric energy. This unit should not be introduced into any new areas, and eventually it should be replaced by the megajoule.

3.3.3.2 *Pressure and Stress*—The SI unit of pressure and stress is the pascal (newton per square metre) and with proper SI prefixes is applicable to all such measurements. Old metric gravitational units for pressure and stress such as kilogram-force per square centimetre (kgf/cm^2) shall not be used. Widespread use has been made of other non-SI units such as bar and torr for pressure, but this use is strongly discouraged. The millibar is widely used in meteorology; this usage will continue for the present in order to permit meteorologists to communicate easily within their profession. The kilopascal should be used in presenting meteorological data to the public.

3.3.4 *Units and Names to Be Abandoned*—A great many metric units other than those of the SI have been defined over the years. Some of these are used only in special fields; others have found broad application in countries that adopted the metric system early. Except for the special cases discussed in the previous sections, non-SI units (as well as special names for multiples or submultiples of SI units) are to be avoided. Various categories of deprecated units are discussed in 3.3.4.1 to 3.3.4.4. The lists are not intended to be complete, but only to indicate more or less prominent examples of each category.

3.3.4.1 *Cgs Units*—All units peculiar to the various cgs systems (measurement systems constructed by using the centimetre, gram, and second as base units) are to be avoided. Among these units are the following, defined for mechanics, fluid mechanics, and photometry: the erg, dyne, gal, poise, stokes, stilb, phot, and lambert. Further use of the cgs units of electricity and magnetism is deprecated. This statement applies to the units designated by the general abbreviations "esu" (for electrostatic cgs unit) and "emu" (for electromagnetic cgs unit), including those units that have been given special names—the gauss, oersted, maxwell, gilbert, biot, and franklin. It also applies to the unit names formed with the prefixes ab- and stat-, for example, the abampere, statvolt, etc.

3.3.4.2 *Decimal Multiples of SI Units*—Those multiples of SI units that cannot be handled by using the SI prefixes are deprecated. Many such examples are covered in subsection 3.3.4.1. An additional example is the angstrom (0.1 nm).

3.3.4.3 *Unit Names to Be Avoided*—Special names for multiples and submultiples of SI units are to be avoided except for the *litre* (3.3.2.4), *metric ton* (3.3.2.5), and *hectare* (3.3.2.3). For example, do not use:

fermi	1 fermi	= 1 fm = 10^{-15} m
micron	1 micron	= 1 μm = 10^{-6} m
millimicron	1 millimicron	= 1 nm = 10^{-9} m
are	1 are	= 1 dam^2 = 100 m^2
gamma (magnetic flux density)	1 gamma	= 1 nT
γ (mass)	1 γ	= 1 μg
λ (volume)	1 λ	= 1 μL = 1 mm^3
mho	1 mho	= 1 S
candle	1 candle	= 1 cd
candlepower	1 candlepower	= 1 cd

3.3.4.4 *Miscellaneous Units*—Other non-SI units that are deprecated include the following:

    calorie
    grade [1 grade = (π/200) rad]
    kilogram-force
    langley (= 1 cal/cm^2)
    metric carat
    metric horsepower
    millimetre of mercury
    millimetre, centimetre, metre of water
    standard atmosphere
      (1 atm = 101.325 kPa)
    technical atmosphere
      (1 at = 98.0665 kPa)
    torr

3.4 *Other Recommendations Concerning Units*:

3.4.1 *Mass, Force, and Weight*:

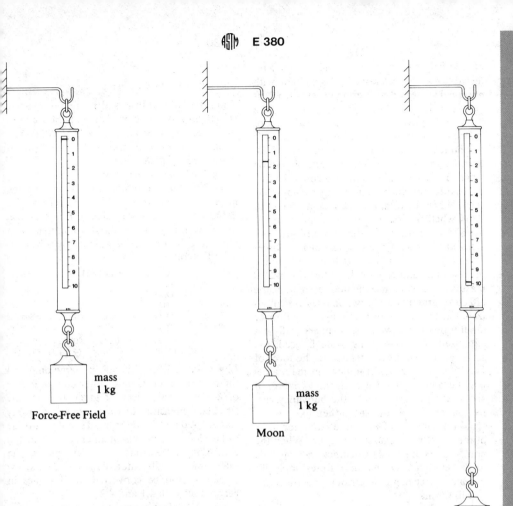

FIG. 1 Illustration of Difference Between Mass (Unit = kilogram = kg) and Force (Unit = newton = N) (see 3.4.1)

3.4.1.1 The principal departure of SI from the gravimetric system of metric engineering units is the use of explicitly distinct units for mass and force. In SI, the name kilogram is restricted to the unit of mass, and the kilogram-force (from which the suffix *force* was in practice often erroneously dropped) should not be used. In its place the SI unit of force, the newton, is used (see Fig. 1). Likewise, the newton rather than the kilogram-force is used to form derived units which include force, for example, pressure or stress (N/m^2 = Pa), energy ($N·m$ = J), and power ($N·m/s$ = W).

3.4.1.2 Considerable confusion exists in the use of the term *weight* as a quantity to mean either *force* or *mass*. In commercial and everyday use, the term *weight* nearly always means mass; thus, when one speaks of a person's weight, the quantity referred to is mass. This nontechnical use of the term weight in everyday life will probably persist. In science and technology, the term *weight of a body* has usually meant the force that, if applied to the body, would give it an acceleration equal to the local acceleration of free fall. The adjective "local" in the phrase "local acceleration of free fall" has usually meant a location on the surface of the earth; in this context the "local acceleration of free fall" has the symbol g (commonly referred to as "acceleration of gravity") with observed values of g differing by over 0.5 % at various points on the earth's surface. The use of *force of gravity* (mass times acceleration of gravity) instead of *weight* with this meaning is recommended. Because of the dual use of the term weight as a quantity, this term should be avoided in technical practice except under circumstances in which its meaning is completely clear. When the term is used, it is important to know whether mass or force is intended and to use SI units properly as described in 3.4.1.1, by using kilograms for mass or newtons for force.

3.4.1.3 Gravity is involved in determining mass with a balance or scale. When a standard mass is used to balance the measured mass, the effects of gravity on the two masses are equalized, but the effects of the buoyancy of air or other fluid on the two masses are generally not equalized. When a spring scale is used, the scale reading is directly related to the force of gravity. Spring scales graduated in mass units may be properly used if both the variation in acceleration of gravity and the buoyancy corrections are not significant in their use.

3.4.1.4 The use of the same name for units of force and mass causes confusion. When the non-SI units are used, a distinction should be made between *force* and *mass*, for example, lbf to denote force in gravimetric engineering units and lb for mass.

3.4.1.5 The term *load* means either mass or force, depending on its use. A load that produces a vertically downward force because of the influence of gravity acting on a mass may be expressed in mass units. A load that produces a force from anything other than the influence of gravity is expressed in force units.

3.4.2 *Temperature*—The SI unit of thermodynamic temperature is the kelvin (K), and this unit is properly used for expressing thermodynamic temperature and temperature intervals. Wide use is also made of the degree Celsius (°C), which is the SI unit used for expressing Celsius temperature and temperature intervals. The Celsius scale (formerly called centigrade) is related directly to thermodynamic temperature (kelvins) as follows:

The temperature interval one degree Celsius equals one kelvin exactly.

Celsius temperature (t) is related to thermodynamic temperature (T) by the equation:

$$t = T - T_0$$

where $T_0 = 273.15$ K by definition.

The International Practical Temperature Scale (IPTS) must be recognized in temperature work of extreme precision. See *ASTM STP 565, Evolution of the International Practical Temperature Scale of 1968.*

3.4.3 *Linear Dimensions:*

3.4.3.1 Nominal dimensions name the item; no SI equivalent is required (see Section 5 for definition of "nominal value"). For example, there is nothing "1 in" about a nominal "1-in pipe," the dimensions of which should be converted as follows:

Nominal Size, inches	Outside Diameter, inches (mm)	Wall Thickness, inches (mm)		
		Sch 40	Sch 80	Sch 160
1	1.315 (33.40)	0.133 (3.38)	0.179 (4.55)	0.250 (6.35)

Likewise, a "2 by 4" is that in name only and refers to the approximate dimensions in inches of a rough-sawn piece of green lumber, the finished dimensions of which are considerably less. A ¼-20 UNC screw thread should continue to be identified in this manner. However, the controlling dimensions of the part, such as the pitch, major, and minor diameters of a screw thread, should be converted to SI values in accordance with 4.1 and 4.2.

3.4.3.2 Surface texture should be expressed in micrometres.

3.4.4 *Energy and Torque:*

3.4.4.1 The vector product of force and moment arm is widely designated by the unit newton metre. This unit for bending moment or torque results in confusion with the unit for energy, which is also newton metre. If torque is expressed as newton metre per radian, the relationship to energy is clarified, since the product of torque and angular rotation is energy:

$$(N \cdot m/rad) \cdot rad = N \cdot m$$

3.4.4.2 If vectors were shown, the distinction between energy and torque would be obvious, since the orientation of force and length is different in the two cases. It is important to recognize this difference in using torque and energy, and the joule should never be used for torque.

3.4.5 *Impact*—Impact strength (actually energy absorption) is measured in terms of energy required to break a standard specimen. The proper unit is joule.

3.4.6 *Pressure and Vacuum*—Gage pressure is absolute pressure minus ambient pressure (usually atmospheric pressure). Both gage pressure and absolute pressure are properly expressed in pascals, using SI prefixes as appropriate. Absolute pressure is never negative. Gage pressure is positive if above ambient pressure and negative if below. Pressure below ambient is often called vacuum; whenever the term *vacuum* is applied to a numerical measure it should be made clear whether negative gage pressure or absolute pressure is meant. See 3.5.5 for methods of designating gage pressure and absolute pressure.

3.4.7 *Dimensionless Quantities:*

3.4.7.1 The values of so-called dimensionless quantities, as for example refractive index and relative permeability, are expressed by pure numbers. In these cases the corresponding SI unit is the ratio of the same two SI units and may be expressed by the number 1.

3.4.7.2 Units such as percent, parts per thousand, and parts per million may also be used.

3.4.7.3 In all cases, the meaning must be unequivocal. Expressions like "The volume fraction of CO_2 in the sample was 1.2 parts per million" or "The mass fraction of CO_2 in the sample was 1.2 parts per million" are permissible, but would not be permissible if the word "volume" in the first expression or "mass" in the second expression were not present.

3.5 *Style and Usage*—Care must be taken to use unit symbols properly, and international agreement provides uniform rules. Handling of unit names varies because of language differences, but use of the rules included here will improve communications in the United States.

3.5.1 *Rules for Writing Unit Symbols:*

3.5.1.1 Unit symbols should be printed in upright type regardless of the type style used in the surrounding text.

3.5.1.2 Unit symbols are unaltered in the plural.

3.5.1.3 Unit symbols are not followed by a period except when used at the end of a sentence.

3.5.1.4 Letter unit symbols are written in lower-case (for example, cd) unless the unit name has been derived from a proper name, in which case the first letter of the symbol is capitalized (for example, W, Pa). The exception is the symbol for litre, L. Prefix symbols use either lower-case or upper-case letters as shown in 2.5.1. Symbols retain their prescribed form regardless of the surrounding typography. For symbols for use in systems with limited character sets, refer to ANSI X3.50 or ANSI/IEEE 260, as applicable. The symbols in ANSI X3.50 are intended for applications in the field of information processing, where unambiguous transmission of information between computers is required. The symbols in ANSI/IEEE 260 are generally consistent with those in ANSI X3.50 and are intended for communication between human beings. The symbols for limited character sets must never be used when the available character set permits the use of the proper general-use symbols as given in this standard.

3.5.1.5 When a quantity is expressed as a numerical value and a unit symbol, a space should be left between them. For example, use 35 mm, *not* 35mm, and 2.37 lm (for 2.37 lumens), *not* 2.37lm.

Exception: No space is left between the numerical value and the symbols for degree, minute, and second of plane angle, and degree Celsius. For example, use 45°, 20°C.

3.5.1.6 When a quantity expressed as a number and a unit is used in an adjectival sense, it is preferable to use a hyphen instead of a space between the number and the unit name or between the number and the symbol. Examples: A three-metre pole... The length is 3 m... A 35-mm film... The width is 35 mm. However, per 3.5.1.5 Exception, a 90° angle... an angle of 90°.

3.5.1.7 No space is used between the prefix and unit symbols.

3.5.1.8 Symbols, not abbreviations, should be used for units. For example, use "A" and not "amp" for ampere.

3.5.2 *Rules for Writing Names*:

3.5.2.1 Spelled-out unit names are treated as common nouns in English. Thus, the first letter of a unit name is not capitalized except at the beginning of a sentence or in capitalized material such as a title.

3.5.2.2 Plurals are used when required by the rules of English grammar and are normally formed regularly, for example, henries for the plural of henry. The following irregular plurals are recommended:

Singular	Plural
lux	lux
hertz	hertz
siemens	siemens

3.5.2.3 No space or hyphen is used between the prefix and unit name. There are three cases where the final vowel in the prefix is commonly omitted: *megohm, kilohm,* and *hectare.* In all other cases where the unit name begins with a vowel both vowels are retained and both are pronounced.

3.5.3 *Units Formed by Multiplication and Division*:

3.5.3.1 With unit names:
Product, use a space (preferred) or hyphen:

   newton metre *or* newton-metre

In the case of the watt hour the space may be omitted, thus:

   watthour

Quotient, use the word *per* and not a solidus:

   metre per second, *not* metre/second

Powers, use the modifier *squared* or *cubed* placed after the unit name:

   metre per second squared

In the case of area or volume, the modifier may be placed before the unit name:

   square millimetre, cubic metre

This alternative is also allowed for derived units that include area or volume:

   watt per square metre

NOTE—To avoid ambiguity in complicated expressions, symbols are preferred over words.

3.5.3.2 With unit symbols:
Product, use a raised dot:

   N·m for newton metre

In the case of W·h, the dot may be omitted, thus:

   Wh

An exception to this practice is made for computer printouts, automatic typewriter work, etc, where the raised dot is not possible, and a dot on the line may be used.

Quotient, use one of the following forms:

   m/s *or* m·s^{-1} *or* $\frac{m}{s}$

In no case should more than one solidus be used in the same expression unless parentheses are inserted to avoid ambiguity. For example, write:

   J/(mol·K) *or* J·mol^{-1}·K^{-1} *or* (J/mol)/K,

but *not*

   J/mol/K

3.5.3.3 Symbols and unit names should not be mixed in the same expression. Write:

   joules per kilogram *or* J/kg *or* J·kg^{-1}

but *not*

   joules/kilogram *nor* joules/kg *nor* joules·kg^{-1}

3.5.4 *Numbers*:

3.5.4.1 The recommended decimal marker is a dot on the line. When writing numbers less than one, a zero should be written before the decimal marker.

3.5.4.2 Outside the United States, the comma is sometimes used as a decimal marker. In some applications, therefore, the common practice in the United States of using the comma to separate digits into groups of three (as in 23,478) may cause ambiguity. To avoid this potential source of confusion, recommended international practice calls for separating the digits into groups of three, counting from the decimal point toward the left and the right, and using a small space to separate the groups. In numbers of four digits on either side of the decimal point the space is usually not necessary, except for uniformity in tables.

Examples:

   2.141 596   73 722   7372   0.1335

Where this practice is followed, the space should be narrow (approximately the width of the letter "i"), and the width of the space should be constant even if, as is often the case in printing, variable-width spacing is used between words. Exceptions: In certain specialized applications, such as engineering drawings and financial statements, the practice of using a space for a separator is not customary.

3.5.4.3 Because *billion* means a thousand million (prefix *giga*) in the United States but a million million (prefix *tera*) in most other countries, this term and others, such as trillion, should be avoided in technical writing.

3.5.4.4 Use of M to indicate thousands, as in MCF for thousands of cubic feet, or in MCM for thousands of circular mils, of MM to indicate millions, of C to indicate hundreds, etc., is deprecated because of obvious conflicts with the SI prefixes.

3.5.5 *Attachment*—Attachment of letters to a unit symbol as a means of giving information about the nature of the quantity under consideration is incorrect. Thus MWe for "megawatts electrical (power)," Vac for "volts ac," and kJt for "kilojoules thermal (energy)" are not acceptable. For this reason, no attempt should be made to construct SI equivalents of the abbreviations "psia" and "psig," so often used to distinguish between absolute and gage pressure. If the context leaves any doubt as to which is meant, the word *pressure* must be qualified appropriately. For example:

"... at a gage pressure of 13 kPa"

or

"... at an absolute pressure of 13 kPa"

Where space is limited, such as on gages, nameplates, graph labels, and in table headings, the use of a modifier in parentheses, such as "kPa (gage)" or "kPa (absolute)," is permitted.

3.5.6 *Pronunciation*—Some recommended pronunciations in English are shown in Table 8.

TABLE 8 Recommended Pronunciation

Prefix	Pronunciation (USA)[A]
exa	ex′ a (*a* as in *a*bout)
peta	pet′ a (*e* as in pet, *a* as in *a*bout)
tera	as in *terra* firma
giga	jig′ a (*i* as in j*i*g, *a* as in *a*bout)
mega	as in *mega*phone
kilo	kill′ oh
hecto	heck′ toe
deka	deck′ a (*a* as in *a*bout)
deci	as in *deci*mal
centi	as in *centi*pede
milli	as in *milli*tary
micro	as in *micro*phone
nano	nan′ oh (*an* as in *an*t)
pico	peek′ oh
femto	fem′ toe (*fem* as in *fem*inine)
atto	as in an*ato*my

Selected Units	Pronunciation
candela	can *dell*′ a
joule	rhyme with *tool*
kilometre	*kill*′ oh metre
pascal	rhyme with *rascal*
siemens	same as *seamen's*

[A] The first syllable of every prefix is accented to assure that the prefix will retain its identity. Therefore, the preferred pronunciation of kilometre places the accent on the first syllable, *not* the second.

4. Rules for Conversion and Rounding

4.1 *General:*

4.1.1 Appendix X3 contains conversion factors that give exact values or seven-digit accuracy for implementing these rules except where the nature of the dimension makes this impractical.

4.1.2 Conversion of quantities should be handled with careful regard to the implied correspondence between the accuracy of the data and the given number of digits. In all conversions, the number of significant digits retained should be such that accuracy is neither sacrificed nor exaggerated. (For guidance concerning significant digits see 4.3.) For example, a length of 125 ft converts exactly to 38.1 m. If, however, the 125-ft length had been obtained by rounding to the nearest 5 ft, the conversion should be given as 38 m; and if it had been obtained by rounding to the nearest 25 ft, the conversion should be given as 40 m.

4.1.3 Proper conversion procedure is to multiply the specified quantity by the conversion factor exactly as given in Appendix X3 and then round to the appropriate number of significant digits. For example, to convert 3 feet 2 9/16 inches to metres: (3 × 0.3048) + (2.5625 × 0.0254) = 0.979 487 5 m, which rounds to 0.979 m. Do not round either the conversion factor or the quantity before performing the multiplication, as accuracy may be reduced. After the conversion, the SI value may be expressed by a multiple or submultiple unit of SI by the use of an appropriate prefix, for example, 979 mm.

4.2 *Accuracy and Rounding*—Accurate conversions are obtained by multiplying the specific quantity by the appropriate conversion factor given in Appendix X3. However, this product will usually imply an accuracy not intended by the original value. Proper conversion procedure includes rounding this converted quantity to the proper number of significant digits commensurate with its intended precision. The practical aspect of measuring must be considered when using SI equivalents. If a scale having division of 1/16 inch was suitable for making the original measurements, a metric scale having divisions of 1 mm is obviously suitable for measuring in SI units. Similarly, a gage or caliper graduated in divisions of 0.02 mm is comparable to one graduated in divisions of 0.001 in. Analogous situations exist in mass, force, and other measurements. Many techniques are used to guide the determination of the proper number of significant digits in the converted values. Two different approaches to rounding of quantities are here described—one for general use and the other for conversion of dimensions involving mechanical interchangeability.

4.2.1 *General Conversion*—This approach depends on first establishing the intended precision or accuracy of the quantity as a necessary guide to the number of digits to retain. This precision should relate to the number of digits in the original, but in many cases this is not a reliable indicator. A figure 1.1875 may be a very accurate decimalization of a noncritical 1³⁄₁₆ that should have been expressed 1.19. On the other hand, the value 2 may mean "about 2," or it may mean a very accurate value of 2 which should have been written 2.0000. It is therefore necessary to determine the intended precision of a quantity before converting. This estimate of intended precision should never be smaller than the accuracy of measurement and should usually be smaller than one tenth the tolerance if one exists. After estimating the precision of the dimension, the converted dimension should be rounded to a minimum number of significant digits (see 4.3) such that a unit of the last place is equal to or smaller than the converted precision. Examples:

1. A stirring rod 6 in long. In this case, precision is estimated to be about ½ in (± ¼ in). Converted, this is 12.7 mm. The converted dimension 152.4 mm should be rounded to the nearest 10 mm, or 150 mm.

2. 50 000 lbf/in² (psi) tensile strength. In this case, precision is estimated to be about ± 200 lbf/in² (± 1.4 MPa) based on an accuracy of ± 0.25 % for the tensile tester and other factors. Therefore, the converted dimension, 344.7379 MPa, should be rounded to the nearest whole unit, 345 MPa.

3. Test pressure 200 ± 15 lbf/in² (psi). Since one tenth of the tolerance is 3 lbf/in² (20.68 kPa), the converted dimension should be rounded to the nearest 10 kPa. Thus, 1378.9514 ± 103.421 35 kPa becomes 1380 ± 100 kPa.

4.2.2 *Special Cases:*

4.2.2.1 Converted values should be rounded to the minimum number of significant digits that will maintain the required accuracy, as discussed in 4.1.2. In certain cases deviation from this practice to make use of convenient or whole numbers may be feasible, in which case the word "approximate" must be used following the conversion. For example:

1⅞ in = 47.625 mm exact
47.6 mm normal rounding
47.5 mm (approx) rounded to preferred number
48 mm (approx) rounded to whole number

4.2.2.2 A quantity stated as a limit, such as "not more than" or "maximum," must be handled so that the stated limit is not violated. For example, a specimen "at least 4 in wide" requires a width of at least 101.6 mm, or at least 102 mm.

4.3 *Significant Digits:*

4.3.1 When converting integral values of units, consideration must be given to the implied or required precision of the integral value to be converted. For example, the value "4 in" may be intended to represent 4, 4.0, 4.00, 4.000, or 4.0000 in, or even greater accuracy. Obviously, the converted value must be carried to a sufficient number of digits to maintain the accuracy implied or required in the original quantity.

4.3.2 Any digit that is necessary to define the specific value or quantity is said to be significant. When measured to the nearest 1 m, a distance may be recorded as 157 m; this number has three significant digits. If the measurement had been made to the nearest 0.1 m, the distance may have been 157.4 m; this number has four significant digits.

4.3.3 Zeros may be used either to indicate a specific value, like any other digit, or to indicate the order of magnitude of a number. The 1970 United States population figure rounded to thousands was 203 185 000. The six left-hand digits of this number are significant; each *measures* a value. The three right-hand digits are zeros which merely indicate the order of *magnitude* of the number rounded to the nearest thousand. The identification of significant digits is only possible through knowledge of the circumstances. For example, the number 1000 may be rounded from 965, in which case only one zero is significant, or it may be rounded from 999.7, in which case all three zeros are significant.

4.3.4 Occasionally data required for an investigation must be drawn from a variety of sources where they have been recorded with varying degrees of refinement. Specific rules must be observed when such data are to be *added, subtracted, multiplied,* or *divided.*

4.3.4.1 The rule for addition and subtraction is that the *answer* shall contain no significant digits farther to the right than occurs in the least precise number. Consider the addition of three numbers drawn from three sources, the first of which reported data in millions, the second in thousands, and the third in units:

```
163 000 000
217 885 000
 96 432 768
477 317 768
```

The total indicates a precision that is not valid. The numbers should first be rounded to *one significant digit* farther to the right than that of the least precise number, and the sum taken as follows:

```
163 000 000
217 900 000
 96 400 000
477 300 000
```

The total is then rounded to 477 000 000 as called for by the rule. Note that if the second of the figures to be added had been 217 985 000, the rounding before addition would have produced 218 000 000, in which case the 0 following 218 would have been a significant digit.

4.3.4.2 The rule for multiplication and division is that the *product* or *quotient* shall contain no more significant digits than are contained in the number with the *fewest significant digits* used in the multiplication or division. The difference between this rule and the rule for addition and subtraction should be noted; the latter rule merely requires rounding of digits that lie to the right of the last significant digit in the least precise number. The following illustration highlights this difference:

Multiplication:
113.2 × 1.43 = 161.876, rounded to 162
Division:
113.2 ÷ 1.43 = 79.16, rounded to 79.2
Addition:
113.2 + 1.43 = 114.63, rounded to 114.6
Subtraction:
113.2 − 1.43 = 111.77, rounded to 111.8

The above product and quotient are limited to three significant digits since 1.43 contains only three significant digits. In contrast, the rounded answers in the addition and subtraction examples contain four significant digits.

4.3.4.3 Numbers used in the above illustrations have all been estimates or measurements. Numbers that are exact counts are treated as though they consist of an infinite number of significant digits. More simply stated, when a count is used in computation with a measurement the number of significant digits in the answer is the same as the number of significant digits in the measurement. If a count of 40 is multiplied by a measurement of 10.2, the product is 408. However, if 40 were an estimate accurate only to the nearest 10, and hence contained but one significant digit, the product would be 400.

4.4 *Rounding Values*[7]:

4.4.1 When a figure is to be rounded to fewer digits than the total number available, the procedure should be as follows:

4.4.1.1 When the first digit discarded is less than 5, the last digit retained should not be changed. For example, 3.463 25, if rounded to four digits, would be 3.463; if rounded to three digits, 3.46.

4.4.1.2 When the first digit discarded is greater than 5, or if it is a 5 followed by at least one digit other than 0, the last digit retained should be increased by one unit. For example 8.376 52, if rounded to four digits, would be 8.377; if rounded to three digits 8.38.

4.4.1.3 When the first digit discarded is exactly 5, followed only by zeros, the last digit retained should be rounded upward if it is an odd number, but no adjustment made if it is an even number. For example, 4.365, when rounded to three digits, becomes 4.36. The number 4.355 would also round to the same value, 4.36, if rounded to three digits.

4.5 *Conversion of Linear Dimensions of Interchangeable Parts*—The use of the exact relation 1 in = 25.4 mm generally produces converted values containing more decimal places than are required for the desired accuracy. It is therefore necessary to round these values suitably and at the same time maintain the degree of accuracy in the converted values compatible with that of the original values.

4.5.1 *General*—The number of decimal places given in Table 9 for rounding converted

---

[7] Adapted from ISO R370 (7).

E 380

toleranced dimensions relates the degree of accuracy to the size of the tolerances specified. Two methods of using Table 9 are given: Method A, which rounds to values nearest to each limit, and Method B, which rounds to values always inside the limits.

In Method A, rounding is effected to the nearest rounded value of the limit, so that, on the average, the converted tolerances remain statistically identical with the original tolerances. The limits converted by this method, where acceptable for interchangeability, serve as a basis for inspection.

In Method B, rounding is done systematically *toward the interior* of the tolerance zone so that the converted tolerances are never larger than the original tolerances. This method must be employed when the original limits have to be respected absolutely, in particular, when components made to converted limits are to be inspected by means of original gages.

Method A—The use of this method ensures that even in the most unfavorable cases neither of the two original limits will be changed by more than 5 % of the value of the tolerance. Proceed as follows:

(*a*) Calculate the maximum and minimum limits in inches.

(*b*) Convert the corresponding two values exactly into millimetres by means of the conversion factor 1 in = 25.4 mm.

(*c*) Round the results obtained to the nearest rounded value as indicated in Table 9, depending on the original tolerance in inches, that is, on the difference between the two limits in inches.

Method B—This method must be employed when the original limits may not be violated, for instance, certain critical mating parts. In extreme cases, this method may increase the lower limit a maximum of 10 % of the tolerance and decrease the upper limit a maximum of 10 % of the tolerance.

(*a*) Proceed as in Method A step (*a*).

(*b*) Proceed as in Method A step (*b*).

(*c*) Round each limit toward the interior of the tolerance, that is, to the next lower value for the upper limit and to the next higher value for the lower limit.[8]

Examples:

A dimension is expressed in inches as .... 1.950 ± 0.016
The limits are .... 1.934 and 1.966

TABLE 9 Rounding Tolerances Inches to Millimetres

Original Tolerance, inches		Fineness of Rounding, mm
at least	less than	
0.000 04	0.000 4	0.0001
0.000 4	0.004	0.001
0.004	0.04	0.01
0.04	0.4	0.1
0.4		1

Conversion of the two limits into millimetres gives .... 49.1236 and 49.9364
Method A—The tolerance equals 0.032 in and thus lies between 0.004 and 0.04 in (see Table 9). Rounding these values to the nearest 0.01 mm, the values in millimetres to be employed for these two limits are .... 49.12 and 49.94
Method B—Rounding toward the interior of the tolerance, millimetre values for these two limits are .... 49.13 and 49.93
This reduces the tolerance to 0.80 instead of 0.82 mm given by Method A.

4.5.2 *Special Method for Dimensions with Plus and Minus Deviations*—In order to avoid accumulation of rounding errors, the two limits of size normally are converted separately: thus, they must first be calculated if the dimension consists of a basic size and two deviations. However (except when Method B is specified) as an alternative, the basic size may be converted to the nearest rounded value and each of the deviations converted toward the interior of the tolerance. This method, which sometimes makes conversion easier, gives the same maximum guarantee of accuracy as Method A, but usually results in smaller converted tolerances.

4.5.3 *Special Methods for Limitation Imposed by Accuracy of Measurements*—If the increment of rounding for the tolerances given in Table 9 is too small for the available accuracy of measurement, limits that are acceptable for interchangeability must be determined separately for the dimensions. For example, where accuracy of measurement is limited to 0.001 mm, study shows that values converted from 1.0000 ± 0.0005 in can be rounded to 25.413 and 25.387 mm instead of 25.4127 and 25.3873 mm with little disadvantage, since neither of the two original limits is exceeded by more than 1.2 % of the tolerance.

4.5.4 *Positional Tolerance*—If the dimen-

---

[8] If the digits to be rounded are zeros, the retained digits remain unchanged.

sioning consists solely of a positional tolerance around a point defined by a nontoleranced basic dimension, the basic dimension must be converted to the nearest rounded value and the positional variation (radius) separately converted by rounding downward.

4.5.5 *Toleranced Dimension Applied to a Nontoleranced Position Dimension*—If the toleranced dimension is located in a plane, the position of which is given by nontoleranced basic or gage dimension, such as when dimensioning certain conical surfaces, proceed as follows:

(*a*) Round the converted reference gage arbitrarily, to the nearest convenient value.

(*b*) Calculate exactly, in the converted unit of measurement, new maximum and minimum limits of the specified tolerance zone, in the new plane defined by the new basic dimension.

(*c*) Round these limits in conformity with the rules in 4.4. For example, a cone of taper 0.05 in/in has a diameter of 1.000 ± 0.002 inch in a reference plane located by the nontoleranced dimension 0.9300 in. By virtue of the taper of the cone, the limits of the tolerance zone depend on the position of the reference plane. Consequently, if the dimension 0.9300 in = 23.6220 mm is rounded to 23.600 mm (that is, a reduction of 0.022 mm), each of the two original limits, when converted exactly into millimetres, must be corrected by 0.022 × 0.05 = 0.0011 mm, in the appropriate sense, before being rounded.

4.5.6 *Consideration of Maximum and Minimum Material Condition*—The ability to assemble mating parts depends on a "go" condition at the maximum material limits of the parts. The minimum material limits, which are determined by the respective tolerances, are often not as critical from a functional standpoint. Accordingly, it may be desirable to employ a combination of Methods A and B in certain conversions by using Method B for the maximum material limits and Method A for the minimum material limits. Alternatively, it may be desirable to round automatically the converted minimum material limits outside the original limits to provide greater tolerances for manufacturing.

4.5.7 While the technique described in 4.5 provides good accuracy of conversion, it will often result in dimensions that are impractical for actual production use. For conversions intended for production, it is usually necessary to round to fewer decimal places and apply design judgment to each dimension to assure interchangeability.

4.6 *Other Units:*

4.6.1 *Temperature*—General guidance for converting tolerances from degrees Fahrenheit to kelvins or degrees Celsius is given below:

Conversion of Temperature Tolerance Requirements

Tolerance, °F	Tolerance, K or °C
2 (±1)	1 (±0.5)
4 (±2)	2 (±1)
10 (±5)	6 (±3)
20 (±10)	11 (±5.5)
30 (±15)	17 (±8.5)
40 (±20)	22 (±11)
50 (±25)	28 (±14)

Normally, temperatures expressed in a whole number of degrees Fahrenheit should be converted to the nearest 0.5 kelvin (or degree Celsius). As with other quantities, the number of significant digits to retain will depend upon implied accuracy of the original dimension, for example:

100 ± 5°F; implied accuracy estimated to be 2°F.
  37.7777 ± 2.7777°C rounds to 38 ± 3°C.
1000 ± 50°F; implied accuracy estimated to be 20°F.
  537.7777 ± 27.7777°C rounds to 540 ± 30°C.

4.6.2 *Pressure or Stress*—As with other quantities, pressure or stress values may be converted by the principle given above. Values with an uncertainty of more than 2 % may be converted without rounding by approximate factors:

1 lbf/in^2 (1 psi) = 7 kN/m^2 = 7 kPa

For conversion factors and values see Appendix X3.

5. Terminology

5.1 To help ensure consistently reliable conversion and rounding practices, a clear understanding of the related nontechnical terms is a prerequisite.

5.2 Certain terms used in this standard are defined as follows:

accuracy (as distinguished from **precision**)—the degree of conformity of a measured or calculated value to some recognized standard or specified value. This concept involves the systematic error of an operation, which is seldom negligible.

approximate value—a value that is nearly

E 380

but not exactly correct or accurate.

coherence—a characteristic of a coherent system as described in X1.9 of Appendix X1. In such a system the product or quotient of any two unit quantities is the unit of the resulting quantity. The SI base units, supplementary units, and derived units form a coherent set.

deviation—variation from a specified dimension or design requirement, usually defining upper and lower limits. (See also **tolerance**.)

digit—one of the ten arabic numerals (0 to 9).

dimension—a geometric element in a design, such as length or angle, or the magnitude of such a quantity.

feature—an individual characteristic of a part, such as screw-thread, taper, or slot.

figure (numerical)—an arithmetic value expressed by one or more digits.

inch-pound units—units based upon the yard and the pound commonly used in the United States of America and defined by the National Bureau of Standards. Note that units having the same names in other countries may differ in magnitude.

nominal value—a value assigned for the purpose of convenient designation; existing in name only.

precision (as distinguished from **accuracy**)—the degree of mutual agreement between individual measurements, namely repeatability and reproducibility.

significant digit—any digit that is necessary to define a value or quantity (see 4.3).

tolerance—the total amount by which a quantity is allowed to vary; thus the tolerance is the algebraic difference between the maximum and minimum limits.

ASTM E 380

SELECTED CONVERSION FACTORS

To convert from	to	multiply by
atmosphere (760 mm Hg)	pascal (Pa)	$1.013\ 25 \times 10^5$
board foot	cubic metre (m³)	$2.359\ 737 \times 10^{-3}$
Btu (International Table)	joule (J)	$1.055\ 056 \times 10^3$
Btu (International Table)/h	watt (W)	$2.930\ 711 \times 10^{-1}$
Btu (International Table)·in./s·ft²·°F (k, thermal conductivity)	watt per metre kelvin [W/(m·K)]	$5.192\ 204 \times 10^2$
calorie (International Table)	joule (J)	4.186 800*
centipoise	pascal second (Pa·s)	$1.000\ 000^* \times 10^{-3}$
centistokes	square metre per second (m²/s)	$1.000\ 000^*\ 10^{-6}$
circular mil	square metre (m²)	$5.067\ 075 \times 10^{-10}$
degree Fahrenheit	degree Celsius	$t_{°C} = (t_{°F} - 32)/1.8$
foot	metre (m)	$3.048\ 000^* \times 10^{-1}$
ft²	square metre (m²)	$9.290\ 304^*\ 10^{-2}$
ft³	cubic metre (m³)	$2.831\ 685 \times 10^{-2}$
ft·lbf	joule (J)	1.355 818
ft·lbf/min	watt (W)	$2.259\ 697 \times 10^{-2}$
ft/s²	metre per second squared (m/s²)	$3.048\ 000^* \times 10^{-1}$
gallon (U.S. liquid)	cubic metre (m³)	$3.785\ 412 \times 10^{-3}$
horsepower (electric)	watt (W)	$7.460\ 000^* \times 10^{-2}$
inch	metre (m)	$2.540\ 000^* \times 10^{-2}$
in.²	square meter (m²)	$6.451\ 600^* \times 10^{-4}$
in.³	cubic metre (m³)	$1.638\ 706 \times 10^{-5}$
inch of mercury (60°F)	pascal (Pa)	$3.376\ 85 \times 10^3$
inch of water (60°F)	pascal (Pa)	$2.488\ 4 \times 10^2$
kgf/cm²	pascal (Pa)	$9.806\ 650^* \times 10^4$
kip (1000 lbf)	newton (N)	$4.448\ 222 \times 10^3$
kip/in.² (ksi)	pascal (Pa)	$6.894\ 757\ ts\ 10^6$
ounce (U.S. fluid)	cubic metre (m³)	$2.957\ 353 \times 10^{-5}$
ounce-force	newton (N)	$2.780\ 139 \times 1^{-1}$
ounce (avoirdupois)	kilogram (kg)	$2.834\ 952 \times 10^{-2}$
oz (avoirdupois)/ft²	kilogram per square metre (kg/m²)	$3.051\ 517 \times 10^{-1}$
oz (avoirdupois)/yd²	kilogram per square metre (kg/m²)	$3.390\ 575 \times 10^{-2}$
oz (avoirdupois)/gal (U.S. liquid)	kilogram per cubic metre (kg/m³)	7.489 152
pint (U.S. liquid)	cubic metre (m³)	$4.731\ 765 \times 10^{-4}$
pound-force (lbf)	newton (N)	4.448 222
pound (lb avoirdupois)	kilogram (kg)	$4.535\ 924 \times 10^{-1}$
lbf/in² (psi)	pascal (Pa)	$6.894\ 757 \times 10^3$
lb/in.³	kilogram per cubic metre (kg/m³)	$2.767\ 990 \times 10^4$
lb/ft³	kilogram per cubic metre (kg/m³)	$1.601\ 846 \times 10$
quart (U.S. liquid)	cubic metre (m³)	$9.463\ 529 \times 10^{-4}$
ton (short, 2000 lb)	kilogram (kg)	$9.071\ 847 \times 10^2$
torr (mm Hg, 0°C)	pascal (Pa)	$1.333\ 22 \times 10^2$
W·h	joule (J)	$3.600\ 000^* \times 10^3$
yard	metre (m)	$9.144\ 000^* \times 10^{-1}$
yd²	square metre (m²)	$8.361\ 274 \times 10^{-1}$
yd³	cubic metre (m³)	$7.645\ 549 \times 10^{-1}$

* Exact

ANNUAL BOOK OF ASTM STANDARDS
Index

Section 1—Iron and Steel Products

SUBJECT	VOLUME
STEEL, STAINLESS STEEL, AND RELATED ALLOYS-	
BARS, CHAIN, AND SPRINGS	01.05
BEARING STEEL	01.05
CASTINGS	01.02
CONCRETE REINFORCING STEEL	01.04
FORGINGS	01.05
PIPING, TUBING, AND FITTINGS	01.01
PRESSURE VESSEL PLATE AND FORGINGS	01.04
RAILS, WHEELS, AND TIRES	01.04
STEEL SHEET AND STRIP	01.03
STRUCTURAL STEEL	01.04
WIRE	01.03
FENCES	01.06
FERROALLOYS AND ALLOYING ADDITIVES	01.02
IRON CASTINGS	01.02
METALLIC COATED IRON AND STEEL PRODUCTS	01.06

This index covers the standards and related material appearing in the 6 volumes contained in Section 1 of the Annual Book of ASTM Standards.

The boldface references are to the ASTM designations; the standards appear in the volume in alphanumeric order. Proposed methods and specifications carry the index reference, "(Proposed)"; they appear, as does other ancillary material, in the Related Material (gray-edged) section in the back of the book. A triangle (Δ) preceding the ASTM designation denotes that Adjunct Material for the standard is published separately; the Adjunct No. is given in the standard. A Combined Index, covering the standards and tentatives appearing in all volumes of the *1984 Annual Book of ASTM Standards,* is issued as Volume 00.01.

Alphabetization in the index is letter-for-letter, with no consideration to punctuation or word division. Initial prepositions of (indented) subentries are ignored for alphabetization.

In the preparation of indexes, every attempt has been made to index standards on three levels: (1) by main subject, using general and specific search terms; (2) by test methods or other significant sections of ASTM Standards; and (3) by cross-references to locate main subject entry terms. (*See also* references are abbreviated as *Sa* and appear under main entry terms.) Specification F 451, for Acrylic Bone Cements (Volume 13.01), illustrates ASTM's method of indexing.

INDEX TERMS FOR MAIN SUBJECT
    ENTRY
Acrylic bone cements,
  spec., **(F 451)**

Bone cements,
  acrylic, spec., **(F 451)**

Cements,
  bone, acrylic, spec., **(F 451)**

INDEX TERMS FOR TESTS
  Compressive strength,
    of acrylic bone cements, test, **(F 451)**
  Doughing time,
    of acrylic bone cements, test, **(F 451)**
CROSS-REFERENCES
  Resins, acrylic
    See **Acrylic bone cements**
  Surgical adhesives
    Sa **Acrylic bone cements**

Index of ASTM Standards, Section 1

A

Abrasion resistant iron castings
  See **Iron castings,** white
Acid cleaning
  See **Cleaning**
ACSR
  See **Aluminum,** electrical conductors
Aerospace applications, materials/tests for
  forgings, of alloy steel (premium quality) blooms and billets, spec., **(A 646) 01.05**
  steel investment castings, reference radiographs, Δ **(E 192) 01.02**
Age-hardened alloys,
  low-carbon nickel-copper-chromium-molybdenum-columbium *and* nickel-copper-columbium, spec. for, **(A 710) 01.04**
Air underwater pressure test
  for seamless and welded austenitic stainless steel tubing (for general service), **(A 269) 01.01**
Air ventilating units
  steel, grille, for detention areas, spec., **(A 750) 01.05**
Alloy castings
  classification, guide, **(A 567) 01.02**
  common requirements, for alloy/steel castings (for general industrial use), spec., **(A 781) 01.02**
Alloys numbering system
  See **Metals and alloys numbering system**
Alloy-steel for nuclear applications
  bolting materials, spec., **(A 540) 01.01**
Alloy tool steel
  See **Steel, stainless steel, and related alloys,** tool steel
Aluminum
  electrical conductors
    steel reinforced (ACSR/AZ), aluminum-coated steel core wire for, spec., **(B 341) 01.06**
    steel reinforced (ACSR), high-strength zinc-coated steel core wire for, spec., **(B 606) 01.06**
    steel reinforced (ACSR), zinc-coated steel core wire for, spec., (for metric see B 498M), **(B 498) 01.06**
Aluminum alloys
  aluminum-zinc alloy-coated steel sheet (by hot-dip process), for storm sewer/drainage pipe, spec., **(A 806) 01.06**
  electrical conductors,
    aluminum-zinc alloy-coated steel overhead ground wire strand for electric power transmission lines), spec., **(A 797) 01.06**
    steel reinforced, high-strength zinc-coated steel core wire for, spec., **(B 606) 01.06**
  wire for electrical use,
    aluminum-zinc alloy-coated steel overhead ground wire strand for electric power transmission lines), spec., **(A 797) 01.06**
Aluminum-coated iron/steel articles
  aluminum zinc-alloy coated steel, fence fabric, spec., **(A 783) 01.06**
  overhead ground wire strand, spec., **(A 797) 01.06**
  steel sheet (coated by hot-dip process), for storm sewer/drainage pipe, **(A 806) 01.06**
  steel sheet (by hot-dip process), spec., **(A 792) 01.06**
  wire rope/fittings (for highway guardrails), spec., **(A 784) 01.06**
  wire strand, spec., **(A 785) 01.06**
  chain-link fences,
    aluminum-zinc alloy-coated steel fence fabric, spec., **(A 783) 01.06**
    fence fabric, spec., **(A 491) 01.06**
    fence fittings, spec., **(F 626) 01.06**
    coating weight, test, **(A 428) 01.06**
    metallic-coated steel wire (for chain link fence fabric), spec., **(A 817) 01.06**
  steel core wire, steel reinforced (ACSR/AZ), for aluminum conductors, spec., **(B 341) 01.06**
  steel wire strand, spec., **(A 474) 01.06**
Aluminum-coated steel core wire
  for aluminum conductors, steel reinforced (ACSR/AZ), spec., **(B 341) 01.06**
Aluminum coatings
  See **Coatings (general),** aluminum coatings
Anchor chains (for ships)
  wildcats, spec., **(F 765) 01.02**
Annealed carbon spring steel
  spec., **(A 684) 01.03, 01.05**
Anti-friction bearings
  See **Steel, stainless steel, and related alloys,** for ball and roller bearings
Apparent porosity
  See **Porosity,** apparent
Arches
  plates and fasteners for, corrugated structural steel, zinc-coated (galvanized), spec., **(A 761) 01.03**
Arc-welded steel pipe/tube
  See **Steel pipe/tube**
Armor tape
  zinc-coated flat steel, spec., **(A 459) 01.06**
Armor wire
  of zinc-coated (galvanized) steel wire, for underground cables (used in communications, control, or power transmission), spec., **(A 411) 01.06**
Atmospheric corrosion resistance
  See **Resistance,** corrosion/corrosion resistance (atmospheric)
Austenitic iron castings
  See **Iron castings**
Austenitic steel
  See **Steel, stainless steel, and related alloys**
Austenitized steel spring wire
  See **Steel wire**
Automotive castings
  See **Iron castings**
Automotive seat spring steel wire
  See **Steel wire**
Automotive steel materials
  structural (high-strength, low-alloy, hot-rolled), plates, of vanadium-aluminum-nitrogen/titanium-aluminum steel, spec. for, **(A 656) 01.04**
Average electron vacancy number
  calculating, **(A 567) 01.02**

Index of ASTM Standards, Section 1

Axles
  of alloy steel, heat-treated, for mass transit and electric railway service, spec., **(A 729) 01.04, 01.05**
  carbon steel (nonheat-treated),
    for export and general use, spec. for, **(A 383) 01.04, 01.05**
  of heat-/nonheat-treated carbon steel, spec., **(A 21) 01.04**
Axle steel bars
  See Steel bars

B

Back connected gages (for ships)
  See Shipbuilding steel materials, gages
Ball bearing steel
  See Steel, stainless steel, and related alloys, for ball and roller bearings
Banded-weld stud-type hanger (for ships)
  See Shipbuilding steel materials
Barbed wire
  of aluminum-coated steel wire, spec., **(A 585) 01.06**
  of zinc-coated (galvanized) steel wire, spec., **(A 121) 01.06**
Barge loading
  of steel products, for domestic shipment, rec. practice, **(A 700) 01.01, 01.03, 01.04, 01.05**
Bars/forging/forging stock
  precipitation-hardening cobalt-containing alloys (UNS R30155/R30816), for high-temperature service, spec., **(B 639) 01.05**
  precipitation-hardening nickel alloys (UNS N07001, N07080, N07500, N07718, N07750), for high-temperature service spec., **(B 637) 01.05**
Bearing quality steel
  See Steel, stainless steel, and related alloys, for ball and roller bearings
Bend testing
  of steel products, **(A 370) 01.01, 01.02, 01.03, 01.05**
Billet steel bars
  See Steel bars
Black plate
  See Tin mill products
Black steel pipe
  See Steel pipe/tube
Board fences
  See Fences/fencing materials
Bolting materials
  Sa Steel bolting materials
  of alloy-steel, for nuclear and special applications, spec., **(A 540) 01.01**
  for nuclear/special applications, requirements, spec., **(A 614) 01.01**
Bolting materials (for shipboard piping)
  See Steel pipe/tube, shipboard piping materials
Bottom connected gages (for ships)
  See Shipbuilding steel materials, gages
Branch connections (for shipboard piping)
  See Steel pipe/tube, shipboard piping materials
  Sa Shipbuilding steel materials

Bridge steel materials
  Sa Automotive steel materials
    Building steel materials
    Generator materials
    Pressure vessel steel materials
    Railroad steel materials
    Ship building steel materials
    Steel bolting materials
    Steel, stainless steel, and related alloys
  general requirements for rolled shapes, plates, sheet piling, and bars, spec., **(A 6) 01.04**
  highway components, steel castings for, carbon steel (70,000 psi (483 MPa)//low-alloy steel (90,000 psi (621 MPa)/120,000 psi (827 MPa)), spec., **(A 486) 01.02**
  sheet piling (for sea walls/cofferdams), spec. for, **(A 328) 01.04**
  structural,
    carbon, for shapes, plates, and bars, spec., **(A 36) 01.04**
    of high-yield-strength, quenched and tempered alloy plate (suitable for welding), spec. for, **(A 514) 01.04**
    spec., **(A 709) 01.04**
  structural (high-strength, low-alloy),
    columbium-vanadium steel shapes, plates, sheet piling, and bars, spec. for, **(A 572) 01.04**
    manganese vanadium steel shapes, plates, and bars, spec. for, **(A 441) 01.04**
    for shapes, plates, and bars, with 50,000 psi (345 MPA) minimum yield point, and 4 in. maximum thickness, spec. for, **(A 588) 01.04**
Brinell hardness
  See Hardness, Brinell
Building constructions/materials
  floor plates (carbon, high-strength low-alloy/alloy steel rolled plates), spec., **(A 786) 01.04**
  windows/window assemblies,
    zinc coatings (on assembled steel windows), spec., **(A 386) 01.06**
Building steel materials
  general requirements for rolled shapes, plates, sheet piling, and bars, spec., **(A 6) 01.04**
  structural,
    carbon, for shapes, plates, and bars, spec., **(A 36) 01.04**
    of carbon steel, with 42,000 psi (290 MPa) minimum yield point, and ½ in. (12.7 mm) maximum thickness, spec. for, **(A 529) 01.04**
  structural (high-strength, low-alloy),
    columbium-vanadium steel shapes, plates, sheet piling, and bars, spec. for, **(A 572) 01.04**
    high-strength/low-alloy structural manganese vanadium steel shapes/plates/bars (for welded/riveted/bolted constructions in welded bridges and buildings), spec., **(A 441) 01.04**
    for shapes, plates, and bars, with 50,000 psi (345 MPa) minimum yield point, and 4 in. maximum thickness, spec. for, **(A 588) 01.04**
Butterfly valve stud bolt assembly (for ships)
  See Steel pipe/tube, shipboard piping materials

Index of ASTM Standards, Section 1

Butt joints (for shipboard piping)
   See **Steel pipe/tube,** shipboard piping materials

C

Cadmium
  electrodeposited coating,
    on steel, spec., **(A 165) 01.06**
  mechanically deposited coatings, on iron/steel metals, spec., **(B 696) 01.04**
  vacuum deposited coating, on iron/steel metals, spec., **(B 699) 01.04**

Calcium-manganese-silicon
  spec., **(A 495) 01.02**

Calcium-silicon
  spec., **(A 495) 01.02**

Calibration/testing
  calibration/testing of instruments for estimating ferrite content of cast stainless steels, spec., **(A 799) 01.02**

Carbon steel
  See **Steel, stainless steel, and related alloys,** carbon steel

Carbon steel sheet, cold-rolled
  for magnetic applications
    See **Steel sheet,** carbon, cold-rolled

Carburizing bearings
  See **Steel, stainless steel, and related alloys,** for ball and roller bearings

Castings
  austenitic alloy castings
    estimating ferrite content in austenitic alloys castings, practice, **(A 799) 01.02**
  evaluation of/specifying textures/discontinuities of steel castings, by visual examination, practice, Δ **(A 802) 01.02**
  investment (steel), for aerospace applications, reference radiographs, Δ **(E 192) 01.02**
  iron-chromium/iron-chromium-nickel alloy (heat-resistant) castings (for general application, spec., **(A 297) 01.02**
  iron-chromium/iron-chromium-nickel/nickel-base alloy (corrosion-resistant) castings (for general application), spec., **(A 743) 01.02**
  nickel/nickel alloys, spec., **(A 494) 01.02**

Cast iron
  See **Iron castings,**

Cast steel products
  See **Steel castings,** for nuclear applications

Cast tool steel
  See **Steel, stainless steel, and related alloys,** tool steel

Centrifugally cast cylinders
  dual metal (gray and white cast iron), spec., **(A 667) 01.02**

Chain
  wildcats (for ship anchor chains), spec., **(F 765) 01.02**

Chain-link fence
  See **Fences/fencing materials**

Chain (steel)
  See **Steel chain**

Charpy impact test
  for cast iron, **(A 327) 01.02**
  of steel products, **(A 370) 01.01, 01.02, 01.03, 01.05**

Charpy V-notch testing
  of structural steel, sampling procedure, spec. for, **(A 673) 01.04**

Chelate cleaning
  of stainless steel parts, equipment, and systems, information on, **(A 380) 01.03**

Chemical analysis
  of steel products, methods, practices, and definitions, **(A 751) 01.01, 01.02, 01.03, 01.04, 01.05**

Chemical descaling
  See **Cleaning,** surface descaling/cleaning

Chests of drawers (for ships)
  chest of drawers (chiffoniers), for ships, spec., **(F 822) 01.02**

Chilled iron castings
  See **Iron castings,** chilled/white iron

Chill testing
  of cast iron, **(A 367) 01.02**
  evaluation of/specifying textures/discontinuities of steel castings, by visual examination, practice, **(A 802) 01.02**

Chromium
  electrodeposition,
    on iron castings, preparation for, practice, **(B 320) 01.02**
  spec., **(A 481) 01.02**

Chromium (12%) alloy steel
  forgings, for turbine rotors and shafts, spec., **(A 768) 01.05**

Chromium alloying additives
  See **Steel, stainless steel, and related alloys,** alloying additives

Chromium/chromium-nickel/chromium-manganese-nickel
  steel plate, sheet, and strip, for fusion-welded unfired pressure vessels, spec., **(A 240) 01.03, 01.04**

Chromium-clad steel
  sheet, strip, and plate, for pressure vessels, chromium-nickel, spec., **(A 264) 01.03, 01.04**
  spec., **(A 263) 01.03, 01.04**

Chromium-manganese-silicon
  steel plates, for pressure vessels, spec., **(A 202) 01.04**

Chromium molybdenum alloy
  pipe,
    branch connections (for use on ships), practice, **(F 681) 01.02**
  steel plates.
    for pressure vessels at elevated temperatures, spec., **(A 387) 01.04**
    quenched and tempered, for welded pressure vessels, spec., **(A 542) 01.04**

Index of ASTM Standards, Section 1

Chromium-nickel alloy
  for spring wire, spec., **(A 313) 01.03, 01.05**
Chromium-silicon steel spring wire
  See **Steel wire**
Chromium-vanadium steel spring wire
  See **Steel wire**
Cladding
  See **Steel, stainless steel, and related alloys, cladding**
Clamp hanger (for ships)
  See **Shipbuilding steel materials**
Classification
  of alloy castings, guide to, **(A 567) 01.02**
  evaluation of/specifying textures/discontinuities of steel castings, by visual examination, practice, Δ **(A 802) 01.02**
  of fence types, **(F 537) 01.06**
  of graphite microstructure in cast irons, Δ **(A 247) 01.02**
Cleaning
  stainless steel parts, equipment, and systems, rec. practice, **(A 380) 01.03**
  surface descaling/cleaning,
    stainless steel parts/equipment/systems, rec. practice, **(A 380) 01.03**
Closed-impression die forgings
  See **Forgings**
Coarse round steel wire
  See **Steel wire**
Coating mass per unit area
  See **Coating thickness**
Coatings
  aluminum,
    on ferrous materials (castings, forgings, fabricated items, structural shapes, tubular products), hot dipped, spec., **(A 676) 01.06**
  electrodeposited,
    cadmium, on steel, spec., **(A 165) 01.06**
    coating thickness, by magnetic field and electromagnetic (eddy current) testing, rec. practice for measuring, **(E 376) 01.06**
    coating thickness by X-ray fluorescence (XRF) techniques, test, **(A 754) 01.06**
  mechanically deposited
    cadmium coatings, on iron/steel metals, spec., **(B 696) 01.04**
    coating thickness, by magnetic field and electromagnetic (eddy current) testing, rec. practice for measuring, **(E 376) 01.06**
    thickness, by X-ray fluorescence (XRF) techniques, test, **(A 754) 01.06**
    zinc coatings on iron/steel metals, spec., **(B 695) 01.04**
  zinc, hot-dip galvanized,
    on assembled steel products, spec., **(A 386) 01.06**
    on assembled steel products, warpage/distortion prevention during galvanizing, rec. practice, **(A 384) 01.06**
    assuring high-quality coatings, rec. practice, **(A 385) 01.06**
    on iron/steel hardware, spec., **(A 153) 01.06**
    on steel fabricated products (rolled/pressed/forged shapes, plates, bars, and strip), spec., **(A 123) 01.06**
    on structural steel products, safeguarding against embrittlement/detection procedures, rec. practice, **(A 143) 01.06**
Coatings (general)
  aluminum coatings,
    coating thickness, by magnetic field and electromagnetic (eddy current) test, rec. practice for measuring, **(E 376) 01.06**
    coating thickness, by X-ray fluorescence (XRF) techniques, test, **(A 754) 01.06**
    on ferrous materials (castings/forgings/fabricated items/structural shapes/tubular products), hot-dipped, spec., **(A 676) 01.06**
  wax coatings,
    assuring high-quality coatings, rec. practice, **(A 385) 01.06**
    on iron/steel hardware, spec., **(A 153) 01.06**
    repairing damaged coatings, practice, **(A 780) 01.06**
    on steel fabricated products (rolled/pressed/forged shapes, plates, bars, and strip), spec., **(A 123) 01.06**
    on structural steel products, safeguarding against embrittlement/dection procedures, rec. practice, **(A 143) 01.06**
Coating thickness
  by magnetic field and electromagnetic (eddy current) testing, rec. practice for measuring, **(E 376) 01.06**
  by X-ray fluorescence (XRF) techniques, test, **(A 754) 01.06**
Coating weight
  of aluminum-coated iron/steel articles, test, **(A 428) 01.06**
  on long-terne sheet (incl. composition), by triple spot test, **(A 309) 01.06**
  of zinc-coated (galvanized) iron/steel articles, test, **(A 90) 01.06**
Cobalt-containing alloys
  precipitation hardening,
    bars, forgings, and forging stock (UNS R30155, R30816), for high-temperature service, spec., **(B 639) 01.05**
Cold forging steel
  See **Forgings**
Cold-headed steel wire
  See **Steel wire**
Cold heading
  of stainless/heat-resisting steel, spec. for, **(A 493) 01.03**
Columbium-vanadium steel
  See **Steel, stainless steel, and related alloys**
Commercial quality steel sheet
  See **Steel sheet**

Index of ASTM Standards, Section 1

Communication/electrical cable
  figure 8-type, of zinc-coated steel wire strand, extra-high strength, concentric-lay, spec., **(A 640) 01.06**
Compression testing
  of cast iron, **(A 256) 01.02**
Concentric lay steel wire strand
  See Steel wire strand
  Sa Electrical conductors
Concrete
  prestressed,
    of uncoated, high-strength steel bar, spec. for, **(A 722) 01.04**
    uncoated seven-wire stress-relieved steel strand for, spec., **(A 416) 01.04**
    uncoated stress-relieved wire for, spec., **(A 421) 01.04**
  reinforcement,
    with axle-steel bars, spec., **(A 617) 01.04**
    with billet-steel bars, deformed and plain, spec., (see also A 615M for metric equivalents), **(A 615) 01.04**
    from cold-worked, deformed steel wire, spec. for, **(A 496) 01.04**
    with fabricated deformed steel bar mats, spec. for, **(A 184) 01.04**
    by low-alloy deformed steel bars, spec. for, **(A 706) 01.04**
    with rail-steel bars, deformed and plain, spec. for, **(A 616) 01.04**
    steel wire (cold-drawn) for, spec., **(A 82) 01.04**
    with steel wire strand (compacted, seven wire, uncoated, stress-relieved), spec., **(A 779) 01.04**
    with welded fabric made from cold-worked deformed steel wire, spec. for, **(A 497) 01.04**
    with welded steel plain bar or rod mats, spec., **(A 704) 01.04**
    with welded steel wire fabric, spec. for, Δ **(A 185) 01.04**
    zinc-coated (galvanized) bars, spec., **(A 767) 01.04**
    zinc-coated (galvanized) steel pipe winding mesh, spec., **(A 810) 01.06**
Concrete pipe
  nonreinforced and reinforced,
    prestressed by hard-drawn steel wire, spec. for, **(A 648) 01.04**
Condenser and heat exchanger tubes
  steel
    See Steel pipe/tube, for high temperature/pressure service
Consumable electrode remelted steel
  See Steel, stainless steel, and related alloys, electrode remelted
Containers
  tank cars/wagons/other shipping containers, tanks—for storing/dispensing combustible/flammable liquids (on commercial/other vessels), 5/10 gal. (20/40 L), spec., **(F 670) 01.02**
Conveyors (laundry/screw)
  assembled, zinc coatings on, by hot-dip process, spec., **(A 386) 01.06**

Copper alloy pipe and tube
  copper-nickel tubing, branch connections, practice for using on ships, **(F 681) 01.02**
Copper-brazed steel tubing
  See Steel pipe/tube, copper-brazed
Copper-clad steel electrical conductors
  composite (copper/copper-clad) concentric-lay-stranded, spec., **(B 229) 01.06**
  concentric-lay-stranded, spec., **(B 228) 01.06**
  wire, hard-drawn, spec., **(B 227) 01.06**
Copper-clad steel wire
  coppered carbon steel wire, spec., **(A 818) 01.06**
  for electrical use, hard-drawn, spec., **(B 227) 01.06**
Copper-clad steel wire strand
  spec., **(A 460) 01.06**
Copper/copper clad steel composite electrical conductors
  concentric-lay-stranded, spec., **(B 229) 01.06**
Copper-copper sulfate-sulfuric acid test
  for detecting susceptibility to intergranular attack in ferritic stainless steel, practice, **(A 763) 01.03**
  for detecting susceptibility to intergranular attack in stainless steels, **(A 262) 01.03, 01.05**
Copper-nickel
  tubing, branch connections, practice for using on ships, **(F 681) 01.02**
Copper sulfate dip
  for locating the thinnest spot on zinc-coated galvanized forgings (the Preece test), test, **(A 239) 01.06**
Corrosion
  corrosion susceptibility testing,
    susceptibility to intergranular attack (in austenitic stainless steels), practice, **(A 262) 01.03, 01.05**
    susceptibility to intergranular attack (in ferritic stainless steels), practice, **(A 763) 01.03**
    susceptibility to intergranular corrosion (in severely sensitized austenitic stainless steel), rec. practice, **(A 708) 01.03**
  intergranular,
    in stainless steel (austenitic, severely-sensitized), susceptibility detection, rec. practice, **(A 708) 01.03**
    in stainless steel (ferritic), susceptibility detection, practice, **(A 763) 01.03**
    in stainless steel, susceptibility detection, rec. practice, **(A 262) 01.03, 01.05**
Corrosion resistance
  See Resistance, corrosion resistance
Corrosion-resistant iron castings
  See Iron castings, white
Corrugating paper machinery
  See Paper machinery
Corrugated steel pipe/tube
  See Steel pipe/tube
Couplings (for shipboard piping systems)
  See Steel pipe/tube, shipboard piping materials
Crane rails
  carbon steel, spec., **(A 759) 01.04**

Index of ASTM Standards, Section 1

Crankshafts
　large, forged,
　　magnetic particle inspection of, spec., **(A 456) 01.05**
　　ultrasonic examination of, spec., **(A 503) 01.05**
Crimp-on weld stud-type hanger (for ships)
　　See **Shipbuilding steel materials**
Culvert pipe/materials
　　See **Drainage materials/systems**
Cupola malleable iron
　　See **Iron castings,** malleable
Cutting test
　for security determination of steel bars, **(A 627) 01.05**
Cyclotrons, steel forgings for
　　See **Nuclear pressure vessels**
Cylinders
　centrifugally cast, dual metal (gray and white cast iron), for pressure-containing parts, spec., **(A 667) 01.02**

D

Data
　shipbuilding/marine product/procedure data sheet (for paints/coatings), Δ **(F 718) 01.02**
Deck treads (steel plate for)
　　See **Steel plate,** rolled floor plate
Definitions of terms relating to
　fencing (chain link), **(F 552) 01.06**
　heat treatment of metals, **(E 44) 01.02**
　iron castings, **(A 644) 01.02**
　steel forgings, **(A 509) 01.05**
Deflection
　of tool steel bars (for security applications), test, **(A 627) 01.05**
Desks
　marine log desks with cabinets, spec., **(F 823) 01.02**
Detention areas
　steel air ventilating units, spec., **(A 750) 01.05**
Detergent materials/systems
　household laundry/dishwasher equipment, assembled equipment, zinc coatings on (by hot-dip process), spec., **(A 386) 01.06**
Die forgings
　　See **Forgings**
Dining room tables (on ships)
　marine mess tables, spec., **(F 824) 01.02**
Discontinuities
　angle-beam ultrasonic examination (of steel plate for pressure vessels), spec., **(A 577/A 577M) 01.04**
　detecting susceptibility to intergranular attack in austenitic stainless steels, rec. practice, **(A 262) 01.03, 01.05**
　detecting susceptibility to intergranular attack in ferritic stainless steels, practice, **(A 763) 01.03**
　detecting susceptibility to intergranular corrosion in severely sensitized austenitic stainless steel, **(A 708) 01.03**
　evaluation of/specifying textures/discontinuities of steel castings, by visual examination, practice, Δ **(A 802) 01.02**
　straight beam ultrasonic examination (of plain/clad steel plate for special applications), spec., **(A 578/A 578M) 01.04**
　straight-beam ultrasonic examination (of steel plate for pressure vessels), spec., **(A 435/A 435M) 01.04**
Distortion prevention
　　See **Warpage/distortion prevention**
Doors (for ships)
　　See **Shipbuilding steel materials,** doors
Double fillet welded joints (for shipboard piping)
　　See **Steel pipe/tube,** shipboard piping materials
Double-jack/-loop chain
　　See **Steel chain**
Double-reduced electrolyte tin plate
　　See **Tin mill products**
Drainage materials/systems
　　Sa **Steel pipe/tube**
　culvert materials (general),
　　steel zinc-coated (galvanized) sheet, spec., **(A 444) 01.06**
　culvert pipe, ductile iron (centrifugally cast), spec., **(A 716) 01.02**
　drain, waste, and vent (DWV) piping systems,
　　pipe (polymeric, pre-coated), galvanized steel, spec., **(A 763) 01.03**
　　sheet (polymeric, pre-coated), zinc-coated (galvanized), spec., **(A 742) 01.06**
　sewer and drain pipe,
　　aluminum-zinc alloy-coated steel sheet (by hot-dip process), for storm sewer/drainage pipe, spec., **(A 806) 01.06**
　　ductile iron, gravity, centrifugally cast, spec., **(A 746) 01.02**
　　installing corrugated steel structural plate pipe (for sewers), practice, **(A 807) 01.06**
　　installing factory-made corrugated steel sewer pipe, practice, **(A 798) 01.06**
　　precoated (polymeric) galvanized sewer/drainage pipe, spec., **(A 762) 01.06**
　　steel zinc-coated (galvanized) sheet for, polymerized precoated, spec., **(A 742) 01.06**
　　structural design of corrugated steel pipe/pipe-arches/arches (for storm/sanitary sewers), practice, **(A 796) 01.06**
　soil pipe (cast iron),
　　rubber gaskets for, spec., **(C 564) 01.02**
　　spec., **(A 74) 01.02**
Drawers
　marine furniture drawers, spec., **(F 825) 01.02**
Drop impact resistance
　　See **Resistance,** impact resistance
Drop tests
　for carbon steel tee rails, **(A 1) 01.04**
　for nonheat-treated carbon steel axles, **(A 21) 01.04**

Index of ASTM Standards, Section 1

Drop tests (*cont.*)
  for railway axles of carbon steel, (**A 383**) **01.04, 01.05**
  for security determination of steel bars, (**A 627**) **01.05**
Dryer rolls (for paper mills)
  ductile iron castings, spec., (**A 476**) **01.02**
Ductile (nodular) iron
  See **Iron castings**, ductile
DWV pipe/tube
  See **Steel pipe/tube**

E

Eddy current examination
  See **Electromagnetic (eddy current) testing**
Electrical conductors
  aluminum,
    steel-reinforced (ACSR/AZ), aluminum-coated steel core wire for, spec., (**B 341**) **01.06**
    steel reinforced (ACSR), high-strength zinc-coated steel core wire for, spec., (**B 606**) **01.06**
    steel-reinforced (ACSR), zinc-coated steel core wire for, spec., (for metric see B 498M), (**B 498**) **01.06**
  aluminum alloys,
    aluminum-zinc alloy-coated steel overhead ground wire strand for electric power transmission lines), spec., (**A 797**) **01.06**
  copper,
    copper/copper-clad steel composite conductors, concentric-lay-stranded, spec., (**B 229**) **01.06**
    copper-clad steel wire, concentric-lay-stranded, spec., (**B 228**) **01.06**
    copper-clad steel wire, hard-drawn, spec., (**B 227**) **01.06**
  steel electrical conductors,
    aluminum-zinc alloy-coated steel overhead ground wire strand for electric power transmission lines), spec., (**A 797**) **01.06**
    copper clad, concentric-lay-stranded, spec., (**B 228**) **01.06**
    copper-clad wire, hard-drawn, spec., (**B 227**) **01.06**
    electrical (flat-rolled), for magnetic applications, spec., (**A 345**) **01.03**
Electrical steel
  See **Steel, stainless steel, and related alloys**, electrical steel
Electric-fusion-/(arc)-welded steel pipe
  See **Steel pipe/tube**
Electric power transmission lines
  of zinc-coated (galvanized) wire strand, spec., (**A 363**) **01.06**
Electric switching equipment
  Sa **Zinc coatings**
  assembled, zinc coatings on, by hot-dip process, spec., (**A 386**) **01.06**
Electrodeposited coatings
  See **Coatings**, electrodeposited coatings

Electrode remelted steel
  See **Steel, stainless steel, and related alloys**, electrode remelted
Electrolytic manganese,
  spec., (**A 601**) **01.02**
Electrolytic tin coated sheet
  See **Coating** (*headings starting with*)
    Steel sheet
    Zinc coatings
Electromagnetic (eddy current) testing
  of coating thickness,
    rec. practice for measuring, (**E 376**) **01.06**
Electron vacancy number
  average, calculating, (**A 567**) **01.02**
Elongation
  conversion from round to flat specimen, (**A 370**) **01.01, 01.02, 01.03, 01.05**
  of steel products, test, (**A 370**) **01.01, 01.02, 01.03, 01.05**
Embrittlement
  of hot-dip galvanized structural steel products, safeguarding against/detecting procedures, rec. practice, (**A 143**) **01.06**
End-quench test
  for hardenability of steel, Δ (**A 255**) **01.05**
Epoxy-coated reinforcing steel bars
  spec., (**A 775**) **01.04**
Etching
  etch structures of stainless steel, classification by oxalic acid test, rec. practice, (**A 262**) **01.03, 01.05**
  macroetching,
    macroetched steel-rating, Δ (**E 381**) **01.05**
    macroetch testing (of consumable electrode remelted steel bars/billets), Δ (**A 604**) **01.03, 01.05**
    macroetch testing (of tool steel bars), Δ (**A 561**) **01.05**
Expansion
  expansion discontinuity,
    evaluation of/specifying textures/discontinuities of steel castings, by visual examination, practice, (**A 802**) **01.02**
External threaded fasteners
  See **Fasteners**
  Sa **Iron bolting materials**
      Nonferrous bolting materials
      Steel bolting materials

F

Farm/field fencing
  See **Fences/fencing materials**
Fasteners, metal
  Sa **Nonferrous bolting materials**
      Steel bolting materials
Fences/fencing materials
  Sa **Guards/guardrails**
  aluminum-zinc alloy-coated steel wire rope fittings (for highway guardrails), spec., (**A 784**) **01.06**

Index of ASTM Standards, Section 1

Fences/fencing materials (*cont.*)
  aluminum-zinc alloy-coated steel wire strand (for guys/messengers/span wires), spec., **(A 785) 01.06**
  chain-link,
    aluminum-coated wire fabric, spec., **(A 491) 01.06**
    aluminum-zinc alloy coated fence fabric, spec., **(A 783) 01.06**
    def. of terms, **(F 552) 01.06**
    fence fittings, spec., **(F 626) 01.06**
    industrial/strength requirements for metal post/rails, spec., **(F 699) 01.04, 01.05**
    installation, practice, **(F 567) 01.06**
    metallic-coated steel wire (for chain link fence fabric), spec., **(A 817) 01.06**
    poly(vinyl chloride) (PVC)-coated, spec., **(F 668) 01.05**
    residential gates/gate posts/accessories, spec., **(F 654) 01.06**
    strength requirements of posts/rails (for industrial fences), spec., **(F 669) 01.06**
    strength requirements of steel posts/rails (for residential chain-link fence), spec., **(F 761) 01.06**
    zinc-coated fabric, for residential use, spec., **(F 573) 01.06**
    zinc-coated fabric, spec., **(A 392) 01.06**
  farm/field,
    of aluminum-coated steel wire, spec., **(A 584) 01.06**
    of zinc-coated (galvanized) iron/steel wire, spec., **(A 116) 01.06**
  post/assemblies,
    for farm/field/line, from hot-rolled steel bars, spec., **(A 702) 01.03, 01.05, 01.06**
    from hot-wrought steel bars, spec. for, **(A 702) 01.03, 01.05, 01.06**
    strength requirements of steel posts/rails (for residential chain-link fence), spec., **(F 761) 01.06**
  poultry netting/fencing (zinc-coated (galvanized) steel), spec., **(A 390) 01.06**
  wooden, design, fabrication, and installation, spec., **(F 537) 01.06**
  woven wire fence fabric (zinc-coated (galvanized)), spec., **(A 116) 01.06**

Ferric sulfate-sulfuric acid test
  for detecting susceptibility to intergranular attack in stainless steels, **(A 262) 01.03, 01.05**

Ferrite content
  calibration/testing of instruments for estimating ferrite content of cast stainless steels, spec., **(A 799) 01.02**
  estimating ferrite content in austenitic alloys castings, practice, **(A 799) 01.02**

Ferritic iron castings
  See **Iron castings**

Ferritic steel
  See **Steel, stainless steel, and related alloys,** ferritic steel

Ferroalloys
  sampling and testing, for size, (before or after shipment), **(A 610) 01.02**

Ferroboron
  spec., **(A 323) 01.02**

Ferrochrome-silicon
  spec., **(A 482) 01.02**

Ferrochromium
  high/low carbon, spec., **(A 101) 01.02**

Ferrocolumbium
  spec., **(A 550) 01.02**

Ferromanganese
  Sa **Electrolytic manganese**
  spec., **(A 99) 01.02**

Ferromanganese-silicon
  spec., **(A 701) 01.02**

Ferromolybdenum
  spec., **(A 132) 01.02**

Ferrosilicon
  spec., **(A 100) 01.02**

Ferrotitanium
  spec., **(A 324) 01.02**

Ferrotungsten
  spec., **(A 144) 01.02**

Ferrous metals/alloys
  Sa **Iron products, general**

Ferrovanadium
  spec., **(A 102) 01.02**

Figure 8-type communication/electrical cables
  of zinc-coated steel wire strand, extra-high strength, concentric-lay, spec., **(A 640) 01.06**

Fillet welded joints (for shipboard piping)
  See **Steel pipe/tube,** shipboard piping materials

Fire escapes
  Sa **Zinc coatings**
  assembled, zinc coatings on, by hot-dip process, spec., **(A 386) 01.06**

Fittings
  cast iron, spec., **(A 74) 01.02**
  for chain-link fences, spec., **(F 626) 01.06**

Flanged joints (for shipboard piping)
  See **Steel pipe/tube,** shipboard piping materials

Flat-link chain
  spec., **(A 466) 01.05**

Floor plate
  See **Steel plate,** rolled plate

Flush doors (for ships)
  See **Shipbuilding steel materials,** doors

Forgings
  Sa **Steel bars**
    Steel bars, forgings, and forging stock
    Steel, stainless steel, and related alloys
  for aerospace/aircraft applications,
    from alloy steel (premium quality) blooms and billets, spec. for, **(A 646) 01.05**
  alloy steel (hot-/hot-cold-/cold-worked),
    and forging billets, Grade 651, for elevated temperature service, spec. for, **(A 477) 01.05**
  austenitic steel,
    ultrasonic examination, rec. practice, **(A 745) 01.05**
  for ball and roller bearings, spec., **(A 295) 01.05**
  carbon/alloy steel,

Index of ASTM Standards, Section 1

Forgings (*cont.*)
    blooms, billets, and slabs, spec. for, **(A 711) 01.05**
    for pipe flanges/fittings/valves/parts, for high-pressure transmission-service pipe, spec., **(A 694) 01.01**
    quenched/tempered/vacuum-treated, for pressure vessels, spec., **(A 508) 01.04, 01.05**
    untreated and heat-treated, for railway use spec., **(A 730) 01.04, 01.05**
    carbon/low-alloy steel,
        components for pressure vessels (with mandatory impact testing requirements), spec., **(A 765) 01.05**
        flanges/fittings/valves for low-temperature service requiring notch toughness testing, spec., **(A 350) 01.01**
    carbon steel,
        for general-purpose piping (fittings/valves), spec., **(A 181) 01.01**
        leaded and resulphurized, for non-welded pressure systems (and general service), spec., **(A 766) 01.01**
        for piping components from −20 to +650°F (−30 to +345°C) with required notch toughness, spec., **(A 727) 01.01**
    cold, of stainless/heat-resisting steel, spec. for, **(A 493) 01.03**
    for corrugating paper machinery,
        forged steel rolls, spec., **(A 649) 01.05**
    crankshaft,
        ultrasonic examination of, spec., **(A 503) 01.05**
        magnetic particle inspection of, spec., **(A 456) 01.05**
    for cyclotrons and related nuclear equipment,
        of carbon steel, spec., **(A 594) 01.05**
    die, closed-impression steel, for general industrial use, **(A 521) 01.05**
    for fabrication of tools,
        from tungsten-/molybdenum-type high speed tool steels, spec., **(A 600) 01.05**
        from wrought carbon steel, spec., **(A 686) 01.05**
    for general industrial use,
        of carbon and alloy steel, untreated and heat-treated, spec., **(A 668) 01.05**
    for generators,
        of alloy steel (for nonmagnetic retaining rings), spec. for, **(A 289) 01.05**
    iron,
        aluminum coatings on, by hot-dip process, spec., **(A 676) 01.06**
    of iron base superalloys, precipitation hardening (hot-cold-finished), for high-temperature service, spec. for, **(A 638) 01.05**
    macroetched steel, rating, Δ **(E 381) 01.05**
    magnetic particle examination of steel forgings, test, **(A 275) 01.05**
    mechanical testing, **(A 370) 01.01, 01.02, 01.03, 01.05**
    for pinions and gears,
        of normalized/tempered carbon steel and quenched/tempered alloy steel, spec. for, **(A 291) 01.05**
    of precipitation hardening cobalt alloys (UNS R30155, R30816), for high-temperature service, spec., **(B 639) 01.05**
    of precipitation hardening nickel alloys (UNS N07001, N07080, N07252, N07500, N07718, N17750), for high-temperature service, spec., **(B 637) 01.05**
    for pressure vessel components,
        of alloy steel, for seamless drums/heads, spec., **(A 336) 01.01, 01.04, 01.05**
        of carbon/alloy steel, for thin-walled vessels, spec., **(A 372) 01.04, 01.05**
        carbon/alloy steel, quenched and tempered, spec., **(A 541) 01.04, 01.05**
        carbon/low-alloy steel, with mandatory impact testing requirements, spec., **(A 765) 01.05**
        carbon steel piping components (flanges/fittings/valves/etc.), for ambient/higher-temperature service, spec., **(A 105) 01.01**
        of carbon steel, spec., **(A 266) 01.01, 01.04, 01.05**
        low-alloy/stainless steel (flanges/fittings/valves/etc.), spec., **(A 182) 01.01**
        of steel, low-alloy, high-strength, quenched and tempered, for fittings and parts, spec., **(A 592) 01.04, 01.05**
    for pressure vessels/isostatic presses/shock tubes/etc.,
        of alloy steel, high-strength, quenched and tempered, spec., **(A 723) 01.04, 01.05**
    for reduction gear rings,
        of carbon and alloy steel, normalized/tempered, and quenched/tempered, spec., **(A 290) 01.05**
    stainless/heat-resisting steel,
        billets and bars, spec. for, **(A 314) 01.05**
        spec., **(A 473) 01.05**
    stainless/heat-resisting steel (age-hardening), spec., **(A 705) 01.05**
    stainless steel (martensitic),
        for high-temperature service, spec., **(A 565) 01.05**
    for steam turbine shafts and rotors, heat stability, test, **(A 472) 01.05**
    steel,
        def. of term, **(A 509) 01.05**
        general requirements, spec., **(A 788) 01.05**
        hardenability test, by end-quench (Jominy) method, **(A 255) 01.05**
        zinc coatings on, assuring high-quality coatings, rec. practice, **(A 385) 01.06**
        zinc coatings on, by hot-dip galvanizing process, spec., **(A 123) 01.06**
        zinc coatings on, safeguarding against embrittlement/detection procedures, rec. practice, **(A 143) 01.06**
        zinc coatings on, warpage/distortion prevention during hot-dip galvanizing process, rec. practice, **(A 384) 01.06**
    of superstrength alloy steel (above 140 ksi (965 MPa)), spec. for, **(A 579) 01.05**

Index of ASTM Standards, Section 1

Forgings (*cont.*)
for turbine generators,
 carbon steel, quenched and tempered, for magnetic retaining rings, spec., **(A 288) 01.05**
for turbine rotor disks and wheels,
 of alloy steel, vacuum-treated, spec., **(A 471) 01.05**
for turbine rotors and shafts,
 of carbon and alloy steel, heat-treated, spec., **(A 293) 01.05**
 of carbon and alloy steel, vacuum-treated, spec., **(A 470) 01.05**
 of chromium (12%) alloy steel, spec., **(A 768) 01.05**
 of steel, vacuum-treated, basic electric, spec., **(A 469) 01.05**
for turbine wheels and disks,
 of alloy steel, heat-treated, spec., **(A 294) 01.05**
ultrasonic examination of,
 turbine and generator steel rotor forgings, **(A 418) 01.05**
 heavy steel, by angle-beam technique, **(A 388) 01.05**
 of vacuum-treated basic electric steel, for generator motors, spec., **(A 469) 01.05**
zinc-coated (galvanized),
 locating the thinnest spot on coating, by Preece test (copper sulfate dip), **(A 239) 01.06**
Fracture appearance transition temperature (FATT)
 See **Temperature tests,** fracture appearance transition temperature (FATT)
Free-machining stainless steel wire
 See **Steel wire**
Free-machining steel bars
 See **Steel bars**
Furniture
chest of drawers (chiffoniers), for ships, spec., **(F 822) 01.02**
marine furniture drawers, spec., **(F 825) 01.02**
marine furniture tops, spec., **(F 826) 01.02**
marine log desks with cabinets, spec., **(F 823) 01.02**
marine mess tables, spec., **(F 824) 01.02**
Furniture spring steel wire
 See **Steel wire**
Fusion
evaluation of/specifying textures/discontinuities of steel castings, by visual examination, practice, **(A 802) 01.02**
Fusion welding
with carbon-manganese-silicon steel plates, for certain atmospheric conditions, spec. for, **(A 573) 01.04**

G

Gages (for ships)
 See **Shipbuilding steel materials,** gages
Galvanized surfaces
 See **Steel pipe/tube**
  Steel sheet
  Zinc-coated (galvanized) surfaces

Gaskets and gasket materials
rubber,
 for cast iron soil pipe and fittings, spec., **(C 564) 01.02**
Gears
pinions/gears, of normalized/tempered carbon steel and quenched/tempered alloy steel forgings, spec., **(A 291) 01.05**
reduction gear rings, of normalized/tempered, and quenched/tempered carbon and alloy steel forgings, spec., **(A 290) 01.05**
Generator materials
 Sa **Automotive steel materials**
  Building steel materials
  Pressure vessel steel materials
  Railroad steel materials
  Ship building steel materials
  Steel bolting materials
  Steel, stainless steel, and related alloys
forgings for nonmagnetic retaining rings, made from alloy steel, spec. for, **(A 289) 01.05**
mechanical testing, **(A 370) 01.01, 01.02, 01.03, 01.05**
steam turbine shafts and rotor forgings, heat stability, test, **(A 472) 01.05**
turbine,
 steel retaining rings, ultrasonic inspection of, rec. practice, **(A 531) 01.05**
turbine rotor disks and wheels,
 of vacuum-treated alloy steel forgings, spec. for, **(A 471) 01.05**
turbine rotors and shafts,
 of heat-treated carbon and alloy steel forgings, spec., **(A 293) 01.05**
 ultrasonic inspection of forgings, **(A 418) 01.05**
 of vacuum-treated basic electric steel forgings, spec. for, **(A 469) 01.05**
 of vacuum-treated carbon and alloy steel, spec. for, **(A 470) 01.05**
turbine wheels and disks,
 of heat-treated alloy steel forgings, spec. for, **(A 294) 01.05**
Girder rails
 See **Railroad steel materials**
Graphite
in iron castings, evaluating microstructure, Δ **(A 247) 01.02**
Grates, steel
 Sa **Zinc coatings**
assembled, zinc coatings on, by hot-dip process, spec., **(A 386) 01.06**
Gray cast iron
 See **Iron castings**
Ground wire strand
 See **Steel wire strand**
Guards/guardrails
 Sa **Fences/fencing materials**
aluminum-zinc alloy coated steel wire rope/fittings (for highway guardrail), spec., **(A 784) 01.06**
woven/welded galvanized steel wire fabric ("hardware cloth"), for windows/screen doors/trees/industrial machines, spec., **(A 740) 01.06**

Guards/guardrails (*cont.*)
   zinc-coated steel wire rope, for highways, spec., **(A 741) 01.06**
Guys (steel wire strand for)
   See **Steel wire strand**

H

Handgrab
   staple/handgrab/handle/stirrup rung, spec., **(F 783) 01.02**
Handles
   staple/handgrab/handle/stirrup rung, spec., **(F 783) 01.02**
Hangers (for ship pipe/tube)
   See **Steel pipe/tube**, pipe hangars
   Sa **Shipbuilding steel materials**
Hardenability
   of steel,
     for bars, subject to end-quench hardenability requirements, spec. for, **(A 304) 01.05**
     by end-quench (Jominy) test, Δ **(A 255) 01.05**
Hardness, Brinell
   of steel products, test, **(A 370) 01.01, 01.02, 01.03, 01.05**
Hardness (indentation)
   of steel products, Brinell, portable, and Rockwell, test, **(A 370) 01.01, 01.02, 01.03, 01.05**
Hardness, Rockwell
   of steel products, test, **(A 370) 01.01, 01.02, 01.03, 01.05**
Hardness, Rockwell (superficial)
   of steel products, test, **(A 370) 01.01, 01.02, 01.03, 01.05**
Hardware (iron/steel)
   See **Iron hardware**
       Steel hardware
Hardware cloth
   spec., **(A 740) 01.06**
Hatch covers
   staple/handgrab/handle/stirrup rung, spec., **(F 783) 01.02**
Heating tests
   heat stability,
     of steam turbine shafts/rotor forgings, test, **(A 472) 01.05**
Heat-resisting stainless steel alloys
   See **Steel, stainless steel, and related alloys**, stainless heat-resisting steel
Heat treatment
   of metals, def. of terms, **(E 44) 01.02**
Heat treatment of metals
   def. of terms, **(E 44) 01.02**
Heavy industrial fence
   See **Fences/fencing materials**
Helical compression springs
   hot-coiled, heat-treated, made of hot-rolled round steel bars, spec. for, **(A 125) 01.05**
High-carbon steel
   See **Steel, stainless steel, and related alloys**, carbon steel (high)

High-fatigue resistant steel spring wire
   See **Steel wire**
High-silicon cast iron
   See **Iron castings**, white
High-speed tool steels
   See **Steel, stainless steel, and related alloys**, tool steel
High-strength, low-alloy structural steel
   See **Steel, stainless steel, and related alloys**, structural
High-stress steel wire
   See **Steel wire**
High tensile strength steel spring wire
   See **Steel wire**
Highway guardrail
   See **Guards/guardrails**
   Sa **Fences/fencing materials**
Hinged doors (for ships)
   See **Shipbuilding steel materials**, doors
HNO_3 test
   See **Nitric acid test procedure**
Hot-dip (galvanized) iron/steel products
   See **Zinc-coated (galvanized) iron/steel articles**
Household laundry and dishwasher equipment
   assembled, zinc coatings on (by hot-dip process), spec., **(A 386) 01.06**
Hydrogen embrittlement
   of hot-dip galvanized structural steel products, rec. practice for detecting, **(A 143) 01.06**

I

Impact testing
   of cast iron, **(A 327) 01.02**
   Charpy impact test,
     for cast iron, **(A 327) 01.02**
     of steel products, **(A 370) 01.01, 01.02, 01.03, 01.05**
     of structural steel - sampling procedures, spec., **(A 673) 01.04**
Inclusions
   evaluation of/specifying textures/discontinuities of steel castings, by visual examination, practice, Δ **(A 802) 01.02**
Inconel alloy
   See **Nickel and nickel alloys**, commercial designations
Indentation hardness
   See **Hardness (indentation)**
Industrial chain-link fence
   See **Fences/fencing materials**
Industrial forgings
   See **Forgings**
Industrial steel materials
   forgings,
     of carbon and alloy steel, untreated and heat-treated, spec., **(A 668) 01.05**

Index of ASTM Standards, Section 1

Installation procedures
  installing corrugated steel structural plate pipe (for sewers), practice, **(A 807) 01.06**
  for thermal insulation piping/machinery (on ships), practice **(F 683) 01.02**
  underground installation,
    installing factory-made corrugated steel sewer pipe, practice, **(A 798) 01.06**
  for wooden fences, spec., **(F 537) 01.06**
Intergranular corrosion
  See **Corrosion**
Interior steel doors (for ships)
  See **Ship building steel materials**
Intermediate hardness carbon spring steel
  spec., **(A 684) 01.03, 01.05**
Internal threaded fasteners
  See **Fasteners**
  Sa **Iron bolting materials**
       Nonferrous bolting materials
       Steel bolting materials
Investment steel castings
  See **Steel castings**
Iron castings
  aluminum coatings on, by hot-dip process, spec., **(A 676) 01.06**
  chilled/white iron, methods of testing, **(A 360) 01.02**
  chill testing, **(A 367) 01.02**
  compression testing, (also consult E 9, Vol. 03.01 for testing of metallic materials), **(A 256) 01.02**
  def. of terms, **(A 644) 01.02**
  ductile,
    austenitic, for heat-/corrosion-/wear-resistant applications, spec., **(A 439) 01.02**
    austenitic, for pressure-containing parts (compressors, expanders, pumps, etc.) for low-temperature (down to −423°F (−234°C)) service, spec., **(A 571) 01.02**
    for culvert pipe, centrifugally cast, spec., **(A 716) 01.02**
    ferritic, for pressure-retaining parts for elevated temperatures, spec., **(A 395) 01.02**
    gravity sewer pipe, centrifugally cast, spec., **(A 746) 01.02**
    for paper mill dryer rolls, spec., **(A 476) 01.02**
    for pressure pipe, spec., **(A 377) 01.02**
    reference radiographs for, guide to, **(E 689) 01.02**
    spec., **(A 536) 01.02**
    electroplating, preparation for, practice, **(B 320) 01.02**
    graphite microstructure in, evaluating, Δ **(A 247) 01.02**
  gray,
    austenitic, for heat-/corrosion-/wear-resistant applications, spec., **(A 436) 01.02**
    for automotive/allied industries, spec., **(A 159) 01.02**
    centrifugally cast dual metal cylinders, spec., **(A 667) 01.02**
    for elevated temperatures for non-pressure containing parts (grate bars, stoker links/parts, ingot molds, etc.), spec., **(A 319) 01.02**
    for flanges/valves/pipe fittings, spec., **(A 126) 01.02**
    for general engineering use, spec., **(A 48) 01.02**
    for pressure-containing parts for temperatures up to 650°F (345°C), spec., **(A 278) 01.02**
    for pressure pipe, spec., **(A 377) 01.02**
    transverse testing, **(A 438) 01.02**
  gray/ductile pipe,
    polyethylene encasement, for application to underground installations, rec. practice, **(A 674) 01.02**
  high-silicon (corrosion-resistant), spec., **(A 518) 01.02**
  impact testing, **(A 327) 01.02**
  malleable,
    ferritic/pearlitic/tempered pearlitic/tempered martensitic grades, for automotive/allied industries, spec., **(A 602) 01.02**
    ferritic, used in general engineering castings, at normal and elevated temperatures, spec., **(A 47) 01.02**
    flanges/pipe fittings/valve parts, for railroad/marine/heavy duty service, at elevated temperatures (up to 650°F (345°C)), spec., **(A 338) 01.02**
    pearlitic, for general engineering use at normal/elevated temperatures, spec., **(A 220) 01.02**
  for the cupola process, spec., **(A 197) 01.02**
  merchant pig,
    for foundry use, spec., **(A 43) 01.02**
  for soil pipe,
    and fittings, spec., **(A 74) 01.02**
    rubber gaskets for, **(C 564) 01.02**
  white,
    abrasion-resistant, for mining/milling/earth-handling/manufacturing industries, spec., **(A 532) 01.02**
    centrifugally cast dual metal cylinders, spec., **(A 667) 01.02**
    high-silicon, for corrosion-resistant service, spec., **(A 518) 01.02**
    statically cast dual metal (chilled white iron-gray iron) rolls, for pressure vessel use, spec., **(A 748) 01.02**
    testing chilled/white iron castings, **(A 360) 01.02**
  zinc-coated (galvanized),
    locating the thinnest spot on coating, by Preece test (copper sulfate dip), **(A 239) 01.06**
Iron-chromium/iron-chromium-nickel alloy
  estimating ferrite content in austenitic alloys castings, practice, **(A 799) 01.02**
  heat-resistant castings (for general application), spec., **(A 297) 01.02**
Iron-chromium/iron-chromium-nickel/nickel-base alloy
  corrosion-resistant castings (for general application), spec., **(A 743) 01.02**
Iron hardware
  zinc coatings on, by hot-dip galvanizing process, assuring high-quality coatings, rec. practice, **(A 385) 01.06**

Index of ASTM Standards, Section 1

Iron hardware (*cont.*)
  safeguarding against embrittlement/detection procedures, rec. practice, **(A 143) 01.06**
  spec., **(A 153) 01.06**
  warpage/distortion prevention during, rec. practice, **(A 385) 01.06**
Iron pipe
  spiral-welded, for conveying liquid, gas, or vapor, spec., **(A 211) 01.01**
Iron products, general
  alloying additives
    See **Steel, stainless steel, and related alloys,** alloying additives
  aluminum coatings on, by hot-dip process, spec., **(A 676) 01.06**
  mechanically deposited coatings on,
    cadmium coatings, on iron/steel metals, spec., **(B 696) 01.04**
    zinc coatings, spec., **(B 695) 01.04**
  reduction of, cold/hot, by wrought alloy steel rolls, spec., for, **(A 427) 01.05**
Iron sheets
  aluminum-coated,
    weight of coating, test, **(A 428) 01.06**
  zinc-coated (galvanized),
    locating the thinnest spot on coating, by Preece test (copper sulfate dip), **(A 239) 01.06**
    weight of coating, test, **(A 90) 01.06**
Isostatic presses
  forgings for, of alloy steel, high-strength, quenched and tempered, spec., **(A 723) 01.04, 01.05**

J

J-band hangar (for ships)
  See **Shipbuilding steel materials**
Joint bars
  carbon steel, low-/medium-high, non-heat treated, spec., **(A 3) 01.04**
  heat-treated carbon steel, spec., **(A 49) 01.04**
Jominy test
  See **End-quench test**

L

Laboratory apparatus
  screens (industrial perforated), for particle size determinations,
    round opening series, spec., **(E 674) 01.03**
    square opening series, spec., **(E 454) 01.03**
Ladders
  fixed vertical steel ladders (for personnel access on ships), spec., **(F 840) 01.02**
Lashings
  staple/handgrab/handle/stirrup rung, spec., **(F 783) 01.02**
Laundry equipment, household
  See **Household laundry and dishwasher equipment**

Leaded carbon steel
  See **Steel, stainless steel, and related alloys,** carbon steel
Left hand doors (for ships)
  See **Shipbuilding steel materials,** doors
Lifting slings
  staple/handgrab/handle/stirrup rung, spec., **(F 783) 01.02**
Light industrial fence
  See **Fences/fencing materials**
Linear changes
  linear discontinuity,
    evaluation of/specifying textures/discontinuities of steel castings, by visual examination, practice, Δ **(A 802) 01.02**
Line-type fences
  See **Fences/fencing materials**
Loading
  steel products for domestic shipment, rec. practice, **(A 700) 01.01, 01.03, 01.04, 01.05**
Log desks (for ships)
  marine log desks with cabinets, spec., **(F 823) 01.02**
Longitudinal wave testing
  of steel retaining rings for turbine generators, as par of ultrasonic inspection, test, **(A 531) 01.05**
Long-terne steel sheet
  See **Steel sheet,** long-terne coated
Low-alloy steel
  See **Steel, stainless steel, and related alloys,** low-alloy steel
Low-carbon steel
  See **Steel, stainless steel, and related alloys**

M

M252 alloy
  See **Nickel and nickel alloys,** commercial designations
Machinery (for ships)
  thermal insulation for, selection/installation, practice, **(F 683) 01.02**
Machine screw steel wire
  See **Steel wire**
Macroetching
  See **Etching,** macroetching
Magnetic alloys
  soft,
    identifying standard electrical grades in ASTM specifications, practice, **(A 644) 01.02**
Magnetic applications
  of flat-rolled electrical steels, spec., **(A 345) 01.03**
Magnetic particle inspection
  of large crankshaft forgings, spec., **(A 456) 01.05**
  of steel forgings, test, **(A 275) 01.05**
Magnetic retaining rings
  for turbine generators, made of quenched and tempered carbon steel forgings, spec. for, **(A 288) 01.05**
Magnetic testing
  of coating thickness,
    rec. practice for measuring, **(E 376) 01.06**

Manganese-chromium-molybdenum-silicon-zirconium, alloy steel
  pressure vessel plats (quenched/tempered), spec., **(A 782) 01.04**
Manganese, electrolytic
  See **Electrolytic manganese**
Manganese-molybdenum-columbium alloy
  low-carbon steel plates, for piping components and welded pressure vessels at moderate/low temperature service, spec., **(A 735) 01.04**
  low-carbon steel plates, shapes, and bars, spec., **(A 699) 01.04, 01.05**
Manganese-molybdenum-/nickel alloy
  plates, for welded boilers, spec., **(A 302) 01.04**
  plates, (quenched and tempered alloy), for welded pressure vessels, spec., **(A 533) 01.04**
Manganese-titanium carbon steel
  plates, for glass or diffused metallic coatings, spec., **(A 562) 01.04**
Manganese-vanadium alloys
  steel plate, for locomotive boiler shells and other pressure vessels, spec., **(A 225) 01.04**
Manholes
  staple/handgrab/handle/stirrup rung, spec., **(F 783) 01.02**
Marine corrosion resistance
  See **Resistance,** corrosion/corrosion resistance
Marine environments, steel for
  general requirements for rolled shapes, plates, sheet piling, and bars, spec., **(A 6) 01.04**
  H-piles and sheet piling, of high-strength, low-alloy steel (for sea walls/dock walls/cofferdams, etc.), spec. for, **(A 690) 01.04**
  sheet piling, of carbon steel, for dock walls/sea walls/cofferdams/etc.), spec., **(A 328) 01.04**
Marine furniture
  doors/marine furniture, spec., **(F 782) 01.02**
Marine steel materials
  See **Ship building steel materials**
Marking
  steel products for domestic shipment, rec. practice, **(A 700) 01.01, 01.03, 01.04, 01.05**
Martensitic stainless steel
  See **Steel, stainless steel, and related alloys**
Meat processing/handling equipment
  assembled, zinc coatings on, by hot-dip process, spec., **(A 386) 01.06**
Mechanically deposited coatings
  See **Coatings,** mechanically deposited
Mechanical spring steel wire
  See **Steel wire**
Mechanical testing
  of steel products, methods and def., **(A 370) 01.01, 01.02, 01.03, 01.05**
Mechanical tubing
  See **Steel pipe/tube**
Medium-carbon steel
  See **Steel, stainless steel, and related alloys**
Merchant pig iron castings
  See **Iron castings,** merchant pig

Mesh (for concrete reinforcement)
  See **Steel wire,** welded fabric
Mesh (for steel pipe/tube)
  zinc-coated (galvanized) steel pipe winding mesh, spec., **(A 810) 01.06**
Messengers (steel wire strand for)
  See **Steel wire strand**
Mess tables
  marine mess tables, spec., **(F 824) 01.02**
Metal-arc-welded steel pipe
  See **Steel pipe/tube,** metal-arc-welded
Metal funnels (for ships)
  See **Shipbuilding materials (general)**
Metallography
  iron castings, graphite microstructure in, evaluating, Δ **(A 247) 01.02**
  macroetched steel, rating, Δ **(E 381) 01.05**
  volume fraction of second phase, by systematic manual point count, rec. practice, **(E 562) 01.02**
Metals and alloys numbering system
  unified (UNS), practice, **(E 527) 01.01, 01.02, 01.03, 01.04, 01.05**
Metals and metallic materials
  heat treatment, def. of terms, **(E 44) 01.02**
Metric practice
  SI (International System of Units), excerpts, **(E 380)** (Related Material—all parts) **01.00**
Microstructure
  calibration/testing of instruments for estimating ferrite content of cast stainless steels, spec., **(A 799) 01.02**
  estimating ferrite content in austenitic alloys castings, practice, **(A 799) 01.02**
  evaluation of/specifying textures/discontinuities of steel castings, by visual examination, practice, Δ **(A 802) 01.02**
Modular gage boards
  See **Shipbuilding steel materials**
Molybdenum alloys
  steel plates, for welded locomotive boiler shells and other pressure vessels, spec., **(A 204) 01.04**
Molybdenum high-speed tool steel
  See **Steel, stainless steel, and related alloys,** tool steel
Molybdenum oxide products
  spec., **(A 146) 01.02**
Music wire
  See **Steel wire,** music wire

N

N-155 alloy
  See **Cobalt alloys,** commercial designations
Negative buoyancy pipe
  Sa **Steel pipe/tube**
  zinc-coated (galvanized) steel pipe winding mesh, spec., **(A 810) 01.06**

Index of ASTM Standards, Section 1

Nelson hanger (for shipboard piping)
    See **Shipbuilding steel materials**
Nickel
    castings, spec., **(A 494) 01.02**
Nickel 205
    See **Nickel and nickel alloys, commercial designations**
Nickel alloys
    clad steel plate, for pressure vessels, spec., **(A 265) 01.03, 01.04**
    steel plates, for welded pressure vessels, double-normalized and tempered, for cryogenic service, spec., **(A 353) 01.04**
        nickel (5%), austenitized, quenched, temperized, reversion-annealed, for service at cryogenic temperatures, spec., **(A 645) 01.04**
        nickel (8–9%), quenched and tempered, spec., **(A 553) 01.04**
        nickel (12%), precipitation-hardening (maraging), spec., **(A 590) 01.04**
        nickel (18%), precipitation-hardening (maraging), spec., **(A 538) 01.04**
        nickel (36%), spec., **(A 658) 01.04**
    spec., **(A 203) 01.04**
Nickel and nickel alloys
    castings, **(A 494) 01.02**
    clad steel plate, spec., **(A 265) 01.03, 01.04**
    precipitation hardening,
        bars, forgings, and forging stock, (UNS N07001, N07080, N07252, N07500, N07718, N07750), for high-temperature service, spec., **(B 637) 01.05**
        plate, sheet, and strip (UNS N07718), for high-temperature service, spec., **(B 670) 01.03**
Nickel and nickel alloys, commercial designations
    Inconel 718 (UNS N07718), spec., **(B 670) 01.03**
    Inconel X750 (UNS N07750), spec., **(B 637) 01.05**
    Nimonic 80 A (UNS N07080), spec., **(B 637) 01.05**
Nickel chromium
    castings, spec., **(A 494) 01.02**
Nickel-chromium-molybdenum
    steel plates, quenched and tempered, for welded pressure vessels, spec., **(A 543) 01.04**
Nickel-clad steel
    plate, for pressure vessels, spec., **(A 265) 01.03, 01.04**
Nickel-cobalt-molybdenum-chromium
    steel plates, quenched and tempered, for welded pressure vessels, spec., **(A 605) 01.04**
Nickel-copper
    castings, spec., **(A 494) 01.02**
Nickel-copper-chromium-molybdenum-columbium alloy steel
    low-carbon, age-hardening plates, for piping components and welded pressure vessels, spec., **(A 736) 01.04**
    with low-carbon, age-hardening process, spec. for, **(A 710) 01.04**
Nickel-copper-columbium alloy steel
    with low-carbon age-hardening process, spec. for, **(A 710) 01.04**

Nickel-molybdenum (UNS N10001, UNS N10665)
    castings, spec., **(A 494) 01.02**
Nickel-molybdenum-chromium
    castings, spec., **(A 494) 01.02**
Nickel oxide sinter
    spec., **(A 636) 01.02**
Nickel steel
    See **Steel, stainless steel, and related alloys, nickel/nickel alloys**
Nimonic alloy 80A
    See **Nickel and nickel alloys, commercial designations**
Nipples, steel, pipe/tube
    See **Steel pipe/tube, pipe nipples**
Nitric acid test
    for detecting susceptibility to intergranular attack in stainless steels, **(A 262) 01.03, 01.05**
Nitric-hydrofluoric acid test
    for detecting susceptibility to intergranular attack in stainless steels, **(A 262) 01.03, 01.05**
Nitriding
    alloy steel bars (for surface hardening by nitriding), spec., **(A 355) 01.05**
Nodular iron castings
    See **Iron castings**
Nomenclature (use of terms/terminology)
    packaging/marketing/loading steel products (for domestic shipments), **(A 700) 01.01, 01.03, 01.04, 01.05**
Nondestructive testing
    electromagnetic (eddy current) testing,
        coating thickness (on metal substrates), rec. practice, **(E 376) 01.06**
    magnetic particle inspection,
        magnetic particle inspection of large crankshaft forgings, spec., **(A 456) 01.05**
        magnetic particle inspection of steel forgings, test, **(A 275) 01.05**
    reference radiographs,
        for ductile iron castings, **(E 689) 01.02**
        for steel castings—heavy walled (2 to 4½ in. (51 to 114 mm) thick), Δ **(E 186) 01.02**
        for steel castings—heavy walled (4½ to 12 in. (114 to 305 mm) thick), Δ **(E 280) 01.02**
        for steel castings—investment (for aerospace applications), Δ **(E 192) 01.02**
        for steel castings (up to 2 in. (51 mm) thick), Δ **(E 446) 01.02**
        for steel welds (fusion), Δ **(E 390) 01.02**
    ultrasonic testing,
        angle-beam ultrasonic examination (of steel plates), spec., **(A 577/A 577M) 01.04**
        of austenitic steel forgings, by straight/angle-beam techniques, practice, **(A 745) 01.05**
        of carbon/low-alloy steel castings, by longitudinal-beam technique, spec., **(A 609) 01.02**
        of carbon steel axles (heat-/nonheat-treated), test, **(A 21) 01.04**
        of heavy-steel forgings, by straight/angle-beam techniques, practice, **(A 388) 01.05**
        of large forged crankshafts, spec., **(A 503) 01.05**

Index of ASTM Standards, Section 1

Nondestructive testing (*cont.*)
longitudinal-beam ultrasonic testing of carbon/low-alloy steel castings, spec., **(A 609) 01.02**
of pressure vessel steel plates, by angle-beam technique, **(A 577/A 577M) 01.04**
of pressure vessel steel plates, by straight-beam pulse-echo technique, spec., **(A 435) 01.04**
of pressure vessel steel plates (rolled carbon/alloy plain and clad), by straight-beam pulse-echo technique, spec., **(A 578) 01.04**
of turbine-generator steel retaining rings, rec. practice, **(A 531) 01.05**
of turbine/generator steel rotor forgings, **(A 418) 01.05**
Nonferrous bolting materials
    Sa **Steel bolting materials**
Nonferrous products
reduction of, cold/hot, by wrought alloy steel rolls, spec. for, **(A 427) 01.05**
Nonmetallic inclusions
evaluation of/specifying textures/discontinuities of steel castings, by visual examination, practice, Δ **(A 802) 01.02**
Nuclear applications, materials for
austenitic stainless steel pipe/tube,
  pipe, spec., **(A 451) 01.01**
  tubing (for breeder reactor core components), spec., **(A 771) 01.01**
for forgings,
  of quenched and tempered vacuum-treated carbon and alloy steel, spec., **(A 508) 01.04, 01.05**
plate,
  steel, special requirements, **(A 647) 01.04**
steel, special requirements,
  alloy steel bolting materials, spec., **(A 540) 01.01**
  bolting materials, **(A 614) 01.01**
Nuclear pressure vessels
of carbon steel forgings (with special magnetic characteristics), for cyclotrons and related nuclear equipment, spec., **(A 594) 01.05**
Nuclear reactor vessels
breeder reactor core components,
  tubing (of austenitic stainless steel alloy), spec., **(A 771) 01.01**
Numbering system, unified (UNS)
for metals and alloys, practice, **(E 527) 01.01, 01.02, 01.03, 01.04, 01.05**
Nuts
    See **Nonferrous bolting materials**
           Steel bolting materials

O

Oceania, steel for
    See **Marine environments, steel for**
Oil-tempered steel wire
    See **Steel wire**
Optical materials/properties/tests
visual examination,
  angle-beam ultrasonic examination (of steel plate for pressure vessels), spec., **(A 577/A 577M) 01.04**

calibration/testing of instruments for estimating ferrite content of cast stainless steels, spec., **(A 799) 01.02**
detecting susceptibility to intergranular attack in austenitic stainless steels, rec. practice, **(A 262) 01.03, 01.05**
detecting susceptibility to intergranular attack in ferritic stainless steels, practice, **(A 763) 01.03**
detecting susceptibility to intergranular corrosion in severely sensitized austenitic stainless steel, **(A 708) 01.03**
evaluation of/specifying textures/discontinuities of steel castings, by visual examination, practice, Δ **(A 802) 01.02**
straight beam ultrasonic examination (of plain/clad steel plate for special applications), spec., **(A 578/A 578M) 01.04**
straight-beam ultrasonic examination (of steel plate for pressure vessels), spec., **(A 435/A 435M) 01.04**
Overhead steel wire strand
    See **Steel wire strand**
Oxalic acid etch test
for classifying etch structures of stainless steels (for susceptibility determination of intergranular attack), **(A 262) 01.03, 01.05**

P

Packaging and packaging materials
steel products for domestic shipment, rec. practice, **(A 700) 01.04, 01.05**
Paints and related coatings
in marine environments,
  shipbuilding/marine product/procedure data sheet, Δ **(F 718) 01.02**
shipbuilding/marine product/procedure data sheet, Δ **(F 718) 01.02**
Panel doors (for ships)
    See **Shipbuilding steel materials,** doors
Panel fences
    See **Fences/fencing materials**
Paper and paperboard
paper mill dryer rolls,
  ductile iron castings for, spec., **(A 476) 01.02**
Paper machinery
corrugating, forged steel rolls for, spec., **(A 649) 01.05**
Paper mill dryer rolls
ductile iron castings, spec., **(A 476) 01.02**
Pearlitic iron castings
    See **Iron castings**
Pearlitic malleable iron castings
    See **Iron castings,** malleable
Peen plating
    See **Coatings,** mechanically deposited
Perforated plate/screens (for particle size analysis)
    See **Plate and screens**
    Sa **Particle size (analysis/distribution)**
Picket fences
    See **Fences/fencing materials**

Index of ASTM Standards, Section 1

Piles
  high-strength, low-alloy steel H-piles/sheet piling (for use in marine environments), spec., **(A 690) 01.04**
  rolled steel plates/shapes/sheet piling/bars, spec. for general requirements, **(A 6) 01.04**
  steel sheet piling, spec., **(A 328) 01.04**
  welded/seamless steel pipe piles, spec., **(A 252) 01.01**

Piles, steel pipe
  See **Steel pipe/tube**, pipe piles

Pinions and gears
  See **Gears**

Pipe-arches
  See **Steel pipe/tube**, pipe-arches

Pipe installation
  See **Installation procedures**

Pipe nipples
  Sa **Steel pipe/tube**, pipe nipples

Pipe (steel)
  See **Steel pipe/tube**

Pipe (steel—for ships)
  See **Steel pipe/tube**, shipboard piping materials

Pipe fittings
  See **Steel pipings fittings**

Plain-carbon steel
  See **Steel, stainless steel, and related alloys**, stainless steel

Plate and screens
  industrial perforated, for particle size determination,
    round opening series, spec., **(E 674) 01.03**
    square opening series, spec., **(E 454) 01.03**

Plate coils
  See **Steel sheet/strip, carbon**, hot-rolled

Plate pipe, steel
  See **Steel pipe/tube**, steel plate pipe

Point count
  systematic manual, for volume fraction of second phase, rec. practice, **(E 562) 01.02**

Polyethylene (PE) plastics
  encasement (for underground installation of gray/ductile cast iron pipe), rec. practice, **(A 674) 01.02**

Polymeric (precoated) steel pipe/tube
  See **Steel pipe/tube**

Polymers
  precoated (polymeric) galvanized steel sewer/drainage pipe, spec., **(A 762) 01.06**

Poly(vinyl chloride) (PVC) plastics (PVC homopolymer, or vinyl chloride copolymer)
  coated steel chain-link fence fabric, spec., **(F 668) 01.06**
  steel chain-link fence fabric (coated), spec., **(F 668) 01.06**

Porcelain enameling
  of steel sheet, spec., **(A 424) 01.03**

Porosity
  evaluation of/specifying textures/discontinuities of steel castings, by visual examination, practice, **(A 802) 01.02**

Posts/assemblies (for chain link fence)
  See **Fences/fencing materials**

Poultry netting/fencing
  of zinc-coated (galvanized) steel wire, spec., **(A 390) 01.06**

Precipitation-hardening cobalt alloy
  See **Cobalt-containing alloys**

Precipitation-hardening nickel alloy
  See **Nickel and nickel alloys, precipitation-hardening**

Preece test
  for locating thinnest spot, on zinc-coated (galvanized) iron/steel articles, **(A 239) 01.06**

Pressure vessel steel materials
  bars,
    ferritic alloy steel, hot-rolled, for elevated temperature or pressure-containing parts suitable for fusion welding, spec., **(A 739) 01.05**
  bars/shapes,
    hot-rolled ferritic alloy, for elevated temperature or pressure-containing parts, spec., **(A 739) 01.05**
  bars/shapes/wire,
    of stainless/heat-resisting steel, spec., **(A 479) 01.04, 01.05**
  castings (carbon/low-alloy steels, and martensitic stainless steels), spec., **(A 487) 01.01, 01.02**
  for cryogenic service,
    of 5 % nickel steel, austenitized, quenched, temperized, and reversion-annealed, spec., **(A 645/A 645M) 01.04**
    of 9 % nickel steel, double-normalized and tempered, spec. for, **(A 353/A 353M) 01.04**
  forgings,
    of alloy steel, high-strength, quenched and tempered, spec. for, **(A 723) 01.04, 01.05**
    of carbon/alloy steel, quenched and tempered, spec., **(A 541) 01.04, 01.05**
    of carbon/alloy steel, quenched and tempered, vacuum-treated, spec., **(A 508) 01.04, 01.05**
    of carbon/low-alloy steel, with mandatory impact testing requirements, spec., **(A 765) 01.05**
    of carbon steel, spec., **(A 266) 01.01, 01.04, 01.05**
    of low-alloy steel, high-strength, quenched and tempered, for fittings and parts, spec., **(A 592) 01.04, 01.05**
  mechanical testing, **(A 370) 01.01, 01.02, 01.03, 01.05**
  plates,
    alloy steel/high-strength low-alloy steel, quenched and tempered, for piping components and welded vessels, spec., **(A 734/A 734M) 01.04**
    alloy steel, high-strength, quenched and tempered, for fusion welded boilers, spec., **(A 517/ A 517M) 01.04**
    carbon-manganese-silicon steel, for fusion welded

pressure vessel steel materials (cont.)
  vessels/structures, spec., (A 537/A 537M) 01.04
  carbon-manganese-silicon steel, heat-treated, for welded vessels at moderate and lower temperature service, spec., (A 738/A 738M) 01.04
  carbon-manganese-silicon steel, for welded vessels at moderate and lower temperatures, spec., (A 612/A 612M) 01.04
  carbon-manganese steel, for welded vessels requiring improved low temperature notch toughness, spec., (A 662/A 662M) 01.04
  carbon-manganese steel, high-tensile strength, for welded vessels, spec., (A 455/A 455M) 01.04
  carbon-silicon steel, for intermediate-/higher-temperature service in welded boilers/etc., spec., (A 515/A 515M) 01.04
  carbon steel, for moderate-/lower-temperature service, in welded vessels with improved notch toughness, spec., (A 516/A 516M) 01.04
  carbon steel, low-/intermediate-tensile strength, for fusion-welded vessels, spec., (A 285/A 285M) 01.04
  carbon steel, manganese-/titanium-bearing, for welded glass-lined vessels, spec., (A 562/A 562M) 01.04
  carbon steel, quenched and tempered, for welded layered vessels not subject to post-weld heat treatment, spec., (A 724/A 724M) 01.04
  carbon steel, with improved low-temperature transition properties, for welded vessels, spec., (A 442/A 442M) 01.04
  chromium/chromium-nickel/chromium-manganese-nickel, (and sheet/strip), for fusion-welded unfired vessels, spec., (A 240) 01.03, 01.04
  chromium-manganese-silicon alloy, for welded boilers, spec., (A 202/A 202M) 01.04
  chromium-molybdenum alloy, for elevated temperature service, spec., (A 387/A 387M) 01.04
  chromium-molybdenum (2-¼ Cr. 1 Mo) alloy, quenched and tempered, for fabrication of welded vessels, spec., (A 542/A 542M) 01.04
  high-strength, low alloy steel, for welded vessels and piping components, spec., (A 737/A 737M) 01.04
  low carbon, age-hardening, nickel-copper-chromium-molybdenum-columbium alloy, for pressure vessels and piping components, spec., (A 736/A 736M) 01.04
  low carbon-manganese-molybdenum-columbium alloy, for piping components and welded vessels at moderate and lower temperature service, spec., (A 735/A 735M) 01.04
  manganese-chromium-molybdenum-silicon-zirconium (quenched and tempered) alloy steel plates, for fusion-welded pressure vessels, spec., (A 782/A 782M) 01.04
  manganese-molybdenum-/nickel alloy, for welded boilers, spec., (A 302/A 302M) 01.04
  manganese-molybdenum-/nickel alloy, quenched and tempered, for welded vessels, spec., (A 533/A 533M) 01.04
  manganese-silicon carbon steel, for welded boilers, spec., (A 299/A 299M) 01.04
  manganese-vanadium alloy, for locomotive boiler shells/boilers for stationary service etc., spec., (A 225/A 225M) 01.04
  molybdenum alloy, for welded locomotive boiler shells/boilers for stationary service/etc., spec., (A 204/A 204M) 01.04
  nickel-alloy, for welded pressure vessels, spec., (A 203/A 203M) 01.04
  nickel-chromium-molybdenum alloy, quenched and tempered, for welded vessels, spec., (A 543/A 543M) 01.04
  nickel (5 %) steel, austenitized, quenched, temperized, and reversion-annealed, for welded vessels at low or cryogenic temperatures, spec., (A 645/A 645M) 01.04
  nickel (8–9 %) steel, quenched and tempered, for welded vessels, spec., (A 553/A 553M) 01.04
  nickel (9 %)-cobalt (4 %)-molybdenum-chromium alloy, quenched and tempered, for welded vessels, spec., (A 605/A 605M) 01.04
  nickel (9 %) steel, double-normalized and tempered, for welded pressure vessels for cryogenic service, spec., (A 353/A 353M) 01.04
  nickel (12 %) precipitation hardening (maraging) alloy, for welded vessels, spec., (A 590) 01.04
  nickel (18 %) precipitation hardening (maraging) alloy, for welded vessels, spec., (A 538/A 538M) 01.04
  nickel (36 %) steel, for welded vessels, spec., (A 658/A 658M) 01.04
  requirements, general, spec., (A 20/A 20M) 01.04
  special requirements for steel plates for nuclear/other special applications, (A 647) 01.04
  ultrasonic examination, by angle-beam technique, spec., (A 577/A 577M) 01.04
  ultrasonic examination, by straight-beam pulse-echo technique, spec., (A 435/A 435M) 01.04
  ultrasonic examination, of rolled carbon and alloy plain and clad steel plates, by straight-beam, pulse-echo technique, spec., (A 578/A 578M) 01.04
seamless drums/heads/etc.,
  from alloy steel forgings, spec. for, (A 336) 01.01, 01.04, 01.05
sheet
  hot-rolled, high-strength, low-alloy steel sheet (for welded layered pressure vessels), spec., (A 812/812M) 01.03, 01.04

Pressure vessel steel materials (*cont.*)
  thin-wall forgings, of carbon and alloy steel, spec., (A 372) 01.04, 01.05
Procedure data sheet
  for shipbuilding/marine paints/coatings, Δ (F 718) 01.02
Procurement and delivery
  of flat-rolled electrical steels, for magnetic applications, procedures for specifying requirements, spec., (A 345) 01.03

Q

Quality assurance
  for hot-dip galvanized zinc coatings, rec. practice, (A 385) 01.06
Quenched and tempered carbon/alloy steel
  See Specific alloys under general heading -
    Steel, stainless steel, and related alloys

R

Railcar loading
  of steel products, for domestic shipment, rec. practice, (A 700) 01.01, 01.03, 01.04, 01.05
Rail fences
  See Fences/fencing materials
Railroad steel materials
  Sa Automotive steel materials
    Bridge steel materials
    Building steel materials
    Generator materials
    Pressure vessel steel materials
    Ship building steel materials
    Steel bolting materials
    Steel, stainless steel, and related alloys
  alloy steel (heat-treated),
    axles, for mass transit and electric railway use, spec., (A 729) 01.04, 01.05
  axles,
    carbon steel, for export and general use, spec., (A 383) 01.04, 01.05
    heat-/nonheat-treated carbon steel, spec., (A 21) 01.04
  bolts/nuts, for railroad tracks,
    heat-treated carbon steel, spec., (A 183) 01.04
  crankshaft forgings,
    magnetic particle inspection of, spec., (A 456) 01.05
    ultrasonic examination of, test, (A 503) 01.05
  forgings,
    carbon and alloy steel, untreated and heat-treated, spec. for, (A 730) 01.04, 01.05
    girder rails, carbon steel (plain/grooved/guard), spec., (A 2) 01.04
  joint bars,
    carbon steel, low-/medium-/high, non-heat treated, spec., (A 3) 01.04
    heat-treated carbon steel, spec., (A 49) 01.04
  leaf springs, heat-treated, made of carbon/alloy steel, spec., (A 147) 01.05
  mechanical testing, (A 370) 01.01, 01.02, 01.03, 01.05
  rails, carbon steel, for cranes, spec., (A 759) 01.04
  screw spikes, spec., (A 66) 01.04
  structural,
    carbon, for shapes, plates, and bars, spec., (A 36) 01.04
  tee rails, spec., (A 1) 01.04
  tie plates,
    low-carbon steel, spec., (A 67) 01.04
  tract bolts/nuts (carbon steel), spec., (A 183) 01.04
  track spikes, soft steel, spec., (A 65) 01.04
  wheels,
    carbon steel, for railway and rapid transit use, spec., (A 551) 01.04
    cast carbon steel, for locomotives and cars, spec. (A 583) 01.04
    cast, for electrical service, spec., (A 631) 01.04
    wrought carbon steel, for locomotives and cars, spec., (A 504) 01.04
    wrought, for electrical service, spec., (A 25) 01.04
  wire, for right-of-way fencing, zinc-coated (galvanized) iron/steel, spec., (A 116) 01.06
Rails (for chain link fence)
  See Fences/fencing materials
Rail-steel bars
  See Steel bars
Rating system
  evaluation of/specifying textures/discontinuities of steel castings, by visual examination, practice Δ (A 802) 01.02
Reactor vessels
  See Nuclear reactor vessels
Reduction gear rings
  See Gears
Reference photographs
  evaluation of/specifying textures/discontinuities of steel castings, by visual examination, practice Δ (A 802) 01.02
Reference radiographs
  for ductile iron castings, guide to, (E 689) 01.02
  steel castings,
    heavy walled, 2 to 4½ in. (51 to 114 mm) thick, Δ (E 186) 01.02
    heavy-walled, 4½ to 12 in. (114 to 305 mm) thick, Δ (E 280) 01.02
    investment, for aerospace applications, Δ (E 192) 01.02
    to 2 in. (51 mm) thick, Δ (E 446) 01.02
  steel welds, fusion, Δ (E 390) 01.02
Register chain
  See Steel chain
Reinforcing steel rods (for concrete)
  See Concrete, reinforcement
Residential chain link fences
  See Fences/fencing materials
Resistance
  corrosion resistance,
    of stainless steel (ferritic), susceptibility detection to intergranular attack, practice, (A 763) 01.0

Index of ASTM Standards, Section 1

Resistance (*cont.*)
   cutting resistance, of tool-resisting steel bars (for security applications), test, **(A 627) 01.05**
   impact resistance,
      of steel products, test, **(A 370) 01.01, 01.02, 01.03, 01.05**
   marine corrosion resistance, of high-strength, low-alloy steel H-piles and sheet piling, spec., **(A 690) 01.04**

Resulphurized carbon steel
   See **Steel, stainless steel, and related alloys, carbon steel**

Retaining rings
   steel, for turbine generators, ultrasonic inspection of, rec. practice, **(A 531) 01.05**

Retaining rings, magnetic
   See **Magnetic retaining rings**

Right hand doors (for ships)
   See **Shipbuilding steel materials, doors**

Right-of-way fencing
   See **Fences/fencing materials**

Rod (steel)
   See **Steel wire rod**

Rolled floor plate
   See **Steel plate,** rolled floor plate

Roller bearings
   See **Steel, stainless steel, and related alloys, for ball and roller bearings**

Rolls
   dual metal (chilled white iron-gray iron), statically cast, for pressure vessel use, spec., **(A 748) 01.02**

Roofing and waterproofing materials
   steel zinc-coated, (galvanized) sheet, coil-coated, spec., **(A 755) 01.06**
   spec., **(A 361) 01.06**

Rope/round wire
   See **Steel wire**

Rotor forgings
   See **Generator materials**

Roughness
   evaluation of/specifying textures/discontinuities of steel castings, by visual examination, practice, Δ **(A 802) 01.02**

Roundness of glass spheres
   in industrial perforated plate and screen, spec., **(E 674) 01.03**

Rubber- or plastic-coated fabric
   steel chain link fence fabric (poly(vinyl chloride) (PVC)-coated), spec., **(F 668) 01.06**

S

S-816 alloy
   See **Cobalt alloys, commercial designations**

Safety chain (steel)
   See **Steel chain**

Sampling
   ferroalloys, for size (before or after shipment), **(A 610) 01.02**

   for impact testing of structural steel, spec. for, **(A 673) 01.04**

Sash chain (steel)
   See **Steel chain**

Scrapless nut steel wire
   See **Steel wire**

Screens
   industrial perforated, for particle size determination,
      round opening series, spec., **(E 674) 01.03**
      square opening series, spec., **(E 454) 01.03**

Screws
   nonferrous bolting materials
      See **Nonferrous bolting materials**

Screws (machine/sheet metal/tapping/wood)
   See **Steel wire**

Screw spikes
   spec., **(A 66) 01.04**

Seamless drums/heads (for pressure vessels)
   of alloy steel forgings, spec., **(A 336) 01.01, 01.04, 01.05**

Seamless/welded steel pipe
   See **Steel pipe/tube**

Security tests
   for tool-resisting steel bars, **(A 627) 01.05**

Seven-wire stress-relieved strand for prestressed concrete
   mechanical testing, **(A 370) 01.01, 01.02, 01.03, 01.05**

Sewer and drain piping systems
   See **Drainage materials/systems**

Shear testing
   of steel retaining rings for turbine generators, as part of ultrasonic inspection, test, **(A 531) 01.05**

Sheet metal screws, steel wire for
   See **Steel wire**

Sheet piling
   See **Steel, stainless steel, and related alloys, sheet piling**

Sheet steel
   See **Steel sheet/strip/plate**

Shipbuilding materials (general)
   product/procedure data sheet (for shipbuilding/marine paints/coatings), Δ **(F 718) 01.02**

Ship building steel materials
   branch connections (for carbon steel/chromium-molybdenum steel pipe and copper-nickel alloy tubing), practice for use of, **(F 681) 01.02**
   chest of drawers (chiffoniers), for ships, spec., **(F 822) 01.02**
   couplings (sleeve-type), of wrought carbon steel, for joining carbon steel pipes, spec., **(F 682) 01.02**
   doors,
      doors/marine furniture, spec., **(F 782) 01.02**
      interior steel doors/frames (for ships), spec., **(F 821) 01.02**
   fixed vertical steel ladders (for personnel access on ships), spec., **(F 840) 01.02**
   gages,
      gage piping assemblies, practice, **(F 721) 01.02**
      modular gage boards, spec., **(F 707) 01.02**

Ship building steel materials (*cont.*)
general requirements for rolled shapes, plates, sheet piling, and bars, spec., **(A 6) 01.04**
marine furniture drawers, spec., **(F 825) 01.02**
marine furniture tops, spec., **(F 826) 01.02**
marine log desks with cabinets, spec., **(F 823) 01.02**
marine mess tables, spec., **(F 824) 01.02**
mechanical testing, **(A 370) 01.01, 01.02, 01.03, 01.05**
pipe hangers (rigid), spec., **(F 708) 01.02**
stainless steel rolled floor plate, spec., **(A 793) 01.02**
staple/handgrab/handle/stirrup rung, spec., **(F 783) 01.02**
structural, for shapes, plates, bars, and rivets, spec., **(A 131) 01.04**
thermal insulation for piping/machinery, practice for selection/installation, **(F 683) 01.02**
welded joints for piping systems, spec., **(F 722) 01.02**
wildcats (for ship anchor chains), spec., **(F 765) 01.02**

Shock tubes
forgings for, of alloy steel, high-strength, quenched and tempered, spec., **(A 723) 01.04, 01.05**

Shrinkage
evaluation of/specifying textures/discontinuities of steel castings, by visual examination, practice, Δ **(A 802) 01.02**

SI (International System of Units)
metric practice,
excerpts, **(E 380)** (Related Material—all parts) **01.00**

Siding materials
steel zinc-coated (galvanized) sheet,
coil-coated, spec., **(A 755) 01.06**
spec., **(A 361) 01.06**

Sieves
perforated plate and screens for,
round opening series, spec., **(E 674) 01.03**
square opening series, spec., **(E 454) 01.03**
wire cloth,
industrial, square opening spec., **(E 437) 01.03**

Silicomanganese
spec., **(A 483) 01.02**

Single-jack/single-loop chain
See **Steel chain**

Single-reduced electrolytic tin plate
See **Tin mill products**

Size
of ferroalloys (before/after shipment), sampling/testing, **(A 610) 01.02**

Sleeve-type pipe couplings (steel)
See **Steel pipe/tube,** couplings

Soil
installing factory-made corrugated steel sewer pipe, practice, **(A 798) 01.06**

Soil pipe
See **Drainage materials/systems,** soil pipe

Span wires
See **Steel wire strand**

Spectroscopy
x-ray fluorescence spectroscopy,
coating thickness determination, test, **(A 754) 01.06**

Spheroidal graphite/nodular iron
See **Iron castings**

Spheroidized carbon spring steel
spec., **(A 684) 01.03, 01.05**

Spiegeleisen
spec., **(A 98) 01.02**

Spikes
screw, spec., **(A 66) 01.04**
track, soft steel, spec., **(A 65) 01.04**

Split cap hanger (for ships)
See **Shipbuilding steel materials**

Springs
steel springs,
from carbon/alloy steel bars, spec., **(A 689) 01.0**
carbon (hard-drawn/galvanized), spec., **(A 764) 01.03, 01.05**
helical (hot-coiled/heat-treated), made from hot-rolled round steel bars, spec., **(A 125) 01.05**
leaf springs (from heat-treated carbon/alloy steel for railway use, spec., **(A 147) 01.05**

Stability
heat stability,
of steam turbine shafts/rotor forgings, test, **(A 472) 01.05**

Stainless steel
See **Steel, stainless steel, and related alloys,** stainless heat-resisting steel

Staple (for ships)
staple/handgrab/handle/stirrup rung, spec., **(F 783) 01.02**

Staterooms
chest of drawers (chiffoniers), for ships, spec., **(F 822) 01.02**

Statically cast metal rolls
dual metal (chilled white iron-gray iron), for pressure vessel use, spec., **(A 748) 01.02**

Stauff twin clamp hanger (for ships)
See **Shipbuilding steel materials**

Steam turbine shafts/rotors
See **Generator materials,** steam turbine shaf

Steel
macroetched, rating, Δ **(E 381) 01.05**
products for domestic shipment, packaging, marking, and loading, rec. practice, **(A 700) 01.01, 01.03, 01.04, 01.05**

Steel air ventilating grille units
for detention areas, spec., **(A 750) 01.05**

Steel assemblies/subassemblies (zinc coatings)
See **Zinc coatings**

Steel ball and roller bearings
See **Steel, stainless steel, and related alloys,** for ball and roller bearings

Steel bar mats/sheets
fabricated, deformed, for concrete reinforcement, spec. for, **(A 184) 01.04**

Index of ASTM Standards, Section 1

Steel bars
 alloy/stainless steel,
  for pressure vessels/valves/flanges/fittings, for high-temperature service, spec., **(A 193) 01.01**
 alloy steel,
  for ball/roller bearings, spec., **(A 295) 01.05**
  cold-finished, spec., **(A 331) 01.05**
  hot-/hot-cold-/cold-worked, for elevated temperature service, spec., **(A 458) 01.05**
  hot-rolled/cold-finished, quenched and tempered, spec., **(A 434) 01.05**
  hot-wrought/cold-finished (quenched/tempered), spec., **(A 434) 01.05**
  hot-wrought (for elevated temperature/pressurized parts), spec., **(A 739) 01.05**
  rolled/forged/strain-hardened, for low temperature service, spec., **(A 320) 01.01**
  standard grades, spec., **(A 322) 01.01**
  subject to end-quench hardenability requirements, specs. for, **(A 304) 01.05**
  for surface hardening by nitriding, spec., **(A 355) 01.05**
 carbon/alloy steel,
  heat-treated, for leaf springs, for railway use, spec., **(A 147) 01.05**
  hot-rolled and cold-finished, general requirements for, spec., **(A 29/A 29M) 01.05**
  hot-rolled, for general purpose springs, spec., **(A 689) 01.05**
 carbon steel,
  bars/shapes/plates (for bridges/buildings), spec., **(A 36) 01.04**
  cold-finished, standard quality, spec., **(A 108) 01.05**
  hot-rolled, merchant quality, spec., **(A 575) 01.05**
  hot-rolled, quenched and tempered, spec., **(A 321) 01.05**
  hot rolled, special quality (include forging/heat-treating/cold-drawing/etc.), spec., **(A 576) 01.05**
  hot-wrought/cold-finished—special quality (for pressure piping components/parts), spec., **(A 696) 01.01, 01.05**
  hot-wrought—special quality (for fluid power applications), spec., **(A 695) 01.05**
  hot-wrought—special quality (for general construction), mechanical requirements, spec., **(A 675) 01.05**
  merchant quality—mechanical requirements, spec., **(A 663) 01.01**
  merchant quality—M-grades, spec., **(A 575) 01.05**
  from standard rail steel, spec., **(A 499) 01.05**
  stress-relieved, cold-drawn, spec., **(A 311) 01.05**
 deformed and plain, for concrete reinforcement,
  axle-steel, spec., **(A 617) 01.04**
  billet-steel, spec., (for metric see A 615M), **(A 615) 01.04**
  epoxy-coated, spec., **(A 775) 01.04**
  rail-steel, spec., **(A 616) 01.04**
  welded steel plain bar/rod mats, spec., **(A 704) 01.04**
 for domestic shipment, packaging, marking and loading methods, rec. practice, **(A 700) 01.01, 01.03, 01.04, 01.05**
 electrode remelted, macroetch testing, Δ **(A 604) 01.03, 01.05**
 epoxy-coated reinforced steel bars, spec., **(A 775) 01.04**
 for fabrication of tools,
  hot-rolled, from tungsten-/molybdenum-type high speed tool steels, spec., **(A 600) 01.05**
  machined flat and square, spec., **(A 685) 01.05**
  from wrought carbon steel, spec., **(A 686) 01.05**
 flat, of austenitic stainless steel, for structural applications, spec., **(A 666) 01.03, 01.04, 01.05**
 hardenability,
  by end-quench (Jominy) method, test, **(A 255) 01.05**
  by end-quench method, spec. for requirements, **(A 304) 01.05**
 hot-/hot-cold-/cold-worked alloys, for elevated temperature service, spec., **(A 458) 01.05**
 hot-rolled and plain, for concrete reinforcement, in mat/sheet form, welded, spec. for, **(A 704) 01.04**
 hot-rolled, for fence posts/assemblies for field and line fencing, spec., **(A 702) 01.03, 01.05, 01.06**
 hot-rolled, round,
  for hot-coiled heat-treated helical compression springs, spec., **(A 125) 01.05**
 hot-wrought (for fence posts/assemblies), spec., **(A 702) 01.03, 01.05, 01.06**
 joint,
  carbon steel, low-/medium-/high, non-heat treated, spec., **(A 3) 01.04**
  of heat-treated carbon steel, spec., **(A 49) 01.04**
 low-alloy, deformed, for concrete reinforcement, spec., **(A 706) 01.04**
 macroetched, rating, Δ **(E 381) 01.05**
 mechanical testing, **(A 370) 01.01, 01.02, 01.03, 01.05**
 for prestressing concrete,
  uncoated, high-strength, spec., **(A 722) 01.04**
 requirements, general, for hot-rolled and cold-finished carbon/alloy steel bars, spec., **(A 29/A 29M) 01.05**
 selecting, according to section/desired mechanical properties, rec. practice, **(A 400) 01.05**
 of stainless/heat-resisting steel (free-machining), hot-rolled/cold-finished, spec., **(A 582) 01.05**
 tool, macroetch testing, Δ **(A 561) 01.05**
 tool resisting, for security applications, spec., **(A 627) 01.05**
 zinc coatings on, by hot-dip galvanizing process, assuring high-quality coatings, rec. practice, **(A 385) 01.06**
  for concrete reinforcement, spec., **(A 767) 01.04**
 safeguarding against embrittlement/detection procedures, rec. practice, **(A 143) 01.06**
 spec., **(A 123) 01.06**

Index of ASTM Standards, Section 1

Steel bars and billets
  electrode remelted, macroetch testing, Δ **(A 604) 01.03, 01.05**
  macroetch testing, inspection, and rating, Δ **(E 381) 01.05**
  stainless/heat-resisting, for forging, spec., **(A 314) 01.05**

Steel bars and shapes
  alloy steel (hot-/hot-cold-/cold-worked), for elevated temperature service, spec., **(A 458) 01.05**
  carbon steel,
    from standard rail steel, spec., **(A 499) 01.05**
  carbon steel (low-/intermediate-tensile strength), structural quality, for general purposes, spec., **(A 283) 01.04**
  flat, tool-resisting, for security applications, spec., **(A 629) 01.05**
  hot-rolled ferritic alloy, for elevated temperature or pressure-containing parts, or both, spec., **(A 739) 01.05**
  of stainless/heat-resisting steel (age-hardening), hot-rolled and cold-finished, spec., **(A 564) 01.05**
  of stainless/heat resisting steel (hot/cold finished), spec., **(A 276) 01.05**
  stainless/heat-resisting steel, spec., **(A 479) 01.04, 01.05**
  zinc coatings on, by hot-dip galvanizing process, assuring high-quality coatings, rec. practice, **(A 385) 01.06**
  safeguarding against embrittlement/detection procedures, rec. practice, **(A 143) 01.06**
  spec., **(A 123) 01.06**

Steel bars/shapes/wires
  carbon steel bars/shapes/plates (for bridges/buildings), spec., **(A 36) 01.04**
  of stainless/heat-resisting steel,
    for cold heading/forging, spec., **(A 493) 01.03**
    spec., **(A 479) 01.04, 01.05**
  of stainless/heat-resisting steel (age-hardening), hot-rolled/cold-finished, spec., **(A 564) 01.05**

Steel bars, forgings, and forging stock
  of iron base superalloys, precipitation hardening (hot-/cold-finished), for high-temperature service, spec., **(A 638) 01.05**
  of stainless steel (martensitic), for high-temperature service, spec., **(A 565) 01.05**

Steel blooms and billets
  of alloy steel (premium quality), for aerospace/aircraft forgings, spec., **(A 646) 01.05**

Steel bolting materials
    Sa **Automotive steel materials**
    Bridge steel materials
    Building steel materials
    Fasteners, metal
    Generator materials
    Iron bolting materials
    Nonferrous bolting materials
    Pressure vessel steel materials
    Railroad steel materials
    Ship building steel materials

alloy/stainless steel,
  for pressure vessels/flanges/valves/fittings, for high-temperature service, spec., **(A 193) 01.01**
alloy steel,
  for nuclear/special applications, spec., **(A 540) 01.01**
  turbine-type, heat-treated, for high-temperature service, spec., **(A 437) 01.01**
  rolled/forged/strain hardened, for low-temperature service, spec., **(A 320) 01.01**
bolts/studs/externally threaded fasteners,
  alloy steel, quenched and tempered, spec., **(A 354) 01.01**
carbon/alloy/martensitic stainless steel,
  nuts (for bolts) for high-pressure/temperature service, spec., **(A 194) 01.01**
fasteners,
  carbon steel, externally/internally threaded, spec. **(A 307) 01.01**
  externally threaded (inc. bolts/studs), of quenched and tempered alloy steel, spec., **(A 354) 01.01**
fasteners and plates, corrugated structural steel, zinc-coated (galvanized), for pipe, arches, and pipe arches, spec., **(A 761) 01.06**
for high-temperature/yield strength (50 to 120 KSI spec., **(A 453) 01.01**
for nuclear/special applications, special requirements, **(A 614) 01.01**
mechanical testing, **(A 370) 01.01, 01.02, 01.03, 01.05**
nuts,
  carbon steel, for general structural/mechanical use, spec., (for metric see A 563M), **(A 563) 01.01**
selecting bolting lengths for (shipboard) piping system flanged joints, practice, **(F 704) 01.02**
track bolts and nuts,
  Sa **Railroad steel materials**
  heat-treated carbon steel, spec., **(A 183) 01.04**

Steel castings
alloy steel,
  heat-treated, for valves/flanges/pressure-bearing parts for high-temperature service, spec., **(A 389) 01.01, 01.02**
austenitic (Hadfield) manganese steel, with alloy modifications, spec., **(A 128) 01.02**
austenitic steel,
  for valves/flanges/pressure-containing parts for low-temperature service, spec., **(A 352) 01.01, 01.02**
  for valves/flanges/pressure-containing parts for high-temperature service, spec., **(A 351) 01.01 01.02**
carbon/alloy,
  high-strength, for structural purposes, spec., **(A 148) 01.02**
carbon/low-alloy,
  heavy-walled, ferritic, for cylinders/valve chests/ steam turbine applications, spec., **(A 356) 01.02**

Steel castings (*cont.*)
    investment, for general applications, spec., **(A 732) 01.02**
    ultrasonic inspection, pulse-echo, by longitudinal-beam technique, spec., **(A 609) 01.02**
  carbon/low-alloy/martensitic stainless steel, normalized/normalized and tempered/quenched and tempered, for pressure-containing parts, spec., **(A 487) 01.01, 01.02**
  carbon steel,
    for fusion welding (of valves/flanges/pressure-containing parts) for high-temperature service, spec., **(A 216) 01.01, 01.02**
    mild-/medium-strength, for general applications, spec., **(A 27) 01.02**
  chromium-nickel alloy, for heat-resisting/elevated-temperature corrosion applications, spec., **(A 560) 01.02**
  chromium (25 %)-nickel (12 %)-iron alloy, for structural/high-temperature service, spec., **(A 447) 01.02**
  classification, of alloy castings, guide, **(A 567) 01.02**
  common requirements, for alloy/steel castings (for general industrial use), spec., **(A 781) 01.02**
  evaluation of/specifying textures/discontinuities of steel castings, by visual examination, practice, Δ **(A 802) 01.02**
  ferritic steel alloy, for valves/flanges/pressure-containing parts, for low-temperature service, spec., **(A 352) 01.01, 01.02**
  heavy-walled,
    2 to 4½ in. (51 to 114 mm) thick, reference radiographs, Δ **(E 186) 01.02**
    4½ to 12 in. (114 to 305 mm) thick, reference radiographs, Δ **(E 280) 01.02**
  for highway bridge components, spec., **(A 486) 01.02**
  investment, for aerospace applications, reference radiographs, Δ **(E 192) 01.02**
  of iron-chromium/iron-chromium-nickel alloys, for heat-resistant/general applications, spec., **(A 297) 01.02**
  iron-chromium/iron-chromium-nickel/nickel-base alloys, for general corrosion-resistant applications, spec., **(A 743) 01.02**
  iron-chromium-nickel-copper alloys, for precipitation hardening treatment, for corrosion-resistant service, spec., **(A 747) 01.02**
  iron-chromium-nickel/nickel-base alloys, for severe corrosion-resistant service, spec., **(A 744) 01.02**
  iron-/cobalt-/nickel-base alloys, for high-strength service at elevated temperatures, spec., **(A 567) 01.02**
  macroetched, rating, Δ **(E 381) 01.05**
  of martensitic stainless/ferritic alloy steel, for pressure-containing parts for high-temperature service, spec., **(A 217) 01.01, 01.02**
  mechanical testing, **(A 370) 01.01, 01.02, 01.03, 01.05**
  qualifications of procedures and personnel for welding of, rec. practice, **(A 488) 01.02**
  reference radiographs for, as applicable to ductile iron castings, guide, **(E 689) 01.02**
  requirements,
    general, for pressure-containing parts, spec., **(A 703) 01.01, 01.02**
    to 2 in. (51mm) thick, reference radiographs, Δ **(E 446) 01.02**
  welding, qualifications of procedures and personnel for, rec. practice, **(A 488) 01.02**
  zinc-coated (galvanized),
    locating the thinnest spot on coating, by Preece test (copper sulfate dip), **(A 239) 01.06**
  zinc coatings on, by hot-dip galvanizing process, assuring high-quality coatings, rec. practice, **(A 385) 01.06**
  safeguarding against embrittlement/detection procedures, rec. practice, **(A 143) 01.06**
  spec., **(A 123) 01.06**
  warpage/distortion prevention of fabricated (by welding) assemblies/subassemblies, rec. practice, **(A 384) 01.06**

Steel chain
  alloy steel, spec., **(A 391) 01.05**
  carbon steel,
    for railroads/construction/industrial uses/spec., **(A 413) 01.05**
    welded machine/coil, spec., **(A 467) 01.05**
    weldless (double jack/loop; single jack/loop; safety; sash; register), spec., **(A 466) 01.05**
  mechanical testing, **(A 370) 01.01, 01.02, 01.03, 01.05**
  wildcats (for ship anchor chains), spec., **(F 765) 01.02**

Steel fasteners
  See **Steel bolting materials,** fasteners

Steel fences
  See **Fences/fencing materials**

Steel flanges/fittings/valves/parts
  alloy/stainless steel,
    bolting materials for, spec., **(A 193) 01.01**
    forged or rolled, for pressure system components, spec., **(A 182) 01.01**
  austenitic stainless steel fittings (as-welded), for general corrosive service at low/moderate temperatures, spec., **(A 774) 01.01**
  bolting materials for
    See **Steel bolting materials**
  carbon/alloy steel,
    flanges, forged, for petroleum/gas pipelines in areas subject to low ambient temperatures, spec., **(A 707) 01.01**
    forged or rolled, for high-pressure transmission-service pipe, spec., **(A 694) 01.01**
    nuts (for bolts), for high-pressure/temperature service, spec., **(A 194) 01.01**
  carbon/low-alloy steel,
    forged or ring-rolled, for low-temperature service

Steel flanges/fittings/valves/parts (*cont.*)
    requiring notch toughness testing, spec., **(A 350) 01.01**
  carbon steel,
    forged, for ambient/higher-temperature service, spec., **(A 105) 01.01**
    forged, for pressure piping systems from −20 to +650°F (−30 to +345°C) with required inherent notch toughness (but testing not required), spec., **(A 727) 01.01**
    forgings, for general-purpose piping, spec., **(A 181) 01.01**
  mechanical testing, **(A 370) 01.01, 01.02, 01.03, 01.05**
  of nickel (8–9 %) alloy steel,
    forged or rolled, for welded pressure vessels at low-temperatures, spec., **(A 522) 01.01**
  for nuclear applications
    See **Nuclear applications,** materials for piping fittings
    See **Steel piping fittings**

Steel forgings
  general requirements, spec., **(A 788) 01.05**
  macroetch testing, inspection, and rating, Δ **(E 381) 01.05**
    See **Forgings**

Steel hardware
  zinc coatings on, by hot-dip galvanizing process, assuring high-quality coatings, rec. practice, **(A 385) 01.06**
  safeguarding against embrittlement/detection procedure, rec. practice, **(A 143) 01.06**
  spec., **(A 153) 01.06**
  warpage/distortion prevention during rec. practice, **(A 385) 01.06**

Steel pipe/tube
  alloy steel,
    electric-resistance-welded, for mechanical tubing, spec., **(A 513) 01.01**
    seamless, for mechanical tubing, spec., **(A 519) 01.01**
    pipe, specialized, general requirements for, spec., **(A 530) 01.01**
  alloy steel, for low-temperature service,
    pipe, seamless and welded, spec., **(A 333) 01.01**
    tubes, seamless and welded, spec., **(A 334) 01.01**
  austenitic alloy,
    heat exchanger tubes, seamless and weldless, with integral fins, spec., **(A 498) 01.01**
    stainless, seamless and welded, for sanitary tubing, spec., **(A 270) 01.01**
    tube, general requirements for, spec., **(A 450) 01.01**
    welded, unannealed tubular products (for low/moderate temperatures and corrosive service), spec., **(A 778) 01.01**
    wrought ferritic and ferritic/austenitic stainless steel piping fittings, spec., **(A 815) 01.01**
  austenitic alloy, for high-temperature central-station service,
    pipe, forged and bored, spec., **(A 430) 01.01**
    pipe, seamless, spec., **(A 376) 01.01**
  austenitic alloy, for high-temperature/corrosive/nuclear pressure service,
    pipe, spec., **(A 451) 01.01**
    tubing (for breeder reactor core components), spec., **(A 771) 01.01**
  austenitic alloy, for high-temperature/pressure service,
    chromium-nickel pipe, electric-fusion-welded, large diameter, spec., **(A 409) 01.01**
    chromium-nickel pipe, electric-fusion-welded, spec., **(A 358) 01.01**
    chromium-nickel tube, seamless, for refinery service, spec., **(A 271) 01.01**
    cold-worked welded austenitic stainless steel pipe, spec., **(A 814) 01.01**
    pipe, centrifugally cast, cold-wrought, spec., **(A 452) 01.01**
    seamless, for boiler/superheater/heat exchanger tubes, spec., **(A 213) 01.01**
    single-double-welded austenitic stainless steel pipe, spec., **(A 813) 01.01**
    stainless pipe, seamless and welded, spec., **(A 31) 01.01**
    stainless tube, seamless and welded, spec., **(A 26) 01.01**
    welded, for boiler/superheater/heat exchanger/condenser tubes, spec., **(A 249) 01.01**
  austenitic stainless steel
    tube, seamless and welded, small-diameter, for general service, spec., **(A 632) 01.01**
    tube, welded, for feedwater heaters, spec., **(A 688) 01.01**
  for ball and roller bearings, spec., **(A 295) 01.05**
  black, hot-dipped (galvanized), welded and seamless,
    in nominal sizes 1/8 to 16 in. incl., for ordinary uses, spec., **(A 120) 01.01, 01.06**
    in nominal sizes 1/8 to 26 in. incl., spec., **(A 53) 01.01**
  butt welds in still tubes, for refinery service, spec., **(A 422) 01.01**
  carbon-molybdenum alloy,
    electric-resistance-welded, for boiler and superheater tubes, spec., **(A 250) 01.01**
    seamless, for boiler and superheater tubes, spec. **(A 209) 01.01**
  carbon-molybdenum alloy (medium-strength),
    seamless, for boiler and superheater tubes, spec., **(A 692) 01.01**
  carbon steel,
    boiler tubes, seamless, for high-pressure service, spec., **(A 192) 01.01**
    cold-drawn buttweld, for mechanical tubing, spec., **(A 512) 01.01**
    electric-resistance-welded, for heat exchanger/condenser tubes, spec., **(A 214) 01.01**
    electric-resistance-welded, for mechanical tubing, spec., **(A 513) 01.01**

steel pipe/tube (*cont.*)
- heat exchanger tubes, seamless and weldless, with integral fins, spec., **(A 498) 01.01**
- pipe, centrifugally cast, for high-temperature/pressure service, spec., **(A 660) 01.01**
- pipe, forged and bored, for high-temperature service, spec., **(A 369) 01.01**
- pipe, seamless, for atmospheric and lower temperatures, spec., **(A 524) 01.01**
- pipe, seamless, for high-temperature service, spec., **(A 106) 01.01**
- pipe, seamless and welded, for water-wells, spec., **(A 589) 01.01**
- pipe, specialized, general requirements for, spec., **(A 530) 01.01**
- tube, electric-resistance-welded, for boilers, spec., **(A 178) 01.01**
- tube, electric-resistance-welded, for boilers/superheaters, spec., **(A 226) 01.01**
- tube, electric-resistance welded, for feedwater heaters, spec., **(A 557) 01.01**
- tube (electric-resistance-welded, metallic-coated carbon steel), spec., **(A 787) 01.01**
- tube, seamless, for mechanical tubing, spec., **(A 519) 01.01**
- carbon steel (cold-formed welded and seamless), structural tubing, for construction of bridges and buildings, spec., **(A 500) 01.01, 01.04**
- carbon steel (hot-formed welded and seamless), structural tubing, for construction of bridges and buildings, spec., **(A 501) 01.01, 01.04**
- carbon steel, for low-temperature service,
  - pipe, seamless and welded, spec., **(A 333) 01.01**
  - tubes, seamless and welded, spec., **(A 334) 01.01**
- carbon steel (low),
  - tubes, seam-welded, round, tapered, for structural use, spec., **(A 595) 01.01, 01.04**
- carbon steel (low-, electric-resistance-welded), for use by chemical industry, spec., **(A 587) 01.01**
- carbon steel (medium),
  - seamless, for boiler and superheater tubes, spec., **(A 210) 01.01**
- carbon steel (wrought),
  - butt-welded pipe fittings, for improved notch toughness, spec., **(A 758) 01.01**
- chromium-molybdenum/chromium-molybdenum-silicon alloy,
  - tube, seamless, cold-drawn, for heat exchanger/condenser tubes, spec., **(A 199) 01.01**
  - tube, seamless, for refinery service, spec., **(A 200) 01.01**
- chromium-nickel alloy,
  - austenitic still tubes, seamless, for refinery service, spec., **(A 271) 01.01**
- cold-drawn,
  - buttweld, carbon alloy, for mechanical tubing, spec., **(A 512) 01.01**
  - tube, seamless, carbon steel, for tubular feedwater heaters, spec., **(A 556) 01.01**
  - tube, seamless, low-carbon steel, for heat-exchanger/condenser tubes, spec., **(A 179) 01.01**

copper-brazed tubing, for automotive/refrigeration/stove industries, spec., **(A 254) 01.01**
corrugated steel,
- aluminum-zinc alloy-coated steel sheet (by hot-dip process), for storm sewer/drainage pipe, spec., **(A 806) 01.06**
- installing corrugated steel structural plate pipe (for sewers), practice, **(A 807) 01.06**
- installing factory-made corrugated steel sewer pipe, practice, **(A 798) 01.06**
- pipe, zinc-coated (galvanized), spec., **(A 760) 01.06**
- plates and fasteners (structural), for pipe, pipe-arches, and arches, spec., **(A 761) 01.03**
- structural design of corrugated steel pipe/pipe-arches/arches (for storm/sanitary sewers), practice, **(A 796) 01.06**

electric-fusion (arc)-welded pipe,
- sizes 4 in/over, spec., **(A 139) 01.01**
- sizes NPS 16/over, spec., **(A 134) 01.01**

electric-fusion-welded,
- pipe, for atmospheric and lower temperatures, spec., **(A 671) 01.01**
- pipe, for high-pressure service at moderate temperatures, spec., **(A 672) 01.01**
- pipe, of carbon and alloy steel, for high-pressure service at high temperatures, spec., **(A 691) 01.01**

electric-resistance-welded,
- pipe, for conveying liquid/gas/vapor, spec., **(A 135) 01.01**
- pipe, for high pressure pipe-type cable circuits, spec., **(A 523) 01.01**
- tube, coiled, for gas and fuel oil lines, spec., **(A 539) 01.01**
- tubing (metallic-coated carbon steel), spec., **(A 787) 01.01**

ferritic alloy,
- tube, general requirements, spec., **(A 450) 01.01**

ferritic alloy, for high-temperature/pressure service,
- heat exchanger tubes, seamless and weldless, with integral fins, spec., **(A 498) 01.01**
- pipe, centrifugally cast, alloy-steel, spec., **(A 426) 01.01**
- pipe, forged and bored, spec., **(A 369) 01.01**
- pipe, seamless, annealed/normalized and tempered, spec., **(A 405) 01.01**
- pipe, seamless, spec., **(A 335) 01.01**
- tube, seamless, for boiler/superheater/heat exchanger tubes, spec., **(A 213) 01.01**

ferritic/austenitic seamless/welded stainless steel pipe (for corrosion-resistant service), spec., **(A 790) 01.01**

ferritic/austenitic seamless/welded stainless steel tube (for corrosion-resistant service), spec., **(A 789) 01.01**

ferritic stainless steel, for general corrosion-resistant/high-temperature service,
- pipe, seamless and welded, spec., **(A 731) 01.01**
- tube, seamless and welded, spec., **(A 268) 01.01**
- tube (welded/unannealed), spec., **(A 791) 01.01**

Steel pipe/tube (*cont.*)
for fire protection systems,
pipe (black, hot-dipped (galvanized), seamless/welded (for fire protection use), spec., **(A 795) 01.01**
fittings
   See **Steel piping fittings**
   Sa **Steel flanges/fittings/valves/parts**
for high-temperature/pressure service (boilers/heat exchangers/condensers/superheaters/etc.),
pipe, austenitic alloy, centrifugally cast, cold-wrought, spec., **(A 452) 01.01**
pipe, austenitic alloy, centrifugally cast, spec., **(A 451) 01.01**
pipe, austenitic alloy, forged and bored, spec., **(A 430) 01.01**
pipe, austenitic alloy, seamless, for central station service, spec., **(A 376) 01.01**
pipe, austenitic alloy, single-/double-welded stainless steel pipe, spec., **(A 813) 01.01**
pipe, austenitic chromium-nickel alloy, electric-fusion-welded, large diameter, spec., **(A 409) 01.01**
pipe, austenitic chromium-nickel alloy, electric-fusion-welded, spec., **(A 358) 01.01**
pipe, austenitic stainless steel, seamless and welded, spec., **(A 312) 01.01**
pipe, carbon/ferritic alloy, forged and bored, spec., **(A 369) 01.01**
pipe, carbon steel, centrifugally cast, spec., **(A 660) 01.01**
pipe, carbon steel, seamless, spec., **(A 106) 01.01**
pipe, austenitic alloy, cold-worked welded stainless steel pipe, spec., **(A 814) 01.01**
pipe, ferritic alloy, centrifugally cast, spec., **(A 426) 01.01**
pipe, ferritic alloy, seamless, annealed/normalized and tempered, spec., **(A 405) 01.01**
pipe, ferritic alloy, seamless, spec., **(A 335) 01.01**
pipe, ferritic stainless steel, seamless and welded, spec., **(A 731) 01.01**
tube, austenitic alloy, welded, for boilers/superheaters/etc., spec., **(A 249) 01.01**
tube, austenitic chromium-nickel steel, for refinery service, spec., **(A 271) 01.01**
tube, austenitic stainless steel, for feedwater heaters, spec., **(A 688) 01.01**
tube, austenitic stainless steel, seamless and welded, spec., **(A 269) 01.01**
tube, carbon/ferritic/austenitic alloys, seamless and weldless, with integral fins, spec., **(A 498) 01.01**
tube, carbon-molybdenum alloy, electric-resistance-welded, spec., **(A 250) 01.01**
tube, carbon-molybdenum alloy, seamless, spec., **(A 209) 01.01**
tube, carbon-molybdenum (medium-strength) alloy, seamless, for boilers/superheaters, spec., **(A 692) 01.01**
tube, carbon steel, electric-resistance-welded, for boilers, spec., **(A 178) 01.01**
tube, carbon steel, electric-resistance-welded, for boilers/superheaters, spec., **(A 226) 01.01**
tube, carbon steel, electric-resistance-welded, for feedwater heaters, spec., **(A 557) 01.01**
tube, carbon steel, electric-resistance-welded, for heat exchangers/condensers, spec., **(A 214) 01.01**
tube, carbon steel, seamless, for boilers, spec., **(A 192) 01.01**
tube, chromium-molybdenum/chromium-molybdenum-silicon alloy, seamless, cold-drawn, for heat exchangers/condensers, spec., **(A 199) 01.01**
tube, ferritic/austenitic alloy seamless, for heat exchanger/corrosion-resistant applications, spec., **(A 669) 01.01**
tube, ferritic stainless steel (for feedwater heater systems), spec., **(A 803) 01.01**
tube, ferritic stainless steel, seamless and welded, spec., **(A 268) 01.01**
tube, iron-chromium-nickel, high-alloy, centrifugally cast, spec., **(A 608) 01.01, 01.02**
tube, low-carbon steel, seamless, cold-drawn, for heat exchangers/condensers, spec., **(A 179) 01.01**
tube, medium-carbon steel, seamless, for boiler superheaters, spec., **(A 210) 01.01**
tube, seamless cold-drawn carbon steel, for feedwater heaters, spec., **(A 556) 01.01**
tube, stainless steel, seamless, for mechanical applications, spec., **(A 511) 01.01**
supplementary requirements, for seamless/electric-resistance-welded carbon steel products, for conformance with ISO boiler construction recommendations, spec., **(A 520) 01.01**
hot-dipped galvanized, welded and seamless,
pipe (black, hot-dipped (galvanized), seamless/welded (for fire protection use), spec., **(A 795) 01.01**
pipe, for ordinary use, spec., **(A 120) 01.01, 01.06**
pipe, spec., **(A 53) 01.01**
installation procedures
installing factory-made corrugated steel sewer pipe, practice, **(A 798) 01.06**
iron-chromium-nickel (high-alloy),
tube, centrifugally cast, for high-temperature/pressure applications, spec., **(A 608) 01.01, 01.02**
low-alloy,
tube, electric-resistance-welded, for pressure containing parts, spec., **(A 423) 01.01**
low-alloy, high-strength,
pipe, welded and seamless, spec., **(A 714) 01.01**
low alloy, high strength, hot-formed welded and seamless,
tube, for bridges/buildings/general structural uses, spec., **(A 618) 01.01, 01.04**
low-carbon/carbon-molybdenum steel,
seamless still tubes, for refinery service, spec., **(A 161) 01.01**

Index of ASTM Standards, Section 1

steel pipe/tube (cont.)
for low-temperature service,
  pipe, carbon and alloy steel, seamless and welded, spec., **(A 333) 01.01**
  tube, carbon and alloy steel, seamless and welded, spec., **(A 334) 01.01**
mechanical testing, **(A 370) 01.01, 01.02, 01.03, 01.05**
mechanical tubing,
  cold-drawn buttweld carbon, spec., **(A 512) 01.01**
  electric-resistance-welded, carbon and alloy steel, spec., **(A 513) 01.01**
  seamless carbon and alloy steel, spec., **(A 519) 01.01**
  stainless steel, seamless, spec., **(A 511) 01.01**
  welded stainless steel, spec., **(A 554) 01.01**
metal-arc-welded pipe, for high-pressure transmission systems, spec., **(A 381) 01.01**
packaging, marking, and loading, for domestic shipment, rec. practice, **(A 700) 01.01, 01.03, 01.04, 01.05**
pipe-arches,
  plates and fasteners for, corrugated structural steel, zinc-coated (galvanized), spec., **(A 761) 01.03**
  structural design of corrugated steel pipe/pipe-arches/arches (for storm/sanitary sewers), practice, **(A 796) 01.06**
pipe nipples,
  austenitic stainless steel, welded and seamless, spec., **(A 733) 01.01**
  carbon steel, welded and seamless, black and zinc-coated (hot-dip galvanized), spec., **(A 733) 01.01**
pipe piles, welded/seamless, spec., **(A 252) 01.01**
for refinery service,
  butt welds in still tubes, spec., **(A 422) 01.01**
  chromium-molybdenum/chromium-molybdenum-silicon, intermediate alloy still tubes, spec., **(A 200) 01.01**
  low-carbon/carbon-molybdenum steel still tubes, spec., **(A 161) 01.01**
  tube, austenitic chromium-nickel steel, spec., **(A 271) 01.01**
requirements,
  general, for carbon/ferritic/austenitic alloy pipe, spec., **(A 450) 01.01**
  general, for specialized carbon/alloy steel pipe, spec., **(A 530) 01.01**
  supplementary, for seamless and electric-resistance-welded carbon steel tubular products for high-temperature service conforming to ISO boiler construction recommendations, spec., **(A 520) 01.01**
sanitary tube, austenitic stainless steel, seamless and welded, spec., **(A 270) 01.01**
for sewers and drainage systems,
  aluminum-zinc alloy-coated steel sheet (by hot-dip process), for storm sewer/drainage pipe, spec., **(A 806) 01.06**
  installing corrugated steel structural plate pipe (for sewers), practice, **(A 807) 01.06**
  pipe, precoated (polymeric) galvanized, spec., **(A 762) 01.06**
  structural design of corrugated steel pipe/pipe-arches/arches (for storm/sanitary sewewrs), practice, **(A 796) 01.06**
shipboard piping materials/systems,
  branch connections, practice for use **(F 681) 01.02**
  couplings (sleeve-type), of wrought carbon steel, for joining carbon steel pipes, spec., **(F 682) 01.02**
  gages—modular gage boards, spec., **(F 707) 01.02**
  gages—piping assemblies, practice, **(F 721) 01.02**
  hangers (rigid), spec., **(F 708) 01.02**
  selecting bolting lengths for flanged joints, practice, **(F 704) 01.02**
  thermal insulation on piping/machinery, selection/installation, practice, **(F 683) 01.02**
  weld joints, spec., **(F 722) 01.02**
spiral-welded pipe, for conveying liquid, gas or vapor, spec., **(A 211) 01.01**
stainless steel,
  pipe (seamless/welded, ferritic/austenitic), for general corrosion-resistant service, spec., **(A 790) 01.01**
  tube, seamless, for high-temperature service, spec., **(A 511) 01.01**
  tube (seamless/welded, ferritic/austenitic), for general corrosion-resistant service, spec., **(A 789) 01.01**
  tube, for water-DWV applications, spec., **(A 651) 01.01**
  tube (welded, unannealed ferritic stainless steel), for high-temperature/corrosive service, spec., **(A 791) 01.01**
  tube, welded, for mechanical applications, spec., **(A 554) 01.01**
steel plate pipe,
  installing corrugated steel structural plate pipe (for sewers), practice, **(A 807) 01.06**
  steel plate pipe, straight-/spiral-seam, electric-fusion (arc)-welded, spec., **(A 134) 01.01**
  structural (hot-formed welded/seamless carbon steel), spec., **(A 501) 01.01, 01.04**
for water-wells, carbon steel pipe, seamless and welded, spec., **(A 589) 01.01**
zinc coatings on, by hot-dip galvanizing process, pipe, corrugated steel, spec., **(A 760) 01.06**

Steel piping fittings
of austenitic stainless steel (as-welded), for general corrosive service at low/moderate rate temperatures, spec., **(A 774) 01.01**
of wrought austenitic stainless steel, spec., **(A 403) 01.01**
of wrought carbon/alloy steel, for low-temperature service, spec., **(A 420) 01.01**
of wrought carbon/alloy steel, seamless and welded, for moderate/elevated temperatures, spec., **(A 234) 01.01**

Index of ASTM Standards, Section 1

Steel pipe fittings (*cont.*)
  of wrought carbon steel, for improved notch toughness, spec., **(A 758) 01.01**
  wrought ferritic and ferritic/austenitic stainless steel piping fittings, spec., **(A 815) 01.01**

Steel plate
  alloy steel,
    alloy steel/high-strength low-alloy plate (quenched and tempered), for pressure vessels, spec., **(A 734/A 734M) 01.04**
    chromium-manganese-silicon, for pressure vessels, spec., **(A 202/A 202M) 01.04**
    chromium-molybdenum alloy plate (for pressure vessels), spec., **(A 387) 01.04**
    chromium-molybdenum (quenched and tempered), for pressure vessels, spec., **(A 542/A 542M) 01.04**
    high-strength/low-alloy/hot-rolled structural steel plate, spec., **(A 656) 01.04**
    high-strength/low alloy pressure vessel steel plates, spec., **(A 737/A 737M) 01.04**
    high-strength, quenched and tempered plate (for fusion-welded boilers/etc. boilers/pressure vessels), spec., **(A 517/A 517M) 01.04**
    high-yield-strength, quenched and tempered plate (for welding), spec., **(A 514) 01.04**
    hot-rolled, high-strength low-alloy carbon/manganese/columbium/vanadium steel plate, spec., **(A 808) 01.04**
    low-carbon age-hardening nickel-copper-chromium-molybdenum-columbium alloy plate (for pressure vessels), spec., **(A 737/A 737M) 01.04**
    low-carbon manganese-molybdenum-columbium alloy plate (for moderate-/lower-temperature service in pressure vessels), spec., **(A 736/A 736M) 01.04**
    low-carbon manganese-molybdenum-columbium alloy plates/shapes/bars (for general applications), spec., **(A 699) 01.04, 01.05**
    manganese-chromium-molybdenum-silicon-zirconium (quenched and tempered) alloy plate, for pressure vessels, spec., **(A 782/A 782M) 01.04**
    manganese-molybdenum/manganese-molybdenum-nickel alloy plate (for pressure vessels), spec., **(A 302/A 302M) 01.04**
    manganese-molybdenum/manganese-molybdenum-nickel plate (quenched and tempered), for pressure vessels, spec., **(A 533/A 533M) 01.04**
    manganese-vanadium alloy, for pressure vessels, spec., **(A 225/A 225M) 01.04**
    molybdenum alloy, for pressure vessels, spec., **(A 204/A 204M) 01.04**
    nickel-5 % (austenitized/quenched/temperized/reversion-annealed) alloy plate, for welded pressure vessels for service at low/cryogenic temperatures, spec., **(A 645/A 645M) 01.04**
    nickel (8/9 %), for pressure vessels, spec., **(A 553/A 553M) 01.04**
    nickel-9 % (double-normalized/tempered), for pressure vessels, spec., **(A 353/A 353M) 01.04**
    nickel-36 % alloy plate (for pressure vessels), spec., **(A 658/A 658M) 01.04**
    nickel alloy, for pressure vessels, spec., **(A 203/A 203M) 01.04**
    nickel-chromium-molybdenum (quenched and tempered), for pressure vessels, spec., **(A 543/A 543M) 01.04**
    nickel-chromium-molybdenum (quenched/tempered), for welded pressure vessels, spec., **(A 543/A 543M) 01.04**
    nickel-cobalt-molybdenum-chromium (quenched and tempered), for pressure vessels, spec., **(A 605/A 605M) 01.04**
    precipitation-hardening (maraging) 12 % nickel plate (for pressure vessels, spec., **(A 590) 01.0**
  austenitic stainless steel,
    sheet/strip/plate/flat bar (for structural applications), spec., **(A 666) 01.03, 01.04, 01.05**
  carbon steel,
    carbon-manganese plate (for moderate-/lower-temperature service in pressure vessels), spec **(A 662/A 662M) 01.04**
    carbon-manganese-silicon plate (heat-treated, pressure vessels, spec., **(A 537/A 537M) 01.0**
    carbon-manganese-silicon plates (with improved notch toughness), spec., **(A 573) 01.04**
    carbon-manganese-silicon steel (heat-treated), for moderate-/lower-temperature service in pressure vessels, spec., **(A 738/A 738M) 01.04**
    carbon steel bars/shapes/plates (for bridges/buildings), spec., **(A 36) 01.04**
    carbon-silicon plate (for intermediate-/higher-temperature service (in pressure vessels), spec **(A 515/A 515M) 01.04**
    high-strength carbon-manganese plate (for pressure vessels), spec., **(A 455/A 455M) 01.04**
    high-strength (for moderate-/lower-temperature service in pressure vessels), spec., **(A 612/A 612M) 01.04**
    low/intermediate tensile strength carbon-silicon steel plates (for machine parts/general construction), spec., **(A 284) 01.04**
    low/intermediate tensile strength plates (for pressure vessels), spec., **(A 285/A 285M) 01.04**
    low/intermediate tensile strength plate/shapes/bars, spec., **(A 283) 01.04**
    manganese-titanium plate (for welded glass-lined pressure vessels), spec., **(A 562/A 562M) 01**
    manganese-silicon alloy plate (for pressure vessels), spec., **(A 299/A 299M) 01.04**
    for moderate-/lower-temperature service (in pressure vessels), spec., **(A 516/A 516M) 01.04**
    precipitation-hardening (maraging) 18 % nickel plate (for pressure vessels), spec., **(A 538/A 538M) 01.04**
    with improved low-temperature transition properties, spec., **(A 442) 01.04**
  carbon steel (high-strength),
    for moderate/lower temperature service in welded pressure vessels, spec., **(A 612/A 612M) 01.04**
  carbon steel (low-/intermediate-tensile-strength),

Steel plate (*cont.*)
    structural quality, for general purposes, spec., **(A 283) 01.04**
    carbon steel (quenched and tempered), for structural applications, spec., **(A 678) 01.04**
    cladding,
        with nickel/nickel alloy, spec., **(A 265) 01.03, 01.04**
    composite, tool-resisting, for security applications, spec., **(A 628) 01.03**
    corrosion-resisting chromium steel clad plate/sheet/strip, spec., **(A 263) 01.03, 01.04**
    corrugated structural, zinc-coated (galvanized), for pipe, pipe-arches, and arches, spec., **(A 761) 01.03**
    general requirements (for pressure vessels), spec., **(A 20/A 20M) 01.04**
    general requirements (for rolled steel plates/shapes/sheet piling/bars), for structural use, spec., **(A 6) 01.04**
    heat-resisting chromium/chromium-nickel stainless steel plate/sheet/strip (for fusion-welded unfired pressure vessels), spec., **(A 240) 01.03, 01.04**
    high-strength/low alloy steel plate,
        high-strength/low-alloy columbium-vanadium steel shapes/plates/sheet piling/bars (for riveted/bolted/welded construction of bridges/buildings/other structures), spec., **(A 752) 01.04**
    high-strength/low-alloy/hot-rolled structural steel plate, spec., **(A 656) 01.04**
    high-strength/low-alloy pressure vessel steel plates, spec., **(A 737/A 737M) 01.04**
    high-strength/low-alloy structural manganese vanadium steel shapes/plates/bars (for welded/riveted/bolted constructions in welded bridges and buildings), spec., **(A 441) 01.04**
    industrial perforated, round opening series, spec. for openings, **(E 674) 01.03**
    installing corrugated steel structural plate pipe (for sewers), practice, **(A 807) 01.06**
    nickel/nickel-base alloy clad steel plate, spec., **(A 265) 01.03, 01.04**
    packaging/marking/loading methods (for steel products for domestic shipment), rec. practice, **(A 700) 01.01, 01.03, 01.04, 01.05**
    rolled floor plate,
        carbon, high-strength low-alloy/alloy steel, spec., **(A 786) 01.04**
        stainless steel, spec., **(A 793) 01.03**
    through-thickness tension testing, **(A 770) 01.04**
    tie plates (low-carbon/high-carbon-hot-worked), spec., **(A 67) 01.04**
    ultrasonic examination,
        angle-beam ultrasonic examination (of steel plate for pressure vessels), spec., **(A 577/A 577M) 01.04**
        straight-beam ultrasonic examination (of plain/clad steel plate for special applications), spec., **(A 578/A 578M) 01.04**
        straight-beam ultrasonic examination (of steel plate for pressure vessels), spec., **(A 435/A 435M) 01.04**
    wildcats (for ship anchor chains), spec., **(F 765) 01.02**
    winding mesh,
        zinc-coated (galvanized) steel pipe winding mesh, spec., **(A 810) 01.06**
    zinc coatings on, by hot-dip galvanizing process,
        assuring high-quality coatings, rec. practice, **(A 385) 01.06**
        safeguarding against embrittlement/detection procedures, rec. practice, **(A 143) 01.06**
        spec., **(A 123) 01.06**

Steel plate pipe
    See **Steel pipe/tube**, steel plate pipe

Steel plate, sheet, and strip
    See **Steel sheet/strip/plate**

Steel posts/rails (for chain link fence)
    See **Fences/fencing materials**

Steel products, general
    aluminum coatings on, by hot-dip process, spec., **(A 676) 01.06**
    chemical analysis, methods, practices, and definitions, **(A 751) 01.01, 01.02, 01.03, 01.04, 01.05**
    electric resistance welded steel shapes, spec., **(A 769) 01.04**
    electrodeposited coatings on,
        cadmium, spec., **(A 165) 01.06**
    flat-rolled, reduction, cold/hot, by wrought alloy steel rolls, spec. for, **(A 427) 01.05**
    macroetched, rating, Δ **(E 381) 01.05**
    mechanically deposited coatings,
        cadmium coatings (on iron/steel metals), spec., **(B 696) 01.04**
        zinc coatings, spec., **(B 695) 01.04**
    mechanical testing, methods and def., **(A 370) 01.01, 01.02, 01.03, 01.05**
    packaging/marking/loading steel products (for domestic shipments), and glossary of terms, **(A 700) 01.01, 01.03, 01.04, 01.05**

Steel rod
    See **Steel wire rod**

Steel rolls
    forged, for corrugating paper machinery, spec., **(A 649) 01.05**
    of wrought hardened alloy, for cold/hot reduction of flat rolled ferrous/nonferrous products, spec., **(A 427) 01.05**

Steel screens
    industrial perforated, round opening series, spec. for openings, **(E 674) 01.03**

Steel screws
    See **Steel bolting materials**
    Sa **Iron bolting materials**
            Nonferrous bolting materials

Steel sheet
    aluminum-coated,
        weight of coating, test, **(A 428) 01.06**
    aluminum-zinc alloy-coated (by hot-dip process),

Steel sheet (*cont.*)
  for storm sewer/drainage pipe, spec., **(A 806) 01.06**
  spec., **(A 792) 01.06**
  carbon, cold-rolled,
    commercial quality (carbon-0.16 to 0.25 max. %), in coils/cut lengths, spec., **(A 794) 01.03**
    commercial quality, in coils/cut lengths, spec., **(A 366) 01.03**
    drawing quality, spec., **(A 619/A 619M) 01.03**
    drawing quality, special killed, for fabricating parts with severe drawing/forming, spec., **(A 620/A 620M) 01.03**
    for structural purposes, spec., **(A 611) 01.03, 01.04**
  carbon steel sheet (for pressure vessels), spec., **(A 414/A 414M) 01.03, 01.04**
  cold-rolled,
    electrolytic zinc-coated, spec., **(A 591) 01.06**
    long-terne coated, spec., **(A 308) 01.06**
    tin-coated by electrodeposition, spec., **(A 599) 01.06**
    type 1 and 2, aluminum-silicon alloy coated (by hot-dip process), spec., **(A 463) 01.06**
  corrosion-resisting chromium steel clad plate/sheet/strip, spec., **(A 264) 01.03, 01.04**
  heat-resisting chromium/chromium-nickel stainless steel plate/sheet/strip (for fusion-welded unfired pressure vessels), spec., **(A 240) 01.03, 01.04**
  hot-rolled, high-strength,
    hot-rolled, high-strength, low-alloy steel sheet (for welded layered pressure vessels), spec., **(A 812/812M) 01.03, 01.04**
  long-terne coated,
    spec., **(A 308) 01.06**
    weight and composition of coating, by triple spot test, **(A 309) 01.06**
  mechanical testing, **(A 370) 01.01, 01.02, 01.03, 01.05**
  packaging/marking/loading methods (for steel products for domestic shipment), rec. practice, **(A 700) 01.01, 01.03, 01.04, 01.05**
  piling, spec., **(A 328) 01.04**
  for porcelain enameling, spec., **(A 424) 01.03**
  stainless chromium-nickel steel clad plate/sheet/strip, spec., **(A 264) 01.03, 01.04**
  zinc-coated (galvanized), by hot-dip process,
    coating thickness, by magnetic field and electromagnetic (eddy current) testing, rec. practice for measuring, **(E 376) 01.06**
    coating thickness, by X-ray fluorescence (XRF) techniques, test, **(A 754) 01.06**
    coil-coated, for roofing and siding, spec., **(A 755) 01.06**
    commercial quality, spec., **(A 526) 01.06**
    for culverts and underdrains, spec., **(A 444) 01.06**
    drawing quality, spec., **(A 528) 01.06**
    drawing quality, special killed, spec., **(A 642) 01.06**
    general requirements, spec., (for metric see A 525M), **(A 525) 01.06**
    high-strength zinc-coated (galvanized) steel sheet spec., **(A 816) 01.06**
    locating the thinnest spot on coating, by the Preece test (copper sulfate dip), **(A 239) 01.06**
    lock-forming quality, spec., **(A 527) 01.06**
    polymeric precoated, for sewer and drainage pipe spec., **(A 742) 01.06**
    for roofing and siding, spec., **(A 361) 01.06**
    structural (physical) quality, spec., **(A 446/A 446M) 01.06**
    weight of coating, test, **(A 90) 01.06**

Steel sheet/strip
  alloy steel, hot-/cold-rolled,
    drawing quality (involving severe cold plastic deformation), spec., **(A 507) 01.03**
    general requirements for, spec., **(A 505) 01.03**
    regular quality, spec., **(A 506) 01.03**
  carbon (15 %), hot-rolled,
    commercial quality, spec., **(A 569) 01.03**
    commercial quality, heavy-thickness coils (formerly designated "plate coils"), spec., (see A 635M for metric equivalent), **(A 635) 01.03**
  carbon (16–25 %), hot-rolled,
    commercial quality, spec., **(A 659) 01.03**
  carbon/high-strength, low-alloy steel,
    hot-/cold-rolled sheet/strip, general requirement spec., (see A 568M for metric equivalents), **(A 568) 01.03**
  carbon, hot-rolled,
    commercial quality, heavy-thickness coils (formerly designated "plate coils"), spec., (for metric see A 635M), **(A 635) 01.03**
    drawing quality, spec., **(A 621/A 621M) 01.03**
    drawing quality, special killed, for fabricating parts with severe drawing/forming, spec., **(A 622/A 622M) 01.03**
    structural quality, spec., **(A 570) 01.03, 01.04**
  columbium/vanadium, high-strength/low-alloy,
    hot-/cold-rolled, for structural purposes, spec., **(A 607) 01.03**
  low-alloy, high-strength, hot-/cold-rolled,
    with improved corrosion resistance, for structural purposes, spec., **(A 606) 01.03**
    for structural/miscellaneous applications requiring improved formability/weldability, spec., **(A 715) 01.03**
  mechanical testing, **(A 370) 01.01, 01.02, 01.03, 01.05**
  packaging, marking, and loading, for domestic shipment, rec. practice, **(A 700) 01.01, 01.03, 01.04, 01.05**
  stainless/heat-resisting,
    chromium-nickel alloy, spec., **(A 177) 01.03**

Steel sheet/strip/plate
  alloy steel,
    hot-/hot-cold-/cold-worked (Grade 651), for high strength at elevated temperatures, spec., **(A 457) 01.03**

Index of ASTM Standards, Section 1

Steel sheet/strip/plate (*cont.*)
  carbon/low-alloy base,
    clad with corrosion-resisting chromium, spec. for, **(A 263) 01.03, 01.04**
    clad with stainless chromium-nickel steel, spec. for, **(A 264) 01.03, 01.04**
    cladding, for pressure vessels,
      with corrosion-resisting chromium, spec., for, **(A 263) 01.03, 01.04**
      with stainless chromium-nickel steel, spec. for, **(A 264) 01.03, 01.04**
    for fabrication of tools,
      from tungsten-/molybdenum-type high-speed tool steels, spec., **(A 600) 01.05**
      from wrought carbon steel, spec., **(A 686) 01.05**
    mechanical testing, **(A 370) 01.01, 01.02, 01.03, 01.05**
    packaging, marking, and loading, for domestic shipment, rec. practice, **(A 700) 01.01, 01.03, 01.04, 01.05**
    stainless/heat-resisting,
      chromium alloy, spec., **(A 176) 01.03**
      chromium-nickel alloy, spec., **(A 167) 01.03**
      chromium/chromium-nickel/chromium-manganese-nickel alloy, for fusion-welded unfired pressure vessels, spec., **(A 240) 01.03, 01.04**
      chromium-nickel-manganese austenitic alloy, annealed/cold-rolled, spec., **(A 412) 01.03**
      flat-rolled, general requirements for, spec., **(A 480/A 480M) 01.03**
    stainless steel (austenitic),
      for structural applications (flat bars included), spec., **(A 666) 01.03, 01.04, 01.05**
    stainless steel (precipitation-hardening), spec., **(A 693) 01.03**

Steel strip
  carbon, cold-rolled,
    spec., (see A 109M for metric equivalent), **(A 109) 01.03**
  carbon, cold-rolled hard,
    untempered spring quality, spec., (for metric see A 680M), **(A 680) 01.03, 01.05**
  carbon, cold-rolled soft,
    untempered spring quality, spec., (for metric see A 684M), **(A 684) 01.03, 01.05**
  carbon/high-strength, low-alloy steel,
    hot-rolled, general requirements, spec., (for metric see A 749M), **(A 749) 01.03**
  carbon, hot-rolled,
    for pressure vessels involving fusion welding or brazing, spec., **(A 414/A 414M) 01.03, 01.04**
  high-carbon (over 25 %), cold-rolled,
    spring quality, general requirements, spec., (see A 682M for metric equivalents), **(A 682) 01.03, 01.05**
  zinc-coatings on, by hot-dip galvanizing process, assuring high-quality coatings, rec. practice, **(A 385) 01.06**
  safeguarding against embrittlement detection procedures, rec. practice, **(A 143) 01.06**
  spec., **(A 123) 01.06**

carbon, cold-rolled, hard,
  untempered high-quality, spec., (for metric, see A 680M), **(A 680) 01.03, 01.05**
carbon, cold-rolled soft,
  untempered high-quality, spec., (for metric, see A 684M), **(A 684) 01.03, 01.05**
carbon, hot-rolled,
  structural quality, spec., **(A 570) 01.03, 01.04**

Steel springs
  from carbon/alloy steel bars, spec. for, **(A 689) 01.05**
  carbon, hard-drawn and galvanized, spec., **(A 764) 01.03, 01.05**
  helical, hot-coiled, heat-treated, made of hot-rolled round steel bars, spec. for, **(A 125) 01.05**
  leaf,
    of heat-treated carbon/alloy steel, for railway use, spec., **(A 147) 01.05**

Steel, stainless steel, and related alloys
  alloying additives,
    ferroalloys, sampling and testing for size (before or after shipment), **(A 610) 01.02**
  alloying additives (boron/columbium),
    ferroboron, spec., **(A 323) 01.02**
    ferrocolumbium, spec., **(A 550) 01.02**
  alloying additives (chromium),
    chromium, spec., **(A 481) 01.02**
    ferrochrome-silicon, spec., **(A 482) 01.02**
    ferrochromium, spec., **(A 101) 01.02**
  alloying additives (manganese),
    electrolytic manganese, spec., **(A 601) 01.02**
    ferromanganese, spec., **(A 99) 01.02**
    ferromanganese-silicon, spec., **(A 701) 01.02**
    silicomanganese, spec., **(A 483) 01.02**
    spiegeleisen, spec., **(A 98) 01.02**
  alloying additives (molybdenum/nickel),
    ferromolybdenum, spec., **(A 132) 01.02**
    nickel oxide sinter, spec., **(A 636) 01.02**
    molybdenum oxide products, spec., **(A 146) 01.02**
  alloying additives (silicon),
    calcium-manganese-silicon, spec., **(A 495) 01.02**
    calcium-silicon, spec., **(A 495) 01.02**
    ferrosilicon, spec., **(A 100) 01.02**
  alloying additives (titanium/tungsten/vanadium),
    ferrotitanium, spec., **(A 324) 01.02**
    ferrotungsten, spec., **(A 144) 01.02**
    ferrovanadium, spec., **(A 102) 01.02**
  alloy steel,
    bars, hot-worked, spec., **(A 322) 01.05**
    blooms, billets, and slabs, for forging, spec., **(A 711) 01.05**
    bolting materials for nuclear/special applications, spec., **(A 540) 01.01**
    bolting materials for pressure vessels/valves/flanges/fittings, for high-temperature service, spec., **(A 193) 01.01**
    bolting materials for pressure vessels/valves/flanges/fittings, for low-temperature service, spec., **(A 320) 01.01**
    bolting materials, turbine-type, heat-treated, for high-temperature service, spec., **(A 437) 01.01**

Steel, stainless steel, and related alloys (*cont.*)
 castings, heat-treated, for valves/flanges/pressure-bearing parts for high-temperature service, spec., **(A 389) 01.01, 01.02**
 castings, high-strength, for structural purposes, spec., **(A 148) 01.02**
 chain, spec., **(A 391) 01.05**
 chromium (12%), forgings, for turbine rotors and shafts, spec., **(A 768) 01.05**
 chromium/chromium-nickel/chromium-manganese-nickel, for fusion-welded unfired pressure vessels, spec., **(A 240) 01.03, 01.04**
 chromium-manganese-silicon, plate, for pressure vessels, spec., **(A 202/A 202M) 01.04**
 chromium-molybdenum, plates, for pressure vessels at elevated temperatures, spec., **(A 387/A 387M) 01.04**
 chromium-molybdenum, quenched and tempered, for welded pressure vessels, spec., **(A 542/A 542M) 01.04**
 forged flanges, for petroleum/gas pipelines, spec., **(A 707) 01.01**
 forged or rolled pipe flanges/fittings/valves and parts, for high-temperature service, spec., **(A 182) 01.01**
 forgings, for magnetic retaining rings for turbine generators, spec., **(A 288) 01.05**
 forgings, for pipe flanges/fittings/valves/parts, for high-pressure transmission-service pipe, spec., **(A 694) 01.01**
 forgings, for seamless drums/heads, and other components for pressure vessels, spec., **(A 336) 01.01, 01.04, 01.05**
 forgings, for thin-walled pressure vessels, spec., **(A 372) 01.04, 01.05**
 forgings, for turbine wheels and disks, spec., **(A 294) 01.05**
 forgings, quenched and tempered, for pressure vessel components, spec., **(A 541) 01.04, 01.05**
 low-carbon, age-hardening nickel-copper-chromium-molybdenum-columbium/nickel-copper-columbium alloy, plates, shapes, and bars, spec., **(A 710) 01.04**
 low-carbon manganese-molybdenum-columbium alloy, plates, shapes, and bars, spec., **(A 699) 01.04, 01.05**
 manganese-molybdenum-/nickel, plates, quenched and tempered, for welded pressure vessels, spec., **(A 533/A 533M) 01.04**
 manganese-molybdenum-/nickel-alloy, plates, for welded boilers, spec., **(A 302/A 302M) 01.04**
 manganese-vanadium, plate, for locomotive boiler shells and other pressure vessels, spec., **(A 225/A 225M) 01.04**
 molybdenum, plate, for welded locomotive boiler shells and other pressure vessels, **(A 204/A 204M) 01.04**
 nickel (5%), austenitized, quenched, temperized, reversion-annealed, for welded pressure vessels at cryogenic temperatures, spec., **(A 645/A 645M) 01.04**
 nickel (8–9%), quenched and tempered, plates, for welded pressure vessels, spec., **(A 553/A 553M) 01.04**
 nickel (9%), plates, double-normalized and tempered, for welded pressure vessels for cryogenic service, **(A 353/A 353M) 01.04**
 nickel (12%), precipitation-hardening (maraging), plates, for welded pressure vessels, spec., **(A 590) 01.04**
 nickel (18%), precipitation-hardening (maraging), plates, for welded pressure vessels, spec., **(A 538/A 538M) 01.04**
 nickel (26%), for welded pressure vessels, spec., **(A 658/A 658M) 01.04**
 nickel-chromium-molybdenum (quenched and tempered) plates, for welded pressure vessels, spec., **(A 543/A 543M) 01.04**
 nickel-cobalt-molybdenum-chromium, quenched and tempered, plates, for welded pressure vessels, spec., **(A 605/A 605M) 01.04**
 nickel, plate, for welded pressure vessels, spec., **(A 203/A 203M) 01.04**
 nuts (for bolts), for high-pressure/temperature service, spec., **(A 194) 01.01**
 nuts, for general structural/mechanical uses, spec., (for metric see A 563M), **(A 563) 01.01**
 pipe, electric-fusion-welded, for high-pressure service at high temperatures, spec., **(A 691) 01.01**
 piping fittings, seamless and welded, for pressure piping/vessel fabrication for service at low temperatures, spec., **(A 420) 01.01**
 rolled floor plates, spec., **(A 786) 01.04**
 alloy steel (heat-treated),
  for axles, for mass transit and electric railway use, spec., **(A 729) 01.04, 01.05**
  forgings, for railway use, spec., **(A 730) 01.04, 01.05**
  forgings, for turbine rotors and shafts, spec., **(A 293) 01.05**
  forgings, for turbine wheels and disks, spec., **(A 294) 01.05**
 alloy steel (high-strength, quenched and tempered), forgings, for pressure vessels/isostatic presses/shock tubes, spec., **(A 723) 01.04, 01.05**
  plate, for use in welded bridges, spec., **(A 514) 01.04**
  plates, for fusion welded boilers and pressure vessels, spec., **(A 517/A 517M) 01.04**
 alloy steel (hot-/hot-cold-/cold-worked),
  forgings and forging billets, Grade 651, for elevated temperature service, spec., **(A 477) 01.05**
  rounds, hexagons, squares, and shapes, for elevated temperature service, spec., **(A 458) 01.05**
 alloy steel (hot-worked), spec., **(A 322) 01.05**
 alloy steel (low),
  castings, heavy-walled, ferritic, for cylinders/valve chests/steam turbine applications, spec., **(A 356) 01.02**
  castings, up to 90,000 psi (621 MPa)/120,000 psi

Index of ASTM Standards, Section 1

Steel, stainless steel, and related alloys (*cont.*)
  (827 MPa), for highway bridge components, spec., **(A 486) 01.02**
 alloy steel (low-, high-strength),
  plates, for piping components and welded pressure vessels, spec., **(A 737/A 737M) 01.04**
 alloy steel (low-, high-strength, quenched and tempered),
  forgings, for fittings and parts for pressure vessel components, spec., **(A 592) 01.04, 01.05**
  plates, for piping components and welded pressure vessels, spec., **(A 734/A 734M) 01.04**
 alloy steel (nonmagnetic),
  forgings, for retaining rings for generators, spec., **(A 289) 01.05**
 alloy steel (normalized and tempered),
  forgings, for reduction gear rings, spec., **(A 290) 01.05**
 alloy steel (premium quality),
  for blooms and billets for aircraft/aerospace forgings, spec., **(A 646) 01.05**
 alloy steel (quenched and tempered),
  bolts, studs, & externally threaded fasteners, spec., **(A 354) 01.01**
  forgings, for pinions and gears, spec., **(A 291) 01.05**
  forgings, for pressure vessel components, spec., **(A 541) 01.04, 01.05**
  forgings, for reduction gear rings, spec., **(A 290) 01.05**
  plates, for piping components and welded pressure vessels, spec., **(A 734/A 734M) 01.04**
 alloy steel (quenched and tempered, vacuum-treated),
  forgings, for pressure vessels, spec., **(A 508) 01.04, 01.05**
 alloy steel (superstrength - above 140 ksi (965 MPa)),
  forgings, spec., **(A 579) 01.05**
 alloy steel (untreated/heat-treated),
  forgings, for general industrial use, spec., **(A 668) 01.05**
 alloy steel (vacuum-treated),
  forgings, for turbine rotor disks and wheels, spec., **(A 471) 01.05**
  forgings, for turbine rotors and shafts, spec., **(A 470) 01.05**
 alloy steel (wrought),
  piping fittings, for moderate/elevated temperatures, spec., **(A 234) 01.01**
 aluminum-coated,
  coating weight, test, **(A 428) 01.06**
 aluminum coatings on
  See **Aluminum coatings**
 austenitic (Hadfield) manganese steel, with alloy modifications, spec., **(A 128) 01.02**
 austenitic steel,
  castings, for valves/flanges/pressure-containing parts, for high-temperature/corrosive service, spec., **(A 351) 01.01, 01.02**
  castings, for valves/flanges/pressure-containing parts, for low-temperature service, spec., **(A 352) 01.01, 01.02**
  estimating ferrite content in austenitic alloys castings, practice, **(A 799) 01.02**
  forgings, ultrasonic examination of, rec. practice, **(A 745) 01.05**
 axle-steel bars, for concrete reinforcement, spec., **(A 617) 01.04**
 for ball and roller bearings,
  carburizing bearing quality billets and products, for anti-friction bearings, spec., **(A 534) 01.05**
  of high-carbon bearing-quality steel products, spec., **(A 295) 01.05**
  of high-carbon chromium-bearing quality steel products, with high-hardenability modifications, spec., **(A 485) 01.05**
  special-quality billets & products, spec., **(A 535) 01.05**
  stainless steel (chromium-carbon bearing quality), spec., **(A 756) 01.05**
 steel bars
  See **Steel bars**
 bolting materials
  See **Iron bolting materials**
  Sa **Nonferrous bolting materials**
  Steel bolting materials
 carbon/low-alloy/martensitic stainless steel,
  castings, normalized/normalized and tempered/ quenched and tempered, for pressure-containing parts, spec., **(A 487) 01.01, 01.02**
 carbon/low-alloy steel,
  castings, investment, for general applications, spec., **(A 732) 01.02**
  castings, pulse-echo ultrasonic inspection, by longitudinal-beam technique, spec., **(A 609) 01.02**
  forged/ring-rolled flanges/valves/fittings, for low-temperature service requiring notch toughness testing, spec., **(A 350) 01.01**
  forgings, for pressure vessel components, with mandatory impact testing requirements, spec., **(A 765) 01.05**
 carbon (low) cold-reduced steel,
  for double-reduced black plate, spec., (for metric see A 650M), **(A 650) 01.06**
  for double-reduced electrolytic tin plate, spec., **(A 626) 01.06**
  for single-reduced black plate, spec., **(A 625) 01.06**
  for single-reduced electrolytic tin plate, spec., **(A 624) 01.06**
 carbon-manganese-silicon steel,
  plates, heat-treated, for welded pressure vessels at moderate/lower temperature service, spec., **(A 738/A 738M) 01.04**
  plates (1 in (25mm) max), for welded pressure vessels at moderate/lower temperatures, spec., **(A 612/A 612M) 01.04**
  plates (1.5 in (38mm) max), for certain atmospheric conditions, spec., **(A 573) 01.04**
  plates (4 in (100mm) max), for fusion welded

Index of ASTM Standards, Section 1

Steel, stainless steel, and related alloys (*cont.*)
    pressure vessels and structures, spec., **(A 537/A 537M) 01.04**
  carbon-manganese steel,
    plates, high-tensile strength, for welded pressure vessels, spec., **(A 455/A 455M) 01.04**
    plates, for welded pressure vessels at moderate/lower temperature service, **(A 662/A 662M) 01.04**
  carbon-silicon steel (low/intermediate tensile strength),
    plates (for machine parts and general use), spec. for, **(A 284) 01.04**
  carbon-silicon steel plates, for intermediate- and higher-temperature service (in pressure vessels), spec., **(A 515/A 515M) 01.04**
  carbon steel,
    axles, nonheat-treated, for export and general use, spec., **(A 383) 01.04, 01.05**
    bars, stress-relieved, cold-drawn, spec., **(A 311) 01.05**
    blooms, billets, and slabs, for forging, spec., **(A 711) 01.05**
    castings, for fusion welding (of valves/flanges/pressure-containing parts) for high-temperature service, spec., **(A 216) 01.01, 01.02**
    castings, heavy-walled, ferritic, for cylinders/valve chests/steam turbine applications, spec., **(A 356) 01.02**
    castings, high-strength, for structural purposes, spec., **(A 148) 01.02**
    castings, mild-/medium-strength, for general application, spec., **(A 27) 01.02**
    castings, up to 70,000 psi (483 MPa) tensile strength, for highway bridge components, spec., **(A 486) 01.02**
    chain, for railroad cars/construction/industrial uses, spec., **(A 413) 01.05**
    chain, welded machine/coil, spec., **(A 467) 01.05**
    chain, weldless, spec., **(A 466) 01.05**
    crane rails, spec., **(A 759) 01.04**
    for fabrication of tools, spec., **(A 686) 01.05**
    fasteners, externally/internally threaded, spec., **(A 307) 01.01**
    forged flanges, for petroleum/gas pipelines, spec., **(A 707) 01.01**
    forgings, for flanges/valves/fittings/parts, for high-pressure transmission-service pipe, spec., **(A 694) 01.01**
    forgings, for piping components (flanges/fittings/valves/etc.), for ambient/higher-temperature service in pressure systems, spec., **(A 105) 01.01**
    forgings, for piping components from −20 to +650°F (−30 to +345°C) with required notch toughness, (but notch testing not required), spec., **(A 727) 01.01**
    forgings, for pressure vessel components, spec., **(A 266) 01.01, 01.04, 01.05**
    forgings, for thin-walled pressure vessels, spec., **(A 372) 01.04, 01.05**
    forgings, leaded and resulphurized, for non-welded pressure systems (and general service), spec., **(A 766) 01.01**
    forgings, quenched and tempered, for pressure vessel components, spec., **(A 541) 01.04, 01.05**
    forgings, with special magnetic properties, for cyclotrons and related nuclear equipment, spec., **(A 594) 01.05**
    for general-purpose piping, spec., **(A 181) 01.01**
    general requirements for rolled shapes, plates, sheet piling, and bars, spec., **(A 6) 01.04**
    girder rails (plain/grooved/guard), spec., **(A 2) 01.04**
    manganese-titanium bearing, for glass or diffused metallic coatings, spec., **(A 562/A 562M) 01.04**
    modular gage boards (for ships), spec., **(F 707) 01.02**
    nuts (for bolts), for high-temperature/pressure service, spec., **(A 194) 01.01**
    nuts, for general structural/mechanical use, spec., (for metric see A 563M), **(A 563) 01.01**
    pipe, electric-fusion-welded, for high-pressure service at high temperatures, spec., **(A 691) 01.01**
    pipe, practice for using branch connections (on ships), **(F 681) 01.02**
    pipe, seamless, for atmospheric and lower temperatures, spec., **(A 524) 01.01**
    pipe, seamless, for high-temperature service, spec., **(A 106) 01.01**
    pipe, welded/seamless, for use in water-wells, spec., **(A 589) 01.01**
    plate (high-strength), for moderate/lower temperature service in welded pressure vessels, spec., **(A 612/A 612M) 01.04**
    plates and bars, with 42,000 psi (290 MPa) minimum yield point, and ½ in. (12.7mm) maximum thickness, spec. for, **(A 529) 01.04**
    plates, carbon-manganese steel, for welded pressure vessels at moderate/lower temperature service, spec., **(A 662/A 662M) 01.04**
    plates, for welded pressure vessels, spec., **(A 516/A 516M) 01.04**
    plates, high-strength carbon-manganese steel, for welded pressure vessels, spec., **(A 455/A 455M) 01.04**
    plates, manganese-silicon, for welded boilers and pressure vessels, spec., **(A 299/A 299M) 01.04**
    shapes, plates, and bars (for bridges), spec., **(A 709) 01.04**
    shapes, plates, and bars (for bridges/buildings/etc.), spec., **(A 36) 01.04**
    sheet piling (for construction of dock walls/sea walls/cofferdams/etc.), spec. for, **(A 328) 01.04**
    tee rails, for railway tracks, spec., **(A 1) 01.04**
    tie plates, low-carbon/high-carbon hot-worked, spec., **(A 67) 01.04**
    tires, for railway and rapid transit use, spec., **(A 551) 01.04**
    tract bolts/nuts spec., **(A 183) 01.04**
  carbon steel (cast),
    for railway locomotive and car wheels, spec., **(A 583) 01.04**

Steel, stainless steel, and related alloys (*cont.*)
  carbon steel (cold-formed welded and seamless),
    structural tubing, for construction of bridges and
    buildings, spec., **(A 500) 01.01, 01.04**
  carbon steel (heat-treated),
    axles, for railway use, spec., **(A 21) 01.04**
    forgings, for turbine rotors and shafts, spec.,
      (A 293) 01.05
    joint bars, spec., **(A 49) 01.04**
  carbon steel (high),
    bearing quality billets/forgings/etc., for ball and
      roller bearings, spec., **(A 295) 01.05**
    chromium-bearing quality billets/forgings/etc.,
      with high-hardenability modifications, for ball
      and roller bearings, spec., **(A 485) 01.05**
  carbon steel (high, hot-worked),
    tie plates, spec., **(A 67) 01.04**
  carbon steel (high-strength, low-alloy), rolled floor
    plates, spec., **(A 786) 01.04**
  carbon steel (hot-formed welded and seamless),
    structural tubing, for construction of bridges and
    buildings, spec., **(A 501) 01.01, 01.04**
  carbon steel (low),
    age-hardening nickel-copper-chromium-molybde-
      num-columbium alloy, plates, for piping com-
      ponents and welded pressure vessels, spec.,
      (A 736/A 736M) 01.04
    age hardening nickel-copper-chromium-molybde-
      num-columbium *and* nickel-copper-colum-
      bium alloys, spec. for, **(A 710) 01.04**
    alloyed with manganese-molybdenum-colum-
      bium, in shapes, plates, and bars, spec.,
      (A 699) 01.04, 01.05
    manganese-molybdenum-columbium alloy,
      plates, for piping components and welded pres-
      sure vessels at moderate/low temperature ser-
      vice, spec., **(A 735/A 735M) 01.04**
    pipe, electric resistance-welded, for the chemical
      industry, spec., **(A 587) 01.01**
    tie plates, spec., **(A 67) 01.04**
  carbon steel (low/intermediate tensile strength),
    plates, for fusion-welded vessels, spec., **(A 285/
      A 285M) 01.04**
    plates (for general use), spec. for, **(A 283) 01.04**
  carbon steel (low, medium, and high, non-heat
    treated),
    joint bars, spec., **(A 3) 01.04**
  carbon steel (nonheat-treated),
    axles, for railway use, spec., **(A 21) 01.04**
  carbon steel (normalized and tempered),
    forgings, for pinions and gears, spec., **(A 291)
      01.05**
    forgings, for reduction gear rings, spec., **(A 290)
      01.05**
  carbon steel (quenched and tempered),
    forgings, for magnetic retaining rings for turbine
      generators, spec. for, **(A 288) 01.05**
    forgings, for pressure vessel components, spec.,
      (A 541) 01.04, 01.05
    forgings, for reduction gear rings, spec., **(A 290)
      01.05**
    plates, spec., **(A 678) 01.04**
    plates, for welded layered pressure vessels, spec.,
      (A 724/A 724M) 01.04
  carbon steel (quenched and tempered, vacuum-
    treated),
    forgings, for pressure vessels, spec., **(A 508) 01.04,
      01.05**
  carbon steel (untreated/heat-treated),
    forgings, for general industrial use, spec., **(A 668)
      01.05**
    forgings, for railway use, spec., **(A 730) 01.04,
      01.05**
  carbon steel (vacuum-treated),
    for turbine rotors and shafts, spec., **(A 470) 01.05**
  carbon steel (wrought),
    couplings (sleeve-type), for joining carbon steel
      pipes, spec., **(F 682) 01.02**
    piping fittings, for moderate/elevated tempera-
      tures, spec., **(A 234) 01.01**
    piping fittings, seamless and welded, for pressure
      piping/vessel fabrication for service at low-
      temperatures, spec., **(A 420) 01.01**
    for railway locomotive and car wheels, spec.,
      (A 504) 01.04
  cast
    Sa **Iron castings**
    Steel castings
    for electrical railway use, spec., **(A 631) 01.04**
  chemical analysis,
    methods/practices/definitions, **(A 751) 01.01,
      01.02, 01.03, 01.04, 01.05**
  chromium-nickel alloy,
    castings, for heat-resisting/elevated-temperature
      corrosion applications, spec., **(A 560)
      01.02**
    heat-resisting stainless steel sheet, strip, and plate,
      for fusion-welded unfired pressure vessels,
      spec., **(A 240) 01.03, 01.04**
  chromium (12%) alloy steel,
    forgings, for turbine rotors and shafts, spec.,
      (A 768) 01.05
  chromium (25 %)-nickel (12 %)-iron alloy,
    for structural/high-temperature service, spec.,
      (A 447) 01.02
  cladding,
    with chromium-nickel steel, spec., **(A 264) 01.03,
      01.04**
    with corrosion-resistant chromium, (sheet, strip,
      and plate), for pressure vessels, spec., **(A 263)
      01.03, 01.04**
    with nickel and nickel alloys, for pressure vessels,
      spec., **(A 265) 01.03, 01.04**
  coating weight
    See **Coating weight**
  for concrete reinforcement
    See **Concrete,** reinforcement
  double-reduced black plate, spec., (for metric see
    A 650M), **(A 650) 01.06**
  electrical steel, flat-rolled,
    for magnetic applications, spec., **(A 345) 01.03**
  electric resistance welded, spec., **(A 769) 01.04**

Steel, stainless steel, and related alloys (*cont.*)
electrodeposited coatings on
  See **Coatings,** electrodeposited
electrode remelted bars and billets,
  macroetch testing, ∆ **(A 604) 01.03, 01.05**
electroplating
  See **Coatings, electroplating**
ferritic alloy steel,
  castings, for pressure-containing parts for high-temperature service, spec., **(A 217) 01.01, 01.02**
  castings, for pressure-containing parts for low-temperature applications, spec., **(A 757) 01.01, 01.02**
  castings, for valves/flanges/pressure-containing parts, for low-temperature service, spec., **(A 352) 01.01, 01.02**
forgings
  See **Forgings**
  Sa **Steel bars, forgings, and forging stock**
general products, chemical analysis of, methods, practices, and definitions, **(A 751) 01.01, 01.02, 01.03, 01.04, 01.05**
general requirements,
  for rolled shapes, plates, sheet piling, and bars, spec., **(A 6) 01.04**
  for stainless/heat-resisting wire, spec., **(A 555) 01.03**
  for stainless/heat-resisting wrought steel products (except wire), spec., **(A 484) 01.05**
for generator applications
  See **Generator materials**
hardenability, by end-quench (Jominy) test, ∆ **(A 255) 01.05**
heat treatment, def. of terms, **(E 44) 01.02**
H-piles and sheet piling,
  of high-strength, low-alloy steel (for use in dock walls/sea walls/cofferdams/etc.), spec. for, **(A 690) 01.04**
intergranular attack, on austenitic stainless steel, detecting susceptibility to, rec. practice, **(A 262) 01.03, 01.05**
intergranular attack in ferritic stainless steels (detecting susceptibility), practice, **(A 763) 01.03**
intergranular corrosion, in severely sensitized austenitic stainless steel, rec. practice, **(A 708) 01.03**
iron base superalloys (precipitation hardening), bars, forgings, and forging stock (hot-/cold-finished), for high-temperature service, spec., **(A 638) 01.05**
iron-chromium/iron-chromium-nickel/alloys,
  castings, for heat-resistant/general applications, spec., **(A 297) 01.02**
iron-chromium/iron-chromium-nickel/nickel-base alloys,
  castings, for general corrosion-resistant applications, spec., **(A 743) 01.02**
iron-chromium-nickel-copper alloys,
  castings, by precipitation hardening treatment, for corrosion-resistant service, spec., **(A 747) 01.02**

iron-chromium-nickel/nickel-base alloys,
  castings, for severe corrosion-resistant service, spec., **(A 744) 01.02**
iron-/cobalt-/nickel-base alloys,
  castings, for high-strength service at elevated-temperatures, spec., **(A 567) 01.02**
for large crankshafts
  See **Crankshafts**
macroetched, rating, ∆ **(E 381) 01.05**
magnetic particle examination of steel forgings, test, **(A 275) 01.05**
manganese-silicon carbon steel,
  plates, for welded pressure vessels, spec., **(A 299/A 299M) 01.04**
for marine environments,
  See **Marine environments,** steel for
martensitic stainless steel,
  castings, for pressure-containing parts for high-temperature service, spec., **(A 217) 01.01, 01.02**
  castings, for pressure-containing parts for low-temperature applications, spec., **(A 757) 01.01, 01.02**
mechanically deposited coatings on
  See **Coatings,** mechanically deposited
mechanical testing, methods and definitions, **(A 370) 01.01, 01.02, 01.03, 01.05**
metallography
  See **Metallography**
nickel (8–9 %) alloy steel,
  forged or rolled flanges/fittings/valves, for welded pressure vessels at low-temperatures, spec., **(A 522) 01.01**
nickel/nickel alloys,
  castings, for corrosion-resistant service, spec., **(A 494) 01.02**
nondestructive testing
  See **Nondestructuve testing**
for pinions and gears
  See **Gears**
pipe and tube
  See **Steel pipe/tube**
plates
  See **Steel plate**
for pressure vessel components
  See **Pressure vessel steel materials**
products for domestic shipment, packaging, marking, and loading methods, rec. practice, **(A 700) 01.01, 01.03, 01.04, 01.05**
products for nuclear/special applications, special requirements,
  alloy-steel bolting materials, **(A 540) 01.01**
  bolting materials, **(A 614) 01.01**
  plates, **(A 647) 01.04**
for railroad materials
  See **Railroad steel materials**
rail-steel bars, for concrete reinforcement, spec., **(A 616) 01.04**
for reinforcing concrete
  See **Concrete,** reinforcement
screw spikes, spec., **(A 66) 01.04**

Steel, stainless steel, and related alloys (*cont.*)
shapes, plates, and bars,
  carbon and high-strength, low-alloy steel (for bridges), spec., **(A 709) 01.04**
  carbon steel (for bridges/buildings), spec., **(A 36) 01.04**
  carbon steel, with 42,000 psi (290 MPa) minimum yield point and ½ in. (12.7 mm) maximum thickness, spec. for, **(A 529) 01.04**
  general requirements, spec., **(A 6) 01.04**
  high-strength, low-alloy manganese vanadium steel (for bridges and buildings), spec. for, **(A 441) 01.04**
  high-strength, low-alloy structural steel, spec. for, **(A 588) 01.04**
  high-strength, low-alloy steel (for welded, riveted, or bolted constructions), spec., **(A 242) 01.04**
  of low-carbon manganese-molybdenum-columbium alloy, spec., **(A 699) 01.04, 01.05**
  of normalized high-strength, low-alloy steel, spec. for, **(A 633) 01.04**
  of quenched and tempered alloy steel (for bridges), spec., **(A 709) 01.04**
  for ship construction, spec., **(A 131) 01.04**
shapes, plates, sheet piling, and bars,
  of columbium-vanadium steel, spec., **(A 572) 01.04**
sheet piling,
  of carbon steel (for construction of dock walls/sea walls/cofferdams), spec. for, **(A 328) 01.04**
  general requirements, spec., **(A 6) 01.04**
sheet piling (and H-piles),
  of high-strength, low-alloy steel (for sea walls/dock walls/cofferdams), spec. for, **(A 690) 01.04**
stainless/heat-resisting steel,
  bar and wire, for cold heading/forging, spec., **(A 493) 01.03**
  bars and shapes, for boilers and other pressure vessel components, spec., **(A 479) 01.04, 01.05**
  bars and shapes (hot/cold finished), spec., **(A 276) 01.05**
  billets and bars, for forging, spec., **(A 314) 01.05**
  bolting materials (bars, bolts, screws, studs, and stud bolts) for pressure vessels/valves/flanges/fittings, for high-temperature service, spec., **(A 193) 01.01**
  calibration/testing of instruments for estimating ferrite content of cast stainless steels, spec., **(A 799) 01.02**
  chromium/chromium-nickel/chromium-manganese-nickel, plate, sheet, and strip, for fusion-welded unfired pressure vessels, spec., **(A 240) 01.03, 01.04**
  chromium-nickel alloy, for spring wire, spec., **(A 313) 01.03, 01.05**
  chromium-nickel alloy, for weaving wire, spec., **(A 478) 01.03**
  cleaning and descaling parts, equipment, and systems, rec. practice, **(A 380) 01.03**
  estimating ferrite content in austenitic alloys castings, practice, **(A 799) 01.02**
  forged or rolled pipe flanges/fittings/valves and parts, for higher-temperature service, spec., **(A 182) 01.01**
  forgings, for general use, and low-/high-temperature service, spec., **(A 473) 01.05**
  general requirements for wrought steel products (except wire), spec., **(A 484) 01.05**
  intergranular attack, detecting susceptibility to, rec. practice, **(A 262) 01.03, 01.05**
  rolled floor plate, spec., **(A 793) 01.03**
  wire, general requirements, spec., **(A 555) 01.03**
  for wire rope, spec., **(A 492) 01.03**
  wire, spec., **(A 580) 01.03**
  wire strand, for guy/overhead ground wires, spec., **(A 368) 01.03**
stainless/heat-resisting steel (age-hardening),
  bars and shapes, hot-rolled and cold-finished, spec., **(A 564) 01.05**
  forgings, for general use, spec., **(A 705) 01.05**
stainless/heat resisting steel (for plate/sheet/strip)
  See **Steel plate**
  Steel sheet
  Steel sheet/strip
  Steel sheet/strip/plate
  Steel strip
stainless/heat-resisting steel (free-machining),
  bars, hot-rolled/cold-finished, spec., **(A 582) 01.05**
  wire, spec., **(A 581) 01.03**
stainless steel (austenitic),
  fittings (wrought), as-welded, spec., **(A 774) 01.01**
  nuts (for bolts), for high-pressure/temperature service, spec., **(A 194) 01.01**
stainless steel (austenitic, severely sensitized),
  intergranular corrosion, susceptibility detection, rec. practice, **(A 708) 01.03**
stainless steel (ferritic),
  detecting susceptibility to intergranular attack, practice, **(A 763) 01.03**
stainless steel (martensitic),
  bars, forgings, and forging stock, for high-temperature service, spec., **(A 565) 01.05**
  castings, for pressure-containing parts for high-temperature service, spec., **(A 217) 01.01, 01.02**
  castings, for pressure-containing parts for low-temperature applications, spec., **(A 757) 01.01, 01.02**
  nuts (for bolts), for high-temperature/pressure service, spec., **(A 194) 01.01**
stainless steel (wrought austenitic),
  piping fittings, spec., **(A 403) 01.01**
steel rolls, forged, for corrugating paper machinery, spec., **(A 649) 01.05**
structural,
  for bridges/buildings, spec., **(A 36) 01.04**
  for bridges (carbon and high-strength, low-alloy steel, and quenched and tempered alloy steel), spec., **(A 709) 01.04**

Index of ASTM Standards, Section 1

Steel, stainless steel, and related alloys (*cont.*)
  carbon, for plates and bars, with 42,000 psi (290 MPa) minimum yield point, and ½ in. (12.7 mm) maximum thickness, spec., **(A 529) 01.04**
  carbon (quenched and tempered) plate, spec., **(A 678) 01.04**
  electric resistance welded, spec., **(A 769) 01.04**
  forgings, of superstrength alloy steel (in excess of 140 ksi (965 MPa)), spec., **(A 579) 01.05**
  low/intermediate tensile strength carbon steel plates (for general use), spec. for, **(A 283) 01.04**
  general requirements for rolled shapes, plates, sheet piling, and bars, spec., **(A 6) 01.04**
  plates and fasteners, corrugated and zinc-coated (galvanized), for pipe, pipe-arches, and arches, spec., **(A 761) 01.03**
  sampling procedure for impact testing, spec. for, **(A 673) 01.04**
  for ship construction, spec., **(A 131) 01.04**
  wildcats (for ship anchor chains), spec., **(F 765) 01.02**
  structural (high-strength, low-alloy),
    columbium-vanadium steel shapes, plates, sheet piling, and bars (for riveted, bolted, or welded bridges/buildings), spec. for, **(A 572) 01.04**
    manganese vanadium steel shapes, plates, and bars (for welded, riveted, or bolted construction), spec. for **(A 441) 01.04**
    for shapes, plates, and bars, with 50,000 psi (345 MPa) minimum yield point, and 4 in. maximum thickness, spec. for, **(A 588) 01.04**
  spec., **(A 242) 01.04**
  structural (high-strength, low-alloy, hot-rolled), vanadium-aluminum-nitrogen/titanium-aluminum, plates, spec. for, **(A 656) 01.04**
  structural (high-yield-strength, quenched and tempered),
    plates (for welded bridges/etc.), spec. for, **(A 514) 01.04**
  structural (normalized high-strength, low-alloy), for shapes, plates, and bars, spec., **(A 633) 01.04**
  tin products
    See **Tin mill products**
  tool-resisting, for security applications,
    air ventilating grille units, for detention areas, spec., **(A 750) 01.05**
    bars and shapes, flat, spec., **(A 629) 01.05**
    bars, spec., **(A 627) 01.05**
    plates, spec., **(A 628) 01.03**
  tool steel,
    alloy tool steel products, spec., **(A 681) 01.05**
    bars, machined flat and square, spec., **(A 685) 01.05**
    bars, macroetch testing, Δ **(A 561) 01.05**
    cast, spec., **(A 597) 01.02**
    tungsten-/molybdenum-type high speed, available in different forms, for fabrication of tools, spec., **(A 600) 01.05**
    wrought carbon, spec., **(A 686) 01.05**
  track spikes, soft steel, spec., **(A 65) 01.04**
  for turbine materials,
    See **Generator materials**
  ultrasonic examination
    See **Nondestructive testing,** ultrasonic testing
  wire
    See **Steel wire**
  wrought,
    for electrical railway use, spec., **(A 25) 01.04**
    welding fittings for nuclear/special applications, requirements, spec., **(A 652) 01.01**
  wrought hardened alloy,
    rolls, for cold/hot reduction of flat rolled ferrous/nonferrous products, spec., **(A 427) 01.05**
  zinc coatings
    See **Zinc-coated (galvanized) iron/steel articles**

Steel strip
  See **Steel strip (following Steel sheet/strip/plate)**

Steel structural strand
  See **Steel wire strand,** structural

Steel structural wire rope
  See **Steel wire rope,** structural

Steel tape
  armoring, flat, used as interlocking/flat armor for electrical cables, spec., **(A 459) 01.06**

Steel, tool steel
  See **Steel, stainless steel, and related alloys,** tool steel

Steel tubing
  See **Steel pipe/tube**

Steel welds
  fusion, reference radiographs, Δ **(E 390) 01.02**

Steel wheels (for railroads/locomotives)
  See **Railroad steel materials**

Steel wire
  alloy, coarse round,
    general requirements, spec., (for metric see A 752M), **(A 752) 01.03**
  alloy, cold-heading quality,
    for hexagon-head bolts, spec., **(A 547) 01.03**
  aluminum-coated,
    aluminum-coated carbon steel wire, spec., **(A 809) 01.06**
    for barbed wire, spec., **(A 585) 01.06**
    chain-link fence fabric, spec., **(A 491) 01.06**
    for farm-field/right-of-way fencing, spec., **(A 584) 01.06**
    weight of coating, test, **(A 428) 01.06**
  aluminum-zinc alloy coated steel,
    chain-link fence fabric, spec., **(A 783) 01.06**
    wire-rope/fittings (for highway guardrail), spec., **(A 784) 01.06**
    wire strand (for guys/messengers/span wires), spec., **(A 785) 01.06**
  for boilers/pressure vessels (of stainless/heat-resisting steel), spec., **(A 479) 01.04, 01.05**
  carbon,
    coppered carbon steel wire, spec., **(A 818) 01.06**

Index of ASTM Standards, Section 1

Steel wire (*cont.*)
   hard-drawn and galvanized, for mechanical springs, spec., **(A 764) 01.03, 01.05**
   metallic-coated steel wire (for chain link fence fabric), spec., **(A 817) 01.06**
   for scrapless nuts, spec., **(A 544) 01.03**
   carbon, coarse round,
     general requirements, spec., (see A 510M for metric equivalents), **(A 510) 01.03**
   carbon, cold-heading quality,
     for machine screws, spec., **(A 545) 01.03**
     for tapping or sheet metal screws, spec., **(A 548) 01.03**
     for wood screws, spec., **(A 549) 01.03**
   hard-drawn,
     for coiled-type upholstery springs, spec., (for metric see A 417M), **(A 417) 01.03, 01.05**
   carbon (low),
     zinc-coated (galvanized), for armoring/protecting submarine/underground cables, spec., **(A 411) 01.06**
   carbon (medium-high), cold-heading quality
     for hexagon-head bolts, spec., **(A 546) 01.03**
   chromium-nickel alloy, for spring wire, spec., **(A 313) 01.03, 01.05**
   chromium-silicon alloy,
     for manufacture of set resistant springs in moderately elevated temperatures, spec., **(A 401) 01.03, 01.05**
   chromium-vanadium alloy,
     for spring wire in moderately elevated temperatures, spec., **(A 231/A 231M) 01.03, 01.05**
     for valve spring wire, spec., **(A 232) 01.03, 01.05**
   cold-drawn,
     for coiled-type springs (upholstery), spec., (for metric see A 407M), **(A 407) 01.03, 01.05**
     for concrete reinforcement, spec., **(A 82) 01.04**
     for mechanical springs/wire forms, spec., **(A 227/A 227M) 01.03, 01.05**
     for zig-zag, square-formed, and sinuous-type upholstery springs, spec., (for metric see A 417M), **(A 417) 01.03, 01.05**
   cold-rolled carbon steel flat wire, spec., **(A 805) 01.03**
   cold-worked, deformed, for concrete reinforcement, spec., **(A 496) 01.04**
     welded wire fabric from, spec., Δ **(A 497) 01.04**
   for concrete, prestressed,
     uncoated, seven-wire, stress-relieved strand, spec., **(A 416) 01.04**
     uncoated, stress-relieved, spec., **(A 421) 01.04**
   for concrete (reinforcement),
     cold-drawn, spec., **(A 82) 01.04**
     cold-drawn, welded wire fabric, spec., Δ **(A 185) 01.04**
     cold-worked, deformed, for welded wire fabric, spec., Δ **(A 497) 01.04**
     cold-worked, deformed, spec., **(A 496) 01.04**
   core,
     aluminum-coated, for aluminum conductors, steel reinforced, spec., **(B 341) 01.06**
     high-strength zinc-coated, for aluminum and aluminum alloy conductors, steel reinforced, spec., **(B 606) 01.06**
     zinc-coated, for aluminum conductors, steel reinforced, spec., (for metric see B 498M), **(B 498) 01.06**
   for electrical use,
     See **Electrical conductors**
   for fabrication of tools,
     from wrought carbon steel, spec., **(A 686) 01.05**
   hard-drawn,
     music spring quality, spec., **(A 228/A 228M) 01.03, 01.05**
     for prestressing concrete pipe, spec., **(A 648) 01.04**
   hard-drawn, high-tensile strength,
     for mechanical springs and wire forms (subject to high static stress), spec., **(A 679) 01.03, 01.05**
   for hexagon-head bolts,
     alloy, cold-heading quality, spec., **(A 547) 01.03**
     medium-high-carbon, cold-heading quality, spec., **(A 546) 01.03**
   high carbon spring,
     for mechanical springs and wire forms (heat treated after fabrication), spec. for, **(A 713) 01.03, 01.05**
   for machine screws, spec., **(A 545) 01.03**
   for mechanical springs and wire forms
     hard-drawn, spec., **(A 227/A 227M) 01.03, 01.05**
     hard-drawn, uncoated, high-tensile strength (and subject to high static stresses), spec. for, **(A 679) 01.03, 01.05**
     for heat treated components, spec., **(A 713) 01.03, 01.05**
     at moderately elevated temperatures, spec., **(A 231/A 231M) 01.03, 01.05**
     oil-tempered, spec., **(A 229/A 229M) 01.03, 01.05**
   mechanical testing, **(A 370) 01.01, 01.02, 01.03, 01.05**
   music wire,
     hard-drawn steel wire, spec., **(A 228/A 228M) 01.03, 01.05**
   oil-tempered,
     for mechanical springs and wire forms, spec., **(A 229/A 229M) 01.03, 01.05**
   oil-tempered carbon,
     for valve springs (and others requiring high-fatigue properties), spec., **(A 230/A 230M) 01.03, 01.05**
   for prestressing concrete pipe, spec., **(A 648) 01.04**
   for scrapless nuts, carbon steel, spec., **(A 544) 01.03**
   of stainless/heat-resisting steel,
     chromium-nickel, for springs, spec., **(A 313) 01.03, 01.05**
     chromium-nickel, for weaving wire, spec., **(A 478) 01.03**
     for cold heading/forging, spec., **(A 493) 01.03**
     free-machining, spec., **(A 581) 01.03**
     general requirements, spec., **(A 555) 01.03**
     rope wire, spec., **(A 492) 01.03**

Steel wire (*cont.*)
spec., **(A 580) 01.03**
strands, for guy/overhead ground wires, spec. for, **(A 368) 01.03**
for tapping/sheet metal screws,
carbon, cold-heading quality, spec., **(A 548) 01.03**
uncoated, stress-relieved, for prestressed concrete, compacted seven-wire strand, spec., **(A 779) 01.04**
seven-wire strand, spec., **(A 416) 01.04**
spec., **(A 421) 01.04**
for upholstery springs,
zig-zag, square-formed, and sinuous-type springs, spec., (for metric see A 417M), **(A 417) 01.03, 01.05**
valve spring quality,
chromium-vanadium alloy, spec., **(A 232) 01.03, 01.05**
oil-tempered carbon steel, spec., **(A 230/A 230M) 01.03, 01.05**
welded fabric, for concrete reinforcement, from cold-worked deformed wire, spec., Δ **(A 497) 01.04**
spec., Δ **(A 185) 01.04**
for wood screws, carbon, cold-heading quality, spec., **(A 549) 01.03**
woven/welded galvanized wire fabric, spec., **(A 740) 01.06**
zinc-coated (galvanized),
barbed wire, spec., **(A 121) 01.06**
carbon steel, for general use, spec., **(A 641) 01.06**
chain-link fence fabric, spec., **(A 392) 01.06**
coating thickness, by magnetic field and electromagnetic (eddy current) testing, rec. practice for measuring, **(E 376) 01.06**
coating thickness, by X-ray fluorescence (XRF) techniques, test, **(A 754) 01.06**
for farm-field/railroad right-of-way fence, spec., **(A 116) 01.06**
high tensile steel, spec., **(A 326) 01.06**
"iron" telephone and telegraph line wire, spec., **(A 111) 01.06**
low-carbon steel, for armoring/protecting submarine/underground cables (used for communications, control, or power purposes), spec., **(A 411) 01.06**
overhead wire strand, for ground/static wires for electric power transmission lines, spec., **(A 363) 01.06**
poultry netting (hexagonal/straight line) and woven poultry fencing, spec., **(A 390) 01.06**
rope and fittings for highway guardrail, spec., **(A 741) 01.06**
tie-wire (for connecting telephone/telegraph line wire to insulators), spec., **(A 112) 01.06**
tying wire, spec., **(A 777) 01.06**
weight of coating, test, **(A 90) 01.06**
woven/welded wire fabric ("hardware cloth"), spec., **(A 740) 01.06**
zinc-coated (galvanized) steel core wire (for aluminum conductors, steel reinforced), spec., (for metric see B 498M), **(B 498) 01.06**

Steel wire rod
alloy, general requirements, spec., **(A 752) 01.03**
carbon, coarse round,
general requirements, spec., (see A 510M for metric equivalents), **(A 510) 01.03**

Steel wire rope
aluminum-zinc alloy coated steel wire,
rope fittings (for highway guardrails), spec., **(A 784) 01.06**
zinc-coated,
for highway guardrail, spec., **(A 741) 01.06**
structural, spec., **(A 603) 01.06**

Steel wire strand
aluminum-coated, for guys, messengers/span wires, spec., **(A 474) 01.06**
aluminum-zinc alloy coated (for guys/messengers/span wires), spec., **(A 785) 01.06**
aluminum-zinc alloy-coated steel overhead ground wire strand for electric power transmission lines, spec., **(A 797) 01.06**
compacted, seven wire, uncoated, stress-relieved (for prestressed concrete), spec., **(A 779) 01.04**
copper-clad, for guys/messengers/span wires, spec., **(A 460) 01.06**
parallel/helical (zinc-coated), spec., **(A 586) 01.06**
of stainless/heat-resisting steel, spec., **(A 368) 01.03**
structural, zinc-coated, spec., **(A 586) 01.06**
zinc-coated (galvanized),
for ground/static wires for electric power transmission lines, spec., **(A 363) 01.06**
for guys/messengers/span wires, spec., **(A 475) 01.06**
zinc-coated, extra-high strength, concentric-lay, for messenger support of figure-8-type communication/electrical cable, spec., **(A 640) 01.06**

Stiles (for ships)
See **Shipbuilding steel materials,** doors

Stirrup rung
staple/handgrab/handle/stirrup rung, spec., **(F 783) 01.02**

Stockade picket fences
See **Fences/fencing materials**

Stranding
See **Steel wire strand**

Strap hanger (for ships)
See **Shipbuilding steel materials**

Strapping (nonmetallic, for shipping containers)
See **Containers,** shipping containers

Strength
strength requirements of steel posts/rails (for residential chain-link fence), spec., **(F 761) 01.06**

Stress-relieved steel strand (for prestressed concrete)
See **Steel wire,** uncoated, stress-relieved

Structural design
structural design of corrugated steel pipe/pipe-arches/arches (for storm/sanitary sewers), practice, **(A 796) 01.06**

Structural iron products
aluminum coatings on, by hot-dip process, spec., **(A 676) 01.06**

Structural rivets
See **Rivets**

Structural steel
    See **Steel, stainless steel, and related alloys,** structural steel
Structural steel products
    Sa **Structural iron products**
    assembled,
        warpage/distortion prevention during hot-dip galvanizing process, rec. practice, **(A 384) 01.06**
        zinc coatings on, assuring high-quality coatings, rec. practice, **(A 385) 01.06**
        zinc coatings on, by hot-dip process, spec., **(A 386) 01.06**
    warpage/distortion prevention during hot-dip galvanizing process, rec. practice, **(A 384) 01.06**
    zinc coatings on, by hot-dip galvanizing process, safeguarding against embrittlement/detection procedures, rec. practice, **(A 143) 01.06**
Structural strand
    See **Steel wire strand,** structural
Stud bolts (for shipboard piping)
    See **Steel pipe/tube,** shipboard piping materials
Studs
    See **Nonferrous bolting materials**
Superficial Rockwell hardness
    See **Hardness, Rockwell (superficial)**
Surface analysis
    angle-beam ultrasonic examination (of steel plate for pressure vessels), spec., **(A 577/A 577M) 01.04**
    detecting susceptibility to intergranular attack in austenitic stainless steels, rec. practice, **(A 262) 01.03, 01.05**
    detecting susceptibility to intergranular attack in ferritic stainless steels, practice, **(A 763) 01.03**
    detecting susceptibility to intergranular corrosion in severely sensitized austenitic stainless steel, **(A 708) 01.03**
    evaluation of/specifying textures/discontinuities of steel castings, by visual examination, practice, Δ **(A 802) 01.02**
    straight beam ultrasonic examination (of plain/clad steel plate for special applications), spec., **(A 578/A 578M) 01.04**
    straight-beam ultrasonic examination (of steel plate for pressure vessels), spec., **(A 435/A 435M) 01.04**
Systematic manual point count
    for volume fraction of second phase, rec. practice, **(E 562) 01.02**

T

Tables
    marine mess tables, spec,. **(F 824) 01.02**
Tailshaft sleeves (for ships)
    See **Shipbuilding materials (general)**
Tank cars/wagons/other shipping containers
    See **Containers,** tank cars/wagons/other shipping containers

Taped lug butterfly valve bolt assembly (for ships)
    See **Steel pipe/tube,** shipboard piping materials
Tapping screws, steel wire for
    See **Steel wire**
Tee rails
    from carbon steel, spec., **(A-1) 01.04**
Telephone/telegraph wire
    zinc-coated (galvanized),
        high-tensile steel, spec., **(A 326) 01.06**
        "iron," spec., **(A 111) 01.06**
        tie-wire (for connecting to insulators), spec., **(A 112) 01.06**
Temperature tests
    fracture appearance transition temperature (FATT), of steel products, **(A 370) 01.01, 01.02, 01.03, 01.05**
Tension (tensile) properties/tests
    of carbon steel axles (heat-treated), **(A 21) 01.04**
    of steel plates (for special applications), for through-thickness tension, test, **(A 770) 01.04**
    of steel products, **(A 370) 01.01, 01.02, 01.03, 01.05**
    tensile strength,
        of steel products, test, **(A 370) 01.01, 01.02, 01.03, 01.05**
Terminology
    packaging/marketing/loading steel products (for domestic shipments), **(A 700) 01.01, 01.03, 01.04, 01.05**
Thermal insulating materials
    for piping/machinery (on ships), selection/installation, practice, **(F 683) 01.02**
Thinnest spot test
    on zinc-coated (galvanized) iron/steel articles, by Preece test (copper sulfate dip), **(A 239) 01.06**
Thin-walled pressure vessels
    See **Pressure vessel steel materials**
Through-thickness tension testing
    See **Tension (tensile) properties/tests**
Tie plates
    steel, low-carbon/high-carbon hot-worked, spec., **(A 67) 01.04**
Tin coated steel sheet
    See **Steel sheet**
Tin coating weights
    for hot-dip/electrolytic tin plate, determination, **(A 630) 01.06**
Tin mill products
    coating thickness,
        by magnetic field and electromagnetic (eddy current) testing, rec. practice for measuring, **(E 376) 01.06**
        by X-ray fluorescence (XRF) techniques, test, **(A 754) 01.06**
    general requirements, spec., (see A 623M for metric equivalents), **(A 623) 01.06**
    mechanical testing, **(A 370) 01.01, 01.02, 01.03, 01.05**
    plate,
        black, cold-rolled, single-/double-reduced, coated

Index of ASTM Standards, Section 1

Tin mill products (*cont.*)
    with chromium and chromium oxide, spec., **(A 657) 01.06**
    black, double-reduced, spec., (for metric see A 650M), **(A 650) 01.06**
    black, single-reduced, spec., **(A 625) 01.06**
    coating weights, determination, **(A 630) 01.06**
    double-reduced, electrolytic, spec., (see also A 626M for metric equivalents), **(A 626) 01.06**
    single-reduced, electrolytic, spec., (see also A 624M for metric equivalents), **(A 624) 01.06**
    packaging, marking, and loading, for domestic shipment, rec. practice, **(A 700) 01.01, 01.03, 01.04, 01.05**
Tin plate
    See **Tin plate mill products**
Tires, steel, for railway use
    See **Railroad steel materials**
Titanium-aluminum steel
    See **Steel, stainless steel, and related alloys**
Tool steel
    See **Steel, stainless steel, and related alloys, tool steel**
Tops
    marine furniture tops, spec., **(F 826) 01.02**
Tracks and track materials
    See **Railroad steel materials**
Track spikes
    of soft steel, spec., **(A 65) 01.04**
Transformer vault frames
    Sa **Zinc coatings**
    assembled, zinc coatings on, by hot-dip process, spec., **(A 386) 01.06**
Transmission lines (electric)
    of zinc-coated (galvanized) wire strand, spec., **(A 363) 01.06**
Transverse testing
    of gray cast iron, **(A 438) 01.02**
Triple spot test
    for coating weight/composition determination, **(A 309) 01.06**
Truck loading
    of steel products, for domestic shipment, rec. practice, **(A 700) 01.01, 01.03, 01.04, 01.05**
Tubular products
    ferrous, aluminum coatings on, by hot-dip process, spec., **(A 676) 01.06**
    steel, packaging, marking, and loading, for domestic shipment, rec. practice, **(A 700) 01.01, 01.03, 01.04, 01.05**
Tungsten high-speed tool steels
    See **Steel, stainless steel, and related alloys, tool steels**
Turbine generators
    See **Generator materials,** turbine
Tying wire (steel)
    See **Steel wire,** zinc-coated (galvanized)

U

U-bolt hanger (for ships)
    See **Shipbuilding steel materials**

Ultrasonic testing
    Sa **Electromagnetic (eddy current) testing**
        Magnetic testing
        Reference radiographs
    of austenitic steel forgings, rec. practice, **(A 745) 01.05**
    of carbon/low-alloy steel castings, by longitudinal-beam technique, spec., **(A 609) 01.02**
    of carbon steel axles, heat-/nonheat-treated, **(A 21) 01.04**
    of heavy steel forgings, rec. practice, **(A 388) 01.05**
    of large forged crankshafts, spec., **(A 503) 01.05**
    of pressure vessel steel plates,
        by angle-beam technique, spec., **(A 577/A 577M) 01.04**
        rolled carbon/alloy plain and clad, by straight-beam pulse-echo technique, spec., **(A 578) 01.04**
        by straight-beam pulse-echo technique, spec., **(A 435) 01.04**
    of steel retaining rings for turbine generators, rec. practice, **(A 531) 01.05**
    of turbine and generator steel rotor forgings, **(A 418) 01.05**
Underwater pressure test
    for seamless and welded austenitic stainless steel tubing (for general service), **(A 269) 01.01**
Unified numbering system (UNS)
    for metals and alloys, practice, **(E 527) 01.01, 01.02, 01.03, 01.04, 01.05**
Upholstery spring steel wire
    See **Steel wire**
UNS (Unified Numbering System)
    for metals and alloys, practice, **(E 527) 01.01, 01.02, 01.03, 01.04, 01.05, 01.06**
UNS A91350 (Wr. alum. alloy, non-heat treatable) wire, for electrical purposes,
    high-strength zinc-coated (galvanized) steel core wire, for aluminum and aluminum alloy conductors steel reinforced, spec., **(B 606) 01.06**
UNS N07001 (nickel-chromium alloy, precipitation hardenable)
    bars, forgings, and forging stock, for high-temperature service, spec., **(B 637) 01.05**
UNS N07080 (nickel chromium alloy, precipitation hardenable)
    bars, forgings, and forging stock, for high-temperature service, spec., **(B 637) 01.05**
UNS N07252 (nickel-chromium alloy, precipitation hardenable)
    bars, forgings, and forging stock, for high-temperature service, spec., **(B 637) 01.05**
UNS N07500 (nickel-chromium alloy, precipitation hardenable)
    bars, forgings, and forging stock, for high-temperature service, spec., **(B 637) 01.05**
UNS N07718 (nickel-chromium alloy, precipitation hardenable)
    bars, forgings, and forging stock, for high-temperature service, spec., **(B 637) 01.05**

Index of ASTM Standards, Section 1

UNS N07718 (*cont.*)
   plate, sheet, and strip, for high-temperature service, spec., **(B 670) 01.03**

UNS N07750 (nickel-chromium alloy, precipitation hardenable)
   bars, forgings, and forging stock, for high-temperature service, spec., **(B 637) 01.05**

UNS R30155 (iron-chromium-nickel-cobalt alloy)
   bars, forgings, and forging stock, for high-temperature service, spec., **(B 639) 01.05**

UNS R30816 (copper-chromium-nickel-niobium-molybdenum-tungsten alloy)
   bars, forgings, and forging stock, for high-temperature service, spec., **(B 639) 01.05**

V

Valves
   See **Steel flanges/fittings/valves/parts**
Valve spring wire
   See **Steel wire**
Vanadium-aluminum-nitrogen steel
   See **Steel, stainless steel, and related alloys**
Vapor degreasing
   of stainless steel parts, equipment, and systems, information on, **(A 380) 01.03**
Voltmeter method
   See **Wattmeter-ammeter-voltmeter method**
Volume
   volume fraction (of a second phase in a multiphase material), by systematic manual point count, rec. practice, **(E 562) 01.02**

W

Warpage/distortion prevention
   during hot-dip galvanizing of steel assemblies/subassemblies (after fabrication), rec. practice, **(A 384) 01.06**
Washers
   Sa **Fasteners**
      Iron bolting materials
      Nonferrous bolting materials
      Steel bolting materials
Water well pipe
   carbon steel (seamless and welded), spec., **(A 589) 01.01**
"Wat'ry main," steel for
   See **Marine environments,** steel for
Weaving wire, steel
   See **Steel wire**
Weight saving steel
   high-strength, low-alloy,
      manganese vanadium steel shapes, plates, and bars, spec., **(A 441) 01.04**
      for shapes, plates, and bars, with 50,000 psi (345 MPa) minimum yield point, and 4 in. maximum thickness, spec. for, **(A 588) 01.04**
      spec., **(A 242) 01.04**

Welded hangers/joints (for shipboard piping)
   See **Steel pipe/tube,** shipboard piping materials
   Sa **Shipbuilding materials**
Welded/seamless steel pipe
   See **Steel pipe/tube**
Welded wire fabric (for concrete reinforcement)
   See **Steel wire,** welded
Welding
   evaluation of/specifying textures/discontinuities of steel castings, by visual examination, practice, **(A 802) 01.02**
Welds
   steel fusion, reference radiographs, Δ **(E 390) 01.02**
Wheels, railway
   See **Railroad steel materials,** wheels
Whelps
   wildcats (for ship anchor chains), spec., **(F 765) 01.02**
Wildcats (for ship anchor chains)
   spec., **(F 765) 01.02**
Winding mesh (for steel pipe/tube)
   zinc-coated (galvanized) steel pipe winding mesh, spec., **(A 810) 01.06**
Windows, steel
   Sa **Zinc coatings**
   assembled, zinc coatings on, by hot-dip process, spec., **(A 386) 01.06**
Wire cloth sieves
   See **Sieves,** wire cloth for
Wire rod, steel
   See **Steel wire rod**
Wire, steel
   See **Steel wire**
Wood and wood products
   fences, design, fabrication, and installation, spec., **(F 537) 01.06**
Wooden fencing materials
   See **Fences/fencing materials**
Wood screws, steel wire for
   See **Steel wire**
Wrought steel products
   See **Steel, stainless steel, and related alloys**

X

X-ray spectroscopy
   See **Spectroscopy,** x-ray spectroscopy

Y

Yield strength and yield point
   of steel products, test, **(A 370) 01.01, 01.02, 01.03, 01.05**

Z

Zinc
   electrodeposition,
      on iron castings, preparation for, rec, practice, **(B 320) 01.02**

Zinc (*cont.*)
  mechanically deposited coatings on iron/steel metals, spec., **(B 695) 01.04**

Zinc alloys
  aluminum-zinc alloy-coated steel overhead ground wire strand for electric power transmission lines), spec., **(A 797) 01.06**

Zinc-coated (galvanized) iron/steel articles
  aluminum-zinc alloy-coated steel sheet (by hot-dip process), for storm sewer/drainage pipe, spec., **(A 806) 01.06**
  bars, for concrete reinforcement, spec., **(A 767) 01.04**
  chain-link fence,
    fabric, residential, spec., **(F 573) 01.06**
    fence fittings, spec., **(F 626) 01.06**
    installation, practice, **(F 567) 01.06**
  high-strength zinc-coated (galvanized) steel sheet, spec., **(A 816) 01.06**
  locating the thinnest spot on coating, by the Preece test (copper sulfate dip), **(A 239) 01.06**
  metallic-coated steel wire (for chain link fence fabric), spec., **(A 817) 01.06**
  pipe, corrugated steel, spec., **(A 760) 01.06**
  pipe, precoated (polymeric) galvanized steel, for sewers and drainage, spec., **(A 763) 01.03**
  plates and fasteners, corrugated structural steel, for pipe, pipe-arches, and arches, spec., **(A 761) 01.03**
  poultry netting/fencing, spec., **(A 390) 01.06**
  repairing damaged coatings, practice, **(A 780) 01.06**
  steel wire strand
    See **Steel wire strand**
  telephone/telegraph line wire, spec., **(A 111) 01.06**
  tying wire, spec., **(A 777) 01.06**
  weight of coating, test, **(A 90) 01.06**
  zinc-coated (galvanized) steel pipe winding mesh, spec., **(A 810) 01.06**

Zinc-coated steel wire
  for aluminum and aluminum alloy conductors, steel reinforced, high-strength zinc-coated, spec., **(B 606) 01.06**
  for aluminum conductors, steel reinforced, spec., (for metric see B 498M), **(B 498) 01.06**

Zinc-coated steel pipe
  See **Steel pipe/tube,** hot-dipped (galvanized)

Zinc coatings
  Sa **Aluminum coatings**
       Coatings (*headings starting with*)
  coating thickness,
    by magnetic field and electromagnetic (eddy current) testing, rec. practice for measuring, **(E 376) 01.06**
    by X-ray fluorescence (XRF) techniques, test, **(A 754) 01.06**
  hot-dip galvanized,
    on assembled steel products, spec., **(A 386) 01.06**
    on assembled steel products, warpage/distortion prevention during galvanizing, rec. practice, **(A 384) 01.06**
    assuring high-quality coatings, rec. practice, **(A 385) 01.06**
    on iron/steel hardware, spec., **(A 153) 01.06**
    on steel fabricated products (rolled/pressed/forged shapes, plates, bars, and strip), spec., **(A 123) 01.06**
    on structural steel products, safeguarding against embrittlement/detection procedures, rec. practice, **(A 143) 01.06**

1984 Membership Application

ASTM
1916 Race Street
Philadelphia, PA 19103

(215) 299-5462
Telex: 710-670-1037

Application is made for membership in ASTM:

☐ Member — An individual or an institution (educational, public library, or a scientific engineering, or technical non-profit society) subscribing to the purpose of the Society provided in the Charter and Bylaws.

☐ Organizational — An individual, business, governmental, research or professional organization, or trade association, or separate facility thereof subscribing to the purpose of the Society provided in the Charter and Bylaws.

PLEASE COMPLETE (Print or type):

NAME (last name first)

JOB TITLE

COMPANY NAME

DEPARTMENT/DIVISION

STREET

CITY, STATE, ZIP CODE

( )
PHONE, EXTENSION

COUNTRY

OFFICIAL REPRESENTATIVE (Organizational Membership Only):

NAME (last name first)

JOB TITLE

MAJOR PRODUCT OR SERVICE PERFORMED BY YOUR ORGANIZATION

IF HOME ADDRESS IS TO BE USED FOR MAILING, PLEASE COMPLETE:

ADDRESS

CITY, STATE, ZIP CODE

Mail to: **ASTM; Attn: Member Services** — See Reverse Side for Benefits and Fees

———— PLEASE COMPLETE INFORMATION ON REVERSE SIDE ————

2/82 Printed in USA

Benefits and Fees for 1984 *Application form on reverse side*

☐ **Member** - Annual Fee $ 50.00   1 January to 31 December
☐ **Organizational** - Annual Fee $350.00   1 January to 31 December

Annual Book of ASTM Standards
- One Free Volume
- One Free Index (1983 or 1984)
- Unlimited Number of Volumes at Member Prices
- Special Quantity Prices
- For Additional Cost-Saving Benefits, Refer to 1984 Publications Catalog

Other ASTM Publications
- One Free Annual Subscription to ASTM Standardization News
- Special Member Prices on Most Publications
- Special Quantity Prices
- For Additional Cost-Saving Benefits, Refer to 1984 Publications Catalog

PLEASE COMPLETE THE FOLLOWING AS IT PERTAINS TO YOU AND YOUR ORGANIZATION:

PROFESSIONAL DATA:
PROFESSIONAL AREAS OF INTEREST: _____

ORGANIZATIONAL DATA (Please check one only):
(a) Book Dealer/Store ☐   (c) Consultant ☐
(b) College/University ☐   (d) Consumer ☐

GOVERNMENT TYPE: (e) Federal ☐   (f) State ☐   (g) City ☐   (h) Local ☐
(i) Hospital/Medical Center ☐

INDUSTRY TYPE: (j) Corporation ☐   (k) Partnership ☐   (l) Properietorship ☐
(m) Labor Union ☐

LIBRARY TYPE: (n) Public ☐   (o) Corporate ☐   (p) Academic ☐
(q) Other _____
(r) Professional Society Association ☐
(s) Other _____